7th International Conference on Wind Turbine Noise (WTN 2017)

Rotterdam, Netherlands
2 - 5 May 2017

ISBN: 978-1-5108-5210-5

Printed from e-media with permission by:

Curran Associates, Inc.
57 Morehouse Lane
Red Hook, NY 12571

Some format issues inherent in the e-media version may also appear in this print version.

Printed by Curran Associates, Inc. (2018)

For permission requests, please contact Institute of Noise Control Engineering/Europe (INCE/Europe) at the address below.

Institute of Noise Control Engineering/Europe (INCE/Europe)
Riverside House
4 Oakland Vale
New Brighton, Merseyside CH45 1LQ
United Kingdom

Phone: +44 (0)151 638 0181
Fax: +44 (0)151 639 5212

Mail@inceeurope.org

Additional copies of this publication are available from:

Curran Associates, Inc.
57 Morehouse Lane
Red Hook, NY 12571 USA
Phone: 845-758-0400
Fax: 845-758-2633
Email: curran@proceedings.com
Web: www.proceedings.com

7th International Conference on Wind Turbine Noise

Day	Start	Session	Subject	Chair
Tuesday	08:30		Registration	
	11:00		Opening	
	11:05	A	Perception and Health Effects - 1	
	13:30	B	Aspects of Source Noise - 1	
	15:40	C	Amplitude Modulation - 1	
	17:15	D	Poster and Exhibition Drinks	
Wednesday	08:30	E	Perception and Health Effect - 2	
	10:15	F	Perception and Health Effect - 3	
	11:40	G	Guest Speaker - 1	
	13:40	H	Shadow Flicker	
	15:50	I	International Perspectives	
	17:15	J	Poster Workshop - Propagation and Background Noise	
Thursday - Hall 1	08:30	K	Assessment and Compliance - 1	
	10:40	L	Assessment and Compliance - 2	
	13:30	M	Low Frequency and Tones	
	15:40	N	Infrasound	
	17:15	O	Seminar: Aeroacoustic investigation of noise sources	
Thursday - Hall 2	08:30	P	Validation of TE Modelling	
	10:40	Q	Combined Modelling	
	13:30	R	Trailing Edge Serrations	
	15:40	S	Source Localisation and Identification	
	17:15	T	Workshop - TBC	
Thursday	19:30		Conference Dinner - De Majesteit	
Friday	08:30	U	Amplitude Modulation - 2	
	10:15	V	Amplitude Modulation - 3	
	11:40	W	Guest Speaker - 2	
	13:20	X	Aspects of Source Noise - 2	
	15:10		Conference Ends	

Details on the following pages

All items on the Programme are open to Delegates

Where the lead author is not presenting the presenter is in (brackets)

INCEEUROPE

7th International Conference on Wind Turbine Noise

Poster Presentations

7th International Conference on Wind Turbine Noise

Tuesday 2nd May		
11:00	**Opening - Gijsjan van Blokland** **President of the Acoustical Society of the Netherlands**	
	Perception and Health Effects - 1	
11:05	Partial masking and the perception of wind turbine noise in ambient sounds.....147	Johansson (Bolin)
11:25	Wind Turbine Noise Dose Response - Comparison of Recent Studies 158	Old
11:45	Impact of noise from suburban wind turbines on human well-being 166	Qu
12:05	Discussion	
12:30	Lunch	
	Aspects of Source Noise - 1	**Jean Tourret**
13:30	Sound power level measurements 3.0.....181	Eilders
13:50	Analysis of sound emission by using amplitude modulation components of wind turbine noise.....191	Okada
14:10	Modelling activities in wind turbine noise aeroacoustics at DTU Wind Energy.....201	Shen
14:30	Long-term experimental campaign on an operating wind turbine for trailing edge serrations verification.....215	Lauret-Ducosson
14:50	Discussion	
15:20	Break	
	Amplitude Modulation - 1	
15:40	Development of an approach to controlling the impact of amplitude modulation in wind turbine noise: exposure-response research, application and implementation.....228	Perkins
16:00	An investigation into correlation between strong wind turbine amplitude modulation and environmental conditions.....241	Halstead
16:20	Extended simulations of wind noise contamination of amplitude modulation ratings.....259	von Hunerbein
16:40	Discussion	
17:15	Introduction of Posters and Exhibitions. Drinks reception for all delegates.	
18:45	Close	

7th International Conference on Wind Turbine Noise

Wednesday 3rd May 2017		
	Perception and Health Effect - 2	
08:30	Low-frequency noise incl. infrasound from wind turbines and other sources 267	Herrmann
08:50	A 'social review' of wind turbine noise.....277	van den Berg
09:10	Wind turbine noise – An overview of current knowledge and perspectives 282	Bauerdorff
09:30	Discussion	
09:55	Break	
	Perception and Health Effect - 3	
10:15	Why do some people believe that they are "made ill" by wind turbine noise 290	Leventhall
10:35	Human response to wind turbine noise: infrasound and amplitude modulation.....300	Feist
10:55	Wind turbines in hilly terrain – Response of residents to sound disturbance related to sound and meteorological measurements.....313	Sjoblom
11:15	Discussion	
	Guest Speaker - 1	
11:40	TBC	
12:40	Lunch	
	Shadow Flicker	
13:40	International Legislation and Regulations for Wind Turbine Shadow Flicker Impact.....317	Koppen
14:00	The development and limits of the German shadow flicker guidelines 329	Ritter
14:20	The visual effects of wind turbines in Japan.....340	Yano
14:40	Presenting insights from shadow flicker compliance monitoring...350	Longbottom (Mackay)
15:00	Discussion	
15:30	Break	
	International Perspectives	
15:50	Addressing a management strategy of Wind Farms Noise Control in Chile 361	Parra
16:10	Comparison of Measured and Modelled Wind Turbine Noise in Indian Terrain.....373	Arivukkodi
16:30	A new characterization of wind turbine noise from Life Cycle Assessment 395	Rivarola
16:50	Discussion	
17:15	Workshop on Poster Subjects. Sound propagation and Background Noise	
18:45	Close	

7th International Conference on Wind Turbine Noise

Thursday 4th May 2017 - Hall 1		
	Assessment and Compliance - 1	
08:30	Experience of reviewing wind farm noise assessments for Scottish local authorities and the implementation of the IOA Good Practice Guide.405	Summers
08:50	Evaluation of wind farm noise in Switzerland – Comparison between measurement and modeling.....415	Desarnaulds
09:10	A case study of how to involve impacted neighbours in measuring and characterizing windfarm noise.....425	Vagene
09:30	A Rigorous Method of Addressing Wind Turbine Noise.....433	Palmer
09:50	Discussion	
10:20	Break	
	Assessment and Compliance - 2	
10:40	Assessment of WTN by separating residual noise without the farm shutdown: Validation of the Italian procedure.....460	Fredianelli (Carpita)
11:00	Characterizing the acoustic noise from wind turbines by using the divergence of the sound pressure in the ambient.....472	Buzduga
11:20	Using long term monitoring for noise assessment of wind farms...482	de Beer
11:40	The Challenges and Benefits of Long-Term Noise Monitoring of Wind Farm Sites.....493	Brush
12:00	Discussion	
12:30	Lunch	
	Low Frequency and Tones	
13:30	Tonal noise mitigation on wind turbines.....505	Marmo
13:50	Origin, Transfer and Reduction of Structure-Borne Noise in Wind Turbines 516	Schneider (Hanus)
14:10	Low-frequency micro-seismic radiation by wind turbines and it's interaction with acoustic noise emission.....525	Gortsas (Kudella)
14:30	Subjective experiments on the perception of tonal component(s) contained in wind turbine noise.....540	Yokoyama
14:50	Discussion	
15:20	Break	
	Infrasound	
15:40	Propagation through a turbulent atmosphere makes blade passage harmonics audible.....551	Richarz
16:00	The assessment of hearing thresholds in the presence of infrasound 561	Zajamsek
16:20	Measurement Techniques for determining Wind Turbine Infrasound Penetration into Homes.....568	Metelka
16:40	Discussion	
17:15	Seminar by TU Delft and GRAS: Aeroacoustic investigation of noise sources: measurements and computations	
18:45	Close	

7th International Conference on Wind Turbine Noise

Thursday 4th May 2017 - Hall 2		
	Validation of TE Modelling	
08:30	Computational Aeroacoustics of Small Vertical Axis Wind Turbines by Applying a Hybrid Approach.....606	Weber
08:50	High Fidelity Airfoil Trailing Edge Noise Predictions via Lattice-Boltzmann Simulations.....615	Manegar
09:10	Small Horizontal Axis Wind Turbine: Aeroacoustic and Aerodynamic Optimization of Airfoil and Blade.....629	Volkmer
09:30	Predicted and Measured Trailing-Edge Noise Emission for a 2.3 MW Wind Turbine.....651	Hornung
09:50	Discussion	
10:20	Break	
	Combined Modelling	
10:40	Coupled wind turbine noise generation and propagation model: A numerical study.....663	Bertagnolio
11:00	A Comprehensive Hamiltonian Ray Tracing Technique for Wind Turbine Noise Propagation under Arbitrary Weather Conditions.....671	McBride (Burdisso)
11:20	Investigation of Amplitude Modulation Noise with a Fully Coupled Noise Source and Propagation Model.....683	Barlas
11:40	Aeroacoustic simulation of multiple wind turbine source interaction 694	Robin
12:00	Discussion	
12:30	Lunch	
	Trailing Edge Serrations	
13:30	Acoustic measurements of a DU96-W-180 airfoil with flow-misaligned serrations at a high Reynolds number in a closed-section wind tunnel 707	Merino-Martinez
13:50	Trailing edge serrations - effect of their flap angle on flow and acoustics 720	Arce Leon
14:10	An Experimental Parametric Study of Airfoil Trailing Edge Serrations 736	Carolus
14:30	Wind Turbine Rotor Noise Prediction and Reduction for Low Noise Rotor Design.....750	Kamruzzaman
14:50	Discussion	
15:20	Break	
	Source Localisation and Identification	
15:40	Modelling and localizing low frequency noise of a wind turbine using an array of acoustic vector sensors.....764	Fernandez Comesana
16:00	Use of the Acoustic Camera to accurately localise wind turbine noise sources and determine their Doppler shift.....774	Bradley
16:20	Wind turbine sound prediction: Modelling and case study on the effect of blade elasticity.....782	Schorle
16:40	Discussion	
17:15	**In Hall 1** - Seminar by TU Delft and GRAS: Aeroacoustic investigation of noise sources: measurements and computations	
18:45	Close	

7th International Conference on Wind Turbine Noise

Friday 5th May 2017		
	Amplitude Modulation - 2	
08:30	The occurrence of nocturnal wind farm rumbling noise.....801	Hansen
08:50	Pre-construction Site Prediction Tool for Wind Farm AM - Do We Now Know Enough?.....812	Birchby
09:10	An Investigation into Short-Term Fluctuations in Amplitude Modulation of Wind Turbine Noise. Preliminary Results.....819	Bonsma
09:30	Discussion	
09:55	Break	
	Amplitude Modulation - 3	
10:15	Application of the UK IOA Method for Rating Amplitude Modulation 841	Coles
10:35	Comparison of the IOA method and Japanese F-S method for quantitative assessment of amplitude modulation of wind turbine noise. A study based on the field measurement results in Japan.....854	Fukushima
10:55	Putting the IOA preferred AM assessment method and the proposed penalty scheme into practice - an outlook for future developments of wind farms in the UK.....864	Lowe
11:15	Discussion	
	Guest Speaker - 2	
11:40	Willem Roose, Professional Miller, will talk about the history of windmills internationally, attempts at generating electricity and his own experience of environmental noise.	
12:20	Lunch	
	Aspects of Source Noise - 2	
13:20	An investigation into the effect of wind shear on the sound emission of wind turbines.....881	Ashtiani
13:40	Variations in measured noise emission of wind turbines due to local circumstances.....891	van der Maarl
14:00	Vertical directivity observations based on statistics of low frequency tonal components measured at downwind and upwind locations.....902	Falourd
14:20	Wind turbine noise measurements in controlled conditions.....913	Boorsma
14:40	Discussion	
15:05	Closing	
15:10	Conference Ends	

7th International Conference
on
Wind Turbine Noise
Rotterdam – 2nd to 5th May 2017

Background Noise Variability Relative to Wind Direction, Temperature, and Other Factors

Tricia Pellerin Tetra Tech
Tricia.Pellerin@tetratech.com

Summary

Characterization of the ambient acoustic environment is often a critical component in the wind energy permitting process. Collection of ambient data may be required during both the pre-construction or post-construction phases to establish a baseline for determining wind turbine sound contribution at different points of reception. The challenge with establishing that baseline is that ambient sound levels continuously fluctuate, affected by multiple factors including existing sound sources, human activity, and meteorological parameters like wind speed and direction. This continuous fluctuation makes both definition and reproducibility of ambient conditions difficult. This paper examines meteorological data collected during surveys at four wind energy facilities throughout the United States. Meteorological data were collected at different heights above ground level using a combination of multiparameter weather sensor stations deployed during survey periods as well as other meteorological towers in proximity to the wind energy facility sites. Relationships between ambient sound level data and various meteorological parameters including wind speed, direction, temperature and relative humidity are reviewed and analyzed.

1. Introduction

Collection of ambient sound data to support the wind energy permitting process can be prompted by a number of different drivers. Most commonly the driver is the applicable noise regulations; however, precedent recently set by other nearby facilities may also influence that decision. Alternatively, the developer may initiate collection of ambient sound data for their internal records and for future comparison with their facilities operational sound levels.

Noise requirements and regulations vary widely by jurisdiction. Standards and criteria are typically either source-based or receiver-based, but for the purposes of environmental permitting compliance, they're typically focused at a noise sensitive area such as a residence, hospital, school, or other location where lower noise levels are of importance to its use. While noise regulations can be presented in a qualitative manner they also often prescribe numerical decibel limits, which can comprise an absolute limit or relative limit or a combination of the two. Absolute limits are specific decibel sound levels that are not to be exceeded at the point of compliance and can be given on an A-weighted decibel (dBA) broadband basis or by unweighted octave band frequency sound levels. Relative limits are those that are set as an allowable incremental increase in decibel level above existing sound levels, which means that existing sound levels would typically need to be established by way of a baseline sound survey.

Generally one of the objectives of a baseline sound survey is to perform it in such a way that it is reproducible; however, ambient sound is inherently variable, making reproducibility challenging.

Many factors including existing sound sources, human activity, and meteorological parameters like wind speed and direction affect the measured ambient sound level. This paper reviews those meteorological parameters and reports an investigation between meteorological conditions and measured ambient data collected at four wind energy facilities throughout the United States

2. Measurement Methodology and Instrumentation

Four wind energy facility sites were selected to represent different climatic regions within the United States including locations in the Midwest (Ohio and Michigan), northwest (Oregon) and southwest (Hawaii). During pre-construction, a combination of concurrent acoustic and meteorological measurements were collected at in proximity to residential receptors situated at typical setback distances used in siting wind turbines (e.g., 1,500 feet or 457 meters). Measurements were conducted in conditions conducive to accurate data collection, avoiding periods of heavy precipitation and elevated winds.

Sound measurements were collected using a Larson Davis 831 real-time sound level analyzer equipped with a PCB model 377B02 ½-inch precision condenser microphone. This instrument has an operating range of 5 decibels (dB) to 140 dB, and an overall frequency range of 8 to 20,000 hertz (Hz) and meets or exceeds all requirements set forth in the American National Standards Institute (ANSI) standards for Type 1 sound level meters for quality and accuracy (precision). The range level error for the Larson Davis 831 is ±0.1 dB relative to the reference range.

The monitoring stations were designed for service as a long-term environmental sound level data-logger measuring devices. Each sound level analyzer used was enclosed in a weatherproof case and equipped with a self-contained microphone tripod. The microphone and windscreen were tripod-mounted at an approximate height of 1.5 to 1.7 meters (4.9 to 5.6 feet) above grade. All sound level analyzer microphones were protected from wind-induced pseudonoise by a 180-millimeter (7-inch) diameter foam windscreen made of specially prepared open-pored polyurethane.

Weather data were collected using a combination of on-site meteorological towers operated by the wind developers and Vaisala WXT520 portable weather transmitters. The Vaisala unit monitors wind speed, wind direction, barometric pressure, temperature and humidity, total rainfall, intensity of rainfall, and duration of rainfall. The WXT520's wind sensor consists of three equally spaced transducers that produce ultrasonic signals. Wind speed and direction are determined by measuring the time it takes for the ultrasonic signal of one transducer to travel to the other transducers. Its stated level of accuracy temperature is ± 0.3˚C (@ 20˚C), 3˚ for wind direction and ± 0.3 m/s or ± 3% (whichever is greater) for wind speeds between 0 and 35 m/s and ± 5% for wind speeds between 36 and 60 m/s.

3. Results

Monitoring at the four energy facilities was conducted at different time periods. At each site, received sound level data were logged continuously for approximately two to three weeks concurrent with meteorological parameters including wind speed, wind direction, temperature, relative humidity and barometric pressure. Data regression analyses were then performed on the sound measurement results to review the observed relationships between ambient sound level and the given meteorological parameters.

3.1 Wind Speed

The relationship between wind speed and ambient sound is well understood. Typically wind speed increased with height and is represented by the following equation:

$$v_z = v_g \cdot (\frac{z}{z_g})^{\frac{1}{\alpha}}, 0 < z < z_g$$

where:

v_z is the wind speed at height z,
v_g is the gradient wind at gradient height z_g,
α is the exponential exponent.

This results in a logarithmic wind profile of speed versus height. The relationship between ambient sound level and wind speed can be examined for wind at any height and generally as wind speed increases so does ambient sound level. At lower wind speeds there is more scatter due to wind-related sound levels being lower and other existing sound sources being more dominant (Lightstone et al, 2010). Figure 1 demonstrates the trend between wind speed and ambient sound level at Site 1 (Michigan).

3.2 Wind Direction

While wind speed changes with height, wind direction generally does not; however, ambient sound level will be influenced by prevailing wind direction depending on the measurement location's position relative to the wind. The International Organization for Standardization (ISO) standard 9613-2 "Attenuation of Sound during Propagation Outdoors" is commonly used in the assessment of wind turbine sound and describes a sound level prediction method under conditions favourable to propagation including the assumption that the receptor or measurement location is downwind of the sound source of interest.

Figure 2 provides an example of the correlation between wind direction and ambient sound levels. The data plotted in Figure 2 reveals that at Site 1 (Michigan) the wind primarily blows from the south and south-westerly direction. This contrasts with Figure 3, which shows a wind more frequently blowing from the north and east directions at Site 2 (Ohio).

3.3 Temperature

Temperature affects sound propagation through absorption and by influencing the speed of sound. Higher temperatures produce a higher speed of sound. The relationship between the temperature of air and the sound speed can be expressed as:

$$V_t = V_0 \sqrt{1 + \frac{t}{273}}$$

where:

t is the Celsius temperature,
Vt is the sound speed at temperature *t*,
V0 is the sound speed at 0°C, which is 331.5 m/s.

Figures 4, 5 and 6 show temperature versus ambient sound data regression analyses for three sites within the midwest, northeast and southwest regions of the United States. Results demonstrate that there is no clear relationship between temperature and ambient sound level, which can most likely attributed to other contributing factors such as relative humidity, which also affects air absorption of acoustic energy. For instance, Figures 4 and 6 show a slight increase in sound level with temperature increase, which could be linked to a decrease in relative humidity.

3.4 Relative Humidity

Typical sound attenuation at a standard temperature of 20°C and a relative humidity of 70% are shown in Table 1.

Frequency (Hz)	**31.5**	**63**	**125**	**250**	**500**	**1,000**	**2,000**	**4,000**	**8,000**
Attenuation (dB/km)	0.0	0.1	0.3	1.0	2.7	5.4	9.7	23.2	74.8

Table 1. Sound Attenuation Rate at T = 20°C and RH = 70 %

Very little attenuation is found for low values of relative humidity and temperature. Monthly and diurnal variation in relation to humidity and temperature can produce sizeable fluctuations in absorptions levels (Chen et al, 2004). Often relative humidity reaches its maximum soon after sunrise and its minimum in the afternoon when temperatures are typically highest. Figure 7 (Site 2, Ohio) demonstrates a slight increase in ambient sound level as relative humidity increases. Results reviewed from other sites showed similar results.

3.5 Barometric Pressure

When the barometric pressure changes from 90 to 110 kPa, the sound speed in air varies from the International Electrotechnical Commission (IEC) standard conditions of 23°C, 101.325 kPa and 50 % RH, by approximately ± 50 ppm. This relatively small variation in sound speed due to barometric pressure changes can be ignored in most acoustical measurements (Wong, 1986). The lack of effect barometric pressure has on ambient sound level is displayed in Figure 8 (Site 2, Ohio).

4. Conclusions

Ambient sound and sound generated by operating wind turbines are affected by metrological conditions. Wind speed if often the primary meteorological parameter considered in the analysis of wind turbine sound impacts; however, other factors such as wind direction, temperature and relative humidity must also be taken into account, but will have a lesser effect on sound propagation.

The results of this study are preliminary but still indicate that review of facility site meteorological may be useful prior to planning an ambient sound monitoring program. Planning could include conducting measurements during a specific season to capture temperature and relative humidity characteristics that could favour sound propagation or positioning certain ambient sound monitoring stations in locations where they will experience a high frequency of downwind conditions.

Similarly, depending on the wind energy facility site, meteorological conditions may also require adjustment within the acoustic modelling analysis. Most often standard engineering assumptions are sufficient in that regard but, in the event, the facility site has year-round consistently high temperatures and relative humidity, increasing those values may better predict actual impacts especially at greater distances from sound sources.

References

Chen, Z. et al (2004) Atmospheric Sound Propagation Considerations for the Birdstrike Project, Montana State University.

Lightstone, A.D. et al (2010) Simultaneous Measurements of Ambient Sound Levels and Wind Speeds, Canadian Acoustics, 23 - Vol. 38 No. 4.

Wong, G.S.K. (1986) Variation of the Characteristic Impedance in Air with Environmental Conditions, The Journal of the Acoustical Society of America 76, 555–559.

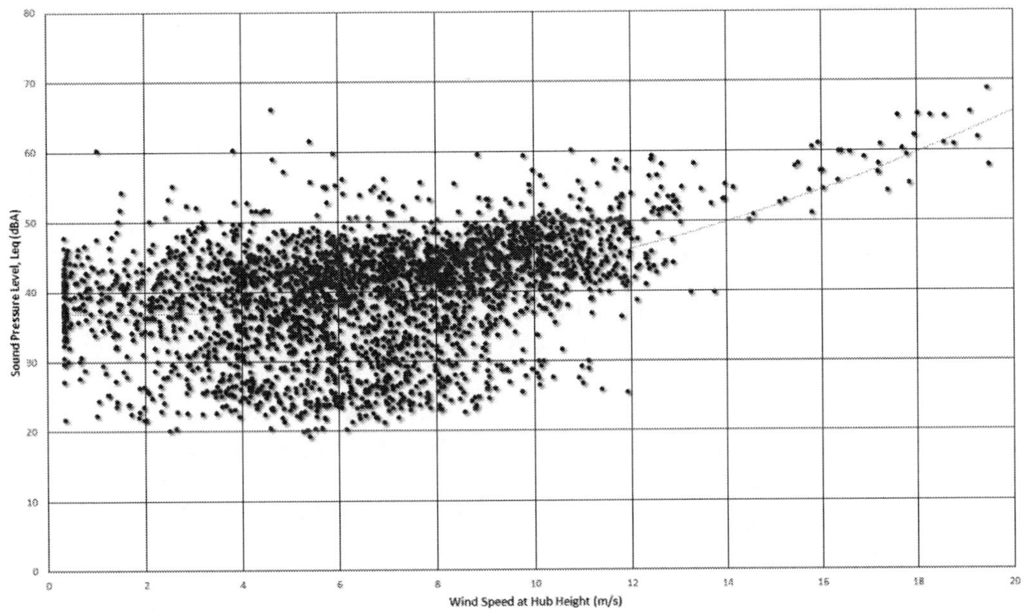

Fig. 1 Ambient Sound Pressure Level versus Wind Speed Data Regression Analysis (Site 1, Michigan)

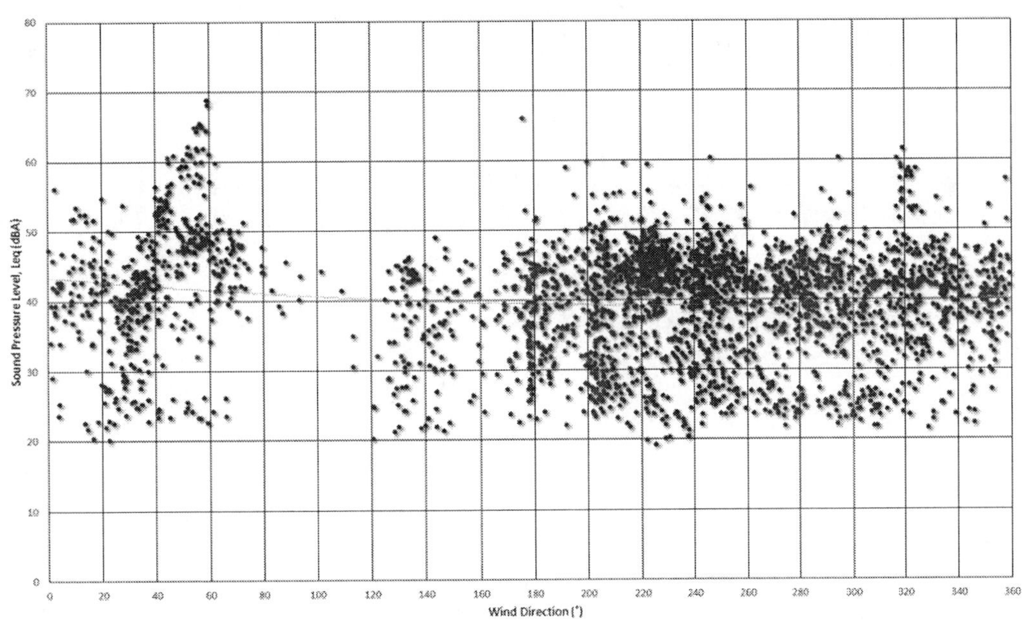

Fig. 2 Ambient Sound Pressure Level versus Wind Direction Data Regression Analysis (Site 1, Michigan)

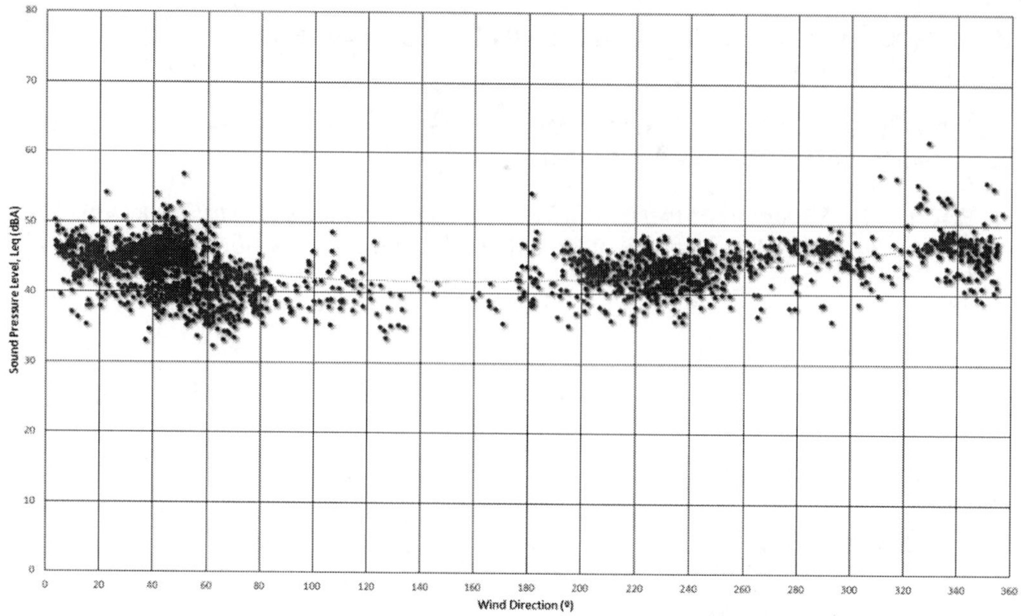

Fig. 3 Ambient Sound Pressure Level versus Wind Direction Data Regression Analysis (Site 2, Ohio)

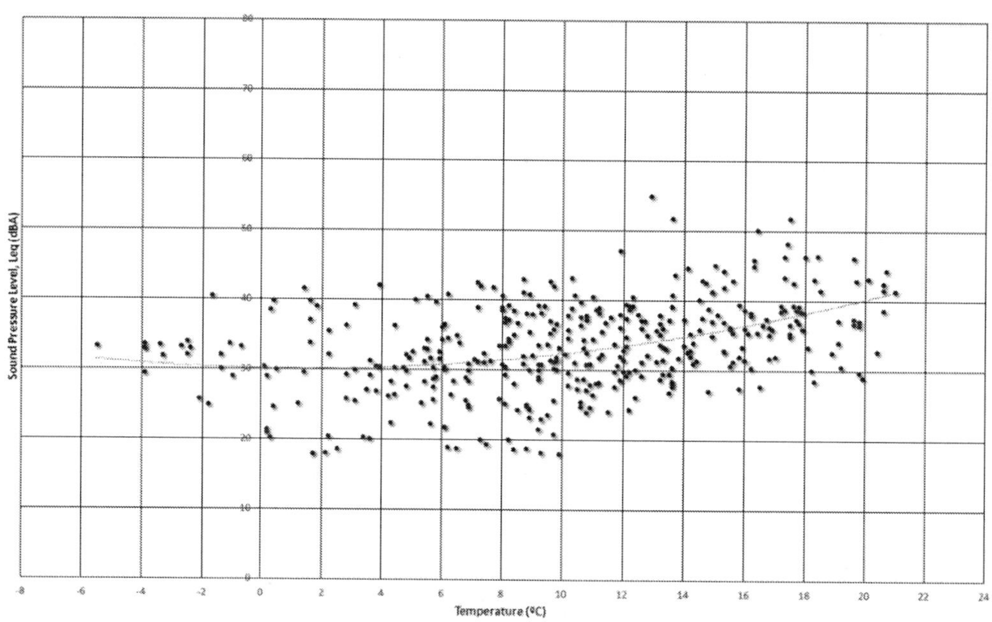

Fig. 4 Ambient Sound Pressure Level versus Temperature Data Regression Analysis (Site 3, Oregon)

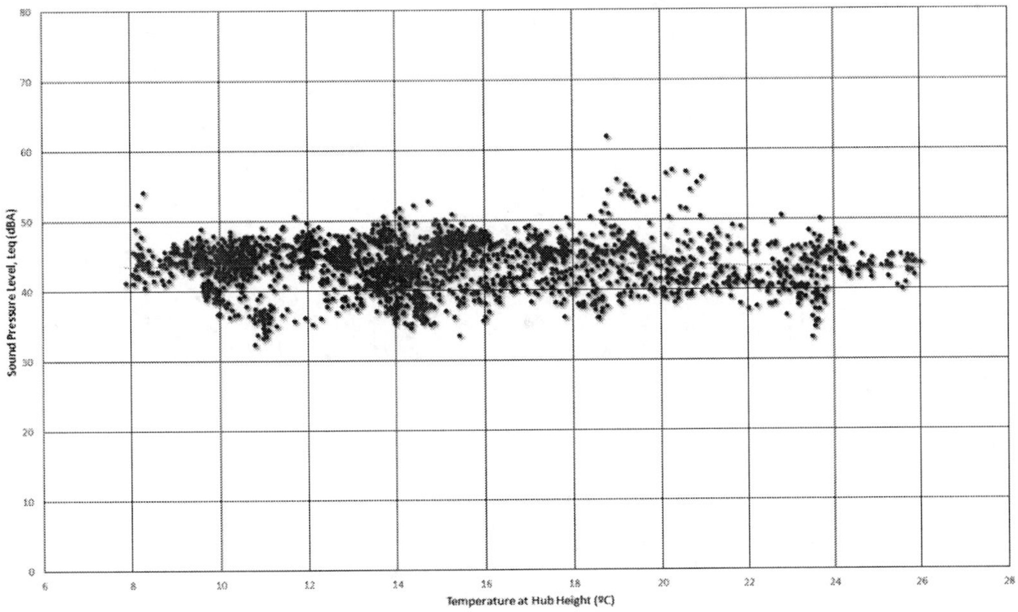

Fig. 5 Ambient Sound Pressure Level versus Temperature Data Regression Analysis (Site 2, Ohio)

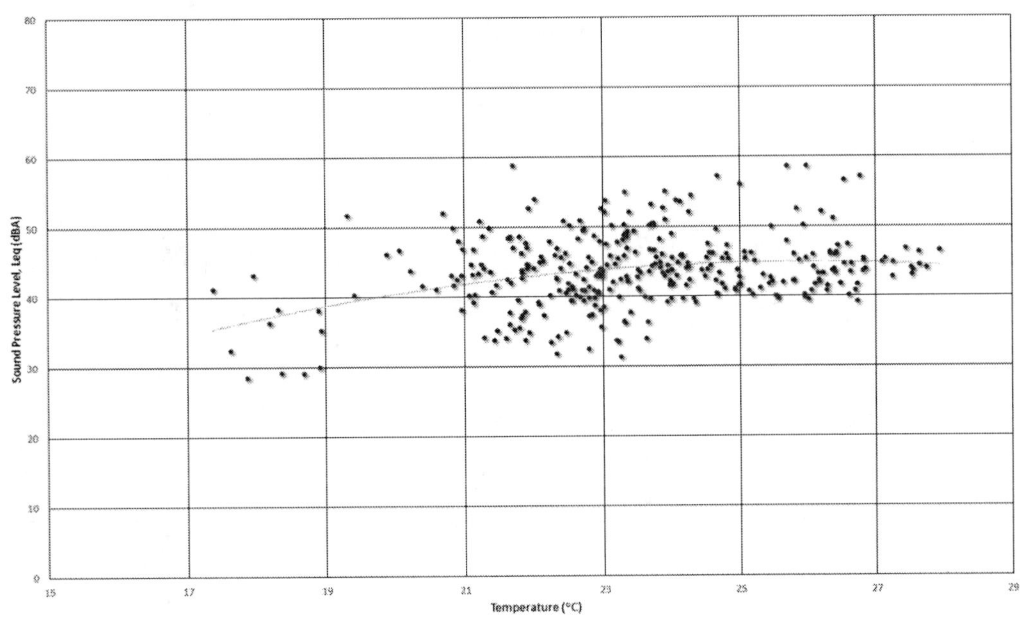

Fig. 6 Ambient Sound Pressure Level versus Temperature Data Regression Analysis (Site 4, Hawaii)

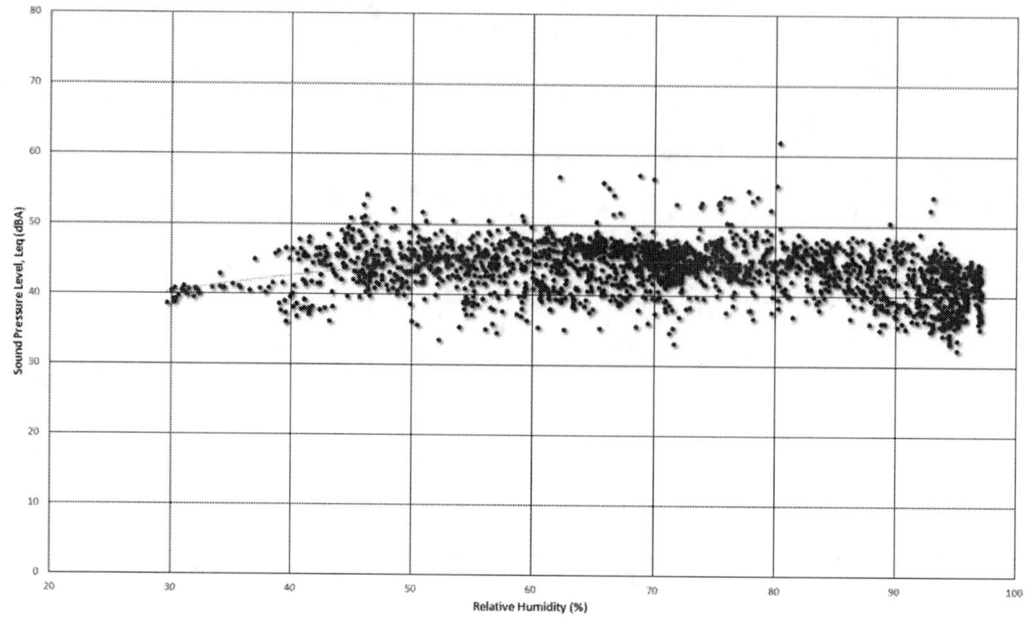

Fig. 7 Ambient Sound Pressure Level versus Relative Humidity Data Regression Analysis (Site 2, Ohio)

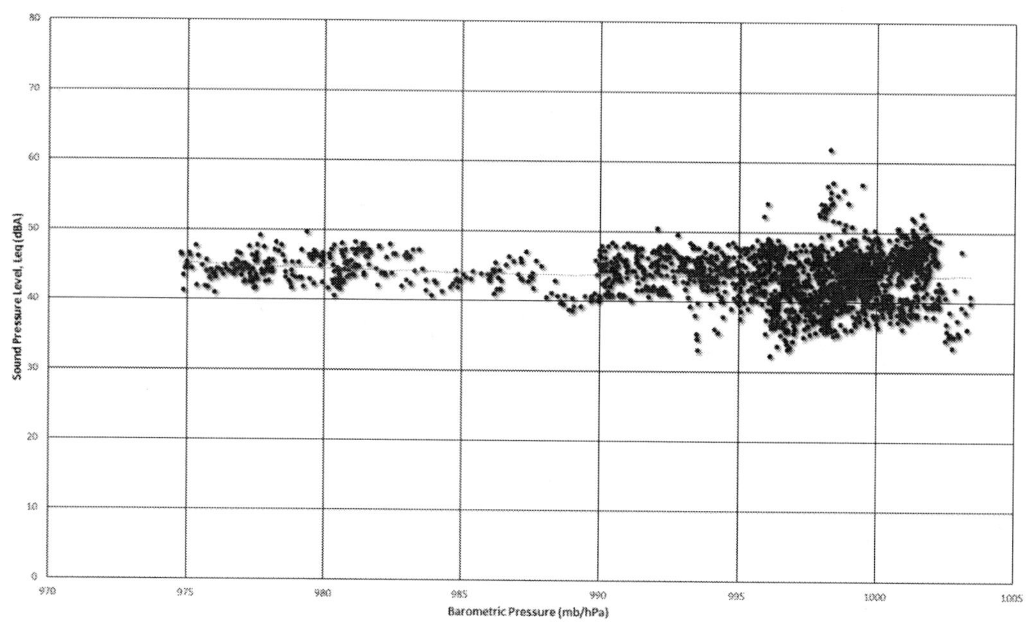

Fig. 8 Ambient Sound Pressure Level versus Barometric Pressure Data Regression Analysis (Site 2, Ohio)

7th International Conference

on

Wind Turbine Noise

Rotterdam – 2nd to 5th May 2017

Efficient tools for assessing the emergence, audibility and masking potential of wind turbine noise by background noise

Dominique Bollinger, HEIG-VD, Yverdon-les-Bains, CH: dominique.bollinger@heig-vd.ch

Xavier Falourd, PRONA SA, Yverdon-les-Bains, CH: x.falourd@prona.ch

Lukas Rohr, PRONA SA, Yverdon-les-Bains, CH: l.rohr@prona.ch

Romain Feuz, PRONA SA, Yverdon-les-Bains, CH: r.feuz@prona.ch

Summary

A complete four seasonal collection of background noise on four planned wind turbine sites combined with the latest knowledge of wind turbine noise emission led to develop efficient tools and site datasheets to assess the masking potential of wind turbine noise by the existing and specific background noise.

For each site, two sonometer and a dosimeter allowed to define and determine sound specificities of each site. Frequency and level analysis of the background noise, combined with statistical correlation analysis and level prediction on each site are performed on the collected data.

Statistical datasheets are established gathering wind directions, background noise and wind speed correlation, 1/3 octaves wind turbine acoustic power related to the wind speed and a superposition of both sonometer levels and wind turbine noise on each 1/3 octave bandwidth. Additional to this, a metrological datasheet completes this characterization of planned sites by displaying wind speed, background noise and temperature distributions and time evolution of wind speeds and background levels for each 1/3 octave bandwidth. The predicted wind turbine noise is then added on these graphics to assess the potential of masking it by the background noise or the risk of emergence.

These tools help planners to assess the risk of emergence and masking potential of wind turbine noise to give a very good way of anticipating possible conflicts and impacts. This case study showed that a reliable prediction of a masking effect cannot be done without a long time onsite measurement of the background noise.

1. Introduction (reprise article WTN 2015)

1.1. A brief history of energy in Switzerland [1.1]

Following the Fukushima accident, the Swiss energy program has been reviewed with a view to reduce and close nuclear plants in Switzerland in the horizon 2050 ([1.5]). This strategy aims to reach a long term energetic supply, integrating first the use of existing resources such as hydraulic energy, which is wide spread in Switzerland, and renewable energy. The second phase would aim to replace the existing encouragement system by an incentive one.

In 2013, the yearly native energy production had a part of 60% of renewable energy, mainly produced by hydraulic energy dams. The energy supplied by solar energy, biomass, biogas waste and wind turbine only reached 3 % of the national production, against 37% for the nuclear plants, the rest being produced by classical thermal plants and long-distance heating. [1.1]

The energy potential for renewable energy has been estimated to 24,2 TWh at the horizon 2050, whose:

- 11,1 TWh for photovoltaic solar energy;
- 4,3 TWh for wind turbine energy;
- 1,2 TWh for biomass energy;
- 4,4 TWh for geothermal energy;
- 3,2 TWh for waste water, incinerators and biogas energy.

1.2. Wind turbines in Switzerland:

Nowadays, Switzerland has 55 wind turbines installed on 32 sites, providing a total of 93,9 GWh in 2014 for a total installed power of 60,3 MW, mainly consisting of individual wind turbines and few parks :

Site types	Nb sites	Wind turb.	2014 prod.
Individual wind turbines	26	26	19,1 GWh
Wind turbine park with 2 wind turbines	3	6	6,0 GWh
Wind turbine park with 3 wind turbines	1	3	12,9 GWh
Wind turbine park with 4 wind turbines	1	4	5,4 GWh
Wind turbine park with 16 wind turbines	1	16	50,5 GWh
Total	**32**	**55**	**93,9 GWh**

Table 1: Wind turbine sites in Switzerland.

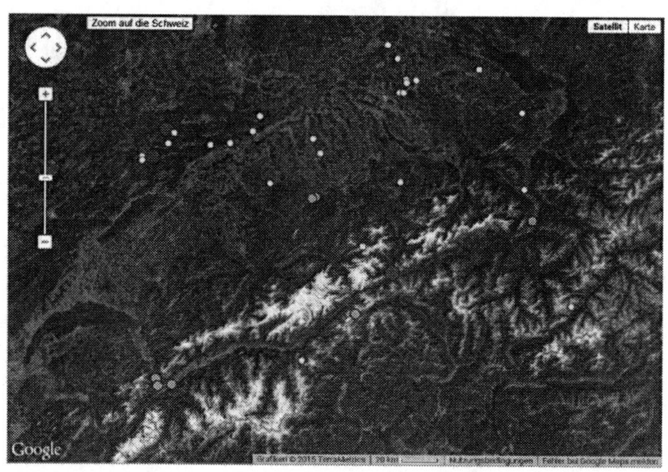

Legende: ○ kleine Windkraftanlage < 300 kW ● grosse Windkraftanlage > 300 kW ● Windpark ○ ehemalige Anlage

Figure 1 : Installed wind turbines in Switzerland [1.4]

The Swiss 2004 wind turbine energy concept gathers 128 potential planning sites with 12 prioritary sites. The average wind speed at 70m over ground level on these potential sites is around 5,6 m/s (min: 4,5 m/s; max: 8,4 m/s). These sites were identified regarding the wind turbine potential, as well as land planning and environmental exlusion criteria.

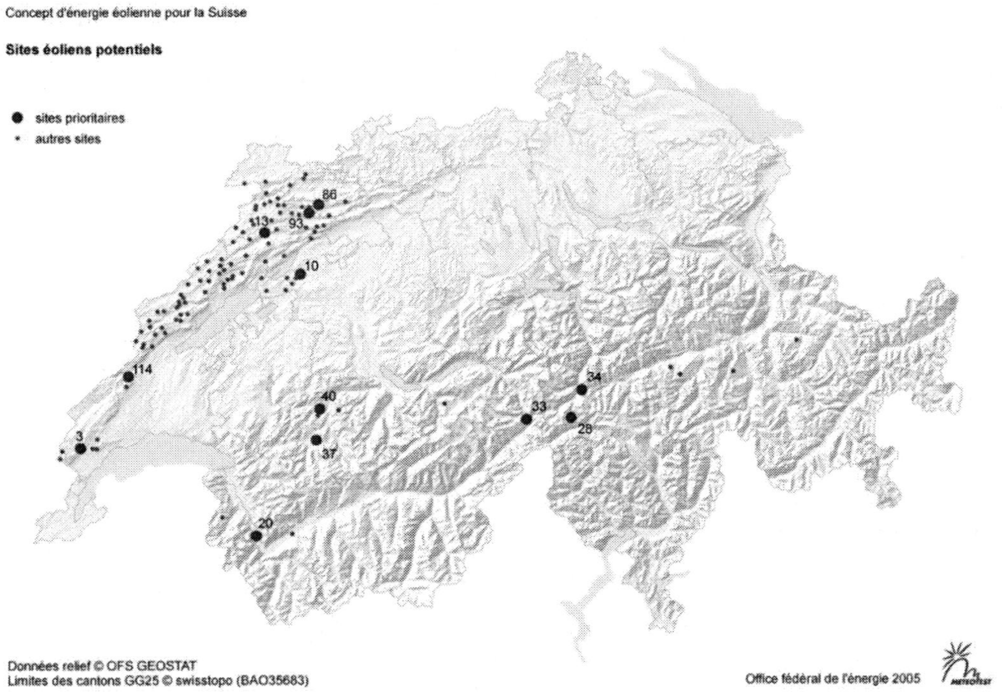

Figure 2 : Potential wind turbine sites [1.4]

1.3. Constraint, limits and the noise problem within legislation and population:

Facing many constraints as well on land planning as on environmental and social acceptance, the wind turbine planning is logically a matter of national planning in a Swiss context which has a strong decentralized planning philosophy. The national planning program and potential sites identification concept supports this fact. To be considered of general interest, hence have a chance of being realized, any wind turbine project needs to ensure in between 5 to 20 MW.

The Swiss democratic system opens possibilities for the population to react through legal procedure and form opposition to the project. Hence, the actual legislation on noise protection follows the precautionary and prevention principles and assimilates wind turbines to industrial noise [1.2]. That means wind turbine immission equivalent level (Leq) should be adapted by a temporal correction and three level correction factors. The main correction factor (K3 = 4) takes into account the impulsiveness components audibility at the immission place. This aspect is not only defined by the purely acoustic impulsiveness such as in typical industrial settings but integrates also well perceptible rhythmic-based discomforts (e.g. noise amplitude modulation)[1.3]. Predicting the noise impact of a whole wind turbine project is nowadays one of the most challenging task for an acoustician: metrics, experience and some good understandings of physical phenomenon at stake miss.

2. The WTN projects

Three different and complementary projects were led by the research partnership HEIG-VD / PRONA SA to study wind turbine noise in its environment. The three focuses were:

- The study of wind turbine noise acoustic emission, its physical phenomenon studied on both modelization and onsite real time measurement (over the 4 different seasons and several different meteorological situations);
- The study of background noise on potential planned sites for wind turbines and potential impacted immission sites;
- A study and realization of an acoustic wind turbine noise installation template to provide regional, local authorities and the population a realistic and genuine reproduction of wind turbine noise for a specific wind turbine and immission site.

2.1 WTN emissions project

The objective of this project is the experimental measurement of noise from a wind turbine ENERCON E-101 for the purpose of validating the usual acoustic models. The understanding of the noise generated by a wind turbine of this type and the knowledge of the performance of acoustic models should allow a controlled noise modeling of future wind farms.

In order to answer questions about the modeling and measurement methodology of noise from a wind turbine, the study is based on a detailed analysis of acoustic data collected for weather and wind different seasons of the year. The secondary objective is to determine to what extent the current Swiss rules effectively assess this type of noise.

The initial project is composed of two distinct parts:

- A series of seasonal measurements of an existing wind turbine.
- The analysis of experimental data and their comparison with modeling.

2.2 Acoustic template

The aim of this project was to reproduce the immission noise of one specific planned wind turbine on a real outdoor environment, for the real distance of the listening site to the planned wind turbine and to simulate shorter distances by different noise reproduction sequences. This project answered some demands of executives of city council to lead to some sensorial experience based on scientific realistic reproduction.

2.3 Background noise measurement and assessment on wind turbine planned sites

The objectives for this project were:

- Determine the background noise characteristics for different wind turbine planned sites and their main immission sites;
- Analyse background noise levels and characteristics to assess its masking potential;
- Recommend a standard procedure for background noise measurement and evaluation on future wind turbine sites;
- Study and analyse the masking, audibility and emrgence phenomenons of the wind turbine noise regarding measured and analysed background noise characteristics and nature, taking into account the sites' topology.

3. Studied sites

3.1 Setting

Each studied site has been equiped on three different locations / measure points as follows:

- The emission or wind turbine location : One dosimeter in the central position, where the planned wind turbine would be built. It has been placed at 10 m high with a meteorological station;
- The predominant background source location : One DUO sonometer with audio recording in a decentralized position close tot he highest presumed background noise source (road, railway, forest border, industrial zone, etc.) at 2 to 10 m high regarding the background noise type ;
- The immission location : One DUO sonometer with audio recording in the most exposed inhabited building (LUS : Locaux à usage sensible / Sensible use buildings), at window height, in accordance to noise legislation (OPB : ordonnance sur la protection contre le bruit / Noise protection legal prescription [2]).

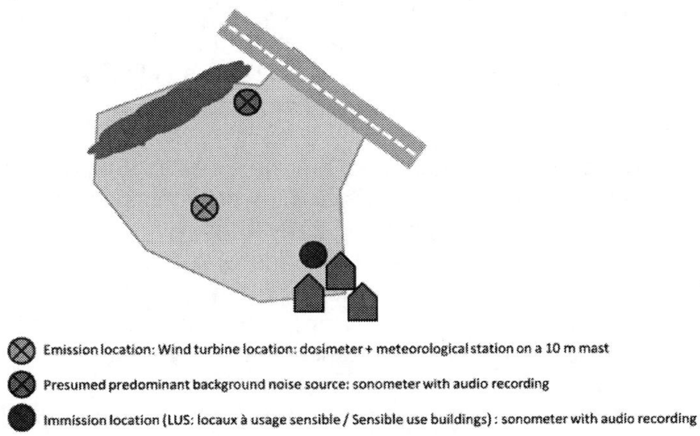

Figure 3: Equipment principles for measured sites

3.2 Focus and objectives

In the wind turbine acoustic phenomenon regarding background noise interaction, three different cases can be identified so we can state on their sonic impact:

- The emergence notion corresponding to a clear and predominant presence of the wind turbine noise over the background noise, thus independently of the respect of legal limit values, and which indicates that the wind turbine noise is clearly identifiable, recognizable and present in the soundscape of considered immission locations;
- The audibility notion corresponding in the possibility to hear the wind turbine noise within the background noise. This wind turbine noise level, if the listeners does not specifically pays attention to it, is mainly lost in the background noise and couldn't be clearly identified or heard. However, when paying attention to it, like traffic noise compared to other installation noise sources f. ex., it would be possible to hear and identify it. It is defined as "audible" but not emerging;
- The masking effect notion corresponding to a complete masking of the wind turbine noise, as well in terms of noise level (dB(A)) as of frequency characteristics. Obviously, this is possible only if the noise phases are synchronous, e.g. happen during the same time and with similar frequency distributions.

These three studied cases base themselves on the studies done in the parallel project which aimed to determine wind turbine noise characteristics, hence giving the input of what the wind turbine noise (emission) on these planned site would be like. These emissions have been used to determine for each measured and analysed site about the probability of masking, audibility or emergence potential of wind turbine noise over the different period of the day and season of the year.

3.3 Some important Swiss legal aspects in noise protection

- The Evaluation Noise Level (Lr: niveau d'évaluation) is used in Switzerland to assess the sonic pollution or the discomfort degree according to the noise legislation.
- A system of three different Exposition Limit Values (VLE : valeurs limites d'exposition) is used in the swiss legislation to assess this "noise pollution":
    - Immission limit values (VLI : valeurs limites d'immission): set to a level so that, according to the state of the science and experiments, immision values lower than these limit values do not significantly annoy population in its well-being. The respect of these values is verified on a regular basis;
    - planning limit values (VP: valeurs de planification): adopted during building zones planning and aiming to ensure a necessary and sufficient protection against new permanent noisy installations (VP < VLI). The respect of these values is a base data for any new project which has to demonstrate that it will be respected;
    - Alarm values (VA: valeurs d'alarme): set to assess the emergency noise sanitation measures to limits and lower immission levels (VA > VLI). These values, as for the VLI, are regularly verified.
- In order to take into account the difference of nuisance level in between day and night, the VLE are generally lower during night (22PM to 6 AM) than day values.
- VLE values depend on the type of noise (road traffic, railroad traffic, plane traffic, etc.) and the relative use zones sensitivity, difined by four different noise sensitivity level:
    - Sensitivity level I (DS I: degré de sensibilité I): use zones needing an increased noise protection, particularly for recreational zone or hospital surroundings;
    - Sensitivity level II (DS II: degré de sensibilité II): use zones with no noisy companies, mostly housing estate, public buildings and public installations;
    - Sensitivity level III (DS III: degré de sensibilité III): use zones with moderately noisy companies, particularly for mixed zones with housing estate combined with craft activities or for agricultural zones;
    - Sensitivity level IV (DS IV: degré de sensibilité IV): use zones containing heavy industries and noisy companies, mostly industrial zones.

Niveau d'évaluation Lr
Jour = 6h-22h; Nuit = 22h-6h

Degré de sensibilité	Valeur de planification		Valeur limite d'immissions		Valeur d'alarme	
	Jour	Nuit	Jour	Nuit	Jour	Nuit
DS I	50	40	55	45	65	60
DS II	55	45	60	50	70	65
DS III	60	50	65	55	70	65
DS IV	65	55	70	60	75	70

Tableau 2 : Emission limit values in Switzerland (VLE) – OPB [2]

4. Measurement and data gathered

4.1 Measurement periods

Four measurement periods have been covered to have a complete view over the four season of the year and their meteorological specificities. These measurement period are shown on table 5. For each site and each measurement period, an hour number of useable measures is stated, knowing that a useable measure, according to the swiss legislation, means a measure without rain and having all acoustic and meteorological data (see table 6).

Each measurement period lasted around two weeks in continuous measures (original sample of 0,125 s). Globally for all sites and all four periods, the hour number of useable measures is 8823 hours.

The following tables base themselves on both Background noise predominant source measure point and Immission point (LUS).

Site	1		2		3		4	
Sonometer	A	B	C	D	E	F	G	H
Period 1	from 4/15 to 4/26	from 4/15 to 4/28			from 6/5 to 6/18	from 6/6 to 6/18	from 6/5 to 6/18	from 6/5 to 6/17
Period 2	from 7/2 to 7/16				from 7/18 to 7/31			
Period 3	from 10/16 to 10/28	from 10/14 to 10/26	from 10/29 to 11/11		from 10/14 to 10/28		from 10/29 to 11/11	
Period 4	from 1/9 to 1/20		from 1/21 to 2/3	from 1/21 to 2/4	from 1/9 to 1/20	from 1/8 to 1/20	from 1/21 to 2/4	

Table 3 : Measurement periods for each site and location

Site	1		2		3		4	
Sonometer	A	B	C	D	E	F	G	H
Period 1	227	274	203	285	308	289	280	255
Period 2	182	182	180	180	306	304	289	289
Period 3	276	267	301	309	288	269	302	307
Period 4	151	150	309	317	150	150	330	314

Table 4 : Hours of usable measures for each site, location and period

4.2 Measured data

Measured data were the following types :

- Sonometers :
  - Leq
  - Lmin/Lmax
  - Partial Leq on 36 1/3 octaves (6.3Hz to 20 kHz)
  - Minimum sample time interval : 125 ms
- Dosimeters :
  - Leq
  - Lmin/Lmax
  - Minimum sample time interval: 1 minute
- Meteorological stations :
  - Wind speed [km/h]
  - Wind direction [°]
  - Temperature [°C]
  - Humidity [%]
  - Pressure
  - Minimum sample time interval : 1 minute

4.3 Analyzed data

A double analysis has been done:

- First based in accordance to swiss noise legislation (OPB), aiming to check the equivalent background noise level compared to legal limit values (aggregated noise level Leq dB(A));
- Secondly based on a frequency acoustic analysis, much more relevant in the case of wind turbine noise. The masking potential of background noise has been then much more specifically and accurately studied.

The following analysis have been done:

- Global and frequency 1/3 octaves Leq [dBA] of the background noise for each site / location;
- Statistical sonometric levels and audio data analysis;
- Correlation analysis to evaluate the possibility of predicting background noise of a given site based on measured parameters such as meteorological values;
- Evaluation level Lr predictions basing ourselves on the parallel wind turbine emission study, particularly studying the relevance of correction factors assessment taking into account the background noise levels and characteristics;
- Building statistical sheets allowing to compare background noise charactersitics and levels in between the "immission site" and the "predominant background noise source", assessing also the masking, audibility or emergence case for each site;
- Building metrological sheets allowing a detailed representation of the background noise and its Lr levels in order to determine eventual frequency and periodic masking potential;
- Making recommendations for studied sites and assessment of correction factors as well as masking, audibility and emergence potentials on immission locations.

5. Background noise levels

5.1 Objective of the analysis

The background noise measure aims to characterize it for each specific site, particularly on its LAeq level but also on its frequency and partial level distributions allowing to evaluate the masking potential of wind turbine noise by the sites' background noise.

We keep in focus here the LAeq level for each site, according to swiss noise legislation. The frequency analysis will be then taken into account later on with the overlay of the background noise for each site in the statistical and metrological sheets.

5.2 Principle

Although all recordings have been done continuously on the four periods for noise levels Leq, globally and per 1/3 octaves to analyse frequency characteristics, we show here only global Leq in order to determine how far from the legal limit values the background level is.

5.3 Results reading and interpretation keys

The results are simply represented by level values Leq(A) for each site, divided in "day" and "night" values. These values allow us to check what are the actual background noise levels compared to the legal limit values.

5.4 Results

The following table shows Leq(A) levels for each site, each of the 4 periods analysed for day and night periods. These values aggregate all measures for the corresponding periods and sites, thus giving a very reliable Leq(A) evaluation according to the swiss noise legislation (OPB [1]).

Mean background noise per day/night perdio for each season

								Period 1		Period 2		Period 3		Period 4	
Location points	LUS	Distance to wind turbine [m]	DS	VP day	VP night	VLI day	VLI night	Leq, day [dB(A)]	Leq, night [dB(A)]	Leq, day [dB(A)]	Leq, night [dB(A)]	Leq, day [dB(A)]	Leq, night [dB(A)]	Leq, day [dB(A)]	Leq, night [dB(A)]
A		315	III	60	50	65	55	46,8	44,4	48,9	45,6	47,1	40,1	43,9	35,6
B	x	561	III	60	50	65	55	49,8	40,4	49,9	44,7	48,1	40,5	44,7	37,3
C	x	576	III	60	50	65	55	51,9	48,5	51,8	57,6	52,6	46,2	48,0	42,6
D		551	III	60	50	65	55	57,6	53,8	58,3	54,3	58,6	53,8	55,0	49,2
E		812	III	60	50	65	55	48,4	45,8	56,0	40,8	49,0	42,6	42,9	37,0
F	x	410	III	60	50	65	55	59,1	54,1	62,6	57,8	60,7	59,0	61,3	55,7
G	x	1206	II	55	45	60	50	52,3	45,8	47,4	41,3	48,9	38,3	53,7	39,7
H		646	III	60	50	65	55	58,3	54,4	58,0	55,0	59,1	54,9	56,0	50,8

Table 5 : Leq (A)of the background noise – field measures

In order to assess the levels of background noise, hence the masking potential of background noise, consideration in relation to legal limit values (VLE: VLI, VP) are highlighted. These considerations do not question in any way the realization of wind turbine projects, they only help to put in parallel the existing soundscape with legal values.

In blue lettering, values are in between the planning values and immission limit values. This means that the actual landscape is already with some noise higher than other sites where the values are under these VP values. Of course, some location show these kind of values at night, since limit values are much lower for nights than days, but especially for one location (location point F) which has some industrial noise sources close to it. This could mean some potential for masking the future wind turbine noise depending on frequency distribution.

In red lettering, values measured are over the immission limit values (VLI), hence over planning values (VP). This means that without even taking into account any new noisy installation, these values are exceeded. Considering the mean Leq(A) levels of wind turbine, masking will be very probable in this situation without the immission levels being increased by wind turbine noise.

In orange lettering, values are highlighted when they are over planning values (VP) and less than 2 dB(A) under immission limit values (VLI). This is the case in between the two latter.

5.5 Conclusions and interpretations

Masking potential cannot be determined for each site only by this analysis, the aim is here to differentiate "legal masking potential", e.g. global Leq(A) cannot imply masking since the frequency distribution is not taken into account. Relative high Leq(A) values indicates a higher background noise, hence a probability of a higher masking potential.

Considering the actual soundscape, we see that location point F shows night values over immission limit values (VLI) as well for location points D, F and H at night for planning values (VP). This would mean, in a purely legal approach according to swiss noise legislation (OPB), that any kind of project respecting planning values (VP), which is a sine qua non condition for new projects, will with strong probability not be determinant for the immission level and present the least emergence risk of the wind turbine noise.

6. Correlation between measured noise vs other measured parameters

6.1 Objective of correlation analysis

Correlation analysis was part of statistical analysis led to assess the dependency in between background noise levels and other parameters, such as meteorological values, day/night period, traffic, etc. thus aiming to better understand the hypothetical dynamics of background noise for each sites in regards to other parameters.

6.2 Principle

A correlation value has been calculated per Leq(A) for each point of location, each season / period for the following parameters:

- Day/night
- Week/week-end
- Traffic
- Wind speed
- Temperature
- Humidity
- Pressure
- Precipitations (rainfall, snowfall)

The correlation calculation has been done on all values, discredited in 1 minute samples to ensure the finest analysis possible of interdependency.

6.3 Results reading and interpretation keys

A correlation factor in between two variables is equal to +1 in the case of one of the variables is an increasing linear function of the other variable. It is equal to -1 in the case of one variable is a decreasing linear function. Intermediate values give information on the degree of linear dependency in between the two variables. The more the correlation factor is close to these extreme values (-1; +1), the stronger the correlation. We simply use the expression “strongly correlated” to express this kind of correlation in between variables. A correlation factor of 0 indicates that the variables aren’t correlated at all.

The correlation factor isn’t dependent on variables units. However, this factor is very sensitive to extreme or outliers (also called deviant values).

- An absolute value of 1 corresponds to a perfect correlation, meaning the possibility to explain and deduct background noise based on the parameter analysed in this correlation analysis;
- An absolute value in between 0,75 and 1 is considered a good correlation;
- An absolute value in between 0,50 and 0,75 (yellow in the following table) is considered as an existing and significative correlation;
- An absolute value in between 0,25 and 0,50 (orange in the following table) is considered a significative but less marked as the latter one;
- An absolute value lower than 0,25 (red in the following table) is considered a weak or invalid correlation.

Negative correlation values follow exactly the same interpretation as positive one, meaning that the values of one parameter are simply inverse correlated.

6.4 Results (sample)

A sample of this analysis results is presented in the following table. The empty cells correspond to periods where data were missing or not measured.

		Period 1	Period 2	Period 3	Period 4	Mean value	Meteo. mean value
Site 1	Day/Night	0.53	0.60	0.66	0.67	0.61	
A	Week/ week-end	0.08	-0.02	-0.11	-0.23	-0.07	
	Traffic	0.37	0.53	0.46	0.44	0.45	
	Wind speed	0.30	0.15	0.25	0.32	0.25	0.19
	Temperature	0.10	0.20	0.34	0.35	0.25	
	Humidity	-0.20	-0.26	-0.47	-0.33	-0.32	
	Pressure		-0.07	-0.20	0.09	-0.06	
	Precipitation		0.13	0.12	-0.03	0.07	
Site 2	Day/Night	0.61	0.55	0.59	0.55	0.57	
B	Week/ week-end	-0.10	-0.20	-0.07	-0.13	-0.12	
	Traffic	0.52	0.54	0.52	0.43	0.50	
	Wind speed	0.09	0.23	-0.15	0.12	0.07	0.11
	Temperature	0.34	-0.12	-0.12	-0.19	-0.02	
	Humidity	-0.32	-0.03	-0.08	-0.26	-0.17	
	Pressure		-0.27	-0.24	0.14	-0.12	
	Precipitation		0.21	0.08		0.15	
Site 3	Day/Night	0.47	0.44	0.22	0.36	0.37	
C	Week/ week-end	0.05	0.14	0.01	0.18	0.10	
	Traffic	0.33	0.29	0.12	0.21	0.24	
	Wind speed	-0.06	-0.03	0.31	0.16	0.09	0.14
	Temperature	0.33	0.13	0.38	0.19	0.26	
	Humidity	-0.27	-0.05	-0.22	-0.19	-0.18	
	Pressure		-0.25	-0.30	0.36	-0.07	
	Precipitation		0.15	0.09	0.10	0.11	
Site 4	Day/Night	0.28	0.44	0.47	0.55	0.44	
D	Week/ week-end	0.10	0.00	0.03	-0.18	-0.01	
	Traffic	0.33	0.44	0.41	0.40	0.40	
	Wind speed	-0.05	0.04	0.05	0.07	0.03	0.12
	Temperature	0.38	0.17	-0.05	-0.24	0.06	
	Humidity	-0.32	-0.04	-0.14	-0.16	-0.16	
	Pressure		-0.42	-0.28	-0.03	-0.25	
	Precipitation		0.17	0.03		0.10	

6.5 Conclusions and interpretations

Looking at the correlation factors, it is obvious that the only good correlation (values above 0,5) are only found for the day/night parameter. Other parameters aren't correlated with the background noise. Some significative correlation can be seen on Site 2, point B for the traffic parameter. This is well explained by the fact that this location was close to a major road where background noise was mainly influenced by traffic noise.

A first conclusion is that background noise level estimation can be done only through a long duration acoustic measurement. Neither meteorological data, nor time period (day/night or week/weekend) can give a sufficient and reliable level prediction.

Analysis of measurement points and field observations help understand a bit better some results:

- Background noise for measurement points close to an industrial installation functioning either permanently or intermittently 24/24 is not really influenced by the hourly period (day/night);
- Background noise is influenced for many measurement locations by the day/night period, because, mainly, by the roads' proximity, also showed by some higher traffic correlation values.
- Wind speed, meteorological data and other local parameters do not influence background noise significantly.

7. Predicted noise levels (Lr)

Based on the study of emission wind turbine noise, predicted noise levels (Lr) have been calculated for each planned wind turbine implantation site, thus leading to values, both global (LAeq) and per frequency 1/3 octaves (partial Leq), for each measurement point.

With these values, the three aspects of masking, audibility and emergence were analyzed on the two following technical sheets: statistic sheet and metrological sheet explained below. These two tools showed powerful ways of anticipating the noise impact of planned wind turbines.

7.1 Conclusions and interpretations

- Globally and taking into account the different sensitivity values (DS: degrés de sensibilité) set for the different use zones, planning noise values (VP) are respected on all sites and all measure locations, as well for day as for night;
- When immission point is under 300 m, the Lr level may exceed planning limit values, but no sensible use buildings were found within these distance to the planned wind turbine sites;;
- On one specific (and quiet) site, the Lr value is above the background noise, respecting though the legal limit values. On this site, we expect an emergence of the wind turbine noise over the background noise;
- Assessing the correction factors according to background noise measured values seems to be more realistic than fixed values.

These different statements show clearly that it is necessary to plan and do long term (4 seasonal) background noise measurement on any planned wind turbine site to assess in advance (anticipate) at best Lr evaluation levels in parallel to meteorological sampling.

8. Statistical sheets

Statistical sheets aim to display data of both immission measurement location, e.g. the immission location of the closest sensible use building, and the presumed predominant background noise source and see if one would be excessively more submitted to background noise. A second aim is to overlay / add the wind turbine noise to estimate also for each measurement location the potential of the three comparative noise cases: masking, audibility and emergence of the wind turbine noise.

8.1 Principle

The following figures show the content of a statistical sheet.

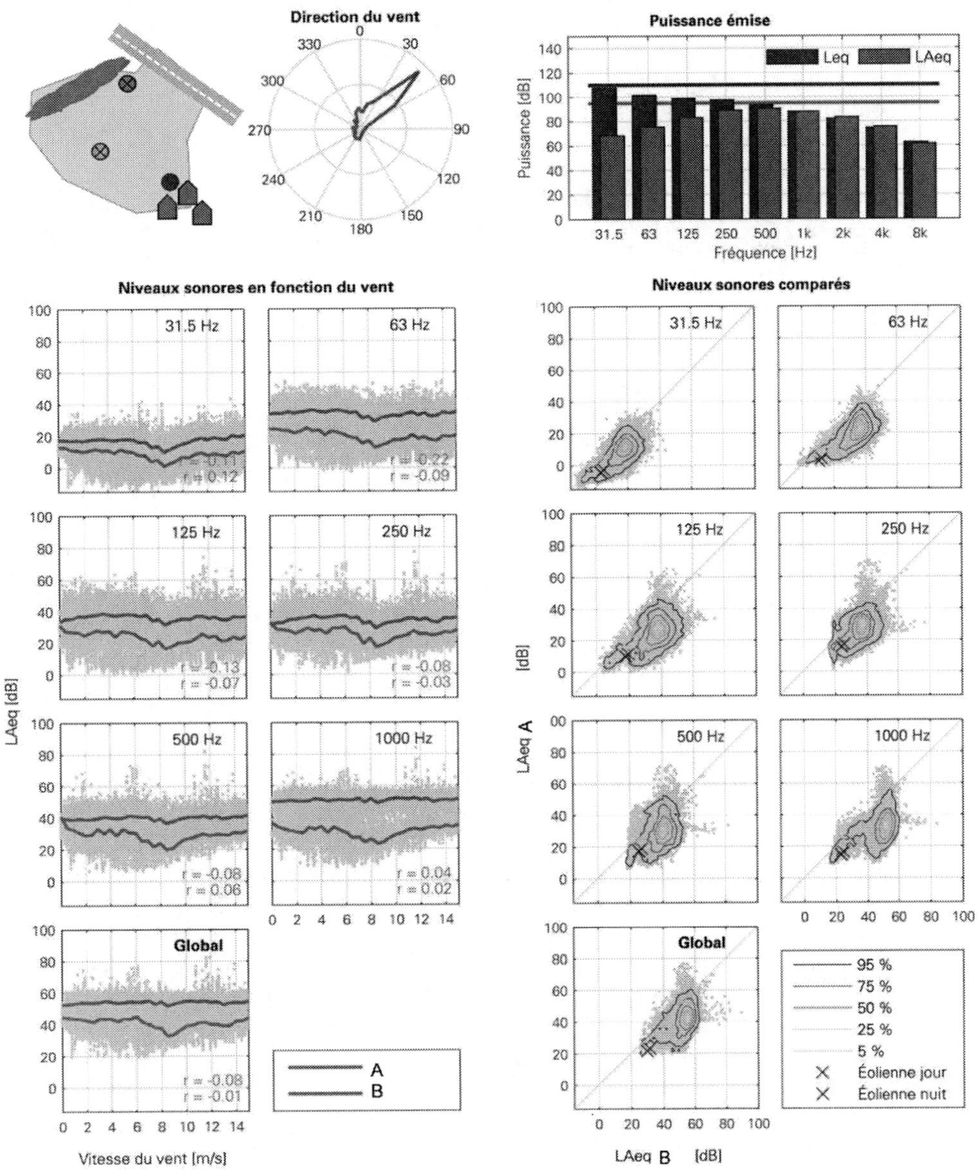

Figure 4: Example of statistical sheet

1. The first element shows the situation of the measurement site with the three different points (wind turbine installation, immission point and predominant background noise source) along with the measured wind direction for the given period.

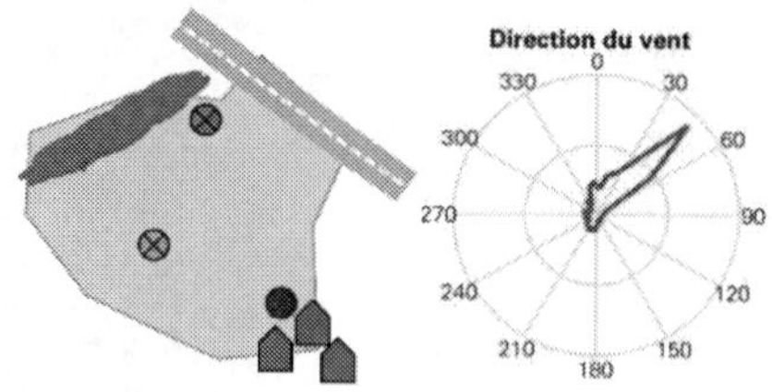

2. The maximum acoustic power of the considered wind turbine is also reported graphically per 1/3 octaves with a mention of the global level (straight line) both for the Z weighting and the A weighting filters.

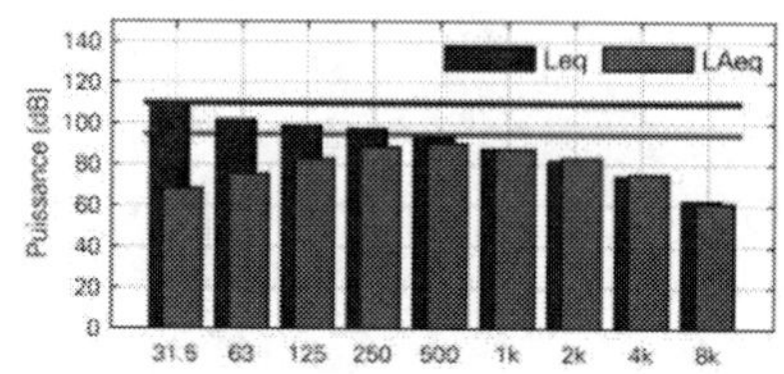

3. The graphics group on the bottom left shows the dependency in between background noise and wind speed at rotor height for each 1/3 octaves as well as for global level. The correlation factor is mentioned for both measurement locations.

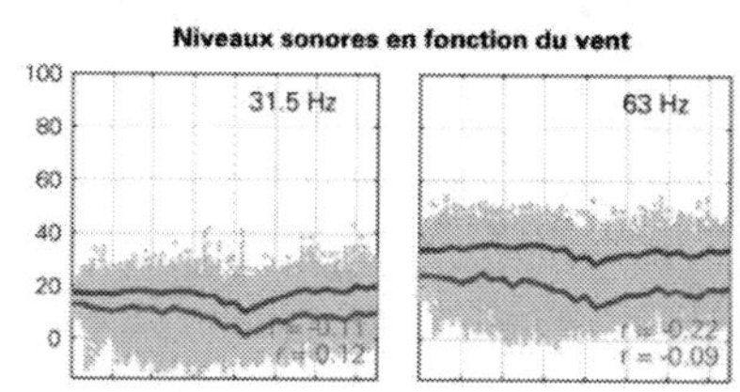

4. The graphics group on the bottom right shows the compared levels for both measurement points of the considered site. The shape of the scatter plot allows to see if some correlation between both measure locations exists (if the scatter plot is symmetrical on the diagonal of the graph). Moreover, the wind turbine acoustic pressure is shown on the graph by two crosses, both for day and night, based on the measured wind speed.

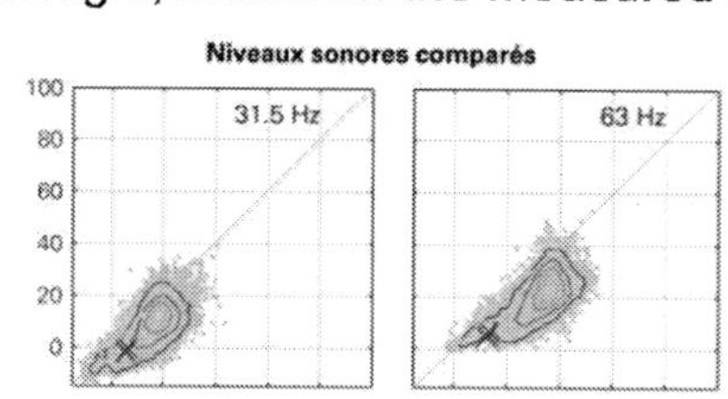

8.2 Reading and interpretation keys

These scatter plot, showing the occurrences of measured levels for both measurement locations, allows to estimate the potentialities of masking effect in regards to wind turbine acoustic pressure level. Four cases can be defined:

- Wind turbine noise masking: when wind turbine LAeq levels are systematically lower to the cloud of points as it can be seen on the following example:

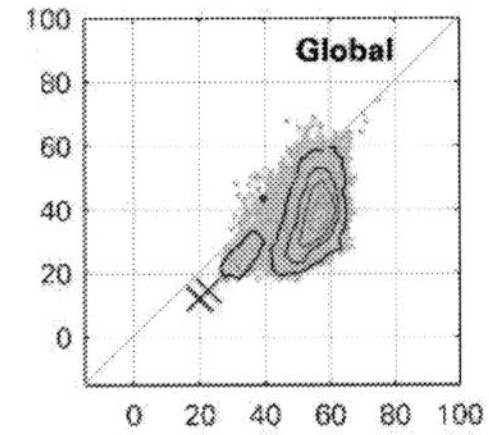

- Wind turbine noise emergence: when we see the inverse situation, e.g. the wind turbine levels are located above the cloud of points. This situation was never observed for any measurement sites.

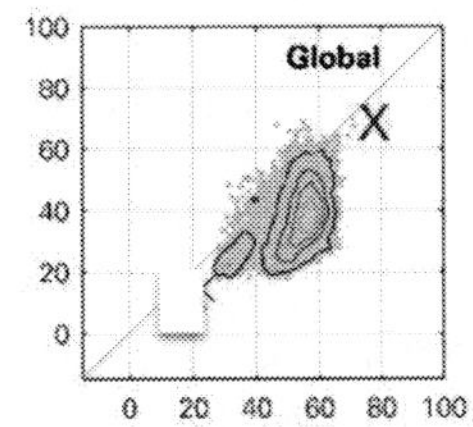

- Wind turbine noise possible and potentially distinct audibility but showing similar noise levels in between background noise and wind turbine noise on the measured locations. This is the case when wind turbine noise levels (both crosses) are "drowned" in the cloud of points, as in the following example. The wind turbine noise is then comparable to the background noise.

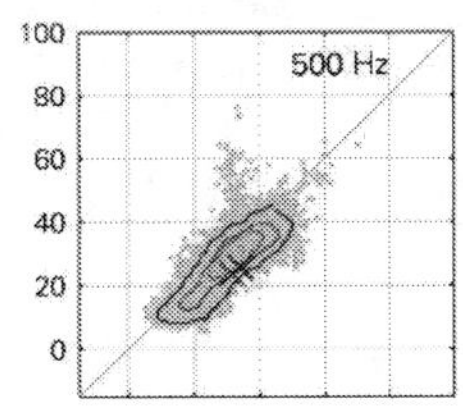

- Undefined case: when wind turbine noise is shown on the side of the cloud of points without being higher (up right, e.g. emergence) neither lower (down left, e.g. masking) of the cloud. It is in this case not possible to interpret on this sole graph the masking or emergence of the wind turbine noise compared to the background noise. This case hasn't been observed in any of the studied sites.

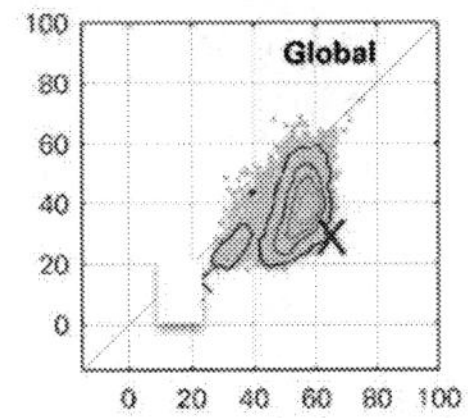

8.3 Results

All results are given by statistical sheets for each measurement site and each period of the year as the example shown above.

8.4 Conclusions and interpretations

8.4.1 Graphical analysis

Thanks to these statistical sheets, the following cases have been observed for the four different sites:

- Site 1 :
  - The wind turbine noise is always (frequency- and global-wise) in the center of the cloud of points: potential audible but no emergence;
  - Potential masking for two seasons;
  - Potential emergence possible for period with very strong winds;
  - The correlation between background noise and wind speed is weak or invalid.
- Site 2 :
  - Masking is nearly permanent except for period 4 (strong winds) where potential audibility is possible (without any emergence);
  - The correlation between background noise and wind speed is invalid.
- Site 3 :
  - The wind turbine noise is always drowned in the center of the cloud of points, e.g. a potential audibility without being emergent;
  - No permanent potential masking, neither potential emergence;
  - The correlation between background noise and wind speed is weak or invalid.
- Site 4 :
  - Masking is very frequent, nearly permanent, particularly for two periods;
  - Potential audibility for period with very strong winds, without any emergence potential;
  - The correlation between background noise and wind speed is invalid.

8.4.2 Note on wind influence

The wind speed at 10 m high is decorrelated from wind speed at rotor height. Yet the influence of wind speed at rotor height on background noise is almost null, unlike the wind speed measured at 10 m height, correlation factors in between wind speed and background noise being systematically higher at 10 m high than at rotor height.

The two following figures show significant correlation factors in between background noise and wind speed at rotor height, but always weak, sometimes null. However, correlation factor are usually higher for low frequencies than high ones.

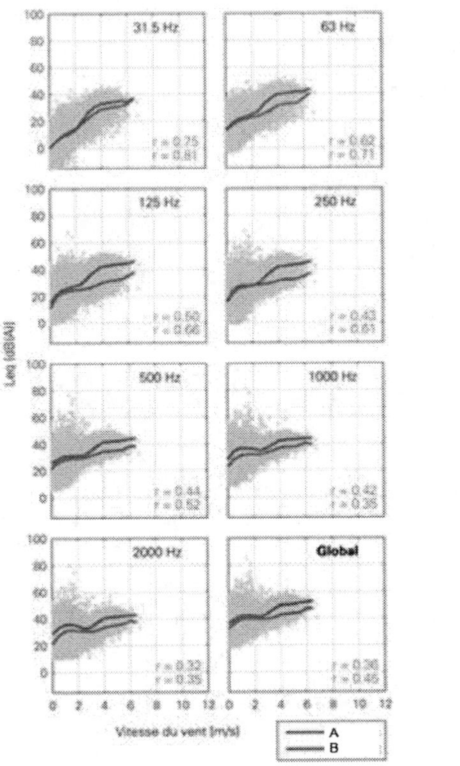

Correlation in between background noise and wind speed at 10 m high.

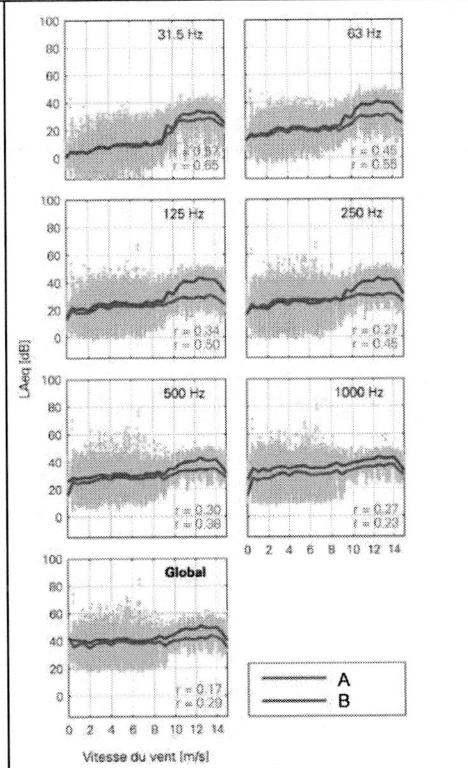

Correlation in between background noise and wind speed at rotor height.

Figure 5: Correlation in between background noise and wind speed at different heights (10 m vs rotor height)

As a result of these two observations, it is not possible to say that wind turbine noise will always be masked by tree leafs sound. This masking case is possible but not certain.

9. Metrological sheets

Each of the measured location point (two per site) has been subject of a metrological sheet per period. These sheets allow to visualize meteorological and acoustic characteristics of each measure point, and appreciate the wind turbine noise emergence, audibility or masking effect in regards to background noise, and this taking into account the frequency distribution.

9.1 Principle

An example of metrological sheet is given in the following figure.

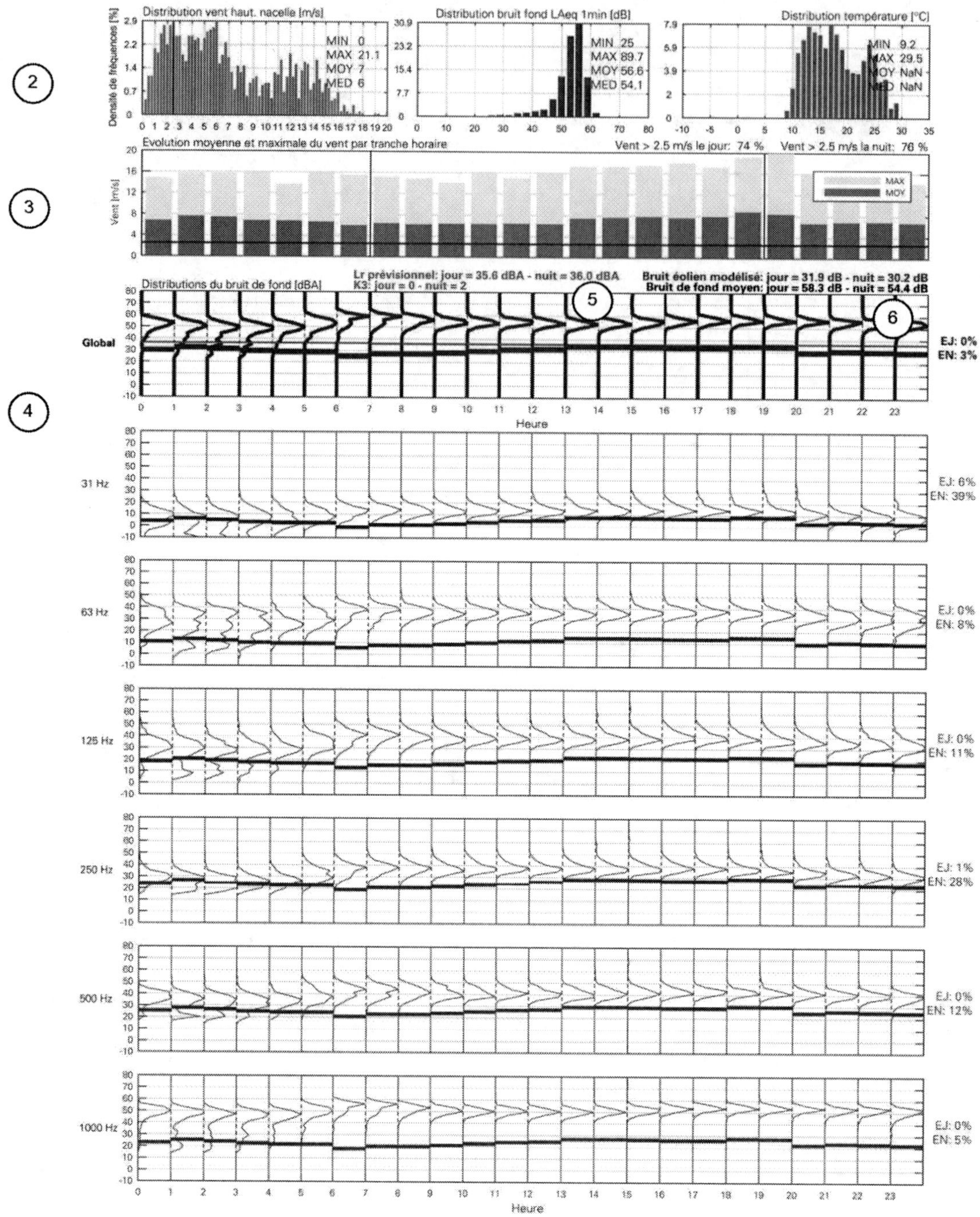

Figure 6: Example of a metrological sheet

The different parts of this metrological sheet are here described:

1. The heading of the sheet (not reproduced here) gives detailed about:
    i. The name of the measured location ;
    ii. The measurement period ;
    iii. The horizontal distance to the emission point (wind turbine installation site) ;
    iv. The sonometer model used;
    v. The XY coordinates ;
    vi. The calculation method used for evaluation levels;
    vii. The wind turbine model planned for the considered site with modelled on builders characteristics and data coming from the "emission" project, the acoustic maximum power and corresponding wind speed.
2. The three first graphics show:
    i. The wind speed distribution at rotor height;
    ii. The background noise distribution (A weighted)
    iii. The temperature distribution

These distributions are obtained on the basis of one minute sampled data. The minimum, maximum, median and mean values are shown on each distribution. On the wind speed distribution, the value of 2,5 m/s is shown with a green vertical line, showing the wind speed under which the wind turbine doesn't spin.

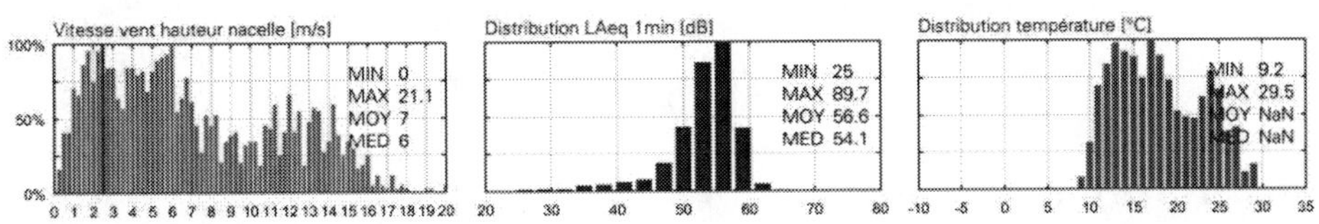

3. The wind speed daily evolution is then represented. Daily and nightly functioning times are calculated (%). The 2,5 m/s wind speed value is also shown here for information purpose.

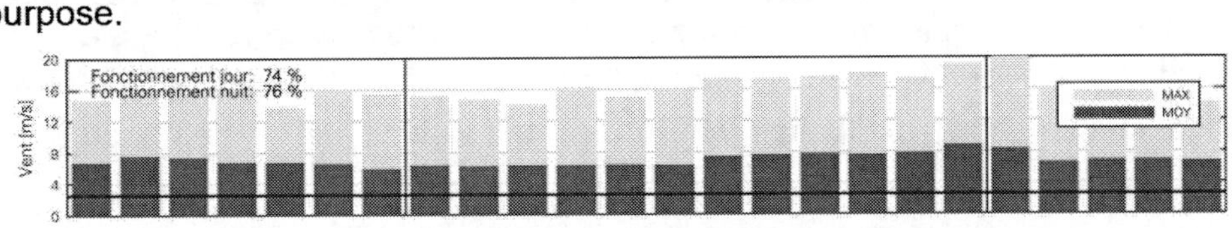

The rest of the metrological sheet allows the visualization of background noise evolution (both global and 1/3 octaves). LAeq noise levels are shown in ordinate and hours on the X-axis. For each time period, wind turbine LAeq 1h acoustic pressure level is facing the background noise level. This comparison allows to indicate which time percentage over which wind turbine noise is higher than background noise.

These values reduced to day/night legal period (OPB [2]), we can calculate two relative values for each global and 1/3 octaves:

- The daily emergence (EJ: émergence jour) in percent;
- The nightly emergence (EN: émergence nuit) in percent.

Figure 6: Example of a metrological sheet

4. In the following example, we can estimate that the global wind turbine noise will be over the background noise 0% of the daily time (EJ : 0%) and 4% of nightly time (EN : 4%).

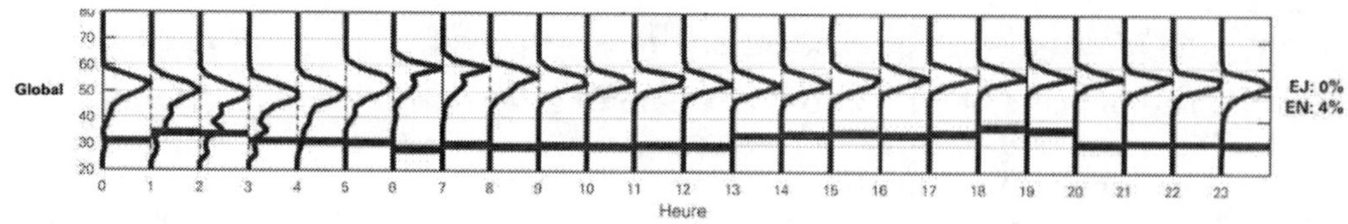

The same analysis are done for each 1/3 octaves.

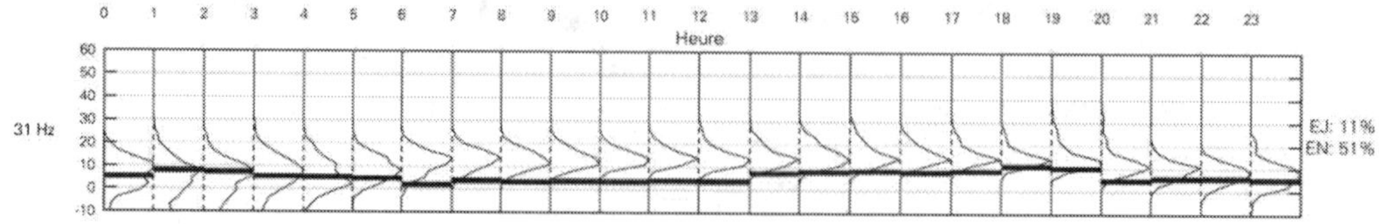

5. The evaluation levels Lr for day and night periods, as well as correction factors K3 are detailed above the different graphics.

Lr jour 36.6 dBA - Lr nuit 38.0 dBA
K_3 jour 0 - K_3 nuit 2

6. Finally the acoustic pressure levels LAeq at the immission point without Ki correction for both day and night periods are given. A mention about the overrun of planning limit values (VP) in regards to sensitivity degrees (DS) is also given.

LAeq,jour: 32.9 dB - LAeq,nuit: 32.2 dB
Dépassement VP DSIII: jour non, nuit non

9.2 Reading and interpretation keys

9.2.1 Wind speed distribution and wind turbine operation

The wind distribution at rotor height allows to estimate the most frequently observed speeds. The 2,5 m/s limit shows the spinning limit of the wind turbine. The more histogram bars are numerous and the more they are high and right to the spinning limit, the more the wind turbine is in operation.

The following example shows a situation where the wind turbine will be nearly permanently in operation.

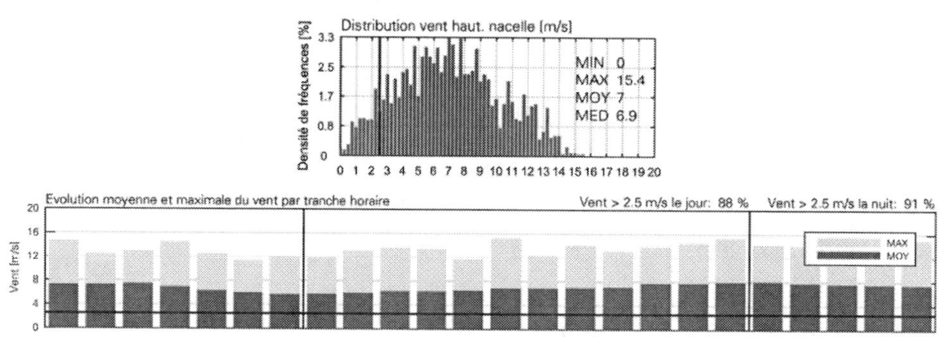

Figure 7: Example of a nearly permanent operation of the wind turbine

(Operation J/N : 88% / 91 %)

The following example shows a wind turbine with low operation time:

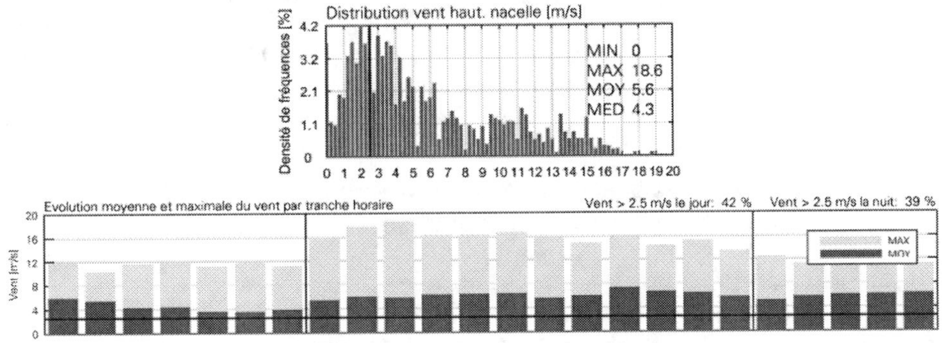

Figure 8: Example of a low operation wind turbine

(Operation J/N : 42% / 39%)

9.2.2 Acoustic pressure level LAeq

The LAeq value corresponds to the wind turbine propagated acoustic pressure level at the immission point. This value, useful to calculate the evaluation level in accordance to the swiss noise legislation (OPB [2]), is submitted to level corrections but also to corrections linked to noise phases, which explains the Lr attenuation when the wind turbine has low operation.

9.2.3 Background noise distribution

The distribution of background noise can be read according to the three different cases already described:

- Masking effect of wind turbine noise: when wind turbine LAeq levels are systematically lower than the preponderant background noise level point, as seen in the following figure for levels at 500 Hz:

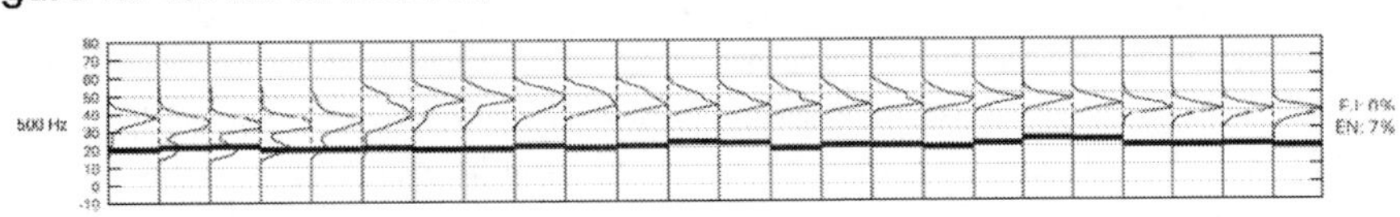

- Wind turbine noise emergence: when the situation is opposite, like the following figure, e.g. the wind turbine noise being above the preponderant background noise level point, here at 500 Hz:

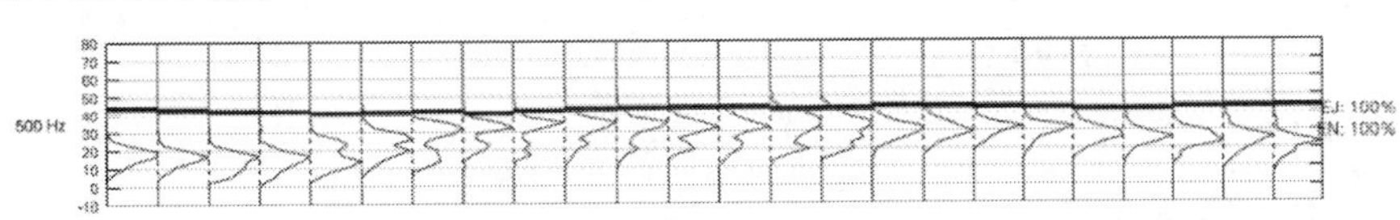

- Wind turbine noise possible and potentially distinct audibility: when both levels do not show such a distinct difference as the two latter cases. This is well shown by the following case where the wind turbine noise is similar to the background noise, here for 125 Hz:

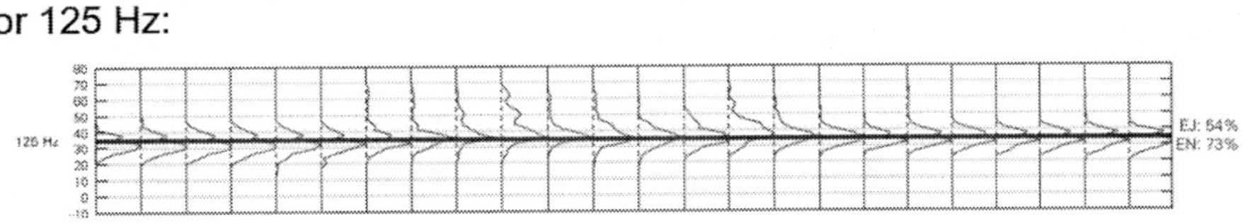

These metrological sheets offer the possibility to assess emergence or masking effect as well for global levels as for 1/3 octaves levels. This emergence can be appreciated on these graphs under the principle that for two similar noises, a value superior of 10 dB(A) becomes distinct, so could be defined as emergent. The emergence is then more reliable when analysed frequency-wise on the 1/3 octaves graphs.

9.3 Results

Metrological sheets have been realized for each measurement location and each seasonal period to provide analysis as described above for each of them.

9.4 Conclusion and interpretations

The following observations can be done :

- Site 1 – location A :
  - Emergence during night, especially in low frequencies (31 Hz) and during strong wind periods in the mid frequencies (500 Hz) ;
  - Very low risk of audibility during day period ;
  - No complete masking effect (neither global, nor frequency-wise).
- Site 1 – location B :
  - Important masking effect, both global and frequency-wise;
  - Possible emergence at night with strong winds.
- Site 2 – location D :
  - Nearly permanent masking effect, both global and frequency-wise, including during strong wind periods;
- Site 3 – location F :
  - Low risk of audibility during night in low frequencies (31 Hz to 125 Hz) ;
  - Possible masking effect for all other cases.
- Site 2 – location C and Site 4 – location G :
  - Nearly permanent masking effect, except during strong wind periods where low risk of audibility can be identified for all frequencies;
- Site 3 – location E and Site 4 – location H :
  - Low distinguishable levels, as well globally as frequency-wise during day periods;
  - Low risk of audibility at night, possible mainly in low frequencies (31 à 125 Hz).

10. Conclusions and perspectives

This background noise study through the different analysis approaches presented in this paper allowed to identify important points here below reproduced:

- Analysis through systematic long term background noise measurement allows a reliable and finer characterization of this background noise for each considered site, in accordance both of legal approach and acoustic properties, including frequency masking potential;
- These analysis allow to assess and determine the actual background noise, estimating also masking effect or emergence potential for each planned site;
- Correlation analysis show a low correlation between background noise and other parameters, except day/night period, this parameter not being sufficient to predict background noise level and timely evolution. This leads to the conclusion that background noise estimation cannot be done without any proper long term measurement;
- Thanks to the parallel study about wind turbine noise emission, a reliable Lr evaluation level prediction propagated on immission sites is possible and relatively easy. This gives

a very good basis to assess masking effect, audibility and emergence potentials for given sites;

- Developing statistical and metrological sheets allows to provide reliable, exhaustive and easy to read information overlaying background and wind turbine noise, this leading to accurate assessment of masking effect, audibility and emergence potentials. Observations can be done both on global levels as well as on 1/3 octaves to provide a finer view of these effects;
- Correlation calculation may show some interesting statement, mainly here discarding some empirical deductions;
- Particularly, it is not possible to state that tree foliage noise will have a masking effect. It may be the case but definitely not certain. Many other noises participate much more to some potential masking effect;
- This study proves that masking effect, audibility and emergence phenomenon can be scientifically determined with a good reliability and realism, taking into account the frequency specificities of both immission sites and wind turbine installations;

The use of the different analysis tools developed in this project and presented here helps anticipate in a reliable manner impacts and masking or emergence potentials for any kind of planned site, adaptable for any kind of wind turbine technology, as long as long term measurement can be done.

These approaches could be used to assess planned wind turbine installation sites and help in the acceptability of these project for the population by offering a transparent approach of assessing a very realistic and scientific anticipation, knowing that trying a background noise prevision through models seems unrealistic, compared to a simple long term measurement period.

References

- [1.1] Message relatif au premier paquet de mesures de la Stratégie énergétique 2050, 13.074, September 4th 2013
- [1.2] Ordonnance sur la protection contre le bruit (OPB 814.41) du 15 décembre 1986 (Etat le 1er février 2015)
- [1.3] Fiche d'information sur le bruit des installations éoliennes, K161-0974, May 5 2011, Confédération suisse, OFEV, Division Lutte contre le bruit
- [1.4] source http://www.wind-data.ch, February 2015
- [1.5] Stratégie énergétique 2050, www.bfe.admin.ch/strategieenergetique2050, 26 August 2014
- [1.6] L'éolien progresse trop lentement en Suisse, Communiqué de Suisse Eole, November 21st 2015
- [1.7] Concept d'énergie éolienne pour la Suisse, Bases pour la localisation de parcs éoliens, OFEN, OFEFP, ARE, Berne, August 2004
- [2] Ordonnance sur la protection contre le bruit (OPB) du 15 décembre 1986
- [3] ISO 9613-2 :1996, Acoustique – Atténuation du son lors de sa propagation à l'air libre – Partie 2 : Méthode générale de calcul.
- [4] to [8] Wind turbine technical sheets and documentation

7th International Conference
on
Wind Turbine Noise
Rotterdam – 2nd to 5th May 2017

Variation of wind induced non-turbine related noise due to position, shelter, wind direction and season

Lars Sommer Søndergaard, DELTA Acoustics – a part of FORCE Technology, Agro Food Park 13, 8200 Aarhus N, Denmark, lss@delta.dk

Rune Egedal, DELTA Acoustics – a part of FORCE Technology, Agro Food Park 13, 8200 Aarhus N, Denmark, rue@delta.dk

Morten Bording Hansen, DELTA Acoustics – a part of FORCE Technology, Agro Food Park 13, 8200 Aarhus N, Denmark, mbh@delta.dk

Summary

Assessment of wind induced non-turbine related noise (here shortly referred to as background noise) in the context of wind turbines in general has two purposes; I) either to perform measurements with as low background noise levels as possible to achieve the best signal-to-noise ratio for optimal determination of a noise level or II) to assess the most appropriate masking noise level to determine audibility of the wind turbine. For both cases knowledge about the nature of background noise and its variation is essential. Variation of wind induced background noise due to position, shelter, wind direction and season has been investigated in this paper, which is based on two case studies – both performed at the same possible wind turbine neighbour, but without nearby wind turbines. The case studies are performed with reference to the Danish wind turbine noise regulations. In both case studies, several synchronized microphones and one synchronized wind mast (10 m height) has been utilized, the first case with focus on low frequencies for an indoor-outdoor relationship both summer and winter, the last case with focus on tonality for different outdoor microphone positions with varying wind directions in autumn. The measurements in general describes a linear relationship between noise level and wind speed. Especially the outdoor measurements show a large variation both between measurement positions, but also for the same position within wind speed bins. When analysed in critical bands, large differences are found (up to 10 dB on average at one wind speed) in the results from the different microphone positions.

1. Introduction

Danish wind turbine noise legislation utilizes calculated noise levels at residents based on sound power level measurements. Compared to emission measurements at residents one of the benefits with the Danish method is ensuring a good signal-to-noise ratio – background noise is generally not a big problem. One of the drawbacks might be that information about audibility of the noise is left out. It is therefore of interest to look further into background noise, here as a case study for a single possible wind turbine neighbour site.

2. Site

A site has been chosen in the western part of Denmark, which is generally flat and with good wind conditions. For the chosen site, there is not much "man-made" noise with nearest neighbour ≈ 260 m away, nearest small town (≈ 200 residents) ≈ 1200 m away and nearest city (≈ 8000 residents) ≈ 3 km away. The focus here is on wind induced non-turbine related noise. For the chosen site two measurement campaigns are available. There are two buildings at the site: The main house and a large barn. The vegetation is mainly foliiferous and surrounding the main house in nearly all directions except south. The main house is a double-storey house from 1964, with brick façade and large window areas.

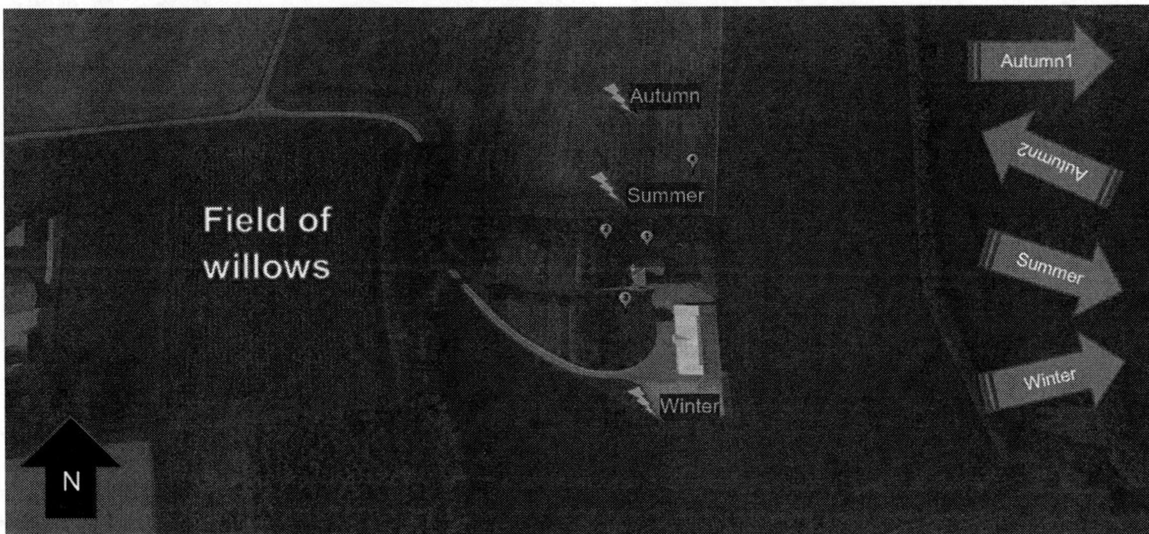

Figure 1 Overall layout of the site, with microphone positions in red and wind mast positions in green. Wind directions for the measurement days are shown with blue arrows in the right side

3. Source data

Source data origins in two case studies performed at the site; both with several synchronized microphones and one synchronized wind mast (10 m height), where one microphone position was reused for both case studies.

3.1 Low Frequency Noise from Large Wind Turbines

The first measurement campaign is performed in 2006-2007 as a part of the large EFP-06 project "Low frequency noise from large wind turbines – quantification of the noise and assessment of the annoyance". This part of the project investigates the relationship between wind speed and background noise produced indoor and outdoor at houses in the country side from other sources than wind turbines, with primary focus on low frequencies. Measurements in 60 second intervals of wind induced non-turbine related noise has been performed for 5 synchronized microphone positions (1 outdoor and 4 indoor microphones).

3.2 Evaluation of tone content in wind turbine noise at receivers

The second measurement campaign is performed in 2015 as a part of a relatively small project performed as reference laboratory for the Danish Environmental Protection Agency to investigate the influence of microphone position on tone evaluation. Measurements in 60 second intervals of wind induced non-turbine related noise has been performed for 4 synchronized outdoor microphone positions.

3.3 Microphone position 1 (only used in 2015 project)

Microphone position 1 (mic1) is the microphone placed closest to the house. The position represents closeness partly to medium high vegetation and partly to a building with eventual turbulent wind inflow. Approximate distance between house and microphone is 4.5 m and a similar distance to relevant vegetation. The surrounding vegetation is primarily trees and bushes, where the trees are approximately 8-10 m high and primarily beech and oak, see Figure 2.

Figure 2 Photo (towards west) of microphone position 1 from 2015

3.4 Microphone position 2 (only used in 2015 project)

Microphone position 2 (mic2) is placed between trees. The position represents a position closely surrounded by medium high vegetation, which could be an overgrown property with much vegetation. The surrounding vegetation is primarily trees and bushes, where the trees are approximately 8-10 m high and primarily consists of beech, oak and chestnut, see Figure 3.

Figure 3 Photo (towards southwest) of microphone position 2 from 2015

3.5 Microphone position 3 (used in both projects)

Microphone position 3 (mic3) is set up on an open lawn in a distance of 15 m from both the house and nearby vegetation.

Figure 4 Photo (towards north-northeast) of microphone position 3. Top photo from December 2006. Bottom photo from October 2015.

The position represents an open position with vegetation in some distance, which could be a property with large open (grass) areas and some vegetation. The surrounding vegetation is primarily trees and bushes in a collective area plus a single chestnut tree (approx. 10 m from the microphone). The trees are approximately 8-10 m high and are primarily wild cherry and oak, see Figure 4 and Figure 5.

Figure 5 Photo (towards south-southwest) of microphone position 3.
Top photo from December 2006. Bottom photo from October 2015.

3.6 Microphone position 4 (only used in 2015 project)

Microphone position 4 (mic4) is placed on an open field relatively long from the house and vegetation (approx. 25 m to nearest vegetation). The position represents a free and open position far from vegetation, which could be a property with very little vegetation. The vegetation is primarily trees and bushes in a hedge, where the trees are approximately 8-10 m high and primarily beech and oak, see Figure 6.

Figure 6 Photo (towards north) of microphone position 4 + 10 m wind mast from 2015

3.7 Wind mast positions

The wind masts have been placed in three different positions. For the 2006 measurement (winter) it was chosen to place it south of the microphone, see Figure 5 (top). For the 2007 measurement (summer) it was chosen to place it just north of the hedge, see Figure 7. For 2015 it was placed far from the vegetation, see Figure 6.

Figure 7 Photo (towards west) of wind mast from 2007

3.8 Site update

In the time between the first measurements in 2006-2007 and the last measurements in 2015 the site of course has changed some. The general layout of buildings and placement of bushes and trees has not changed. But bushes and trees has of course grown in the meantime, and some trees has been cut down, especially west of the house, which can be seen in Figure 4. The largest change has been for the area marked "Field of willow" in Figure 1. For 2006-2007 this area was a regular field with regular crops like grass or wheat. For 2015 this area was a forest of young willows (to be used for fuel) with an approximate height of 8 m. Similarly, the small forest area south of the willow field was newly planted in 2006-2007, where it in 2015 has an approximate height of 6 m. As a general result the site in general are more sheltered in 2015 than in 2006-2007.

4. Results

In total data was available for four days; two from each project. By comparing the data new information can be extracted. Overall differences for the four measurement days are listed in the table below (all measurements were performed in daytime).

	Year	**Month**	**Season**	**Wind direction**
Summer	2007	August	Summer	West-Northwest
Winter	2006	December	Winter	West-Southwest
Autumn day 1	2015	October	Autumn	West
Autumn day 2	2015	October	Autumn	East

4.1 Variation due to season

One of the measurement positions for the two measurement campaigns was nearly identical and it is therefore interesting to compare the measurements in this position. The microphone was positioned approximately 15 m from the house in a height of 1m (2006-2007 measurements) and 1.5 m (2015 measurement), see photos in Figure 4 and Figure 5.

For the measurements in October the vegetation was still green and no sign of "autumn colours" were seen and only a few of the leaves had fallen of the trees. So, in that sense the vegetation was probably comparable to summer.
All measurements were averaged over 60 seconds. See Figure 8 for an overview of number of measurements for each measurement day. As can be seen most measurements was obtained for Autumn Day 1 and Summer.

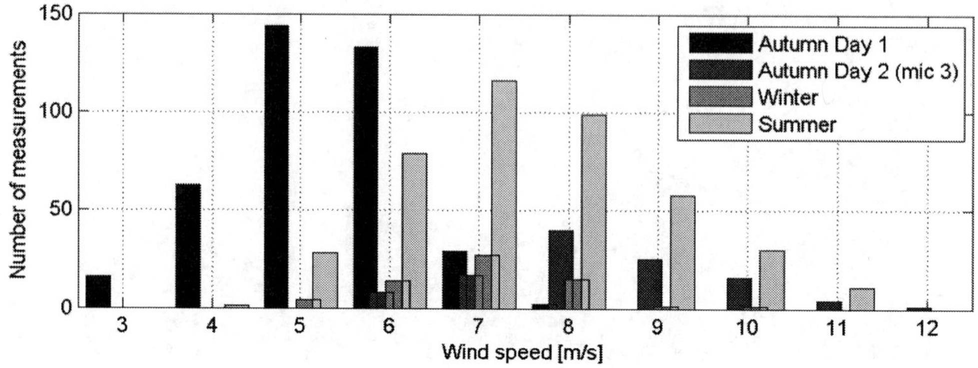

Figure 8 Overview of number of 60 seconds' measurement periods for the four measurement days

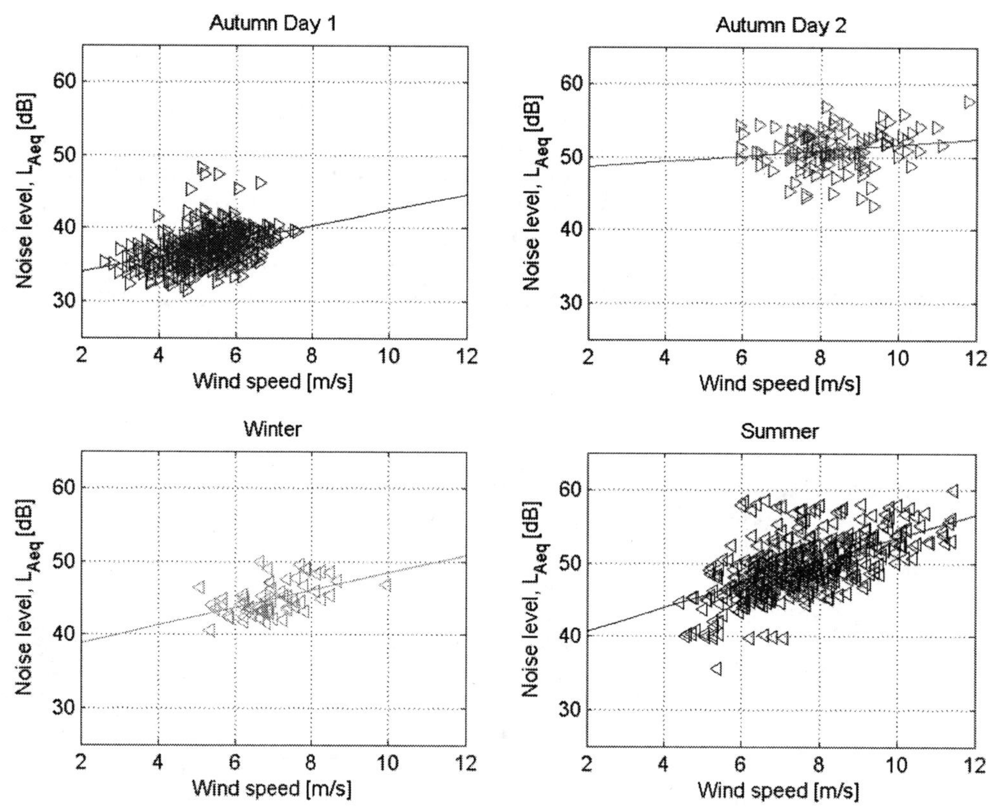

Figure 9 Noise level versus wind speed for the four different measurement days

In Figure 9 the relationship between noise level ($L_{Aeq,60seconds}$) and wind speed is shown for the four measurement days together with a linear regression line. It can be seen there is a fairly linear relationship between wind speed and noise level. It can also be seen that the noise level at similar wind speeds are not necessarily the same for the measurement days – for example at 7 m/s the average noise level changes between approximately 40 to 50 dB.
Assuming the amount of leaves on the trees has a relationship with the noise level, the logical assumption would be that the winter situation would have the lowest noise level, then the autumn and then summer. Instead the lowest noise level is found for the first autumn measurement day. The average frequency distribution for the obtained wind speeds for the different measurement days is shown in Figure 10, where it can be seen that for the summer measurement the frequency distribution for the different wind speeds are approximately parallelly displaced where it for Autumn Day 1 and Winter measurement days the change primarily is limited to only some frequency areas. For Autumn Day 2 the change in frequency spectra is very small which fits very well with the nearly horizontal slope of the regression line in Figure 9.

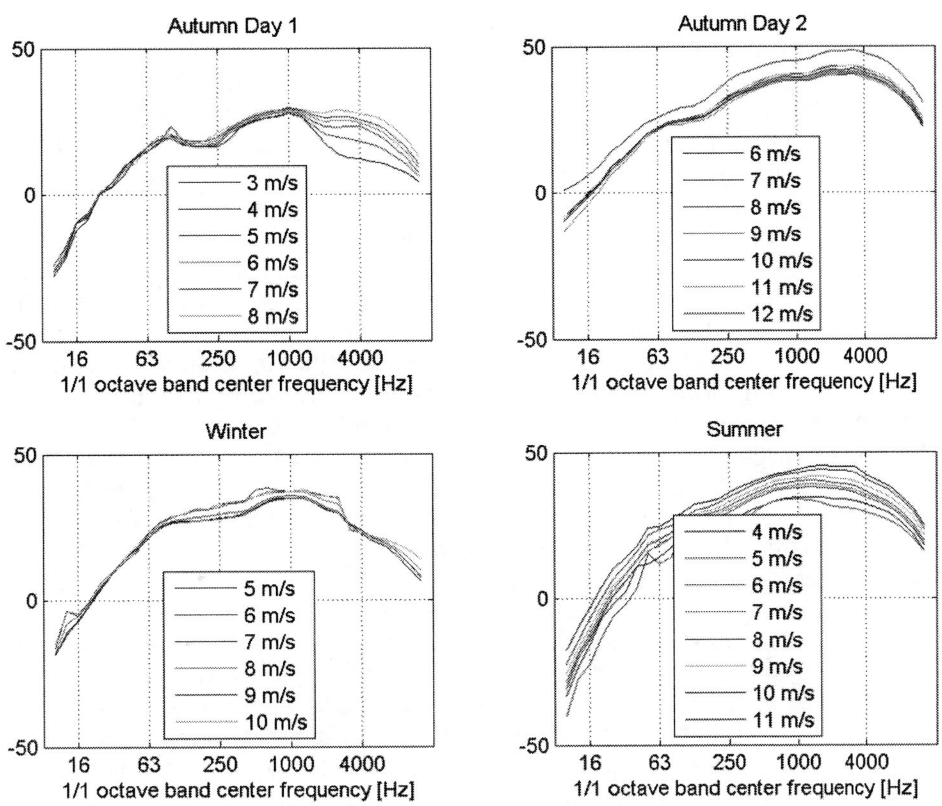

Figure 10 Average noise levels for the obtained wind speeds for the different measurement days

For easy comparison, the average noise levels for the wind speed 7 m/s is shown in Figure 11.

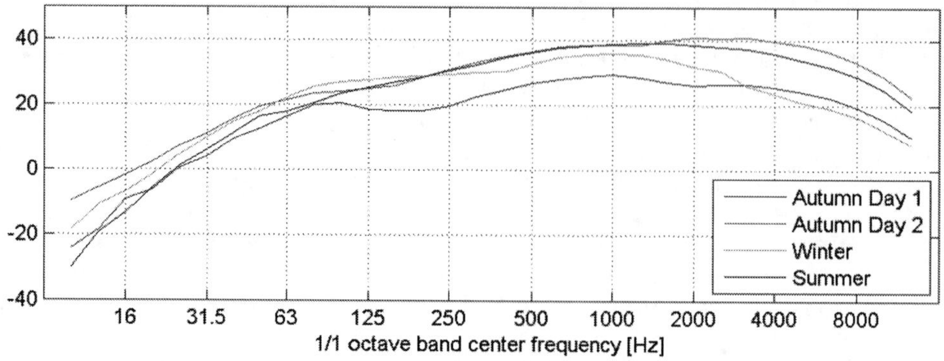

Figure 11 Average noise level for 7 m/s for the different measurement days

4.2 Variation due to position, wind direction and shelter

It is also interesting to look at the different microphone positions for the 2015 measurement campaign. Data was recorded for 4 synchronized microphones for the different days (a couple of days passed between measurement day 1 and 2) which had very different wind directions and also a change in wind speed. Both measurement days was in October but vegetation was still green and the leaves hadn't started falling of the trees. Measurement day 1 had wind speeds from 3 to 8 m/s and the wind direction was from west. Measurement day 2 had wind speeds from 6 to 12 m/s and the wind direction was from east-southeast. Noise level versus wind speed (average time 60 seconds) is shown in Figure 12 for the four microphone positions together with a linear regression curve. It can be seen that there is a decent linear relationship, where the vegetation noise rises with wind speed. It can also be seen that there is a large variation in noise level with the same wind speeds within microphone positions in the magnitude of 10-15 dB. The explanation for this is probably that there are approximately 100 m between microphone and wind mast, and the wind speed in the vegetation (the noise source) and the 10 m wind mast is probably not that well correlated.

For determining audibility for a wind turbine neighbour the distance between the turbines and residents in Denmark will typically be down to 500-600 m (for modern large wind turbines) and even though both the noise source (the wind turbine(s)) and the masking noise (the vegetation noise) are correlated with the same source (the wind) they might not be correlated (in time?) possibly leading to periods with audible wind turbine noise.

Comparing Day 1 and Day 2 there seem to be a jump in noise level of approximately 3-5 dB from Day 1 to Day 2. The possible reason for this is most likely that even though the wind mast is placed in the same unsheltered and free position for both days the wind direction is nearly opposite. At Day 1 the wind direction is west. The microphone positions and vegetation close to the microphone position is relatively sheltered when the wind direction is from west both due to that the terrain west of the site is slightly elevated and there is a lot of vegetation west of the site sheltering the house. Oppositely with wind from the east-southeast the house, microphones and vegetation close to the microphone is relatively unsheltered – especially microphone 3.

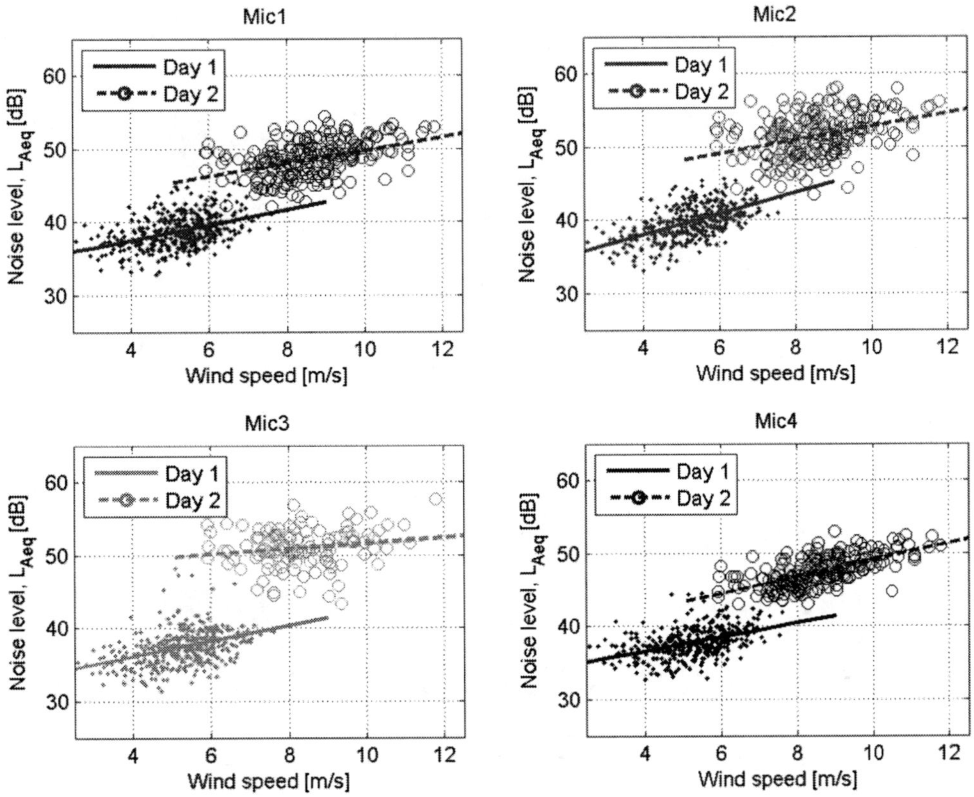

Figure 12 Noise level ($L_{Aeq,60seconds}$) versus wind speed for the four microphones together with a linear regression curve

To reduce the influence of distance between the wind mast and the vegetation close to the microphones the difference between the synchronized microphones is investigated by subtracting the 60 seconds' noise level of each of microphone position 2-4 with microphone position 1. The result of this is shown in Figure 13 together with a linear regression curve. The difference between microphone position 4 and 1 are in average quite small which is interesting since microphone position 1 is relatively sheltered while microphone 4 is unsheltered. The largest difference is seen between microphone position 3 and 1. All the chosen microphone position are valid positions – meaning that it is positions which could have been chosen for a "real" noise assessment. Due to the large variation, it is quite clear that it is very important to choose microphone position with great care.

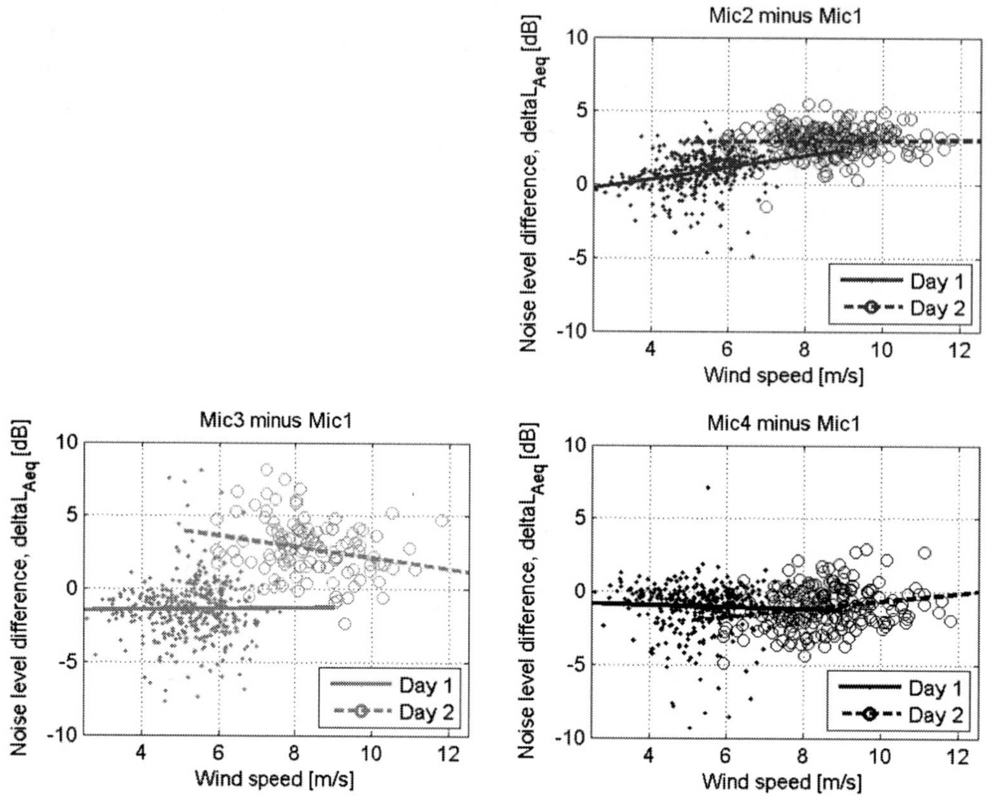

Figure 13 Difference in noise level ($L_{Aeq,60seconds}$) between microphone 2-4 and microphone 1 versus wind speed together with a linear regression curve

4.3 Microphone position variation influence on tonal audibility

When assessing tonal audibility, the level of masking noise is equally important to tonality level. In overall the audibility of a tone can be described as the difference between the tone level and the masking noise level, see an example in Figure 14. To investigate the influence of microphone position on assessment of tonal audibility the noise data has been processed through objective tonal audibility software (noiseLAB) by successively adding an artificial tone at the 1/3 octave band centre frequencies to determine the masking noise level at the critical bands centred at the 1/3 octave band centre frequencies.

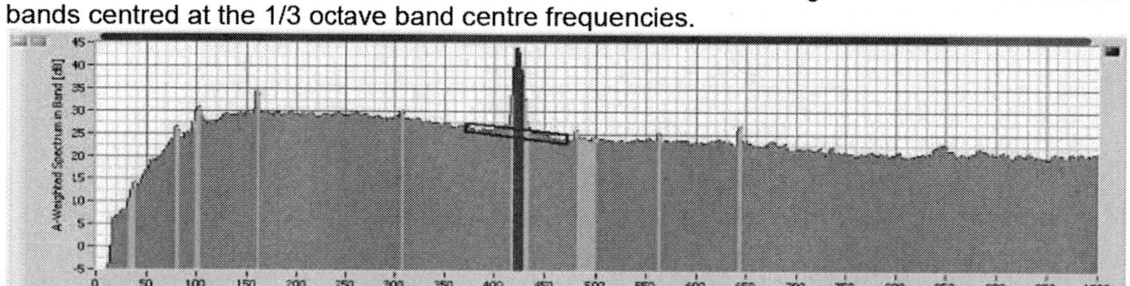

Figure 14 Example of tonal audibility evaluation. The investigated tone is shown in red and the masking noise is shown in blue. Masking noise level is shown in the rectangular black box

In Figure 15 the variation in masking noise level between microphone positions is shown for different wind speeds and measurement days. The largest tonal audibility will be found in the position with the lowest masking noise level, but as it can be seen none of the microphone positions have the lowest masking noise level for all measurement days, wind speeds and frequencies. For example, at Day 1 with a wind speed of 6 m/s (top left plot) microphone position 3 has the lowest masking noise level for all frequencies, but for Day 2 with wind speeds 6 and 8 m/s microphone position 3 has the highest masking noise level. Microphone position 4 (the microphone on the field) could be a sensible choice since it is generally one of the lowest ones – but only for frequencies above 250 Hz. The probable explanation for this is that this microphone position is the least in shelter/most exposed to the wind, and the rise in masking noise level at frequencies below 315 Hz is due to wind induced noise (even with both primary and secondary wind screen). This large variation between (valid) microphone positions could have a significant impact on tonality assessment and resultantly also both on the wind park and residents. As an example, consider a tone at 125 Hz assessed in microphone position 3 at a wind speed of 8 m/s to a tonal audibility of -1 dB. If the same assessment had been done in microphone position 1 the resulting tonal audibility could easily be +4 dB.

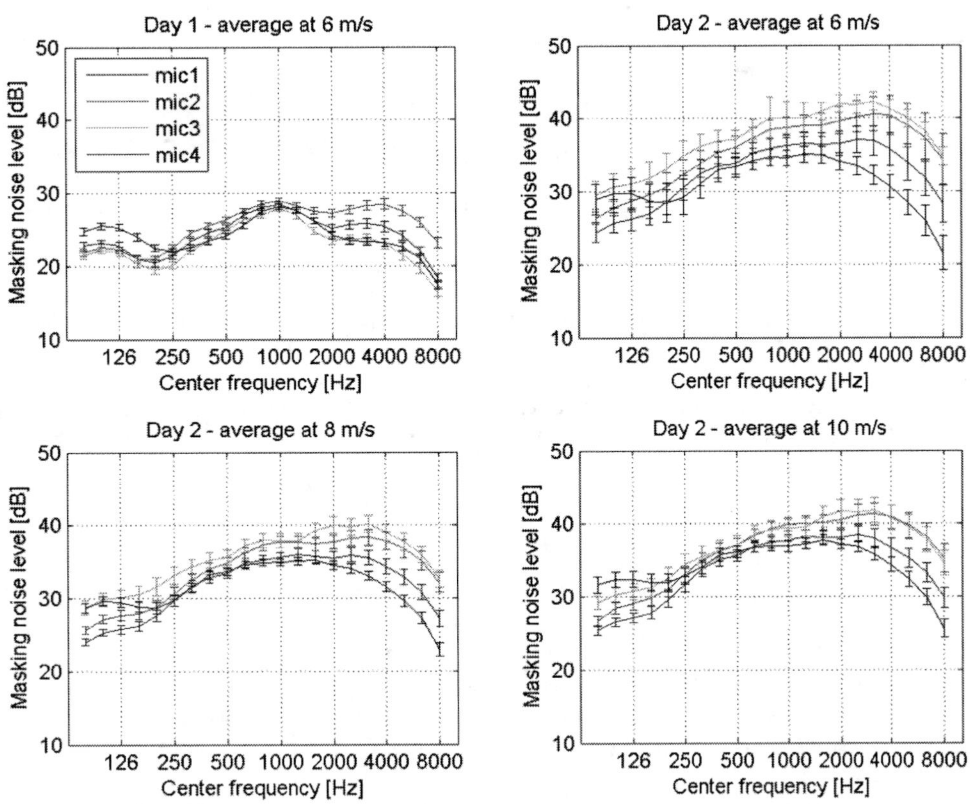

Figure 15 Average masking noise level and 95 % confidence interval for different microphone positions, measurement days and wind speeds

Finally, it is interesting to look at the variation within wind speeds. Figure 16 shows the masking noise levels for a wind speed of 8 m/s (7.5 – 8.5 m/s) for measurement day 2 for the four microphone positions. A microphone position is shown in each of the four plots containing both all masking noise levels within 8 m/s wind speed together with the average masking noise level, 95 % confidence levels, maximum level and minimum level. As can be seen there is a large variation of 10-20 dB for the different frequencies, with the largest variation for the highest frequencies.

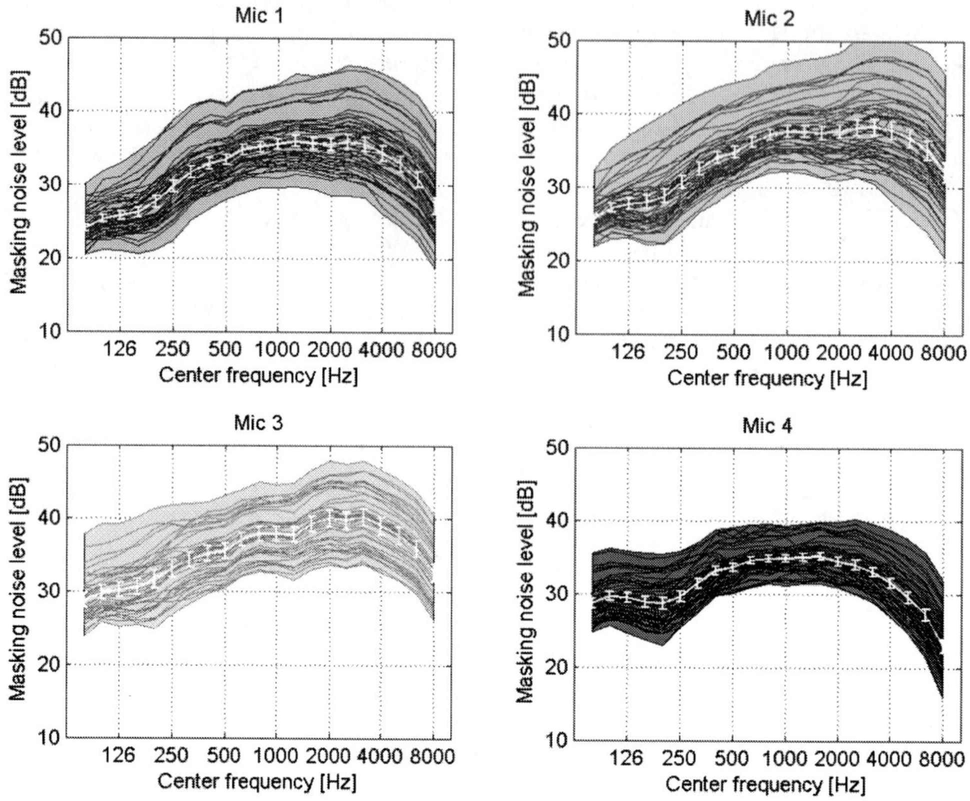

Figure 16 Masking noise levels (colour line), average masking noise level (white), 95 % confidence interval (white), maximum and minimum levels (coloured area) at 8 m/s for Day 2 for the four microphone positions

5. Conclusions

Two measurement campaigns performed at the same site has been further investigated and intercompared in order to gain knowledge about variation of wind induced non-turbine related noise in a Danish context.
Based on the results there are a linear relationship between wind induced non-turbine related noise and wind speed – the more wind the more noise.
Further it seems (as expected) that wind induced non-turbine related noise level is dependent on wind direction, wind exposure/shelter, season and measurement position.

Assessment of wind induced non-turbine related noise in the context of wind turbines in general has two purposes; I) either to perform measurements with as low wind induced non-turbine related noise levels as possible to achieve the best signal-to-noise ratio for optimal determination of a noise level or II) to assess the most appropriate masking noise level to determine audibility of the wind turbine.

If the purpose of measurement is to achieve as low as possible wind induced non-turbine related noise it is assumed that the key is to both obtain a sheltered microphone position and a position with low vegetation noise; to obtain the latter, it is important to be either:

- Far from vegetation (which most often means an unsheltered position)
- With sheltered vegetation (is often possible, but condition might change with wind direction/season)

- Shielded/sheltered from a building, which can cause shielding and/or reflections of wind turbine noise

No one of the used microphone positions had the lowest noise level for all situations. In future work, it could be interesting to test further positions for example ground microphone positions (like IEC 61400-11) or mounted on the wall. If possible it can for now be suggested to use several microphone positions for tonal assessment. Alternatively, if the tone frequency is known in advance one microphone should suffice, and a free field position (like microphone position 4) should be used for tone frequencies above 250 Hz and a sheltered position (like microphone position 1) should be used for tone frequencies below 250 Hz.

In order to choose the most appropriate masking noise level the results can be used as guidance depending on context.

As future work it could of course be interesting to measure over a longer period with varying wind speed, wind direction and season to achieve a good description of the influence on wind induced non-turbine related noise.

References

Søndergaard, Lars S and Søndergaard, Bo (2008) *Low Frequency Noise from Large Wind Turbines – Background Noise Measurement and evaluation,* DELTA AV 138/08

Søndergaard, Lars S and Pedersen, Torben H (2015) *Vurdering af toneindhold i vindmøllestøj hos naboer (Evaluation of tonal content in wind turbine noise at residents)* Reference laboratory for Noise measurements for the Danish Environmental Protection Agency/DELTA, Report nr. 28 (in Danish)

Danish Environmental Protection Agency, *Støj fra vindmøller (Noise from wind turbines)* – Guideline from the Danish Environmental Protection Agency no. 1, 2012 (in Danish).

Statutory Order on Noise from Wind Turbines; Statutory Order no. 1284 of 15. December 2011 (in Danish. English translation can be seen at http://eng.mst.dk/media/mst/66206/engelsk_vindmoellebekendtgoerelse.pdf).

IEC 61400-11 Ed. 2.0 IDT *Wind turbine generator systems – Part 11: Acoustic noise measurement techniques*

Danish Environmental Protection Agency, *Guidelines for Measurements of Environmental Noise*, 6/1984 (in Danish), Nov. 1984.

7th International Conference
on
Wind Turbine Noise
Rotterdam – 2nd to 5th May 2017

Evaluation of Wind Turbine Noise in Japan

Mimi NAMEKI[1]; Akira SHIMADA[1]; Nobuo MACHIDA[2]; Hideki TACHIBANA[3]
1:Ministry of the Environment, Japan (MOEJ) , oto@env.go.jp
2: Nihon University,Japan, machida@eme.cst.nihon-u.ac.jp
3: The University of Tokyo,Japan, pon-t@iis.u-tokyo.ac.jp

Summary

While increasing in size and number, wind turbines in Japan are often located in quiet rural areas due to the country's wind energy availability. Since the noise of a wind turbine is more noticeable in a quiet environment, this location of wind turbines and their unique acoustical character such as amplitude modulation, sometimes raise complaints about noise by neighborhood residents even if the noise generated by wind turbines is not very loud compared to other environmental noises.

The Ministry of the Environment of Japan (MOEJ) set up an expert committee to discuss issues related to wind turbine noise (WTN) in 2013. In November 2016, the expert committee published a report on the investigation, prediction and evaluation methods of WTN. The report compiles recent scientific findings on WTN, including the results of nationwide field measurements in Japan and the results of a review of the scientific literature related to the health effects of WTN. The report sets out a methodology for investigation, prediction and evaluation as well as case examples of countermeasures. With regard to WTN evaluation, the report proposes that WTN should not be more than 5dB above the background noise where background noise levels are above 35-40dB.

MOEJ plans to develop a noise guideline on WTN and a detailed technical manual for WTN investigation based on the methodology presented by the report.

1. Introduction

Among renewable energy sources, wind power generation is an important energy sources that emits neither air-polluting substances nor greenhouse gases and can also contribute to energy security because the power can be generated by a natural resource readily available in Japan. The Basic Energy Plan of Japan (Cabinet decision in April, 2014) regards wind power generation as an energy source that can be made economically viable because its generation cost could be as low as that for thermal power generation if it could be developed on a large scale.

The number of wind power facilities installed in Japan started to increase around 2001, and 2,034 units were installed by 2014 (as of the end of March, 2015) (1).

According to the Supplementary Materials for the Long-term Energy Supply and Demand Outlook issued by the Agency for Natural Resources and Energy in July, 2015, approximately 10 million kW of wind power is expected to be installed by 2030, which represents a nearly four-fold increase from the existing installed wind power capacity of approximately 2.7 million kW (2).

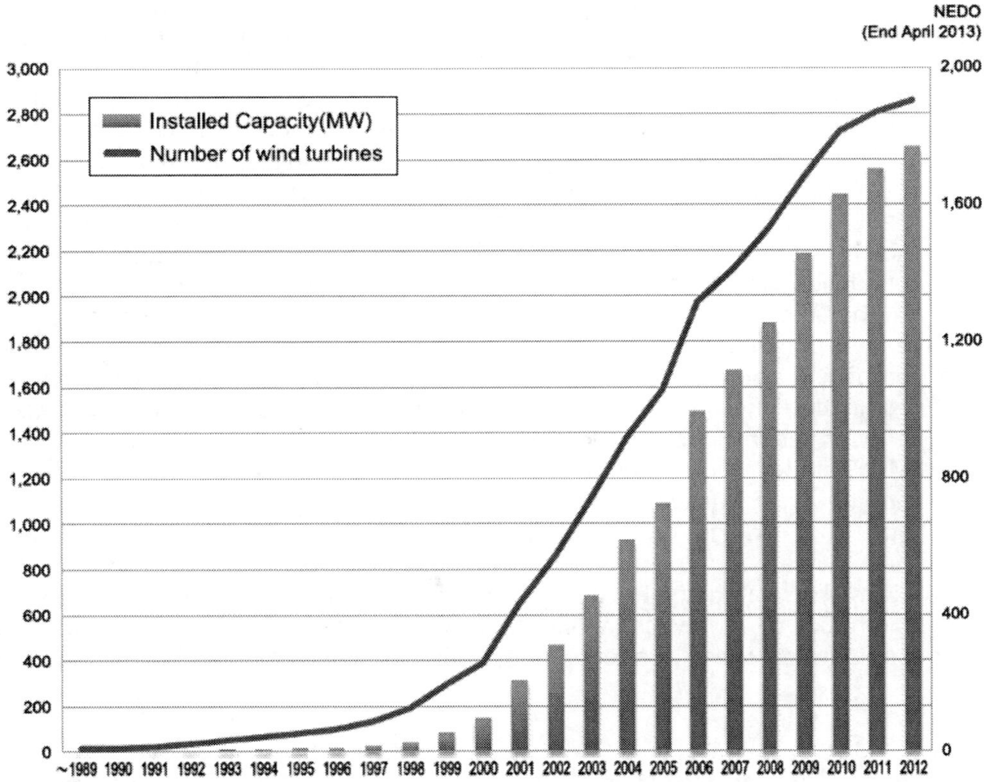

Figure 1. Installed capacity and number of wind turbines in Japan (Source: NEDO)

Wind power facilities emit a certain amount of noise due to their power generation mechanism in which blades rotate by catching wind to generate power. While the noise level is normally not significantly large, there are cases where even a relatively low level of noise causes complaints as wind power facilities are often constructed in agricultural/mountainous areas that have suitable weather conditions including wind direction and velocity that were originally quiet. There have not only been noise complaints but also complaints of inaudible sound of a frequency of 20 Hz or less.

Against such a backdrop, as a result of the amendment of the Order for Enforcement of the Environmental Impact Assessment Act in October, 2012, the establishment of wind power stations came to be classified as relevant projects under the Act and discussions on the environmental impact assessment of wind power facilities have taken place.

In assessing the impact of noise resulting from the installation of a facility, the procedure of environmental impact assessment performed before installation examines "the extent to which such noise can be feasibly avoided or reduced" and, if applicable, "whether it is intended to be consistent with standards or criteria given by the Japanese government or local municipalities from the perspective of environmental protection." For the former examination, the extent to which the impact of noise resulting from the implementation of the relevant project can be feasibly avoided or reduced is assessed by comparing multiple countermeasures in terms of the structure, layout, output, the number of units, and technical noise reduction measures in accordance with the maturity of the project plan. The assessment can also be performed by examining to what extent more feasible technology can be incorporated, etc. Specifically, assessment is made from such viewpoints as whether the local noise level will not be significantly raised, whether the layout plan for the project secures a sufficient distance between the facility and residences, etc.

The Environmental Quality Standards for Noise are generally used for the Environmental Impact Assessment. However, there are acoustic characteristics peculiar to noise generated from wind power facilities (hereinafter, "wind turbine noise"). It is thus necessary to develop methods relevant to the investigation, prediction, and evaluation of wind turbine noise based on the latest scientific findings.

The Ministry of the Environment of Japan (hereinafter, "MOEJ") has set up an expert committee and examined ideas and issues about methods for investigating, predicting, and assessing wind turbine noise from 2013 to 2016. The expert committee published a report on the investigation, prediction and evaluation methods of wind turbine noise in November 2016. During the development of the report, the MOE started a one-month public comment period. All comments were considered, and changes were made to the report where appropriate. The report compiles recent scientific findings on wind turbines in terms of noise, including the results of nationwide field measurements in Japan and the results of review of the scientific literature related to the health effects of wind turbine noise. The report sets out methodology for investigation, prediction and evaluation as well as case examples of countermeasures. This report summaries the report by the expert committee.

2. Outline of the report

The report by the expert committee consists of three parts. The first part explains key findings from past researches, namely the field survey measuring wind turbine noise in Japan and a literature review on wind turbine noise and human health. The second part proposes methods for investigating, predicting and evaluating wind turbine noise. A guideline on wind turbine noise is proposed in this part. The third part states the actions recommended by the expert committee. The following chapters summarize those three parts of the report.

3. Key findings

3.1 Findings from the field survey

Field surveys measuring wind turbine noise conducted in Japan from 2010 to 2012 revealed the following.

- In terms of spectral characteristics, wind turbine noise generally has a spectral slope of -4 dB per octave. It has a 1/3 octave band sound pressure level in all parts of the super-low frequency range, which means 20 Hz or lower, is below the ISO threshold of hearing for pure tones and the criterion curve for the evaluation of low frequency noise proposed by Moorhouse et al. (Fig. 2). Super-low frequency range components of wind turbine noise are at imperceptible levels. Therefore, wind turbine noise is not an issue caused by super-low frequency range.

- In regard to the audible frequency range, in the range from about 40 Hz and above, the 1/3 octave band sound pressure level is above the said criterion curve and the threshold of hearing defined by ISO 389-7. Therefore, wind turbine noise should be regarded as "audible" sound (noise) in discussing it.

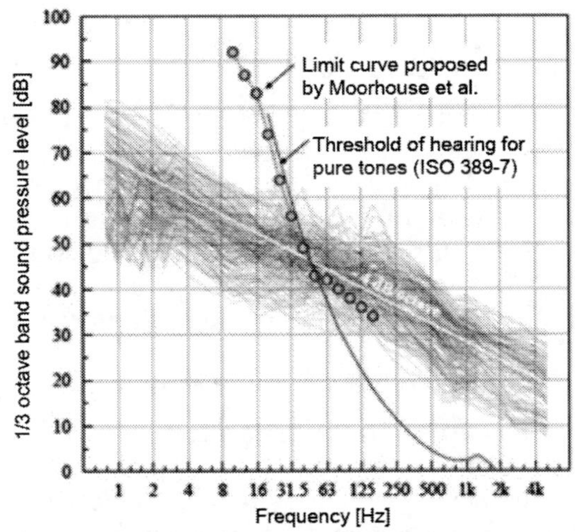

Figure 2. Results of the analysis of frequency characteristics of wind turbine noise (at 164 locations in the vicinity of 29 wind power facilities in Japan)

- Noise exposure levels of nearby residents from wind power facilities are distributed in the range of 26–50 dB in time-averaged A-weighted sound pressure levels. While this implies that wind turbine noise is not significantly higher than other types of environmental noise, it can cause serious annoyance to those living residential areas in the vicinity of wind power facilities located in extremely quiet agricultural/mountainous areas.
- Low-frequency components of wind turbine noise obtained from field measurements were within the range of those of other environmental sounds.
- In Japan, it is known that the following relation holds between L_{Aeq}, which properly excludes non-relevant noise, and L_{A90}: $L_{Aeq} \fallingdotseq L_{A90}+2$ dB

It is also generally said that acoustic isolation is not always effective for noise from wind power facilities because it contains more low-frequency components. In a quiet environment with little noise of other types, it is relatively more easily heard than ordinary noise is.

3.2 Findings from the literature review on health effects

After careful assessment of the evidence obtained from peer reviewed research results from around the world, it has been concluded that wind turbine noise has likely no negative effects on human health.

However, amplitude modulation and the tonal sounds of wind turbine noise tend to increase annoyance. Existing research results indicate that wind turbine noise over 35 – 40 dB raises annoyance and that the risk of sleep disturbance may increase accordingly.

No clear association is seen between infrasound or the low-frequency noise of wind turbine noise and human health.

Some research results have suggested that wind turbine noise related annoyance is also affected by other issues such as visual aspects or economic benefits.

4. Methods for investigating and predicting wind turbine noise, a perspective for its evaluation, and responses against it

In light of the findings described in Section 2, the issue of wind turbine noise should be taken not as one of super-low frequency sound below 20 Hz but as one of "audible" sound (noise), and it should be basically measured at the A-weighted sound pressure level. We here summarize matters to be noted in conducting an investigation and/or the prediction of noise before and after installing wind power facilities and a perspective for wind turbine noise evaluation.

4.1 Investigation and prediction before installation

3.1.1 Matters to be noted upon an investigation

In selecting a method for investigation, it is necessary to collect various kinds of information in light of business and regional characteristics in order to conduct prediction and evaluation appropriately. Particularly with regard to wind turbine noise, it is important to distinguish and discuss three major issues:

(1) Sound source characteristics

It is necessary to pay attention to:

- information on the wind power facility concerned, including its specifications, manufacturer, model number, hub height, rotor diameter, rated wind velocity, and power generation;
- the sound power level of the generated noise;
- the A-weighted overall value and frequency characteristics (including the 1/3 octave band sound power level) of the sound power level at the rated (maximum) output (to grasp the situation of maximal environmental impact);
- A-weighted overall values and frequency characteristics (including the 1/3 octave band sound power level) of sound power levels under different wind velocities;
- pure tonal frequency components (to be determined in accordance with IEC 61400-11:2012); and
- existing data pertaining to the same model in operation.

(2) Propagation characteristics

In Japan, wind power facilities are often installed in agricultural/mountainous areas. Sound waves emitted from a wind power facility installed in an agricultural/mountainous area are affected by various factors before propagating to a sound receiving point (assessment point), in comparison with one installed on a large, flat piece of land such as a plain or desert. Its noise level and frequency characteristics tend to change due to phenomena including reflection, absorption, transmission, refraction, and diffraction. It is therefore necessary to pay attention to:

- phenomena such as the reflection, absorption, or diffraction of wind turbine noise due to undulating terrain or ridges,
- the state of the ground surface (including rivers and lakes), and
- meteorological information such as wind conditions including wind direction, velocity, and frequency.

(3) Information on a sound receiving point (assessment location)

With regard to locations where an investigation is conducted, focusing on the daily life and activities of residents in the vicinity of a wind power facility, it is necessary to pay attention to:

- the configuration of establishments particularly requiring consideration for environmental conservation such as schools and hospitals and the outline of housing configuration (including the structure of each house), and
- the state of the acoustic environment (degree of quietness) of the area in question.

(4) The specific method for investigation

In measuring residual noise in a given area, it is necessary to pay attention to the following.

a. Sound to be excluded

Sounds of the types given below should be excluded. Since wind power facilities operate when wind is blowing, noises caused by wind such as the sound of rustling leaves are not excluded. ("Wind noise" generated by wind hitting a sound level meter's microphone is excluded, however.)

i) transitory noise such as the sound of automobiles passing nearby and aircraft noise
ii) artificial sound not occurring regularly such as sound generated by accidents/incidents, vehicles driven by hot-rodders, emergency vehicles, etc.
iii) natural sound not occurring regularly such as sound generated by natural phenomena including rain and defoliation, animals' cries, etc.
iv) sound incidental to measurement such as the voice of a person talking to a measurer, sound of tampering with measuring instruments, etc.

b. Surveying and other equipment

As the wind is generally strong in areas around wind power facilities, it is important to use a windbreak screen in order to avoid the effects of wind noise to the extent possible when measuring residual noise. Several kinds of urethane spherical windbreak screens of different diameters are commercially available. In general, the larger the diameter of such a screen is, the less likely a sound level meter inside the screen will be affected by wind noise. Installing a windbreak screen can reduce the impact of wind noise up to a wind velocity of around 5 m/s.

c. Survey areas and locations

Considering the propagation characteristics of wind turbine noise, the survey targets areas susceptible to an environmental impact by wind turbine noise, such as residential areas in the vicinity of a wind power facility (generally within a radius of about 1 km from a wind turbine). An area in which a quiet environment should be conserved such as hospital premises may be included in these target areas. In selecting specific survey locations in the survey areas, in addition to locations where a wind power generation facility is planned to be installed, such locations are to be selected that are immune to local impacts of particular sound sources where the average level of noise in the relevant area can be assessed, including residential areas around the wind power generation facility. Measurement is to be performed at an outdoor location 3.5 m or more distant from a reflective object, excluding the ground.

d. Survey period and hours

In order to grasp conditions throughout the year accurately, a survey is to be conducted in each period of the year for different typical meteorological conditions under which a wind turbine operates (for instance, each season if meteorological conditions vary greatly by seasons).

The period of a single survey should be appropriately determined in consideration of the time variation of noise due to the impact of meteorological conditions and other elements. As measurement values may be unstable depending on wind conditions, a survey should be performed for three or more consecutive days in principle. The survey should be conducted both during the day (6:00–22:00) and at night (22:00–6:00) hours.

3.1.2 Matters to be noted in prediction

As mentioned above, in Japan, wind power facilities are often installed in agricultural/mountainous areas. In comparison with cases where such a facility is installed on a large, flat piece of land such as a plain or desert, sound waves emitted from a wind power facility installed in a mountainous area diffuse in a more complicated manner as they propagate due to the influence of geological states, vegetation, meteorological conditions such as wind conditions, etc. In addition, it should be noted that the propagation of wind turbine noise is extremely complicated as it is subject to attenuation by distance, reflection and absorption by the ground surface, reflection and diffraction by acoustic obstructions, attenuation by atmospheric absorption, etc.

Among the prediction methods used, while "ISO 9613-2 : 1996" allows incorporation of more detailed conditions, the prediction calculation becomes rather complex. Furthermore, there is the problem of how the reflection rate should be calculated in cases where the effect of reflection by the ground surface becomes an issue, as is the case with a wind turbine installed on a ridge.

The New Energy and Industrial Technology Development Organization (hereinafter, "NEDO") published a prediction method for the environmental impact assessment of wind power generation in July, 2003 (revised as the second version in February, 2006). This models wind power facilities as sound source points and uses sound power levels provided by manufacturers of wind power generators. This method takes into account distance attenuation due to sound diffusion in the propagation process and attenuation by atmospheric absorption. While this method can be used easily, it is difficult to consider meteorological effects, etc.

It is necessary to pay attention to such characteristics of methods in making predictions.

4.2 Survey after the installation of a wind turbine

As stated in Section 3.1, predicting wind turbine noise involves elements with large uncertainty such as emission characteristics of noise from the source and effects of meteorological conditions as well as the terrain and structures in the propagation process. Predicted values before the installation of a wind turbine and measured values after installation may sometimes differ greatly.

We here summarize matters to be noted in a survey after the installation of a wind turbine.

(1) Conditions of measurement

It is necessary to grasp the conditions of measurement and other relevant local matters that may impact the propagation of noise. At least, one should grasp the wind direction and velocity at the nacelle height, the variation of power output, and meteorological data required for calculating the attenuation by atmospheric absorption (wind direction and velocity, temperature, and humidity).

(2) Survey method

Wind turbine noise varies greatly according to the wind conditions, and a wind turbine often starts and suspends operation repeatedly. Therefore, measurement should be performed in appropriate hours considering the state of operation of the wind power facility in question. For example, a method is conceivable that measures the average level in a 10-minute period in which wind turbine noise is stable (10-minute equivalent noise level: $L_{Aeq,}$ 10 min) and regards it as the representative value. If the relevant wind power facility operates steadily for many hours, it is effective for obtaining robust data, for instance, to measure noise for 10 minutes every hour on the hour and calculate the average energy over the entire period of time.

For measurement locations, period, etc., refer to what is noted for a survey before the installation.

(3) Survey Results

The representative value of a survey after the installation of a wind power facility should be taken as the A-weighted equivalent sound pressure level measured over a period of time in which the effect of wind turbine noise is at its maximum and in which the effect of background noise is low (e.g. during night time). It is also required to confirm whether there is any pure tonal component.

The equivalent noise level during operation can be estimated by adding around 2 dB to the noise level exceeded for 90% of the measurement period (L_{A90}).

4.3 Evaluation of wind turbine noise

With regard to the evaluation of wind turbine noise, the expert committee proposed the development of a new guideline. Detailed proposals on the new guideline are as follows:

- The guideline should be applied when a wind power facility will be newly built or a wind power facility will be retrofitted to add a power generation facility.
- As a guideline value, "residual noise + 5dB" is proposed.
- Residual noise should be measured when wind is steady.

- In low noise environments, a lower limit for wind turbine noise should be set since there is no acoustic benefit. WTN should be limited to 35dB in the areas where background noise is lower than 30 dB and where some noise sensitive locations exist. For other areas, 40 dB should be set as the lower limit of wind turbine noise.
- To apply the guideline, locations where WTN might affect residents' daily activities (e.g. nearest dwellings) should be selected.
- To conserve the indoor environment, evaluation should be made based on outside noise data (both day and night).

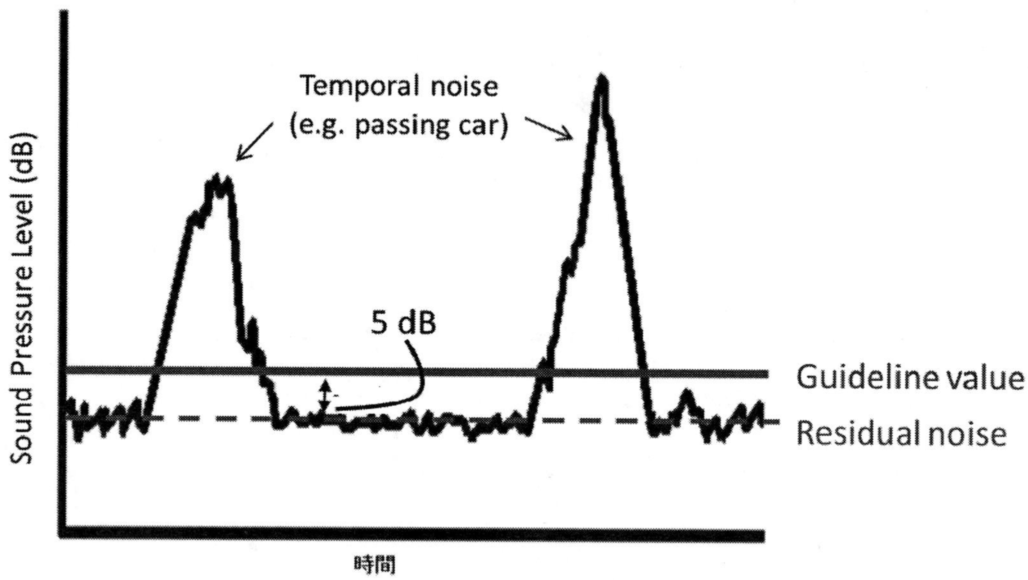

Figure 3. Image of relationship between residual noise and guideline value

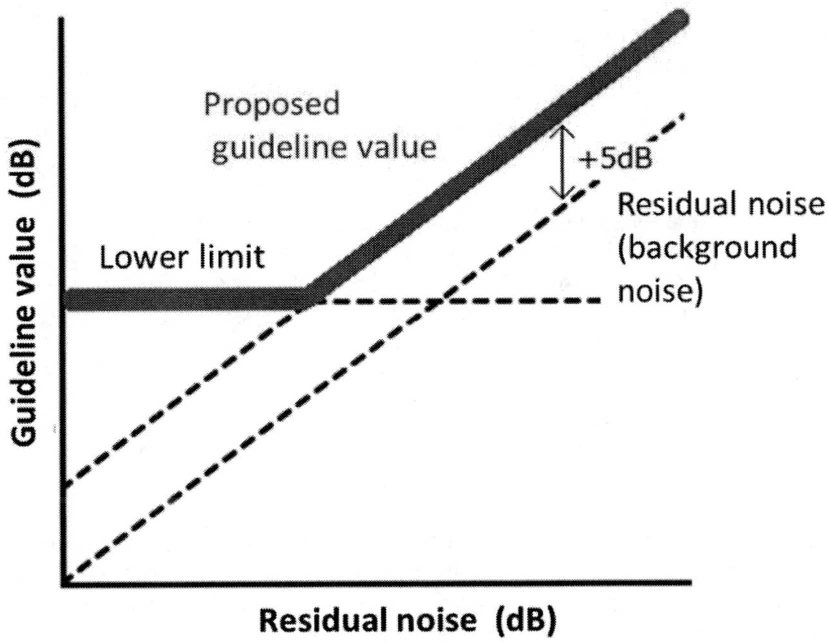

Figure 4. Image of guideline value

5. Next Actions

5.1 Actions to be taken by operators and manufacturers of wind power facilities

Operators and manufacturers will continue to be expected to accumulate survey data after the installation of wind power facilities, implement technical measures, such as developing low noise blades or implementing additional soundproofing measures, and maintenance measures intended to reduce noise, etc. Furthermore, they are also expected to examine and develop technology supporting the broad promotion of efforts for noise control including the examination of an aerodynamic sound propagation prediction model reflecting locational conditions.

5.2 Actions to be taken by administrative agencies (the government of Japan and local municipalities)

5.2.1 Collecting and sharing information on wind power facilities, raising awareness

It is necessary to develop and improve manuals for appropriately responding to complaints concerning wind power facilities. At the same time, it is necessary to examine a framework for sharing knowledge of technological countermeasures implemented by operators that can be applied to other facilities, to administer education and training programs to enhance local municipality officials' expertise further, to promote understanding by local residents through the dissemination of precise information on the auditory impression of wind turbine noise and similar matters as well as raising their awareness of such information, etc.

It is possible that not only the magnitude and properties of sound but also visual elements are related to complaints about noise from wind power facilities. It is necessary to continue to gather knowledge on the impact of elements other than noise and examine responses.

5.2.2 Perspective for the evaluation of wind turbine noise

It is necessary to consider all facilities, not limited to wind power facilities, located therein. It is also necessary to examine what methods for investigating, predicting, and assessing the sound environment in quiet areas in Japan should be like while surveying examples in other countries.

5.3 Actions to be taken by all parties concerned

When it comes to wind turbine noise, it is important to facilitate communication among relevant stakeholders including operators of wind power facilities, manufacturers, local municipalities, and local residents, in light of issues unique to sensory pollution. It has been reported that annoyance caused by wind turbine noise is low among residents who perceive wind turbines positively so that receptivity to the installation of a wind turbine facility is an important factor. There are cases where actions for maintaining a favorable relationship with local residents reduced complaints. Such actions include a wind power facility operator's holding briefing sessions, creating an optimal business plan based on a comprehensive analysis of the distance separating residences and the relevant establishment in conjunction with the installation and layout of a wind power facility, continuing to deal with complaints, and concluding an agreement with local residents and municipalities. It is necessary to enhance communication among the parties concerned in this light.

6. Conclusion

This paper summarizes the basic ideas and methods proposed by the report published by the expert committee on wind turbine noise in November 2016. MOEJ plans to develop a noise guideline and a detailed technical manual for wind turbine noise investigation based on the methodology presented by the report in 2017.

References

1. New Energy and Industrial Technology Development Organization (NEDO). NEDO offshore wind energy progress Edition II. [Internet] 2013. p.5. Available from: http://www.nedo.go.jp/content/100534312.pdf?from=b
2. Agency for Natural Resources and Energy, Ministry of Economy, Trade and Industry. [Internet] 2015. p.47. Available from: http://www.enecho.meti.go.jp/committee/council/basic_policy_subcommittee/mitoshi/011/pdf/011_07.pdf

7th International Conference
on
Wind Turbine Noise
Rotterdam – 2nd to 5th May 2017

Noise guidelines for countries where no national regulations are available. A discussion of the IFC EHS Guidelines for Wind Energy

Pierre Dutilleux – Evergy Engineering GmbH – Malsenstrasse 84, 80638 Munich Germany, pierre.dutilleux@evergy.de

Summary

The acoustic assessment of a wind farm project is performed according to the national regulations of the host country. Experience gained in developed wind markets shows that a detailed regulatory framework is necessary in order to achieve reliable and stable results. Some countries however have not yet adopted such regulations. In this case, the International Finance Corporation (IFC) Environment, Health and Safety (EHS) Guidelines can be applied as an alternative standard.

1. Introduction

The development and the siting of wind-farm projects must take into consideration the local environmental constraints and health protection regulations of the country of application. In developed and industrial countries, the intricate sets of regulations and standards which have been developed over the past decades for the industry and the infrastructure projects have been adapted or extended to address the wind-energy specific issues [1].

In many countries around the world however, no such comprehensive sets of regulations and standards are available yet. In the lack of local specific regulations, the project developer might design the project on the basis of his own expertise or according to his own standards but, once financing institutions (banks and/or investors) get involved in the project; these might impose further standards.

"The Equator Principles is a risk management framework, adopted by financial institutions, for determining, assessing and managing environmental and social risk in projects. It is primarily intended to provide a minimum standard for due diligence to support responsible risk decision-making. The Principles are a framework to require the implementation of good international practice in relation to the environmental and social issues arising in projects that Equator Principles Financing Institutions (EPFIs) are financing or advising." The Equator Principles refer either to the International Finance Corporation (IFC) Performance Standards and the World Bank Group Environmental, Health and Safety Guidelines (EHS) [3] or, if relevant and applicable, to local or national law relating to environmental and social matters [2].

2. The IFC EHS General and the Wind Energy Guidelines, 2007

The World Bank Group has defined the Environmental, Health, and Safety (EHS) Guidelines which are structured in a general document applicable for all projects as well as several specific documents, one of which is dedicated to wind energy. The General Guidelines set Noise Level Guidelines in their Section 1.7 [3].

Table 1.7.1- Noise Level Guidelines[54]		
Receptor	One Hour L_{Aeq} (dBA)	
	Daytime 07:00 - 22:00	Nighttime 22:00 - 07:00
Residential; institutional; educational[55]	55	45
Industrial; commercial	70	70

[54] Guidelines values are for noise levels measured out of doors. Source: Guidelines for Community Noise, World Health Organization (WHO), 1999.

[55] For acceptable indoor noise levels for residential, institutional, and educational settings refer to WHO (1999).

Table 1: IFC EHS Noise Level Guidelines

The 2007 edition of the IFC EHS General Guidelines states that the "noise impacts should **not** exceed the levels presented in [Table 1], **or** result in a maximum increase in background levels of 3 dB at the nearest receptor location off-site".

The IFC EHS Wind Energy Guidelines [4] recall that "noise impacts should **not** exceed the levels presented in the General EHS Guidelines, **nor** result in a maximum increase in background levels of 3 dB at the nearest receptor location." Both formulations, although similar, are obviously not identical.

The French version of the IFC EHS Guidelines for Wind Energy says: « Le bruit généré par les éoliennes **ne doit pas** excéder les niveaux indiqués dans les Directives EHS générales. Il **ne doit pas non plus** entraîner une augmentation de l'intensité du bruit de fond supérieure à 3 dB à l'emplacement du récepteur le plus proche ». This wording is again different from both of the English versions. We conclude that the different versions of the 2007 Guidelines are somewhat ambiguous and therefore leave room for interpretation.

The ambient noise level is the sum of the background noise level and of the wind-farm noise level. In some countries, the basic rule says that the higher the background noise level, the louder the wind-farm is allowed to be. In other countries such as Germany, the maximum wind-farm noise level is limited irrespective of the background noise level [14]. The following diagram illustrates the planning domains A, B and C which could be exploited depending on the formulation and/or the applicable regulatory framework. We refer to a few selected countries for illustration purposes, only.

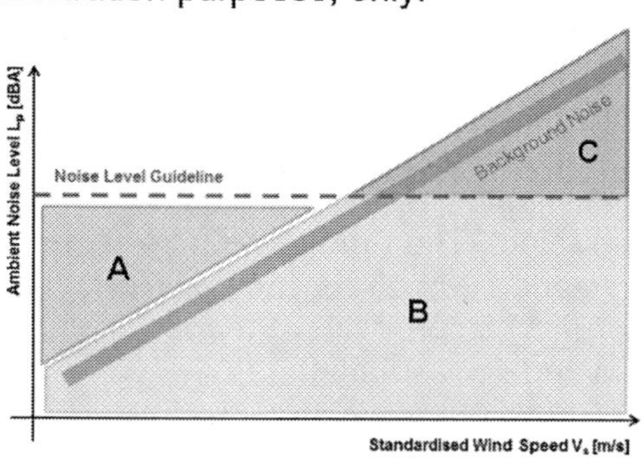

Planning allowed in domain:	A	B	C
Germany	Yes	Yes	No
France	Yes if <= 35dBA	Yes	Yes
IFC EHS Wind Energy, French Version, 2007	No	Yes	No
UK	Yes if <= 35 to 43 dBA	Yes	Yes

The domain A corresponds to quiet areas which can be particularly protected in some countries such as Austria ("schutzwürdige Gebiete" [17]) and in France (if ambient noise is beyond 35 dBA). The domain B is usually allowed in most countries. The domain C must be avoided in some countries such as Germany; in this case the maximum output of the wind turbines is curtailed so that the wind-farm noise level does not increase during strong wind periods.

The dwellers would be best protected under application of the IFC EHS Wind Energy Guideline in its French flavour of 2007 but its application would lead to substantial operational restrictions during high wind-speed periods. The production losses at high wind speeds might be acceptable in regions with moderate average wind speeds but they would heavily affect the profitability of the projects which are developed in regions with very high wind-energy potential.

3. Guidelines for Wind Energy 2015

An updated version of the Guidelines for Wind Energy has been published in 2015 [6].

The 2015 Guidelines for Wind Energy are currently available in the English, Spanish and Arabic languages (we have reviewed the English version, only). The Noise Level Guidelines are formulated in the General Guidelines, only. The ambiguity discussed in Section 2 has been avoided and the allowed planning domain is now very broad as shown in the following diagram:

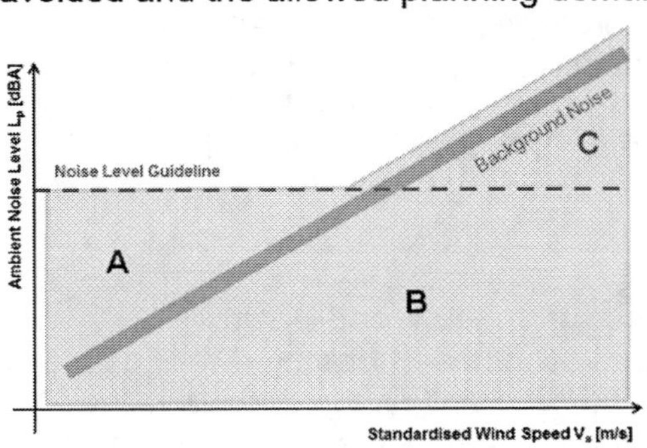

Compared to the Guidelines of 2007, the new version has been significantly extended by addressing some implementation details. The 2007 version simply mentioned that "The applicability of specific technical recommendations should be based on the professional opinion of qualified and experienced persons" and provided a list of basic general reference publications. The 2015 version now calls for Good International Industry Practice (GIIP) and refers to a lengthier list of more specific national publications. We are of the opinion that this list of references is fairly unbalanced. For example, reference is made to the French regulations [7] which link to the French preliminary standard on the assessment of wind-farm noise [8]. On the other hand we did not find any reference to the German regulations on wind turbine noise although the extended planning and operating experience of this country would be worth mentioning. The probably more consistent set of national references is provided by the ETSU-R-97 report [9] and its Institute Of Acoustics (IOA) Good Practice Guide [10].

The IFC stresses the point in its Guidance Note 3 that the noise impact of the wind-farm project must be carefully assessed [13]. Whereas very few details were given in the 2007 version of the Guidelines about the noise impact assessment method, the outline of a method can be found in the 2015 version of the Guidelines for Wind Energy which is largely inspired from the ETSU-R-97 report [9] and the IOA Good Practice Guide [10].

"The [party responsible for implementing and operating the project] will refer to the EHS Guidelines or other internationally recognized sources, as appropriate, when evaluating and selecting resource efficiency and pollution prevention and control techniques for the project. The EHS Guidelines contain the performance levels and measures that are normally acceptable and applicable to projects. When host country regulations differ from the levels and measures presented in the EHS Guidelines, [parties responsible for implementing and operating the project] will be required to achieve whichever is more stringent. If less stringent levels or measures than those provided in the EHS Guidelines are appropriate in view of specific project circumstances, the client will provide full and detailed justification for any proposed alternatives through the environmental and social risks and impacts identification and assessment process. This justification must demonstrate that the choice for any alternate performance levels is consistent with the objectives of this Performance Standard" [12].

"During the project life-cycle, the [party responsible for implementing and operating the project] will consider ambient conditions and apply technically and financially feasible resource efficiency and pollution prevention principles and techniques that are best suited to avoid, or where avoidance is not possible, minimize adverse impacts on human health and the environment. The principles and techniques applied during the project life-cycle will be tailored to the hazards and risks associated with the nature of the project and consistent with good international industry practice (GIIP), as reflected in various internationally recognized sources, including the World Bank Group Environmental, Health and Safety Guidelines (EHS Guidelines)" [12].

"Good International Industry Practice (GIIP) is defined as the exercise of professional skill, diligence, prudence, and foresight that would reasonably be expected from skilled and experienced professionals engaged in the same type of undertaking under the same or similar circumstances globally or regionally. The outcome of such exercise should be that the project employs the most appropriate technologies in the project-specific circumstances" [12].

4. Discussion

The Noise Level Guidelines are based on recommendations from the World Health Organisation (WHO) and are usually suitable for developed countries where the dwellings are built according to minimum standards and where the partitions as well as the doors and the windows are able to provide a minimum acoustic insertion loss. The figures shown in Table 1 are in the range of what is practised in Europe where the industry and the infrastructures have already reached a high level of development. The large number of now operating wind farms in some European countries should not be used as an argument for the general suitability of such Noise Level Guidelines. Several cases are documented where the dwellers complain although the immission levels are well within the applicable noise level guidelines [15]. In countries with lower construction standards, the outdoor noise level guidelines might have to be lowered in order to ensure that the indoor noise levels comply with the WHO recommendations for indoor levels. As an example, special attention should be paid when the people are living in compounds with patios where they usually keep the doors and the windows open during the night.

The evasive formulations of the 2007 guidelines left a large interpretation freedom to the project developers. The outline of the assessment procedure as introduced in the 2015 wind-energy guidelines should pave the way for detailed and qualified investigations and assessments. Experience gained in the UK shows however that lengthy discussions often go along with its application [10].

The acoustic indicator to choose for the acoustic assessment is not specified in the IFC EHS Guidelines. This leaves the project developer the possibility to choose an indicator which is compatible with the otherwise applicable national regulations. The choice of an acoustic indicator and/or of an associated project-specific assessment procedure must however remain consistent with the objectives of the Performance Standards [12]. The outline of the noise impact assessment procedure as proposed by the 2015 Guidelines is inspired by the ETSU-R-97 report and one of the outcomes of this report is that, after extensive and valuable investigations, the $L_{A90,10min}$ acoustic indicator has been found particularly suitable. This gives a hint about the level of expertise and scrutiny which is expected by the Performance Standards.

When the projects are developed at sites with very high wind-energy potential, the risk is high that the acoustic measurements are heavily affected by the wind-induced noise at the microphones. By referring to the method proposed by ETSU-R-97, the 2015 Guidelines also open the possibility to account for this factor [10][16].

5. Conclusions

The IFC EHS Guidelines set minimum standards for the acoustic assessment of wind farm projects which should be adhered to unless more stringent host country regulations are in place.

The Guidelines do not provide all the implementation details and leave room for alternative methods if they can be justified to be more appropriate for the project. Expert knowledge should be sought in order to implement solutions in line with Good International Industrial Practice. The outline of the assessment method which is proposed by the 2015 version of the Guidelines for Wind Energy sets high standards.

During the course of the project development, over to the construction, the commissioning and the operation phases, the regulatory framework often evolves and the feasibility of the project has to be reassessed whenever a significant change is introduced. These changes can be perceived as threats by the project developer, while the scientific, technological and economic progress might offer new opportunities for optimisation and improvement of the project.

References

[1] Mark Bastasch (2011) *Criteria*. In Dick Bowdler and Geoff Leventhall *Wind Turbine Noise, Chapter 8*

[2] Equator Principles Association (2016) *Governance Rules*, January

[3] World Bank Group, IFC (2007) *Environmental, Health, and Safety (EHS) Guidelines. General EHS Guidelines: Environmental Noise Management,* 30.04.2007

[4] World Bank Group, IFC (2007) *Environmental, Health, and Safety (EHS) Guidelines for Wind Energy,* 30.04.2007

[5] World Bank Group, IFC (2007) *Directives environnementales, sanitaires et sécuritaires pour l'énergie éolienne,* 30.04.2007

[6] World Bank Group (2015) *Environmental, Health, and Safety Guidelines for Wind Energy,* 07.08.2015

[7] *Arrêté du 26 août 2011 relatif aux installations de production d'électricité utilisant l'énergie mécanique du vent au sein d'une installation soumise à autorisation au titre de la rubrique 2980 de la législation des installations classées pour la protection de l'environnement*

[8] NF S 31-114 *Mesurage du bruit dans l'environnement avec et sans activité éolienne*

[9] ETSU for the Department of Trade and Industry (DTI) (1997) *The Assessment and Rating of Noise from Wind Farms,* Report ETSU-R-97

[10] Institute of Acoustics (IOA) (2013) *A Good Practice Guide to the Application of ETSU-R-97 for the Assessment and Rating of Wind Turbine Noise*

[11] Institute of Acoustics (IOA) (2014) *A Good Practice Guide to the Application of ETSU-R-97 for the Assessment and Rating of Wind Turbine Noise, Supplementary Guidance Note 4: Wind Shear*

[12] World Bank Group, IFC (2012) *Performance Standards 3, Resource Efficiency and Pollution Prevention.* 01.01.2012

[13] World Bank Group, IFC (2012) *Guidance Note 3, Resource Efficiency and Pollution Prevention.* 01.01.2012

[14] Pierre Dutilleux, Joachim Gabriel (2008) *Assessment of the acoustic noise issues of wind farm projects in the light of the experience gained in Germany*. DEWEK, November

[15] Joachim Gabriel, Steffen Vogl, Thomas Neumann, Gundula Hübner, Johannes Pohl (2013) *Amplitude Modulation and Complaints about Wind Turbine Noise*. Wind Turbine Noise, Denver

[16] Justin Adock, Christophe Delaire, Daniel Griffin, Mark Jiggins (2015) *Study of secondary wind shield performance in the field*. Wind Turbine Noise, Glasgow

[17] Judith Lang (1998) *Anforderungen an schalltechnische Projekte*. R-157, Umweltbundesamt, Wien

7th International Conference
on
Wind Turbine Noise
Rotterdam – 2nd to 5th May 2017

Environmental Impact Assessment(EIA) of Wind Power Generation Plan in Korea

Hyosung Sun, Korea Environment Institute, hssun@kei.re.kr

Summary

The Environmental Impact Assessment(EIA) of a wind power generation plan is performed to evaluate the environmental effect prediction and prepare the reduction measures for minimizing the environmental damage. Among the various environmental effects by performing a wind power generation plan, this paper introduces the environmental impact assessment concerned with the wind turbine noise.

1. Introduction

In Korea, the Environmental Impact Assessment(EIA) of a development project is performed in order to evaluate the prediction results of the environmental effect and prepare the reduction measures for minimizing the environmental damage. In the case of a wind power generation plan, the environmental effect of constructing and operating a wind turbine is evaluated in order to establish the environment-friendly plan. The important items of the environmental effect by a wind turbine consist of geographical damage, ecological effect, noise including low frequency, scenic influence. Especially, the noise problem from a wind turbine is more important, because most of the civil complaints from a wind turbine are caused by the wind turbine noise with low frequency. Therefore, this paper is focused on the introduction of the environmental impact assessment concerned with the wind turbine noise including low frequency.

2. EIA Contents of Wind Turbine Noise

The Environmental Impact Assessment(EIA) contents of a wind turbine noise consist of present-condition investigation, noise affect prediction, noise decreasing method, noise monitoring. In the stage of present-condition investigation, the noise measurement is performed in the residential facilities around the installation region of a wind turbine in order to understand the background noise distribution before the installation and operation of a wind turbine. In the step of noise affect prediction, the noise effect distribution is predicted in the construction time and operation time of a wind turbine by the appropriate noise prediction methods. In the stage of noise decreasing method, when the noise values exceed the noise limit, the various noise decreasing methods are applied in order to satisfy the noise limit for minimizing the noise damage from development plan performance. In the step of noise monitoring, when a wind turbine is constructed or operated, the noise measurement is performed in the residential facilities around the construction and operation region of a wind turbine.

3. Prediction of Wind Turbine Noise

The prediction of the wind turbine noise according to the construction and operation of a wind turbine plays an important role in the environmental impact assessment of a wind power generation plan. In the case of the construction time of a wind turbine, the noise of the various construction equipments applied to the construction of a wind turbine are calculated. After the individual noise values of the various construction equipments are composed, the noise values in the residential facilities around the construction region of a wind turbine are computed by a point-source equation of attenuation in distance. In the case of the operation time of a wind turbine, the noise value including low frequency according to the operation of a wind turbine is calculated. After the maximum noise power level according to wind speed change is chosen, the noise values in the residential facilities around the operation region of a wind turbine are computed by a wind turbine noise propagation prediction method. When the multiple wind turbines are operated, the individual noise values of a single wind turbine are computed and the final noise values at the residential facilities are calculated by composing the individual noise values.

4. Decreasing Method of Wind Turbine Noise

After the noise prediction according to the construction and operation of a wind turbine is performed, the noise prediction results are compared with the noise limits. When the noise values of the residential facilities exceed the noise limit, the noise reduction measures are established in order to minimizing the noise damage from the construction and operation of a wind turbine. In the case of the construction time of a wind turbine, when the construction noise value exceed the construction noise limit, the various noise reduction measures such as sound-proofing wall installation, low-noise construction equipment application, etc. are adopted according to the characteristics of the construction region and surrounding area of a wind turbine. In the case of the operation time of a wind turbine, when the operation noise value exceed the operation noise limit, the various noise reduction measures such as low-noise wind turbine application, separation distance from a residential facility, etc. are adopted in order to solve the civil complaints from the noise problem.

5. Wind Turbine Noise Monitoring

Because the environmental impact assessment of a wind power generation plan predicts the future circumstances, it has the limitation of including the uncertainty of the future circumstances. The wind turbine noise monitoring is performed as the solution plan of the above limitation of environmental impact assessment. When the construction and operation of a wind turbine is actually performed, the noise measurement of the residential facilities around the construction and operation of a wind turbine is executed. When the noise values from the noise monitoring exceed the noise limit, the additional noise reduction measures are applied in order to minimize the wind turbine noise damage and solve the civil complaint from a wind turbine noise.

6. Conclusions

This paper introduces the environmental impact assessment concerned with the wind turbine noise. The EIA contents of a wind turbine noise is composed of present-condition investigation, noise affect prediction, noise decreasing method, noise monitoring. Among them, the prediction of the wind turbine noise according to the construction and operation of a wind turbine is performed by a source noise power level information and noise propagation method. When the noise prediction results exceed the noise limit, the various noise reduction measures are established to minimize the noise damage and civil complaint. Also, the wind turbine noise monitoring is performed to overcome the limitation of the uncertainty of the environmental impact assessment.

7th International Conference
on
Wind Turbine Noise
Rotterdam – 2nd to 5th May 2017

Assessment of the error between measured and predicted noise levels from wind farms

Luc Schillemans – Tractebel Luc.Schillemans@Tractebel.Engie.com
Marc Van Caillie – Laborelec Marc.VanCaillie@Engie.com
Simon Courret – Engie Simon.Courret@Engie.com
Colin Le Bourdat – Engie Colin.LeBourdat@GDFSuez-FE.com

Summary

Literature studies show that for wind turbines there can exist non negligible differences between the measured and the predicted noise pressure values at resident's locations.

For this present study, measurements have been selected from three French ENGIE wind farms with several weeks of available data. The wind farms have at least 4 wind turbines of 2 to 3MW. Noise measurements were selected at about 4 locations at each wind farm.

Simulations are done using the basic ISO 9613-2 method and the advanced Nord2000 method for the determination of the transfer function, being the difference between the noise power level at the hub height of the wind turbines and the noise pressure level at each resident location. The wind turbines operate in standard and different curtailment modes. The relation between the wind speed at hub height and the noise power level is determined according to IEC 61400-11. The wind speed, direction, turbulence and shear were measured by means of a ground based Lidar system. Sensitivity analysis regarding the transfer function for the advanced Nord 2000 calculation code were also investigated.

Comparisons with measurements were done based on the sound pressure level that is exceeded for 50% of the time during a 10 minutes interval (occurrences). As the overall measurements were clearly influenced by residual noise, the 500Hz octave band value was selected showing the best signal to noise ratio.

Comparison was done at moments with the largest observed difference in noise level between on and off condition of the wind turbine, resulting in reliable measured specific levels for a very small number of occurrences. The method was then extended for different wind speeds and directions bins. The minimum value was retained for each bin, most likely to be the least influenced by other noise sources. A correction of the residual noise was performed.

Finally, calibration values are proposed.

1. Introduction

A correct wind turbine noise propagation modelling is essential both for wind farm design, taking into account the noise impact on residents, as well as for wind farm control, in order to assess noise curtailment strategies, once the wind farm is in operation.

Wind turbine noise propagation in a complex non homogeneous medium, as it is the case for operating wind turbines, is a non trivial issue. Many physical processes are involved like noise refraction and noise diffusion, due to the wind and thermal stratification of the atmosphere in which the turbine is operating.

This study about wind turbine noise propagation modelling covered the following tasks:

- A bibliographical analysis: a general overview of the problematic, the main mathematical resolution techniques of the acoustic wave equation, the main standards (normalised calculation schemes) and their implementation in software packages, and an investigation of the return of experience.

- A testing of different acoustic propagation codes and comparison of the results by assessing their predictions versus measurements for some selected wind farms. This comparison is presented in details in this paper, and more dedicated to the Radenac case study.

2. Summary on wind turbine noise propagation modelling issues and selection of ISO 9613-2 and Nord 2000 calculation codes

Three different situation may occur, i.e a positive wave front speed gradient (downwards refraction condition), a negative wave front speed gradient (upwards refraction condition), and a zero stratification (i.e an homogeneous case). The next table summarizes the origins of the wave front speed gradients and the consequences on acoustic propagation.

Negative gradient – upwards refraction	Positive gradient – downwards refraction	Zero gradient
• **Thermal origin**: negative temperature profile, day conditions, unstable thermal conditions • **Wind origin**: wind direction in opposite direction with the wave front direction • **Unfavourable** to wave front propagation • Possibility of a shadow zone	• **Thermal origin**: positive temperature profile, night conditions (thermal inversion conditions), stable thermal conditions • **Wind origin**: wind direction aligned with direction with the wave front direction • **Favorable** to wave front propagation	• No wind and no temperature profile (neutral thermal conditions at high wind speeds), begin and end of day or cloudy conditions • Opposite contributions between thermal and aerodynamic effects, like : -opposite wind speed propagation compared to wave front propagation at night -aligned wind speed compared to wave front propagation during a sunny day • Homogeneous conditions

Figure 1: wave font gradients (origins and consequences on noise propagation)

Examples of combination of mean wind/temperature vertical profile effects on wave front refraction is illustrated at the next figure: no stratification (A); negative mean temperature profile – typical day situation (B); positive temperature mean profile – typical night situation (C); uniform positive mean wind speed profile (D); combination of mean wind and thermal profiles.

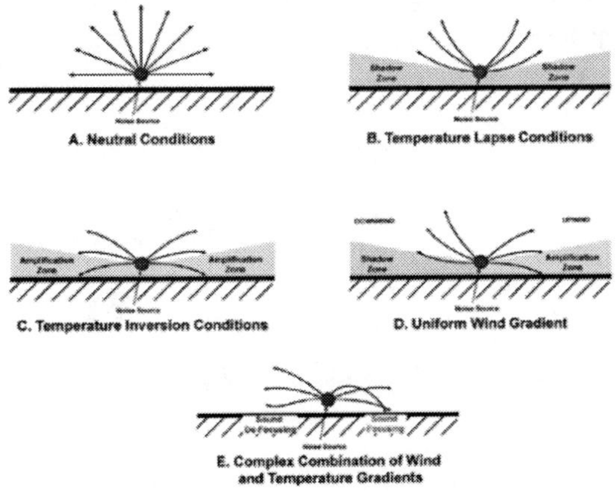

Figure 2: illustration of wave front gradients and their impact on noise propagation

Furthermore, fluctuating values of the wind or the thermal field is at the origin of diffusive scattering processes and potential noise contamination of the so called shadow zone.

Due to the height of the wind turbine (i.e the acoustic source), the distance between the source and the reception point, the topography of the site as well as the presence of obstacles have to be considered when considering refraction and diffusion effects on the wind turbine noise propagation.

In order to be able to model the acoustic propagation of the wave fronts, the acoustic wave equation in complex medium (heterogeneous and in presence of convection) has to be resolved.

The so called analytical ray method is directly inspired from optics and is purely geometric.The goal is to follow the incremental temporal evolution of the location of wave front elements, created by a punctual source, in other to capture trajectories, called "rays". Straight or curved rays occur respectively in homogeny or heterogeneous medium. In a second time, the acoustic pressure is calculated. The analytical ray method, combined with heuristic models in order to take into account processes involved in noise propagation, is currently implemented in most engineering noise propagation standards.

Ray based standards with reduced complexity such as ISO 9613-2 or CONCAWE are widely used for wind farm noise predictions, despite the fact that these standards were not developed for wind turbine noise propagation modelling.

Other standards are characterised by varying degrees of complexity, where the more complicated standards based on curved rays, such as Nord2000 or Harmonoise, could give the most accurate results for prediction of wind turbine noise.

On the other hand, numerical methods (in tome or frequency domain) like the linearized Euler Equation method (LEE), Parabolic Equations (PE) approach, Boundary Element Method (BEM) or Fast Field Program (FFP), the Greens Function Parabolic Equation (GFPE) are not directly useful in engineering prediction methods, due to the excessive calculation times required and/or limitations (related to the presence of the boundaries and/or the stratification of the propagation medium).

In a software package the calculation scheme can either be standardized or proprietary (proper development of the software manufacturer).

In current analysis the Windpro package from EMD International in Denmark is used. Under consideration is the standardized ISO 9613-2 calculation scheme and the more proprietary NORD2000 calculation scheme.

3. Wind farm selection

For the purpose of this study, three different sites were investigated all located in France, namely Radenac, Landes de Couesme and Hambers.

At all sites the wind speed and turbulence was measured using a Lidar.

The next table presents the main characteristics of the 3 sites and the main results of the wind measurement campaigns.

Item	RAD Radenac	LDC Landes de Couesme	HAM Hambers
	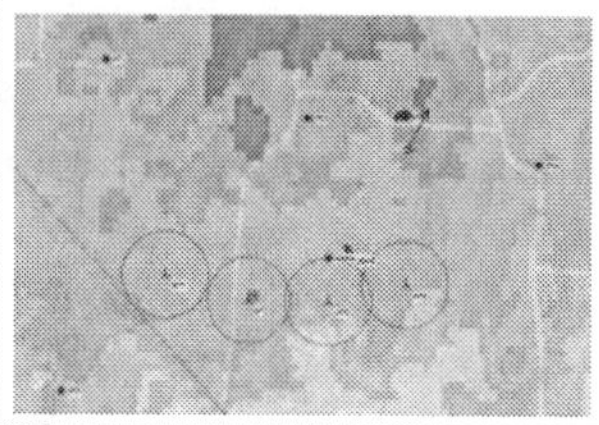	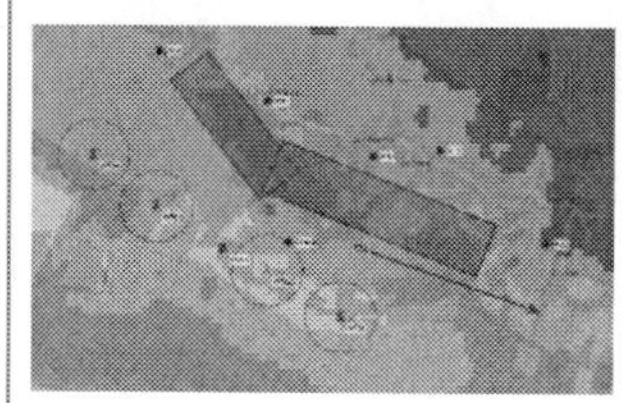	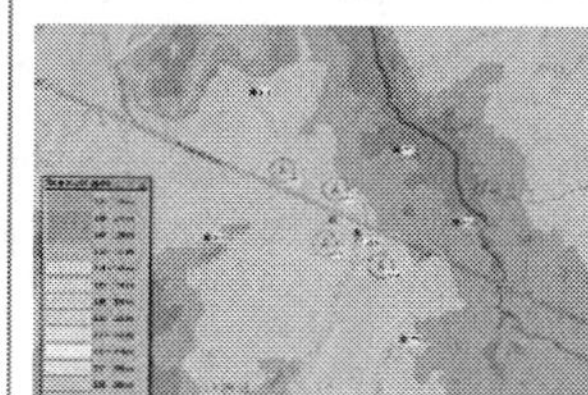
Turbines	2MW Senvion MM92 4 wind turbines	3MW Alstom Wind ECO 110 XX	2MW Senvion MM92 4 wind turbines
Hub height	100m	90m	100m
terrain type	flat	hilly	Hilly
Measurement	14/08-16/09/2014	18/10-31/10/2013	05/12-11/12/2015
wind rose	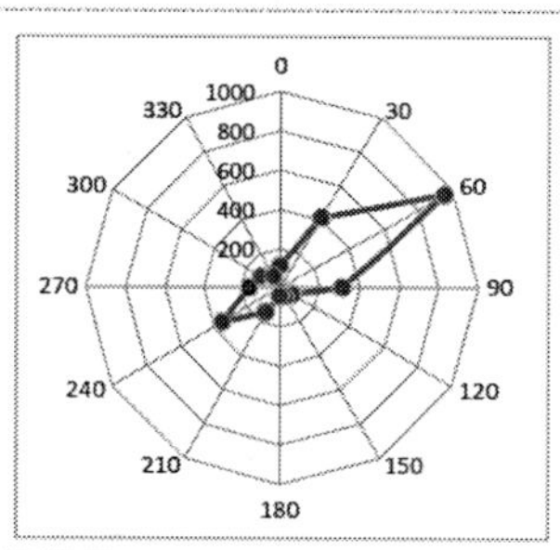	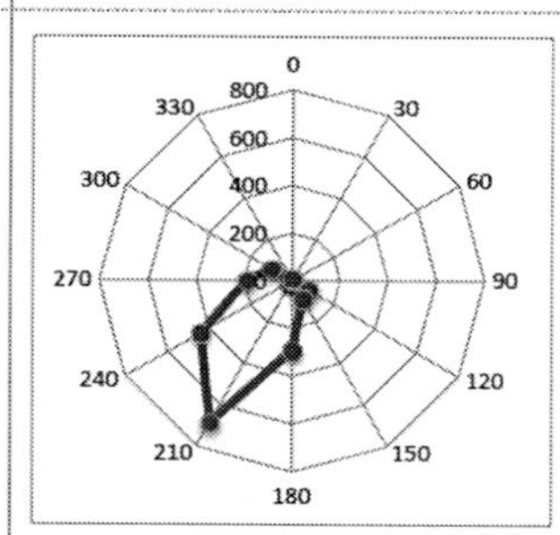	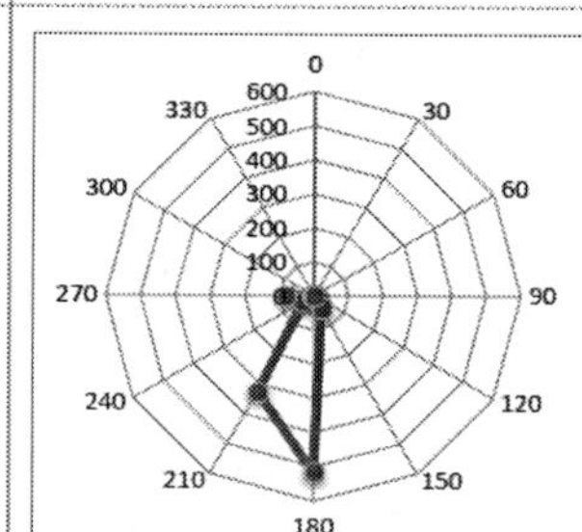
wind speed	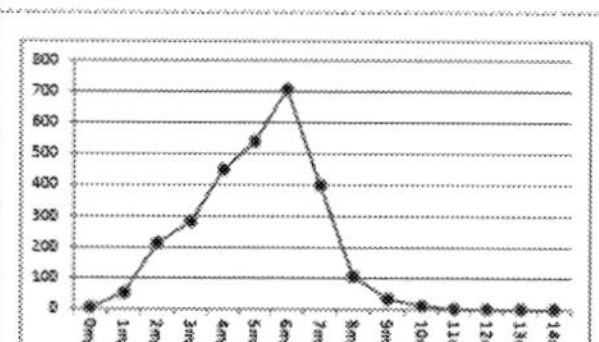	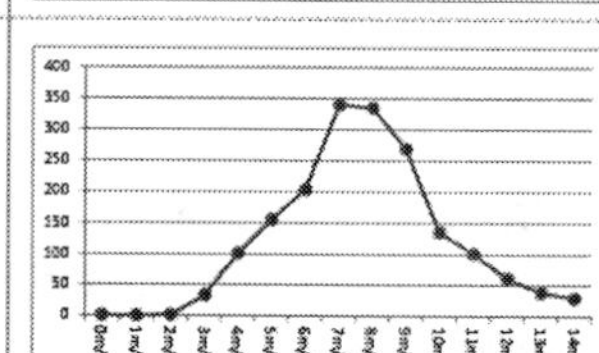	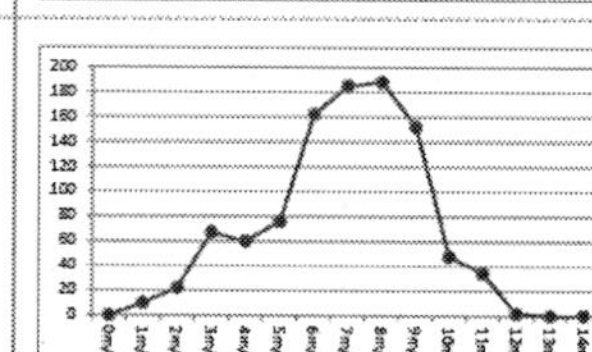

wind shear	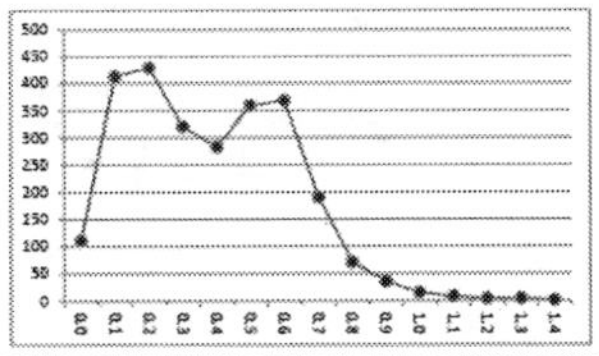	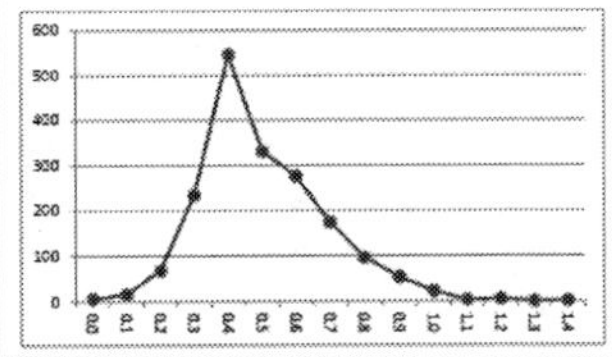	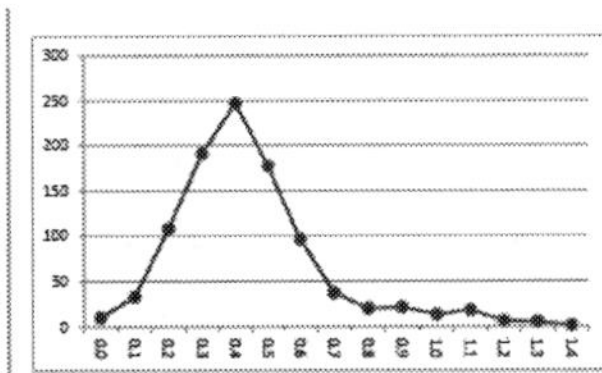
immission points	upwind	downwind	Mixed
wind park wake effects	no	no	Yes

Figure 3: the selected wind farms and results of the wind measurement campaign

The distances between the residents and the wind turbines varies between 500 to 1000m.

4. Wind turbine sound power level and spectral content

Sound power levels are used to characterize the noise generated by the wind turbine, are function of the wind speed as well as of the curtailment strategy. As an example, and for the Senvion MM 92 wind turbine (with a hub height at 100.0 m and total height of 146.3 m), the used sound power levels given by the manufacturer are illustrated hereafter.

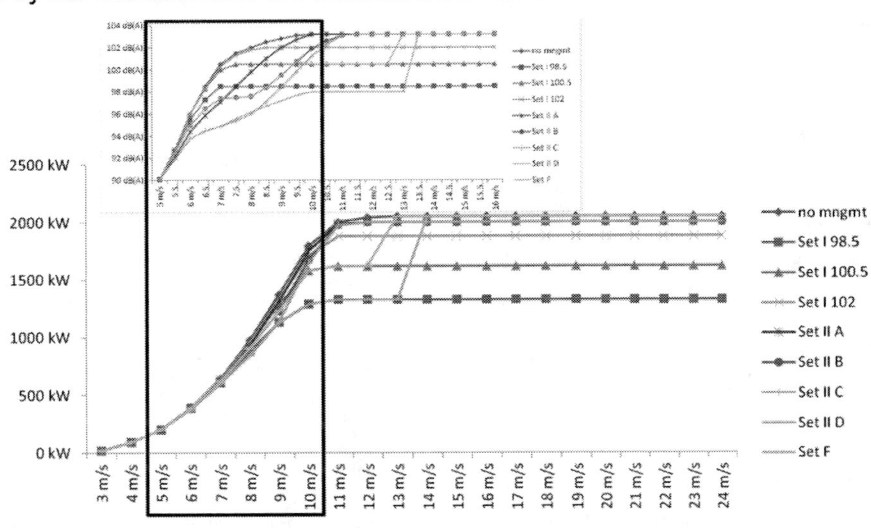

Figure 4: power curves and sound power levels (Senvion MM92)

For moderate wind speeds lower than 8m/s, which occur a large portion of time, the curtailment strategy has a large effect on the emitted noise however with a limited impact on the produced electric power.

The wind turbine sound power also depends on the frequency. The figure shows the A-weighted spectrum for the MM-92 wind turbine.

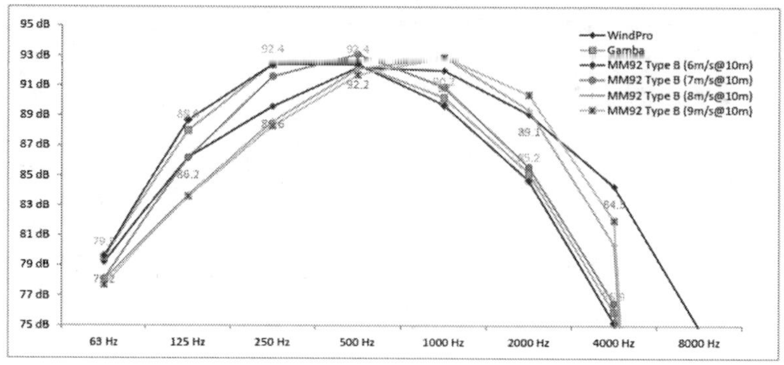

Figure 5: normalized sound power spectra (Senvion MM92)

All shown spectra are normalised to the same overall value of 97.5dB(A). The standard spectrum that is used by Windpro is also shown.

The spectra are function of the wind speed itself. At lower velocities the lower frequencies become more dominant. The manufacturer specification sheets measured by independent certifying organisations show this frequency shift. For further analysis and comparison at the 500Hz 1/1 octave band, a wind speed dependent correction should be applied between the overall level used and the spectral band value. A constant value is used as there is not much difference between the spectra at this frequency band.

5. Data analysis

5.1 Windturbines transient conditions

When a wind turbine is changing its operational mode as for instance going from normal mode to stop or restarting again or shifting to any curtailment mode, this happens during a 10 minute period of the stored data. As in that period of 10 minutes the operating conditions are changing, this period cannot be used for any further analysis. This is called a transient condition.

Therefore, the data in a 10' interval will only be retained and used for further analysis when the operational modes of all wind turbines are not different from the 10' before or after.

The 24 hours of a day is divided into the following parts:

Night	End of the night	Day	End of the day	Beginning of the night
0h-5h	5h-7h	7h-20h	20h-22h	22h-24h

5.2 Acoustic parameter

The acoustic measurements are provided as 1/3rd octave band values from 50Hz up to 10kHz at the different points. These are given as equivalent values Leq but also as the 50% statistical value L50. These two indices are also at hand for the overall A-weighted sound pressure level.

In was observed from the data that the signal to noise ratio is optimal for the 500Hz band as the residual noise tends to be louder at higher frequencies and the specific noise of the wind turbine is maximal at this frequencies. It is also the frequency which is the least compensated by the A-weighted sensitivity of the human ear. Also, particular for the Radenac campaign, it was noted that at unaccountable 80 to 100Hz components were also present form the wind turbines which actually should not be present.

Consequently, the values retained are the 10' 1/3rd octave band L50 values. Then, these values are being A-weighted and followed by a logarithmic sound pressure level summation in order to obtain the 500 Hz 1/1 octave band A-weighted 10' L50 value.

6. Sound transfer function

The difference between the acoustic sound power level L_w (SWL) and the sound pressure level L_p (SPL) at a specific point in the neighbourhood is called the sound transfer function TF (or also the transmission loss).

TF = Lw – Lp

A lower TF means that the noise impact to the environment will be higher for a constant noise power at the source.

More in detail this would be in the case of ISO 9613-2 equal to:

TF = (Adiv + Aatm + Agr + Abar + Amisc) - K - Dc + Cmet

with

Lw (SWL): Sound power level

K: Pure tone

Dc: Directivity correction

Adiv: the attenuation due to geometrical divergence

Aatm: the attenuation due to atmospheric absorption

Agr: the attenuation due to ground effect

Abar: the attenuation due to a barrier

Amisc: the attenuation due to miscellaneous other effects

Cmet: Meteorological correction

In ISO 9613-2 the ground factor G-value used is 0.5, called mixed conditions. For Nord2000 the corresponding acoustic hardness is considered as type D which stands for a crop field in spring or autumn or grass.

The transfer function are observed to be different for different calculation schemes which are based on different methods for calculating the transfer of the sound to the environment.

Notice that the transfer of sound is more frequency dependant in the Nord2000 scheme than in the ISO 9613-2 scheme. Spectral values can differ by more than 2 dB between schemes. In general, the transmission loss is smaller for the Nord2000 than the ISO 9613-2 method, leading to higher sound pressure levels for one given sound power level.

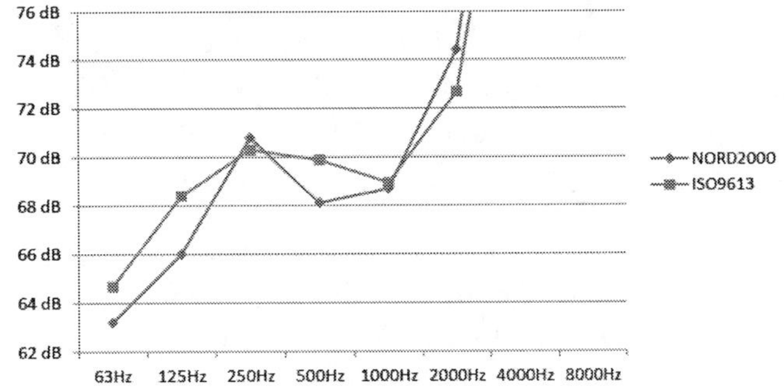

Figure 6: example of transfer function between one turbine and one reception point (Radenac)

When one applies a typical noise power spectrum, then the impact at each reception point can be calculated for both calculation schemes both for the 500 Hz octave band and for the overall level, and the relative comparison between Nord 2000 and ISO 9613 can be done.

In general, the difference for the results between the Nord2000 and the ISO 9613-2 method is smaller for the overall value than for the 500Hz octave band.

Nord2000-ISO 9613-2	500 Hz octave band	Overall value
Pt02/E1	2.5	1.3
Pt02/E2	2.2	0.9
Pt02/E3	3.0	0.9
Pt02/E4	3.0	1.1
Pt03/E1	-2.9	-2.8
Pt03/E2	1.1	0.8
Pt03/E3	1.8	1.0
Pt03/E4	2.8	0.9

Figure 7: difference between the modelled sound pressure levels in Nord 2000 and Iso 9613 for different configurations (wind turbine E1-4 / reception points 02-03) at Radenac

.

7. NORD2000 Transfer Function: sensitivity analysis

It is known that for ISO 9613-2 the transfer function is only dependant on the geometrical dispersion and the ground factor. Here, the magnitude of the wind speed or the wind direction nor the wind shear have no impact, neither have the meteorological condition of clear or cloudy sky nor a day or night condition.

As presented at §2, in particular situations, the impact of curved rays can influence the levels behind barriers. This could also be behind topographical elements as ridges or hills. There the shadow zone can be impacted.

The scheme of Nord2000 holds these effects more into account. It is now examined for the site of Radenac to which extent this holds, as illustrated hereafter.

Period of day	Meteo	Ovl	63 Hz	125Hz	250Hz	500Hz	1kHz	2kHz	4kHz	8kHz
Day	Clear sky	69.0	63.2	66.1	70.8	68.2	68.7	74.3	92.1	142.8
Day	Cloudy sky	69.0	63.2	66.0	70.8	68.1	68.7	74.4	92.5	143.6
Night	Clear sky	69.0	63.2	66.0	70.8	68.0	68.7	74.4	92.7	143.9
Night	Cloudy sky	69.0	63.2	66.0	70.8	68.1	68.7	74.4	92.5	143.6

Figure 8: example of the sensitivity of the transfer function for one wind turbine/reception point case (Nord 2000)-Radenac

Notice that the results are very much independent of the day/night conditions or clear/cloudy sky for this specific case (less than 1 km between the wind turbine and the reception point and flat terrain).

On the other hand, there is an impact of the wind direction. Between 0 and 180° the 250Hz band transfer function changes +0.4dB, in the band 500Hz this is -1.3dB and in the 1kHz the difference is +0.7dB.

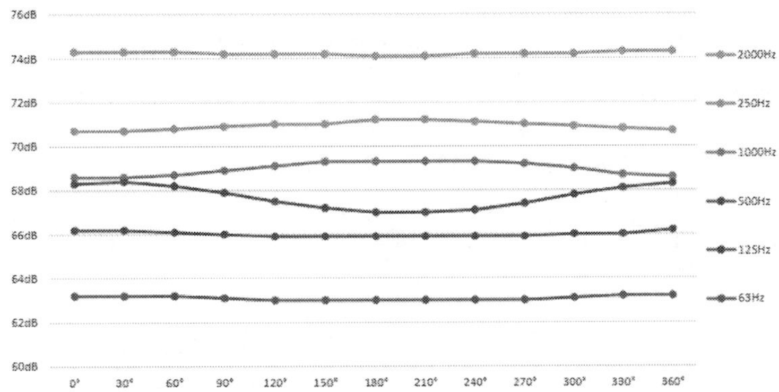

Figure 9: transfer function vs wind direction for one wind turbine/reception point (Nord 2000) - Radenac

The overall level impact of course depends on the shape of the spectrum of the noise power level but for the wind turbine used here and using a spectrum at maximum wind speed it is seen that these (moderate) spectral changes are rather counteracting resulting in a small overall change in the immission sound pressure level of -0.4 dB.

Simulations at different wind speeds show an influence on some spectral components of the TF of maximum 0.1dB, negligible for present site.

Also, simulations with a wind shear of 0.15 and of 0.6 show no impact on the transfer function.

At larger distances (no comparison available with measurements for this study), the sensitivity of the transfer function to wind direction and other parameters can occur. As an illustration, the following figure shows the noise impact towards the environment of the wind turbine 3 at a wind speed of 8m/s at hub height coming from the 60° direction (clockwise from the vertical) and with a wind shear component of 0.6.

Notice that indeed the immission levels vary considerably but only at larger distances (above 1km). The levels downstream are higher than upstream where a shadow zone is created.

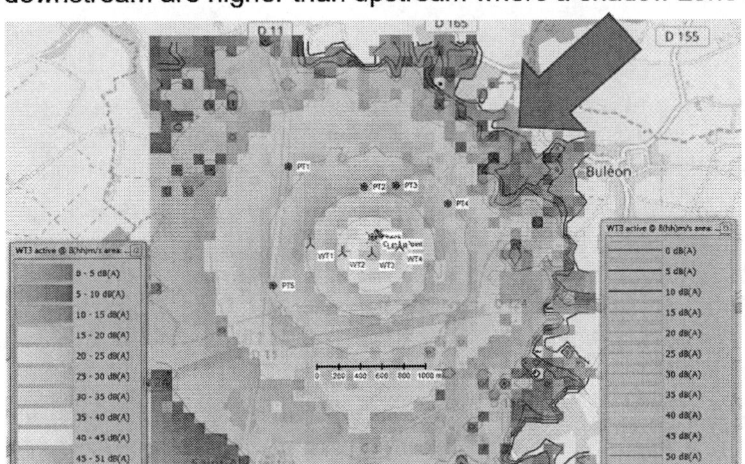

Figure 10: noise map (wind turbine 3 operating, 8m/s at hub height, wind shear 0.6, day and clear sky conditions) (Nord 2000)

This analysis was done with clear sky conditions during daytime. We should also expect an impact of day/night, clear/cloudy conditions at lager distances. Again, no comparison with measurements was possible at large distances due to lack of measurements.

8. Towards the objective of the determination of calibration values

The 23 days of the measurement campaign results in about 3312 occurrences of 10 minute time steps for Radenac. The wind turbines were in different operational modes and in different combinations of wind turbine in operation and in off mode.

Using the noise power level of the wind turbines as function of the wind velocity and in combination with the transfer function to different points in the neighbourhood it is possible to calculate, for each time step, the expected noise impact.

Plotting measurements as function of the wind speed and for each time stamp shows a large (vertical) spread of the noise level due to other noise sources present. The aim is to extract from these measurements values without any other polluting noises than the noise from the wind turbine itself.

The next figure shows the measured ambient noise (i.e. the wind turbine and the residual noise levels) and the calculated specific wind turbine noise levels (here using ISO 9613-2) at point 2 at the Radenac wind farm. It shows all the moments where at least one wind turbine is in operation.

It is clear that not only the analysis should be further be done for each set of different operational conditions of the wind turbines but also using wind speed bins where higher levels due to residual noise can be excluded.

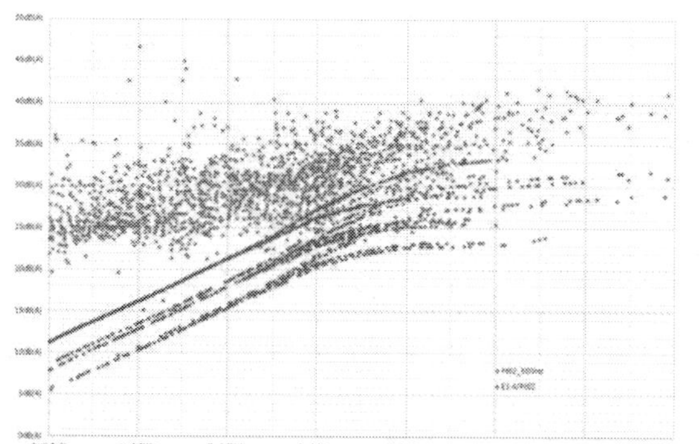

Figure 11: measured ambient (red dots) and modelled (ISO 9613) wind turbine noise levels (blue dots) versus hub height wind speed at one point for the Radenac case, all operating conditions of the wind turbines (standard and curtailed modes)

The measured noise is the ambient noise levels, including residual noise sources. At first sight it is easy to see that the higher values at a given hub height wind speed are probably due to other sources different from the wind turbines. One searches for the minimum value in a bin. However, setting the bin to small will result is few or no values. On the other hand, when setting the bin too big, the effect of the noise power level becomes important as this can change substantially with a small wind speed variation at these low speeds.

In the next figure, the noise level at point 2 with all wind turbines only in standard mode (no curtailment) is shown. At speeds between 3 and 5.5m/s the difference between the ambient measured noise and the calculated values becomes larger due to the residual noise which seems to be function of the wind speed. As the parameter used is L50, these values are at least present during 50% of the time. The lower value is then probably originated in wind induced noise. The higher values can be due to the combined presence of a number of other noise sources.

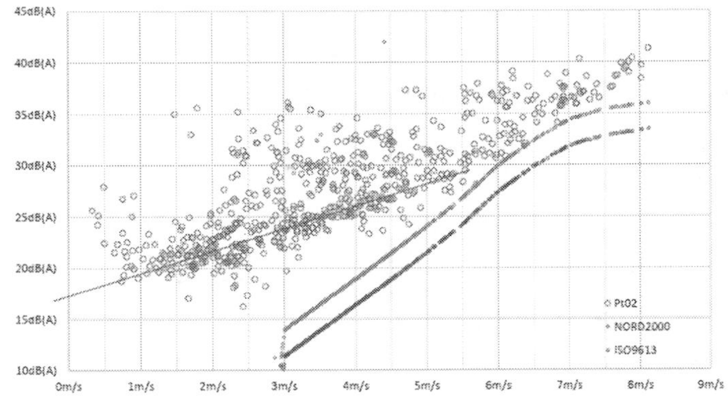

Figure 12: measured ambient (red dots) and modelled (ISO 9613, Nord 2000) wind turbine noise levels (blue and grey dots) versus hub height wind speed at one point for the Radenac case, all turbines in standard mode operation

At first glance it seems that the calculated values are in the neighbourhood of the lowest measured values. However, analysis should be made per wind turbine contribution if these scenarios are present. The next figures show results when only the wind turbine 2 or 3 are in operation. There the difference between the measured and calculated value becomes more apparent.

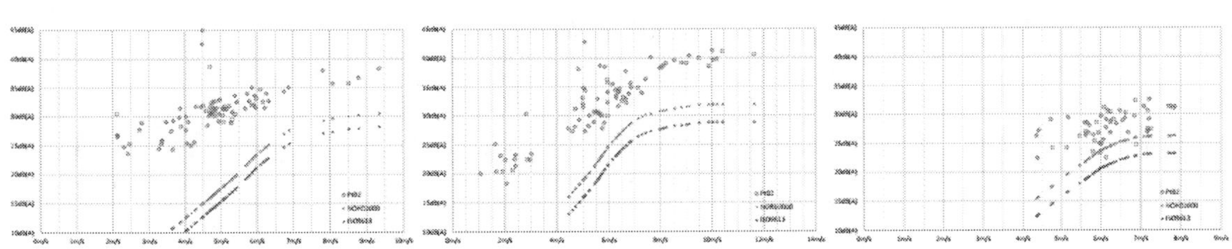

Figure 13: measured ambient (red dots) and modelled (ISO 9613, Nord 2000) wind turbine noise levels (blue and grey dots) versus hub height wind speed at one point for the Radenac - case with only wind turbine 2 (left) and 3 (middle) in standard mode, and with only wind turbine 3 in F mode (right)

Based on all data available, it is now the challenge to establish the error that is made using ISO 9613 and Nord 2000 calculation schemes with respect to the measurement campaign and to determine the calibration values. The methodology proposed hereafter will be based on the above mentioned measured minimum value and wind speed bin approach.

9. Procedure for the determination of calibration values

In order to derive calibration values taking into account considerations presented at previous paragraph, the following procedure is proposed:

1. Add wind turbines electrical power data (1' minute value conversion to 10' values)
2. Add noise measurements 10' 1/3 octave band LA50 values

3. Add Lidar data number format conversion
4. Add part of day (day, evening, night)
5. Add curtailment timing (noise control modes)
6. Calculate measured octave band values at 500Hz
7. Model in Windpro the site model (insert topography, locate turbines, locate measuring points, determine ground factor, ...)
8. Perform numeric simulation for ISO 9613-2 and NORD2000 method
9. Extract sound transfer function TF and add it to calculation sheet
10. Apply a typical spectrum for this type of turbine
11. Determine and exclude the transient intervals
12. Classify wind speed and direction (into bins)
13. Determine the acoustic power for turbines as function of curtailment mode and wind speed
14. Calculate in each immission point the contribution for each wind turbine (with transfer function and typical spectrum) and this for all turbines
15. Make pivot tables for the minimum sound pressure Lp value measured and also calculated in each situation of wind speed and direction and operational mode
16. Analyse results (errors) and evaluate reliability of different scenarios
17. Retain calibration values

10. Illustration of the procedure for the Radenac case

The next figure shows the measured ambient and residual noise levels at point 2 during the daytime for the 500Hz 1/1 octave band value. The minimum measured value is presented for a combination of the wind turbine operational modes and this for the velocity ranges from 4 to 7 m/s in steps of 1 m/s and for the wind directions of 30 to 90° in steps of 30°.Transient conditions as discussed before are here not considered.

n of Pt02_50(		E1_Sta	E2_Sta	E3_Sta	E4_Sta						
		STAN	STAN	STAN	STAN	STOP	STOP	STOP	STOP	STOP	STOP
		STAN	STAN	STOP	STOP	STAN	STAN	STOP	STOP	STOP	STOP
		STAN	STOP	STAN	STOP	STAN	STOP	STAN	STAN	STOP	STOP
Vhh_cla	Vdir_Cla	STAN	STOP	STAN	STAN	STOP	STOP	STAN	STOP	STAN	STOP
4	30	23.3				26.9	24.3	25.2			25.7
	60	23.6	26.8	26.8	25.0	27.4	25.0	35.1			24.2
	90	25.8	28.3		27.4	27.2	28.3	27.4	27.8		26.6
5	30	28.8			25.5			27.3	28.9		23.5
	60	24.2	29.3	30.0	27.8	28.4	28.5	26.7	29.3	32.6	26.0
	90		28.2			26.9	29.5	30.2	27.3		26.8
6	30	30.8		31.5	26.5			31.4	27.8	30.8	25.0
	60	30.6	32.6		27.6		31.6	29.6	31.6	28.8	26.4
	90	32.1	29.7				31.4	33.0			29.0
7	30				30.8			34.5	32.5		29.6
	60	33.1	35.0		33.8			32.5	32.2		28.5
	90	35.9									32.3

Figure 14: measured ambient and residuals noise levels (minimum values) classified in wind speed-direction bins, and for different operational conditions of the 4 wind turbines (standard "stan"-stop modes)

The minimum value is taken from all measurement values measured with the same environmental conditions for the wind speed and direction and also for the operational modes of the wind turbines.

As the conditions are identical then only the lowest measured value is likely to have the least effect of residual noise from other sources. As such, this measured ambient noise value is the closest to the specific noise generated by the wind turbines. In any case, a correction will be added related to the level of the residual noise compared to the ambient noise.

This is also done when all turbines are stopped. In that condition the measured value is the residual value generated by other sources besides the wind turbines. As the minimum of this series is taken, it is assumed that no lower value can be attained and that this value is continuously present. This value is retained as representative when the turbine are operating.

For the same environmental and operational conditions a second pivot table shows the number of intervals, or occurrences, that are present to retain the minimum value from. This is informative important because a minimum value of a large number of occurrences would seem more reliable.

ronotimbre [(		E1_Sta	E2_Sta	E3_Sta	E4_Sta						
		STAN	STAN	STAN	STAN	STOP	STOP	STOP	STOP	STOP	STOP
		STAN	STAN	STOP	STOP	STAN	STAN	STOP	STOP	STOP	STOP
		STAN	STOP	STAN	STOP	STAN	STOP	STAN	STAN	STOP	STOP
Vhh_cla	Vdir_Cla	STAN	STOP	STAN	STAN	STOP	STOP	STAN	STOP	STAN	STOP
4	30	5				1	1	2			2
	60	15	7	1	2	5	6	1			13
	90	7	5		3	1	6	2	1		9
5	30	3			3			5	1		4
	60	11	7	1	11	5	14	10	11	2	16
	90		5			1	9	1	3		15
6	30	1		2	8			3	9	1	3
	60	16	3		10		5	11	8	1	24
	90	19	7				7	1			9
7	30				1			4	2		2
	60	14	1		4			2	7		7
	90	10									5

Figure 15: number of occurences for the table presented at figure 14

Based on the ambient and the residual noise, then the correction value to the measured ambient noise can be calculated to finally determine the specific noise contribution coming from the wind turbines. This is done with the formula:

$$C(\Delta)=\Delta-10*\log(10^{\Delta/10}-1)$$

Here C is called the correction value and Δ is the arithmetic difference between the ambient and the residual noise.

Specific noise correction		E1_Stat	E2_Stat	E3_Stat	E4_Stat						
		STAN	STAN	STAN	STAN	STOP	STOP	STOP	STOP	STOP	STOP
		STAN	STAN	STOP	STOP	STAN	STAN	STOP	STOP	STOP	STOP
		STAN	STOP	STAN	STOP	STAN	STOP	STAN	STAN	STOP	STOP
Vhh_class	Vdir_Class	STAN	STOP	STAN	STAN	STOP	STOP	STAN	STOP	STAN	STOP
4	30					6.2					
	60		3.5	3.5	7.7	2.8	7.7	0.4			
	90		4.9		7.7	8.9	4.9	7.7	6.2		
5	30	1.5			4.3			2.3	1.5		
	60		2.7	2.2	4.7	3.7	3.6	8.3	2.7	1.1	
	90		5.6			16.4	3.3	2.7	9.6		
6	30	1.3		1.1	5.3			1.1	3.2	1.3	
	60	2.1	1.2		6.2		1.6	2.8	1.6	3.7	
	90	2.9	8.3				3.7	2.2			
7	30				6.2			1.7	3.1		
	60	1.8	1.1		1.5			2.2	2.4		
	90	2.5									

Figure 16: correction factor to be applied to the minimum measured ambient noise levels of the figure 14

Only C values are retained for moderate correction value (less than 4) because these measurements are the least influence by the residual noise. This is true when the residual noise is more than 2 dB lower than the ambient noise.

The measured ambient noise is corrected with the C values to result in the specific noise contribution from the wind turbines as illustrated at the next figure.

Turbine specific noise Measurement - C)		E1_Stat	E2_Stat	E3_Stat	E4_Stat						
		STAN	STAN	STAN	STAN	STOP	STOP	STOP	STOP	STOP	STOP
		STAN	STAN	STOP	STOP	STAN	STAN	STOP	STOP	STOP	STOP
		STAN	STOP	STAN	STOP	STAN	STOP	STAN	STAN	STOP	STOP
Vhh_class	Vdir_Class	STAN	STOP	STAN	STAN	STOP	STOP	STAN	STOP	STAN	STOP
4	30										
	60		23.3	23.3		24.6		34.7			
	90										
5	30	27.3						25.0	27.4		
	60		26.6	27.8		24.7	24.9		26.6	31.5	
	90						26.2	27.5			
6	30	29.5		30.4				30.3	24.6	29.5	
	60	28.5	31.4				30.0	26.8	30.0	25.1	
	90	29.2					27.7	30.8			
7	30							32.8	29.4		
	60	31.3	33.9		32.3			30.3	29.8		
	90	33.4									

Figure 17: corrected specific noise levels of the wind turbines based on the ambient levels and the correction factors presented at figures 14 and 16

The specific values coming from the wind turbines are now predicted by calculation, at first in ISO 9613-2. For each wind turbine the overall noise power level for each 10 minute occurrence is determined. Further, for each immission point the contribution of each wind turbine is superimposed holding into account the sound power level and the transfer function between source and receiver. Also, a correction is made for the 500Hz 1/1 octave band value with respect to the overall level as explained at §4. Then, as for the measurements, also here the minimum value of the sound pressure level at the receiver is retained for further analysis.

Min of E1:4/l		E1_Sta	E2_Sta	E3_Sta	E4_Sta						
		STAN	STAN	STAN	STAN	STOP	STOP	STOP	STOP	STOP	STOP
		STAN	STAN	STOP	STOP	STAN	STAN	STOP	STOP	STOP	STOP
		STAN	STOP	STAN	STOP	STAN	STOP	STAN	STAN	STOP	STOP
Vhh_clas	Vdir_Clas	STAN	STOP	STAN	STAN	STOP	STOP	STAN	STOP	STAN	STOP
4	30	13.5				11.9	8.6	12.1			-68.8
	60	13.8	10.8	16.3	10.9	12.2	8.7	16.4			-68.8
	90	13.8	12.9		11.9	12.6	7.7	12.9	13.1		-68.8
5	30	18.8			18.9			17.5	18.3		-68.8
	60	18.5	19.0	22.3	16.0	16.4	12.9	15.5	14.4	14.5	-68.8
	90		15.3			17.2	12.9	15.7	13.8		-68.8
6	30	24.6		22.5	21.1			22.6	17.2	19.2	-68.8
	60	24.4	23.5		22.0		17.9	20.9	19.6	22.9	-68.8
	90	24.2	21.8				19.3	22.0			-68.8
7	30				28.4			28.2	24.3		-68.8
	60	30.7	25.6		27.1			27.1	23.2		-68.8
	90	29.8									-68.8

Figure 18: predicted noise levels of the wind turbines according to the classification of figure 14 (ISO 9613-2)

Finally, comparing the measured specific noise level with the calculated noise levels represents the error.

M=CALC-MEAS		E1_Stat	E2_Stat	E3_Stat	E4_Stat						
		STAN	STAN	STAN	STAN	STOP	STOP	STOP	STOP	STOP	STOP
		STAN	STAN	STOP	STOP	STAN	STAN	STOP	STOP	STOP	STOP
		STAN	STOP	STAN	STOP	STAN	STOP	STAN	STAN	STOP	STOP
Vhh_class	Vdir_Class	STAN	STOP	STAN	STAN	STOP	STOP	STAN	STOP	STAN	STOP
4	30										
	60		-12.5	-7.0		-12.4		-18.3			
	90										
5	30	-8.5						-7.5	-9.1		
	60		-7.6	-5.5		-8.3	-12.0		-12.2	-17.0	
	90						-13.3	-11.8			
6	30	-4.9		-7.9				-7.7	-7.4	-10.3	
	60	-4.1	-7.9				-12.1	-5.9	-10.4	-2.2	
	90	-5.0					-8.4	-8.8			
7	30							-4.6	-5.1		
	60	-0.6	-8.3		-5.2			-3.2	-6.6		
	90	-3.6									

Figure 19: errors between predicted and measured wind turbine noise levels according to the classification of figure 14 (ISO 9613-2)

The next step is then to calculate the noise impact but with a different transfer function determined by the calculation scheme of Nord2000, leading to the errors illustrated at the next figure.

M=CALC-MEAS		E1_Stat	E2_Stat	E3_Stat	E4_Stat						
		STAN	STAN	STAN	STAN	STOP	STOP	STOP	STOP	STOP	STOP
		STAN	STAN	STOP	STOP	STAN	STAN	STOP	STOP	STOP	STOP
		STAN	STOP	STAN	STOP	STAN	STOP	STAN	STAN	STOP	STOP
Vhh_class	Vdir_Class	STAN	STOP	STAN	STAN	STOP	STOP	STAN	STOP	STAN	STOP
4	30										
	60		-10.2	-4.2		-9.8		-15.3			
	90										
5	30	-5.8						-4.5	-6.2		
	60		-5.2	-2.6		-5.7	-9.8		-9.2	-14.0	
	90						-11.1	-8.8			
6	30	-2.3		-5.1				-4.7	-4.4	-7.3	
	60	-1.4	-5.6				-10.0	-2.9	-7.4	0.8	
	90	-2.3					-6.2	-5.8			
7	30							-1.6	-2.1		
	60	2.1	-6.0		-2.4			-0.2	-3.6		
	90	-0.9									

Figure 20: errors between predicted and measured wind turbine noise levels according to the classification of figure 14 (Nord 2000)

After this "semi-automatic" approach, individual errors are analysed more into details in order to assess if they have to be conserved or excluded. As an example, some minimum ambient levels can be still polluted with other sources than the wind turbines and cannot be retained. In other cases, the electrical power do not match with the observed wind speed, these events have also to be excluded.

11. Results

The calibration M value stands for the difference between the calculated and the measured value ($L_{p,calc}$ - $L_{p,meas}$). A negative value indicates that the measured values are higher than the calculated values. Retained calibration values are based on deep analysis of the individuals errors, as explained above.

The ISO 9613-2 simulations are performed with a recommended G=0.5 value so with 50% absorption present which is the preference value according to British recommendations.

If one would take higher G-values (G=1), thus more absorptive surroundings (like grass and pasture), then the calculated value would be 4dB lower at 500Hz thus making the error even bigger with -4dB.

For comparison, the Nord2000 soil classes result in a -2dB to +2dB range for 500Hz, as illustrated hereafter.

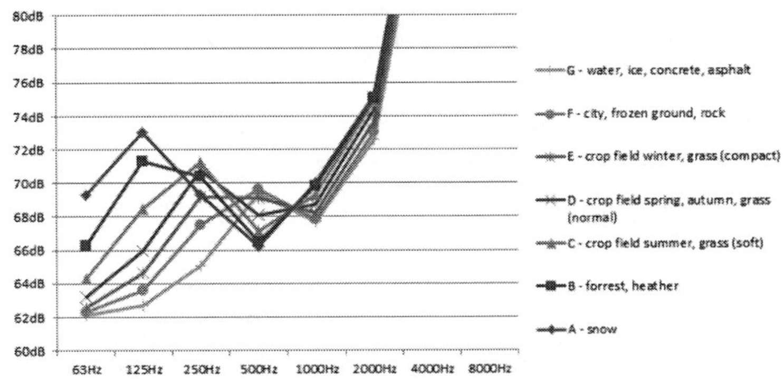

Figure 21: Nord2000 transfer function for turbine 3 to point 3 vs soil condition

11.1 Radenac

At night the residual noise is the lowest so the ambient value is very much equal to the specific wind turbine noise.

The proposed resulting calibration values are based on analysis of the individual errors (not presented here).

M = Calculated - Measured	Daytime	Nighttime
ISO 9613-2 (with G=0.5)	-5 dB(A)	-6 dB(A)
NORD2000	-3 dB(A)	-4 dB(A)

Figure 22: calibration values (500Hz), Radenac

As such, it is seen that the Nord2000 results are more accurate than the ISO 9613-2 results.

It is the assumption with ISO 9613-2 that sound transfer is always calculated as downwind, thus generating the highest noise values at the receiver. This is done independent of the actual wind direction or the orientation between source and receiver. The transfer values from this scheme are only made up for downwind condition. In present case the measurements are dominant upwind (the houses are standing upstream with respect to the wind turbines). So, again, the calculation would even be overestimating the actual value and the opposite is true because the calculation shows an underestimation.

Nord2000 shows that the impact of upstream or downstream conditions is about 1.3dB at 500Hz and 0.4dB overall.

ISO 9613-2 offers the possibility to hold into account that not 100% of the time downwind conditions are present by presenting the meteorological factor C_{met} for calculating the long term averaged value. This effect ranges between 0 and C_0. This last value is often taken equal to 2 dB as a recommended maximum value. However, as long as the levelled distance between source and receiver is smaller than 10 times the summed height of source and receiver then the correction value C_{met} equals zero. This is the case here as this distance is here about 915 meter.

This factor is often used but, again, this lowers the absolute result which is not preferable as the result is already an underestimation.

11.2 Landes de Couesme

The number of transients are high due to a complex noise control scheme and due to recurrent occurrences where no or partial data is at hand thus limiting the number of available analysis data.

During the day there is one most common noise control scheme for wind speeds of 8, 9 and 10m/s (Modes E, E, D and D for wind turbines 1, 2, 3 and 4).

In these cases the ambient noise is considerable higher than the residual noise so that the correction values C are low.

The calibration values at these higher speeds are weighted according the number of occurrences. The calibration values for point 1 are about 1 dB greater than for point 2.

During the night no data is present at higher wind speeds and for the lower wind speeds the number of occurrences is also too low to make any general conclusions. However, for what it's worth the (averaged between two points) result for the 8m/s wind speed indicates also that at night the calibration value is about 1dB worse than during daytime.

M = Calculated - Measured	Daytime	Nighttime
ISO 9613-2 (with G=0.5)	-3 dB(A)	-4 dB(A)
NORD2000	0 dB(A)	-1 dB(A)

Figure 23: calibration values (500Hz), Landes de Couesme

11.3 Hambers

The number of occurrence is the highest in the direction of 180° and for the wind speeds of 8, 9 and 10m/s. This renders for the daytime an average calibration value of about 3 dB(A) for ISO 9613-2 and this reduces to 0 dB(A) for Nord2000.

At point 4 the residual noise was such that large correction values would be needed to attain the specific values which consequently are much less reliable.

Point 5 has a more quiet residual effect and results in about -3 dB(A) for the highest speeds for ISO 9613-2. For Nord2000 this results also in about 0 dB(A) but with a larger spread among values.

During the night the deviations in point 2 are large, up to about 10 dB(A) for ISO 9613-2 and 7 dB(A) for Nord2000. It is assumed that a noise control method is applied which not hold true taking into account the measurement values.

At point 5 the wind speeds of 8 to 9m/s delivers, most likely, calibration values of -4 for ISO 9613-2 and -3 for NORD2000.

M = Calculated - Measured	Daytime	Nighttime
ISO 9613-2 (with G=0.5)	-3 dB(A)	-4 dB(A)
NORD2000	0 dB(A)	-3 dB(A)

Figure 24: calibration values (500Hz), Hambers

12. Conversion for overall correction values

Previous analysis was done on the 500Hz octave band values leading to calibration values valid at these frequencies. This was done primarily because of the quality of the measurements with interfering residual noise values. To eliminate as much as possible these effects, only the octave band that is dominant for the wind turbine noise in the surroundings is used.

However, it is the aim to establish calibration values which can be applied to the calculated overall value.

As the measured overall values in general can be of poor quality, a comparison with these measurements can only be done for isolated and selected cases.

This was done in a first part of this research project resulting in a high level of confidence in the overall measured levels of the wind turbines, however for a very limited number of selections and around transient states.

Based on the errors on the overall levels observed for selected cases, the next table show the 500 Hz and overall calibration values (noted with square brackets).

Item			RAD Radenac	LDC Landes de Couesme	HAM Hambers
Calibration value	ISO 9613-2	day	-5 [-3]	-3 [-2]	-3
		night	-6 [-4]	-4 [-3]	-4
	NORD2000	day	-3 [-2]	0 [0]	0
		night	-4 [-3]	-1 [-1]	-3

Figure 25: calibration values (500 Hz, and overall levels (brackets))

13. Conclusions

In this study three sites were investigated.

The wind turbines are in the 2MW range with a hub height at 90 to 100m.

The immission points are typical between 500 and 1000m from the wind turbines.

Only measurements with a clear line of sight are considered.

As the noise level from the wind turbines is low at the receivers, the measurements are often very much influenced by other sources. For more accurate determination of the calibration values, a high quality of noise measurements is necessary where the residual noise is minimal.

Of course, uncertainty in the sound power level will be influence the final calibration value. For this study the sound power levels used were provided by the wind turbine manufacturer.

Two different calculation schemes ware investigated being the very common ISO 9613-2 method and the very advanced NORD2000 method.

The former uses transfer functions, between source and receiver, which only depend on distance, ground factor and screening. The latter can also hold into account the wind direction and wind shear and works with curved rays. It can also take into account thermal conditions (day/night, clear/cloudy sky)

It has been found in the current project that the transfer functions for Nord2000 are not sensitive to the actual wind speed and wind shear, nor to the choice of day or night or clear or cloudy sky, at short distances (less than 1 km, where the measurement points were located).

The effect of the wind direction on the results, which can be determined with Nord2000 and not with ISO 9613-2, can be seen on larger distances. We also can expect impact of day/night clear/cloudy sky at lager distances. No comparison with measurement is possible for this study at larger distances.

From the transfer functions in general it is seen that the spectral and overall differences of the transfer functions between ISO 9613-2 and Nord2000 are more pronounced at shorter distances (lower transfer function values) than at greater distances.

The lower transfer functions value for Nord2000 at 500Hz octave band will provide higher calculated noise levels from the wind turbines at smaller distances. The calculation results for Nord2000 approaches more the measured values than the ISO 9613-2 results.

A methodology was developed in order to analyse the error between the calculation and the measurements, leading to calibration factors. This methodology is based on wind speed/direction bins and on the minimal value of the measured ambient and residual noise levels for given bins.

From the results of the 3 measurement campaigns, we observe that the calculations underestimate as well as for ISO 9613 than for Nord 2000 the real measured noise values.

The next table shows the final retained calibration values to be applied to the overall calculated values.

Item			RAD Radenac	LDC Landes de Couesme	HAM Hambers
terrain type			flat	hilly	hilly
immission points			upwind	downwind	mixed
wind park wake effects			no	no	yes
Calibration value	ISO9613	day	**-3**	**-2**	-2
		night	**-4**	**-3**	-3
	NORD2000	day	-2	0	0
		night	-3	-1	-3

Figure 26: calibration values (overall levels)

Calibration values are valid when for ISO 9613-2 a 50% ground factor is taken (G=0.5). For higher ground factor values (G=1), like grass and pasture, the calculated value would underestimate even more the real noise impact with a supplemental deviation of 4dB.

ISO 9613-2 is said to calculate noise effects in the downwind conditions. A correction factor on this assumption taking into account different wind directions is possible but has very limited effect for distances smaller then about 1000m (current case).

As noise levels are only in conflict with regulations at short distances the ISO 9613-2 method is regarded as sufficiently accurate when using an appropriate calibration value. At larger distances, or at shorter distances but in presence of obstacles in the line of sight, the Nord 2000 method should be recommended.

Finally, it worth to mention that this methodology should be trained and fed with data coming from different wind farm configurations, in order to be able to generalize the universality of the current proposed calibration values.

References

Jones, A and McKenna T (2010) A new method for statistical analysis: the two part reverse interrogation of sparse data Jnl Reverse Statistics 55(2), 65-219

First international meeting on wind turbine noise: perspectives for control Berlin 17th and 18th October 2005 – S Garriges

A review of the use of different noise prediction models for windfarms and the effects of meteorology- G Parry- Acoustics 08 Paris

Propagation acoustique en milieu extérieur complexe. Caractérisation expérimentale in-situ des conditions micro météorologiques -éléments méthodologiques et métrologiques- B Gauvreau&al, 2009

Propagation acoustique en milieu extérieur- B Gauvreau (power point)

Propagation acoustique en milieu extérieur : application de l'équation parabolique rapide au couplage d'effets météorologiques et de topographies complexes. Thèse de doctorat de l'université du Maine- F Edern Aballéa, 2004

Analyse de la propagation acoustique à basse altitude par l'équation parabolique tridimensionnelle. Thèse de doctorat de l'école centrale de Lyon – Y Delrieux, 1991

A good practice guide to the application of etsu-r-97 for the assessment and rating of wind turbine noise, institute of acoustics, 2013

Wind shear and its effects on wind turbine noise assessment, Sgurr – D Mc Laughlin mioa

Neglect of wind shear in assessing long range propagation of wind turbine noise- M Wtoft

Comparison of predicted and measured wind farm noise levels and implications for assessments of new wind farms – T Evans & al, Acoustics Australia, 2012

The propagation of noise from petroleum and petrochemical complexes to neighbouring communities prepared by cj manning for concawe's special task force on noise propagation

Iso 9613-2 acoustics – attenuation of sound during propagation outdoors – part 2: general method of calculation

Nordic environmental noise prediction methods, Nord 2000, summary report, delta, 2001

Nord 2000, comprehensive outdoor sound propagation model part 1 : propagation in an atmosphere without significant refraction, Delta, 2000

Nord 2000, comprehensive outdoor sound propagation model part 2 : propagation in an atmosphere with refraction, Delta, 2000

Comparison of noise calculation methods, F Probst, 2011

7th International Conference
on
Wind Turbine Noise
Rotterdam – 2nd to 5th May 2017

Sound propagating from wind turbines in winter conditions

Kristina Conrady, Department of Earth Sciences, Uppsala University, Sweden, kristina.conrady@geo.uu.se
Anna Sjöblom, Department of Earth Sciences, Uppsala University, Sweden, anna.sjoblom@geo.uu.se
Conny Larsson, Department of Earth Sciences, Uppsala University, Sweden, conny.larsson@met.uu.se

Summary

Simultaneous acoustic and meteorological measurements, as well as daily snow observations, at a wind farm in northern Sweden are analysed for one snow season. The purpose is to examine the meteorological influence on sound propagating from the wind turbines, and to evaluate if the site is representative for other sites as well. Measurements of these types are crucial, since significant knowledge gaps exist in the implementation of the complexity of the atmospheric boundary layer (ABL) in sound propagation models, especially for conditions in cold climates. It is known that processes in the ABL in cold climates differ from those in warmer zones, for instance due to a snow cover. It has been shown previously that there is an effect of snow on sound propagating from wind turbines, as well as other sound sources, but neither the effect of different snow conditions nor the impact of snow on trees has been fully investigated yet. The results show that there is a difference between different snow conditions, and that snow has an attenuating or amplifying effect on sound from wind turbines. Furthermore, the vertical wind speed and temperature gradient influence the effect. Hence, the impact of snow on sound propagation cannot be generalised as just a damping effect, and has to be taken into account when planning and maintaining wind farms.

1. Introduction

Environmental noise pollution and noise regulation are two keywords becoming more and more relevant to the discourse. To give stakeholders the opportunity to define regulations, knowledge of sound propagation and processes influencing it are essential. The knowledge of how sound propagates inside the atmospheric boundary layer (ABL) can be used for planning wind farms or other sources of environmental noise in order to reduce the noise received by humans. Several sound propagation models are already used for these purposes, however they are not yet very sophisticated and do not take into account the complexity of conditions in cold climates (NPL, 2014).

Since the vertical temperature gradient in cold climates is typically positive, the sound waves' paths are usually bend towards the surface. This bending is called "refraction" and is caused by sound speed gradients. Downward bending, in turn, leads to an increased interaction between sound waves and the surface. Therefore, the knowledge of the acoustical properties of the surface are crucial in order to determine the surface's effect on the sound. In cold climates, a snow cover is present during several months per year and, additionally, snow might cover trees and constructions – the so-called "upplega". However, neither the effect of different snow qualities nor the effect of upplega are completely understood yet.

2. Measurements

In the vicinity of a wind farm with twelve wind turbines in northern Sweden, meteorological and acoustic measurements were simultaneously conducted (Fig. 1). Ten-minute average values of temperature, relative humidity, atmospheric pressure, wind speed and wind direction at several heights, as well as sound levels were collected between 05.11.2013 and 30.04.2014. Furthermore, snow observations were conducted on a daily basis. The wind turbines are located on a hill approximately 270 m above sea level. Around 1 km northeast of the wind farm, a sound level meter and a microphone are located on a slope within a forest. An 18-m meteorological mast was erected directly on the edge of that forest. Additional data was used delivered by two meteorological towers 15 km northwest and 10 km southeast of the wind farm. The few roads in the area are rarely used and the little number of houses in the vicinity of the 18-m mast lead to a fairly low background level.

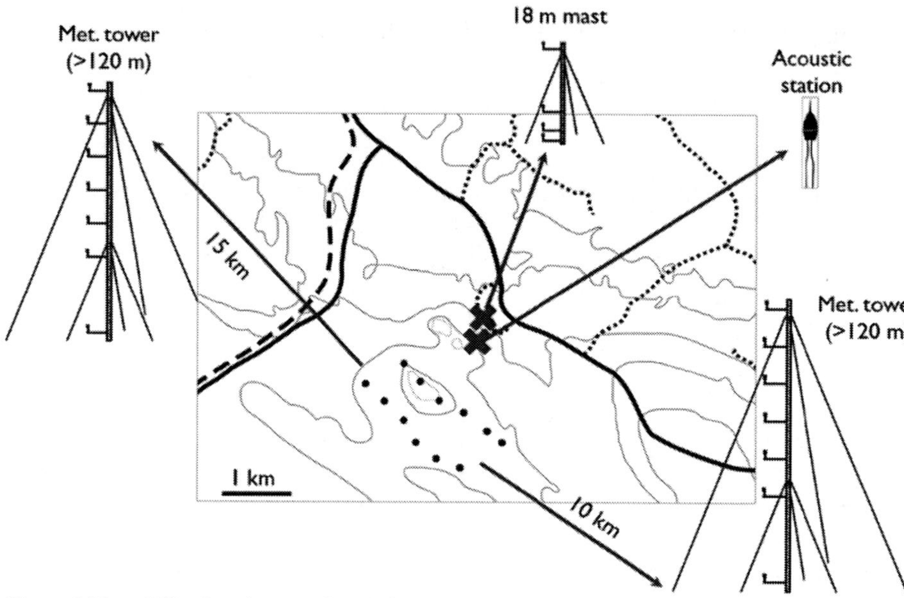

Figure 1 Map of Site A and measuring stations.

3. Results and discussion

3.1 Impact of snow quality

Based on the snow observations, the snow was classified and separated in the following classes: dry, damp, frozen and wet snow. To analyse the snow qualities' effects on sound propagation, a frequency distribution for each snow quality was made (Fig. 2). The distributions of the relative sound pressure level, ΔL, for damp and frozen snow are rather normally distributed and broadly similar, with averages of 3.4 dBA and 2.5 dBA, respectively. In contrast, the distribution for dry snow is skewed to negative values. It consists of two parts – one with a maximum around 2 dBA and a second one with a clearly weaker maximum around -11 dBA. An explanation could be the existence of two different types of dry snow, which was shown by Albert (2003). Flat grained snow has different acoustic properties than spherical grains, however, both could be combined in the dry-snow class. A second explanation could be meteorological conditions which coincide with dry snow.

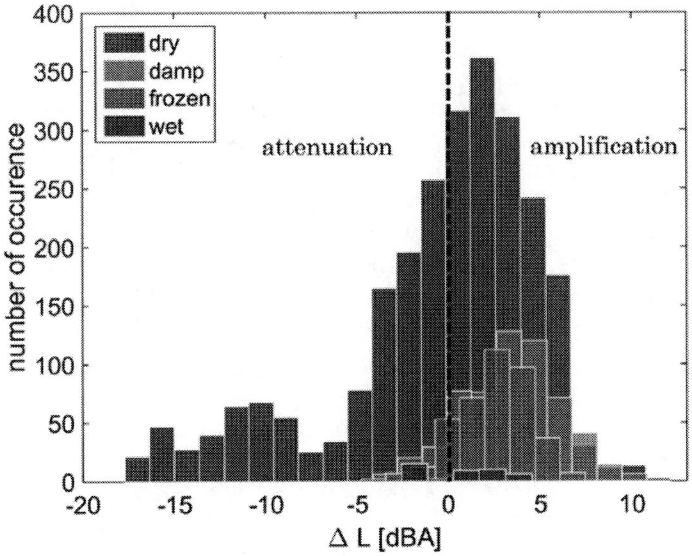

Figure 2 Distribution of sound pressure level, ΔL, for cases with dry (blue), damp (green), frozen (orange) and wet snow (purple) (Conrady et al., in prep.).

3.2 Impact of upplega

Similar to Figure 2, frequency distributions of ΔL were made in order to compare conditions were upplega was observed with those when no upplega was observed (Fig 3). In the latter case, positive ΔL dominate and ΔL is rather normally distributed with an average of 2.7 dBA. The distribution for upplega is skewed to negative ΔL and shows one maximum around -2 dBA and a second but weaker one around -12 dBA. The left part of the distribution for upplega conditions is broadly similar to the distribution for dry snow. That might be caused by a coincidence of dry snow and upplega.

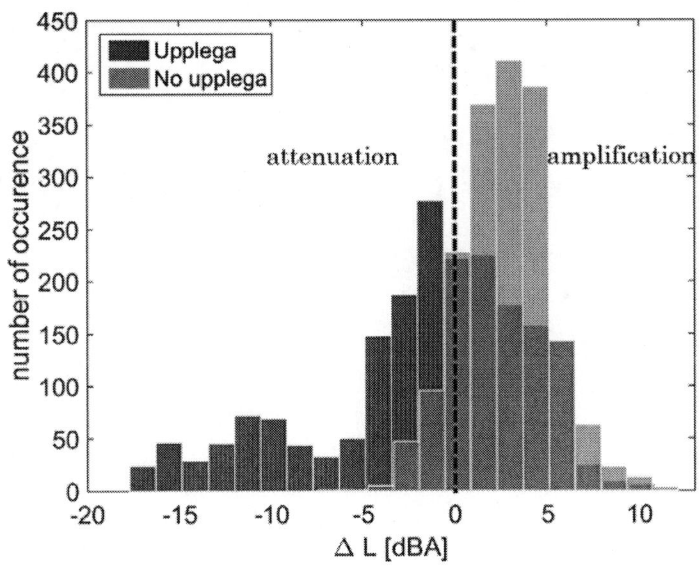

Figure 3 Distribution of sound pressure level, ΔL, for cases with upplega (purple) and no upplega (green) (Conrady et al., in prep.).

4. Conclusions

Sound pressure levels in the vicinity of a wind farm in northern Sweden were analysed for one snow season in 2013/14. The aim was to investigate the impact of snow on sound propagating from the wind turbines. A classification of snow was introduced to compare the effects of different snow qualities, namely dry, damp, frozen and wet snow. One main finding is, different snow qualities affect sound differently. On average, dry snow dampens sound best. However, maximum ΔL are similar for dry and damp snow, and lower for frozen snow. Another main finding is, upplega leads on average to attenuation and clearly lower ΔL compared to conditions without upplega. Maximum ΔL are similar for conditions with and without upplega.

References

Albert, D G (2003) *Observations of acoustic surface waves in outdoor sound propagation* J. Acoust. Soc. Am. 113(5), 2495-2500

Conrady, K, Sjöblom, A and Larsson C (in prep.) *Impact of snow on sound propagating from wind turbines*

NPL (2014) *Guide to Predictive Modelling for Environmental Noise Assessment*

7th International Conference
on
Wind Turbine Noise
Rotterdam – 2nd to 5th May 2017

Wind turbine noise prediction using Olive Tree Lab Terrain

Alexis BIGOT, SIXENSE Environment: alexis.bigot@sixense-group.com
Panos ECONOMOU, PEMARD: panos@pemard.com
Costas ECONOMOU, PEMARD: costas@pemard.com

Summary

Calculation of noise propagation from wind turbines is complex, and large variations of sound immission levels are commonly observed at the same wind speed. These variations are influenced by sound emission (aerodynamic noise), but also by meteorological parameters such as temperature gradient, wind speed profile, wind direction, and turbulences. Commonly used models (ISO 9613-2 and Nord2000 for example) generally predict the average sound pressure levels adequately under downwind conditions, but often fail to predict noise levels in upwind conditions. In this paper we present the results of a collaborative research between SIXENSE Environment (ex SOLDATA Acoustic) and P.E Mediterranean Acoustics Research & Development (PEMARD), using on site experience on more than 350 French windfarms, and Olive Tree Lab - Suite v4.0 software which uses wave based geometrical acoustics to calculate sound propagation, including atmospheric refraction. The goal is to combine both approaches and introduce and test key parameters for wind turbine noise prediction. Calculation results are compared to long term noise & meteorological measurements. A good correlation is shown between calculation and measurements even in case of complex meteorological situations.

1. Introduction

Although models for outdoor sound propagation in a homogeneous atmosphere, where the speed of sound is constant, have been studied and developed extensively in the past decades the results are accurate only for short ranges of approximately under 200 m. At higher ranges the variation of the speed of sound due to wind and temperature stratification needs to be taken into account. Modelling the propagation of sound through such a non-homogeneous atmosphere is one of the most difficult tasks in outdoor acoustics due to the multiple physical phenomena that need to be accounted for such as turbulent scattering, creeping waves, caustics and many others (Attenborough, et al., 1995). A good historical review of sound propagation in moving media can be found in (Bateman, 1914) (Ostashev, 1997) (Piercy, Embleton, & Sutherland, 1977) and (Delany, 1977). Despite extensive research over the past decades there is no practical engineering model that can take into account all of the phenomena simultaneously.

This paper investigates the capabilities of Olive Tree Lab – Suite v4.0 (OTL-Suite), in performing long range sound propagation calculations. In general, OTL-Suite incorporates in its calculation engine, various types of models for the calculation of phenomena such as spherical wave reflection coefficients, multiple diffractions, atmospheric refraction and turbulence, and atmospheric absorption. The models used in the software engine are methodologies which provide accuracy and reasonable calculation times. In the case of refraction two separate models are being used, the model of (L'Espérance, Herzog, Daigle, & Nicolas, 1992) for downwind and upwind refraction and the model by (West, Walkden, & Sack, 1989) for shadow zone calculations, explained later on in this paper. OTL-Suite is a unique acoustic simulation software utilizing wave based geometrical acoustics (WBGA) (Lam, 2005) which preserve the wave nature of sound propagation. It is capable of modelling sound transmission in a non-homogenous atmosphere with linear sound speed profiles or by linear approximations of logarithmic sound speed profiles.

The paper begins with a brief historical review, followed by the theoretical model implemented by OTL-Suite. Subsequently benchmark cases results developed by (Attenborough, et al., 1995) and (WP2 Team, 2002) are compared to OTL-Suite calculation results, followed by a section where OTL-Suite calculation results are compared to measurements available for wind turbine noise. It is argued that a scatter plot of dB(A) values for a range of atmospheric parameters is a much better way to validate numerical models of long range sound propagation due to the dynamic nature of atmospheric conditions. Finally conclusions are presented.

2. A brief Historical Survey

A good historical review of sound propagation in moving media can be found in (Bateman, 1914) (Ostashev, 1997) (Delany, 1977) and (Piercy, Embleton, & Sutherland, 1977). In particular, the introduction in (Bateman, 1914) provides an excellent account of the early qualitative observations and mathematical formulations of the problem, while (Ostashev, 1997) has a detailed account of the investigations which occurred in the interwar period. What follows is a brief overview of the development of this field over the past decades.

Early modern analytical prediction schemes for atmospheric acoustics were developed during the early period after WWII. These schemes would approximate the sound speed with linear profiles and then graphically combine them for the cases of stratified mediums. The advantage of assuming a linear sound speed profile in a medium is that it allows for a closed form solution to the wave equation. The widespread adoption of the computer also led to the development of numerical algorithms which could tackle more general problems in ocean acoustics. Later in the 1980s these methods would also be implemented for the field of atmospheric acoustics.
One of these numerical methods was the Fast Field Program which was originally developed for underwater acoustics and was later implemented for atmospheric acoustics in the mid-1980s. The intention was to make the fastest possible algorithm that could carry out propagation predictions in real time. This method is capable of calculating the sound pressure of a monopole source above a flat ground and immersed in a layered atmosphere. Complicated wind and temperature profiles can be approximated by dividing the atmosphere into multiple horizontal layers with constant wind and temperature profiles. The FFP was originally designed as a two dimensional formulation for an axisymmetric atmosphere but was later generalized to three dimensions (Nijs & Wapenaar, 1990) (Wilson, 1993).

The Parabolic Equation method was applied to the field of ocean acoustics in the late 1970s and atmospheric acoustics in 1988 after being successfully used in such diverse fields as electromagnetic wave propagation, seismic waves, quantum mechanics and many others (Attenborough, et al., 1995). Whereas the Fast Field Program can only model horizontal layers of the homogenous atmosphere and homogenous grounds, the PE method is capable of modelling arbitrary terrains and atmospheric conditions including range-dependent phenomena such as turbulence. Two solutions are popular, the finite difference Crank-Nicholson Parabolic Equation method and the Green's Function Parabolic Equation method. The CNPE has been shown to be more accurate in situations with large sound speed gradients while the GFPE is more efficient. Like the FFP both of these methods were originally developed for a two dimensional axisymmetric atmosphere although a three dimensional GFPE method was later developed (Salomons E. M., 2001).

All the above numerical methods can be considered to be wave models and they successfully model arbitrary cases of inhomogeneous atmospheres and terrains. However they are still too computationally expensive to be used for practical engineering purposes. This is why there is an interest to expand the classical ray model from geometrical acoustics to deal with inhomogeneous mediums. Although the ray model is considered to be only a high frequency approximation of the wave model it does have the advantages that computational times tend to be faster while also providing an easy visual interpretation of wave propagation (Salomons E. M., 1994).

Rayleigh was the first to tackle the ray model for moving inhomogeneous mediums in his 1896 treatise. The model was further developed to be able to include phenomena such as caustics and range-dependent sound speed profiles. These models were still too complicated to implement for engineering purposes as the ray paths in an inhomogeneous medium need to be calculated numerically. It was in the early 1990s that a more practical model was proposed by (L'Espérance, Herzog, Daigle, & Nicolas, 1992). This model used the fact discussed by (Embleton, Thiessen, & Piercy, 1976) that the rays in downwind conditions are grouped in 4 rays for each order of reflection greater than 1. The model included the effects of turbulence, atmospheric absorption, geometrical spreading, the ground effect and refraction for linear sound speed profiles. Salomons developed a model to include logarithmic and power profiles (Salomons E. M., 1994) and also combined the ray model with theories of caustics (Salomons E. M., 1998).

In the case where the receiver is in the shadow zone in an upward sound propagation atmosphere and ray modelling fails to reach the receiver, the ray model can easily be combined with the residual method first treated by Pierce in his classic textbook (Pierce, 1994) and later implemented by many researchers who finally improved the method to be able to calculate the sound pressure level anywhere in the shadow zone (Berry & Daigle, 1988) (West, Walkden, & Sack, 1989). A limiting assumption of the residual series method is that it assumes a linear sound speed profile. The above methods do not take into account the effect of turbulence scattering sound into the shadow zone, a phenomenon that increases the SPL in the high frequencies considerably (Salomons E. M., 2001). A more recent paper presents an alternative analytical solution that includes turbulent scattering in the shadow zone (Lam, 2009).

Starting in the late 1990's these models were eventually implemented in engineering prediction schemes. Between 1996 and 2000 DELTA developed the Nord2000 prediction scheme which was capable of predicting various industrial noise sources and included the heuristic model by (L'Espérance, Herzog, Daigle, & Nicolas, 1992) although it only implemented a single bounce version of the model (Attenborough, Li, & Horoshenkov, 2007) (Plovsing, B; Kragh, J, 2006) (Plovsing, B; Kragh, J, 2006). Harmonoise, a European project, was developed in 2002 to offer a state of the art prediction scheme for which other prediction schemes could base themselves on. The Harmonoise scheme also has an improved method for linearly approximating a logarithmic sound speed profile which was later also implemented in the Nord2000 prediction scheme (Salomons, Maercke, Defrance, & deRoo, 2011) (Plovsing, B; Kragh, J, 2006).

3. Theoretical Background

What follows is a brief description of the models used by OTL-Suite. Further details can be found in the references cited.

3.1 Theory of the propagation of sound in a non-homogenous atmosphere inside the bright zone

For cases of downward or upward refraction OTL-Suite implements the heuristic model originally developed by (L'Espérance, Herzog, Daigle, & Nicolas, 1992). The advantage of the model is that it is simple to implement and takes into account multiple bounces of rays in cases of strong downward refraction instead of just the two rays of the single bounce model. The model does this by taking advantage of the fact that in the case of a positive gradient there is one direct path, three paths with one order of reflection and four paths for each successive order of reflection. Thus the intersection points of each path with the ground can be found by finding the roots of a fourth order polynomial equation for each reflection order.

Once the rays are found their path lengths and times are calculated using geometrical parameters described in the original paper. The model also takes into account atmospheric absorption and turbulence.

3.2 Theory of propagation of sound in the shadow zone

The heuristic model predicts that in cases of negative sound speed gradients and where the receiver is located in a shadow zone, no rays will reach the receiver and the sound pressure level will be 0. In reality there is a creeping wave which propagates above the ground and diffracts acoustical energy into the shadow zone (Pierce, 1994). In order to predict the sound pressure level in the shadow zone, OTL-Suite combines the heuristic model with a residual method outlined in (West, Walkden, & Sack, 1989). This involves expressing a Z-dependent Green's function in a residual form whose solutions are Airy functions. The pressure at the receiver is then calculated using the Hankel function.

3.3 Approximating a logarithmic sound speed profile with a linear sound speed profile

The input parameters required to model a logarithmic sound speed profile in OTL-Suite are: the Temperature at ground level (T), the temperature at a height z defined by the user, the wind speed $u(z_u)$ at a height z_u, the roughness constant (z_0) and the wind direction (φ) defined in OTL-Suite as the clockwise angle from the North with the downwind condition blowing from south to north. Figure 1 below shows how these parameters are entered in OTL-Suite.

In cases of a logarithmic sound speed profile the sound speed is described with the following equation:

$$c(z) = A\ln\left(\frac{z}{z_0}+1\right) + Bz + c_0$$

Where A and B are given by:

$$A = \frac{u(z_u)\cos\theta}{\ln\left(\frac{z_u}{z_0}+1\right)} \qquad B = \frac{dT}{dz}\frac{10.025}{\sqrt{T+273.15}}$$

This time θ is the wind direction relative to the propagation of sound between the source and receiver and $\frac{dT}{dz}$ is the linear temperature gradient.

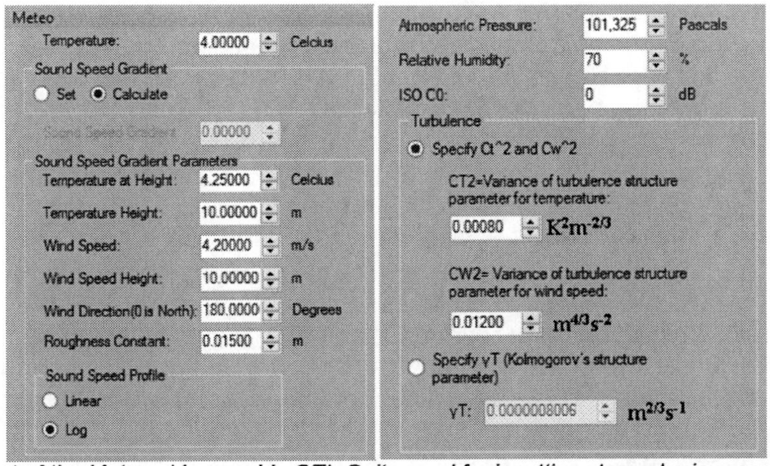

Figure 1: Screenshot of the Meteo side panel in OTL-Suite used for inputting atmospheric parameters for the modelling of atmospheric refraction, turbulence and absorption. These particular parameters were used for DELTA Cases 3 and 4 described in Section 6.

Since both the heuristic model and the residual method need to approximate any general sound speed profile with a linear profile, OTL-Suite employs the method by Harmonoise to approximate a logarithmic profile (Plovsing, B; Kragh, J, 2006). This involves finding the radius of curvatures of the logarithmic (A) and the linear (B) parts of the profile and combining them as follows:

$$r_{A,B} = \frac{1}{\frac{1}{r_A} + \frac{1}{r_B}}$$

Where

$$r_A = sign(A)\frac{R}{8}\sqrt{\frac{2\pi c_0}{|A|}} \qquad r_B = sign(B)\sqrt{\left(\frac{c_0}{|B|}\right)^2 + \left(\frac{R}{2}\right)^2}$$

The effective linear sound speed gradient can then be found using:

$$a = \frac{1}{r_{A,B}\cos\varphi}$$

Where φ is given by:

$$\varphi = \sin^{-1}\left(\frac{\sqrt{R^2 + (z_r - z_s)^2}}{2r_{A,B}}\right) + \tan^{-1}\left(\frac{z_r - z_s}{R}\right)$$

R is the horizontal range between the source and receiver while z_s and z_r are the source and receiver heights respectively.

4. Outline of Benchmark Cases

For the present study the results of OTL-Suite were compared against the benchmarks cases in (Attenborough, et al., 1995) which we will refer to as the 1995 benchmark cases, and some of the benchmark cases in (WP2 Team, 2002) which we will refer to as the Harmonoise benchmark cases. The 1995 benchmark cases include analytical solutions for linear sound speed profiles but they are only done for monochromatic frequencies. The Harmonoise benchmarks cases are done in a 1/3 Octave frequency resolution and include comparisons to the modern engineering prediction scheme Nord2000. They also include logarithmic sound speed profiles thus allowing us to test the capabilities of OTL-Suite in linearly approximating logarithmic sound speed profiles.

What follows is an outline of the benchmark cases used.

4.1 1995 Benchmark Cases

The 1995 benchmark cases consist of four cases corresponding to different atmospheres: a homogenous atmosphere with uniform sound speed (Case 1), a non-homogenous atmosphere with a strong positive linear sound speed gradient of 0.1 (Case 2), a non-homogenous atmosphere with a strong negative linear sound speed gradient of -0.1 (Case 3) and a composite sound speed profile (Case 4) which was not used as it exceeds the capabilities of OTL-Suite. Full details and descriptions of the cases can be found in (Attenborough, et al., 1995). In the original study only the analytical, FFP and PE methods of all the cases were presented.

The intention of the original paper was to develop benchmark cases of extreme atmospheric conditions but without the inclusion of effects such as turbulence, rough ground or uneven terrain. This would allow simple versions of new numerical methods to be tested against these benchmarks before being expanded to include other physical phenomena.

In the three cases considered calculations were performed for Source-Receiver ranges of up to 10000 m. The calculations were performed for three monochromatic frequencies: 10, 100 and 1000 Hz. Here we present the results for Case 2 and Case 3 at a range of 10000 m and a frequency of 100 Hz. The receivers were separated by 25 m.

The ground impedance was described using the Delany and Bazley 1 parameter model with a Flow resistivity of 205000 Pa s m^{-2} as opposed to the 4-parameter model used in the benchmark paper. The parameters used in the model are summarised in **Erreur ! Source du renvoi introuvable.**.

Table 1: Parameters used for 1995 Benchmark Cases

Parameter	Value
Density of air (ρ_0)	1.205 kg/m^3
Atmospheric Pressure	1 atm
Relative Humidity (RH)	70 %
Temperature (T_0)	22 °C
Ground Flow Resistivity (σ) (D&B)	205000 Pa s m^{-2}
Source Height (h_s)	5 m
Receiver Height (h_r)	1 m
Range (R)	10000 m
Frequency (f)	100 Hz

4.2 Harmonoise Benchmark Cases

OTL-Suite was compared against Case 1.1 of the Harmonoise benchmark cases. This case consists of a flat ground with uniform impedance for different Source-Receiver heights and Ranges. In total there are 144 different subcases. The atmospheric conditions used in the particular subcases under investigation are summarized **Table 2** below:

Table 2: Atmospheric conditions used in the Harmonoise Benchmark subcases

Index m	Atmospheric condition	Sound speed profile
m = 2	Linear sound speed Profile, no turbulence	$a = 0.05\ s^{-1}$, $c(z) = c_0 + az$
m = 3	Logarithmic sound speed profile, no turbulence	$b = 1\ ms^{-1}$, $c(z) = c_0 + b\ln(1+z/z_0)$
m = 5	Logarithmic sound speed profile, no turbulence	$b = -1\ ms^{-1}$, $c(z) = c_0 + b\ln(1+z/z_0)$

Due to the large number of subcases in case 1.1 the subcases were narrowed down to the ones consisting of a non-homogenous atmosphere, the ones that did not include atmospheric turbulence (thus the ones which have an index m = 2, 3 and 5), the subcases consisting of a locally reacting ground (grass), a range of 2000 m and a source/receiver height combination of $h_s = 0.5$ m with $h_r = 1.5$ m and $h_s = 5$ m with $h_r = 4$ m. These particular source/receiver height combinations were chosen to test the linear approximation of a logarithmic profile when the sources and receivers are close to the ground and far from the ground.

Thus the list of subcases considered are: C11_2132m and C11_3232m where the index m corresponds to the atmospheric conditions m = 2, 3 and 5. The parameters used for all the subcases are outlined in **Table 3** below:

Table 3: Modelling Parameters used in the Harmonoise Benchmark Cases

Parameter	C11_21322	C11_21323	C11_21325	C11_32322	C11_32323	C11_32325
Source height (m)	0.5	0.5	0.5	5	5	5
Receiver height (m)	1.5	1.5	1.5	4	4	4
Range (m)	2000	2000	2000	2000	2000	2000
Speed of sound (ms^{-1})	340	340	340	340	340	340
Roughness constant (m)	0.1	0.1	0.1	0.1	0.1	0.1
Ground Flow Resistivity ($Pa\ s\ m^{-2}$)	200000 (Grass)	200000 (Grass)	200000 (Grass)	200000 (Grass)	200000 (Grass)	200000 (Grass)

OTL-Suite is compared to Basic and Engineering models, which are given below with their acronyms. The Basic models are, the Crank-Nicholson Parabolic Equation method (CPE TNO), the Green's Function Parabolic Equation method (GPE CST), the Fast Field Program (FFP CST) and for the subcases with a linear sound speed profile the Meteo-BEM (MBE CST) model. The engineering models, are the Nord2000 propagation model (N20 DEL) and the CRAYL model (CRA DEL). The distinction between Basic and Engineering models was made in (WP2 Team, 2002) and it applies to the rest of the paper. More details of these models can be found in (WP2 Team, 2002).

5. Results

5.1 1995 Benchmark Cases

Good agreement was found between OTL-Suite and the FFP, PE and analytical solutions used in the 1995 Benchmark Cases. In Case 2, the downward refracting atmosphere, OTL-Suite follows the trend quite well although the minima and maxima are significantly sharper than the 1995 Attenborough Case, especially at large ranges. Nevertheless in a more realistic scenario these minima and maxima would most likely be smoothed out by turbulence. In Case 3 there is a discontinuity present at a range of about 400 m indicating that the receiver is now in the shadow zone where the Transmission Loss drops sharply. Figure 5 in section 6.2 shows some of the sound ray paths from the source to receivers located at a range of 5000-7000 m.

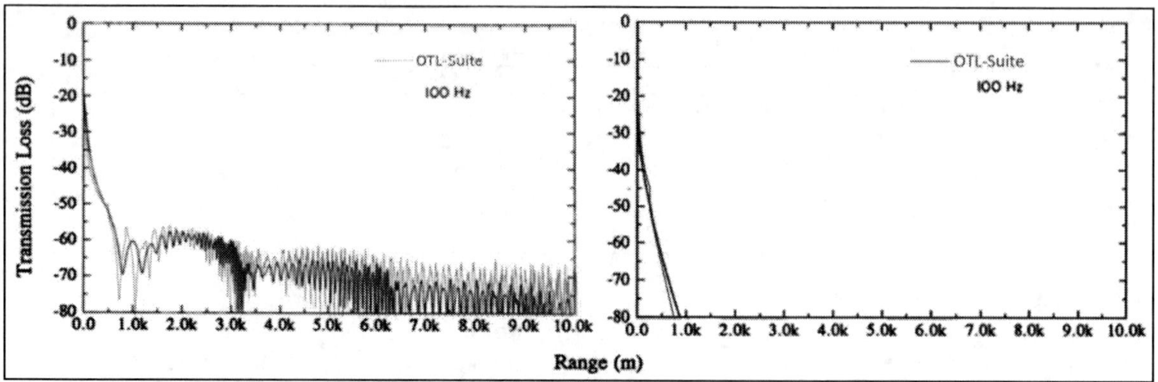

Figure 2: Comparison between OTL-Suite calculations (red line) and 1995 Benchmark Cases (black line). Left graph is for the case of a strong positive linear sound speed gradient of 0.1 s^{-1} while the right graph is for the case of a strong negative sound speed gradient of -0.1 s^{-1}. Both curves show transmission loss vs distance at 100Hz. Calculated graphs are superimposed on published data.

5.2 Harmonoise Benchmark Cases

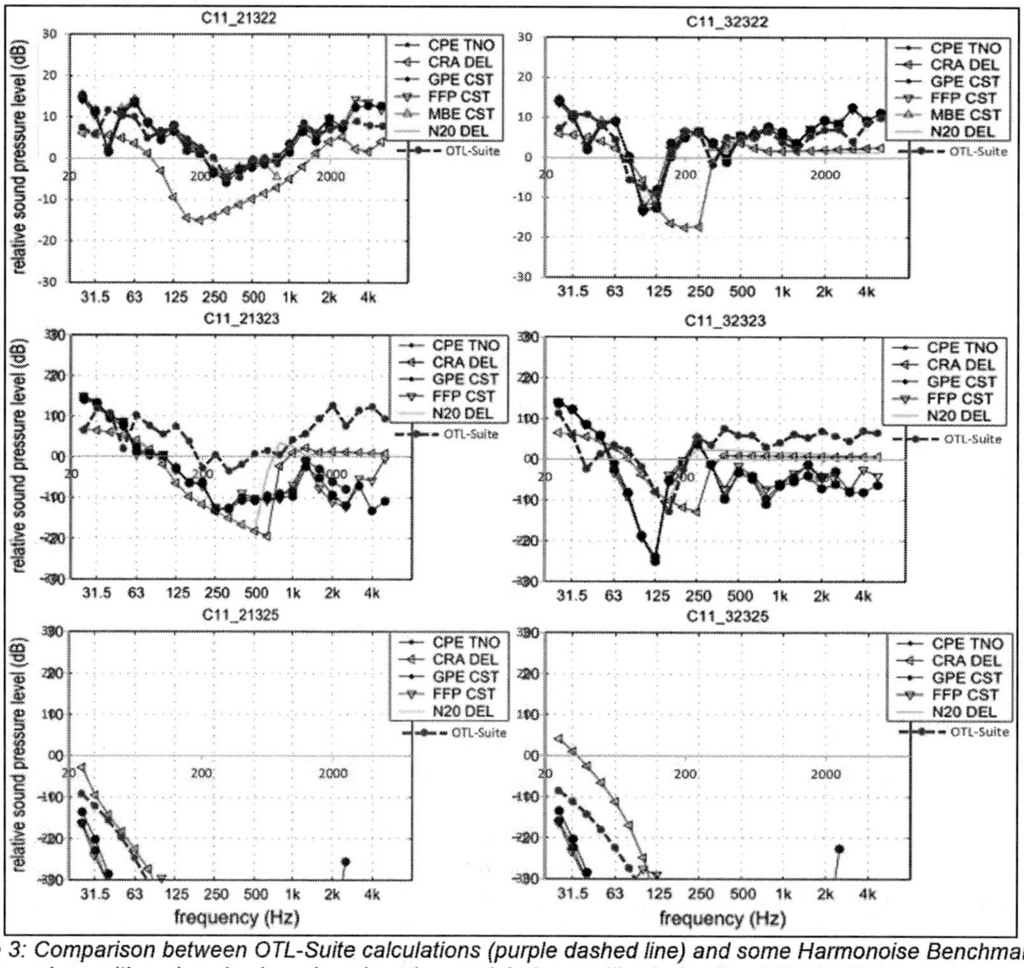

Figure 3: Comparison between OTL-Suite calculations (purple dashed line) and some Harmonoise Benchmark subcases done with various basic and engineering models (see outline in Section 4.2). For the graphs on the left the source and receiver heights are 0.5 and 1.5 m respectively while for the graphs on the right the source and receiver heights are 5 and 4 m respectively. The graphs on the top are for a linear sound speed gradient of the 0.05 s^{-1}, the middle graphs have a positive logarithmic coefficient of 1 ms^{-1} while the bottom graphs have a negative logarithmic coefficient of -1 ms^{-1}. Calculated graphs are superimposed on published data.

The results for the Harmonoise Benchmark subcases are shown in Figure 3 above. For the subcases where the source and receiver are close to the ground (0.5 m and 1.5 m respectively) a good agreement with both basic models and engineering models is found for the case with linear refraction (Subcase C11_21322). Once a logarithmic profile is assumed the results of OTL-Suite and the engineering models deviate from the basic models (Subcase C11_21323) significantly. There is always a frequency shift between the interference minima. This is to be expected because since OTL-Suite and the engineering models use a linear approximation for the logarithmic profile, the path length and time differences will be different leading to a shift of the interference minima.

For the subcases where the source and receivers are further away from the ground (5 m and 4 m respectively) there is a better agreement between OTL-Suite and the basic models for the logarithmic cases (subcases C11_32323) in the low frequencies although there are still high deviations. This is to be expected because of to the shape of the logarithmic curve. As the source and receiver move away from the ground the linear approximation will better match the logarithmic one.

There is also a discrepancy between OTL-Suite and the engineering models in all of the subcases. This can be explained by the fact that the engineering models are single bounce models that only take into account two paths whereas OTL-Suite implements a multiple bounce model. The discrepancy occurs because at long ranges there will be a significant amount of paths for downward refractions which the single bounce models of the engineering models do not take into account.

For subcases C11_21325 and C11_32325 where the receivers are in the shadow zone there is a large deviation between the engineering models and the basic models with OTL-Suite displaying a closer agreement with the basic models.

6. Comparison with noise measurements

For the case of wind turbine noise, the comparisons between OTL-Suite calculations and measurements was done in 2 steps.

In the first step we used the loudspeaker measurements which were made in the framework of the validation of Nord2000 (Plovsing & Kragh, 2009). This step is interesting because the loudspeaker was positioned at a height of 50m, which is comparable to the height of the noise sources of a wind turbine. The parameters used for these cases are detailed in **Table 4**.

In the second step OTL-Suite calculations were compared to noise measurements around a wind farm consisting of 6 wind turbines (hub height 80m). This test case was chosen because in some meteorological configurations (high wind shear in stable atmospheric conditions) the background noise is more than 10 dB lower than the WTN noise, even at ranges of 500m from the wind turbines. High wind shear also has the advantage that it results to a low wind speed near the ground reducing the wind disturbance on the microphone.

Due to the unpredictable range of atmospheric parameters in any given situations we propose a scatter plot of dB(A) values vs the atmospheric parameters for validating atmospheric acoustics.

6.1 Comparison with loudspeaker measurements by DELTA

In this test case the loudspeaker was placed at a height of 50m. The noise source's amplitude and directivity was known enabling us to calculate the excess propagation effect (the difference between the total sound level and direct sound) in 1/3 octave frequency resolution. Although (Plovsing & Kragh, 2009) used a ground Flow Resistivity of 200000 $Pasm^{-2}$ there is great uncertainty about the ground modelled therefore the value of the Flow Resistivity was adjusted to 400000 $Pasm^{-2}$ for DELTA Case 1 and 50000 $Pasm^{-2}$ for DELTA Cases 2, 3 and 4 in order to match the first interference minimum.

Table 4: Parameters used for the DELTA Validation Cases. Input data taken from (Plovsing & Kragh, 2009) **or extrapolated from their graphical representations of the sound speed profiles. The ground Flow Resistivities were adjusted from DELTA's 200000 $Pasm^{-2}$.**

Parameter	Delta Case 1	Delta Case 2	Delta Case 3	Delta Case 4
Source Height (m)	50	50	50	50
Receiver Height (m)	2	2	2	2
Ranges (m)	456	1020	412	912
Temperature at Ground (°C)	4	4	4	4
Temperature Height z (m)	10	10	10	10
Temperature at Height z (°C)	4.25	4.25	4.25	4.25
Wind Speed Height z_u (m)	10	10	10	10
Wind Speed at Height z_u (ms^{-1})	4.2	4.2	4.2	4.2
Wind Direction relative to Sound Propagation Direction (degrees)	0 (downwind)	0 (downwind)	180 (upwind)	180 (upwind)
Roughness Constant (m)	0.015	0.015	0.015	0.015
Ground Flow Resistivity ($Pasm^{-2}$)	400000	50000	50000	50000

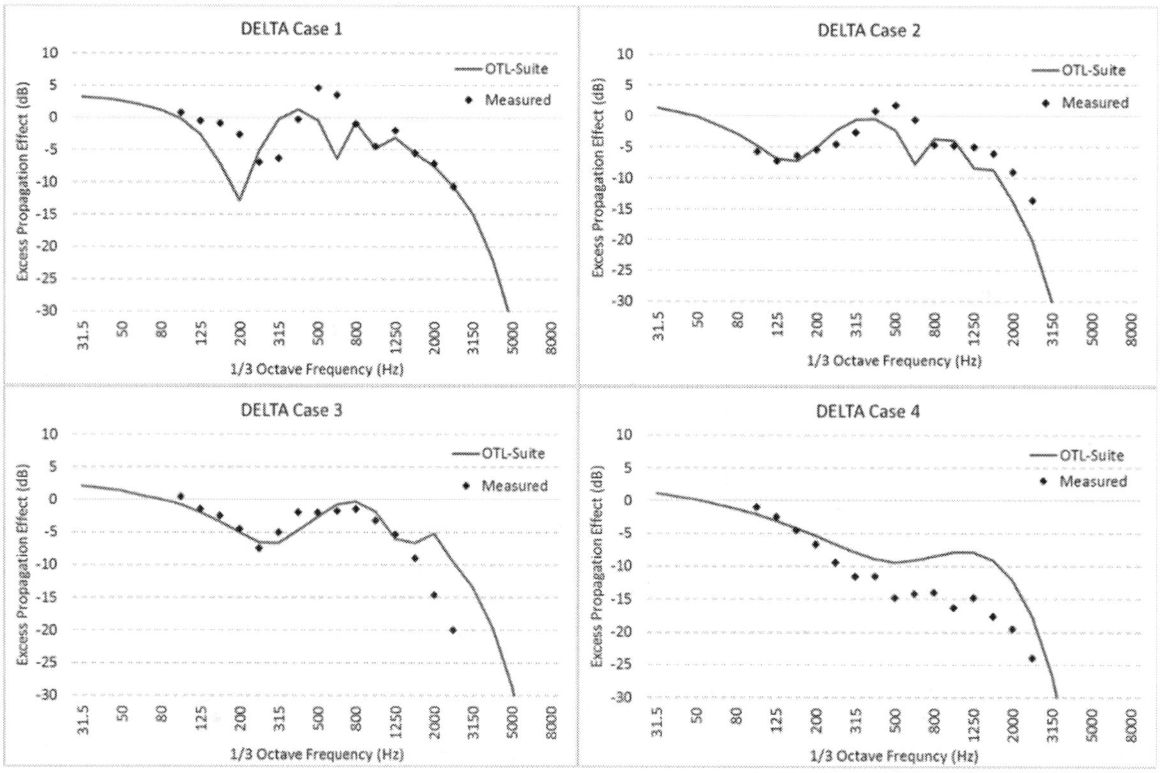

Figure 4: Measured and predicted excess propagation effect. Delta Cases 1 and 2 are for downwind conditions while Delta Cases 3 and 4 upwind conditions. The source receiver horizontal range is approximately 500 m for Cases 1 and 3 and approximately 1000 m for Cases 2 and 4.

Figure 4 above presents the results for two ranges (approximately 500m and 1000m) for both downwind and upwind conditions. There is a good agreement for downwind propagation, and a more or less good agreement in the upwind propagation. This is consistent with the comparisons with the Harmonoise benchmark cases described in Section 5.2.

It is difficult to analyse this case further, because of the reliability of the input data: some of the parameters (like temperature and roughness) had to be extrapolated from the graphical sound speed profiles available in (Plovsing & Kragh, 2009).

6.2 Comparison with noise measurements around a wind farm

The wind farm that was investigated consisted of 6 wind turbines (hub height 80m, rotor diameter 90m). The meteorological measurements recorded were: wind speed and wind direction at heights of 2m, 10m and the hub height of 80m; temperature, humidity and atmospheric pressure at heights of 2m and 10m.

The microphones were positioned at a height of 1.5m and at horizontal ranges of 150m and 500m from the wind turbines; measurements were done in a 1/3 octave band frequency spectrum and full audio spectrum for some locations.

Noise measurements are presented in L_{eq} for a horizontal range of 150m from the wind turbines, and L_{50} for large ranges.

The wind turbine is modelled as a point source. The sound power level of the source is available from measurement reports. There were three cases taken into consideration with the parameters outlined in **Table 5** below:

Table 5: Parameters used for OTL-Suite model to compare against WTN measurements

Parameters	WTN Case 1	WTN Case 2	WTN Case 3
Source Height (m)	80	80	80
Receiver Height (m)	1.5	1.5	1.5
Range (m)	150	500	500
Temperature at Ground (°C)	10.7	4.1	3.6
Temperature Height z (m)	10	10	10
Temperature at Height z (°C)	10.732	4.382	3.757
Wind Speed Height z_u (m)	10	10	10
Wind Speed at Height z_u (ms^{-1})	6.8	5.0	4.4
Wind Direction relative to Sound Propagation Direction (degrees)	Downwind	Downwind	Upwind
Roughness Constant (m)	0.05 (shear factor 0.16)	0.91 (shear factor 0.28)	1.33 (shear factor 0.31)
Ground Flow Resistivity ($Pasm^{-2}$)	225000	225000	225000

The Excess Attenuation is first calculated in narrow frequency bands starting at 20 Hz in steps of 5 Hz until 500 Hz, where the steps switch to 20 Hz until at 10000 Hz. The frequencies of the Excess Attenuation are then combined into the centre frequencies of a 1/3 Octave spectrum ranging from 25 Hz to 10000 Hz. The direct sound (which includes the source characteristics) is then added to the Excess Attenuation in the 1/3 Octave spectrum to obtain the Sound Pressure Level. This is then also combined into a 1/1 Octave spectrum and then given in dB(A) values.

OTL-Suite allows users to calculate the Excess Attenuation at extremely high resolutions (from 0,001 to 100.000 Hz at 0,001 Hz increments, in constant frequency steps or constant percentage steps). The resolution chosen here is a compromise between accuracy and performance.

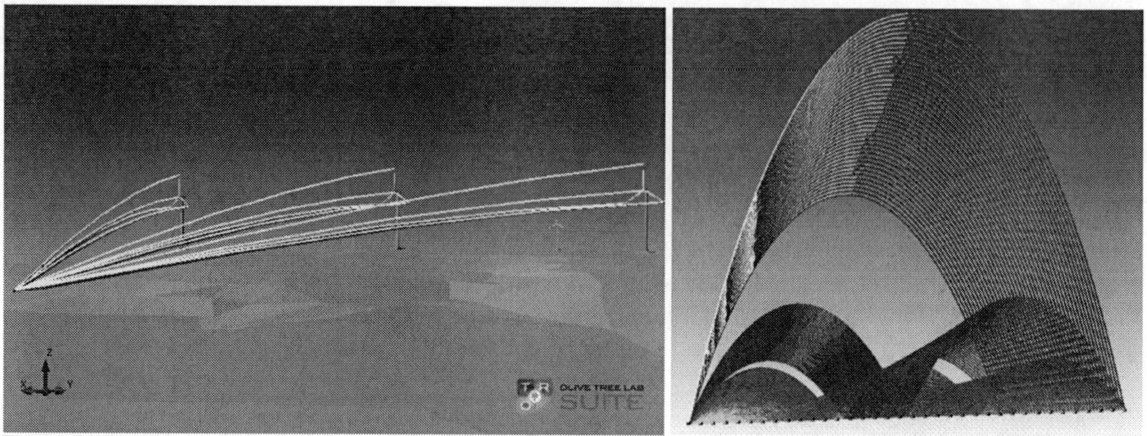

Figure 5: On the Left: View of the 3D model for 3 wind turbines. On the Right: Some of the sound ray paths between source and 80 receivers for the 1995 Benchmark Case 2. The receivers are located at a range of 5000 – 7000 m.

A first comparison is presented in Figure 6 below for WTN Case 1, in a 1/1 Octave band frequency spectrum and 1/3 Octave band frequency spectrum in downwind conditions.
We can see a good agreement between calculations and measurements at a range of 150m of the wind turbine, with almost the same interference minimum at about 125 Hz.
This means that the sound power level taken as input data and the propagation model works fine, even in the point source approximation.

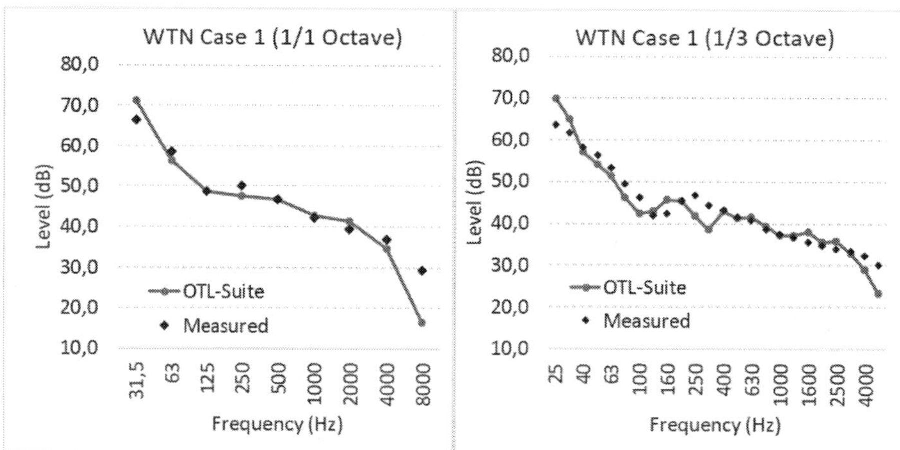

Figure 6: Measured and predicted noise level for WTN Case 1 in 1/1 band and 1/3 band. Downwind conditions with a range of 150m.

Figure 7 below presents the results for WTN Case 2, at a range of 500m from a wind turbine in downwind conditions. Calculations are presented in 1/3 Octave band frequency spectrum, and for a set of 10-minute meteorological data in dB(A).

We can see a good agreement in the 1/3 Octave frequency spectrum with some small differences in the low frequencies under 40 Hz, which were also visible at a range of 150m. There is a very good agreement on the dB(A) scatter plot.

Figure 8 below presents the results for WTN Case 3, at a range of 500m from a wind turbine in upwind propagation.

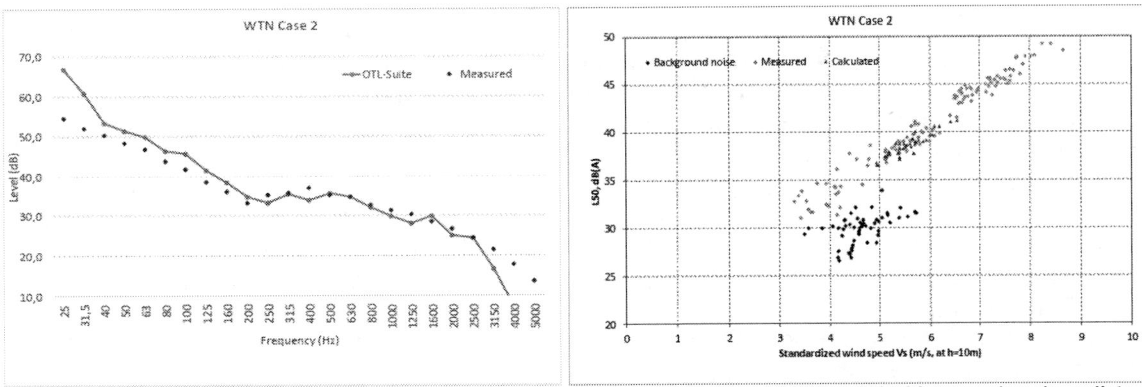

Figure 7: Left graph: Measured and predicted noise levels for WTN Case 2. Right graph: Measured and predicted Noise Levels for Case 2 in dB(A) for a set of 10-minute meteorological data. Downwind conditions with a range of 500m.

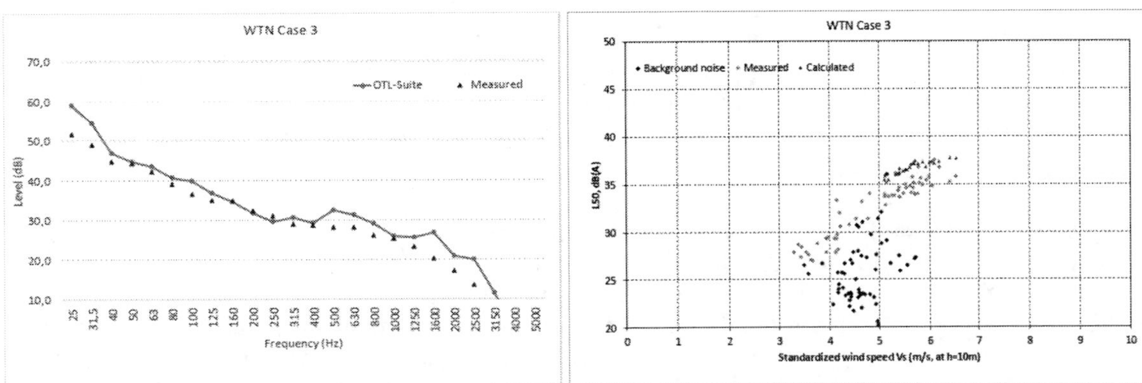

Figure 8: Left graph: Measured and predicted noise levels for Case 3. Right graph: Measured and predicted Noise Levels for Case 3 in dB(A) for a set of 10-minute meteorological data. Upwind conditions with a range of 500m.

We can see some differences between measurement and calculation in the spectrum calculations, but a quite a good agreement in the dB(A) scatter plot. However the calculation results in dB(A) seems to correspond to the maximum of the measured values.

This is consistent with the comparisons between OTL-Suite and the benchmark cases in Sections 5.1 and 5.2. OTL-Suite seems to overestimate the high frequencies at long ranges compared to the basic models. It should also be noted that if the receiver is located in the shadow zone for upwind conditions OTL-Suite will use the default xy plane as the ground and ignore the imported rough ground model. A rough ground would most likely further attenuate the Sound Pressure Levels. Nevertheless the calculated results are within an acceptable range to the measured ones.

7. Conclusion

This paper is a result of a collaborative research between SIXENSE Environment and PEMARD. The calculation results of OTL-Suite are compared with benchmark cases and long term noise & meteorological measurements taken especially for this paper, at a wind turbine farm. Comparison of calculation results with benchmark cases is good. In the case of wind turbine noise there is a very good agreement in the downwind cases and acceptable results in upwind condition. The key point of the calculation is the knowledge of full meteorological data, including wind speed profile and temperature gradient. This can easily be assessed with two meteorological stations at 2m and 10m height, as presented in a previous paper at WTN 2015 (Bigot, Slaviero, Mirabel, & Dutilleux, 2015). A good way of presenting calculation results is to compute one calculation for each 10mn sample of meteorological data, and present the scatter plot of the dB(A) values.

The paper shows that the complexity of atmospheric dynamics cannot be fully represented by a single practical engineering model. This is already demonstrated in the Harmonoise validation reference. The main source of discrepancy between measured and predicted data in ray models is the approximations used in calculating sound speed profiles. However, for engineering purposes accuracy has to be traded with calculation time. This being said, the ray models, implemented with multiple reflection paths, seem to be better suited as a compromise between accuracy and calculation time. Furthermore, sound ray paths allow for the visualisation of sound propagation.

Future work could include the study of subsonic noise propagation in OTL-Suite, which allows calculations of infrasound. In future development of OTL-Suite, noise sources could be modelled as moving dipole and quadrupoles sources (instead of monopoles) allowing for more realistic calculations including the calculation of modulation effects. It would also be worthwhile to compare more measurements with further developments of WBGA to include phenomena such as the semi-analytical model for full logarithmic sound speed profiles (Salomons E. M., 1994), the effects of caustics (Salomons E. M., 1998) and the more recent model of the effect of turbulent scattering of acoustical energy into the shadow zone (Lam, 2009).

References

Attenborough, K., Li, K. M., & Horoshenkov, K. (2007). *Predicting Outdoor Sound.* New York: Taylor & Francis.

Attenborough, K., Taherzadeh, S., Bass, H. E., Di, X., Raspet, R., Becker, G. R., . . . Hoof, H. A. (1995). Benchmark Cases for outdoor sound propagation models. *Journal of the Acoustical Society of America, 97*(1), 173-191.

Bateman, H. (1914). The Influence of Meteorological Conditions on The Propagation of Sound. *Monthly Weather Review*, 258-265.

Berry, A., & Daigle, G. A. (1988). Controlled experiments of the diffraction of sound by a curved surface. *Journal of the Acoustical Society of America, 83*(6), 2047-2058.

Bigot, A., Slaviero, D., Mirabel, C., & Dutilleux, P. (2015). Influence of vertical temperature gradient on background noise and on long-range noise propagation from wind turbines. *6th International Meeting on Wind Turbine Noise.*

Delany, M. E. (1977). Sound Propagation in the Atmosphere: A Historical Review. *Acustica, 38*(10), 201-233.

Embleton, T. F., Thiessen, G. J., & Piercy, J. E. (1976). Propagation in an inversion and reflections at the ground. *Journal of the Acoustical Society of America, 59*(2), 278-282.

Lam, Y. W. (2005). Issues for computer modelling of room acoustics in non-concert hall settings. *Acoustical Science and Technology Journal, 26*(2), 145-155.

Lam, Y. W. (2009). An analytical model for turbulence scattered rays in the shadow zone for outdoor sound propagation calculation. *Journal of the Acoustical Society of America, 125*(3), 1340-1350.

L'Espérance, A., Herzog, P., Daigle, G. A., & Nicolas, J. R. (1992). Heuristic Model for the Outdoor Sound Propagation Based on an Extension of the Geometrical Ray Theory in the Case of a Linear Sound Speed Profile. *Applied Acoustics, 37*(2), 111-139.

Nijs, L., & Wapenaar, C. P. (1990). The influence of wind and temperature gradients on sound propagation calculated with the two-way wave equation. *The Journal of the Acoustical Society of America, 87*(5), 1987-1998.

Ostashev, V. E. (1997). *Acoustics in Moving Inhomogeneous Media.* London: E & FN Spon.

Pierce, A. D. (1994). *Acoustics: An Introduction to Its Physical Principles and Applications.* New York: Acoustical Society of America.

Piercy, J. E., Embleton, T. F., & Sutherland, L. C. (1977). Review of noise propagation in the atmosphere. *Journal of the Acoustical Society of America, 61*(6), 1403-1418.

Plovsing, B., & Kragh, J. (2009). *Validation of the Nord2000 propagation model for use on wind turbine noise, Report AV 1236/09.* Hørsholm: DELTA Acoustics and Electronics.

Plovsing, B; Kragh, J. (2006). *Nord2000. Comprehensive Outdoor Sound Propagation Model. Part 1: Propagation in an Atmosphere without Significant Refraction.* Hørsholm: DELTA Acoustics & Vibration Report AV 1849/00.

Plovsing, B; Kragh, J. (2006). *Nord2000. Comprehensive Outdoor Sound Propagation Model. Part 2: Propagation in an Atmosphere with Refraction.* Hørsholm: DELTA Acoustics & Vibration Report AV 1851/00.

Salomons, E. M. (1994). Downwind propagation of sound in an atmosphere with a realistic sound-speed profile: A semianalytical ray model. *Journal of the Acoustical Society of America, 95*(5), 2425-2436.

Salomons, E. M. (1998). Caustic diffraction fields in a downward refracting atmosphere. *Journal of the Acoustical Society of America, 104*(6), 3259-3272.

Salomons, E. M. (2001). *Computational Atmospheric Acoustics.* London: Kluwer.

Salomons, E., Maercke, D. v., Defrance, J., & deRoo, F. (2011). The Harmonoise Sound Propagation Model. *Acta Acustica united with Acustica, 97*(1), 62-74.

West, M., Walkden, F., & Sack, R. A. (1989). The Acoustic Shadow Produced by Wind Speed and Temperature Gradients Close to the Ground. *Applied Acoustics, 27*(3), 239-260.

Wilson, D. K. (1993). Sound field computations in stratified, moving medium. *Journal of the Acoustical Society of America, 94*(1), 400-407.

WP2 Team. (2002). *Task 2.2. State of the art of modeling.* Delft: HAR22TR-020222-TNO03.

WP2 Team. (2002). *Task 2.3. Benchmark calculations and modeling approximations .* Delft: HAR23TR-020222-TNO10.

Annual investigation of sound propagation from a boreal wind park

Karl Bolin. KTH-Royal Institute of technology, Aeronautical and Vehicle Engineering, Marcus Wallenberg Laboratory for Sound and Vibrations research, Teknikringen 8, SE-100 44, Stockholm, kbolin@kth.se

Ilkka Karasalo. KTH- Royal Institute of technology, Aeronautical and Vehicle Engineering, Marcus Wallenberg Laboratory for Sound and Vibrations research, Teknikringen 8, SE-100 44, Stockholm, ilkkak@kth.se

Esbjörn Olsson SMHI- Swedish Meteorologic and Hydrologic Institute, Folkborgsvägen 17, SE-601 76 Norrköping, Sweden, Esbjorn.Olsson@smhi.se

Summary

This paper presents a study of sound propagation of a wind park of two turbines in the boreal forest belt in southern Sweden. Three different sound propagation algorithms have been used in order to compare results and to assess similarities and dissimilarities for propagation distances of 1 km and 2 km. All three sound propagation codes belong to the parabolic equation family and are therefore capable of incorporating realistic sound speed profiles and alternating terrain. Annual sound propagation have been calculated every 3rd hour using weather forecast data (HiRLAM) using sound speed profiles depending wind velocity, temperature, air humidity and atmospheric pressure. The ground conditions have been using the Delany and Bazely impedance model combined with information on the local ground conditions and altitudes from geographical databases. Results show distributions with slightly more loss than spherical spreading and large deviations from spherical spreading are observed, especially for the longest propagation distance.

Introduction

Sound propagation from wind turbines have become an increasingly important topic as the growth of turbine sizes are accompanied with increasing sound powers and sound is thus perceived at longer distances. Moreover, the relative costs of a proper acoustic assessment of wind parks have decreased and can now be considered marginal compared to the total investment of a wind park.

This paper uses state-of-art sound propagation algorithms to investigate sound spreading from a wind park of two turbines in boreal terrain over one complete year. Three propagation codes are reported in this paper, these are numerical solvers of the parabolic equation and are consequently capable of using arbitrary and range dependent meteorological input from the pan-European weather prediction code HiRLAM to assess the sound propagation. Ground conditions are simulated by using flow resistivity derived from the European Corine-database covering the investigated area with a grid size of 25 m. Local altitude differences are taken into account in two of the propagation codes by digital maps with 50 m grid resolution. Consequently both changing terrain and atmospheric properties are considered in the reported propagation calculations which.

Meteorological data from the mesoscale HiRLAM used by several European countries meteorological institutes are used to calculate the sound speed gradient in this paper. Data are available for every 3rd hour over one year and this dataset have been used to investigate changes in sound exposure due to changing weather. The results are examined for the three codes and compared against each other in order to assess the robustness of the three different solvers. Moreover, the seasonal, daily and hourly changes are analyzed and therefore a more thorough understanding of when high sound levels from wind turbine noise can be expected to occur for boreal sites. This information can be of importance for mitigating wind turbine noise annoyance at the same time as maximizing the amount of power produced by turbines by active control of the turbines according to prevailing meteorological conditions.

Method

Sound propagation algorithms

This paper utilizes tree sound propagation algorithms which are briefly described in the text below. Comparable results have thus been created and this paper focuses on comparing differences and similarities of the calculated results. All three solvers of the wave equation are based on the Helmholtz equation and use the approximation of the Helmholtz equation giving a parabolic equation, the interested reader is referred to Salomons (2002) for a derivation of the parabolic equation from the Helmholtz equation.

The family of parabolic equations is considered suitable by the authors to assess propagation of wind turbine sound due to their capacity to use almost arbitrary sound speed profiles and changing terrain conditions. These factors are considered attractive for wind power applications as it has been shown by (van den Berg 2002 and Öhlund & Larsson, 2015) among others that both meteorology and ground conditions are important factors when assessing wind turbine noise. Moreover, the ability to include arbitrary sound speed profiles facilitates the use of real meteorological data and eliminates the need to truncate these as would normally be the case for many other solvers, such as ray tracers and normal mode models. The parabolic solvers are also computationally fast enough to allow long time periods and long propagation distances to be assessed on ordinary (as of 2017) computers without the need to use clusters that would be the case for Finite Difference Time Domain solvers. One restriction with parabolic equation solvers are their narrow angle approximation which restrict near field accuracy, this approximation would make parabolic equations unsuitable to deal with emission measurements of a wind turbine, however with the source heights and propagation distances reported in this paper this approximation is not violated for any of the reported results.

Jeltsch Energy-conserving Parabolic Equation

The Jeltsch Energy-conserving Parabolic Equation computational method is based on a parabolic equation derived from the Helmholtz equation for frequency-domain soundfields. It is based on formulating the PE as an ordinary differential-algebraic system of equations by discretizing in space, and solving the system by a fourth-order, strongly damped scheme introduced by Jeltsch (1977), for further details of the algorithm see Karasalo & Sundström (1996). This propagation algorithm was originally developed for underwater acoustic applications and later adapted to atmospheric sound propagation by introducing a local artificial absorption layer Salomons (2002) as an approximation of the non-reflecting condition at the upper boundary. Irregular terrain is handled by using orthogonal boundary-adapted curvilinear coordinates described in Abrahamsson (1991), which simplifies the formulation of stable boundary conditions at the upper and lower boundaries. An adaptive range-step controlled by a local error estimate is used for marching in range in order to yield suitable lengths of each range step.

Greens Function Parabolic Equation

The Greens Function Parabolic Equation (GFPE) is a marching algorithm which computes a vertical pressure distribution at each new range step. It was developed by Gilbert and Di (1993) and later improved Gilbert (2015). The method is a Fourier split-step algorithm designed for atmospheric sound propagation and can use range-steps in the order of 10 wavelengths (λ), considerably longer than conventional parabolic equation methods such as the Crank Nicholson using 0.1 λ range steps. GFPE is suitable in the present application because of its computational efficiency. The parameters of the GFPE method were chosen by suggestions in the references Gilbert & Di (1993) and Salomons (2002). Thus, the horizontal and vertical step sizes ($\Delta r=10\lambda_0$ and $\Delta z=0.1\lambda_0$) depend on the wavelength at the ground, λ_0.

Greens Function Parabolic Equation with terrain

The original GFPE algorithm is formulated for flat ground and a cylindrical coordinate system. A direct solution to introduce a curvilinear lower boundary representing changing altitudes in the terrain is considered by the authors as a complicated operation. However, Parakkal et al (2012) suggested to use a refraction factor corresponding to changing terrain in the closely related Fourier based Beilis-Tappert Parabolic Equation. An implementation by the authors of the present paper of the same refraction factor as suggested by Parakkal et al (2012) in the GFPE algorithm shows to give comparable calculations of sound propagation and is therefore implemented and reported in the present paper.

Site description

Ryningsnäs is a test site for wind turbines in boreal terrain in the South of Sweden and is situated as shown on the map in figure 1. The two 2.5 MW wind turbines have 80 m respectively 100 m hub height. A detailed map of the site can be seen in figure 2 as can be seen the turbines are positioned on a plateau mainly covered by boreal forest. Southwest of the park a dale with a river can be observed which marks the change from high ground to lower altitudes. Not seen on the map in figure 2 is a cutting area in the vicinity of the turbines as seen on the aerial photograph shown in figure 3.

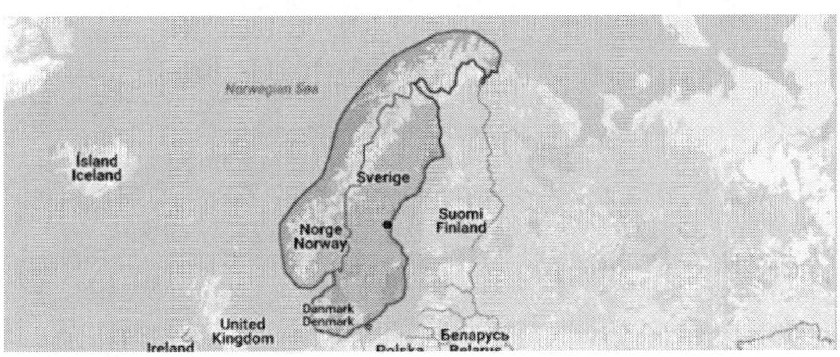

Figure 1: Map of Scandinavia, the location of the Ryningsnäs wind park is shown as a red dot.

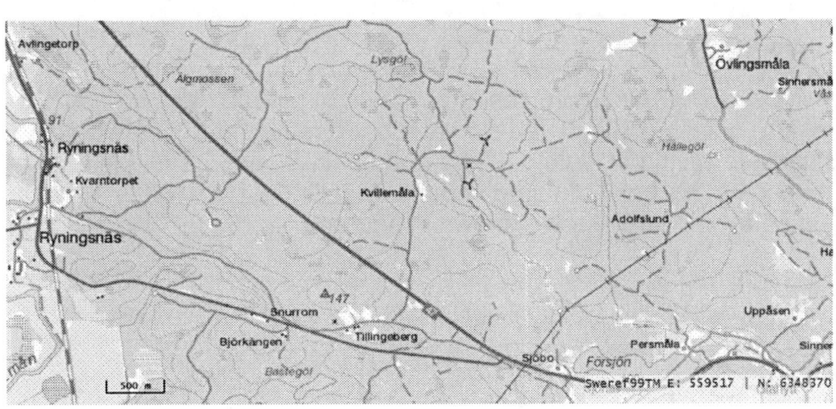

Figure 2: Detailed map of the investigated site, the wind turbines are marked as three blades and are seen in the middle of the map. The map's length of 500 m distance can be seen in the lower left corner of the figure.

Figure 3: An aerial photography of the two wind turbines. The picture is taken from the South, thus the large clear cut is around the southern wind turbine.

Meteorological Input

Meteorological information is available to the authors from the Swedish Meteorological and Hydrological Institute. These data originate from the meso-scale HIRLAM model which gives height profiles of wind velocity, temperature and humidity with grid size 5x5 km at every 3 hours. Although the model is not optimized for conditions at the boundary layer or near boundary layer the data have seasonal and daily variations and it is thus interesting to examine variations of the sound propagation in different directions from the wind turbine. For the calculations reported in this paper the time period is from June 2010 until May 2011, the reason for using these old meteorological data (as of 2017) is that the sound propagation calculations could be compared to sound measurements reported in Öhlund & Larsson (2015) but this comparison is not reported in the current paper.

Ground conditions

Geographical databases can provide information of ground height and ground cover in Europe. The data used in this article have been retrieved from the Swedish land survey database "Metria" covering all the territory in Sweden. Altitude data can be used to model the ground inclination and are specified by a 50x50 m grid and the area around the wind turbines are shown in figure 4.

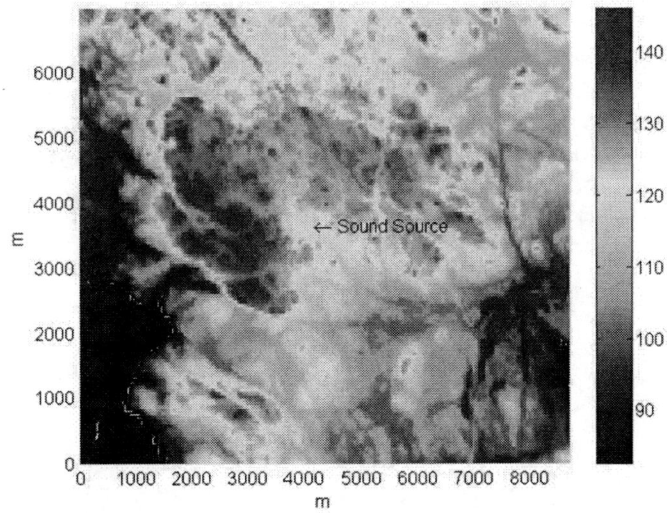

Figure 4: Altitude map (in meters above sea level) of the area around the wind farm. The assumed source position is shown in the figure.

Ground coverage data is publicly available from the Swedish Environmental Protection agency in accordance to the pan-European Corine-database organizing ground into around 80 ground classes. These data are available in a grid size of 25x25 m for Sweden. The ground coverage of the site is shown in figure 5. As can be seen the Southwest corner of figure 2, 4 and 5 all show the farmland and the area is a plateau mainly covered by coniferous forest with several lakes around the site, shown as deep red areas in figure 5.

Figure 4: Terrain type map of the area around the wind farm. The assumed source position is shown in the figure.

Ground coverage data are not directly flow resistivities but measurements coupling ground classes to ground impedance and flow resistivity by Sohlman et al (2004) have been used in this paper in order to assess the ground. The model suggested by Delany & Bazley (1970) has been used to compute the ground impedance in the sound propagation calculations in the present paper. Figure 6 shows the data obtained from the databases for the western propagation and the corresponding flow resistivity.

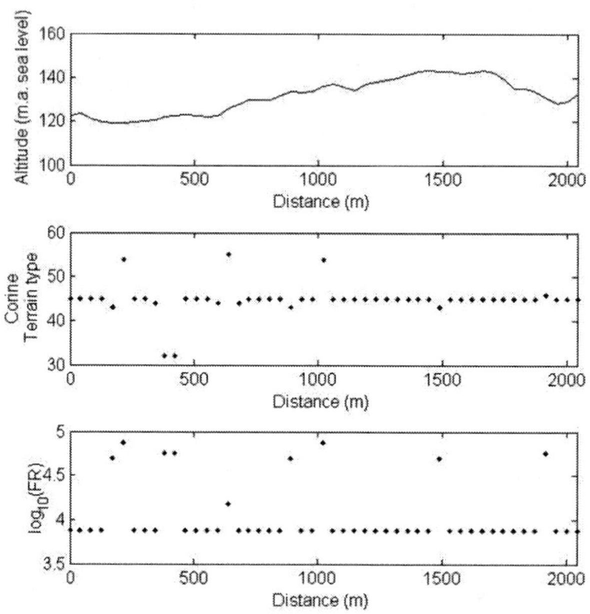

Figure 6: Ground conditions for the western propagation direction. The upper graph contain the height data, the middle the terrain type according to the Corine database and the lower the resulting flow resistivity (FR) used.

Results

All sound transmission results in this paper are presented as transmission losses (TLs) and defined as the loss of acoustic energy from 1 m distance of a spherical source to the reception points and reported in dB. 80 Hz narrowband frequency has been used in order to decrease the computational time that increases with increasing frequencies for parabolic equations. The source position is assumed to be the southern wind turbine at position WGS84 decimal position (N 57,274348°,E 15,986999°). Propagation directions have been chosen to the cardinal directions (North, South, East and West) and propagation distances 1 km and 2 km from the midpoint between the two wind farms have been chosen.

Jeltsch Energy-conserving Parabolic Equation

Resulting TL from the JEPE computations are shown in figure 7. The dominating wind direction is Southwest and tendencies for higher TLs, i.e. lower sound levels, can be observed for the Southern and Eastern direction compared to the Northern and Western directions. This trend seems to increase with increasing distance as the 2 km TLs have more obvious differences in TL depending on direction than the 1 km calculations. Noteworthy is that no statistical analysis such as ANOVA have been performed in order to evaluate if these observations are statistically significant or not.

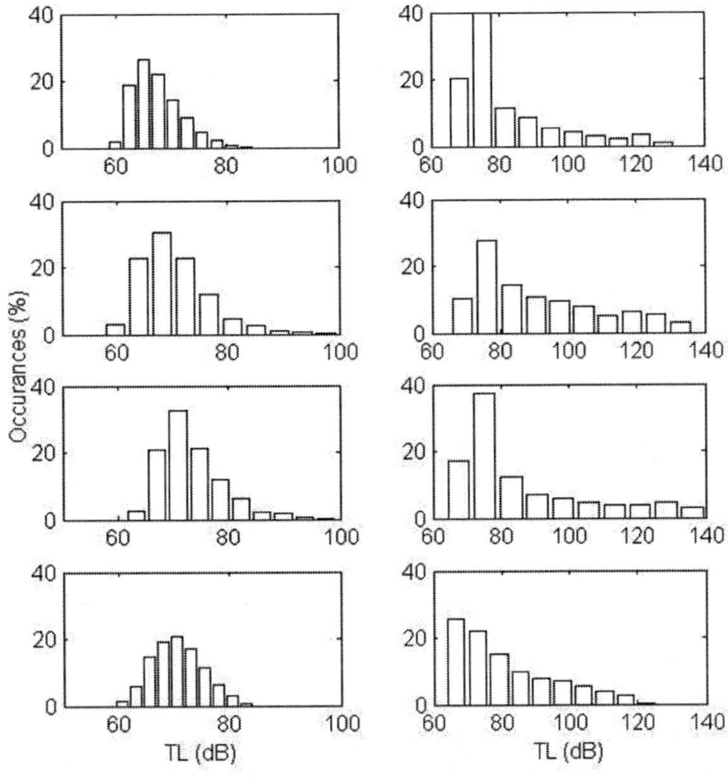

Figure 7: Histograms of the TL for the JEPE calculation are shown in the figure. Left column shows the TL at 1 km distance and the right column shows the TL for 2 km distance. The rows show each cardinal direction in the order: North (upper), South, East and West (lowest).

Greens Function Parabolic Equation

Resulting TL from the GFPE computations are shown in figure 8. The TLs seems to be comparable to the spherical spreading values of TL= 60 and 66 dB for 1 km and 2 km with somewhat higher TLs probably due to the atmospheric absorption. Differences of the TL distributions depending on direction of propagation is visible but the calculations show less trends of low TLs downwind of the turbine compared to the JEPE results. Moreover the dynamics of the TL have decreased showing less occasions of high TL, i.e. sound shadow due to an upward refraction compared to the JEPE. The dynamics of the TL for 2 km distance is higher than for the 1 km distance which is plausible as increasing propagation distances would increase the importance of refraction.

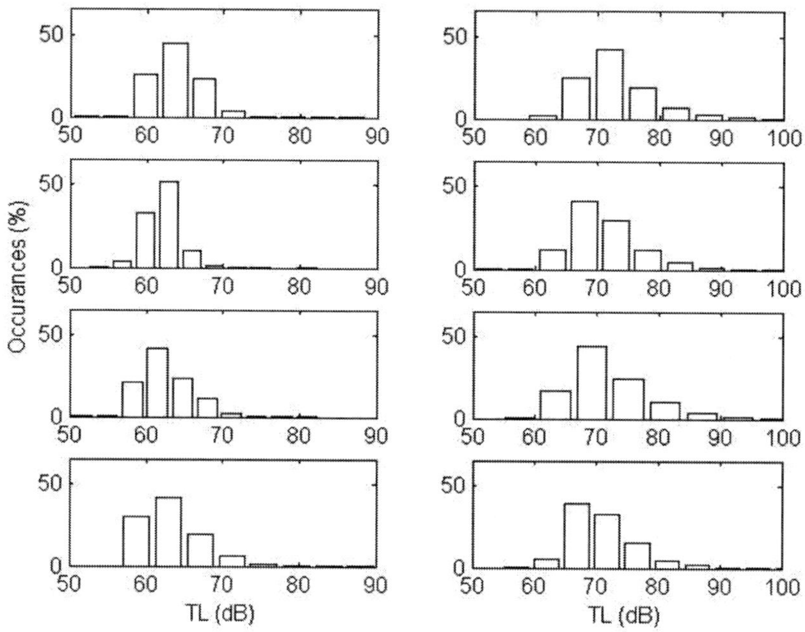

Figure 8: Histograms of the TL for the GFPE calculation are shown in the figure. Left column shows the TL at 1 km distance and the right column shows the TL for 2 km distance. The rows show each cardinal direction in the order: North (upper), South, East and West (lowest).

Greens Function Parabolic Equation with terrain

The TL for the GFPE calculation including alternating altitude can be seen in figure 9. By comparing the distributions of GFPE calculations with and without terrain it can be observed that the distribution patterns are similar but far from identical which gives a cue that the effect of terrain might have to be taken into account when investigating sound propagation from wind turbines at 1 km and 2 km distance.

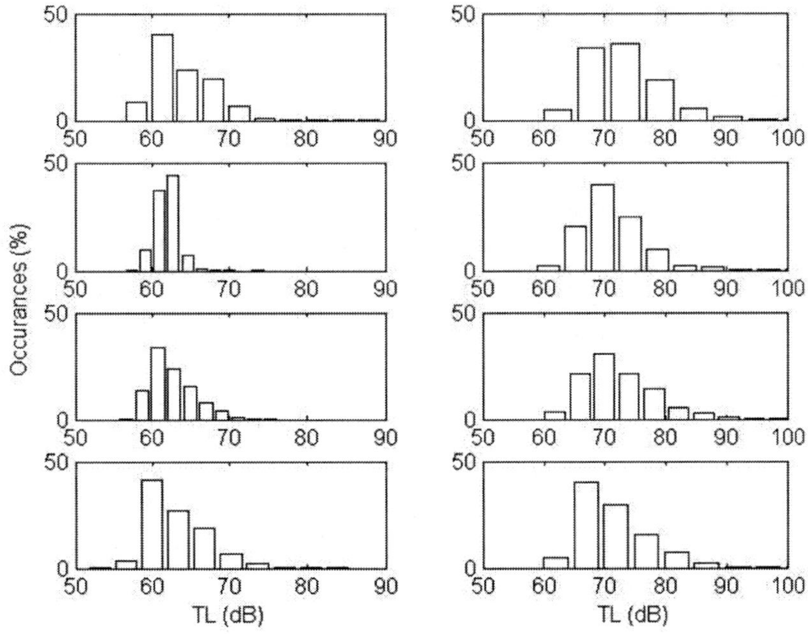

Figure 9: Histograms of the TL for the GFPE calculation including terrain are shown in the figure. Left column shows the TL at 1 km distance and the right column shows the TL for 2 km distance. The rows show each cardinal direction in the order: North (upper), South, East and West (lowest).

Discussion and Conclusion

The results from this paper show that geographical and meteorological databases can be used to describe the environment with some detail around a wind farm. This information enables investigations of spatial and temporal changes in sound propagation from wind turbines. Sound energy focuses in different directions and noise will thus receive sound at different times and in different quantities. This could be used both when planning wind farms to assess local noise doses and also when managing wind farms to minimize noise and maximize power output.

The shown results in this paper focus on the time distributions of different PE algorithms and differences between the different numerical schemes are observedmover the analyzed year, especially at the longest propagation distance of 2 km and less for the shorter distance of 1 km.

A detailed analysis of specific times and propagation directions are not reported in this paper but could be performed. This could be an interesting approach in order to compare the different numerical models in more detail and evaluate which method, if any, that is currently most trustworthy, regarding this issue the authors want to stress that numeric sound propagation calculations include several parameters that have to be adequately adjusted in order to acquire reliable results and thus comparison to long term measured results is considered essential before applying these algorithms in actual wind farms.

Acknowledgements

This paper has been performed in the SWIP project (New innovative solutions, components and tools for the integration of wind energy in urban and peri-urban areas) which received funding from the EU-FP7 Programme for research, technological development and demonstration under grant agreement No. 608554.

References

Abrahamsson L. (1991). *Othogonal grid generation for two-dimensional ducts*, J Comp Appl Math, **34**, p. 305-314

Delany M. E. & Bazley E. N (1970). *Acoustical properties of fibrous absorbent materials*, Appl. Acoust. **3**, p. 105–116

Gilbert K. E. (2015) "A numerically stable formulation of the Green's function parabolic equation: Subtracting the surface-wave pole."J Acoust Soc Am. **137**,EL8-14.

Gilbert, K. E., and Di, X. (1993). "A fast Green's function method for one-way sound propagation in the atmosphere," J. Acoust. Soc. Am. **94**, p. 2343–2352.

Jeltsch R. (1977). *Multistep methods using higher derivatives and damping at infinity,* Math. Comp. **31**, p. 124-138.

Sohlman M, Jonasson H & Gustafsson A, (2004), Using Satellite data for the determination of the acoustic impedance of ground, Report to the Swedish National Space Board 2004-12 by Metria, Sweden

Karasalo I. and Sundström A. (1996), *JEPE – a high order PE-model for range dependent fluid media*, Proc of the 3rd European Conference on Underwater Acoustics, Heraklaion, Crete, p. 189-194

Parakkal, S and Gilbert, K. E. and Di, X. (2012), "Application of the Beilis and Tappert parabolic equation to long-range sound propagation over irregular terrain" The Journal of the Acoustical Society of America, **131,** p.1039-46

Salomons, E.M. (2002) Computational Atmospheric Acoustics (Kluwer Academic Publishers, Dordrecht, Netherlands, 335 pp.

van den Berg G. P. (2004) *Effects of the wind profile at night on wind turbine sound,* Journal of Sound and Vibration **277**, p. 955–970.

Öhlund & Larsson (2015), Meteorological effects on wind turbine sound propagation, Applied Acoustics, **89**, p. 34-41

7th International Conference
on
Wind Turbine Noise
Rotterdam – 2nd to 5th May 2017

Comparison of measured and calculated noise levels in far distances of wind turbines

Ulf Kock, DNV GL - Energy, Sommerdeich 14 b, 25709 Kaiser-Wilhelm-Koog, Germany
e-mail: ulf.kock@dnvgl.com

Dr. Ivan Asin Cruz, DNV GL - Energy, Sommerdeich 14 b, 25709 Kaiser-Wilhelm-Koog, Germany
e-mail: ivan.asin.cruz@dnvgl.com

Arno Trautsch, DNV GL - Energy, Sommerdeich 14 b, 25709 Kaiser-Wilhelm-Koog, Germany
e-mail: arno.trautsch@dnvgl.com

Summary

The prediction of the noise caused by the wind turbines is one of the most relevant items during the planning phase of any wind park. Deviations of the predicted noise levels with respect to the measurements performed after the construction and the commissioning of the wind park might lead to noise reduced operation modes or in extreme cases to a complete shutdown of the turbines. The aim is the investigation if the calculation according ISO 9613-2 does underestimate the noise in the far field for wind turbine noise and if, to which extend.

In order to evaluate the accuracy of the predictions models DNV GL has measured the noise levels and calculated the predicted levels in far distances from the wind turbine on behalf of the German federal state of Schleswig-Holstein. This study is performed on various wind turbine models representing the current state of the art situation of the market. For this purpose, some wind turbine models with at least 3 MW of rated power, rotor diameters larger than 100 m and hub heights of 90 m or larger have been studied at different times of the year and different weather conditions to get a better insight in the possible influences on the sound propagation and to evaluate the deviation of the propagation calculation. The effect of the roughness length is taken into account by considering different site conditions located in the north of Germany. These locations comprise moorlands, coastal areas and hilly land.
The software application CadnaA has been used to predict the noise levels produced by the wind turbines.
DNV GL will present the results of the project and compare the findings with the research project performed by Uppenkamp und Partner on behalf of the "Landesamt für Natur, Umwelt und Verbraucherschutz" in the state North Rhine-Westphalia in 2012. The results of this study are presented both in the audible frequency range (20 Hz to 20 kHz) as well as in the low frequency spectra (1 Hz to 200 Hz).

1. Introduction

The renewable energy production has experienced a large expansion in the last 25 years in parallel to the development of the technology. Nowadays wind turbine generator systems (WTGS) with nominal power values of several MWs, rotor diameters in the order of 100 m or more, and hub heights up-to 150 m cannot only be found in prototype testing locations but also in commercial wind farms. Larger WTGS can produce higher noise emissions both from aerodynamic and machine sources.

The wind farms are usually located near already existing infrastructures (roads, electric network, etc) in order to reduce the construction and installation cost. This is not an issue in the case of offshore wind parks. But in regions with high population density such as continental Europe can lead to criticism and complaint from the local citizens. In parallel to the technology development, local authorities have developed different environmental and health protection regulations.

A major concern during the planning of any wind farm project is the sound power level (SPL) caused by the WTGS in residential areas in the vicinity. An accurate prediction of the SPL may have a big impact on the final park layout or operation conditions, i.e.: a more compact design of the park, bigger WTGS model, nominal or reduced operation modes, etc.
After the construction and commissioning of the turbines some local administrations require a noise measurement to confirm the validity of the predictions and fulfil the corresponding directives.

In this study DNV GL has investigated if the calculation according ISO 9613-2 [1] underestimates or overestimates the noise in the far field for wind turbine noise. The possible effect of the roughness length has been accounted for by considering 3 different sites located in Schleswig-Holstein (north region of Germany). The predictions and measurements are compared at different distances from the WTGS up-to 1000 m, and similarly differences between the audible (20 Hz to 20 kHz) and low frequency (1 Hz to 200 Hz) are evaluated.

2. Measuring objects, prediction and experimental setup

In the following the measurement objects, i.e. the WTGS, the prediction tools as well as the measurement setup are presented.

2.1 Wind turbine generator system and location

In order to present the current status of the market several WTGS models have been evaluated. Each WTGS is located at different site. The variety of locations allows for the analysis of the effects of the roughness lengths in the final result.
The following table presents a summary of the characteristics of each WTGS model and locations.

Table 1: Summary of the WTGS characteristics

Manufacturer	Siemens	Senvion	Vestas Wind Systems A/S
Model	SWT-3.6-120	3.0M122	V112-3.3MW
Rotor diameter [m]	120	122	112
Hub height [m]	90	139	119
Distance tower centre to rotor plane [m]	3.8	4.2	4.5
Nominal power [kW]	3600	3000	3300
Location	Galmsbüll (Germany)	Holtsee (Germany)	Norderheistedt (Germany)
Type of location	Coastal area	Hilly land	Moorland

2.2 Predictions

The predictions are based on the software application CadnaA and the regulation ISO 9613-2 [1] implemented on it. The prediction software characterises the WTGS as a point-size noise source located at the tower centre and at a hub height with respect to the ground. In first approximation, the noise propagation is calculated taking into account only on the attenuation factors corresponding to the 500 Hz and without any ground effect.
The SPL is calculated at each house or residence known to be inhabited. Additionally, the equal SPL points (isolines) are represented.

2.3 Experimental setup

The noise measurement are performed according the German technical guideline for wind turbines [1], which is based on the IEC-guideline Edition 2 [3]. The measurement is not performed according the actual IEC Ed. 3 guideline as the FGW-guideline is the reference for building and operation permits according to the German Federal Immission Control Act, ("BImSchG").
The equipment consists in a Class 1 noise level meter placed on an acoustically hard board, a meteorological mast and WTGS data acquisition system. The data is collected by a central data-acquisition system with a 1 Hz frequency. The position of the measurement follows the recommendations from [1], i.e. in the downwind direction and from the WTGS tower centre at a distance R_0 equal to the total WTGS height.
In addition to the primary measurement point R_0, additional noise level meters are placed at 4 m-height masts. These positions are located at various distances ranging between R_0 and 1000 m approximately. The variety of selected measuring positions provides a general overview on any deviations due to the distance to the WTGS.
The exact locations are selected based on meteorological and geographical conditions occurring during the measurement. A summary of each positions is given in the following tables.
The measurement equipment is calibrated on a regular basis.

Table 4: Summary of the measurement conditions in Norderheistedt

Location	**Norderheistedt (Germany)**		
WTG Model	Vestas V112-3.3MW		
Type of location	Moorland		
Roughness length [m]	0.05		
Measurement dates	2016-11-18/19	2016-12-01/02	2017-01-10/11
R0 [m]	160	175	170
Alternative positions [m]	322 510 694	500 700 980	500 700 1000
Measured wind-speed range at 10m height [m/s]	4.4 – 10.7	6.0 – 11.5	4.6 – 10.4
Predominant wind direction	S – SW	NNW	SSW
Temperature [°C]	2.0 – 3.8	0.3 – 6.8	2.7 – 3.4
Air pressure at 2 m height [hPa]	999 – 1001	1015 – 1018	1006 – 1011
Atmospheric conditions	Clear and dry	Clear and dry	Cloudy, foggy, high humidity (>95%)
Turbulence intensity at 10 m height	11.7 %	15.3 %	12.8 %

3. Analysis of the measured data

The sound pressure level was continuously recorded during the whole measured period. Any abnormal noise sources, such as airplanes or road traffic, were marked during the measurement and removed from the posterior analysis. In addition, time periods in which the WTGS was in operation (total) or shutdown (background) were also marked, allowing for an easier noise source classification during the analysis.

In the following we present the methodology to determine the sound pressure and power level arising from the wind turbine only after subtraction of the effects due to the background.

3.1 Sound pressure level

The sound pressure level is ordered as a function of the wind speed and a polynomial function is fitted through all remaining data points after the removal of spurious values.
The wind speed of background data points corresponds to the values measured at the 10 m-height mast. The wind speed values corresponding to the total noise are derived either from the measured power values and their relation to the power curve of the operational mode, or from the nacelle anemometer corrected to 10 m height. For further details please refer to [2]
From the regression curves the noise pressure levels are obtained for integer wind speed values. According to the regulation [2] the wind speed bin size is 1 m/s centred at integer wind speed values, i.e.: [5.5, 6.5) m/s, [6.5, 7.5) m/s, [7.5, 8.5) m/s, etc.
The sound pressure level due to the wind turbine only is calculated as the energetic difference between the total and background noise levels and can be represented as

$$L_i = 10\log\left(10^{0.1L_{T,i}} - 10^{0.1L_{B,i}}\right)$$

where

- $L_{T,i}$ refers to the total noise pressure level at the i-th wind speed bin
- $L_{B,i}$ is to the background noise pressure level at the i-th wind speed bin
- L_i represents the noise pressure level at the i-th wind speed bin due to the WTGS only.

3.2 Sound power level

The sound power level of the WTGS is calculated for each integer wind speed value from the background subtracted sound pressure level. An isotropic spherical propagation of noise from the rotor centre to the microphone position (R_1). The SPL at the i-th wind speed ($L_{W,i}$) bin relates to the sound pressure level ass

$$L_{W,i} = L_i - 6 + 10\log\left(\frac{4\pi R_1^2}{S}\right)$$

with S representing the reference surface area of the board (1 m^2), and 6 dB is the contribution due to coherent interference at the acoustically hard board.

4. Comparison between ISO 9613-2 and FGW Measurements

In the following the results comparing the predictions according to the ISO 9613-2 [1] guideline and the FGW measurement procedures [2] are presented.

Figure 1: Comparison of measured (red) and calculated (green) noise levels

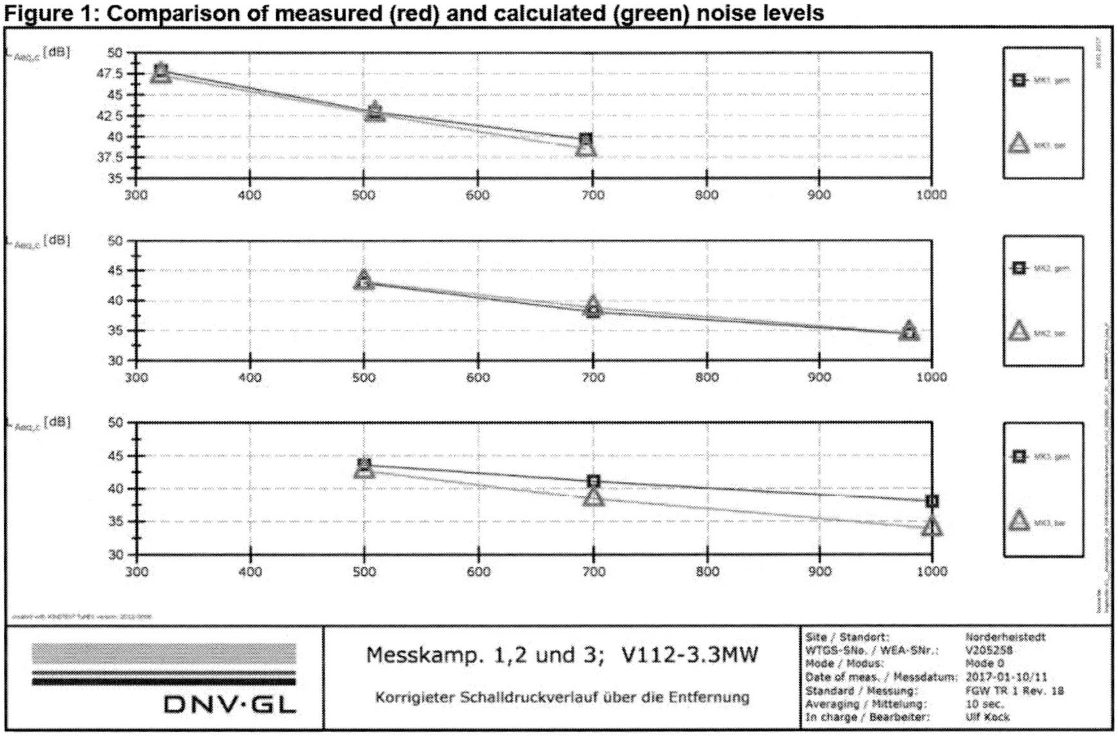

Figure 2: Deviations between measured and calculated noise levels of the three performed measurements

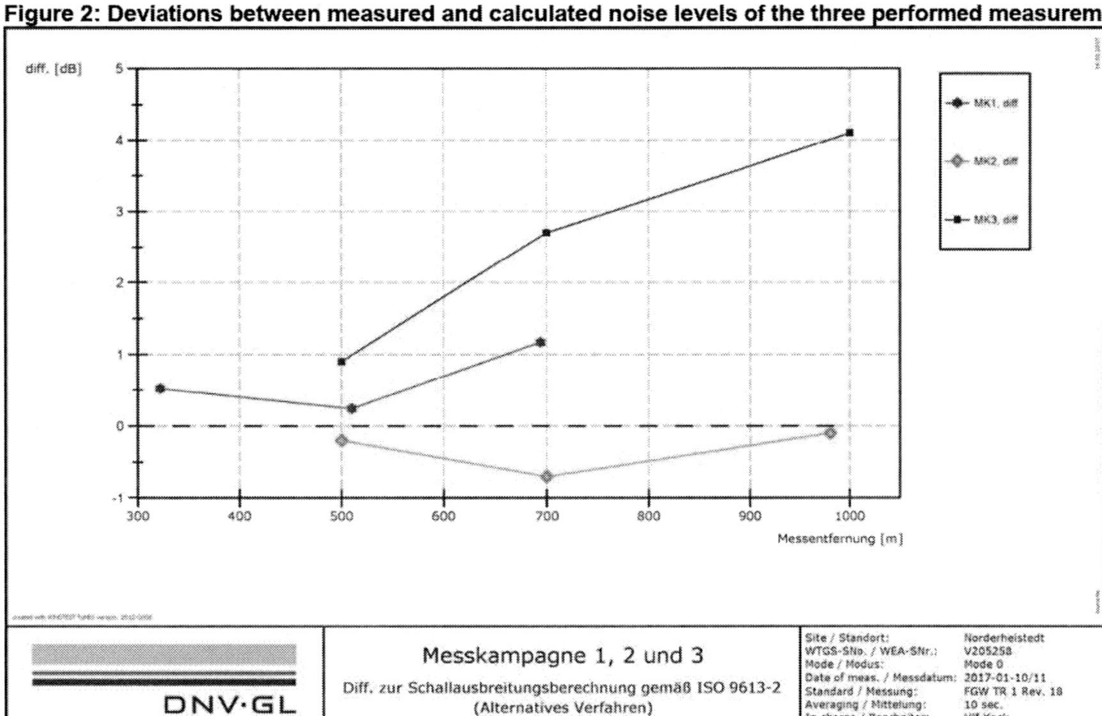

5. Conclusions

The first results suggest that predicted noise pressure levels are lower than the measured ones. This observed deviation increases with the distance to the wind turbine generator system.

At this stage, only three of the 15 scheduled measurements where performed. Hence the statistical observation of the deviations between calculated and measured levels is not possible. The measurement with the highest deviation was performed at weather conditions with very high humidity. Further measurements will be performed to verify which links between the weather- and winds-conditions and the atmospheric layers.

References

[1] ISO 9613-2:1996
Acoustics -- Attenuation of sound during propagation outdoors -- Part 2: General method of calculation
1996-12-01

[2] Technische Richtlinien für Windenergieanlagen, Teil 1, Rev. 18
(Technical guideline for wind energy generator systems, part 1, revision 18)
Herausgeber: Fördergesellschaft Windenergie e. V.,
Stresemannplatz 4, 24103 Kiel
2008-02-01

[3] IEC 61400-11 Ed. 2.1
Wind Turbine Generator Systems, Part 11: Acoustic Noise Measurement Techniques
2006-11-28

[4] IEC 61400-11:2012,
Deutsche Fassung EN 61400-11:2013, Windenergieanlagen - Teil 11: Schallmessverfahrens
(German edition EN 61400-11:2013, Win turbine generator systems – Part 11: Acoustic Noise Measurement Techniques)

[5] SFQS-DE-WI-076
Kalibrierung und periodische Überprüfung (Calibrations and Periodic Tests)
2013-06-21

[6] SVAC-G-PR-003
Messung der Schallemission an Windenergieanlagen – Summenpegelmethode (Noise Emission Measurements on Wind Turbines - Sum Level Method)
2014-03-31

[5] DIN 45645-1
“Ermittlung von Beurteilungspegeln aus Messungen - Teil 1: Geräuschimmissionen in der Nachbarschaft“ (Determination of the noise rating level through measurements – Part 1: Noise immissions in the proximity)
1996-07

[6] DIN 45681
“Bestimmung der Tonhaltigkeit von Geräuschen und Ermittlung eines Tonzuschlages für die Beurteilung von Geräuschimmissionen“ (Determination of the noise tonality and evaluation of an additional tonality for assesment of nosie immissions)
2005-03

7th International Conference
on
Wind Turbine Noise
Rotterdam – 2nd to 5th May 2017

Comparison of Sound Propagation Models for Offshore Wind Farms

Guangsheng (Sam) Du, Valcoustics Canada Ltd.: sam@valcoustics.com
A.D. Lightstone, Valcoustics Canada Ltd.: al@valcoustics.com
Joseph Doran, Valcoustics Canada Ltd.: joseph@valcoustics.com

Summary

Propagation of sound from offshore wind turbines over water is different from propagation from land-based wind turbines. Prediction of noise from offshore wind turbines generally involves propagation of noise over large distances meaning small inaccuracies or uncertainty in the prediction models can become significant. It has been shown that under downward refracting atmospheric conditions, sound can propagate for extended distances over water. Different meteorological conditions that occur over water may attenuate or enhance sound propagation.

A number of noise propagation prediction models developed or used by authorities in different countries, as well as published, numerical (theoretical), methods, to predict receptor sound levels, are summarized. The numerical methods which can calculate sound pressure levels by including various meteorological conditions and are widely accepted as an accurate estimation tool for long range outdoor sound propagation are also investigated. Comparison of the advantages and limitations of each model are given. In addition, the status of each of the sound propagation models are described. This includes the commercial status and commercial application of each model as well as the available application software packages.

1. Introduction

Wind power is a renewable energy source that has the potential to contribute significantly to meeting energy needs around the world. As the sites on land with good wind potential become less available, an alternative approach is to locate wind farms offshore. Offshore wind resources may be stronger and more reliable than wind over land. Thus, offshore wind farms are being constructed all over the world especially in Europe.

Sound wave propagation from offshore wind turbines over water is different from propagation from land-based wind turbines. For offshore wind turbines, it generally involves large distances, temperature inversion, refracting atmospheric conditions, effect from rough sea surface, meteorological conditions over water, phenomenon of cylindrical propagation, etc. Thus, noise prediction models commonly used for land-based wind turbines may not be suitable for offshore wind turbines. Various studies have shown that under downward refracting atmospheric conditions, sound can propagate for extended distances over water.

The existing models to predict sound level at a given distance and atmospheric conditions have various degree of accuracy, complexity and computing speed. Several noise prediction models developed or used by authorities in different countries, as well as published, numerical (theoretical), methods, to predict receptor sound levels are discussed briefly. Some of them are empirical in nature and do not include the real atmospheric conditions. Others are more sophisticated and complex and include extended atmospheric parameters and therefore, require significant computation time. The numerical methods such as the parabolic equation methods can calculate sound pressure levels by including various meteorological conditions and are widely accepted as an accurate estimation tool for long range outdoor sound propagation.

2. ISO 9613-2

2.1 Summary

The ISO 9613-2 standard "Acoustics – Attenuation of Sound During Propagation Outdoors – Part 2, General Method of Calculation" describes an empirical engineering method for prediction of environmental noise outdoors at a given distance from a variety of sources of sound [1]. It calculates the attenuation of sound outdoors over the distance between the source of sound and the point of reception. The result of this method is the equivalent A-weighted sound pressure level of a known source under meteorological conditions favourable to propagation (e.g., downwind). The ISO 9613-2 method has a stated tolerance of ±3 dB for a source height of up to 30 m and a distance from 100 m up to 1000 m). For distance greater than 1000 m, the accuracy is not given in the standard.

The ISO 9613-2 model accounts for downwind conditions as well as a moderate temperature inversion over ground with wind speeds ranging from 1 to 5 m/s measured at a height between 3 and 11 m above ground. The sound pressure level resulting from this method is considered to be a level that is seldom exceeded.

The model accounts for the various input parameters including geometrical divergence, atmospheric absorption, ground attenuation, reflection from surfaces and sound barrier attenuation. However, it does not account for other meteorological conditions which could result in greater sound propagation, such as temperature inversion over water. This has led to lower predicted sound pressure levels with propagation over water than those observed.

The method also predicts a long-term average A-weighted sound pressure. The long-term average A-weighted sound pressure level encompasses levels for a wide variety of meteorological conditions.

The method consists of octave-band algorithms (with a nominal mid-band frequency from 63 Hz to 8 kHz) for calculating the attenuation of sound which originates from a point sound source or an assembly of point sources. The source (or sources) may be moving or stationary. It does not apply to sound from aircraft in flight, or to blast waves from mining, military or similar operations.

To apply the method of this part of ISO 9613, several parameters need to be known with respect to the geometry of the source and of the environment, the ground surface characteristics and the source strength in terms of octave-band sound power levels for directions relevant to the propagation.

2.2 Model Application

Although, the standard is explicitly not intended for calculating sound propagation for inversion conditions over water (in addition to the distance and source height limitations), it has been used to assess noise impact from offshore wind turbines [2, 3, 4, 5].

Kelsall concluded that ISO 9613-2 may be suitable for predicting noise propagation from wind turbines over water at the distances of up to 9 kilometres (km), based on sound level measurements. There was good agreement between the measured results and ISO 9613-2 modelling results at 31.5, 63, 125 and 250 Hz octave frequency bands. At long distances (e.g., 3 km or greater), the wind turbine noise is dominated by frequencies of 500 Hz or below due to significant air absorption of the frequencies above 500 Hz [6].

The noise assessment report for the Atlantic Array Offshore Wind Farm, dated June 2013, Revision A, prepared by Channel Energy Limited [7], was undertaken using the octave band method of ISO 9613-2 but with cylindrical spreading at 2000 m and beyond, instead of spherical (point source) spreading.

The ISO 9613-2 model has been explicitly adopted by the environmental approval authorities in various jurisdictions including British Columbia, New Brunswick, Nova Scotia, Manitoba, Alberta and Ontario in Canada; Flanders and Wallonia in Belgium; Finland; Germany; Netherlands; the United Kingdom; New Zealand; as well as South Australia, Queensland, Western Australia; Victoria; Tasmania and New South Wales in Australia, for wind turbine noise prediction as well as for industrial/commercial noise assessment.

A major advantage of ISO 9613-2 is that it has been widely and successfully used for the analysis and assessment of a large variety of sources, including land-based wind turbines. It has been implemented in most commercially available modelling software such as CadnaA, SoundPLAN and windPRO for a number of years [8, 9, 10].

The major limitations of the ISO 9613-2 model are:

- The standard warns that temperature inversions over water are not covered, only inversion conditions over land. This may result in lower predicted sound levels over water than those observed.
- The stated estimates of accuracy are for a very limited range of source heights and distances. Accuracy of ±1 dB is indicated for source heights between 5 and 30 m and distances of less than 100 m, and ±3 dB for heights up to 30 m and distances between 100 and 1000 m. Both land based and offshore wind farms will have greater source heights (currently up to 100 m) and greater distances to many receptors. Offshore wind farm distances can be expected to significantly exceed the apparent distance limit of ISO 9613-2, as is the case for many land-based wind farms.

Thus, the typical wind farm, whether land-based or off-shore, violates some of the apparent limitations of the model. However, it should be noted that exceeding the apparent limitations does not necessarily mean the results are invalid – just that the model has not been formally validated for those conditions and there is no stated estimate of accuracy when the given limits are exceeded.

There are other limitations to the model, such as when there is ground attenuation and other than flat terrain. The derivation and theoretical basis of the ground attenuation equations are not given and the algorithms are not intuitive and not documented by any specific publications. However, these aspects are largely irrelevant to offshore wind farms, except perhaps where there may be relevant in-land receptors at a non-trivial distance from the shore line, opposite to an off-shore wind farm.

The evidence appears to be that ISO 9613-2 can under-predict sound levels propagated for long distances over water, but adjustments (as in the Swedish model below) can be made.

3. Swedish Model

3.1 Summary

The Swedish Environmental Protection Agency (SEPA) has issued an alternative engineering method to the ISO 9613-2 procedure for sound propagation from distant off-shore wind turbines. For ranges up to 1000 m, hemispherical spreading is used for both land and water. For distances greater than 1000 m (break point), cylindrical spreading is used. The SEPA method assumes hard ground and a standard atmospheric attenuation [11, 12, 13].

Initially in 2002, the Swedish model set the break point distance at 200 m. This was later corrected to 700 m and finally set to 1000 m.

The Swedish model considers the frequency spectrum from 63 Hz to 4 kHz and not up to the usual 8 kHz.

The Swedish model is only valid for downwind conditions.

3.2 Model Application

The major difference between the Swedish Model and ISO 9613-2 is that the Swedish Model uses cylindrical spreading when the distance between source and receiver is greater than 1000m. The Swedish model assumes hard ground and atmospheric attenuation at a temperature of 0°C and 70% relative humidity according to ISO 9613-1 [11, 12, 13]. In addition, the Swedish model only considers the frequency spectrum from 63 Hz to 4 kHz.

Thus, the Swedish model has basically the same set of advantages and disadvantages as described for ISO 9613-2, except that propagation of sound beyond 1000 m over water (and potentially temperature inversions over water) has been addressed.

Similar to ISO 9613-2, the Swedish Model makes minimal assumptions about the atmospheric conditions such as temperature gradients above the sea and wind conditions (which are difficult to implement into the modelling).

Notwithstanding the limitations, this method has been used to assess noise impact from offshore wind farms in the UK [48, 49, 50, 51 and 52].

The Swedish Model has only been approved for use in Sweden. However, the model was used in projects in the UK where it was assumed to show a worst case propagation over longer distances. The Swedish model was also used to compare sound prediction results for three floating test wind turbines in the US, together with other models. However, it is unknown whether the Swedish model was an approved model in the US for permitting purposes. The model has been implemented in the commercial software packages CadnaA and WindPro [9, 14].

4. Danish Model

4.1 Summary

In 1991, the Danish Ministry of Environment published a method for determination of noise from wind turbines. The Danish model, assuming hard ground, overestimates the levels of noise propagating over ground, but gives reasonable results offshore for limited distances up to 500 m for the overall A-weighted sound pressure level. In the octave band version, the reliable distance extends to 2 to 5 kilometres. The models fail at large distances because of multiple reflections from the sea surface building up and leading to cylindrical spreading of the sound energy [11, 15, 16, 17].

The Danish model gives reasonable result at small distance for air absorption, but may result in considerable error over large distances.

The Danish model is only valid for downwind conditions.

Moreover, the Danish authorities have developed a method to calculate indoor sound levels for low frequencies from wind turbines.

In accordance with the Danish statutory order issues on January 1, 2012, due to the increased coherence between direct and reflected sound at low frequencies, a more specific and detailed approach was chosen to avoid underestimation of the noise levels in the frequency range from 10 to 160 Hz, independent of distance and height of the wind turbine. For land based wind turbines, the ground correction is +6 dB at 10 Hz and decreases to 0 dB at 160 Hz. For off-shore wind turbines, the "ground" correction is +6 dB at 10 Hz and decreases to +4 dB at 160 Hz.

4.2 Model Application

The Danish model, assuming hard ground, overestimates the levels of sound propagating over ground, but gives reasonable results offshore for limited distances up to 500 m, for the overall A-weighted sound level [11, 15, 16, 17]. In the octave band version, the reliable distance extends to 2 to 5 kilometres. The model fails at large distances because of multiple reflections from the sea surface building up and leading to cylindrical spreading of the sound energy. If a turbine has a pure tone component, a penalty of +5 dB is applied.

The model has been implemented in the commercial software package WindPro [14].

5. CONCAWE Model

5.1 Summary

The CONCAWE model dates back to 1981; the method is focused on the propagation of noise from petroleum and petrochemical complexes to neighbouring communities. The model takes into account not only significant topographical features, but also the meteorological conditions prevailing at the site. The latter feature allows the prediction of long term equivalent continuous sound levels and long term statistical sound levels, in addition to probable maxima and minima, on the basis of the statistical distribution of wind velocity and Pasquill Stability for the area [18, 19].

The CONCAWE model has based many of the algorithms on experimental data. This was done for the ground attenuation and all the meteorological effects.

The CONCAWE model enables octave band sound pressure levels to be calculated at a receiver point for a given meteorological scenario. The CONCAWE model considers the range of octave bands from 63 Hz to 4 KHz.

5.2 Model Application

The CONCAWE model takes into account geometrical spreading (spherical divergence), atmospheric absorption, ground attenuation, meteorological attenuation, barrier shielding, source/receiver height correction and in-plant screening. The algorithms governing the meteorological conditions prevailing at the site allows the prediction of long term equivalent continuous sound levels and long term statistical sound levels, in addition to probable maxima and minima, on the basis of the statistical distribution of wind velocity and Pasquill Stability for the area. While theoretically, it is advantageous to be able to include a wide range of meteorological conditions in the modelling analysis, in practice, the needed input information may not be available, negating this potential advantage.

The CONCAWE model is based on empirical data from land-based petrochemical complexes for sound propagation over ground. The ground effect calculation produces results that are similar to ISO 9613-2 in that it under-predicts sound levels propagating over water.

The CONCAWE model has also been used for offshore wind farm projects in the UK [20, 21, 22].

The CONCAWE model has been adopted in Alberta, Canada; the United Kingdom; Hong Kong; South Australia, Queensland, Western Australia and New South Wales in Australia, for wind turbine noise prediction as well as for industrial/commercial noise assessment.

The CONCAWE model has been implemented in the commercial software package CadnaA and SoundPLAN [9].

6. Nordic Prediction Model (Nord2000)

6.1 Summary

Nord2000 is a calculation model developed as a joint project between the Nordic countries, Denmark, Norway, Sweden, Iceland and Finland [23, 24, 25, 26]. Nord2000 considers the influence of wind (direction, speed, gradient), temperature, ground absorption and screening. It is also possible to choose different wind speed and temperature gradients. Nord2000 is suitable for calculations over hilly terrain as it takes varying topography into account. It also takes into consideration the acoustic characteristics of water surface and therefore is appropriate for calculation of sound propagating over water.

The propagation model is based on analytical solutions - geometrical ray theory and theory of diffraction. The model calculates one-third octave band attenuation from 25 Hz to 10 kHz for homogeneous or inhomogeneous atmosphere conditions. A disadvantage is that all of the input data, such as temperature gradient variations, associated standard deviations and standard deviation of variation in wind speed may not be readily available.

The input variables that may be taken into account are:

- The terrain profile defined by start and end coordinates of the straight-line segments and the ground flow resistivity and roughness (unevenness) of each segment;
- Height of source and receiver above the first and last terrain point, respectively;
- Aerodynamic roughness length of the ground (used to define the wind speed profile);
- The average wind speed component in the direction of propagation and the height at which the wind speed is specified;
- The standard deviation of variations in wind speed component;
- Temperature along the propagation path near the ground;
- Standard deviation of temperature gradient variations;
- Turbulence strength parameters due to wind and temperature, respectively; and
- Relative humidity of the air.

The Nord2000 model allows calculation of short-term levels for specified weather conditions such as short-term (less than 30 minutes or one hour) equivalent sound pressure levels or maximum levels. Long-term noise levels (e.g., yearly average of day, evening and night sound level) can be obtained by combining short-term noise levels calculated by Nord2000 with meteorological statistics. In practice, short-term level calculations are made for a limited set of meteorological classes, and the long-term levels are the weighted average of these results. This approach makes it possible to calculate long-term levels such as maximum sound levels for longer periods, or even complete statistical distributions of sound levels.

The model is particularly accurate at small distances. The model has only been validated by measurements at distances up to 200 m where good accuracy has been found (deviations within ±2 dB of overall A-weighted sound pressure levels in most cases). The method has been validated by comparison with measurements and with other prediction methods, such as Parabolic Equations (PE), which are believed to be more accurate. See Section 8 below.

6.2 Model Application

Nord2000 model has been used to assess noise from offshore wind farms as well as other sources of noise.

Nord2000 has been validated by more than 500 propagation cases based on measurements on land for various non-wind turbine sources as well as by reference results obtained from accurate numerical prediction methods [27].

The propagation model has been widely used to assess noise from offshore wind farms in the Nordic countries as well as in the UK [28, 29, 30].

The model has been implemented in commercial software packages such as CadnaA, SoundPlan, exSound2000 and SPL2000 [9, 10]. It may not be as easy to use as the ISO 9613-2 model since it requires more input parameters than ISO 9613-2.

Compared to ISO 9613-2 model, Nord2000 is relatively complex in that all calculations in Nord2000 are done in one third octave bands and it takes into account a lot more input variables than ISO 9613-2. This means it has the potential for better accuracy. However, correctly selecting the input variables, such as temperature gradient variations, associated standard deviations and standard deviation of variation in wind speed, is not necessarily practicable as they may not be readily available. This may be a major disadvantage in practice.

The Nord200 model has been explicitly adopted by the environmental approval authorities in Denmark and Norway for wind turbine noise prediction as well as for industrial/commercial noise assessment.

7. Harmonoise P2P Model

7.1 Summary

The Harmonoise model is the result of a co-operation between a number of European countries [31, 22, 33]. It is an engineering model for predicting environmental noise levels. This prediction model is based on solutions and concepts close to those found in the Nord2000 model. It predicts the sound pressure level at the receiver position in one-third octave bands from 25 Hz to 10 kHz from the sound power level of the source. The effects of various factors are calculated separately and subtracted from the source sound power level. These factors include spherical divergence, air absorption, reflections from ground, diffraction at sound barriers, energy losses during side reflections, and effects of scattering zones.

7.2 Model Application

A series of measurements was conducted in the Collie Basin, W.A., to provide reliable measurements of actual noise levels under various meteorological conditions [34, 35]. It involved a loudspeaker source producing 1/3 octave bands of filtered pink noise, with measurements at distances from approximately 1000 m to 3000 m and simultaneous monitoring of meteorological conditions using a tethered balloon. Attenuations between the loudspeaker and the measurement locations were recorded for a total of 37 measured 1/3 octave attenuation spectra. These measurement data points were compared with the predictions from Harmonoise P2P model based on the measured meteorological data (i.e., wind speed at 10 m height and temperature gradient between 10 m and 30 m). It is concluded that agreement between the reference model and experimental results ranges from excellent in flat terrain situations down to fairly good in more complex configurations (hilly, viaduct) [33]

The Harmonoise model has not been used for wind farm noise studies.

The model has been implemented in the commercial software package CadnaA [9].

8. Partial Differential Equation Based Methods

8.1 Summary

The parabolic wave equation is frequently used in acoustic engineering to estimate long range sound propagation. The method essentially calculates the sound pressure level in the direction of propagation by solving an approximate form of the Helmholtz equation. This partial differential equation can be discretized using various numerical methods such as the Crank-Nicholson Parabolic Equation (CNPE), the Green's Function Parabolic Equation Method (GFPE) and the Extended Finite Element Method (XFEM). A brief discussion of each of these methods is summarized below.

8.2 Crank-Nicholson Parabolic Equation (CNPE) Method

In the CNPE method, the sound speed and the ground conditions can vary with range and height. Axial symmetry is assumed. Parabolic Equation (PE) [36] methods can be used in three dimensions as well, though it would lead to quite time consuming calculations [11, 37, 38].

The advantage of this method compared to the engineering methods in the previous sections is that surface impedance and a sound speed profile, atmospheric turbulence and surface roughness can all be included in the calculations. For example, the sound speed profile can be obtained from wind and temperature profiles measured using weather balloons or similarity scaling theory.

The CNPE is obtained by applying a finite difference discretization to the above equation; in the solution, the pressure at each range step is obtained from that at the previous range. The CNPE is especially suited for calculation of low frequencies. The CNPE method is limited to quite small propagation angles (±15°), giving restrictions on the relation between the source and the receiver height. Later a so-called wide angle PE was developed, which increased the possible propagation angle, but it is still restricted to around (±30°).

8.3 Green's Function Parabolic Equation (GFPE) Method

The GFPE method is a Fourier, split-step algorithm designed for atmospheric sound propagation and can use range-steps in the order of 10 wavelengths, considerably longer than conventional

Parabolic Equation (PE) methods such as the CNPE. GFPE is suitable in the present application because of its computational efficiency and because it has been shown to give reasonably good agreement to measurements over a water surface [39, 40, 41, 42, 43].

The method is a marching algorithm which computes a vertical pressure distribution at each new range step.

The GFPE can deal with complex ground impedance, arbitrary wind and temperature vertical profiles, and atmospheric turbulence.

8.4 Fast Field Program (FFP)

The Fast Field Program (FFP) technique was developed for prediction of underwater sound propagation and has been adapted to propagation in the atmosphere by several authors. Four such adaptations are called the CERL-FFP, CFFP, SAFARI, and FFLAGS [44, 45, 46, 47].

The basis of the FFP method is to work numerically from exact integral representations of the sound field within a layered atmosphere. By taking the Hankel transform (i.e., the weighted sum of an infinite number of Bessel functions) of the wave equation it is possible to obtain a height-dependent transformed wave equation for the sound pressure. This forms the starting point of the FFP.

Various numerical difficulties follow from the truncation of the integral to a finite sum and from the behaviour of the integrand. Different methods of dealing with these difficulties are used in CERL-FFP, CFFP, SAFARI, and FFLAGS.

9. Conclusions

Propagation of sound from offshore wind turbines over water is different from propagation from land-based wind turbines because with offshore wind turbines propagation of sound over much larger distances is generally involved and because the terrain conditions are much different. Small inaccuracies or uncertainty in the prediction models can become significant. Each of the noise prediction models described herein has their advantages and limitations. Some are empirical in nature and are easy to implement. Most have been implemented in various commercially available software packages. Others are based on parabolic wave equations which are complicated to implement and require intensive computation time. Notwithstanding the limitations of some of the models, they have been adopted by various approval authorities for wind turbine noise prediction. Even in cases where the stated limitations in the model have been exceeded (e.g., source height maxima or distance limits), the model [albeit with some custom "tweaking" (such as cylindrical propagation instead of spherical spreading)] appear to have been used successful for actual offshore wind turbine projects.

References

[1] ISO 9613-2, Acoustics - Attenuation of sound during propagation outdoors - Part 2 General method of calculation. Genève: International Organisation for Standardisation,1996

[2] Bolin, K. et al, Long Term Estimations of Low Frequency Noise Levels over Water from an Off-shore Wind Farm, J. Acoust. Soc. Am. 135(3), pp 1106-1114, March 2014

[3] Aker, P. et al, Offshore Wind Turbines: Estimated Noise from Offshore Wind Turbine, Monhegan Island, Maine, U.S. Department of Energy, November 2010

[4] Aker, P. et al, Offshore Wind Turbines: Estimated Noise from Offshore Wind Turbine, Monhegan Island, Maine, Addendum 2, U.S. Department of Energy, March 2011

[5] Block Island Wind Farm and Block Island Transmission System In Air Acoustic Report, Tetratech EC, Inc., September 2012
[6] Kelsall, T., Measurements of Propagation of Sound Over Water at Low to Mid Frequencies Applicable to Wind Turbine Noise Calculation, Canadian Acoustics, Vol. 40(3), pp 58-59, 2012
[7] Carter, P., Environment Statement: Volume 3: Annex 18.1 – Airborne Noise Propagation Models, Channel Energy Limited, May 17, 2013
[8] DataKustik website, http://www.datakustik.com/en/products/cadnaa/modeling-and-calculation/calculation-standards/
[9] SoundPLAN website, http://www.soundplan.eu/english/soundplan-acoustics/soundplan-modules/industrial-noise-indoors-outdoors/
[10] EMD International A/S website, http://www.emd.dk/windpro/windpro-modules/environment-modules/decibel/
[11] Mylonas, L., Assessment of Noise Prediction Methods over Water for Long Range Sound Propagation of Wind Turbines, Uppsala University (Campus Gotland), April 2014
[12] Boué, M. Long-range Sound Propagation over the Sea with Application to Wind Turbine Noise, TRITA-AVE 2007:22 ISSN 1651-7660
[13] Swedish Environmental Protection Agency, Measurement and Calculation of the Sound from Wind Turbines, June 10, 2013
[14] Greater Gabbard Offshore Wind Farm Environmental Statement, Project Management Support Services, October 2005
[15] The Danish Ministry of the Environment, Statutory Order on Noise from Wind Turbines, Translation of Statutory Order no. 1284, December 15, 2011
[16] Søndergaard, B. and Plovsing, B., Noise from Offshore Wind Turbines, Environmental Project No. 1016, the Danish Environmental Protection Agency, 2005
[17] Kragh, J. et al, Noise Immission from Wind Turbines, Final report of project JOR3-CT95-0065, DELTA Acoustics & Vibration, AV 1509/98, 1998
[18] Manning, C.J., The Propagation of Noise from Petroleum and Petrochemical Complexes to Neighbouring Communities, AT 931, Acoustic Technology Limited, May 1981
[19] Tonin, R., Estimating Noise Levels from Petrochemical Plant, Mines and Industrial Complexes, Acoustics Australia, Vol. 12 No. 2, pp 59-67, 1985
[20] Teesside Offshore Wind Farm Environmental Statement, EDF Energy (Northern Offshore Wind) Ltd, March 2014
[21] Wikipedia Webpage: https://en.wikipedia.org/wiki/Teesside_Wind_Farm
[22] Beatrice Offshore Wind Farm Environmental Statement: Section 17 - Wind Farm Airborne Noise, Arcus Renewable Energy Consulting Ltd., April 2012
[23] Kragh, J. et al, Nordic Environmental Noise Prediction Methods, Nord2000, Summary Report, General Nordic Sound Propagation Model and Applications in Source-Related Prediction Methods, DELTA Acoustics & Vibration, December 21, 2001, revised June 30, 2002
[24] Plovsing, B., Nord2000, Comprehensive Outdoor Sound Propagation Model, Part 1: Propagation in an Atmosphere without Significant Refraction, DELTA Acoustics & Vibration, December 31, 2000, last revised March 31, 2006
[25] Plovsing, B., Nord2000, Comprehensive Outdoor Sound Propagation Model, Part 2: Propagation in an Atmosphere with Refraction, DELTA Acoustics & Vibration, December 31, 2000, last revised March 31, 2006
[26] Plovsing, B., Nord2000, Comprehensive Model for Predicting Outdoor Sound Propagation, Experience from Validating the Model, Inter-Noise, 2001
[27] DELTA website, http://www.madebydelta.com/delta/Business_units/TC/Services+by+technology/Acoustics/Low+frequency+noise/Nord2000.page

[28] Hornsea Offshore Wind Farm, Project One - Environmental Statement: Volume 4, Annex 4.3.1 - Airborne Noise Technical Report, Report Number UK04-050200-REP-0051, Smart Wind Limited, July 2013
[29] Wigham, S. et al, Horns Rev 3 Offshore Havmøllepark: Offshore Noise Emission, Document No. 3621200091, Version 01, Orbicon A/s, February 2014
[30] Appelqvist, P., Suurhiekka Wind Farm – Calculation of Noise from Wind Turbines, Ingemansson, Report No. 30-06056-08112000, November 2008
[31] Salomons, E. et al, The Harmonoise Sound Propagation Model, Acta Acoustica United with Acustica, Vol. 97 pp 62 to 74, 2011, doi:10.3813/aaa.918387
[32] Defrance, J. et al, Outdoor Sound Propagation Reference Model Developed in the European Harmonoise Project, Acta Acoustica United with Acustica, Vol. 93 pp 213 to 227, 2007
[33] Bullen, R., The Harmonoise noise prediction algorithm: Validation and Use under Australian Conditions, Australian Acoustical Society, Proceedings of Acoustics 2012
[34] Herring Storer Acoustics / Wilkinson Murray, Collie Basin Acoustic Study, Report 8731-3-07076, 2008
[35] Environmental Protection Authority, Perth, Western Australia, Bluewaters Power Station Expansion – Phase III and Phase IV, Collie: Report and Recommendations of the Environmental Protection Authority, Report 1349, March 2010, ISSN 1836-0483
[36] Craddock, J. N. and White, M. J., Sound propagation over a surface with varying impedance: A parabolic equation approach, J. Acoust. Soc. Am. 91(6), pp 3184-3191, June 1992
[37] Johansson, L., Sound Propagation Around Off-shore Wind Turbines: Parabolic Equation Calculations for Baltic Sea Conditions, Stockholm 2003, ISRN-KTH-BYT-/R-03/192-SE, ISSN 1651-5536
[38] Gauvreau, B. et al, Traffic noise prediction with the parabolic equation method: Validation of a split-step Pade´ approach in complex environments, J. Acoust. Soc. Am. 112(6), pp 2680-2687, December 2002
[39] Bolin, K., Wind Turbine Noise and Natural Sounds: Masking, Propagation and Modeling, Doctoral Thesis, Stockholm 2009
[40] Van Renterghem, T. et al, Airborne sound propagation over sea during offshore wind farm piling, J. Acoust. Soc. Am. 135 (2), pp 559--609, February 2014
[41] Bolin, K. et al, Long range sound propagation over a sea surface, J. Acoust. Soc. Am. 126(5), pp 2191–2197, November 2009
[42] Oliveira, A. P., The effect of wind and turbulence on sound propagation in the atmosphere, https://fenix.tecnico.ulisboa.pt/downloadFile/395144345753/resumo.pdf
[43] Daigle, G. A. et al, Propagation of noise above ground having an impedance discontinuity, J. Acoust. Soc. Am. 77(1), pp 127–138, January 1985
[44] Daigle, G. A., Outdoor Propagation: Review of Physical Mechanism and Computational Models, Journal of the Canadian Acoustical Association, Vol. 22, No. 3, 1994
[45] Attenborough, K. et al, Benchmark cases for outdoor sound propagation models, J. Acoust. Soc. Am. 97(1), pp 173-191, January 1995
[46] L'Espérance, A. et al, Outdoor Sound Propagation in Presence of Atmospheric Turbulence: Experiment and Theoretical Analysis with the FFP Algorithm, J. Acoust. Soc. Am. 98, pp 570-579, 1995
[47] Gabillet, Y. et al, Sound Propagation in a Wind tunnel: Comparison of Experiments with the FFP and a Residue Solution, Appl. Acoust. 43, pp 321-331, 1994
[48] Burbo Bank Extension Offshore Wind Farm, Environment Statement: Volume 2: Chapter 11 – Offshore Noise, Done Energy Burbo Extension (UK) Ltd., March, 2013
[49] Burbo Bank Extension Offshore Wind Farm, Environment Statement: Volume 1: Chapter 6 – Project Description, Dong Energy Burbo Extension (UK) Ltd., March, 2013

[50] Burbo Bank Extension Offshore Wind Farm, Environmental Statement: Annex 11 – Offshore Noise, Dong Energy Burbo Extension (UK) Ltd., March, 2013

[51] Application for the Burbo Bank Extension Offshore Wind Farm Order Letter, Department of Energy and Climate Change, 26 September 2014

[52] Hughes, D. et al, Noise Assessment for the proposed Walney Extension (WOW03) wind farm, Subacoustech Environmental Report No. E341R0508, Subacoustech Environmental Ltd May 8, 2013

7th International Conference
on
Wind Turbine Noise
Rotterdam – 2nd to 5th May 2017

Verification and Validation of the "PNoise" Airfoil Trailing-Edge Noise Prediction Module inside "QBlade".

Joseph Youssif Saab Junior, Escola de Engenharia Maua: saab@maua.br
Marcos de Mattos Pimenta, Escola Politecnica da USP: marcos.pimenta@poli.usp.br
José Roberto Castilho Piqueira, Escola Politecnica da USP: diretoria@poli.usp.br
David Marten, TU-Berlin: david.marten@tu-berlin.de
Georgios Pechlivanoglou, TU-Berlin: george@pechlivanoglou.com
Christian Navid Nayeri, TU-Berlin: christian.nayeri@tu-berlin.de
Christian Oliver Paschereit, TU-Berlin: oliver.paschereit@tu-berlin.de
Alexandre Marturscelli Faria, Escola Politecnica da USP: martuscellifaria@usp.br

Summary

A 2D airfoil trailing-edge noise module "PNoise" was developed under a Poli-USP and TU-Berlin collaboration project and integrated into TU-Berlin wind turbine design environment "QBlade", available under General Public License. From the release of the v0.95 integrated version up to this date, more than 11,500 downloads have been made of the code. Even if a small percentage of the users intend to use the new TE-noise assessment feature, this would stress the importance of dissemination of the PNoise code verification and validation information.
The TE noise module is based on a modified BPM TE noise model, with turbulent boundary layer data provided by the integrated XFLR5 hybrid solver.
The calculation result validation and code verification process was successfully accomplished within the original limitations and datasets of the BPM model. However, an effort to extend Reynolds number validation to a range more representative of the flow over large WT main airfoils was impaired by the lack of robust TE noise spectral and 1/3 octave experimental data at higher Reynolds numbers.
The code is a work-in-progress intended to allow future assessment of all airfoil sources at early development stages and also complete rotor noise assessment. However, the practical usability of the PNoise module has been already demonstrated with the recent design of new airfoils that have the potential for significantly reduced TE noise levels when compared to traditional airfoils, at high Reynolds number flows.

1. Introduction

The PNoise TE noise code is based on a modified BPM TE noise model (Brooks, Pope, & Marcolini, Airfoil Self-Noise and Prediction, 1989) with the XFLR5 (Drela, Youngren, & Deperrois, 2009) providing the turbulent boundary layer data, both integrated inside the unique wind-turbine-design, graphical interface and user-friendly environment provided by the QBlade software (Pechlivanoglou, Marten, Weinzierl, Moesus, & Wendler, 2009), (Marten, Extension of an Aerodynamic Simulator for Wind Turbine Blade Design and Performance Analysis, 2010), (Marten & Wendler, Qblade Guidelines v0.6, 2013), (Marten, Qblade Short Manual V0.8, 2014).

Other self-noise sources as well as inflow noise models will be added in the future as part of the collaboration scope. Also, a "quasi-3D rotor" noise prediction tool is already specified.
The 2D TE noise module was developed and integrated in a betta version into QBlade V0.8, when it was thoroughly verified and validated. However, the module was later integrated into the newer QBlade V0.95 for public release. During this re-integration process, some improvements were made to the output graphs and files and also to the internal structure of the code, prompting a new validation and verification procedure, which is the object of this text.

2. Verification and Validation Range

The PNoise code and solution verifications were accomplished within the original limitations of the BPM model (Brooks, Pope, & Marcolini, Airfoil Self-Noise and Prediction, 1989).
The validation of the results was accomplished against the original BPM experimental spectra provided in that same seminal BPM paper.
The use of the model for assessing the TE noise of a generic airfoil geometry at large Reynolds and Mach number flows has become a practical reality with the current integration of the BPM model to the XFLR5 and QBlade functionalities, however, as defined by Oberkampf and Roy (Oberkampf & Roy, 2012), the use of a model beyond the original validation scope is called a *prediction* and, by definition, implies that it shall be made at the user own responsibility and risk, particularly in the case of absolute noise value assessment.
For improved performance, when using TBL displacement thickness reading over the TE from a XFLR5 output file, a recommendation is made for the data to be taken at 98% chord station (Saab Jr & Pimenta, Displacement Thickness Evaluation for Semi-Empirical Airfol Trailing-Edge Noise Prediction Model, 2016) as a *compromise station* among fully turbulent and transition flows, but the number is provided as a default value that may and should be altered at the discretion of the user, according the nature of the specific flow. The same reasoning applies to the default eddy-convection Mach number ($0.8{\cdot}M$) and other default input data, like the observer distance from the source and the directivity angles, for instance.

2.1 Model validity range and scope.

The BPM model (Brooks, Pope, & Marcolini, Airfoil Self-Noise and Prediction, 1989) is based on previous experimental work (Brooks & Hodgson, 1981), (Brooks & Marcolini, 1985), (Brooks & Marcolini, Airfoil Trailing-Edge Flow Measurements, 1986), summarized in table 1.

Table 1 - The experimental cases which provide the database for the BPM 2D TE noise model.

	Reference	Chord-Based Reynolds Number Range	Mach Number Range	AOA (α)	TU	Type of flow	TE Type
1	(Brooks & Hodgson, 1981)	$9.5{\times}10^5 < Re_C < 2.5{\times}10^6$	< 0.19	$0^0, 5^0, 10^0$	N/A	Tripped	From blunt to sharp variations
2	(Brooks & Marcolini, 1985)	$4.8{\times}10^4 < Re_C < 2.5{\times}10^6$	≤ 0.208	0^0	$< 0.05\%$	Tripped and untripped	Very Sharp
3	(Brooks & Marcolini, 1986)	$Re_C < 3.0{\times}10^6$	≤ 0.208	Up to 19.8^0	$\sim 0.03\%/ < 0.54\%$ Uniform flow / TE	Tripped and untripped	Very Sharp
4	(Brooks, Pope, & Marcolini, Airfoil Self-Noise and Prediction, 1989)	$Re_C \leq 1.5{\times}10^6$	≤ 0.208	Up to 19.8^0	Low turbulence	Tripped and untripped	Very Sharp

In the seminal BPM paper (Brooks, Pope, & Marcolini, Airfoil Self-Noise and Prediction, 1989), the TE noise model was introduced and validated for turbulent (tripped) flow up to $Re_C \leq 1.5\times10^6$, $M < 0.21$ and 19.8^0 AOA (angle-of-attack). All experiments and thus, the resulting model validation, were made for the NACA 0012 airfoil, based on the acoustic spectra measured in this range. For further details, see page 51 of the BPM report. Also, the BPM authors state that:

> *"For the turbulent-boundary-layer-trailing-edge noise and separation noise sources, an accurate and generally applicable predictive capability is demonstrated, especially for the important conditions of high Reynolds numbers and low to moderate angle of attack"*

> *"The unique prediction capability presented should prove useful for the determination of broadband noise for helicopter rotors, wind turbines, airframe noise and other cases where airfoil shapes encounter low-to-moderate speed flow"*

A later NREL validation study for the model (Moriarty & Migliore, 2003) showed good agreement of the BPM prediction with data taken from a series of wind tunnel tests performed at the NLR, The Netherlands (Oerlemans, 2003). The comparison was made at $M = 0.21$ and AOA ranging from 0^0 to 13.1^0. The agreement was good for frequencies near 3 kHz but for lower frequencies (~800 Hz) the differences found were up to 6 dB. The study did not expand the validation range of the model.

More recently, Doolan and Moreau (Doolan & Moreau, 2013) have plotted SPL spectra as a function of Strouhal number for some experiments, against BPM predictions. For the case of the IAG Wind tunnel data (Herrig & Würz, 2008) at $Re_C{\sim}2.9\times10^6$, it has shown good agreement with BPM prediction at $M = 0.20$, for peak Strouhal number and higher frequencies. However, by verifying the IAG Wind Tunnel Data made available by (Herrig & Würz, 2008), it seems that the higher chord-based Reynolds number of the experiment was close to $Re_C{\sim}2.4\times10^6$.

Further attempts to extend the Reynolds number validation of the PNoise tool integrated into QBlade, using the research of (Devenport, et al., 2010) based on data from the Virginia Tech Aeroacoustic Tunnel, were not successful and will be reported in a follow-on paper.

3. Results

A detailed set of operating instructions for the PNoise code may be downloaded online along with the QBlade code (Saab Jr, et al., 2016).

A preliminary modification had to be made to the XFLR5 output routines embedded into the QBlade, in order to save displacement thickness (δ^*) data along with each polar operational point object. This information was not previously saved to file, which became necessary since it is employed as the transversal turbulence scale at the TE noise model. It is also employed as a turbulence scale for other self-noise sources that are intended to be implemented in the future. This preliminary modification did not impart any abnormalities to the code (Saab Jr, et al., 2016).

3.1 Test Cases

The verification and validation process involved analyzing the results of six different combination cases, displayed in table 2.

Table 2 - Case numbers (flow data sources and angles of attack combinations) for the verification and validation process.

Case numbers	NACA0012 Sharp TE airfoil	
	Transversal turbulence scale source	
Flow Angle	BPM correlations	XFLR5 data
Zero AOA	1	4
Below Switching angle	2	5
Above Switching angle	3	6

The calculation procedure verification for all six cases was accomplished against step-by-step calculations carried out in spreadsheets for each one of them. The verification cases requiring XFLR5 output data were run in a non-integrated fashion, with displacement thickness data calculated, exported to file, linearly interpolated to the desired chord station and then inputted in the spreadsheet for the remainder of the BPM calculation.
The numeric validation of the reference spreadsheets themselves had been previously accomplished against peak frequency, peak level and roll-off behavior, compared to graphical TE noise spectra, both experimental and calculated, provided in the original BPM paper (Brooks, Pope, & Marcolini, Airfoil Self-Noise and Prediction, 1989).

For all code verification cases, the procedure adopted was to subtract the code SPL calculations for 1/3 octave bands, from the values calculated with the aid of the correspondent reference spreadsheet. For brevity, the differences found are shown for a few selected frequencies only, 50 Hz, 1,000 Hz and the peak frequency, plus for the overall Sound Pressure Level.

3.1 Code verification with BPM correlations for TBL data, for zero AOA - Case 1.

The typical input screens for the PNoise code are illustrated in figures 1 and 2 below. For the selection of the original BPM correlations and type of flow (fully turbulent or transition), the second screen is used.

Case 1 is the baseline case displayed in figure 11 of the BPM original paper (Brooks, Pope, & Marcolini, Airfoil Self-Noise and Prediction, 1989), with the following flow and geometric features:

- NACA 0012, Sharp Trailing Edge.
- Reynolds: 1,500,000
- Mach: 0.21
- Tripping: @15% chord, both sides
- Chord: 0.3048 m
- Wetted TE span: 0.4572 m

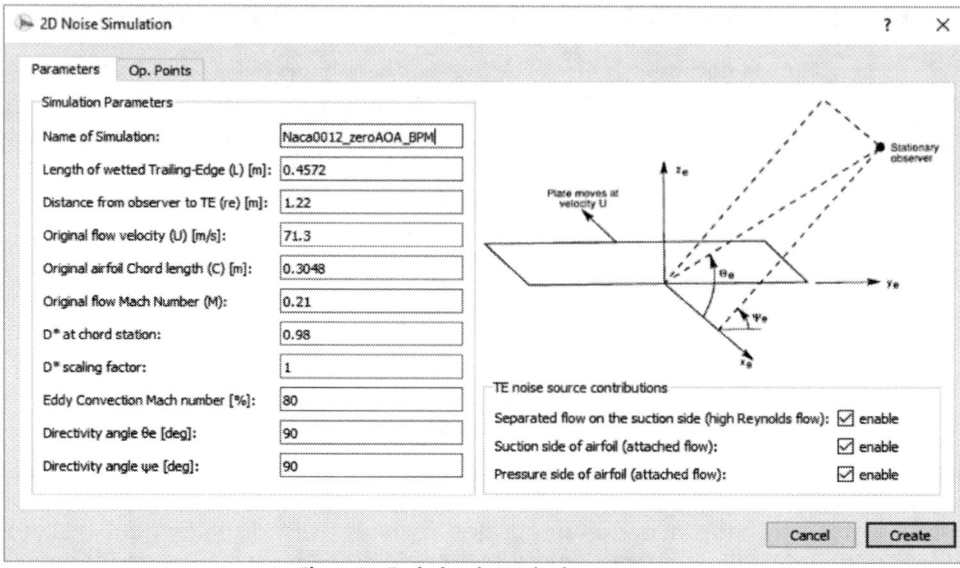

Figure 1 – Typical main PNoise input screen.

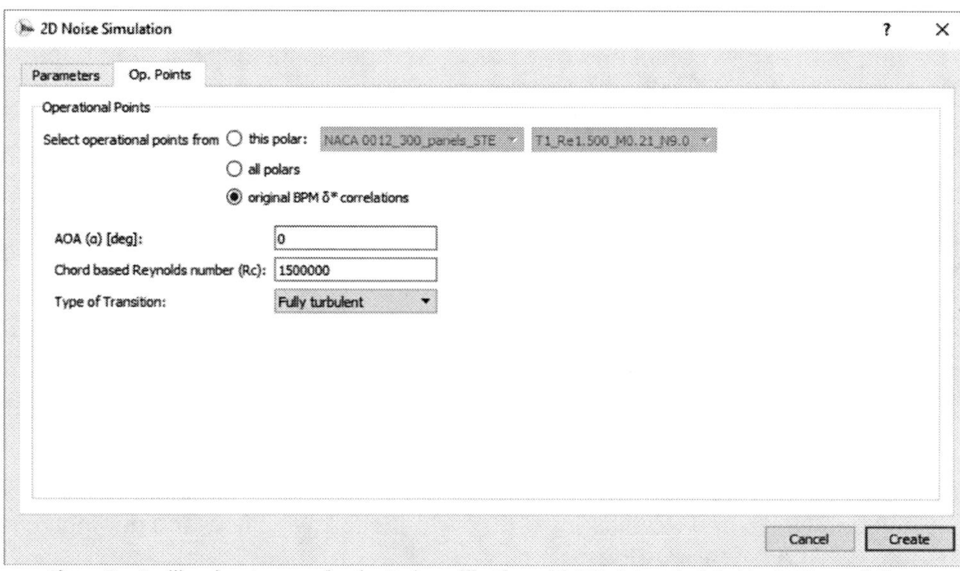

Figure 2 – Auxiliary input screen for the Code verification with TBL data calculated by BPM correlations.

When the Original BPM correlations for displacement thickness are selected, the dialog opens three new fields for the user to enter AOA data, chord-based Reynolds Number and to select among Transition or Fully Turbulent Flow.

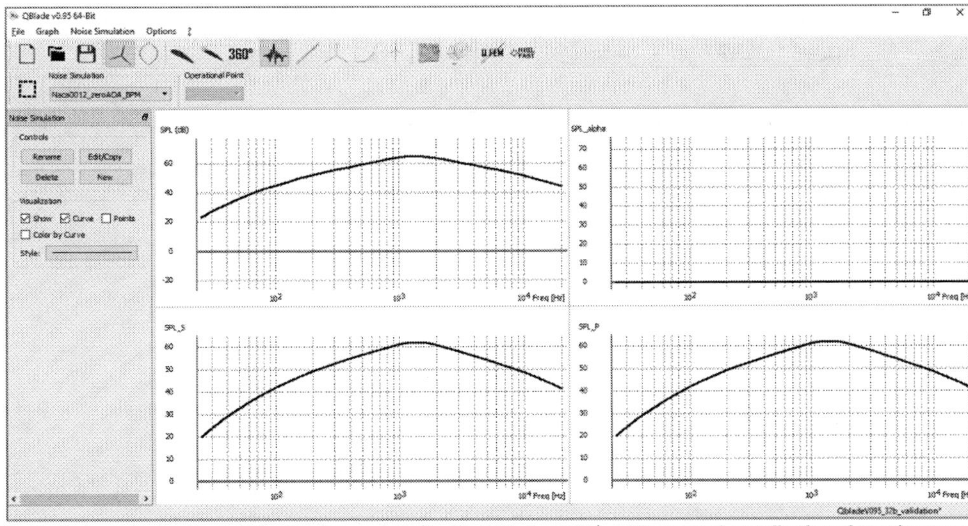

Figure 3 – Typical spectral output of the PNoise module (inside QBlade v0.95), with spectral contributions from the pressure side, suction side and AOA (zero contribution for case 1), plus overall spectra.

The numerical verification was made with the aid of the calculation spreadsheet, fed with the BPM δ* correlations (δ = 2.66E-3 m for this particular case), in order to reproduce the original model 1/3 octave band spectrum. The differences among the SPL data calculated by the code and the spreadsheet may be seen in table 3.

Table 3 – Numerical validation of the code calculation for Case 1, against spreadsheet calculation.

Frequency (Hz)	Source	SPL_alpha (dB)	SPL_S (dB)	SPL_P (dB)	SPL (dB)	Diff.(dB)
50	Spreadsheet	-1138.39	31.19	31.19	34.20	+1.29
	Code	-1136.90	32.48	32.48	35.49	
1,000	Spreadsheet	-938.72	59.63	59.63	62.64	+1.28
	Code	-938.24	60.91	60.91	63.92	
1,250	Spreadsheet	-936.48	60.61	60.61	63.62	+1.27
	Code	-935.20	61.88	61.88	64,89	

The peak frequency of 1,250 was correctly predicted by the code and the overall SPL resulted in 73.2 dB against 71.9 dB for the spreadsheet, a difference of 1.25 dB. The same kind of systemic difference is seen on all frequencies compared.

3.2 Code verification with BPM correlations for TBL data, for AOA = 4°- Case 2

The BPM model has different calculation procedures for AOA below and above the "switching angle" defined as the angle above which the angle-of-attack SPL contribution becomes dominant. The switching angle calculated to match the experimental conditions for the 0.2286 m–chord BPM airfoil at a Reynolds number flow of 1,120 Million, is 9.5°. Thus, one verification was made below the switching angle (@4°) and another one above it (@12.5°).

Case 2 was run with the following flow and geometric features:

- NACA 0012, Sharp Trailing Edge.
- Reynolds: 1,120,000
- Mach: 0.21
- Tripping: @15% chord, both sides

- Chord: 0.2286 m
- Wetted TE span: 0.4572 m

The typical graphic output for the AOA=4° airfoil attitude is shown in figure 4 and the comparison table is seen in table 4.

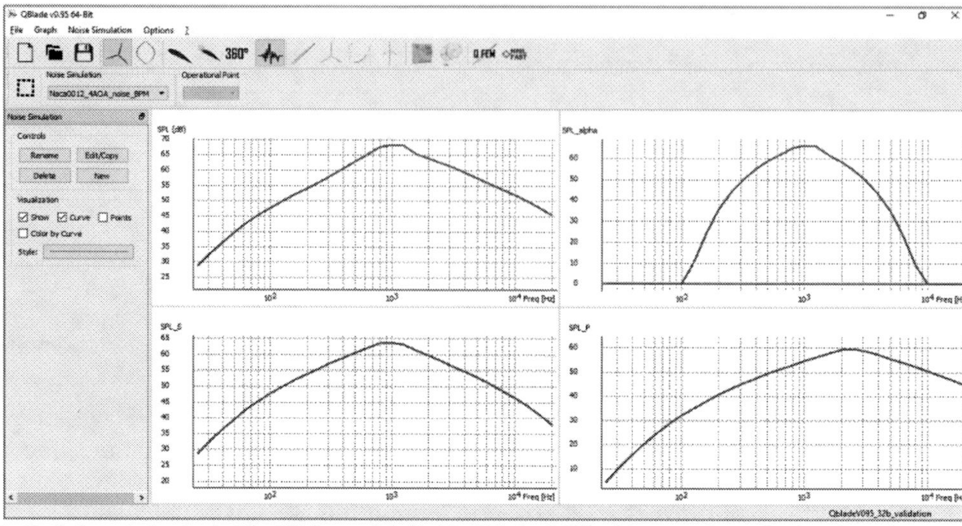

Figure 4 - Typical spectral output of the PNoise module (inside QBlade v0.95), with spectral contributions from the pressure side, suction side and AOA, plus overall spectra.

Table 4 - Results from PNoise code (inside QBlade v0.95) against verification spreadsheet, for 4° AOA, BPM correlations.

Frequency (Hz)	Source	SPL_alpha (dB)	SPL_S (dB)	SPL_P (dB)	SPL (dB)	Diff.(dB)
50	Spreadsheet	-85.55	38.20	18.76	38.25	+1.29
	Code	-84.10	39.49	20.06	39.54	
1,000	Spreadsheet	64.90	62.44	53.18	67.03	+1.28
	Code	66.17	63.72	54.46	68.31	

The peak frequency of 1,000 is correctly predicted by the code and the overall SPL is 75.8 dB against 74.4 dB for the spreadsheet, a difference of 1.4 dB. A systemic difference close to +1.3 dB is seen for both frequencies compared.

3.3 Code verification with BPM correlations for TBL data, for AOA=12,5° - Case 3.

Case 3 was run with the following flow and geometric features:

- NACA 0012, Sharp Trailing Edge.
- Reynolds: 284,400
- Mach: 0.21
- Tripping: @15% chord, both sides
- Chord: 0.0254 m
- Wetted TE span: 0.4572 m

The typical graphic output for the AOA above the switching angle was recovered correctly as shown in figure 5, i.e. the sole significant contribution to the TE noise level in this flow attitude is the angle-of-attack or detached boundary layer over the suction side contribution, known to the model as SPL-α contribution.

Figure 5 - Typical spectral output of the PNoise module (inside QBlade v0.95), with spectral contributions from the AOA, only.

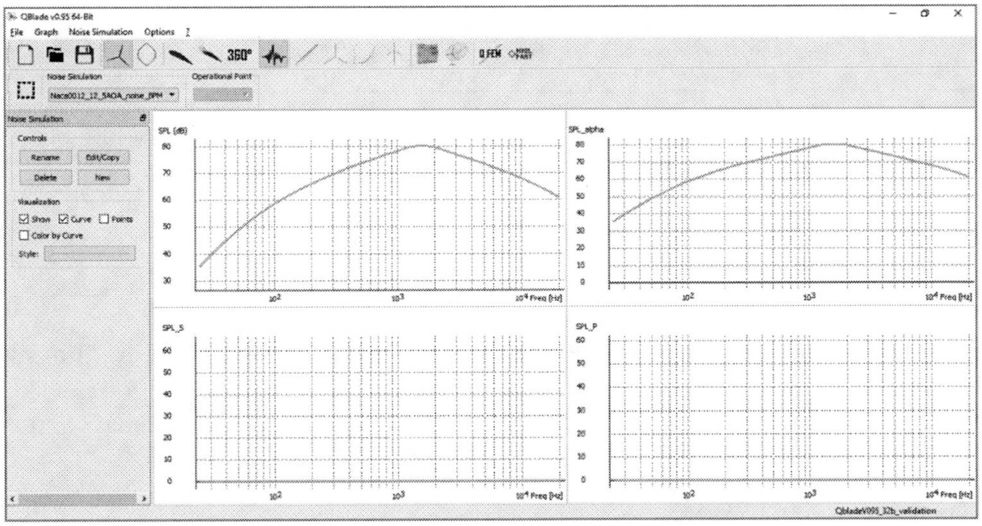

Table 5 - Results from PNoise code (inside QBlade v0.95) against verification spreadsheet, for 12.5° AOA, BPM correlations.

Frequency (Hz)	Source	SPL_alpha (dB)	SPL_S (dB)	SPL_P (dB)	SPL (dB)	Diff.(dB)
50	Spreadsheet	48.75	-∞	-∞	48.75	-0.22
	Code	48. 53	-2.1E+9	-2.1E+9	48.53	
1,000	Spreadsheet	78.41	-∞	-∞	78.41	-0.20
	Code	78.21	-2.1E+9	-2.1E+9	78.21	
1,600	Spreadsheet	80.42	-∞	-∞	80.42	-0.18
	Code	80.24	-2.1E+9	-2.1E+9	80.24	

The OASPL is 88.6 dB for the spreadsheet and 88.4 dB for the code calculation, a -0.2 dB difference and the peak frequency in in the band of the 1,600 Hz central frequency for both calculations, a shown in table 5.

3.4 Code verification with TBL data calculated by XFLR5 for zero AOA - Case 4.

The flow and geometric and flow input parameters for cases 4, 5 and 6 are the same employed for the cases 1, 2 and 3, respectively, except that the original BPM displacement thickness correlations are replaced by XFLR5 flow calculation results.

The results for zero AOA (case 4) are shown in table 6 below.

Table 6 - Results from QBlade v0.95 code calculation against verification spreadsheet, for zero AOA, XFLR5 flow data.

Frequency(Hz)	Source	SPL_alpha(dB)	SPL_S (dB)	SPL_P (dB)	SPL (dB)	Diff.(dB)
50	Spreadsheet	-1252.71	19.64	19.74	22.70	+1.2
	Code	-1252.15	20.93	21.04	24.00	
1,000	Spreadsheet	-948.83	53.60	53.65	56.64	+1.3
	Code	-947.54	54.88	54.93	57.92	
2,500	Spreadsheet	-938.72	58.32	58.33	61.34	+1.3
	Code	-937.46	59.59	59.61	62.60	

The differences are systemic and around 1.3 dB, which was considered acceptable. The peak frequency is within the 2,500 Hz band for both spectra.
The overall unweighted sound pressure level is 69.6 dB for the Spreadsheet and 70.8 dB for the Code, a 1.2 dB difference over prediction by the code.

3.5 Code verification with TBL data calculated by XFLR5 for AOA below the switching angle (4°) – Case 5.

The code output for case 5, with 4° AOA (below the switching angle) is illustrated in figure 6 and the calculated results for selected frequencies are shown in table 7.

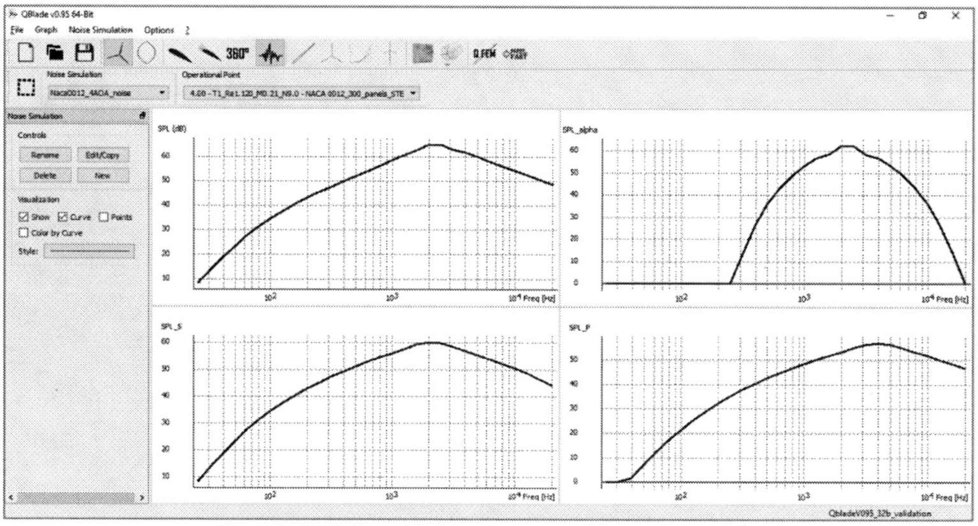

Figure 6 – Typical graphical output screen of the PNoise module

Table 7 - Results from PNoise code (inside QBlade v0.95) against verification spreadsheet, for 4° AOA, XFLR5 flow data.

Frequency (Hz)	Source	SPL_alpha (dB)	SPL_S (dB)	SPL_P (dB)	SPL (dB)	Diff.(dB)
50	Spreadsheet	-243.38	21.83	5.67	21.93	+1.3
	Code	-241.81	23.12	6.97	23.23	
1,000	Spreadsheet	51.88	54.77	47.13	57.04	+1.3
	Code	53.18	56.05	48.42	58.33	
2,500	Spreadsheet	61.10	58.89	52.10	63.47	+1.4
	Code	62.38	59.96	54.92	64.82	

The differences are systemic and of the order of 1.3 dB, which was considered acceptable. The peak frequency is contained within the 2,500 Hz band for both spectra.
Also the OASPL is 71.1 dB for the Spreadsheet and 72.2 dB for the code or a 1.1 dB difference.
The user should be warned that for negative AOA, the upper surface, initially a suction side, becomes a pressure side and the lower surface, initially a pressure side, becomes a suction side. Since the graphs are labelled "pressure" and "suction" sides, not "upper" and "lower" sides of the airfoil, the output will appear overlapped (unchanged) in the graphical display when symmetrical angles are simultaneously seelcted for calculation (e.g., +4° and -4° AOA).

3.6 Code verification with TBL data calculated by XFLR5 (inside QBlade) for AOA above the switching angle (12.5°) - Case 6.

The results for some frequencies at 12.5° AOA (above the switching angle), are shown in table 8.

Table 8 - Results from PNoise code (inside QBlade v0.95) against verification spreadsheet, for 12.5° AOA, XFLR5 flow data.

Frequency (Hz)	Source	SPL_alpha (dB)	SPL_S (dB)	SPL_P (dB)	SPL (dB)	Diff.(dB)
50	Spreadsheet	1.78	-∞	-∞	1.78	+0.3
	Code	2.03	-2.1E+9	-2.1E+9	2.03	
1,000	Spreadsheet	57.67	-∞	-∞	57.67	0.0
	Code	57.67	-2.1E+9	-2.1E+9	57.67	
8,000	Spreadsheet	72.92	-∞	-∞	72.92	-0.1
	Code	72.82	-2.1E+9	-2.1E+9	72.82	

SPL_α should be the sole effective noise source contributor for an angle above the switching angle, which is exactly the behavior displayed.
The OASPL is 80.7 dB for the spreadsheet and 80.6 dB for the code or a 0.1 dB difference.

4. Conclusions

The calculation procedure verification for the code was made for 6 cases, covering the zero, below and above switching angle conditions, with turbulent boundary layer data provided both by the original BPM correlations and the XFLR5 embedded into the QBlade code.
The code calculation verification procedure was made for all cases against spreadsheet reference calculation, prepared as per the original model and previously verified against the original BPM-calculated spectra.
The code calculations displayed a systemic, positive overprediction of about 1.3 dB for the cases below the switching angle [0°, 4°]. This applies to sample frequencies selected for comparison, which included the peak frequency for each case and also to the overall SPL.
The code calculations displayed a closer adherence to reference calculation in the cases for AOA above the switching angle (12.5°), where the differences found ranged in the [-0.2,+0.1] dB interval, for selected frequencies and for the overall SPL.
The +1.3 dB systemic difference perceived in some of the cases was considered acceptable for the first release of the PNoise module, but improvements in calculation accuracy will be made for follow-up releases.
All correspondence and suggestions for improvements are welcome and should be addressed to the first author, which is the sole responsible for any bugs in the current version.

Acknowledgements
The first author would like to acknowledge the invaluable support received from the Technische Universität Berlin Hermann Föttinger Institute (HFI) Group, Germany; Escola Politecnica da USP,

Brazil and Escola de Engenharia Maua, Brazil. Also, the authors are indebted to programmers Ricardo Marques Augusto and Nikolai Moesus for their help during the coding phase of the project.

References

Brooks, T. F., & Marcolini, M. A. (1986). Airfoil Trailing-Edge Flow Measurements. *AIAA Journal*, pp. 1245-1251.

Brooks, T. F., Pope, S., & Marcolini, M. A. (1989). *Airfoil Self-Noise and Prediction.* NASA. Langley: NASA.

Brooks, T., & Hodgson, T. H. (1981). Trailing Edge Noise Prediction From Measured Surface Pressures. *Journal of Sound and Vibration 78(1)*, pp. 69-117.

Brooks, T., & Marcolini, M. (1985). Scaling of Airfoil Self-Noise Using Measured Flow Parameters. *AIAA Journal (23) 2*, pp. 207-213.

Devenport, W., Burdisso, R., Camargo, H., E., C., Remilieux, M., & Rasnik, M. a. (2010). *Aeroacoustic Testing of Wind Turbine Airfoils.* NREL SR-500-43471.

Doolan, C., & Moreau, D. (2013). Review of NACA 0012 Turbulence Trailing Edge Noise Data at Zero Angle of Attack. *5th International Meeting on Wind Turbine Noise* (pp. 1-10). Denver, Co.: INCE Europe.

Drela, M., Youngren, H., & Deperrois, A. (2009). *XFOIL/XFLR5.* MIT, Cambridge, Ma.

Herrig, A., & Würz, W. K. (2008). New CPV-Results of Naca 0012 Trailing-Edge Noise . *International Conference on methods of Aerophysical Research, ICMAR 2008*, (pp. 1-9).

Marten, D. (2010). *Extension of an Aerodynamic Simulator for Wind Turbine Blade Design and Performance Analysis.* Technischen Universität Berlin. Berlin: TU Berlin.

Marten, D. (2014). *Qblade Short Manual V0.8.* Berlin: TU Berlin.

Marten, D., & Wendler, J. (2013). *Qblade Guidelines v0.6.* Berlin: TU Berlin.

Moriarty, P., & Migliore, P. (2003). *Semi-Empirical Aeroacoustic Noise Prediction Code for Wind Turbines.* NREL, Golden, CO.

Oberkampf, W., & Roy, C. (2012). *Verification and Validation in Scientific Computing.* Cambridge: Cambridge University Press.

Oerlemans, S. (2003). *Wind Tunnel Aeroacoustic Tests of Six Airfoils for Use on Small Wind Turbines.* Golden, Co.: NREL SR-500-34470.

Pechlivanoglou, G., Marten, D., Weinzierl, G., Moesus, N., & Wendler, J. (2009). *QBlade.* Technische Universität Berlin, Berlin.

Saab Jr, J. Y., & Pimenta, M. M. (2016). Displacement Thickness Evaluation for Semi-Empirical Airfol Trailing-Edge Noise Prediction Model. *Journal of the Brazilian Society of Mechanical Sciences and Engineering, V.38, N.2*, pp. 385-394.

Saab Jr, J. Y., Pimenta, M. M., Piqueira, J. R., Marten, D., Pechlivanoglou, G., Nayeri, C., & Paschereit, C. O. (2016). *Verification and Validation of the Airfoil Trailing-Edge Noise Prediction Module (PNoise) inside QBlade V0.95.* Sao Paulo, Brazil.

7th International Conference
on
Wind Turbine Noise
Rotterdam – 2nd to 5th May 2017

Partial masking and perception of wind turbine noise in ambient noise

Anders Johansson MWL, KTH- Royal Institute of Technology, Stockholm, Sweden: aebjo@kth.se
Jesper Alvarsson School of Social Sciences, Södertörn University, Stockholm, Sweden: jesper.alvarsson@sh.se
Karl Bolin MWL, KTH- Royal Institute of Technology, Stockholm, Sweden: kbolin@kth.se

Summary

The level of perceived annoyance is often linked to the type of sound heard and an increase in loudness often means an equally higher level of annoyance. Wind turbine noise has been reported as being more annoying than other common environmental noise sources, especially for low levels below 50 dB(A). Commonly the increase in annoyance is attributed to amplitude modulations of the turbine sound, which facilitates detection in a constant background noise. Therefore it is natural to consider the ambient sounds surrounding wind turbines and how they affect the perceived annoyance. For instance, could annoyance and noise guidelines be different in city dwellings and near highway-areas compared to forest landscapes? This paper addresses this question by conducting a listening experiment. Adopting the method of magnitude production, 20 listeners produced sound levels, according to predefined levels of annoyance shown on the Borg CR100-scale. Recordings from three wind turbines of different size were rated, both heard alone and in the presence of ambient noise from either: a deciduous forest, busy road or city street. Results show that rated alone, only a small difference exists between the wind turbine sounds. However, clear differences in annoyance could be distinguished between different ambient sounds, where the deciduous forest stands out as the overall poorest masker. The findings suggest that the type of ambient surrounding could be taken into consideration when mitigating noise annoyance from wind turbines.

1. Introduction

Moving towards a future solemnly dependent on renewable energy it is natural to contemplate on the layout of this landscape. Wind energy has proven to be a great source of energy and the development has multiplied over the last decade with numbers increasing in both size and quantity. The natural development site for wind energy has been in remote areas, for example large parks are planned and built offshore or in sparsely populated areas. The one reason for this remote deployment, besides the prospect of windy weather and low establishment costs, is to limit the negative impacts of wind turbines such as visual and acoustical annoyance. Out of these aspects the acoustical emissions are identified as the most annoying (Van den Berg, Pedersen et al. 2008).

Comparing dose-response curves (Miedema and Oudshoorn 2001), (Miedema 2004) for wind turbine noise and other noise sources, wind turbines appear to be even more annoying than road or aircraft-noise and especially for relatively low sound levels below 50 dB(A). A common theory attributes the increase in annoyance to the fluctuating character of the turbine sound, which would facilitate detection in a constant background noise. This links annoyance to the detectability of the sound but does not address how the sound character of the wind turbine or background sound affects the level of annoyance.

It has also been shown that the perception of noise can be significantly influenced by other psychological factors and moderate the level of annoyance associated with the noise, making it more or less disturbing. For instance, a Dutch survey (Pedersen, Berg et al. 2009) showed that people who benefited economically from the wind turbines reported less annoyance compared to people who did not. The same survey also confirms the correlation of annoyance and other negative attitudes towards wind turbines, which had been found in earlier studies (Pedersen and Waye 2004), (Pedersen, Hallberg et al. 2007).

With this knowledge in mind it is natural to consider the ambient sounds surrounding wind turbines and how they affect the perceived annoyance. It is easy to assume that an environment which already contains unwanted noise would not suffer greatly from additional sonic pollutions. For example, could annoyance and noise guidelines be different in city dwellings and near highway-areas compared to forest landscapes? Previous research indicates that positive effects in reducing the level of annoyance have been achieved by masking from natural sounds (Bolin 2010) However, an epidemiological study showed no decreased annoyance of wind turbine sound by masking road noise (Pedersen, van den Berg et al. 2010).

To evaluate the perception of wind turbine sound and the potential of masking from ambient sounds, the listening test is considered a proven method and a good compromise between validity and reliability. In this experiment the method of magnitude production was adopted with the belief that this procedure facilitate a natural pace as the listener is given direct control of the test progression and can makes judgments when ready. We also think this test procedure is more engaging for the listener, who otherwise easily fallsh into listening fatigue.

The purpose of the study is to extend the research of masking of wind turbine sounds, by performing a listening test which also include urban ambient sounds in a comparison model where levels of annoyance are produced according to the Borg CR100-scale (Borg 2001).

2. Method

The layout of the listening test was divided into two parts where levels of annoyance were first produced for wind turbines alone. Background sound was then added, and levels of annoyance were produced for the partially masked wind turbine sounds. Figure 1 shows the third octave band spectra $L_{A,eq}(f)$ of the three types of wind turbines and the three different background sounds. In addition, pink noise was also included as a reference source to compare short time annoyance. In figure 1 all levels have been normalized so that the third octave levels of each sound equals 0.

2.1 Wind turbine sounds

In order to get a variation in sound character and frequency content of the sounds, three binaural recordings of wind turbines of different size and rated power were selected. In the following, these sounds will be referred to as: Small-, Medium- and Large-WT (wind turbine).

The wind turbine referred to as Small-WT is a 12kW vertical axis turbine with hub-height 6 meters, turbine radius 3 meters and blade-length 5 meters. The Medium-WT is of ordinary horizontal axis type and almost four times as big, with rated at 48 kW power. The rotor diameter is 14.6 meters and hub height 21 meters. The Large-WT, also of horizontal axis type, is an additional four times as big, with 2 MW rated power and 95 meters hub height. The recording distance for the small- and Medium-WT was 200 meters, while the Large-WT was recorded at 500 meters. The Average blade passage frequency was estimated from the recordings to be around 3, 1 and 0.6 Hz, for the three wind turbines.

The sound characters of the Small-WT is described by the authors as high pitched and energetic, while the Medium-WT has a more low pitched sound resembling that of a washing machine with a "thumping" rhythm. The sound of the Large-WT is similar in character to the Medium-WT, except that it has more low frequency content. A faint whistling sound emanating from the gearbox can be heard in this recording.

For use in the listening test all sounds were normalized with its A-weighted equivalent sound level and set to have a dynamic range from 0 to 87 dB(A), measured at the ear drum.

2.2 Background sounds

For background sound ambience, recordings from a deciduous forest, city-street and a busy road, were used. These sounds were selected to represent surroundings of possible development sites for wind power. Similar to the selection of the wind turbine sounds, care was taken to select sounds with distinguished sound characteristics, as described below.

The road traffic sound is a binaural recording measured at a 50 meter distance from a four-lane highway with constant passing traffic and speed limit of 70 km/h. Both larger trucks and passenger cars can be discerned in the recording. The pace of the cars and trucks passing is very calm and is not perceived as stressful by the authors.

The deciduous forest recording is named Woodland Atmosphere and is a binaural recording from BBC. The recording contains the sounds of rustling leafs at a distance, which has a clear resemblance to white noise as seen in the spectrum (figure 1). When gusts of wind blows through the trees slow amplitude variations up to ca 10 dB are generated. There is no birdsong and few impulsive sounds such as squeaking or falling branches present in the recording.

The city ambience is also a BBC recording named City Traffic. It features a busy street in a seemingly large city with a variety of vehicles stopping and starting. The traffic appears slow and tranquil and footsteps or occasional chats from pedestrians can be heard. Amplitude variations are typically 5dB and some impulsive sounds generate peaks up to 10 dB above the mean level.

In order to determine suitable levels for the background sound, ten listeners (who did not perform in the main test) participated in a procedure where they adjusted the sound level to a level they perceived as natural. The same equipment as in the main listening test was used and the results showed an inter-individual average about 60 dB(A). This level was used for all background sounds in the listening test.

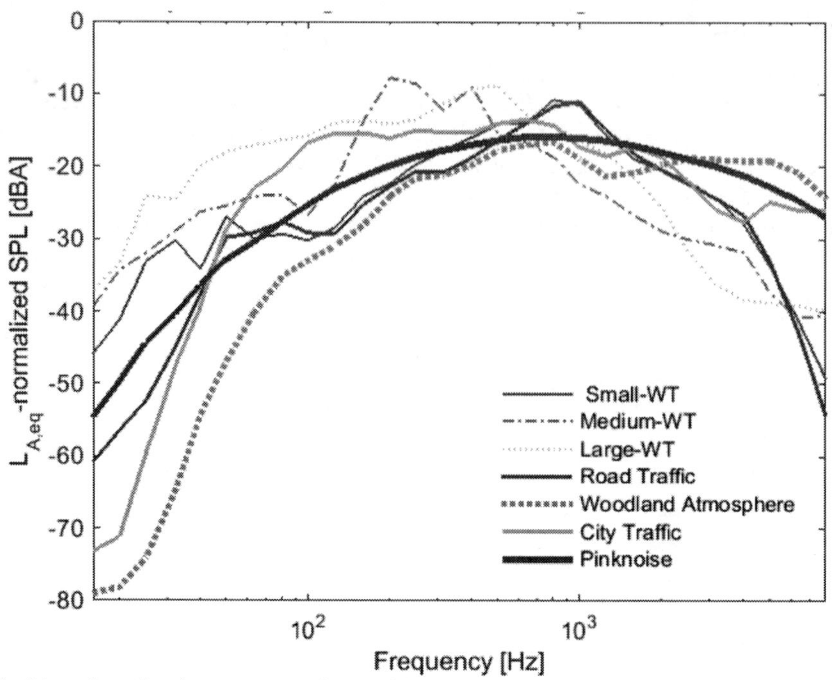

Figure 1: Showing the $L_{A,eq}$-normalized third octave spectrum for all experimental sounds. Levels have been normalized so the third octave levels of each sound equals 0 dB.

2.3 Procedure

The participant was seated in front of a laptop and the software played the sound through a pair of headphones. Prior to the experiment the listener read through instructions, which included a thorough description of the Borg CR100-scale and examples on how to interpret the different levels as perceived annoyance. Before the experiment started the listener had a chance to practice and familiarize with the test interface and sounds. The test leader was present during this initial phase to answer possible questions.

The listening test consisted of two sessions, rating of single sounds and rating of partially masked sounds. The first session contained four sounds; Small- Medium- and Large-WT plus pink noise. Subjects were instructed to produce estimates of six different magnitudes, namely "just noticeable", "very weak", "weak", "moderate", "strong" and "very strong". All sounds and levels were presented in random order different for each test subject.

Which magnitude the participants should produce was shown on the laptop screen by a blue arrow on the current magnitude, see figure 2 for the case where the subject is instructed to adjust the sound to a "just noticeable" level. The volume control was manipulated by the participant via the rotatable knob, which was pressed when the participant perceived the appropriate sound level.

In the second session, each wind turbine sound was played simultaneously with one of the three background sounds: road traffic, forest and City Traffic, resulting in a total of nine sound-combinations. The background sound had a fixed level and the listener was instructed, as previously, to produce levels of annoyance but to only focus on the wind turbine sound. Estimates were produced for four magnitudes from "Just noticeable" to "Moderate".

Apart from randomized presentation order of both annoyance levels and sounds, randomization was also applied to start time and sound level for each excerpt. Changing the onset sound level was performed to minimize the risk of a habituated behaviour when adjusting the sound level. Each listener performed four repetitions for each level and sound, which resulted in 96 production-tasks per session. One session took in average 45-60 minutes to complete, including two scheduled 5 minutes breaks. A longer break (~15 minutes) was also included between the two sessions.

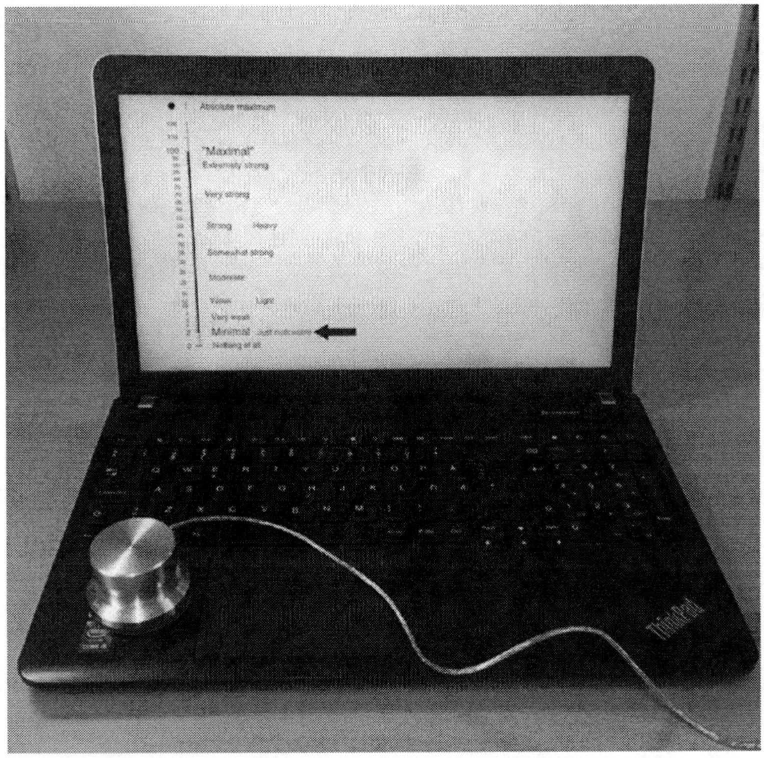

Figure 2: Overview of the experimental-setup, showing the Borg CR100-scale and arrow, pointing towards one of the verbal descriptors describing the level of annoyance the listener should produce by adjusting the sound level via the rotatable knob, also shown in the picture.

2.4 Recording equipment

The wind turbine recordings were conducted using a Bruel and Kjear (BK) Head and Torso Simulator 4128D (HATS), with one BK 4190 microphone in each ear. The microphone signals were fed to a BK type 2690 Nexus Conditioner. The conditioner was set at 1V/Pa, high pass filter at 20 Hz and low pass filter at 22400 Hz. The amplified signal was fed to a Sound Devices 788T recorder. Calibration of the microphones was done with a Larson Davis model CAL200, 1 kHz, 94 dB tone calibrator. While recording, the HATS and the measurement microphone centre were positioned 1.5 meters above the ground, 1 meter apart. The recordings were sampled at 44100 Hz with a bit depth of 24.

2.5 Listening test equipment

The sound-program was reproduced using Bose QC15 noise cancelling headphones for which the frequency response had been measured and equalized flat from 30-8000 Hz with the use of a Mini-DSP control board. The headphones and DSP-board were connected to the laptop via an Asus XONAR U7 soundcard for which the transfer amplitude at 1 kHz were independently measured and normalized for both right and left channels. Sound level adjustments were done via a Griffin Technology PowerMate rotatable knob. The test software was written in MATLAB R2014b using the GUI-interface and all post processing of measurement-data were also performed in MATLAB. The listening test was conducted in a quiet office space.

2.6 Listeners

The 20 persons, 8 women and 12 men, who participated in the listening experiment were all students or personnel at KTH at an age between 19 and 36 years. All participants were asked prior to the test if they suffered from any hearing impairment or considered themselves to have a reduced hearing ability. Participants were informed that the experiment posed no risk for the hearing system and that the experiment could be aborted at any time if the participant felt uncomfortable. As a gesture of gratitude from the researches, the participants received two cinema tickets.

3. Results

To correct for outliers the analysed data are excluding adjustments which deviated more than 10dB from the individual median of the repetitions of a sound, this elimination was performed for 96 judgments per individual, counting 20 persons in the test. For each individual the geometrical mean was then calculated for every level and sound.

3.1 Single sounds ratings

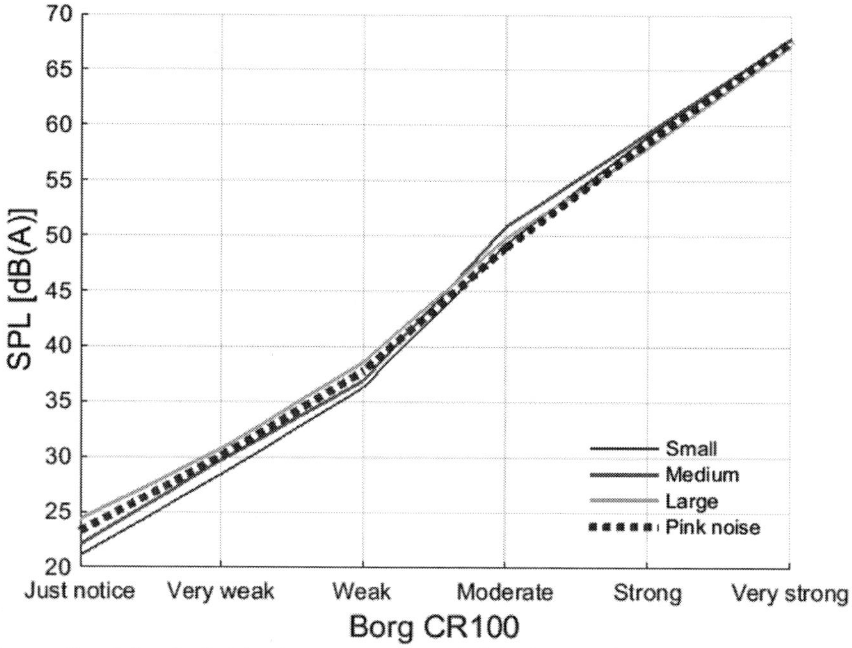

Figure 3: Shows the inter-individual average A-weighted sound level produced for annoyance levels "just noticeable" to "very strong", for sounds small-, medium- and large-WT plus pink noise.

Figure 3 shows the mean produced sound volume for single sounds as a function of annoyance level indicated on the Borg CR100-scale. For the three lowest intensities it can be seen that the produced level is lower for the small wind turbine. For instance, at "Just noticeable" which exhibit the greatest differences between sounds, the sound pressure level for the large wind turbine is allowed to be 3 dB higher and still be perceived as equally annoying.

The range from "Just noticeable" to "Very strong" is approximately 50 dB and the average step between two intensities is about 10 dB, which correlates well with the design of the Borg

CR100-scale where a doubling in perceived stimuli intensity equals a step between two adjacent verbal anchors (Fastl 1999, Borg 2001). In general the total range can be interpreted as the sound level between a whisper and a noisy vacuum cleaner at 1 m. distance.

When calculating the inter-individual standard-error it is found to be varying between 1 - 2 dB with an even distribution among sounds but with a slight increase for higher intensity levels. This could be an indication that it becomes more difficult to determine the sound level for higher levels of annoyance, which probably also leads to a decrease in correlation between subject ratings. Nevertheless, the overall standard error is to be considered relatively low, indicating that individuals tend to produce similar experiences. However, when performing a repeated measures ANOVA with annoyance level and sound type as independent variables, the rated sounds show not to be different at 5% significance level ($F(3,58) = 1.347$, $P < 0.268$).

3.2 Partial masking ratings

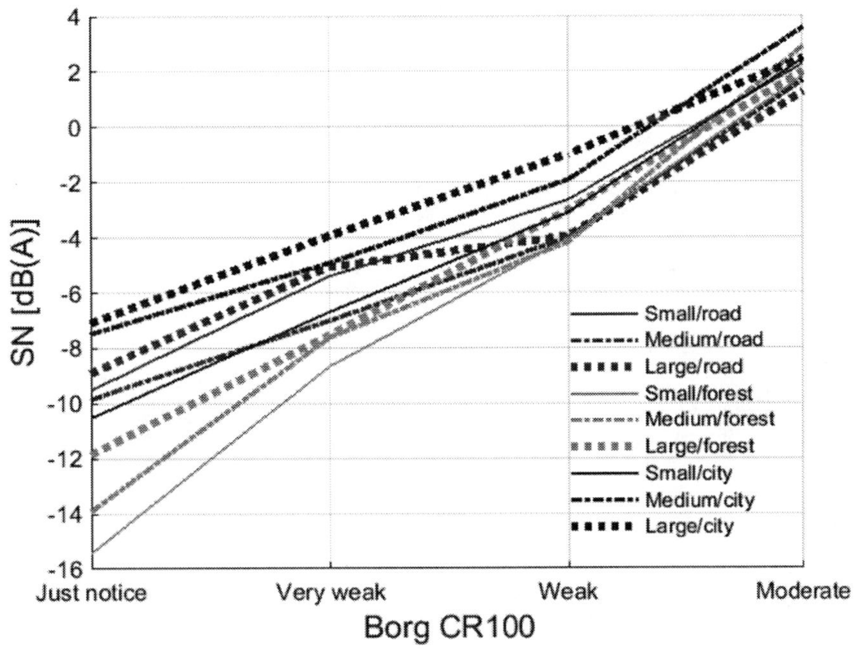

Figure 4: Shows the inter-individual average A-weighted signal-to-noise ratio as a result of produced annoyance levels for the small-, medium- and large-WT in the presence of the road, forest and city ambient background sounds. Estimates was produced for four magnitudes from "just noticeable" to "moderate",

Figure 4 shows the signal-to-noise ratio between the wind turbines and background sounds as a function of annoyance level. The result, plotted as the inter-individual mean indicate that the forest background is the poorest masker compared to the two other background sounds. This is clearly visible in figure 2 where the different sound combinations are ordered in groups after background sound type. It can also be seen that the wind turbines follow a similar pattern and that the largest wind turbine is rated as the least annoying sound within these groups.

In comparison to the single sounds the rated difference between the partially masked sounds is more pronounced (8 compared to 4 dB, at "just noticeable"). This result is somewhat expected, as combinational effects are likely to occur between masker and sound. For example, a good masker will probably be successful in reducing an already less annoying sound, while a less efficient masker will have trouble reducing a sound which was already annoying from the beginning. It can be seen that this effect is at its strongest for the lower intensity levels, where partial masking is high. For higher levels of annoyance the sounds begin to be rated more equally. This converging pattern is an indication that it is the sound intensity that dominates the level of annoyance (Berglund 1990).The overall range of the ratings is about 20 dB, signal to noise-ratio, with an average step of 5 dB, between adjacent intensity levels. This suggests that a doubling in annoyance level for the partially masked sounds does not translate to a 10 dB difference in signal intensity (Fastl 1999). The average value the "just noticeable" level is about -10 dB which correlates well with other research regarding the audibility of wind turbine sounds (Bolin 2010). The inter-individual standard-error was found to fluctuate around 1 dB for every sound and level.

The more prominent differences between ratings are reflected in the statistical properties where a repeated measures ANOVA with a Greenhouse-Geisser correction show statistically significance between the rated sound combinations ($F(4.096, 77.827) = 7.557$, $P < 0.000$) and the effect size is large (eta2 = .285). Performing a Post hoc test using a Bonferroni correction further reveals that significant differences lay between sounds having different backgrounds i.e. within a masker group the sounds are not significantly different. This result indicates that the main effect of perceived annoyance from wind turbines is largely dependent on the environmental background sound.

To get a clearer view on the relations between the wind turbine and backgrounds sounds and the effect on perceived annoyance, the data was rearranged and averaged over wind turbine type and background sound category.

3.3 Partial masking averages

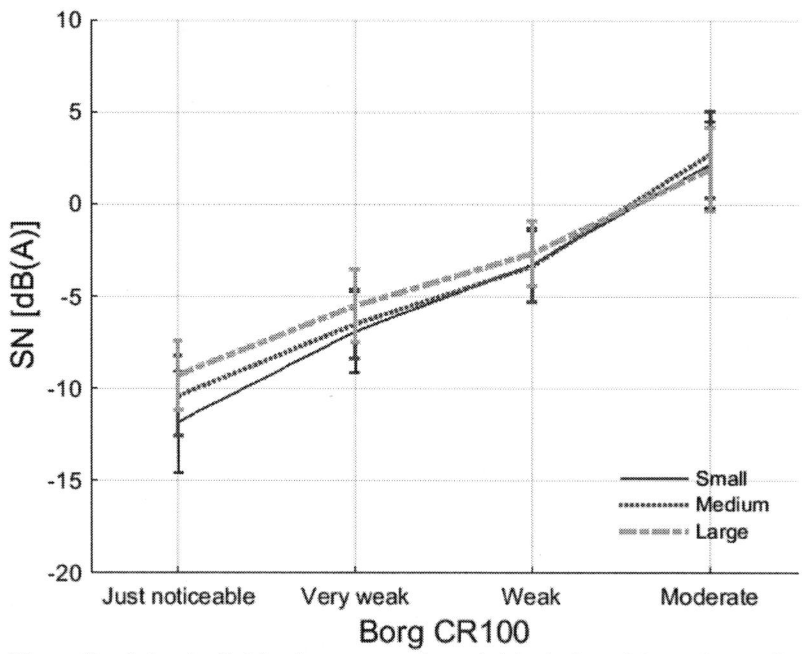

Figure 5: Show the inter-individual average A-weighted signal-to-noise ratio as a result of produced annoyance levels for the partially masked wind turbine sounds; small-, medium- and large-WT in the presence of the road, forest and city ambient background sounds. The result is calculated as an average over all background sounds for each wind turbine sound. Error-bars shows the 95% confidence interval.

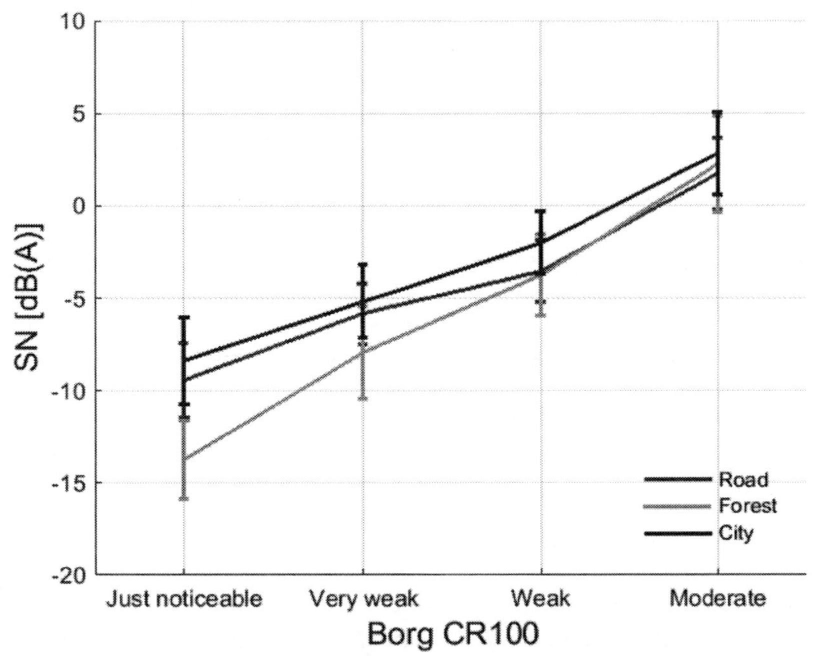

Figure 6: Show the inter-individual average A-weighted signal-to-noise ratio as a result of produced annoyance levels for the partially masked wind turbine sounds; small-, medium- and large-WT in the presence of the road, forest and city ambient background sounds. The result is calculated as an average over all wind turbine sounds for each background sound. Error-bars shows the 95% confidence interval.

Figure 5 and 6 show these averages, and as the plots are derived from the same data-set, they also share approximately the same statistical properties with 1 dB standard error and a 15 dB range. In figure 3 it is seen that the rating-order for wind turbine type follow the single sounds pattern, with the small wind turbine rated as the most annoying. However, the statistical analysis show no significant difference between sounds ($F(2, 38) = 2.081$, $P < 0.139$). The pattern of background sounds, as shown in figure 4, are a confirmation of previous observations, where the forest sound is rated as the least efficient masker, by far. The repeated measures ANOVA with applied Greenhouse-Geisser correction also show that mean differences are statistically significant ($F(1.447, 27.499) = 16.251$, $P < 0.000$) and that the effect size is very large (eta2 = .461). Post hoc analysis further confirms that rated differences are significant between all background sounds.

The above analysis of figure 3 and 4 reflects what was seen in figure 1 and 2 i.e. the main effect in reducing the level of annoyance is contributed to the environmental background sounds. What differs between the two compilations it that the effects seem to be more prominent when presented as averages over a range of conditions. This is reflected in the p-value and effect size of both measures.

4. Conclusions

In this experiment the annoyance level for wind turbines of different size and sound character was rated, alone and together with background sound from a deciduous forest, busy city and road traffic. Results reveal that the wind turbines alone did not differ in perceived annoyance level. However, when the background sound was added, significant differences depending on these sounds appeared. It was shown that the city background sound was twice as efficient at reducing the annoyance level compared to the deciduous forest. In sound level figures this translates to 5 dB difference. These numbers are true for the threshold of detection (around -10 dB signal-to-noise ratio). For lower masking values the effect decreases. At typically zero signal-to-noise ratio the effect has ceased and the wind turbine sounds were rated as equally annoying.

Whether the masking-efficiency is attributed to the frequency content or temporal structure and sound quality of the background sound remains unclear. Though, it is noted that the city background sound, which was proven the most efficient masker, has an elevated range between 50 and 500 Hz compared to the other background sounds. Vice versa, the forest background sound, rated as the least efficient masker is seen to have the lowest energy in this range.

The main reason for adopting a magnitude production test-model was to allow for the listener to complete the test in a pace that felt natural and not stressful, this in order to get closer to the subjective experience. An indication that this method was successful was that the listeners adapted to the test procedure and spent longer time on the judgment of partially masked sounds to compensate for the more demanding task.

Naturally, it is important to take into consideration that since this test was carried out with headphones and in a laboratory environment under a restricted time-period, comparisons to the effect of long-term exposure of sound should be precautious. Nonetheless, it is our belief that the differences seen in this experiment will translate to the real world. A sound perceived as more annoying than the other in laboratory conditions, is likely to stay so under prolonged exposure, though the levels seen here may of course not be absolute.

In conclusion, this experiment highlights the effects of masking and show that there are benefits to explore these effects as they are shown to have a considerable effect in reducing the annoyance level. This is especially true for the threshold level of detectability, were we also saw the largest effects. What you cannot hear, does not bother you. This aspect can be beneficial to consider when planning for wind energy development near or in urban landscapes.

Acknowledgements

This paper has been performed in the SWIP project (New innovative solutions, components and tools for the integration of wind energy in urban and peri-urban areas) which received funding from the EU-FP7 Programme for research, technological development and demonstration under grant agreement No. 608554.

References

Berglund, B. P., A. Rankin, K. (1990). "Relationship between loudness and annoyance for ten community sounds." Environment International **16**.

Bolin, K. N., Mats E. Khan, Shafiquzzaman (2010). "The Potential of Natural Sounds to Mask Wind Turbine Noise." Acta Acoustica united with Acustica **96**(1): 7.

Borg, G. B., E (2001). "A new generation of scaling methods: level anchored ratio scaling." Psychologica **28**: 30.

Fastl, H. Z., W (1999). Psycho-acoustics Facts and Models,. Heidelberg, Germany, Springer: 203-233.

Miedema, H. M. and C. G. Oudshoorn (2001). "Annoyance from transportation noise: relationships with exposure metrics DNL and DENL and their confidence intervals." Environmental Health Perspectives **109**(4): 409-416.

Miedema, H. M. E. V., Henk (2004). "Noise annoyance from stationary sources: relationships with exposure metric day-evening-night level (DENL) and their confidence intervals." The Journal of the Acoustical Society of America **116**(1): 9.

Pedersen, E., et al. (2009). "Response to noise from modern wind farms in The Netherlands." The Journal of the Acoustical Society of America **126**(2): 634-643.

Pedersen, E., et al. (2007). "Living in the Vicinity of Wind Turbines — A Grounded Theory Study." Qualitative Research in Psychology **4**(1-2): 49-63.

Pedersen, E., et al. (2010). "Can road traffic mask sound from wind turbines? Response to wind turbine sound at different levels of road traffic sound." Energy Policy **38**(5): 2520-2527.

Pedersen, E. and K. P. Waye (2004). "Perception and annoyance due to wind turbine noise—a dose–response relationship." The Journal of the Acoustical Society of America **116**(6): 3460-3470.

Van den Berg, F., et al. (2008). "Project WINDFARMperception: Visual and acoustic impact of wind turbine farms on residents. Final report, FP6-2005." Science-and-Society **20**.

7th International Conference
on
Wind Turbine Noise
Rotterdam – 2nd to 5th May 2017

Wind Turbine Noise Dose Response – Comparison of Recent Studies

Isaac Old. RSG, 55 Railroad Row, White River Junction, Vermont, U.S.A..
isaac.old@rsginc.com
Kenneth Kaliski. RSG, 55 Railroad Row, White River Junction, Vermont, U.S.A..
ken.kaliski@rsginc.com

Summary
In the past few years there have been three major cross-sectional studies that have studied human response to sound from operating wind power projects. These are the Swedish and Danish studies compiled by Janssen, Eisses, and Perderson [1], the Japanese research study conducted by Kuwano, Yano, Kageyama, Sueoka, and Tachibanan [2], and the Canadian research study of Michaud, Feder, Voicescu, Marro, Guay, Denning, McGuire, Bower, Lavign, Murray, Weiss, and Van Den Berg [3]). While each of these studies has resulted in a dose-response curve, wind turbine sound levels have been represented with a different metric in each survey. Differences in these metrics has led to some confusion when the dose response curves have been compared or cited. The following paper uses knowledge from the literature on the relationship between different sound propagation modeling algorithms and sound level measurements, to compare all three curves.

Introduction
There have been three major wind turbine noise dose-response studies performed in the last 15 years.[1] Each of these studies, has compared either measured or predicted sound levels from modern upwind pitch-controlled wind turbines to the annoyance level of nearby residents, using similar measures of annoyance. The difficulty is that each study has used different sound level metrics and sound propagation modeling parameters to predict the sound level that residents are exposed to.

In trying to compare the results of these studies for use in practical purposes (e.g. in looking at appropriate noise limits), the different methodologies can lead to misunderstandings of what sound levels correspond to different levels of annoyance under a particular metric or averaging time.

[1] The study by Pawlaczyk-Luszczynska et al is not being included, due to a lack of included sound propagation modeling information.

This paper will look at each of the three studies, describing methods used to derive sound exposure, and the results. The results will then be normalized using a common ISO 9613-2 sound propagation modeling algorithm, assuming mixed ground porosity (G=0.5), which has been shown to yield accurate results compared to measurement results of hourly equivalent average sound levels ($L_{Aeq(1\text{-}hour)}$) over flat or constant gradient ground ([4] and [5]).

Dose Response Study Descriptions

Swedish and Dutch Studies

The Swedish and Dutch dose response studies were reported by Eja Pedersen along with other authors. There were three studies in total.

The first was conducted in Sweden in the year 2000 [6]. The study area was flat, rural and agricultural. Sound levels were predicted using an older version of the Swedish Environmental Protection Agency (SEPA) method.

The second study was conducted in Sweden in 2004 [7]. In this case the study areas were more varied in both terrain and population density. The SEPA method was also used to predict sound levels at residences.

The third study was conducted in 2007 in the Netherlands [8]. The modeling methodology was different, using ISO 9613-2, with fully absorptive ground (G=1) and a receiver height of 5 meters. The study compared modeling results in the previous two studies and found that the average difference between the SEPA method and ISO 9613-2 was 0.3 dB. Results from all three studies were combined to compare with dose-response curves of transportation noise. To make the comparison, predicted sound levels were converted to Day-Evening-Night equivalent average sound levels (L_{DEN}), using work by Van Den Berg predicting annual average wind turbine sound levels [9].

Janssen et al. have since derived dose-response equations for the three studies [1].

Japanese Study

A dose response study from Japan used a different approach for sound propagation modeling [2]. Instead of using an established modeling algorithm, the study derived an equation from sound level measurements, performed with a 0.2 meter height microphone. The derived equation does not take into account sound propagation from all turbines in the study area, but only the closest turbine. The metric measured at the site was the "$L_{Aeq,n}$", which is the average of 10-minute periods measured throughout the nighttime with the turbine at full power. The equation works out to a sound level decay rate of 6.7 dB per doubling of distance and a typical turbine sound power of about 100 dBA. No mention is made about whether measurements were made upwind, downwind, or crosswind of the residences, so it is assumed that no distinction was made based on wind direction.

Health Canada Study
Health Canada conducted a dose-response study that compared modeled sound levels to both subjective and objective responses. To predict resident sound exposure, the study used ISO 9613-2, with G=0.7 ground factor (70% porous ground) and measured wind turbine sound powers (as opposed to using manufacturer-published sound powers) [3]. Residences were modeled at a 4-meter height.

Analysis Procedures
To compare results from all three studies, attempts were made to normalize the sound levels used for resident sound exposure to a common metric. The metric used is the A-weighted hourly equivalent average sound level ($L_{Aeq(1\text{-hour})}$), as is modeled in ISO 9613-2 with G=0.5 ground attenuation with 4-meter height receivers. This set of metrics has been shown to accurately model 1-hour equivalent average sound levels over flat or constant gradient terrain [5] [4]. These parameters are often employed for the pre-construction modeling of wind projects.

As noted earlier, one of the Swedish/Danish and the Canadian studies used ISO 9613-2 and two of the Swedish/Danish studies use the SEPA algorithm for the prediction of sound levels at each survey respondent. Janssen et al., in derivation of the dose response curves did not try and recalculate results obtained by the SEPA method with ISO 9613-2, but the differences between the two methods was found to be rather small at 0.3 dB [8].

The differences in modeling results between these studies is due to the particular ground factors and receiver heights used in ISO 9613-2. The Swedish/Dutch studies assumed a receiver height of 5 meters and ground factor of G=1 (100% porous) while Health Canada assumed a receiver height of 4 meters and ground factor of G=0.7.

To determine the difference in sound level predictions between these studies and our reference parameters (G=0.5, 4-meter receiver height), an ISO 9613-2 sound propagation model was set up for a large distributed wind farm, assuming flat ground similar to most of the research study areas. The model contained 203 wind turbines with 266 receivers which ranged from 360 meters to 5,700 meters from the nearest wind turbine. We compared the model results using the Canadian and Swedish/Danish parameters to the results using our reference parameters. We then took the median of the difference between models over all of the receivers. Our results showed that modeled levels in the Swedish and Dutch studies would be 1.6 dB lower than our reference model. The Health Canada modeled results would be 0.8 dB lower than our reference model.

The Japanese study used a different sound calculation method than the other studies. Nighttime sound level measurements were taken, from which average results were correlated with the distance to the closest turbine. Emissions from multiple turbines are not differentiated and upwind/downwind effects were not taken into account. Measurements were made with a microphone height of 0.2 meters, minimizing the influence of wind. Again, using our reference model, we found that the modeled difference between receiver heights of 0.2 meters and 4 meters was a median of 2.6 dB. The lack of wind direction effects would contribute to lower predicted sound levels. Using data published by Fukushima et al. on the directionality of wind turbine sound, we determined that sound levels will be 2.8 dB lower than the assumption of a downwind-only direction, with a difference of 8 dB between full downwind and cross-wind conditions [10]. Since the other studies, using ISO 9613-2 and SEPA calculation algorithms, assumed downwind propagation, the Japanese algorithm's assumption will result in lower overall predicted sound levels.

For the Japanese study, we use the dose response curve found in Figure 12 of Reference [2].

A comparison of the SEPA, ISO 9613-2, and Kuwano et al sound propagation equations is shown in Figure 2.

For the Japanese and Canadian studies, response to sound level was reported in 5 dB bins, whereas Janssen et al.'s dose response curves allow calculation of does at any level. For comparison, the midpoint of each sound exposure bin for the Japanese and Canadian studies was considered the reference sound level. So for the 40 to 45 dBA bin, the reference sound level was 42.5 dBA.

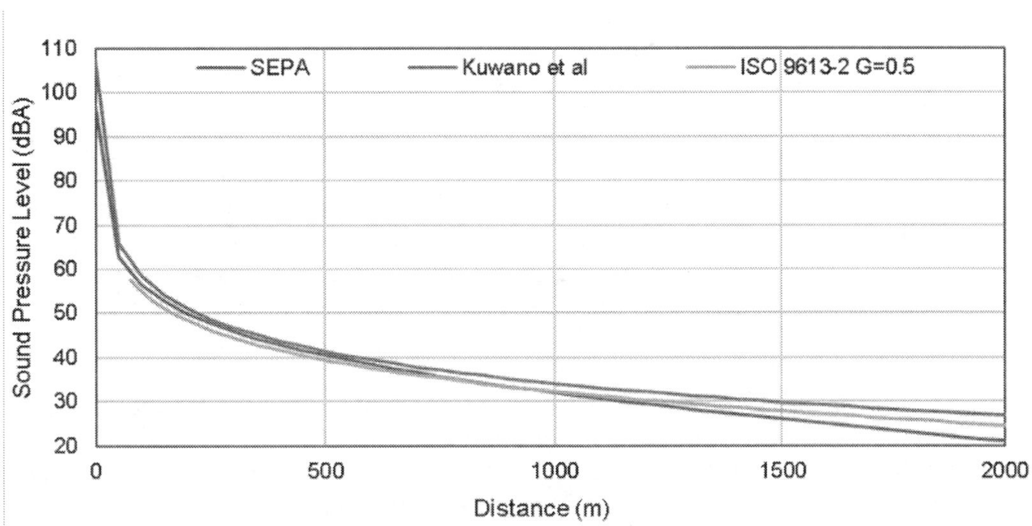

Figure 1: Comparison Between Decay Rate of Kuwano et al., ISO 9613-2, and SEPA Sound Propagation Equations (Assuming 105 dBA Sound Power for SEPA and ISO 9613-2 equations)

Results

Dose response curves from all three studies, with a normalized sound metric are shown in Figure 1. The normalized metric is a one-hour Leq, modeled using the ISO 9613-2 methodology with G=0.5 and a 4-meter receiver height.

For both the Swedish/Dutch and Canadian studies, results are given for both indoor and outdoor annoyance. The results for these studies show good agreement. The Japanese study shows higher levels of annoyance than the others at lower sound levels and lower levels of annoyance at higher sound levels.

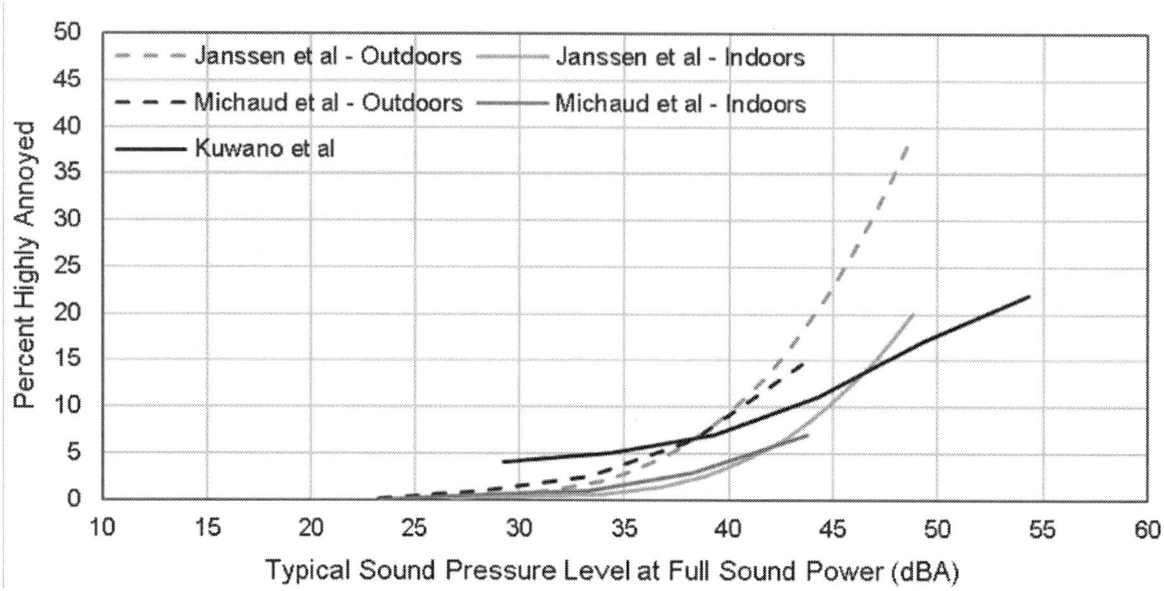

Figure 2: Wind Turbine Noise Does Response Curves Normalized to 1-hour L_{eq}, G=0.5, 4-meter height

Discussion

Our results show that the Swedish/Dutch and Canadian normalized dose-response curves show good agreement, while the normalized Japanese curve has higher overall levels of annoyance at low sound levels, and then a lower level of annoyance at higher sound levels. There are several possible reasons for the differences, centering on the method used for sound level prediction and the study areas that were used.

The sound exposure equation used by the Japanese study is based on measurements that were performed near individual turbines in the study. For generation of the dose-response relationship, the expected sound level dose is then derived from the distance from a residence to the closest turbine. Both the Swedish/Dutch study and Health Canada study modeled sound levels from all nearby turbines. Most of the sites featured multiple turbines, so this approach will tend to slightly overestimate sound levels at single-turbine sites and underestimate sound levels at multiple-turbine sites, particularly relative to the other studies, which assume downwind propagation from all turbines, even for "wind park" turbine layouts.[2] Taking into account multiple turbines would likely shift the curve to the right. Background sound could also have influenced the Japanese study's background sound level measurements. This would have the effect of inflating

[2] Even though the measurements were conducted within a wind farm, most of the measurements were collected close to the turbine under test, so sound from the closest turbine would dominate.

wind turbine sound levels, particularly at lower turbine-only levels. Removal of all background sound would have the effect of compressing the curve.

The Canadian study was conducted in flat to rolling terrain. The primary study area, in the Province of Ontario, largely contains distributed wind farms, so residences may be located with turbines in multiple directions. The secondary study area, the Province of Prince Edward Island, is largely flat and the turbine layouts are both distributed and centralized. One of the Swedish and the Dutch studies were located in primarily flat terrain [6] [7] and the second Swedish study was located in mixed terrain [8]. In contrast, most of the areas in the Japanese study were mountainous and several of the areas had the turbines located on ridgetops [11]. Mountainous areas can have lower wind speeds at receivers. Ridgetop projects are also more likely to have arrays where all turbines are either up- or down-wind of a given residence. This will result in higher maximum sound levels, but lower long-term sound levels [12].

Many of the factors that contribute to wind turbine noise annoyance such as: attitude to the noise source, noise sensitivity, and concern over landscape littering may not be static between cultures, contributing to varying responses. This is at least partially demonstrated in the Canadian study, where results are segmented between provinces. Those results showed that that wind turbine noise was considered more annoying on average in Ontario than on Prince Edward Island.

Conclusions

This paper aligned results from three recent dose-response studies of wind turbine noise to a common sound level metric in the attempt to both allow comparison of results with short term sound level metrics and allow comparisons between studies. The three studies included:

- The Swedish and Dutch studies performed by Pedersen et al and Janssen et al.'s dose response curves derived from results of these studies;
- Health Canada's Canadian study, described in Michaud et al; and
- The Japanese Study, described by Kuwano et al.

Using information known about the relative sound levels predicted by different sound propagation modeling algorithms, results from all three studies were normalized to sound propagation modeling results using ISO 9613-2, with mixed ground (G=0.5) and a 4-meter receiver height. These parameters have been shown to be accurate in predicting a one-hour equivalent average sound level from wind turbines.

Our conclusions can be summarized as follows.

- The normalized dose-response curves of the Swedish/Dutch and Canadian studies showed good agreement,

- The normalized Japanese dose-response curve showed higher overall annoyance at lower turbine-only sound levels, with annoyance increasing at a slower rate than the other studies.
- This difference may be due to the Japanese study's use of a sound level prediction algorithm that did not take into account spectral turbine sound power levels or the number of turbines near a residence, but rather only on the distance from a residence to the closest turbine.

Bibliography

[1] S. Janssen, H. Vos, A. Eisses and E. Pedersen, "A Comparison Between Exposure-response Relationships for Wind Turbine Annoyance and Annoyance Due to other Noise Sources," *Journal of the Acoustical Society of America,* pp. 3746-3753, 2011.

[2] S. Kuwano, T. Yano, T. Kageyama, S. Sueoka and H. Tachibana, "Social Survey on Wind Turbine Noise in Japan," *Noise Control Engineering Journal,* pp. 503-520, 2014.

[3] D. Michaud, K. Feder, S. Keith, S. Voicescu, L. Marro, J. Than, M. Guay, A. Denning, D. McGuire, T. Bower, E. Lavigne, B. Murray, S. Weiss and F. Van Den Berg, "Exposure to Wind Turbine Noise: Perceptual Responses and Reported Health Effects," *Journal of the Acoustical Society of America,* pp. 1443-1453, 2016.

[4] RSG, "Massachusetts Study on Wind Turbine Acoustics," Massachusetts Clean Energy Center, 2016.

[5] T. Evans and J. Cooper, "Comparison of Predicted and Measured Wind Farm Noise Levels and Implications for Assessments of New Wind Farms," *Acoustics Australia,* pp. 28-36, 2012.

[6] E. Pedersen and K. Persson-Waye, "Perception and Annoyance Due to Wind Turbine Noise - a Dose-response Relationship," *Journal of the Acoustical Society of America,* pp. 3460-3470, 2004.

[7] E. Pedersen and K. Persson-Waye, "Wind Turbine Noise, Annoyance and Self-Reported Health and Well-being in Different Living Environments," *Occupational and Environmental Medicine,* pp. 480-486, 2007.

[8] E. Pedersen, F. Van Den Berg, R. Bakker and J. Bouma, "Response to Noise From Modern Wind Farms in the Netherlands," *Journal of the Acoustical Society of America,* pp. 634-642, 2009.

[9] F. Van Den Berg, "Criteria for Wind Farm Noise: Lmax and Lden," in *Acoustics,* Paris, 2008.

[10] A. Fukushima, K. Yamamoto, H. Uchida, S. Sueoka, T. Kobayashi and H. Tachibana, "Study on the Amplitude Modulation of Wind Turbine Noise: Part 1 - Physical Investigation," in *Internoise*, Innsbruck, Austria, 2013.

[11] H. Tachibana, H. Yano, A. Fukushima and S. Sueoka, "Nationwide Field Measurements of Wind Turbine Noise in Japan," *Noise Control Engineering Journal,* vol. 62, no. 2, pp. 90-101, March-April 2014.

[12] K. Kaliski and E. Duncan, "Calculating Annualized Sound Levels for a Wind Farm," in *Proceedings of Meetings on Acoustics*, Baltimore, Maryland, 2010.

[13] D. Michaud, K. Feder, S. Keith, S. Voicescu, L. Marro, J. Than, M. Guay, A. Denning, T. Bower, P. Villeneuve, E. Russel, G. Koren and F. Van Den Berg, "Self-reported and Measured Stress Related Responses Associated with Exposure to Wind Turbine Noise," *Journal of the Acoustical Society of America,* pp. 1467-1479, 2016.

[14] M. Bliss, "Environmental Noise Assessment of Proposed Wind Farms Using Annual Average Ldn," in *Fifth International Conference on Wind Turbine Noise*, Denver, Colorado, 2013.

[15] S. Keith, K. Feder, S. Voicescu, V. Soukhovtsev, A. Denning, J. Tsang, N. Broner, T. Leroux, W. Richarz and F. Van Den Berg, "Wind Turbine Sound Pressure Level Calculations at Dwellings," *Journal of the Acoustical Society of America,* pp. 1436-1442, 2016.

[16] K. Kaliski and E. Duncan, "Propagation Modeling Parameters for Wind Power Projects," *Journal of Sound and Vibration,* pp. 12-15, December 2008.

7th International Conference
on
Wind Turbine Noise
Rotterdam – 2nd to 5th May 2017

Impact of noise from suburban wind turbines on human well-being

Fei Qu School of Architecture, University of Sheffield, Sheffield, S10 2TN, UK. fei.qu@sheffield.ac.uk
Aki Tsuchiya School of Health and Related Research, University of Sheffield, S1 4DA, UK. a.tsuchiya@sheffield.ac.uk
Jian Kang School of Architecture, University of Sheffield, Sheffield, S10 2TN, UK. j.kang@sheffield.ac.uk

Summary

This study aims to investigate the relationship between exposure to wind turbine noise and human well-being. While most existing studies have focused on rural settings, this study concerns people living in the vicinity of suburban wind turbines, based on a field study around three suburban wind farms in the UK. The main sample (n=261) was asked to: rate their general health and well-being; identify annoyance by environmental nuisances including wind turbine noise; and attribute health concerns to the noise. A separate control group (n=96) recruited in the same areas was asked similar questions, but without attributing health problems to wind turbines. Noise-mapping techniques were used to obtain A-weighted wind turbine sound pressure levels (SPLs) at each respondent's dwelling. It was found that the odds ratio of being annoyed by wind turbine noise increased with each dB increase in SPLs (OR=1.2; 95% CI: 1.11 to 1.40), with no significant difference between the two groups. Respondents who were annoyed by wind turbine noise were significantly more likely to report difficulty falling asleep and sleeping less deeply at night. Positive associations were found between SPLs and adverse health effects including nausea and dizziness. However, the main sample reported significantly less health problems and better self-assessed general health. More health problems were found to be related to SPLs in the control group (whose health and well-being were measured without reference to wind turbine noise). Moderating factors, including noise sensitivity, sustainability in life, negative attitude to wind turbines, and visibility, were found to have significant impacts on annoyance and well-being. To conclude, dose-response relationships are found between wind turbine noise and annoyance in suburban environments. Respondent's knowledge of the motivation of the survey leads to more optimistic responses regarding adverse health impact, a methodological finding which should be noted in future research.

1. Introduction

The potential adverse impacts of wind turbine noise have attracted substantial attention. Dose-response relationships between exposure to wind turbine noise and the percentage of annoyed residents have been found in five studies conducted in Sweden, the Netherlands, Poland and Canada, respectively (Michaud et al., 2016; Pawlaczyk-Łuszczyńska, Dudarewicz, Zaborowski, Zamojska-Daniszewska, & Waszkowska, 2014; Pedersen, van den Berg, Bakker, & Bouma, 2009; Pedersen & Waye, 2004, 2007). In addition, a dose-response relationship between self-reported sleep disturbance and noise exposure was found in three studies (Bakker et al., 2012; Pawlaczyk-Łuszczyńska et al., 2014; Pedersen, 2011). Other health-related effects such as psychological distress were found to be associated with wind turbine noise, with noise annoyance as a mediator (Pedersen, 2011; Shepherd, McBride, Welch, Dirks, & Hill, 2011).

Much of the existing research has focused on rural settings and there is a gap in the research for wind turbines in urbanised environments. While recent studies have addressed the future potential wind energy projects in urban environments in terms of wind behaviour and energy yield (Cooney, Byrne, Lyons, & O'Rourke, 2017; Ishugah, Li, Wang, & Kiplagat, 2014; Millward-Hopkins, Tomlin, Ma, Ingham, & Pourkashanian, 2013), very little work has been carried out on the noise impact on humans in urban residential areas. Two studies looking at both rural and suburban environments have found that the risk of being disturbed by wind turbine noise is pronounced in quiet areas compared to noisy areas (Bakker et al., 2012; Pedersen et al., 2009). The question remains whether the absence of significant relations between noise exposure and annoyance is because noisier environments better mask the wind turbine noise, or because people living in noisier areas have adapted more (Bakker et al., 2012). There is a need to investigate the effect of wind turbine noise on the residents in urbanised environments with high population density and different background noise intensity.

In terms of the methods used in previous field surveys, similar questionnaires were used across four studies to assess the resident's responses to wind turbine noise (Pawlaczyk-Łuszczyńska et al., 2014; Pedersen et al., 2009; Pedersen & Waye, 2004, 2007). This questionnaire included a number of questions on wind turbines, and thus it is possible that respondents could see the objective of the surveys. In other studies, the questionnaire was explicit that its objective was to investigate adverse health effects potentially associated with wind turbines (Michaud et al., 2016; Nissenbaum, Aramini, & Hanning, 2012). However, such questionnaires may introduce self-reporting bias into the survey results, and there is a need for questionnaires that minimise self-reporting bias.

The aim of this study is therefore to investigate the relationships between exposure to wind turbine noise, resident's response to the noise, and their health and well-being. A-weighted sound pressure levels (SPLs) at individual respondents' dwellings were calculated using noise mapping, and were linked to questionnaire responses to examine whether the dose-response relationships also exist in urbanised settings. Self-reporting bias was minimised by using two variants of the questionnaire: one with, and another without, specific questions on wind turbines. These two variants of the questionnaire allowed an investigation on whether knowledge of the motivation of the survey affects respondents' reporting of health impacts.

2. Method

The survey investigated the relationship between exposure to wind turbine noise and human well-being. Paper questionnaires were delivered to select residents of three sample sites across the UK in the vicinity of large wind turbines in suburban-urban settings. A-weighted sound pressure levels (SPLs) were calculated using noise mapping techniques, for the most exposed façade of each target dwelling. The relationships between SPLs and human health and well-being were investigated through quantitative analysis of the questionnaire data. Possible self-reporting bias associated with asking people for their perceived causes of health problems was minimised by recruiting a separate control group without any focusing on wind turbine noise. Differences between the main and control groups in relation to reported health and well-being were examined.

2.1 Questionnaire

The questionnaire asked about responses to wind turbine noise, sleep disturbance, the prevalence of health-related problems, and general health. It also measured socio-economic status and attitudes to environmental causes. Generally, the questionnaire had three sections in the following order: (i) well-being and health, (ii) evaluation of the neighbouring environment (including wind turbine noise), and (iii) socio-demography and their dwelling. Most questions were drawn from established national surveys of health and well-being such as the British Household Panel Survey (BHPS), with several modifications to fit this survey.

a) Questionnaire variants

In order to minimise potential self-reporting bias caused by the knowledge of the motivation of the survey, two variants of the questionnaire were generated. In each sample site, one group received "Questionnaire Variant 1", which included explicit questions on the impacts of the local wind turbines on the respondent's well-being, such as: rating their general health and well-being; identifying annoyance by environmental nuisances including wind turbine noise; reporting health problems they experienced; describing the sound of wind turbines and indicating their attitude to wind turbines. This allowed respondents to directly attribute any well-being concerns they have to the presence of the local wind power project. A separate control group recruited in the same areas received "Questionnaire Variant 2", which focused on well-being and health, but without attributing any problems to wind turbines. Reference to wind turbine was removed, except in one question on awareness of and annoyance with various environmental nuisances, one of which was wind turbine noise. Other questions that did not mention wind turbines were kept identical in Variants 1 and 2.

b) Outcome variables

The main outcome variables included awareness of and annoyance with wind turbine noise, self-reported sleep disturbance, prevalence of specific health problems, general health, and subjective well-being. In this questionnaire, annoyance was assessed in four questions. The first question (Q1) was adapted from a previous survey (Pedersen & Waye, 2004), in which respondents were requested to state their responses to a seven environmental nuisances, including wind turbine noise. Respondents were asked to first indicate whether they noticed any of the nuisances, and if yes, to rate their degree of annoyance on a 5-point scale from "not at all" to "extremely". In Variant 1, the annoyance with wind turbine noise was further examined in three questions, addressing annoyance overall (Q2), outdoors and indoors (Q3), as well as in specific circumstances (Q4). Sleep disturbance in this survey was measured without making reference to noise and was kept identical in Variants 1 and 2. The question had a number of items on sleep disturbance, including "hard to fall asleep", "sleep less deeply", "lie awake for a while" and so on. Respondents were required to choose all the statements that described their sleep. The purpose was to identify the relationship between wind turbine noise and different degrees of sleep disturbance, which have been reported to be affected by environmental noise in various studies (Muzet, 2007), but have not been examined in existing wind turbine noise studies. Further, participants were asked to indicate whether they experienced the listed ten physiological and psychological health problems during the past week, including headache, dizziness, ear discomfort, cardiovascular disease, tension and edginess, lack of concentration. In Variant 1, respondents were then given the opportunity to indicate whether they felt wind turbine noise might be the cause. The response scale was configured as "yes", "possibly", "no", and "I don't know". In addition, all respondents were asked to self-assess their general health on a 5-point scale from poor to excellent.

c) Moderating variables

As moderating variables, the survey included questions on socio-demographic, personal/attitudinal, and architectural factors. Firstly, socio-demographical factors such as age, sex, longstanding illness, and household income that were found to influence noise annoyance and health in previous studies were assessed (Bluhm, Nordling, & Berglind, 2004; Dolan, Peasgood, & White, 2008; Fields, 1993; Frijters & Beatton, 2012). Second, personal/attitudinal questions addressing personal noise sensitivity, environmental sustainable lifestyle, and attitude to the noise source were added in line with previous studies (Weinstein 1980; Guski 1999; Job 1999). Noise sensitivity was measured in one question with two items: a) "I find it hard to relax in a place that's noisy" and b) "I get used to most noises without much difficulty", assessed on a 6-point scale from agree strongly to disagree strongly. The question was drawn from an established 21-items noise sensitivity questionnaire (Weinstein, 1978), shortened in this questionnaire following (Benfield et al., 2014). A question on attitudes to environmentally sustainable lifestyles (sustainability for short) was adapted from two questions in the BHPS. Respondents' attitude to wind turbines was also assessed using eight antonym adjectives to describe wind turbines, drawn from a previous study (Pedersen & Waye, 2004). A question identified the respondent's financial stakes in the wind farm. Furthermore, three questions were on architectural factors, such as the number of bedrooms in the dwelling, the housing type, and the orientation of the dwelling, which were found to have effects on resisting the wind turbine noise in previous studies (Qu, Kang, & Tsuchiya, 2015). Furthermore, dwelling-related questions measured visibility of the wind turbine, length of residency, and ownership of the dwelling.

2.2 Study sites and sample

The target population of the survey was defined as residents who lived within two kilometres of modern wind turbine(s) in suburban areas in the UK. Three wind farm sites were selected as study areas. Participants were selected using multi-stage sampling, and questionnaires were mailed or door dropped to the sampled individuals in late 2014. A small number (N=51) of the questionnaires were delivered face-to-face.

a) Study area

Three typical suburban wind farm sites were selected to concentrate the sample in three clusters of households for further sampling. Data on operational onshore wind energy projects in the UK were obtained from the UK wind energy online database (UKWED, n.d.) and a map of each wind farm site on Google Earth. The selection of study sites was mainly based on two criteria: (1) Each wind turbine on the site should be a modern large turbine with power capacity more than 2MW. (2) The wind farm site should have a sufficient number of residents living within two kilometres, ideally in a suburban context with densely populated residences. The features of the selected sites are summarised in Table 1: one is in a suburban area in East Midland (site A), another is in the suburb of a large city in Scotland (site B), and the third is in a town by the eastern coast of England (site C). The wind turbines were large and modern, with tower heights between 80 and 85m. All sites could be classified as suburban-urban with high population densities (2000-4000/km^2).

Table 1. Characteristics of the study sites

Site	Characteristics	Turbine model	Location
A. Newthorpe Sewage Treatment wind farm	• Surrounded by 3 suburban areas • A highway and a small railway are present • Highly visible • Semi-detached dwellings	1 turbine Nordex N100 3.4 MW / 80m Year 2014	Nottinghamshire, Midlands of England
B. Michelin Tyre Factory wind farm	• Inside industrial area in the city • Proximity to suburban residential areas • Relatively low traffic noise	2 turbines Enercon E82 2.3 MW / Year 2006	City of Dundee, Scotland
C. Lowestoft wind farm	• At seaside with strong wind • Surrounded by highly populated urban area • Long terrace dwellings • Occasionally shut down	1 turbine Vestas2 NM923 2.75 MW / 80m Year 2005	Lowestoft, Suffolk, East England

Source: UKWED, site visits

b) Study sample

To ensure that residents exposed to different levels of noise were adequately represented in the sample, disproportionate stratified sampling was applied with wind turbine noise levels as the strata. Preliminary noise modelling was carried out for the three sites to predict the distributions of wind turbine noise, considering different wind turbine models and terrain conditions. Each site was then divided into four noise strata with 5 dB(A) intervals. Addresses were randomly selected from the edited version of the electoral register, by stratum, for each of the two questionnaire variant. Where there were several adults at the same address, one individual was selected at random. A total of 2971 individuals were sampled (2238 for Variant 1 and 733 for Variant 2), where the sample size of each noise stratum was balanced across the three sites. A detailed description of the sampling strategy is available from the authors by request.

2.3 Noise exposure calculations

To examine the spatial distribution of wind turbine noise levels in studied sites, noise maps were calculated using the software package CadnaA (DataKustik GmbH, 2006). The map and topographical information of the study sites were obtained from the EDINA Ordnance Survey Digimaps in the UK (Ordnance Survey, 2013). A-weighted sound pressure levels (SPLs) on the most exposed façade (maximum façade exposures) of target buildings were obtained. The calculation in the software was based on the ISO 9613 (International Standards Organisation, 1996) sound propagation standard. Noise emission from the wind turbine was calculated under downwind conditions. In line with the IEC 61400-11 standard (IEC 61400-11, 2012), the wind turbine was simulated as a point source at hub height. The spectrum of the point source was set based on that given by the manufacturer, where the SPLs are relatively high at low-frequencies and attenuate with octave. The ground absorption was set to 0.5 in accordance with the Good Practice Guide in the UK (Perkins et al 2013). Temperature was set to 10 ℃, relative humidity to 70% for atmospheric absorption, consistent with common practice (Keith et al 2016). The reflection order by buildings was set to 3, based on a previous study (Kang, 2006). The façade-receiver distance was set to 0.05m. The calculations using the above method have been verified by field measurements focusing on wind turbine noise attenuation around buildings. The detailed validation process is available by request.

2.4 Statistical analysis

Statistical analyses were performed using SPSS version 22.0 (Statistics, 2009). Descriptive statistics were provided for the characteristics of the participants. Response to wind turbine noise was presented as proportions of the number of respondents in each 5 dB(A) stratum with 95% confidence intervals (CI). Annoyance measured on verbal and ordinal scales were dichotomised, with slightly annoyed to extremely annoyed classified as "annoyed". In the analysis of questions with multiple items, such as sleep disturbance which had six, each item was treated as a variable such as "difficulty in falling asleep", "sleep less deeply" and "lie awake". In the analysis of variables with two questions, such as sensitivity and sustainability, a derived variable was created on a 6-point ordinal scale computed by the numeric sum of the two original variables. Oblique rotated principle axis factor analysis was employed to extract the oblique factor underlying the 14 inter-related adjectives for the respondents' attitudes to wind turbine noise.

Differences in distribution of observations and respondent characteristics between Variants 1 and 2 were tested using Pearson's chi-square for categorical variables, and *t*-test for continuous variables, with *p*-values below 0.05 considered statistically significant. Differences between the two variants in outcome variables with ordinal scales (e.g. general health) were tested with the Mann-Whitney's *U* test. Differences in distribution of respondent characteristics across four sound categories were also examined using Pearson's chi-square for categorical variables or analysis of variance (one-way ANOVA) for continuous variables.

Binary logistic regression was applied to analyse the effects of noise on annoyance, sleep disturbance, and adverse health problems. The main explanatory variable, noise exposure, was represented by A-weighted SPL, calculated for the most exposed façade of a dwelling. Preliminary regression analyses were carried out to select the variables for the final models presented in this paper by exploring the influence of subjective factors, where possible moderating factors were added to the regression model one-by-one, always keeping A-weighted SPL in the model. Though the site dummies did not have any influence in some preliminary regressions, these variables were included in the analyses to exclude bias from social and acoustic differences between areas. Odds ratios (ORs) were reported for each variable with 95% confidence intervals (CIs), with *p*-value below 0.05 considered statistically significant. The Nagelkerke psudo-R^2 was applied as a measure of explained variance. Hosmer-Lemesow goodness-of-fit [$p_{(H-L)}$] was presented for each logistic regression model, with *p*-value >0.05 indicating no statistically significant difference between the modelled and the observed data.

3. Results

3.1 Descriptive statistics of respondents and noise exposure

a) Response rate and noise exposure

The numbers of respondents of the two questionnaire variants were 261 and 96, respectively, with a total of 357. The overall response rate was 12.0%, similar across variants 1 and 2. Overall, there were fewer respondents with noise exposures over 40 dB(A) (15%), while the proportions of respondents in the other three noise intervals were similar (26-29%). There was no statistically significant association between the four noise intervals and the two variants (X^2=3.649, p=.302). A Mann-Whitney *U* also indicated that the distribution of respondents in four noise groups was the same across questionnaire variants [$U(n_1$=259, n_2=91)=10962.5, p=.304].

b) Study group characteristics

The mean age in the study population was 56 (SD = 17.7), and 49% were male. Most of the respondents were employed (43%) or retired (41%). Over half (55%) of the respondents reported to be sensitive to noise based on the two questions on sensitivity. Overall, 49% of the respondents lived in detached or semi-detached houses, while 34% of the respondents lived in mid-terrace or end-of-terrace dwellings. In total, 68% of the respondents privately owned their accommodation, while the remaining lived in rented dwellings.

The characteristics of the respondents were similar across the two variants. No statistically significant differences were found in age, gender, education, household income, noise sensitivity, or housing type. A statistically significant difference was found in long standing illness (X^2=4.826, p=.036), with 39% in Variant 1 and 48% in Variant 2. In addition, significantly more respondents in Variant 2 had no qualification (X^2=9.479, p=.050).

Since the two variants look reasonably similar, in the following analysis, effects of wind turbine noise on response to the noise, sleep, health problems and well-being will be examined by pooling the data across the two variants. Variables regarding long-standing illness and highest qualification that are significantly different across variants will be controlled for in regressions.

c) Characteristics related to wind turbine noise

There was a significant difference between different age across the four sound categories [one-way ANOVA F (3, 352)=9.879, p=.000], with a significant quadratic trend [F (1, 352)=19.601, p=.000], indicating that the respondents in the lowest and highest sound categories were significantly older than those in the middle sound categories. Significantly more respondents in the higher sound categories were not employed or retired (X^2=22.275, p=.008). Respondents in the lower sound categories had more bedrooms [F (3, 343)=10.512, p=.000] and were more likely to be in detached or semi-detached houses, while those in the higher sound categories were more likely to be in a terrace house or a flat (X^2=37.246, p=.000). A statistically significant correlation was found between sound categories and ownership of the accommodation (X^2=30.163, p=.003), which decreased with higher sound categories. No statistically significant differences in gender, long-standing illness, education, or noise sensitivity were found across sound categories.

3.2 Response to wind turbine noise

Overall, 16% of the respondents (*n*=59) noticed the wind turbine noise and 11% of the respondents (*n*=39) were annoyed by it. Of the Variant 1 respondents, 17% (*n*=45) indicated they were annoyed outdoors and 10% (*n*=25) were annoyed indoors. Of those annoyed outdoors, 56% (*n*=25) were slightly annoyed, 38% (*n*=17) were annoyed moderately and very, only 7% were extremely annoyed. The adjectives that were agreed to by the most respondents were environmental friendly (71%), followed by efficient (41%), necessary (38%), and harmless (37%). "Ugly" was the most often selected among the negative adjectives (23%), while "pretty" was selected much less often (6%). The most frequently reported nuisances that were noticed by respondents were traffic noise (40.6%) and noise from neighbours (38.4%); while wind turbine noise was noticed by 16.6% of the respondents. The least noticed among all nuisances included odor, bugs/pests/vermin, pollution/grime/dust, and the noise from traffic and neighbours. Similar proportions were observed when assessing annoyance with environmental nuisances. Wind turbine noise was less frequently reported as annoying than all other listed environmental nuisances.

Dose-response relationships were found between categories of wind turbine noise and percentage of respondents who noticed and were annoyed by them. Table 2 shows the proportion of respondents who are aware of and annoyed with wind turbine noise by categories of noise exposures. Taking both Variants 1 and 2, the proportion of respondents who noticed wind turbine noise increased from 5% (n = 5, 95%CI: 1%-11%) at the SPL category below 30 dB(A) to 47% (n = 25, 95%CI: 33%-61%) at the SPL category above 40 dB(A). The proportion of those annoyed by wind turbine noise also increased with sound category, from 3% (n=3, 95%CI: 0%-6%) in the lowest to 30% (n=16, 95%CI: 17%-43%) in the highest. In terms of response to wind turbine noise in different circumstances, more respondents noticed and were annoyed by the noise when the wind was strong or at night, while fewer noticed and were annoyed when they were inside the dwelling with the windows closed. Chi-square tests show that responses to wind turbine noise were significantly different between sound categories.

Table 2. Awareness of and annoyance with wind turbine noise related to sound exposures shown as proportion within each SPL category with 95% CI.

Respondent [Percentage (95% CI)]	Total	Maximum sound pressure levels at dwelling [dB(A)] <30 (*n*=114)	30-35 (*n*=102)	35-40 (*n*=90)	>40 (*n*=53)	Chi-square test of difference between noise categories
Variant 1+2						
Noticed among other nuisances	16 (13-20)	5 (1-11)	12 (6-19)	20 (12-29)	47 (33-61)	X^2=45.056, *p*=**.000**
Annoyed among other nuisances	11 (8-15)	3 (0-6)	8 (3-14)	13 (7-21)	30 (17-43)	X^2=24.598, *p*=**.000**
Variant 1						
Annoyed overall	12 (8-16)	1 (0-4)	9 (3-16)	20 (10-31)	25 (12-39)	X^2=20.042, *p*=**.000**
Annoyed outdoors	16 (12-21)	4 (0-9)	14 (6-23)	22 (12-32)	35 (20-50)	X^2=20.950, *p*=**.000**
Annoyed indoors	9 (6-13)	3 (0-7)	5 (0-10)	15 (7-25)	23 (10-37)	X^2=16.255, *p*=**.001**
Noticed when wind is strong	12 (8-17)	3 (0-7)	11 (4-18)	12 (4-21)	35 (20-51)	X^2=31.776, *p*=**.000**
Noticed when inside with window closed	3 (1-5)	0	2 (0-5)	2 (0-6)	13 (3-24)	X^2=15.778, *p*=**.001**
Noticed when at night	10 (6-14)	3 (0-7)	9 (3-17)	12 (4-21)	23 (11-37)	X^2=15.067, *p*=**.002**
Annoyed when wind is strong	9 (5-12)	1 (0-4)	9 (3-16)	7 (2-14)	25 (11-40)	X^2=24.735, *p*=**.000**
Annoyed when inside with window closed	2 (0-4)	0	2 (0-5)	2 (0-6)	8 (0-17)	X^2=7.871, *p*=**.049**
Annoyed when at night	6 (3-9)	1 (0-4)	6 (1-12)	8 (2-16)	13 (3-24)	X^2=9.381, *p*=**.025**

A series of bivariate binary logistic regression was used to examine the influence of subjective factors on being annoyed by wind turbine noise. It was found that the odds of being annoyed by wind turbine noise were not statistically different between questionnaire variants and sites. Gender, income, long-standing illness, noise sensitivity, and architectural factors were not significantly associated with being annoyed. Of respondent's attitude to wind farm project measured on 14 adjectives, five factors (utility, appearance, necessity, efficiency, and environmental impact of the wind turbine) were identified and added to the regression model one by one. Only negative attitudes to environmental impact (i.e. not environmental friendly, dangerous, ugly) were significantly associated with annoyance.

Results of the final regression models are shown in Table 3, where Model 1 explains the variance in being annoyed by wind turbine noise using the whole data and Model 2 explains the same variables using the sample from Variant 1. For both models, annoyance with wind turbine noise were positively associated with SPLs. Age was positively associated with annoyance in a diminishing downward slope. Having higher qualifications beyond O-levels significantly decreased the probability of being annoyed. Annoyance with wind turbine noise was not significantly different between variants and sites. In Model 2, effects of visibility and attitudes related to wind turbines were examined using Variant 1. Visibility of the wind turbine did not significantly change the odds of being annoyed when controlling for other covariates. Having negative attitudes to the environmental impact of wind turbines was found to increase the odds of annoyance by more than four times.

Table 3. Association between annoyed by wind turbine noise, SPLs, and covariates

Model	Variables	*p*-value	Odds Ratio (OR)	95% CI for OR
	Annoyed by WTN *[n=356, R^2=0.264, $p_{(H-L)}$=0.308]*			
1	SPL	<0.001	**1.18**	(1.08-1.28)
(Variant 1+2)	Age	<0.05	**1.24**	(1.05-1.47)
	Age squared	<0.01	**0.81**	(0.69-0.94)
	Highest qualification (ref: O-level)			
	- No qualification	0.153	0.49	(0.18-1.31)
	- A-level	<0.1	0.29	(0.07-1.19)
	- Higher education below degree	<0.1	0.31	(0.08-1.14)
	- Degree level	<0.05	**0.25**	(0.06-0.98)
	- Other (professional certificate)	0.602	1.51	(0.32-7.22)
	Site (ref: Site C)			
	- Site A	0.928	0.94	(0.30-2.93)
	- Site B	0.242	1.77	(0.67-4.68)
	Variant 2	0.799	0.89	(0.38-2.11)
	Annoyed by WTN *[n=254, R^2=0.339, $p_{(H-L)}$=0.331]*			
2	SPL	<0.05	**1.12**	(1.00-1.26)
(Variant 1)	Age	<0.05	**1.24**	(1.03-1.48)
	Age squared	<0.05	**0.80**	(0.67-0.96)
	Highest qualification (ref: O-level)			
	- No qualification	0.167	0.40	(0.11-1.48)
	- A-level	<0.1	0.21	(0.04-1.17)
	- Higher education below degree	<0.05	**0.22**	(0.05-0.93)
	- Degree level	<0.1	0.25	(0.06-1.14)
	- Other (professional certificate)	0.634	1.69	(0.20-14.41)
	Visibility of the WT (ref: can't see any from home)			
	- See WT from window	0.249	2.43	(0.54-10.98)
	- See WT from garden	0.851	0.82	(0.10-6.80)
	- See WT from both window & garden	<0.1	4.81	(0.93-24.95)
	Negative attitude to the environmental impact of WT (no/yes)	<0.001	**4.84**	(1.84-12.73)
	Site (ref: Site C)			
	- Site A	0.599	0.69	(0.17-2.78)
	- Site B	0.962	1.03	(0.29-3.68)

Statistically significant correlations in boldface.

3.3 Sleep disturbance

Respondents in both Variants 1 and 2 indicated sleep disturbances without referring to noise. There was no significant difference between the variants regarding the prevalence of each sleep problem. Overall, 13.4% of the respondents did not have their sleep disturbed. The most often chosen problems were "sleep less deeply" (33.1%) and "lie awake for a while" (32.6%). Moreover, 18.4% of the respondents reported "hard to fall asleep" and 4.7% reported they were "taking sleeping pills to fall asleep". No associations between wind turbine noise levels and sleep were found.

However, sleep disturbance was significantly associated with overall annoyance with wind turbine noise and annoyance with the noise indoors, tested in logistic regression analyses controlling for age, gender, long-standing illness, noise sensitivity, and sites. The measured sleep problems were not associated with annoyance of wind turbine noise outdoors and at night. Taking both Variants 1 and 2 into account, one unit increase of the annoyance with wind turbine noise significantly increased the odds of sleeping less deeply by 1.54 (95% CI: 1.06-2.25). Using the sample from Variant 1, the odds of sleeping less deeply was increased by 1.83 (95% CI: 1.11-3.03) with one unit increase of annoyance overall, controlling for the influence of negative attitude and visibility to the wind turbines. Annoyance with wind turbine noise indoors was found to increase significantly the odds of difficulty falling asleep by 1.33 (95% CI: 1.01-1.76). Of subjective factors, age and having a longstanding illness significantly increased the odds of sleeping less deeply. Visibility of the wind turbine from both window and garden significantly increased the odds of "sleep less deeply" by 2.78 (95% CI: 1.20-6.42) than those who only see it from a window. Annoyance with wind turbine noise did not significantly affect not being disturbed, lying awake or taking sleep pills.

3.4 Health problems and general health

No significant difference in self-reported general health was found between variants. However, it was found that the prevalence of each reported health problem was statistically significantly different between the variants. The percentage of respondents reporting a given health problem was generally 1.5 – 2 times higher in Variant 2 than in Variant 1. The largest difference concerned difficulty in intellectual activities (12% in Variant 1 versus 33% in Variant 2). In addition, the respondents in Variant 1 were asked for the perceived health impact of wind turbine noise before identifying health problems, where 89% of the respondents indicated that wind turbine noise had no effect on their health. Only 1% of respondents reported that wind turbine noise had effects on health sometimes, while 8.4% of respondents indicated "I don't know". Moreover, when respondents in Variant 1 were invited to attribute the cause, very few indicated their health problems were caused by wind turbine noise, less than 7% for each health problem. Therefore, the less lower prevalence of reported health problems in Variant 1 might be that some respondents who subjectively perceived no health impact of the noise, knowing the motivation of the survey was to link their reported health problems to wind turbine noise exposure, hesitated to report any health problems. On the other hand, the higher prevalence of health problems in Variant 2 is also in line with the significantly higher proportion of respondents having long-standing illness in Variant 2 than in Variant 1.

Logistic regression analysis was carried out using each health problem experienced as binary outcome variable and SPL at respondent's dwelling as an explanatory variable, controlling for known moderating factors and sites. As there was significant difference in self-reported health problems between the variants, separate regression models were carried out for each health problem, using the whole data, data of Variant 1, and of Variant 2, respectively. No significant association was found between noise exposure and health problems among respondents in Variant 1. More significant associations between SPLs and prevalence of health problems were found using Variant 2. The results of the regressions with significant associations are shown in Table 4. Wind turbine noise levels were found to be related to more physical symptoms (i.e. nausea, dizziness, ear discomfort), than psychological distresses (i.e. stress, mood swings). As shown in Table 4, wind turbine SPLs were found to be positively associated with nausea ($p<0.05$) and dizziness ($p<0.1$) using the whole data. Among respondents of Variant 2, wind turbine SPLs were found to significantly increase dizziness and ear discomfort ($p<0.05$), as well as nausea and difficulty in intellectual activities, though with weaker significance ($p<0.1$). Higher noise sensitivity was found to increase the odds of nausea and dizziness in Variant 2. Age, gender and having a long-standing illness were also associated with health problems. It is worth noting that annoyance with wind turbine noise had no significant influence if added to the model, and the associations between SPLs and health problems in Variant 2 remained significant. These statistically significant relationships between wind turbine noise and health could be due to random chance in statistical tests, especially for the small sample size of Variant 2. Nevertheless, the results indicated a significant difference between self-reported health across questionnaire variants, where the control group (to whom the motivation of the research was masked) reported more health problems.

Table 4. Association between health problems, wind turbine SPLs, and covariates.

Model (sample)	Variables	*p*-value	Odds Ratio (OR)	95% CI for OR
1	***Nausea*** *[n=348, R^2=0.290, $p_{(H-L)}$=0.119]*			
(Variant 1+2)	SPL	<0.05	**1.09**	(1.01-1.17)
	Age	<0.001	**0.95**	(0.92-0.97)
	Female	<0.001	**5.39**	(2.25-12.93)
	Longstanding illness (no/yes)	<0.001	**4.39**	(1.87-10.35)
	Sensitivity to noise (scale 1-6)	0.763	0.96	(0.74-1.25)
	Site A	0.949	1.03	(0.39-2.70)
	Site B	0.237	0.74	(0.19-1.50)
	Variant 2	<0.05	**2.72**	(1.22-6.08)
2	***Dizziness*** *[n=348, R^2=0.218, $p_{(H-L)}$=0.099]*			
(Variant 1+2)	SPL	<0.1	1.06	(0.99-1.13)
	Age	<0.01	**0.97**	(0.95-0.99)
	Female	<0.01	**2.91**	(1.42-5.94)
	Longstanding illness (no/yes)	<0.001	**4.39**	(2.10-9.16)
	Sensitivity to noise (scale 1-6)	0.281	1.14	(0.90-1.45)
	Site A	0.594	0.79	(0.33-1.90)
	Site B	0.546	0.76	(0.32-1.83)
	Variant 2	<0.01	**2.65**	(1.30-5.37)
3	***Nausea*** *[n=94, R^2=0.496, $p_{(H-L)}$=0.742]*			
(Variant 2)	SPL	<0.1	1.16	(0.99-1.34)
	Age	<0.01	**0.94**	(0.90-0.98)
	Female	<0.05	**5.87**	(1.14-30.30)
	Longstanding illness (no/yes)	<0.05	**5.53**	(1.17-26.11)
	Sensitivity to noise (scale 1-6)	<0.05	**2.21**	(1.10-4.44)
	Site A	0.338	0.37	(0.05-2.87)
	Site B	0.105	0.15	(0.02-1.49)
4	***Dizziness*** *[n=94, R^2=0.387, $p_{(H-L)}$=0.914]*			
(Variant 2)	SPL	<0.05	**1.17**	(1.03-1.34)
	Age	0.100	0.97	(0.94-1.01)
	Female	<0.05	**3.62**	(1.01-12.99)
	Longstanding illness (no/yes)	<0.1	2.93	(0.87-9.89)
	Sensitivity to noise (scale 1-6)	<0.05	**1.98**	(1.15-3.41)
	Site A	0.605	1.62	(0.26-9.92)
	Site B	0.345	0.39	(0.06-2.73)
5	***Ear discomfort*** *[n=94, R^2=0.291, $p_{(H-L)}$=0.536]*			
(Variant 2)	SPL	<0.05	**1.15**	(1.02-1.30)
	Age	<0.05	**1.05**	(1.01-1.09)
	Female	0.317	1.83	(0.56-6.02)
	Longstanding illness (no/yes)	0.586	1.38	(0.43-4.45)
	Sensitivity to noise (scale 1-6)	<0.1	1.47	(0.95-2.27)
	Site A	0.627	0.66	(0.12-3.55)
	Site B	0.333	0.44	(0.08-2.32)
6	***Difficulty in intellectual activities*** *[n=94, R^2=0.320, $p_{(H-L)}$=0.929]*			
(Variant 2)	SPL	<0.1	1.11	(0.99-1.23)
	Age	0.943	0.99	(0.97-1.03)
	Female	0.163	2.13	(0.74-6.15)
	Longstanding illness (no/yes)	<0.001	**5.94**	(2.03-17.36)
	Sensitivity to noise (scale 1-6)	0.174	1.13	(0.89-1.88)
	Site A	0.786	0.81	(0.18-3.65)
	Site B	0.240	0.40	(0.09-1.84)

Statistically significant correlations in boldface.

Table 5. Association between health problems and annoyance with WTN, shown as *p*-value and odds ratio with 95%CI tested with logistic regression using Variant 1 sample.

Health problems (outcome variable)	Explanatory variables (in Variant 1)					
	Annoyed overall		Annoyed outdoors		Annoyed indoors	
	p-value	OR (95% CI)	*p*-value	OR (95% CI)	*p*-value	OR (95% CI)
a) Headache	<0.1	2.49 (0.98-6.35)	0.436	1.39 (0.61-3.18)	0.524	1.03 (0.96-1.10)
b) Nausea	**<0.05**	**4.48 (1.15-17.49)**	0.100	2.70 (0.83-8.83)	0.663	0.69 (0.13-3.73)
c) Dizziness	0.250	2.01 (0.60-6.96)	0.666	0.76 (0.22-2.63)	0.658	1.35 (0.36-5.02)
d) Ear discomfort	0.968	0.98 (0.31-3.11)	0.622	1.29 (0.47-3.55)	0.524	0.64 (0.16-2.55)
e) Cardiovascular disease	0.425	2.23 (0.31-15.94)	**<0.05**	**8.19 (1.38-48.63)**	0.998	0.00 0.00
f) Stress	0.227	1.72 (0.71-4.17)	0.514	1.30 (0.60-2.82)	0.288	1.67 (0.65-4.33)
g) Tension and edginess	**<0.01**	**3.63 (1.47-8.93)**	**<0.05**	**2.63 (1.20-5.78)**	**<0.05**	**3.34 (1.26-8.86)**
h) Difficulty in intellectual activities	**<0.05**	**3.37 (1.07-10.59)**	0.200	1.93 (0.71-5.28)	<0.1	3.00 (0.95-9.46)
i) Mood swings	0.220	1.79 (0.71-4.55)	0.521	1.32 (0.57-3.04)	0.675	1.24 (0.45-3.44)
j) Lack of concentration	**<0.05**	**2.53 (1.04-6.17)**	0.207	1.68 (0.75-3.76)	0.426	1.49 (0.56-3.94)

Adjusted for WTN, age, sex, longstanding illness, noise sensitivity, visibility of WT, and site. n=249-250, R^2=0.165-0.289.

Table 6. Association between general health, wind turbine SPL, and covariates.

Model (sample)	Variables	*p*-value	B	95% CI for B
1	***General health (1 poor - 5 excellent)*** *(n=345, R^2=0.396)*			
(Variant 1+2)	SPL	0.251	0.009	(-0.006, 0.023)
	Age	0.400	-0.002	(-0.007, 0.003)
	Female	0.905	0.010	(-0.159, 0.180)
	Household income (ref: < £20,000)			
	£20,000 - £29,999	0.973	0.005	(-0.257, 0.266)
	£30,000 - £49,999	<0.01	**0.385**	(0.129, 0.641)
	more than £50,000	<0.1	0.329	(-0.035, 0.692)
	I don't know / missing	0.208	0.155	(-0.086-0.396)
	Longstanding illness (no/yes)	<0.001	**-0.992**	(-1.178, -0.806)
	Sensitivity to noise (scale 1-6)	<0.05	**-0.061**	(-0.117, -0.006)
	Sustainable lifestyle (scale 1-6)	<0.001	**0.120**	(0.056, 0.185)
	Site A	<0.05	**0.263**	(0.037, 0.489)
	Site B	0.364	0.089	(-0.104, 0.283)
	Variant 2			
2	***General health (1 poor - 5 excellent)*** *(n=252, R^2=0.385)*			
(Variant 1)	SPL	<0.01	**0.024**	(0.007, 0.042)
	Age	0.417	-0.003	(-0.009, 0.004)
	Female	0.683	0.042	(-0.161, 0.245)
	Household income (ref: < £20,000)			
	£20,000 - £29,999	0.596	0.084	(-0.228, 0.396)
	£30,000 - £49,999	<0.01	**0.421**	(0.112, 0.731)
	more than £50,000	<0.05	**0.453**	(0.017, 0.889)
	I don't know / missing	<0.05	**0.292**	(0.001, 0.583)
	Longstanding illness (no/yes)	<0.001	**-0.931**	(-1.155, -0.707)
	Sensitivity to noise (scale 1-6)	0.116	-0.054	(-0.120, 0.013)
	Sustainable lifestyle (scale 1-6)	<0.001	**0.130**	(0.054, 0.205)
	Site A	<0.1	0.219	(-0.040, 0.479)
	Site B	0.216	0.165	(-0.097, 0.427)
3	***General health (1 poor - 5 excellent)*** *(n=93, R^2=0.536)*			
(Variant 2)	SPL	<0.05	**-0.030**	(-0.059, -0.002)
	Age	0.393	-0.004	(-0.014, 0.005)
	Female	0.623	-0.080	(-0.405, 0.244)
	Household income (ref: < £20,000)			
	£20,000 - £29,999	0.529	-0.160	(-0.663, 0.344)
	£30,000 - £49,999	0.216	0.285	(-0.170, 0.741)
	more than £50,000	0.823	-0.076	(-0.746, 0.594)
	I don't know / missing	0.773	-0.064	(-0.500, 0.373)
	Longstanding illness (no/yes)	<0.001	**-1.121**	(-1.463, -0.780)
	Sensitivity to noise (scale 1-6)	0.122	-0.079	(-0.180, 0.021)
	Sustainable lifestyle (scale 1-6)	0.165	0.091	(-0.038, 0.221)
	Site A	0.240	0.303	(-0.206, 0.813)
	Site B	0.103	0.413	(-0.086, 0.911)

Statistically significant correlations in boldface.

Using Variant 1, the relationship between health problems and annoyance due to wind turbine noise was tested with logistic regression controlling for SPL, visibility of the wind turbine and other covariates as above (see Table 4). The results are shown in Table 5. It was found that more psychological distresses were associated with being annoyed by wind turbine noise. Tension and edginess was positively associated with being annoyed by wind turbine noise overall, outdoors and indoors. Nausea, difficulty in intellectual activities, and lack of concentration were associated with annoyance overall. Cardiovascular disease was associated with being annoyed by wind turbine noise outdoors. Visibility of the wind turbine was not associated with health problems.

Respondents in Variant 1 also gave more optimistic responses regarding general health. Table 6 shows the results of ordinary least squares (OLS) regression models for wind turbine noise and general health, controlling for the effects of subjective factors. No significant association was found between wind turbine SPL and general health using the whole data (Model 1). SPL was found to be associated with general health using data of Variant 1 and of Variant 2 separately, where the effects were different between variants (Model 2 vs 3). Higher level of wind turbine noise was significantly associated with better general health of respondents in Variant 1. Higher degree of sustainability in life and higher household income significantly increased the level of general health among Variant 1 respondents. Of the respondents in Variant 2, on the contrary, higher wind turbine noise was significantly associated with poorer general health. Having a long-standing illness significantly decreased general health. The reason for the difference between variants, however, could arguably be because differences in socioeconomic factors across the variants were not fully controlled for. Although income, long-standing illness, sensitivity to noise and sustainability in life were found to have significant impacts on self-reported general health in relation to variants, the inter-collinearity between factors might change the direction of the coefficient of noise.

4. Conclusions and discussions

This study extends the existing basis for health impact of wind turbine noise by further exploring the dose-response relationship in densely populated suburban-urban settings. It was found that A-weighted SPL at the dwelling was positively associated with noise annoyance of the respondent. The odds ratio of being annoyed by wind turbine noise increased with each dB increase in SPLs, controlling for the effect of moderating factors. However, the proportions of respondents annoyed (moderately, very and extremely) by wind turbine noise in this study were lower than those reported in the Swedish (Pedersen & Waye, 2004, 2007) and Dutch (Pedersen et al., 2009) studies in rural areas. The percentages of very annoyed respondents were also much lower than those reported in the Polish (Pawlaczyk-Łuszczyńska et al., 2014) and Canadian (Michaud et al., 2016) studies. Compared to other environmental nuisances, respondents noticed and were annoyed the least frequently by wind turbine noise. This was the opposite to the results of previous studies which suggested wind turbine noise was the most frequently assessed as annoying amongst a similar set of nuisances (Pawlaczyk-Łuszczyńska et al., 2014). One could argue that in suburban-urban areas wind turbine noise is less prominent than other environmental nuisances. On the other hand, it has been proposed that peoples' beliefs about the importance of the source of the noise can decrease annoyance (Fields, 1993), and the different degree of annoyance might be due to the different socio-economic characteristics and attitudes between the suburban-urban respondents of this study and the rural respondents of previous ones. For example, this study has found that education and attitudes to wind energy projects moderate annoyance with wind turbine noise significantly.

Wind turbine noise was found to influence health and well-being. Noise levels was not associated with sleep, but degree of noise annoyance significantly increased the possibility of sleep disturbance including sleeping less deeply and difficulty falling asleep. Visibility of the wind turbine from both window and garden significantly increased the odds of a less deep sleep. Positive associations were found between wind turbine SPL at a dwelling and adverse health problems, including nausea and dizziness. Dizziness and ear discomfort were found to be related to SPL in the control group (Variant 2). It was further found that among the Variant 1 respondents, annoyance with the noise significantly differentiated the prevalence of health problems: psychological problems such as tension and edginess were significantly and positively associated with being annoyed by wind turbine noise, but not with the noise level by itself. This is consistent with those reported in four previous studies using Swedish, Dutch and Polish samples, where annoyance with wind turbine noise (but not noise level itself) was consistently associated with feeling tense or stressed (Pawlaczyk-Łuszczyńska et al., 2014; Pedersen, 2011). It should be noted that a significant relationship between noise annoyance and health should not be taken as evidence of a causal pathway from the noise to health, as the study method did not establish causality between variables, e.g., adverse health problems might cause annoyance, in the reverse direction.

An important finding of the study lies in the difference between the main and the control groups. The associations between noise levels and the prevalence of health problems were not significant in the main sample (Variant 1), who reported significantly less health problems and better general health. Moreover, respondents in Variant 1 gave more optimistic responses regarding general health, with higher SPL significantly associated with *better* self-reported general health, the opposite to the relationship found in Variant 2 (control). The reason could be related to the effect of questions of the two variants. Unlike Variant 2, where the purpose of the research was masked, it was clear to participants in Variant 1 that their health data would be analysed in relation to wind turbine noise. This might have led to less health problems being reported by Variant 1 respondents. The results suggest that respondents' knowledge of the motivation of the survey in the main variant lead to more optimistic responses regarding adverse health impact of wind turbines, a methodological finding which should be noted in future research.

Acknowledgements

This research has been funded by the University of Sheffield CCDRI network scholarship.

References

Bakker, R. H., Pedersen, E., van den Berg, G. P., Stewart, R. E., Lok, W., & Bouma, J. (2012). Impact of wind turbine sound on annoyance, self-reported sleep disturbance and psychological distress. *Science of the Total Environment*, *425*, 42–51.

Benfield, J. A., Nurse, G. A., Jakubowski, R., Gibson, A. W., Taff, B. D., Newman, P., & Bell, P. A. (2014). Testing noise in the field: A brief measure of individual noise sensitivity. *Environment and Behavior*, *46*(3), 353–372.

Bluhm, G., Nordling, E., & Berglind, N. (2004). Road traffic noise and annoyance-an increasing environmental health problem. *Noise and Health*, *6*(24), 43–49.

Cooney, C., Byrne, R., Lyons, W., & O'Rourke, F. (2017). Performance characterisation of a commercial-scale wind turbine operating in an urban environment, using real data. *Energy for Sustainable Development*, *36*, 44–54.

DataKustik GmbH. (2006). *Cadna/A Manual Version 3.6*. Greifenberg: DataKustik GmbH.

Dolan, P., Peasgood, T., & White, M. (2008). Do we really know what makes us happy? A review of the economic literature on the factors associated with subjective well-being. *Journal of Economic Psychology*, *29*(1), 94–122.

Fields, J. M. (1993). Effect of personal and situational variables on noise annoyance in residential areas. *The Journal of the Acoustical Society of America*, *93*(5), 2753–2763.

Frijters, P., & Beatton, T. (2012). The mystery of the U-shaped relationship between happiness and age. *Journal of Economic Behavior and Organization*, *82*(2–3), 525–542.

Guski, R. (1999). Personal and social variables as co-determinants of noise annoyance. *Noise and Health*.

IEC 61400-11. (2012). *Wind Turbine Generator Systems - Part 11: Acoustic Noise Measurement Techniques*. Geneva: International Electrotechnical Commission.

International Standards Organisation. (1996). ISO 9613-2 1996 Acoustics - Attnuation of sound during propagation outdoors (part 2).

Ishugah, T. F., Li, Y., Wang, R. Z., & Kiplagat, J. K. (2014). Advances in wind energy resource exploitation in urban environment: A review. *Renewable and Sustainable Energy Reviews*, *37*, 613–626.

Job, R. F. S. (1999). Noise sensitivity as a factor influencing human reaction to noise. *Noise & Health*, *3*, 57–68.

Kang, J. (2006). *Urban Sound Environment*. CRC Press.

Keith, S. E., Feder, K., Voicescu, S. A., Soukhovtsevallison, V., Tsangnorm, D., Keith, S. E., … Leroux, T. (2016). Wind turbine sound pressure level calculations at dwellings Wind turbine sound pressure level calculations at dwellings, *1436*.

Michaud, D. S., Feder, K., Keith, S. E., Voicescu, S. A., Marro, L., Than, J., … Weiss, S. K. (2016). Exposure to wind turbine noise : Perceptual responses and reported health effects. *Jasa*, *139*(March), 1443–1454.

Millward-Hopkins, J. T., Tomlin, A. S., Ma, L., Ingham, D. B., & Pourkashanian, M. (2013). Assessing the potential of urban wind energy in a major UK city using an analytical model. *Renewable Energy*, *60*, 701–710.

Muzet, A. (2007). Environmental noise, sleep and health. *Sleep Medicine Reviews*, *11*(2), 135–142.

Nissenbaum, M. A., Aramini, J. J., & Hanning, C. D. (2012). Effects of industrial wind turbine noise on sleep and health. *Noise & Health*.

Ordnance Survey. (2013). EDINA Digimap Ordnance Survey Service. Retrieved August 26, 2014, from http://digimap.edina.ac.uk

Pawlaczyk-Łuszczyńska, M., Dudarewicz, A., Zaborowski, K., Zamojska-Daniszewska, M., & Waszkowska, M. (2014). Evaluation of annoyance from the wind turbine noise: a pilot study. *International Journal of Occupational Medicine and Environmental Health*, *27*(3), 364–88.

Pedersen, E. (2011). Health aspects associated with wind turbine noise - Results from three field studies. *Noise Control Engineering Journal*, *59*(1), 47–53.

Pedersen, E., van den Berg, F., Bakker, R., & Bouma, J. (2009). Response to noise from modern wind farms in The Netherlands. *The Journal of the Acoustical Society of America*, *126*(2), 634–43.

Pedersen, E., & Waye, K. P. (2004). Perception and annoyance due to wind turbine noise--a dose-response relationship. *The Journal of the Acoustical Society of America*, *116*(6), 3460–70.

Pedersen, E., & Waye, K. P. (2007). Wind turbine noise, annoyance and self-reported health and well-being in different living environments. *Occup Environ Med*, *64*(7), 480–486.

Perkins, R., Acoustics, H. L., Bass, J., Lewis, T., Mackay, J., Mcgovern, R., & Mckenzie, A. (2013). A Good Practice Guide to the Application of ETSU-R-97 for the Assessment and Rating of Wind Turbine Noise, (1), 1–40.

Qu, F., Kang, J., & Tsuchiya, A. (2015). Effects of built environment morphology in residential areas on resisting wind turbine noise on building façades. In *6th International Meeting on Wind Turbine Noise*. Glasgow.

Shepherd, D., McBride, D., Welch, D., Dirks, K. N., & Hill, E. M. (2011). Evaluating the impact of wind turbine noise on health-related quality of life. *Noise and Health*, *13*(54), 333–339.

Statistics, I. S. (2009). IBM SPSS Statistics 22 Core System User's Guide. *SPSS Inc.*

UKWED. (n.d.). UK wind energy database. Retrieved December 1, 2014, from http://www.renewableuk.com/en/renewable-energy/wind-energy/uk-wind-energy-database/index.cfm/maplarge/1

Weinstein, N. D. (1978). Individual differences in reactions to noise: a longitudinal study in a college dormitory. *The Journal of Applied Psychology*, *63*(4), 458–466.

Weinstein, N. D. (1980). Individual differences in critical tendencies and noise annoyance. *Journal of Sound and Vibration*, *68*(2), 241–248.

7th International Conference
on
Wind Turbine Noise
Rotterdam – 2nd to 5th May 2017

Sound power level measurements 3.0

ir. L.M. Eilders, Peutz bv: l.eilders@peutz.nl
ing. E.H.A. de Beer, Peutz bv: e.debeer@peutz.nl

Summary

The apparent sound power level of a wind turbine is determined in accordance to the international standard IEC 61400-11 edition 3.0. This paper focuses on the challenges introduced in this version and how to deal with them. The main difficulties are k-factor dependent wind speed, disturbances leading to incomplete datasets, continuous full analysis necessary for determination of a complete dataset.

The standard requires 10 measurements (30 if tonal) per 0.5 m/s wind speed bin for both the total noise and the background noise. The wind speed measured at 10 meter height, used for the background measurements is now calibrated in-situ with a k-factor dependent on Turbine ON data, which means that the normalized wind speeds can change bins until the last recorded measurement. This can have a major impact if measurements stretch over multiple days due to varying meteorological conditions. Furthermore, disturbances during the measurements have to be filtered from the dataset. Manual filtering afterward can result in an incomplete dataset of sound power levels related to the required wind speed range. In cases where the difference in the sum of the 1/3-octave bands of the estimated sound power level based on the total and background noise for a given wind speed bin is less than 3 dB, the result is not reported. This can only be monitored by doing the complete analysis per 1/3-octave band on site since a measurement with enough data points can in the end still result in an incomplete dataset.

Constraints on the measuring positions of both wind speed and sound pressure level also require special attention to limit downtime during rapidly changing wind directions.

Extensive automation with a cloud based database is developed in light of aforementioned issues as a possible solution and is presented in this paper.

1. Introduction

The apparent sound power level of a wind turbine is determined in accordance to the international standard IEC 61400-11 Wind turbines – Part 11: Acoustic noise measurement techniques edition 3.0 which replaced edition 2.1. Detailed evaluation of the differences between Edition 3.0 and Edition 2.1 have been made before. [Jozwiak, R. et all, 2015]

Jozwiak, R. et all, name the most significant instrumentation changes as:

- 1/3 Octave Band centre frequencies extended down to 20 Hz;
- A resolution of 1 to 2 Hz for the entire frequency range for tonal analysis;
- Rotor RPM is now a mandatory parameter to be logged;
- A minimum sample rate of 1 Hz for turbine parameters and wind speed.

Other significant changes are:

- The implementation of 10 second averages for each interval;
- The apparent sound power level is now related to hub height instead of 10 m height;
- K-factor correction method for nacelle anemometer and 10 m anemometer

The use of the k-factor method to relate the wind speed measured at 10 m height and nacelle to hub height is on itself minor. This change however has some major implications for monitoring the progress during sound power level measurements.

2. Measurement procedure

The apparent A-weighted sound power levels, spectra, and tonal audibility at bin centre wind speeds at hub height and 10 m height of an individual wind turbine are assessed by sound power level measurements conform IEC 61400-11 edition 3.0. The standard describes where and how the sound pressure level and the wind speed at 10 m height are to be measured as well as which parameters of the turbine should be logged during the measurements. Based on the recorded data the apparent A-weighted sound power levels, spectra, and tonal audibility at bin centre wind speeds at hub height and 10 m height can be calculated.

2.1 Measurement locations

2.1.1 Microphone location

The direction of the positions shall be within ±15° relative to the downwind direction of the wind turbine at the time of measurement. The downwind direction can be derived from the yaw position. The horizontal distance (R_0=H + D/2) from the wind turbine tower vertical centreline to each microphone position shall be as with a tolerance of ±20%, max ±30m. The allowed location for the reference microphone is shown in figure 1.

2.1.2 10 m anemometer location

For measurement of background noise an anemometer mounted on a met mast of at least 10 m height shall be used. The position of the met mast should be relatively undisturbed and represent the free wind at the turbine position. The allowed location for the 10 m high anemometer is shown in Figure 1. Here the grey highlighted areas to the left and the right of the turbine show the possible location of the 10 m anemometer, the grey highlighted area directly downstream of the wind turbine indicate the allowed area where the reference microphone has to be located. If during measurements any of the two measurement positions falls outside of their allowed range the measurement position is invalid and has to be excluded from the measurement set.

2.2 Wind speed range

The wind speed range is related to the specific wind turbine. As a minimum it is defined as the hub height wind speed from 0.8 to 1.3 times the wind speed at 85% of maximum power rounded to wind speed bin centres.

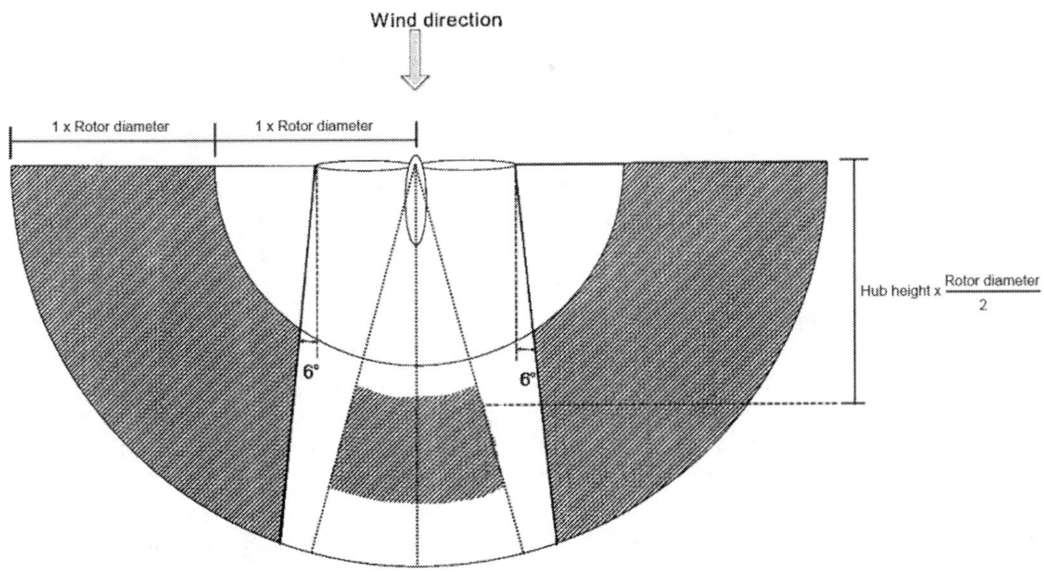

Figure 1 Allowable range for 10 m anemometer and reference microphone position.

2.3 Background noise

With the wind turbine shut down, and using the same measurement set-up, the background noise shall be measured immediately before or after each measurement series of wind turbine noise and during similar wind conditions. When measuring background noise every effort shall be made to ensure that the background sound measurements are representative for the background noise that occurred during the wind turbine noise emission measurements. It is necessary to measure the background noise several times during the measurement period to cover the same wind speed range as for the total noise.

2.4 Number of measurements

At least 180 measurement shall be made overall for both total noise and background noise covering corresponding wind speed ranges (see section 2.2). At least 10 measurements shall be made in each wind speed bin for both the total noise and background noise.

Additional noise measurements may be needed to determine the audibility of an identified tone. When a tone is found in a wind speed bin the number of required measurements depends on the number of measurements containing a tone of the same origin. 30 measurements may be required to assess the audibility of the tone if less than 6 measurements contain the tone of the same origin.

2.5 Required wind turbine parameters

During the measured several parameters of the wind turbine have to be logged. These include but are not limited to:

- power output;
- generator speed;
- rotor speed;
- nacelle wind speed;
- yaw angle.

2.6 Required site parameters

During measurements the air temperature and atmospheric pressure have to be measured.

3. Keeping track of measurements

In theory, for a wind speed range of 7.5 m/s to 12.5 m/s at hub height a minimum measurement time of 60 minutes is required to obtain enough measurements for both the total noise and the background noise (30 minutes each).

In practice however there are several factors that can influence the number of measurements inside a wind speed bin. The most important factors are discussed below.

3.1 Total and background noise

For a given wind speed interval of 0.5 m/s both the total and background noise have to be measured. Therefore the turbine must be switched off and on during a measurement. The standard states that every effort shall be made to ensure that the background sound measurements are representative and measured during similar wind conditions as the total noise. After switching the turbine on or off it typically takes about 5 minutes for the rotor to be completely stopped or sped up to normal rotational speed and pitch.

Switching the turbine state then leads to an exclusion of circa 5 minutes of measuring time. Since, with changing wind conditions, a certain wind speed interval might only occur once during the measurement day. And it might be crucial to make an accurate assessment on when to switch turbine states to ensure both total and background noise are sufficiently measured.

3.2 K-factor

The wind speed measured at the nacelle and the wind speed measured at 10 meter height, used for the background measurements is now calibrated in-situ with a k-factor. The k-factor for both the total noise and background noise is dependent on Turbine ON data. The k-factor is defined as the average value of the ratio of the wind speed derived from the power curve (V_P) and the measured wind speed (V_{nac} or V_{10m}).

For the total noise measurements the k-factor is applied to the measured wind speeds for the data points with power levels outside the allowed range of the power curve to derive the normalised wind speed at hub height. This means that the normalized wind speeds outside the allowed range of the power curve can change bins until the last recorded measurement for the total noise measurements.

With edition 3.0 of the standard the background wind speed is now calculated to hub height using the k-factor. This implies that the 10 meter high anemometer used for background measurements is dependent on the Turbine ON data. The amount of measurements inside the background wind speed bins can only be known by calculating the k-factor. Therefore the normalized wind speeds of all measurements can change bins until the last recorded measurement for the background noise.

The use of a k-factor can have a significant impact on the measurement results if measurements stretch over multiple days due to varying meteorological conditions.

3.3 Disturbances

The nature of the apparent sound power level measurements means that there are always outside influences, for example disturbing noise from birds, planes, passing cars, working farmers, rain etc.

Edition 2.1 of the standard incorporates the use of a fourth order regression best fit through the measured sound pressure levels to determine the sound pressure level at each integer wind speed. In edition 3.0 the average per 1/3 Octave (which might not be at wind speed bin center) is used instead of a best fit. Measured disturbances can significantly raise the average sound power level and influence the measurement results.

Disturbances during the measurements have to be filtered from the dataset. Manual filtering afterward can result in an incomplete dataset of sound power levels related to the required wind speed range if the amount of measurements in a bin are just above the required 10 (or 30 in order to assess the audibility of a tone).

3.4 Calculation criteria

In cases where the difference in the sum of the 1/3-octave bands of the estimated sound power level based on the total and background noise for a given wind speed bin is less than 3 dB, the result is not to be reported. This can only be monitored by doing the complete analysis per 1/3-octave band on site as a measurement with enough data points can without real time analysis afterwards still result in an incomplete dataset.

3.5 Other parameters

Constraints on the measuring positions of both the wind speed measured at 10 meter height and the sound pressure level also require special attention to limit downtime during rapidly changing wind directions. Measurements with invalid measurement locations because of ±15° constraint for the microphone position or 10 m nacelle position outside the specified range become excluded from the calculations. When the wind direction is changing a measurement position may soon become invalid. One can anticipate and move a measurement location prior to the change in wind direction to limit downtime and possible miss measuring a certain wind speed.

3.6 Drawbacks of post processing measurement data

Measurements based on post processing data according to IEC 61400-11 edition 3.0 have the following drawbacks:

- The in-situ calibration of the k-factor leads to possible shifting of measurements between wind speed bins;
- Due to a change of wind direction the measurement position could afterwards be found outside the allowable range;
- Disturbance correction is done on an energy basis instead of a statistical basis (edition 2.0) leading to a higher chance of exclusion of background measurements during specific wind speeds;
- Third octave band analysis may even show that specific bands should be excluded because of narrow banded disturbance noise;
- If after post analysis missing wind speed bins are found, additional measurements need to be done, leading to delay because one should wait for the specific wind conditions to occur;
- Additional measurements lead to increased lead time of (for instance) prototype testing which may be considered inefficient.

In chapter 4 a measurement procedure and monitoring system is described in which the aforementioned drawbacks are negated.

4. Solutions

As discussed in Chapter 3 there are several factors that have to be known in order to assess if a measurement dataset is complete. Being able to accurately monitor the measurement process is key in decision making in the field which will not only reduce measurement time but improve the quality of the measurements.

In order to achieve accurate monitoring and be in control at the site about all measurement constraints extensive automation is required. The solution of the authors is presented below.

4.1 Cloud database

Data collected from the turbine, 10 m anemometer and sound pressure levels are stored in a central database based on a 1 Hz sample rate. The microphone position is, depending on wind turbine hub height and rotor diameter, up to approximately 200 m downwind of the turbine and can frequently change due to changing wind direction during a measurement. Wireless transfer of the data is therefore the most practical solution. Therefore all logged acoustic, wind speed, and wind turbine parameter values are independently and real-time sent to a remote cloud database. Measurement data and audio files are stored locally for redundancy. Due to the large size of the raw audio files, tonal analysis is performed post-measurement. Multiple microphone positions can be used simultaneous to determine the effect of directivity on the sound power levels.

4.2 Real time computing

With all the measurement data collected in a central storage, monitoring software continuously runs full data processing in accordance to the international standard IEC 61400-11 edition 3.0. In short the following steps are performed real-time:

- time synchronisation;
- 10 s averaging of 1 Hz measurement data;
- Analysing noise measurement data to exclude disturbances;
- Analysing wind turbine measurement data to determine turbine status;
- Checking for invalid measurement positions;
- Normalizing measured wind speeds based on k-factors;
- Sort measurement data to wind speed bins;
- Calculate values at wind speed bin centers;
- Calculate total noise and background noise;
- Perform background correction;
- Calculate total and 1/3-Octave band apparent sound power levels;
- Calculate uncertainties.

Logged variables and calculation results can be viewed real time via a web portal: "Peutz monitoring portal".

4.2.1 Wind turbine status (ON/OFF)

Based on the collected data from the wind turbine (Figure 2), the wind turbine state (ON/OFF) is automatically determined. By analysing the power output and the generator speed the current state of the wind turbine is deducted. Changes in wind turbine state are also accurately determined and automatically excluded from calculations (see Figure 3).

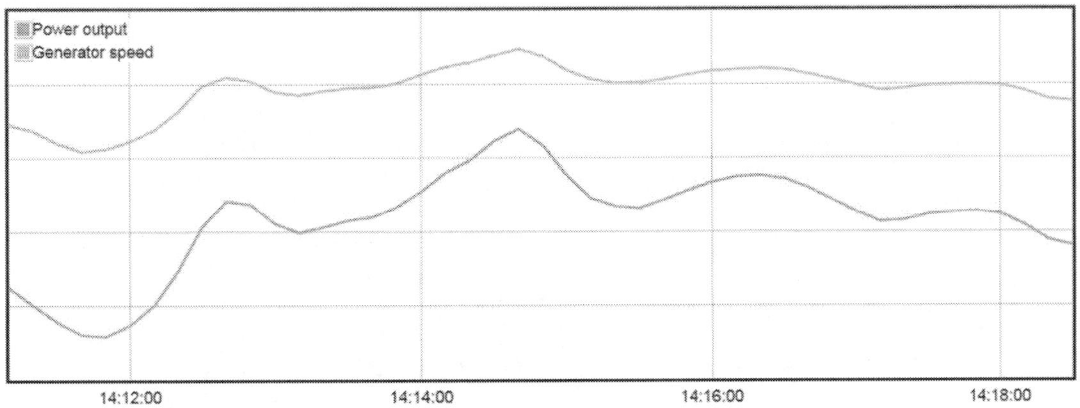

Figure 2 logged generator speed [rpm] and power output [kW] of last 10 minutes

4.2.2 Acoustic disturbances

The acoustic data is processed and analysed and obvious disturbances of birds etc. are automatically filtered out. The equivalent sound pressure level during a 10 second interval is compared to the peak sound pressure level that occurred inside that 10 second interval. Disturbances that are short and high in intensity can easily be distinguished as the difference between the peak level and the equivalent sound pressure level will be high. Disturbances can also be set manually (see Figure 3).

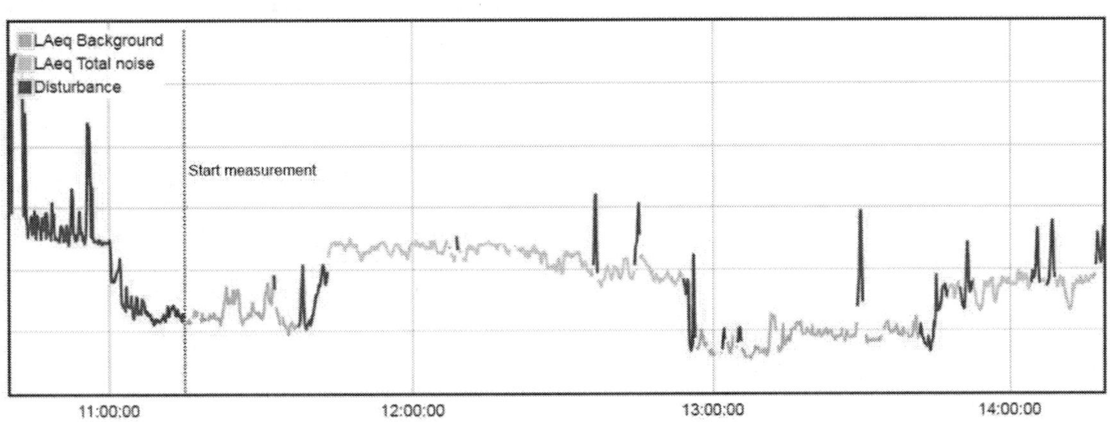

Figure 3 Recorded sound pressure level at the reference microphone position [dB(A)]. With distinction between total noise, background noise and disturbances.

4.2.3 Measurement positions

The location of the reference microphone and the 10 m anemometer is monitored by the software. Their location is on placement determined with GPS and stored in the database. The yaw angle is projected on the preferred measurement location and is monitored and checked real time. The valid locations are dynamically calculated based on the wind turbine parameters stored in the database. Measurements where the measuring locations are outside the valid area are excluded from the calculations. When the wind direction is changing and a measuring position might soon become invalid, one can anticipate and move a measurement location prior to limit downtime and possible miss measuring a certain wind speed. Not only is the validity of a measurement position easily distinguishable, the possible new location can be easily picked, with an interactive web tool located at the monitoring portal, as the projected wind direction can be manually changed to be able to take into account terrain obstacles such as waterways, ditches, tree lines, buildings, etc.

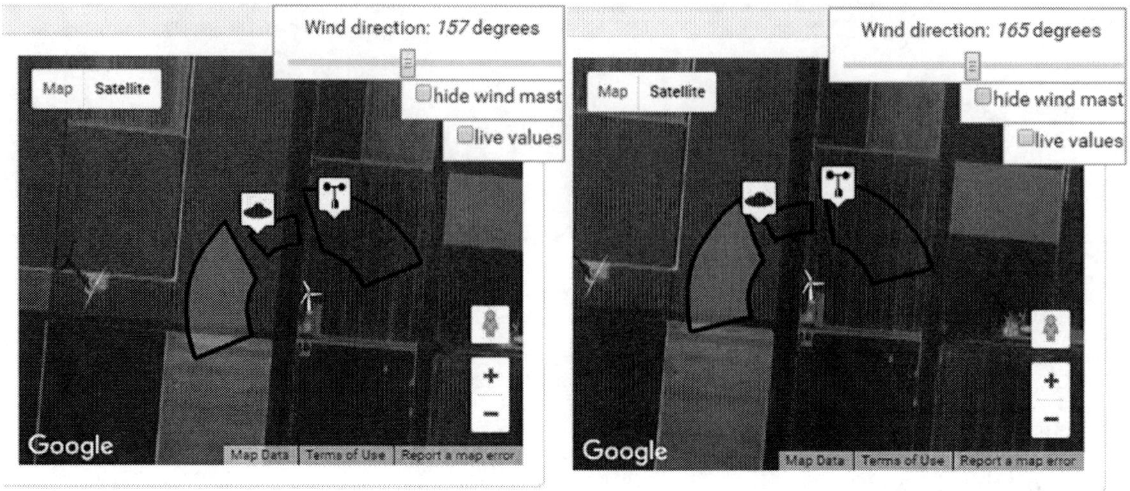

Figure 4 current measurement positions and their allowable range. (left: all valid, right: microphone position not valid)

4.2.4 K-factor

All measured wind speeds are known and can be monitored (Figure 5). The k-factors for both the nacelle wind speed and the 10 m anemometer wind speed are continuously calculated. The normalized 10 m and nacelle wind speed are continuously updated.

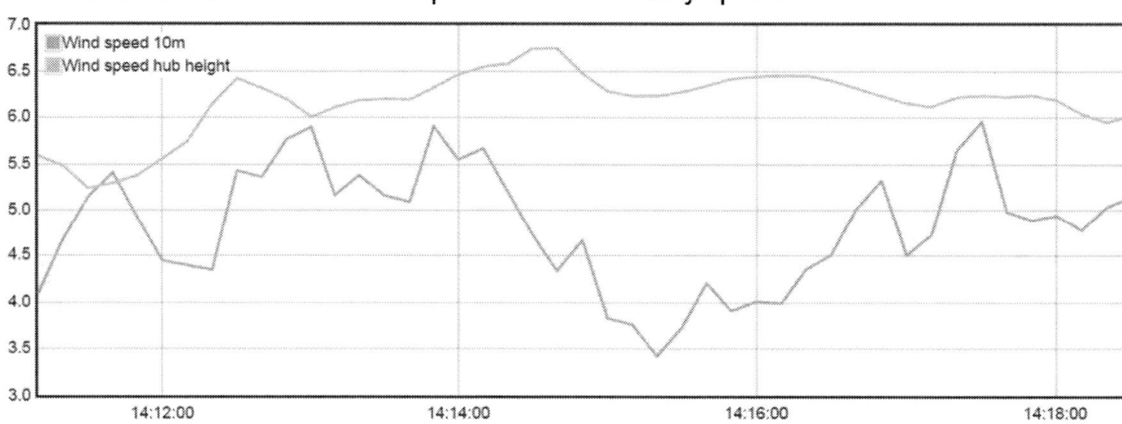

Figure 5 Measured wind speed at 10 m height and hub height (not normalized)

4.2.5 Number of measurements

An overview of the amount of valid measurements in wind speed bins for the total noise and background noise is continuously updated (Figure 6). The required wind speed range is always shown. If wind speed measurements are recorded outside the required wind speed interval the table is extended. For each wind speed bin the number of measurements excluding disturbances and invalid measurement positions are given. If the minimum required amount of measurements to assess the apparent sound power level (10 measurements) is reached the value turns orange. If enough measurements have been recorded to assess the audibility in case of a recorded tone the number turns green. The total number of measurements in the required wind speed range is given as a total which will turn green if it meets the measurement requirement of 180 measurements.

Wind speed	4.5	5.0	5.5	6.0	6.5	7.0	7.5	8.0	8.5	9.0	9.5	10.0	10.5	11.0	11.5	12.0	12.5	Total
Count total noise	0	4	20	58	87	47	43	60	91	72	31	20	4	0	0	0	0	321
Count background noise	6	14	29	43	59	53	31	30	26	19	25	12	11	4	0	0	0	158

Figure 6 Overview of the number of valid measurements in a wind speed bin for the measured total and background noise

4.2.6 Apparent sound power level

If an apparent sound power level can be calculated for a given wind speed bin the total LWA and the 1/3-octaves are shown in the monitoring portal. Marked with * where the difference in the sum of the 1/3-octave bands of the estimated sound power level based on the total and background noise for a given wind speed bin is between 6 and 3 dB or [ ] for the cases where a background correction of 3 dB is applied for a given 1/3-octave.

Cases are not reported where the difference in the sum of the 1/3-octave bands of the estimated sound power level based on the total and background noise for a given wind speed bin is less than 3 dB. The calculated apparent sound power levels are also shown graphically for easy assessment at a glance (Figure 7).

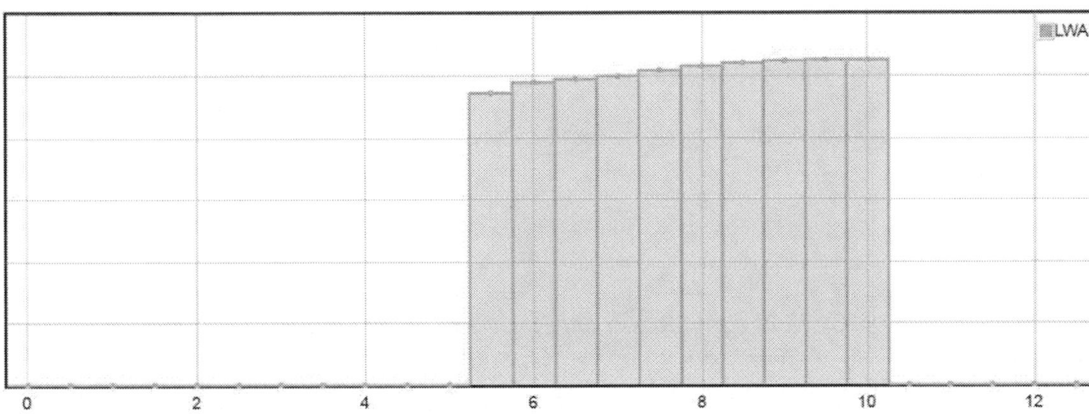

Figure 7 Apparent sound power level [dB(A)] per wind speed bin [m/s]

5. Conclusions

As discussed in Chapter 3 there are several factors that have to be known in order to assess if a measurement dataset is complete. Post processing the measurement data has major drawbacks (see section 3.6).

Switching often between wind turbine on or off will ensure that the background measurements are performed under similar conditions as prescribed by the standard. Each switch will however take some time in which measuring is not possible. Being able to monitor the measurement process real-time greatly assists in decision making in the field.

To be in control during measurements a real-time monitoring portal ("Peutz monitoring portal") is implemented wherein all measurement data is processed real time according to the standard including correction for disturbances and third octave band calculations.

The location of the reference microphone and the 10 m anemometer in relation to the yaw angle (wind direction) is shown real time in the monitoring portal. Measurements with invalid measuring locations are excluded from the calculations. When the wind direction is changing and a measuring position might soon become invalid. One can anticipate and move a measurement location prior to limit downtime and possible miss measuring a certain wind speed.

An overview of the amount of valid measurements in wind speed bins for the total noise and background is provided in the monitoring portal. The status of the measurement progress is visible at a glance.

If an apparent sound power level can be calculated for a given wind speed bin the total LWA and the full 1/3-octaves are shown in the monitoring portal. Possible points of interests can be immediately assessed.

The system can be easily used in the field and extended with multiple microphone locations. With enough microphones unmanned measurements can be performed as the whole range can be covered. Noise monitoring at residents can also be added.

The system is currently in the progress of being accredited for wind turbine measurements conform IEC 61400-11 edition 3.0 and has been used to perform several acoustic measurements on prototype wind turbines.

The system allows for efficient measurements which will not only reduce measurement time but improve the quality of the measurements as the ability to limit measurements to one day will ensure equal environmental conditions.

Results are readily available during and immediately after a measurement.

References

Jozwiak, R. et all. (2015) *Field comparison of IEC 61400-11 Wind turbines – Part11: Acoustic noise measurement techniques: Edition 3.0 and Edition 2.1* 6th International Meeting on Wind Turbine Noise Glasgow 20-13 April 2015

IEC 61400-11:2006 edition 2.1 Wind turbines – Part 11: Acoustic noise measurement techniques.

IEC 61400-11:2012 edition 3.0 Wind turbines – Part 11: Acoustic noise measurement techniques.

7th International Conference
on
Wind Turbine Noise
Rotterdam – 2nd to 5th May 2017

Analysis of sound emission by using amplitude modulation components of wind turbine noise

Yasuaki Okada, Faculty of Science and Technology, Meijo University
1-501 Shiogamaguchi, Nagoya 468-8502, Japan, E-mail: okada@meijo-u.ac.jp
Shinya Hyodo, Faculty of Science and Technology, Meijo University
1-501 Shiogamaguchi, Nagoya 468-8502, Japan, E-mail: 1204390089@ccalumni.meijo-u.ac.jp
Koichi Yoshihisa, Faculty of Science and Technology, Meijo University
1-501 Shiogamaguchi, Nagoya 468-8502, Japan, E-mail: yosihisa@meijo-u.ac.jp
Teruo Iwase, Professor Emeritus, Niigata University
8050 Igarashi 2-nocho, Niigata 950-2181, Japan, E-mail: iwase@cc.niigata-u.ac.jp

Summary

The noise generated from operating wind turbines is classifiable as aerodynamic or mechanical sound as a whole. To investigate the radiation characteristics of wind turbine noise under various wind conditions, field measurements of noise emitted from a single wind turbine have been performed over long periods. The six receiving points were set circularly around the wind turbine with a rated power of 1.5 MW. Wind turbine operational data were also collected at 1 s intervals along with corresponding acoustic data. The distinguishable sound directivity was revealed and a simple empirical formula for the directivity correction was proposed, assuming the wind turbine to be a point source with combined bi- and omnidirectional patterns. Furthermore, it was found that the relationship between wind turbine noise and the rotor rotational speed is extremely strong compared with that with the wind speed. However, it is generally difficult to obtain the actual rotor speed at several seconds intervals during noise measurements. Therefore, we focused on the amplitude modulation of sound generated from blades of the wind turbine and compared the rotor speeds estimated from the amplitude modulation with the actual mean values. The results showed that the rotor rotational speed can be identified with a reasonable accuracy by using the amplitude modulation components obtained around the wind turbine and the apparent A-weighted sound power levels set based on the estimated rotor speeds agree well with those based on the actual speeds. Additionally, the calculated strengths of the amplitude modulation depth in emission areas are presented.

1. Introduction

Regarding the directional characteristics of wind turbine noise (WTN), some studies based on aerodynamic sound theories and experiments have been carried out [1-4]. According to several studies, it was demonstrated that the overall sound pressure levels L_A of the tailing edge noise in the crosswind direction are 4 – 6 dB lower than those in the up- and downwind direction, although those measurement data were collected under limited wind conditions.

We have also performed field measurements around single wind turbines under various wind conditions, to examine the dependence of wind turbine operational conditions on the radiation characteristics of WTN, the horizontal directivity for estimating the noise emission levels, and the tonal components included in WTN [5,6]. In this study, the results of the rotor speed dependence of the apparent A-weighted sound power levels of WTN and a simple empirical formula for the directivity correction derived from the measurement data are presented.

Additionally, the rotor rotational speed was estimated by using the amplitude modulation (AM) components contained in WTN and the power levels set based on the estimated rotor speeds were compared with those based on the actual mean values. Furthermore, the strength of the AM depth in emission areas was calculated. The F-S method proposed by Fukushima and Tachibana was used as a reference for determining the AM components [7-9].

2. Field measurements

Noise measurements were performed around a single wind turbine over eight days [5]. The turbine has an upwind rotor diameter of 70.5 m and a hub height of 65 m. Its rated wind speed, output power, and rotor rotational speed are 12 m/s, 1.5 MW, and 20 rpm, respectively. To assess the directional characteristics of WTN, 6 receiving points were circularly arranged in the range of 210° around the wind turbine, as shown in Fig. 1. The points were set at a horizontal distance of 50 m except for one point (distance of 40 m). An all-weather-type wind screen with a diameter of 20 cm was installed on each microphone and acoustic signals measured using the A-frequency weighting were recorded on PCM recorders (48 kHz sampling, 16 bits).
After field measurements, the A-weighted sound pressure levels in 1/3-octave bands for the frequency range from 50 Hz to 5 kHz were analyzed at 100 ms intervals ($L_{p,100ms}$) from the recordings. To eliminate periods with intruding intermittent specific noise (e.g. road vehicle and birds) and wind-induced noise at a microphone, instantaneous changes in the 1/3-octave band spectrum owing to them were checked carefully while reproducing the sound. Then, the apparent A-weighted sound power levels (L_{WA}) were calculated using the time-averaged sound pressure levels over 10 s ($L_{peq,10s}$). In addition, with the wind turbine stopped forcibly, the residual sound levels ($L_{A95,T}$ in 1/3 octave bands) under various wind conditions were checked.

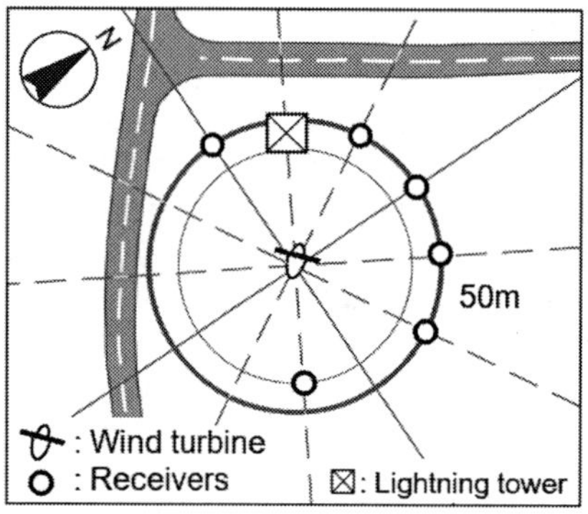

Rated power	1.5 MW
Hub height	65.0 m
Rotor diameter	70.5 m
Cut-in wind speed	3.0 m/s
Rated wind speed	12.0 m/s
Rated rotor speed	20.0 rpm

Figure 1 – Arrangement of receiving points around a wind turbine.

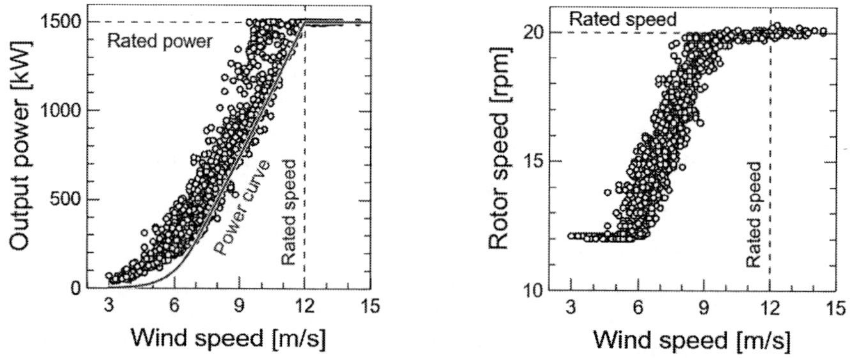

Figure 2 – Operational conditions of the wind turbine during noise measurements.

2.1 Wind turbine operational conditions

Meteorological and wind turbine operational data (wind speed at hub height, nacelle direction, output power, and rotor speed) were collected at 1 s intervals. Figure 2 shows the mean values of them calculated over 10 s during the measurements. The number of data was 1,842. The noise measurements were made over the range from the cut-in wind speed (3 m/s) to the rated speed (12 m/s). The rotor rotational speed reaches the rated speed (20 rpm) at a mean wind speed of about 9 m/s, whereas the rated wind speed is 12 m/s.

2.2 Dependences of wind speed and rotor rotational speed on WTN

Figs. 3 and 4 show the apparent A-weighted sound power levels plotted against the wind speed at hub height and rotor rotational speed, respectively. The circles in the figures represent the measured L_{WA} within ±30° relative to the up- and crosswind direction. The measuring position was determined from the angle between each receiving point and the nacelle direction. The solid lines indicate the calculated mean levels at 1 m/s and 0.5 rpm intervals, respectively.

The apparent A-weighted sound power level of WTN increases with increasing wind speed, as presented in Fig. 3. However, the correlation between L_{WA} and the wind speed at hub height is only moderate, because the change in the wind speed is rapid as well as irregular under actual meteorological conditions. On the other hand, the correlation between L_{WA} and the rotor speed is extremely strong compared with that with the hub height wind speed, as presented in Fig. 4. In addition, L_{WA} increases almost linearly up to the rated rotor speed, regardless of the direction of measuring position relative to the nacelle.

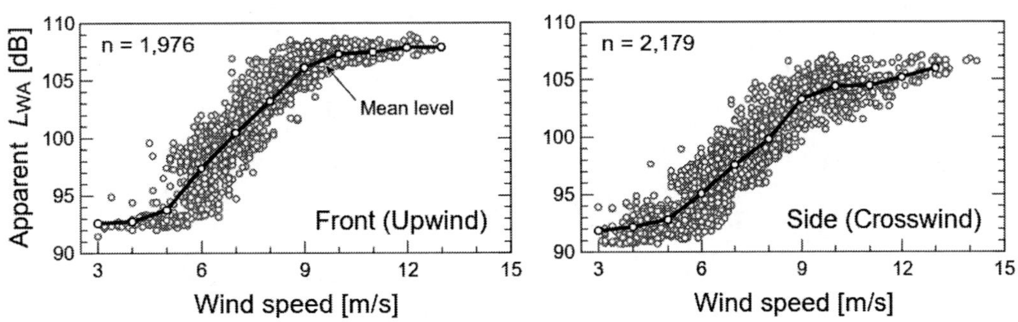

Figure 3 – Wind speed dependence of apparent A-weighted sound power level. Measuring positions are within ±30° relative to the up- and crosswind directions (n: number of data).

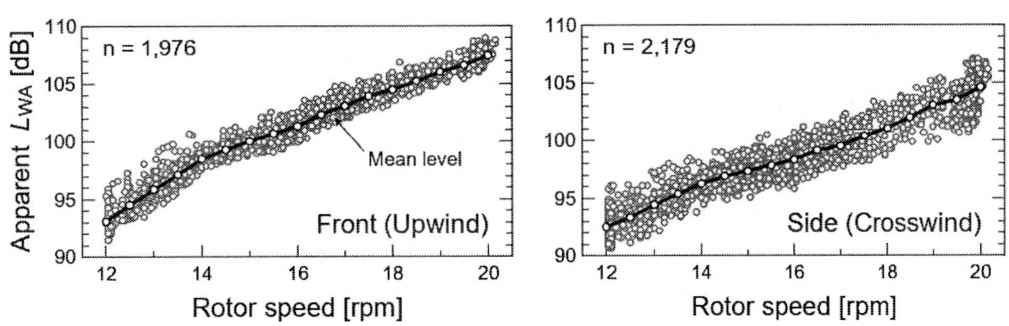

Figure 4 – Rotor speed dependence of apparent A-weighted sound power level. Measuring positions are within ±30° relative to the up- and crosswind directions (n: number of data).

2.3 Empirical formula for horizontal directivity of WTN

Several calculation methods for the horizontal sound directivity of WTN have been proposed on the basis of aerodynamic sound theories or semi-empirical prediction methods [1-3]. To simplify the modeling of the directivity pattern of WTN, we have focused on the difference between the sound pressure levels $L_{peq,10s}$ in the up- or downwind direction and those in the other directions, which was obtained through field measurements. A simple regression formula was applied, assuming the directivity pattern of aerodynamic and mechanical sound to be bi- and omnidirectional, respectively. The directivity correction $\Delta L_{dir,\theta}$ is expressed by combining both directional patterns as follows:

$$\Delta L_{dir,\theta} = 10\lg\left(\frac{1+a\,|\cos^b\theta|}{1+a}\right) \quad 0 \le \theta \le 360^\circ, \tag{1}$$

where θ is the direction of the measuring position relative to the wind turbine and a and b are coefficients for the sound directivity. Those coefficients (a, b) were derived from the measured A-weighted sound pressure levels in 1/1-octave bands, which are the dominant frequency components of WTN [6].

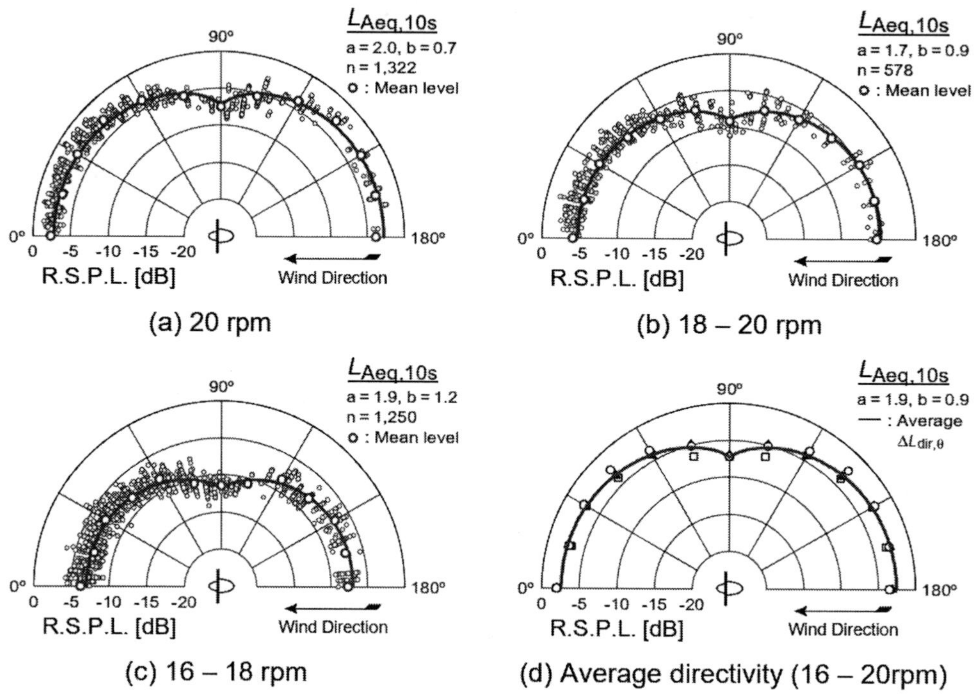

Figure 5 – Distribution of A-weighted sound pressure level and mean level at 15° intervals (—: $\Delta L_{dir,\theta}$ calculated by Eq.(1), a, b: coefficients for $\Delta L_{dir,\theta}$, n: number of data).

Table 1 – Calculated level difference $\Delta L_{dir,90^\circ}$ relative to the up- and downwind directions of the wind turbine (a, b: coefficients for $\Delta L_{dir,\theta}$).

Rotor speed	$L_{Aeq,10s}$		250 Hz		500 Hz		1 kHz	
[rpm]	$\Delta L_{dir,90^\circ}$	a, b	$\Delta L_{dir,90^\circ}$	a, b	$\Delta L_{dir,90^\circ}$	a, b	$\Delta L_{dir,90^\circ}$	a, b
20	-4.8	2.0, 0.7	-6.7	3.7, 0.8	-6.3	3.3, 0.8	-3.4	1.2, 0.6
18 – 20	-4.3	1.7, 0.9	-5.7	2.7, 1.0	-5.6	2.6, 0.9	-3.6	1.3, 0.7
16 – 18	-4.6	1.9, 1.2	-5.9	2.9, 1.5	-6.1	3.1, 1.2	-3.8	1.4, 1.2
Average	-4.6	1.9, 0.9	-6.0	3.0, 1.1	-6.0	3.0, 1.0	-3.4	1.2, 0.9

Figure 5 shows the measured A-weighted sound pressure levels (○) and the mean values calculated (○) at 15° intervals. The measured data were divided into 3 groups in consideration of the rotor speed dependence of the emitted noise L_{WA}. The solid lines indicate the directivity correction $\Delta L_{dir,\theta}$ derived from the results under operational conditions at rotor speeds of 20 rpm, 18 – 20 rpm, and 16 – 18 rpm. The coefficients a and b are shown in Table 1.

The calculated $\Delta L_{dir,\theta}$ agrees reasonably well with the measured sound pressure levels, whereas the coefficients a and b depend on the frequency band as well as the rotor rotational speed. The difference $\Delta L_{dir,90°}$ between sound levels in the up-/ downwind direction and those in the crosswind direction are within 4 – 5 dB for $L_{Aeq,10s}$, and those at 250 Hz and 500 Hz are within 6 –7 dB, as shown in Table 1. This tendency is similar to the measurement results at other wind turbine sites [1,2].

In addition, to grasp the average horizontal directivity of WTN, the average $\Delta L_{dir,\theta}$ was derived using the mean level differences under each operational condition. Figure 5(d) shows the average directivity of the A-weighted sound pressure level at speeds of 16 – 20 rpm. The coefficients a and b are 1.9 and 0.9, and the average $\Delta L_{dir,90°}$ is about -5 dB, respectively.

In order to validate this average directivity $\Delta L_{dir,\theta}$, it was compared with the results measured around another wind turbine with a rated power of 2.0 MW [6]. Figure 6 shows a comparison between the directivity patterns of the A-weighted sound pressure level in 1/1-octave bands (250 Hz, 500 Hz and 1 kHz) for different wind turbines. The average directivity $\Delta L_{dir,\theta}$ of the A-weighted sound pressure level is qualitatively similar to that obtained at another wind turbine, whereas the frequency dependence of the decrease in the sound level in the crosswind direction is different for individual wind turbines.

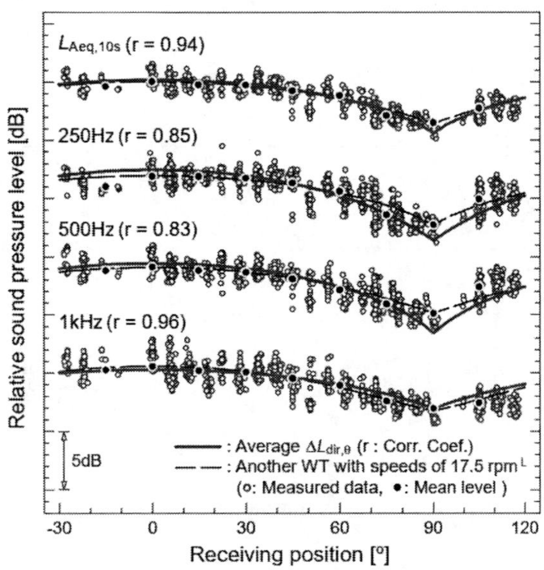

Figure 6 – Comparison of horizontal directivity of the A-weighted sound pressure level in 1/1-octave bands for different wind turbines (number of data: 1,312).

3. Examination focused on amplitude modulation

As mentioned previously, the radiation characteristics of WTN depend strongly on the rotor rotational speed. However, it is generally difficult to obtain the actual rotor speed at several seconds intervals during noise measurements. Therefore, we focused on the periodic fluctuations (AM) of sound generated from blades of a wind turbine, and the rotor speed was estimated from the blade-passing-frequency (BPF), which can be detected by calculating fourier spectrum of the AM components contained in WTN. Then the estimated rotor speeds were compared with actual values. In addition, the AM depths D_{AM} [7-10] which evaluate the strength of the AM components, in emission areas under various operational conditions were calculated.

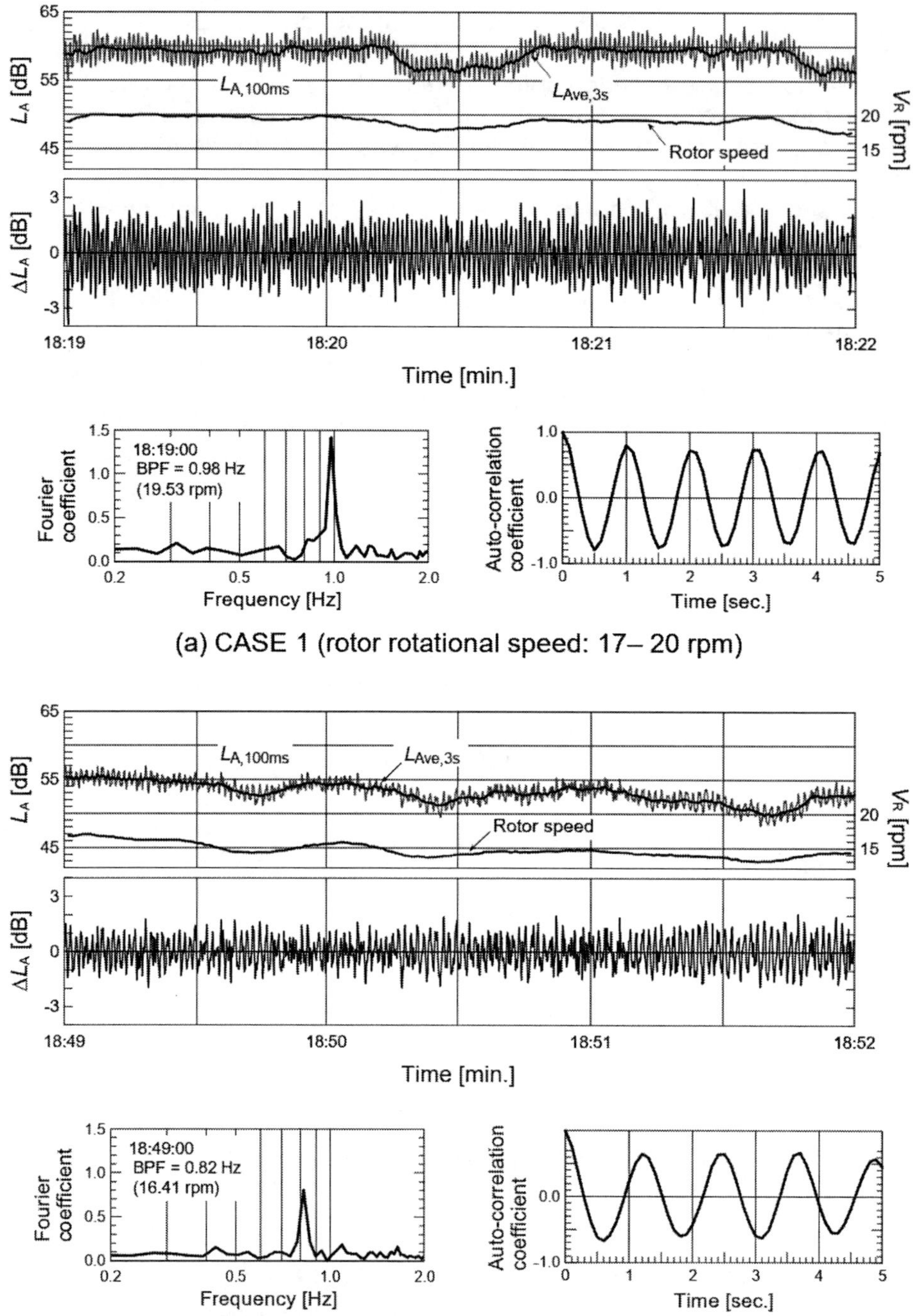

(a) CASE 1 (rotor rotational speed: 17– 20 rpm)

(b) CASE 2 (rotor rotational speed: 13 – 17 rpm)

Figure 7 – Examples of A-weighted sound pressure level $L_{A,100ms}$, moving average value $L_{Ave,3s}$, level difference ΔL_A and fourier and auto-correlation coefficients of ΔL_A in the first 30 s (V_R: rotor rotational speed at 1 s intervals).

3.1 Estimation of rotor rotational speed

The measurement data of WTN in emission areas contain periodic fluctuations (AM) due to blade passing as well as irregular gradual changes caused by increase or decrease in the radiated sound power. In this study, the F-S method was used as a reference, in which the difference ΔL_A between the A-weighted sound pressure level with the FAST time-weighting $L_{A,F}$ and that with the SLOW time-weighting $L_{A,S}$ is calculated in order to extract the periodic fluctuation of AM by removing the gradual change. Then, the AM depth D_{AM} is obtained by calculating the 90% range of ΔL_A. The effectiveness of this method has been theoretically proved [9], while we used a moving average value of $L_{A,100ms}$ instead of $L_{A,S}$ to simplify the calculation procedure, as follows:

$$\Delta L_A = L_{A,100ms} - L_{Ave,3s} , \quad (2)$$

$$D_{AM} = \Delta L_{A,5} - \Delta L_{A,95} , \quad (3)$$

where $L_{Ave,3s}$ is the 3-seconds moving average value of $L_{A,100ms}$ (blade-passing-interval for the target wind turbine: 1.7 – 1.0 s) and $\Delta L_{A,5}$ and $\Delta L_{A,95}$ are the 5% and 95% point on the cumulative distribution of ΔL_A.

Figure 7 shows the time-traces of $L_{A,100ms}$, $L_{Ave,3s}$ and level difference ΔL_A between them, when the wind turbine operated at around the rated speed (CASE1, 17 – 20 rpm) and low speed (CASE2, 13 – 17 rpm). The A-weighted sound pressure level of WTN changes gradually with the rotational speed V_R and those variations could be expressed by the moving average value $L_{Ave,3s}$. The extent of the periodic fluctuation ΔL_A which represents the AM component of WTN was apt to increase with increasing rotor speed. In addition, the rotor speed varies about 2 rpm in only a few minutes. Thus, in this study, the rotor speed was estimated by using the AM component ΔL_A over 30 s or 1 min. The examples of the fourier spectrum and auto-correlation function of ΔL_A in the first 30 s are presented in Fig. 7. It can be seen that the BPF of ΔL_A is apparently detected as a fundamental frequency of 0.98 Hz or 0.82 Hz, corresponding to a periodicity of 1.0 s or 1.2 s, and the estimated rotor speeds (19.53/ 16.41 rpm) were within the actual speed range (19 – 20/ 16 – 17 rpm), respectively.

Next, the BPFs were detected by using ΔL_A calculated from all of the A-weighted sound pressure levels ($L_{A,100ms}$, $L_{Ave,3s}$) obtained in a measuring period of 8 days, and the rotor speeds calculated from the BPFs were compared with the actual mean values.

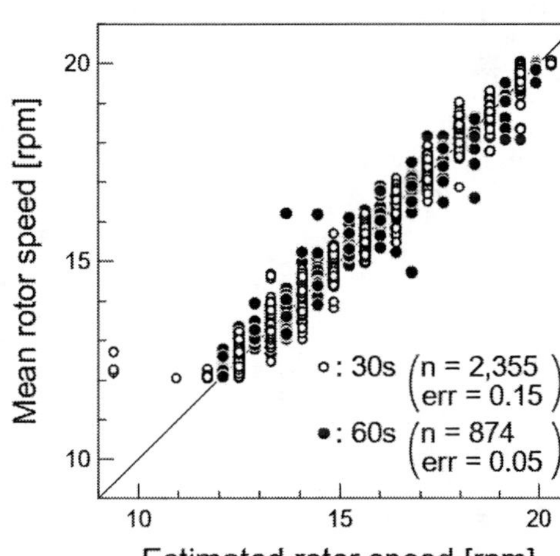

Figure 8 – Comparison between rotor speed estimated by using AM components of WTN and actual mean value (Analysis time: 30, 60 s, n: number of data).

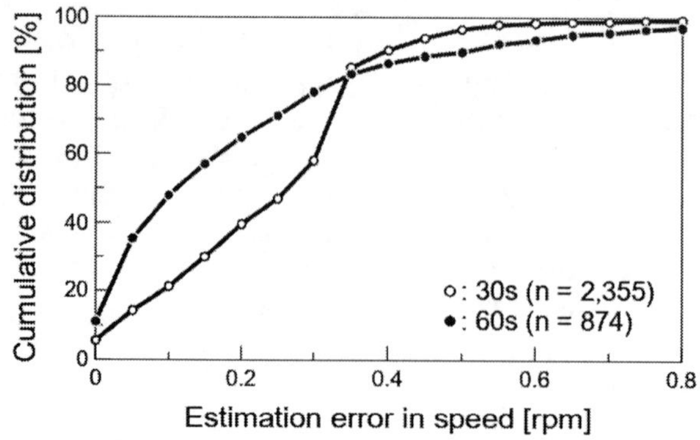

Figure 9 – Cumulative distribution of estimation error in rotor speed (Analysis time: 30, 60s, n: number of data).

Figs. 8 and 9 show comparison results in the case of using ΔL_A for 30 s and 1 min, and the cumulative distribution of difference between the estimated rotor speeds and the actual values. The number of data were 2,355 for 30 s and 874 for 1min, respectively. The estimated rotor speed agree reasonably well with the actual mean values, whereas the difference of more than 1.0 rpm between them are seen in a few cases. In the case of using data for 30 s, the 90% of all the estimated rotor speeds are within ±0.4 rpm of the actual mean values. Thus, the rotor rotational speed could be identified accurately by using the AM components of sound pressure levels measured around the wind turbine.

3.2 Radiation characteristics of WTN on the basis of estimated rotor speed

The average A-weighted sound power levels set based on the estimated rotor speed were compared with those based on actual mean speed. Figure 10 shows those results in the case of using data for 30 s. The number of data used to calculate the mean power levels at 1.0 rpm is given in parentheses. The unfilled circle indicates the mean value calculated using the power levels over 10 s, as presented in Fig. 4. The power levels on the basis of the estimated speed are good agreement with those based on the actual values. Thus, in case any rotor rotational speed is not be obtained, the radiation characteristics of the emitted noise can be determined by using the estimated speed from the measurement data (AM) around the wind turbine. Note that the sound pressure level is different by the direction of the measuring position relative to the nacelle, as mentioned in Section 2.

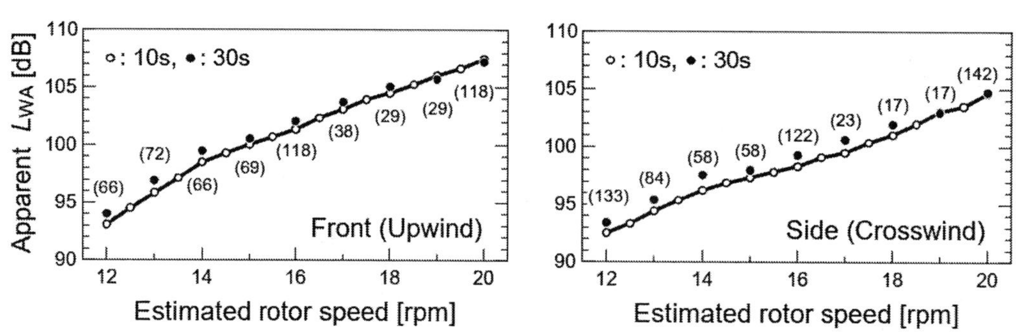

Figure 10 – Comparison between apparent A-weighted sound power levels set based on estimated speed and those based on actual mean value which is shown in Fig. 4 (Analysis time: 30s, (n): number of data).

3.3 AM depth in emission areas

The AM depths D_{AM} in emission areas were calculated from 90% range of the level difference ΔL_A for 30 s and 1 min, respectively. At first, to validate the extraction method of ΔL_A from the measurement data, the calculated AM depths D_{AM} were compared with those values in accordance with F-S method, as shown in Fig. 11. The AM depths D_{AM} calculated from both methods were almost the same. The mean difference between both values was -0.16 dB and the standard error was less than 0.5 dB, whereas the differences were almost 1 dB in some cases.

Figure 12 shows the AM depth D_{AM} calculated from ΔL_A for 30 s and 1 min, respectively. The strengths of D_{AM} in 30 s and 1 min were almost the same and distributed primarily within a range of 1.5 – 4.0 dB. In the future, we intend to investigate the dependence of rotor rotational speed on the strength of AM components and the horizontal directivity of them on the basis of noise measurement data in emission areas.

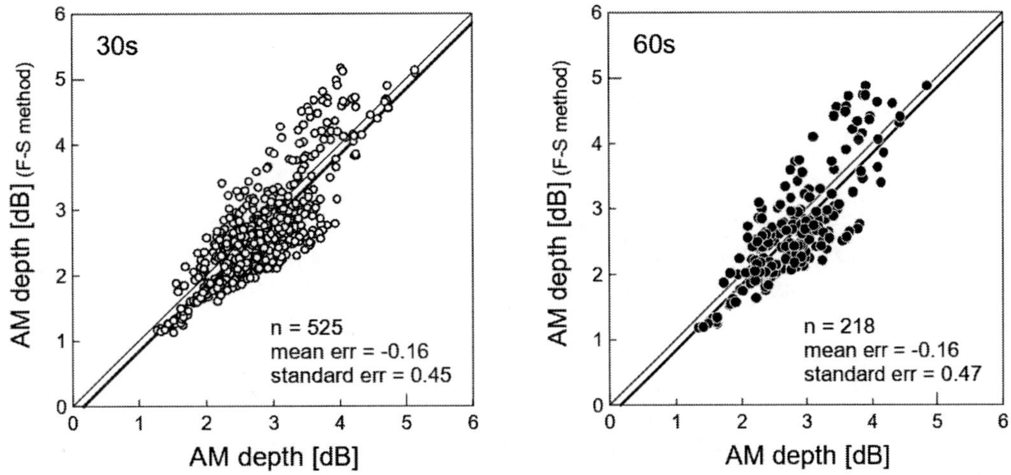

Figure 11 – Example of comparison with AM depth D_{AM} calculated by F-S method (Analysis time: 30, 60s, n: number of data).

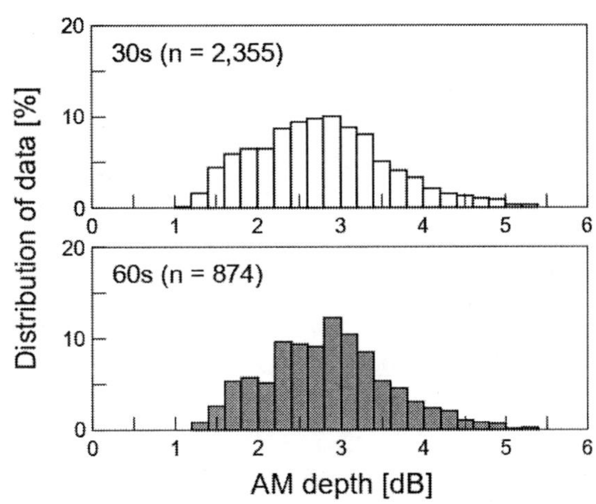

Figure 12 – Distribution of AM depth D_{AM} in emission area (Analysis time: 30, 60s, n: number of data).

4. Conclusions

Field measurements have been performed at a single wind turbine site over long periods to examine the radiation characteristics of WTN at the ground level under various operational conditions. The distinguishable sound directivity is revealed, whose pattern can be expressed by a simple empirical formula, assuming the wind turbine to be a point source with combined bi- and omnidirectional patterns. The A-weighted sound pressure levels in the crosswind direction are 4 – 5 dB lower than those in the up- and downwind directions. The tendency is similar to the results at other wind turbine sites in the previous measurements [1,2]. Additionally, the noise emission radiated from wind turbines depend strongly on the rotor rotational speed, which can be estimated from the AM components contained in WTN.

References

[1] S. Oerlemans *et al.* (2009) *Prediction of wind turbine noise and validation against experiment*, Aeroacoustics **8**(6), 555-584

[2] M. Friman (2011) *Directivity of sound from wind turbines: A study on the horizontal sound radiation pattern from a wind turbine*, Department of Aeronautical and Vehicle Engineering

[3] S. Lee *et al.* (2013) *Numerical modeling of wind turbine aerodynamic noise in the time domain*, J. Acoust. Soc. Am. **133**(2), 94-100

[4] H. H. Hubbard *et al.* (1991) *Aeroacoustics of large wind turbines*, J. Acoust. Soc. Am. **89**(6), 2495-2508

[5] Y. Okada *et al.* (2015) *Radiation characteristics of noise generated from a wind turbine*, Acoustical Science and Technology **36**(5), 419-427 (doi:10.1250/ast.36.419)

[6] Y. Okada *et al.* (2016) *Horizontal directivity of sound emitted from wind turbines*, Acoustical Science and Technology **37**(5), 239-246 (doi:10.1250/ast.37.239)

[7] H. Tachibana *et al.* (2014) *Nationwide field measurements of wind turbine noise in Japan*, Noise Control Eng J. **62**(2), 90-101

[8] A. Fukushima *et al.* (2013) *Study on the amplitude modulation of wind turbine noise: Part 1– Physical investigation*, Proc. 42nd Inter-noise 2013

[9] A. Fukushima *et al.* (2015) *Practical measurement method of wind turbine noise*, Proc. 6th Wind Turbine Noise 2015

[10] UK Institute of Acoustics (2016) *A Method for Rating Amplitude Modulation in Wind Turbine Noise*, Amplitude Modulation Working Group Final Report

[11] Y. Okada *et al.* (2016) *Study on empirical formula for the sound directivity around a wind turbine*, Proc. 45th Inter-noise 2016

7th International Conference
on
Wind Turbine Noise
Rotterdam – 2nd to 5th May 2017

Modelling Activities in Wind Turbine Aeroacoustics at DTU Wind Energy

Wen Zhong Shen, Department of Wind Energy, Technical University of Denmark: wzsh@dtu.dk
Wei Jun Zhu, Emre Barlas, Harald Debertshäuser, Jens Nørkær Sørensen, Department of Wind Energy, Technical University of Denmark

Summary

This paper gives a comprehensive overview on the modelling activities in wind turbine aeroacoustics at DTU Wind Energy in the last 20 years, and it gives also a summary of the state-of-the-art wind turbine noise prediction models used in DTU's software WindSTAR. Various noise generation models have been developed at DTU, which include Acoustic analogy, Flow-acoustics splitting technique (Splitting technique), Amiet's model, HAWC2-Noise model and BPM-FLEX-CAA model. For long distance acoustic propagations, the novel wind turbine noise propagation model using CFD and Parabolic Equation (PE) method has been developed. As a feature of the propagation package in WindSTAR, the rotating wind turbine noise source is modelled with an unsteady moving source located in one or a number of PE computational domains. In order to take the wind turbine wake, atmospheric turbulence and wind shear effects into account, the noise propagation model is combined with CFD/RANS or CFD/LES computations.

1. Introduction

In order to alleviate the global warming, many countries set the governmental goal of 100% renewable energy consumption by 2050. This means a vast increase of wind turbine installations will happen in the next 30 years with increased rotor size and number of wind farms, which will give certain environmental impacts. One of the major impacts is the noise generation from wind turbines and wind farms. For example, the newly updated Danish wind turbine noise regulation [1] is very restricted for both broadband and low frequency wind turbine noise radiation. Unfortunately, the energy capture and noise generation are two competing factors. Therefore it needs to develop sophisticated design tools to fulfil the requirements of high power performance and low noise emission for wind turbine and wind farm design.

Either single wind turbine or wind farm noise generation consists of basic aerodynamic nature, which is caused by turbulent flows interacting with wind turbine blades. There exist different noise prediction models, most of which are based on semi-empirical scaling laws. The development in high-performance computing (HPC) technology provides many possibilities to perform computational fluid dynamics (CFD) and computational aero-acoustics (CAA) simulations for noise generated from wind turbine blades.

The following sections provide a review of DTU's in-house developed noise generation and propagation models, which form the basis in the software WindSTAR (Wind turbine Simulation Tool for AeRodynamic noise). The topics cover researches in noise generation from airfoil noise to wind turbine, and its long range propagation under various flow conditions. The paper is organized as follows: Section 2 introduces the wind turbine flow solver which is the platform of the acoustic solver; Section 3 discusses the in-house developed wind turbine noise generation models in WindSTAR-Gen (Gen means generation); In Section 4, the noise propagation model and the coupling of noise generation and propagation models in WindSTAR-Pro (Pro means propagation) is discussed. Conclusions are presented in Section 5.

2. Wind turbine flow solver

Flow simulations are often needed before the step of CAA computations where the flow is governed by the three-dimensional Navier-Stokes equations. For turbulent flow induced noise, different turbulence scales are responsible for the noise generation. In Large Eddy Simulation (LES), the large scales are directly computed but the small scales are modelled with an eddy viscosity based sub-grid-scale (SGS) model.

At DTU, the EllipSys3D code [2, 3] was developed for both CFD RANS and LES computations. The solver is based on structured grid, with a multi block / cell-centred finite volume discretization. The Navier-Stokes equations can be solved either steady or unsteady using pressure-velocity coupling technique where the predictor-corrector method is used. In the predictor step, a second-order backward differentiation scheme is used for time discretization and a second-order central difference scheme is used in spatial discretization, except for the convective terms that are discretized by the QUICK upwind scheme. To avoid numerical oscillations from the velocity-pressure decoupling, the improved Rhie-Chow interpolation developed by Shen et al. [4] is used during the corrector step. Instead of using the SIMPLE algorithm, the improved SIMPLEC scheme for collocated grids [5] is implemented as well such that the solution is independent of the relaxation parameter and time-step. The solver is based on a five-level multi-grid technique that improves the convergence speed quite well. The EllipSys3D code is programmed with a multi-block topology, and therefore it is parallelized relatively easily using Message Passing Interface (MPI). When LES is performed, the mixed scale turbulence model of Ta Phuoc [6] is used.

3. Wind turbine noise generation models

Many noise generation models have been developed at DTU which include flow-acoustic splitting technique [7-11], acoustic analogy [12-14], Amiet's model [15], HAWC2-noise model [16-20], and BPM-FLEX-CAA model [21-22]. Since the BPM-FLEX-CAA model is new and consists of an engineering model BPM and a CAA model splitting technique, we only present the BPM-FLEX-CAA model here. Moreover, other engineering models, such as Amiet's model, HAWC2-noise model, can also be used to replace BPM in BPM-FLEX-CAA. For the other models, the reader is referred to [23].

3.1 BPM-FLEX-CAA model

This section introduces the hybrid model developed by Debertshäuser et al. [21-22] that combines a semi-empirical model and an advanced CAA method. In the CAA method, the acoustic simulations are coupled with the LES/AL (Actuator Line technique) simulations at the same time. In the case of flexible blades, the aero-elastic code FLEX5 is an optional build-in tool.

Semi-empirical BPM model

In the semi-empirical BPM model, the modelling of turbulent inflow noise is based on Lawson's scaling law [24]

$$\mathrm{SPL}_{inflow} = 10log_{10}\left[D\rho^2 c_0^2 L \frac{\Delta l}{r^2} M^3 I^2 \hat{k}^3 \left(1+\hat{k}^2\right)^{-7/3}\right] + C \qquad (1)$$

where D is the sound directivity, ρ is the air density, L is the typical atmospheric turbulence length scale, Δl is the airfoil section semi-span, M denotes the Mach number, $\hat{k}$ is the normalized wavenumber, and I is the turbulence intensity. The rotor self-noise calculation is based on the 2D airfoil self-noise model [25]. A steady BEM method is often combined with the airfoil self-noise model which consists of scaling laws for several noise mechanisms. The general form of the airfoil self-model reads,

$$\mathrm{SPL} = 10log_{10}\left(\frac{\delta M^n lD}{r^2}\right) + G_1(St) + G_2(Re) + G_3(\delta) + C. \tag{2}$$

In Eq. (2), it is shown that the airfoil noise is related to a thickness parameter δ, either relating to the boundary layer thickness or blunt trailing-edge (TE) thickness. The length parameter l represents either the length of airfoil span or the length of blade tip. The other functions, G_1, G_2 and G_3 are related to the Strouhal number, Reynolds number and thickness parameter. The major self-noise mechanisms are depicted in Figure 1, where TE noise, TE blunt noise, stall noise and tip noise are individually shown.

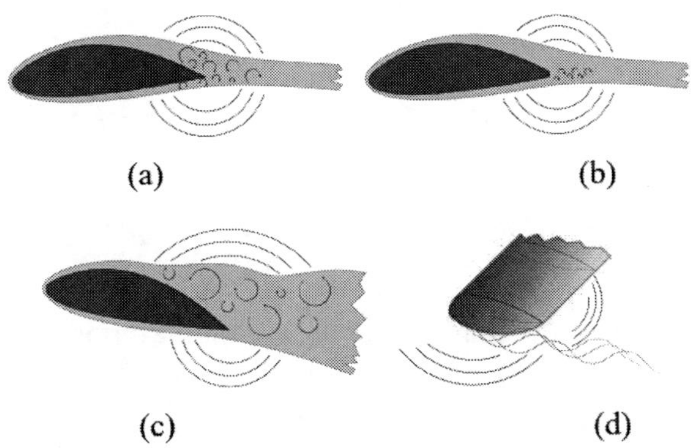

Figure 1: Wind turbine blade self-noise mechanisms.

The wind turbine noise prediction model directly applies the airfoil noise model to the rotating wind turbine blades [26]. The total wind turbine aerodynamic noise is summed up from each of the blade elements such as

$$SPL_{total} = 10log_{10}\left(\sum_i^n 10^{0.1SPL_{total}^i}\right) \tag{3}$$

where the total noise is the sum of n blade elements and SPL_{total}^i is the total noise from each airfoil element.

AL model

The development of the BPM noise prediction tool requires rotor aerodynamic computations as the basic input to the model. The most popular method used for this purpose is the BEM method that directly computes the local velocity and angle of attack at the blade elements. A new approach, using the CFD based Actuator Line (AL) technique [27] is introduced by Debertshäuser et al. [21-22]. The AL technique breaks the limit of the BEM method and the outputs of the aerodynamic simulation are time dependent, fluctuating flow field. The complex flow conditions can thus be modelled, such as turbulent inflow, wind shear and yaw, etc. The time dependent flow parameters extracted from CFD computations are fed into the BPM model. The approach to compute the flow field over the wind turbine blades is done by adding the volume force to the momentum equation

$$\frac{\partial U_i}{\partial t} + \frac{\partial(U_iU_j)}{\partial x_j} = -\frac{1}{\rho_0}\frac{\partial P}{\partial x_i} + \nu\frac{\partial^2 U_i}{\partial x_j^2} + \boldsymbol{f}, \tag{4}$$

$$\boldsymbol{f} = \frac{1}{2}\rho V_{rel}^2 c(C_L, C_D). \tag{5}$$

The force $\boldsymbol{f}$ is computed iteratively with the blade element method combined with tabulated airfoil characteristics. As shown in Figure 2, the elements of the rotating blades are represented with a body force. At each time-step, the EllipSys3D flow solver gives a velocity field. The relative velocity V_{rel} at each blade segment is calculated by identifying the blade position at the current time instant. To take the blade flexibility into account, the AL approach is coupled with the in-house developed aero-elastic code FLEX5 that handles more complex structural deformations and thus influences noise generations.

CAA model

The Splitting method is based on the flow and acoustics splitting approach [7-8] and implemented in the EllipSys code as the CAA solver [9-11]. The compressible Navier-Stokes equations are decomposed into an incompressible flow part and an acoustic perturbation part. The final set of acoustic equations consists of density equation, pressure correlation equation and auxiliary correlated velocity fluctuation equations

$$\frac{\partial \rho^*}{\partial t}+\frac{\partial f_i}{\partial x_i}=0, \tag{6}$$

$$\frac{\partial p^*}{\partial t}-c^2\frac{\partial \rho^*}{\partial t}=-\frac{\partial \bar{P}}{\partial t}, \tag{7}$$

$$\frac{\partial f_i}{\partial t}+\frac{\partial}{\partial x_j}\left[f_i\left(\bar{U}_j+u_j^*\right)+\rho_0\bar{U}_i u_j^*+\left(p^*+\frac{2}{3}\rho^* k\right)\delta_{ij}\right]=\frac{\partial}{\partial x_j}\left[\rho^* v_t\left(\frac{\partial \bar{U}_i}{\partial x_j}+\frac{\partial \bar{U}_j}{\partial x_i}\right)\right]. \tag{8}$$

The unknowns with a superscript (*) indicate acoustic variables and the capital letters U_i and P are the resolved incompressible flow variables. The auxiliary variables are defined as $f_i=\rho u_i^*+\rho^*\bar{U}_i$. The system is closed with a speed of sound equation that is valid for real gases,

$$c^2=\frac{\gamma\left(\bar{P}+p^*\right)}{\rho_0+\rho^*} \tag{9}$$

The acoustic computation may start after the flow field is fully established. The incompressible flow parameters form the input to the acoustic equations. The flow and acoustic simulations can have different meshes and different time-steps determined from their CFL numbers, which can greatly accelerate the computations. In the following CAA calculations, the incompressible Navier-Stokes equations are always solved by the second-order finite volume EllipSys code, but the acoustic equations have the option to use high-order wavenumber optimized schemes, ranging from 2nd-order to 10th-order schemes [11].

Application of the BPM-FLEX-CAA model to the Nordtank 500 kW turbine

Figure 2 (a) is the user interface of the BPM noise prediction tool. The input of blade geometry as well as airfoil data is required for the BEM model. Turbulence length scale and turbulence intensity are inputs to the inflow noise model. Trailing-edge geometry details, such as bluntness and serration sizes are applied for the purpose of more specific modelling. The A-weighted sound pressure spectra are calculated and compared with measurements at wind speeds of 6, 8, and 10m/s in Figure 2(b). The model is seen in good agreement with the experiment, especially at wind speeds of 8 and 10m/s.

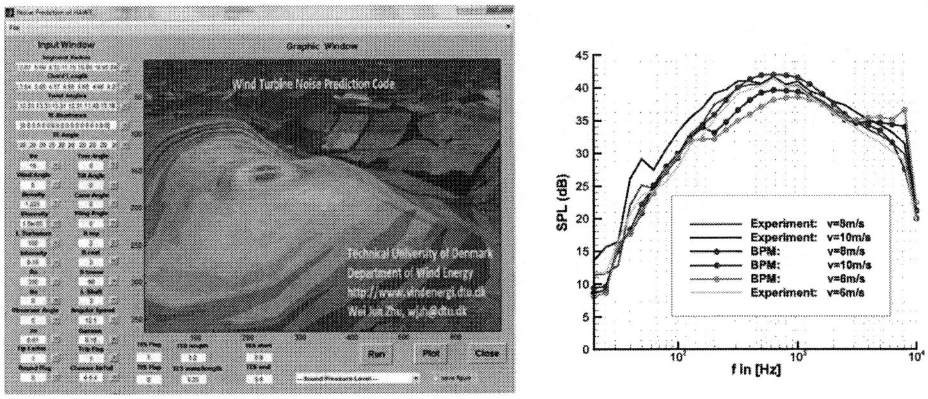

(a) (b)

Figure 2: (a) User interface of the noise generation model. (b) Predictions using the BPM model and compared with the DTU field measurements for a Nordtank 500 kW turbine at free-stream wind speeds of 6m/s, 8m/s and 10m/s.

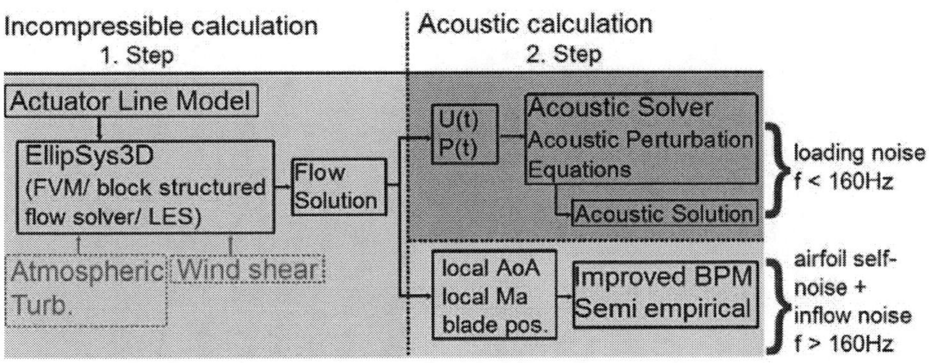

Figure 3: Flow chart of the coupled BPM-FLEX-CAA model.

The noise generation under the atmospheric turbulence and wind shear condition is complicated. To solve this problem, a coupling approach is proposed in Figure 3 where *Step1* contains the LES/AL flow simulation and *Step2* contains the CAA splitting method and BPM hybrid models. The low frequency noise created from the atmospheric turbulence and wind shear is captured by the CAA model using the splitting technique. To apply the CAA approach only for the low frequency calculation can greatly save the computational time. The high frequency component is calculated with the BPM model and the flow inputs obtained from LES. The time dependent angle of attack at each blade element is computed from the AL technique. In Figure 4 the stream-wise velocity field at a wind speed of 10m/s is shown. The wind turbine centre is located at Z = 5R, Y = 1.8R. The atmospheric turbulence is introduced at Z = -1R and thus influences the incoming flow facing the wind turbine, as well as on the wake development. In particular, it has two influences: (1) on the local angle of attack and local relative velocity that are used for the BPM noise calculation; (2) on the noise radiation in the computational domain.

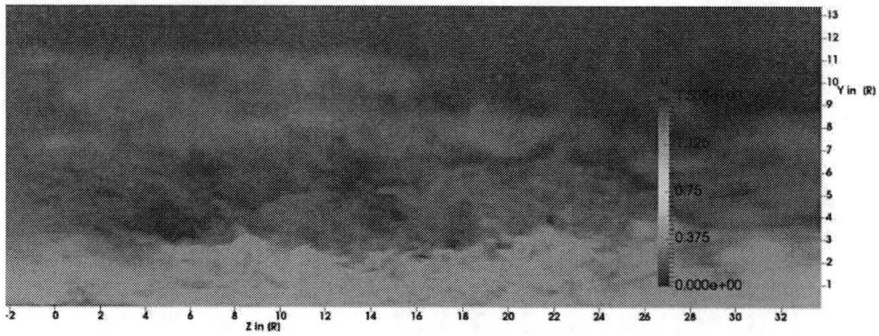

Figure 4: Normalized steam-wise velocity calculated with the actuator line method where the rotor centre is positioned at z = 5R, and y = 1.8R.

At a wind speed of 10m/s, the wind turbine generated aerodynamic noise field is seen in Figure 5. As seen in the figure, there is a high level of sound pressure generated at the rotor position. The propagation of wind turbine noise is influenced by the atmospheric turbulence as well as the wake turbulence. As a result of the time history sound pressure, Figure 6 contains the overall sound pressure level recorded during a time period of about two minutes at the three wind speeds. The fluctuations of the sound pressure level are due to the inflow turbulence, wake and wind shear.

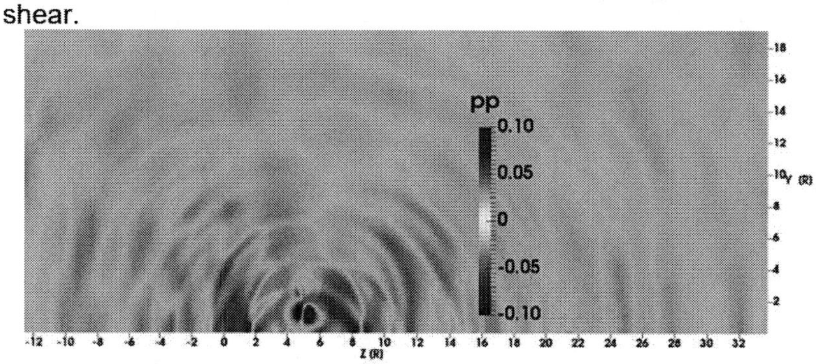

Figure 5: Acoustic pressure calculated with the flow-acoustic splitting technique for the Nordtank 500 kW turbine at a wind speed of 10m/s; Turbine centre is placed at Z = 5R, Y = 1.8R.

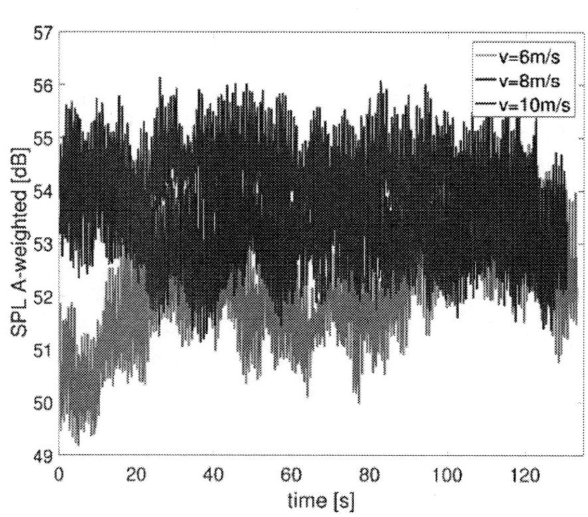

Figure 6: A-weighted overall sound pressure levels at a distance of 45m for the Nordtank 500kW turbine at wind speeds of v = 6, 8, 10m/s over a short time period.

4. Modelling of wind turbine noise propagation

The noise propagation model used in the WindSTAR-Pro software is presented here.

4.1 Noise propagation model

Various outdoor sound propagation modelling techniques are available that can be adopted for predicting wind turbine and wind farm noise propagations (see Berengier et al. [28] for a brief review). Due to the nature of wind turbine noise source characteristics, most of the methods cannot be directly used as a wind turbine noise propagation tool. For far field noise propagations, we choose the Parabolic Equation (PE) model as a compromise between numerical accuracy and computational effort. The PE method is a solution of the wave equation with the approximations of harmonic wave propagations with a finite angle. The solution of each PE simulation yields a steady solution at each frequency. For rotor noise propagations, if a time dependent solution is desired, multiple PE simulations can be carried out successively in order to capture the SPL time evolution.

Following the notation in Blanc-Benon et al. [29], the PE methods can be divided into two groups: scalar and vector PE. The scalar PE is the conventional approach where the moving atmosphere is represented by a hypothetical frozen medium with a fixed effective sound speed $c_{eff} = c + v_x$ where v_x is the wind velocity component along the direction of propagation between source and receiver. For the vector PE, the vector properties of the velocity field are maintained, which can be obtained from a flow simulation. Thus the propagation equation contains new terms. The scalar PE method is expected to perform well when the source and receiver are close to the ground because the sound propagation direction is nearly horizontal and the scattering angle (angle between the sound wave and a wave scattered by the turbulence) do not exceed a certain limit. For both methods the derivation can be carried out from the wave equation for the sound field P' in an inhomogeneous moving medium, as shown in Ostashev et al [30]. With the assumption of uniform density, the equation reads

$$\left[\nabla^2 + k^2(1+\epsilon) - \frac{2i}{\omega}\frac{\partial v_i}{\partial x_j}\frac{\partial^2}{\partial x_i \partial x_j} + \frac{2ik}{c_o}\boldsymbol{v}\cdot\nabla\right]P'(r) = 0 \tag{10}$$

where $k = \omega/c_0$, ω is the radian frequency of the sound, c_0 is the reference speed of sound, $P'(r)$ is the monochromatic sound pressure field, and $\epsilon = (c_0/c)^2 - 1$. Note that Eq. (10) is reduced to the Helmholtz equation if $\boldsymbol{v}$ = 0, i.e. without any ambient flow. More details of mathematical manipulation in order to reduce the equation to one way parabolic equation is referred to Ostashev et al [30].

The derived equations can be solved with various numerical techniques. The code is developed with multiple options, either using finite difference methods or FFTs (Green's Function Parabolic Equation Gilbert [31]). Depending on the selected PE method, the treatment of undulating terrains may vary. Two different methods are implemented, namely domain decomposition or terrain following coordinates. First, one treats the complex terrain as a succession of flat domains. After each flat domain, the coordinate system (*x*, *z*) is rotated so that the *x*-axis remains parallel to the ground (see Aballéa [32]). The second method is based on the terrain following coordinate transformation of the Helmholtz equation by Sack and West [33]. Different PE methods implemented in the solver are summarized in Table 1:

PE Methods	Velocity Treatment	Terrain Treatment	Numerical Technique	Required Resolution (1/lambda)	Additional Note
Wide Angle (WAPE)	Scalar	Domain Decomposition	Finite Diff.	$\frac{\lambda}{8}$ - $\frac{\lambda}{10}$	Tridiagonal Solver
Mean-Wind (MW-WAPE)	Vector	Domain Decomposition.	Finite Diff.	$\frac{\lambda}{8}$ - $\frac{\lambda}{10}$	Penta-diagonal

					Solver
Turbulent-Wind (TW-WAPE)	Vector	Domain Decomposition	Finite Diff.	$\frac{\lambda}{8}$ - $\frac{\lambda}{10}$	Penta-diagonal Solver
Green's Function (GFPE)	Scalar	Domain Decomposition.	FFTs	$5\,\lambda - 40\,\lambda$	Not relevant
Generalized Terrain (GTPE)	Scalar	Terrain Follow	Finite Diff.	$\frac{\lambda}{8}$ - $\frac{\lambda}{10}$	Tridiagonal Solver

Table 1: List of numerical methods developed at DTU for wind turbine noise propagations.

In addition to the terrain geometry, all these PE methods require a set of noise source and flow inputs. These inputs can be obtained from various methods that vary in computational time depending on the complexity and accuracy. These include simple or complex source models, simple analytical flow models, low fidelity or high fidelity flow solvers or field experiments. The source models and flow inputs are listed below:

1. **Starter function:** a starter function is always needed to begin the PE calculation. The initial SPL value from a wind turbine is desired such that this function mimics a wind turbine as a sound source as accurately as possible.
   (a) *Single Point Source;*
   This is the conventional approach for a steady single point source representing a monopole as the PE starting function. An adjustment for capturing wind turbine far field directivity [26] can be carried out by weighting the starter function with the source strength obtained at the corresponding observer azimuthal angle. This results in a similar SPL field shown in Figure 7. This approach applies a relatively simpler monopole sound source and combined with wind turbine noise directivity characteristics.

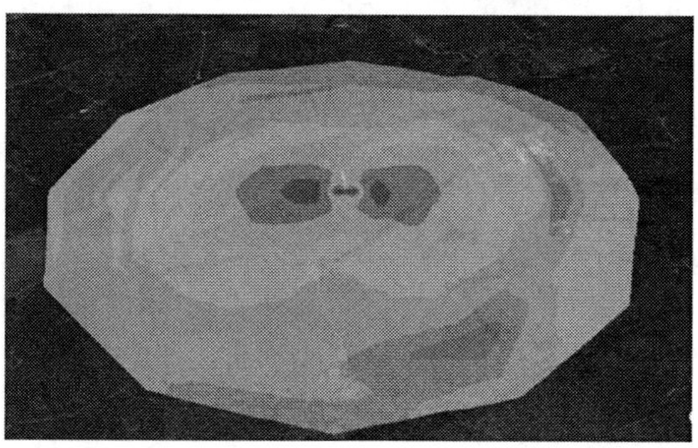

Figure 7: Single wind turbine noise modelled with a single point source at hub height.

   (b) *Moving source along the rotor vertical direction:*
   A more realistic source approach can be achieved by taking into account the blade rotating phenomena with a source moving in the rotor vertical direction. Since the wind turbine aerodynamic noise is mainly located in the outer part of blade, a wind turbine can be treated as lumped sources located near the blade tips, and different source locations result in dynamic sound pressure level in the frequency domain as well as sound directivity change. To reduce the complex 3D source to three incoherent sources rotating with the blades, only a specific time step corresponding to the true blade position in the vertical line is used. As shown in Figure 8, a wind turbine rotating in and out of the 2D PE domain intersects at the bottom and top tip heights. Thus we can model it via three point source that are translated either up or down at each time step by taking the turbine rotational speed into account.

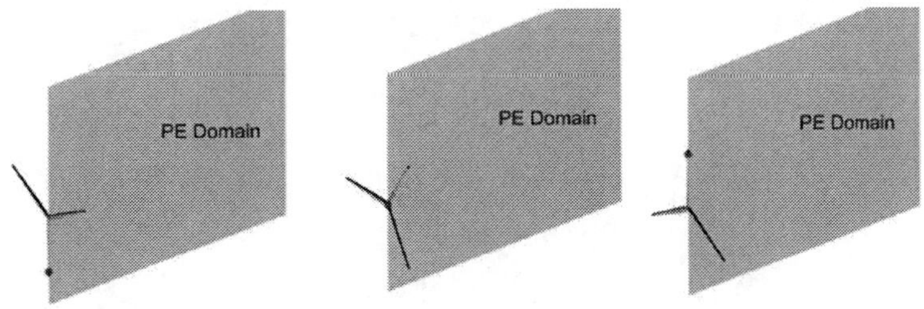

Figure 8: Snapshots of source locations during a rotation.

(c) *Coupling with unsteady source model:*
A more sophisticated and computationally demanding source model is to couple the PE model with the time varying sound source obtained from a generation model. With this approach the source power level at each time step is extracted from a noise generation model (for example, the BPM model) and then fed into PE for accurate propagation calculations. Repeating this multiple time steps yields more realistic wind turbine noise time signals at far field.

2. **Background flow field**: as aforementioned, the conventional approach uses a combined value from speed of sound and wind speed. In order to do this, we need the temperature distribution as well as the wind field projected in the 2D plans from source to receiver. These are obtained from:
   (a) 'Monin Obukhov Similarity Theory' + 'Analytical Wake Model'. The idea is based on the superposition of an atmospheric wind profile with an analytical wake model.
   (b) 'Linearized Flow Solver' + 'Analytical Wake Model': The background flow field is obtained from the linearized flow solver WAsP Engineering developed at DTU. Since the turbine effect is not modelled with the flow solver, the superposition with an analytical wake model is carried out. Additionally, the synthetically generated turbulence is superposed.
   (c) 'Reynolds averaged NS (RANS)' + 'Actuator Disc (AD): A flow solver that solves Reynolds Averaged NS equations where the turbine is modelled with an Actuator Disc method. In this approach, the initial background flow and turbine generated wake are simulated together.
   (d) 'LES' + 'AL': Large Eddy Simulation where the turbine is modelled with an Actuator Line technique. This yields a time dependent flow output, thereby at each time step the background flow of PE is updated. This is so far the most accurate and time consuming method.
   (e) Field flow measurements from a single or multiple met masts: This method is reserved for the validation of the code.

3. **Ground impedance**: The ground characteristics are needed to determine the boundary conditions for PE.
   (a) Model of Delany Bazley [34]: An empirical model for the calculation of ground impedance, obtained from fibrous absorbing materials.
   (b) Model of Attenborough [35]: A theoretical model for the calculation of ground impedance where the ground is approximated as a semi-infinite porous medium, or as a porous layer with a rigid backing.

4.2 Wind turbine noise propagation modelling

Using various combinations of the PE inputs, the single wind turbine noise propagation on flat terrain is studied extensively by Barlas et al. [36] [37]. In these studies, the effects of wind shear, turbulence levels and source modelling techniques are investigated. The simulated wind turbine wake under different turbulence intensity (TI) is seen in Figure 9. The inflow turbulence intensity clearly influences the wake structure which further affects the sound propagation in the wake.

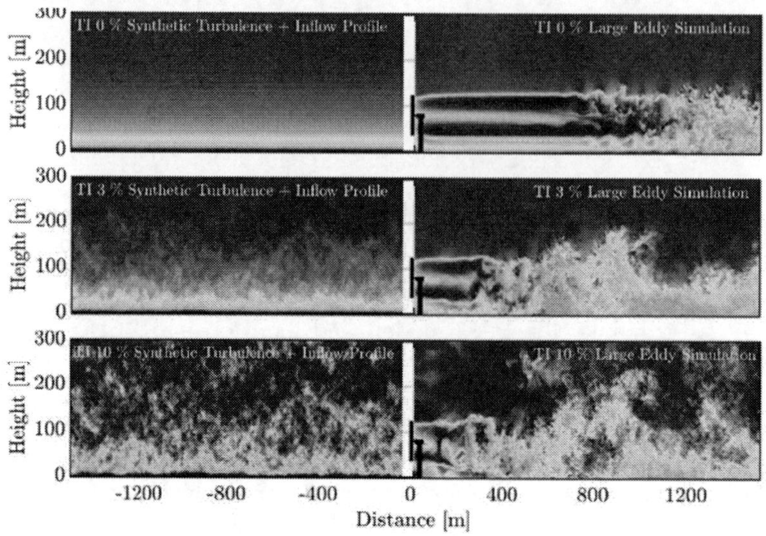

Figure 9: Ambient flow and wake behind a single wind turbine.

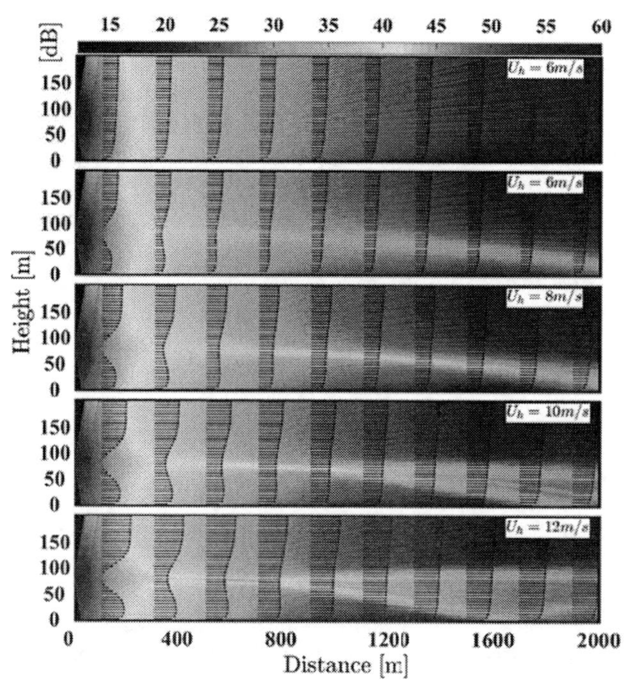

Figure 10: SPL at a hub-height wind speed of 6m/s without wake effect (upper figure), and at wind speeds of 8m/s,10m/s,12m/s with wake effect.

To investigate the influence from the wind turbine wake, PE simulations are carried out at a 6m/s inflow wind speed under the neutral atmospheric stability condition. In Figure 10, the horizontal wind speed is shown as vectors and the colour graphs are the sound pressure level generated from a wind turbine at a hub height of 80m. The differences in SPL are clearly seen for the cases with and without wake effect. Including the effect of wake, the wind turbine noise propagation is also influenced by incoming wind speed such as shown in Figure 10 for different inflow wind speeds. The effect from the wind speed is mainly due to the change of the wake profile that refracts the sound waves in different directions.

Considering a receiver is located at 2 meters height above ground level, we can have a close look at the sound pressure loss along fixed propagation paths. In Figure 11, the change of SPL in function of the downstream distance up to about 2500m is shown. For the case at a wind speed of 10m/s without wake effect, the line with triangles indicates a simple logarithmic decay of SPL. The other three cases show the sound propagation under the effect of wind turbine wake at three wind speeds. For these three cases, it is observed that there is some increase of SPL up to 12dB compared to the case without wake effect. From such study, it is realized that the wind turbine noise propagation is strongly affected by the wake itself. Using a logarithmic decay of wind turbine sound propagation is a not a good assumption.

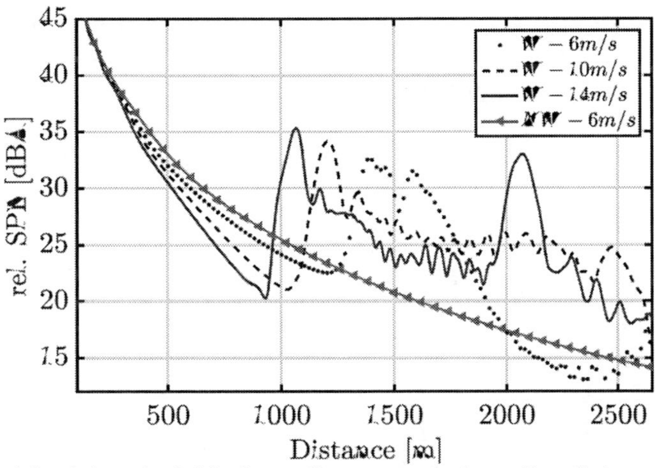

Figure 11: SPL at 2 meters height above the ground along the distance up to 2500 meters in a neutral atmospheric boundary layer. Case with wake (W) and without wake (NW).

Other effects, such as wind shear, also play an important role. Figure 12 contains the information of TI and wind shear effects. The SPL contours clearly show that increasing TI and shear factor leads to different SPL distributions. As shown in the figure, the computational domain also includes sound propagations in the upstream direction. The wind turbine rotor centre is located at x=0m and z=80m. The wind comes from left to right such that the sound waves in the upstream direction are bent upwards. After the rotor, the sound waves propagate towards the ground, which creates a noise problem to the receiver. As plotted in the figure, in each row the wind shear increases from 0.14 to 0.45 and in each column, the turbulence intensity increases from 0% to 10%. Thus, the plot shows a matrix of wind shear and turbulence effects. From the figure, it is seen that the inflow turbulence level is an important factor for wind turbine noise propagations in the downstream direction. In the upstream direction, increasing the wind shear results a reduced sound pressure level.

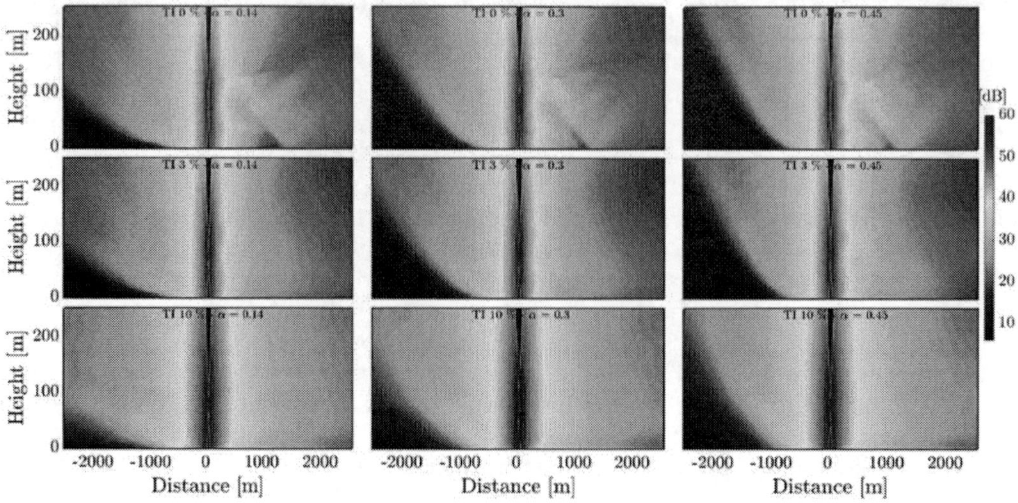

Figure 12: Time-averaged SPL under various flow conditions. Source is located at 0m distance and the flow comes from left to right. From top to bottom: turbulence intensity is 0%, 3% and 10%; from left to right: wind shear factor is 0.14, 0.3 and 0.45.

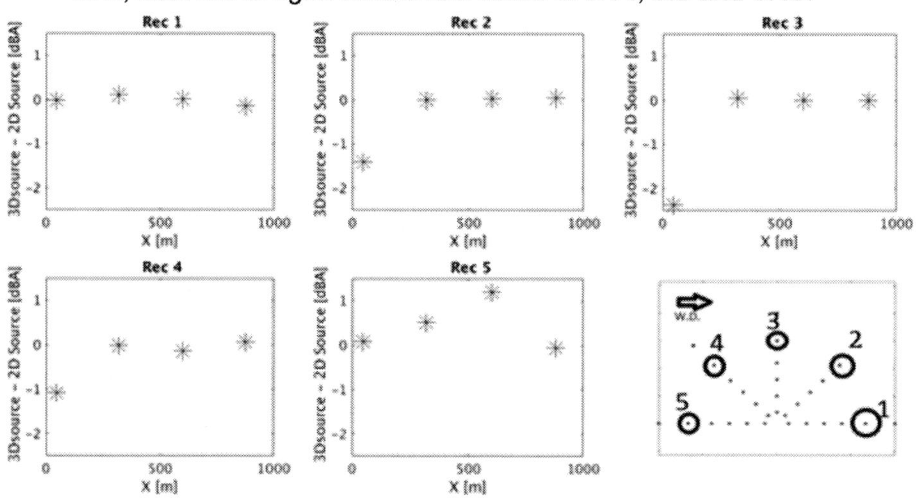

Figure 13: Evaluation of the source models (b) and (c) in function of direction and distance to the turbine.

In order to check the differences in the source modelling, sound propagations with the source models (b) and (c) are computed. In Figure 13, the differences of the 3D source model (c) and the 2D source model (b) are plotted in function of distance to the turbine and direction to the wind direction. From the figure, it is seen that the most affected area is in the near field below 300m and in the directions (2-4), but the differences are within 1dB.

5. Conclusions

In this paper, the wind turbine noise generation and propagation models developed at DTU are presented and summarized. The current review is only focused on the part of numerical modelling. The applications of these models depend on the computational resource and the purpose of study. The Navier-Stokes based CAA methods are computationally heavy which is suitable for understanding detailed noise generation mechanisms. The engineering models are more applicable for airfoil and rotor design purposes because of the advantage of little computational effort. Wind turbine noise generation models can be coupled with the wind turbine noise propagation model, such that a complete wind turbine noise simulation tool is developed. It is found that the atmospheric flow condition can be well-coupled with the PE method. The simulations shown that wind turbine noise propagations are largely influenced by the ambient flow as well as the wind turbine created wake. The PE model can be naturally coupled with CFD generated flow fields to simulate wind turbine noise propagation in complex terrain and complex atmospheric conditions.

References

[1] Søndergaard B (2014), *Noise and Low frequency noise from Wind Turbines*. Inter. Noise, Melbourne, Australia, 1-12.

[2] Michelsen JA (1992), *Basis3D - a Platform for Development of Multiblock PDE Solvers*. Technical Report, AFM 92-05, Technical University of Denmark, Denmark.

[3] Sørensen NN (1995), *General Purpose Flow Solver Applied to Flow over Hills*. Risø-R-827-(EN), Risø National Laboratory, Roskilde, Denmark.

[4] [Shen WZ, Michelsen JA and Sørensen JN (2001), *An improved Rhie-Chow interpolation for unsteady flow computations*. AIAA Journal, 39: 2406-2409.

[5] Shen WZ, Michelsen JA, Sørensen NN and Sørensen JN (2003), *An Improved SIMPLEC Method on Collocated Grids for Steady and Unsteady Flow Computations*. Numerical Heat Transfer, Part. B, 43: 221-239.

[6] Ta Phuoc L (1994), *Modèles de sous maille appliqués aux écoulements instationnaires décollés*. Proceedings of the DRET Conference: Aérodynamique Instationnaire Turbulente-Aspects Numériques et Expérimentaux, Paris, France: DGA/DRET editors.

[7] Shen WZ and Sørensen JN (1999), *Acoustic modeling of low-speed flows*. Theoretical and Computational Fluid Dynamics, 13: 271-289.

[8] Shen WZ and Sørensen JN (2001), *Aero-acoustic modelling of turbulent airfoil flows*. AIAA Journal, 39: 1057-1064.

[9] Shen WZ, Mikkelsen JA and Sørensen JN (2004), *A collocated grid finite volume method for aeroacoustic computations of low-speed flow*. Journal of Computational Physics, 196: 348-366.

[10] Shen WZ, Zhu WJ and Sørensen JN (2009), *Aero-acoustic computations for turbulent airfoil flows*. AIAA Journal, 47: 1518-1527.

[11] Zhu WJ, Shen WZ, Sørensen JN (2011), *High-order numerical simulations of flow induced noise*. Journal of numerical methods in fluids, 66: 17-37.

[12] Zhu WJ, Shen WZ (2015), *Airfoil Trailing Edge Noise Generation and Its Surface Pressure Fluctuation*. Journal of Power and Energy Engineering, 3:14-19.

[13] Zhu WJ, Shen WZ (2015), *Numerical simulation of airfoil trailing edge serration noise*. The 6th International Meeting On Wind Turbine Noise, Glasgow, UK.

[14] Zhu WJ, Shen WZ (2016), *LES tests on airfoil trailing edge serration*. The Science of Making Torque from Wind, Journal of Physics: Conference Series 2016, 753(3): 022062.

[15] D'Amico S (2016), *Validation and improvement of a noise prediction model for wind turbines*. Thesis work at DTU Wind Energy-M-0118.

[16] Fischer A (2011), *Experimental Characterization of airfoil boundary layers for improvement of aeroacoustic and aerodynamic modeling*. PhD thesis, DTU Wind Energy Department. Roskilde, Denmark.

[17] Bertagnolio F, Fischer A, Zhu WJ (2014), *Tuning of turbulent boundary layer anisotropy for improved surface pressure and trailing-edge noise modelling*. Journal of Sound and Vibration, 333: 991-1010.

[18] Fischer A, Bertagnolio F, Madsen HA, (2016), *Improvement of TNO type trailing edge noise model*. European Journal of Mechanics - B/Fluids, In Press.

[19] Bertagnolio F (2015), *Experimental investigation of stall noise towards its modelling. In 6th international conference on wind turbine noise*. Conference proceedings, Glasgow, UK.

[20] Madsen HA (2010), *Low frequency noise from wind turbines mechanisms of generation and its modelling*. Journal of Low Frequency Noise Vibration and Active Control, 29(4): 239-251.
[21] Debertshäuser H, Shen WZ, Zhu WJ (2015), *Development of a high-fidelity noise prediction and propagation model for noise generated from wind turbines*. The 6th International Meeting On Wind Turbine Noise, Glasgow, UK.
[22] Debertshäuser H, Shen WZ, Zhu WJ (2016), *Aeroacoustic Calculations of Wind Turbine Noise with the Actuator Line/Navier-Stokes Technique*. Proceedings of the 34th Wind Energy Symposium, American Institute of Aeronautics & Astronautics, San Diego, CA.
[23] Zhu WJ, Shen WZ, Barlas E, Bertagnolio F, Sørensen JN (2017), *Wind turbine noise generation and propagation modeling at DTU Wind Energy: A review*. Submitted to Renewable and Sustainable Review.
[24] Lowson MV (1993), *Assessment and Prediction of Wind Turbine Noise. 1.Basic Aerodynamic and Acoustic Models*. Flow Solutions Report 1993/06, W/13/00317/00/00, 1–46.
[25] Brooks TF, Pope, DS and Marcolini MA (1989), *Airfoil Self-Noise and Prediction*. NASA Reference Publication 1218, National Aeronautics and Space Administration, USA.
[26] Zhu WJ, Heilskov N, Shen WZ and Sørensen JN (2005), *Modeling of aerodynamically generated noise from wind turbines*. Journal of Solar Energy Engineering 127: 517-528.
[27] Sørensen JN, Shen WZ (2002), *Numerical computations of wind turbine wakes*. Journal of Fluids Engineering 124: 393-399.
[28] Berengier MC, Gauvreau B, Blanc-Benon Ph, Juve D (2003), *Outdoor sound propagation: A short review on analytical and numerical approaches*. Acta Acustica 89: 980 – 991.
[29] Blanc-Benon Ph, Dallois L, Juve D (2001), *Long range sound propagation in a turbulent atmosphere within the parabolic approximation*. Acta Acustica 87: 659 – 669.
[30] Ostashev V, Juve D, Blanc-Benon Ph (1997), *Derivation of a wide-angle parabolic equation for sound waves in inhomogeneous moving media*. Acta Acustica, 83(3): 455–460.
[31] Gilbert KE (2014), *A numerically stable formulation of the Green's function parabolic equation: Subtracting the surface-wave pole*. JASA Express Letters, Published Online
[32] Aballéa F (2004), *Propagation acoustique en milieu extérieur: Application de l'équation parabolique rapide au couplage d'effets météorologiques et de topographies complexes*. PhD dissertation, Université de Maine, France.
[33] Sack RA, West M (1995), *A parabolic equation for sound propagation in two dimensions over any smooth terrain profile: The generalised terrain parabolic equation (GT-PE)*. Applied Acoustics, 45(2):113-129.
[34] Delany ME, Bazley EN (1970), *Acoustical properties of fibrous absorbent materials*. Appl. Acoust. 3:105-116.
[35] Attenborough K (1985), *Acoustical impedance models for outdoor ground surfaces*. Journal Sound Vibration 99: 521-544.
[36] Barlas E, Zhu WJ, Shen WZ, Kelly M, Andersen SJ (2017), *Effect of wind turbine wake on atmospheric sound propagation*. Applied Acoustics (In revision).
[37] Barlas E, Zhu WJ, Shen WZ, Andersen SJ (2016), *Wind turbine noise propagation modelling: An unsteady approach*, The Science of Making Torque from Wind, Journal of Physics: Conference Series, (ISSN:1742-6596) (DOI: http://dx.doi.org/10.1088/1742-6596/753/2/022003), vol: 753.

7th International Conference
on
Wind Turbine Noise
Rotterdam – 2nd to 5th May 2017

Long-term experimental campaign on an operating wind turbine for trailing edge serrations verification

Irene Lauret-Ducosson, EDF EN: Irene.Lauret@edf-en.com
Albert Alarcon, EDF DTG: albert.alarcon@edf.fr
Isabelle Schmich Yamane, EDF DTG: isabelle.schmich-yamane@edf.fr

Summary

From an environmental point of view, a wind farm operation results from the compromise between electrical production and integration into its environment, notably its acoustic environment. By adopting a continuous improvement process to increase its wind farms efficiency, EDF aims at reducing its wind farms noise impact: if the source sound power is reduced preserving power performance, the power production and noise integration compromise is lightened. In this context, wind turbine manufacturers develop solutions to limit the wind turbine noise generation where one of the most widespread examples is the use of trailing edge serrations. These solutions can be retrofitted to operating wind turbines. However, there is a lack of industrial feedback about the efficiency of the above-mentioned solution in real operating conditions. To this end, EDF retrofitted trailing edge serrations on an operating wind turbine and conducted an in-situ experimental campaign in order to validate the manufacturer's noise performance claim. The measurement campaign was twofold: while a standard IEC 61400-11 measurement campaign was carried out to match the industry standard verification procedure, a long-term measurement campaign was conducted in order to validate the trailing edge serration proficiency in various meteorological conditions. To fulfil the experimental long-term campaign a specific measurement system was designed consisting of a combined standalone power supply solution and a real-time remote data delivery. This paper presents the originality of the measurement campaign and of the measurement system and the data analysis conclusions.

1. Introduction

Wind is a clean and practically inexhaustible source of energy. However, wind farm noise can limit the widespread adoption of wind energy, since wind farm operation is a compromise between electrical production and integration into its acoustic environment.

Regulations are enforced to ensure this integration. In some cases, compliance requires wind farm curtailment in order to limit the noise impact of the wind farm on its close environment.

By adopting a continuous improvement process to increase wind farm efficiency, EDF aims at increasing wind farm output production over their lifetime, while maintaining the acoustic ambiance of the site and compliance with noise regulations. This initiative has commenced with existing assets in order to maximise the renewable energy delivered to the grid, yet solutions found can also be used for future projects in order to offer a low cost of energy.

2. Continuous improvement process for noise optimisation

2.1.Optimisation potential

An initial diagnosis was conducted on EDF's French operating wind farms. The inventory of noise-curtailed wind farms revealed several GWh/year lost production, with 70% of these losses being borne by two wind turbine types (see Figure 1).

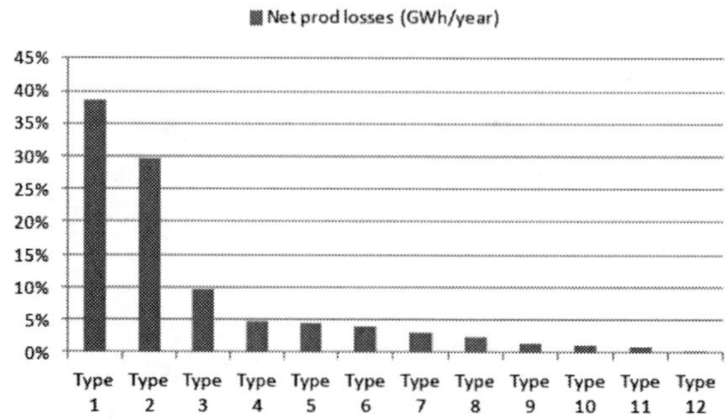

Figure 1 - Net production losses due to noise curtailment for EDF's French operating wind farms

2.2.Root causes & possible solutions

There are several causes to the limitation of production due to noise curtailment. The fishbone diagram shown in Figure 2 is used to identify the root causes for the net production of a wind farm, with regards to noise environment integration.

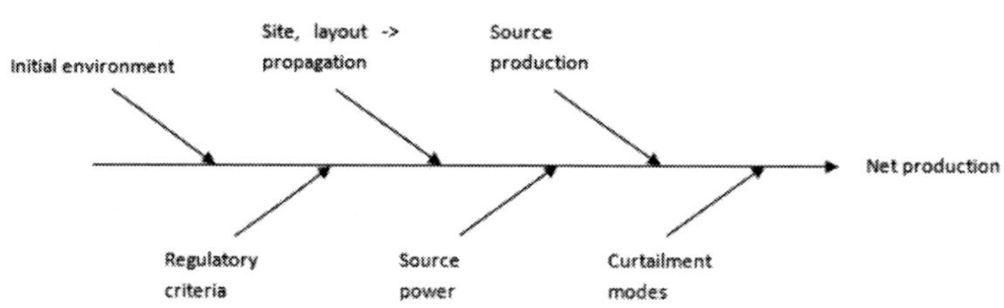

Figure 2 - Wind farm net production fishbone

Considering already operational wind farms, some root causes are fixed: initial environment, layout and turbine type, regulatory limits. However, other root causes can still be acted upon, e.g. the source sound power – the focus of this study.

2.3.Characterise & optimise

A sound power reduction of 1-2.5dB(A) results in a few percent estimated production regain, depending on the site. This degree of reduction is considered an achievable objective.

In this context, wind turbine manufacturers develop solutions to limit the wind turbine noise generation, where one of the most widespread examples is the use of trailing edge serrations.

Several studies [1] [2] showed that in laboratory conditions, trailing edge serrations could reduce noise generation by this order of magnitude. At lower sound power levels (i.e. low wind speeds) a lower noise reduction is expected, whereas a higher noise reduction can be expected at maximum sound power level.

Trailing edge serrations can be retrofitted to operating wind turbines.

2.4. Test & verification

However, there is a lack of industrial feedback about the efficiency of the above-mentioned solution in real operating conditions. In order to assess the in-situ implementation of the solution, EDF retrofitted trailing edge serrations on an operating wind turbine and conducted an experimental campaign in order to validate the manufacturer's noise performance claim.

3. In-situ experimental campaign

3.1. Context

The main objective for the measurement campaign was measuring the impact of trailing edge serrations, for which the expected sound power reduction is 1-2.5dB(A). This was achieved by comparing the noise emitted by an operating wind turbine before and after the installation of this add-on device.

To this end, the measurement campaign is twofold. A standard IEC 61400-11 measurement campaign was carried out to match the industry standard verification procedure. At the same time, a long-term measurement campaign was also conducted in order to explore a wide range of operating conditions and validate the trailing edge serration proficiency in various meteorological conditions.

3.2. Experimental setup

3.2.1. Measurement site

The test site was selected considering the following limitations:

- Low background noise
    - Secluded test wind turbine or neighbouring wind turbines shut down
    - No trees, little vegetation
    - No noise source in close vicinity: roads, etc
- Flat ground area over a few hundred meters
- 3G network coverage available for data download
- Safe location for the measurement units

The selected test site displayed a single row layout and the test turbine was the last in the row. The next closest wind turbine was shut down during the measurements. The crops adjacent to the test turbine were harvested prior to testing. The test site was located in a quiet rural area, with low ambient noise, in spite of nearby low traffic roads.

3.2.2. Measurement points location

The seven measurement points were located around the test turbine (see Figure 4) according to the historical wind rose (see Figure 3) at the site, in order to secure a downwind microphone position regardless of the wind direction. The measurement points were located at the distance specified in IEC standard 61400-11 (hub height + half rotor diameter).

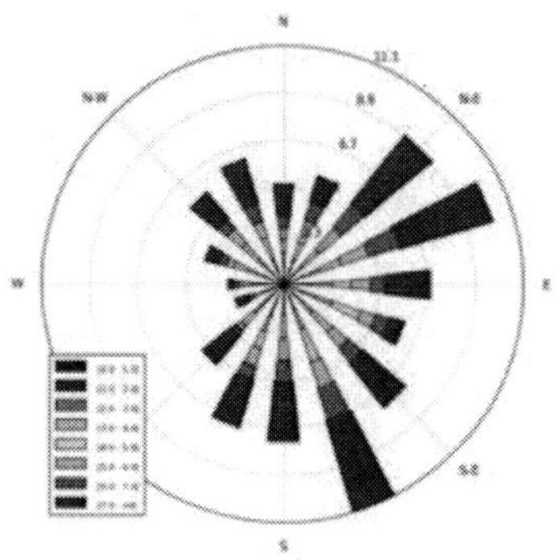

Figure 3 - Historical September wind rose (2010-2015)

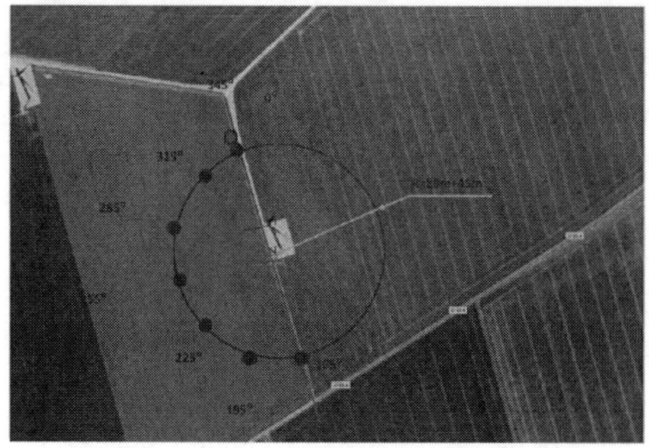

Figure 4 - Measurement points location around test turbine

3.2.3. Test turbine operation

The proficiency of the trailing edge serrations was evaluated for the full production mode as well as two reduced operation modes. Consequently, an operation plan (see Table 1) was designed for the test turbine in order to combine the three modes together, with regular periods of shut down to record background noise.

Day 1	00:00 - 01:00	01:00 - 02:00	02:00 - 03:00	03:00 - 04:00	04:00 - 05:00	05:00 - 06:00	06:00 - 07:00	07:00 - 08:00	08:00 - 09:00	09:00 - 10:00	10:00 - 11:00	11:00 - 12:00
	Mode A		Shut down	Full prod mode		Mode A		Full prod mode		Shut down	Mode A	
	12:00 - 13:00	13:00 - 14:00	14:00 - 15:00	15:00 - 16:00	16:00 - 17:00	17:00 - 18:00	18:00 - 19:00	19:00 - 20:00	20:00 - 21:00	21:00 - 22:00	22:00 - 23:00	23:00 - 00:00
	Full prod mode	Shut down	Full prod mode	Mode A			Shut down	Mode A			Shut down	Full prod mode

Day 2	00:00 - 01:00	01:00 - 02:00	02:00 - 03:00	03:00 - 04:00	04:00 - 05:00	05:00 - 06:00	06:00 - 07:00	07:00 - 08:00	08:00 - 09:00	09:00 - 10:00	10:00 - 11:00	11:00 - 12:00
	Mode B		Shut down	Full prod mode		Mode B		Full prod mode		Shut down	Mode B	
	12:00 - 13:00	13:00 - 14:00	14:00 - 15:00	15:00 - 16:00	16:00 - 17:00	17:00 - 18:00	18:00 - 19:00	19:00 - 20:00	20:00 - 21:00	21:00 - 22:00	22:00 - 23:00	23:00 - 00:00
	Full prod mode	Shut down	Full prod mode	Mode B			Shut down	Mode B			Shut down	Full prod mode

Table 1 - Test turbine operation plan during the measurement period

The operation plan was implemented for 18 days before and 14 days after the installation of the trailing edge serrations. The total measurement duration was 39 days, including a 7 day interim period for installation work.

3.2.4. Measurement equipment

To fulfil the specific demands of the experimental long-term campaign, a measurement system was designed consisting of a combined standalone power supply solution and a real-time remote data delivery. The seven recording units were autonomous and could be remotely controlled via a 3G connection.

Each recording unit consisted of (see Figure 5, Figure 6 and Table 2):

- 1 microphone with rain protection placed on a hard board (diameter 1.2m) on the ground
- 1 Brüel & Kjær sound level meter, class 1, type 2270
- 1 modem connected to the 3G network
- 1 Lithium-Iron-Phosphate battery
- 1 solar panel
- 2 GPS tracking security boxes

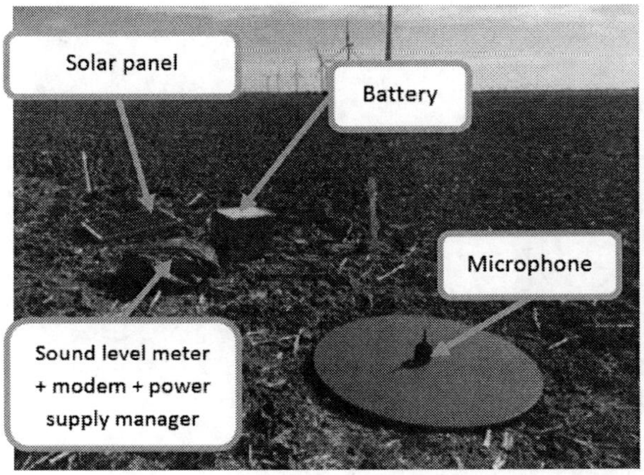

Figure 5 - Recording unit and 3G modem

Figure 6 - Sound level meter, modem and power supply manager

Solar panel	Battery
Power 60 W	Weight 13,6 kg
No load voltage 20 V	Volume 5 dm3
Short-circuit current 4,02 A	Voltage 12,8 V
Maximum voltage 16,6 V	Capacity 100 Ah
Maximum output current 3,59 A	Energy 1280 Wh
Weight 7 kg	Theoretical acquisition duration 25d
Dimensions 80x60x10 cm	

Table 2 - Technical specification for the solar panel and battery

Figure 7 shows the general scheme of the power supply manager which handles the energy flow between the battery charge, the solar panel production and the energy call of the measurement, recording and communication systems.

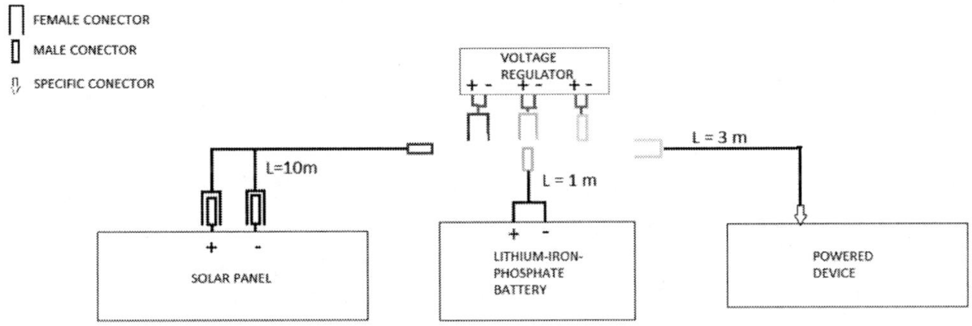

Figure 7 - General scheme of the power supply manager

One major issue with the design of the measurement system was equipment safety, as it was left exposed for several weeks in open fields. An electronic surveillance beacon, equipped with a GPS tracking system and motion detector was placed on each unit. The beacon sends a warning signal via the 3G network in case a movement is detected on a unit and a local team can then check on the equipment. The system can be set up and monitored remotely via a web interface shown in Figure 8.

Figure 8 - Web interface for set up and monitoring of the surveillance system

4. Analyses and results

4.1. IEC 61400-11 analysis

A standard IEC 61400-11 measurement campaign and data analysis was carried out as per industry best practise, in order to validate the manufacturer's noise performance claim for the sound power level reduction resulting from the use of trailing edge serrations.

The measurements and analysis were done for all three tested operational modes but the results are presented here for the full production mode only.

Wind bin	2m/s	3m/s	4m/s	5m/s	6m/s	7m/s
Measured sound power reduction	0.9	0.0	1.3	1.7	1.3	2.0
Manufacturer's sound power reduction			1.3	1.3	1.2	1.5
Samples before installation	*48*	*375*	*544*	*492*	*228*	*79*
Samples after installation	*62*	*225*	*280*	*52*	*24*	*27*
Uncertainty U_c before installation	*1.7*	*1.6*	*1.3*	*1.0*	*0.9*	*0.9*
Uncertainty U_c after installation	*1.4*	*1.4*	*1.4*	*1.1*	*1.0*	*0.9*

Table 3 - Sound power level reduction after installation of trailing edge serrations (analysis acc. IEC 61400-11)

The results show a reduction of the sound power levels (see Table 3), following the installation of the trailing edge serrations. This was found for all three tested operational modes as shown in Figure 9.

- Reduction of the sound power levels for wind speeds at 10m ≥ 3m/s
  - Full production mode: maximum reduction of up to 2dB(A)
  - Mode A: reduction of 0.5-1.5dB(A)
  - Mode B: reduction of 1-1.8dB(A)
- Negligible reduction of sound power levels for wind speeds at 10m ≤ 3m/s.

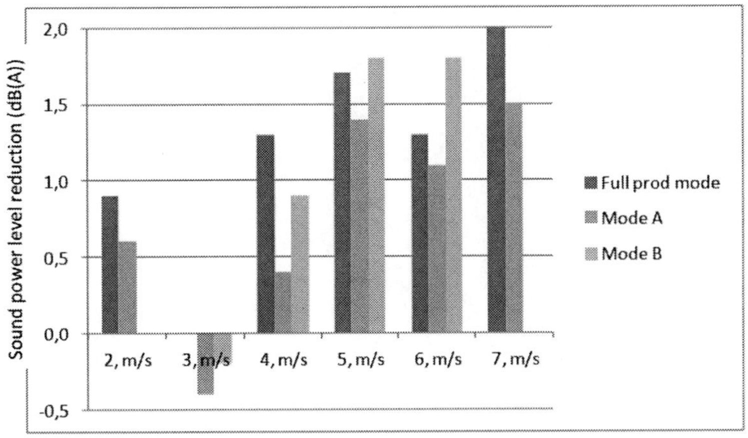

Figure 9 - Sound power level reduction after installation of trailing edge serrations (analysis acc. IEC 61400-11)

4.2.Spectral analysis

This analysis complements the previous section by focusing on the third-octave band spectra (in the audible frequency range (20Hz-10kHz)) for the noise emitted by the wind turbine before and after the installation of the trailing edge serrations. The analysis is presented for the full production mode and for two wind speed bins: 6m/s and 10m/s (see Figure 10 and Figure 11) at hub height. Similar conclusions can be drawn for the other measured operational modes.

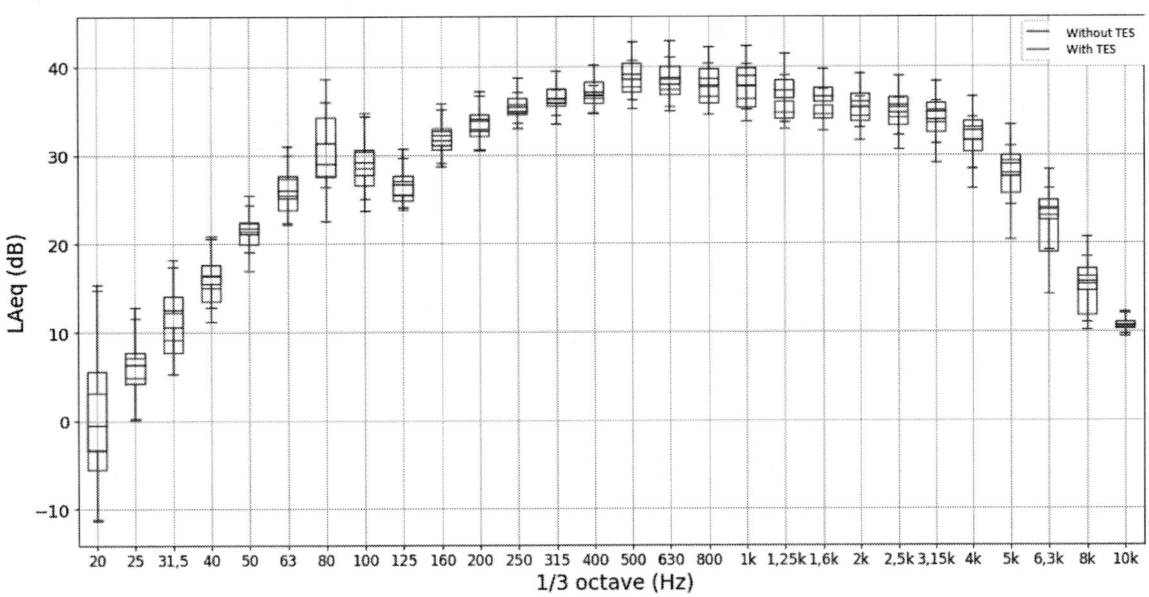

Figure 10 - Box plot for the third-octave band spectra, before (blue) and after (red) the installation of TES. Data for the full production mode at 6m/s at hub height.

While the data acquired before and after the installation of the trailing edge serrations appear to have a similar signature in the low frequency range (20Hz to 400Hz), slight differences can be observed in the upper range up to 2.5kHz.

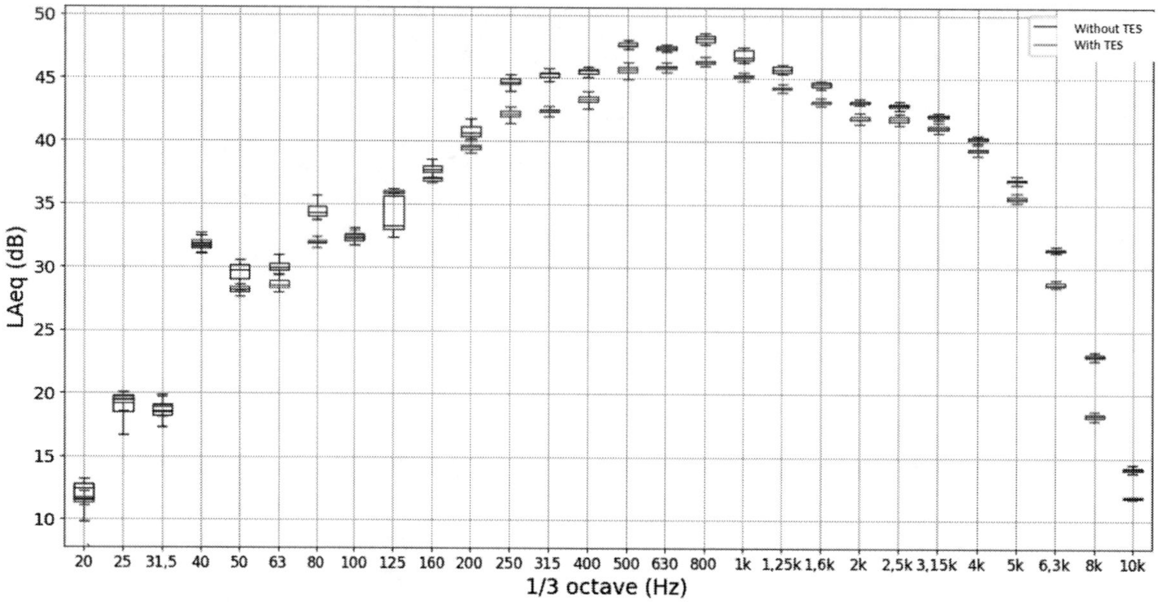

Figure 11 - Box plot for the third-octave band spectra, before (blue) and after (red) the installation of TES. Data for the full production mode at 10m/s at hub height.

At higher wind speeds, the differences in the data acquired before and after the installation of the trailing edge serrations are more marked. While the spectra in the low frequency range remain similar, they differentiate in the upper frequency range, from 315Hz to 10kHz.

The analysis of the mean spectra difference before and after the installation of the trailing edge serrations confirms the proficiency of the system, which acts in the medium range frequency (from 315Hz upwards) with a sound power level reduction in the order of 1-2dB(A). The system does not have an impact in the lower frequency range.

4.3. Directivity analysis

This analysis complements the previous sections by focusing on the directivity of the noise emitted by the wind turbine before and after the installation of the trailing edge serrations. The results are presented for the full production mode. Similar conclusions can be drawn for the other measured operational modes.

The directivity is presented in Figure 12 as measured sound pressure levels $L_{AEq,1min}$, before and after the installation of the trailing edge serrations, in 30 degree sectors around the wind turbine. It shows the usual features of wind turbine noise directivity: higher levels in the downwind sector (270-90° through 0°) than in the upwind sector (90-270° through 180°), with the maximum level directly downwind of the wind turbine (0°) and minimum levels in the crosswind directions (90° and 270°).

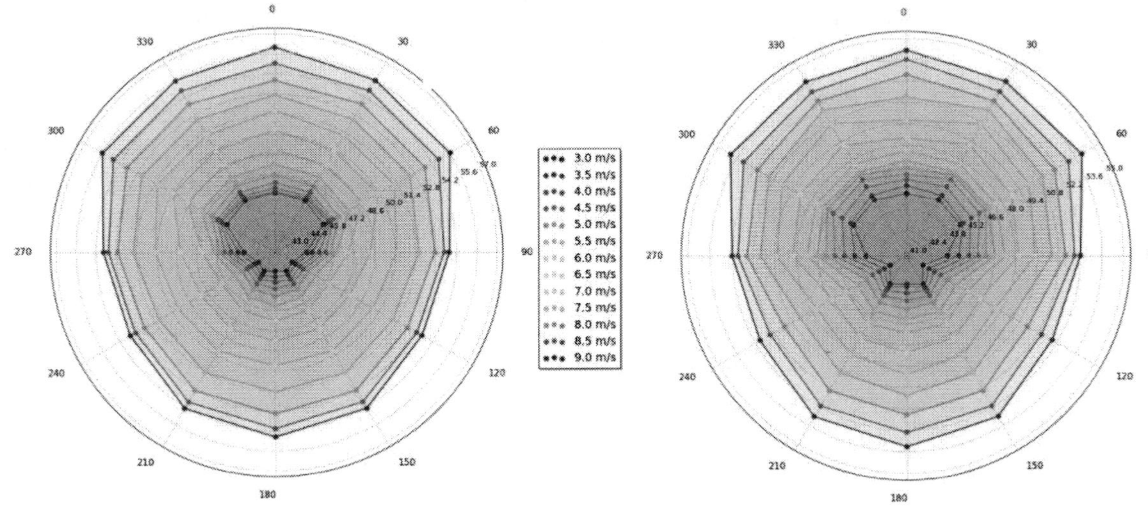

Figure 12 - Directivity diagram for mean measured $L_{AEq,1min}$ (dB(A)), before (left) and after (right) the installation of TES. Data for the full production mode.

In accordance with the previous analyses, the noise reduction tends to increase as wind speed increases. A reduction of up to 2dB(A) is measured in the downwind direction (0°) and upwind direction (180°). Finally, the noise reduction is non-existent or more limited for the low wind speeds and in the crosswind directions (90° and 270°). It is important to emphasise that no general change in wind turbine noise directivity occurred due to the installation of the trailing edge serrations.

4.4. Statistical analysis

The statistical analysis [3] complements the previous sections as the long-term measurement database can be analysed in order to validate the proficiency of trailing edge serrations in various meteorological and operational conditions.

The main advantage of this method lies in taking into account the variance of the datasets acquired before and after the installation of trailing edge serrations. The representativeness of the datasets variance is ensured by the duration of the measurements. Thus, the statistical method aims at being more robust than a comparison of shorter-term measurements based on mean values. Moreover, the techniques used for the statistical analysis are based on non-parametric methods, which do not require the population normality, even though the nature of wind turbine noise requires that the data is analysed according to the wind speed.

The general approach of this analysis is based on the use of $L_{A50,1min}$ as noise indicator and the use of the wind speed at hub height and wind direction with respect to the rotor position (0° is directly downwind). The data is not filtered as the use of $L_{A50,1min}$ ensures the exclusion of particular events (such as traffic or plane noise).

For each dataset before and after the installation of the trailing edge serrations, the data is sorted into 10x 1m/s wind speed bins (3-12m/s, centred around the integer value) and 8x 45° wind direction bins (0-315°, centred around 0°). The data is then compared for each sector using box plots and statistical values (mean, median, standard deviation). The analysis is presented for the full production mode. Similar conclusions can be drawn for the other measured operational modes.

For the full production mode, 35270 samples before and 27503 samples after the installation of the trailing edge serrations were sorted according to the methodology above. Figure 13 gives an overview of the noise levels $L_{A50,1min}$ distribution according to wind direction and wind speed.

A primary comment is the difference in data coverage between the datasets before and after the installation of trailing edge serrations. Attention must be paid to the data count per bin when comparing the datasets. Sectors 0°, 90° and 180° are the most relevant to be looked at, considering the directivity analysis observations above and the data coverage for each dataset.

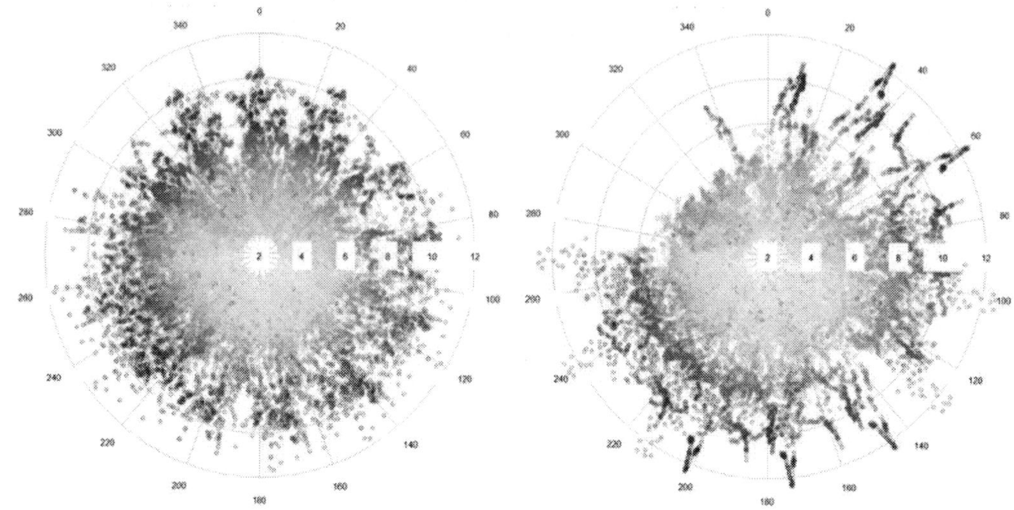

Figure 13 - Polar diagram for mean measured $L_{A50,1min}$ in dB(A) with shading ranging from 44dB(A) in yellow to 51dB(A) in red, before (left) and after (right) the installation of TES. Data for the full production mode.

4.4.1. Sector 0° (downwind)

The box plot for measured $L_{A50,1min}$ in sector 0° (Figure 14) shows a population difference between the datasets before and after the installation of the trailing edge serrations. The reduction is low for wind speeds up to 6m/s but sets in for wind speeds beyond. Although the data count in the 9m/s and 10m/s wind speed bins is lower, a reasonable variance of these samples allows remaining positive about the proficiency of the trailing edge serrations.

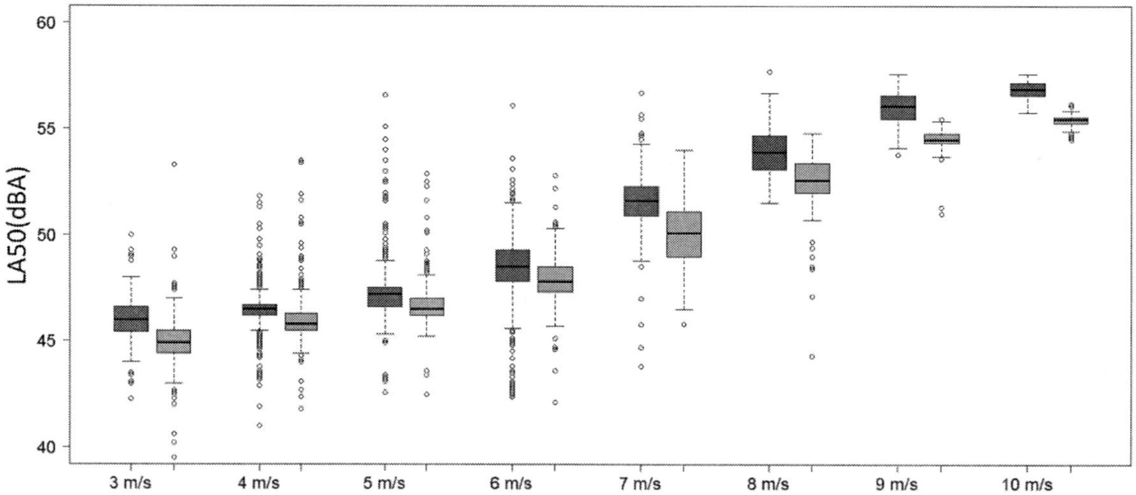

Figure 14 - Box plot for LA50,1min in sector 0°, before and after the installation of TES. Data for the full production mode.

Table 4 below shows the mean difference per wind speed bin between the datasets. The mean difference measures the notion of average gain due to the trailing edge serrations. The same observations can be made: while the noise reduction remains minor for the low wind speeds, it shows a consistent reduction of around 1.5dB(A) above 6m/s.

Wind speed bin	TES	mean	median	std dev	data count	mean difference
3 m/s	with	**44.8**	**44.8**	**1.6**	**179**	**1.1**
	without	**45.9**	**45.9**	**1.3**	**183**	
4 m/s	with	**46.0**	**45.7**	**1.2**	**351**	**0.5**
	without	**46.4**	**46.5**	**0.8**	**899**	
5 m/s	with	**46.7**	**46.5**	**1.1**	**460**	**0.5**
	without	**47.2**	**47.1**	**1.1**	**962**	
6 m/s	with	**47.8**	**47.7**	**1.1**	**508**	**0.6**
	without	**48.5**	**48.5**	**1.6**	**743**	
7 m/s	with	**49.9**	**49.8**	**1.4**	**138**	**1.6**
	without	**51.5**	**51.5**	**1.1**	**700**	
8 m/s	with	**52.3**	**52.6**	**1.6**	**130**	**1.4**
	without	**53.7**	**53.5**	**1.1**	**334**	
9 m/s	with	**54.5**	**54.5**	**0.7**	**80**	**1.6**
	without	**56.0**	**55.9**	**0.7**	**204**	
10 m/s	with	**55.4**	**55.5**	**0.4**	**105**	**1.4**
	without	**56.8**	**56.8**	**0.5**	**54**	

Table 4 - Characteristic quantities of sample population for sector 0° and the full production mode.

For the 10m/s wind speed bin, the dataset after the installation of the trailing edge serration, in spite of a reduced data count, seems to be consistent in terms of variance with the dataset before the installation of the trailing edge serration. The comparison of the datasets up to 10m/s is therefore justified.

4.4.2. Sector 90° (crosswind) and 180° (upwind)

In Figure 15 and Figure 16 below, the same box plot comparison for the 90° (crosswind) and 180° (upwind) sectors is presented for the full production mode.

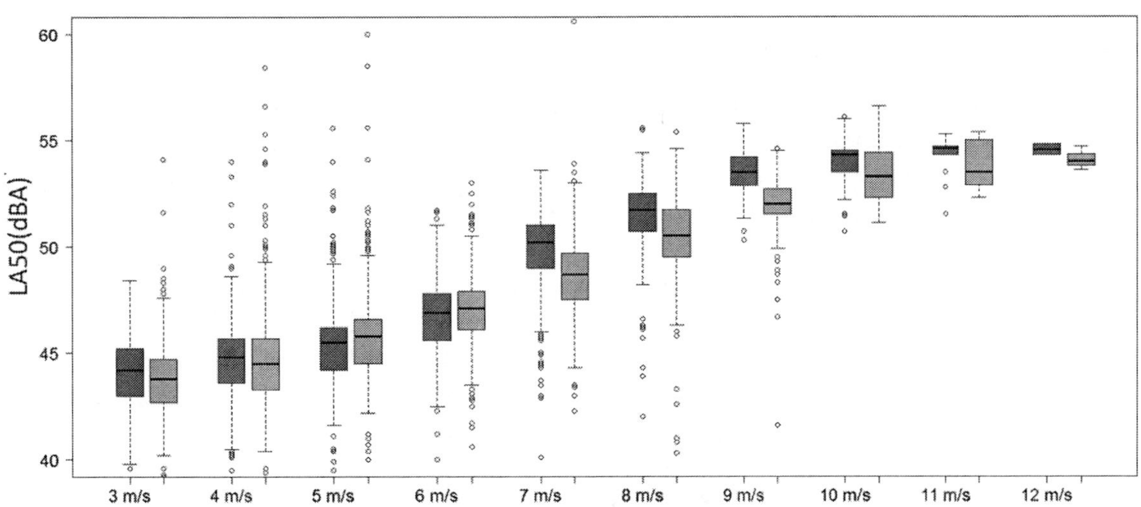

Figure 15 - Box plot for $L_{A50,1min}$ in sector 90°, before and after the installation of TES. Data for the full production mode.

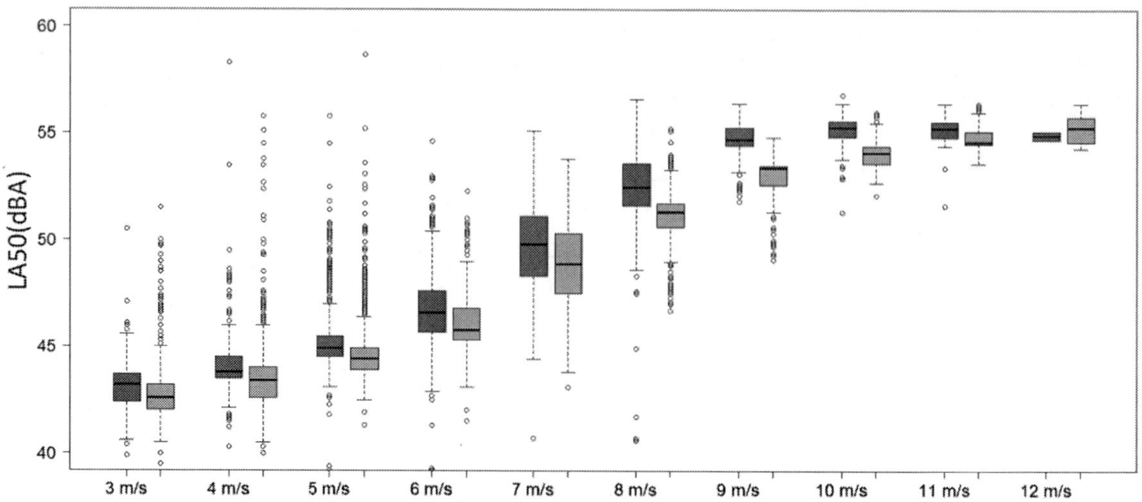

Figure 16 - Box plot for $L_{A50,1min}$ in sector 180°, before and after the installation of TES. Data for the full production mode.

In the upwind direction, the results remain fairly similar to those in the downwind sector, with a slightly lower noise reduction (around 1dB(A)) from the 7m/s wind speed bin. The reduction for the crosswind direction remains more moderate, often less than 1dB(A).

5. Conclusion

This paper presents the continuous improvement process EDF is undertaking to increase wind farm efficiency through a noise optimisation approach. This process is developed in a first stage via the implementation of trailing edge serrations.

These aerodynamic blade add-ons allow reducing the wind turbine noise at the source. The objective of this study was verifying the performance of such a noise reduction system in real operating conditions. To achieve this, a comparison of sound power levels measured in an in-situ experimental campaign before and after the implantation of the trailing edge serrations was performed.

From a scientific point of view, the strength of this work lies in:

- The specific design of a sound power measurement system, autonomous for five weeks and allowing remote data transfer.
- The insight gained through field experience of the proficiency of trailing edge serrations in real life operational and meteorological conditions.
- The full scope analysis following three axes:
  - The IEC standard procedure combined with a statistical analysis showed the reduction of total sound power levels for wind speeds at hub height above 6 m/s. The maximum achieved noise reduction is 1.5-2dB(A) in full production mode (lower reduction for lower operational modes). Regarding the statistical method, it is important to point out that while it requires little effort to implement and for data quality check, it naturally introduces a variance dimension in the data comparison, resulting into an in-depth analysis.
  - The spectral analysis showed that the reduction occurs in the 300Hz-3000Hz frequency range. On the one hand, this is in line with the experimental lab results

presented by the scientific community; on the other hand, this is to be considered for the wind turbine noise propagation over a long distance.

- The directivity analysis showed the highest reduction in the downwind and upwind directions while a more limited reduction was observed crosswind. This effect attenuates the wind turbine directivity. At no time did trailing edge serrations increase the noise.

From an operational point of view, this work confirms trailing edge serrations as an effective solution for wind turbine noise reduction. As a result, it can be deployed to the wind farms of the same turbine type with the certainty that the overall acoustic ambiance of the site is safeguarded, the noise regulatory compliance is maintained, and more renewable energy is delivered to the grid.

References

[1] Oerlemans, S., Fisher, M., Maeder, T., Kögler, K. (2009) *Reduction of wind turbine noise using optimized airfoils and trailing-edge serrations* AIAA Journal, Vol. 47, pp. 1470-1481.

[2] J Hurault, A Gupta, E Sloth, NC Nielsen, A Borgoltz, P Ravetta (2015) *Aeroacoustic wind tunnel experiment for serration design optimisation and its application to a wind turbine rotor* International conference on Wind Turbine Noise 2015

[3] A. Alarcon, I. Schmich-Yamane, M. Alayrac, F. Juncker (2016) *Meteorological factors characterization in atmospheric noise propagation using a variance analysis approach* 45th International Congress and Exposition of Noise Control Engineering INTER-NOISE 2016

7th International Conference
on
Wind Turbine Noise
Rotterdam – 2nd to 5th May 2017

Development of an approach to controlling the impact of amplitude modulation in wind turbine noise: exposure-response research, application and implementation

PERKINS, Richard A, WSP | Parsons Brinckerhoff: perkinsr@pbworld.com
LOTINGA, Michael J (co-lead), WSP | Parsons Brinckerhoff: michael.lotinga@pbworld.com
BERRY, Bernard, Berry Environmental Ltd
GRIMWOOD, Colin J, CJG Environmental Management Ltd
STANSFELD, Stephen A, Queen Mary University of London

Summary

WSP | Parsons Brinckerhoff was commissioned by the UK Government Department of Energy and Climate Change to undertake a review of research into the effects of and response to the acoustic character of wind turbine noise known as amplitude modulation (AM), or more specifically an increased level of modulation of aerodynamic noise as perceived at neighbouring residential dwellings, with a view to securing protection, where it might be justified within the planning regime. This paper describes how the literature review was undertaken and the key findings on the state of knowledge of the effects of AM on people, and the response relationships that have been identified. It highlights the gaps in the knowledge base and the risks of bias in the studies reviewed. The paper also describes potential methods to control AM, the recommended penalty scheme method, how this might be applied in practice, and feedback received following publication of the main research report.

1. Introduction

Amplitude modulation (AM) has been one of the most controversial issues in evaluating community response to wind turbine noise (WTN). This characteristic has been shown to exacerbate the annoyance people feel in response to hearing it (Pedersen & Waye, 2004), and may contribute to the greater negative response to perception of WTN compared with other types of environmental noise at similar exposure levels (Janssen, Vos et al, 2011).

This paper outlines recent research funded by the UK Government, which reviewed available evidence on human perception and response to AM in WTN, with the aim of recommending a suitable method for control to use in planning wind farms. Also discussed are current limitations, issues related to implementation of the control, and identification of further research needs.

There have been previous attempts to develop a planning-stage control for AM. These have typically been highly technical in nature, either concerned with restricting specific numbers of AM event occurrences that meet a set of conditions (Planning Inspectorate, 2009), or a character-penalty scheme (RenewableUK, 2013). There have also been simpler schemes such as NZS 6808:2010, which, in addition to proposing an 'interim' measurement method, advises a flat 5 dB character penalty, which in principle could be assigned from a subjective assessment.

2. Project in Brief

In addition to the main objectives of reviewing the evidence on the human response to AM, and providing a recommendation for a control, the project required close cooperation with the UK Institute of Acoustics (IOA) AM Working Group (AMWG). The AMWG conducted a concurrent independent research study aimed at developing a robust objective method for detecting and rating AM in real WTN signals (Bass, Cand et al, 2016).

The exposure-response review was led by researchers at WSP | Parsons Brinckerhoff, supported by external independent noise and health specialist consultants. The research team reported to a Steering Group comprising DECC, the Dept for Environment, Food and Rural Affairs, the Dept for Communities and Local Government, Public Health England, and representatives for the Devolved Authorities.

3. Review of Human Response to AM

3.1 Approach

A systematic approach to the review was adopted and is described in detail within the final report (WSP | Parsons Brinckerhoff, 2016). To minimise potential publication bias, searches were carried out in peer-reviewed ('black' literature) science and health research databases as well as other sources such as relevant conference proceedings and industry or government-funded research ('grey' literature). The initial search yields were sifted by examination of titles and abstracts. The resultant database was categorised according to the study type: category 1 comprised publications on scaled responses to quantified AM WTN; category 2 comprised other potentially relevant sources, including AM complaint case-studies, exposure-response studies of non-WTN AM, epidemiological field studies of WTN, planning issues relating to AM WTN, and any other useful studies of AM in WTN. This initial longlist included 134 papers. Further examination of each paper was undertaken and a relevance rating assigned, on the basis of which a shortlist of 69 papers was compiled for the full review, including 15 papers in category 1. The potential effects of selection bias (due to the application of relevance ratings and the categorisation processes) are considered unlikely to be significant, mainly due to the relatively small number of studies into the AM WTN exposure-response relationship, i.e. category 1. Category 2 material mainly provided supporting and contextual information and so any effect in terms of outcomes for the research is not expected to be critical. Suspected duplicates were retained, due to the relatively small number of papers addressing AM exposure-response.

A systematic template was used to review the shortlist, prompting reviewers to extract equivalent information from each paper and to consider the robustness and risks of bias. Each category 1 study was assigned to two reviewers to ensure consistency; differences were resolved by discussion. The responses received from reviewers were synthesised and conclusions to be drawn were considered by the research team. During drafting of the output report, two further studies were published that would have met the category 1 selection criteria, and these were also given limited reviews, included as annexes. The initial study recommendations were subject to an external independent peer review, feedback from which was incorporated into the final version of the published report.

3.2 Outcomes

The primary negative health effect consistently identified in relation to exposure to WTN is annoyance; sleep disturbance and stress are also highlighted in the literature, but the evidence in most cases indicates that these effects typically arise as a result of the annoyance experienced – the evidence for a direct connection between WTN exposure at typical levels (eg 25-45 dB $L_{Aeq,T}$ outdoors) and sleep disturbance is not consistent, whereas significant relationships between noise-related annoyance and disturbed sleep are consistently observed

(Pedersen & Persson Waye, 2007; Bakker, Pedersen et al, 2012). The outcome for an individual whose sleep is affected may be damaging to health in either case, but this suggests that it would be sufficient to develop a control for AM from the evidence base for annoyance; by reducing annoyance, the associated indirect pathway effects would be expected to be similarly reduced.

The evidence reviewed further indicates that, of the acoustic factors contributing to the annoyance response to AM in WTN, the time-averaged overall level and depth of modulation appear to be most important in determining response. Results from the exposure-response experiments reported by Lee, Kim et al (2011) and von Hünerbein, King et al (2013) are shown in Figures 1 and 2a[1] ($N = 30$ and 20 respectively). Annoyance ratings are observed to increase with both time-average level ($L_{Aeq,T}$) and modulation depth (MD, ΔL), with the former dominating the relationship [NB. in Figure 2a, the original MD values have been re-scaled to the AMWG rating method developed by Bass, Cand et al (2016) – see Appendix].

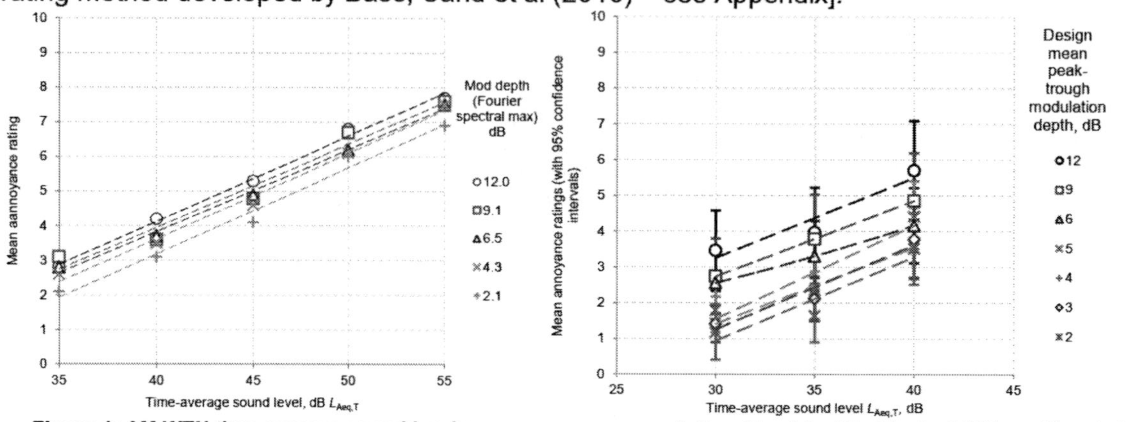

Figure 1: AM WTN time-average sound level exposure-response relationships identified by (a, left) Lee, Kim et al (2011); (b, right) von Hünerbein, King et al (2013)

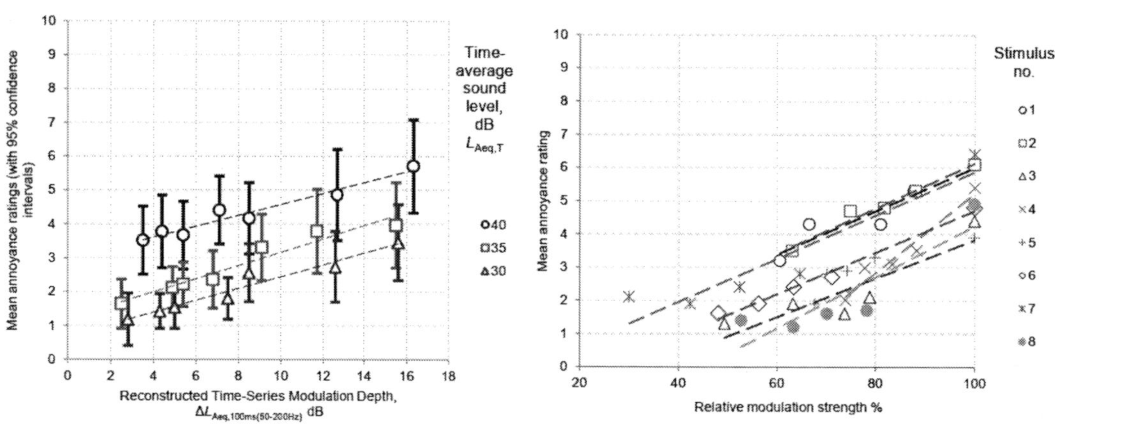

Figure 2: AM WTN modulation depth exposure-response relationships identified by (a, left) [adjusted from] von Hünerbein, King et al (2013); (b, right) Ioannidou, Santurette et al (2016)

The results for $L_{Aeq,T}$ show statistically significant ($p<0.05$) relationships with annoyance; the results for MD are less strong, and significance is typically only found when comparing high and low depth values from the ranges. Further evidence for the influence of MD has been reported in lab study results by Ioaniddou, Santurette et al (2016), shown in Figure 2b (in which the MD for each stimulus is normalised to its maximum value), which shows a significant effect of relative modulation strength on rated annoyance ($N = 19$). Results from a field-based case-

[1] NB. Linear regression lines are shown in the figures to aid visibility of broad trends and potential relationships, but have not been tested as robust statistical models for parametric relationships.

study at a site with historical noise complaints also indicated a significant relationship between measured WTN AM and reported annoyance (Bockstael, Dekoninck et al, 2012). In general across these studies, differences in individual subjective ratings and associated uncertainty tend to expand with increasing MD.

A threshold for perception of the fluctuations in a modulating WTN-like sound has been studied by Yokoyama, Kobayashi et al (2015); the lab results shown in Figure 3 (re-scaled as before), indicate that around 40-50% of participants (N = 17) perceived fluctuation at MDs of 2 dB, increasing to 95-100% at 3 dB; $L_{Aeq,T}$ appears related to the likelihood of detection, as expected according to the psychoacoustic model developed by Fastl (1982). It can be said from this that fluctuation in broadband WTN-like sounds can be sensed by most people with normal hearing at approximately 2 to 3 dB depth, with the latter being roughly the definite detection threshold.

Studies have also examined the subjective equivalence of a modulating WTN sound compared with its steady-amplitude counterpart. This has been examined using a method of paired comparison adjustment, in which one of the signals is modified in level until the participant judges both sounds to be equivalent. In the results reported by von Hünerbein, King et al (2013), the steady signal was adjusted relative to the AM, and the target response for equalisation was 'annoyance', while in those of Yokoyama, Kobayashi et al (2015), the AM signal was adjusted, and the judgement required was of perceived 'noisiness'. Nonetheless the experiments were very similar, and the results are shown together in Figure 4a; both datasets have again been re-scaled to the AMWG MD ratings. On average, the equivalence between the AM and negligible-AM WTN sounds used is approximately in the range 0-4 dB.

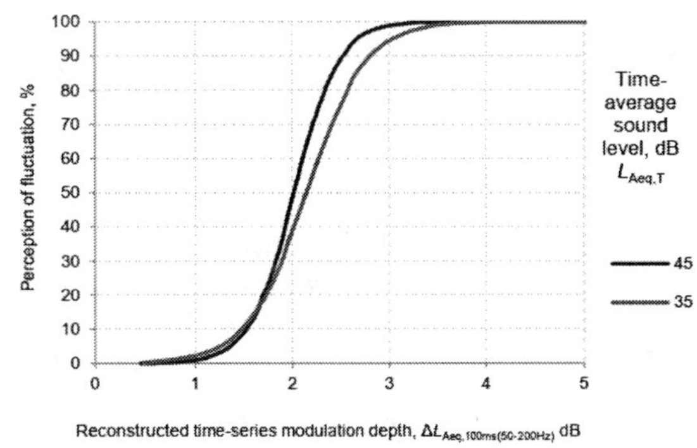

Figure 3: AM WTN detection threshold identified by [and adjusted from] Yokoyama, Kobayashi et al (2015)

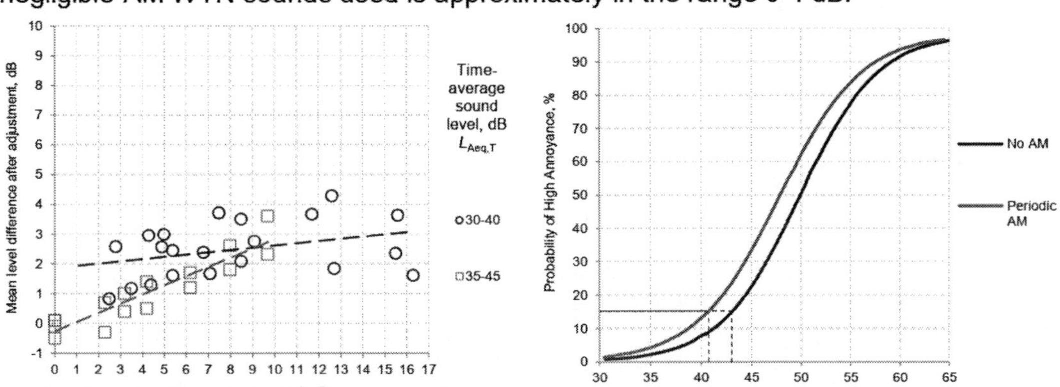

Figure 4: AM WTN equivalent response exposure-response relationships (a, left) identified by [and adjusted from] (○) von Hünerbein, King et al (2013) and (□) Yokoyama, Kobayashi et al (2015); (b, right) Schäffer, Schlittmeier et al (2016)

The results of an experiment with a larger sample (N = 60) have been used to develop a logistic regression model for the probability of high annoyance associated with WTN sounds exhibiting i) no significant AM and ii) periodic AM with a varying MD in the range of around 6 to 9 dB $\Delta L_{pA,F}$ (Schäffer, Schlittmeier et al, 2016)[2]. This model, shown in Figure 4b, indicates an equivalent annoyance for AM of around 2-3 dB for levels in the range 35-50 dB $L_{Aeq,T}$.

[2] As noted by Bass, Cand et al (2016), use of $L_{pA,F}$ to determine MD tends to produce lower values than using $L_{Aeq,100ms}$, which is the metric employed in the AMWG method.

Human sensitivity to periodic AM in broadband noise has been shown to peak over the range 2-8 Hz (Fastl, 1982; Bradley, 1994; Bengtsson, Persson Waye et al, 2004; Moore, 2013); 4 Hz is often-quoted due to its use in the definition of the model for 'fluctuation strength' outlined by Fastl, 1982. The modulation frequency (f_m) of WTN AM has likewise been shown to have an effect on lab ratings of annoyance. The results reported by Ioaniddou, Santurette et al (2016) and shown in Figure 5a indicate annoyance increasing over the f_m range 0.5 to 2 Hz, but not significantly so. Sensitivity test results presented by von Hünerbein, King et al (2013) indicate an increase in annoyance when comparing f_m of 0.8 Hz and 1.5 Hz (which was not tested statistically due to the small sample size: $N = 11$). The results are shown in Figure 5b, which, for the sake of comparability, have been subject to adjustments to account for differences in $L_{Aeq,T}$ and MD between stimuli, and so should be interpreted with caution (the adjustments are detailed in the Appendix). In both cases (Figure 5a and Figure 5b), it can be seen that differences between the stimuli, such as spectral content, can have a larger influence on ratings than f_m over the ranges considered.

In another lab experiment, Schäffer, Schlittmeier et al (2016) found that AM WTN was rated more annoying at the same $L_{Aeq,T}$ than AM road traffic noise (RTN), in broad agreement with the field-study analysis of Janssen, Vos et al (2011). This was considered potentially attributable to the f_m range in which WTN AM tends to vary, and the closer proximity of this range to that of peak fluctuation sensitivity (in contrast with the lower f_m of the RTN stimuli used). Accordingly, it is surmised that f_m has an effect on response, but, within the context of modern large-scale commercial wind turbines (with f_m in the range 0.5 to 1.5 Hz, and typically around 1 Hz), this appears to be relatively slight.

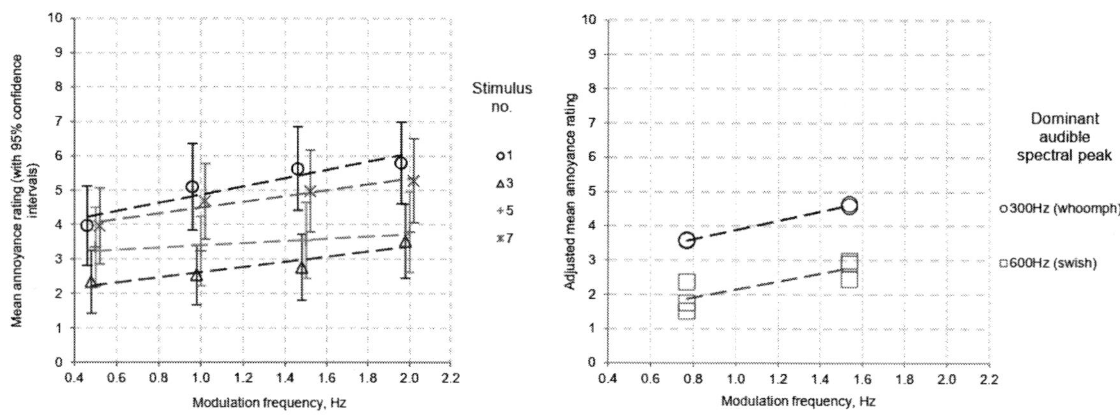

Figure 5: AM modulation frequency exposure-response relationships identified by (a, left) Ioannidou, Santurette et al (2016); (b, right) [adjusted from] von Hünerbein, King et al (2013)

The spectral content of AM WTN has been highlighted as a further factor in the negative experiences that have been reported by affected residents (van den Berg & Bowdler, 2011), with descriptive terms such as 'swish' and 'whoomph' used to differentiate between the emphasis of mid and lower frequency ranges respectively. This has been examined in several studies: Yokoyama, Kobayashi et al (2015) showed that the onset of perception of fluctuation sensation in WTN began at 80-100 Hz for WTN recordings made at typical receptor ranges. The tests reported by von Hünerbein, King et al (2013), and Lee, Kim et al (2011) used stimuli with peaks at lower (whoomph-like) and upper (swish-like) frequency ranges, which tended to produce higher annoyance ratings for the latter. However in both these cases the MD for the upper frequency stimuli was greater, which is likely to have confounded the relationship – the indicative results in Figure 5b, for which the influence of MD has been adjusted for (see Appendix), suggest the lower frequency range could have been rated higher if the MDs were controlled for. On the other hand, the study by Ioaniddou, Santurette et al (2016) examined the effect of exposure to intermittent periods of lower spectral frequency AM (with greater MD), interspersed within periods of 'normal' AM (upper spectral frequency, and smaller MD).

Counterintuitively, the results did not show any significant modification of annoyance by intermittent low-frequency AM; responses instead appeared to be determined by the AM character of the 'normal' periods. Reports from field research suggest that impulsive AM with low-frequency character ('thumping') is particularly disturbing to wind farm neighbours, and could contribute to increased noise complaints (van den Berg, 2004; Bowdler, 2008; Large & Stigwood, 2014). Further work would be beneficial to more fully understand the influence of the spectrum of AM WTN on responses: while it seems likely to be a factor in determining response, its influence compared with $L_{Aeq,T}$ and MD is uncertain.

As summarised by Perkins, Lotinga et al (2016), there have been several studies, both field and lab-based, which have shown that a wide range of non-acoustic factors have a significant influence on the annoyance responses attributed to WTN, including: noise sensitivity, turbine visibility, colour and flicker, attitude to wind energy and turbine aesthetics, exposure to wind energy-related media, neighbourhood land-use, economic involvements with wind turbines, association of sound with wind turbines, and general health (also outlined in WSP | Parsons Brinckerhoff, 2016).

The effects of diurnal variation in AM (ie time of day, occurrence frequency, and prolongation/intermittency) are not well documented in the evidence reviewed, although there are several field reports of increased impacts occurring at evening, night or early morning (van den Berg, 2009; Gabriel, Vogl et al, 2013; Stigwood et al, 2014). Van den Berg (2005) has shown how atmospheric and wind conditions more frequently encountered at night are likely to increase risk and severity of AM occurrence. Further probable factors are increased sensitivity and sense of intrusion during the night-time, and lower levels of other background sounds (Pedersen, Hallberg et al, 2007). The influence of AM exposure duration on expected responses has not yet been studied in detail.

3.3 Robustness and Limitations of the Evidence

Some potential risks of bias in the evidence have already been highlighted. In general, the results from laboratory-based exposure-response studies are limited by small samples typically recruited from somewhat unrepresentative populations (eg university students and staff); while the sample recruited by Schäffer, Schlittmeier et al (2016) was larger and with a broader representation, including a wider age group and a majority of rural residents (52%), none of the participants were living near turbines. The lab exposures used are also relatively brief, between 10s and 30s; while this may not significantly affect the short-term response ratings expected within the experimental setup, this cannot be expected to be closely representative of the responses that might be expected from those exposed within sensitive settings, for longer durations, and in which the expectation of cessation of the exposure (ie respite) may be uncertain.

The field studies on the other hand, involve WTN-exposed populations but carry risks of selection bias, especially where problematic situations involving WTN have developed. Typically they do not feature control cases for comparison. All the field studies identified are cross-sectional, preventing examination of changes in the measured responses over time (and establishment of causal relationships). Very few field studies identified directly compared quantifiable AM with scaled responses, which limited their value against the aims of this research project.

The above issues aside, one of the main knowledge gaps identified in the review in relation to the aims of this research is the effects of variation in AM exposure: duration, intermittency and prolongation.

4. AM Planning Control

4.1 Penalty Scheme

The evidence reviewed above indicates that AM increases annoyance, and that the expected response to a given occurrence could be broadly quantifiable relative to an equivalent period of WTN without significant AM; this supports the proposition for a character penalty scheme control, which can be imposed at planning stages, to be activated in reaction to complaints about possible AM occurrences.

4.2 Aim and Principle

The aim of the control is to reduce the additional impact of AM, ie its severity and occurrence; if AM is not reduced, the overall penalised level of WTN must be reduced to compensate (ie to meet the limit).

The principle of the proposed penalty scheme is that AM increases the annoyance caused by WTN, and that this increase can be characterised by adding a value to the overall WTN level, to equalise it with a negligible-AM WTN sound (in essence this principle is the same as for character adjustments used in other standardised methodologies, including BS 4142:2014 and ISO 1996-1:2016).

4.3 Formulation

Based on the studies considered above, it is proposed that a combination of the time-averaged level and MD parameters is a reasonable objective expression for the expected annoyance response to AM WTN. A more detailed expression could include spectral content and modulation frequency, but the current evidence appears to be less clear on the strength of these parameters. One approach to addressing the issue of spectral content is inherent within the method developed by Bass, Cand et al (2016), which employs frequency filtering to ensure the signal is evaluated for the range that produces the maximum AM rating; this metric is believed to be a robust and effective approach to detecting and evaluating real AM in WTN. The proposed threshold for the penalty is 3 dB MD, rated using this metric; this is the level at which detection can be confidently expected, and adverse responses may start to increase significantly. Based on the equivalent response evidence, the magnitude for the penalty is a variable 3 to 5 dB over the MD range 3 to 10 dB (and 5 dB thereafter). The AM character penalty scheme as proposed is shown in Figure 6a, with relevant data from the supporting evidence. The result of application of this penalty scheme to the absolute response data shown in Figure 1b is illustrated in Figure 6b – it should be noted that this is a separate response dataset from the equivalent response data used to inform the penalty shown in Figure 6a. As would be expected, the average responses are significantly (p<0.01) correlated with $L_{Ar,T}$ dB, with Pearson r-value of 0.872 (compared with 0.684 for the separated parameters $L_{Aeq,T}$ dB and 0.693 for $\Delta L_{Aeq,100ms(50\text{-}200Hz)}$ dB).

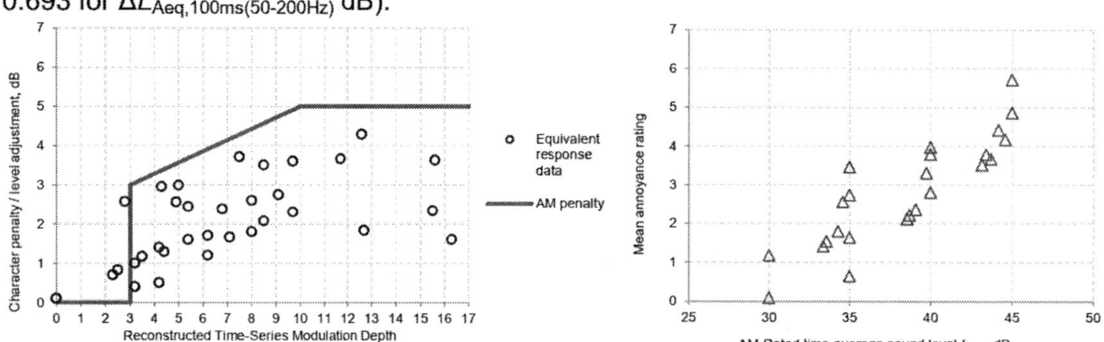

Figure 6: AM penalty scheme (a, left) value to be applied and equivalent response datasets [adjusted from] von Hünerbein, King et al (2013) and Yokoyama, Kobayashi et al (2015); (b, right) application to absolute annoyance response dataset from von Hünerbein, King et al (2013)

4.4 Application

Steps to be taken in applying the proposed scheme are as follows:

Instatement: It would be added within a planning condition attached to new development consent for large-scale commercial wind turbines (not small domestic turbines falling outside the scope of the method – discussed below).

Activation: The scheme would be considered in reaction to complaints about AM in WTN received by the local authority.

Action: Monitoring of WTN would be required under the scheme, including the specification of equipment suitable for obtaining measurements to produce ratings of AM in accordance with the AMWG Reference Method, which gives values for individual 10-minute periods (Bass, Cand et al, 2016).

Rating: The ratings produced would be considered against the penalty scale shown in Figure 6. The corresponding penalty values would be added to the WTN levels measured using existing methodologies for compliance testing as set out in ETSU-R-97 and the IOA Good Practice Guide (2013, 2014) for integer wind speeds to derive a rated equivalent level $L_{Ar,T}$ – implementation of this step is discussed further below.

Assessment: The rated levels would be compared with the overall noise limits set out in the planning consent, with the condition that, if both of the following two clauses are met, the difference between day/night limits at that wind speed will also be added to the rated level (the purpose of this condition is to ensure that AM impacts do not increase during the night-time):

1. A higher (less stringent) noise limit is in place for night-time at the wind speed being considered; and
2. An AM penalty is assigned at the same wind speed (ie the AM rating is ≥3 dB).

Enforcement: Limit exceedances demonstrating a breach of the condition could be enforceable by the local authority, in which case the specific wind speeds in which limits are breached should frame the mitigation requirements (eg a breach of the condition at 7ms^{-1} wind speed should not entail action to be taken to mitigate at 9 ms^{-1}, but at 7ms^{-1} – this point is mainly relevant to designing operational mode mitigation strategies, as opposed to engineering solutions such as blade treatments) – this may be formalised by a 'mitigation scheme to be agreed and implemented' clause, or similar, in the condition.

Mitigation: This should address a reduction so that the overall rated level consistently meets the limits; there are two pathways to achieve this: i) reduce AM in the WTN; ii) reduce the time-average level of WTN.

4.5 Assumptions and Limitations

Practical implementation of the above application remains an area that requires further technical development. In particular, a number of issues are highlighted within the WSP | Parsons Brinckerhoff (2016) report, and have also been raised during feedback on the scheme:

i. How to apply the penalty to the derived WTN levels for compliance assessment – the current UK practice set out by Cand et al (2013, 2014) is to derive averaged WTN levels for each wind speed, subtract averaged background sound and compare with the limits. The penalties for AM should be calculated from the AM ratings for individual 10-minute periods, not from a rating averaged over a longer period. For compatibility with current practice, one approach would be to then average the penalties over the assessment period, and apply this to the average level (in effect this is the same approach taken to tonal character penalisation). However the application of the AM penalty should avoid a situation in which averaging could dilute the effectiveness of the scheme in achieving its aim of controlling AM impact, which could require that AM-penalised periods be assessed separately to non-penalised periods. This is an important issue that requires further consideration.

ii. The determination of non-WTN background noise levels and how this will relate to the penalty scheme implementation in terms of deriving the rated levels.

The scheme is limited by the exclusion of non-acoustic influences on responses, such as visibility of turbines – future work might consider how these interrelated issues could be combined, but in terms of providing a workable control, this is a necessary limitation.

It is also underpinned by the assumption of modulation frequencies not too dissimilar to those used in the research supporting the scheme, and the range that forms the basis for the practical application of the AMWG rating metric, ie up to 1.6 Hz. Where higher f_m are expected, such as with small turbines, it is possible the scheme could underestimate the impact of AM character. Further research could assist in extending the applicability of the control.

4.6 Discussion and further work

Duration of impact

The review highlighted a lack of robust evidence on which to determine the effect of prolongation of exposure on expected responses; that is, how much is too much when it comes to AM? This is an area that has not been well explored and is difficult to study and quantify, yet is highly relevant in determining the application of the penalty scheme. It does not seem reasonable to suggest that brief, sporadic or occasional occurrences of 10-minute periods of AM >3 dB ΔL constitutes justifiable grounds for imposing corrective measures, yet it seems clear that frequent and prolonged exposure to AM >3dB (where the sound is audible) should be avoided; in between these extremes an effective, practical and sustainable approach must be established. However, this issue is not easily resolved. The current difficulty facing UK practitioners is in applying an AM control in a way that is i) practical, ii) effective, and iii) lawful. Approaches to addressing point i) could be relatively straightforward to achieve, but in combination with ii) and iii) the problem is considerably less tractable. An arithmetic averaging approach has the benefit of compatibility, practicality, is straightforwardly understood, and removes the need to make further judgements of duration and intermittency of impacts, as these considerations are automatically (naturally) incorporated. However, this implies that periods without AM are amalgamated in the total with penalised periods, and may lead to underestimation of the impacts in the AM periods. An alternative approach might address the magnitude and the duration of periods of AM separately, which could avoid the risk of impact underestimation, but implies judgement – difficulties arise with how to aggregate and interpret the overall impact with regards to frequency of occurrence and prolongation of AM.

One possible avenue to assessing the impact of intermittent AM could involve a type of cumulative 'dose' value formed from the 10-minute AM penalties. This could be another way to account for the magnitude of impacts alongside the durations and intermittency of occurrences, but would require further research to develop.

Alternative approaches

The partial foundation of the proposed penalty scheme on the 'equivalent annoyance' experimental data obtained by von Hünerbein, King et al (2013) has been questioned by Bowdler [quoted in Pease (ed.), 2016] on the basis that the data use fixed $L_{Aeq,T}$ as the control between paired samples. An alternative scheme based on L_{A90} differences presented in the same study was proposed, which is suggested to be more appropriate to the ETSU-R-97 method, itself based on L_{A90} assessment. However, it has been demonstrated that the L_{A90} scheme proposed could not be derived from the cited dataset without bias, due to the lack of control for L_{A90} in the paired comparisons (Lotinga, Perkins et al, 2017) – a problem that was accepted by the original researchers in their report; the same concern is also raised by Bass, Cand et al, (2016). Moreover, this proposal implies a direct connection between the trough depth in the artificial stimuli, and the long-term $L_{A90,10min}$ level used to assess WTN within ETSU-R-97, measures which are not directly comparable. According to Lotinga, Perkins et al (2017), the main issue arising from this alternative proposal with relevance to the penalty scheme is whether the assumption in ETSU-R-97 of an approximately constant relationship between $L_{Aeq,10min}$ and $L_{A90,10min}$ within WTN remains broadly valid during periods of real AM occurrence. This could be the subject of further research work, but in the absence of evidence to the contrary, the penalty should be added to the WTN $L_{A90,10min}$, which is employed as a reasonable and practical proxy for WTN $L_{Aeq,10min}$ in ETSU-R-97.

The use of objective noise exposure metrics to measure or predict subjective responses has limitations, especially in the case of WTN in which non-acoustic factors, such as visual impacts and attitude to wind energy, have a considerable influence on the expected variance in noise-related annoyance (Pawlaczyk-Łuszczyńska, Dudarewicz, et al, 2014). In the future, multi-dimensional quality-of-life measures may be a more effective way to evaluate and articulate the impacts and sustainability of wind farm developments, in which objective noise exposure and AM rating could play a part in a holistic approach, within a wider assessment framework (Shepherd, Welch et al, 2013). Further research aimed towards integrating approaches to assessing environmental and health impacts of noise-generating developments, and innovative tools and measures reflecting the need to consider health and wellbeing would be welcome. Stansfeld (2016) has also identified a particular need for use of objective measures of health effects in wind turbine noise research.

Non-periodic AM

All of the measurement, rating and assessment information presented above assumes that the AM is generally periodic. There remains the potential issue of the impact of non-periodic (or random-fluctuation) AM, which may occur in complex turbine interaction situations. Interestingly, the study by Schäffer, Schlittmeier et al (2016) suggests that the response to AM in WTN may not be directly related to its periodicity, and random-fluctuation AM with an 'f_m' varying within a similar range could be equal in impact to periodic AM (ie the modelled response curve was almost identical to that shown for periodic AM in Figure 4b) – this would be very difficult to detect and assess using current methods, indicating a potential area for further research to address.

5. Conclusions

The research review has concluded that i) WTN contributes to annoyance, ii) AM increases that contribution, and iii) that existing evidence provides a reasonable base for an objective control framework for AM based on a principle of 'equivalent response' with negligible-AM WTN, by including the main acoustic exposure factors thought to affect response. A suitable character penalty control has been proposed and is recommended for application in planning wind farm developments. However, significant questions remain regarding the extent and prolongation of impacts, and how this can be addressed within a practical, effective and lawful implementation of the proposed scheme.

It is hoped that the proposed control will lead to the development of more proactive approaches to prediction and mitigation of AM on the part of wind farm operators and developers.

References

Bakker, R H, Pedersen, E, van den Berg, G P, Stewart, R E, Lok, W and Bouma, J, 2012. "Impact of wind turbine sound on annoyance, self-reported sleep disturbance and psychological distress". In: *Science of the Total Environment*, 425, pp 42-51.

Bass, J, Cand, M, Coles, D, Davis, R, Irvine, G, Leventhall, G, Levet, T, Miller, S, Sexton, D and Shelton, J, 2016. *IOA Noise Working Group (Wind Turbine Noise) Amplitude Modulation Working Group Final report: A method for rating amplitude modulation in wind turbine noise.* Institute of Acoustics.

Bengtsson, J, Persson Waye, K and Kjellberg, A, 2004. "Sound characteristics of low frequency noise and their relevance for the perception of pleasantness". In: *Acta Acustica united with Acustica*, 90, pp 171-180.

Bockstael, A, Dekoninck, L, Can, A, Oldoni, D, De Coensel, B and Botteldooren, D, 2012. "Reduction of wind turbine noise annoyance: an operational approach". In: *Acta Acustica united with Acustica*, 98, pp 392-401.

Bowdler, D, 2008. "Amplitude modulation of wind turbine noise: a review of the evidence". In: *Acoustics Bulletin*, 33 (4), pp 31-35.

Bradley, J S, 1994. "Annoyance caused by constant-amplitude and amplitude-modulated sounds containing rumble". In: *Noise Control Engineering Journal*, 42 (6), pp 203-208.

BSI, 2014. *BS 4142:2014 Methods for rating and assessing industrial and commercial sound*. London: British Standards Institution.

Cand, M, Davis, R, Jordan, C, Hayes, M and Perkins, R, 2013. *A Good Practice Guide to the application of ETSU-R-97 for the assessment and rating of wind turbine noise.* Institute of Acoustics.

Cand, M, Davis, R, Jordan, C, Hayes, M and Perkins, R, 2014. *A Good Practice Guide to the application of ETSU-R-97 for the assessment and rating of wind turbine noise. Supplementary Guidance Note 5: Post completion measurements.* Institute of Acoustics.

Energy Technology Support Unit Working Group on Noise from Wind Turbines, 1996. *ETSU-R-97. The assessment and rating of noise from wind farms.* Department of Trade and Industry.

Fastl, H, 1982. "Fluctuation strength and temporal masking patterns of amplitude-modulated broadband noise". In: *Hearing Research* 8, pp 59-69.

Gabriel, J, Vogl, S, Neumann, T, Hübner, G and Pohl, J, 2013. *Amplitude modulation and complaints about wind turbine noise.* 5th International Meeting on Wind Turbine Noise, Denver, 28-30 August 2013.

Ioaniddou, C, Santurette, S and Jeong, C-H, 2016. "Effect of modulation depth, frequency, and intermittence on wind turbine noise annoyance". In: *Journal of the Acoustical Society of America* 139 (3), pp 1241-1251.

ISO, 2016. *ISO 1996-1 Acoustics — Description, measurement and assessment of environmental noise — Part 1: Basic quantities and assessment procedures.* Geneva: International Standards Organisation.

Janssen, S A, Vos, H, Eisses, A R, and Pedersen, E, 2011. "A comparison between exposure-response relationships for wind turbine annoyance and annoyance due to other sources". In: *Journal of the Acoustical Society of America* 130 (6), pp 3746-3753.

Large, S and Stigwood, M, 2014. *The noise characteristics of 'compliant' wind farms that adversely affect its neighbours.* 43rd International Congress and Exposition on Noise Control Engineering (Inter-noise), Melbourne, 16-19 November 2014.

Lee, S, Kim, K, Choi, W and Lee, S, 2011. "Annoyance caused by amplitude modulation of wind turbine noise". In: *Noise Control Engineering Journal*, 59 (1), pp 38-46.

Moore, B C J, 2012. *An introduction to the psychology of hearing.* 6th edition. Bingley: Emerald Group Publishing.

NZS 6808:2010. *Acoustics – wind farm noise*. Standards New Zealand.

Pawlaczyk-Łuszczyńska, M, Dudarewicz, A, Zaborowski, K, Zamojska-Daniszewska, M and Waszkowska, M, 2014. "Annoyance related to wind turbine noise". In: *Archives of Acoustics*, 39 (1), pp 89-102.

Pease, J (ed.), 2016. "News from the IOA 'ETSU – is it time to move on?' meeting: Agreement still elusive". In: *Noise Bulletin*, December 2016.

Pedersen, E and Persson Waye, K, 2004. "Perception and annoyance due to wind turbine noise – a dose-response relationship". In: *Journal of the Acoustical Society of America* 116 (6), pp 3460-3470.

Pedersen, E and Persson Waye, K, 2007. "Wind turbine noise, annoyance and self-reported health and well-being in different living environments". In: *Occupational and Environmental Medicine*, 64 (7), pp 480-486.

Pedersen, E, Hallberg, LR-M and Persson Waye, K, 2007. "Living in the vicinity of wind turbines – a grounded theory study". In: *Qualitative Research in Psychology*, 4 (1-2), pp 49-63.

Perkins, R A, Lotinga, M J (co-lead), Berry, B, Grimwood, C J and Stansfeld, S A, 2016. *A review of research into the human response to amplitude-modulated wind turbine noise and development of a planning control method*. 45th International Congress and Exposition on Noise Control Engineering (Inter-noise), Hamburg, 21-24 August 2016. [online]

Planning Inspectorate (UK), 2009. Appeal decision notice. Appeal Ref: APP/Q1153/A/06/2017162. Land to the south east of North Tawton and the south west of Bow [online] last accessed 09/02/2017.

RenewableUK, 2013. *The development of a penalty scheme for amplitude modulated wind farm noise*, [online] last accessed 09/02/2017.

Schäffer, B, Schlittmeier, S J, Pieren, R, Heutschi, K, Brink, M, Graf, R, and Hellbrück, J, 2016. "Short-term annoyance reactions to stationary and time-varying wind turbine and road traffic noise: a laboratory study". In: *Journal of the Acoustical Society of America* 139 (5), pp 2949-2963.

Shepherd, D, Welch, D, Dirks, K N and McBride, D, 2013. "Do quiet areas afford greater health-related quality of life than noisy areas?". In: *International Journal of Environmental Research and Public Health*, 10, pp 1284-1303.

Stansfeld, S A, 2016. *WHO Environmental noise guidelines for the European region – what is new? 2. New evidence on health effects from environmental noise and implications for research*. 45th International Congress and Exposition of Noise Control Engineering (Inter-noise), Hamburg, August 21-24, 2016.

Stigwood, M, Stigwood, D and Large, S, 2014. *Initial findings of the UK Cotton Farm Wind Farm long term community noise monitoring project*. 43rd International Congress and Exposition on Noise Control Engineering (Inter-noise), Melbourne, 16-19 November 2014.

Van den Berg, F and Bowdler, D, 2011. "Amplitude Modulation". In: Leventhall, G and Bowdler, D (eds.). *Wind turbine noise*, Chapter 5. Brentwood: Multi-Science Publishing Co.

Van den Berg, G P, 2004. "Effects of the wind profile at night on wind turbine sound". In: *Journal of Sound and Vibration*, 277, pp 955-970.

Van den Berg, G P, 2005. "The beat is getting stronger: the effect of atmospheric stability on low frequency modulated sound of wind turbines". In: *Journal of Low Frequency Noise, Vibration, and Active Control*, 24 (1), pp 1-24.

Van den Berg, F, 2009. *Why is wind turbine noise noisier than other noise?* 8th European Congress and Exposition on Noise Control Engineering (EuroNoise), Edinburgh, 26-28 October, 2009.

Von Hünerbein, S, King, A, Piper, B and Cand, M, 2013. *Wind turbine amplitude modulation: research to improve understanding as to its cause and effect. Work package B(2): development of an AM dose-response relationship*, [online].

WSP | Parsons Brinckerhoff, 2016. *Wind turbine AM review: Phase 2 report. 3514482A V3.* Department for Business, Energy and Industrial Strategy.

Yokoyama, S, Kobayashi, T, Sakamoto, S and Tachibana, H, 2015. *Subjective experiments on the auditory impression of the amplitude modulation sound contained in wind turbine noise*. 6th International Meeting on Wind Turbine Noise, Glasgow, 20-23 April 2015.

Appendix: Data Adjustments

Re-scaling of original modulation depth values to the AMWG metric has been carried out using the methods detailed by Bass, Cand et al, (2016). The open-source Python software provided by the AMWG for calculating AM ratings is available at https://sourceforge.net/projects/ioa-am-code/?source=directory. Re-scaled results and derivatives thereof are presented in Figures 2a, 3, 4a and 6. In Figures 4a and 6a, the original response data of Yokoyama, Kobayashi et al (2015) have been opposite-signed, inverting the dB values for comparative purposes.

Three adjustments are made to the data used in Figure 5b: i) for differences in $L_{Aeq,T}$, ii) for differences in MD, iii) for differences between the sensitivity study and the main test results. The adjustments for i) have been derived from figure A1a, which shows the same data as Figure 1b, grouped over all MDs. The gradient of the linear regression line for 35-40 dB $L_{Aeq,T}$

has been used (broadly corresponding to the range of $L_{Aeq,T}$ in the data addressing f_m in Figure 5b) as a rough estimation of the annoyance rating per dB increase in $L_{Aeq,T}$, and this value (0.32175) has been used as an approximation to normalise the ratings in Table 8.5 of the report by von Hünerbein, King et al (2013) to an $L_{Aeq,T}$ of 35 dB. Similarly, an adjustment has also been made for MD, using the values for 35 dB dB $L_{Aeq,T}$, as shown in Figure A1b (scaled to the original design mean peak-trough MD $\Delta L_{Aeq,100ms}$, for consistency with the original data); the value of the gradient (0.2437...) has been used as an approximation to normalise ratings to a MD of 5 dB. The data points from the original test for 1 dB MD have been omitted as they are outside the interpolation range for this adjustment. Finally, the ratings have been equally translated so that the mean annoyance rating for the reference stimulus with MD 5 dB $\Delta L_{Aeq,100ms}$, time-average level 35 dB $L_{Aeq,T}$ and f_m 0.8 Hz is the same as that obtained in the main test. The output of the adjustments is shown in Figure 5b.

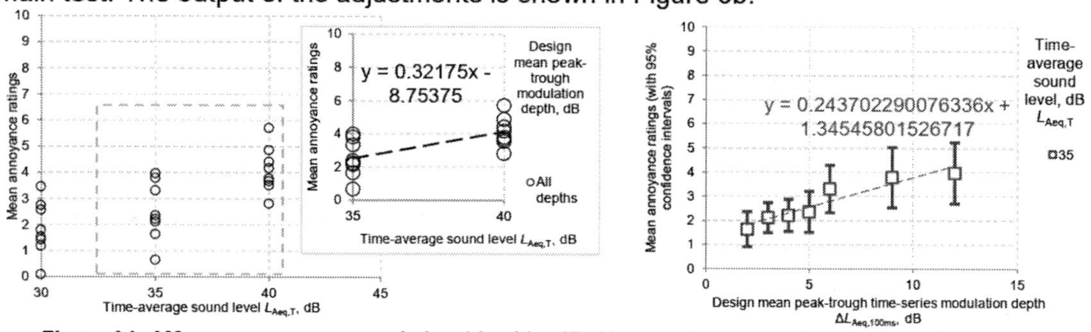

Figure A1: AM exposure-response relationships identified by von Hünerbein, King et al (2013) and used to derive adjustment values for data in Figure 5b; (a, left and inset) time-average sound level; (b, right) modulation depth

7th International Conference
on
Wind Turbine Noise
Rotterdam – 2nd to 5th May 2017

AN INVESTIGATION INTO CORRELATION BETWEEN STRONG WIND TURBINE AMPLITUDE MODULATION AND ENVIRONMENTAL CONDITIONS

Duncan Halstead, Aercoustics Engineering Ltd.: DuncanH@aercoustics.com
Adam Suban-Loewen, Aercoustics Engineering Ltd.: AdamS@aercoustics.com
Payam Ashtiani, Aercoustics Engineering Ltd.: PayamA@aercoustics.com

1 SUMMARY

This paper presents the results of an improved detection algorithm for wind turbine amplitude modulated sound used in conjunction with meteorological measurements to investigate possible correlations between high wind turbine amplitude modulated noise and different environmental conditions. The study was conducted using measurement data taken at nearby residences to three different wind farms of varying size and turbine model in Ontario, Canada. Long term measurements were taken at locations roughly 450m to 900m from the nearest turbines. The data available for analysis represents over 1000 hours of sound recordings, acquired between September 2015 and December 2016.

An algorithm, based on work from the Institute of Acoustics and modified by Aercoustics, was used to detect and determine the modulation depth of noise from wind turbines at these three wind farms. The algorithm was found to be effective at automatically identifying periods of wind turbine amplitude modulation, despite the presence of noise from other sources in the ambient environment.

Modulation depths were compared with meteorological data collect at 10-meter and hub height at each measurement location. Wind speed, wind direction, temperature, atmospheric pressure, and relative humidity were all analysed with respect to the calculated modulation strength from the wind turbine noise.

It was found that the driving factor behind the wind turbine modulation strength at each measurement location was the local ambient masking noise, rather than the turbine noise emission. This was evidenced by the clear relationship between 10-meter wind speeds and detected modulation strength. Correlations with hub height wind speeds showed similar results, but not as clean. Wind direction analysis did not provide anything conclusive, and more analysis on a greater dataset is required before any trends can be seen.

The effects of other meteorological variables were not deemed to have a noticeable impact on detected modulation strength.

2 INTRODUCTION

Amplitude Modulated (AM) noise from wind turbines has been the subject of much study lately, including Institute of Acoustics [1], RenewableUK [2], and other independent researchers [3] [4] [5].

One of the prevailing methods for wind turbine AM detection is the application of an FFT[1] on a calculated level-vs-time signal to get information on the frequency at which the level is modulating. Modulation frequencies around the wind turbine blade-pass frequency are attributed to wind turbine AM. The most recent work using this method was published in 2016 by the Institute of Acoustics' (IOA) working group on wind turbine amplitude modulated sound [1].

That method was used as a basis for this study. Some changes have been implemented to improve the rejection of ambient sound, as well as preserve modulation depth information between time and frequency domains. A rating metric was also developed to assess the relative strength of modulation between periods in the same dataset. This algorithm was applied to three measurement datasets that were acquired by long-term monitoring rigs installed at different wind farms in Ontario over the past year and a half.

This study is intended to evaluate the ability of the modified algorithm to detect wind turbine AM, and look for any correlations between the environmental conditions during periods of wind turbine AM to see if there are any common factors.

3 MEASUREMENT SETUP AND SITE CONDITIONS

The measurement data used for this study was obtained from three long-term monitoring rigs installed in-field for different periods during the last two years. These measurements were conducted to audit the acoustic compliance of the nearby wind farm – a process required by the farm's operating permit – rather than a response to specific noise complaints.

The monitoring rigs are built to synchronously measure acoustic and meteorological (MET) data at heights of 4.5 meters and 10 meters, respectively. The microphone is fitted with a primary and secondary windscreen to reduce the effect of wind generated self-noise. The weather station is programmed to measure wind speed, wind direction, temperature, atmospheric pressure, and relative humidity. The system is designed to operate remotely, measuring every night from 10pm to 5am. Measurement data is logged in 1-minute averages, and the signal recordings are stored for post-processing. A picture of a typical measurement apparatus is shown in Figure 1.

For the compliance audits, these measurement rigs are installed near residences around the wind farm. The sites used in this study range from 450m to 900m from the nearest turbines.

All the measurements used for analysis in this study were taken between September 2015 and December 2016. Each measurement campaign ranged from 6 weeks to 5 months, and represents roughly 1000 hours of measurement data. All measurements were unattended, so it was important that any algorithm developed is capable of sifting through large amounts of data automatically.

Figure 1: Measurement apparatus

[1] Fast Fourier Transform – numerical implementation of the Discrete Fourier transform

4 METHODOLOGY

The acoustic measurement data was processed in three steps, starting from the raw signal recordings. The first step was to determine the modulation time signal and its frequency content. The second step was to use this data to isolate the modulation energy attributable to wind turbine AM. The third and final step was to take the wind turbine AM information and examine any relationship with operational and meteorological parameters.

4.1 Step 1 – Determining the Modulation Spectra

The algorithm used to detect AM is based on the latest work by the Institute of Acoustics, published in September 2016 [1]. Specific optimizations and simplifications were added by Aercoustics to improve the algorithm's ability to preserve energy equivalence of the calculated modulation depths between the time and frequency domains. The procedure is outlined as follows:

1. Log the 1/3rd octave bands in 4 ranges: 50-200Hz, 200-400Hz, 200-800Hz, and 200-1250Hz (inclusive) in 100ms intervals. Calculate the overall A-weighted level by energy summing the spectra from each 1/3rd octave range. The result is a 10Hz time domain modulation signal that is limited in bandwidth to the frequency range selected (see Figure 2 for an example). The subsequent analysis is performed on each range.

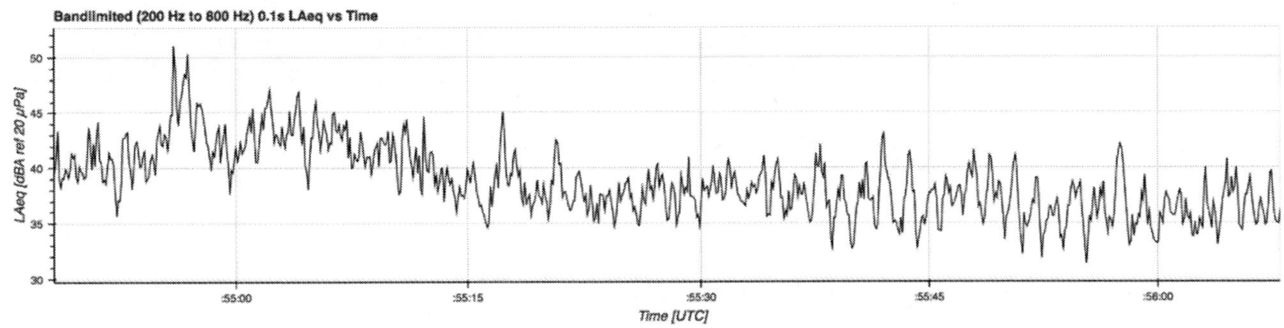

Figure 2: A-weighted, bandlimited 0.1s time domain signal containing wind turbine AM starting just over halfway through the recording.

2. Apply a high pass filter with a cut-off frequency of 0.2Hz to the time domain signal. This is an alternative to the method of "detrending" the time signal with a 3rd order polynomial that is used by the IOA. It was found to be a simpler implementation, removing sources of error attributed to over/under-fitting of the polynomial to varying length intervals of the time signal.

3. Analyze the filtered, band-limited 10Hz time domain signal by calculating rolling, short duration FFTs and then reducing to a single 1-minute spectrum using an L10 function. Frequencies below 5Hz, in the region of the wind turbine blade-pass frequency (BPF) and its harmonics are of interest. Frequency resolution, windowing, and reduction parameters are chosen to conserve accurate peak-to-peak amplitude of worst-case modulation depth in the frequency domain. A statistical reduction approach is used to reject outliers.

The spectra obtained using this methodology are used to build modulation spectrograms for each measurement campaign. A 1-minute interval was chosen to provide a high temporal resolution. A 10-minute interval could also be chosen, which matches the format common to most turbine SCADA systems. Two sample spectra, one clean and one contaminated, are shown in Figure 3

The method used by IOA to rate wind turbine AM does not directly match the use case of correlation assessment with meteorological parameters. In particular, the BPF and harmonic bandwidths are assumed in the IOA method to be constant over time for every wind turbine, and the assessment of masking modulation is limited to only a few spectral lines. A custom metric to rate relative wind turbine AM strength is created to address this.

To help isolate wind turbine AM from modulations caused by ambient sources, rotational speed from the turbine SCADA, provided in 10-minute intervals, is used to identify the specific BPF for each measurement interval. While this may not be necessary for some turbines, it is found to be particularly important for turbines whose rotational speed may vary significantly.

Relative modulation strength is assessed for each modulation spectrum, providing a single value for each 1-minute interval, matching the sampling frequency of meteorological data. The value of the relative modulation strength metric is normalized such that 0 represents little to no AM, 1 represents max AM for the site, and negative values represent cases of contaminated modulation spectra. This metric can be used to sort and filter the data for AM detection; at which point the modulation depth values from the modulation spectra can then be used as a physical quantification of the level of AM.

4.2 Step 3 – Analyzing Periods of High Wind Turbine Amplitude Modulation

The calculated relative modulation strength was compared to the measured MET parameters to see if any correlations were evident. MET data was used from the weather stations on the monitoring rig as well as the anemometer from the closest wind turbine (mounted at hub height). Specifically, parameters assessed in this study are the ambient temperature, atmospheric pressure and relative humidity at 10-meters, as well as wind directions and wind speeds at both the turbine hub height and 10-meter measurement locations. Turbine power output was used to filter out periods where the turbines were not operational.

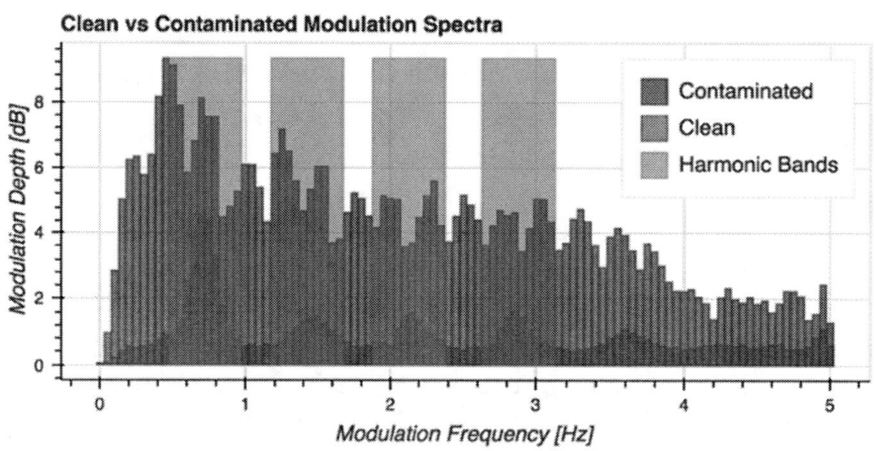

Figure 3: Clean spectrum with wind turbine AM (green) vs. a noisy spectrum with no obvious AM (blue). Harmonic bands are overlaid at expected harmonic frequencies.

5 SITES ASSESSED

Measurement data obtained from three different long term monitoring rigs was analysed using the methodology described in Section 4. Each rig was installed between 450 and 900 meters from the closest turbines. Each site varied in wind turbine make, site layout, and surrounding topography. Wind farm size ranged from less than 10 to over 100 wind turbines. All three wind farms in this study comprised turbines having power ratings over 1.5 MW and blade diameters over 80 meters. Details regarding each site are as follows:

5.1 SITE ONE

The first site in this study is a wind farm having only two turbines in the immediate vicinity, both south/southeast of the monitoring location roughly 650m away. The remaining turbines in the farm are located further to the south. The topography of the surrounding area is flat farm land, with no trees near the measurement position. The turbines in this wind farm have the largest blade lengths of the three sites analysed.

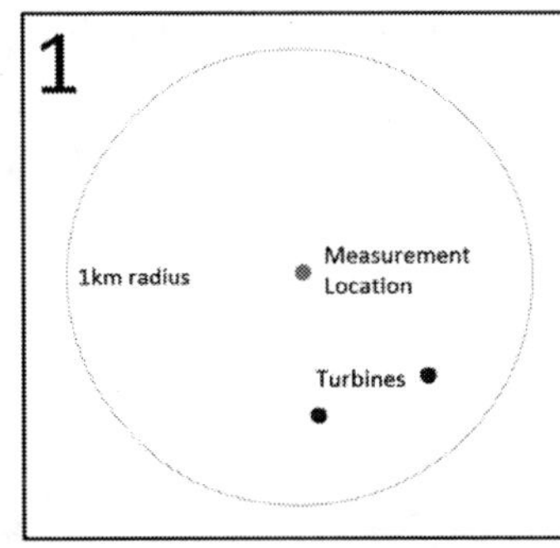

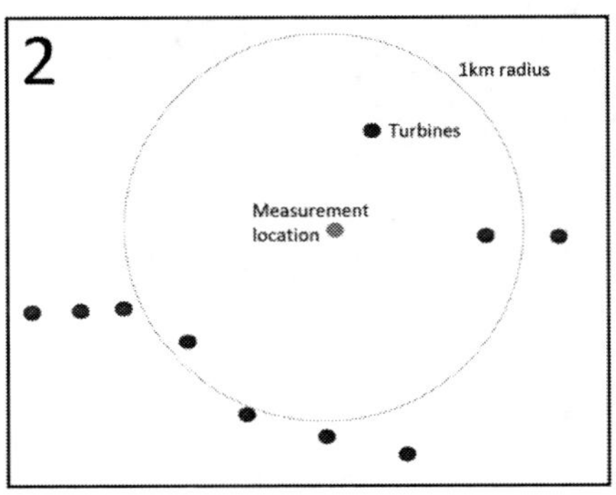

5.2 SITE TWO

This wind farm is made up of stall controlled wind turbines, having the shortest blade lengths of those in the study. The measurement position is roughly 450m southwest of the nearest turbine. However, within a radius of 1,000m there are turbines in nearly all directions. The area surrounding the measurement position is flat farmland, with a narrow line of trees immediately southwest of the receptor.

5.3 SITE THREE

This site lies in a forest, with the surrounding topography punctuated by ridges and valleys. The monitoring rig itself is in a valley, with the nearest turbines located on a ridge 900 meters to the northwest. There is a vertical difference of roughly 230m from the measurement position to the hub of the nearby turbines. Additional turbines are located southwest, nearly 2km away.

The ridge, combined with the forest, makes for site conditions where the measured wind speeds at 10m are consistently low when the hub height wind speeds are high. The low wind speeds at 10m translate to low ambient levels.

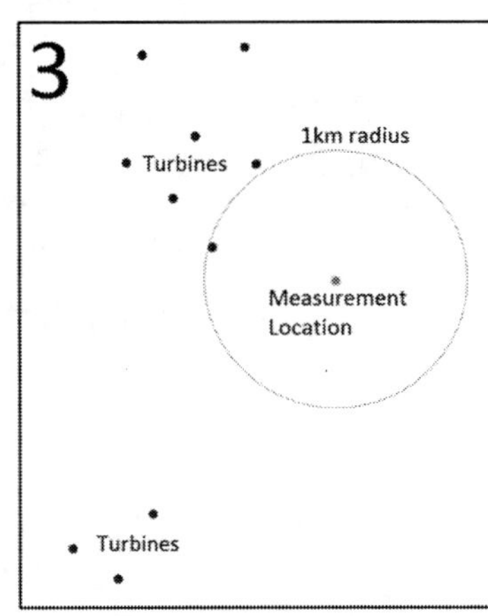

6 RESULTS

Analysis results are separated into sections detailing the performance of the detection algorithm, the correlations between wind conditions, and correlations with the remaining environmental factors.

6.1 Detection of Wind Turbine Amplitude Modulation

The aim of the detection algorithm was to have a high success rate of automatically detecting wind turbine AM, while minimizing false positives that are generated by ambient noise sources that modulate near the turbine BPF.

6.1.1 Modulation Spectrograms

Overall modulation spectrograms are plotted below, one for each site. Each spectrogram in this section comprises the entire measurement campaign. Measurements were taken nightly between 10pm-5am, however, for clarity the data has been concatenated into one continuous plot, with no gaps between measurement periods. The x-axis for each plot is in hours.

Figure 4: Modulation spectrogram, entire measurement campaign - Site One

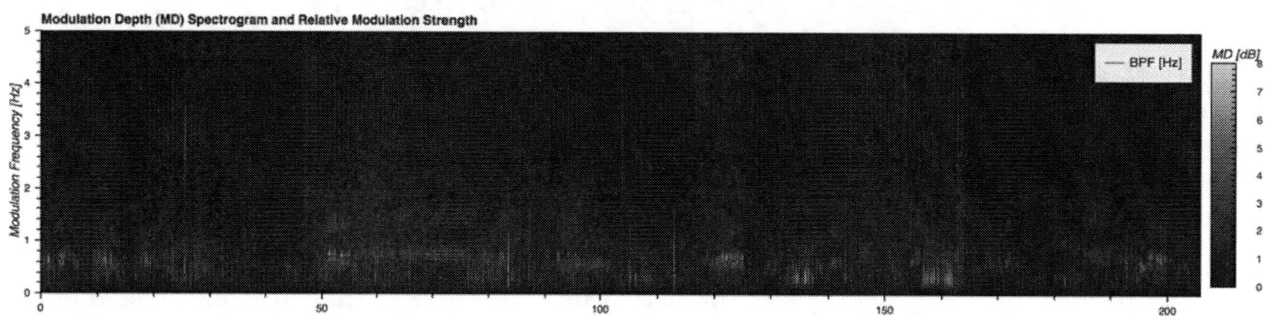

Figure 5: Modulation spectrogram, entire measurement campaign - Site Two

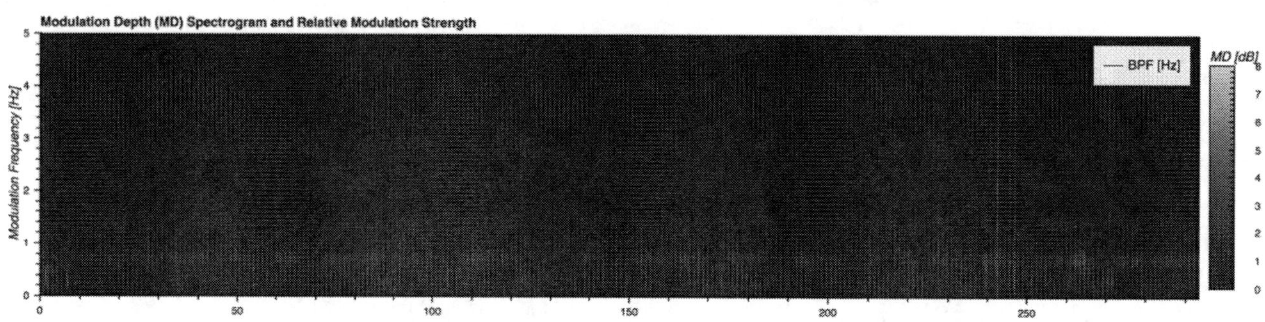

Figure 6: Modulation spectrogram, entire measurement campaign - Site Three

The colour of each data point corresponds to the calculated modulation depth, in a sliding scale from 0-8dB. The upper limit is capped, and all modulation depths greater than or equal to 8dB are plotted in yellow. The BPF of the turbine is overlaid on each curve, in red. For Site One and Three, the BPF was determined using the actual rotational speed of the nearest turbine, obtained from the wind farm SCADA. The BPF in site 2 is assumed to be constant, due to the nature of the turbines.

The modulation spectrograms show multiple periods where the amplitude modulation in the region of the turbine BPF is visually apparent. Generally, the modulations track quite well with the turbine BPF, with a couple exceptions.

> In the second half of Site One, strong modulations are apparent at frequencies below the BPF; this is likely due to wind-induced noise causing lower frequency modulations that mask the wind turbine AM.
>
> Site Three shows some periods where the actual modulation frequency deviates from the turbine BPF; this is likely a result of the SCADA information available from only a single turbine, rather than the cluster of turbines near the measurement position.

Overall, the modulation spectrograms appear to provide an effective means of visually identifying periods of strong AM in a measurement dataset. This provides a useful method of manually selecting periods for further analysis.

6.1.2 Quantifying Modulation Depth

A shortcoming of previous methods of AM detection were the inability of the algorithms to accurately replicate actual modulation depths observed in the time domain levels. Modifications to the algorithm were made specifically to address this issue. The resulting modulation depths determined from the modulation spectra were found to have a much better agreement with the actual modulation depths observed in the time domain levels. An example section is provided below in Figure 7.

Figure 7: Comparison of the modulation depth from the FFT with time signal

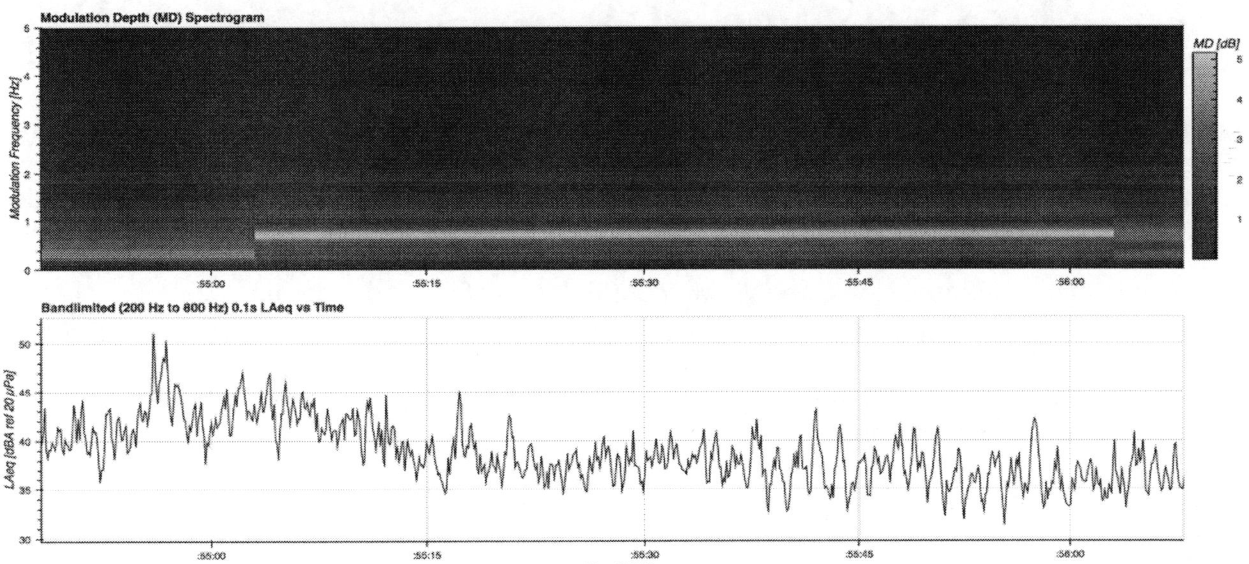

In Figure 7, the time domain signal in the lower plot corresponds to the same time interval as the modulation spectrogram above it. The spectrogram shows a modulation depth of about 5.2dB, which appears to agree with the time domain signal below it.

6.1.3 Effect of Bandwidth Selection on Data Quality

The modulation spectrograms were generated for each of the four frequency ranges outlined in Section 4.1. In general, it was found that the 200-1250Hz bandwidth provided the best signal-to-noise ratio (SNR) for all sites. All further analysis was done on the data calculated from this frequency range. A comparison of the effect of different frequency ranges on the modulation spectrograms is presented below for one site.

Figure 8: Frequency bandwidths, clockwise from the top left, 50-200Hz; 100-400Hz; 200-800Hz; 200-1250Hz

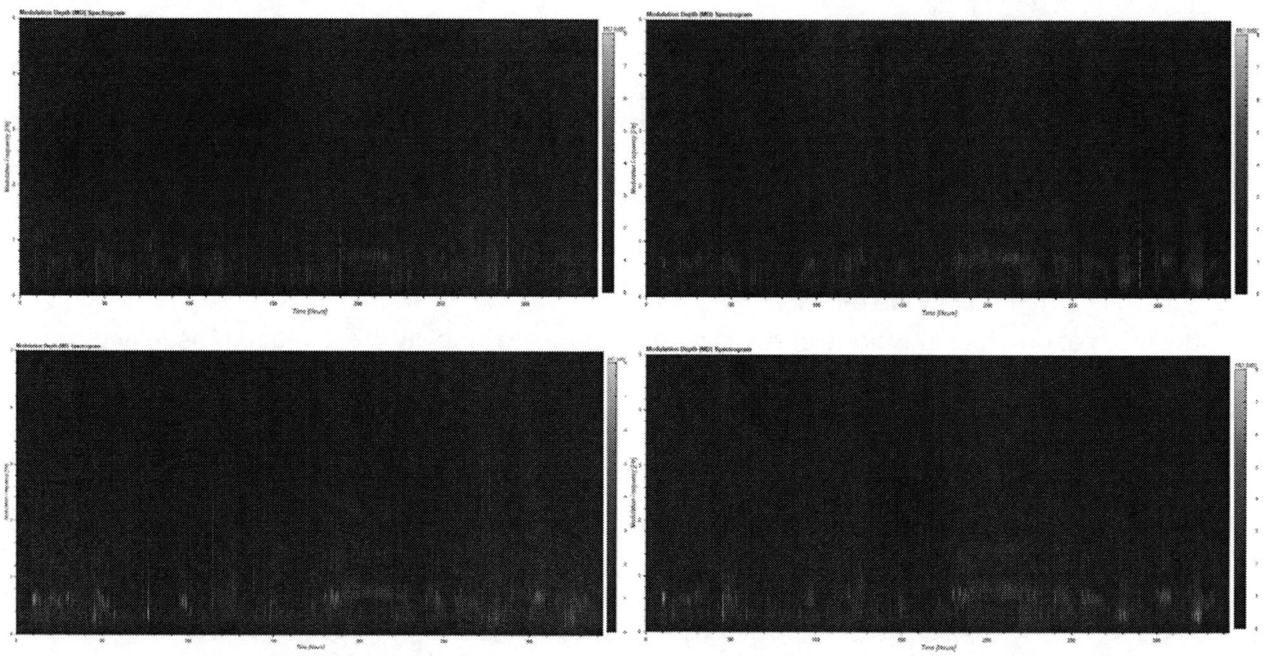

Of all the frequency ranges used, 200-800Hz and 200-1250Hz resulted in the best SNR. The 200-800Hz range had less individual contaminating events, visible as vertical lines in the plots, but the 200-1250Hz had a better SNR from modulation energy to average background energy. For this reason, the 200-1250Hz bandwidth was selected for further analysis.

6.1.4 Comparison of Modulation Strength Metric with Modulation Spectrogram

Periods of wind turbine AM are visually distinguishable as "hot" regions that occur around the wind turbine BPF in the modulation spectrogram. The strength metric for wind turbine AM developed for this paper is compared with the modulation spectrograms to assess how well the metric correlates with actual observed modulation depths.

In Figure 9, the modulation spectrogram for a single site is plotted above the calculated modulation strength as a function of time. The time axes for both plots are equivalent, and so the modulation strength at a given point can be directly compared with the corresponding region of the spectrogram directly above it.

Figure 9: Comparison of modulation strength metric with modulation spectrogram

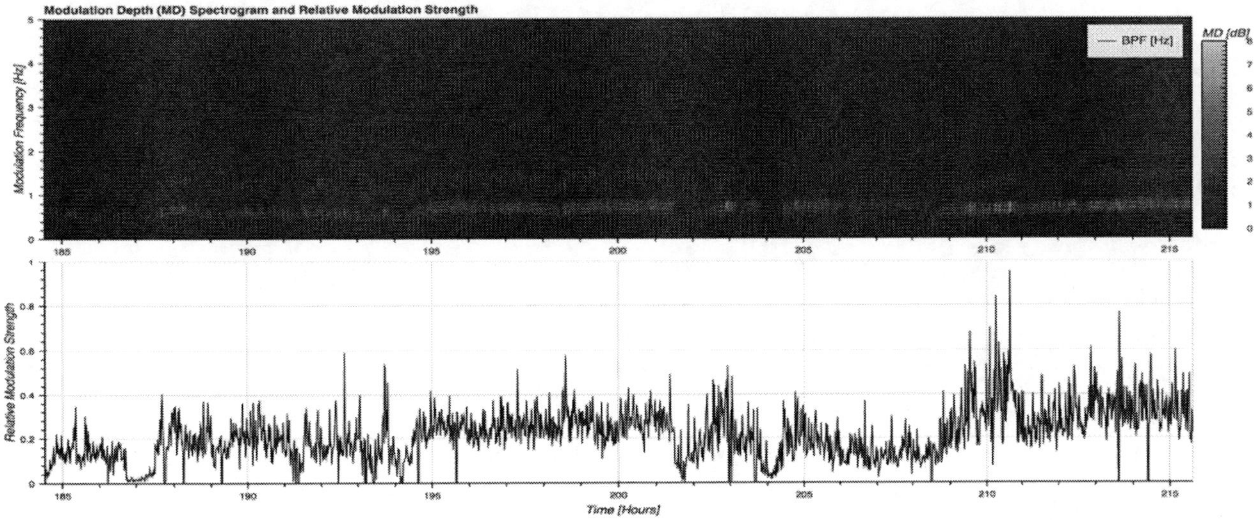

It can be seen in Figure 9 that the metric for modulation strength tracks quite well with the actual modulation shown in the corresponding spectrogram. The measurement starts with weak modulation around the BPF, followed by a period of no modulation, after which the modulation comes back at a higher strength. Throughout the whole section, the modulation strength metric appears to show a close relationship with the visible modulation at the wind turbine BPF. The metric even drops to zero during the period where the modulation in the spectrogram appears to disappear. Modulation strength is highest near the end of the plot, where the corresponding region of the spectrogram shows the hottest colours.

6.2 Modulation Strength and Wind Speed

After validation, the modulation strength metric was plotted against various meteorological parameters to see if there were any trends to be seen in the data. This section presents the results of the analysis with respect to wind speed.

Periods of high modulation strength were found to occur more often when there was an appreciable difference between the hub height and 10-meter wind speeds. High hub height winds indicate the turbine is generating closer to its maximum sound power, and low 10-meter winds usually correspond to low ambient sound levels.

There are a few periods of high modulation that occur below the wind turbine BPF, shown in the second half of the spectrogram. These low frequency modulations occur during periods when the 10-meter wind speeds are high, and are therefore attributed to wind-induced ambient noise, rather than wind turbine AM.

Figure 10 shows, over the course of an entire measurement campaign, the variation in wind speed and wind direction measured at both the 10-meter anemometer (from the monitoring rig) and the hub-height anemometer (from the nacelle of the nearest turbine). Hub height measurements are plotted in green and 10-meter height measurements are plotted in blue.

Figure 10: Plot of the variation in 10-meter and hub-height wind speed and direction compared with modulation strength over the duration of the monitoring campaign

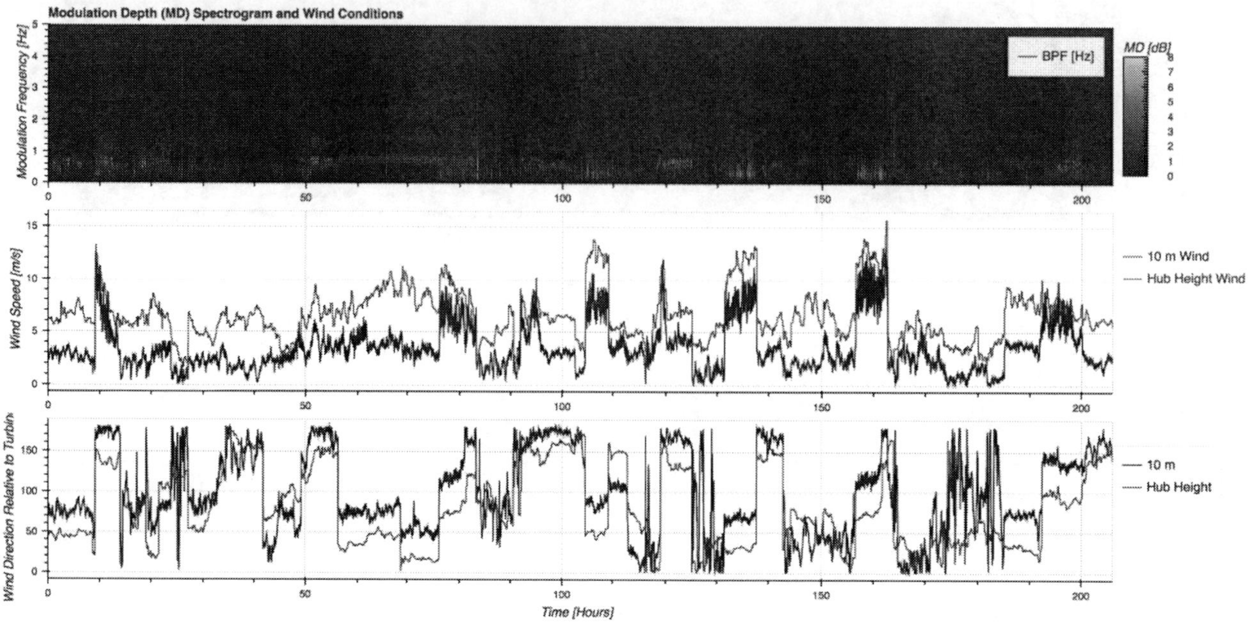

To further investigate the relationship between wind speed and wind turbine AM, the modulation strength was plotted as a function of measured wind speed for each data set. The relationships between modulation strength and 10-meter wind speed as well as hub height wind speed are shown in Figure 11 and Figure 12, respectively.

Figure 11: Modulation strength as a function of 10-meter wind speed

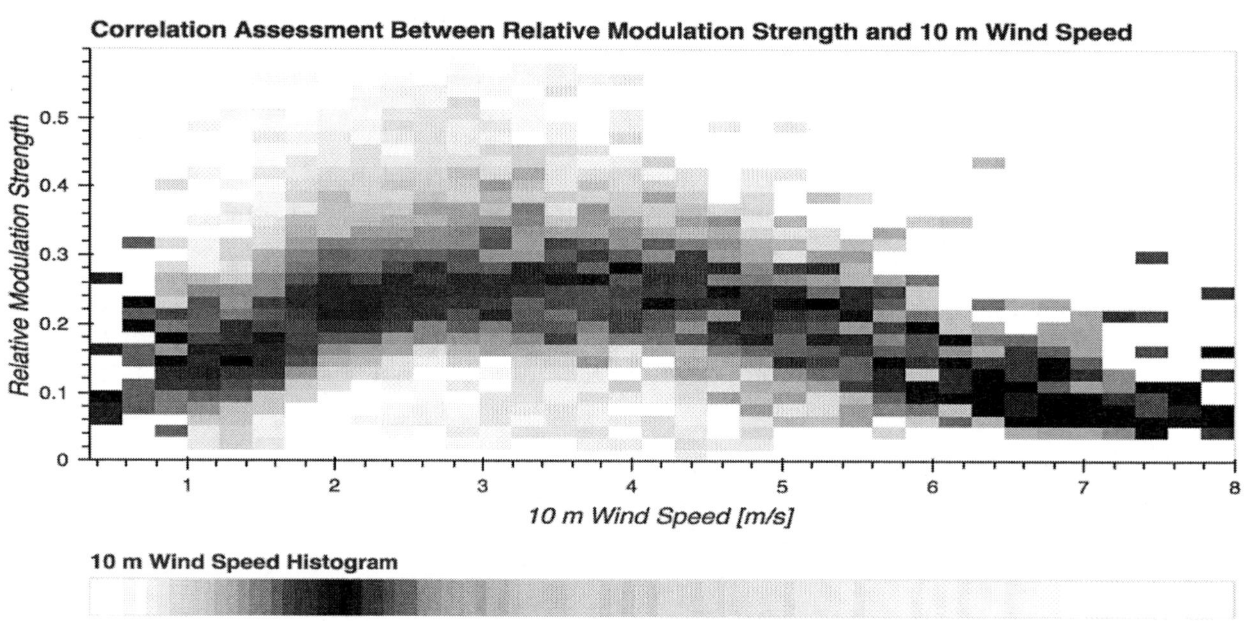

Both plots are coloured relative to the individual probability densities in each wind bin (0.2m/s wide), rather than the whole dataset. The aim is to remove the effect of data concentration in specific wind bins biasing the colour plot. The histogram of total data per wind bin is presented below each plot. These plots are generated with the measurement data from Site Two.

Figure 12: Modulation strength as a function of hub-height wind speed

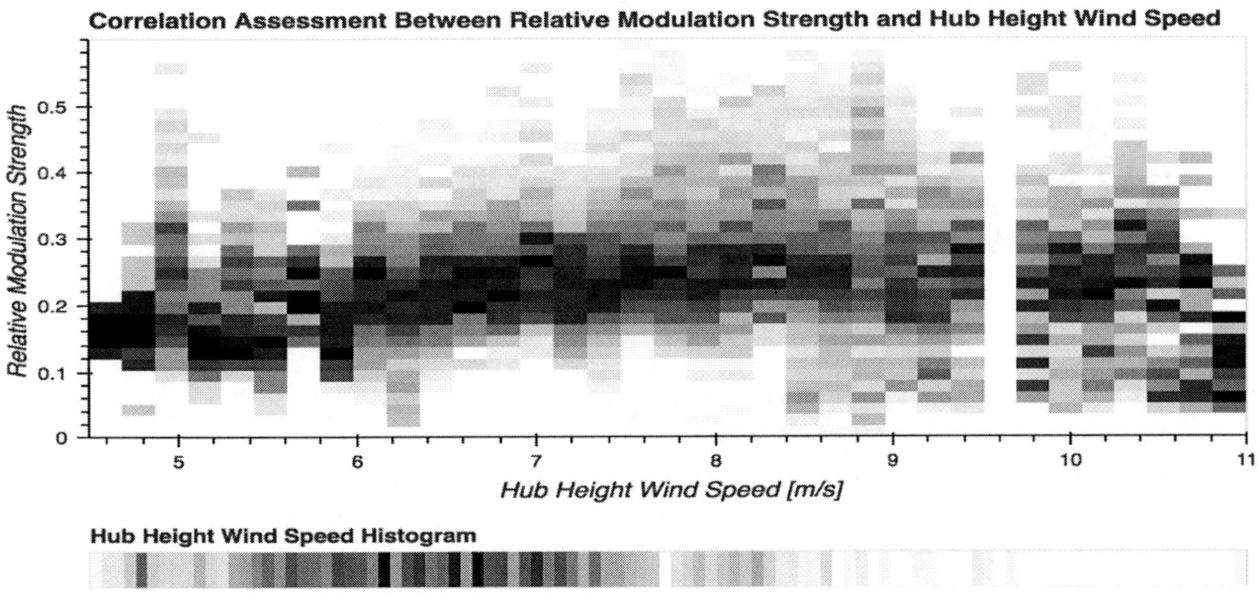

From these results, it appears that there is a more clearly defined relationship to modulation strength with 10-meter wind speed than there is with hub height wind speed. This can be explained intuitively, as surface wind speed has a more direct correlation with the measured ambient sound level. Below a certain ground-level wind speed, usually 3-4m/s, the wind-induced noise in the ambient is negligible. However, above that threshold, it becomes a driving factor in the ambient sound level. This can be seen in Figure 11, as the 10-meter wind speed increases above 4 m/s, the modulation strength decreases, due to increased ambient noise that masks the wind turbine AM. Below 3m/s, the modulation strength also decreases, likely due to the hub height winds dropping below rated power, reducing the wind turbine sound emission.

The same trend can be seen in Figure 12, however the trend is not as clear. It can be concluded from these plots that the driving factor in detecting wind turbine AM is not the hub-height wind speed, but rather the 10-meter wind speed. It follows from this observation that sorting wind speeds by surface wind speeds may be a more effective method of isolating wind turbine sound, and that the driving factor to the audibility of wind turbines is the ambient sound level.

6.3 Modulation Strength and Wind Direction

Similar to the plots of wind speed in the previous section, modulation strength was also plotted as a function of wind direction at 10-meter height in Figure 13. The results of Site One are plotted below. Site One has turbines in only one direction, and is therefore the best candidate to examine any relationship between wind turbine AM and wind direction.

Figure 13: Modulation strength as a function deviation from downwind at 10-meter height

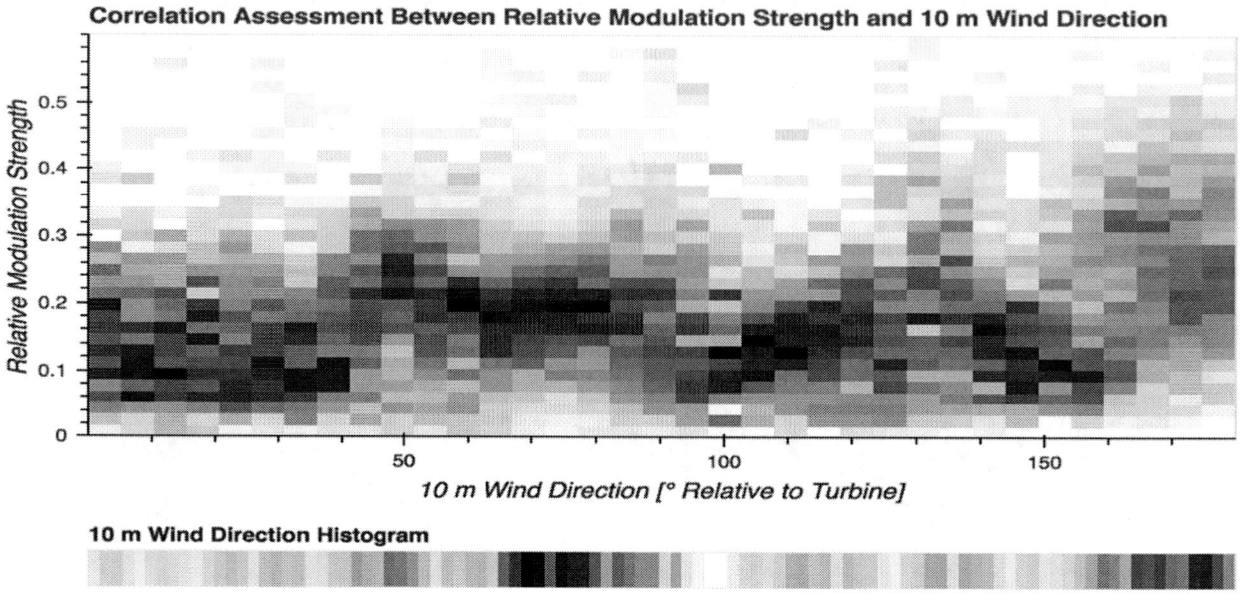

The values in the wind direction plot are presented as *absolute* deviation from downwind direction, where 0° corresponds to the receptor downwind of the nearest turbine and 180° corresponds to the receptor upwind of the nearest turbine.

The data plotted in Figure 13 does not appear to show any significant trends. While there does seem to be higher modulation strengths in crosswind conditions, the differences are slight and the data spread is large. Analysis in greater detail, and on a larger dataset is required before any definitive conclusions can be made with respect to wind direction.

6.4 Other Meteorological factors

The other meteorological variables were examined as part of this study, including the relative humidity, atmospheric pressure, and ambient temperatures measured at 10-meters. Plots of each variable vs. the calculated wind turbine AM are presented below for each site. Modulation strength is on the y-axis, and the specific MET variable is on the x-axis for each plot.

6.4.1 Site One

Humidity

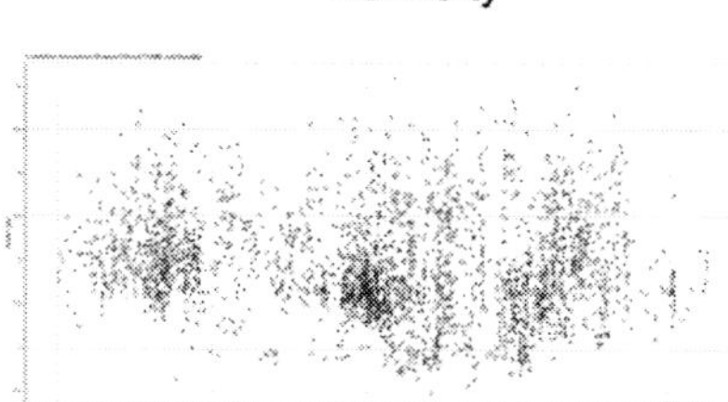

Atmospheric Pressure

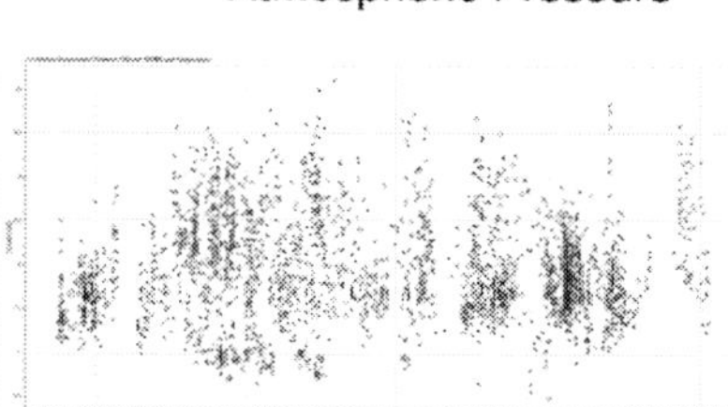

Temperature

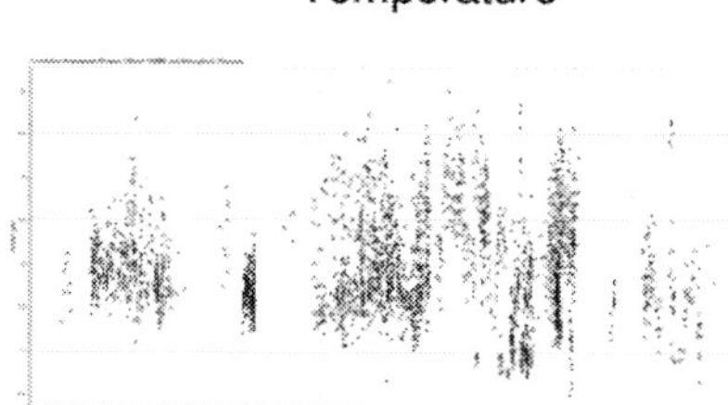

6.4.2 Site Two

Humidity | Atmospheric Pressure | Temperature

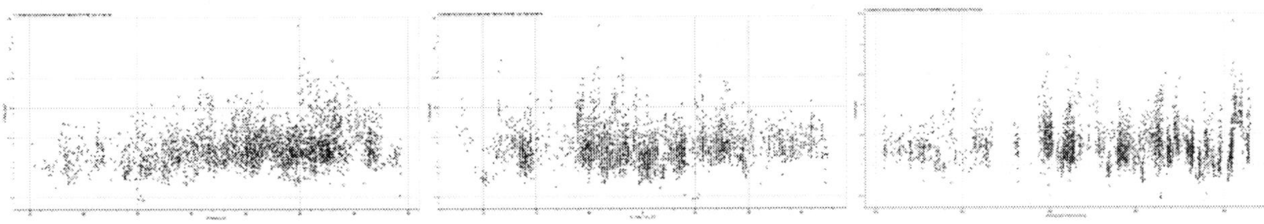

6.4.3 Site Three

Humidity | Atmospheric Pressure | Temperature

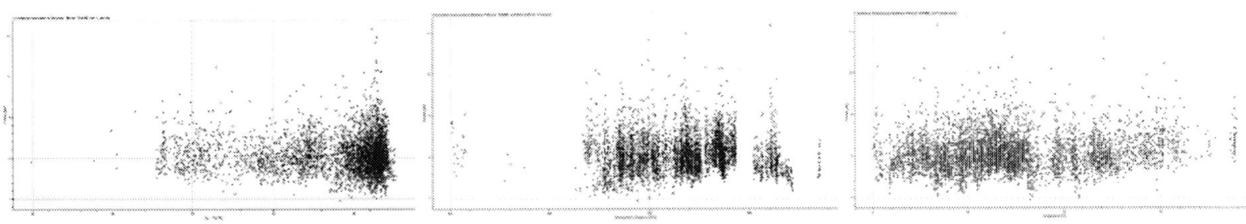

Evident from the plots above, there does not appear to be any significant correlation between temperature, pressure, or humidity and wind turbine AM strength.

7 DISCUSSION

In this study, there are two main points of discussion, namely (1) the effectiveness of the algorithm at automatically and reliably detecting periods of wind turbine AM and (2) the impact of any of the assessed meteorological variables on the prevalence and strength of wind turbine AM.

7.1 THE DETECTION ALGORITHM

The detection algorithm was found to be good at preserving the actual peak-to-peak amplitudes from the time signal. Comparisons between the modulation depth spectra and the actual peak-to-peak values in the time domain signal indicate a good agreement, including cases where there is strong but intermittent wind turbine AM.

A shortcoming of this method is that intervals where the turbine rotational speed changes may not be accurately captured, due to the changing BPF during the time interval. This problem is exacerbated by the fact that most SCADA systems provide data in 10-minute averages.

7.2 THE RATING METRIC

The metric by which the AM is rated in this study provides a relative measure of the AM strength over the course of a measurement campaign. This rating can be used in conjunction with the calculated amplitudes from the modulation spectra to provide quantitative values of the absolute modulation depth corresponding to periods of relatively strong AM. This step was beyond the scope of this study.

The method by which the strength of the wind turbine AM is judged is still a topic of much debate, but the results of this paper show that an objective and automated metric for quantifying wind turbine AM is quite possible. Further study is required.

7.3 Meteorological Analysis

Wind speed was found to influence the detected AM, due mainly to the relationship between 10-meter wind speed and the ambient sound level. Modulation strength was found to increase with 10-meter wind speed to a point, after which wind-induced noise increased the ambient levels such that the wind turbine noise was masked. For flat topography, it was found that AM was most detectable up to 10-meter wind speeds of around 4 m/s, above which the ambient sound masking reduced the detectability of the wind turbine AM.

Different wind directions appeared to result in differing modulation strengths. There were, however, no statistically significant trends in any of the three data sets analysed in this study. The difficulty in analysing the data for wind direction was found to be the huge disparity between the amount of available data at different wind directions. Most sites will have a prevailing wind pattern, which would always create a bias in the dataset towards those conditions. Generally, it is accepted that wind direction affects the propagation of sound, and so one would expect that it would affect the measured strength of wind turbine AM at a receptor. More analysis on a greater dataset would be required in this area to determine with confidence whether a specific wind direction consistently produces greater measured modulation strengths.

There was no significant correlation between wind turbine AM and temperature, relative humidity, or atmospheric pressure. While it cannot be totally ruled out that these meteorological parameters influence the measured AM, there was no significant correlation – positive or negative – to modulation depth found in this study.

8 REFERENCES

[1] Institute of Acoustics, "A Method for Rating Amplitude Modulation in Wind Turbine Noise," Institute of Acoustics, 2016.

[2] RenewableUK, "Wind Turbine Amplitude Modulation: Research to Improve Understanding as to its Cause and Effect," RenewableUK, London, 2013.

[3] H. Tachibana, "Study on the amplitude modulation of wind turbine noise: Part 1 - Physical investigation," in *INTER-NOISE*, Innsbruck, 2013.

[4] J. McCabe, "Detection and Quantification of Amplitude Modulation in Wind Turbine Noise," in *Wind Turbine Noise Conference*, Rome, 2011.

[5] C. Di Napoli, "Long Distance Amplitude Modulation of Wind Tubine Noise," in *Wind Turbine Noise Conference*, Rome, 2011.

APPENDIX – ALL PLOTS

Modulation spectrograms, wind speed, wind direction, and correlation plots, organized by site.

Site One – Wind turbines with large rotor diameters located south of the measurement location

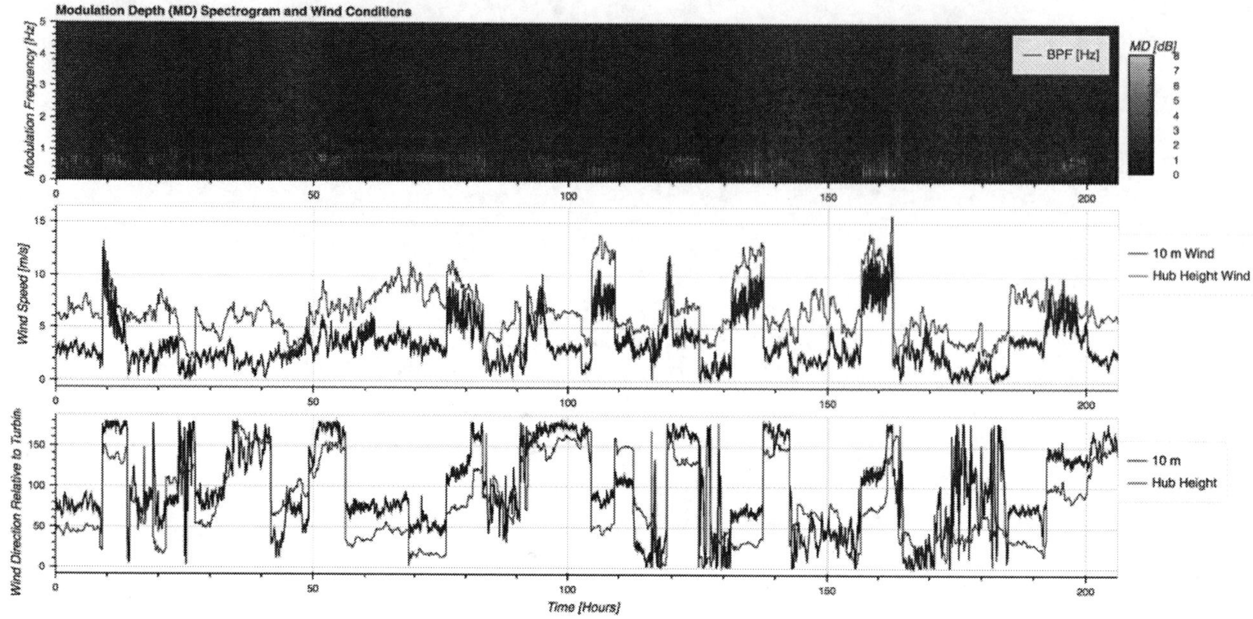

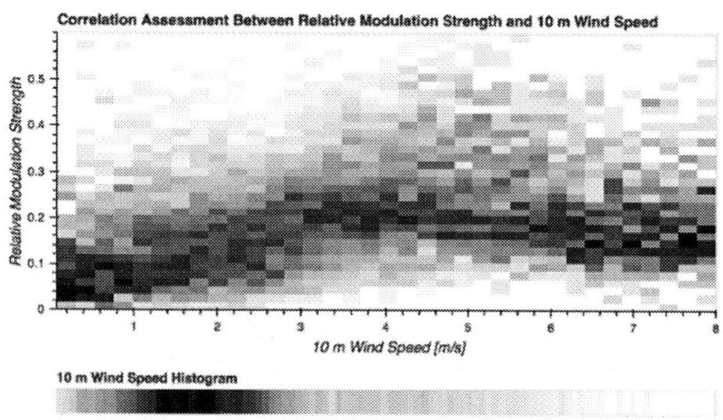

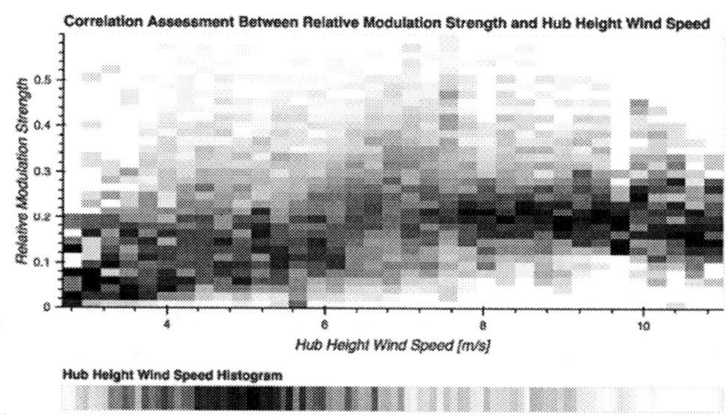

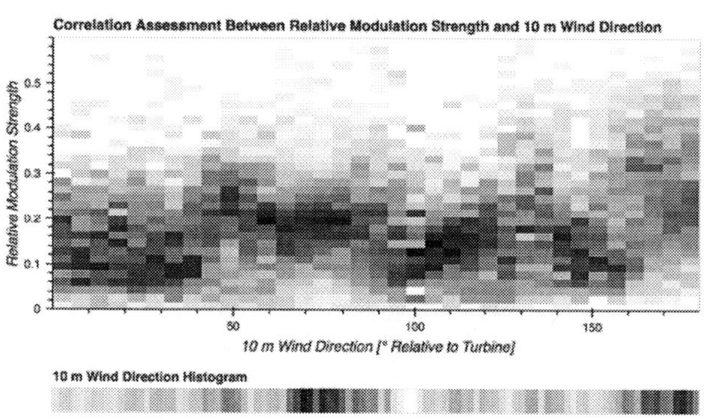

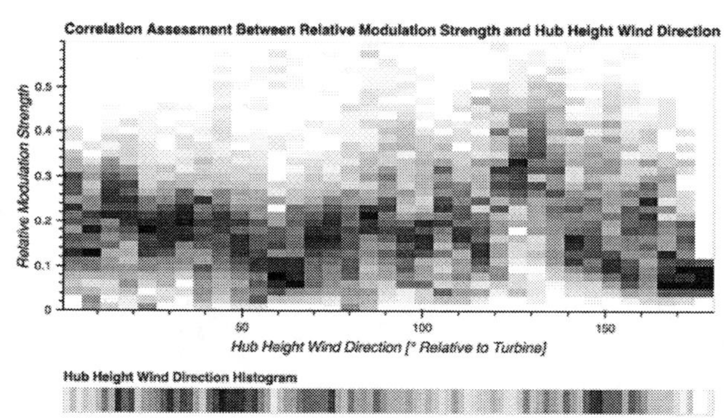

Site Two – Stall controlled turbines surrounding the measurement location

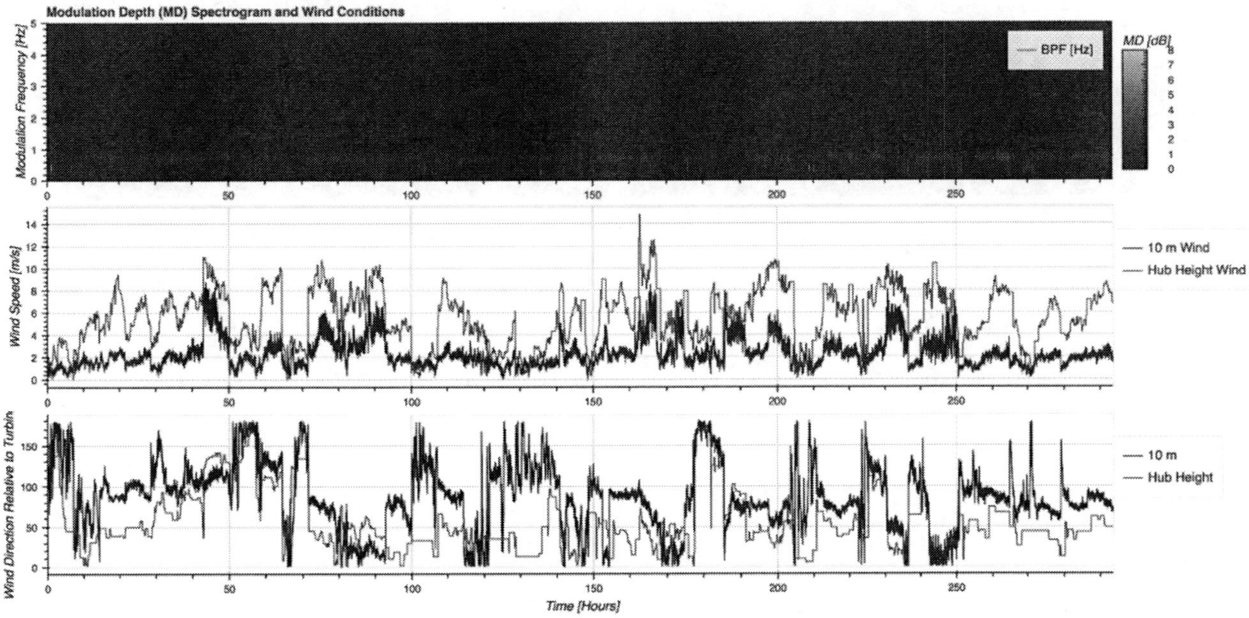

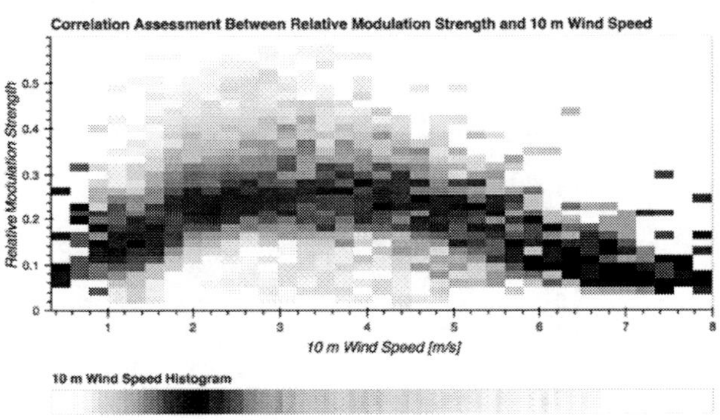

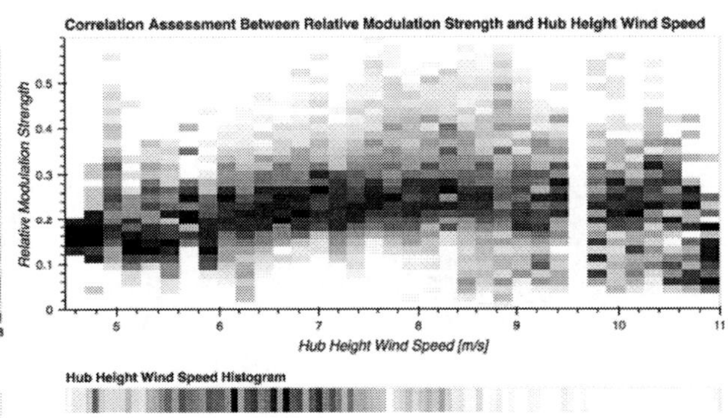

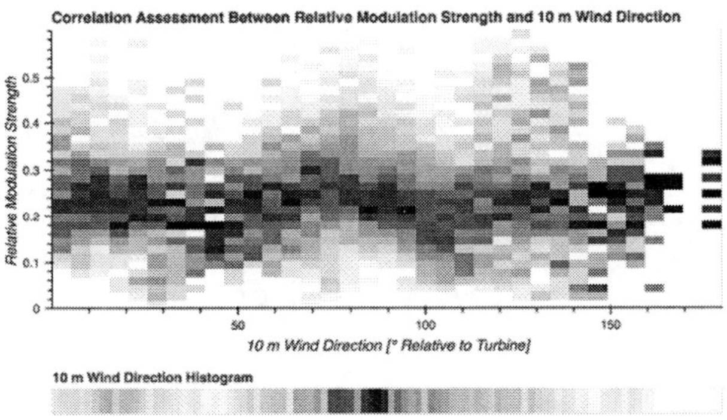

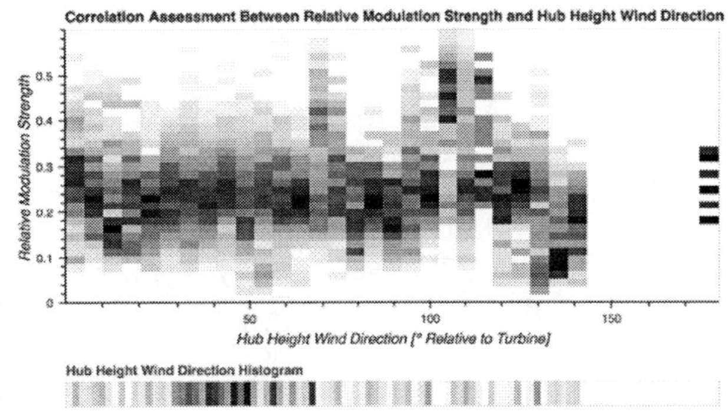

Site Three – Wind turbine on a ridge, measurment location in a forested valley

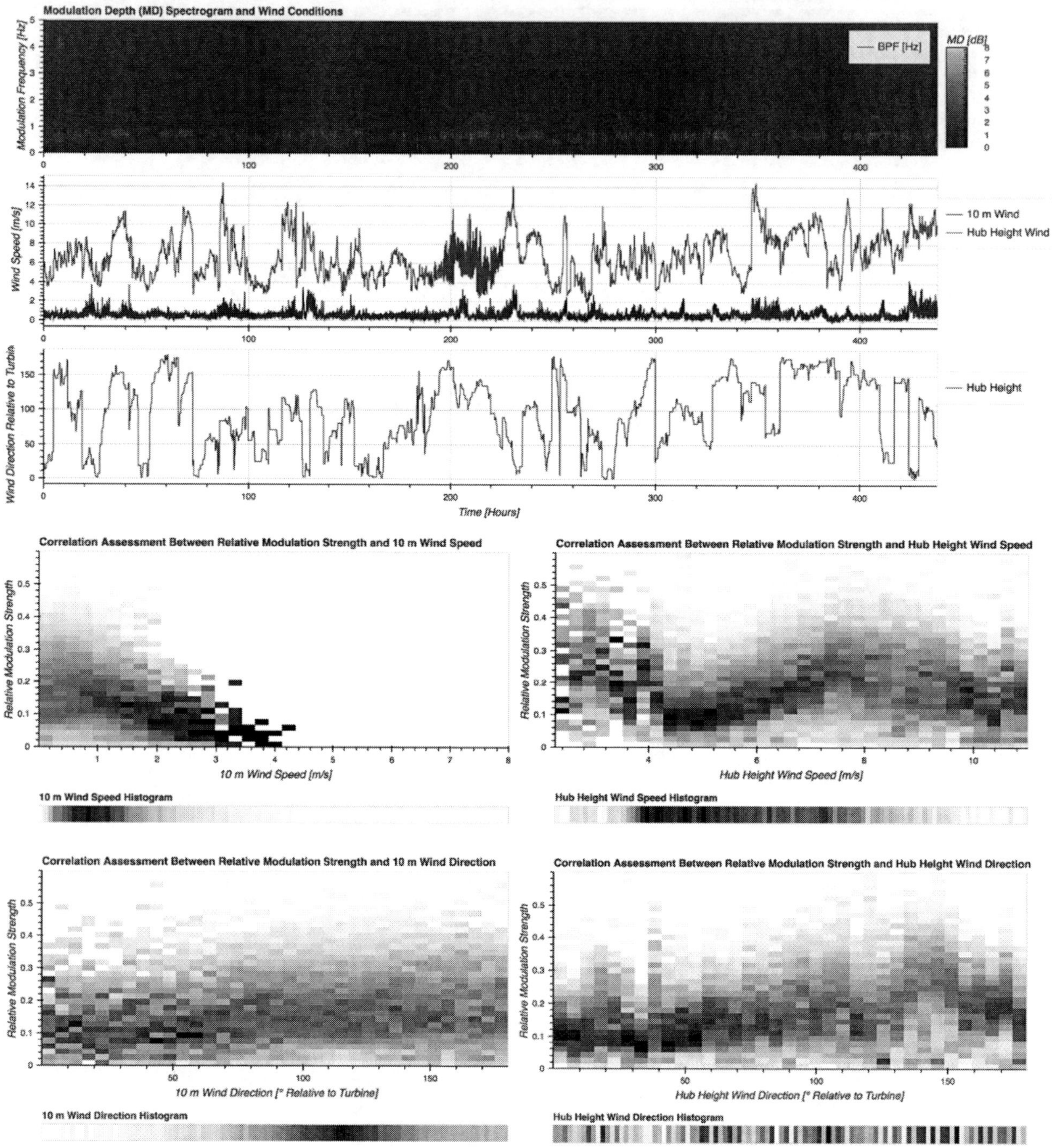

7th International Conference
on
Wind Turbine Noise
Rotterdam – 2nd to 5th May 2017

Extended simulations of wind noise contamination of amplitude modulation ratings

Sabine von Hünerbein, University of Salford, Salford M5 4WT UK, s.vonhunerbein@salford.ac.uk
Paul Kendrick, University of Salford, Salford M5 4WT UK, P.Kendrick@salford.ac.uk
Trevor Cox, University of Salford, Salford M5 4WT UK, T.J.Cox@salford.ac.uk

Summary

Microphone wind noise can corrupt outdoor measurements and recordings and especially the rating of Amplitude Modulation (AM) depth. In a previous study simulations of synthesised wind turbine sounds in wind noise have shown that even relatively at low wind speeds of 2.5 m/s AM rating errors of over 4 dBA can result. Microphone wind noise is intermittent, and consequently one solution is to analyse only uncorrupted parts of the recordings which has be shown to be an efficient way of recovering the true AM metrics. In the current study more wind noise only stimuli have been used to evaluate the probability of false positives in AM ratings. In wind turbine noise between 42% and 61% of the stimuli were rated at 0 dBA modulation depth in the presence of wind noise. Where wind noise only was used about 11% of the stimuli resulted in valid AM ratings. Further work is required to clarify points on perception and the nature of real recordings with increased variability.

1. Introduction

In recent years the AM characteristics of wind turbine noise has been claimed to be responsible for increased annoyance as compared to other environmental noise sources. Measures have been requested to quantify and mitigate AM in wind turbine noise (e.g. Cand *et al.*, 2013). In an attempt to develop a reliable measurement method for AM noise the Institute of Acoustics founded the Amplitude Modulation Working Group (AMWG). They launched a consultation on three proposed AM rating methods (Bass et al., 2015) which, after considerable data analysis and consolidation of input resulted in the final report by Bass *et al.*, 2016. It recommends to use a method employing both frequency domain and time domain signal processing to arrive at accumulated 10 minute ratings of AM depth measured in dBA. This method, the authors claim, delivers robust results given the contamination of sound measurements by ambient noise and the variable nature of wind turbine noise.

One particular type of noise known to affect many outdoor recordings and measurements is microphone wind noise even where the common foam windshields with a diameter of about 10 cm are used. Based on the consultation document simulations of synthesised wind turbine noise including microphone wind noise have been published (Kendrick, von Hünerbein and Cox, 2016). The results suggested that the ratings would be affected by wind noise. In these simulations wind turbine noise at various AM depths with a constant averaged sound level of 40 dBA were 'contaminated' by microphone wind noise following an algorithm by Van den Berg (2006). The wind noise was based on a data base of high resolution wind measurements (Fritts *et al.*, 2001).

Because any method proposed for a regulatory framework needs to be trusted by all stakeholders, the current work applies the final proposed method to the sounds used in the previous simulation. This was to quantify the extent to which the new algorithm is still affected by wind noise. Pure microphone wind noise was then rated by the AM method to explore the potential for "false positive ratings".

2. Simulation design

The design of the simulation followed three steps. In a first step, the sound samples were synthesised, in a second step the samples were frequency filtered and down-sampled to the format required by the AM rating method. This was then run with the recommended settings and the resulting ratings analysed.

2.1 Design of wind turbine noise combined with microphone wind noise

A selection of 20 second synthesised AM stimuli used in the von Hünerbein *et al.,* (2013) study was chosen. The microphone wind noise was generated using the same duration as the test files, by repeatedly selecting at random velocity time histories from the CASES99 wind data (Fritts, *et al.,* 2001). For each test signal, 35 examples were generated, each with mean wind speeds from nominally 0 m/s to 10.3 m/s in 0.25 ms steps. The relative wind noise levels are representative for the 40dB stimuli as described in Kendrick *et al.,* 2016 and Jackson *et al.,* 2014 both of which provide further details on stimuli design. The resulting stimuli are saved as audio files at a sampling rate of 41 kHz. The sound files were then frequency filtered into the frequency bands 50 – 200 Hz, 100 – 400 Hz and 200 – 800 Hz and down-sampled to the 100 ms sampling rate required by the AMWG rating method.

2.2 AM rating of wind noise only

For the microphone wind noise stimuli 346 stimuli of 20 s length and covering the wind speeds from 0 m/s to 10.3 m/s in 0.25 ms steps were again generated as audio files. The sound files were then frequency filtered into the frequency bands 50 – 200 Hz, 100 – 400 Hz and 200 – 800 Hz and down-sampled to the 100 ms sampling rate required by the AMWG rating method. The resulting data are 10 s non-overlapping blocks.

2.3 The AMWG rating method

Of the three AM metrics proposed in the AMWG consultation document, the final report recommends an adapted version of the method combining frequency domain analysis with time domain rating:
First the AM rating method performs a Fast Fourier Transform of the band-limited 100 ms time series data. The fundamental modulation frequency and the corresponding first two harmonics are thus detected. After checking the prominence of the respective frequency peaks to determine whether there is sufficient energy in those, the time series is recreated from the significant elements of the Fourier transform. The method assumes that the frequency filters and peak selection will minimise the ambient noise contributions so that the AM contributions dominate. This works if contaminating noise has a) not a similar periodicity and b) has no significant energy contributions in the frequency bands of interest. These assumptions will therefore be most likely fulfilled for sounds like high pitched bird noises and broad band traffic noise.

Wind noise however, is often modelled as a superposition of eddies with sizes ranging from several hundred meters to a few millimeters. The energy distribution of these eddies follows the so called Kolmogorov energy cascade. According to that theory low frequency eddies in the frequency range of wind turbine noise amplitude modulation are expected to contain significant energy contributions and are therefore likely to affect the rating method adversely. False

positive ratings from wind noise only might result. Or the wind noise could mask the AM of the wind turbine sounds thereby reducing the rated AM depth.
The AM rating method for continuous measurements then goes on to accumulate ratings in 10 minute intervals and assign an AM depth where at least 50% of the samples were rated to contain valid AM. If wind noise affects the 10 s ratings then a skewed rating of the 10 minute intervals could result if enough values are affected.

3. AM ratings

The following results show AM ratings for wind turbine noise affected by microphone wind noise to demonstrate the concern about the reliability of the currently proposed AM metric. It goes on to investigate the ratings and statistics for "false-positive ratings" of microphone wind noise stimuli without any turbine signal.

3.1 Amplitude modulated wind turbine noise plus wind noise

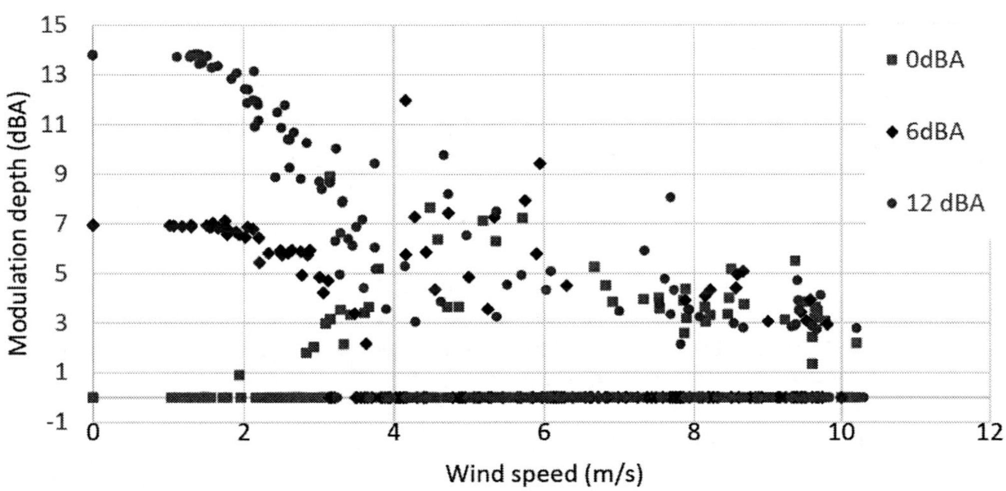

Fig. 1 AM ratings using the 50 – 200 Hz filter for AM modulated wind turbine noise including wind noise on the microphone. The design modulation depths are 0 dBA (blue squares), 6 dBA (black diamonds) and 12 dBA (red circles). AM ratings are affected by wind noise from wind speeds as low as 2 m/s assuming a 10 cm ∅ wind shield.

Figure 1 shows that as the wind speed increases the ratings of the modulation depth start to be affected by the wind noise. The frequency filter used for this graph was the 50 -200 Hz band where the wind noise energy is highest. When there is little AM in the wind turbine signal, microphone wind noise will tend to increase the modulation depth and at high modulation depths the effect is an overall reduction. This is because the wind noise can often contain gusts which will act to increase the modulation depth when the background level is fairly stationary, but where periodic modulations exist, they will be masked by the wind-noise. The simulations show that errors can occur even at very low wind speeds. This is because the metric is computed from the broadband A-weighted signal and microphone wind noise is dominated by large amplitude low frequency components. Errors appear to become significant above 2 m/s.

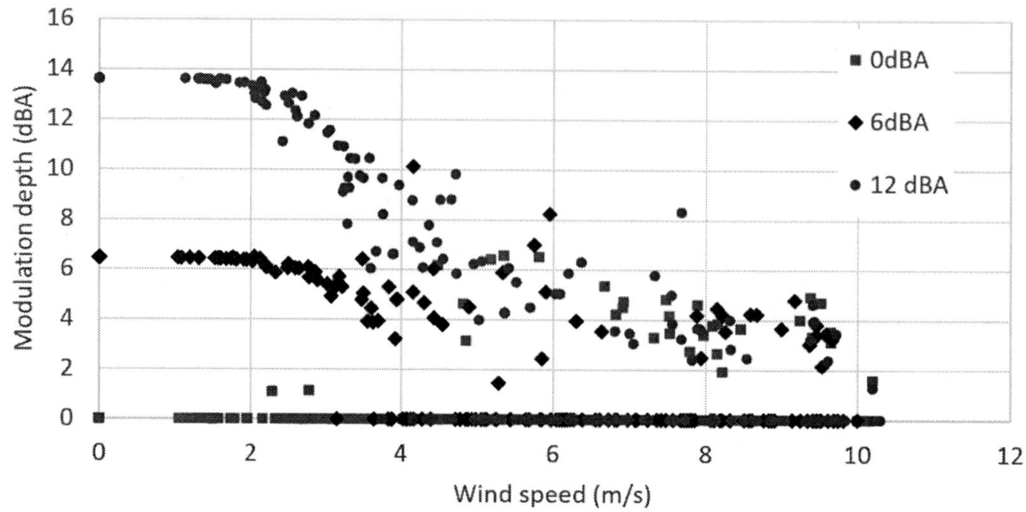

Fig. 2 AM ratings using the 200 – 800 Hz filter for AM modulated wind turbine noise including wind noise on the microphone. The design modulation depths are 0 dBA blue squares, 6 dBA black diamonds and 12 dBA red circles. AM ratings are affected by wind noise from as wind speeds as low as 2 m/s assuming a 10 cm ∅ wind shield.

Figure 2 shows the same data but using the band limited signal between 200 and 800 Hz. This shows that the metric is more robust, where errors begin to become significant at around 2.5 m/s. This improvement is due to the band limiting to higher frequency, where wind noise contains less energy and is therefore slightly less dominant.

It is notable in both figures that many ratings for the 6dBA and 12 dBA stimuli at higher wind speeds are 0 dBA. This is also evident from the data in Table 1 where between 42% and 61% of the data are rated at 0dB for high AM wind turbine noise.

Frequency band, Hz	50-200	100-400	200-800
Design AM depth, dBA			
6	61%	51%	52%
12	50%	46%	42%

Table 1: The percentage of high AM stimuli that were rated 0dBA in the presence of wind noise

The reason for these high percentages is the fundamental difficulty of distinguishing signal from noise at common listening distances from wind turbines. The high number of 0 dBA ratings will therefore possibly affect the cumulative 10 minute rating by reducing the detected number of intervals affected by AM.

It could be speculated that the occurrence of AM is thus potentially underestimated. However, it is also important to mention that the quoted wind speeds relate to the microphone position which at the recipient location is typically at a height of about 1.5 m above the ground where higher wind speeds are rare. The assumption of a 10 cm ∅ wind shield could also contribute to the overestimation of the problem because secondary wind shields are commonly used for the evaluation of environmental noise and it could be argued that those measurements will not be affected by wind speeds as low as 2-3 m/s. It is currently unclear which minimum wind speeds affect the measurements. Several members of the AMWG working group have voiced their surprise at these results, stating that they do not agree with their experience in the field. (AMWG, 2016). It is curious though that high AM does not seem to occur at high wind speeds (Coles *et al.*, 2016)

3.2 Wind noise only stimuli

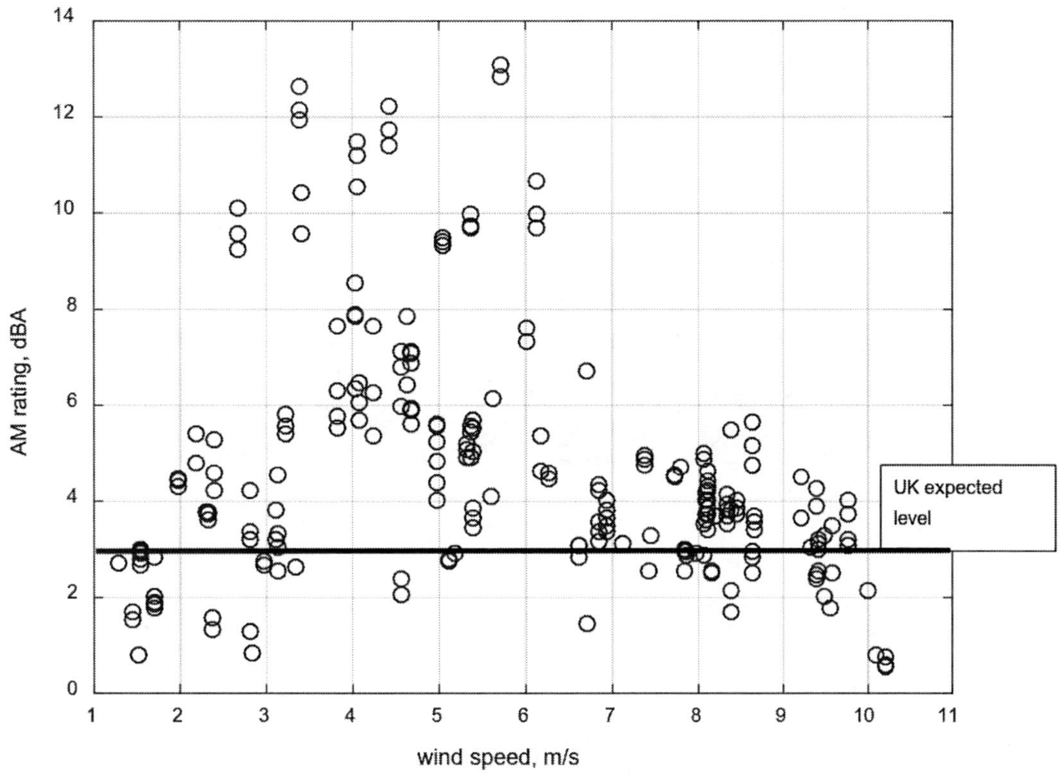

Fig 3. All incorrect AM ratings from wind noise. The displayed 222 data points represent 11% of the wind noise samples. Therefore, 89% of the data have been rated at 0 dBA AM depth.

Where the recordings only contain microphone wind noise, the AM rating method produced false positives in 11% of the stimuli at ratings between 0.5 and 13 dBA. 76% of the false positives were rated at above 3dBA, the normal expected level of AM from a wind turbine according to UK guidelines ETSU-R-97 (1996). The majority of ratings are between 2 and 8 dBA with higher ratings being only 12% of all ratings. This agrees with the results from Section 3.1 which indicated that AM ratings for stimuli with high AM depth would generally be rated lower in the presence of wind turbine noise and might explain some of the findings of Coles *et al.* (2016).

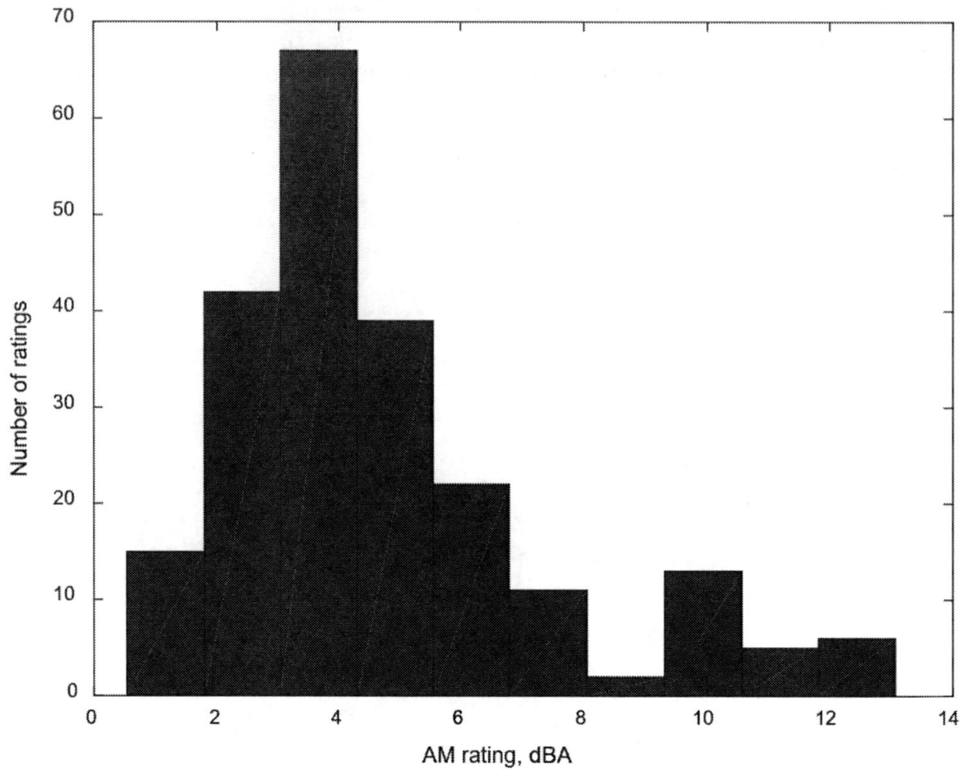

Fig 4. Histogram of the microphone wind noise AM ratings. The majority of the ratings is lower than 8 dBA which agrees with the results from Figs 1 and 2.

In addition to AM rating and peak prominence the AMWG rating method also provides the fundamental modulation frequency at which the rating was calculated. Figure 5 shows that the ratings from wind noise show a range of fundamental frequencies between 0.4 and 0.9 Hz which is by design within the range that AM modulation from wind turbines occurs. However, consistency checks might be devised where an unrealistic deviation from the blade passing frequencies is detected in the analysis and wrong ratings might be thus removed. Further research is necessary and this might only work for sites with a small number of audible wind turbines as the interference between different audible turbines might produce a range of modulation frequencies similar to the one observed from wind noise.

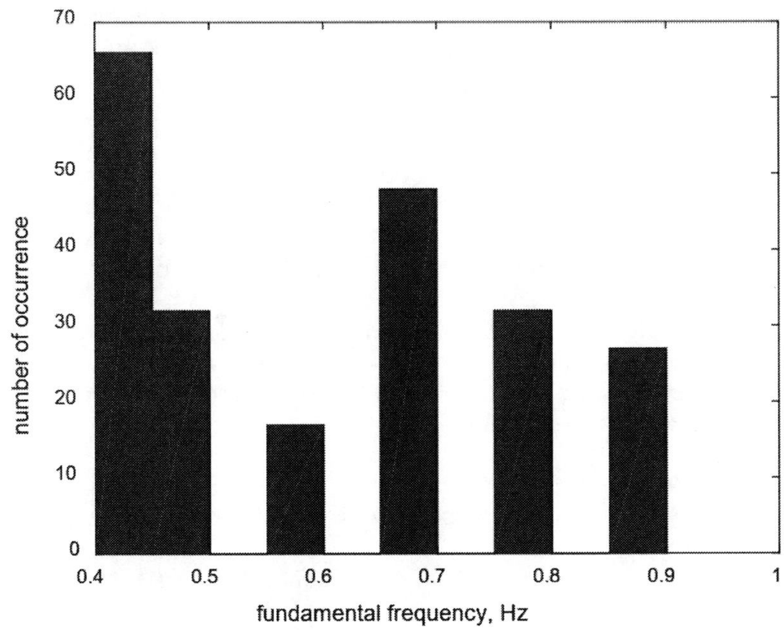

Fig 5. Histogram of identified fundamental AM frequencies from wind noise. The identified AM from wind noise shows a relative wide spread of fundamental AM frequencies which does not necessarily agree with the AM frequencies of audible wind turbine noise and might provide a mechanism to exclude more erroneous ratings from wind noise.

4. Conclusions

The recently proposed AM rating method was tested for microphone wind noise with and without wind turbine noise.
Where a combination of wind turbine noise and microphone wind noise is recorded a considerable effect on the ratings can be expected with over 40% of the 10 s periods rated at 0 dBA thus reducing AM detection. Where there is no wind turbine noise but microphone wind noise only, about 11% of the microphone wind noise has been rated as AM.
There are several limits to the scope of the understanding at this stage. The most important one is that we do not know how human perception reacts to the combination of wind noise and wind turbine noise. It could be speculated that in similar fashion to the microphone sensor the ear and brain will experience reduced amplitude modulation where wind noise in the ear is present. Common sense and experience suggest that wind noise can be heard even in the absence of much other environmental noise. The ear might be better at distinguishing between directional wind turbine noise and surrounding wind noise though. The authors are not aware of any scientific evidence for or against this hypothesis.

The second limitation is the knowledge on the typical wind conditions around the microphones that record wind turbine noise at the recipients' location as these are not commonly monitored. It has been suggested that typical conditions might be at low wind speeds especially at sheltered locations. Better wind shields than assumed for this study might also contribute to a reduction of the rating problems.

Successful removal of wind noise using a machine learning algorithm has been demonstrated in a previous paper for the artificial stimuli at sound pressure levels from 40 dB and for recordings with a minimum sampling rate of 41 kHz for the sound recordings (Kendrick *et al.*,

2016). If microphone wind noise turns out to be a persistent problem this approach might have to be evaluated in more realistic conditions with signals including variations in frequency content, masking levels, modulation frequencies among others. Real recordings need to be used to evaluate the method.

References

AMWG, (2016) Personal communications at Gdansk EWEA conference.

Bass, J., Cand, M., Coles, M., Davis, R., Irvine, G. Leventhall, G., Levet, T., Miller, S. Sexton, D. and Shelton, J. (2015) *Methods for rating amplitude modulation in wind turbine noise - IoA Discussion Document,* Institute of Acoustics, Amplitude Modulation Working Group. http://www.ioa.org.uk/, [3/21/2016].

Bass, J., Cand, M., Coles, M., Davis, R., Irvine, G. Leventhall, G., Levet, T., Miller, S. Sexton, D. and Shelton, J. (2016) IOA Noise Working Group (Wind Turbine Noise), Amplitude Modulation Working Group, Final Report, A Method for Rating Amplitude Modulation in Wind Turbine Noise, Institute of Acoustics, Amplitude Modulation Working Group. http://www.ioa.org.uk/, [13/2/2017].

Cand, M, Bullmore, A, Smith, M, von Hünerbein, S and Davis, R (2013) *Wind turbine amplitude modulation: research to improve understanding as to its cause & effect.* Renewable UK, http://www.renewableuk.com/, [4/11/2016].

Coles, D, Cand, M., Level, T (2016) *Application of the UK IOA AM Rating Method*, EWEA, Gdansk, November 2016.

ETSU-R-97 (1996), *The Assessment and Rating of Noise from Wind Farms*, Final Report for the Department of Trade & Industry, The Working Group on Noise from Wind Turbines, September 1996

Fritts, D., Poulos, G., and Blumen, B. (2001). *Cooperative atmosphere-surface exchange study—1999*, http://www.eol.ucar.edu/projects/cases99/ [21/3/2016].

Jackson, IR, Kendrick, P, Cox, TJ, Fazenda, BM, and Li, FF (2014). *Perception and automatic detection of wind-induced microphone noise*, J. Acoust. Soc. Am. 136, 1176–1186.

Kendrick, P, von Hünerbein, S and Cox, TJ (2016), The effect of microphone wind noise on the amplitude modulation of wind turbine noise and its mitigation , The Journal of the Acoustical Society of America (JASA), 140(1), EL79.

von Hünerbein, S, King, A, Piper, BJ and Cand, M (2013) Wind Turbine Amplitude Modulation: Research to Improve Understanding as to its Cause and Effect - Work Package B(2): Development of an AM Dose-Response Relationship, RenewableUK, http://www.renewableuk.com/, [4/11/2016].

7th International Conference on Wind Turbine Noise Rotterdam – 2nd to 5th May 2017

Low-frequency noise incl. infrasound from wind turbines and other sources

Lorenz HERRMANN[1]; Ulrich RATZEL[2]; Otto BAYER[3], Klaus-Georg KRAPF[1], Martin HOFFMANN[3], Janosch BLAUL[1], Clemens MEHNERT[3]
[1] Wölfel Engineering GmbH + Co. KG, Germany
[2] Umweltministerium Baden-Württemberg, Germany
[3] LUBW Landesanstalt für Umwelt, Messung und Naturschutz Baden-Württemberg, Germany

Summary

In recent years, the issue of low-frequency noise – especially infrasound – has aroused great interest not only among experts, but also among the general public. The reason for this is probably the discussion about the expansion of wind power. In the years 2013-15, LUBW Landesanstalt für Umwelt, Messungen und Naturschutz Baden-Württemberg and company Wölfel Engineering performed extensive measurements of low-frequency noise (incl. infrasound of 1 Hz and higher) in the immediate vicinity of six wind turbines, in urban and rural areas as well as in areas that are explicitly dominated by road traffic. The aim of the project was to collect comparable data about the occurrence of infrasound and low-frequency noise in the vicinity of wind turbines and other sources. The measurements on wind turbines with a capacity of 1.8 to 3.2 MW were performed simultaneously with different distances to the respective wind turbine. The infrasound emitted by wind turbines could be measured very well in the close vicinity of the turbines. Here, the sound intensity is below the human perception threshold. The large amount of data was documented in different evaluations (e.g. linear third-octave band levels, narrow-band spectra, G-rated overall sound pressure levels depending on wind speed or time of day). The measurement method and the main results are presented in this paper.

Keywords: Infrasound, Wind turbine, Measurement, Low-frequency noise
(See . http://www4.lubw.baden-wuerttemberg.de/servlet/is/262445/?shop=true&shopView=6647 .)

1. Introduction

In the years 2013-2015, the State Institute for Environment, Measurements and Nature Conservation (LUBW) and company Wölfel Engineering performed extensive measurements of low-frequency noise (incl. infrasound of 1 Hz and higher) in the immediate vicinity of six wind turbines. Furthermore, measurements were performed in urban and rural areas (without source reference) as well as in areas dominated by road traffic and measurements on technical equipment in residential buildings. The aim was to collect current comparable data about the occurrence of infrasound and low-frequency noise of wind turbines and other sources.

[1] Herrmann@woelfel.de
[2] Ulrich.Ratzel@um.bwl.de
[3] Windenergie@lubw.bwl.de

2. Methods

Various criteria had to be taken into consideration when selecting the measuring locations. For example, roads with heavy traffic, forests and industrial plants had to be strictly avoided. Different types of wind turbines (WT) were selected with a capacity between 1.8 and 3.2 MW. Parallel and synchronous measurements were performed at three measuring points each, at different distances to the respective WT. The nearest measuring point was determined in accordance with FGW guideline and IEC 61400-11 depending on hub height and rotor diameter. If possible, the more distant measuring points were positioned at distances of about 300 m and 700 m with identical set-up. To record the low frequencies, measuring microphones (G.R.A.S.) calibrated to measure low frequencies were used with a lower limit frequency of approx. 1 Hz. The microphones were mounted on sound-reflecting plates and provided with double windscreens. Placement of the measuring microphone in a hole in the ground in relevant measuring situations did not lead to additional level reductions of the wind-induced background noise in the infrasound range. When measurements on a WT were performed, possible neighboring turbines were stopped. All measuring points were evaluated by analogy to the FGW guideline and IEC 61400-11 (1). In addition, G-levels and narrow-band spectra were determined from the recorded audio files subsequent to the measurements.

3. MEASUREMENT RESULTS

3.1 Wind Turbines

In total, measurements on six wind turbines were carried out (Enercon E-66, E-82 and E-101, REpower MM92 and 3.2M114 as well as Nordex N117). As an example, the results of the measurement on a 2.4 MW Nordex N117 will be presented in the following. The results for the other turbines are basically comparable and described in detail in the project report (2).

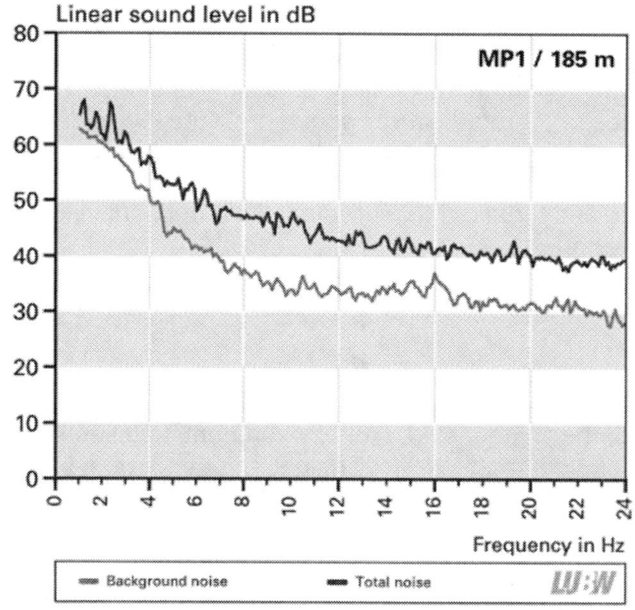

Figure 1: Narrow-band spectrum (resolution 0.1 Hz). Total noise (violet, upper curve) with WT in operation and background noise (green, lower curve) with WT out of operation, measured in near sound field at a distance of 185 m to the turbine. The discrete frequencies below 6 Hz are easily visible (upper curve).

Figure 1 shows the narrow-band spectrum from 1 to 24 Hz with a resolution of 0.1 Hz, which was measured at the reference point at a distance of 185 to the turbine. In the period of time chosen for this representation, the average measured wind speed in the operating noise was approx. 7.6 m/s, in the background noise approx. 6.9 m/s. If possible, periods with similar wind speed and little gustiness were selected for the evaluation. In the ordinate, the measured linear sound pressure level without frequency weighting is shown. With the WT in operation, discrete maxima in the frequency range below 6 Hz are clearly visible. They correspond to multiples of the rotor passage frequency (here visible at approx. 0.6 Hz: 1.2 Hz, 1.7 Hz, 2.3 Hz, 2.9 Hz, etc.) For the far field (Figure 2), there is hardly any difference noticeable between operating noise and background noise. The discrete maxima that are clearly visible in the near field do no longer exist.

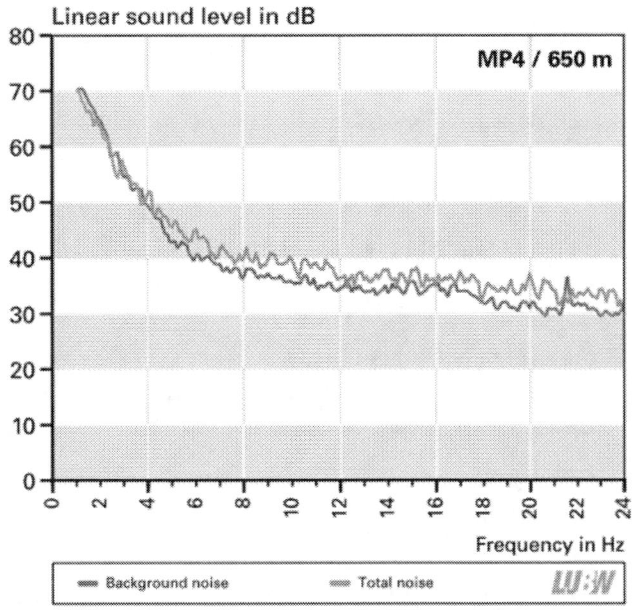

Figure 2: 0.1-Hz narrow-band spectrum of total noise (orange, upper curve – with WT in operation) and background noise (green, lower curve – with WT out of operation) measured at a distance of 650 m.

Figure 3 (upper diagram) shows the measured average A-weighted and G-weighted overall sound pressure levels (according to ISO 7196) for the entire measuring period, at a distance of 185 m to the plant. Each data point corresponds to a 10 s equivalent continuous sound pressure level. The levels are plotted against the corresponding wind speeds (at a height of 10 m and rounded to 0.1 m/s). Measurement times with disturbing noise were excluded from the analysis. The G-levels are shown red and green respectively - red with the turbine in operation (total noise), green with the WT out of operation (background noise). The level difference between operating and background noise in the G-weighting is clearly visible. The A-levels of operating noise are additionally specified (violet). At a distance of 650 m (Figure 3, lower diagram) no significant differences between operating and background noise are visible any more in the G-weighting. The broader distribution of measured values in the A-level can be explained by the stronger influence of the background noise.
In Figure 4, the measured sound pressure levels of a representative period of time for the operating noise at the three measurement points at different distances are compared as linear third-octave levels. For each measurement, a period of time was chosen in which the wind speed was as constant as possible. Here the average measured wind speed was 7 m/s.

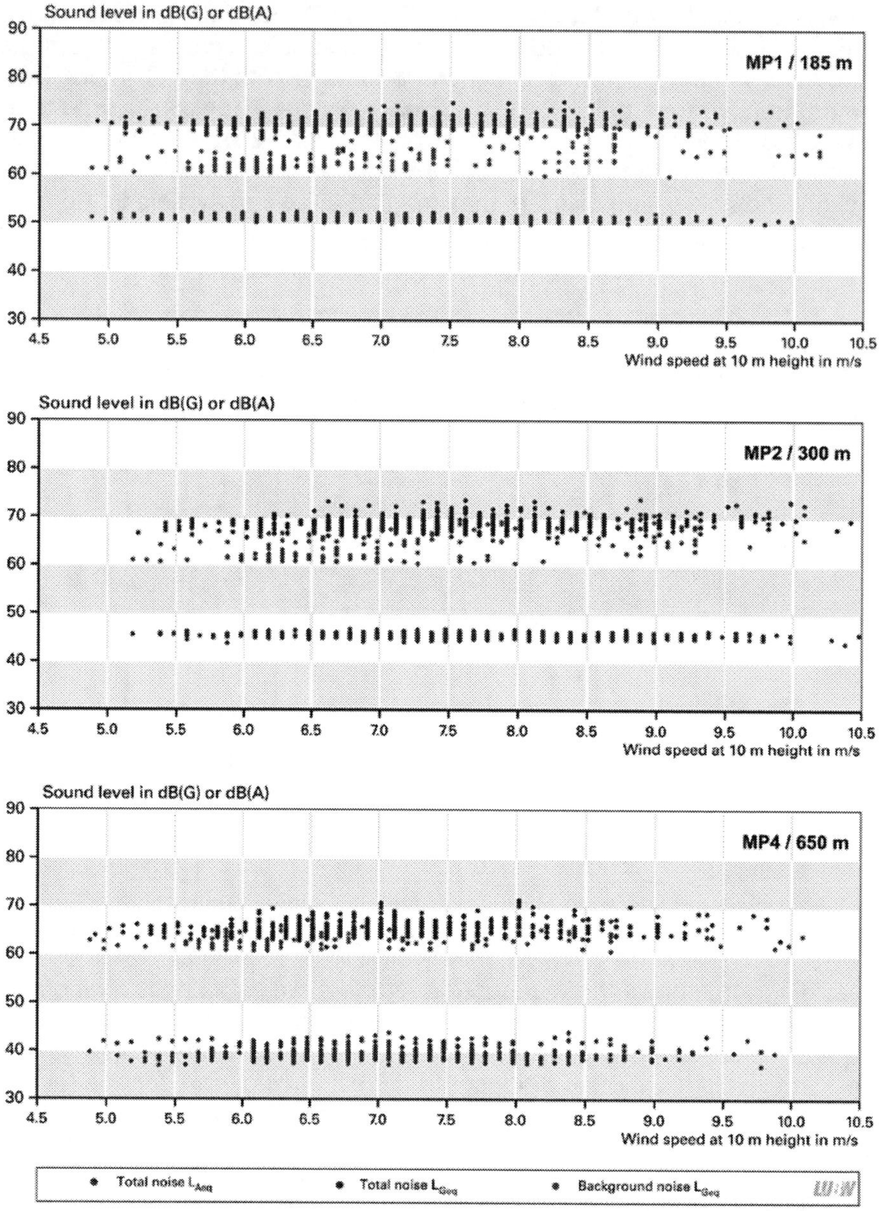

Figure 3: A-level (violet dots) and G-level (red dots with WT in operation, green dots with WT out of operation). At a distance of 185 m (upper diagram), significant differences in the G-level between WT in and out of operation can be seen. At a distance of 650 m, there are only minor differences (lower diagram).

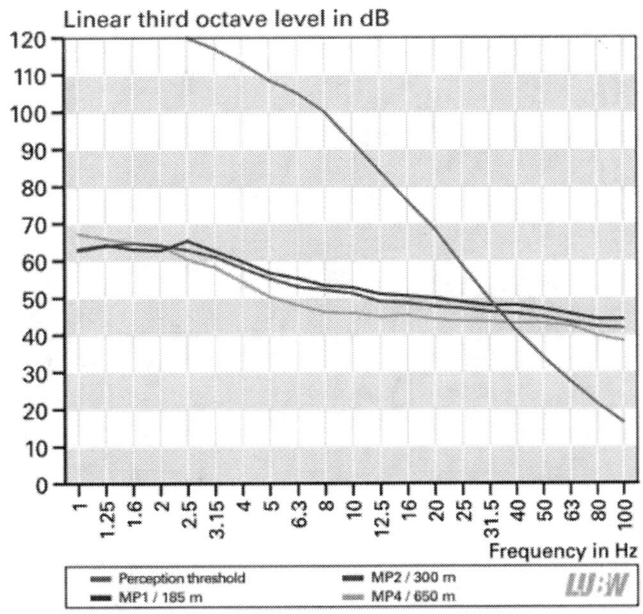

Figure 4: Third-octave band spectra of the total noise measured at distances of 185 m, 300 m and 650 m to the WT. For comparison purposes, the perception threshold was additionally included in the figure (grey).

In addition, the perception threshold was shown according to the draft of DIN 45680:2013 (3). Below 8 Hz, it was supplemented by literature values (4). In the figures, the background noise, as it is e.g. generated by surrounding vegetation at these wind speeds, for example, is included in the measured total noise. From about 32 Hz upwards, the levels are below the perception threshold for all measuring points. In case of infrasound, they are even very far below the perception threshold, by 20 up to more than 50 dB below the threshold. Comparable results were found in the other measurements that were carried out. At distances of 120 m to 190 m, the G-levels were between 60 and 80 dB(G) in all turbines, incl. wind noise. According to a Polish study, values of about 89 dB(G) were measured in the center of a wind farm with 25 Vestas V80 turbines. At the edge of the wind farm, about 67 dB(G) were measured (5).

3.2 Road traffic

During the project, intensive measurements of road traffic noise also have been performed, both outside and inside buildings. Figure 5 shows linear third-octave band spectra (hour average sound pressure level) throughout the total frequency spectrum, measured on the immission side at an inner-city street in the afternoon between 16:00 and 17:00. Between 0:00 and 1:00, they are 10-15 dB lower. The resonance in the low-frequency range, here with a maximum of about 40 Hz, is striking.

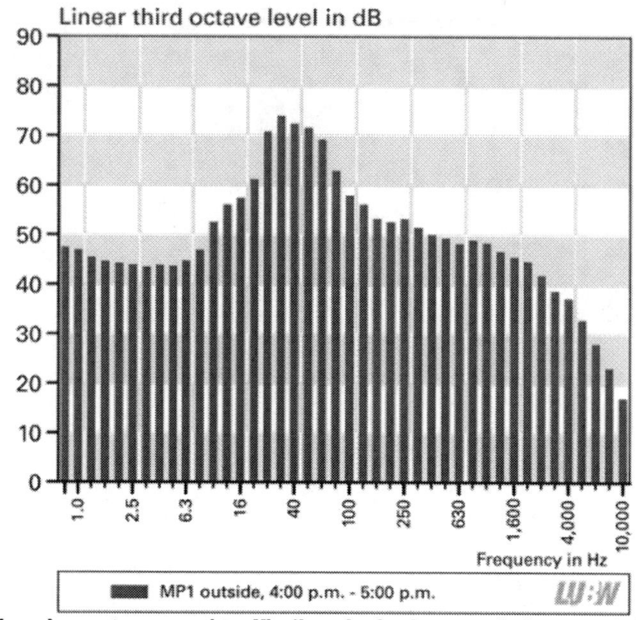

Figure 5: Third-octave band spectrum road traffic (immission), recorded on weekdays between 16:00 and 17:00. The traffic intensity is 14,000 vehicles/24h with a percentage of trucks of 3 %.

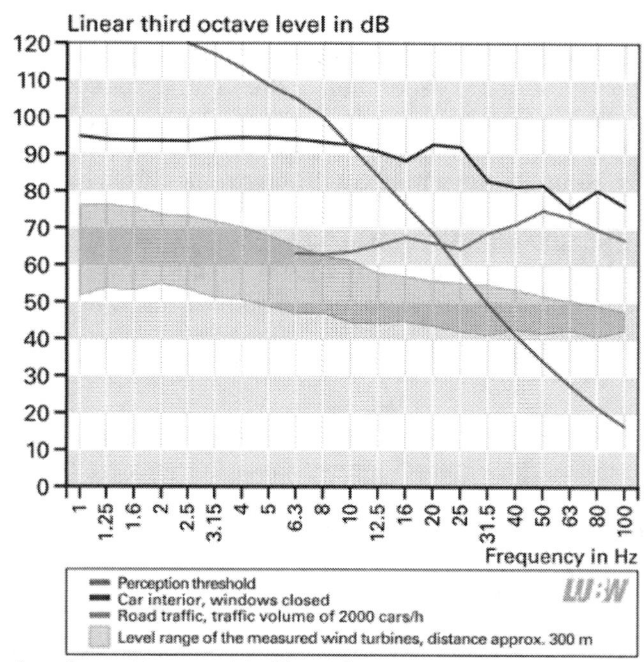

Figure 6: Third-octave band spectrum road traffic at the permanent measuring station Reutlingen (blue curve center) with 2,000 vehicles/h. Results of the WT measurements (red) and level inside a running passenger car with closed windows (upper blue curve). Perception threshold (grey).

Further data were recorded by the permanent measuring station Reutlingen, Baden-Württemberg (6) and are shown in Figure 6. For comparison purposes, the results of the wind turbine measurements were added, as well as the levels measured inside a running passenger car (130 km/h, with closed windows). The latter are significantly higher.

3.3 Urban and rural Background

The measurements in the urban background without source reference were performed on a rather quiet square without direct influence of traffic noise. They could be performed with no or low wind. As expected, the levels fall respectively, from day all through the evening until night (Figure 7).

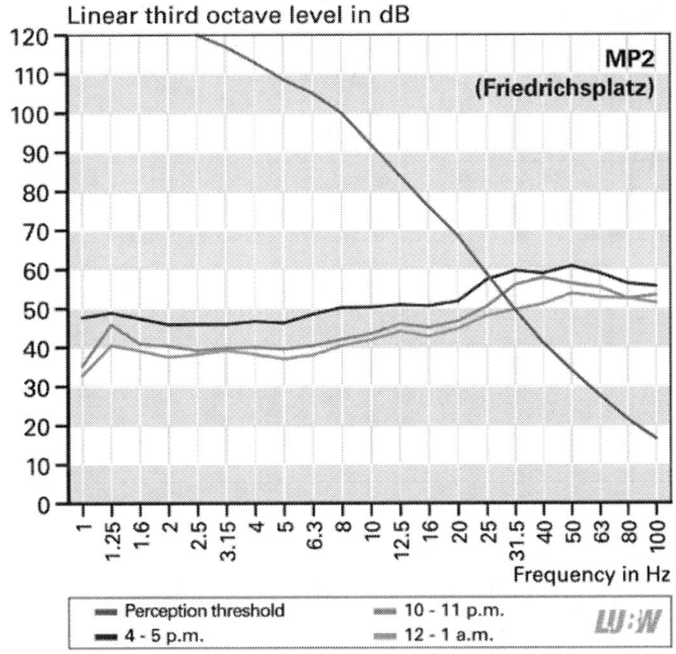

Figure 7: Third-octave band spectra (green) inner city of Karlsruhe on a rather quiet square at different times. Perception threshold (grey).

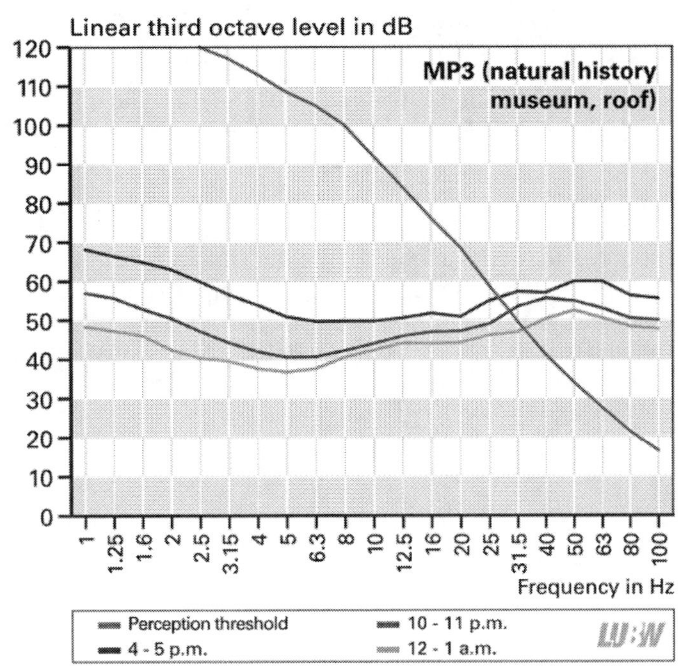

Figure 8: Linear third-octave band spectrum inner city of Karlsruhe, measured on a roof at different times. Perception threshold (grey).

The values in the infrasound range measured on the roof of an adjacent building tend to be higher than those measured at the square, which might be associated with interferences from more distant sources (Figure 8).
The measurements in the rural background were performed without wind turbines around. Figure 9 shows the results in a meadow, in the free field, with values measured at wind speeds of 5 resp. 10 m/s. They are in a similar range as for measurements in the immediate vicinity of WT. The levels measured at the edge of the forest and in the forest were lower.

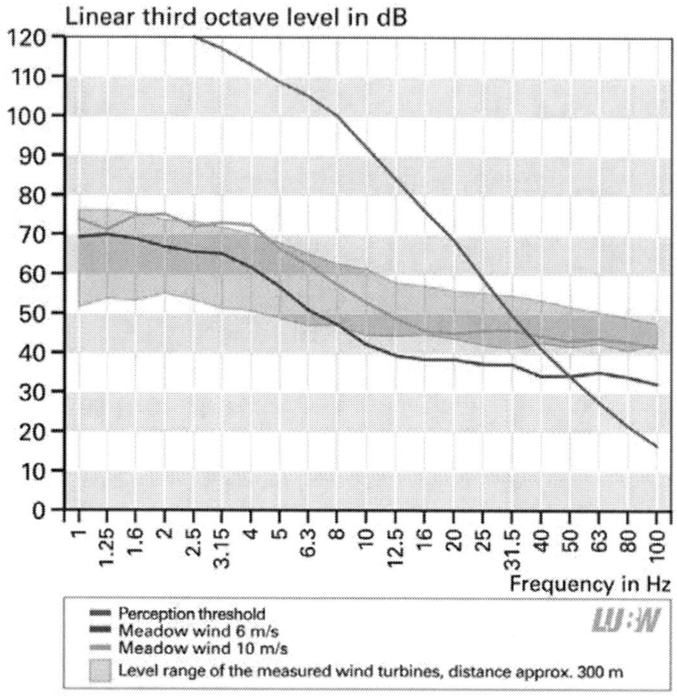

Figure 9: Comparison of the noise situation in the free field (green – without WT in the vicinity) with the level range of a WT at a distance of 300 m (red). Perception threshold (grey).

3.4 Sound sources in the household

As regards the sound sources in the household, oil-fired heating and washing machine are noticeable sound sources in the low-frequency range (Figure 10). Both of them contain significantly low-frequency portions (oil burner, spinning) during operation, whereas the gas-fired heating works relatively quiet.

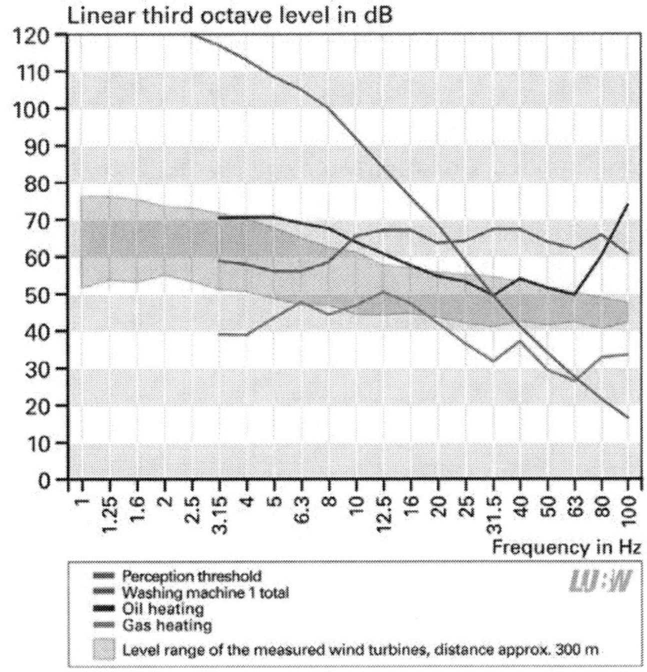

Figure 10: Third-octave band spectra of oil-fired heating (violet), washing machine (blue) and gas-fired heating (lowest curve). For comparison WT (red), perception threshold (grey).

4. Conclusions

Infrasound is produced by a large number of different technical and natural sources. It is an ordinary part of our environment that is present everywhere.

The contribution of wind turbines, however, is insignificant. The infrasound level produced by wind turbines is significantly below the human perception threshold. In our investigations in the vicinity of the plants, the levels could well be measured, the natural frequencies below approx. 6 to 8 Hz could also be determined. At the more distant measuring points, however, they could not be detected any more. The differences between wind turbines in and out of operation were negligible or non-existent.

There is no scientifically substantiated evidence of adverse health effects of infrasound in this level range.

The comprehensive project report which provides much more information and data is readily available on the Internet (2).

References

IEC 61400-11: Acoustic noise measurement techniques, Ed. 2.1, date of issue 2006-11 – German version: DIN EN 61400-11: Wind turbines – Part 11: acoustic noise measurement techniques, date of issue 2013-09

Low-frequency noise incl. infrasound from wind turbines and other sources – Report on results of the measurement projects 2013-2015, LUBW 2016
URL: http://www4.lubw.baden-wuerttemberg.de/servlet/is/262445/?shop=true&shopView=6647

DIN 45680: Draft: Measurement and assessment of low-frequency noise immissions (September 2013), date of issue 2013-09, with regard to the perception threshold identical with draft 2011-08

Møller H. & Pedersen C.: Hearing at low and infrasonic frequencies, Noise & Health, Vol. 6, Issue 23, S. 37-57, 2004

Ingielewicz, R. & Zagubien, A.: Infrasound Noise of Natural Sources in the Environment and Infrasound Noise of Wind Turbines, Pol. J. Environ. Stud. Vol.23. No.4.1323-1327, 2014

Bayer, O. et al.: Messung von Straßenverkehrslärm 2013 - Zusammenfassende Darstellung der Messergebnisse für die stationären Verkehrslärmmessstationen Karlsruhe und Reutlingen, LUBW 2014
URL: http://www4.lubw.baden-wuerttemberg.de/servlet/is/253556/

7th International Conference
on
Wind Turbine Noise
Rotterdam – 2nd to 5th May 2017

A 'social review' of wind turbine noise

Frits van den Berg, GGD Amsterdam Public Health Service, fvdberg@ggd.amsterdam.nl
John Bolte, RIVM National Institute for Public Health and Environment and Smart Sensor Systems Group, The Hague University of Applied Sciences

Summary

In order to better understand the conflicts arising in areas with plans for wind farms, the Dutch Ministry of Infrastructure and the Environment in 2014 set up a pilot for a Knowledge Platform on Wind Energy. In this pilot, wind turbine sound (WTS) was the only topic to be addressed. The Platform had a project team of five persons from knowledge institutions and a Quartering Party of stakeholders in wind energy consisting of persons from national and regional authorities, from a national association of residents and the national wind energy association. There were two discussion meetings with a large number of persons from a number of organizations, providing different views on wind turbine noise and its effects.

The product of the pilot was a Knowledge Communication including three subtopics: WTS as such, health effects of WTS, and public perception of WTS. This was written by a team of four writers with these three areas of expertise. The content was based on scientific publications and on additional other material mentioned in the discussions. The stakeholders involved could give their comments on a first and second draft so as to provide a Communication that incorporated the different views on the desired information.

A formal first version of the Knowledge Communication was published in June 2015. Following this the pilot was evaluated by a management consultancy. It was concluded that there were some misunderstanding about the project goals and their implementation, but there was a broad support for continuation of the pilot, though perhaps in another form. In March 2016 the Ministry announce that the follow-up will be twofold: a national dialogue on wind energy as part of a national platform on energy and environment; and a network of expertise that will provide scientific expertise on wind energy.

1. Introduction

As it is in many countries, the Netherlands strive for a sustainable energy supply. A characteristic for the deployment of sustainable energy is the visibility of decentralised installations such as wind turbines. The resulting confrontation of 'sustainable' and 'nearby' can be difficult for citizens and administrators. The realization of wind energy generation leads to questions and worries and a need for clarity. The questions may refer to the effect of wind turbines on neighbouring residents, the environment, landscape and nature. It may concern the potential effects of noise and other nuisances, a possible decrease in house prices, the balance between the energy and climate performance and costs, and the potential contribution to the economy and 'green growth'.

Initiated by Thijs Aarten of Aarten Wind Solutions (now AES), the Ministry of Infrastructure and the Environment decided to start a pilot in 2014 to explore the merits of the exchange of knowledge and experience between all parties involved in the planning and realization of wind energy projects. This was dubbed the pilot 'Knowledge Platform Wind Energy' (KP WE) in line with other such platforms concerning 'hot topics' in society. A possible continuation of this platform was to be decided after evaluation of the pilot phase in 2015. The pilot itself was restricted in time (half a year) and content (health effects of wind turbine noise) and was to serve two main activities: to offer a platform for effective interactions or 'dialogue' between society, scientific community, the business community and authorities, and providing reliable and impartial information based on the collection, analysis and communication of experiences and knowledge on wind energy (in the pilot: effects of wind turbine noise).

In the KP WE action plan it was stated that authorities, entrepreneurs and citizens acknowledged the wind energy perspective but at the same time experienced problems in practical realisation. A growing critical attitude was seen with respect to the development of new wind projects. This was considered an urgent matter in view of the 4 GW onshore wind power capacity to be added in the near future to the 2,5 GW present at that time (2014).

2. Activities in the pilot Knowledge Platform Wind Energy (KP WE)

The first aim -a platform for effective interactions- was realized in two ways. One way was the structure of the pilot management, the other the consultation of societal groups. The second aim - providing reliable and impartial information- was realized by drafting a 'Knowledge Communication' including contributions from all parties involved.

2.1 Providing a dialogue

Wind energy, in the pilot restricted to wind turbine noise and its effects, was to be taken as a problem in society and not just an environmental or technical problem. Stakeholders were distinguished in four groups: (organizations representing) neighbouring residents, enterprises, authorities, and institutes for research and education (in Dutch the "four O's" as all names begin with an O).
The Project Team structured and coordinated the pilot project and consisted of representatives of research institutes (RIVM National Institute for Public Health and the Environment, ECN Energy research Centre of the Netherlands, EAE/RUG Energy Academy Europe/University of Groningen), the Dutch Association of Community Health Services and Regional Medical Emergency Offices (GGD-GHOR), and was chaired by Aarten from AWS.
The Quartering Party was a group of eight persons advising the Project Team and included representatives of all types of stakeholders. In this party residents were represented by an organization of residents who were neighbours to wind farms (in Dutch abbreviated as NLVOW). They had been critical of wind energy projects, but formally represented all neighbours, including those negative or positive about wind energy.

Early in the pilot a number of persons from all types of stakeholders were invited to attend a roundtable exploration meeting' in November 2014. At this meeting five topics/points of view were determined, based on the stakeholder typology (the "four O's"), and these were to be discussed at five tables. Each participant was to attend two tables. The topics were:
- Audible noise perceived by residents, including noise limits.
- Low frequency noise affecting residents.
- The type of research we need
- The perspective of authorities.
- The perspective of wind farm developers and entrepreneurs

About 40 persons attended this meeting.

The results were analysed and could be categorized in four themes:

1. Exposure to noise and noise reduction (objective as well as subjective);
2. Effects (health effects and public acceptation);
3. Decision making and procedural aspects (legislation and regulation, communication, administrative responsibilities, distribution of costs and benefits, transparency, trust);
4. Knowledge and skills (scientific and non-scientific) and technical solutions.

Decision making and procedural aspects (item 3) were stressed as very relevant, also in the discussion and perception of noise and it effects. This was an important reason to include this in the Knowledge Communication.

A second general meeting was held later in the project, in March 2015. In this meeting a draft of the Knowledge Communication on the (health) effects of wind turbine sound was presented. There was also an evaluation of the project so far and, anticipating a permanent KP WE, a discussion on important themes concerning wind energy. About 50 persons attended this meeting. There was a general appreciation for the open discussion between stakeholders. Comments on the Knowledge Communication were diverse and have not been analysed as were the comments in the first meeting.

2.2 Drafting the Knowledge Communication

According to the action plan the Knowledge Communication was to clarify the need in society to understand wind turbine sound and its effects. To this end those with practical knowledge and experts were to meet to evaluate the knowledge available. This was the start for a working document that was to be presented to all involved.

The results of the first meeting were analysed by a small number of persons (the 'Knowledge Forum') from the institutions taking part in the project and they proposed a table of contents to the writers team. The writers team consisted of four persons from four institutions (GGD, RUG, ECN and RIVM), coordinated by one of them. At the start it was decided that the writers would not be mentioned as authors, but the institutions would stand up for the content.

There were three main chapters: one on the noise itself, one on the health effects of wind turbine noise and a third on the public opinion on wind farms. To add possibilities for action, each chapter was supplemented with a number of items of interest.

Several versions were drafted and each was put to the members of the Project Team and Quartering Party. They provided comments on each version and the writers team coordinator replied to each comment similar to the procedure in a scientific peer review process. This process showed that a 'social review' did not quite follow the rules of a scientific review. One important point was the use of understandable language even though not all content will be understandable to a lay person; e.g. explaining what L_{den} is. A second point was social relevance in contrast to scientific relevance. E.g. a noise limit cannot be judged as scientifically correct or incorrect as it is the outcome of a political discussion. However, from a social point of view it is legitimate to discuss the arguments for setting a specific limit.

In the process a summary and an introduction were added to the three chapters. Also the items of interest were put in a separate chapter and more detailed knowledge with respect to wind turbine sound and its effects were included in an appendix.

In the end all participating organizations accepted the final text except the NLVOW representing the residents. In their opinion the lack of an independent check or review of the text was an important drawback. The final text was published as a first version of the Knowledge Communication and titled "Geluid van windturbines versie 1.0"("Sound of wind turbines, version 1.0").

3. Evaluation of the pilot

After the conclusion of the project the Ministry of Infrastructure and the Environment ordered a management consultancy to evaluate the pilot. The consultancy (Twynstra Gudde) consulted all documents, spoke with 17 members of 10 organizations involved in the pilot project and reported their findings in late 2015.

3.1 Project implementation

The conclusion with respect to the dialogue was that it was well organized. Participants experienced a safe and open way to talk with each other. However, the effect and value of the dialogue had to show in practice and at the moment of evaluation was unclear.
With respect to the Knowledge Communication the conclusion was that, even though the team of writers had worked hard and tried to make an objective communication, some participants thought the final text was not entirely objective, either because of the content or because of the interests of some participating organizations. The final version was subscribed by all participants except one, much to the regret of many of those interviewed. Because of this there was no broad support for the text.

3.2 'Lessons learned'

For a possible permanent Knowledge Platform on Wind Energy, a number of recommendations were given based on the experiences of the pilot project. With respect to the dialogue they can be summarized as:

- Commitment of all participants to the approach and aims of the Knowledge Platform; this was not achieved in the pilot project as most of the approach, aims and structure of the pilot was decided at the start or early in the project by the Ministry and/or the project team. Also the pilot was restricted in time and scope.
- For similar reasons the participants did not share a common definition of the dialogue nor the relation between the results of the dialogue and the Knowledge Communication.
- Building trust is very valuable and relies on continuity of those taking part and the feeling that what has been said will lead to results. To this end agreements must be reached on how and by whom the consequences of the dialogue will lead to an outcome.

Recommendations with respect to the Knowledge Communication can be summarized as follows:

- Participants all expressed the need for an independent party to collect, interpret and share available knowledge. The objectivity of the writers team with persons from renowned institutions is an element that should be preserved.
- One participant did not support the Knowledge Communication because it was not reviewed by an outside independent party, not involved in the knowledge Platform. Even though others did not endorse this view, the fact that this was not decided unanimously at the start was an obstacle to the full support of the Communication.
- A lack of clarity was also expressed about the reactions to the comments of participants with respect to a draft of the Communication: a reaction was addressed to the participant involved and not the others.
- The dialogue did not end when the Knowledge Communication was published. The evaluation report recommends more planning at the start or during a project with respect to who would be using the results and how they could be helped best. Little attention was paid to the distribution and use of and communication about the KC, or how to link it to practical performance. These 'practicalities' were not foreseen in the project budget or time.

4. Conclusion and outlook

The pilot project shows the difficulties that occur when an attempt to get stakeholders together for a true dialogue is restricted in scope, time and budget. In turn this has influenced the approach in and structure of the project. There was no unanimous agreement at the start on a number of issues and this led to different expectations in a later stage.
The need to interpret research results appears to be different for different groups, reflecting their interests at stake. Perhaps the dialogue in a Knowledge Platform can acknowledge these differences and bridge knowledge gaps, but this will certainly take more time.

In March 2016 the Ministry of Infrastructure and the Environment (that had financed the pilot) announced a follow-up of the pilot Knowledge Platform Wind Energy to parliament. Together with the Ministry of Economic Affairs a national dialogue on energy issues and a network of expertise wil be set up. Both initiatives aim to give more attention to spatial quality and environmental management in energy projects. A national platform on Energy and Environment has recently been announced and this will join authorities, the energy industry, civil organizations and knowledge institutions. Wind energy will be one of the topics. Also a network of expertise is now in the make that will focus on the spatial planning aspects of energy projects.

References

The Introduction is an adaptation of text on a web page of Aarten Energy Solutions (www.aarten-es.nl/specifieke-klant-oplossingen/kennisplatform-windenergie, consulted 12 Feb. 2017) announcing the start of the Knowledge Platform pilot project.
Further text in this paper is based on documents produced in the pilot project (such as the action plan and meeting reports, and the Knowledge Communication, all in Dutch, may be downloaded at www.kennisplatformwindenergie.nl (consulted 15 February 2017), and the evaluation report of the project.

7th International Conference
on
Wind Turbine Noise
Rotterdam – 2nd to 5th May 2017

Wind turbine noise – an overview of current knowledge and perspectives

Andrea Bauerdorff, German Environment Agency : andrea.bauerdorff@uba.de
Steffen Körper, German Environment Agency : steffen.koerper@uba.de

Summary

The issue of noise pollution caused by wind turbines is an important part of the implementation of the Energy Turnaround (Energiewende). In 2014, 28.2 % of the total electrical energy was generated from renewable energy sources within the European Union. In Germany, the number of these installations has continuously increased in the last years. Within the framework of the importance of wind energy, the protection of the people from wind turbine noise has to be taken into consideration. Especially low-frequency noise, infrasound and amplitude-modulated sound are in the focus on public discussion.
The German Environment Agency is currently working on two research projects. In 2015 started the project "Noise Effects of Infrasound Exposure". The results of the study are expected in 2018. A key objective of the investigations of this study is to clarify which sound pressure level generated by infrasound from technical sources can be relevant in respective to impairment perception. For this purpose a laboratory study under controlled conditions is carried out. Within this study a detailed questionnaire is worked out and accompanying infrasound measurements will be conducted. Moreover, in a further research project the impact of wind turbines noise on health and quality of life of the population is reviewed. Although there are different studies on individual questions, a holistic view of this topic is missing. The annoyance potential of wind turbines will be investigated with a large-area annoyance survey. Another focus will be laid on the amplitude modulation. A "Handbook for affected persons" based on the results of the research project shall be developed.

1. Introduction

Many people are exposed to high levels of noise that adversely affect their health and quality of life. Noise is now experienced virtually every-where and around the clock, in towns and in the country, day and night. For this reason many people are exposed to noise, which can have a negative effect on their health and their quality of life.To a certain degree, noise is a pollutant, which has only a localised effect unlike other pollutants but which can be found virtually everywhere as there are so many areas in Germany that are affected by it. Not all sound can be automatically classified as noise. Various factors unrelated towards the level of sound generated also play a role in classifying sound as noise. Thus, e.g. also personal attitudes and dispositions or even information, which are related to the noise source, have an effect.
Wind turbines are of great importance to meet the demand for electricity with renewable sources of energy. In recent years, the number of wind turbines has continuously increased in Germany. In 2016, there are 27270 wind turbines with an installed wind power capacity of 45911 MW [1]. The environmental policy of the German Government aims to promote the share of wind energy use in power supply. This goal is still widely accepted in society [2]. On the other side, there are numerous complaints of citizens concerning the noise generated by wind turbines.

The binding immission values for noise immissions are specified by the "Technical Instructions on Noise Abatement - TA Noise" in Germany [3]. Wind turbines are one example of noise sources that can cause annoyance to the residents, although these installations comply with the binding immission values. Studies have shown that the periodicity of noise generated by wind turbines is one main cause for the annoyance. For instance, a study conducted in 2014 by the Martin Luther University in Halle-Wittenberg concluded that noise modulations are one of the reasons of annoyance [4].
The State Office for the Environment, Measurement and Nature Conservation of the German Federal State of Baden-Wuerttemberg carried out a measurement project entitled “Low-frequency noise incl. infrasound from wind turbines and other sources”. In this project infrasound and low-frequency noise was measured in the vicinity of several modern wind turbines. This measurement campaign showed that the measured noise levels are below the human hearing or perception threshold [5].

2. Research project on behalf of the German Environment Agency: “Noise Effects of Infrasound Exposure”

Wind turbines produce low-frequency noise and infrasound. The acknowledged rule of technology for the determination and assessment of low-frequency noise is the German standard DIN 45680 “Measurement and assessment of low frequency noise immission” [6] with its accompanying Supplement 1 “Measurement and assessment of low-frequency noise immissions in the neighbourhood – Guidelines for the assessment for industrial plants” [7]. The DIN 45680 contains a standardized hearing threshold level.
The research project “Feasibility study on infrasound effects, development of research designs to assess the impact of infrasound on humans by different sources” on behalf of the German Environment Agency [8] has shown that DIN 45680 as well the international standard ISO 7196 “Acoustics - Frequency-Weighting Characteristic for Infrasound Measurements” still have deficits [9].The DIN 45680 is currently under revision [10], [11]. Within this activity a noise perception threshold is developed. This threshold and the hearing threshold are illustrated in figure 1 together with the sound pressure levels of low-frequency noise measurements of a 2 MW wind turbine in 250 m distance. The results are similar to the above-mentioned measurement results of the Federal State of Baden-Wuerttemberg [5].

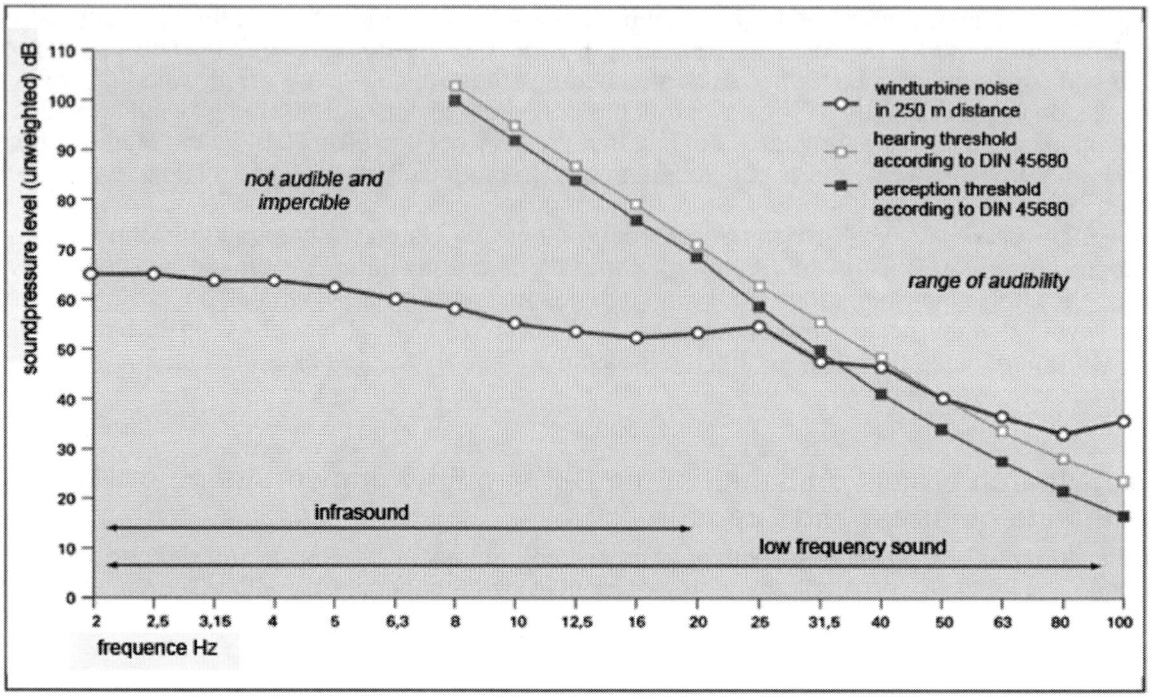

Fig 1: Hearing and perception thresholds of DIN 45680 [7], [11] and sound pressure level at low-frequency range of a 2 MW wind turbine in 250 m distance
Source: Bavarian Environment Agency [12]

Research project details

Based on the results of the research project "Feasibility study on infrasound effects" the German Environment Agency commissioned a follow-up research project "Noise Effects of Infrasound Exposure" in spring 2015. It is carried out by a consortium under leadership of "Möhler + Partner Ingenieure AG". This investigation shall clarify the connection between infrasound immissions beyond the perception and hearing threshold and the impact on human beings. To reach this goal a laboratory study is carried out. 30 test person take part in this study. The sound pressure levels will be controlled during the experiment. Before, while and after the sound exposure the electrocardiography (ECG), the electroencephalography (EEG) and the blood pressure as well as the coordination function of each subject are recorded. Additionally a questionnaire will provide specific information about the momentary perceived annoyance.

On the one hand, a quiet residential environment forms the usual impairment situation for the relevant people and on the other hand, an unnaturally environment, such as a low-reflection laboratory or windowless space, could possibly have an unspecifiable effect on the reaction of the subjects. For this reason, a living space with standard furnishings was chosen (Fig. 2).

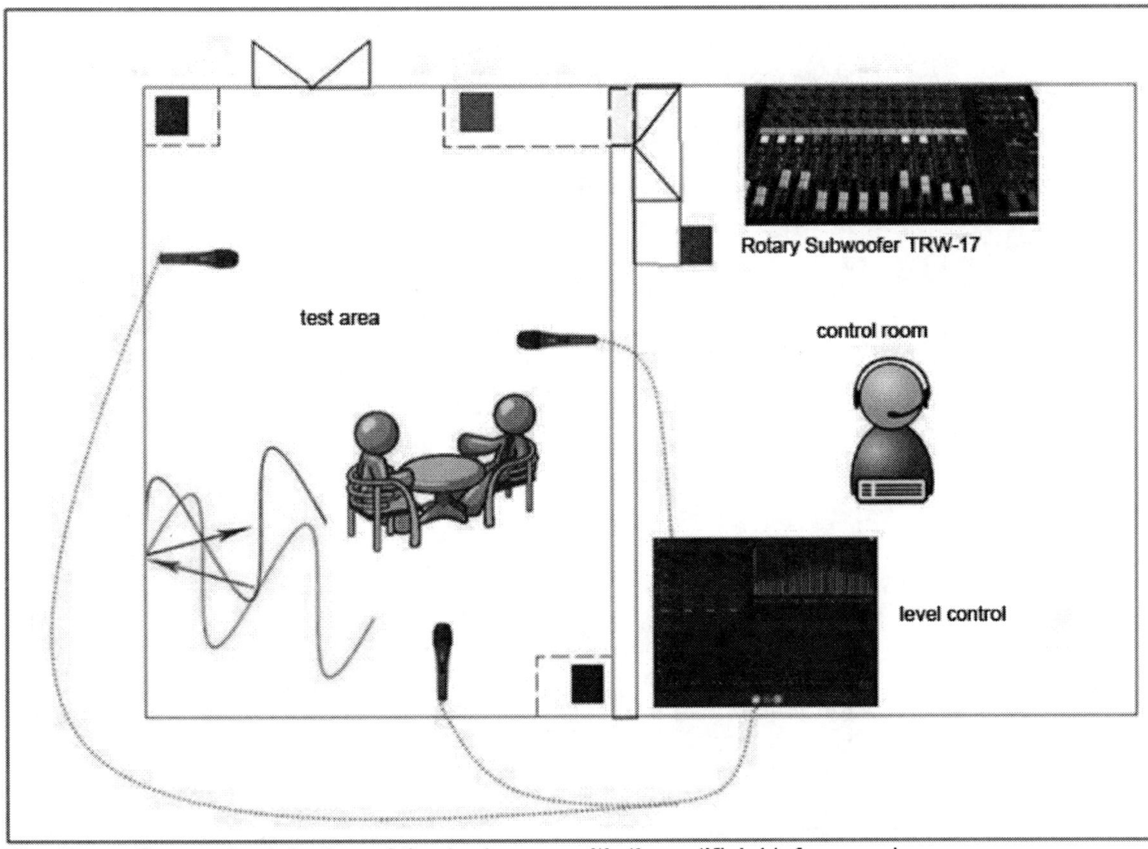

Fig. 2: Conceptual design of the test room with the artificial infrasound source
Source: Interim report on Noise Effects of Infrasound Exposure [13]

For the study an innovative loudspeaker concept of the Rotary Subwoofer TRW-17 from "Eminent Technologies" is used. With the Rotary Subwoofer it is possible to generate signals in a frequency range from 1 to 30 Hz and sound pressure levels up to 120 dB (Fig. 3).

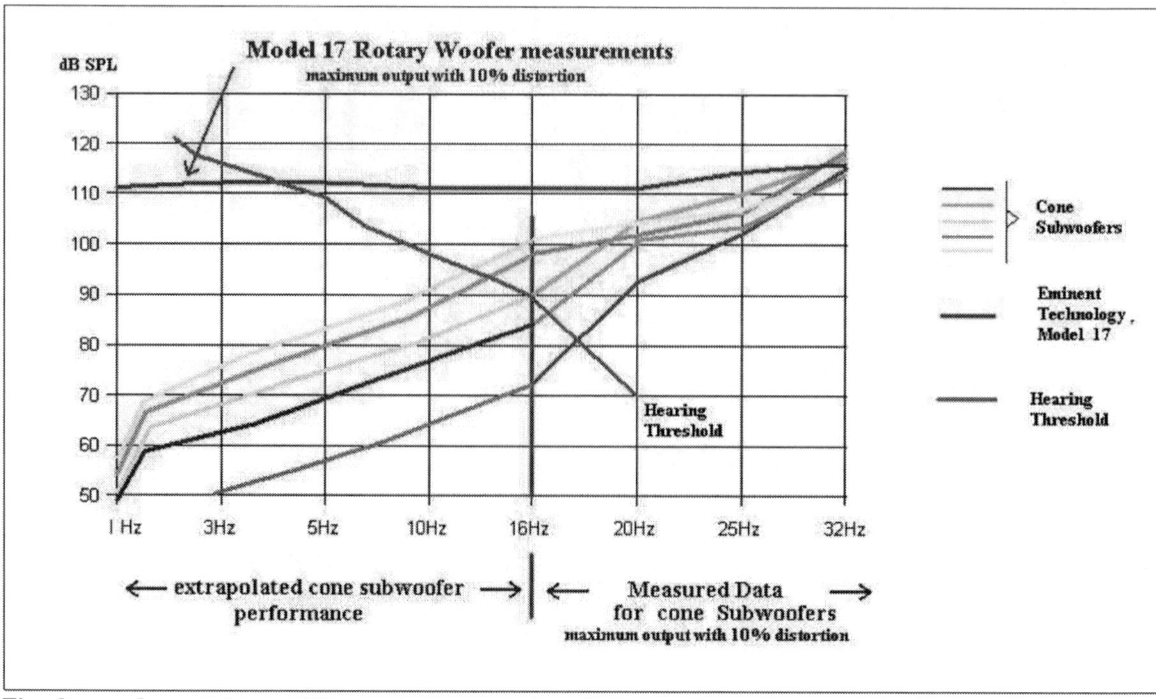

Fig. 3: Comparison of the emission spectra of cone subwoofers and the Rotary Subwoofer TRW-17 infrasound source
Source: Interim report on Noise Effects of Infrasound [13]

The operating mode of a rotating Subwoofer is based on the following principle: rotating wings of an electric motor generate the sound pressure. The amplitude and frequency of the airborne sound radiation are varied by tilting the wings (Fig. 4).

Fig. 4: Infrasound loudspeaker - Rotary Subwoofer TRW-17
Source: Interim report on Noise Effects of Infrasound [13]

3. Planned Research project on behalf of the German Environment Agency: "Noise effects by Wind Turbines onshore"

In Germany, wind turbines with a height of more than 50 m, are submitted to licensing pursuant to the German Federal Immission Control Act [14]. Accordingly installations should be constructed and operated in such a way that does not cause harmful effects on the environment or other hazards, considerable disadvantages and considerable nuisance to the general public. Moreover, precaution should be taken to prevent harmful effects on the environment. One important element of the licensing process deals with the protection against noise.
At the moment the licensing process based on the reglementation of the TA Noise (TA Lärm) [3] is reconsidered. Measurements of "uppenkamp and partner" [15] assigned by the federal state of Nordrhein-Westfalen showed a systematic exceedance of the calculated noise rating levels. Additionally the annoying effects of amplitude-modulation are not yet considered within the assessment of wind turbine noise, because of their imprecise possibility of forecast and prediction.
As consequence of the latest developments many federal states in Germany initiated additional investigations and measurements. In this context a wide range of questions are reviewed, including the appearance, propagation and abatement of noise from wind turbines as well as the impact on the health and quality of life of the population. The on-going studies all focus on specific acoustic subjects, but a holistic overview of the developments and results is missing. For this reason the German Environment Agency plans a further study which show give a comprehensive overview of new research results on these topics. Moreover, a large-area annoyance survey on the impacts by wind turbine noise is also part of this investigation. In this context the effects of amplitude-modulated noises on humans will be also analyzed. The results of the study will be presented and discussed on a symposium by the German Environment Agency. In addition to this the reach results will be summarized in a publication addressed to politicians and the general public to support the on-going discussion about licensing procedure.

4. Conclusions

Wind turbines are of great importance to meet the demand for electricity with renewable sources of energy. Especially in Germany, the number of these installations has continuously increased in the last years. In this context it particularly important to preserve the protected goods of environment and nature as well as health. The German Environment Agency supports this aim with information, analyses, and assessments including research projects commissioned by the Agency.

References

[1] Status of land based wind energy development in Germany, Deutsche Wind Guard GmbH, Varel, Germany 2016

[2] Umfrage zur Akzeptanz der Windenergie an Land Frühjahr 2016 (Survey of Wind Energy Acceptance onshore spring 2016), Fachagentur zur Förderung eines natur- und umweltverträglichen Ausbaus der Windenergie an Land e.V., Juni 2016

[3] Technische Anleitung zum Schutz gegen Lärm, TA Lärm (Technical Instructions on Noise Abatement - TA Noise), Sixth General Administrative Provision to the Federal Immission Control Act of 26 August 1998 (Joint Ministerial Gazette (GMBl.) No. 26/1998 p. 503)

[4] Dr. Johannes Pohl, Dipl. Ing. Joachim Gabriel, Prof. Dr. Gundula Hübner, Abschlussbericht zum Projekt „Untersuchung der Beeinträchtigung von Anwohnern durch Geräuschemissionen von Windenergieanlagen und Ableitung übertragbarer Interventionsstrategien zur Verminderung dieser“ (Study of the annoyance of neighbors by noise emissions from wind turbines and the derivation of transferable strategies for the reduction of these), DBU, 2014

[5] Ratzel, U. et al., Tieffrequente Geräusche und Infraschall von Windkraftanlagen und anderen Quellen. Bericht über Ergebnisse des Messprojektes 2013-2015 (Low-frequency noise incl. infrasound from wind turbines and other sources. Report on results of the measurement project 2013-2015), LUBW, 2016

[6] DIN 45680: Messung und Bewertung tieffrequenter Geräuschimmissionen in der Nachbarschaft (Measurement and assessment of low-frequency noise immissions in the neighbourhood), issue March 1997, Beuth Verlag

[7] DIN 45680, Beiblatt 1: Messung und Bewertung tieffrequenter Geräuschimmissionen in der Nachbarschaft (Measurement and assessment of low-frequency noise immissions in the neighbourhood – Guidelines for the assessment for industrial plants) issue March 1997, Beuth Verlag

[8] Krahé, D.; Schreckenberg, D; Möhler, U., et al.: Machbarkeitsstudie zu Wirkungen von Infraschall, Entwicklung von Untersuchungsdesigns für die Ermittlung der Auswirkungen von Infraschall auf den Menschen durch unterschiedliche Quellen (Feasibility study on infrasound effects, development of research designs to assess the impact of infrasound on humans by different sources), research project on behalf of the German Environment Agency, Texte 40/2014

[9] ISO 7196: Acoustics - Frequency-weighting characteristic for infrasound measurements, issue September 1995

[10] E DIN 45680: Draft - Messung und Bewertung tieffrequenter Geräuschimmissionen (Measurement and assessment of low-frequency noise immissions) issue April 2013, Beuth Verlag

[11] E DIN 45680, Beiblatt 1: Draft - Messung und Bewertung tieffrequenter Geräuschimmissionen - Hinweise zur Beurteilung - Technischer Anlagen (Measurement and assessment of low-frequency noise immissions - Guidelines for the assessment – Technical equipment) issue April 2013, Beuth Verlag

[12] Windenergieanlagen - beeinträchtigt Infraschall die Gesundheit? (Does infrasound from wind turbines affects health?): Umwelt Wissen - Klima und Energie / Bavarian Environment Agency [editor] – 4th updated edition, Augsburg: 2014, Germany

[13] Krahé, D.; ´Müller, U.; Schreckenberg, D; et al.: Interim report on Noise Effects of Infrasound Exposure, research project on behalf of the German Environment Agency

[14] Bundes-Immissionsschutzgesetz (Federal Immission Control Act) in the version promulgated on 17 May 2013 (BGBl. I p. 1274) as last amended by Article 1 of the Act of 20 November 2014 (BGBl. I p. 1714)

[15] Engelen, J; Wenzel, P: Schalltechnischer Bericht der erweiterten Hauptuntersuchung zur messtechnischen Ermittlung der Ausbreitungsbedingungen für die Geräusche von hohen Windenergieanlagen zur Nachtzeit und Vergleich der Messergebnisse mit Ausbreitungsberechnungen nach DIN ISO 9613-2 (Acoustic report on the expanded main investigation on measurements for the determination of sound propagation conditions of high of wind turbines and comparsion with measurement results of sound propagation calculations according to DIN ISO 9613-2), Landesamt für Natur, Umwelt und Verbraucherschutz Nordrhein-Westfalen, Essen 2014

7th International Conference on Wind Turbine Noise Rotterdam – 2nd to 5th May 2017

Why do some people believe that they are "made ill" by wind turbine noise

Geoff Leventhall Consultant UK geoff@activenoise.co.uk

Summary. Wind turbine noise is either inaudible or is at a low level, but there are on-going claims that it is a cause of illness. It is necessary to consider adverse mechanisms and whether these might be direct or indirect. A direct mechanism is an effect such as hearing loss, which requires very high long-term levels. There is no confirmed direct mechanism by which the low levels of noise from wind turbines affects our health, although there has been much speculation about effects of infrasound. An indirect mechanism is when reaction to a noise leads to stress-related responses. After a period of time these stress responses may lead to the symptoms of somatoform illness. Thus, whilst there is no confirmed direct physical mechanism for effects of wind turbine noise, there are psychological routes to adverse outcomes. These routes are heavily influenced by publicity, modification of beliefs, feelings of resentment, etc. to the point when "the truth" may be obscured, or a "new truth" develops.

1. Introduction Some people become ill in the vicinity of wind turbines and an assumption is that this is due to energy radiating from the turbines to the person. The most common radiation to the person is noise, with electrical fields second. But a "visual radiation" is also important. A "response radiation" from the person back to the turbine is not normally considered, but does occur, mainly developing from how a person responds to the wind turbine (Fig 1). An affected person may send a distress response back to the turbine, as they see the turbine as the focus of their problems. Or they may feel that the turbine is a beautiful construction, which they enjoy looking at. The physical input which comes from the turbine is returned as an emotional output, such as annoyance, pleasure or indifference.

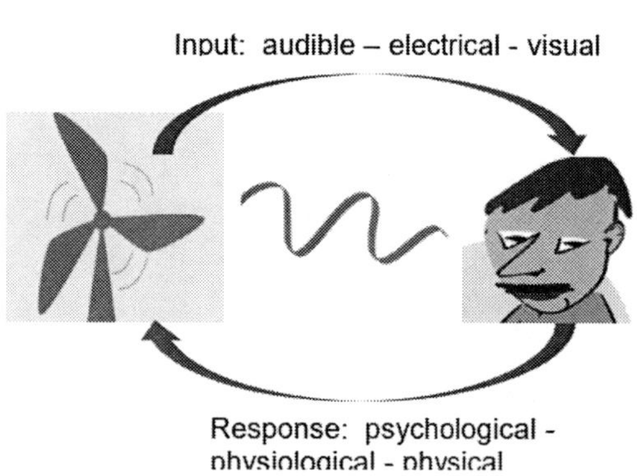

Fig 1 Turbine and human interaction

One explanation of the route to onset of illness is that the noise, visual effects or electric field from the turbines cause direct physiological changes in the person, leading to illness. Contained within this is a common belief that infrasound from the turbines disturbs balance system functioning within the inner ear, leading to feelings of nausea. According to Pierpont and the Wind Turbine Syndrome the source of problems is infrasound (Pierpont 2009). Pierpont has assiduously propagated this idea, leading to it being taken up by others and spread further throughout the internet. An objector web page (www.quixoteslaststand.com) gives links to over 2000 anti-wind groups world-wide, which are outputting negative views on wind turbines.

The way in which infrasound has been driven as the cause of problems from wind turbines is interesting. The present writer's opinion is that this is partly because the higher audio frequency range is well documented and has well established criteria, with which wind farms, in general, comply, whilst infrasound and lower frequencies do not have specific associated criteria. This opened up the opportunity for considerable negative publicity about infrasound and its effects (Illusory Truth Effect, see later) whilst at the same time objectors, having developed concerns in others, then adopted the blocking measure of demanding more work on infrasound before further developments were permitted. When it was shown, and accepted, that infrasound levels were low and that the blade frequency components were well below hearing threshold, objectors had to find a fall-back position and declared that it was the pulsed nature of the emissions which was the problem. They referred to "pulsed infrasound" as if this was a special and harmful type of infrasound.[1] Work is currently in progress on human response to pulses similar to wind turbine pulses, so far indicating that pulses of the type generated by blade tower interaction are unlikely to be a problem (Walker and Celano 2015, Tonin, Brett et al. 2016). Of course, there is a further fall back position, related to the duration of exposure to the blade-tower pulses.

2. The Illusory Truth Effect The Illusory Truth Effect, also known as the Truth Effect, describes the influence of repetition on beliefs. (Dechene, Stahl et al. 2010, Henkel and Mattson 2011, Polage 2012). The Illusory Truth Effect is well known as a main component of advertising and propaganda and has been increasingly investigated in recent years, possibly stimulated by the power of the internet as a source of information. The basis of the Illusory Truth Effect is:

- Repetition of false statements leads to development of belief in them.
- Repetition of similar statements from different sources strengthens this belief.
- The path to belief is made easier by each previous repetition.

We all also have our preferred beliefs. When there is a choice, we tend to believe what we wish to believe, so stabilising our personal equilibrium, which is supported by our beliefs. We feel comfortable when our existing beliefs are confirmed and, if we have become antagonistic towards wind turbines, we are receptive to negative statements about them.

---

[1] The pulses and the components are different ways of describing the same phenomenon and occur together. It requires only about five or six harmonic components to develop a pulse.

Prior knowledge does not necessarily protect from the Illusory Truth Effect (Fazio, Brashier et al. 2015). Fazio et al concluded from their experiments that "participants demonstrated knowledge neglect, or the failure to rely on stored knowledge, in the face of fluent processing experiences". This means that our existing knowledge and beliefs may be ignored, or abandoned, following constant exposure to illusory truth.

The Illusory Truth effect is, of course, related to the nocebo effect in its end result. See Chapman et al for the nocebo effect (Chapman, Joshi et al. 2014)

3. The Role of Stress Existing stress concepts in relation to noise have been reviewed, (Benton and Leventhall 1994) describing how stress may be divided into three broad categories:

- Cataclysmic stress (sudden physical events that are widespread and devastating – earthquakes, war).
- Personal stress (e.g. separation or loss).
- Background stress (persistent, low intensity).

The impact of cataclysmic and personal stress events may differ in terms of scale, but support groups are often available for those affected, leading to personal and/or social support. The impact and the support reduce with time, as will the associated symptoms. Background stressors, such as an unwelcome new windfarm, might not intrude into full awareness and may, after onset, form an integral part of the individual's daily environment. Background stressors do not normally lead to a complete failure of personal coping, so individuals are likely to continue with full daily routines. Individuals lose control through a gradual erosion of their coping capacities and this loss of control is central to the impact of environmental stressors. An individual's perception of a stimulus's 'controllability' Is a key factor in its impact as a stressor. Decrease in the amount of perceived control leads to an increase in stress whilst, in a few cases, the stress may rise from background towards a harsher level.

Different circumstances arise when pre-stressing occurs through the action of objector groups (Illusory Truth Effect). The stress then arises through anticipation of an adverse outcome and may continue throughout the planning, construction and operation of the turbines. Those made anxious have, in effect, been primed to present adverse reactions to the concept of turbines.

4. Individual Differences Criteria are normally developed through a statistical correlation between a cause and its effect, moderated by a legislative decision on numerical limits e.g. 40dBA. Criteria do not relate to specific individuals, but are a compromise, determined by a population average. Consequently, criteria do not protect everybody and, as they are a compromise between requirements, are not intended to do so. However, it is individuals who complain and who are treated as individuals by their medical doctor, although their individuality is lost within criteria. Individual differences, and the resulting dissatisfaction of a small minority, remain a challenge for all noise sources and their criteria.

There are a number of illustrations of individual differences, but the following is a simple one (Fig 2) (Leventhall 1998). The inputs (sound) pass through Detection (outer-middle-inner ear) to the auditory cortex, where they trigger Perception. These first two stages are well defined, but the third stage of Response occurs in the

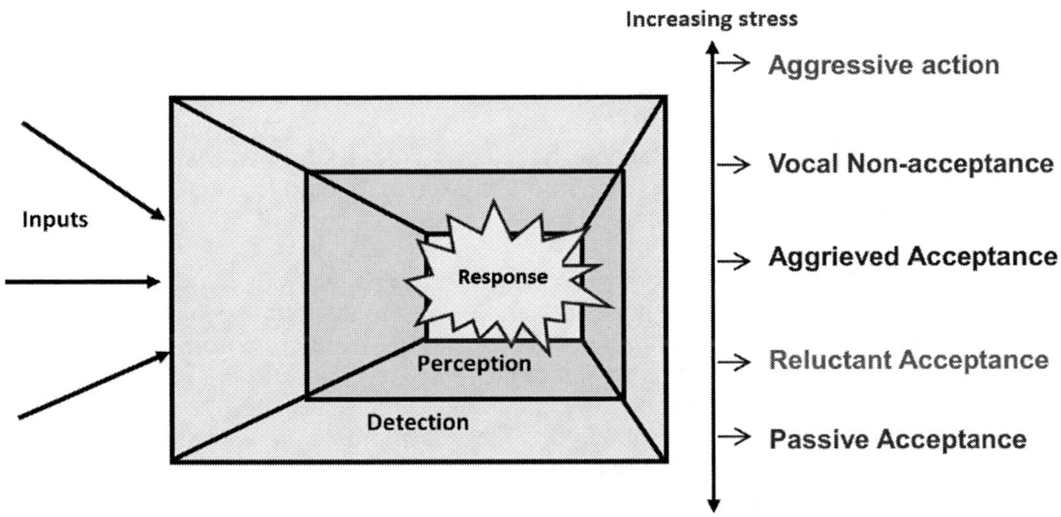

Fig 2 Individual differences in response to noise

emotional part of the brain (limbic system) and has considerable individual differences, ranging from a passive acceptance of the situation (I can hear it, but it does not bother me) to the need to complain or take some action (This noise is ruining my life). Each step up the response scale is accompanied by an increasing level of stress and the associated problems which this brings.

5. Some medical opinions of the Wind Turbine Syndrome (WTS). Farboud surveyed the literature on infrasound and the ear and concluded that, contrary to Pierpont's claims, there was no evidence for effects on the vestibular system at the levels from wind turbines.(Farboud, Crunkhorn et al. 2013). Farboud followed up the initial study with investigations of a vocal WTS complainant, who had initiated legal action. Farboud concluded that the emerging opinion demonstrated no clear evidence of the existence of a WTS or that the low frequency noise causes "significant physiological effects". Additionally, "the symptoms of WTS are vague and can be attributed to psychosomatic factors and annoyance due to the proximity to low frequency whirring, rather than a true physiological impairment." (Farboud and Trinidade 2014).

Harrison also looked at WTS from a medical perspective (Harrison 2015). He considered excitation of the vestibular system by sound, which requires very high levels, in excess of 100dB, and cautiously concluded that there is no known physiological mechanism which might lead to WTS at the levels of noise from wind turbines.

Rubin et al approached the Wind Turbine Syndrome as psychologists (Rubin, Burns et al. 2014) showing that there are ample psychological factors to explain the responses of some people to wind turbines. This is supported by earlier work on attitudes to electromagnetic fields (Witthoft and Rubin 2013).

As infrasound is an acoustic wave It is, of course, to be expected that the ear and auditory cortex respond to infrasound, as has been shown. (Dommes, Bauknecht et al. 2009)

The edifice which Pierpont has built around the Wind Turbine Syndrome is unstable, with little evidence to support it. However, it is an interesting example of the power of the Illusory Truth Effect. Pierpont's WTS has established itself only through repetition, not through science. Additionally, the Illusory Truth Effect has given wind turbine neighbours information on what symptoms to expect, leading to consistency of complaints. Such consistency reinforces belief.

6. Noise, Stress and Health. Exposure to an unwanted noise might develop psychological stress and, in a small number of people, lead to symptoms which cannot be distinguished from those of persons who claim to be suffering from the Wind Turbine Syndrome. Symptoms similar to those of the Wind Turbine Syndrome predate the claims of Pierpont. (Nagai, Matsumoto et al. 1989), (Leventhall 2002, Pedersen, Møller et al. 2008). These symptoms are well known to occur amongst the group known as "Hum Sufferers", who are affected by noises which have not been definitely located or even objectively detected (Demming 2004, Bommer, Young et al. 2016, Frosch 2016) A difference between Hum Sufferers and those responding to wind turbines is that, for the wind turbine group, the source is assumed known, which gives a focus for complaints. However, symptoms are similar, as shown in Table 1, in which the first column describes symptoms of WTS as listed by Pierpont, whist the second column is a compilation of symptoms attributed to other, often unknown, noise sources. The similarity is remarkable, leading to the conclusion that what has been described as a new syndrome is a well-established result of unmanaged stress. Not all persons exhibit all symptoms, but sleep problems are common. Similar effects to those of Table 1 may also occur in

WTSyndrome (Pierpont)	**Noise stress (e.g. the HUM)**
sleep disturbance headache tinnitus ear pressure dizziness vertigo nausea visual blurring tachycardia irritability problems with concentration and memory panic episodes associated with sensations of internal pulsation or quivering "which arise while awake or asleep"	insomnia headache pressure in the ears or head dizziness nausea eye strain fatigue distraction nose bleeds feeling vibration muscle spasms palpitations skin burning stress tension etc

Table 1 Comparison of symptom compilation from Wind Turbine Syndrome and Noise Stress

the absence of noise, providing that sufficient other stressors are acting. Perhaps the most common effects are nausea and sickness in anticipation of stressful events (e.g. job interviews) and trembling due to fear or anger (loss of control).

Psychological stress explains many of the adverse effects claimed for wind turbine noise. Extreme cases may result in physiological effects, such as disturbed sleep. However, recent work has shown that sleep disturbance is similar both near to and remote from wind turbines.(Jalali, Nezhad-Ahmadi et al. 2016, Michaud, Feder et al. 2016).

The responses of people are very complex and those who have developed an aversion to a specific noise respond strongly whenever that noise exceeds their hearing threshold. It is possible to reduce sensitivity to noise through a course of Cognitive Behavioural Therapy, so improving the participants' quality of life. (Leventhall , Robertson et al. 2012). It is probable that negative statements from objectors, which develop concerns in residents near to proposed wind farms, have caused a greater degree of adverse effects than might have otherwise occurred.

7. **An Illustration. The Cape Bridgewater Wind Farm Study.** This is an intriguing study of how Illusory Truth can divert attention from the most probable cause of a problem to the most improbable. That is, focus of attention on the illusory truth of the effects of infrasound leads to neglect of other well-established subjective processes. The study was a collaboration between the developer (Pacific Hydro), the consultant (Steven Cooper) and the residents at three complaint locations.(Cooper 2014). Complaints had continued for about six years without resolution, leading to very stressed and unhappy complainants. One of the properties had been vacated. Cooper confirmed that the wind farm complied with its design criteria.

Although noise and vibration were experienced, the main complaint was of "sensation" which, according to Cooper, includes "headache, pressure in the head, ears or chest, ringing in the ears, heart racing, or a sensation of heaviness". These symptoms should be compared with Table 1 above, where they will also be found.

Cooper claimed that the residents' distress was due to infrasound from the turbines, but the appendices to the Cape Bridgewater report, which include residents' diaries of their perceptions, showed that the turbines were normally audible. In fact, hotspots of noise were noted in and around the residences. Cooper was determined to promote infrasound as the cause of problems, presumably because he had developed a prior belief in this, and so failed to take into account the broad research which has been conducted in the area of human response to audible noise. This literature, containing extensive relevant information on human response to noise, and to other unwanted elements in the environment, was ignored. Thus, the potential that resident responses are linked to ***audible*** noise was neglected in favour of inaudible infrasound, although it is clear from the Cape Bridgewater Report that the wind farm was audible at the residences, as shown by many indications of audibility in residents' diaries. It is also clear from descriptions in the Report that some of the turbines were visible from the residences.

The Cape Bridgewater Report concedes that the measured levels of vibrations are below the levels for human perception, and vibration is accordingly dismissed as a source of impact. However, for consistency, the same conclusion could be made in relation to inaudible infrasound, but it was not.

The extensive research, which has been carried out over many years on effects of noise, shows that the main effect of low levels of unwanted audible noise, from any source, is creation of hostile reactions and negative thoughts, leading to stress and to the adverse effects which might follow from stress. Negative attitudes to the noise source intensify reactions.

Therefore, the response of the residents at Cape Bridgewater is unlikely to be due to effects from low levels of inaudible infrasound, as the Cape Bridgewater Report claims. Responses are likely to be due to the audible noise which is, of course, produced together with the infrasound. Both are generated by the rotating blades, occur at the same time and cannot be separated.

Stress from low levels of audible noise is associated with a number of somatic sensations, particularly of the heart and stomach. Stress from wind turbines, if it arises, is often low level but, in a very small number of persons, the reactions may become intense and overpowering, so that neighbouring wind turbines dominate their lives.

Reaction to noise, especially to low level noise, is largely conditioned by attitudes to the noise and its source. It is known that noise level contributes only about 20% of the total annoyance from noise, whilst it is feelings and opinions about the noise and its source which shape many of our responses, influencing tolerance levels. (Job 1988) (Job 1996, Hatfield, Job et al. 2001) (Miedema and Vos 1999). Negative emotions give an additional impact to an unwanted stimulus. The attitudes of nearby residents towards wind turbines is a major factor in the effects that turbines may have on them (Rubin, Burns et al. 2014).

It has been shown that sham exposures to infrasound (Crichton, Dodd et al. 2013) Crichton, Dodd et al. 2014) or to sham electric fields (Witthoft and Rubin 2013) produce adverse symptoms in those who have been primed to expect these. The Cape Bridgewater Report takes no account of this work, which goes back over 45 years (Jonsson and Sorensen 1970).

The Cape Bridgewater Report is deficient in that it focusses on only one, unproven and speculative causal hypothesis (i.e. due to infrasound) and, in doing so, has neglected all other hypotheses, including those based on a wide scientific field of subjective acoustics, which is sufficient to explain the symptoms experienced by the residents at Cape Bridgewater. That is, there is a well-established route to distress through psychological reactions to long term, unwanted, low-level, audible noise. Assumptions of effects from inaudible infrasound are misleading and unnecessary.

8. Conclusions It is proposed that adverse effects of wind turbine noise on residents occur in the same manner as effects from other noise sources, of which there are many examples, and which are well known to those with "street level" experience of environmental noise problems. The effect mechanism progresses through detection to stress to reactions and is especially vigorous in those who have been primed to believe that an effect will occur. Psychological stress explains many of the adverse effects claimed for wind turbine noise. Extreme cases may result in physiological effects, whilst negative emotions give an additional impact to an unwanted stimulus. The attitudes of nearby residents towards wind turbines is a major factor in the effects that turbines may have on them, although it is only a very small number of those exposed who present the most extreme responses. Criteria do not cover the needs of this group.

There is no convincing evidence that the infrasound from wind turbines, at the levels at residences, is a problem, but it will be useful if those who claim to be affected, for example those who become nauseous, are investigated further in a controlled manner in an infrasonic chamber. Currently, their claims remain anecdotal.

The fact that distress from nearby wind turbines appears to have a psychological origin, rather than a direct physiological cause, does not imply that those affected are in any way misrepresenting the effects upon them. Their distress is real.

References

Bommer, A., et al. (2016). "Mysterious and Annoying: Hums, Tones, and Other Hard-to-Detect Residential Noise Complaints " Noise-Con 2016.

Chapman, S., et al. (2014). "Fomenting sickness: nocebo priming of residents about expected wind turbine health harms." Frontiers in Public Health: doi: 10.3389/fpubh.2014.00279.

Cooper, S. (2014). "The results of an acoustic testing program at Cape Bridgewater wind farm." http://www.pacifichydro.com.au/english/our-communities/communities/cape-bridgewater-acoustic-study-report/?language=en.

Crichton, F., et al. (2013). "The Power of Positive and Negative Expectations to Influence Reported Symptoms and Mood During Exposure to Wind Farm Sound." Health Psychology .

Dechene, A., et al. (2010). "The Truth About the Truth: A Meta-Analytic Review of the Truth Effect." Pers Soc Psychol Rev **14**: 238-258.

Demming, D. (2004). "The Hum: an Anomalous Sound Heard Around the World." Journal of Scientific Exploration **18**(4): 571-595.

Dommes, E., et al. (2009). "Auditory cortex stimulation by low-frequency tones—An fMRI study." Brain Research **1304**(December): 1 2 9 - 1 3 7.

Farboud, A., et al. (2013). "'Wind Tuirbine Syndrome':fact or fiction." The Journal of Laryngology & Otology, **127**(3): 222-226.

Farboud, A. and A. Trinidade (2014). "Otological Effects of low-Frequency noise: Case of Wind Turbine Syndrome from a Successful Plaintiff." Otolaryngology - Head and Neck Surgery **Vol 151**(Supplement): P-221.

Fazio, L. K., et al. (2015). "Knowledge Does Not Protect Against Illusory Truth." Journal of Experimental Psychology: General **144**(5): 993-1002.

Frosch, F. G. (2016). "Manifestations of a low-frequency sound of unknown origin perceived worldwide, also known as "the Hum" or the "Taos Hum"

" International Tinnitus Journal **20**(1): 59-63.

Harrison, R. V. (2015). "On the biological plausibility of Wind Turbine Syndrome

" International Journal of Environmental Health Research **25**(5): 463-468.

Hatfield, J., et al. (2001). "The influence of psychological factors on self-reported physiological effects of noise." Noise and Health **3**: 1-13.

Henkel, L. A. and M. E. Mattson (2011). "Reading is believing: The truth effect and source credibility." Consciousness and Cognition **20**: 1705-1721.

Jalali, L., et al. (2016). "The impact of psychological factors on self-reported sleep disturbance among people living in the vicinity of wind turbines." Environmental Research **148** 401-410.

Job, R. F. S. (1988). "Community response to noise: A review of factors influencing the relationship between noise exposure and reaction." J. Acoust. Soc. Am. **83 (3),** : 991 - 1001.

Job, R. F. S. (1996). "The influence of subjective reactions to noise on health effects of noise." Environment International **22**(1): 93-104.

Jonsson, E. and S. Sorensen (1970). "Relation between annoyance reactions and attitude to source of annoyance." Public Health Reports **85**(12): 1070 - 1074.

Leventhall , G., et al. (2012). "Helping sufferers to cope with noise using distance learning cognitive behaviour therapy." J Low Frequency Noise Vibration and Active Control **31**(3): 193-204.

Leventhall, H. G. (1998). "Making noise comfortable for people." ASHRAE Transactions Vol 104, pt1,: 896 - 900.

Leventhall , H. G. (2002). "35 years of low frequency noise - " Stephens Medal Lecture. Proc IoA 24, Proceedings CD.

Michaud, D. S., et al. (2016). "Effects of wind turbine noise on self-reported and objective measures of sleep." Sleep **39**(1): 97-109.

Miedema, H. and H. Vos (1999). "Demographic and attitudinal factors that modify annoyance from transportation noise." JASA **105**: 3336 - 3344.

Nagai, N., et al. (1989). "Process and emergence of the effects of infrasonic and low frequency noise on inhabitants." Jnl Low Freq Noise Vibn **8**(3): 87-89.

Pedersen, C. S., et al. (2008). "A detailed study of low-frequency noise complaints." Jnl Low Freq Noise Vibn and Active Control **27**: 1-34.

Pierpont, N. (2009). "Wind Turbine Syndrome." K-Selected Books.

Polage, D. C. (2012). "Making up History: False Memories of Fake News Stories." Europe's Journal of Psychology **8**(2): 245–250.

Rubin, G. J., et al. (2014). "Possible psychological mechanisms for “wind turbine syndrome”. On the windmills of your mind." Noise and Health **16**(69): 116-122.

Rubin, J. G., et al. (2014). "Possible psychological mechanisms for “wind turbine syndrome”. On the windmills of your mind." Noise and Health **16**(69).

Tonin, R., et al. (2016). "The effect of infrasound and negative expectations to adverse pathological symptoms from wind farms." Jnl Low Freq Noise Vibn Ac Cntrl **35**(1): 77-90.

Walker, B. and J. Celano (2015). "Progress Report on Synthesis of Wind Turbine Noise and Infrasound " Proc WInd Turbine Noise 2015, Glasgow.

Witthoft, M. and G. J. Rubin (2013). "Are media warnings about the adverse health effects of modern life self-fulfilling? An experimental study on idiopathic environmental intolerance attributed to electromagnetic fields (IEI-EMF)." Journal of Psychosomatic Research **74**: 206-212.

Witthoft, M. and J. G. Rubin (2013). "Are media warnings about the adverse health effects of modern life self-fulfilling? An experimental study on idiopathic environmental intolerance attributed to electromagnetic fields (IEI-EMF)." Jnl Psychosomatic Research **74**: 206-212.

7th International Conference
on
Wind Turbine Noise
Rotterdam – 2nd to 5th May 2017

Human response to wind turbine noise: infrasound and amplitude modulation

Chris Feist, University of Minnesota, St. Anthony Falls Lab: feist020@umn.edu

Peggy Nelson, University of Minnesota, Speech-Language- Hearing Sciences: nelso477@umn.edu

William Herb, University of Minnesota, St. Anthony Falls Lab: herb0003@umn.edu

Matthew Lueker, University of Minnesota, St. Anthony Falls Lab: luek0004@umn.edu

Noah Stone, University of Minnesota, Dept. of Mechanical Engineering: stone700@umn.edu

Summary

An ongoing project at the University of Minnesota seeks to characterize infrasound and amplitude modulated audible wind turbine noise and the human perceptual responses to these acoustic signals. The goal is to inform state agencies that regulate and license wind turbine farms in Minnesota, USA with advancements in turbine noise monitoring, analysis, and physiological impacts.

Acoustic noise data from two wind turbine sites have been collected, including an intensive noise monitoring effort from a University-owned, single turbine site with an array of 36 audible-range microphones and three infrasound microphones. This facility also includes a 130 meter metrological tower. Additional data have been collected from the single turbine site and a multi-turbine wind farm using one audible microphone, one infrasound microphone, and a LiDAR wind profiler. These data are analyzed to establish bounds on the frequency range of the noise, levels of amplitude modulation, and characteristics of infrasound noise over a range of operating conditions. This will inform the design of stimulus for the human response testing. Preliminary analysis indicates that infrasound levels are below published human thresholds, but the component of infrasound attributable to the turbine blade passing frequency varies substantially over time and operating conditions.

Human response testing is performed at the Center for Applied and Translational Sensory Science at the University of Minnesota. Infrasound and audible stimuli are presented separately and in combination with varying amounts of amplitude modulation and spectral "peakiness". Response measurements include a force plate sensor to measure human sway response to the noise sources. Preliminary data suggests that measures of sway are stable and reproducible, and that some persons may show changes in stance in the presence of combinations of audible and infrasound stimuli. We present results of both the wind turbine noise analysis and the human response pilot testing.

Project funding provided by customers of Xcel Energy through a grant from the Renewable Development Fund.

1. Introduction

A project ongoing at the University of Minnesota seeks to characterize infrasound and amplitude modulated audible wind turbine noise and the human perceptual response to these noise sources. The research team consists of the St. Anthony Falls Laboratory (SAFL), Department of Speech-Language-Hearing Sciences (SLHS), and the Minnesota Department of Commerce and is funded by Xcel Energy's Renewable Development Fund. Wind turbine acoustic data has been collected at a single turbine research site and multi-turbine wind farm. The data is analyzed to establish levels of amplitude modulation and characterize infrasound during turbine operation which will help shape the human response testing. The results from human response testing and thorough characterization of noise from a single and multi-turbine site could help inform state agencies in establishing more advanced regulations and licenses procedures. The project will also provide more information and understanding on health and annoyance concerns citizens have vocalized about wind turbines regarding noise.

The project uses a new amplitude modulation methodology, "A Method for Rating Amplitude Modulation in Wind Turbine Noise," recently published in August, 2016 by IOA Amplitude Modulation Working Group (AMWG, 2016). This method reflects the most current research on amplitude modulation and is the closest to a standard amplitude modulation calculation for wind turbines. The metric is defined as an amplitude modulation (AM) depth and will be a parameter in human response testing. This project will help in beginning to bridge the gap between AM depth and human response.

1.1 Literature Review

A literature review guided the research plan for the current project. The goals of the review was to provide insight into measurement techniques of wind turbine noise, current analysis methods, any shortcomings with respect to measurement techniques and analysis, and identify key areas of research interest. IEC 61400-11 (International Electrotechnical Commission, 2012) was used as the standard for wind turbine acoustic noise measurement techniques. Measurement systems and techniques were designed to meet this standard as much as possible but were modified or expanded upon in areas of measurement positions, amplitude modulation, and infrasound measurements.

Previous work shows that infrasound measurement requires specialty audio measurement equipment and secondary windscreens are required to reduce wind-induced microphone noise (Hansen, Zajamsek, & Hansen, 2014). The likely source of infrasound is a rapid change in angle of attack as the turbine blade passes through the perturbed flow upwind of the tower (Hansen, Zajamsek, & Hansen, 2014). Measured infrasound levels are generally below the threshold of human perception; however, thresholds are not as well defined as those in the audible range.

Amplitude modulation, and specifically enhanced amplitude modulation, has been identified as a cause of noise complaints and annoyance. We use methods outlined in literature to analyze AM, most notably "A Method for Rating Amplitude Modulation in Wind Turbine Noise," and determine annoyance thresholds of human test subjects. There are a number of speculated causes of EAM and we investigate correlation between atmospheric conditions wind shear to AM modulation depth.

Overall, literature shows that both infrasound and the modulation frequency of AM are closely linked to the blade passing frequency and greatly affected by atmospheric conditions.

2. Field Measurements

Field measurements for this project come from two acquisition systems and two field sites. The two field sites are a single turbine research site and a multi-turbine industrial wind farm. The data acquisition systems include a 36 microphone directivity array system deployed at the single turbine site in 2012 and a two microphone, mobile, single-point system deployed at both the single turbine and wind farm sites in 2016.

2.1 Single Turbine Research Site (Eolos)

The research site includes a single 2.5 MW Clipper C96 wind turbine (80 m hub height, 96 m rotor diameter), additional blade and tower monitoring equipment, a 130 m tall meteorological tower located 160 m south of the turbine, and a deployable WindCube LiDAR system.

The wind turbine SCADA information is continuously logged. SCADA is recorded at a frequency of 1 Hz with the ability to log 20 Hz on demand. The wind turbine is also fitted with strain gauges throughout the wind turbine blades and at the base of the wind turbine tower. Accelerometers are also installed in the rotor blades.

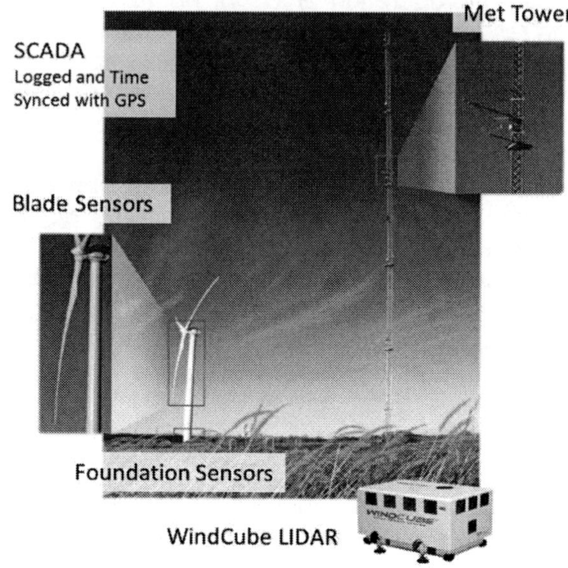

Figure 2.1: Eolos single turbine wind energy research site

The research grade meteorological tower spans the entire swept area of the turbine blades and has instruments installed on boom arms at ten different heights on the tower. Sonic anemometers are located on four of the boom arms and are located at heights of 10, 30, 80, and 129 m. The remaining boom arms have cup and vane anemometers and temperature and relative humidity sensors at locations from 7 to 126 m. The LiDAR system is capable of measuring wind speed and direction from 40 m up to 200 m.

The turbine is controlled by researchers so background noise samples can be collected at the same location as noise measurements by placing turbine in standby and turning off ancillary noise sources such as fans.

2.2 Wind Farm

Pleasant Valley Wind Farm is a wind farm consisting of 100 Vestas V100 2.0 MW wind turbines (100 m rotor diameter, 95 m hub height). The dominant wind directions at the site are from the south and northwest. As a result of this, the wind turbines are predominantly oriented in northeast-southwest rows. The topography is flat and the turbines are surrounded by agricultural fields.

The mobile WindCube LiDAR system is placed upwind of the turbine of interest to provide incoming meteorological data and a mobile field weather station provides wind information local to the acoustic measurement location. Factors such as crops, trees, and other obstacles may limit measurement locations.

The turbine(s) are not able to be turned off so background noise samples are collected at a designated location with similar topography and vegetation ~5.5 km from the closest wind turbine.

2.3 2012 Measurement Equipment

In June of 2012, United Technology Research Center (UTRC) in collaboration with the University of Minnesota – St Anthony Falls Laboratory (SAFL) performed wind turbine acoustic measurements at the Eolos wind research site. These measurements were conducted during daylight hours from June 7th to June 19th. This system was only deployed at the single turbine site.

A total of 75 data channels were collected for this measurement campaign, of which 36 were an evenly spaced circular directivity microphone array located 102 m from the turbine. This directivity array data from this array was analyzed for this current project.

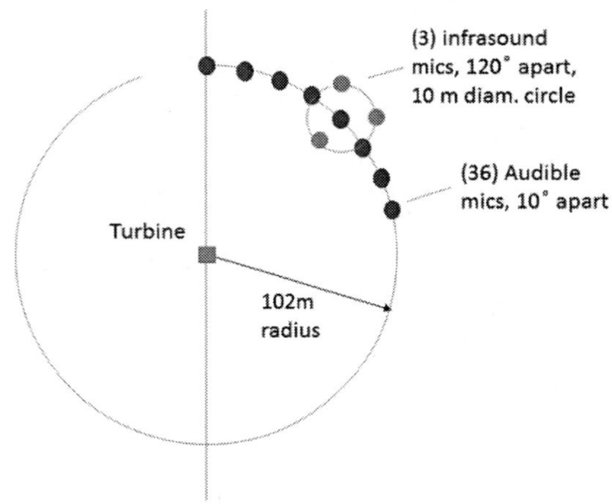

Figure 2.2: 2012 directivity array at Eolos site

The Brüel and Kjær Type 4958 ¼" microphones (flat response from 10 Hz to 20 kHz) were sampled at 32,768 Hz and high passed filtered at 20 Hz. The microphones were placed on glass disks with windscreens to meet IEC 61400-11 requirements. A number of turbine conditions were collected including normal operation, background, and cooling fans on and off for both conditions.

The additional channels included three infrasound microphones surrounding the 40° mic in a 10 m diameter circle, microphones and accelerometers located in the nacelle and base of turbine (power conversion equipment), hub tachometers, timing signal, and meteorological conditions for the the Eolos met tower

2.4 Current Measurement Equipment

A mobile data acquisition system was constructed for the current project. It measures noise using both an audible and infrasound microphone sampled at 50 kHz. The system was used at both the Eolos site and the wind farm. Distance from the turbine of interest varied from 300 to 600 meters at a variety of orientations. Data collection occurred from June to December 2016 and included day and night measurements. Ancillary measurements included meteorological data from a mobile weather station located near the audio equipment and GPS for heading and distance from the turbine of interest. The system was constructed to comply with IEC 61400-11. System component are listed below.

Figure 2.3: 2016 measurement equipment at Eolos site

- Audible microphone: Brüel and Kjær Type 4191 (3 Hz to 22.4 kHz)
- Infrasound microphone: Brüel and Kjær Type 4193 with UC0211 adapter (0.1 Hz to 5 kHz)

- 1 m plywood measurement boards
- ACO Pacific 7 inch diameter primary windscreen
- Custom 20 inch diameter secondary windscreen with insertion loss characterized
- MetOne weather station located at 3 meters
- INFILTEC microbarometer INFRA20
- GPS location
- Brüel and Kjær Nexus amp/power supply
- National Instruments A/D board – 9239

The WindCube LiDAR was deployed upwind of the turbine at both the Eolos site and the wind farm for all current field measurements.

2.5 Wind Speed Measurements and Reporting

Wind speed used for reporting was taken from either the LiDAR data or the MetOne weather station for both sites. Wind speed reported from the LiDAR is measured wind speed at hub height with a time delay (unless others stated) using the distance from the turbine divided by a 10 minute average convective wind speed to estimate wind speed at the turbine location. The convective wind speed is determined from hub-height wind speed measurements. Wind speed reported from the MetOne weather station located near the acoustic equipment measures wind speed at three meters and is adjusted to the 10 meter standard reference height according to IEC 61400-11. The MetOne reported wind speed represents the wind speed at the observer and there is no time delay for the wind speed reported. This wind speed is termed $U_{10m_{met1}}$.

In addition to wind speed reported at hub height from LiDAR measurements, a quasi wind shear calculation is performed by taking the difference between wind speed at the top-tip and bottom-tip elevations measured by the LiDAR. For the Eolos turbine, the top-tip elevation is 128 m and the bottom-tip elevation is 32 m. The multi-turbine site had top-tip and bottom-tip elevations of 150 m and 50 m, respectively. This wind speed difference is termed $U_{\Delta tip}$.

3. Noise Data Analysis

Audible noise data and wind data from 2012 and 2016 were analyzed to characterize the overall noise levels, correlations to wind conditions, frequency content, and amplitude modulation as a function of wind conditions, and orientation/distance from the turbines. The 2012 data set from the Eolos site, with synchronized measurements of turbine noise and wind profiles, gave the opportunity to examine the directionality of turbine noise and the relationships between turbine noise production and incoming wind conditions. The 2016 data included both day and night measurements for a range of distances from the turbine(s). The infrasound data were analyzed to 1) characterize the overall levels of infrasound under different wind conditions, 2) characterize the variability in spectral content of the infrasound signals and 3) distinguish between ambient infrasound and turbine-generated infrasound. In particular, the variability in the prominence of the blade-passing frequency and harmonics in the frequency spectra was characterized to give bounds for the infrasound waveforms used in human response testing.

3.1 Analysis Methods

Audible Signals

For analysis of the audible noise signals, the raw microphone data were processed to give 100 ms L_{eq} time series in 10 s segments. Unweighted 10 s L_{eq} noise data were analyzed to determine how well the noise measured at the microphones was correlated to wind measured at the met tower. For each 10 s L_{eq} measurement, a corresponding mean wind velocity was calculated from the met tower. The length of the averaging window for wind data was varied, and a time delay between the noise data and wind data was also introduced to take into account the distance from the met tower to the turbine tower. For each wind speed increment, the variable delay (τ) was calculated as:

$$\tau = x\cos(WD)\,/U, \qquad (3.1)$$

where U is the average wind speed for a given averaging window and measurement height composition, WD is the wind direction, and x is the distance from the met tower to the turbine tower (160 m).

Amplitude modulation analysis was also performed based on 10 s L_{eq} values. For the purposes of this project, it was desirable to choose a method that separates AM due to the turbine itself from other fluctuation in background noise. The AM analysis method recently published by the Institute of Acoustics (IOA) was chosen, is summarized below, and is detailed in (AMWG, 2016):

1) The raw microphone signal is processed using an A-weighting filter, followed by bandpass filters with frequency ranges of 50-200 Hz, 100-400 Hz, and 200-800 Hz.
2) The 100 ms L_{eq} is calculated for each filtered signal, for a 10 s segment of data
3) A Fourier transform is taken of each 10 s segment
4) The prominence of the blade-passing frequency (BPF) and its harmonics are assessed. Measurements lacking a clear peak of the fundamental BPF are discarded.
5) A time domain L_{eq} signal is reconstructed, based on the spectral peak heights at the BPF and the 2nd and 3rd harmonics (if applicable), and the peak-peak amplitude of the AM is measured using the 95th and 5th percentiles of the reconstructed L_{eq}.

The IOA method then assesses the number of 10 s segments with significant amplitude modulation within 10 minute windows, to find periods of consistently significant AM. In the present study, this final step was omitted, to compile statistics on AM depth over a range of wind conditions.

Infrasound Signals

The measured infrasound signals were assumed to be composed of infrasound generated by the turbine overlaid with ambient background infrasound. To capture a sufficient number of low frequency oscillations in a sample segment, the infrasound signals were analyzed in 30 s segments. The algorithms used in the AM analysis for identifying and quantifying the prominence of the blade-passing frequency (BPF) and harmonics were also used for the infrasound analysis. Fourier analysis (FFT) was used to generate a frequency spectra for each 30 s segment. Multiple frequency spectra were normalized to the BPF and averaged in bins of similar wind conditions, to obtain characteristic spectra of turbine-generated infrasound for different wind speeds.

3.2 Analysis Results

Wind Correlation Results

Wind correlation analysis was performed using data from June 10, 2012, and focused on data from four fixed microphones (0, 90, 180, and 270 deg.). For normal turbine operation an L_{eq} averaged over these microphones was calculated and strong correlations were found between the met tower wind velocities and the measured noise levels. Met tower wind measured at hub height (80 m) and 130 m were the best predictors of L_{eq}, with correlation coefficients of up to 0.84. Taking the mean of all 10 anemometers on the met tower gave slightly higher correlation coefficients (up to 0.86) compared to the individual wind speeds. Introducing a time delay (Eq. 3.1) between the wind averaging window and the L_{eq} measurement gave significantly better correlations – for a 60 s averaging window on hub height wind speed, the correlation coefficient increases from 0.79 to 0.84 using the delay. The wind speed measured on top of the turbine nacelle, which is part of the SCADA data set, was a relatively poor predictor of turbine noise (Fig. 3.1) compared to the met tower wind speed. This is probably due to interference from the turbine blades. Turbine power output was a relatively good predictor of turbine noise, particularly for longer averaging times (Fig. 3.1), with correlation coefficients of up to 0.85.

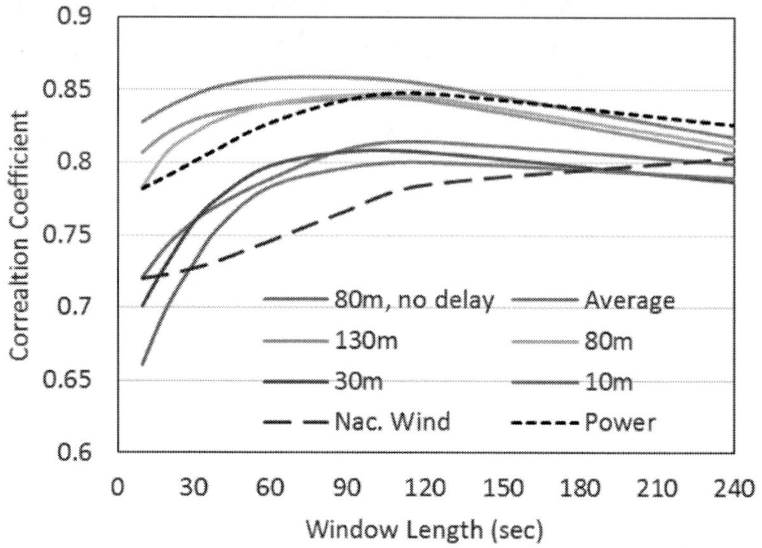

Figure 3.1: Correlation coefficient of 10 s L_{eq} to wind speed for various anemometer heights and varying wind averaging window lengths. Eolos data, June 10, 2012, normal turbine operation. Data are also given for wind measured on the nacelle and for turbine power output. Average equates to the mean wind speed over all ten anemometers. A time delay (Eq. 3.1) was used, except where noted.

Amplitude Modulation Results

The AM analysis was performed both on the 2012 and 2016 data sets from the Eolos and wind farm sites. AM results from the 2012 Eolos measurements are included here, although there are concerns that the measurements were taken only ~100 m from the turbine, and therefore may be in the near-field region and not representative of AM at typical residential distances. (Oerlemans, 2014).

There was significant variability in the AM modulation depth in both the 2012 and 2016 data sets, as summarized in Table 3.1. Although background (turbine off) noise samples only contained about 10% with measureable AM, the median modulation depth of background noise was slightly higher (2012 data) or similar to (2016 data set) turbine-on samples. Maximum measured AM depth that was audible and recognizable as wind turbine noise was 3.7 dB. Literature indicates AM depth can be up to 10 dB (AMWG, 2016), however, this seem to be a rare event and was not captured during our measurements.

For all three data sets, no systematic relationships were found between AM depth and wind conditions. Examples of AM depth plotted over varying $U_{\Delta tip}$, and indicator of wind shear, are given in Figure 3.2 for 2016 measurements at both the Eolos and multi-turbine sites. $U_{\Delta tip}$ was calculated for the same 10 s period as the AM depth analysis ($U_{\Delta tip}$ is a 10 s average here). AM depth results include measurements ranging from 300 to 600 m at multiple orientations from the turbine of interest and results are for the frequency bandwidth of 100 – 400 Hz.

Table 3.1. Summary of amplitude modulation results for the three noise data sets, based on the IOA methodology. These results are for the middle frequency band (100-400 Hz), which had the highest fraction of samples with AM.

	2012		2016 EOLOS		2016 Wind Farm	
	Normal Operation	Back-ground	Normal Operation	Back-ground	Normal Operation	Back-ground
# Segments Analyzed	3416	1528	2040	720	3048	1080
% with AM	32%	10%	27%	8%	28%	8%
Median AM depth	2.5 dB	3.1 dB	2.2 dB	2.4 dB	2.3 dB	2.4 dB
Standard Deviation	0.9 dB	3.0 dB	0.7 dB	1.2 dB	0.6 dB	1.2 dB
90th percentile	4.7 dB	8.7 dB	3.1 dB	4.3 dB	3.0 dB	4.5 dB

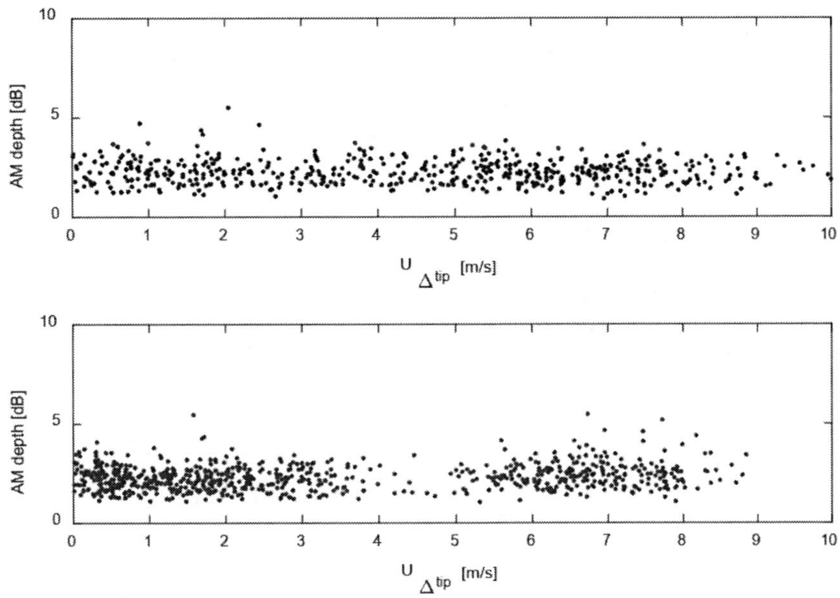

Figure 3.2: Amplitude modulation depth plotted against $U_{\Delta tip}$ for the Eolos site (top) and multi-turbine site (bottom). AM depth corresponds with the bandpassed frequency range of 100-400 Hz and 10 s segment length

Infrasound Results

The measured infrasound levels were well below published human response thresholds (Watanabe and Moller, 1990) with a minimum difference of 15 dB at 20 Hz. Infrasound level differences between human response thresholds and measured infrasound increased with decreasing frequency (60 dB difference at 8 Hz).

Table 3.2 shows infrasound prominence and peak infrasound quantities at the 3rd harmonic for 2016 Eolos and wind farm measurements. The 3rd harmonic was chosen because it was the most prominent with a maximum prominence of 10-12 dB. The 3rd harmonic of the wind farm prominence is not reported because the wind farm 3rd harmonic isless defined. This is likely due to presence of multiple turbines causing multiple, slightly different BPF's occurring at different phases at the specific location of the acoustic measurements.

Figure 3.3 (left) gives averaged frequency spectra for the infrasound signals measured during turbine operation at the Eolos site in 2016 for varying wind speed bins (1 - 7 m/s), with the blade passing frequency and harmonics highlighted. Wind speed bins are determined by using

$U_{10m_{met1}}$ with a 30 s average window. The 30 s data segments were resampled such that the BPF was equal to 1 Hz. For turbine-on conditions, spectral peaks are visible at harmonics 2 – 7 of the BPF, but not at the fundamental. The fundamental BPF peak is believed to be masked by high ambient background noise in this frequency range. The averaged frequency spectra show a clear increasing in infrasound level with increasing wind speed. Additionally, the infrasound peaks associated with the BPF harmonics are very clear at ground level wind speeds below 4 m/s but are not seen at wind speeds above 4 m/s.

Figure 3.3 (right) shows the infrasound peak value at the 3rd harmonic for the resampled 30 s data segments vs. $U_{10m_{met1}}$. There is a strong correlation coefficient, 0.7, between infrasound level and wind speed. However, this strong correlation is not observed when infrasound is plotted with hub-height wind speed.

As with the audible noise, both background and turbine-on infrasound levels increased systematically with wind speed (Figure 3.3 and 3.4). Figure 3.4 gives averaged frequency spectra for the infrasound signals for both the turbine-on and background measurement conditions. The infrasound levels are very similar at frequencies below the 2nd harmonic. At frequencies above the 2nd harmonic (about 1.5 Hz) the turbine-on measurements show increased infrasound levels over the background measurements. However, this difference decreases with increasing wind speed.

Figure 3.5 gives the average, resampled, infrasound spectra for the Eolos and multi-turbine site measurements. The 30 s data segments included in this average were taken from segments where the 3rd harmonic had a peak prominence that was greater than the 95th percentile for the resampled data set. The average spectra for the Eolos data show strong peaks at harmonics 2-7. The wind farm spectra has fewer and less defined peaks (harmonics 3-6). This is likely caused by multiple turbines and is explained above.

Table 3.2. Summary of the BPF 3rd harmonic infrasound characteristics for 2016 measurements

	2016 EOLOS Turbine on	2016 Wind Farm Turbines on
Number of 30 s Segments Analyzed	4180	4826
Median Level @ 3rd Harmonic (dB)	59.5	62.6
Standard Deviation @ 3rd Harmonic (dB)	5.2	6.0
90th percentile @ 3rd Harmonic (dB)	65.5	70.3
Max Prominence @ 3rd Harmonic (dB)	10-12	Not Reported

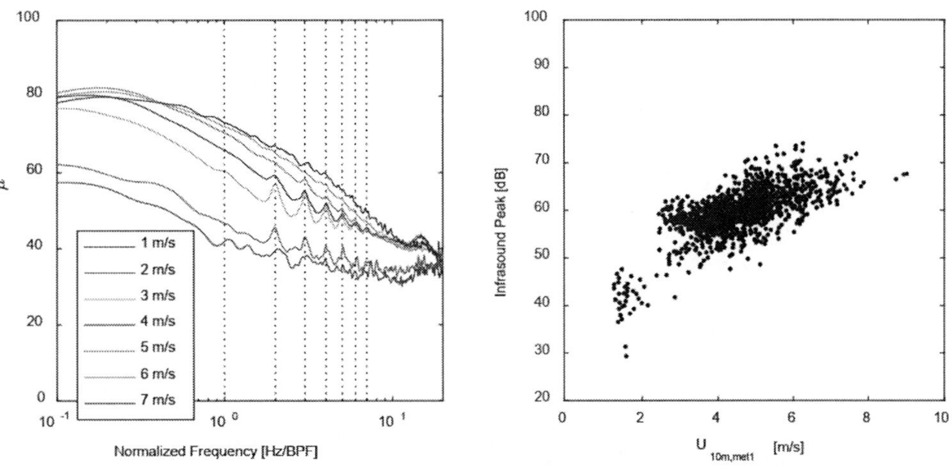

Figure 3.3: (left) Resampled infrasound spectra for the Eolos site, turbine-on, 2016 measurements. Wind speed used for binning is $U_{10m_{met1}}$. Only data segments where a BPF was determined are used in the averaging. (right) Infrasound peak values from the 3rd harmonic plotted against $U_{10m_{met1}}$

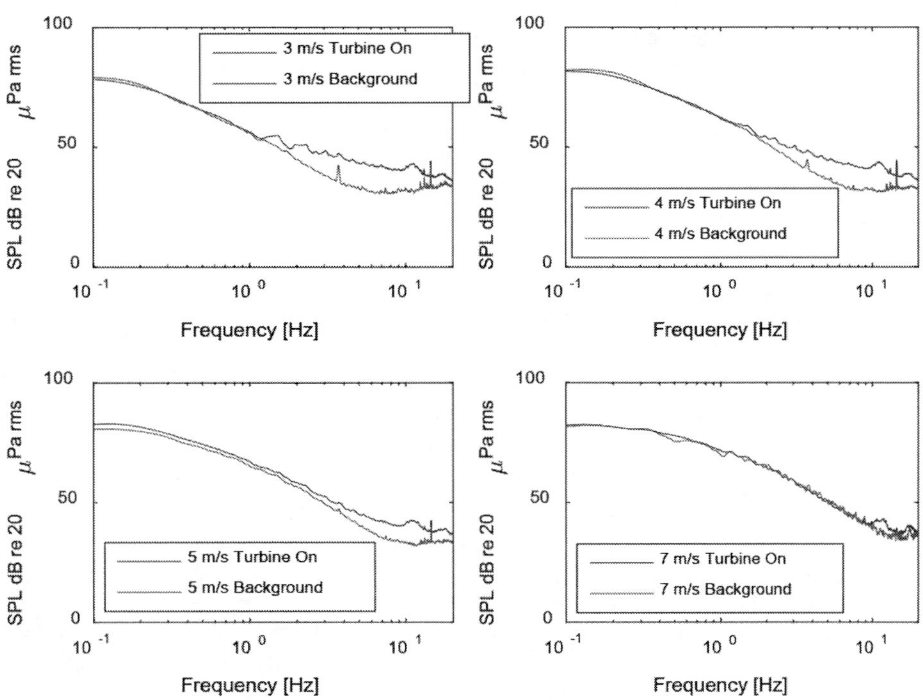

Figure 3.4: Background and turbine-on average infrasound spectra for the Eolos site, 2016 measurements. The spectra are binned by $U_{10m_{met1}}$ wind speed. Averaging is performed on all 30 s data segments.

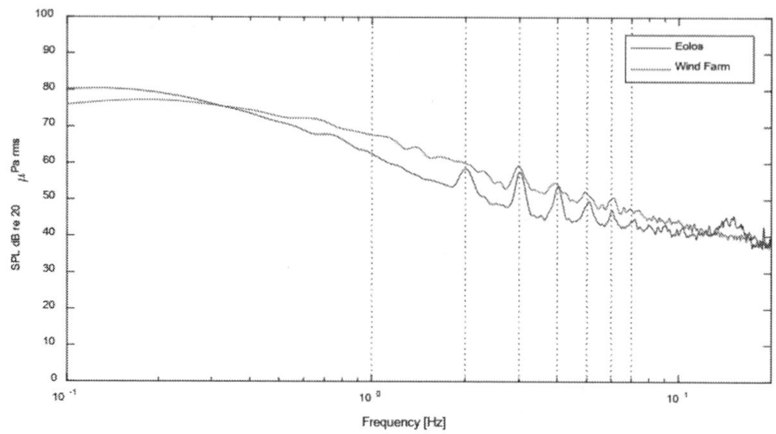

Figure 3.5: Average infrasound spectra for Eolos and wind farm sites using 95th percentile and greater as a threshold for data segments included in the average calculation.

4. Human Response Testing

Human response testing will consist of a target of ~100 healthy people individually subjected to a combination of amplitude modulated audible and infrasound noise files for a short duration in a controlled laboratory environment. Results will consist of postural stability, self-reported detection and rating of intensity of amplitude modulated audible sound and infrasound from turbines, and self-reported symptoms such as nausea. They also will complete pre- and post-testing surveys for symptoms measurements of postural stability. Subjects of final testing will be blind to the knowledge that the stimuli were recorded or derived from acoustic measurements near wind turbines.

As of the writing of this paper, pilot testing has been completed. This included ten healthy adults subjected to amplitude modulated audible and infrasound signals recorded 300 meters from the Eolos turbine. The facilities are the same for pilot testing and the full testing. The stimuli, measurements, and analysis described below were used for pilot testing and are a close representation of what will be used for final testing. The pilot testing results described below are similar in format to what will be presented for final testing.

4.1 Facilities

Each individual will be tested at the Center for Applied and Translational Sensory Science (CATSS) at the University of Minnesota, with testing protocol approved by the University of Minnesota Institutional Review Board. The testing room is a 6' by 15' by 8' room with reproduced audible and infrasound recordings obtained from the Eolos and wind farm recordings.

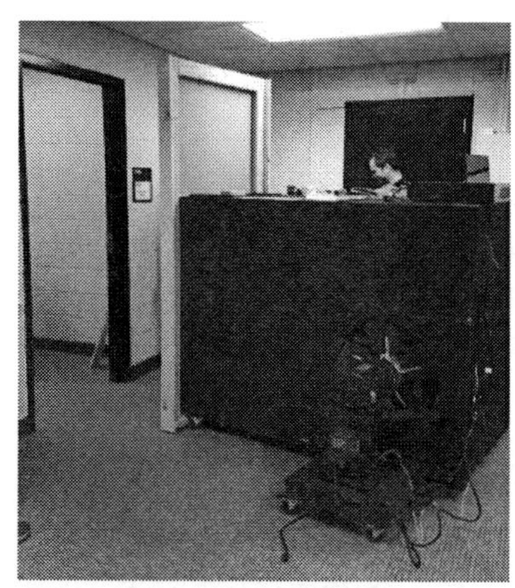

Figure 4.1: Infrasound and audible subwoofer enclosure and testing room entrance. Enclosure pushed up to entrance after subject enters.

Postural stability and sway is measured by having individuals stand on an AMTI balance forceplate which measures left-right and front-back sway continuously.

Infrasound stimuli are generated using an Eminent Technologies© rotary subwoofer with a frequency range of .01 to 30 Hz. Audible stimuli are played through a custom subwoofer with a frequency range of 50 - 800 Hz simulating the audible turbine noise at 300-500 meters.

Audible and infrasound signals were recorded from the Eolos single turbine research site and the wind farm described above. They are recreated in the CATSS lab and noise levels are calibrated using an infrasound microphone (Brüel and Kjær Type 4193 with UC0211 adapter), Infiltec microbarometer, and Brüel and Kjær 2250 sound level meter.

4.2 Pilot Testing Stimuli

Three types of infrasound stimuli were presented: a) no infrasound, b) unaltered recorded infrasound obtained from the wind turbine; and c) spectrally peaky infrasound artificially generated to enhance spectral peaks.

Three types of audible stimuli were presented: a) no audible sound, b) steady-state recorded audible turbine noise, and c) the audible turbine noise with sinusoidal amplitude modulation at the BPF superimposed on the signal with a modulation depth typical of the most modulated recordings from the turbine.

All exposures were randomized and presented with eyes open and eyes closed. Each stimuli was repeated and there are additional two baseline conditions. Each stimulus lasted 40 s, including a 5 s startup, 30 s of recording, and a 5 s ramp down.

4.3 Pilot Testing Human Response Measurements

Measurement of human responses include measurements from the balance forceplate and self-reporting from the individuals. During exposure, postural sway is measured for front-back and left-right and analyzed for the area of the Center of Pressure (CoP) which indicates the amount of movement induced by the stimulus. After each exposure condition, listeners were asked to indicate (Yes/No) whether they detected the acoustic noise of interest and asked to rate the pleasantness/unpleasantness of the sounds using a sliding visual scale from Very Unpleasant to Very Pleasant. Following the full procedure, participants filled out a survey of any symptoms they experienced during the session.

4.4 Pilot Testing Human Response Analysis

The results of the postural sway (CoP) are analyzed using several factors:

- Pre- vs. post- test baseline change in CoP indicating whether the postural stability of the individual significantly changed overall after exposure to the range of stimuli.
- Comparison of CoP during amplitude modulated versus unmodulated audible sounds
- Comparison of CoP during infrasound present versus absent

The results of the self-reporting will be used to for descriptive statistics of detection and symptoms during final testing.

4.5 Pilot Testing Results

Pilot testing postural stability (CoP) for the eyes-closed, front-back stability measure is described below. This was the measure with greatest variability among subjects. The results in the table are preliminary and represent the type of the results reported for actual testing.

Table 4.1. Pilot testing mean front-back Center of Pressure area (cm^2) (postural stability) with eyes closed

Condition	CoP area mean	CoP SD
Pre-test baseline	1.05	*0.18*
Post-test baseline	1.02	*0.16*
Stimuli with AM	1.40	*0.25*
Stimuli with no AM	1.32	*0.21*
Stimuli with infrasound	1.41	*0.25*
Stimuli with no infrasound	1.26	*0.19*

Overall the initial results for pre- versus post- baseline conditions show no effect of stimulus over the full testing procedure. Within the pilot study, little effect of amplitude modulated audible sound versus no AM on sway was observed. Sway with and without infrasound present show a potential trend for the presence of infrasound to increase response. More analysis and increased confidence will be possible with the final testing which has a larger sample size. In the exit questionnaire, two subjects indicated symptoms. In each case they reported mild (rating of 1 on a scale from 0 to 4) experience of fatigue, difficulty focusing, nausea, ear fullness, and general discomfort. No ratings above 1 were experienced, and all other ratings were 0 (not at all.) In the pilot testing, three out of ten subjects indicated that they detected a signal when there was infrasound present.

5. Conclusions

Measurements were recorded at the Eolos single turbine site and a mult-turbine wind farm. Blade passing frequencies were normalized to 1 Hz in order to collapse data into a single dataset. Amplitude modulation calculated using the Institute of Acoustics method (AMWG, 2016) showed no relation to wind shear in the data collected. Infrasound peak levels at BPF harmonics were prominent but well below human perception thresholds and the BPF was fully masked by background infrasound.

Stimulus signals used for final human response testing will be based upon the measurements collected at the Eolos and wind farm site. The characteristics of these recreated signals are based up the levels and prominence of BPF harmonics for infrasound and modulation frequency and depth of a 50 to 800 Hz carrier frequency. The Eolos data on Figure 3.5 is representative of the peaks and prominence of the BPF harmonics that will be used in human testing. Maximum measured AM depth that was audible and recognizable as wind turbine noise was 3.7 dB. The AM depth used for human response testing is yet to be determined.

As of the time of writing, only pilot testing has been completed and full testing will begin in late spring 2017.

Acknowledgements
The authors would like to acknowledge Dan Shannon of United Technologies Research Center who led the 2012 data collection campaign at the Eolos research site, Meredith Adams and Michael Sullivan for input into experimental design of human response testing procedures, and the project technical advisory panel for guidance and feedback.

Project funding is provided by customers of Xcel Energy through a grant from the Renewable Development Fund.

References

AMWG, A. M. (2016). *A Method for Rating Amplitude Modulation in Wind Turbine Noise.* Institute of Acoustics.

Hansen, K., Zajamsek, B., & Hansen, C. (2014). Identification of low frequency wind turbine noise using secondary windshields of various geometries. *Noise Control Engineering Journal.*

International Electrotechnical Commission. (2012). *IEC 61400-11 Wind Turbine Generator Systems—Part 11: Acoustic Noise Measurement Techniques.*

Oerlemans, S. (2014). Effect of wind shear on amplitude modulation of wind turbine noise. *International Journal of Aeroacoustics*, 751-728.

Watanabe, T., & Moller, H. (1990). Low Frequency Hearing Thresholds in Pressure Field and in Free Field. *Journal of Low Frequency Noise Vibration and Active Control, 9*(3), 106-115.

7th International Conference
on
Wind Turbine Noise
Rotterdam – 2nd to 5th May 2017

Wind turbines in hilly terrain – response of residents to sound disturbance related to sound and meteorological measurements

Anna Sjöblom, Uppsala University, Sweden: Anna.Sjoblom@geo.uu.se, Conny Lasson, Uppsala University, Sweden, Conny.Larsson@met.uu.se, Kristina Conrady, Uppsala University, Sweden, Kristina.Conrady@geo.uu.se

Summary

How people perceive sound and noise annoyance from wind turbines is investigated in the project "Human perception of sound from wind turbines in hilly terrain related to sound measurement" running between October 2015 and December 2017. The project will be presented in more detail at the conference, together with the first results from the measurements with a focus on the connection between different weather types and sound intensity.

1. Introduction

Expansion and drift of modern wind turbines in a forest environment are important issues from both a business and a political energy policy perspective. Noise is one of the most important environmental factors for planning and drift of wind turbines and previous a relatively high proportion of residents are disturbed by wind noise, compared to, for example, interference from traffic noise (e.g., Pedersen and Waye, 2007). However, relatively little is known on how sound from wind turbines is perceived in hilly terrain.

2. The project

The project "Human perception of sound from wind turbines in hilly terrain related to sound measurement" is part of the programme Vindval of the Swedish Environmental Protection Agency and is financed by the Swedish Energy Agency between October 2015 and December 2017. Sound diaries coupled to state-of-the-art meteorology and acoustic measurement methods are used along with complimentary modelling approaches to determine how terrain, amplitude modulation, and lee from the wind, affect disturbance by noise and sleeping disorders of local population. How mediating factors, such as personal attitude and sound sensitivity, shape the overall acoustic experience are also examined.

3. Meteorological and acoustical measurements

Three locations in Sweden have been selected, all located in hilly terrain: southern, middle and northern Sweden. In southern Sweden measurements were taken in the winter 2015/2016 while the two other sites focus on the winter 2016/2017.

At all three locations, a weather station with measurements of temperature, wind and humidity at three levels (Figure 1) was placed in the vicinity of the wind farms together with acoustical measurements. Furthermore, the measurements were taken close to residents potentially affected by disturbing sound from the wind farms. The effect of weather on the sound propagation can hence be studied, especially the gradients of wind and temperature. The sound data is filtered in order to exclude background sound, using the same method as Larsson and Öhlund (2014), including filtering out large variations and high frequency sound.

Figure 1. Meteorological measurements in southern Sweden.

4. First results

The equivalent sound pressure level (L_{eq}) from the acoustical measurements in southern Sweden winter 2015/2016 are shown in Figure 2. The displayed data comes from 10 minutes averages from all data (marked as all) and the filtered data (marked as selected). It is clear that the sound level exceeds 40 dbA quite regularly, which require further analysis.

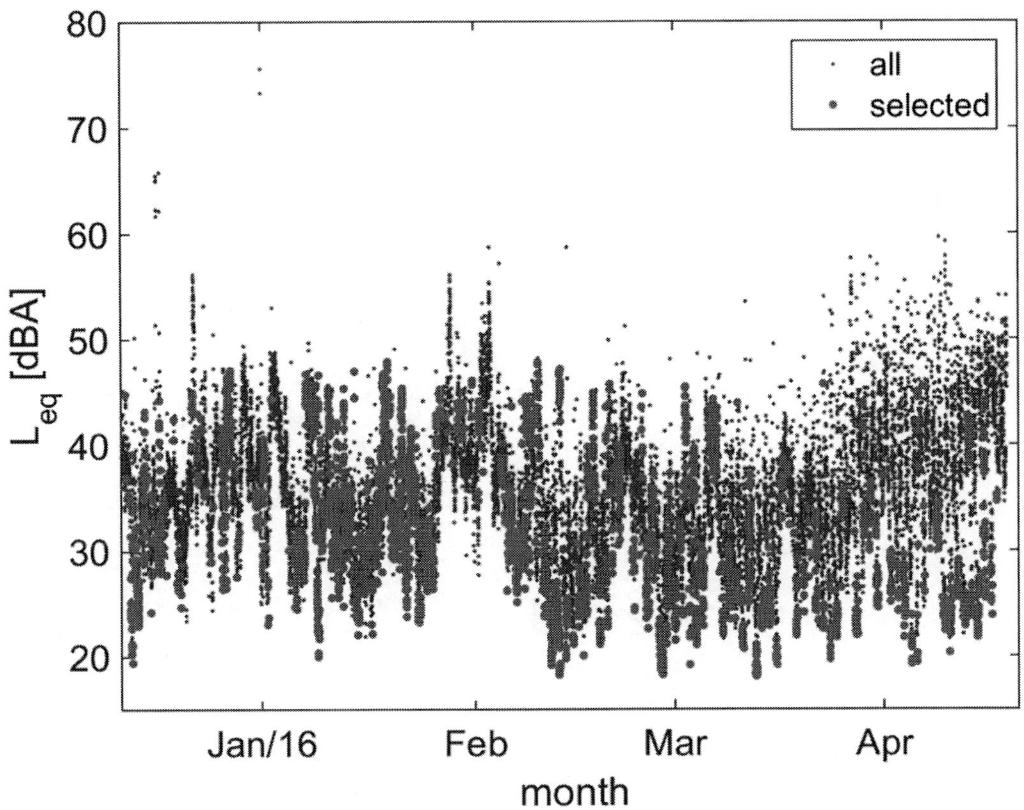

Figure 2. Equivalent sound pressure level (L_{eq}) from the measurements in southern Sweden winter 2015/2016; all data (all) and filtered data (selected).

5. Conclusions

Meteorological and acoustical measurements have been taken at three places in Sweden. Further analysis and results will be shown at the conference, and the influence of weather will be discussed.

References

Larsson, C and Öhlund O (2014) *Amplitude modulation of sound from wind turbines under various meteorological conditions*. The journal of the Acoustical Society of America 135(1).

Pedersen, E and Kerstin Persson Waye K P (2007) *Wind turbine noise, annoyance and self-reported health and well-being in different living environments*. Occupational and Environmental Medicine 64.7, 480-486.

7th International Conference
on
Wind Turbine Noise
Rotterdam – 2nd to 5th May 2017

International Legislation and Regulations for Wind Turbine Shadow Flicker Impact

Erik Koppen, Arcadis Netherlands: erik.koppen@arcadis.com
Mahesh Gunuru, Arcadis Netherlands (formerly): mahesh.mahes@yahoo.in
Andy Chester, Arcadis United Kingdom: andrew.chester@arcadis.com

Summary

When a wind farm is being developed, citizens are often concerned about the effects of shadow flicker which is caused as a result of the rotating turbine blades periodically blocking the sun light. Shadow flicker impacts are often limited by regulations which require the wind turbine is shut down at critical periods when the effects of shadow flicker occur for too long. This may lead to energy production losses depending on the specific situation.

This study presents the results of a comparative study into shadow flicker regulations in a number of countries. The results show not all countries have guidelines or regulations for assessing and limiting shadow flicker impacts. Of those countries that do have regulations or guidelines for shadow flicker impact assessment, most countries have based their regulations on the German Guideline "Hinweise zur Ermittlung und Beurteilung der optischen Immissionen von Windenergieanlagen (WEA-Schattenwurf-Hinweise)" *(Guideline for Identification and Evaluation of the Optical Emissions of Wind Turbines)*. This guideline states a limit value of 30 hours per year and 30 minutes per day for the astronomical maximum possible shadow (worst case). When a shadow flicker control module is used, the German guideline states the real shadow impact must be limited to 8 hours per year. However, there are differences in the exact implementation, like the consideration of only the worst case, only the real case or both the worst and the real case shadow impact. Other common differences are the exact definition of shadow flicker sensitive receptors and the zone of influence which has to be considered. This can lead to considerable differences in energy production losses by a shadow flicker control module. Denmark and the Netherlands have their own specific limit values. The Dutch legislation is most deviating since the limit value comprises a combination of days per year and minutes per day.

1. Introduction

In sunny conditions wind turbines cast a shadow on the neighbouring area. Shadow flicker is the flickering effect caused by the rapid periodic occurrence of shadow by the rotating turbine blades. The impacts of shadow flicker impact vary with time and place depending on several factors such as the position and height of the sun relative to the wind turbines and the receptors, the wind turbine hub height and its rotor diameter, cloud cover and wind direction.

Shadow flicker may cause annoyance depending on how long and how often the effect occurs, the flicker frequency and the contrast. The annoyance mostly occurs inside buildings, where the shadow flicker is perceived through a window opening. Shadow impacts are often limited by regulations stating the wind turbine is shut down at critical periods when the effects of shadow flicker occur for too long. This may lead to energy production losses depending on the specific situation.

This paper is an attempt to identify and compare existing government legislation and guidelines regarding the impacts of shadow flicker. The information is gathered from government websites, government documents, policies, guidelines, and wind farm shadow flicker assessment reports. Since not all information was available in English some details might be lost in the translation. Overall, this paper is believed to be accurate.

2. Shadow Flicker Assessment

When assessing shadow flicker impacts, the worst case and/or real case impacts are determined.

Worst case impact

The worst case shadow flicker impact - the astronomical maximum possible shadow flicker duration - is defined as the shadow flicker duration which occurs when the sun is always shining during daylight hours, the sky is always clear, the wind turbine is always rotating and the rotor is always perpendicular to the receptor areas.

Real case impact

The real case shadow flicker impact – the really expected shadow flicker duration – is the shadow flicker duration when taking into account average sunshine hour probabilities and wind statistics of the particular region.

3. Legislation and Guidelines Governing Wind Turbine Shadow Flicker

3.1 Overview

To give the reader a sense of disparity of wind turbine shadow flicker regulations, an overview is presented in Table 1 summarising the shadow flicker regulations and acceptable threshold limits as published by different countries and their respective jurisdictions.

Most countries that have regulations or guidelines for the impacts of shadow flicker and their assessment have based their regulations on the German Guideline "Hinweise zur Ermittlung und Beurteilung der optischen Immissionen von Windenergieanlagen (WEA-Schattenwurf-Hinweise)" *(Guideline for Identification and Evaluation of the Optical Emissions of Wind Turbines)* [1]. This guideline is described in paragraph 3.2.

The subsequent paragraphs describe the shadow flicker regulations in a selected number of countries in more detail: Australia, Belgium, Denmark, the Netherlands and the United Kingdom. The regulations in the other considered countries are less distinguishing and only listed in the table below.

Table 1: Overview of Shadow Flicker Limit Values and Receptor Locations

Country	Shadow Flicker Limit Values	Receptor Locations	Legislation / Guideline
Australia – National Level [2]	- Worst case: 30 hours/year - No daily limit - Real case: 10 hours/year (only required if worst case exceeds 30 hours/year)	Each dwelling 50m from its centre within distance of 265 x maximum blade chord	Guideline, no legislation at national level
Australia – Queensland [3]	- Worst case: 30 hours/year and 30 min./day - Real case: 10 hours/year (only required if worst case exceeds 30 hours/year)	Each dwelling 50m from its centre within distance of 265 x maximum blade chord	Guideline
Australia – Tasmania [4]	Refers to national guideline	Refers to national guideline	Guideline
Australia - New South Wales [5]	30 Hours/year	Dwellings within 2km distance	Guideline
Australia - Western Australia [6]	Set back distance of 1km	Noise-sensitive buildings not associated with the wind farm	Guideline
Australia – Victoria [7]	30 Hours/year	Dwellings, including garden fenced areas of dwellings	Guideline
Australia - South Australia [8]	Refers to national guideline	Refers to national guideline	Guideline
Austria [9]	Worst case: 30 hours/year and 30 min./day	Sensitive buildings, zone of influence approximately 2000m-2500m	No legislation
Belgium – Flanders Region [10] [11]	Real case: - 8 hours/year and 30 min./day - On industrial sites, with the exception of	Dwellings, hospitals, nursing homes, school buildings, office buildings etcetera	Legislation

Country	Shadow Flicker Limit Values	Receptor Locations	Legislation / Guideline
	dwellings, 30 hours/year and 30 min./day		
Belgium – Walloon Region [12]	Worst case: 30 hours/year and 30 min./day	Dwellings, hospitals, nursing homes, school buildings etcetera	Legislation
Brazil [13]	Worst case (recommended): 30 hours/year and 30 min./day	Sensitive buildings	No legislation, EHS guideline for wind energy World Bank Group
Canada [14]	Worst case: 30 hours/year and 30 min./day	Sensitive buildings	No legislation or guideline, but common practice
Denmark [15]	Real case: 10 hours/year	Dwellings	Guideline
Germany [1]	- Worst case: 30 hours/year and 30 min./day - Real case: 8 hours/year (only required if shadow flicker control system is used)	Living rooms, lounges, bedrooms, classrooms in school buildings, offices, laboratories and workplaces within a distance in which rotor blade covers at least 20% of the sun disk	Guideline adopted by many Federal States
India [16]	Worst case: 30 hours/year and 30 min./day	Dwellings	No legislation or guideline, but common practice
Ireland [17] [18]	Maximum 30 hours/year recommended	Dwellings within 10 rotor diameters distance	Guideline
Japan [19]	30 Hours/year	Dwellings	No legislation, only for EIA purposes
Netherlands [20]	Maximum 17 days per year more than 20 minutes' real case shadow flicker	Dwellings, school buildings hospitals, nursing homes, day-care centres etcetera within a distance of 12 times the rotor diameter	Legislation
Poland [21]	30 Hours/year	Dwellings	No legislation, but common practice

Country	Shadow Flicker Limit Values	Receptor Locations	Legislation / Guideline
Serbia [22]	30 Hours/year and 30 min./day	Dwellings and offices and dwellings within 500m distance	Guideline
Sweden [23]	- Worst case: 30 hours/year and 30 min./day - Real case: 8 hours/year	Sensitive buildings	Guideline
UK – England, Wales [24] [25] [26]	No set limit value, but common practice is maximum 30 hours/year, and 30 minutes/day	Dwellings within zone of 10 rotor diameters from each turbine and between 130 degrees either side of north (relative to each turbine)	Guideline and common practice
UK – Scotland [27]	No set limit value, but as a general rule at distance 10 rotor diameters shadow flicker is not expected to be a problem	Dwellings	Guideline
USA – National Level [28]	30 Hours/year and 30 min./day	Occupied buildings	Guideline
USA - Connecticut [29]	30 Hours/year	Occupied buildings	Legislation
USA – Wisconsin [30]	- 30 Hours/year - Reasonable shadow flicker mitigation when experiencing 20 hours or more per year of shadow flicker	Dwellings and community buildings	Legislation

3.2 Germany

Germany has a detailed guideline for calculating and assessing the impacts of shadow flicker. This guideline "Hinweise zur Ermittlung und Beurteilung der optischen Immissionen von Windenergieanlagen (WEA-Schattenwurf-Hinweise) " *(Guideline for Identification and Evaluation of the Optical Emissions of Wind Turbines) [1]*, was issued by the´Länderausschuss für Immissionsschutz' *(States Committee for Pollution Control)* in 2002. It has since been adopted by many federal states and is considered common practice for wind turbines and wind farms in Germany.

The German guideline states shadow flicker must be considered up to the distance where at least 20% of the sun disk is covered by the rotor blade. At larger distances the shadow flicker will be too diffused to cause annoyance. Further, the shadow flicker is assessed only for sun angles over the horizon of at least 3 degrees. For lower angles the shadow flicker is neglected due to the less bright sun light and screening by vegetation and buildings.

The German guideline considers the following as sensitive rooms:

- living rooms including lounges;
- bedrooms, including overnight rooms in lodges and bedrooms in hospitals and sanatoriums;
- classrooms in school buildings, colleges and similar institutions;
- offices, laboratories, workplaces, training rooms and similar workplaces.

Outdoor areas such as terraces and balconies, adjacent to buildings are considered sensitive areas between 6 a.m. and 10 p.m.

Geographical areas which have been designated for future developments with sensitive rooms shall be assessed at the most critical spots at a height of 2 meter above ground level.

For indoor rooms the assessment height is the window center. For outdoor areas the assessment height is 2 meter above ground level.

The limit values for the worst case - the astronomical maximum possible - shadow flicker impact are:

- 30 minutes per day, and;
- 30 hours per year.

If a shadow flicker control system is used which automatically stalls the wind turbine at the times shadow flicker is expected to occur, the real case shadow flicker impact must be limited to 8 hours per year.

3.3 Australia

National Government

Australia has no national legislation for the impacts of shadow flicker from wind turbines, but in 2010 the Environment Protection and Heritage Council (EPHC) issued a (draft) guideline [2]. This guideline recommends an exposure limit of 30 hours/year modelled (i.e. worst case). There is no limit for daily exposure duration. In most circumstances where a dwelling experiences a modelled level of shadow flicker less than 30 hours per year, no further (real case) investigation is required. In cases where the modelled impacts of shadow flicker are more than 30 hours/year, then the measured shadow flicker (i.e. real case) must be determined. The limit value for the measured shadow flicker is 10 hours/year.

The maximum zone of influence is defined as 265 x maximum blade chord. This means no assessment is required for dwellings beyond this distance. The shadow flicker is assessed only for sun angles over the horizon of at least 3 degrees. The assessment method requires reporting of the maximum value of shadow flicker duration within 50 m of the centre of a dwelling. Depending on jurisdictions, shadow flicker assessment may not be required for participating landowners.

Queensland
The Australian State of Queensland issued planning guidelines in 2016 [3]. This guideline recommends the same limit values and maximum zone of influence as the national guideline.

Tasmania
The Australian State of Tasmania has no legislation or guideline for shadow flicker, but refers to the national guideline [4].

New South Wales
The Australian State of New South Wales also has no legislation for shadow flicker, but did issue a guideline. The impact of shadow flicker should be assessed for dwellings within a 2km distance from a turbine. The shadow flicker duration should not exceed 30 hours per year [5].

Western Australia
The State of Western Australia has no legislation or a guideline for shadow flicker, but recommends a distance of 1km between the turbine and receptors [6].

Victoria
The Australian State Victoria has no legislation for shadow flicker, but did issue guidelines [7]. Victoria recommends a setback distance of 1km from the turbine, unless evidence is provided that the owner of the dwelling has consented in writing to the location of the turbine. The shadow flicker experienced surrounding the area of a dwelling (garden fenced area) must not exceed 30 hours per year.

South Australia
The State of South Australia has no legislation for shadow flicker, but a guideline that refers to the national guideline [8].

3.4 Belgium

Flanders
The Flanders region of Belgium has legislation for regulating shadow flicker impact [11]. The current legislation was implemented in 2012 [10], but was revised in 2016 regarding receptors on industrial sites.

The legislation states a wind turbine should be equipped with an automatic shadow flicker control system if a shadow flicker sensitive receptor is present within an zone experiencing 4 hours per year of expected shadow flicker. The operator is required to keep a log book per wind turbine with the relevant data to determine shadow flicker and for each turbine and relevant sensitive receptors a shadow flicker calendar with the astronomical maximum possible shadow flicker duration. For at least the first two years of operation the operator will draft a report showing the effective shadow flicker for each relevant object per year and detailing the mitigating measures that have been taken.

For dwellings and all other relevant shadow flicker sensitive receptors the limit value is a maximum of 8 hours' effective shadow flicker per year and 30 minutes per day. The only exceptions are shadow flicker sensitive receptors other than dwellings on industrial sites. For these receptors the limit value is a maximum of 30 hours' effective shadow flicker per year and 30 minutes per day.

In order to understand the legislation, expected shadow flicker is the real case shadow flicker impact and effective shadow flicker is the number of hours of shadow flicker at the sensitive object as determined from measurements or the log book of the turbines.

The explanatory memorandum defines a shadow flicker sensitive receptor as an inner space where shadow flicker can cause nuisance. This includes but is not limited to receptors such as dwellings, hospitals, nursing homes, school buildings and office buildings. Further, it states that the expected shadow flicker will be calculated for sun angles over the horizon of more than 3 degrees assuming a standard window size of 5-meter-wide and 2-meter-high at 1 meter above ground level.

The wind turbines have to be automatically halted when they cause an excess of shadow flicker at sensitive receptors, unless it is shown that due to physical reasons no nuisance can occur (e.g. sun blinds installed, screening by receptors or vegetation etc.). Also, the turbines do not need to be stopped if during the shadow flicker period no persons will be present or if individual agreements with private persons can be reached.

Wallonia
The Walloon Region of Belgium has legislation for regulating shadow flicker impact, implemented in 2014 [12]. The astronomical maximum possible shadow flicker is limited to 30 hours per year and 30 minutes per day for dwellings and other sensitive receptors.

3.5 Denmark

Denmark has no legislation on the impacts of shadow flicker, but does have guidance to limit the impact [15]. The Ministry of Environment recommends that the real case shadow flicker impact on dwellings should not exceed 10 hours per year. If this is threatened to be exceeded an automatic shadow flicker control system has to be installed to limit the impact.

3.6 Netherlands

The Netherlands has legislation for regulating the impacts of shadow flicker [20]. The current legislation was implemented in 2007. The legislation states the wind turbine shall be equipped with automatic shadow flicker control system which stalls the turbine if shadow flicker occurs at sensitive receptors and the distance between the turbine and the sensitive receptor is less than 12 times the rotor diameter and if on average shadow flicker occurs more than 17 days per year for more than 20 minutes per day. Shadow flicker is only considered relevant if a sensitive receptor has windows at the side where shadow flicker occurs.

The legislation considers sensitive receptors such as dwellings, school buildings, hospitals, nursing homes, mental institutions, day-care centres etcetera. Receptors like office buildings and hotels are not considered to be sensitive receptors.

3.7 United Kingdom

England and Wales

In England and Wales planning policy for onshore wind turbines is contained in a number of documents, principally the Government's National Planning Policy Framework (NPPF) [24], the National Policy Statement for Renewable Energy Infrastructure [25], and online planning practice guidance for renewable and low carbon energy. Local authorities may also contain policies on onshore wind development in up-to-date local planning policy for a particular area.

The NPPF does not specifically provide guidance on shadow flicker; however, guidance is included within the Planning Practice Guidance for Renewable and Low Carbon Energy [26] document originally published in July 2013. This states that "Only properties within 130 degrees either side of north, relative to the turbines can be affected at these latitudes in the UK".

According to the National Policy Statement for Renewable Energy Infrastructure, in England and Wales the maximum potential number of hours that shadow flicker could occur at each affected occupied building should be calculated, using industry good practice. However, there are no standards set for acceptable exposure limits. Best practice guidance on the interpretation of the significance of effects as a result of shadow flicker on receptors generally references European best practice. As described in paragraph 3.2, Germany references two methods for setting limits as follows [1]:

- An astronomic worst case scenario limited to a maximum of 30 hours per year and 30 minutes on the worst affected day, and;
- A realistic scenario including meteorological parameters limited to a maximum of 8 hours per year.

A significant effect is therefore generally considered to occur where the proposed wind turbine will affect the receptor over substantial parts of the day and/or over the year. This is assumed to be over 30 hours a year, and 30 minutes per day.

Within the UK, there are no nationally set separation distances between wind turbines and housing. Appropriate distances should be maintained between wind turbines and sensitive receptors to protect amenity, and the two main impact issues that determine the acceptable separation distances are visual amenity and noise. The arrangement of wind turbines should be carefully designed within a site to minimise effects on the landscape and visual amenity while meeting technical and operational siting requirements and other constraints. The National Policy Statement for Renewable Energy Infrastructure, in England and Wales sets out that shadow flicker assessment should be undertaken where wind turbines have been proposed within 10 rotor diameters of an existing occupied building".

Some local councils have determined setback distances within their Local Plan's, however as set out in the Department for Communities and Local Government document, Renewable and Low Carbon Energy, local planning authorities should not rule out otherwise acceptable renewable energy developments through inflexible rules on buffer zones or separation distances. Other than when dealing with setback distances for safety, distance of itself does not necessarily determine whether the impact of a proposal is unacceptable.

Scotland

The Scottish Government's document 'Onshore Wind Turbines' states where shadow flicker could be a problem, developers should provide calculations to quantify the effect. In most cases when a separation between wind turbines and nearby dwellings is provided (as a general rule 10 rotor diameters) shadow flicker is not expected to be a problem [27].

4. Conclusion

The study shows not all countries have guidelines or regulations for assessing and limiting shadow flicker impacts. Most countries that do have regulations or guidelines for shadow flicker impact assessment have based their regulations on the German Guideline "Hinweise zur Ermittlung und Beurteilung der optischen Immissionen von Windenergieanlagen (WEA-Schattenwurf-Hinweise)" (Guideline for Identification and Evaluation of the Optical Emissions of Wind Turbines). In countries lacking regulations for shadow flicker the German guideline is often applied as best practice.

The German guideline states a limit value of 30 hours per year and 30 minutes per day for the astronomical maximum possible shadow duration (worst case scenario). In case a shadow flicker control module is used the expected shadow impact (real case scenario) must be limited to 8 hours per year.

There are a number of differences in the exact implementation of the shadow flicker regulations. Some countries and jurisdictions only consider the worst case scenario, sometimes both the impact per year and per day and sometimes just the impact per year. Relatively few countries consider also the real case impact. Also, there are differences in the definition of sensitive receptors and the relevant zone of influence.

All countries and jurisdictions that consider the worst case scenario have set a limit value of 30 hours per year. Those that also consider the impact per day have all set a limit of 30 minutes per day. In the relatively few cases where the real case impact is regulated the limit value for dwellings is 8 hours per year, with the exception of Australia, Denmark and the Netherlands. Australia and Denmark have a recommended limit value of 10 hours per year. The Dutch legislation is most deviating since the limit value comprises a combination of days per year and minutes per day. In the Netherlands, an automatic shadow flicker control system which stalls the turbine is required if on average shadow flicker occurs for more than 17 days per year for more than 20 minutes per day within a zone of 12 times the rotor diameter from the wind turbine.

It must be noted that is not always clear that those countries that have a limit value of 30 hours per year refer to the worst case scenario. Therefore, we cannot exclude that some countries might apply this limit value to the real case impact instead of the worst case impact as intended by the German guideline.

The differences in the exact definition of shadow flicker sensitive receptors and the zone of influence which has to be considered have impacts on the results. This can lead to considerable differences in production losses by a shadow flicker control module. For example, in Germany and the Flanders Region of Belgium office buildings and workplaces are considered sensitive receptors, whilst in a number of other countries like for example the Netherlands these are not considered sensitive. This means that a turbine close to an office building or another workplace can in one country lead to a considerable production loss while in another country there would be no loss at all.

References

[1] Länderausschuss für Immissionsschutz (2002) *Hinweise zur Ermittlung und Beurteilung der optischen Immissionen von Windenergieanlagen (WEA-Schattenwurf-Hinweise)*

[2] Environment Protection and Heritage Council (EPHC) (2010) *National Wind Farm Development Guidelines – Draft*

[3] State of Queensland Department of Infrastructure, Local Government and Planning (2016) *Wind Farm State Code Planning Guideline*

[4] Environmental Protection Authority Tasmania (2017) *Wind Farms*, website http://epa.tas.gov.au/regulation/industrial-activities/regulation-of-industrial-activities/wind-farms, accessed on 4 February 2017

[5] State of New South Wales Department of Planning & Infrastructure (2011) *Draft NSW Planning Guidelines Wind Farms*

[6] Western Australian Planning Commission (2004) *Guidance for Wind Farm Developments* Planning Bulletin 67

[7] Victoria State Government Department of Environment, Land, Water & Planning (2016) *Policy and Planning Guidelines for Development of Wind Energy Facilities in Victoria* 2015, 12, 20, 31, 41

[8] Central Local Government Region of South Australia (2014) *Wind Farm Development Guidelines for Developers and Local Government Planners*

[9] Gschiel, E, Amt der Burgenland Landesregierung (2014) *Umweltmedizinische Begutachtung am Beispiel Windkraft*

[10] Vlaamse overheid - Departement Leefmilieu, Natuur en Energie (2012) *Toelichtingsnota nieuwe milieuvoorwaarden voor windturbines*

[11] Vlaamse overheid - Departement Leefmilieu, Natuur en Energie (2017) *VLAREM II*, Art. 5.20.6.2.1-5.20.6.2.3

[12] Gouvernement wallon (13 February 2014) *Arrêté du Gouvernement wallon portant conditions sectorielles relatives aux parcs d'éoliennes d'une puissance totale supérieure ou égale à 0,5 MW*

[13] World Bank Group (7 August 2015) *Environmental, Health and Safety Guidelines Wind Energy*, 12-13

[14] Natural Forces Wind Inc (June 2013) *Gaetz Brook Wind Farm Shadow Flicker Assessment Report*

[15] Danish Government - Miljøministeriet Naturstyrelsen (2015) *Vejledning om planlægning for og tilladelse til opstilling af vindmøller,* 19-20

[16] India Centre for Science and Environment (2013) *EIA Guidelines Wind Power Sector*, 10

[17] Ireland Government Department of Environment (2013) *Wind energy Development Guidelines*

[18] Irish Wind Energy Association (2017) *Environmental Impacts*, website http://www.iwea.com/index.cfm/page/environmentalimpacts#q72 accessed on 5 February 2017

[19] Japanese Government Ministry of the Environment (2010), *Summary Report of Working Group Regarding Basic Concept on Environmental Impact Assessments Related to Wind Power Generation Facilities*, 11

[20] Nederlandse overheid - Ministerie van Volkshuisvesting, Ruimtelijke Ordening en Milieubeheer (2017) *Regeling algemene regels voor inrichtingen milieubeheer*, Art. 3.12

[21] GEO Renewables S.A. (2014) *Korytnica Wind Farm Non - Technical Summary*, 20

[22] United Nations Development Programme Serbia & Ministry of Environment and Spatial Planning of the Republic of Serbia (2010) *Guidelines on the Environmental Impact Assessment for Wind Farms,* 24

[23] Boverket (2009) *Vindkraftshandboken Planering och prövning av vindkraftverk på land och i kustnära vattenområden*, 36-38

[24] UK Government Department for Communities and Local Government (March 2012) *National Planning Policy Framework*

[25] UK Government Department of Energy and Climate Change (July 2011) *National Policy Statement for Renewable Energy Infrastructure (EN-3)*
[26] UK Government Department for Communities and Local Government (July 2013) *Planning practice guidance for renewable and low carbon energy*
[27] Scottish Government (24 October 2012) *Onshore wind turbines*
[28] National Association of Regulatory Utility Commissioners (NARUC) Grants & Research Department (January 2012) *Wind Energy & Wind Park Siting and Zoning Best Practices and Guidance for States*
[29] State of Connecticut Regulation of Connecticut Siting Council (February 2010) *CSC Wind Regulations*, 9-12
[30] Public Service Commission of Wisconsin *(December 2012) Wind Energy Systems, PSC 128.15 Shadow flicker*

7th International Conference
on
Wind Turbine Noise
Rotterdam – 2nd to 5th May 2017

The development and limits of the German shadow flicker guidelines

Peter Ritter pr@emd.dk; emd Deutschland GbR, Kassel, Germany
Timo Mertens t.mertens@cube-engineering.com, CUBE Engineering GmbH, Kassel, Germany

Summary

Shadow flicker from wind turbines is an immission for residents living near wind farms. The residents have to be protected against health critical immissions. Therefore the assessment of potential shadow flickering is part of the environmental impact assessment of a wind farm. A detailed guideline was established in Germany between 1997 and 2006 to avoid critical health disturbances which are already adopted in other countries. The guideline is well proven and since applying the guideline in Germany no complaint with significant health effects occurred. Mitigation measures which have been taken at operational wind farms such as turbine shut down strategies, have been proved to be very successful, so that shadow flicker cannot be considered to be a major issue anymore. From this point of view the guideline is a success story for establishing residential friendly recommendations. This paper describes the background of the development of the guideline and practical experiences with shadow flicker from wind turbines.

Introduction

Shadow flicker from Wind turbine generators (WTG) may occur when the rotating blades of a wind turbine pass through the sun's rays seen from a specific location. This creates a moving shadow with a flicker effect. This impact varies spatially and temporally and depends on a number of environmental conditions such as weather, topography, or the distance between the turbine and the receptor. Residents in case of the shadow flickering effect are the indoor and outdoor areas of houses and other buildings, where the people stay for a while.

Figure 1: Shadow from a wind turbine generator

With the growth of the rotor diameter to more than 40 m and appropriate hub heights the first complaints to the local authorities on shadow flicker occurred in 1997. Before this time the distance from WTGs to residents was determined by the noise immissions of WTGs because shadow flicker effects were not relevant. Due to the fact that larger WTGs create disturbing shadow flickering in distances where the noise immissions are within the limits, the shadow flicker was then identified as a relevant immission from wind turbines and was included in the permitting process for wind farms.

During this phase an expert group was founded in cooperation with the enviornmental agency of Schleswig Holstein in 1997 and started to discuss and define guidelines for shadow flicker. To identify the degree of daily and annual disturbance two research studies were initiated. In the first study several residents influenced by shadow flicker from wind farms were interviewed to estimate the disturbance in comparison to prognoses of the shadow flicker. In the second study the continuously shadow flicker influence on the human productivity was analysed to assess the acceptable daily duration.

Based on these two studies the German shadow flicker guideline was developed and fixed in the expert group over a period of more than 10 years. With the input from various experts from research, industry and public authorities the guideline "WEA-Schattenwurf-Hinweise" for prognoses and permitting was defined including recommendations and limits of shadow flicker. Finally, all federal states in Germany adopted the guideline for the wind farm permission process.

1. Physical basics of shadow flicker from wind turbines

For a better understanding of the shadow flicker effect, some physical basics about the formation of shadow will be described.

1.1 Sun radiation

The sun radiation is depending on the position of the earth to the sun. During one year with small deviation (leap years) the earth rotates in an ecliptic orbit around the sun. The rotating

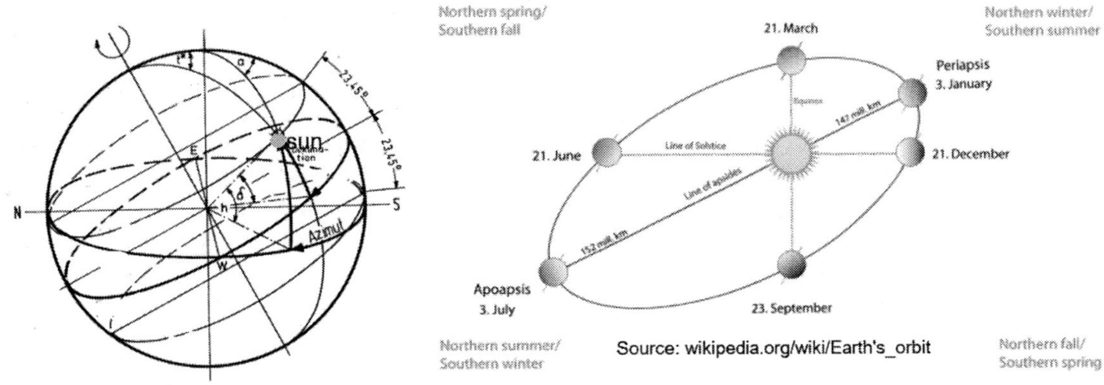

Figure 2: Sun radiation and earth orbit

axis of the earth is aslope to the orbit and the distance of the to the sun varies between 147 mill. and 152 mill. kilometres.
Caused by the aslope axis and the rotation of the earth around the sun the change of the seasons exists at the 21th of March, 21th of June, 23th of September and 21th of December. The position of the sun from the earth view is normally described with sun height (declination) and the cardinal direction (azimuth). The location of the earth and the radiation depending on location, date and time is well known and described in mathematical formulas in several publications and is programmed in software tool libraries.

1.2 Shadow flicker

Shadow flicker from WTG for the residents begins, if the WTG rotor is located in between the sun and the building. The flicker varies spatially and temporally and depends on environmental conditions like the position and height of the sun, wind speed, wind direction and cloudiness. The probability of shadow flickering occurrence and the extent of its effect on the residents depend on the factors such as the direction of the windows relative to the turbine, the distance from the turbine, the turbine hub height and the rotor diameter, the speed of blade rotation, the time of year and the time of day.

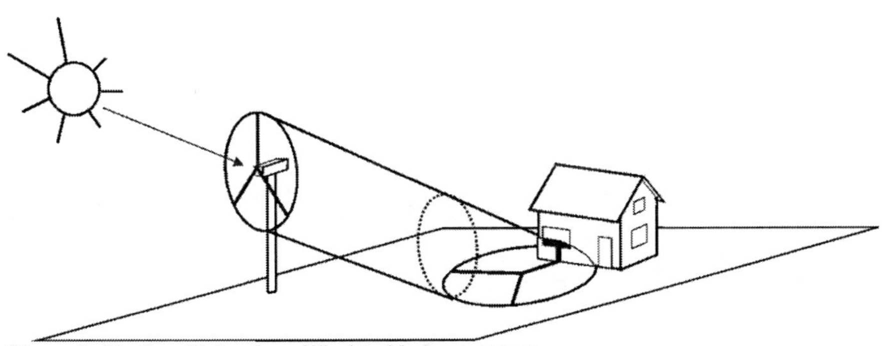

Figure 3: Shadow flicker from Wind turbine generators

The frequency of the shadow flicker depends on the rotation speed of the turbines. Normally the tip speed at the blades of the WTG is more or less similar. So with higher rotor diameter the rotor speed will decrease. E.g. the Enercon E-40-6.44 with 800kW operates with 18- 34 rpm and the new Enercon E141 with 4 – 11 rpm. The shadow flicker frequency for the E-40 are between 0,9-1,7 Hz and the E-141 between 0,2 to 0,5 Hz. The frequency is below the critical frequency of 2.5 Hz for epileptic seizures.

1.3 Shadow flicker intensity

The shadow intensity is defined by the part of the sun which will be covered by the blade. Based on the definition the expert group agreed that a relevant shadow influence area exists if more than 20% of the sun is covered from the average blade width. The average blade width is defined as medium value of the maximum blade width and the blade depth at 90% of the rotor length. Smaller WTG e.g. Enercon E-40 from the beginning in 1998 had a shadow influence area of 835m. New WTG e.g. the Enercon E141 have a shadow influence area of 1,835m.

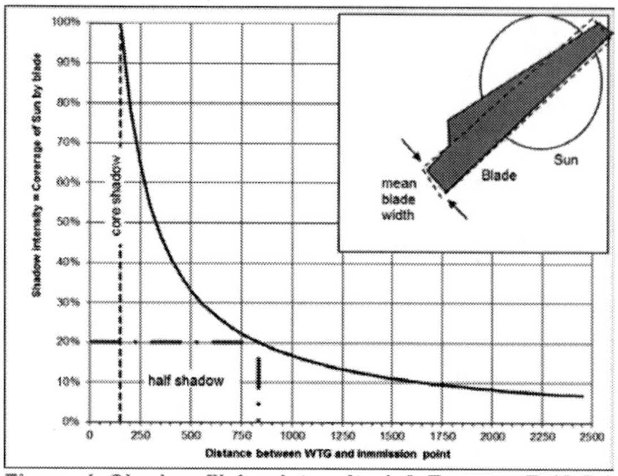

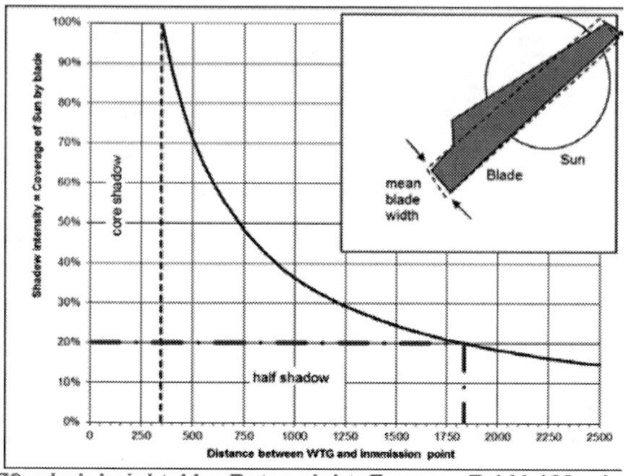

Figure 4: Shadow flicker intensity, left Enercon E-40-6.44 78m hub height 44m Rotor, right: Enercon E-141 129m hub height 141m Rotor

1.4 Cumulated shadow flicker and relevant shadow area

Depending on the geometric constellation of the WTG to the immission point the duration and frequency of the flicker varies a lot. Some areas around the WTG are not affected by the shadow flicker at all, e.g. south of the WTG in northern parts of the earth, because the sun in this regions never shines from the north. But also in the winter time at the middle of the day the shadow distance is limited by the fact that the sun rises to a minimum high.

Additionally the cumulated period of shadow flickering of more than one WTG is relevant to assess the shadow flicker disturbance around a wind farm. The cumulated shadow flicker comprises the counted shadow flicker minutes over one year at one immission point(shadow receptor). The result of annually counted shadow flicker can be shown in a map with iso-lines of the shadow impact.

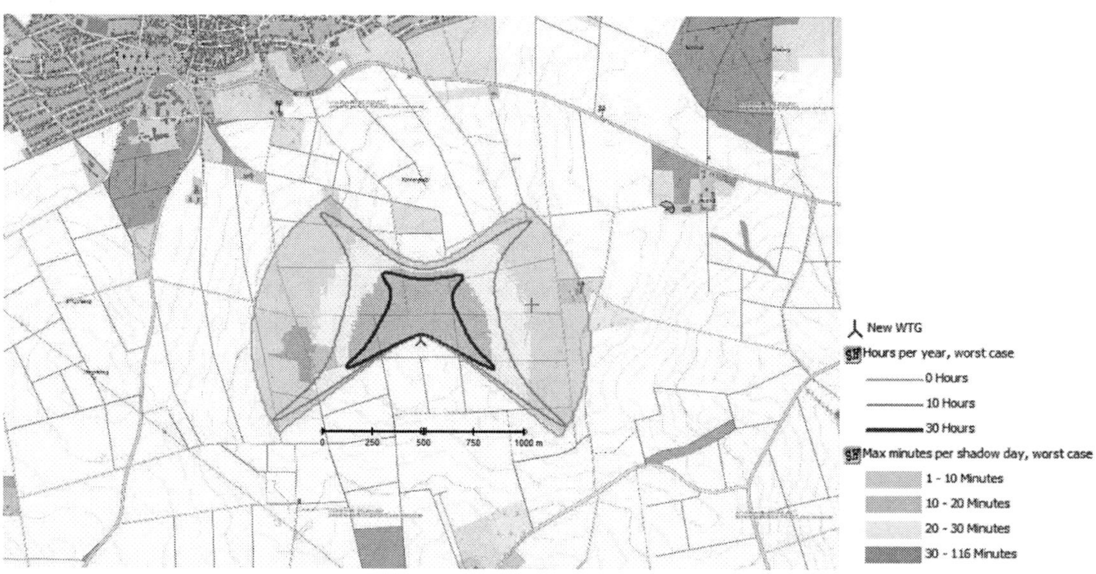

Figure 5: relevant Shadow areas around a WTG Enercon E-40-6.44 78m hub height, 44m rotor diameter

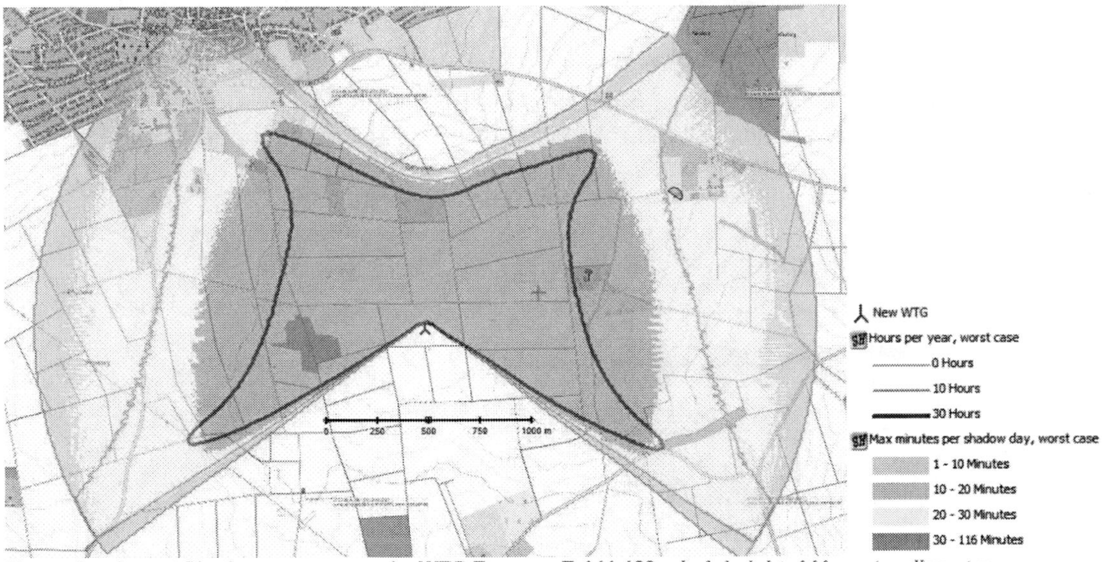

Figure 6: relevant Shadow areas around a WTG Enercon E-141 129m hub height, 141m rotor diameter

2. The guideline

Germany was the first country which defined a guideline for shadow flicker of WTGs.

2.1 Development guideline

Caused of first complaints from residents the engage with the disturbance of shadow flicker of wind turbines were started. The local authority of the city Schleswig in the northern part of Germany started to group some experts to analyse and to discuss the shadow flicker in 1997. The group defined the goal to generate practical requirements for the protection of the residents around WTGs. Similar to other immissions such as noise- and odour nuisance, it was planned to establish a guideline for shadow flickering. The main topics discussed were:

- Significant and relevant critical shadow flicker,
- Limit for shadow flickering over a year,
- Limit for shadow flickering over a day,
- Relevant shadow flickering conditions (shadow intensity, shadow flicker distance, weather conditions for flicker,
- Protected areas and buildings (shadow receptors),
- Comparable prognoses, requirements and physics.

Over several years the practical use of the guideline was approved and additional recommendations and further tests were established. Since the erection of the first WTG in Germany more than 25 years ago, no complaint with health effects is known. Therefore no adjustments on the guideline were necessary since 2006 and the guideline was adopted in further countries.

2.2 Working group for shadow flicker

In 1997 the expert group for envioronmental impacts of WTG "Arbeitskreis Umwelteinwirkungen von Windenergieanlagen" was founded. The members of the group belonged to research institutes, authorities, manufacturers and software companies, meteorological experts, consultants and medical scientists. The group met several times over a period of several years until all topics were addressed and the guideline was well proven and finally adjusted in 2006.

2.3 Studies and analyses for shadow flicker in Germany

To clearly determine the stress caused by shadow flicker the Psychologic Institute of the University of Kiel was assigned to make detailed interviews with residents (number of 223) around wind turbines[i]. The main focus was to identify the level of disturbance and the significant critical limit, above which the health of the people was affected during one year.

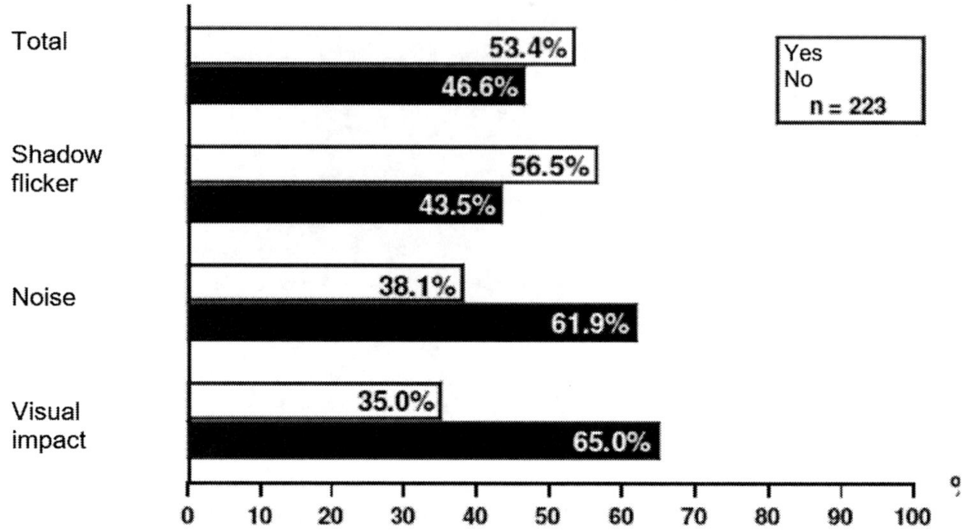

Figure 7: Feedback of disturbance of WTG caused of different effects

Based on the results of the detailed questioning significant critical and less critical locations where identified. The annual possible shadow flicker was calculated by modelling and simulations. The comparison of both methods showed a correlation between the significant disturbance and the calculated shadow flicker.

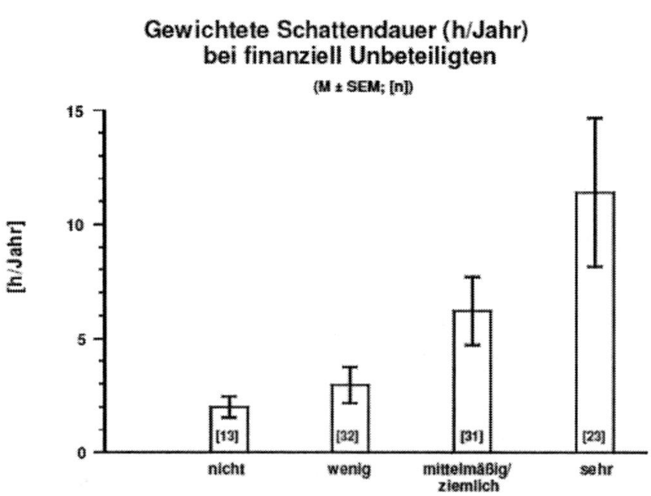

Figure 8: correlation of disturbance and shadow flicker

Based on these results, the limit of 30h annual shadow flicker, which correspond to 8h/year real shadow flicker, was defined by the expert group.

To define the second limit for the daily maximum shadow flicker a second study of the same institute was conducted[ii]. In this study the same share of men and women from a group of students (number of 32 with a mean age of 23 years) and other adults (number of 25 with a mean age of 47 years) were studied. Half of the group was exposed to shadow flicker while the other half only sat in a laboratory for a time of 60 minutes. The test persons had to work and fulfil different tasks (e.g. calculations and visual search tasks) at a PC on a desk exposed to shadow flicker. Their working power and mental and physical condition, the cognitive stress processing and the vegetative nervous system (heart rate, blood pressure, skin conductivity and finger temperature) were controlled and analysed during a total period of 120 minutes. In the group exposed to shadow flicker the shadow flicker was activated after 40 minutes for a period of 60 minutes.

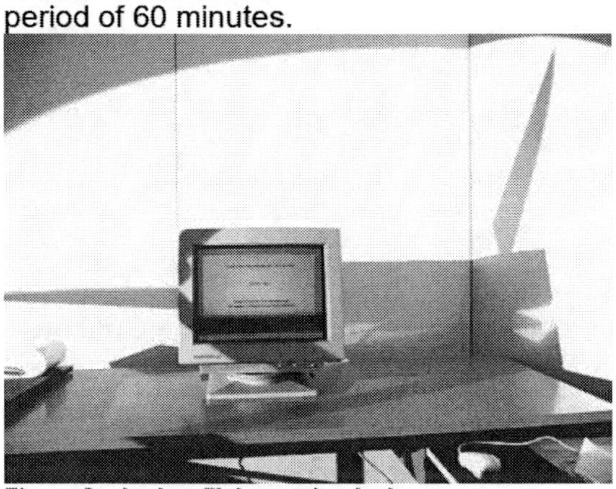

Figure 9: shadow flicker at the desk

Table 1: main result of shadow flicker in the laboratory

	1 period No shadow flicker	2th period Shadow flicker	3th period Shadow flicker	4th period, After shadow flicker
	1.-20. min	21.-40. min	41.-60. min	
disturbance students professionals		 reduced constant	 constant constant	
Cognitiv stress handling students professionals	 Slide moderate	 Slide strong	 slide moderate	
feeling students professionals	 declined	 improved declined	 improved	 improved
performance students professionals	 declined declined		 improved improved	 declined
physiological reactions students professionals	 increased	 increased increased	 specific increased increased	 specific increased increased

The results gave an indication about the stress factor resulting due to shadow flickering. The study showed that under specific conditions periodic shadow did not constitute a significant disturbance. However, the documented increased demands on mental and physical energy indicated that cumulative long-term effects might meet the criteria of a significant nuisance.

Young people could better cope with this stress and could improve their work performance in stress situations. Older persons were not able to compensate this stress and their performance was reduced.

Based on these results the expert group decided to set the maximum daily limit of shadow flickering.

2.4 Assumptions

To handle, validate and compare the shadow flicker effects and calculations some assumptions were defined in the expert group:

- Sensible difference in brightness of the shadow flicker is 2,5%;
- There is no shadow flicker with sun heights over the horizon less than 3 degree;
- The assessment of the shadow flicker has to be made on worst case conditions. Worst case means, the sun is always shining during the daylight hours, the rotor is always spinning and the rotor circle is always vertical to the sun radiation. The calculated immissions have to be calculated for one point, if normal windows are affected from the shadow flicker.
- relevant shadow flicker exist only with sunshine at radiation of more than 120W/m².

2.5 Limits

The limits of shadow flickering impact for a neighbour to a wind farm according to the German guidelines are:

- a maximum of 30 hours per year of astronomical maximum possible shadow (worst case);
- a maximum of 30 minutes per day of astronomical maximum possible shadow (worst case);
- if an automatic shadow regulation is installed, the real shadow impact must be limited to 8 hours per year.

If one of these thresholds is exceeded, mitigation methods such as turning off turbines during critical times must be considered. Some wind farm operators offer a zero shadow immission strategy. They switch the turbines always off during shadow flicker on residential areas.

2.6 Comparison of worst case and real case limits (8 hours per year)

To compare the worst case shadow flicker values with the real shadow flicker the German Weather Service DWD estimated the mean real sunshine hours out of the annual limit of 30h. Approximately 30% will be hours with relevant radiation intensity for flickering.

2.7 Protected areas for shadow flicker

To define relevant shadow receptors, the expert group decided that short-term used rooms and areas without any relaxing phases are not necessarily to be protected against critical shadow flicker.

The following relevant immission areas are definedas protected areas:

- living rooms,
- sleeping rooms,
- class rooms or similar rooms for education,
- office, working places and places for commerce,
- relaxing areas like balconies or terraces lying directly close to the building.

3. Shadow flicker prognosis

As part of the development and planning process for a wind farm, computer models are used to predict and quantify the impact of shadow flickering around the wind farm. The results of the calculation can be included in the environmental assessment of the wind farm. One of the leading computer packages which are used in the industry to model the shadow flicker is windPRO with its module SHADOW. To get permission for erecting a wind turbine a shadow prognosis has to be delivered to the permitting authority. A detailed approach to generate the shadow prognosis is given in the above described guideline.

3.1 Recommendations

To make a shadow flicker prognosis, local conditions have to be known. Besides the exact coordinates and sizes of the WTG (hub height, rotor diameter, rotor blade width), the residential areas around the planned WTG have to be known. The consultant visits the site to document each critical effected area and building. The basis for such site visit is a calculated shadow map of the site which shows all the relevant shadow flicker areas. In combination with local detailed maps the critical buildings and areas will be evaluated and documented by photos.

3.2 Software tools

The SHADOW flicker module of windPRO calculates the occurrence of shadow flicker impact time and intervals from one or more turbines for receptors at given locations. In addition, a map of the cumulated annual and daily shadow can be generated. Beside the calculation of the shadow flickering for a certain point (as the German guideline demands), a calculation for defined windows, rooftops or facades can be conducted. The worst case scenario or a real case scenario based on weather statistics can be calculated. The software tools simulate for each hour in the year the shadow constellation, so for full areas a large number of calculations have to be done and it need some time for calculation.

If the weather condition is overcast or calm, or if the wind direction forces the rotor plane of the WTG to be parallel to the line between the Sun and the neighbour, the WTG will not produce shadow impacts, but the impact will still appear in the calculations. In other words, the calculation is a worst-case scenario, which represents the maximum potential risk of shadow impact. A calendar can be printed for any specific point of observation, which indicates the exact days, and time periods where shadow impact may occur.

Apart from calculating the potential shadow impact at a given neighbour, a map presenting the iso-lines of the shadow impact can also be printed. This printout will render the amount of shadow impact for any location within the affected area.

3.3 Typical results of shadow flicker prognoses

The result of a typical shadow flickering prognosis is a table with the cumulated maximum shadow flicker over one year and the maximum minutes per day at the different receptors (immission points) (Figure 10).

Calculation Results

Shadow receptor

No. Name	Shadow, worst case: Shadow hours per year [h/year]	Shadow days per year [days/year]	Max shadow hours per day [h/day]	Shadow, expected values: Shadow hours per year [h/year]
A Immission point, shadow receptor	9:07	34	0:24	1:29
B Immission point, shadow receptor	30:43	55	0:43	5:23

Total amount of flickering on the shadow receptors caused by each WTG

No. Name	Worst case [h/year]	Expected [h/year]
1 ENERCON E-141 EP4 4200 141.0 !-! hub: 129,0 m (TOT: 199,5 m) (27)	39:50	6:52

Figure 10: results of a shadow calculation with windPRO

Additionally, a graphical shadow calendar will be presented (Figure 11) to get a perfect overview of the shadow flicker.

SHADOW - Calendar, graphical

Calculation: Receptors E-141

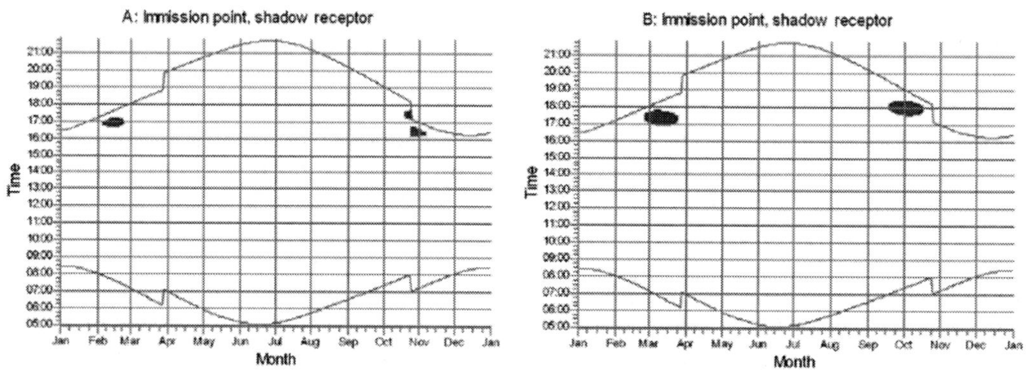

Figure 11: graphical calendar of shadow flicker at results (windPRO)

3.4 IT technologies in the WTG and wind farms to avoid Shadow flicker

To avoid critical significant shadow flicker of more than 30 Minutes per day and 8h per year the WTG could be equipped with a shadow control system. All manufacturers offer either their own shadow control management system or a system in cooperation with an external provider . The shadow control management takes care of the switch off of a WTG via a light sensors and an integrated timer calendar. The light sensors recognize a defined sunlight brightness level and the internal clock checks the time. If limit values are exceeded, the wind power generator is stopped for the necessary period of time. Instead of the automatically switch off system in some cases it is possible to equip concerned neighbours with a remote control to switch of the turbine in the critical shadow flickering periods. This avoids unnecessary stops when the concerned people are not at home or in the affected rooms during the shadow flicker.

4. Conclusion

Shadow flicker of wind turbines is a relevant issue for the development and operation of a wind farm. As a part of the environmental assessment a shadow analysis has to be made for the permission. With the requirements of the German guideline, the practical experience over several years, the existing software tools and shadow flickering management systems in wind turbines the challenge of shadow flicker could be handled in a clear and practical way. Depending on the local residential situation, the annual yield reduction caused by shadow flicker control management is in most cases very low and does not influence the feasibility of a wind project.

References

P. Brinckerhoff, A. Court, W. A. Drive, Department of Energy and Climate Change, *Update of UK Shadow Flicker Evidence Base*,

[i] Dr. J. Pohl, Dr. F. Faul, Prof. Dr. R. Mausfeld (1999), *Belästigung durch periodischen Schattenwurf von Windenergieanlagen, Feldstudie*, Universität Kiel
[ii] Dr. J. Pohl, Dr. F. Faul, Prof. Dr. R. Mausfeld, 2000, *Belästigung durch periodischen Schattenwurf von Windenergieanlagen, Laborpilotstudie*, Universität Kiel

7th International Conference
on
Wind Turbine Noise
Rotterdam – 2nd to 5th May 2017

The visual effects of wind turbines in Japan

Takashi Yano, Kumamoto University, yano@gpo.kumamoto-u.ac.jp
Sonoko Kuwano, Osaka University, kuwano@see.eng.osaka-u.ac.jp
Hideki Tachibana, The University of Tokyo, pon-t@iis.u-tokyo.ac.jp

Summary

The impact of wind turbines has been a serious social problem in Japan since 2000s. The Ministry of the Environment initiated a nationwide survey on the effects of wind turbine noise from 2010 to 2012 across Japan. In total 747 responses were obtained. The distance from the nearest wind turbines to respondents ranged from 90 to 1466m and the noise exposure ranged from 26 to 50 dB $L_{Aeq,n}$. The aural effects of wind turbines on people have mainly been investigated and published in international journals [S. Kuwano et al. (2014), T. Kageyama et al. (2016)]. However, wind turbines affect people around wind farms both aurally and visually. This study focuses not only aural but also visual effects of wind turbines on communities. In the social survey, visibility, visual annoyance, and shadow flicker were asked. The relations between the distance and the visual effects were analysed with logistic regression analysis. The most significant relation was found between the distance and the shadow flicker in the garden. The distance corresponding 10% of people who claimed shadow flicker in the garden was 860 m (95% CI; 770-1020). The distance corresponding 10% very annoyed by wind turbine noise were 780 m (95%CI; 660-1060). These findings imply that the setback distance should be considered from aural and visual points of view. The effects of factors moderating shadow flicker and noise annoyance are also discussed.

1. Introduction

For the global sustainability, green energy has strongly been required in the world. As a useful candidate, the construction of wind power plants has strongly been promoted in Japan since 2000. More than 2000 wind turbines had been constructed by 2014. As wind turbines produce not only electricity but also emit noise, people living in the vicinity of wind farms complain against noise.

Therefore many studies have been conducted across the world, mainly in developed countries, to investigate the effects of wind turbines on communities. Accumulating the findings, several reviews have also published [1-4]. They have reached similar conclusions. For example, McCunney et al. [3] critically reviewed and concluded that epidemiological studies have shown the association between living near wind turbines and annoyance and that infrasound and low frequency sound do not present unique health risks.

To properly perform the wind turbine noise policies in Japan, a research project entitled "Research in the evaluation of human impact of low frequency noise from wind turbine generators" had been conducted for three years from 2010 to 2012, funded by a grant from the Ministry of the Environment, Japan. A series of physical measurements, laboratory psychological experiments, and social surveys were conducted and the findings were published in journal articles [5-10]. The social survey found the association of wind turbine noise (abbreviated as WTN hereafter) with annoyance and sleep effects but not with health effects. Michaud et al. [11] recently conducted a large scale survey on the health effects of WTN in Canada. They also found the association of WTN with annoyance but not with sleep and health effects. Both studies have reached the almost same conclusions as McCunney et al.

In the Japanese study the effects of WTN were mainly discussed [9, 10]. However, wind turbines affect communities not only aurally but also visually. Knopper et al. [2] reviewed scientific papers on wind turbines and human health, and summarized the risk of seizures by shadow flicker (abbreviated as SF hereafter). They suggested that SF does not induce seizures in people with photosensitive epilepsy. Even if SF does not yield serious health effects on people, it may bother, disturb or annoy them. Pedersen and Larsman [13] constructed structural equation models showing how visual attitudes affect noise annoyance.

In this study, the association between WTN and annoyance is discussed further than the former paper [9]. That is, simple relationships between distance and noise annoyance or visual effects are shown, and the setback distance is discussed from aural and visual point of view, based on the data set obtained in the aforementioned survey.

2. Method

2.1 Social survey

Social surveys were conducted at 34 sites near wind turbine generators (wind turbine site) and at 16 control sites which have similar characteristics to wind turbine sites but not affected by wind turbine generators across Japan from Hokkaido to Okinawa Prefecture. Most of the sites were rural areas where the houses were sparsely located. The interviewers visited each house and, if they could get the informed consent from the residents, asked each question with face-to-face interview method. The questionnaire is based on that proposed by Namba et al. [13] and modified for this survey. It contains 12 questions on length of residence, satisfaction with residential area, annoyance caused by noise, sleep quality, demographic variables, attitude to wind turbine, and health effects measured by THI (Total Health Index) [14]. The items on residential areas are convenience of shopping, convenience of transportation, amount of greenery, clean air, quietness, and public facilities.

Annoyance caused by road traffic, aircraft, Shinkansen train, conventional train, factory, construction, and wind turbine noises was evaluated by ICBEN (or ISO) 5-point verbal scale [15, 16], phrased as "Thinking about the last 12 months or so, when you are here at home, how much does each noise listed below bother or annoy you? Please choose the appropriate number."

1. not at all
2. slightly
3. moderately
4. very
5. extremely
9. inaudible

The visual effects of wind turbines were asked with the following questions:

"Can you see any wind turbine generators from your home?

1. yes
2. no

"If you can see the wind turbine generators from your house, are they disturbing the landscape?"
1. no problem
2. disturbing

"Does the shadow flicker reflect in your house?"
1. yes, in the garden
2. yes, in the house
3. no

When the respondents select "2" in the last question, it is considered that the SF also reflects in the garden.

2.2 Noise measurement and estimation

During the same period as social surveys, noise recordings were performed at the 34 sites for successive five days with sound level meters specially designed to measure sound pressure from 1Hz to 20 kHz. The microphones were put at 20cm high from the ground with double wind screens. The rated generation power of the wind turbines under investigation was from 400 kW to 3,000 kW, mainly larger than 1,500 kW.

Since the effects of WTN on people were serious particularly during night time as shown in section 3.2, the time-averaged A-weighted sound pressure level of WTN under a rated operation condition during night time from 22:00 to 6:00 ($L_{Aeq,n}$) was obtained as the energy-mean of the time-averaged A-weighted sound pressure level over 10 min of every hour.

The noise reduction in distance was obtained by logarithmic regression equation formulated with noise levels measured at eight points each site and the WTN exposure levels of individual residences were estimated from the regression curve using the distance from the nearest wind turbine. In such a way, 651 data in total were obtained, ranging from 26 to 50 dB.

3. Results

3.1 Demographic variables

The number of respondents per site was about 21 on average, ranged from 3 to 42. The total number of respondents was 747 at the wind turbine sites and 332 at control sites. The responses could be obtained from 49 % at wind turbine sites and 45 % at control sites.

The respondents were 357 males and 387 females (3 no inscription) at wind turbine sites. Males (48%) and females (52%) were well balanced. About 80 % were aged their fifties or above. About 85 % of the respondents had been living for more than 10 years at the same places where they lived when the survey was conducted and about 25 % were engaged in agriculture, fishery or forestry, about 40 % had no job or housewives, and the rest were university students, owners of their own shops or company employees, etc.

In order to investigate the visual effects, the corresponding physical factor should be the visual angle of wind turbines from residents. However, as it is difficult to identify, the distances from the residents to the nearest wind turbines were used in this study. Table 1 shows the distributions of sex and age in the four distance categories: less than 400, 400-599, 600-799, 800 or above. The distributions of sex and age were almost consistent among the categories. The majority is older people: more than 60% is 60s or older.

Table 1 Frequency of demographic variables (relative frequency)

Distance	Sex		Age					
(m)	Male	Female	-20s	30s	40s	50s	60s	70s-
-399	63 (44)	81 (56)	6 (4)	7 (5)	19 (13)	21 (15)	48 (33)	43 (30)
400-599	107 (52)	99 (48)	8 (3)	14 (7)	21 (10)	37 (18)	56 (27)	70 (33)
600-799	98 (49)	102 (51)	7 (4)	15 (8)	15 (8)	30 (15)	65 (33)	68 (34)
800-	89 (46)	105 (54)	7 (4)	12 (6)	16 (8)	36 (19)	48 (25)	75 (39)
Total	357 (48)	387 (52)	28 (4)	48 (6)	71 (10)	124 (17)	217 (29)	256 (34)

3.2 Satisfaction with residential areas and noise annoyance

The distributions of satisfaction with residential environment are typical in Japanese rural areas and similar between wind turbine and control sites: people satisfied natural environment such as greenery, clean air, and quietness but complained somewhat to social environment such as shopping, transportation, and public facilities.

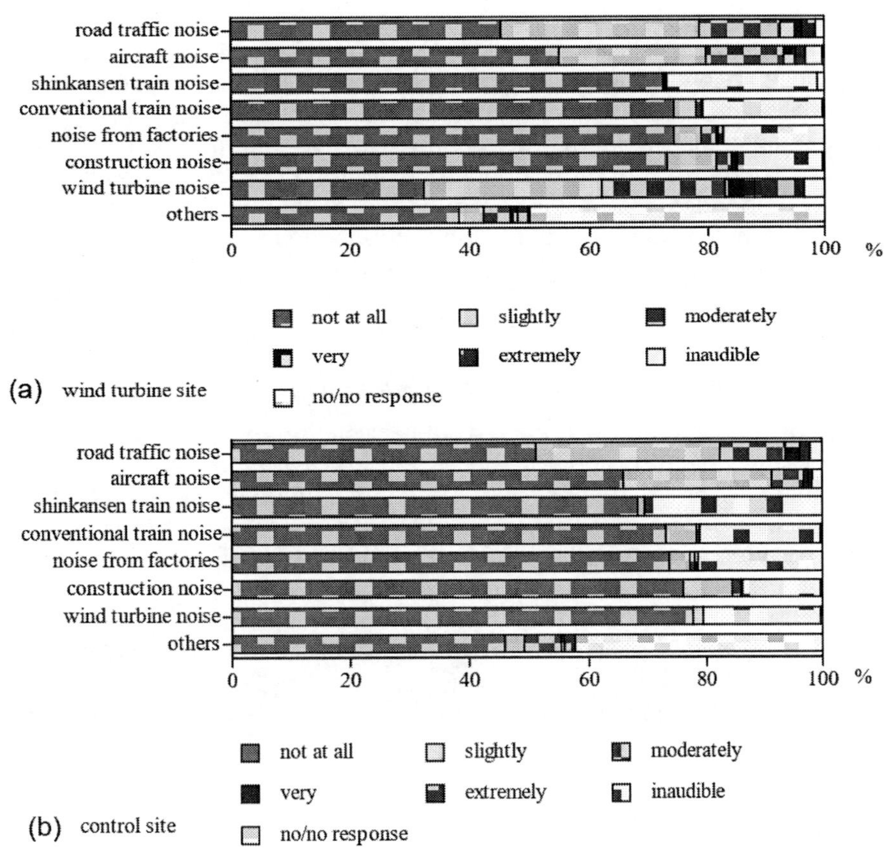

Fig.1 Annoyance of respective noises at wind turbine and control sites [9]

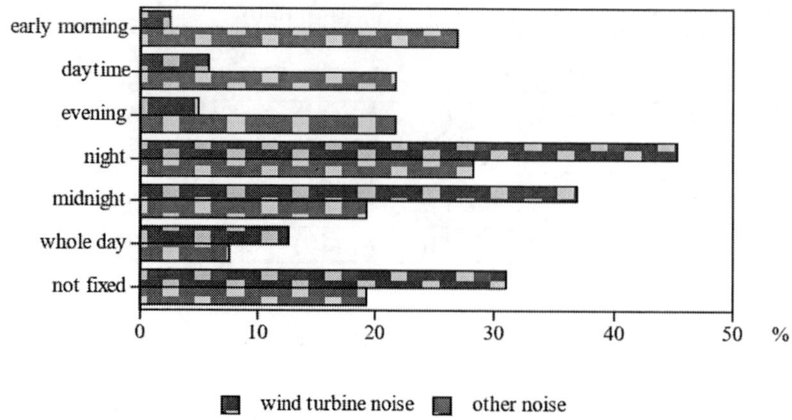

Fig.2 Comparison of annoyed period between respondents who chose WTN as most annoying and those who chose the other noise [9]

Fig. 1 shows the distributions of annoyance response to various environmental noises. Except WTN the distribution patterns are similar between wind turbine sites and control sites: people were not annoyed by Shinkansen, conventional railway, factory, and construction noises but somewhat annoyed by road traffic and aircraft noises. This indicates that road traffic and aircraft noises were the main components of background noise. At wind turbine sites more than 30 % of respondents were moderately or more annoyed by WTN. Fig. 2 compares the annoyed period in a day between respondents who answered that WTN was most annoying and those who answered that the other noise was most annoying. Though the respondents who chose the other noise had no specific period annoyed by noise, those who chose WTN were annoyed by noise during night and midnight. That is why $L_{Aeq,n}$ is used as the noise index for WTN in this study [9].

3.3 Relationships between exposure/distance and noise annoyance

The exposure-response relationships were obtained with logistic regression analysis hereafter. Table 2 and Fig. 3 show the results of logistic regression analysis and the relation between $L_{Aeq,n}$ and prevalence of annoyance, respectively. Percent extremely annoyed (abbreviated as % EA), percent extremely or very annoyed (%VA) and percent extremely or very or moderately annoyed (%MA) were used as the noise annoyance indices. All the coefficients of $L_{Aeq,n}$ were statistically significant ($p<0.05$).

Table 3 and Fig.4 show the results of logistic regression analysis and the relation between distance and prevalence of annoyance, respectively. All the coefficients of distance were statistically significant ($p<0.0001$). Distance was much more associated with noise annoyance than noise exposure.

Table 2 Parameter estimate for logistic regression models between $L_{Aeq,n}$ and prevalence of noise annoyance

Model	Parameter	Estimate	SE	*p*	95% Lower CI	95% Upper CI
EA	Intercept	-6.0363	1.2052	<0.0001	-8.4769	-3.7421
	$L_{Aeq,n}$	0.0942	0.0293	0.0013	0.0377	0.1530
VA	Intercept	-4.4640	0.9461	<0.0001	-6.3622	-2.6477
	$L_{Aeq,n}$	0.0689	0.0234	0.0032	0.0237	0.1154
MA	Intercept	-2.2044	0.6712	0.0010	-3.5346	-0.9004
	$L_{Aeq,n}$	0.0425	0.0169	0.0121	0.0095	0.0759

Table 3 Parameter estimate for logistic regression models between distance and prevalence of noise annoyance

Model	Parameter	Estimate	SE	*p*	95% Lower CI	95%Upper CI
EA	Intercept	-0.7905	0.3491	0.0236	-1.4793	-0.1080
	Distance	-0.0028	0.0006	<0.0001	-0.0041	0.0016
VA	Intercept	-0.7731	0.2840	0.0065	-1.3331	-0.2182
	Distance	-0.0018	0.0005	<0.0001	-0.0028	0.0009
MA	Intercept	0.1749	0.2079	0.4001	-0.2314	0.5842
	Distance	-0.0013	0.0003	<0.0001	-0.0019	-0.0007

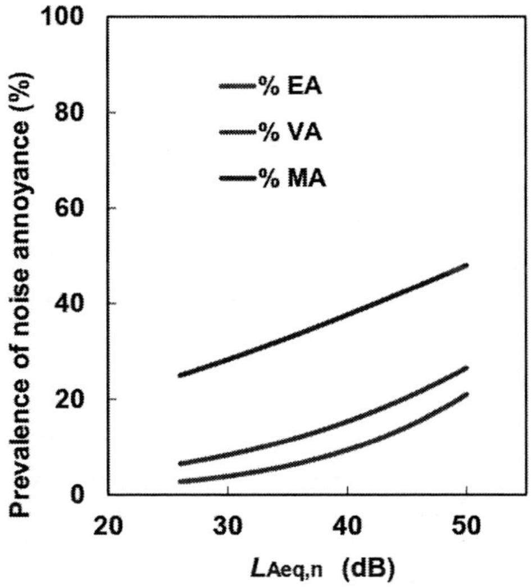

Fig.3 Relationships between $L_{Aeq,n}$ and prevalence of annoyance [9]

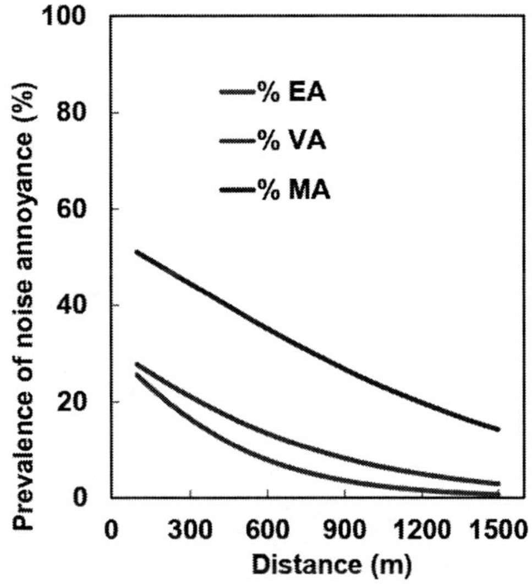

Fig.4 Relationships between distance and prevalence of annoyance [9]

3.4 Relationships between distance and visual effects

Table 4 and Fig.5 indicate the results of logistic regression analysis and the relation between distance and prevalence of visual effects, respectively. The coefficient of distance for visual annoyance was not significant ($p>0.05$) but it was strongly significant for SF outdoor ($p<0.0001$) and significant for SF indoor ($p<0.05$). Though visual annoyance and SF indoor gradually decreased with distance, SF outdoor sharply decreased. It is reasonable that SF is much more noticeable outdoor than indoor and that visual annoyance may be influenced by not only distance but also the topography and the other factors.

Table 4 Parameter estimate for logistic regression models between distance and visual effects of wind turbines

Model	Parameter	Estimate	SE	*p*	95% Lower CI	95% Upper CI
Visual annoyance	Intercept	-1.4699	0.3147	<0.0001	-2.0959	-0.8604
	Distance	-0.0009	0.0005	0.0621	-0.0019	0.000028
SF outdoor	Intercept	0.4085	0.2588	0.1144	-0.0957	0.9199
	Distance	-0.0030	0.0004	<0.0001	-0.0039	-0.0022
SF indoor	Intercept	-1.7533	0.4191	<0.0001	-2.5904	-0.9435
	Distance	-0.0018	0.0007	0.0118	-0.0032	-0.0004

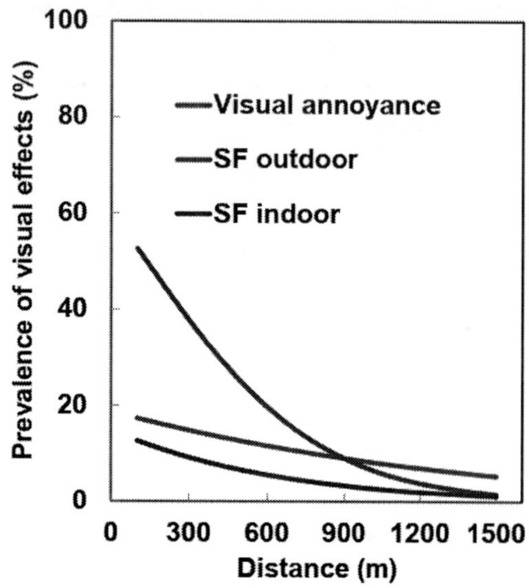

Fig.5 Relationships between distance and prevalence of visual effects

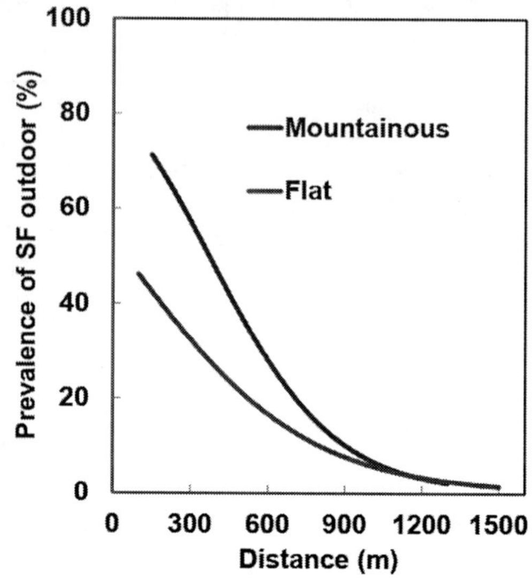

Fig.6 Comparison of distance-SF outdoor relationships between mountainous and flat areas

3.5 Multiple logistic regression analysis of SF outdoor

Multiple logistic regression analysis was applied using SF outdoor as the dependent variable and distance, sex, age, topography (mountainous or flat), and interaction between distance and topography as independent variables. The classification of topography is based on researchers' observation: 10 sites are mountainous and others are flat. The results are shown in Table 5. The coefficients of distance and topography are statistically significant, $p<0.0001$ and $p=0.0036$, respectively. Sex and age are not significant. The interaction between distance and topography is not significant and this suggests that the relation between distance and SF outdoor has the similar trend between mountainous and flat areas. However, since topography is significant, the relation between distance and SF outdoor should be considered separately between mountainous and flat areas as shown in Fig. 6. When the distance is shorter, people in mountainous area notice SF more easily than those in flat area. The odds ratio of SF outdoor between mountainous and flat areas was significant only in the range from 400-599 m (0.22, 95%CI: 0.10-0.45) and thus there seems to be no difference in SF outdoor between the two areas more than 600 m away from wind turbines.

Table 5 Parameter estimate for multiple logistic regression models of SF outdoor

Model	Parameter	Estimate	SE	*p*	95% Lower CI	95%Upper CI
SF outdoor	Intercept	1.3332	0.5361	0.0129	0.2839	2.3904
	Distance	-0.0033	0.0005	<0.0001	-0.0043	-0.0024
	Sex (M:0, F:1)	0.1712	0.1951	0.3802	-0.2104	0.5554
	Age	-0.0682	0.0672	0.3102	-0.1985	0.0654
	Topography (Mountain:0, Flat:1)	-0.6508	0.2238	0.0036	-1.0915	-0.2114
	Distance * Topography	0.0012	0.0010	0.2159	-0.0007	0.0031

3.6 Multiple logistic regression analysis of WTN annoyance

Multiple logistic regression analysis was applied to "very annoyed or not (VA)" as the dependent variable with stepwise backward elimination of independent variables: $L_{Aeq,n}$, sex, age, topography, and visual effects (visibility, visual annoyance, SF outdoor, or SF indoor), and interaction between $L_{Aeq,n}$ and topography or visual effects. The results are shown in Table 6.

Table 6 Parameter estimate for multiple logistic regression model of VA with stepwise backward elimination

Model	Parameter	Estimate	SE	*p*	95% Lower CI	95%Upper CI
VA	Intercept	-7.0686	1.3343	<0.0001	-9.7954	-4.5499
	$L_{Aeq,n}$	0.1350	0.0347	0.0001	0.0690	0.2056
	Topography (Mountain:0, Flat:1)	-1.5050	0.3313	<0.0001	-2.1878	-0.8804
	Visual annoyance (not:0, disturbed:1)	1.9412	0.2845	<0.0001	1.3858	2.5041
	SF outdoor (not:0, yes:1)	0.7498	0.2787	0.0071	0.1985	1.2943
	$L_{Aeq,n}$ * Topography	0.1668	0.0632	0.0083	0.0455	0.2941

The remaining significant variables are $L_{Aeq,n}$, topography, visual annoyance, SF outdoor, and the interaction between $L_{Aeq,n}$ and topography. Visual annoyance is much correlated with noise annoyance (Pearson's Chi square test, $p<0.0001$) and thus the significance of visual annoyance is very reasonable. SF outdoor is very significant for noise annoyance. The main effects of visibility, SF indoor, and all interactions between $L_{Aeq,n}$ and visual effects are not significant. The last one indicates that $L_{Aeq,n}$-%VA relationships are not affected by the visual effects. The interaction between $L_{Aeq,n}$ and topography was significant. This means that $L_{Aeq,n}$ affected annoyance more in flat areas than in mountainous areas.

4. Discussion

It is frequently reported that WTN is more noticeable during night time than day time as shown in Fig 2. This is because background or residual noise level is lower during night time than day time. The background or residual noise level is very critical for the assessment of WTN. In some countries the regulation or guideline values for WTN is determined relative to residual noise levels based on the classification of areas [17]. For example, the guideline value in New Zealand is higher one of 35 dB or residual noise level plus 5 dB in rural area and of 40 dB or residual noise level plus 5 dB in the other areas. Japanese government will establish the similar guidelines for WTN [18].

Japanese environmental quality standard for road traffic noise during night time is 45 dB in residential area. Kuwano et al. [9] compared L_{dn}-%Highly Annoyed (%HA) relationships between WTN and road traffic noise. WTN is more annoying than road traffic noise and there is 9 to 6 dB penalty for WTN in the range from 10 to 20 %HA. This finding may provide a reasonable evidence to show the consistency among the guidelines for environmental noises, particularly between WTN and road traffic noise in Japan.

In Fig. 6 the effect of SF outdoor is greater in mountainous areas than in flat areas. Since wind turbines are usually constructed at peaks in mountainous areas, the altitudes are higher in mountainous areas than flat areas. Thus the shadows are clearly reflected on the ground in mountainous areas. That may be why the SF outdoor is the most associated with distance in Table 4.

When the distance corresponding to 10% SF outdoor in Fig.5 is calculated, it is 860 m (95% CI: 770-1020). When the distances in mountainous and flat areas are separately calculated from Fig. 6, they are 900 m (95% CI: 800-1100) and 800m (95% CI: 690-1040), respectively. Since both curves are consistent each other in the range more than 900 m in Fig. 6, totally and separately calculated distances are almost the same.

As well as SF outdoor, when the distance corresponding 10 % VA is calculated from Fig.4, it is 780 m (95% CI: 660-1060). In the same way the distance corresponding 20 % SF outdoor and 20 % VA are 590m (95% CI: 530-670) and 340 (95% CI: 60-460), respectively. The INCEJ report [19] summarizes not only noise limit but also setback distances in various countries and states in detail. The latter is diverse, ranging 300-1000 m. The present study supports the setback distance in various countries and states. However, which noise annoyance index and visual effect should be used and how much prevalence is taken as criteria should be discussed further.

From the multiple logistic regression analysis, $L_{Aeq,n}$, topography, visual annoyance, SF outdoor, and the interaction between $L_{Aeq,n}$ and topography are significant. The topography itself is reflected on the noise exposure, and the effect of topography is considered to be that of residual noise level. That is, the residual noise level may be higher in flat areas than mountainous areas because of more transportation noise and wave sound near coasts. The effects of topography are considered to be caused by residual noise. Since $L_{Aeq,n}$–noise annoyance relationships are significantly affected by topography, the effects of residual noise on WTN annoyance should be systematically investigated further to enforce the evidence of guidelines for WTN. SF outdoor also significantly affected noise annoyance. This indicates that wind turbines may be regulated not only aurally but also visually.

5. Conclusions

This study re-analysed the data from the nationwide survey on the effects of wind turbines noise in Japan from aural and visual points of view and concluded as follows:

1) Comparing exposure-annoyance relationships between WTN and road traffic noise provide a reasonable evidence for their guidelines.
2) Shadow flicker outdoor is the most associated with distance of the visual effects.
3) Shadow flicker outdoor is more prominent in mountainous area than flat area because of higher latitude.
4) The distance corresponding to 10 % shadow flicker outdoor is 860 m. This is almost the same as the distance corresponding 10 % VA, 780m.
5) Though visibility and shadow flicker indoor did not affect noise annoyance significantly, visual annoyance and shadow flicker outdoor significantly affect noise annoyance.
6) This study supports the setback distances in different countries and states. However, which noise annoyance index and visual effect should be used and how much prevalence is taken as criteria should be discussed further.

6. Acknowledgements

The authors appreciate Drs. Makoto Morinaga and Shigenori Yokoshima for their valuable comments on the statistical analysis. This study was financially supported by Environment Research and Technology Development Fund (S2-11) by the Ministry of the Environment, Japan.

References

[1] Jeffery, RD and Krogh, CME (2014) Industrial wind turbine and adverse health effects Can J Rural Med 19(1), 21-26

[2] Knopper, LD, Ollson, CA, McCallum, LC, Aslund, MLW, Berger, RG, Soueine, K, and McDaniel, M (2014) Wind turbine and human health Frontiers in Public Health 63, 1-20

[3] McCunney, RJ, Mundt, KA, Colby WD, Dobie R, Kaliski, K, and Blais, M (2014) Wind turbine and health: A critical review of the scientific literature JOEM 56(11), e108-e130

[4] Schmidt, JH and Klokker, M (2014) Health effects related to wind turbine noise exposure: A systematic review PLOS one

[5] Tachibana, H, Yano, H, Fukushima, A, and Sueoka, S (2014) Nationwide field measurements of wind turbine noise in Japan Noise Control Engr. J. 62(2), 90-101

[6] Yokoyama, S, Sakamoto, S, and Tachibana, H (2014) Perception of low frequency components in wind turbine noise Noise Control Engr. J. 62(5), 295-305

[7] Sakamoto, S, Yokoyama, S, Yano, H, and Tachibana, H (2014) Experimental study on hearing threshold for low frequency pure tones Acoust. Sci. & Tech. 35(4), 213-218

[8] Yokoyama, S, Sakamoto, S, Tsujimura, S, Kobayashi, T, and Tachibana, H (2015) Loudness experiment on general environmental noise considering low-frequency components down to infrasound Acoust. Sci. & Tech. 36(1), 24-30

[9] Kuwano, S, Yano, T, Kageyama, T, Sueoka, S, and Tachibana, H (2014) Social survey on wind turbine noise in Japan Noise Control Engr. J. 62(6), 503-520

[10] Kageyama, T, Yano, T, Kuwano, S, Sueoka, S, and Tachibana, H (2016) Exposure-response relationship of wind turbine noise with self-reported symptoms of sleep and health problems: A nationwide socioacoustic survey in Japan Noise & Health 18, 53-61

[11] Michaud, DS et al (2016) Exposure to wind turbine noise: Perceptual responses and reported health effects J. Acoust. Soc. Am. 139(3), 1443-1453

[12] Pedersen, E and Larsman, P (2008) The impact of visual factors on noise annoyance among people living in the vicinity of wind turbines J. Environmental Psychology 28, 379-389

[13] Namba, S et al. (2010) Proposal of fundamental items for social survey on noise problems Acoust. Sci. Tech. 31(2), 124-128

[14] Aoki, S, Suzuki, S, Yanai, H (1991) Development of a new health questionnaire, the Total Health Index, as a tool for quantitative evaluation of perceived physical and mental health "Method and Application of mental Health Surveys: the Todai Health Index" edited by Suzuki, S and Roberts, S Tokyo University Press, 59-87

[15] Fields JM et al. (2001) Standardized general-purpose noise reaction questions for community noise surveys–research and a recommendation J. Sound Vib., 242(4), 641-679

[16] ISO TS 15666 (2003) Acoustics-Assessment of noise annoyance by means of social and socio-acoustic surveys

[17] Tachibana, H (2015) Guidelines for wind turbine noise in foreign countries J. Acoust. Soc. Jpn., 71(4), 198-205 (in Japanese)

[18] Nameki, M, Shimada, A, Machida, N, Tachibana, H (2017) Evaluation of wind turbine noise in Japan Proc. of Wind Turbine Noise 2017

[19] INCEJ (2016) Survey report on future sound environment and method of noise and vibration regulation: the effects of wind turbine noise (in Japanese)

7th International Conference
on
Wind Turbine Noise
Rotterdam – 2nd to 5th May 2017

Presenting insights from shadow flicker compliance monitoring.

Peter Longbottom, Visualwind: peter@visualwind.co.uk
James Mackay, TNEI: james.mackay@tnei.co.uk

Summary

At-property shadow flicker monitoring campaigns have not historically been implemented at wind farm sites due to the understanding that shadow flicker issues can be resolved in the site design / planning stage and through adoption of shadow flicker mitigation strategies. However, complaints relating to shadow flicker have persisted and led to the need for such compliance monitoring. Technology for monitoring shadow flicker at receptor properties has only recently been developed and deployed. Results from these campaigns is discussed and the primary causes of unexpected flicker are presented.

1. Introduction

Shadow flicker from wind turbines has received significantly less attention than wind turbine noise. It is relatively easy to predict when shadow flicker will occur, given the location and dimensions of properties and nearby turbines. In theory, therefore, using shutdown modules, flicker can be eliminated.

However, despite careful predictions and mitigation measures, unexpected flicker is still being observed. Flicker is occurring at unexpected times, being observed from turbines that are significant distances away, and in some cases, is being unexpectedly observed at low light levels.

In 2014 the first at-property shadow flicker monitoring campaign was undertaken, using the WindComply system's high resolution light sensors and wide angle video cameras. Since then, flicker has been monitored at several sites in the UK and northern Europe. Comparison of observed results with predictions reveals a complex picture involving multiple factors such as the accuracy of prediction model input data, the impact of distance between turbine and property on flicker intensity, the relationship between light levels and flicker intensity, and complexities in shutdown logic.

An accurate record of flicker times is an important step in resolving issues. However, there is a further challenge; although some national guidance specifies an upper limit on hours of flicker per year, it lacks discussion about what constitutes unacceptable flicker. What light intensity levels and variation mark the boundary between acceptable and unacceptable flicker? Analysis of sensor and video data is adding to a growing understanding of the relationship between light intensity and the nature of shadow flicker.

In this paper, insights into the primary causes of unexpected flicker and thoughts on establishing the boundary between acceptable and unacceptable flicker will be discussed, with some additional thoughts on how the methods used in long-term cloud-connected flicker campaigns can be applied in the field of noise monitoring.

2. Developing technology for flicker monitoring

There are three major objectives for long term remote monitoring. These are capture of qualitative and quantitative light data, implementation of a power system for long-term independent operation and automated data transfer to a cloud server. The challenges of delivering each of these is discussed below.

2.1 Capturing quantitative and qualitative light data

Monitoring shadow flicker involves recording the variation in light intensity over time. Constant monitoring of light levels enables a full record of all shadow flicker events. Shadow flicker is observable as a regular pattern in the light intensity times series plot (see Fig. 1).

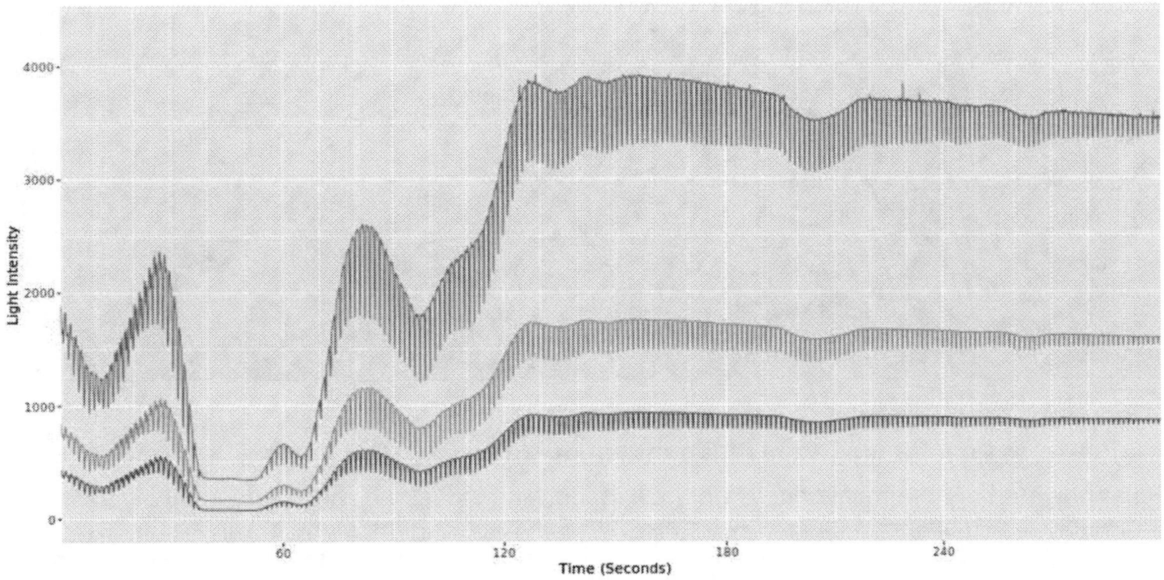

Figure 1 – Variation in light intensity over time

A combination of three light sensors with varying filter opacities was used to provide useful results across a full range of light levels. The light sensors were polled at a high frequency (80Hz) to ensure that all flicker events were captured even when the tip of a blade passes in front of the sun for a fraction of a second.

Although the light level sensors provide a full record of all shadow flicker events, they do not identify which wind turbine is causing the shadow flicker. To support the quantitative light sensor data, a video camera was used to capture qualitative data. The video data was used for identifying which turbines were responsible for which flicker event as well as giving a qualitative indication of the nature event and giving evidence of correct turbine shutdowns.

2.2 Power system for long-term independent operation

Solar powered remote monitoring is not new in the wind industry having been used in meteorological monitoring for over two decades. But achieving the same in wind turbine environmental monitoring has been more difficult due to the increased power consumption of the devices used.

Achieving autonomous operation with batteries charged from photovoltaic panels was a requirement for WindComply flicker monitoring campaigns. In some cases, a full year of uninterrupted operation was required. The winter months with their short days and often cloudy weather presented the biggest challenge. Significant time was invested in component selection and optimising device power regimes in order to bring power consumption to an absolute minimum without compromising the collection and transfer of data.

2.3 Cloud connected monitoring

Remote access to data is a luxury that has not typically been available in the field of wind turbine environmental monitoring. However, for a remote shadow compliance monitoring campaign, immediate access to data was essential. If a complaint is received during the monitoring campaign, the site operator needs to be able to respond with accurate and up-to-date results from the field. Moreover, immediate knowledge of data enables multiple (sometimes minor) changes to be made to the flicker shutdown program during the monitoring campaign rather than waiting till the end of a long campaign.

In recent years, the significant speed improvements and cost reduction of mobile data as well as the availability of multiple network roaming data contracts have made cloud connected remote monitoring cost effective. Multi-protocol modems, VPN technology and multiple-network SIMs were used to maximise the WindComply unit's communications reliability in remote locations. Light sensor and video data is compressed by the WindComply unit before being sent at short intervals to the WindComply server for client access via a web login.

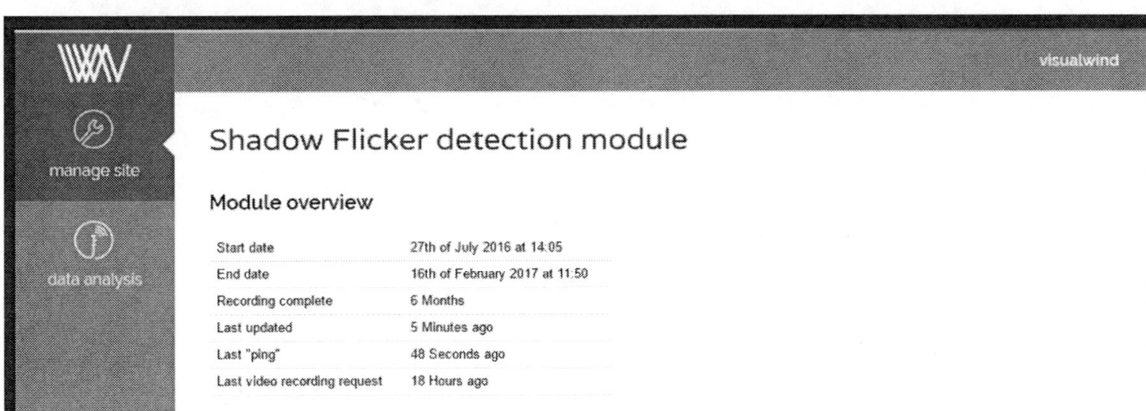

3. Identifying the causes of unexpected flicker

In some cases, where at-property flicker monitoring has been required, there had been no previous investigation into shadow flicker and the monitoring campaign provided the first step in addressing flicker issues. However, in most cases, the monitoring was used to investigate, quantify and resolve cases of unexpected flicker. Typically, a shadow flicker impact assessment has been used to predict flicker times and shutdown modules have been installed on the relevant turbines with the expectation that no flicker would be observed by residents.

When a complaint or complaints are received during the first year of operation, it is necessary to have accurate information to know which turbines are causing flicker at what times and dates so that the flicker can be mitigated. In each case, the WindComply team at Visualwind and TNEI has worked closely with the site operator to analyse light level data and video data for comparison with turbine the shutdown schedule to identify the causes of unexpected flicker. The causes of unexpected flicker can be categories into three main groups as outlined below.

3.1 Timing inaccuracies with shutdown schedules

In some cases, issues in the setup and implementation of the shutdown module have led to timing inaccuracies. Incorrect location information for the property or wind turbine creates an incorrect shutdown schedule. At one site, the Windcomply monitoring campaign found that the shutdown module had been incorrectly set to allow a fixed number of hours of flicker in each year before implementing shutdown.

In other cases, there are no obvious inaccuracies in the input data but flicker is observed a few minutes either side of the shutdown schedule. In such cases, it is very difficult to establish whether the discrepancy between the predicted flicker and observed flicker is caused by inaccuracies in model input data or inaccuracies in the computer models used to generate the shutdown schedules. Whilst analysis has found some common modelling mistakes (such as not adjusting the model to consider the variation of shadow flicker over time (as the suns orbit changes slightly every year) a significantly larger dataset would be required to start to draw conclusions on the accuracy of the various prediction models.

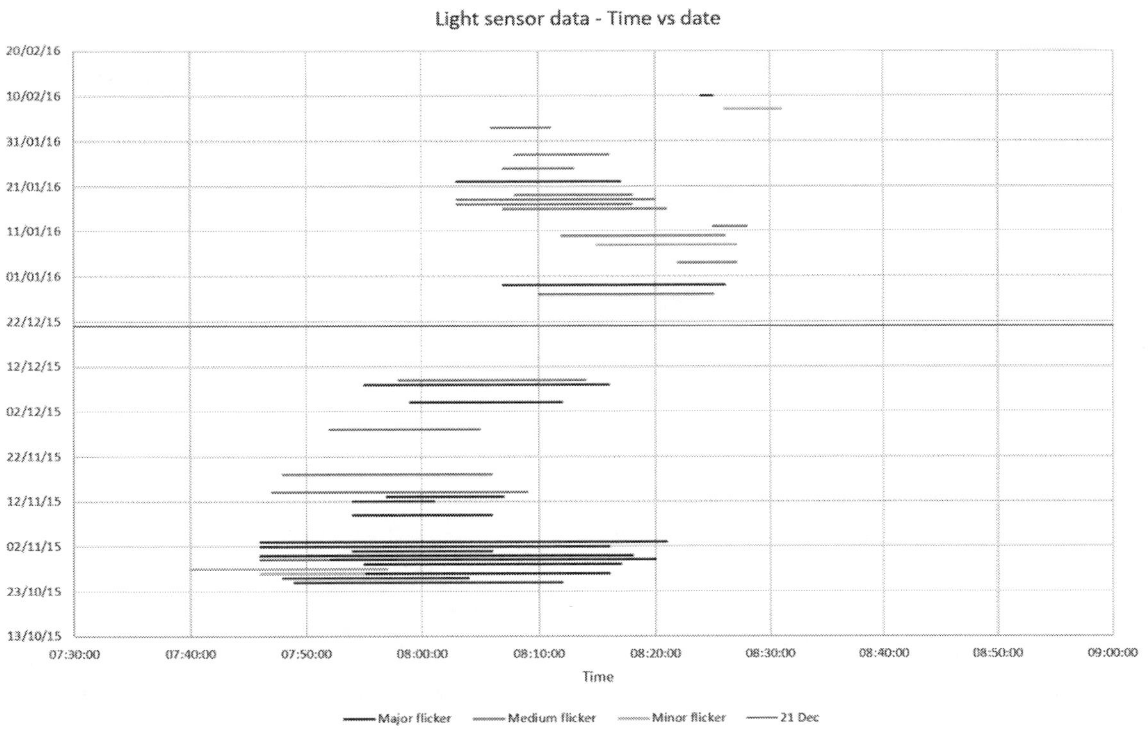

Figure 2 – Example of shadow flicker classification

3.2 Flicker experienced at light levels below shutdown schedules

The flicker mitigation modules installed in wind turbines use light sensors installed on the turbines to establish the light levels. When the turbine enters a shadow flicker time window, a shutdown command is only sent to the turbine controller if the light levels are above a certain threshold. There is no guidance as to what light level should be used as a threshold below which the impact of shadow flicker is deemed to be no longer an issue (see Section 5 below).

During WindComply monitoring campaigns, flicker complaints have occurred at light levels below the shutdown threshold. In such cases, the flicker shows in the WindComply sensor data as well as being clearly visible in the video files. How the operator responds in such cases is considered in Section 4.1 below.

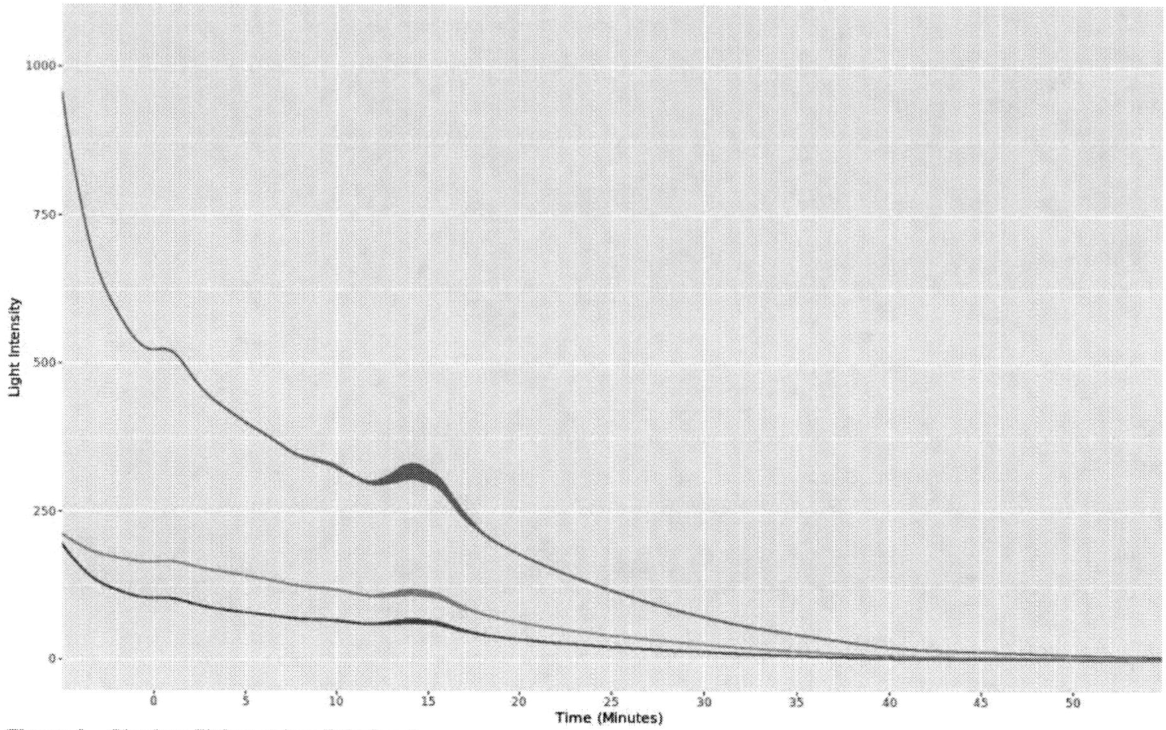

Figure 3 – Shadow flicker at low light levels

3.3 Flicker caused by turbines beyond 10 rotor diameters

Historically in the UK, it has been standard practice to assume that the impact of shadow flicker from turbines beyond 10 rotor diameters will not be significant enough to trigger complaints. As such most impact assessments only included turbines within a radius of 10 rotor diameters from receptor properties and correspondingly, flicker shutdown modules were installed on turbines within this radius.

In practice the 10 rotor diameter study area is a guideline not an absolute threshold at which shadow flicker can occur. If the right calculation parameters are selected shadow flicker modelling software can predict shadow flicker out to distances as great as 25 rotor diameters, it is however usually assumed that shadows at such large distances are likely to be too diffuse to have an impact. The lack of an absolute cut off at which shadow flicker would cease to be noticeable and / or annoying was recognised in a 2011 report written by Parsons Brinckerhoff (PB) for the Department of Energy and Climate Change (now the Department of Business, Energy & Industrial Strategy). The PB report, titled *'Update of UK Shadow Flicker Evidence Base'* noted:

"The current recommendation ... to assess shadow flicker impacts within 130 degrees either side of north is considered acceptable, as is the 10 rotor diameter distance from the nearest property. It is acknowledged that this is a 'one size fits all' approach that may not be suitable depending on the latitude of the site."

During WindComply monitoring campaigns, complaints have been received which upon investigation are shown to originate from turbines outside 10 rotor diameters. How the operator responds in such cases is considered in section 4.1 below.

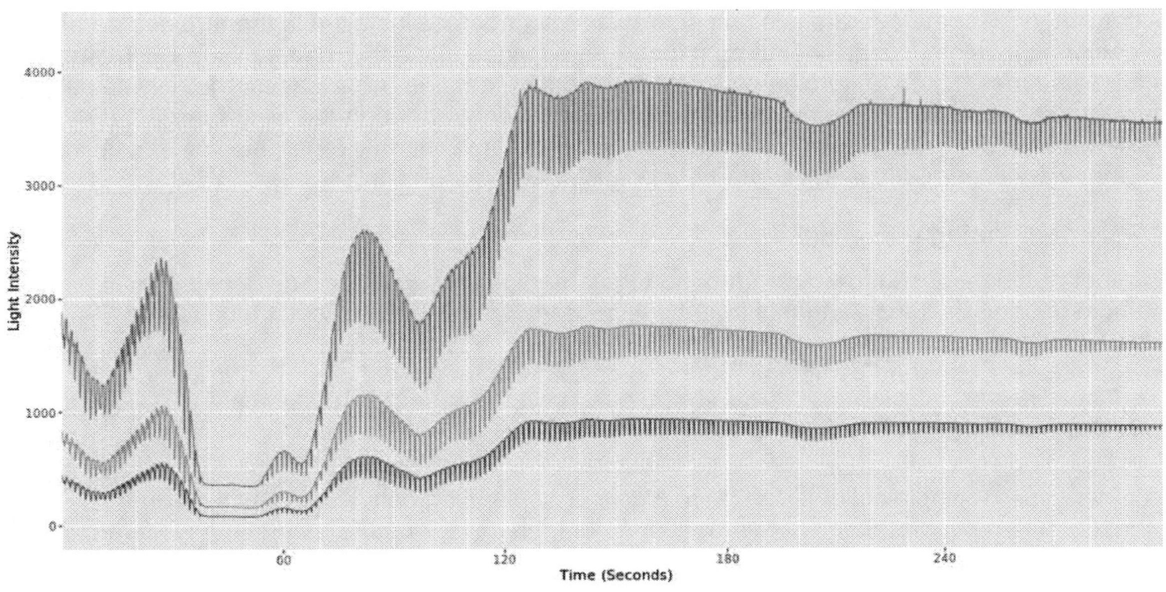

Figure 4

4. Outcomes

The rationale for implementing at-property shadow flicker compliance monitoring varies from site to site but the deliverable outcomes for most wind operators fit into two main categories. These are discussed below;

4.1 Resolving unexpected flicker through fine tuning of shutdown schedules, light level thresholds and flicker management of turbines beyond 10 rotor diameters

Based on recent experience of deploying the Windcomply shadow flicker module, where unexpected flicker occurs, and especially where it leads to complaints, most wind operators wish to resolve the issues by adjusting shutdown schedules and/or light level thresholds to eliminate the unexpected flicker.

WindComply data is used to adjust the schedule to match the exact timings of recorded flicker. This requires a degree of interpolation and judgement because in any monitoring campaign there are several days where cloudy weather conditions mean that no data is gathered. In some cases, where the number of bright days is especially low, a second season of monitoring is helpful to further fine tune the shutdown schedule.

Fine tuning of light threshold levels is a more difficult question to resolve given its subjective nature. Wind operators have typically been willing to increase the sensitivity (reduce the light level threshold) in response to complaints but finding an appropriate threshold for general use is a wider question that may warrant further attention and investigation (see Section 5).

To date, based on Visualwinds recent experience, operators have been hesitant to retrofit flicker mitigation modules on turbines beyond 10 rotor diameters despite complaints being received from such turbines, given that in this area the guidance is more clear. Most flicker from such distant turbines is classed by WindComply as low intensity and complaints about flicker form such turbines is rare, but further work is needed to understand why in some cases, flicker from turbines more than 10 rotor diameters from a property has led to more moderate flicker (see Section 5).

A challenge for some operators is that some shadow modules installed in the turbines do not allow for schedules or light level thresholds to be adjusted since the module itself calculates shutdown times based on turbine and property location data. In these cases, the only solution currently available is to write a separate shutdown schedule through the wind farm SCADA system which shuts the turbine down during the flicker time window regardless of light levels. Further work is currently being carried out in this area to resolve this issue.

4.2 Demonstrating compliance

Another outcome of shadow flicker monitoring campaigns for wind farm owners is the demonstration to a Local Authority that the wind farm complies with planning conditions.

In some cases a second campaign has been carried out for the wind farm operator to demonstrate that adjustments made during a previous monitoring campaign have successfully led to flicker mitigation.

In other cases, complaints of flicker have triggered a planning condition requiring some monitoring to be undertaken but the monitoring campaign demonstrates that the wind farm is compliant with the planning conditions.

5. Understanding the impact of light levels and distance on flicker intensity

The intensity of a shadow will be governed by a number of factors including:

A. The proportion of total received light at a location being obscured by the object casting the shadow (e.g. shadow flicker in a room where sunlight coming from the window is the only light source will be more significant than flicker in an identical room with lights on); and
B. The proportion of the sun that is being covered by the rotating blade (and the type of shadow that is therefore cast).

Consideration of variable A can be hard to quantify as it will vary from room to room depending on artificial light levels and the presence of other windows / openings that may also let light in. The potential impact of variable B can however be considered by calculating the proportion of the suns disc being obscured using information on the relative locations of the sun, the turbine and the receptor.

Figure 5 below illustrates the difference between umbra, penumbra and antumbra shadows. Whilst the terms are usually used when discussing celestial bodies their use can be helpful to quantify shadows cast by turbines too.

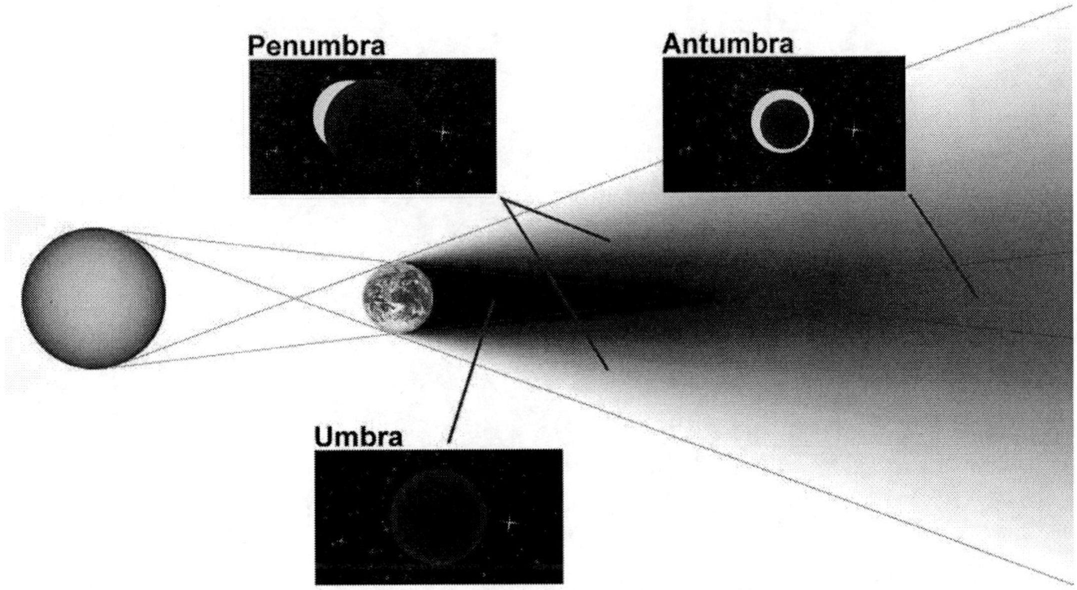

Figure 5 – Shadow types

Whilst turbine blades can be several meters wide at their widest point the, umbra region of any shadow is only cast a relatively short distance before at least some of the sun becomes visible either side of the turbine blade. As distance from the turbine increases the proportion of the suns disc which is covered decreases, the proportion of the suns light being blocked is reduced and the intensity of the shadow decreases.

Notwithstanding the ability to calculate the proportion of the suns disc which is covered and the impact on the resulting light levels as measured at a receptor, there is limited published information available regarding peoples response to changes in light level. Subjective responses may be influenced by both the variation in light level and the absolute levels of light (i.e, is a change of 25% in light level on a bright day more or less annoying than a 25% change in light level on an overcast day, at what point does the variation become imperceptible).

It is anticipated that collection and analysis of data collected by the Windcomply module can help feed into and further study into human response to help inform future guidance both in relation to the size of appropriate study areas to be used for modelling and potentially to assist with complaints investigations.

6. Applying principles of cloud-connected monitoring to noise monitoring

Many of the challenges discussed in Section 2 are also relevant to wind farm noise monitoring and analysis. Access to the most up to date noise data collected at a site can be very helpful, particularly when wind turbine shut downs are required to collect background noise data. The ability to provide live updates on the data collected can be very useful to ensure that the appropriate level of data is collected whilst ensuring that revenue losses associated with unnecessary turbine shut downs are minimised.

7. Conclusions

The collection of accurate at-property quantitative and qualitative flicker data has delivered specific outcomes for wind farm owners – resolving unexpected flicker and demonstrating compliance – but it also contributes to a wider industry understanding of wind turbine shadow flicker, specifically understanding why unexpected flicker is occurring and the impact of light levels and turbine distances on the nature and intensity of shadow flicker experienced at residential properties. The correlating of light level data with subjective recipient data will enable progress towards finding a suitable light level threshold for turbine shutdown as well as a greater understanding of the impact of distance on flicker intensity.

References

To follow

7th International Conference
on
Wind Turbine Noise
Rotterdam – 2nd to 5th May 2017

"Addressing a Management Strategy of Wind Farms Noise Control in Chile"

[*]José David Parra,[**] Igor Valdebenito,[***] Víctor Hugo Lobos
Division of Air Quality and Climate Change
Department of Standards and Policies
Section of Environmental Noise
Ministry of Environment
Government of Chile
San Martín 73, Santiago, Chile.
[*] jparra@mma.gob.cl;[**] ivaldebenito@mma.gob.cl;[***] vlobos@mma.gob.cl

---

Summary

During the last years, wind energy has grown exponentially around the world. Non-conventional renewable energies (NCRE) have had an important impact on society, seeking to establish itself as one of the main sources of energy in many countries.

One of the most important challenges in the development of wind energy is to promote the use of NCRE, avoiding the environmental impacts to the communities. To generate an assessment strategy and guidelines that allow the development of clean energy, in all its aspects, is the challenge that today stands out and addresses wind energy.

One of the goals committed in the Chilean National Energy Policy is that by 2050 at least 70% of the national electricity generation comes from renewable energies, for which, within the guidelines of the aforementioned policy, is to promote a high penetration of renewable energies in the electrical matrix. Likewise, in relationship to environmental effects, the National Energy Policy recognizes that the country's progress will require improving the regulatory framework through programs to review and elaborate new regulations, environmental management instruments and standards of environmental sustainability reaching the energy sector, with coordination between different government entities that can contribute to the sustainable development of the NCRE.

The present work aims to perform an evaluation of standards and tools of noise management of wind farms, generating also an analysis of the current situation of the System of Environmental Impact Assessment (SEIA) in Chile and the practices that must be improved for the management of the noise of this type of projects. It is intended to establish future work paths related to the sustainable development of wind projects in the country, presenting the guidelines for the development of a management strategy of wind farms noise (WFN) control in Chile.

Key words: Wind Farms Noise Assessment, Management Strategy, Wind Farms Noise Control, Government Alliance.

---

1. Introduction

Noise is one of the most important impacts generated by wind farms and is also a complex issue to tackle, so research, generation of knowledge and noise reduction of wind turbines are necessary for the progress of this renewable energy sector (5).

Wind power in Chile by December 2015, has 906 MW operating and has a high potential to develop, especially in the regions of Antofagasta, Coquimbo, Biobío and Los Lagos. Thus, at the same date, there was 5,706 MW environmentally approved in the SEIA and 1,700 MW in the environmental qualification process. The northern part of the country has large tracts of land where projects have been set up without any impact on the community. However, as they approach the central and southern areas of the country, project areas tend to present rural communities due to development of agricultural and livestock activities in these areas.

Lately in Chile, noise from wind farms has caused negative reactions from some communities, with some projects already in operation being questioned. This has generated concern in the energy sector, especially in view of future projects, as the development of wind energy will continue to increase, which will require a proper study of noise phenomena that could have an impact on the community.

1.1 Renewables and Wind Energy in Chile

In recent years, there has been a large growth of non-conventional renewable energy (NCRE) in Chile. In 2005, there were 286 MW of NCRE capacity installed, while a total of 2,135 MW was achieved in September 2015, to constitute 11.43% of the country's electricity generation in October 2015.

One of the important characteristics of wind energy is its variability condition, as it depends on the atmospheric conditions. This leads to the need for wind measurements to determine the yield related to the capacity factor of the resource in the country.

Thus, the Ministry of Energy in Chile has developed important analyzes of the north, center and south of the country, establishing areas with an interesting wind potential, highlighting the significant wind profiles in the north and south of the country, mainly near coastal and mountainous areas. To date it has been possible to identify areas with great wind potential, reaching values close to 40,000 MW, among which are areas with a presence of population under schemes of different population density. In these areas, the objective is to promote the development of wind energy projects in a harmonious and compatible way with human activity, in order to move towards a sustainable energy and to be able to fulfill in this way with the goal established in the National Policy of Energy for the year 2050 that establishes that at least 70% of the national electricity generation must come from renewable energies.

Due to the development of wind energy, there is a need to generate, together with the Ministry of Environment, a Program for the Review and Development of new regulations and instruments for environmental management that affect the energy sector. In this process, the Ministry of Energy in coordination with the Ministry of Environment will address the environmental management issues that are considered necessary to include in the regulatory review. In this context, the current situation of evaluation of wind farms in the country should be modified, establishing new management instruments, particularly for the management of WFN control in Chile.

1.2 Energy Policy of Chile

The Energy Policy of Chile (1), "Energy 2050", proposes a vision of the energy sector by 2050 that corresponds to a reliable, sustainable, inclusive and competitive sector. This view is based on a systemic approach, which proposes as its main objective to achieve and maintain the reliability of the entire energy system, while meeting sustainability and inclusion criteria and contributing to the competitiveness of the country's economy. In short, through these attributes, it is established as an objective to move towards a sustainable energy in all its dimensions.

To achieve this vision by 2050, the Energy Policy is based on 4 pillars: Security and Quality of Supply, Energy as a Development Engine, Compatibility with the Environment and Energy Efficiency and Education. On these bases, the various measures and plans of action proposed up to the year 2050 must be developed.

The Energy Policy 2050 recognizes that the country's progress will require the improvement of the regulatory framework on a regular basis, through programs for the revision and elaboration of new regulations and environmental management instruments and sustainability standards for the energy sector, in coordination between different government entities. These programs should reflect the interests of society, ensuring the maximization of social welfare, as well as progressively reducing the gaps between the existing environmental regulations in Chile and those in other countries of the Organization for Economic Co-operation and Development (OECD). The foregoing will also involve revising such regulations and standards to keep the country's regulatory framework updated with the best international practices.

Thus, at the end of 2016, an alliance is created between the Ministry of Energy and the Ministry of Environment, generating an agreement that will allow to develop a work for the sustainable development of wind energy, particularly addressing a management strategy of WFN control in Chile. The creation of this alliance aims to study the current regulatory framework for WFN and the management strategies of this source at the international framework, seeking to generate the management instruments that allow both the developer and the project consultant, an implementation in accordance with the current international standards, avoiding to generate impacts to the community in future projects. The strategy is sustained in three key stages: Diagnosis, Design and Implementation. Figure 1 shows the particular activities in a general view of each stage.

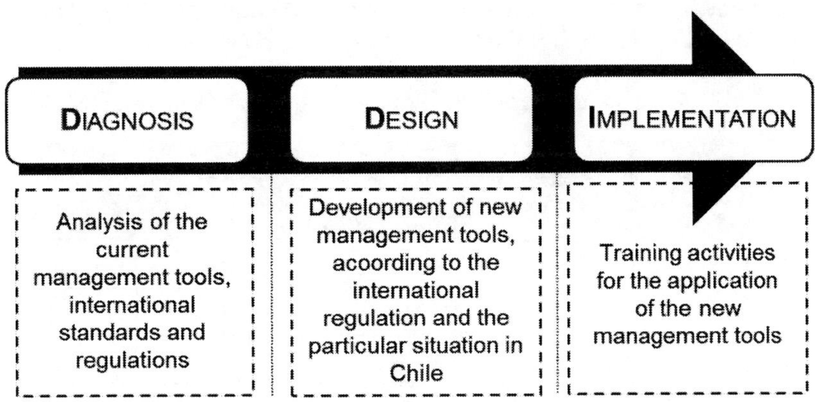

Figure 1: Stages of the strategy.

2. Environmental Impact Assessment of Wind Farms in Chile

The Law 19.300 of General Bases of the Environment (4), creates the System of Environmental Impact Assessment (SEIA) introducing the environmental variable in the projects analysis. The Environmental Assessment Service (EAS) is responsible for implementing the process of environmental evaluation of projects through a public on-line platform accessible to the community.

2.1 System of Environmental Impact Assessment (SEIA)

One of the main instruments to prevent the environmental deterioration of the country is the SEIA, being administrated by the Environmental Assessment Service (EAS). This instrument allows to introduce the environmental dimension in the design and execution of the projects and activities that are carried out in the country. EAS evaluates and certificates that the initiatives, both in the public sector and in the private sector, can comply with the environmental requirements that apply to them (3).

SEIA generates a series of processes, which must demonstrate through an Environmental Impact Statement (EIS) or and Environmental Impact Assessment (EIA) to comply with current regulations in the country with a proper assessment and mitigation strategies of the environmental impacts produced by the project. Likewise, environmental authorities must certify and ratify compliance with current regulations, qualifying the mitigation and control measures proposed by the owner, and finally establishing a positive or negative environmental rating.

Each wind project that enters must present a study of acoustic impact to the EAS. This is because one of the environmental regulations in Chile, corresponds to a standard that regulates the noise of productive activities (stationary sources).

2.2 Noise regulation in Chile

The main noise regulation in Chile (2) corresponds to the Supreme Decree No. 38/11 (DS 38/11) of the Ministry of Environment. This regulation aims to control the noise of productive activities, characterized in chilean legislation as stationary sources. The norm regulates the noise of several of those sources, among them, energy facilities, defined as energy generation, distribution or storage facilities.. Therefore, wind farm projects are being evaluated by this normative since they qualify as energy facilities.

This noise regulation presents the procedures to perform noise measurements. Measurements must be made at the receiver location with specific certificated equipment, presenting a detailed report of the location and sound pressure levels obtained in the control inspection.. The following table presents the noise limits, defined for five zones, according to the General Ordinance of Urbanism and Constructions (GOUC) in Chile.

Table 1: Maximum permissible noise levels according to DS 38/11.

Maximum permissible corrected noise levels (CNL) in dB(A)		
Zone	**7 to 21 hours**	**21 to 7 hours**
I	55	45
II	60	45
III	65	50
IV	70	70
Rural	Minimum between background noise level + 10 dB(A) or Zone III	Minimum between background noise level + 10 dB(A) or Zone III

2.3 Historic comparative analysis of environmental impact assessment of wind farms in Chile

In this section, results of a comparative analysis are presented. A review of every project under evaluation by the SEIA was made, since the first wind project in Chile back in 2006.

A total of 106 projects were evaluated according a defined criterion. 83 projects are in approved status and 23 projects are in qualification status.

The analysis criterion was defined by the following points.

- Land-use
- Background noise analysis
- Noise prediction methodologies
- Noise monitoring proposals
- Applied mitigation measures

Results of the analysis are presented below.

2.3.1 Land-use

This type of project is usually located in rural areas, far away from the residential areas. Due to the distribution of the receivers in each project, some of them qualify as an urban receiver; however, in general, wind projects are evaluated according to the limits established in DS 38/11 for rural areas. In this way, the following table presents the limits that have been applying for wind farms in Chile.

Table 2: Maximum permissible levels applied in the WFN assessment in Chile.

Day limit	Night limit
Background Noise Level + 10 dB(A) or 65 dB(A)	Background Noise Level + 10 dB(A) or 50 dB(A)

Because the source remains in the time, the limits that must be fulfilled in the assessment corresponds mainly to the limit established for night time, as shown in table 2 (BNL[1] + 10 dB(A) or 50 dB(A)).

2.3.2 Background Noise Levels (BNL)

DS 38/11 establishes a background noise measurement methodology:

"The countinous equivalent sound pressure level (Leq) should be measured continuously, until the reading stabilized, recording the value of Leq every 5 minutes. Reading shall be understood as stabilized when the arithmetic difference between two consecutive registers is less than or equal to 2 dB(A). The considered level will be the last of the recorded levels. In any case, the measurement will not be extended for more tan 30 minutes."

It has been observed that the reviewed projects correctly apply this criterion, identifying each receiver and each source that contributes to BNL. Measurement campaigns are carried out from 1 to 2 days regularly, measuring in day and night periods. Once the BNL are established for each receiver, the maximum permitted noise levels are defined according to table 2, presented above.

On the other hand, special attention has been paid to the cumulative noise impact, carrying out assessment that considers the project under evaluation and the neighboring projects that could produce impact to the community.

Within the shortcomings of the assessment has been revealed the lack of consideration of the operating conditions of wind turbines as one of the primary factors for the baseline survey of background noise, this is, wind speeds between 4 and 25 m/s. The current regulation in Chile do not present requirements for wind analysis in measurements. Therefore, correlation between measured noise levels and wind speed are not considered in the assessment.

[1] BNL: Background Noise Level

2.3.3 Noise prediction techniques

The trend of the use of certain acoustic prediction techniques has been observed. Particularly ISO 9613 (9) appears as the most widely used propagation model for the prediction of wind turbine noise in Chile. Other techniques have been applied by some consultants and some assessments are based on theoretical assumptions and international observations when a wind farm is far removed from the communities and would not require a modeling assessment.

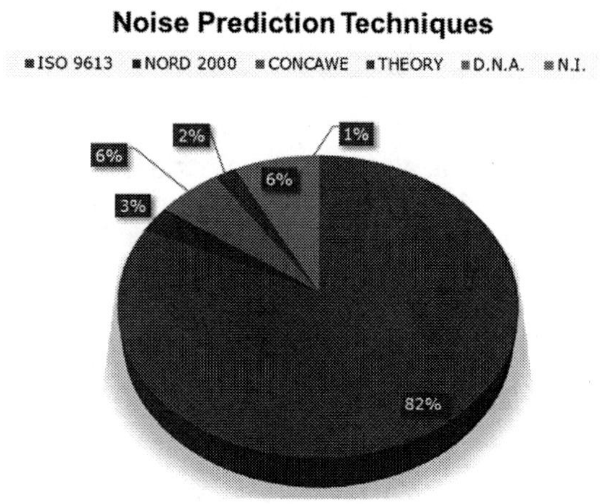

Theory: Theoretical methodologies
D.N.A.: Does Not Apply
N.I.: No Information

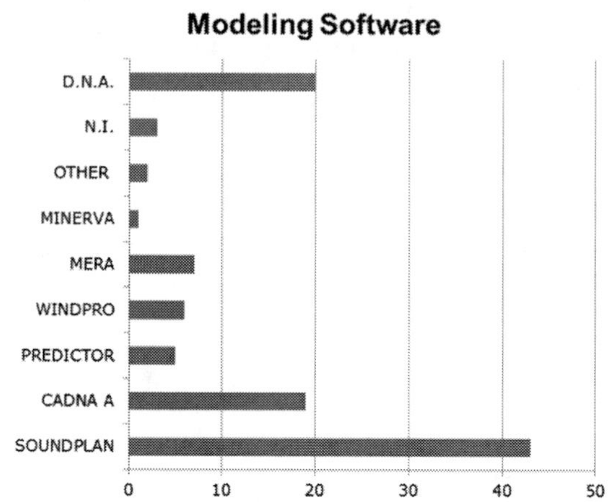

OTHER: Alternative codes developed for modeling according to ISO 9613, for example.

Figure 2: Noise prediction techniques used in Chile.

ISO 9613 applies with the limitations of the model. The input parameters for the code (climatic conditions, considerations that could present errors, source data according to technical reports, etc.) are not correctly stated. On the other hand, the evaluator of the project has evidenced the limitations of ISO 9613, reason why recently in the country, has been requested the implementation of more advanced techniques or international ISO considerations that present a valid theoretical foundation in a certain range of action.

2.3.4 Post-installation noise monitoring

It is identified that some projects present noise monitoring plans, once it has been implemented, usually 2 monitoring by the first year and an annual monitoring from the second year. However, the monitoring plans presents lack of information by not presenting the considerations that were used to establish the methodology. It is important to note that the variability of atmospheric conditions is a very important factor when noise measuring is made, so the monitoring plan should contain an analysis of the variability of the climatic conditions to decide the number of measuring periods and how long will be the measure campaign, in order to characterize in the best way the variation of the noise levels during a year of monitoring, for example.

2.3.5 Noise control measures

When corresponds, noise control measures are applied to specific wind turbines in a wind farm, the one closest to the affected receiver. It is regularly proposed to vary the mode of operation to a quieter mode. However, because the maximum levels allowed in the country are not strict, the projects comply with the regulations at the stage of operation. This generates that no mitigation strategies are presented recurrently. This could lead to future disputes, since without having a demanding entrance limit; a receiver could present future complaints and would have to incur expenses to control the noise emissions once the wind turbine has already been installed. This situation has been seen in Chile during the last years. On the other hand, this generates a poor development of noise control strategies for wind turbines noise in the country, since there is no defined management model for these cases.

3. Noise Regulation in Chile and International Standards

3.1 International Regulation

In relation to the international noise regulation, different criteria are identified which allow to establish evaluation references. The criteria varies from the measurement techniques (equipment, atmospheric conditions considered, measuring time, quantity), prediction techniques, until the definition of the maximum permissible noise levels. Different practices in the international regulation that could be adopted in the development of a management instrument have been evidenced. Emphasizing the wind analysis in the measurements by establishing the appropriate correlations with the background noise as well as extending the measurement period to generate a better characterization of the variation of noise levels in relationship to the atmospheric conditions could be important guidelines for the development of a management strategy.

The development of an acoustic model that can represent the evaluation scenario is very important, since the noise level to be evaluated will be the one that the modeling software or the prediction technique presents. There is no international standard describing a specific method for modeling and the prediction of wind turbines noise. However, it is accepted by the UK acoustic consultants that noise from a wind farm is calculated according to the ISO 9613 regulation (9), "Acoustic – Attenuation of sound during propagation outdoors". Although there are other sound propagation techniques, ISO standards are considered as very useful and efficient tools in the calculation of sound emission levels. However it is known that ISO 9613 has important limitations for the prediction of wind turbine noise (9, 10). In some of the countries the use of more advanced prediction techniques such as NORD 2000 or CONCAWE is recommended, which could generate better predictions due to the more specific considerations in calculations.

On the other hand, one of the important decisions is the definition of the separation distances between the wind turbines and the receivers. It is important to remark that in many cases there is no statutory separation distances stipulated in legislation. Recommendations or suggestions for separation are made generally through planning policy and guidance. The range of distances varies between 350m and 2km (14). Also, a separation distance according to the rotor diameter or hub height is established by many countries.

In the environmental noise regulation, one of the most important factors to study is the maximum noise level allowed in the receiver. For this type of sources have been evidenced noise levels that vary according to the levels of background noise, wind speeds or simply fixed limits that apply for certain situations. The following table (Table 3) shows maximum permitted noise levels for wind farms in rural and residential areas for different countries.

Table 3: Noise limits for wind farms in different countries (13).

Country	Descriptor	Rural Area	Residential Area
Germany	Leq dB(A)	**Day: 45 Night: 35 (Sensitive area)**	**Day: 50 Night: 35**
Belgium (Flanders)	Leq dB(A)	**Day: 45 Night: 43**	**Day: 44 Night: 39**
Belgium (Wallonia)	Leq dB(A)	**45**	**45**
Canada (Ontario)	Leq dB(A)	**40 – 4 m/s 45 – 8 m/s 51 – 10 m/s**	**45 – 4 m/s 45 – 8 m/s 51 – 10 m/s**
Denmark	Leq dB(A)	**42 – 6 m/s 44 – 8 m/s**	**37 – 6 m/s 39 – 8 m/s**
US (Indiana, Tipton Country)	Leq dB(A)	**45**	
Finland	Leq dB(A)	**Day: 45 Night: 40**	
France	Leq dB(A)	**Day: BN + 5 Night: BN + 3**	
Netherlands	Lden dB - Ln dB	**Lden: 45 Ln: 41**	**Lden: 45 Ln: 41**
New Zealand	L90,10 dB(A)	**35 ó BN + 5**	**40 ó BN + 5**
Norway	Lden dB	**45**	
UK	L90, 10 dB(A)	**BN + 5 ó 40 BN + 5 ó 43**	
Sweden	Leq dB(A)	**35**	**40**
South Australia	L90, 10 dB(A)	**35 or BN + 5**	**40 or BN + 5**

3.2Comparison of noise limits with Chile

From the above table it is possible to note that one of the less strict levels corresponds to 45 dB(A). The most demanding level corresponds to a level of 35 dB(A). According to the maximum levels allowed in Chile (Table 2), it can be observed that establishing a comparison between the less strict level for the international regulation, according to Table 3, and the strictest level in Chile, there is a difference of 5 dB(A), which places the country with one of the lowest international noise limit levels, being 50 dB(A) for the night. In comparison to the less demanding level in Chile, there would be a difference of 15 dB(A). Chile presents a day limit of 65 dB(A) for WFN. It is possible to note that if the level of 65 dB(A) is compared to the strictest level presented in table 3, there will be a difference of 30 dB(A). In addition, according to those countries that adopt the BNL analysis, Chile is at an undemanding level in the international framework, applying the criterion of BNL + 10 dB(A) for WFN. It can be concluded that noise regulation in Chile for WFN is undemanding.

4. Development of a management strategy

4.1 Scope

The scope of the strategy seeks to present guidelines for the implementation of new wind farm projects in the country, avoiding the generation of noise impact in the community. The alliance between the Ministry of Energy and the Ministry of Environment in Chile, is committed to act generating a set of proposals that allow addressing the problem in the short, medium and long term, ensuring the sustainable development of this energy sector

4.2 Considerations

4.2.1 Noise limits and descriptors

The study of noise limits is one of the key factors to promote the development of noise management of wind farms. The international regulation presents an important reference for noise limits. It is necessary to study the characteristics of noise limits that could be implemented in the country and whether or not they would depend on background noise and wind speeds. Also, the analysis of the most appropriate noise descriptor to apply the limits should be studied.

4.2.2 Noise modeling and measurement techniques

One of the important considerations of the work is to study sound prediction techniques. ISO 9613 appears as an important reference, however, due to its limitations, it is necessary to study its application range together with the implementation of new methods included in acoustic modeling software. The work proposal will also include acoustic measurements of wind turbines according to the international technical standard IEC 61400-11 (12), which will allow the generation of acoustic models according to measured wind turbine models commonly implemented in the country.

4.2.4 Land-use planning

The General Ordinance of Urbanism and Constructions (GOUC) in Chile, contains the statutory provisions of the law, regulates the administrative procedures, the urban planning process, land urbanization, construction and technical standards of design and construction required in urbanization and construction in the country. These provisions allow to establish the permitted areas and limits of construction for different buildings and facilities.

Land-use planning is a key factor to future wind projects, as it would allow establishing the distance boundaries between the source and the receiver, thus preventing the approach of wind projects to the community.

4.2.5 Wind turbines noise perception

The study of the perception of the noise is very important, since as it is known, the characteristics of this type of sources are very particular, in comparison to other sources. The perception of sound and people's reaction to noise is highly variable and subjective (6-8). Because of this variability, it is difficult to generalize about the impacts of wind farms. In this sense, a perception analysis of WFN will be very important to raise information regarding the current situation of the country in relation to the noise of wind turbines.

4.3 Strategy proposals

The management tools visualized for the development of a WFN control strategy in Chile is detailed below. Figure 3 presents a summary of the strategy proposals detailed in the following paragraphs.

4.3.1 SEIA noise assessment guide

One of the proposals to start generating information considers the inclusion of "wind turbine noise" theme in the noise assessment guide that is being developed by SEIA. This would allow the establishment of references for consultants in short-term, including recommendations for modeling and noise measurement along with establishing international references for the analysis and WFN assessment in relation to the improvement space identified in this work.

4.3.2 Good practice guideline for wind turbines noise assessment

A second proposal considers the development of a guide of specific good practice guidelines that will allow establishing in a more detailed way the factors to be considered in WFN assessment, also including international references and recommendations to carry out the assessment.

4.3.3 GOUC Modification

A modification of the GOUC will be one of the key factors in the development of the strategy. To define a specific distance between the receiver and the source will avoid future community problems with the project developers. This is a precautionary measure that could have a major positive impact on the sustainable development of wind energy in the country. In addition, there is the possibility of establishing agreements between the community and companies to create protected spaces to avoid the installation of future receivers in a certain area of land belonging to a private.

4.3.4 DS 38/11 Modification

In the long-therm, a modification of DS 38/11 could be important for national regulation of WFN. Excluding the source from DS 38/11 and regulate WFN with and international regulation or even the development of an specific standard for WFN in Chile, would allow establishing the necessary guidelines for the assessment, according to the definition of new maximum permitted noise levels, studied adequately for the national territory.

4.3.5 Social Information

. One of the needs is to “demystify” and clean up the information about the impacts of wind projects and making it more reliable that finally generates a rejection of this type of energy due mainly to noise issues. It is necessary to inform the community about the noise issues of wind farms properly and also to inform the development of the work done in view of the problems. The results will be that it will weigh the issue to a level of high importance in the country, also generating reliable information for those affected.

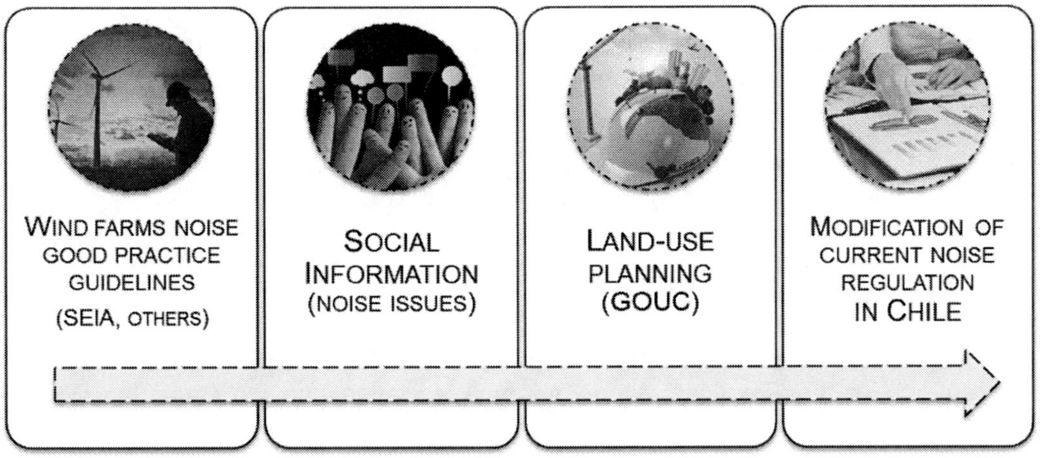

Figure 3: Strategy proposals.

5. Conclusions

It has been possible to highlight very important areas of improvement in the current regulation for Chile as presented in the proposals mentioned above, which allows the design of a set of elements that will be part of the strategy being applied in different time periods. The Chilean regulation is not very demanding and the guidelines for the evaluation of the noise of wind farms have not yet been defined, so the country is nowadays in a key space for the development of wind energy, particularly in the strategies of WFN control.

Guidance to project consultants and evaluators in the framework of the SEIA is fundamental for the generation of good practices in the WFN assessment. Short-term strategies can perfectly address the guidelines that will allow an adequate analysis of this type of sources in the country.

The review of international practices for measurement, modeling and maximum permissible noise levels, together with the perception phenomena and effects of wind turbine noise, are key factors for the development of a WFN management strategy in the country.

It is important to note that sustainable development of wind energy must consider in equal and congruent parts the growth of the economy, the well-being of the people and the care of the environment. In this way the reduction of the cost of energy and the low emission of noise, as far as wind projects are concerned, is essential for the development of wind energy.

Future works correspond to the design of the strategy and the implementation of each of the actions studied. It is expected that by 2018 the guidelines will be fully defined along with the implementation of those elements that are sought to address in the short-term.

Acknowledgement

This work has been supported by the Ministry of Energy and Ministry of Environment of the government of Chile.

References

1. Energy Policy of Chile, Ministry of Energy, Government of Chile (2016).
2. DS 38/11, Noise Emission regulation for stationary sources, Ministry of Environment of Chile (2011).
3. Environmental Assessment Service, System of Environmental Impact Assessment, Government of Chile, http://www.sea.gob.cl)/ (2017).
4. Law 19,300 of General Bases of the Environment, Government of Chile (1994).
5. R. Bowdler and G. Leventhall, Wind turbine noise, Multi-Science Publishing (2012).
6. Colby, D., R. Dobie, G. Leventhall, D. Lipscomb, R. McCunney, M. Seilo, and B. Sondergaard. Wind Turbine Sound and Health Effects: An Expert Panel Review, A report prepared for the American Wind Energy Association and the Canadian Wind Energy Association, 85 pp. (2009).
7. Draft Programmatic Environmental Impact Statement on Wind Energy Development on BLM-Administered Lands in the Western United States. U.S. Department of the Interior, Bureau of Land Management, FES 05-11 Vol 1:398 pp. (2004).
8. Rogers, A. L., J. L. Manwell, and S. Wright. Wind turbine acoustic noise: a white paper. Renewable Energy Research Laboratory, Center for Energy Efficiency and Renewable Energy, Department of Mechanical and Industrial Engineering, University of Massachusetts at Amherst, Amherst, MA. 26 pp. (2004).
9. ISO 9613:1996 "Acoustics – Attenuation of sound during propagation outdoors" (1996).
10. Kaliski, K., D. K. Wilson, S. Vecherin, and E. Duncan, Improving Predictions of Wind Turbine Noise Using PE Modeling, NOISE-CON 2011, Institute of Noise Control Engineers, Portland, OR (2011).
11. General Ordinance of Urbanism and Constructions (GUOC), Ministry of housing and urbanism, Government of Chile (1992).
12. IEC International Standard 61400-11 (2002).
13. E. Koppen and K. Fowler, International Legislation for Wind Turbine Noise, EuroNoise 2015, Maastricht, Netherlands (2015).
14. S. Cave, Wind Turbines: Planning and Separation Distances, Northern Ireland Assembly, Ireland (2013), NIAR 767-13.

7th International Conference
on
Wind Turbine Noise
Rotterdam – 2nd to 5th May 2017

Comparison of Measured and Modeled Wind Turbine Noise in Indian Terrain

G.Arivukkodi, A.E, National Institute of Wind Energy, Chennai,
Email Id : arivukkodi.niwe@nic.in
Dr.S.Gomathinayagam, Director General, National Institute of Wind Energy, Chennai. Email Id: dg.niwe@gov.in
Dr. S.Kanmani, Professor& Director, Centre for Environmental Studies, Anna University, Chennai. Email Id: skanmani@annauniv.edu

Summary

Today, India is in fourth position in terms of installed wind power capacity, cumulatively with 28.14 GW. Towards achieving Indian set target of 60 GW by 2022, wind farms are likely to be close to habitats of human and other living beings, which may get affected by noise of operation of WTG (Wind Turbine Generator). First attempt on noise study from wind turbines in India was carried out in Kayathar, Tuticorin district, Tamilnadu to study the noise propagation from wind turbines with different installed capacity and make of WTG, in flat terrain. Emphasis was given on the reliable experimental procedure and estimation of the real field noise impact in both high (Wind speed 4-13 m/s) and moderate (Wind speed 0.7 – 8.6 m/s) seasons. Acoustical parameters and non-acoustical parameters were measured simultaneously according to IEC 6 1 4 0 0 -11. The onsite measurements were compared with empirical noise prediction models ISO 9613-2 to get an idea for accuracy on ambient noise propagation, this showed a difference of ± 5dB (A) in the noise perception for every increment in the distance (x'.m), at an average air density of 1.161 kg/m3, temperature of 27.78 ◦C and pressure of 995 mb. In order to identify the major source of noise emission, individual blade noise was investigated using noise prediction models and were verified by developed MATLAB codes for different regions in blade. Aerodynamic blade noise associated with the passage of air over the blade were 15.5% (600kW) and 27.83 %(2MW) in the overall sound pressure level (SPL) perceived at a distance of 1 m, which showed a milder contribution of noise to the total noise level. Based on the study, an equation was derived to obtain SPL (dB (A)) at the base of the tower for wind turbines and at varying distances of measurement. Thus, at this point the consorted acoustical research would facilitate social acceptance of wind farms and identify their source of noise production for mitigation at design stage.

Keywords: Wind Turbine, Sound Pressure level, IEC 61400 – 11, ISO 9613-2, Aerodynamic Noise , Noise Prediction Models , Decay Constant

1. Introduction

Acoustical noise produced by wind turbine is an important field of investigation because of the growing interest in renewable energies and because of the impact that operation of these structures have on the environment. In recent years, there has been a spectacular increase in wind power installations worldwide.India ranks 4th in the wind energy with largest installed capacity of 28419.40 MW as on 30.11.2016[1]. However, despite the fact that wind energy comprises of a technologically mature, economically competitive and environmentally friendly energy source, it is often accompanied by concerns related with noise impact, usually encountered in the direct vicinity of wind farms. The total noise generated by a wind turbine is made up of several components, broadly grouped as mechanical and aerodynamic noise. Whenever the wind speed is below "cut-in", the blades rotate very slowly without electric power generation and consequently minimal noise is generated. When the turbine is operating between the winds speeds of approximately 4m/s and 30m/s the sound power level monotonically increases.

Wind turbine noise is due to a combination of both mechanical and aerodynamic operations. In general, mechanical noise is caused from the interaction of turbine components (generated by the turbine's mechanical and electrical parts). Aerodynamic noise is caused by blades passing through the air and it is generally broadband in nature which can have a swishing character [2, 3].

Generally as it should be expected the level of noise and its diffusion are influenced by the type of terrain and its meteorological conditions of respective site. The sound from wind turbine noise can either be reduced or enhanced depending on certain factors such as the area type (ie. rural or urban), domiciles residing near the wind farms and the type of communities affected such as residents, industrial, tourists. [4-6]. Different acoustic sound properties not fully described by the equivalent A-Weighted level, can be related to the perception and annoyance for wind turbine noise, also depending on the operating conditions of the wind farm. This hypothesis was supported by Persson Waye and Ohrstrom [7]. It was suggested that the presence of the sound characteristics subjectively described as lapping, swishing and whistling was responsible for the differences in perception and annoyance between the sounds.

The perception of wind turbine noise, in fact could be covered by wind generated noise. However usually the wind turbines have a stable, constant, variable rotor speed, that results in a quite steady state noise emission even if the wind speed and the background level is low [8]. Acknowledging the importance of the noise impact concerning the further adoption of wind energy, in the current study authors present and evaluate a set of real field noise measurements, acquired from three wind turbines of different installed capacity, make, height, rotor diameter, year commissioned (ageing of the turbine), located at the Indian topography. With respect to the identification of noise source from wind turbines, aerodynamic blade noise was considered since the wind turbine mechanical noise can be reduced by some simple mitigation measures such as using quitter gearbox, periodic maintenance, provision of acoustic shields, changing of some mechanical parts which can affect the emitted noise.

Generally there seems a complex propagation path and additional noise sources that interferes between the source of acoustical noise (wind turbines) and receptor of acoustical signal (surrounding area). Thus, to get a clear picture on noise propagation from wind turbines, sound pressure level (SPL) from the source and receptor are compared with background SPL [9.10] Emphasis was first given on the development of an accurate and reliable experimental procedure according to IEC 61400:11 2002-12 - Acoustic noise measurement techniques and secondly on the estimation of the real field noise impact of the existing wind turbines disassociated from the background noise for several wind speed values, in both high and moderate wind seasons.

Thirdly the blade noise propagation from the wind turbines was identified. Finally wind turbine noise emission was predicted at varying distances of measurement by comparing the simulated results derived by the application module of standard software tool ISO 9613-2 with the available onsite measurements. At the present time, there are no common international noise level standards or regulations for sound pressure levels due to differences in regulatory measurements and community expectations world over.Though Europe and U.S, countries follow the fixed noise limits, in most countries however, noise regulations define upper bounds for the noise to which people may be exposed. These limits depend on the country and may be different for daytime and night-time [11]. In India, Ministry of Environment, Forest and climate change has given notification on "The Noise Pollution (Regulation and Control Rules, 2000) stating the noise limits (dB (A) Leq) with respect to day time and night time for various category of area/ zone (industrial, commercial, residential and silence zone), while no specific characterization and control levels of wind turbine noise in the habitats have been specified, since noise emanates from different sources [12]. The study also compares the simulated results with real acoustic noise emission measurements which gives an idea for obtaining the accuracies, uncertainties of wind turbine noise prediction models.

2. Description and background study

The methodology involved for carrying out the objectives of the study starts with selecting the wind turbines in flat terrain with respect to different installed capacity, Height, Rotor Diameter and Commissioned Date located at Kayathar, Tuticorin District in Tamilnadu, India located at latitude N 8° 57' 44.05", longitude E 77° 43' 12.73". Emphasizing the studies on noise survey carried out over the globe, an attempt was made to study the noise propagation. The terrain of the site was flat and gently sloping from the turbines location towards the western direction surrounded by cluster of trees, which were 10m high, and 800m wide and around 950m away from the turbine under study. The type of wind turbine, their respective gradients, direction and relative boundary under study are illustrated in Figure 2.1

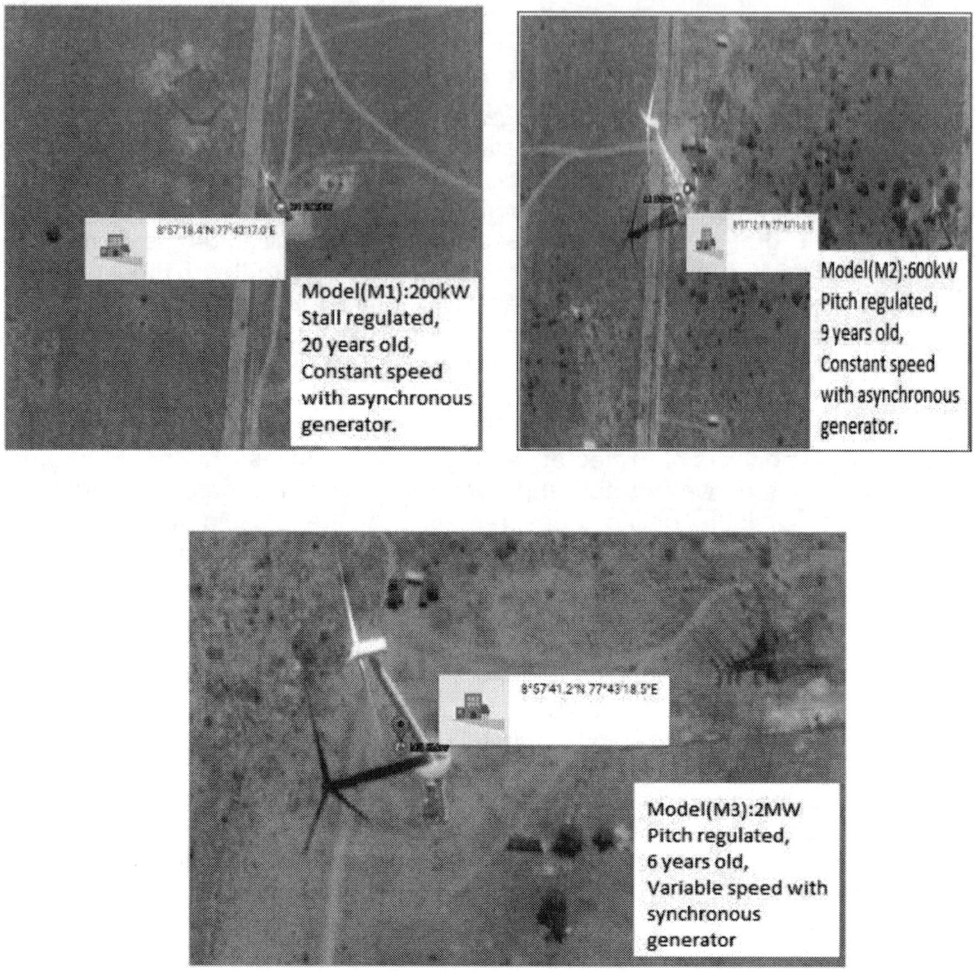

Figure 2.1 Maps showing the wind turbines gradient, direction and relative boundary

Since India has enormous wind potential ,It is essential to study the wind characteristics in the field since it is closely related to wind energy potential, wind farm design, operational management of wind energy forecasting and conversion systems. As shown in Figure 2.2 the measurements at the wind monitoring station were carried out according to the standards.

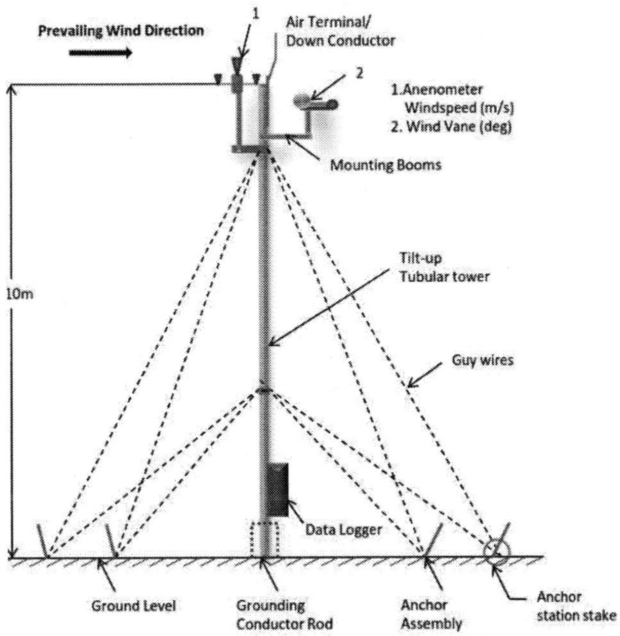

Figure 2.2 Wind Monitoring Station

With the help of wind data collected, assessment on wind characteristics were carried out to understand the general wind potential of the site and its variation with respect to meteorological conditions. Indian monsoon is the most prominent of the world's monsoon systems. It blows from the northeast during cooler months and reverses direction to blow from the southwest during the warmest months of the year. High wind speed was perceived during the month of May to August. Moderate wind speed prevailed during the month of September to April. As stated by Indian meteorological centre, at early October, variable winds are very frequent everywhere. At the end of the month, the entire Indian region is covered by northerly air and the winter monsoon takes effective. The surface flow is deflected by the Coriolis force and becomes a north-easterly flow thereby decreasing the regional wind speed. Considering the wind speed variation and environmental aspects, noise propagation in the study area was also carried and compared with the modeled results to find the source of noise production from the wind turbines, as it propagates distances.

2.1 Methodology

The methodology adopted during the acoustic measurement procedure is described in the following:

Noise measurements test procedures from wind turbines at the noise receptor location were conducted according to IEC 61400-11. Prerequisite preparation of the site included cleaning and marking, foundation, installation of weather monitoring station, fixation of data logger (NRG) and respective sensors (NRG). Acoustical Measurements were taken in the study area comprising a cumulative installation of 10MW capacity wind farm (shown in Figure 2.3 with varied wind turbine make, capacity, model, year of commissioning). The measurement setup is shown in Figure 2.4

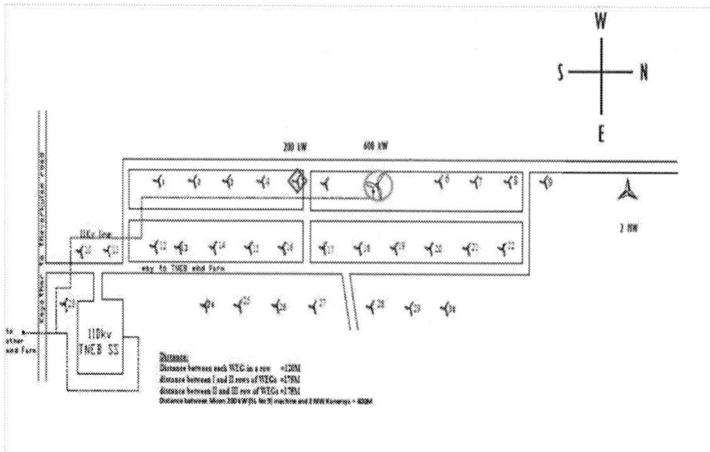

Figure 2.3 Layout Diagram for the study area

The anemometer (to measure wind speed) and the wind vane (to measure direction) were placed before the wind turbine at a height of 10m above the ground level and at a distance greater than two (2D) and less than four times (4D) the diameter (D) of the wind turbine rotor [13]. The complete setup for the measurement at site is shown in Figure 2.4

Figure 2.4 Onsite measurement settings according to IEC 61400-11

Wind turbines have aerodynamic and aero acoustic behaviours with unique characteristics that make their prediction more challenging. The wind turbine aerodynamic noise shown in Figure 2.5 is very important to be investigated because of its high emission levels in addition to the difficulty of its control compare with the mechanical noise [14]. Correspondingly semi-empirical relations were used to predict the blade noise from the overall sound pressure level perceived. The noise from wind turbines with installed capacity of 600kW and 2MW were investigated to find the typical amount of noise source contributed during the operational period of wind turbines in the field.

Figure 2.5. Dominant sound source –aerodynamic noise

2.2 Onsite measurement settings according to IEC 61400-11

Using the measurement procedure and methodology guided by the standard, the measurements of the wind speed and noise level at the wind turbine site were measured using Modular Precision Sound Analyzer Investigator type 2260. Non acoustical measurements viz., local meteorological conditions (temperature, direction and barometric pressure) was recorded throughout the measurements period in the data-logger. Based on the wind turbine dimension, the measurement points were selected along a side with the distances in a radial direction. This is seen in Figure 2.6 and Table 2.1 as per the standard where the horizontal distance between measurement point and the tower of the wind turbine recommended being equal to the hub height plus the radius of the rotor with a tolerance of ±20 %. The inclination angle between the measurement point and hub height determined to be in the range of 25°- 40° and the direction of downwind measurement point ±15° of the wind turbine [15].

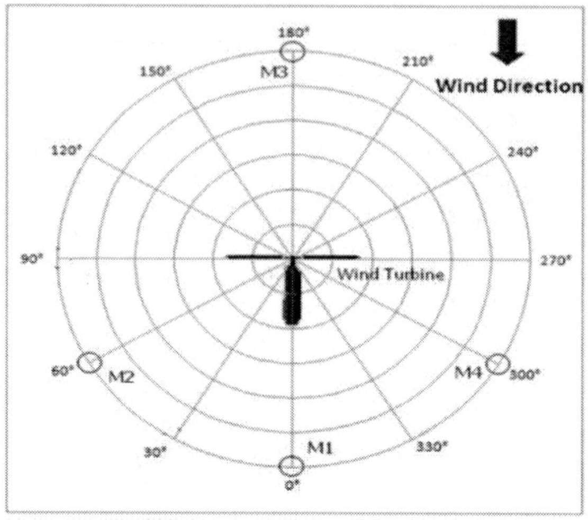

Figure 2.6 Study Area- Measurement Points

Table 2.1 Wind Turbine Noise measurements

Capacity	**200 kW**	**600 kW**	**2 MW**
Model	Model 1	Model 2	Model 3
Hub Height(H)	30m	75m	80m
Diameter of rotor (D)	24m	52m	82m
Reference Distance (Ro)	42m	101m	121m
Mast range b/w 2D-4D (m)	48-96	104-208	164-328

Acoustical measurements were carried out in the downward direction under two stages for three wind turbines (M1-200 kW, M2-600 kW, M3-2MW), i.e. first when the wind turbines are in "on condition" and second "off condition" i.e., immediately after the time that the wind turbines are stopped. In this way ambient background noise was accounted under the same weather conditions. The wind speed (at 10 m from the ground) generally recommended being around 8 m/s (1 minute average). Table 2.1 describes the installed capacity, make, hub height, rotor diameter, reference distance and mast height for three wind turbines (M1-200 kW, M2-600 kW, M3-2MW).

In this context, one of the main problems when measuring the noise emission of wind turbines is the influence of the background noise ie. the wind at the microphone, the wind acting at the adjacent trees, shrubs and structures, traffic on nearby roads and rail tracks, aircrafts and industries, animal and human activities. Such a background noise was measured at the site under turbine off condition[15,16]. It is noted that all sensors and instruments of the acoustic and non-acoustic measurement equipment (refer Appendix A) were calibrated and certified in verified laboratories according to National and International Standards.

Sound characteristics as described here could be of relevance for perception and annoyance, especially at low background noise levels. The perception of wind turbine noise, in fact, could be covered by wind generated noise. The polarizing issue of wind-turbine noise is often framed one of two ways: Turbines are either harmless, or they tend to have powerful adverse effects, especially for sensitive individuals [17-19].To understand the source of noise propagation from the wind turbines, theoretical study on the blade noise was carried out. What distinguishes wind turbine noise emission from more conventional sources of sound is that it tends to increase with increasing wind speed. Concurrently, the ambient noise environment at neighbouring locations will also often change with wind speed [19].

In our present work, using different prediction model stated by Lowson, overall sound pressure and sound power level was calculated and analysed for wind turbines of different installed capacity at varying distances. These models require simple input parameters. Lowson classified sound prediction models into three categories depending on the mechanisms causing wind turbine noise [20.21]. The following Table 2.2 shows the semi empirical Equations 2.1 to 2.4 for different categories.

Table 2.2 Semi empirical equations for different categories

Category I models	Equations
Lowson	$\mathrm{LwA} = 10\log_{10}\mathrm{PWT} + 50$ Eq.2.1
Hau	$\mathrm{LwA} = 22\log_{10}\mathrm{D} + 72$ Eq.2.2
Hagg	$\mathrm{LwA} = 50\log_{10}\mathrm{V_{TIP}} + 10\log_{10}\mathrm{D} - 4$ Eq.2.2
Modified hagg	$\mathrm{LpA}=\mathrm{C1}\log_{10}\mathrm{V_{TIP}} + \mathrm{C2}\log_{10}\mathrm{nB}\frac{\mathrm{Ab}}{\mathrm{Ar}} + \mathrm{c3}\log_{10}\mathrm{cT} + \mathrm{c3}\log_{10}\frac{\mathrm{D}}{\mathrm{r}} - \mathrm{c5}\log_{10}\mathrm{D} - \mathrm{C6}$ Eq.2.4

The above mentioned equations require inputs such as rated power of wind turbine (PWT) in Watts, rotor diameter (D) in meters, tip speed at rotor blade (VTip) in m/s, number of blades (nB), blade area (Ab), rotor area (Ar), axial force coefficient (cT), rotor diameter (D), the distance between the rotor hub and the observer (r), and a few constants (C1-C6).

Individual blade noise as represented in Figure 2.7 viz., Inflow turbulence noise, The turbulent boundary layer trailing edge(TBL-TE) , Laminar boundary layer vortex shedding (LBL-VS) noise, Stall Separation Noise, The Tip Vortex Formation (TIP-VF) was found using Grosveld and Brooks, Pope, Morcolini (BEM) aerodynamic model. In this model turbine noise is divided into segments, each segment has its own chord, span, angle of attack, free stream velocity, and hence each segment has its own contribution on the total sound level emitted[22-25].

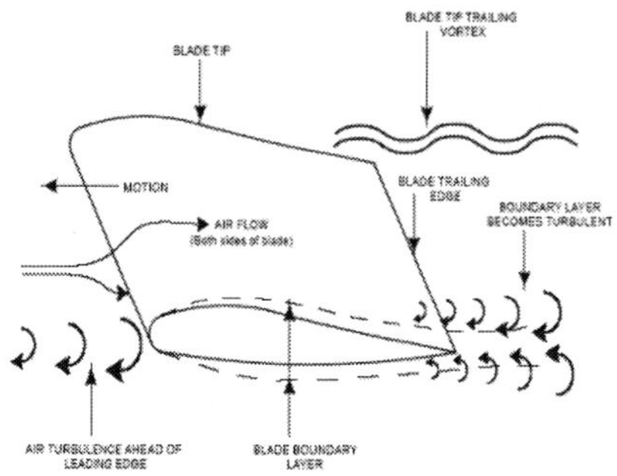

Figure 2.7 Aerodynamic noise sources associated with wind turbine blade

The overall sound power level, sound pressure level and individual blade noise results are illustrated using different wind turbine noise prediction models. The values of different input parameters used and numerical results so obtained from different models were analyzed both manually and using developed MATLAB codes.

3. Results and Discussion

Wind turbine noise studies during high and moderate wind seasons. By applying the methodology as per IEC 61400-11(edition 2) previous section, a number of sets of noise measurements for different wind turbines were obtained under varying wind speeds i.e., during high (4 to13m/s) and moderate wind (0.6 to 8.6m/s) seasons. An analysis of the results obtained with the wind turbines in operation, indicates the recorded sound power level (Lw) and sound pressure level (SPL).

As analyzed by Bill Dawson, Neil Mackenzie [26] wind farm noise regulations stipulate wind speed dependent criteria (referenced to wind speed at the hub height of the turbines), under the assumption that during high-wind speed conditions (when wind turbines generate higher noise levels), there would be a corresponding high wind speed and masking noise level at nearby receivers. Following, another finding as shown in Figure 3.1 was quite different that the sound pressure level (58.64dB (A),

58.02 dB (A), 58.38 dB (A),) measured for three wind turbines (200 kW, 600 kW, 2 MW) increased as the wind speed (4-13m/s) increased relatively in the high windy season.

As per the measured data's the maximum sound pressure level (SPL) obtained for three wind turbines M1-200 kW, M2-600 kW and M3-2 MW during high wind season were 61.21dB(A), 63.42dB(A) and 66.73dB(A). The minimum SPL were 57.42dB (A) 53.18dB(A) and 50.94dB(A) respectively . Among the three wind turbines, stall regulated wind turbine (M1-200kW) showed the maximum noise emission. Under moderate wind season sound pressure level measured for M3-2 MW was 55.10 dB(A) at the wind speed range (0.9-8.6m/s), whereas there was no significant sound pressure level from wind turbines of capacities M1-200 kW and M2-600 kW because of low wind speed perceived during the season. Figure 3.2 shows the scatter of noise data and background data measured at downwind direction for the 3 wind turbines (M1-200 kW, M2-600 kW, and M3-2 MW). Number of scatter of noise data and background data recorded for M1, M2, and M3 at high wind season and moderate wind season were (101, 212, 72- high wind season) (62, 193, 59 moderate wind season).

According to IEC 61400-11, Noise measurements for the operating wind turbine (wind turbine plus background noise) are correlated with background noise measurements at standardized wind speeds. The noise measurements are then corrected for background noise using following Equation 3.1.

$$Ls = 10 \log [10(Ls + n/10) - 10(Ln/10)] \quad \text{Eq.3.1}$$

Where,

L_s is the equivalent sound pressure level (dB) of the wind turbine operating alone,

L_{s+n} is the equivalent sound pressure level (dB) of wind turbine plus background noise,

L_n is the equivalent sound pressure level (dB) of the background noise.

The Figure 3.1 shows the apparent sound power level, maximum and minimum turbine sound pressure and background noise level for three wind turbines (M1-200 kW, M2-600 kW and M3-2 MW).

The sound pressure level measurement data obtained for three wind turbines in both high and moderate wind season were corrected for background noise using equation 3.1. At downwind direction, the background noise for the three model wind turbines at distinct wind speeds showed difference in a range of 3.98-4.92 dB(A) which was less than 6 dB(A) but more than 3 dB(A) higher than the background level, *Ls+n* was corrected by subtraction of 1.3 dB(A), but according to standard IEC 61400-11 the corrected data were not used for the determination of the apparent sound power level or directivity[27-29].

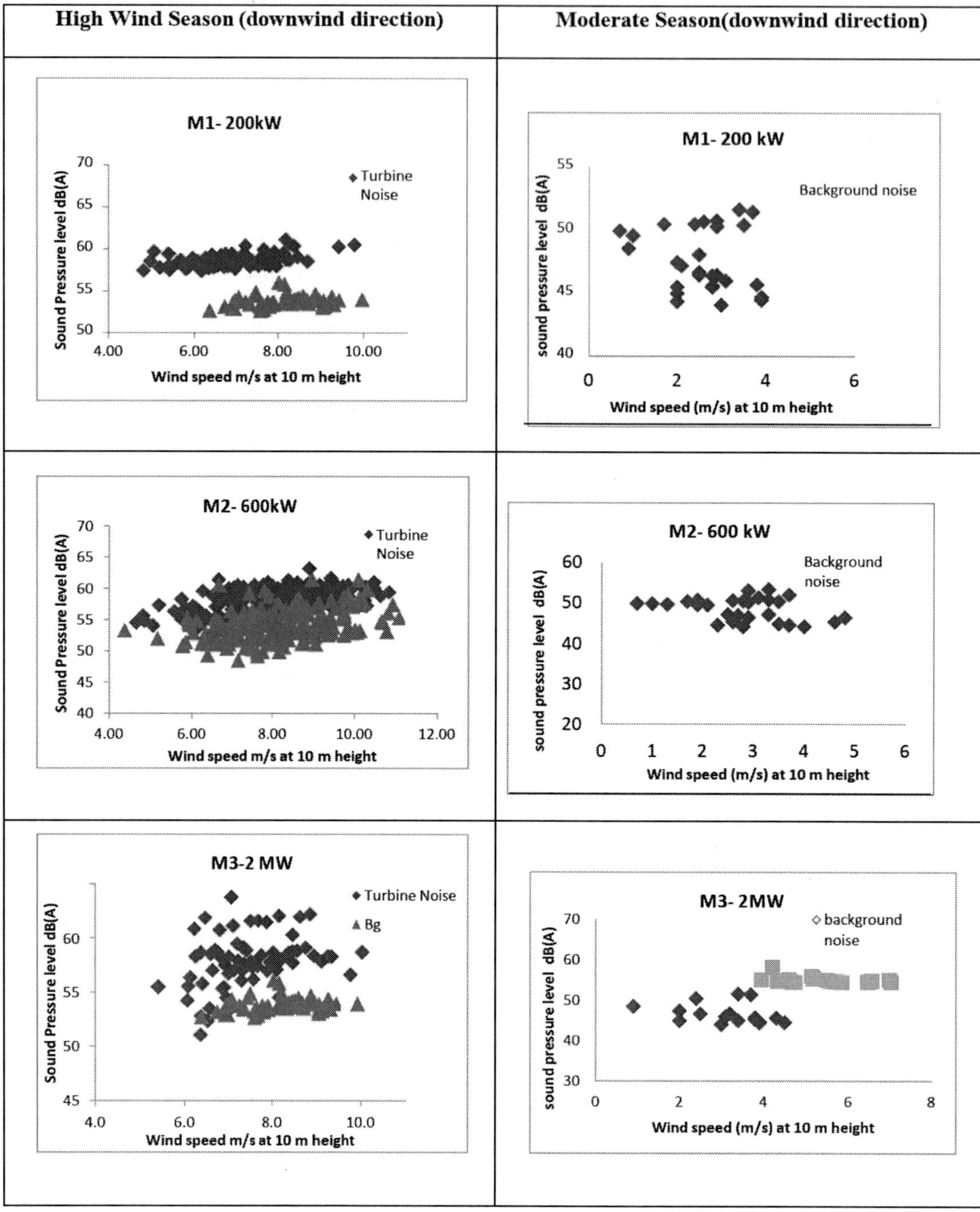

Figure 3.1 Wind Turbine Noise Studies – High Wind Season and Moderate Wind Season (200 kW, 600 kW and 2 MW)

3.1 Propagation of Noise from Wind Turbine

For assessing the noise propagation from wind turbines, an atmospheric propagation model is used to derive the expected noise level at the surroundings from the sound power level at the wind turbine at varying distances. The noise prediction software WindPRO –DECIBEL can carry out calculation based on eight models. The model ISO 9613-2 is a general and internationally accepted tool to estimate far field noise levels under conditions favourable to the propagation of noise with respect to both performance and in the project planning stage [30-32]. The WindPRO module DECIBEL for noise impact calculations is based on the noise emission data at 10m height or at hub height of the wind turbine [33]. The distance to the limit can be calculated to find the close proximity of dwelling location. In this context, necessary settings are required and are shown in Figure 3.2. Then, by taking into account the topography, the roughness of the area, potential obstacles and coefficients related to local meteorological conditions, the noise generated by the wind turbines at the ground or at a specific height for a given wind speed was calculated.

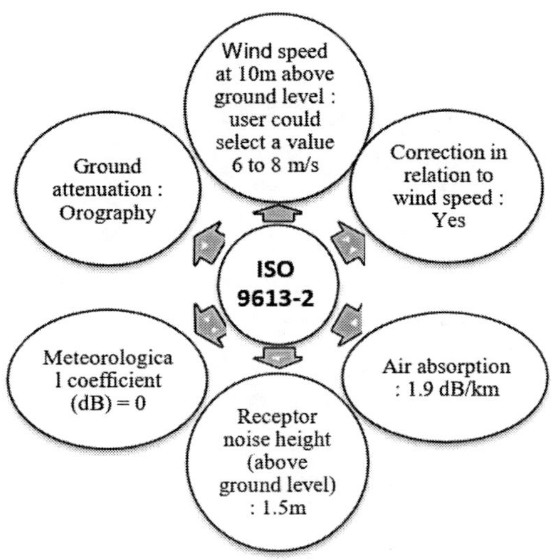

Figure 3.2 Settings of Software tools

3.2 Propagation of blade Noise

Model considers that the noise source noise is represented as a point located at the hub height. This model is valid only for low frequencies and where the turbulence length scale is large compared to the blade chord. Using the illustrated input parameters, the total sound pressure level was found. Relative velocity of the blade element for the respective wind turbines were found by subtracting natural wind speed with blade tip speed. The individual blade parameters were calculated at the blade length section of 0.7. In order to assess the noise impact, the sound pressure level of the individual wind turbine needs to be estimated. The following Figures 3.3 and 3.4 shows the individual noise contribution in the blade region for wind turbines of varying installed capacity and aerodynamic parameters.

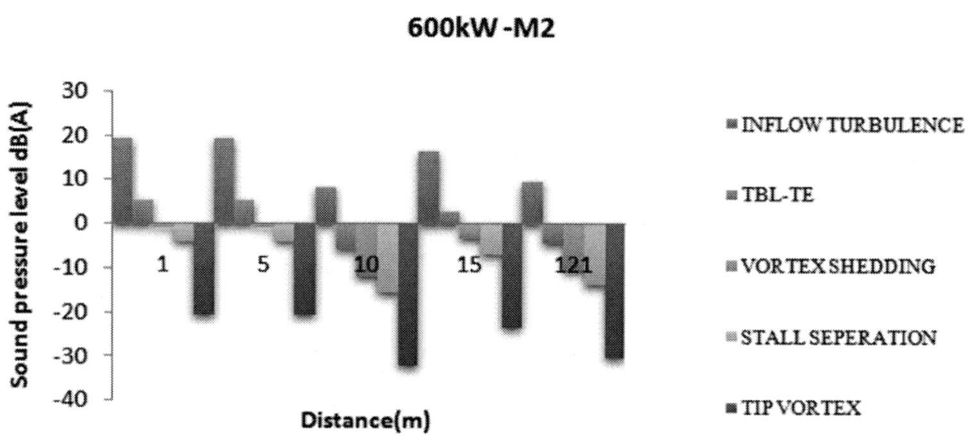

Figure 3.3 Blade noise contribution of 600kW at varying distances

Sound pressure level was measured in the study area (Kayathar, Tamilnadu) at varying distances as per IEC 61400-11 (measured 10m from the ground at 8m/s taking 1 minute average).

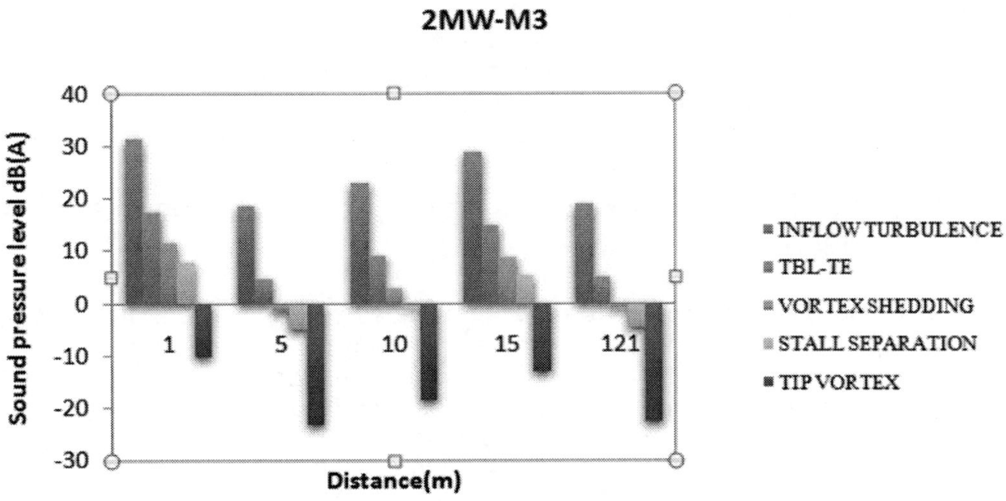

Figure 3.4 Blade noise contribution of 600kW at varying distances

The overall sound pressure level measured for 600kW (M2) and 2 MW(M3) are 95.02dB(A) and 97.02 dB(A) respectively. Theoretically by using semi-empirical equations individual blade region noise found to be varying at different distance of measurement (distance between observer and receiver). Figure 3.1 and 3.2 indicate the variation in SPL perception with respect to varying distance. For wind turbine 600kW (M1), the aerodynamic noise associated with wind turbine blade ie., inflow turbulence, turbulent boundary layer trailing edge TBL-TE, laminar boundary layer vortex shedding (LBL-VS), stall separation , tip vortex formation are 19.34 dB(A), 4.04 dB(A), 8.12 dB(A), 16.61 dB(A), 9.60 dB(A) and for 2MW aerodynamic noise obtained are 31.46 dB(A), 18.53 dB(A), 22.92 dB(A), 28.80 dB(A), 19.05 dB(A). Equivalent noise of blade was found using equation 3.2.

$$Eq = 10log_{10}\{\frac{1}{n}+[10^{L1}/_{10} + 10^{L2}/_{10} + 10^{L3}/_{10} + 10^{L4}/_{10} + 10^{L5}/_{10}]\} \qquad \text{Eq 3.2}$$

Where Eq is the equivalent noise, n is the number blade regions,L1 is inflow turbulence, L2 is turbulent boundary layer trailing edge TBL-TE, L3 is laminar boundary layer vortex shedding (LBL-VS), L4 is stall separation,L5 is tip vortex formation.Fractions of 15.5% and 27.83 %contribution to the overall noise perceived from 600kW (M2) and 2MW(M3), were estimated. The results showed decrease in sound pressure level as the distance of measurement increased. It also showed aerodynamic noise perceived from blade is less compared to the overall SPL. This may be due to distance effect, air absorption, ground and meteorological effects, attenuation in blade region and surrounding areas.

Aerodynamic noise associated with the passage of air over the blades is typically the most important component of wind turbine acoustic emissions. A large number of complex flow phenomena occur, each of which generate sound in particular frequency bands. Aerodynamic sound level generally increases with rotor speed.

4. COMPARISON BETWEEN ONSITE NOISE LEVEL MEASUREMENTS AND MODELED RESULTS

Figure 4.1 presents the comparison between measured and model output (o/p) of noise levels with varying distance recorded, at wind speed range from 4 – 13m/s during high wind season. The intensity of noise from turbine primarily depends on the distance between the source and the point. As expected the distance between the source and the receiver increased, the sound pressure level decreased. At this point, the noise level of three wind turbines (M1-200 kW, M2-600 kW, and M3-2MW) which may be perceived by a receptor was measured with varying distances away from the respective turbine. Propagation path is another factor deciding the noise intensity. Thus, in the case of flat terrain assuming a hemispherical path for noise propagation without obstructions, the sound pressure level (SPL) at a distance R from a wind turbine radiating noise at an intensity of Lw is obtained by equation 4.1:

$$\mathrm{Lp} = \mathrm{Lw} - 10\log(2\pi\mathrm{R}^2) - \alpha\mathrm{R} \qquad \text{Eq.4.1}$$

Where Lp is sound pressure level (dB(A)),

Lw is sound power level (dB(A)),

R is the distance between the source and receiver (m),

α is the absorption coefficient 0.005dB(A) /m.

Comparative graph showing modeled and measured SPL at the wind speed range from 4 to 13 m/s (high wind season) are presented in Figure 4.1. For the three wind turbines M1-200kW, M2-600 kW and M3-2MW initial distance of measurement were taken according to standard. The SPL recorded for wind turbines M1, M2, M3 were 59 dB(A) at 42 m , 58 dB(A) at 101m and 58 dB(A) at 121m. Using the software WindPRO-DECIBEL close proximity distances for measurement of noise emission were found for all the three wind turbines M1, M2 and M3. Further using the equation 4.1, SPL for the above mentioned wind turbines were calculated for the respective distances obtained from the software. The measured and model SPL for M1, M2 and M3 were 35 dB(A) at 425 m, 36 dB(A) and 35 dB(A) at 630 m, 32 dB(A) and 35 dB(A) at 840 m respectively. This showed a fairly good agreement between the measured data (at study area) and modeled data (using relevant software WindPRO module DECIBEL – ISO 9613-2 general international standard) at varying distance and respective wind speed. The results of the ISO model indicate a value of SPL at the wind speed between 4 and 13m/s with a difference of ±5 dB (A) for every consecutive increment in distance(x'.m) formulated as mentioned earlier. This variation may be due to local meteorological effects around the study area.

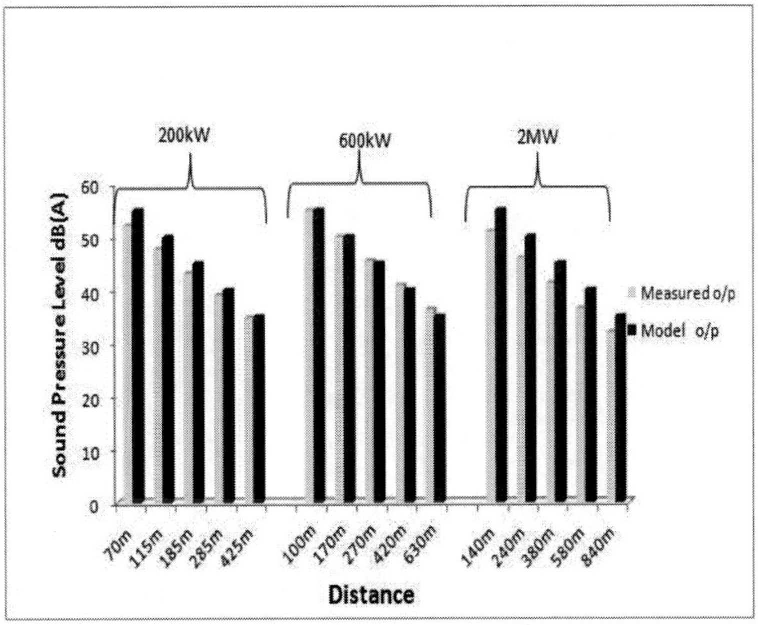

Figure 4.1 Comparison of Wind Turbine Noise Propagation (M1-200kW, M2-600kW and M3-2MW Capacity Machines)

Fig 4.2 shows the decay constant (β) for both measured and model o/p derived for three wind turbines, (M1-200 kW, M2-600 kW M3-2MW) at varying distance of measurement by dividing measured SPL to that of respective distances. It ranged between 0.04-0.79 dB (A)/ m. For wind turbines of varying installed capacity (M1-200 kW, M2-600 kW, M3-2MW) with increase in their hub height of 30m, 75m, 80m at respective distance of measurement 70m, 100m, 140m it indicated a gradual decrement in decay constant with respective to the above mentioned conditions (0.75 dB(A)/m, 0.55 dB(A)/m and 0.36dB(A)/m)).

Alternatively by measurement and model o/p fitting a simple exponential, decay equation can be derived from the following equation 4.2,

$$\mathrm{Lp}(\mathrm{x}) = \mathrm{Lpk}.\, \mathrm{e}^{-\beta \mathrm{x}} \qquad \text{Eq 4.2}$$

Where,

Lp (x) is the SPL (dB (A)) at respective distance (m),

Lpk is the SPL at the base of the tower (dB (A)),

β is the decay rate (dB (A)/m),

x is the distance of measurement (m).

SPL at the base of the tower can be obtained at varying distance by using the derived equation 4.2. Example for M1 -200kW at a distance of 70m, the measured SPL and derived (titled) β were 52.2 dB (A) and 0.75 respectively. Thus, sound pressure level at the base of the tower at the respective distance obtained by using Eq 4.2 was 56.45 dB (A). Similarly Lpk can be derived for different installed capacity wind turbines at varying distances of measurement.

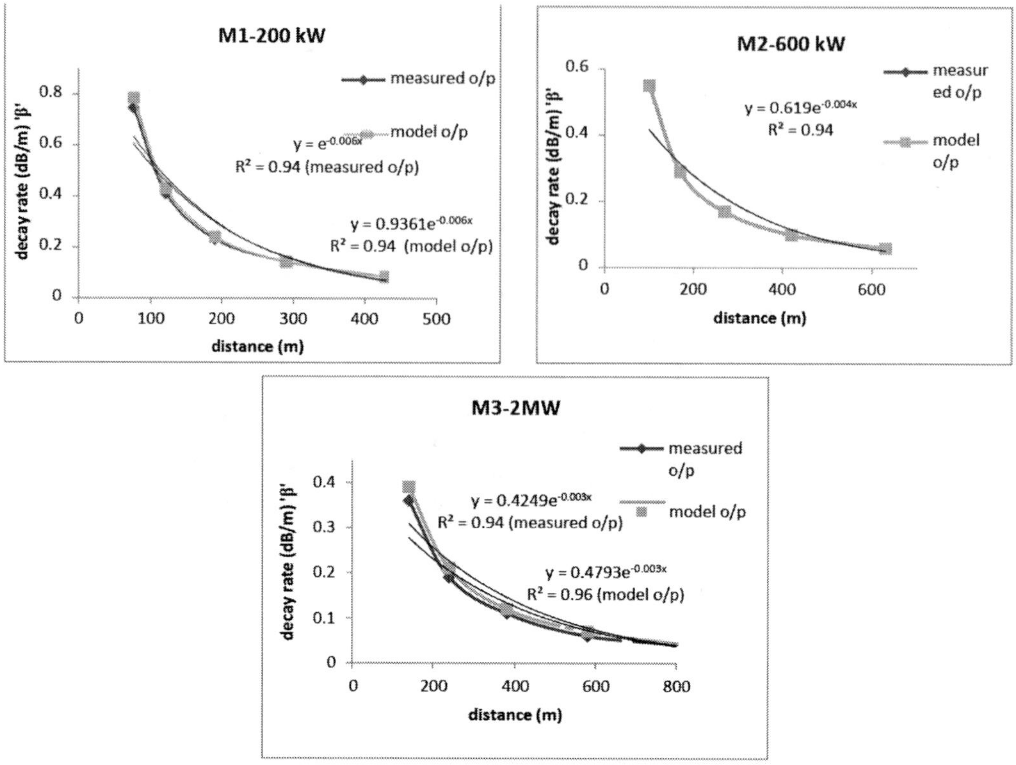

Figure 4.2 Decay rate for measured and model o/p for M1-200kW, M2-600 kW, M2-2MW

5. Conclusion:

In the current set of experiments a first pilot study has been made in India pertaining to the estimation of noise measurements produced by the operation of wind turbines located in a medium sized wind farm (less than 10MW) in flat terrain. Comprehensive onsite measurements were carried out at different measuring points, where the distance was derived from the standards for the most habitually prevailing wind speed range in the area. In the analysis of results special emphasis was given to both the evaluation of the noise level at varying measuring distances and the existing background noise. The SPL was derived by the "on operation" status of wind turbine and the existing background noise was determined on the basis of measurements during the "off" condition of the wind turbine. It should be mentioned that the noise measurements under the same wind speed conditions for both seasons i.e. high wind season at wind speed 4 - 13m/s and moderate season at wind speeds 0.7- 8.6 m/s for the three wind turbines (M1-200 kW, M2-600 kW, M3-2 MW) indicated an uncertainty range of ±5 dB(A) as the distance of measurement increased. At this point it is noted that there was no SPL variation recorded for 600 kW and 200 kW wind turbines since the wind speed was relatively low. Hence, according to the results obtained, the sound pressure level increased with increase in wind speed and there was a decrease in the same when the distance of measurement increased. An attempt was made to derive a equation to obtain sound pressure level at the base of the tower for different installed capacity wind turbines at varying distances with the measured SPL or model SPL and the respective decay rate of noise level.

The increased use of wind turbines develops the need to assess their impact on the environment where they are going to be installed since one of the most important aspect of wind turbines environmental impact is its noise. In order to assess their noise impact, the sound power level and sound pressure level of the individual wind turbine needs to be estimated. There are number of models dealing with this issue. In the current paper the individual noise from blade regions were investigated for respective wind turbines at varying distances of perception. With the help of the semi empirical models depending on available input parameters one can predict the individual blade region noise contribution to the equivalent noise. In summarizing the increased use of wind turbines develops the need to assess their impact on the environment where they are going to be installed since one of the most important aspect of wind turbines environmental impact on community/ habitat is its noise.

Nevertheless, to attain a clear trend of the noise emission spectrum from wind turbine, the ongoing specific research will be continued to a greater wind speed variation range as well as multiple reference points of measurements during the day and night of moderate and windy seasons in Indian flat and complex terrain to hardness clean wind energy. Finally, the availability of detailed onsite measurements in flat terrain allowed for comparisons to be made with modeled results at varying distances. Generally, an incremental validation of the prediction models was provided by observing a fairly good agreement between experimental and modeled results, for practical applications following international standards of best practices.

6. Acknowledgement

The authors acknowledge National Institute of Wind Energy (NIWE) under Ministry of New and Renewable Energy, Government of India, Chennai, Tamilnadu, for extending technical support and research infrastructure for the study. The authors also acknowledge the technical assistance rendered by Ms.A.Madhumitha as a project research assistant.

7. References

1. Ministry of New and Renewable Energy (MNRE), Government of India, available at http://www.moef.gov.in (assessed on January 2017).

2. Sung Soo Jung, Wan-Sup Cheung, Cheolung Cheong and Su-Hyen Shin., Experimental Identification of Acoustic Emission Characteristics of Large Wind Turbines with Emphasis on Infrasound and Low- Frequency Noise. J. of the Korean Phy. Soc., 2008, 53, 1897 – 1905.

3.Wagner, S., Bareib, R.and Guidati, G., Wind Turbine Noise. Springer.,1996, DOI:10.1007/978-3-642-88710-9.

4. Mark Bastasch., Jeroen van Dam., Bo sondergaard.and Anthony Rogers., Wind turbine Noise – An overview. Canadian Acoustic., 2006, 34, 7-15.

5. John, K, Kaldellis. and Zafirakis, D., The wind energy (r) revolution: A short review of a long history. Renewable energy., 2011, 36, 1887-1901.

6. Environmental Impact Assessment (EIA) study for a Proposed 100MW Wind Energy Project. 2012, prepared by Kurrent Technologies Ltd., Kenya., Kajiado District, Kenya for Kipeto Energy Limited, Report No. KT/1000/002/11.

7. Persson Waye ,K. and Agge ,A., Experimental quantification of annoyance unpleasant and pleasant wind turbine sounds. Proc. of Internoise., 2000, 6, 3994-3997.

8.Hubbard ,H.H., Grosveld, F.E. and Shepered, K.P., Noise characteristics of large wind turbine generators. Noise Con. Eng. J., 1983, 21, 21-29.

9. Erich and Hau. Wind Turbines Fundamentals, Technologies Application, Economics., Springer-Verlag Berlin Heidelberg, 2006, DOI 10.1007/3-540-29284-5.

10. ETSU-R-97, The Assessment and Rating of Noise from Wind Farms., 1996, UK Department of Trade and Industry.

11. Anthony ,L., Rogers , James ,F., Manwell., Sally Wright, M.S. PE .,Wind Turbine Acoustic Noise. A white paper 2006, Prepared by the Renewable Energy Research Laboratory Department of Mechanical and Industrial Engineering, University of Massachusetts at Amherst.

12.The Environment (Protection) Act, 1986,amended 1991 – S.O.123(E),14/2/2000 – Noise Pollution (Regulation and Control) Rules, 2000, under the gazette notification of the Government of India in the Ministry of Environment, Forests and Climate change.

13. IEC 61400:11 2002-12 Wind turbine generator systems – Part 11 : Acoustic noise measurement techniques, Second edition. International Technical Commission.

14. John, K, Kaldellis. and Zafirakis, D., The wind energy (r) revolution: A short review of a long history. Renewable energy., 2011, 36, 1887-1901.

15. Pederson ,E., Hallberg ,L.R. and Persson Waye, K.P.,Living in the vicinity of wind turbines- A- Grounded Theory Study. Q. Res. in Psy., 2007, 1,49-63.

16. Kaldellis ,J.K., Garakis, K. and Kapsali, M., Noise impact assessment on the basis of onsite acoustic noise emission measurements for a representative wind farm, Rene. Ene. 2012, 41, 306-314.

17. Claudio Guarnaccia, Nikos.E. Mastorakis. and Joseph Quartieri., Wind Turbine Noise- Theoretical and experimental study., 2011 , Int. J. of Mech.,5, 129-137.

18. Claudio Guarnaccia, Nikos.E. Mastorakis. and Joseph Quartieri., Wind Turbine Noise- Theoretical and experimental study., 2011 , Int. J. of Mech.,5, 129-137.

19. H. H. Hubbard, F. W. Grosveld, K. P. Shepherd, Noise characteristics of large wind turbine generators, Noise Control Engineering Journal, Vol. 21, 1983, pp 21–29

20.Examination Of The Significance Of Noise In Relation To Onshore Wind Farms, Seai Wilton Park House, Wilton Place DublinPrepared by: Marshall Day Acoustics., 2013

21.Lowson, M. V., "Assessment and Prediction of Wind Turbine Noise," Flow Solutions Report 92/19, ETSU W/13/00284/REP, 1992, pp. 1-59.

22. Hau, E., Langenbrinck, J and Palz, W., "WEGA Large Wind Turbines," Springer-Verlag, Berlin, 1993, pp. 1-143. http://dx.doi.org/10.1007/978-3-642-52129-4.

23. Hagg, F., "Aerodynamic Noise Reduced Design of Large Advanced Wind Turbines," ECWEC'90, Proc. of the European Community Wind Energy Conference, Madrid, Spain,pp. 384-388, September 1990.

24. Brooks, F. T.; Pope, D. S. and Marcolini, M. A. ,"Airfoil Self-Noise and Prediction,". NASA RP-1218, 1989, pp. 1-137.

25. Grosveld, F. W., "Prediction of Broadband Noise from Horizontal Axis Wind Turbines," Journal of Propulsion and Power. 1985, Vol. 1, No. 4, pp. 292-299.

26. Bill Dawson and Neil Mackenzie., Meteorological stability impacts on wind turbine noise assessments.,2013, Proc.of Acou., Victor Harbor, Australia.

27. Pierre Dutilleux and Joachim Gabriel., Assessment of the acoustic noise issues of wind farm projects in the light of the experience gained in Germany, DEWI GmbH, Ebertstrasse 96, 26382 Wilhelmshaven.

28. Tom Evans and Jonathan Cooper ., Effects of different meteorological conditions on wind turbine noise. Proc. of Acou., 2013, Victor Harbor, Australia.

29. Rhys Brown., Andrew Mitchell and Luis Najera Medina., Forecasting low frequency noise from wind farms. Proc.of Acou.,2013, Victor Harbor, Australia.

30. Dutilleur ,P. and Gabriel, J., Recommendations for improved acceptance of wind farm projects in France with regard to acoustic noise., 2007, Proc.of the 2nd int. meeting on wind turbine noise., ISBN: 978-1-62276-390-0 ,Lyon, France.

31. Klug, H.A., Review of wind turbine noise 2005.,Proc. of first int. conf. on wind turbine noise, Berlin, Germany.

32. Frontier Power Systems .,Noise Impact Assessment of the Proposed Hermanville Wind Farm, PEI Energy Corporation., 2013,Alberton, PE.

33. WindPro 2.6-DECIBEL Module, online user manual of Windpro Software, Version 2.6, 2010.

7thInternational Conference
on
Wind Turbine Noise
Rotterdam – 2nd to 5th May 2017

A new characterization of wind turbine noise from Life Cycle Assessment.

Andrea Rivarola. Instituto Nacional de Tecnología Industrial INTI. Universidad Tecnológica Nacional-Facultad Regional Mendoza. Argentina. andreari@inti.gob.ar
Pablo Arena. Universidad Tecnológica Nacional-Facultad Regional Mendoza. Argentina. aparena@frm.utn.edu.ar
Héctor Mattio. Wind Energy Independent Consulting, Chubut, Argentina. hmattio@gmail.com

Summary

Purpose
At present, Life Cycle Assessment (LCA) does not include noise as an obligatory impact category. Noise is not a material emission, which makes difficult to quantify the impact caused by noise, and to compare it with other impact categories. The aim of this work is the development of a methodology for including Noise Impact Indicators (NII) based on Site-Specific Characterization Factors (SSCF) in LCA studies.

Methods
First, the Noise Indicator Model (NIM) presents SSCF calculus that includes a fate analysis and site characteristics (temperature, relative humidity, wind speed, terrain, etc). Then, the NII is calculated using these characterization factors. In this part, the model considers source characteristic, noise emissions of the source (wind turbine noise emission) and site background noise, both coming from the inventory stage. The visualization of data and final results has been done using a Geographical Information System (GIS). We propose to apply the NIM in five regions of Argentina, evaluating a wind turbine as stationary noise source.

Conclusions and Discussions.
The calculated NII, enabled us to identify which sites presented annoying noises, and relate them to an hourly and seasonal meteorological condition. The results thus obtained were more informative than those obtained from other models that not include the refined analysis proposed by this. Another advantage was that the measuring unit employed in the results was simple to interpret them and in accord with the recommended units for noise indicators (WHO 2003).

Outlook
The proposed NIM allows the quantification of noise impact through the NII which are coherent with processes related to outdoor sound propagation, as well as with meteorological data variability related to each study site and with the LCA methodology. It is through spatial and temporal differentiation that a greater sensitivity in the results can be reached by the NIM compared to other general models (ISO/DIS 9613-21989). Finally, although some NII calculus

related factors may be improved, the model seems promising for noise impact analysis within the LCA framework.

Keywords: Site-dependant characterization factor, Midpoint noise indicator, Wind turbine.

1. Introduction

Noise is considered an optional impact category within LCA, and is excluded in most of the LCA studies (Guinée et al 2002). The development of environmental impact indicators in this category is an arduous task, due to the lack of consensus about the parameters to include, and the complexity of the involved environmental mechanisms, which are highly dependent on local conditions.

This situation leads, to an absence of reliable Noise Indicator Model (NIM) fully compatible with LCA (Müller -Wenk 2004; Benetto et al. 2006). The most significant attempts to include this category in LCA were made in assessing noise coming from road traffic (Lafleche et al. 1997, Müller -Wenk 2002, 2004, Meijer et al 2006, Althaus et al. 2009, Cucurachi et al 2012, 2014). There are some discussions on the best way for inclusion of noise impacts in LCA (Heijungs et al 2016, Ongel 2016).

On the other hand, wind turbine noise is a controversial subject among manufacturers and installers of wind parks, inhabitants living close to the parks and environmentalist groups (Pedersen 2007). Employing LCA compatible noise indicators could provide objective information about the problem, since evaluates quantitatively the impact caused by noise in different places. In this way a Noise Indicator could help to identify sensitive areas whose inhabitants could suffer the annoying noises.

This article describes a proposed method for a quantitative assessment of noise impact in LCA, through a model that develops a midpoint category indicator. The model has been applied for assessing the impact caused by the noise coming from the operation of wind turbines located in different sites of Argentina, taking into account seasonal and hourly variations.

2. Methods

2.1 Development of a characterization factor for noise category

The model analyzes noise as a local impact category, through characterization factors that include a spatial and temporal analysis. To include a spatial analysis, the characterization model includes a fate analysis of the pollutant (noise emission) taking into account the environmental mechanisms involved. The temporal differentiation is evaluated according to the disturbance provoked by the contaminant. This varies according to the moment of the day (day or night) and to the season of the year (winter or summer).

The environmental mechanism (outdoor sound propagation) together with the environmental intervention that originates the noise emission (noise source) are integrated into the characterization model through the selection and combination of the factors related to the physical phenomenon concerned.

2.2 Development of an impact indicator for noise category

Choosing indicators at a mid-point level is commonly considered as the best practice. (Guinee et al. 2001) but the decision making process is easier using end-point indicators due to their straightforward interpretation. The choice of an indicator at different levels will also require special attention to the consistency of the impact framework as a whole related to the type of indicator chosen (Leske et al. 2003). The model proposed in this work produces a midpoint indicator that evaluates quantitatively the alteration of the site environmental quality at a level close to the original environmental intervention.

At the inventory phase, one must gather information from the exact site where the study is taking place. The required information includes noise emission data from the source, data from the site background noise level and climactic characteristics.

2.3 Proposed Model

Outdoor noise propagation is a complex phenomenon depending on a number of factors that are sometimes difficult to estimate and model.

A thorough study of the most significant site factors which influence outdoor noise propagation from wind turbines has been performed, and they have been included in the Noise Indicator Model (NIM). For a stationary noise source, like the wind turbine, these factors are: a) Geometrical Divergence Factor, b) Atmospheric Absorption Factor, c) Ground Factor, d) Meteorological Effect Factor (Attenborough 2007, 2008; Crocker 2007; Rossing 2007; Van den Berg 2003, 2006). In this model, a plain terrain has been considered since it is a frequent topography where wind farms are installed (e.g. Patagonian region in Argentina, off-shore wind farms etc), and it is the easiest situation to analyze. The presence of artificial barriers and vegetation can be linked to additional mechanisms that only arise in specific situations.

The Characterization Factor (NCF) was calculated according to the equation 1:

$$NCFi = GDF + \mathrm{AAFi} + \mathrm{GFi} + \mathrm{MEFi} \ldots \quad (1)$$

GDF represents the Geometrical Divergence Factor, which is a function of the source-receptor distance, and it is independent of site characteristics. It was calculated from equation 2, considering that for a stationary source the noise propagates spherically (Attenborough et al 2007).

$$GDF = \left[20 \times Lg\left(\frac{d}{d_o}\right) + 11 \right] \quad (2)$$

In equation 2, d represents the source-receptor distance, while do represents a reference distance.

AAFi points out the Atmospheric Absorption Factor. As sound propagates through the atmosphere its energy converts gradually into heat, through several molecular processes known as air absorption. AAFi depends on the source-receptor distance (d) and on the air absorption coefficient (α), which in turn depends on the air temperature, relative humidity and pressure, and on the noise frequency (Blackstock 2000). AAFi was calculated through the following expression (ISO/DIS 9613-1, 1990):

$$AAFi = \alpha \times \frac{d}{1000} \quad (3)$$

GFi represents the Ground Factor. It is considered that above the reflecting ground, sound reaches a receptor (R) from a source (S) in two ways: directly through a direct ray (dr) trajectory, and by the reflection from the ground through a reflected ray (rr) trajectory. Thus, the GFi represents attenuation due to the ground as the result of interference between direct and reflected sound, greatly depending on the type of ground surface, on the grazing angle, the difference in length of the runs (distance rr - distance dr) and on the sound frequency. The selected calculation method for the ground factor is valid for both short (less than 100 meters) and long distances, for plain terrain, and assuming favorable atmospheric conditions for propagation, meaning that the ray from the source to the receptor is refracted downwards (ISO/DIS 9613-2, 1996). The GFi was determined by the following equation:

$$GFi = SZF + IZF + RZF \quad (4)$$

SZF: represents the source zone factor, IZF considers the intermediate zone between the source and the receptor, and RZF represents the receptor zone factor. First, we selected the Ground Type for each zone. The employed Ground Type classification, according to an acoustic perspective, is indicated in Table 1.

Table 1. Ground classification from acoustic perspective. Source ISO/DIS 9613-2

Ground Type (GT)	Hard Ground (HG)	Soft Ground (SG)	Mixed Ground (MG)
Characteristics	Pavement, asphalt, concrete, water. Slightly porous	With grass, trees or vegetation. Porous	Hard and soft areas
Values	GT = 0	GT = 1	0< GT* >1

*Equal to the soil proportion, which is soft.

Second, we calculated the SZF, IZF and RZF by the chosen Ground Type for each region using the expressions proposed by ISO/DIS 9613-2, 1996.

The last factor considered in equation 1 is the Meteorological Effects Factor (MEF), estimated from the algorithms proposed in the CONCAWE model (Manning 1981).

This model considers the influence of the wind speed, solar radiation and the stability of the atmosphere (Pasquill 1961)
Finally, the Noise Impact Indicator is calculated according to the following expression:

$$NII = \Sigma(NEi - NCFi) - BNF \quad (5)$$

NEi is the Noise Emission or Sound Power Level for each frequency (i), in dB(A); NCFi represents the Noise Characterization Factor for each frequency (i), in dB(A) and BNF is the existing Background Noise Factor in the considered region, in dB(A).
NII represents the sound muffling capacity of a region exposed to a certain level of noise, in the predominant sound propagation direction. The NII is determined for specific meteorological conditions and for the moment of day (day or night) in which the environmental perturbation occurs. From this indicator we can identify those sites that present annoying noises and the conditions under which they are generated during the year. The indicator obtained is measured in dB(A) units, the recommended unit for noise impact indicators, where A refers to the perceived noise by the human hear (0 dB -hearing threshold- to 100 dB – threshold of pain) (WHO 2003).
The Table 2 shows the proposed classification to evaluate noise indicator values.

Table 2 Proposed classifications for NII

Proposed ranks for NII [dB(A)]	*Noise classification*	*Site muffling capacity*
< 8	Not annoying noise	Favorable
8	Annoying noise	Unfavorable
>>8	Very annoying noise	Very unfavorable

In order to give further insights into this model, a sensitivity analysis has been performed to determine the most significant contributors to noise impact. The four factors involved in the calculation of the noise indicator are dependent on the distance to the source. The Absorption

Factor, Ground Factor and Meteorological Effect Factor depend also on other variables like atmosphere temperature and humidity, source-receptor distance, noise frequency, etc. The results suggest that the temperature and atmospheric humidity have a relatively minor influence on the indicator, which could increase at higher frequencies and greater distances. At lower frequencies, noise propagates more efficiently, being less influenced by air absorption. On the other hand, the Ground Factor seems to be higher for hard ground, and has a strong influence on the indicator. To sum up the Ground Factor and Meteorological Effect Factor suggest the most relevant influence on the Noise Impact Indicator.

3. Example: Noise Impact of Wind Energy in Argentina

In order to show the application of the Noise Indicator Model, a case study has been performed for different locations with high wind energy potential in Argentina. Several scenarios with different meteorological conditions have also been considered for evaluating the influence of extreme climatic conditions on outdoor noise propagation.

3.1 Evaluated Sites

The model was run for five sites of Argentina: 1) Cutral Có in Neuquén, 2) Arauco in La Rioja, 3) Comodoro Rivadavia in Chubut, 4) Trelew in Chubut, 5) Bahía Blanca in Bs. As. They have been chosen by their significant eolic potential which have been analyzed by Argentinean Wind Energy Geographic Information System and some of them have just installed wind turbines.

3.2 Tools and Data employed

The Argentinean Wind Energy Geographic Information System (CREE et al 2006) was used to obtain information about the wind energy potential of the chosen sites. We evaluated the speed frequency distribution and the predominant wind directions. A V47-660 kW wind turbine was selected as noise source which is highly efficient in the vast majority of wind conditions and has been installed widely in Argentina. We analyzed the power curve of the selected wind turbine, verifying the functioning of this equipment in the site. In Table 3 we show the data of noise emission or data of sound power level (SPL) for frequency used in the calculi.

Table 3 Measurement Data at a Sound Power Level (SPL) corresponding to V47-660 kW. Source: VESTAS 2012

Frequency in 1/1 of octaves (Hz)	*SPL, ref [dB(A)]*
63	78,2
125	86,1
250	89,8
500	95,2
1000	97,0
2000	92,9
4000	87,9
8000	69,2
Weighted total value (A)	100,8

We employed statistical data of the wind speed, wind direction, temperature, relative humidity and cloudiness (SMN 2012) with the aim of evaluating the meteorological conditions of the specified sites. Mean monthly solar radiation was determined for the different sites through the Argentine Solar Atlas (Grossi Gallegos et al 2007). Arc View GIS was used to assist visualization of Noise Impact Indicators in different areas of Argentina. It has allowed to identify sensitive zones around the source easily.

In Argentina there are no available background noise measurements to be used as Background Noise Factor. However, it is possible to estimate the BNF value according to site characteristics following existing standards (IRAM 4262 2001).
To analyze the influence of meteorological data variability during the year, different scenarios, shown in Table 4, have been proposed. They also take into consideration the time of the day when the data had been collected, with the aim of evaluating the influence of extreme climatic conditions on outdoor noise propagation (mean annual data vs. monthly data).

Table 4 Characteristics of possible scenarios

Scenarios	*Climactic Conditions*	*Temporal differentiation*
S-Aa	A- Situation employing mean annual data	a- Day
S-Ab		b- Night
S-Ba	B- Situation employing mean monthly data	**a- Day - June - Winter**
S-Bb		**b- Night - June - Winter**
S-Ca	C - Situation employing mean monthly data	a- Day - January - Summer
S-Cb		b- Night - January – Summer

The application of these scenarios to the chosen sites allows us to introduce the temporal and spatial differentiation into the analysis.

4. Results and Discussion

In Table 5, we have pointed out the chosen relevant scenarios (S-Ba and S-Bb), for each analyzed site. These scenarios have presented the most significant NII values according to the proposed classification for NII in Table 2.

Table 5 Summarize the Noise Indicators Results

Site	*Sceneries*	*Pasquill's Category*	*Meteorological Category*	*Ground Type*	*Source Distance (m)*	*Dominant Propagation Direction*	*NII (dB(A))*
Cutral Có Neuquén	S-Ba	D	6	0,5	300	South-West	8
	S-Bb						
Arauco La Rioja	S-Bb	D	6	0	400	South	9
C. Rivadavia Chubut	S-Ba	D	6	0	400	West	9,2
	S-Bb			0,5	300		8,2
Trelew Chubut	S-Bb	C	6	0	400	West	9,5
	S-Ba			0,5	300		8,5
Bahía Blanca Bs. As.	S-Bb	D	6	0,5	300	North-West	8,5
	S-Ba						

For Noise Impact Indicator, the above results indicate that the most unfavorable condition for outdoor noise propagation occurs generally at night and during the winter season, especially in June. This suggests that variables like solar radiation and cloudiness degree have a strong influence in the meteorological factor. Besides, the results indicate that the presence of a hard ground (GT = 1) gives greater NII than a mixed ground (GT = 0,5).

The obtained results reflect the influence of spatial (different sites) and temporal (seasonal and hourly) differentiation, proposed by the model on outdoor sound propagation. These emphasize the importance of including local characterization factors for quantitative impact evaluation.

These results also justifies employing monthly instead of yearly meteorological data, since the last one ignores different climactic conditions on sound propagation.
The Figure 1 shows the Noise Indicators according to Table 5, calculated at maximum distance from the source.

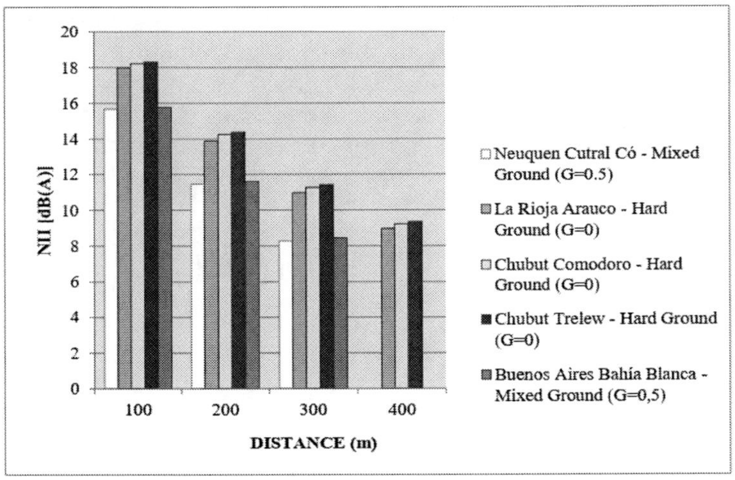

Fig. 1 Sites with Annoying NII vs. Source distance

Then, the Figure 2 shows how the noise propagates on the different sites according to the analyzed variables.

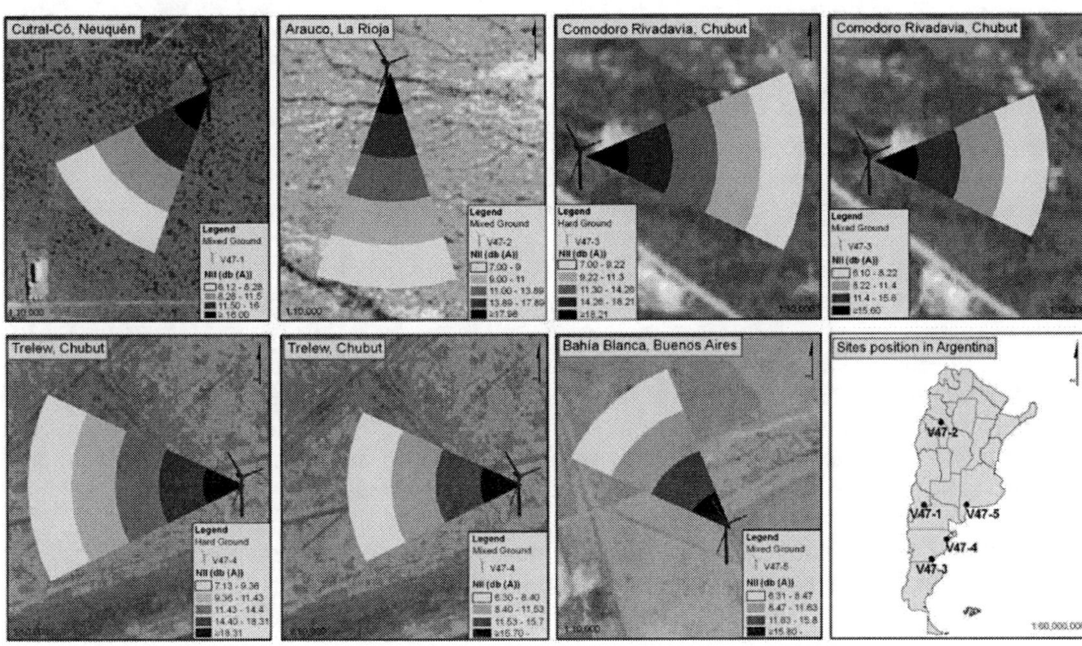

Fig. 2 Location map of study area showing the sensitive areas around the studied sites and the dominant direction of propagation.

As a final part of the analysis, the Noise Impact Indicator results are compared to those obtained from the ISO 9613-2 model, widely adopted as the basis for outdoor noise propagation calculus, but that does not consider the Meteorological Effects Factor. Apart from that, there is some discussions about the limitations of applying this calculation method to noise emission sources of great height (Gonzalez et al. 2011)

This model was applied to La Rioja-Arauco and Neuquén- Cutral-Co sites, which presented the highest values for Noise Impact Indicator. ISO model determines noise level at different distances from the source, which has been used as an input for the calculation of Noise Impact Indicator employing equation 5. In Figure 3 the NII results for the sites calculated from both models has been compared. We observed that the sensitive zones to annoying noise calculated according to the ISO Model are smaller than those calculated from our model. This is due to the absence of the meteorological factor in the ISO calculi, which manifests itself only at 200 m measured from the noise source and in the predominant direction of noise propagation.

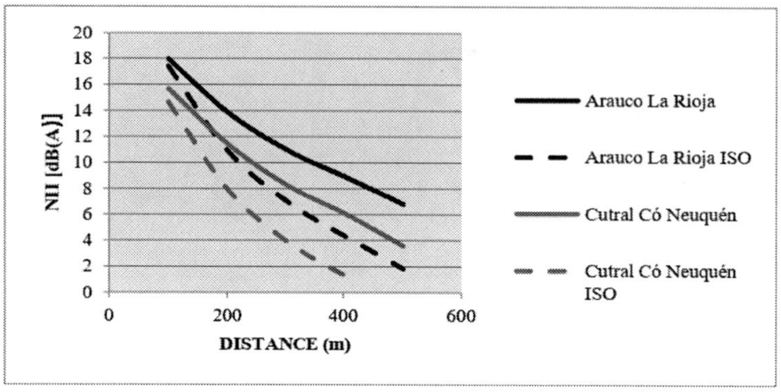

Fig.3 ISO Model vs. NIM Model

5. Conclusions and Outlook

The Noise Indicator Model represents a simple and complete way to study outdoor noise propagation. It has been applied in different sites of Argentina. Through this model, we could establish objective comparisons between sites employing Noise Impact Indicators.

NIM employs a simple measure unit (dB (A)). It is simple to interpret since it has a clear physical meaning and it is in agreement with the units recommended by noise indicators.

With the analysis refinement proposed by the model presented in this work, sensitive zone can be detected dependent on the variability of meteorological data employed.
Although, MIN model seems to be suitable to noise study from a life cycle perspective, it would be also interesting to develop further studies to advance in the analysis of the meteorological factor and the consideration of the topography of complex terrains.

The proposed model calculates a Noise Impact Indicators at a midpoint level. However, the model could be used for the development of an end point indicator.

References

Attenborough, K (2008) Sound Propagation in the Atmosphere. In: Crocker M J(ed) Handbook of Noise and Vibration Control (ed M. J. Crocker), John Wiley & Sons, Inc., Hoboken, NJ, USA. doi: 10.1002/9780470209707.ch5

Attenborough K, Ming Li K, Horoshenkov K (2007) Prediction Schemes. In: Taylor & Francis e-Library (ed) Predicting Outdoor Sound, pp 380-406.ch12

Althaus, H. J., De Haan, P., & Scholz, R. W. (2009). Traffic noise in LCA: part 2: analysis of existing methods and proposition of a new framework for consistent, context-sensitive LCI

modeling of road transport noise emission. International Journal of Life Cycle Assessment, 14(7), 676–686.
Benetto E, Dujet C, Rousseaux P (2006) Fuzzy-Sets Approach to Noise Impact Assessment. Int J LCA 11 (4) 222 – 228.
Blackstock D T (2000) Fundamentals of Physical Acoustics. Copyright 2000 John Wiley & Sons, Inc 541 pp. ISBN 0-471-- 31979-1
C.R.E.E. et al (2006) Centro Regional de Energía Eólica. Sistema de Información Geográfico Mapa Eólico Nacional. Ministerio de Planificación Federal, Inversión y Servicios Públicos 2006. Available on website: https://www.minem.gob.ar/www/833/25683/mapas.html (last accessed date Febrery 12, 2017)
Crocker M J et al. (2007) Handbook of Noise and Vibration Control Edited by Crocker M J. John Wiley & Sons, Inc. ISBN 978-0-471-39599-7. DOI: 10.1002/978047029707.
Cucurachi, S., Heijungs, R., & Ohlau, K. (2012). Towards a general framework for including noise impacts in LCA. International Journal of Life Cycle Assessment, 17(4), 471–487.CrossRef (http://dx.doi.org/10.1007/s1136701103774)
Cucurachi, S., & Heijungs, R. (2014). Characterisation factors for life cycle impact assessment of sound emissions. Science of the Total Environment, 468, 280–291.CrossRef (http://dx.doi.org/10.1016/j.scitotenv.2013.07.080)
Grossi Gallegos H, Righini R (2007) Atlas de energía solar de la República Argentina. Publicado por la Universidad Nacional de Luján y la Secretaría de Ciencia y Tecnología, Buenos Aires, Argentina, 74 páginas + 1 CD-ROM, mayo de 2007 (ISBN 978-987-9285-36-7), Registro de la Propiedad Intelectual No. 554247 del 15 de marzo de 2007.
Guinée JB, Gorrée M, Heijungs R, Huppes G, Kleijn R, de Koning A,van Oers L, Wegener Sleesewijk A, Suh S, Udo de Haes HA, de Brujin H, van Duin R, Huijbregts MAJ, Lindeijer EW, Roorda AAH, van der Ven BL, Weidema BP (2001) Life cycle assessment: an operational guide to the ISO Standards. Final Report, May 2001, Ministry of Housing, Spatial Planning and Environment (VROM) and Centrum voor Milieukunde (CML), Rijksuniversiteit, Den Haag and Leiden.
González, E., Rezzano Tizze, N., Bianchi Falco, F Algunas limitaciones de la Norma ISO 9613 – Parte 2 para el estudio de propagación de ruido de aerogeneradores de gran porte. Reunión Regional de Acústica. AUA 2011. RRA – 1105.
Guinée J B, Gorrée M, Heijungs R, Huppes G, Kleijn R, Koning A de, Oers L van, Wegener Sleeswijk A, Suh S, Udo de Haes H A, Bruijn H de, Duin R van, Huijbregts M A J (2002) Handbook on life cycle assessment. Operational guide to the ISO standards. I: LCA in perspective. IIa: Guide. IIb: Operational annex. III: Scientific background. Kluwer Academic Publishers, ISBN 1-4020-0228-9, Dordrecht, 2002, 692 pp.
Heijungs, R. & Cucurachi, S. Life Cycle Assessment of Noise Emissions: Comment on a Recent Publication Environmental Model Assessment (2016). doi:10.1007/s10666-016-9545-z
IRAM 4062. Norma Argentina. 3° Edición 2001. Ruidos Molestos al vecindario. Método de medición y clasificación. Annoying noises for the neighborhood. Method for their measurement and classification
ISO/DIS 9613-1, International Organization for Standardization (1990): Acoustics-Attenuation of Sound during Propagation Outdoors, Part 1, Calculation of the Absorption of Sound by the Atmosphere, CH-1211 Geneva 20, Switzerland.
ISO/DIS 9613-2, International Organization for Standardization (1989): Acoustics-Attenuation of Sound during Propagation Outdoors, Part 2, A General Method of Calculation, CH-1211 Geneva 20, Switzerland.
Lafleche V, Sacchetto F (1997) Noise Assessment in LCA-A Methodological Attempt. Int J LCA 2 (2) 111-115.
Leske T, Buckley C (2003) Towards the development of a salinity impact category for South African environmental Life-cycle assessments: Part1- A new impact category. ISSN 0378-4738=Water SA Vol. 29 N° 3. Available on website http://www.wrc.org.za (last accessed date July 6, 2012)

Manning, J C (1981) CONCAWE: The propagation of noise from petroleum and petrochemical complexes to neighboring communities-Phase I"AT 674. Final Report issued to stitching.
Meijer A, Huijbregts MAJ, Hertwich E, Reijnders L (2006): Including Human Health Damages due to Road Traffic in Life Cycle Assessment of Dwellings. Int J LCA 11. Special Issue 1, 64 – 71.
Müller-Wenk R (2002) Attribution to road traffic of the impact of noise on health. Environmental series no. 339. Swiss Agency for Environment, Forest and Landscape (SAEFL), Bern.
Müller-Wenk R (2004) A method to include in LCA road traffic noise and its health effects. Int J Life Cycle Assess 9:76–85.
Nielsen PH, Laursen JE (2000) Integration of External Noise Nuisance from Road and Rail Transportatiom in lifecycle assessment.18pp [online]: http://www.asdu.ait.ac.th/faculty/FacultyByID.cfm?FacultyID=459.
Ongel, A. (2016). Inclusion of noise in environmental assessment of road transportation. Environmental Modeling and Assessment, 21, 181–192.
Pasquill, F. 1961. The Estimation of the dispersion of windborne material. Meteorological Magazine. 90: 33-49.
Pearsons K, Bennett R, (1974) Handbook of Noise Ratings, NASA CR-2376.
Pedersen E (2007) Thesis PhD Human response to wind turbine noise: perception, annoyance and moderating factors.
Piercy T, Embleton, Sutherland L (1977) Review of Noise Propagation in the Atmosphere, JASA 61 (6), pp 1403.
Rossing T D, (2007). Springer Handbook of Acoustics. ISBN 978-0-387-30446-5
SMN (2012) Servicio Meteorológico Nacional Argentino Available on website: http://www.smn.gov.ar/ (last accessed date July 6, 2012)
Van den Berg F G P (2003) Effects of wind profile at night on wind turbine sound. Journal of sound and vibration. Doi:10.1016/j.jsv.2003.09.050
Van den Berg F G P (2006) Thesis Phd: The sound of high winds: the effect of atmospheric stability on wind turbine sound and microphone noise.
WHO (2003) technical meeting on noise and health indicators. Second meeting - Results of the testing and piloting in Member States. Bonn Germany, 18th, 19th December 2003 Summary report. Avalaible for download through. www.euro.who.int/document/NOH/Nohindicmeetrptdec03.pdf
VESTAS (2012) V47-660 Kw. Available on website: http://www.iufmrese.cict.fr/concours/2002/CG_2002STI_lycee/Pour_en_savoir_plus/Vestas_V47.pdf (last accessed date July 6, 2012)

7th International Conference
on
Wind Turbine Noise
Rotterdam – 2nd to 5th May 2017

Experience of reviewing wind farm noise assessments for Scottish local authorities and the implementation of the IOA Good Practice Guide to the Application of ETSU-R-97 for the Assessment and Rating of Wind Turbine Noise

Steve Summers, ACCON UK Ltd, Citibase, 95 Ditchling Road, Brighton, BN1 4ST, United Kingdom. steve.summers@accon-uk.com
Graham Parry, ACCON UK Ltd, Unit B Fronds Park, Frouds Lane, Aldermaston, Berks RG7 4LH, United Kingdom. graham.parry@accon-uk.com

Summary

ACCON UK has been carrying out reviews of noise assessments for onshore wind farms submitted in support of planning applications to two Scottish local authorities. This work has enabled us to review the similarities and differences between the approaches of a number of acoustic consultants. This paper considers the approaches used for background noise surveys, noise prediction modelling, the selection of noise criteria and limits and the assessment of cumulative noise. Most of the assessments considered have been carried out since the publication by the UK's Institute of Acoustics of their Good Practice Guide. An overview of the success of the Good Practice Guide in standardising the implementation of noise assessments based on ETSU-R-97 is therefore also presented. Case studies are discussed which highlight the approach to cumulative assessments and the different cumulative assessment results which can be obtained for common receptors when assessed for neighbouring wind farm developments.

1. Introduction

The 1996 report 'The Assessment and Rating of Noise from Wind Farms', ETSU-R-97[1], sets out the methodology for assessing noise from wind turbines that is approved by the UK Government and the Scottish Government. Following concerns that there was a lack of consistency in how wind farm noise assessments were being carried out, the UK's Institute of Acoustics (IOA) published 'A Good Practice Guidance to the Application of ETSU-R-97 for the Assessment and Rating of Wind Turbine Noise' (the GPG)[2] in 2013. The GPG resulted from a request by the UK Department of Energy and Climate Change (DECC), to develop recommendations from the Hayes McKenzie Partnership Report[3] on 'Analysis of How Noise Impacts are considered in the Determination of Wind Farm Planning Applications'. The GPG provides a comprehensive guide to the implementation of the measurement, prediction and assessment procedures given in ETSU-R-97. A framework for setting noise limits is given in ETSU-R-97 as well. Implementation of noise limits is also covered by the GPG, along with model planning conditions to address operational noise from wind turbines.

As the IOA GPG has become an established reference document, adoption of its recommendations within the United Kingdom has become a widely accepted method of demonstrating that a noise assessment for a new wind farm has followed best practice. ACCON on behalf of a number of Scottish Local Authorities has carried a review of noise assessments submitted in support of planning applications for over twenty wind farms in Scotland. This work started in 2014 for East Ayrshire Council and has also included wind farms in the South Ayrshire Council area since 2016. This work has enabled us to review and compare the approaches of different acoustic consultants. Most of the assessments considered have been carried out since the publication of the GPG. As a result, we have been able to carry out a review that assesses the similarities and differences between the approaches to the noise assessments submitted for different wind farms. Specifically, the implementation of the various recommendations of the GPG have been investigated by completing a summary matrix considering the key requirements of a wind farm noise assessment. This investigation has considered the approaches used for background noise surveys, use of turbine sound power data, noise prediction modelling, the selection of noise criteria and limits and the assessment of cumulative noise.

Two case studies are presented. The first shows how the assessments of noise from neighbouring wind farm proposals resulted in distinctly different assessment findings due to differing approaches to obtaining background noise data. The second case study shows the necessity of taking account of all relevant planning information from nearby wind turbine developments as part of the consideration of cumulative noise issues.

2. Review of Implementation of Good Practice Guide

This review has considered nineteen noise assessments for noise assessments submitted in support of planning applications submitted between August 2013 and October 2016. The GPG includes a number of summary boxes, numbered SB1, SB2 etc. These provide succinct summaries of key good practice recommendations and several of these are referenced in the review set out below.

2.1 Background Noise Surveys

Our analysis shows that for all the assessments the consultants indicated that the Local Planning Authority was consulted over the approach to the noise surveys, including the selection of monitoring locations. This demonstrates that practitioners are generally following the GPG SB5 recommendation which indicates that the Environmental Health Department of the local authority should be consulted about the surveys and invited to attend the installation of monitoring equipment. The number of noise monitoring locations used ranged from 1 to 7 for the assessments where noise surveys were carried out. The average number of monitoring locations was 4. In the vast majority of the assessments in the study, the spread of survey locations was considered sufficient to meet the recommendation of SB4 which states that the study area and monitoring locations should be chosen *'with the objective of collecting sufficient data to enable the background noise levels at each noise-sensitive receptor within the study area to be characterised'*. Two applications did not include any baseline noise surveys. In one application the predicted turbine noise levels were below 35 dBL_{A90} and therefore noise surveys were not required under the ETSU simplified assessment methodology. The other assessment that did not include noise measurements relied on an assessment that demonstrated that the predicted turbine noise levels from the proposed wind farm would be at least 10 dB below the conditioned limits set at receptor locations in common with two existing/consented wind farms.

GPG recommendation SB3 states *'Any contribution to background noise levels of noise from an existing wind farm must be excluded when assigning background noise and setting noise limits for a new development.'* In three assessments directional filtering was applied to the survey data to remove contributions from existing wind turbines. In one of these assessments it was indicated that the predicted noise levels from the existing turbines were also used to correct the background data. If the approach was described correctly this would have excluded the contributions of the existing turbines twice and therefore was not a correct application of the GPG recommendations.

In relation to the requirements of SB9, wind speed data was in all cases obtained concurrently with the noise survey data. For all assessments, apart from one, this was measured using a meteorological mast or SODAR system such that hub height wind speeds could be correctly derived. Only in one assessment was a 10 m high met mast used for wind speed measurements and this was for a development of two turbines. The GPG states that the duration of the background noise survey should be determined by the need to acquire enough valid data and the number of data points recommended is summarised in recommendation SB12. The GPG states that the requirements are unlikely to be met in less than two weeks. The majority of surveys took place for between 2 weeks and 2 months. Only one assessment used a shorter duration of 10 days. A further assessment used a long survey period of 3 months in order to obtain sufficient valid data.

Recommendations SB14, SB15 and SB16 address the need to exclude the following background noise data from the analysed data respectively: uncommon or atypical noise sources, the dawn chorus and data affected by rainfall. Whilst the vast majority of assessments indicated that atypical data was excluded as well as data points affected by rainfall, the dawn chorus was rarely mentioned. Presumably this was treated as part of general atypical data. It would be helpful if all noise assessment reports were to provide fuller description of these data exclusions in order to more clearly demonstrate that the appropriate exclusions have been made. A useful way of illustrating the approach, as adopted by some consultants, is to indicate the excluded data points on the scatter graphs, colour coded for rain exclusions and other types of exclusions.

2.2 Turbine Sound Power Level Data

The GPG recognises that predictions of turbine noise levels normally '*consider a "candidate turbine" at the planning stage, which is representative of the range of turbines which may be installed at the site*'. Most of the assessments have adopted a single candidate turbine model. However, four assessments have adopted an "envelope" approach. This involves considering a range of turbine models and using the highest sound power level at each wind speed from the manufacturers' data for each model.

All of the assessments included an allowance for uncertainty within the turbine sound power input data used for the noise predictions. In all cases apart from one a correction of between 1.0 dB and 2.0 dB was added as the sound power was either manufacturers' specified data or tested sound power data. In one case warranted data from test reports was available and the consultant found that a margin of 1.645 σ was apparent between the tested and warranted values and therefore the warranted values could be used directly in the noise model. All of these approaches are in accordance with paragraph 4.3.6 of the GPG.

All of the assessments except for one have used octave band frequency data and adopted the G=0.5 ground factor in the ISO 9613-2 calculations as recommended in the GPG. In one assessment frequency data was not used. In the absence of spectral data GPG para 4.3.3 recommends '*instead of applying equation (10) of the standard, a conservative calculation should be made using Agr = -3 dB (effectively hard ground), and the air absorption rate corresponding to the 250 Hz octave band.*' The consultant did not follow this procedure. They carried out predictions using G=0.5, assuming the sound power level applied to the 250 Hz octave band. We tested this methodology by applying the spectral data from a similar turbine model and repeated the calculations with G=0.5. The two sets of predictions showed agreement within +/-1 dB(A), indicating that the methodology was acceptable despite deviating from the GPG recommendation. It is understood that adopting *Agr = -3 dB* does result in worst case predictions.

2.3 Noise Prediction Methodology

The ISO 9613-2 prediction methodology has been used by all of the assessments with the reports also indicating that the settings and modifications given in the IOA GPG were used. In the vast majority of assessments the methodology sections explain that the topographic screening corrections have been limited to 2 dB. Many also indicate that the + 3 dB correction has been added for propagation across concave ground profiles. However, very few of the assessment reports set out detailed breakdowns that show how these adjustments have been applied to the predictions for each receptor. In the majority of cases it is therefore unclear whether the required modifications to the ISO 9613-2 procedures have actually been fully implemented.

In all but one of the assessments an all-purpose proprietary noise modelling package such as CadnaA or SoundPlan has been used for the noise predictions. In one assessment the wind industry modelling software Resoft was utilised. It is noted that this assessment was for a development with only two turbines and no detailed predictions of cumulative noise were carried out.

2.4 Determining the ETSU-R-97 Limits

The procedure given in ETSU-R-97 is to determine noise limits from a combination of a lower fixed limit and a limit derived from the prevailing background noise +5 dB where this value exceeds the lower fixed limit. The following extract from the GPG sets out the required approach which includes providing justification for the selected limits:

'3.2.2 The day amenity noise limits have been set in ETSU-R-97 on the basis of protecting the amenity of residents whilst outside their dwellings in garden areas. The daytime amenity noise limits are formed in two parts: Part 1 is a simple relationship between the prevailing background noise level (with wind speed) with an allowance of +5 dB; Part 2 is a fixed limit during periods of quiet. ETSU-R-97 describes three criteria to consider when determining the fixed part of the limit in the range of 35 dB to 40 dB L_{A90}, all of which should be considered. They are:

1) the number of noise-affected properties;
2) the potential impact on the power output of the wind farm; and
3) the likely duration and level of exposure.

3.2.3 The rationale for a choice of this limit, or factors which would assist the determining authority in this respect should be set out in the assessment. It is beneficial to the decision maker to display both sets of limits to illustrate the range available and/or the noise limit for the development if agreed previously with the LPA.'

This GPG recommendation is based directly on the factors discussed on page 65 of ETSU-R-97. Paragraph 3.2.5 of the GPG goes on to state that assessing the above three factors '*represents a relevant consideration when determining applicable noise limits*'. As part of the EIA process limits must be set in order to assess the noise impacts. On this basis limits should be selected and the rationale for the selection stated.

Two thirds of the assessments reviewed have either provided reasoning for the selected limits or adopted the lower end of the ETSU daytime fixed limit range and hence no further justification was necessary. Six of the assessments have simply presented assessments of the predicted turbine noise levels against two sets of limits, one using the 35 dB fixed lower limit and one using the 40 dB fixed lower limit. This approach fails to follow the GPG recommendation of choosing a limit and setting out the reasons for the choice.

It is not possible to provide a concise summary of the noise limits adopted across the range of assessments due to the variation in approach in addressing noise from both the development alone and cumulative noise in combination with other wind turbines. However, most assessments have considered the proposed wind farm in isolation as the first stage of the noise assessment. For these assessments, daytime noise limits using a fixed lower limit of 35 dB L_{A90} have been adopted in all instances except where both 35 dB and 40 dB L_{A90} limits have been used. Night-time assessments have generally considered the standard ETSU-R-97 43 dB L_{A90} lower fixed limit. However, two assessments have utilised a limit of 40 dB L_{A90} and one has used 38 dB L_{A90}.

2.5 Assessment of Cumulative Noise Effects

The noise limits in ETSU-R-97 apply to the noise from all wind turbines as summarised by the following quote from page 58 of ETSU, *'...absolute noise limits and margins above background should relate to the cumulative effect of all wind turbines in the area which contribute to the noise received at the properties in question...'* Therefore the noise assessment for a wind farm must take account of existing wind turbines, those consented but not yet built, as well as submitted planning applications for turbine developments, in addition to the turbines of the proposed development itself. All the assessments have addressed cumulative noise. There are two principal approaches that may be taken in the assessment of cumulative effects, as set out below.

1. Detailed cumulative predictions not required

One approach does not require detailed noise predictions of cumulative noise levels. This involves demonstrating either of the following:

a) The predicted noise levels at the receptors from the proposed wind farm are at least 10 dB above the combined contribution from all other turbines. On this basis the contribution from other turbines will be insignificant.

b) The predicted noise levels at the receptors from the proposed wind farm are at least 10 dB below the existing conditioned noise limits for a development that has already been consented. On this basis there will be no significant contribution from the newly proposed turbines to the noise levels at the receptors.

2. Detailed cumulative predictions required

The second approach is to carry out cumulative nose predictions taking account of all relevant turbines and to compare these with the ETSU derived noise limits. The GPG gives detailed guidance on assessing cumulative noise and on approaches to sharing or apportioning limits between nearby wind farms should this be necessary.

All of the assessments reviewed have addressed cumulative noise. Three assessments have adopted approach 1. The remaining assessments have generally followed approach 2 and included a detailed numerical assessment of cumulative noise levels. The majority of these assessments have included a breakdown showing the contribution of each wind farm to the total predicted levels. This is good practice and found to be useful when reviewing the assessments on behalf of the LPA. However, the provision of such a breakdown is not explicitly required by the GPG.

A full analysis of the approaches adopted for cumulative noise assessments is beyond the scope of this paper due to the differences in the number and location of existing and proposed wind turbines in relation to each application and the resulting varying approaches to assessing cumulative noise. However, around 30% of the noise assessments considered did adequately consider cumulative noise issues as part of the initial noise assessment submitted. Once any difficulties with the initial assessment were communicated with the relevant acoustic consultant, the issues have generally been rectified by further assessments submitted as Further Environmental Information (FEI). A recurring problem with cumulative assessments is where they have omitted consideration of another wind farm that has only recently been submitted as a planning application. This issue is perhaps unsurprising considering the long timescales that usually apply to the environmental assessment of wind farms and the associated timeframes between submission and approval of planning applications.

3. Case Studies

3.1 Differences in Background Noise Levels – Glenouther and Blair Wind Farms

Two applications for adjacent wind farms showed that different approaches to obtaining background noise data could lead to significant differences in background noise levels and derived noise limits applied to the same receptors. Figure 1 provides a plan showing the two proposed wind farms: Glenouther wind farm and Blair wind farm. There are a number of complications to obtaining background noise data in this locality. Firstly, the presence of the large Whitelee wind farm located to the south east; and secondly, the potential influence of road traffic noise from the M77 motorway. The two noise assessments followed distinctly different approaches to obtaining background noise data, although both followed GPG guidance.

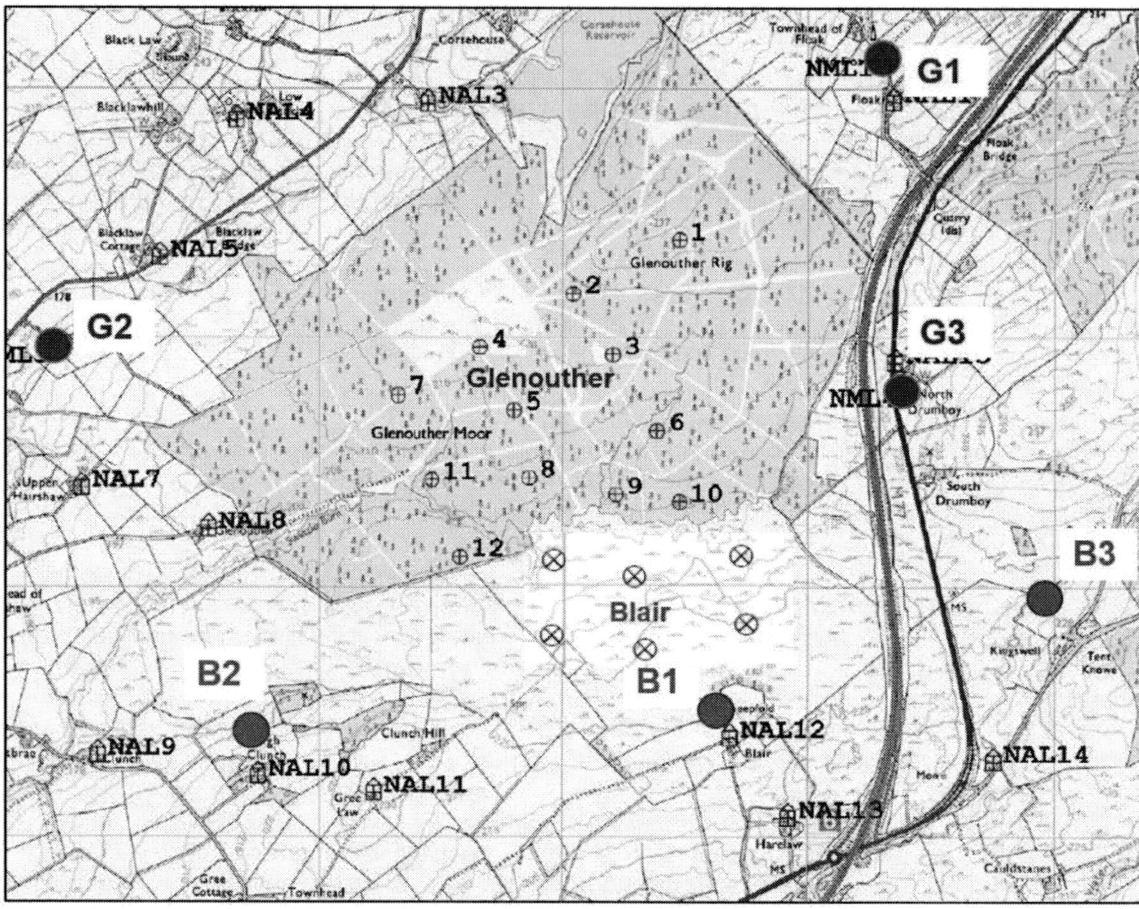

Fig 1: Plan of Blair and Glenouther Wind Farms and associated Monitoring Locations

Glenouther

The consultants for this development chose to adopt the background noise survey data from an earlier application at the same site which carried out noise surveys in 2006, before Whitelee came into operation. This is equivalent to the method given in GPG paragraph 5.2.3 of utilising data from the ES of an existing wind farm. The noise survey data was re-analysed taking account of GPG recommendations. For monitoring location G1 directional filtering was used to remove the contribution of traffic noise from the M77. This follows advice given in GPG SB19. Filtering out of the M77 noise is appropriate because the receptor is upwind of the motorway, whereas it would be downwind of the proposed wind farm in the predominant prevailing wind condition. Directional filtering was not applied to the data for G3 as the receptor would be downwind of both the wind farm and the M77 during prevailing wind conditions.

Blair

The consultants for the Blair wind farm chose to carry out background noise surveys and to utilise directional filtering to remove the contribution of noise from existing turbines. Monitoring location B1 used directional filtering to exclude noise from the M77. This will have also filtered out noise from the Whitelee wind turbines. For location B3, used to represent receptors west of the M77 motorway, data for wind directions 0° to 180° was filtered out to remove noise from Whitelee. There was concern that the daytime noise measurements were influenced by noise from a boiler flue. This led to the night-time background noise data being used in place of the daytime readings to ensure a conservative approach was taken. For monitoring location B2, directional filtering was applied for a narrow angle of westerly winds to remove any contribution from a single turbine approximately 2 km away.

Comparisons

The background noise data for locations G1 and B1 was used in each assessment for receptors located to the south of both wind farms and to the west of the M77. Data from locations G2 and B2 was adopted for receptors to the south west and west of both wind farms. Data from locations G3 and B3 was used was for receptors to the east of the M77. Graphs of the resulting background noise levels against wind speed are shown in Figures 2, 3 and 4. It can be seen that the results for the Blair assessment for B1 are 3 to 4 dB higher than those for the Glenouther assessment G1 up to wind speeds of 11 m/s. Similarly, comparing B2 and G2, the background noise levels obtained for Blair are around 5 dB higher than those derived for Glenouther. Conversely, in the case of the data used for locations east of the M77, the results obtained for the Blair assessment for B3 are 2 to 8 dB lower than those obtained for location G3. Interestingly, if the actual daytime data obtained at B3 is considered (rather than the night-time data adopted in the assessment), a very close agreement can be identified between the B3 and G3 measured levels.

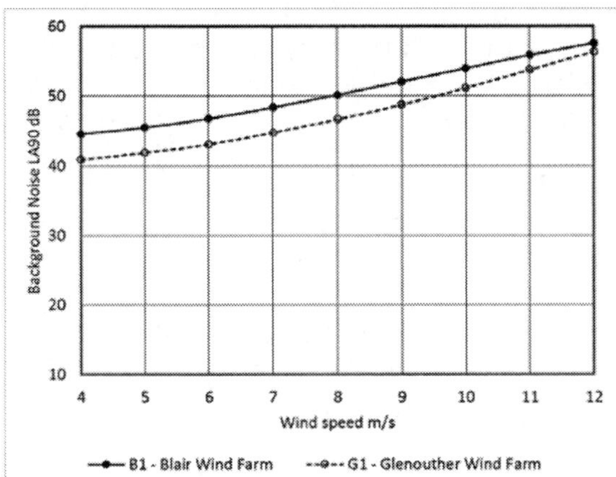

Fig 2: Daytime background B1 and G1

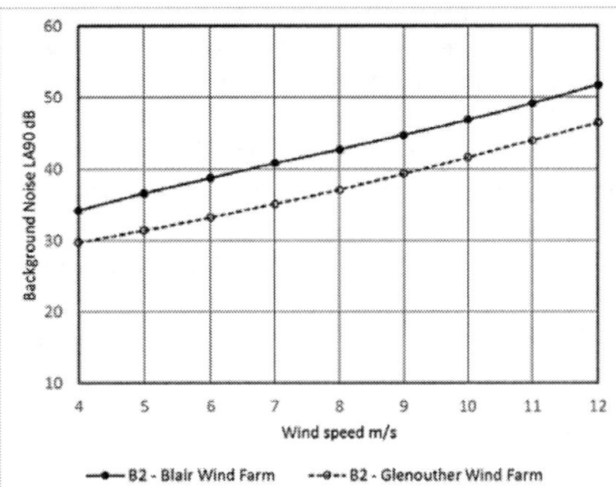

Fig 3: Daytime background B2 and G2

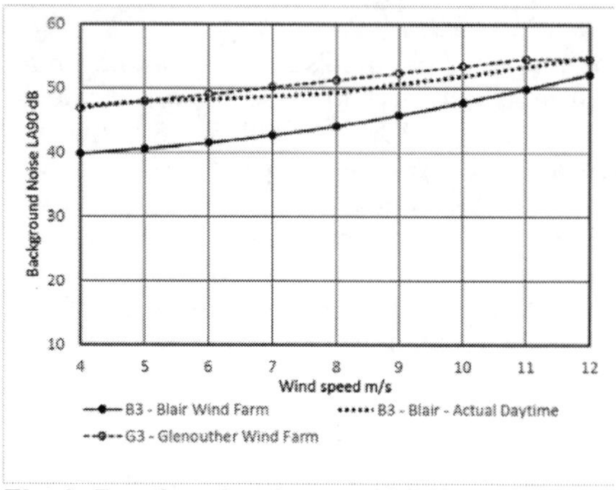

Fig 4: Daytime background B3 and G3

It is appropriate to consider the effects of wind shear correction differences between the two data sets Paragraph 5.2.4 of the GPG states the following:

'If the developer wishes to utilise previously presented background noise level data, care should also be taken with respect to any differences in wind speed conditions between the original and proposed site. The underlying principle of ETSU-R-97 requires that the background noise levels at any given location must be correlated with the wind speeds measured on the wind farm site of interest. Where a systematic difference exists between the wind conditions on the two sites, then a correction will need to be applied, meaning that the derived background noise curves for the two sites will be different.'

It is understood that the met mast used during the 2006 survey was to the north of the Glenouther site. The Blair met mast was located centrally within the Blair site. Examination of the ground topography indicates that the elevations of both met masts were likely to have been similar and therefore wind speeds measured at both locations are likely to be similar for the same wind conditions. A difference between wind speeds at the two survey sites is therefore considered unlikely to explain the large differences between the B1/G1 and B3/G3 results.

There is also a difference between the hub heights of the proposed turbines for the two wind farms. The proposed hub heights for Blair were 59 m and those for Glenouther were 80 m. As the background noise levels are standardised to a 10-metre height wind speed, based on the hub height wind speed, this may account for some of the differences between the background noise data. However, the magnitude of the differences is larger than could be explained by the correction to the noise levels due to the standardisation for different hub heights. It is noted that none of the noise monitoring locations applied to similar receptors were carried out at the same properties and therefore some differences between the two sets of results would be expected. There was also an eight year difference between the two sets of measurements. However, as the data analysis in both cases was designed to exclude noise from the key major new noise source in the area, Whitelee wind farm, there are no clear reasons as to why the two assessments obtained background noise levels with such large differences.

The planning application for Blair wind farm was dismissed following an appeal. However, it is clear that if both the Blair and Glenouther wind farms had been consented it would have been necessary to agree common background noise levels in order to apportion noise limits between the two developments. The acoustic consultants acting for Glenouther did carry out their own review of the differences in adopted background noise levels between the two proposals. The consultants concluded that they should continue to use the 2006 data as the background noise levels were generally lower and therefore resulted in a worst-case assessment.

3.2 Cumulative Noise - Loudoun Community Wind Farm

This application was for two turbines proposed to be constructed to the west/south-west Sneddon Law wind farm (15 turbines) and Whitelee Wind Farm. A plan showing the location of the Loudoun proposal and the exiting developments is given in Figure 5. Whitelee is the largest wind farm in the UK comprising 215 turbines. Sneddon Law wind farm has a cumulative noise condition in place which sets a limit on the total combined noise immissions from Sneddon Law and Whitelee wind farms at the properties nearest to the Sneddon Law development. This condition was put in place following a planning inquiry. The reporter held that given the large number of Whitelee turbines, the operators of Sneddon Law would not be in a position to arrange a shutdown of Whitelee Wind Farm should a noise complaint investigation be required. On this basis it would not be possible for noise immissions at the closest properties to Sneddon Law to be measured without noise from Whitelee Wind Farm. It was thus necessary for the consented noise limits to be based on the total cumulative noise levels allowed from both wind farms.

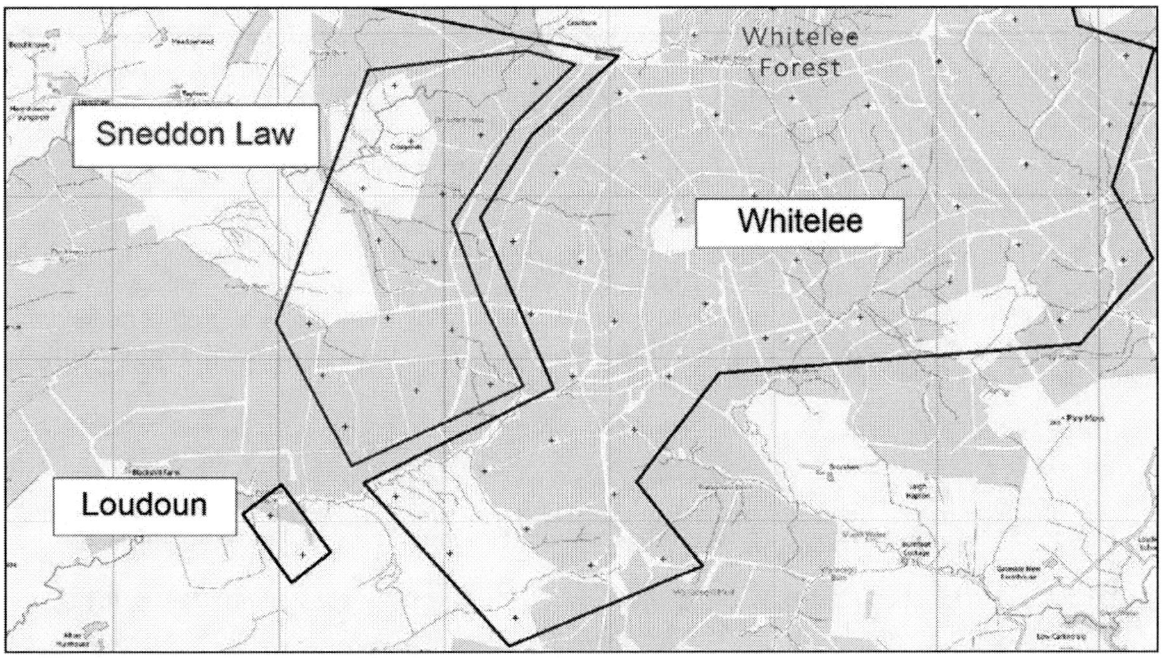

Fig 5: Loudoun Community Wind Farm Location Plan

An initial examination of the cumulative noise assessment included in the ES for the Loudoun development indicated that this assessment had been carried out appropriately. Predictions of cumulative noise immissions were carried out and these had been assessed against ETSU derived noise limits based on the background noise survey carried out for the project. However, closer examination revealed two difficulties with the approach taken. Firstly, the background noise levels and resulting derived noise limits were much higher than those in the Sneddon Law consent (for the receptors in common). Secondly, given that the planning inquiry had found that it was necessary to impose a cumulative noise condition on Sneddon Law, it would follow that a similar condition would most likely be necessary for Loudoun Community Wind Farm.

An additional noise assessment was subsequently submitted for Loudoun utilising noise limits set 10 dB below the Sneddon Law cumulative noise limits. Adoption of such limits followed the approach identified in the Hayes McKenzie Partnership (HMP) 2011 report referenced in the GPG. The HMP report states that *'If an existing wind farm has permission to generate noise levels up to ETSU-R-97 limits, planning permission noise limits set at any future neighbouring wind farm would have to be at least 10 dB lower than the limits set for the existing wind farm to ensure there is no potential for cumulative noise impacts to breach ETSU-R-97 limits (except in such cases where a higher fixed limit could be justified)'.* This meant that it was no longer necessary to provide detailed cumulative noise predictions and an assessment against derived ETSU limits.

The revised assessment indicated that a substantial reduction in noise levels would be required from the standard operational settings of the candidate wind turbine. Although this turbine model does have noise reduced settings, at the time of writing it remains unclear that sufficient noise reduction could be obtained using the candidate turbine model.

4. Conclusions

The investigation has shown that of the sample of nineteen noise assessments reviewed it was generally found that the recommendations in the IOA Good Practice Guide for implementing ETSU-R-97 were being followed. This provides evidence that the GPG has improved the quality and consistency of wind farm noise assessments. This research has demonstrated that the main area where there are ongoing difficulties with the noise assessments is the treatment of cumulative noise issues. Cumulative noise considerations are often complex both in terms of technical aspects of noise assessment and the implications for planning and consent conditions. This perhaps indicates there is a need for a further review of good practice guidance on cumulative noise.

One of the case studies highlighted how assessments for neighbouring wind farms applied differing background noise levels and therefore noise limits for common receptors. This was despite both assessments following GPG recommendations for obtaining background noise data. The other case study highlighted the need to consider noise conditions of nearby wind farms as a complex cumulative noise situation applied in this situation/case.

References

1. The Working Group on Noise from Wind Turbines (1996) ETSU-R-97, *The Assessment and Rating of Noise from Wind Farms*, ETSU for the DTI (Department of Trade and Industry).

2. Institute of Acoustics (2013), *A Good Practice Guidance to the Application of ETSU-R-97 for the Assessment and Rating of Wind Turbine Noise.*

3. Hayes McKenzie Partnership Report, *Analysis of How Noise Impacts are considered in the Determination of Wind Farm Planning Applications*, Ref HM: 2293/R, 6th April 2011.

Acknowledgements

The authors would like to acknowledge the assistance of James Mackay of TNEI Services Ltd.

7th International Conference on
Wind Turbine Noise
Rotterdam – 2nd to 5th May 2017

Evaluation of wind farm noise in Switzerland – Comparison between measurement and modeling

Desarnaulds Victor, EcoAcoustique SA Lausanne : desarnaulds@ecoacoustique.ch
Fécelier Ronan, EcoAcoustique SA Lausanne : fecelier@ecoacoustique.ch
Magnin Dimitri, EcoAcoustique SA Lausanne : magnin@ecoacoustique.ch

Summary

The different evaluation-methods for the wind farm noise in Switzerland – computer models (for new wind farm projects) or on-site measurements (for existing wind farms) are often discussed by the concerned authorities and organizations. In order to improve the evaluation of the wind farm noise, this research project aims to compare the current Swiss calculation method with the results of in-situ measurements of a specific wind park.

The measurement's results allow to validate some elements concerning material and instrumentation, the duration to cover the different meteorological conditions, the parameters to be recorded, the relevant periods, as well as the measurement positions. These results show also the limits of the measurement's method. Given the particularity of the site (situation on a ridge with strong wind exposure), it is not possible to extract exactly the wind turbine noise from the background noise, even if the noise of the wind turbine is partially audible in the audio recordings. In this configuration, the measured sound level represents the noise of the wind turbine mixed with the background noise, even after the suppression of other interfering noises. The different methods tested (periods with high audibility of the wind turbine noise, third-band-analysis, statistical analysis) do not allow to isolate clearly the wind farm noise.

Concerning the calculation's results, the Swiss method recommended to determine the wind farms noise is comparable to the ones used in neighbouring countries. All these methods are based on a simplified approach of the noise propagation, which mainly does not take into account the meteorological effects. Due to the application of a special ground-connection-factor, the results of the mandatory Swiss method (ISO-norm 9613-2 – modified according to EMPA recommendation) are usually 1 to 3 dB(A) higher than those obtained with the commonly used international norm (ISO 9613-2).

The comparison between the results of the measurement and the modeling shows that the average global sound level (annual averaged L_{Aeq} for daytime) obtained from the measurements is 7 dB(A) higher than the values obtained by the modeling. If one takes into account the statistical index L_{A90}, the difference is about 4 dB(A). With increasing wind speed ($v > 7$ m/s) the difference between measurement and modeling is particularly marked. This important discrepancy between measurement and calculation results is mainly due to the fact that the measured wind turbine noise is overrated by the presence of background noise (especially from wind in the vegetation).

In order to optimize the methods of measurement and calculation, it would be necessary to perform a more detailed frequency analysis (FFT). The data should also be completed with complementary measurements in several positions (also in the areas less exposed to wind) while the wind turbine is interrupted (« stop-and-go », procedure unfortunately not possible in the frame of this project) and to extend the procedure to several wind parks.

1. Introduction

The different evaluation-methods of the wind farm noise in Switzerland – computer models (project of new wind farms) or in-situ measurements (as for existing wind farms) are often discussed by the concerned authorities and organizations (federal and cantonal public authorities, Suisse Eole ...). For modeling purposes, the FOEN (Federal Office for the Environment) recommends a method based on the EMPA report "Lärmermittlung und Massnahmen zur Emissionsbegrenzung bei Windkraftanlagen" [1]. In Switzerland, there isn't yet any official measurement method for the evaluation of the wind farm noise.

In order to improve the evaluation of the wind farm noise, this research project aims to compare the current Swiss calculation method with the results of one-site measurements of a wind park. This research project is funded by the Swiss Federal Office of Energy - SFOE (project SI/501150).

2. Measurement

2.1 Methodology

The main objective of the measurement method proposed in this research project is to remain simple (equipment, parameter...) and efficient. In order to take into account all the interfering noises (from rain, wind at the microphone and in the trees) and to improve the representativeness of the results, we use a statistical approach over a long measurement period.

The measurement method has also been defined with a view of being as close as possible to ISO 1996-2: 2007 "Description, measurement and assessment of environmental noise -- Part 2: Determination of environmental noise levels" [2] and NF S 31-010 1996 " Acoustics - Environmental noise characterization and measurement — Special measuring methods » [3]:

- wind with an angle relative to the direction of the receiving source of ± 60 ° during daytime and ± 90 ° during nighttime
- wind speed (measured at a height between 3 and 11 m) between 2 and 5 m/s during daytime and with 0.5 m / s more for the nighttime.
- no strong negative temperature gradient close to the ground.
- no disturbing condition, in particular close to the microphone. It is advisable to avoid making measurements when the wind speed is greater than 5 m/s, or in case of heavy rain.

Before performing on-site measurements, a series of laboratory tests were carried out to validate the equipment, in particular performance according to wind speed of various windscreen models.

For on-site measurement, the two locations are selected on both sides of a wind turbine at a distance of approximately 200 m in the direction of the prevailing winds (South-West, North-East). These positions, however, dictated by local constraints (plot boundary, presence of isolated trees), fulfill the ISO 1996-2: 2007 [4] requirements (in the direction of the prevailing winds) and remain relatively distant from disturbing noise sources as forests and other wind turbines present in the area. Moreover, choosing a position relatively close to the wind turbine allows to reduce the uncertainties related to long distance propagation and the influence of background noise (increase of S/N ratio).

The on-site measurement performed over one month (May-June 2015) covers varied weather conditions, which are representative of those usually found in this area.

2.2 Results

Laboratory

5 windscreen models (windshields available on the market) have been tested in laboratory :

- Standard windscreens (included with the sound level meter Norsonic Nor 140), 60 mm diameter (Nor 1451).
- Standard windscreens 90 mm diameter (Nor 1434).
- Microphone Outdoor Protection Kit (Windshield Nor 1212, 50 mm diameter)
- Large (200 mm) diameter windscreen (Outdoor Microphone Protection System Rion WS 03)
- Double windscreen (Nor 1216 + CA 4575)

The results (see Figures 1 and 2) show that the most efficient systems to limit airflow noise at high speeds (> 5 m/s) are the double windscreen (Nor 1216 + CA 4575) followed by the large diameter windscreen (Rion WS 03). All systems tested are quite equivalent for air velocities lower than 5 m/s.

For typical noise spectrum of a wind turbine, the overall sound level correction (L_{Aeq}) due to frequency response of the double windscreen (Nor 1216 + CA 4575) is +0.3 dB(A), mainly due to high frequencies. No correction is needed for the other windscreen.

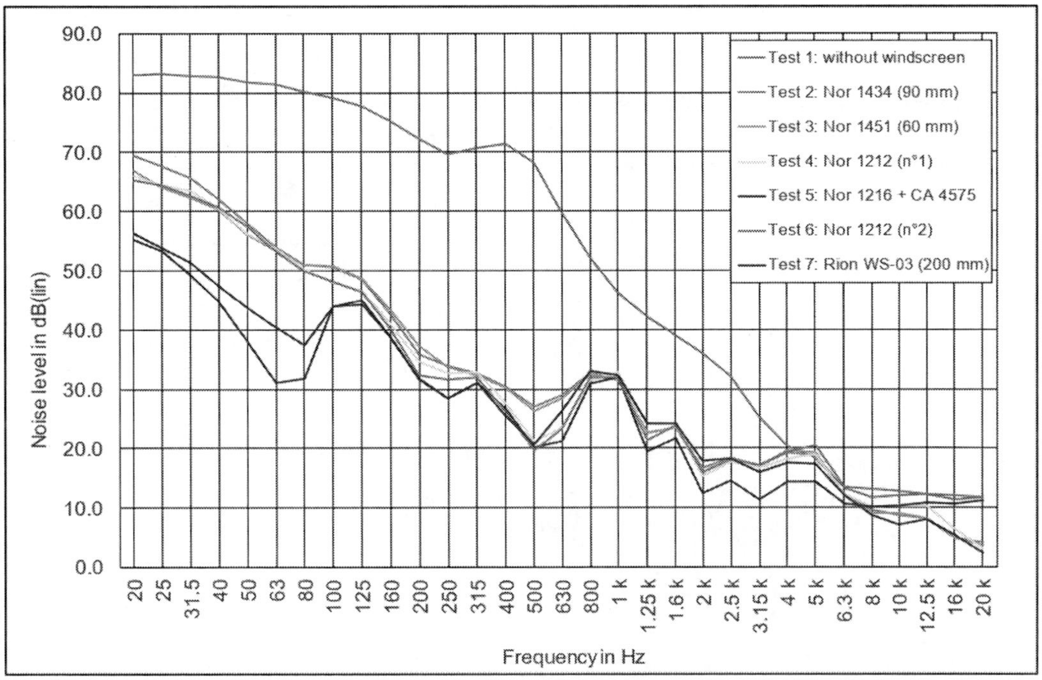

Figure 1 : Frequency level for various windscreens at 4.7 m/s wind speed

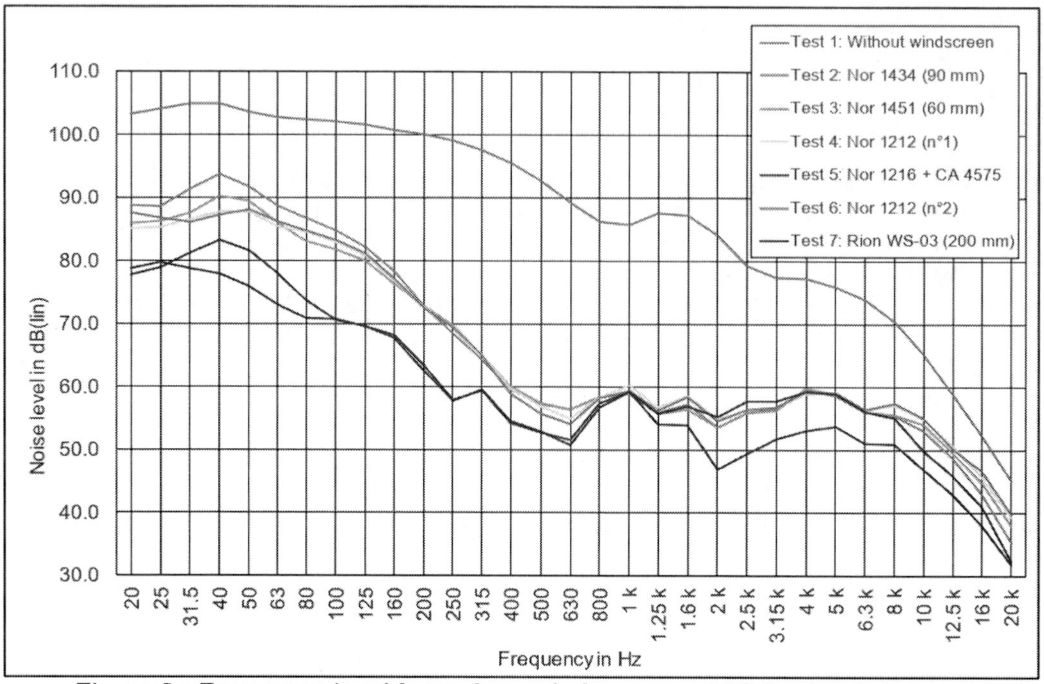

Figure 2 : Frequency level for various windscreens at 12.8 m/s wind speed

On-site

In a first stage, short term on-site measurements are grouped in a single figure for each measuring position (see Figure 3 for position 1). Measured sound levels take into account all sound events that occur near the microphones.

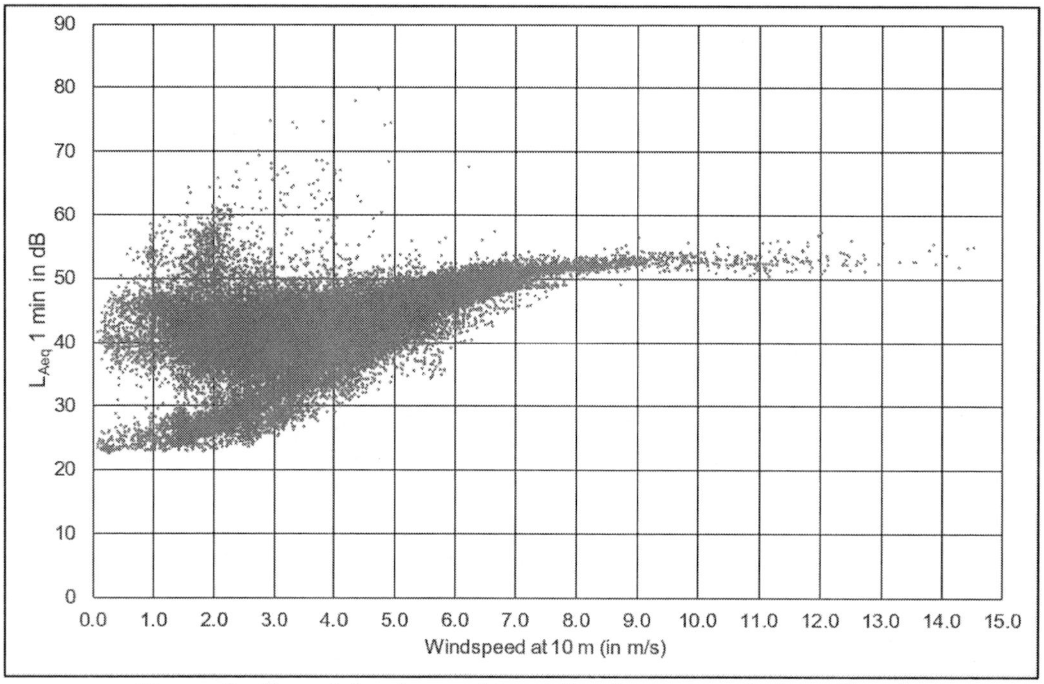

Figure 3 : Raw data of sound level (L_{Aeq}) measured at position 1

In a second stage, in order to extract the useful information, it is necessary to remove disturbing noises due to:
• Unfavorable weather condition (rain, wind gust at the microphone)
• Human activities (tractor, forestry work, construction sites in the vicinity)
• Noise from animals (cow bells, birds, crickets)

Based on post processing analysis of audio recordings, the samples containing such disturbing noises have been removed. The suppression of disturbing noise led us to consider only the measurement results during nighttime (from 22h00 to 4h00) for the rest of this research project (see Figure 4).

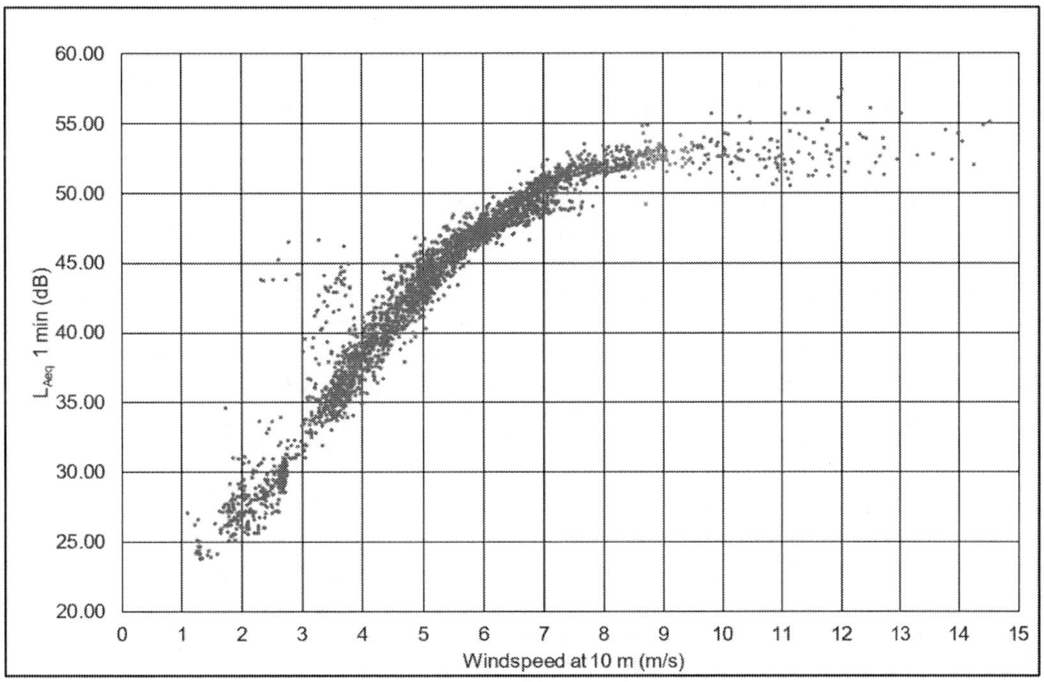

Figure 4 : Selected data (after suppression of disturbing noise) of sound level (L_{Aeq}) measured during nighttime (22h-4h) at position 1 for main wind direction

Based on these selected data, trend curves (third order polynomial) are plotted in order to determine the sound levels for each wind class. Then, the annual average sound levels for day and night periods are calculated based on these trend curves and according to the different wind classes' occurrences (see Table 1).

Windspeed m/s	Sound Level L_{A90}	Day occurrence %	Weighted Day L_{A90}	Night Occurrence %	Weighted Night L_{A90}
<4.5	--	42.9	--	31.0	--
4.5-5.5	43	11.4	34	10.9	34
5.5-6.5	48	9.3	37	10.8	37
6.5-7.5	50	8.2	39	10.1	40
7.5-8.5	52	7.0	40	8.7	41
8.5-9.5	53	6.1	41	8.1	42
>9.5	54	15.1	46	20.4	47
Total		100	**L_{day}=49**	100	**L_{night}=50**

Table1 : Day and night annual average level calculation (L_{Aeq} from sound levels and wind occurrence in each wind speed class, position 1)

Results show that the annual average level is 1 dB(A) higher for nighttime than for daytime. Even with an average occurrence (15 -20%), the highest wind speed class (>9.5 m/s) represents the essential contribution to noise (50%).

However, a detailed analysis of the different periods shows that wind noise at the microphone is always significant when the wind turbine operates at high speed. The only periods when wind noise is low are of course limited to weak wind periods when the wind turbine operates with a relatively low power or is off and thus with reduced noise emissions. The values obtained in Table 1 therefore constitute the measured noise levels of wind turbine noise combined with wind noise at the microphone and residual background noise (especially from wind in the vegetation, even at long distance). Unfortunately, it is not possible, in our specific situation, to extract noise data only due to wind turbine (without any disturbance from background noise) even if such noise is audible in the audio recordings.

According to the methodology adopted in some countries, the use of statistical indicators such as L_{A90} (which represents the A-weighted sound level exceeded for 90% of the measurement period) makes it possible to deduce part of the remaining disturbing noises.

The same calculation is then carried out on the basis of statistical indices L_{A90} level to determine the annual average sound levels for day and night periods (see Table 2).

Windspeed m/s	Sound Level L_{A90}	Day occurrence %	Weighted Day L_{A90}	Night Occurrence %	Weighted Night L_{A90}
<4.5	--	42.9	--	31.0	--
4.5-5.5	39	11.4	31	10.9	31
5.5-6.5	43	9.3	34	10.8	35
6.5-7.5	46	8.2	36	10.1	37
7.5-8.5	49	7.0	38	8.7	39
8.5-9.5	50	6.1	39	8.1	40
>9.5	51	15.1	43	20.4	44
Total		100	**L_{day}=46**	100	**L_{night}=47**

Table 2 : Day and night annual average statistical L_{A90} level calculation (Position 1)

Results with L_{A90} lead to same conclusion as for L_{Aeq} but with values 3 dB lower.

3. Calculation

3.1 Methodology

The application of the internationally recognized noise propagation model ISO 9613-2 [2] to wind turbine noise is problematic because of the important height of such noise sources. For that reason, the FOEN recommends in Switzerland to use the ISO 9613-2 standard with certain adaptations, in particular concerning the ground effect [1]. Various computation models for noise propagation are compared for our particular wind turbine situation (ISO 9613-2 [2], Ljud från vindkraftverk [5], CNOSSOS [6], Nord 2000 [7], Harmonoise [8]).
Except for the simplified Swedish method ([5]), they are all quite similar to the ISO 9613-2 method.
Computer modeling is also carried out using CadnaA software (version 4.2) with 3D terrain model including vegetation and wind turbines as omnidirectional noise sources.

3.2 Results

Calculation results according to the various models and parameters (Ground factor G from 0 to 1) are summarized in Table 3 and Figure 5. The average annual sound levels of the various calculation methods is 42 dB(A) ± 2 dB(A), except for the Harmonoise max method (Class S5) with slightly higher results (45 dB(A)).

Wind speed m/s	Day Occurrence %	ISO 9613 (EMPA)	ISO 9613 (G=0)	ISO 9613 (G=0.5)	ISO 9613 (G=1)	Ljud från vindkraftverk (G=1)	CNOSSOS (G=1)	Nord 2000 (G=1)	Harmonoise (G=1)
4.5-5.5	11.4	38.7	40.8	39.3	37.9	40.3	37.0	39.2	39.5 -> 41.9*
5.5-6.5	9.3	42.8	45.0	43.5	42.0	44.5	41.1	43.4	43.7 -> 46.1*
6.5-7.5	8.2	44.7	46.9	45.4	43.9	46.3	43.0	45.3	45.4 -> 48.8*
7.5-8.5	7.0	45.4	47.6	46.1	44.6	47.1	43.7	46.0	46.1 -> 49.5*
8.5-9.5	6.1	45.5	47.7	46.2	44.7	47.2	43.8	46.1	46.2 -> 49.6*
>9.5	15.1	45.1	47.3	45.8	44.4	46.8	43.4	45.7	45 -> 48.3
Annual LAeq daytime		**42**	**44**	**42**	**41**	**43**	**40**	**42**	**42 -> 45**

Table 3: Annual average noise level calculated for position 1 with the various models and parameters (Ground factor G from 0 to 1)

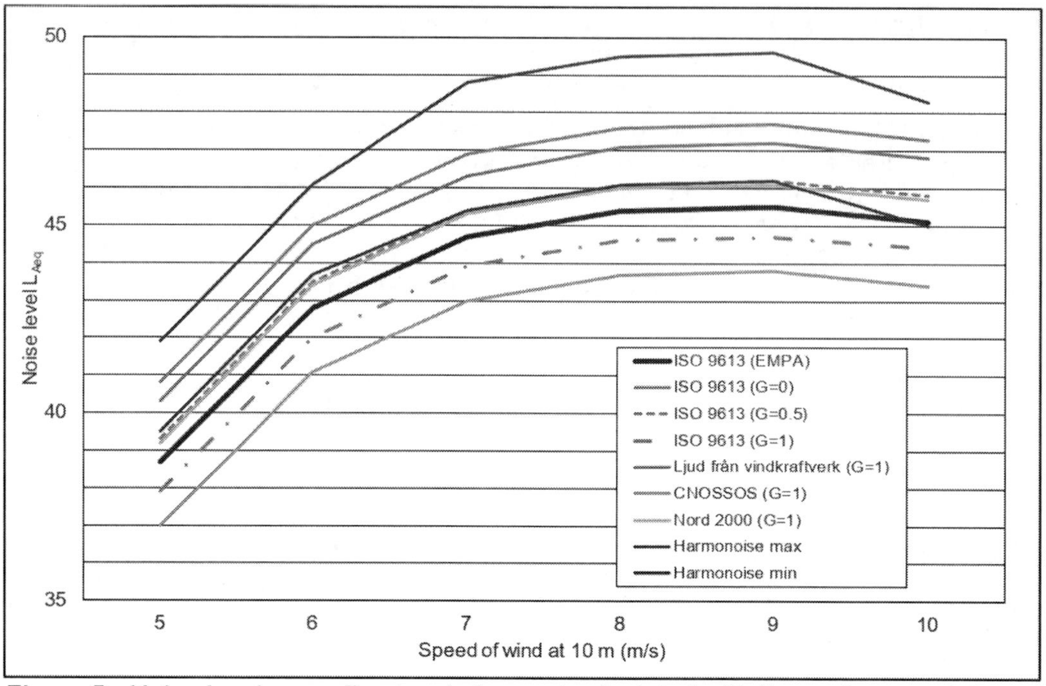

Figure 5 : Noise level according to wind speed calculated with the various models and parameters (Ground factor G from 0 to 1)

The modelling's results show that:

- Most of the calculation methods tested do not take into account meteorological effects (except for the Harmonoise method). The noise propagation is therefore considered to be independent of wind speed and direction.
- The results of the different calculation methods are within ± 2 dB(A) (except for Harmonoise Class S5). This range is relatively small compared to the uncertainties associated with this type of calculation (between -6 and +3 dB (A) according to the EMPA [1]).
- The Swiss method (EMPA) differs from other models, using a single correction factor for the ground effect (+1 dB).

4. Comparison between measurement and calculation

The comparison between measurements and modeling is illustrated in Figure 6 and Table 4.

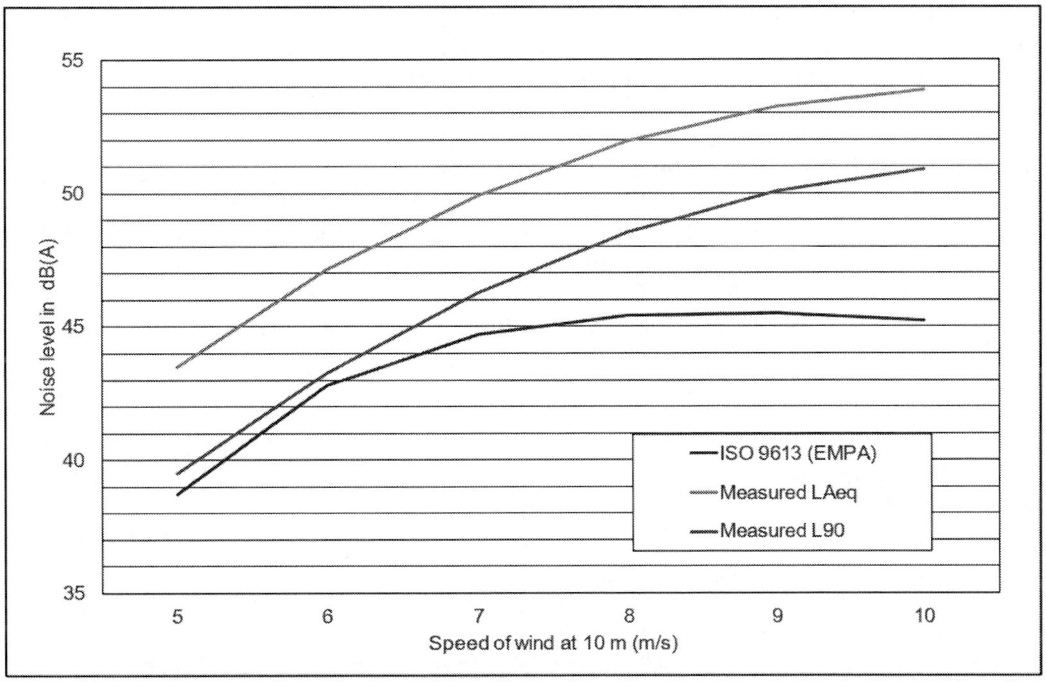

Figure 6 : Calculated and measured sound level according to wind speed and direction (position 1)

Daytime, Position 1	dB(A)
Measurement, annual L_{Aeq}	49
Measurement, annual L_{A90}	46
Calculated L_{Aeq} (ISO 9613-2 including Empa correction)	42

Table 4 : Summary of measured and calculated annual average levels (daytime, position 1)

The average sound levels (annual L_{Aeq} for daytime) obtained by measurements are 7 dB(A) higher than the calculated results. When taking into account the L_{A90}, the difference is only 4 dB(A). The difference between measurements and modeling increases with the wind speed and becomes very significant at high wind speed (v> 7 m/s).

This large discrepancy between measured and calculated results is mainly due to the fact that the measurements results include not only wind turbine noise but also some residual background noise (mainly due to wind noise in the vegetation), which cannot be extracted.

5. Conclusions

The comparison between the results of the measurement and the modeling shows that the average global sound level (annual averaged L_{Aeq} for daytime) obtained from the measurements is 7 dB(A) higher than the values obtained by the modeling. If one takes into account the statistical index L_{A90}, the difference is about 4 dB(A). With increasing wind speed (v > 7 m/s) the difference between measurement and modeling is particularly marked.

This important discrepancy between measurement and calculation results is mainly due to the fact that the measured wind turbine noise is overrated by the presence of background noise (especially from wind in the vegetation). In order to optimize the methods of measurement and calculation, it would be necessary to perform a more detailed frequency analysis (FFT). The data should also be completed with complementary measurements in several positions (also in the areas less exposed to wind) while the wind turbine is interrupted (« stop-and-go », procedure not possible in the frame of this project) and to extend the procedure to several wind parks.

References

[1] EMPA, Untersuchungsbericht Nr. 452'460, Lärmermittlung und Massnahmen zur Emissionsbegrenzung bei Windkraftanlagen, Dübendorf, 22. Januar 2010.

[2] Norme ISO 9613-2 (2007) *Description, measurement and assessment of environmental noise -- Part 2: Determination of environmental noise levels.*

[3] Norme francaise NF S 31-010 (1996), *Environmental noise characterization and measurement — Special measuring methods.*

[4] Norme ISO 1996-2:2007, *Méthode générale pour le mesurage du bruit à l'extérieur.*

[5] Ljud från vindkraftverk, Naturvårdsverket, rapport 6241, décembre 2011.

[6] Common Noise Assessement methods in Europe (CNOSSOS-EU), report EUR 25379 EN, 2012.

[7] Nord 2000, Comprehensive outdoor sound propagation, Model Part1: Propagation in an atmosphere without significant refraction, revised 31 march 2006.

[8] SALOMONS Erik, VAN MAERCKE Dirk, DEFRANCE Jérôme, DE ROO Foort, *The Harmonoise sound propagation model*, 22 January 2010.

7th International Conference
on
Wind Turbine Noise
Rotterdam – 2nd to 5th May 2017

A case study of how to involve impacted neighbours in measuring and characterizing windfarm noise

Sveinulf Vågene (Petroleum geophysicist) Stavanger, Norway.
E-mail: svagene@getmail.no

Summary

In order to resolve the frequent noise-related conflicts between windfarm neighbours and windfarm operators it is necessary to thoroughly characterize and understand the noise-problem from the perspective of the impacted residents. Valuable data can be acquired and used to understand issues if longitudinal studies are designed to capture the neighbors own experience of the noise. Data collection methods and learnings from the Lista Windfarm in the south of Norway are presented. It is shown how to involve windfarm neighbors through self-reported noise diaries to measure and observe the noise from the wind turbines near their dwellings.

1. Introduction

There are few studies to be found on self-reported long-term data recorded by people who are impacted by noise from windfarms. Thus, the impacted neighbours own experience, a very important dimension in our understanding of the noise, is missing. A problem that is not being properly understood can rarely be adequately solved. Understanding this dimension is therefore a prerequisite for addressing and moving towards a solution of the problems with windfarm noise.

This paper is a case study on establishing successful collaboration between researchers and neighbours impacted by windfarm noise. Suggestions are given on how to avoid pitfalls and efficiently continue collaboration over longer time periods with the objective of collecting time series of recorded and self-reported noise data of high quality. The basis for this study is project learnings from a study carried out during the years from 2013 through mid-2016 at the Lista Windfarm in south Norway (Vågene, S and Larsen, W (2015), and (2016)). This study benefited from cooperation with a highly motivated and diligent windfarm neighbour who recorded the noise data.

The value of time series of noise data recorded by the people who experience the noise are many, among which are:

- A thorough understanding of how impacted neighbours actually experience the wind farm noise. The neighbours own experience constitute a large source of information which today, in most studies, has been ignored or only casually addressed by simple questionnaires.
- Ability to link time series of recorded and self-reported noise and wind data.
- Daily observation and logging of noise unrelated to the windfarm, and which should be removed from the dataset. This allows more efficient and reliable editing and removal of unwanted noise from the data.

- In some cases near real-time editing and removal of unwanted sound data.
- Detection of unexpected noise phenomena that may occur. Such data may give new insights.
- Allowing the impacted residents to participate in the process of collecting the data may also give more ownership and buy-in to the conclusions of the studies performed.

The key challenges in working with neighbours include convincing them to perform the rigorous and often tedious task of gathering and recording the data on a daily basis over longer time periods. Issues of data integrity also need to be considered. Strategies to mitigate these challenges are discussed in this paper with the objective that they may help other researchers to move more quickly towards setting up such a study and getting buy-in from neighbours to assist in collecting new data.

The term **self-reported** data is here used broadly and denotes recording of data which are purely subjective, such as how annoying the noise is on any day, as well as data that are of a more observational nature such as auditory assessment of which wind turbine emits noise during the day.

2. Collecting the Data

The Lista project started as an informal attempt to find out more about the wind farm noise level at a neighbour's home. Very little relevant information was found on how to perform this type of self-reported long-term study. Our data collection method therefore needed to be designed and then evolved as we went along. This required a number of iterations of experimentation during the initial phase of the project until a methodology was found that worked well within the project constraints. This chapter describes the method and some of the experiences made.

2.1 Selecting Neighbours to record the Data

At Lista one person did all the recording of data, but a family member living in the same household assisted from time to time by making observations when the primary person (the observer) was unavailable. The observer initially wanted only to understand what the actual noise levels at his home were. With that objective he had acquired professional sound and wind recording equipment. Technical support was solicited from a friend, a Health Environment and Safety specialist trained in assessing noise. The HES specialist support was very valuable to ensure correct use and calibration of the sound recording instrument. Normally, this role may have to be filled by the project leader or a project technician.

2.2 Selecting a Database Tool to capture the Data

The database tool in which the data are to be entered need to be easy to use, and preferably, familiar to the users in order to minimize need for database training. Excel was found to be an excellent tool for this: simple in its basic layout, but with possibilities for sophisticated data analysis. All the data were therefore recorded in one single large spreadsheet with 30 columns resulting in more than 10.000 data-points covering 15 months of systematic recording in our final version of the database spreadsheet. Approximately half of the 30 columns needed to be populated for each data entry. Regular data backups are, of course, recommended.

2.3 Instrumental Recording of Noise Data

Data were downloaded from the sound level meter to a laptop by the neighbour. Values to be recorded were noted on a paper log before they were transferred to an Excel spreadsheet. During the early phase we experimented with different period averages, including the whole day, but in the end we decided to log LAeq (and wind data) averaged over six hour periods.

Average values reduce the data volume while enabling study of how the noise varies through the 24 hour day. Most of the highly sampled raw noise data were also kept.

The observer edited the instrumentally recorded sound data for unwanted noise, such as farming activity, lawn mowers and noise from large flocks of migrating birds, on a daily or weekly basis. This allowed for near removal of unwanted sounds in the data. This would in most cases not be practical to do on long-term data series after project completion.

Figure 1: View towards east from the observer's dwelling. Four wind turbines are partly visible in this direction.

The sound meter microphone and wind gauge were placed in open, undisturbed areas with no permanent disturbing sound-sources such as brooks or trees with rustling leaves nearby, but not too far away from the house to have easy access.

2.4 Instrumental Recording of Wind and Weather Data

Measurement of local wind speed and direction at the receiver location was a requirement as one of the study objectives was to define how pronounced the wind shadow was. Initially a simple wind gauge was utilized, but as the wind speed readings were low an advanced wind gauge was acquired. Both wind gauges were run together for a few months without showing any substantial difference and the first batch of data from the simpler gauge were accepted for use.

Initially other weather data like temperature, clouds and type of precipitation also were recorded. However, we found that recording this data locally added little value as the data was nearly identical to the data we could freely download from the Lista meteorological station eight kilometres away. This reduced the amount of work for the observer.

2.5 Self-reported Noise Data

The following self-reported data were recorded based on the neighbour's auditory observations and own experience:

- Daily noise emission from eight surrounding wind turbines (table 1).
- Observation of audible noise during each of the four six-hourly intervals (table 2).
- Daily perceived strength of total noise from the surrounding wind turbines.
- Daily perceived level of annoyance from the noise.
- Days with observed noise - at any time during the day.
- Nights when noise was so annoying it was not possible to sleep in regular bedroom.
- General comments on the character of the noise and notable changes in environmental conditions.

Table 1: Example of daily audible noise emission from eight wind turbines surrounding the observer dwelling. The numbers indicate that noise was observed at the receiver location at least during one of the six-hourly intervals and do not necessarily imply that the noise was present at all times during the day.

Time of measurement			Audible noise from the wind turbine = 1							
Date	From	To	Gjeldal, Turbine T 11	Gjeldal, Turbine T 10	Borgåsen, Turbine T 13	Sudland, Turbine T 12	Åsen, Turbine T 9	Dalennen, Turbine T 6	Rudjordfjellet, Turbine T 3	Almås, Turbine T 4
11-Mar-15	00:00	06:00		1				1	1	1
11-Mar-15	06:00	12:00								
11-Mar-15	12:00	18:00								
11-Mar-15	18:00	00:00								
12-Mar-15	00:00	06:00	1	1	1	1				
12-Mar-15	06:00	12:00								
12-Mar-15	12:00	18:00								
12-Mar-15	18:00	00:00								
13-Mar-15	00:00	06:00	1	1	1	1	1			
13-Mar-15	06:00	12:00								
13-Mar-15	12:00	18:00								
13-Mar-15	18:00	00:00								
14-Mar-15	00:00	06:00		1						
14-Mar-15	06:00	12:00								
14-Mar-15	12:00	18:00								
14-Mar-15	18:00	00:00								
15-Mar-15	00:00	06:00	1	1	1	1				
15-Mar-15	06:00	12:00								
15-Mar-15	12:00	18:00								
15-Mar-15	18:00	00:00								
16-Mar-15	00:00	06:00	1	1						
16-Mar-15	06:00	12:00								
16-Mar-15	12:00	18:00								
16-Mar-15	18:00	00:00								
17-Mar-15	00:00	06:00	1	1						
17-Mar-15	06:00	12:00								
17-Mar-15	12:00	18:00								
17-Mar-15	18:00	00:00								

Table 3: Example of data recording of self-reported daily noise strength and observation of audible noise during six hour intervals

Time of measurement			Self Reported noise strength (daily)			Audible noise heard during six hour periods of the day			
Date	From	To	Strong	Medium	Weak	Night	Morning	Afternoon	Evening
21-Feb-15	00:00	06:00		1					
21-Feb-15	06:00	12:00							
21-Feb-15	12:00	18:00						1	
21-Feb-15	18:00	00:00							1
22-Feb-15	00:00	06:00		1					
22-Feb-15	06:00	12:00					1		
22-Feb-15	12:00	18:00						1	
22-Feb-15	18:00	00:00							1
23-Feb-15	00:00	06:00		1		1			
23-Feb-15	06:00	12:00					1		
23-Feb-15	12:00	18:00						1	
23-Feb-15	18:00	00:00							1
24-Feb-15	00:00	06:00	1			1			
24-Feb-15	06:00	12:00					1		
24-Feb-15	12:00	18:00						1	
24-Feb-15	18:00	00:00							1
25-Feb-15	00:00	06:00			1				
25-Feb-15	06:00	12:00					1		
25-Feb-15	12:00	18:00							
25-Feb-15	18:00	00:00							
26-Feb-15	00:00	06:00		1		1			
26-Feb-15	06:00	12:00					1		
26-Feb-15	12:00	18:00						1	
26-Feb-15	18:00	00:00							1

Observation of audible noise from as many as eight wind turbines means that the observer will need to assess accurately the direction from which the noise arrives. Fortunately, the human ear is a fairly good directional instrument and experience with specific environments will improve this skill further. We found that in most cases it was relatively easy to determine from which direction the sound from individual wind turbines originated. Being able to see part of the turbine made determination slightly easier, but the two turbines which were not visible were almost as easy to identify in the local soundscape.

2.6 Other Self-reported Data

A column in the spreadsheet was assigned to note unusual noise events. This enabled capture of data normally missed such as abnormal character of sound, special sound conditions and peculiar changes in sound related to weather. Also, other mechanical or biological sounds of sufficient strength and duration to influence the average values of the sound data were noted.

3. Discussion and Suggestions for a Successful Project

Long-term studies combining recorded and self-reported data by non-scientific personnel have often been considered impractical, largely due to the challenges of getting people to participate. When self-reported data are collected, validity issues and avoiding biased or incomplete data must, of course, be considered. These issues can be managed through good preplanning, selection and efficient communication with the volunteer neighbours participating in the project. In the following some suggestions for maximising success of a neighbour-involved noise assessment project are given.

3.1 Selecting Neighbours to Record the Data

Recording of a time series dataset is a meticulous task which requires motivation, dedication, stamina and attention to detail, even if the project lasts only for a month. Needless to say, it is not a task which anyone will be willing or suited to perform. Wind farm neighbours who are annoyed by the noise from the farm are, however, a highly motivated group of people who may be willing to do a lot of work to document their experience of the noise. Trust is in this case an important factor, and experience from working with neighbours suggest that a study commissioned or financed by the owner of the windfarm may not be as likely to get motivated participants as one carried out by an independent research institution.

Careful selection of the people for data collection is essential to the success of such a project. This requires that the project scientist gains good knowledge of the neighbours he will work with. Selecting a small group of neighbours at different locations and encouraging them to work together to exchange knowledge and experience is advantageous. Participating neighbours should live at quiet places without too much traffic or farm noise as was the case with the neighbour in the Lista study. A map of the windfarm showing the location of impacted and annoyed neighbours (Figure 2) may be used to select potential participants to approach.

People with past experience with data recording and data management, are to be preferred. However, such experience may be rare. These are, however, skills that can be acquired. The most important qualities to look for, in the author's experience, are the motivation and tenacity of the participants being asked to join the project.

One may think it is best to find people who are "neutral" to the windfarm to participate in a study. This may be a valid approach in some cases. However, this will become a pitfall if the study objective is to understand impacted neighbours experience of the noise. People who do not complain may be living too far away where the noise is weak, or at protected locations where the noise is not very dominant. Others may also be hearing impaired, or they may be receiving substantial economic benefit from the windfarm owner. A group of “neutral” observers will not give much insight in the impacted neighbour’s experience of the noise. Careful thought needs to be given to the objective of the study if such participants are to be included.

It is recommended that the project manager sets aside some individual time for each participant to teach them a few simple tricks (like splitting screens if Excel is being used) on how to enter data while minimizing error in large spreadsheets.

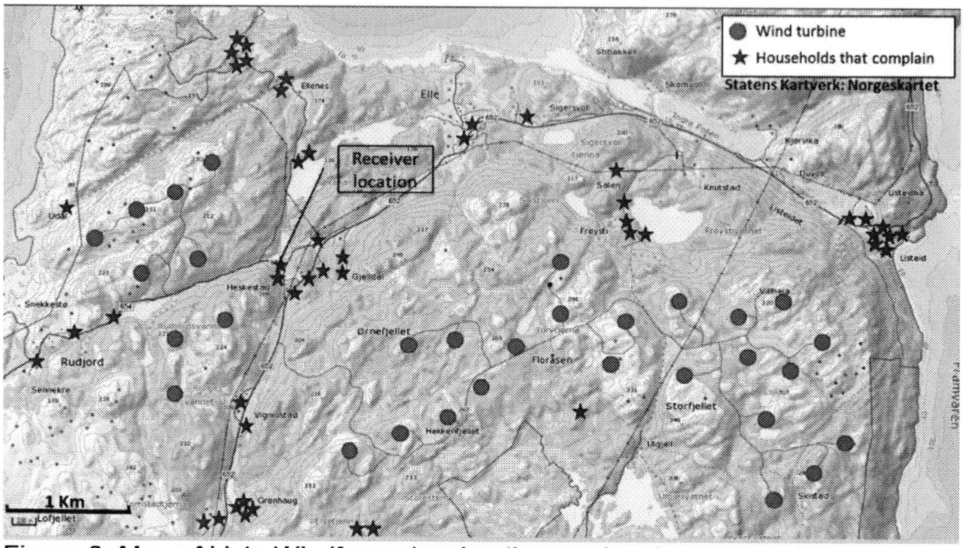

Figure 2: Map of Lista Windfarm showing the receiver location and the neighbouring dwellings that find the noise annoying (red stars). Blue dots are the wind turbines.

3.2 Maintaining the Momentum of the Study

Recording noise data over longer periods can become a tedious task if one is left alone to do the work for longer periods. It is therefore a key success factor to regularly stay in touch with the recording neighbours. This ensures an effective exchange of learnings as well as exchange of ideas and suggestions that can be used to make the study method more efficient. For example, one may find it advantageous to record an additional parameter, or that one of the chosen parameters may be of limited value or too much work to capture. Frequent contact is good for motivation as it gives an opportunity for the project manager to demonstrate that there is high interest in the work that is being done. Patience, tenacity and people skills are also good qualities for the project manager to bring to the project.

3.3 Quality Control of Data

Our data set consisted of more than 10.000 data points recorded over a time period of 15 months. Such large datasets require efficient cleaning and quality control to ensure data integrity. Space entries in cells making them look blank, misplaced or missing commas and numbers mistakenly plotted in the wrong cells must be corrected. Most of these can easily be identified and fixed with reasonable Excel skills. Excel functions like column averages, max value and global replaces were very useful functions for this task. Particularly useful to identify bad data points is plotting or cross plotting of data columns and examining the plots for anomalous values. Having the paper log as backup was in a few instances useful in correcting data-points that had been incorrectly entered. It is also good to retain as much as possible of the initial raw data collected.

3.4 Preventing Compromised Data

Many of the people recording the data will be neighbours who are very annoyed by the noise and therefore not very inclined to like the windfarm. It is therefore important to consider the risk of the data being biased in one direction or other. This risk can be mitigated by taking the following actions:

- Prior to the startup of the project it is recommended to hold a meeting to inform the participating neighbours. Among other things one may mention that the consequence of a compromised dataset is that it cannot be used. It may also be good to describe how the quality control process will be performed and point out that error in the data can be detected through various comparative methods of analysis.
- Reminding the participants, from time to time, of the importance of accurately and diligently recorded data.
- During and after the study plotting or cross plotting self-reported noise data against recorded noise and weather data, and other analysis, may be performed to verify consistency of data.

Most validity concerns that have been raised in literature on self-reported data have been related to responses to single questionnaires sent to people. Responding through a questionnaire is a relatively low threshold of effort, thus making it more exposed to biased contributions. Experience from the Lista long-term data collection effort suggests that this is a small problem, if at all one, for long-term studies. "Natural selection" will, in most cases, result in any less serious participants, due to the high effort, dropping out of long-term studies. Thus only the motivated and dedicated will stay to complete the work. Also the large size of the resulting datasets will make systematically biased entries very challenging and less likely due to the sheer effort of such an attempt.

3.5 Unexpected Data and Knowledge

It is not uncommon that unexpected observations are made in a study. As such events or information may be unknown at the stage of study design there may not initially be a specific study parameter assigned to capture such information and it may consequently be lost. Space should therefore be set aside in the recording log to note such events and the participating observers should be encouraged to note anything unusual. It is also recommended to frequently stay in contact with the recording neighbours to hear how things are going and to ask if they have made any interesting observations regarding the noise. At the beginning of the study it may be useful to ask the recording neighbours to describe their own past experience of the noise.

Some examples of unexpected information gained in the Lista study are:

- The noise is felt to be at its strongest when snow is falling. And, very surprisingly, noise is often experienced to be stronger when snow is on the ground.
- Substantial changes in noise levels may occur when moving laterally distances as short as 10 m at some receiver locations.
- The wind-turbines, at relatively moderate winds, often emit loud noises which some neighbours describe as sudden "roars" which can make them jump both when outside or inside the house.
- During strong storms some of the wind turbines emitted a protracted wailing, screeching sound which some neighbours found very annoying.

Such observations may, if they are not one-time occurrences, be systematically logged over time to get statistically significant data to assess their frequency of occurrence. In the case of Lista this data may be useful in understanding for example nightly wake-up frequencies and annoyance levels among the windfarm neighbours.

4. Conclusions

Windfarm neighbours annoyed by the noise from the windfarms constitute a highly motivated group of people who often may be willing to perform substantial work for research in order to shed more light on their own experiences with the noise. These people constitute a large untapped resource for research. Neighbour involved projects to record digital and self-reported noise data from windfarms can be performed with success if they are well planned and executed. Being able to link physical data, like sound and wind speed, with self-reported data can be a very useful tool in understanding noise from windfarms.

Experience from the Lista project suggests that validity concerns regarding self-reported data, as known from other fields such as medicine, should be viewed as a smaller risk in studies where longer time-series of data are being recorded. Risk of biased data can be managed by good preplanning.

Good communication is a key success factor. Regular and frequent follow-up and involvement/discussion of the project with its participants is prerequisite, particularly in the early phase, to identify and handle any problems, misunderstandings and drops in motivation that may occur.

References

Vågene, S and Larsen, W (2015) Impact of wind-turbine noise on local residents in mountainous terrain at Lista Windfarm, south Norway 6th International Meeting on Wind Turbine Noise, Glasgow 20-13 April 2015 (Conference CD)

Vågene, S and Larsen, W (2016) Wind turbine noise as experienced by impacted residents ICA2016, Proceedings of the 22nd International Congress on Acoustics, Buenos Aires

7th International Conference on Wind Turbine Noise Rotterdam – 2nd to 5th May 2017

A Rigorous Method of Addressing Wind Turbine Noise

William K.G. Palmer P. Eng. TRI-LEA-EM trileaem@bmts.com

Summary

A rigorous and repeatable method is proposed to monitor and assess wind turbine noise. The method had to consider that an effective monitoring system must take into account more than just averaging sound power levels over a long term. The method recognizes that humans are bothered by the changes and annoying characteristics that occur, as well as long term averages. Others describe this as the need to determine how the special characteristics of sound quality may impact quality of life. To verify this approach, assessments were conducted using the method at two wind power developments. Use of this rigorous method permitted gathering evidence of the presence of characteristics described as annoying by residents. The evidence produced by this method is clear: the method itself is repeatable and it considers the requirements of a more comprehensive system. In contrast, compliance methods currently in use have not demonstrated the capability of verifying non-conformance of the same wind power developments, even though in one case the monitoring has been in progress for eight years. Use of the proposed method will permit others to gather quality evidence in a similar manner.

1. Introduction

Until as late as the 1960's, noise from most any source was considered only as an annoyance. In 1970, though, the perception of noise changed, with the publication by Karl D. Kryter of the seminal work, titled, *"The Effects of Noise on Man."* Even now, nearly 50 years later, as one reads through Kryter's work, themes often discussed today keep reappearing:

- Masking, Loudness, and Auditory Fatigue
- Equal Loudness Contours
- Perceived Noisiness (Annoyance)
  - Loudness Versus Noisiness
  - Influence of Cognitive Values
- Judged Perceived Noisiness and Perceived Noise Level
- Background Noise in Real Life
- Effective Perceived Noise Level
- Laboratory Versus Field Test Conditions
- Relative Accuracy of Physical Units for Predicting Judged Received Noisiness
- Community Reactions to Noise
- Indoor versus Outdoor Listening – Relative Judgements
- Non-Auditory System Responses to Noise
  - Health
- General Physiological Responses to Noise
  - Stress and Health
  - Sleep and Health
- Effects of Noise on Mental and Motor Performance

Kryter went on in other papers such as, *"Non-Auditory Effects of Environmental Noise,"* published in 1972, to outline a need for a measurement basis other than just direct physical measurements. As he noted, "The most direct, and perhaps most valid, insight into the possible presence and magnitude of stress reactions in general living environments is probably that which has been obtained from attitude surveys and real-life behaviour of people." The fact that people are reacting is more important than a simple measure. Although Kryter noted that "it appears" people adapt to noise, he acknowledged that, "This conclusion is deduced from a relatively small amount of research and incompletely tested concepts."

Yet, only 50 years later, a noise source unimagined by Kryter, wind turbines, some nearly 200 metres tall, can be found across the countryside in many countries. Some wind turbines, are located as close as 400 to 500 metres to homes. Yet, the work by Kryter has continued through many who have followed him, such as Klaus Genuit, and André Fiebig, of HEAD acoustics, in Germany, who noted in, *"Psychoacoustics and its Benefit for the Soundscape Approach,"* published in *Acta Acustica United* in 2006, "The increase of complaints about environmental noise shows the unchanged necessity of researching this subject. By relying on sound pressure levels averaged over long time periods and suppressing all aspects of quality, the specific properties of environmental noise situations cannot be identified because annoyance caused by environmental noise has a broader linkage with various acoustical properties such as frequency spectrum, duration, impulsive, tonal and low-frequency components, etc. than only with SPL [Sound Pressure Level]. In many cases these acoustical properties affect the quality of life."

Others, such as Mathias Basner and Wolfgang Babisch have furthered the work of Kryter at conferences such as the International Conference on the Biological Effects of Noise (ICBEN) publishing works such as:

- *"ICBEN review of research on the biological effects of noise 2011-2014",* concluding, "These reviews demonstrate that noise is a prevalent and often underestimated threat for both auditory and nonauditory health and that strategies for the prevention of noise and its associated negative health consequences are needed to promote public health,"
- *"Auditory and non-auditory effects of noise on health," Lancet,* 2014
- *"Cardiovascular effects of environmental noise,"* 2014.

Still, the regulation of wind turbines is largely based on simple parameters such as a 40 dBA L_{eq} limit, often averaged over long time intervals, and aspects of the quality of the sound are largely ignored, such as the difference from the natural environment of this omnipresent source that may be 15 or more dB greater than ambient, and often cyclical or tonal as opposed to random. Compliance protocols have been established in some jurisdictions that are so complex, that today, 10 years after wind arrays commenced operation, it has been impossible to complete reports based on the protocol to show compliance. Meanwhile, the turbines continue to operate, while real-life behavioral changes (as projected by Kryter) are occurring, such as walking away from family homes, leaving them unsold, after complaints to the operator and regulator resulted in no remediation.

Numerous papers, including some by this author, have identified what are dismissed with disdain as "anecdotal reports" of adverse impacts that occurred with the start up of wind turbines in the environment of those impacted. However, there is a solid basis for presenting such lists. It mirrors the approach taken by most medical doctors when a patient first presents himself or herself with a new adverse health complaint. Taking a patient "history" is the way most doctors begin. Similarly, engineers and problem solvers often begin to address a new problem by looking for changes that have occurred. Yet, some maintain there is no proof that the start up of the turbines was the change that caused the impact, even though the conditions diminish when the person vacates the area, and recur when the person returns. They may attribute it to the stress self-generated by refusing to accept a change. Ignoring those suffering will not result in solving the problem predicted by Kryter of people making real-life behavioral changes. The rigorous method established in this paper permits measuring the physical emissions (noise) from wind turbines, and confirming some aspects of the quality of the noise that are identified as problematic to demonstrate evidence of the cause for the suffering.

2. Predicting Noise from a Wind Turbine Array

Wind turbines are licensed on the basis of a predicted sound pressure level that will occur at a receptor after the array is put into service. Part of the rigorous approach in this paper is to be able to duplicate those calculations of predicted sound pressure levels, and to understand some of their limitations. The intent is to describe a method that can be simply replicated without requiring the use of complex computer models. Understanding the problem without having to revert to mystical (and usually expensive) "black box" algorithms that return inexplicable results is the goal. This prediction method is based on the International Standards Organization (ISO) standard 9613-2 *"Acoustics – Attenuation of sound during propagation outdoors. Part 2 – General method of calculation."* It is not the newest standard used for this purpose, but it is still widely used to generate a first approximation. There are numerous limitations of the code such as:

- It recommends that it be used for distances not over 1000 metres, while we use it to predict attenuation out to 2 or 3 kilometres.
- It assumes a point source of the sound, while for wind turbines, the predominant noise source is in the region of the blade tips, so may follow a locus equal to the rotor diameter, and the distance to the receptor may only be a few (perhaps 3 to 10) times that distance, so the source is certainly not equivalent to a point source.
- The code specifies it is for use with ground based sources such as road or railways so that the distance from the source to the receptor is many times the height of either, while wind turbines with noise emitters up to 200 metres overhead really are not ground based when the distance to receptors may only be a few times the height of the source.
- It only considers frequencies down to 63 Hz, while for wind turbines the low frequencies may be a predominant factor.
- It is based on generally soft ground from the source to the receptor, while in winter, frozen ground conditions, or during inversion conditions over water, the code is limited, particularly when single values of ground attenuation are chosen.

Still, even with these limitations, an estimate based on ISO 9613-2 gives at least a first approximation, and it will be used in this paper.

2.1 Determining Distance to Turbines Within Area of Interest

As a general rule, turbines to be considered will be bounded by a circle with a radius not over 5 or 6 times the distance to the closest turbine. Beyond that, the predominating effect of the closest turbine will be so dominant that calculating the effect of more distant turbines is of limited value. The simplest method of determining the distances to applicable wind turbines is to use a scaled ruler on a map showing the turbine locations centred about the point of interest. If more than a few cases will need to be calculated, a template of scaled concentric circles is prepared as shown in Figure 1.

The map might be available from the developer's public filings, or if that is not readily available, even a printout from "Google Maps" can be used. From the figure it is possible within a few minutes to estimate the distance from point of interest, R145 to all turbines within 3 km as shown in the table below the figure.

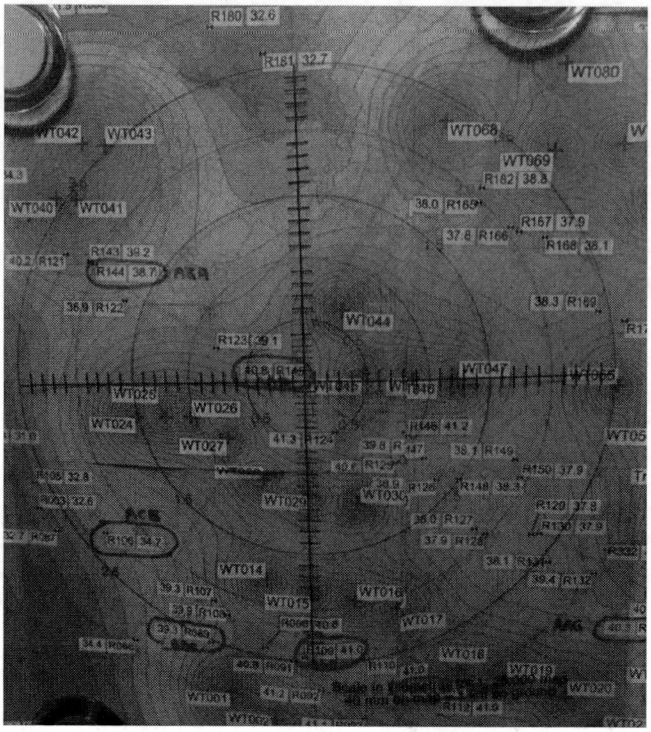

FIGURE 1 – A Simple Map Tool

WT 045 – 460m	WT 044 – 650m		WT 027 – 800m	WT 046 – 850m
WT 028 – 950m	WT 029 – 1000m		WT 030 – 1200m	WT 047 – 1250m
WT 014 – 1800m	WT 015 1900m		WT 016 – 2000m	WT 041 – 2400m
WT 040 – 2600m	WT 043 – 2600m		WT 042 – 2700m	WT017 – 2200m
WT018 – 2800m	WT001 – 2900m		WT055 – 2500m	WT056 – 2900m
WT068 – 2300m	WT 069 – 2600m			

If one wants a more precise set of distances to turbines, (as were used for the calculations in this paper) and listings of the coordinates of each turbine and point of interest are available, either from a developer's documentation, or from a field trip with a hand held GPS unit, then a more rigorous calculation can be performed by calculating the results from:

Distance = Square Root [$|X\ coordinate_1 - X\ coordinate_2|^2 + |Y\ coordinate_1 - Y\ coordinate_2|^2$]

In this case, the coordinates of R145 and WT 045 are given in the developer's records as:
$X_R = 459854$ $Y_R = 4907073$ $X_{WT} = 460305$ $Y_{WT} = 4907113$
Solving, Distance = Square Root [$|459854\text{-}460305|^2 + |4907073\text{-}4907113|^{2}$] = 453 (m)

Accordingly solving for all the turbines identified by the 3000-metre template (rounding up the suggested 6 x closest turbine distance) gives results of:

WT 045 – 453m	WT 044 – 632m		WT 027 – 818 m	WT 046 – 840m
WT 028 – 942m	WT 029 – 988m		WT 030 – 1153m	WT 047 – 1216m
WT 014 – 1771m	WT 015 – 1861m		WT 016 – 1961m	WT 041 – 2532m
WT 040 – 2536m	WT 043 – 2487m		WT 042 – 2626m	WT 017 – 2125m
WT 018 – 2829m	WT 001 – 2881m		WT 055 – 2494m	WT 056 – 2851m
WT 068 – 2263m	WT 069 – 2601m			

While these are more precise, the difference is small enough that for a quick calculation, one must consider that the visual method described first needed about 5 minutes, while looking up the table values and doing the calculations individually took over 2 hours, and the end result will have very little difference.

2.2 Calculating Sound Pressure Level at Point of Interest

An Excel spreadsheet was prepared to calculate the sound pressure levels at any receptor. It is a simple spreadsheet, yet includes all the relevant aspects of ISO 9613-2. Inputs to the spreadsheet include:

- The Sound Power Level for the turbines used, as provided by the manufacturer. In some cases it may be necessary to interpolate between given values to determine a Sound Power Level for a particular turbine output, or wind shear.
- The distances between all relevant turbines and the point of interest for which the sound pressure level will be calculated.
- Details such as turbine hub height, residence heights, and environmental condition (weather) specifics.

The full details of the spreadsheet will not be given in this paper for brevity, but copies of the relevant data entry and results pages of spreadsheet are included as Figures 2 and 3. Interested individuals may contact the author for more information.

Wind Turbine Sound
ISO 9613-2 Calculator

For Vestas V82 rated at 1.65 MW
turbine use, with assumptions for propagation given in
MOE Oct 2008 Guidlines for Wind Farms

Based on MOE Assumptions
Oct. 2008

Wind Development: Enbridge Underwood
Turbine Type: Vestas V82 1.65 MW

Distance Source to Receptor	575	m			
Source (hub) Height	80	m	Source Zone (30h)	575	m
Source Base elevation	0	m	Source G	0.7	
Top of Source Height	80	m	Receiver Zone (30h	135	m
Receptor Height	4.5	m	Receiver G	0.7	
Receptor Base Elevation	0	m	Middle Zone	0	m
Top of Receptor Height	4.5	m	Middle G	0.7	
Temperature	10	C	Receptor SPL (dBA)	34.7	
Relative Humidity	70	%	Receptor SPL (dBC)	46.1	

	63 Hz	125 Hz	250 Hz	500 Hz	1000 Hz	2000 Hz	4000 Hz	8000 Hz	All Freq
Source PWL dB	108.2	106.1	104.6	99.3	96.1	92.8	91.7	80.2	108.2
A Weighted PWL dBA	82.0	90.0	96.0	96.1	96.1	94.0	92.7	79.1	102.5
Att.Divergence	66.2	66.2	66.2	66.2	66.2	66.2	66.2	66.2	
Att.Ground (Source Zone)	-1.5	-0.5	-0.5	-0.5	-0.5	-0.5	-0.5	-0.5	
Att.Ground(Middle Zone)	0.0	0.0	0.0	0.0	0.0	0.0	0.0	0.0	
Att.Ground(Receptor Zone)	-1.5	2.0	0.5	-0.4	-0.4	-0.5	-0.5	-0.5	
Att.Atmosphere	0.1	0.2	0.6	1.1	2.1	5.6	18.9	67.8	
Att.Atmosphere dB per 1000m	0.1	0.4	1.0	1.9	3.7	9.7	32.8	117.9	
Receptor SPL (unweighted)	44.9	38.1	37.8	32.9	28.7	21.9	7.5	-52.9	46.7
A Weighting Factor	-26.2	-16.1	-8.6	-3.2	0.0	1.2	1.0	-1.1	
Receptor SPL (A Weighted)	18.7	22.0	29.2	29.7	28.7	23.1	8.5	-54.0	34.7
C Weighting Factor	-0.8	-0.2	0.0	0.0	0.0	-0.2	-0.8	-3.0	
Receptor SPL (C Weighted)	44.1	37.9	37.8	32.9	28.7	21.7	6.7	-55.9	46.1

Instructions for use: First load list of relevent turbines and distances onto sheet 2 and hub height particulars above.
1. Insert unweighted source PWL in dB into orange cells above for the hub wind speed estimate is to be made for.
2. Insert distance from Turbine (Source) to Receptor in the pink cell above for each turbine.
3. Transfer data for Receptor unweighted SPL dBZ (yellow) and for A-weighted SPL dBA (blue) to Sheet 3.

Figure 2 – Wind Turbine Sound Calculator – Input Sheet 1

We will look further at the outputs from the wind turbine noise calculator when we discuss measurements taken in Section 3

2.3 Adjusting for Seasonal Impacts

Although the calculator is based on average conditions of 10°C and 70% relative humidity, changing these parameters results only in minor propagation changes, but do not show the significant effect in wind turbine output (and sound generation, hence Sound Power Level) as air density changes. Blade condition, including dirt (insects), wear, or minor icing also result in an increase in turbine Sound Power Level, which are not inputs to the calculation.

Wind Turbine Sound
ISO 9613-2 Calculator

Calculated for Receptor
R285

Based on MOE Assumptions
Oct. 2008

	Wind Development:Enbridge Underwood									
	Turbine Type: Vestas V82									
Turbine	Distance	Val(dBA)	Val 63Hz	Val 125Hz	Val 250Hz	Val 500Hz	Val 1kHz	Val 2kHz	Val 4kHz	Val 8kHz	Val(dBZ)
WT 108	575	34.7	44.9	38.1	37.8	32.9	28.7	21.9	7.5	-52.9	46.7
WT 107	900	30.0	41.0	33.9	33.5	28.4	23.6	14.9	-7.0	-95.1	42.6
WT 106	1011	28.8	40.0	32.9	32.4	27.2	22.2	12.8	-11.7	-109.2	41.6
WT 101	1089	28.0	39.4	32.2	31.7	26.4	21.2	11.4	-14.9	-119.0	40.9
WT 109	1177	27.1	38.7	31.4	30.9	25.5	20.2	9.9	-18.4	-130.1	40.2
WT 090	1228	26.6	38.3	31.1	30.5	25.1	19.7	9.0	-20.5	-136.5	39.8
WT 110	1297	26.0	37.8	30.5	30.0	24.5	18.9	7.9	-23.2	-145.1	39.3
WT 089	1605	23.6	35.9	28.6	27.8	22.0	16.0	3.0	-35.2	-183.2	37.4
WT 100	1615	23.5	35.9	28.5	27.7	22.0	15.9	2.9	-35.5	-184.5	37.3
WT 091	1618	23.5	35.9	28.5	27.7	21.9	15.8	2.8	-35.6	-184.8	37.3
WT 092	1919	21.5	34.3	26.9	25.9	19.9	13.2	-1.6	-47.0	-221.8	35.7
WT 088	2004	21.0	34.0	26.5	25.5	19.4	12.5	-2.8	-50.2	-232.2	35.3
WT 087	2417	18.7	32.3	24.7	23.4	16.9	9.4	-8.4	-65.3	-282.5	33.6
WT 093	2652	17.6	31.6	23.8	22.4	15.7	7.8	-11.5	-73.8	-311.0	32.8
WT 094	2856	16.7	31.1	23.2	21.7	14.8	6.4	-14.0	-81.1	-335.6	32.3
WT 041	6772	5.5	24.8	14.6	10.7	0.3	-15.1	-59.0	-216.6	-804.4	25.3
Overall SPL (dB)		38.9	50.2	43.1	42.5	37.3	32.3	24.0	7.8	-52.9	51.8

Instructions for use:
1. Load list of turbines and distance into first two comumns.
2. Copy and Paste first distance to Sheet 1
3. Copy dBA and Octave Values from Sheet 1 and Paste Sheet to 2 for the turbine. Use "Paste Special" - "Values"
4. Repeat for other turbines.

Figure 3 – Wind Turbine Sound Calculator – Output Sheet 2

2.4 Adjusting for Close Turbine Spacing

The Sound Power Levels provided by manufacturers that form the basis of the calculation are based on measurements performed on single turbines on a test site. As will be discussed in the measurements section, the results obtained by measurement do not always match the predicted case, particularly in an environment where turbines impact each other due to close spacing, resulting in additional turbulence, which raises the turbine Sound Power Level.

3. Measuring the Actual Noise Levels

The brief overview of potential influences on the turbine Sound Power Level, and propagation, lead to the need to conduct measurements to determine if the actual measurements match the predictions. We will discuss the key parameters of a basic measurement system.

3.1 System Requirements

Measuring sound pressure levels with a sound level meter is really not adequate to be able to determine annoyance. A calibrated recording system is critical to be able to determine the quality characteristics of the sound, and to select time segments for analysis that are free from extraneous influences such as vehicles, wildlife, humans, and environmental conditions of rain or heavy wind. In reality it is not difficult to listen to recordings and to select relatively "clean" sound signatures of the desired parameter independent of extraneous influences.

This paper will describe one possible system. There is no claim made that this is the "only" manner of doing the job, nor should the mention of any particular manufacturer be considered as an exclusive endorsement. It is simply that this works for us, and outlines some of the specifications to consider in setting up a wind turbine recording system.

Characteristics of the recording system used:

- Microphone – a good quality, omnidirectional microphone with a wide frequency band, and a relatively low noise floor. Typical measurement microphones are condenser type, which require some sort of power supply for polarization. Systems used to prepare this report include:
    - ACO Pacific 7046 free field microphone capsule with 4012 companion preamp, and PS9200 9V battery operated external power supply for 200V polarization.
        - Frequency response ± 2 dB 3 Hz to 20 kHz
        - Sensitivity 50 mV/Pa
        - Noise Floor 10 – 12 dBA
    - Earthworks M30BX omnidirectional measurement microphone, with internal 6 V battery for polarization
        - Frequency response specification 9 Hz to 30 kHz +1/-3dB (although observed to be wider)
        - Sensitivity 30 mV/Pa
        - Noise Floor 22 dBA equivalent

Figure 4 – Microphone Wind Screen and Mounting

Primary and Secondary windscreen

- Outdoor measurements typically require a primary and secondary windscreen to reduce the effects of ambient breezes passing over the microphone capsule

- For us, the Ontario wind turbine measurement protocol requires use of a 90 mm primary windscreen and a concentric 450 mm secondary windscreen. The secondary one we use has 25 mm of reticulated open cell foam, with 8 pores per 10 mm. It is fabricated from two 16 inch (40.6mm) diameter open wire metal flower baskets, with a cylindrical 16 inch (40.6mm) central open mesh section to produce an oblong wind screen suitable to enable the microphone with its primary windscreen used to have the cartridge at the centre of the outer wind screen.

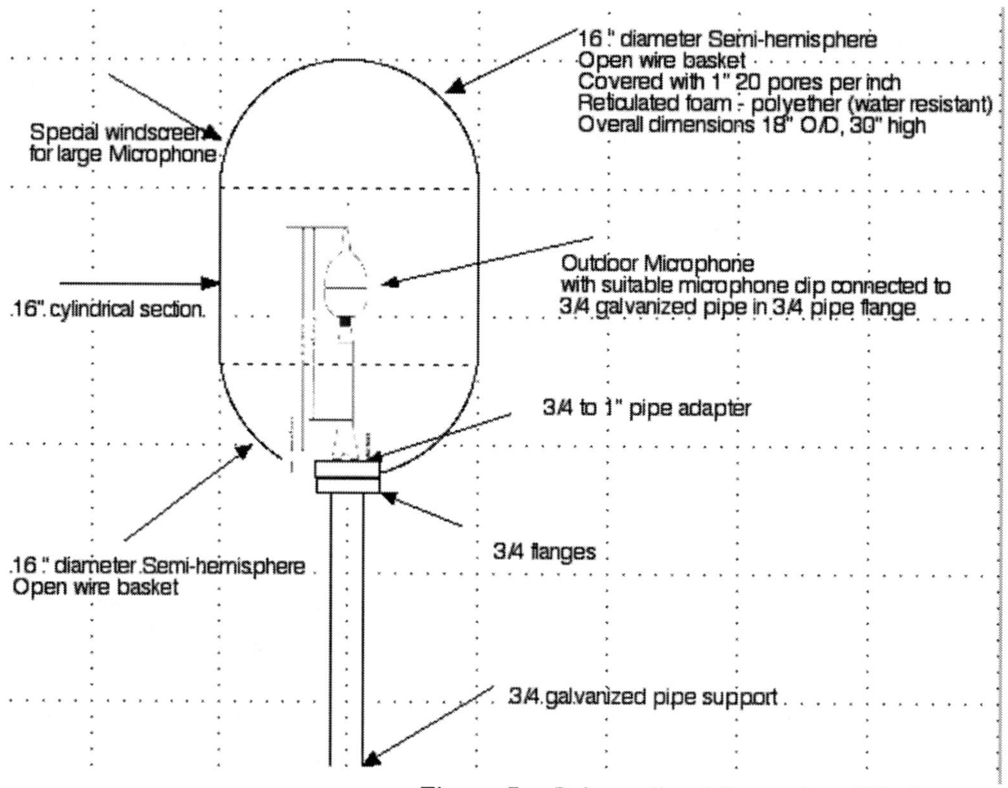

Figure 5 – Schematic of Secondary Windscreen

- USB Digitizer
  - We use a M-Audio Fast Track Audio Interface
    - Accepts two microphone inputs (to enable simultaneous indoor and outdoor measurements to be conducted).
    - Can supply phantom power to microphones if required, but we avoid this by using the microphones with integral battery power supplies.
    - Up to 24 bit, 48 kHz operation (but generally used at the standard 44.1 kHz sampling frequency).
    - The interface can be powered from the USB bus of the computer it is plugged into, or in the case of some models from an external power supply. In our usage, we have found that the newer "Fast Track Pro" models of the interface are prone to AC contamination that generates 60 Hz and harmonic contamination of the produced digital signal if the

computer that the interface is connected to by the USB bus is plugged into AC power, while the older first and second generation Fast Track interfaces ones were not prone to AC contamination.

- Recording Software
  - We've had good success using Audacity for the Macintosh. For a free software application, it is very versatile, and permits separation of a two track "stereo" recording (as generated when making a simultaneous indoor and outdoor recording) into separate "monaural" tracks for individual calibration and processing.

- Recording Platform
  - Here we are using an antique Macintosh iBook G4 computer, to run Audacity. It gives us about 6 to 8 hours of "unplugged" recording capability, remote from "mains" power. Could we upgrade? Certainly, but we are also of the opinion that "if it works and it isn't broke, don't fix it." Consistent use of the same system eliminates a source of concern for change when comparing two sets of results. We've also used newer models of the Macintosh. No doubt "that other platform" might also be used, but we cannot comment.

- Signal Processing Software
  - Audacity by itself will meet most of the user's needs for signal processing. The Audacity program permits saving files in .wav format, for later processing, and permits doing Fast Fourier Transform (FFT) analysis. This breaks the signal into equal slices of frequency width, which permits identification of special frequency concerns such as tonality. Audacity also provides a signal generator to generate noise (white, pink, or Brownian), or tones (of various nature). With a bit of effort, simultaneous signals can be overlaid to produce a multi-featured signal that can replicate measured conditions for controllable listening tests. A "Poster" presentation will be used at the conference to demonstrate some of these replica signals for an audio "jury" listening test, but they will not be part of the conference presentation which will focus on the assessment method, and evaluation of the results it generated.
  - An alternative versatile audio signal-processing program used is the Faber Acoustics Electroacoustics Toolbox. An especially convenient feature it has is to enable calibrator traces to be input so that all subsequent recordings from the same recording campaign are automatically recorded in a calibrated manner.

- Calibrator
  - You will require a 1000Hz calibrator to calibrate your microphone system before and after each set of recordings (between set up and teardown).
  - We use a Lutron Model SC-941 sound calibrator that generates a 94 dB 1000 Hz signal in compliance with an ISO-9001 quality management system.
  - It must be periodically tested against a traceable national standard.

3.2 Conducting Measurements at K2 Wind Power Development

Ontario presents perhaps a unique situation for wind turbine transient monitoring. The Ontario electrical grid has typical daily demands ranging from 12,000 to 15,000 MW at night, and from 18,000 to 22,000 MW in the daytime. The typical contributors to that Ontario electrical grid are from 10,000 to 12,000 MW of base load nuclear, 3000 to 7000 MW of hydraulic generation, and 1000 to 7000 MW of natural gas fired generation. On top of that "dispatchable generation" (can be called on to increase or decrease generation on demand), Ontario has installed some 6,800 MW of "variable generation" (4,600 MW of wind generators and 2,200 MW of solar) that generate depending on the availability of their natural resource. Often, when the wind blows best, as it is wont to do when the system load is smaller than the baseload generation, the Ontario Independent System Operator has the authority to "curtail" wind generators (stop accepting wind generation, while they are still being paid as if generating). The result is a common occurrence of having wind power on and off at short notice. Figure 6 is a typical output chart for the K2 wind power development for 5 days in January 2017 showing many occasions when the turbine output changed even though the monitoring system shows it was capable of (and paid for) higher outputs. The chart is generated from data of the Ontario Independent Electrical System Operator with "capability" provided by specially installed wind test towers at the wind power development, and "output" from the revenue metering system. The chart shows the values for the average condition for the hour preceding and not the actual hour ending value.

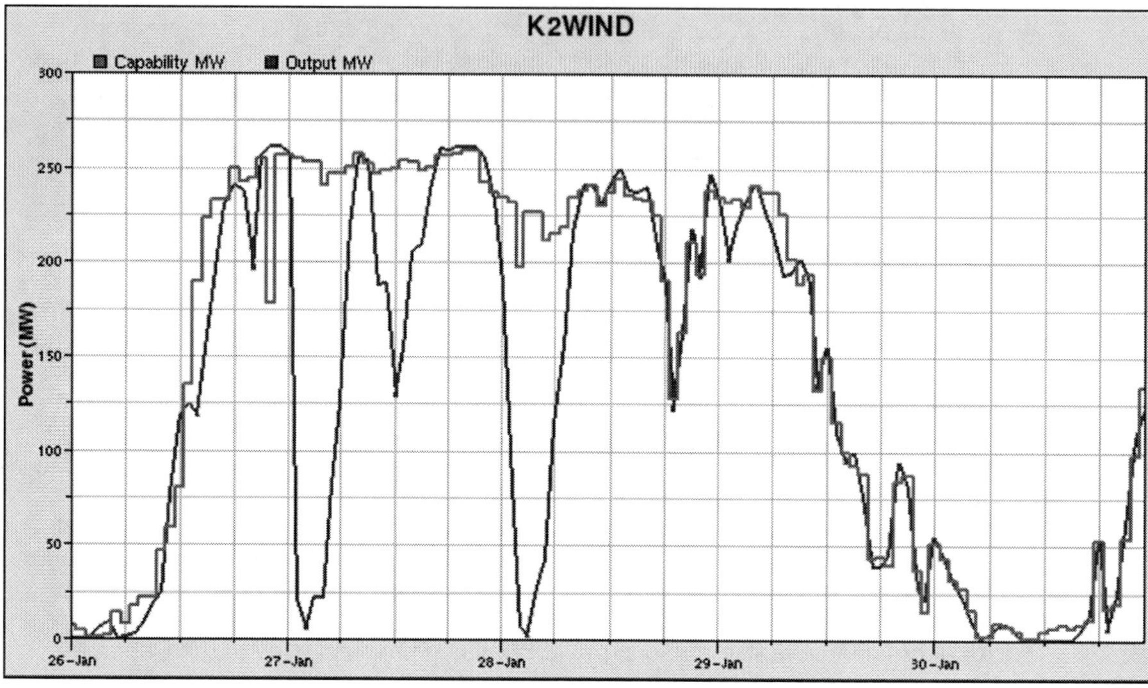

FIGURE 6 – K2 Wind Capability and Output January 26 to 30, 2017

As a result there are lots of opportunities to carry out monitoring as turbines go from intermediate or high power to low power even though the wind conditions may be relatively unchanged.

A family who live in this K2 Wind power development, presented to the Multi-Municipal Wind Turbine Working Group, comprised of elected and municipally appointed citizen representatives from about 14 municipalities in Bruce, Grey, and Huron Counties. It tries to address citizen issues related to wind project operation. The residents reported these turbines to be tonal, emitting a "sickening" sound, ever since operation started about a year ago. "We have complained to the operator and the regulator, and nothing has improved. Anything you can do to help," they asked, "would be appreciated."

The working group visited their home, and found that the Ministry of the Environment and Climate Change had installed a short term monitoring station at the home for 9 days. The residents could press a button when they believed the turbines were problematic to start a 10-minute recording for later analysis by the MOECC staff. The residents were also asked to make recordings of times when the turbines were not problematic, and any comment they had at the time any recording was initiated. At the end of the monitoring period, the Ministry staff provided the residents with a USB stick containing the twenty-seven 10 minute recording files as .wav documents (twenty-five initiated by the residents plus the initial and final test initiated by the Ministry staff). The Ministry reported that it could find no problems with the turbines. In the majority of the cases (19 of the 25 initiated by the residents) the staff reported that no assessment of the data could be made as the recording also showed indications of wind in trees, or wildlife. The Ministry staff found the turbines were compliant in the 6 cases they did evaluate. The Ministry report identified their assessment of the dBA rating for each recording, and the Ministry comments. Looking at the Ministry report and listening to the sound files provided some interesting insights when preparing this paper, as shown in Figure 7.

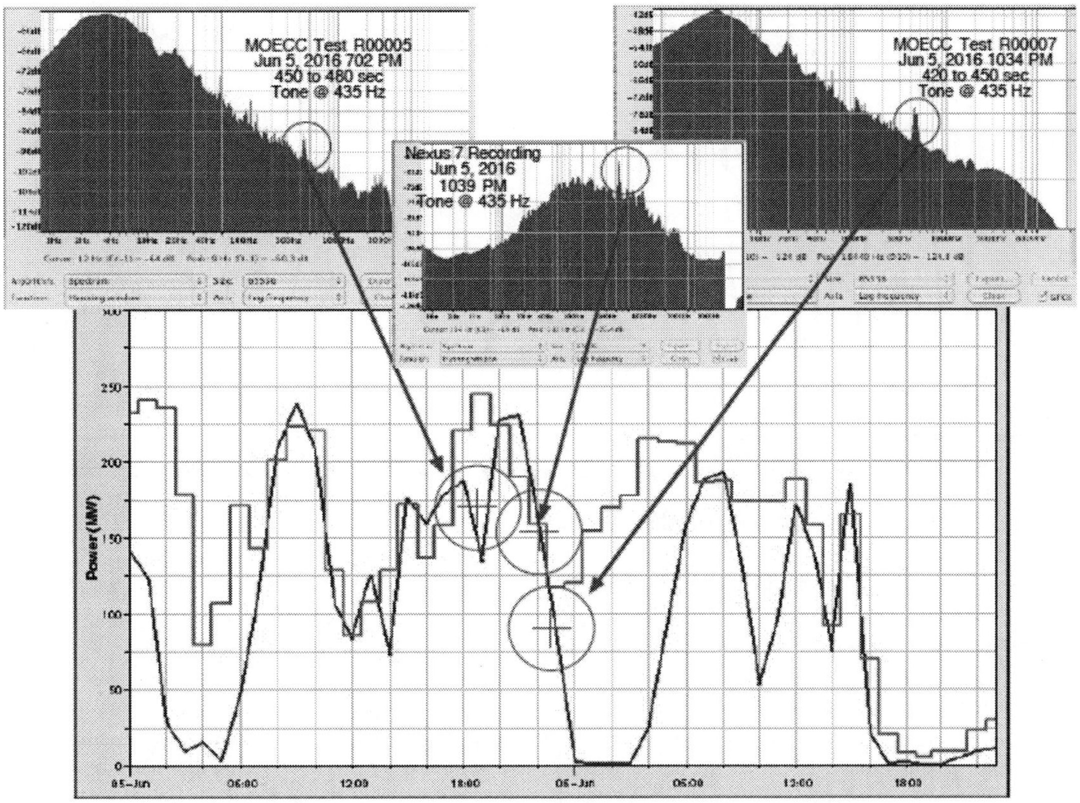

FIGURE 7 – Initial Assessment of Recordings and Turbine Output for June 5

The problematic periods seemed to be occurring as the turbines were curtailed, and the FFT charts generated by Audacity from the Ministry recordings for this initial look certainly seemed to show indicators of tonality at about 450 Hz that matched listening to the recordings. This tonality did not correlate to either tree noise or wildlife, as proven by generating a broad tonal test signal centred at 450 Hz as shown in the FFT of the sound sample and doing a listening test of that test tone. The residents provided an additional recording done on a hand held Nexus 7 tablet. While not of measurement protocol standards, it too revealed a very obvious tonal character.

Monitoring was set up by the author at this home, making simultaneous recordings indoor (in a vacant bedroom using the ACO Pacific microphone) and outdoor (using the Earthworks microphone) connected to the recording system described previously.

The wind power development capability and output for K2 Wind for one of the monitoring periods on Nov 10 and 11 is shown in Figure 8.

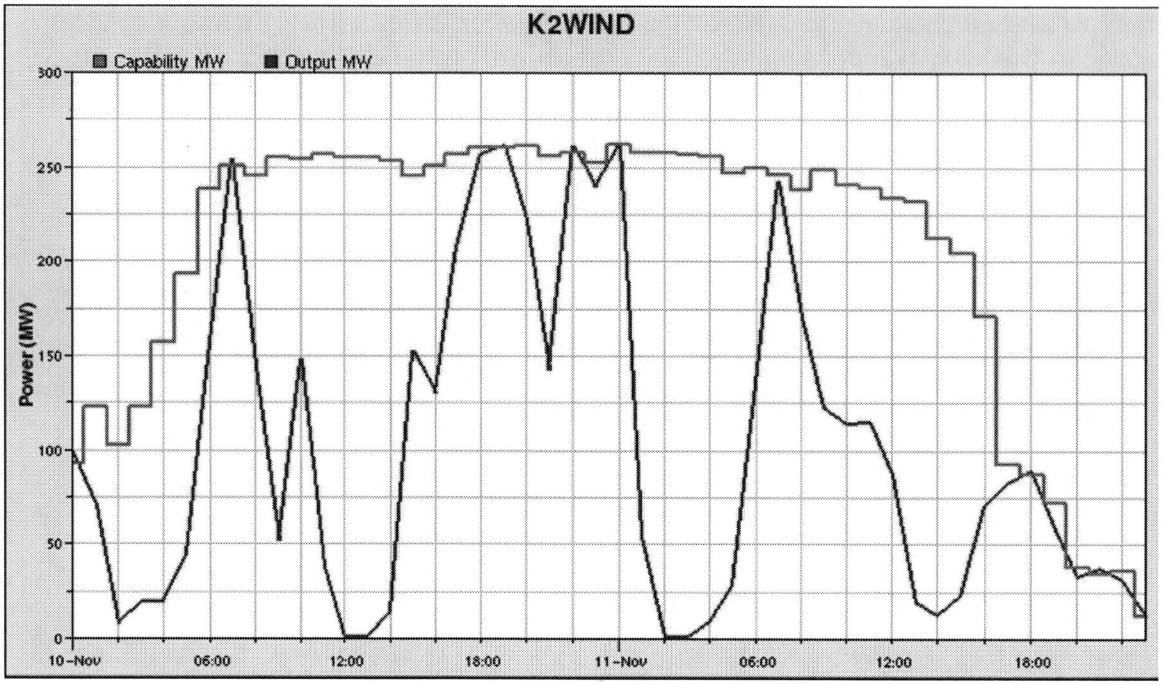

Figure 8 – K2 Wind Output and Capability on Nov 10 & 11, 2016

Evaluation of the outdoor recordings, considering the extremes of high power operation (at full output of 262 MW) and curtailed operation (when generating 21 MW while capable of 262 MW) produced the FFT shown in Figure 9.

While the sound pressure level was higher during the high power operation; when the turbines were curtailed, a very clear tonal peak centred about 450 Hz was seen. This tonal peak was about 10 dB in magnitude above the baseline sound present at the time and would be clearly noticeable. (The author personally observed it at times during the monitoring period.)

Then, attention was turned to compare the conditions indoors and outdoors for the same times, as shown in Figure 10.

The most obvious observation was that while the higher frequencies are attenuated when passing through the house walls (this is a well insulated house with thermal windows), the low frequencies below about 100 Hz are nearly as high as outdoors, and the indoor sound FFT shows much more "roughness" with variation in the order of 10 dB at frequencies about 200 Hz, and at lower frequencies. This was similar to the observations made previously by the author and presented to the Acoustical Society of America at the ASA 168th meeting in Indianapolis, titled, "Room modes – a predictor of wind turbine annoyance." That paper arose after a study at a different home in another wind power development with a different turbine type showed that in rooms where annoyance was felt, the frequencies flagged by room mode calculations and the low frequency spikes observed from the wind turbine measurements coincided.

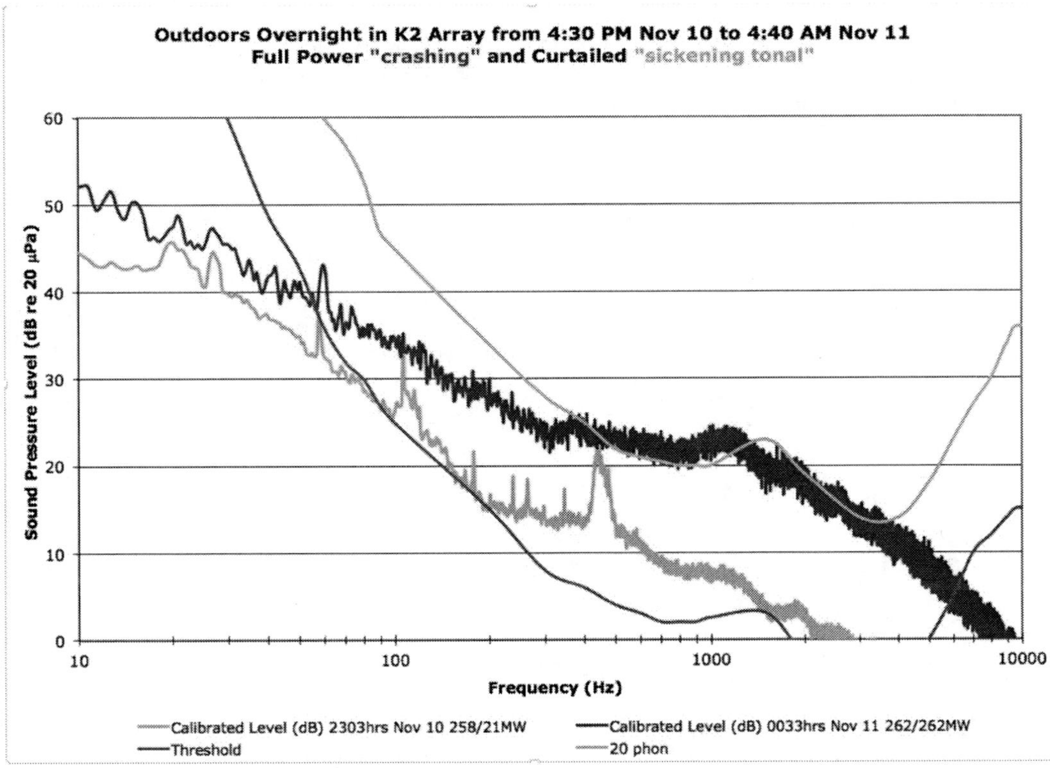

FIGURE 9 – Outdoors Overnight in K2 array

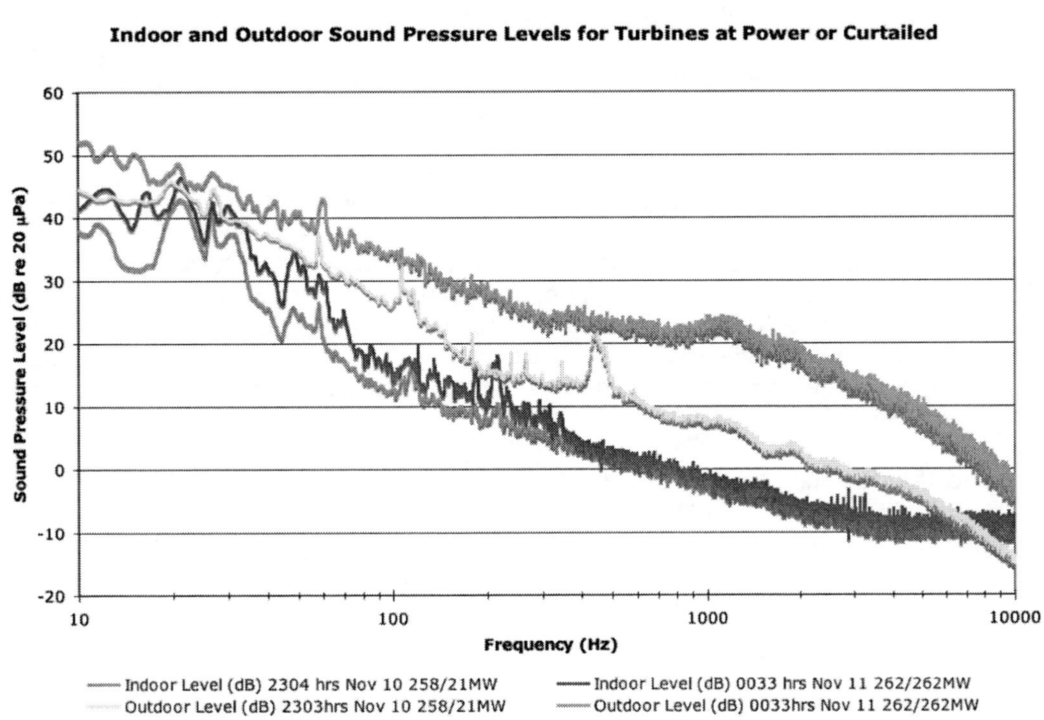

FIGURE 10 – Indoor and Outdoor Comparisons at the Same Times in K2 Array

Carrying out a more in depth analysis of the Ministry of the Environment and Climate Change provided sound samples from June 7 to 11 permitted more observations to be made. Figure 11 shows the K2 Wind Capacity and Output for June 11 and 12.

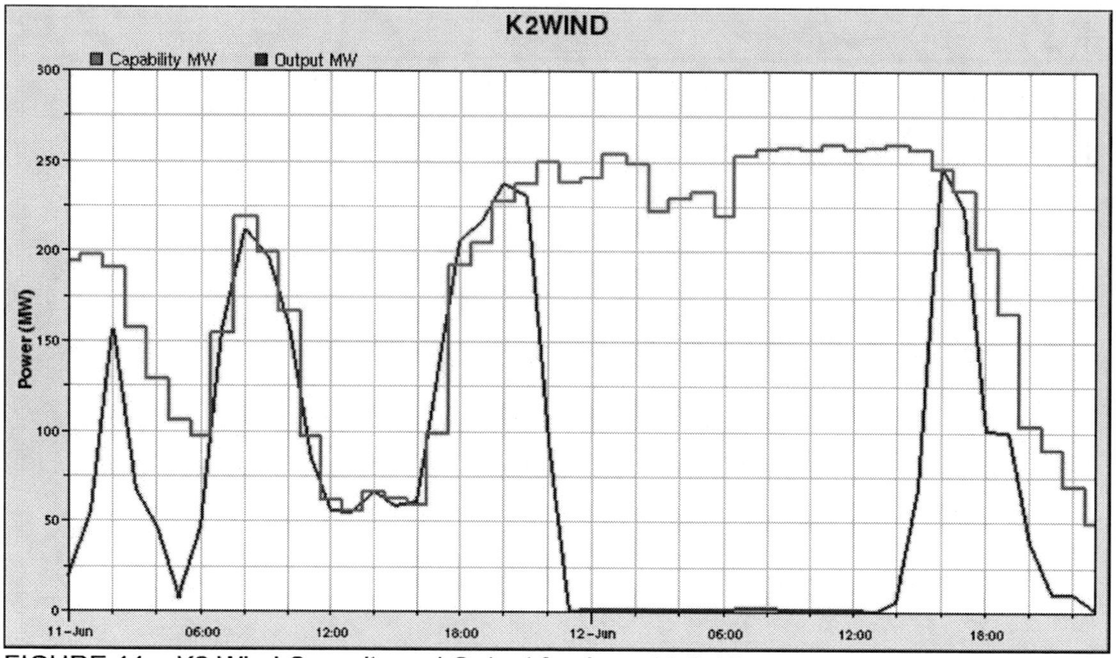

FIGURE 11 – K2 Wind Capacity and Output for June 11 & 12

Although the Ministry did not provide calibration files for their sound recordings they did provide in their report their assessment of the sound pressure level for each sample. Using the Electroacoustics Toolbox, and working backwards to set the given sound pressure level for a number of the recordings provided as the calibration level, permitted a "Quasi Calibration" of the Ministry data, and from that a calibrated FFT analysis was made. The result of that analysis is presented in Figure 12.

Again, it was seen that when the residents described adverse effects in their comments filed with their initiation of recordings, FFT analysis of the sound recordings taken at those times clearly show a tonal condition occurring at about 450 Hz. Reference to the Output curve in Figure 11, shows that the tonality occurred just before the turbine output was curtailed. As noted earlier, the output curves derived from the Independent Electricity System Operator give the average output for the hour preceding each hourly data point, and do not necessarily show the exact time of the change. However, it was clear that the tonal condition again corresponded to the onset of curtailment.

In Figures 9 and 12, the traces representing the "Threshold of Audibility" from ISO 226:2003 and the associated "20 phon" threshold have been included as an indicator that the sound pressure levels seen were well above the thresholds. However, adding data from another source to a chart of FFT results presents a problem. As one who has experience with FFTs can testify, the value shown on an FFT chart is not as important as the indication of frequencies they give. In fact, as FFT's are prepared with larger sample sizes (of smaller width) the frequency resolution improves, but the indicated sound pressure level falls. This is shown in Figure 13.

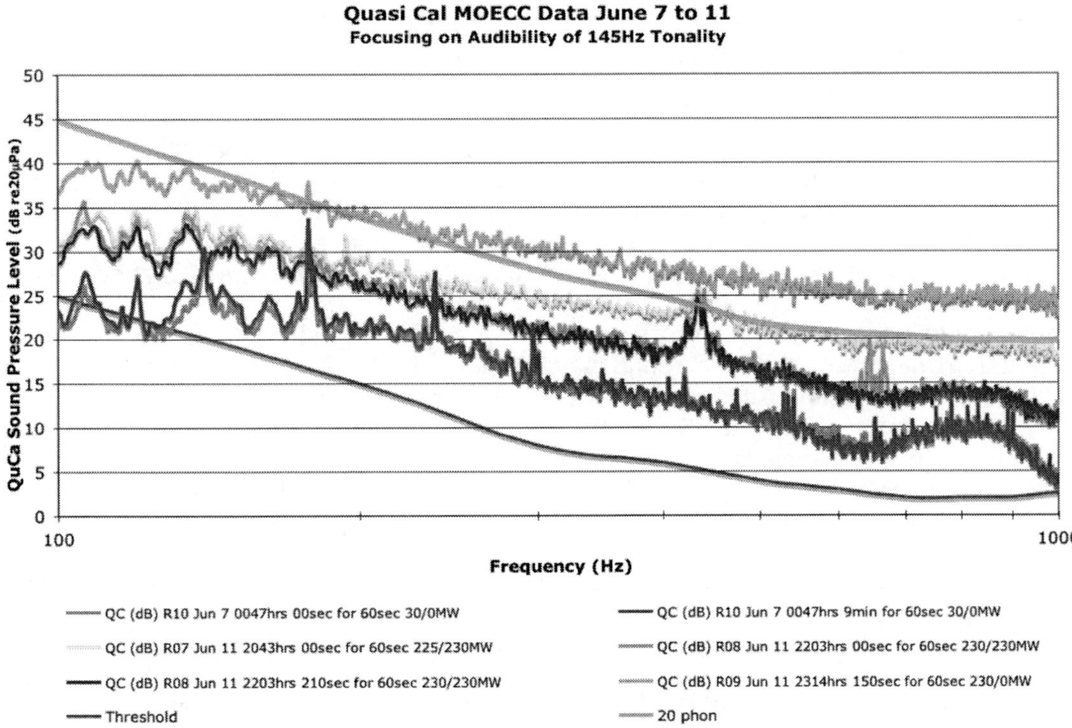

FIGURE 12 – Quasi-Calibrated MOECC Data for June 7 through 11

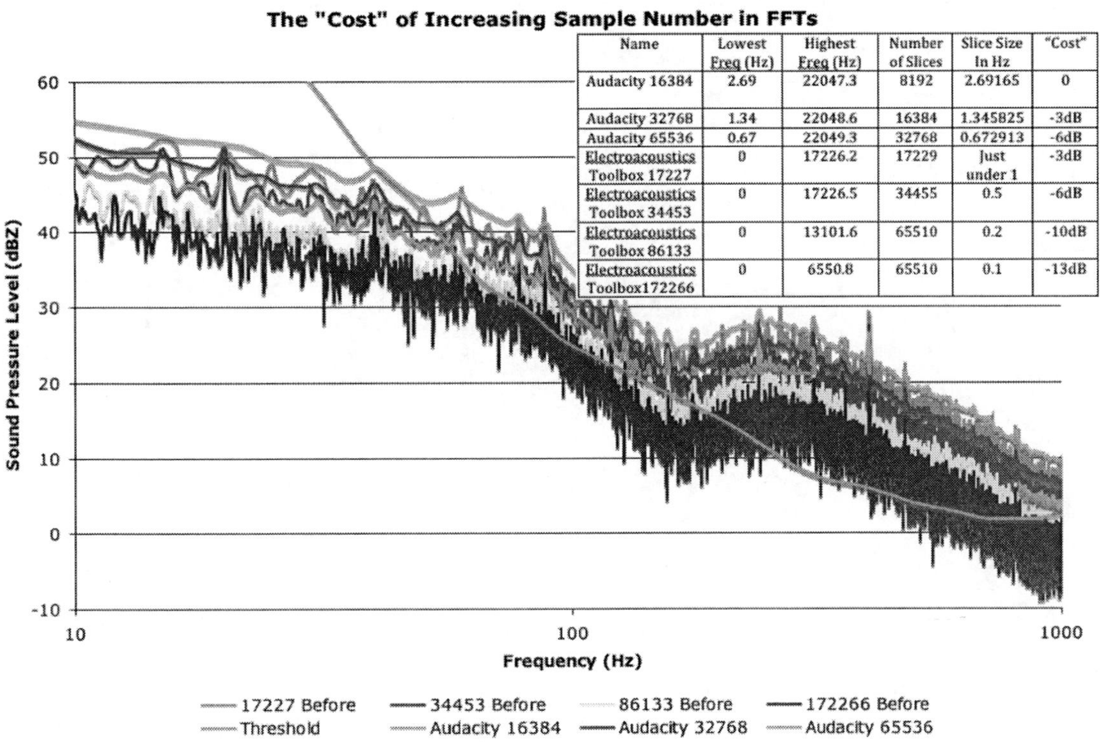

Name	Lowest Freq (Hz)	Highest Freq (Hz)	Number of Slices	Slice Size In Hz	"Cost"
Audacity 16384	2.69	22047.3	8192	2.69165	0
Audacity 32768	1.34	22048.6	16384	1.345825	-3dB
Audacity 65536	0.67	22049.3	32768	0.672913	-6dB
Electroacoustics Toolbox 17227	0	17226.2	17229	Just under 1	-3dB
Electroacoustics Toolbox 34453	0	17226.5	34455	0.5	-6dB
Electroacoustics Toolbox 86133	0	13101.6	65510	0.2	-10dB
Electroacoustics Toolbox172266	0	6550.8	65510	0.1	-13dB

FIGUE 13 – A Caution When Plotting FFTs

Figure 13 would seem at first glance to show that the 7 traces displayed must show very different sound samples, as some are well below the Threshold of Audibility while others are well above it. Yet, the FFT's were prepared for exactly the same sound sample.

Four of the seven FFTs of the same sample were performed by the Electroacoustics Toolbox and three by Audacity, with different sample sizes. A table on Figure 13 shows the "cost" of increasing the number of samples. As an example, doubling the number of samples results in decreasing the indicated Sound Pressure Level by about 3 dB. In fact the chart shows that over the different FFT's, there is a reduction in the indicated Sound Pressure Level of about 13 dB.

The caution here is that displaying the "Threshold of Audibility" on a FFT display may not be an accurate determination of whether a sound is audible or not. The FFT's in this report were prepared generally using the Audacity tool, with 65,536 sample lines, for a sample slice width of 0.69 Hz. This presents an indicated Sound Pressure Level some 6 dB below the first tool in the list, which uses 16,384 sample lines, for a sample slice width of 2.69 Hz. Thus, showing the "Threshold of Audibility" trace might be misleading, as for example in this case, the sample was very definitely audible as it showed the case of all turbines in an array in service surrounding a home only 453 m from the nearest turbine. Yet, the 65,536 Audacity sample shows that the indicated Sound Pressure Level was only slightly above the Threshold of Audibility. The Sound Pressure Level presented on an FFT can only be considered as an indication for comparison purposes, while the strength of an FFT is showing frequency specifics such as tonality.

3.3 Conducting Measurements at the Underwood Wind Power Development

Measurements made on November 7 & 8, 2016 will be used as an example. On these days, the wind turbine output and capability is shown in Figure 14.

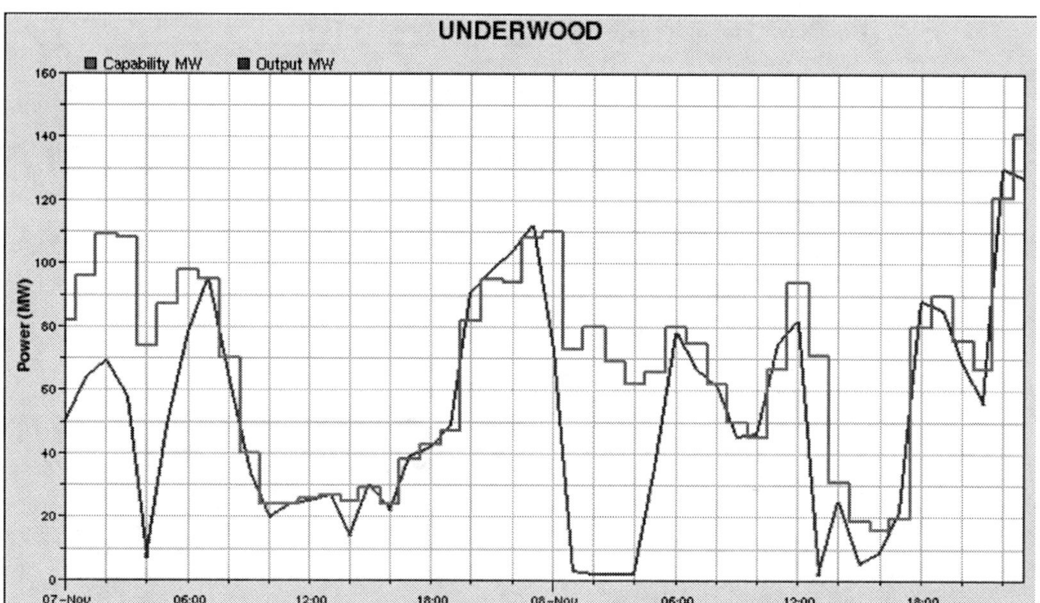

FIGURE 14 – Underwood Wind Power Development Output and Capability Nov 7 & 8

The figure shows that near midnight on Nov. 7, the turbines changed from about 110 MW to near zero. In fact, what physical presence in the field conducting monitoring showed was that the turbines continued at an unchanged power level until about 0030 hours on Nov 8, when the turbines were heard to stop quickly, with a very abrupt transient as all turbines in the array received a stop signal. Monitoring was carried out outdoors at 4 sites soon before the turbines were shut down, at one of the 4 sites during the transient, and soon after at all 4 sites again.

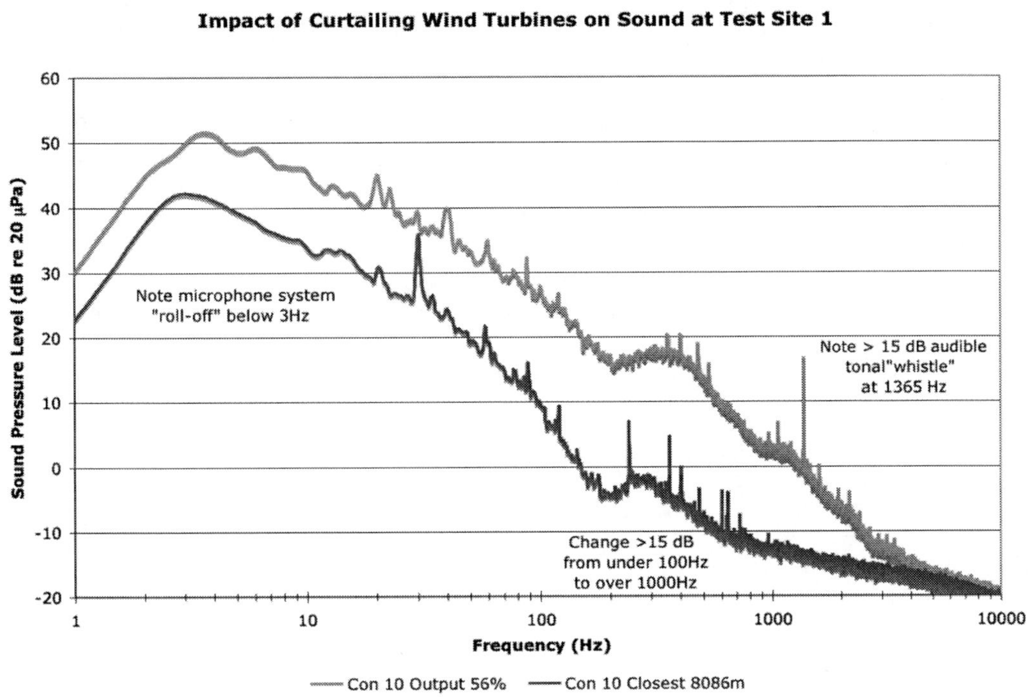

FIGURE 15 – Impact of Curtailing Wind Turbines on Sound at Test Site 1

The FFTs of the Sound Pressure Levels before and after the turbine operation was curtailed, at the first site on Bruce Concession 10, a roadway with little nighttime traffic, are shown in Figure 15. The Figure shows the microphone "roll-off" below 3 Hz, that there was a change of about 15 dB from under 100 Hz to over 1000 Hz with the wind turbines shut down, and that when the turbines were operating, the FFT shows a very clear tonal "whistle" at about 1365 Hz indicated as 17 dB higher than the Sound Pressure Levels at frequencies just below and just above the tonal condition.

A premise of the Ministry of the Environment and Climate Change wind turbine monitoring protocol is that monitoring to show compliance must be conducted over a long period. The protocol requires the initial acoustic monitoring by residents to produce at least a 10-minute sample for each complaint period, and the final compliance protocol requires a minimum of 120 one-minute measurement intervals for each integer of wind speed. During each of those one-minute intervals there must be no changes in wind speed or direction. A further 60 samples are required for each integer wind speed with the turbines not operational. So far data collection has taken years to obtain a sufficient number of samples, and in at least one array, initial reports showed that over 90% of samples taken were discarded as non-compliant. All samples are logarithmically combined to determine the Leq produced by the facility, which eliminates any short-term change effects. This appears to be precisely the sort of monitoring that was cautioned against by Genuit and Fiebig described in Section 1 when they noted, "By relying on sound pressure levels averaged over long time periods and suppressing all aspects of quality, the specific properties of environmental noise situations cannot be identified, because annoyance caused by environmental noise has a broader linkage with various acoustical properties such as frequency spectrum, duration, impulsive, tonal and low-frequency components, etc. than only with SPL [Sound Pressure Level]. In many cases these acoustical properties affect the quality of life." The annoyance aspects that impact the quality of life of impacted residents are not being assessed.

For this facility as an example, where the turbines first went into operation in November 2008, and citizen complaints occurred soon after, it has not yet been possible to complete a report to demonstrate compliance. The monitoring is still in progress, over 8 years later, with the turbines continuing in operation, and residents continuing to complain. The hypothesis is that individual samples are not representative due to variation. As a test of this hypothesis, two test samples were taken in the first minute of a 3-minute monitoring sample and in the last minute of the 3-minute test record, and the FFTs were compared to see if there was indeed any correspondence. The two traces for this first location are shown in Figure 16. They would appear to be very nearly identical, and the differences would not be adequate to dismiss either one as unrepresentative. Similar compliance was seen at another site when tested. The rigorous testing method described in this paper is showing indications of some of the special acoustical properties that are affecting the quality of life, as the testing method independently verifies that the conditions described by residents do indeed exist.

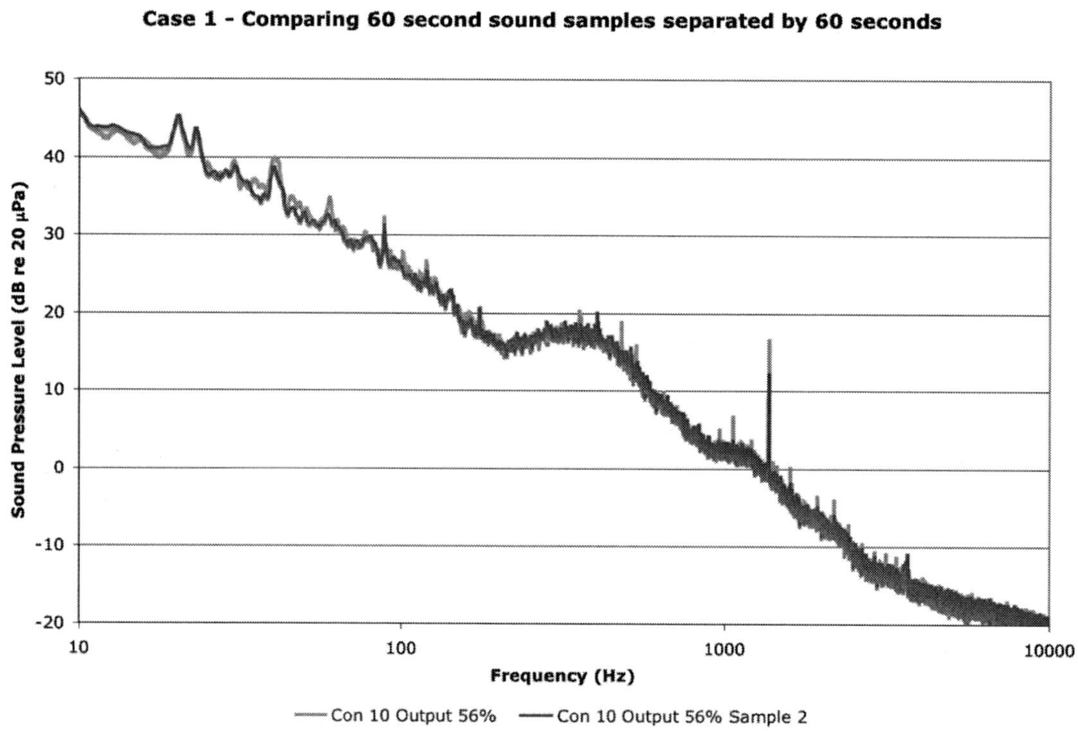

Figure 16 – Comparing Two Samples at First Monitoring Site shows Remarkable Correlation.

Figure 17 shows the comparison between the first two test sites, which are separated by about 1.5 km, for the case of the readings with the turbines in service and the turbines shut down. The samples at each location are separated by some 30 minutes, yet still show a very similar pattern. The troubling conditions are not only localized to one turbine, but are distributed through the array with minor variation in amplitude.

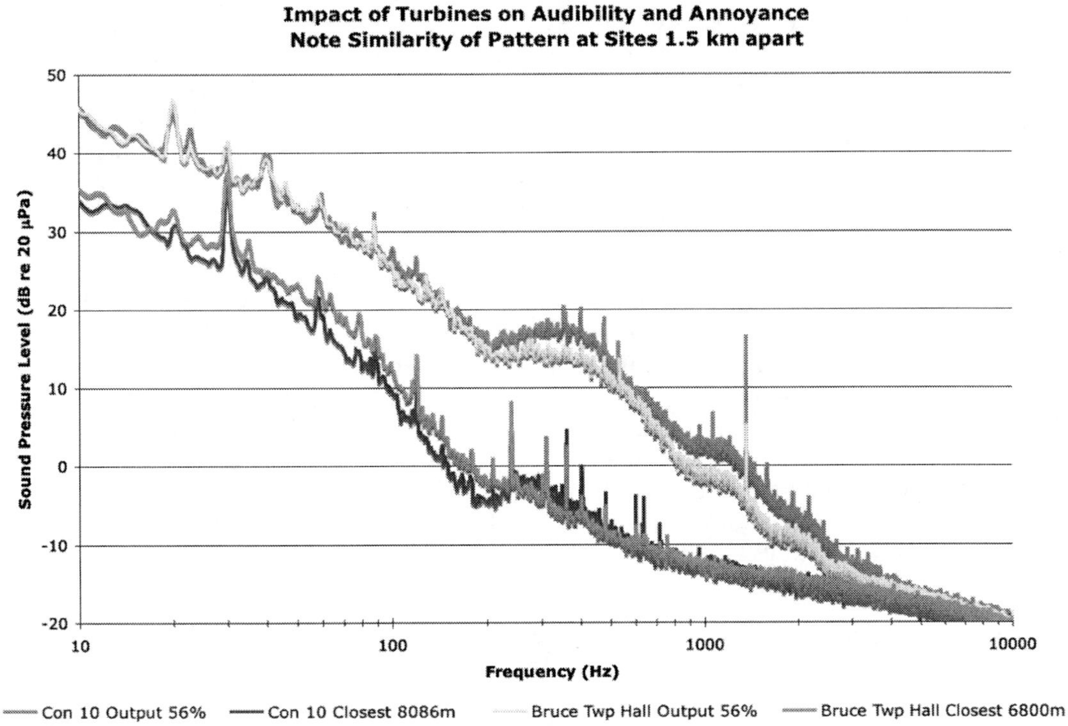

FIGURE 17 – Comparing Two Test Sites 1.5 km Apart Shows Similarity

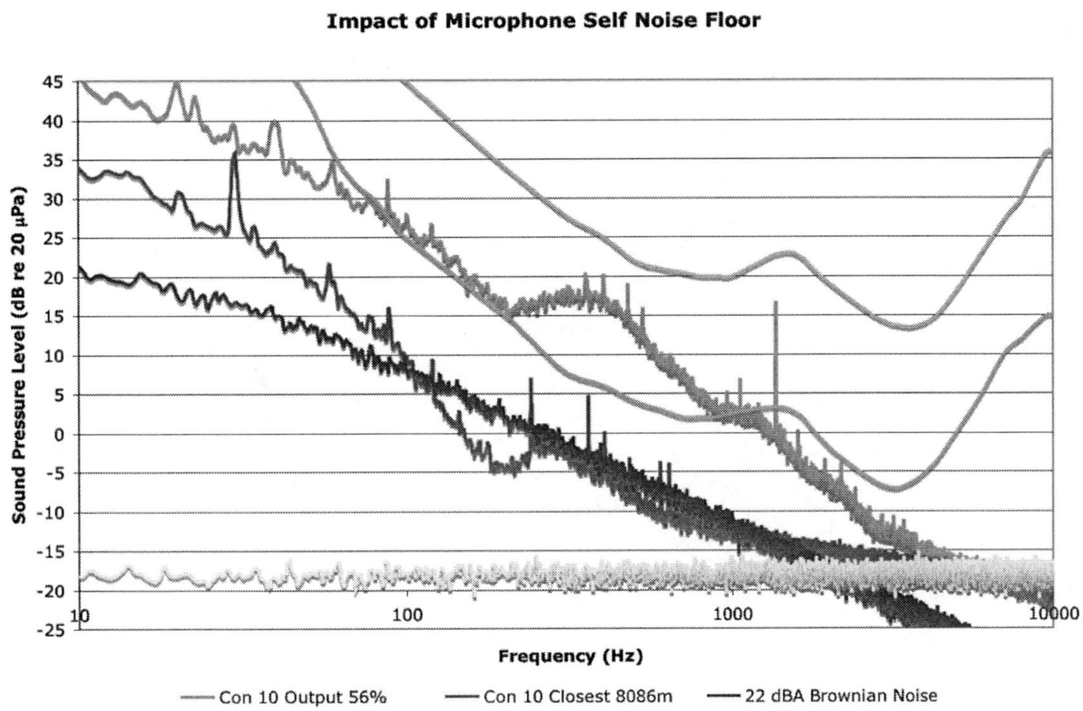

FIGURE 18 – Demonstration of the Impact of Microphone Self Noise Floor

As a further consideration, the FFT Sound Pressure Level for the first test site was compared in Figure 18 to the manufacturer's suggested noise floor of 22 dBA for the Earthworks microphone using both a 22 dBA White Noise trace, and a 22 dBA Brownian Noise trace. The MSc Thesis in Acoustics by Benjamin Russo at The Pennsylvania State University (2013) shows that some components in a microphone system will display a fairly flat noise response with frequency, (like white noise) while others will show a 1/f characteristic (like Brownian noise). Figure 18 shows that for either case, neither the noise floor nor the threshold of audibility will prevent the microphone from being effective as the sound pressure level is above these limits, even if it shown low on the FFT analysis.

The results before and after the turbines are shut down at the second test site, outside the Bruce Township Hall (R285) are shown in Figure 19. The difference from before to after is seen to be in the order of 20 dB, and again an audible tonal signal is displayed on the FFT at 1365 Hz. Ontario regulations require that if tonality of wind turbines is detected as it has been at the K2 turbines and at the Underwood turbines a 5 dB penalty is to be applied. The MOECC issued approval for the K2 turbines in their "Renewable Energy Approval" issued July 23, 2013 on the basis of an application that said the turbines were not tonal, and similarly the "Certificate of Approval Air" for the Underwood turbines was issued July 4, 2007 on the basis that they were not tonal. The rigorous monitoring method demonstrates evidence that both are indeed tonal.

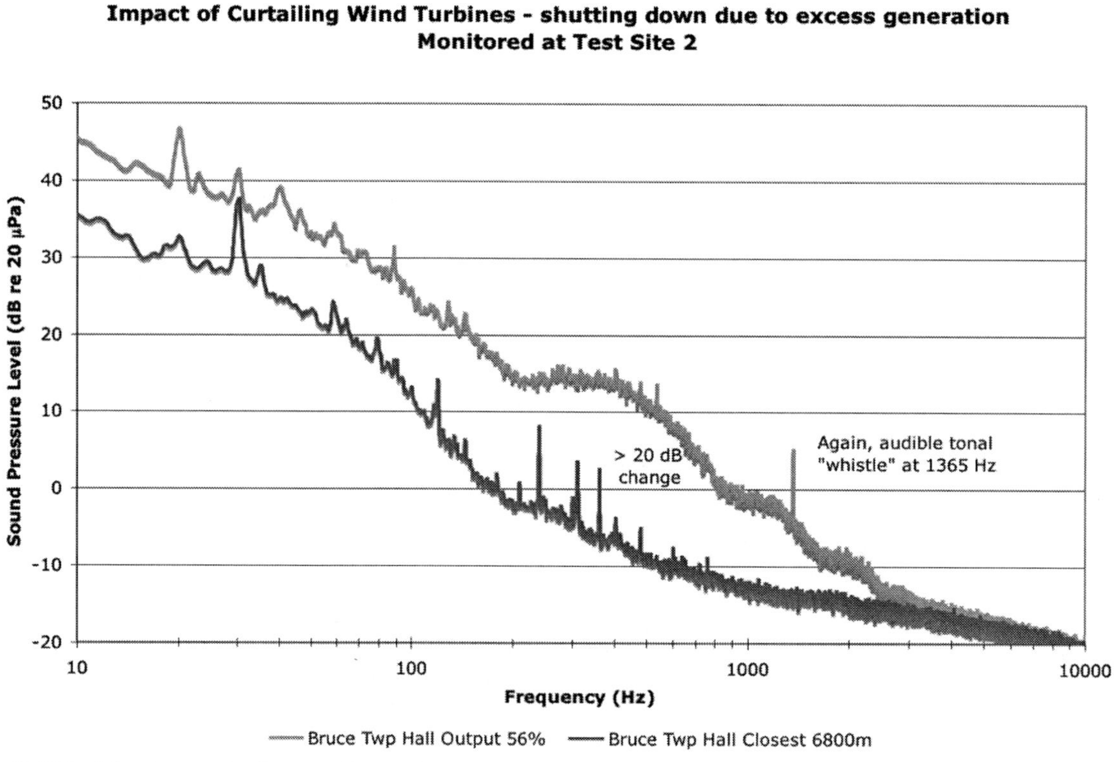

FIGURE 19 – Impact of Curtailing Wind Turbines at Test Site 2 – Bruce Township Hall

Figure 20 shows the case of the monitoring before and after curtailing of operation at the third test site. Again, with the turbines operational, the Sound Pressure Level displayed by the FFT (recognizing that it may well be low) shows to be very near to the 20 phon annoyance level. When the turbines are not in operation, the sound level is reduced by some 20 dB. Again, a very significant tonality is detected. In this case, there is some residual tonality detectable even with the local turbines shut down. While carrying out the monitoring with the turbines shut down, it was possible to hear a repetitive slapping sound, as if cables in the turbine tower were slapping the tower, exciting it to ring. This may be a cause of the indicated tonality even when shut down.

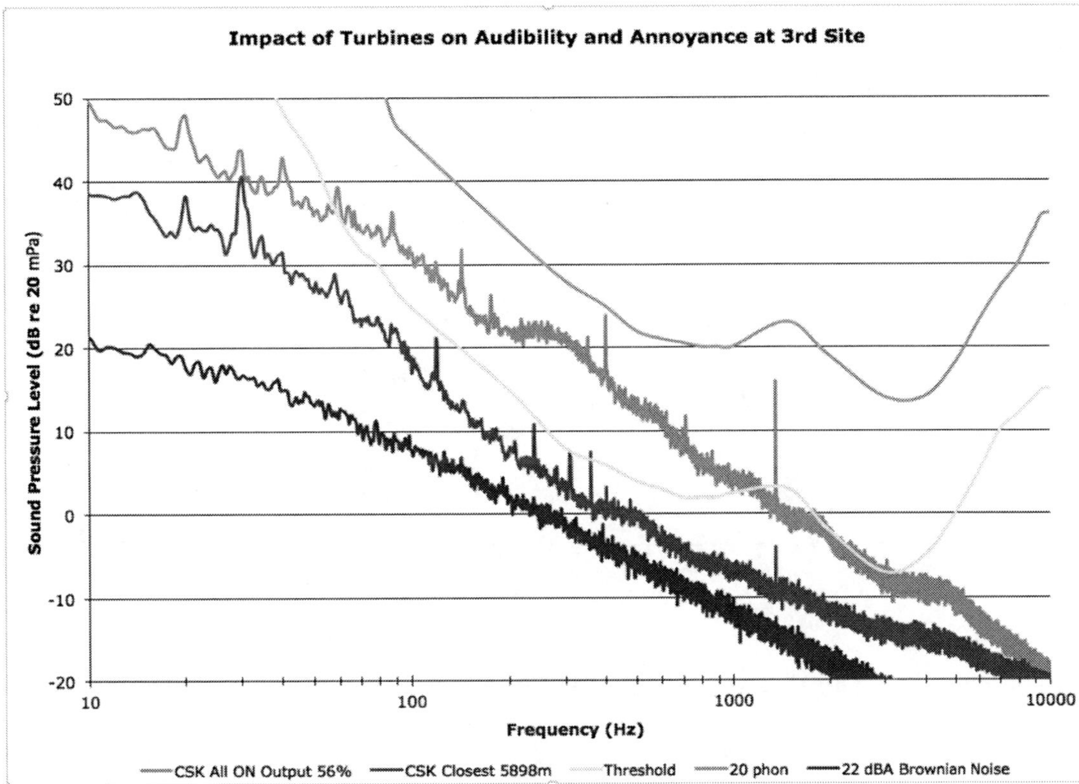

FIGURE 20 – Impact of Turbine Audibility and Annoyance at Test Site 3

At the fourth monitoring site, monitoring was set up and in service when the turbines shut down at about 00:30 AM. However, at this site, the closest turbine to the home, WT045, continued in operation. There is a sound monitoring site near this wind turbine, which may be a factor in why this turbine continued in operation, even when the others were shut down so that a lower sound recording could be made. At this site shown in Figure 21, the condition both before and after the transition remained above the threshold of audibility. Here too, tonality was detected at 1365 Hz.

The shut down transient itself generated a significant impact, as all of the surrounding turbines but this one received a stop signal, and the blades rotated to the feathered position from full speed operation. The recording of the transient itself will be available for the poster based audio listening test to give an indication of the sort of transient that is occurring routinely day after day, with the curtailment of wind turbines, that often occurs about midnight, just after residents may have gotten to sleep as the electrical load drops and the need to reduce generation manifests itself.

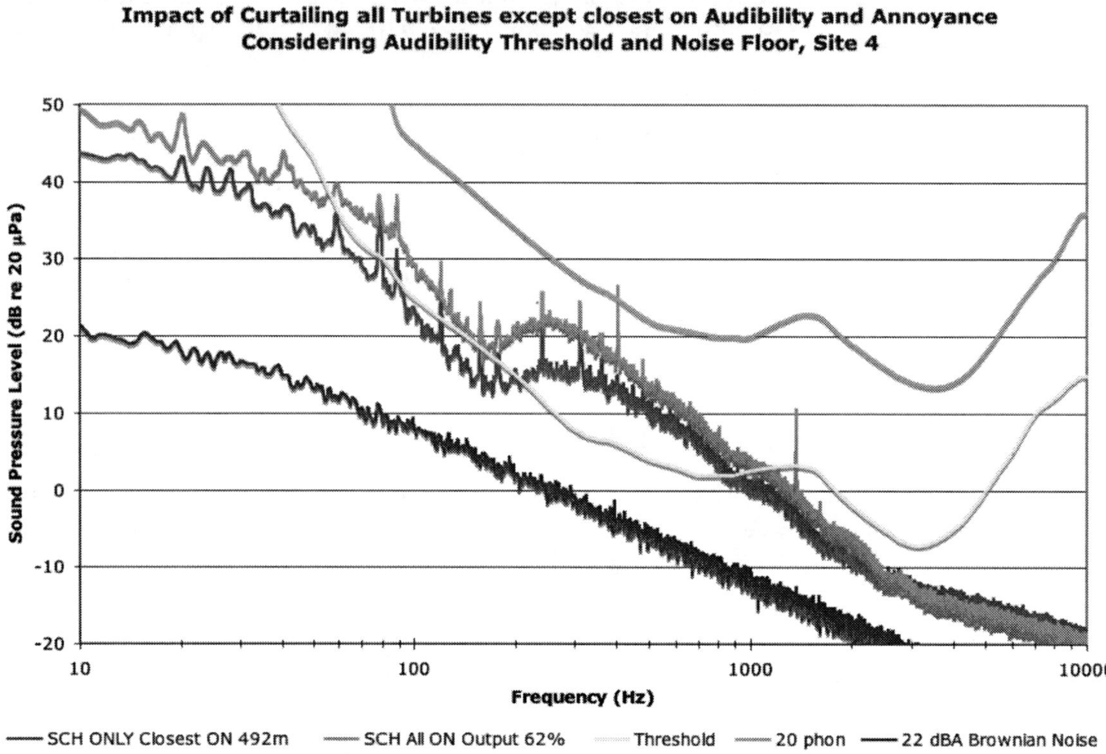

FIGURE 21 – Impact of Curtailing All Turbines But One (the Closest) at Monitoring Site 4

3.4 Comparing Measured Vs Predicted Sound Pressure Levels at Two Sites

Earlier in Section 2.2, the calculation of the Predicted Sound Pressure Levels was described. Following the measurements made in Section 3.3, it was possible to compare the measurement results to the predictions. To do this, a calculation of the third-octave sound pressure levels was carried out by "binning" the outputs of the FFT performed on the sound recordings and logarithmically adding them together as relevant to each third-octave. The overall A-weighted and Z (unweighted) sound pressure levels were also calculated. It is recognized that using the FFT results does allow the same vulnerability as described before for the FFT display, in that the higher the resolution of the FFT, the lower is the indicated sound pressure level. Thus, the indicted sound pressure levels may be lower than exist in reality.

The measured vs predicted Sound Pressure Levels for R145 is shown in Figure 22 (the 4th monitoring site where one turbine was left in operation) and for R285 in Figure 23, (the 2nd monitoring site at Bruce Township Hall).

The sound measurements shown in Figure 22 and Figure 23 were derived by separating the FFT results into one-third octave bins as described above, and then these were converted to octaves. The predicted results were from the Wind Turbine Sound Calculator in Section 2.2.

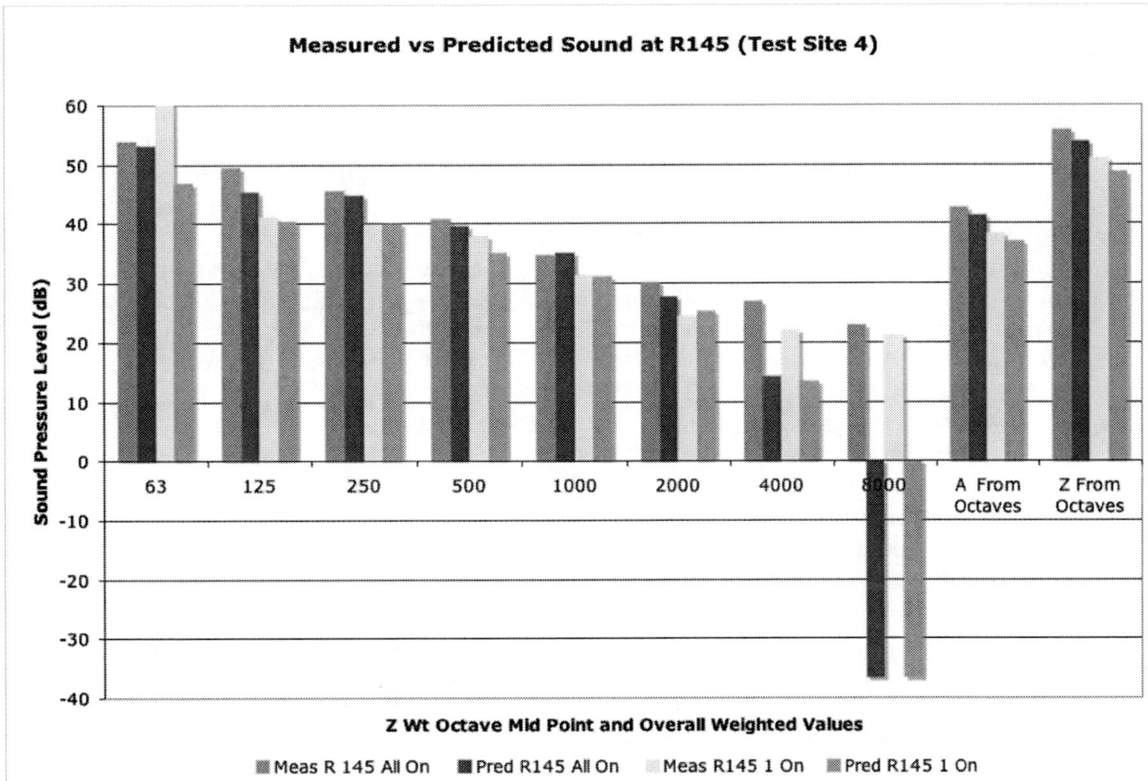

FIGURE 22 – Measured vs Predicted Sound at R145, the 4th Test Site

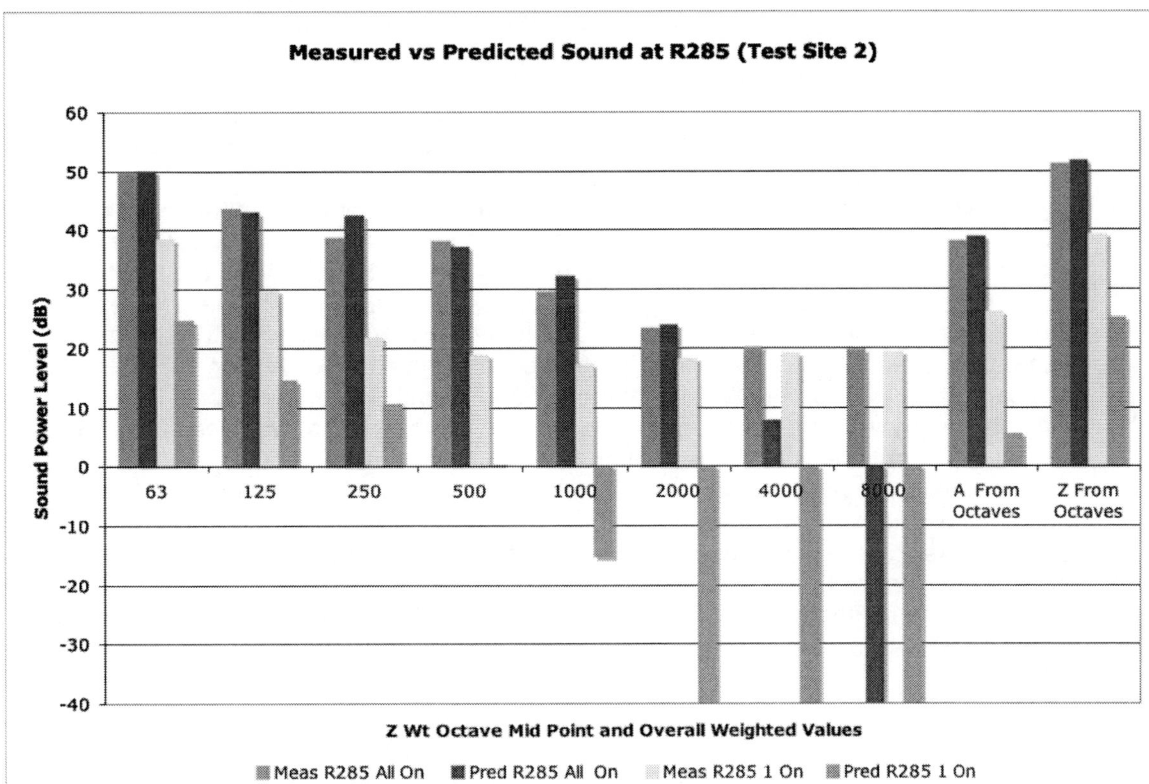

FIGURE 23 – Measured vs Predicted at R285, the 2nd Test Site

Figures 22 and 23 make several observations possible.

- At Test Site 4, for the all turbines "on" monitoring condition, the overall dBA measured sound pressure level exceeded the predicted value from the calculator in section 2.3 by 42.8 dBA vs. 41.4 dBA. For all but one octave (at 1000 Hz) the measured sound pressure level was greater than the predicted level. This may be accepted for the 4000 and 8000 Hz octaves, where the atmospheric attenuation of the higher frequencies from the turbines reduces them well below ambient, to make the predicted value very low. However, when the measured value is greater than the predicted value for octaves 2000 Hz and lower it suggests that the Sound Power Level for the turbines is higher than the value provided by the manufacturer, or the attenuation is less than predicted by the ISO 9613-2 code.
  - As the code itself is generally well verified for the attenuation factors, the error would appear to be in the turbine Sound Power Level. A number of possible conditions for this were identified earlier. These turbines have been in service for 8 years now, and the blades are wearing, and dirtier than new. The turbines are not individually located as at a test site, but are spaced at about 5 rotor diameters apart, so may influence each other. The temperature when the monitoring was conducted was below 10 °C, so the greater air density may have impacted each turbine output. However, it is not the requirement of the monitoring program to identify the actual cause of the measured sound being over the predicted sound, it is only to be able to show that the actual sound pressure level at the receptor was above the licence value. These turbines were licensed on the basis that the sound pressure level would be 39.2 dBA at R145 when the wind speed 10 metres above ground was 6 m/sec. On the night of the monitoring, the wind speed at R145 was well below 6 m/s, but the measured sound pressure level exceeded 40 dBA by nearly 3 dB when calculated from the octaves from 63 to 8000.
- At Test Site 4, for the 1 turbine "on" monitoring condition, the overall measured dBA exceeded the predicted value by 1.3 dBA at 38.4 vs. 37.1. A possible cause for the reduction from the excess in the all turbine state would be that as the other turbines were stopped and producing less turbulence than when operating, there would not have been the same inter turbine interaction.
- At Test Site 2, the predicted sound pressure for the all turbines "on" monitoring condition at 38.9 dBA, exceeded the measured value at 38.1 dBA. The probable cause for this can be seen from the wind output chart in Figure 11. For both the R145 and the R285 site, the predicted value was based on the maximum shown output before the turbines were shut down at 110 MW, while the chart shows that the output was actually rising in the time before the shut down, so it may have been less than 110 MW when the monitoring was carried out at Test Site 2. Thus, the predicted sound power level would have been estimated high, while the measured value would have been what was actually occurring.
- At Test Site 2, for the 1 turbine operating state, the sole turbine WT045, at 6772 metres distant was well beyond the propagation estimation specifications for ISO 9613-2. The fact that the measured sound pressure level exceeded the predicted value is not remarkable since the predicted contribution from the distant turbine (at 5.5 dBA) to the ambient was minimal. The measured value at 26.2 dBA is an expected ambient condition at night, showing the significant excess above ambient caused by the operating wind turbines.

4. A Reproducible Manner of Producing Listening Tests

In the "poster" presentation for this paper, a repeatable model for the prediction of the cyclic sound of a wind turbine will be demonstrated. Further, a demonstration of a repeatable model for modelling the tonal characteristics observed at both the K2 array and the Underwood array will be demonstrated. The demonstrations will help to understand that even if sounds have the same A-weighting, they do not have the same annoyance. The listening test will show that supressing special characteristics of sound quality in some current acceptance criterion can fail to identify real problems faced by residents in the area of a wind power development. Figure 24 gives a brief overview of how the signal generator function of the Audacity program can be used to create a replica of a modulated cyclical signal using the envelope tool to modify a basic Brownian noise signal. In a similar manner, the tone generator function on Audacity was used to overlay onto a modulated cyclical signal to simulate a tonal wind turbine. Samples such as these were demonstrated to a number of residents with experience living near wind turbines. Interesting remarks were made such as, “That is exactly what it sounds like!”

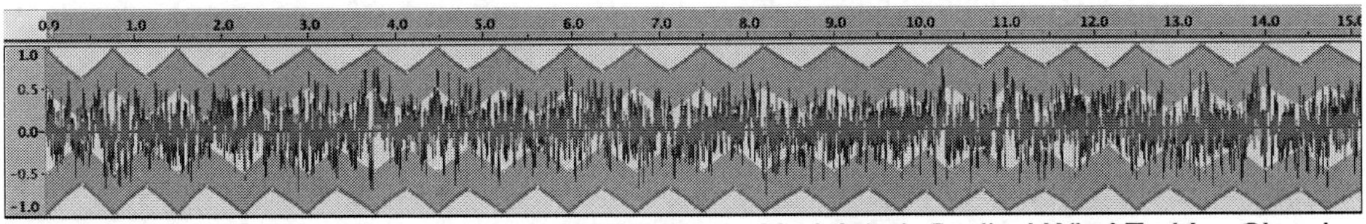

FIGURE 24 – A Demonstration of how to Simulate a Modulated, Cyclical Wind Turbine Signal

5. Conclusions

This paper has demonstrated a method for rigorous monitoring of wind turbine sound. The goal of the method was to establish evidence for the condition noted by Karl D. Kryter: “The most direct, and perhaps most valid, insight into the possible presence and magnitude of stress reactions in general living environments is probably that which has been obtained from attitude surveys and real-life behaviour of people.” Behaviours such as walking away from an unsold loved home to live at the home of a family member, or when normal people become activists in trying to communicate their concerns provide such valid insights. The rigorous method had to consider the present acceptance criterion for wind turbines, in light of the insight given by those who study the quality of noise and its relation to annoyance. Those who study the subject identify that, “Current acceptance criterion relying on sound pressure levels averaged over long time periods and suppressing all aspects of quality cannot identify the specific properties of environmental noise situations.”

A repeatable and transparent method of predicting the expected sound pressure level was presented. A rigorous method of monitoring the actual sound conditions was described. This was used to conduct assessments at two different wind power developments with two different turbine types. Using the method it was possible to generate reproducible evidence of some of the special acoustical properties that are affecting quality of life. Thus it could verify that conditions identified by residents as troublesome do exist, when the current acceptance criterion was unable to detect problems.

The paper outlines a method of preparing reproducible sounds to permit a “jury-test” at the poster session in a repeatable manner of special acoustical qualities such as modulated cyclical sound, or tonality. The demonstration will show evidence that two sounds with the same A-weighting, in the absence of consideration of the special characteristics of the sound, are not equal in annoyance.

6. Acknowledgements

Many have contributed to the development of this monitoring protocol, and this list of acknowledgments will be incomplete, but representative of some who helped, even without knowing of their influence:

- Residents permitted the intrusion of allowing monitoring inside and around their home, both occupied and vacant, and visits at any hour of the day or night to collect data. The locations reported in this paper are only indicative of many where residents have permitted monitoring at their homes. In some cases residents even got up during the night to plug in the monitoring computer to ensure the battery would be recharged when needed. Thanks for your tolerance, and for discussions to outline what your observations were as to what you found most bothersome. They were invaluable at making sense of a mystery.
- Werner Richarz provided the initial thinking for an Excel spreadsheet used to evaluate the third-octave contributions from FFT results.
- Discussions with Kristen Persson Waye were instrumental in clarifying the issue of cyclical sound as bothersome.
- Discussions with Jo Solet were useful in helping to understand sleep disturbance issues.
- Speaking face to face with Mathias Basner and Wolfgang Babisch at a special Acoustical Society of America presentation honouring the work of Karl D. Kryter was a big factor in focusing on "The Effects of Noise on Man".
- Discussions with, and listening to papers presented by Klaus Genuit, André Fiebig, and Brigitte Schulte-Fortkamp were very helpful in understanding the issue of the quality of noise and it's relation to annoyance.
- The understanding of my wife, Jean, to permit me to focus almost single-mindedly on a subject for over 10 years allowed this work to happen.

References

Karl D. Kryter, *The Effects of Noise on Man*, published in 1970 by Academic Press, 1970. Kindle Edition available from Amazon.com.

Karl D. Kryter, *Non Auditory Effects of Environmental Noise*, published in Environmental Noise, A.J.P.H., March 1972

Klaus Genuit and André Fiebig, *Psychoacoustics and its Benefit for the Soundscape Approach,* published in Acta Acustica United, 2006, Vol 92, (2006) 1-1

Mathias Basner, Wolfgang Babisch, Arian Davis, et al, *Auditory and non-auditory effects of noise on health*, Lancet, vol 383, April 12, 2014

Thomas Munzel, Tommaso Gori, Wolfgang Babisch and Mathias Basner, *Cardiovascular effects of environmental noise exposure*, European Heart Journal (2014) 35, 829-836.

Mathias Basner, et al, *ICBEN review of research on the biological effects of noise 2011-2014*, published in Noise & Health, March-April 2015, vol 17:75, 57-82

Benjamin Russo, *Thermal Noise in Condenser Microphone Back Volumes*, MSc Thesis in Acoustics at Pennsylvania State University, Graduate School, College of Engineering, August 2013.

William Palmer, *Wind Turbine Annoyance – a clue from acoustic room modes*, ASA 168th conference paper, 2014.

William Palmer, *Wind Turbines – A Changed Environment*, 6th International Meeting on Wind Turbine Noise, Glasgow, April 2015.

William Palmer, *Understanding the human impact caused by the sound of wind turbines*, ASA 169th conference paper, May 2015.

William Palmer, *Considerations Regarding an Acoustic Criterion for Wind Turbine Acceptability*, Acoustics Week in Canada Conference Paper, 2016, Canadian Acoustics, 254-971, P8

7th International Conference
on
Wind Turbine Noise
Rotterdam – 2nd to 5th May 2017

Assessment of WTN by separating residual noise without the farm shutdown: validation of the Italian procedure.

Luca Fredianelli. IPCF-CNR, Via G. Moruzzi, 1 - 56124 Pisa, Italy.
E-mail: luca.fredianelli@gmail.com

Stefano Carpita. Department of Physics - University of Pisa,
Largo B. Pontecorvo, 3 - 56127 Pisa, Italy.
E-mail: carpitastefano@gmail.com

Paolo Gallo. Environmental Protection Agency of Tuscany Region,
Department of Pisa – Via N.Porpora, 22 – 50144 Firenze, Italy.
E-mail: p.gallo@arpat.toscana.it

Diego Palazzuoli. Environmental Protection Agency of Tuscany Region,
Department of Pisa – Via N.Porpora, 22 –50144 Firenze, Italy.
E-mail: d.palazzuoli@arpat.toscana.it

Gaetano Licitra. Environmental Protection Agency of Tuscany Region,
Via Marradi 700, 57100 Livorno, Italy.
E-mail: g.licitra@arpat.toscana.it

Summary

In 2013 a procedure for the assessment of noise impact of operational wind farms has been published by the Italian Institute for Environmental Protection and Research (ISPRA) and the Environmental Protection Agency of Tuscany Region (ARPAT). By means of measurement campaigns of specific noise and weather parameters at the receivers lasting at least 2 weeks and through iterative steps, the method provides the evaluation of noise impact produced by operational wind farms, without stopping the energy production for measurement purpose. A validation of the procedure is presented in the paper, after focusing on the issues related to it.

1. Introduction

The ISPRA-ARPAT Italian procedure for wind turbine noise assessment [1] estimates the immission and residual components of noise from the environmental levels measured at the receiver during a measurement campaign lasting at least 2 weeks.

The extrapolation of immission and residual levels is achieved through a data analysis procedure (DAP) not requiring the farm shutdown and it is mainly founded on the the following basic ideas:

- the residual noise is correlated to the wind speed measured at ground near the receiver (v_{gr});
- the immission levels are correlated to a new parameter, the equivalent blades rotational speed N_{eq} representing the rotational speed of a virtual turbine producing the overall wind farm immission. The N_{eq} is calculated as an average of the single turbines rotor speeds weighted according to the different propagation paths and conditions.
- the use of a long term measurement campaign allows to select the 10 min intervals when the immission levels are negligible respect to the background noise. By calculating the activation threshold, an initial rough estimate of the residual noise can be evaluated.

The procedure estimates the immission curve as a function of N_{eq}, and the residual curve as a function of v_{gr}.

A numerical evaluation of the uncertainty of the outputs has been presented at the

WTN conference of 2015 [2], however the validation on virtual and real scenarios and a sensitivity analysis of the input parameters are mandatory steps in order to evaluate the effectiveness of the procedure for wind turbine noise assessment. Sensitivity analysis results have been illustrated in [3]. This paper presents an overview of the validation results obtained with different methods.

2. Validation methods

The validation is a complex task, needing the comparison of the outputs levels with reference values properly estimated. The immission levels can be simulated with noise software, with their due uncertainties, but the residual noise evaluation is the major problem, it being related to the exact time and place of evaluation. The most effective way to validate the procedure would be to compare the residual noise from the procedure with measured levels in a specific measurement campaign with the wind farm shutdown. However, this conflicts with the economic and technical problem of stopping the farm for long periods in order to allow the characterization of all meteorological and v_{gr} conditions. In a similar way, a valid alternative would be obtained using a wind farm under construction, where the residual can be measured before the installation. Also in this case, there would be the problem of finding two periods with exact meteorological and vegetation condition to be compared, especially among different seasons or years. Moreover, new farms are becoming very rare in Italy and to date no plant managers allowed the required shutdown.

The most common methods in literature for residual measurement in the WTN assessment have spatial or temporal flaws, related to measurements in different sites or periods, that lead to not sufficiently reliable results [4].

1. In the *"proxy method"* the spatial coherence between the residual estimated at the proxy position and the actual residual at the receiver is not guaranteed.
2. In the *"ante-operam residual assessment"* temporal flaws are the primary disadvantages.
3. In the *"shielding method"*, the total exclusion of the noise source is not guaranteed, especially for low frequencies. Also spatial flaws could occur.
4. The *"turbines shutdown method"* has a big economic impact and may also

present temporal flaws.

The *Proxy Method* involves the use of a "proxy" background sound level monitoring location, located far enough away from the receiver site that turbine noise is negligible. The sound environment should be similar to monitoring location(s) near the turbines (without turbines operational). This requires matching locations for flora, fauna, meteorological conditions, nearby roadways with similar traffic, residential noise sources, and commercial/industrial noise sources [5].

Limitations of the proxy method are mostly due to difficulties in finding suitable proxy locations, especially in a mountainous region, where meteorological conditions, land use, vegetation and roads change rapidly. To determine appropriate proxy locations, measurements need to be performed in advance of project operations, thus becoming the *ante-operam residual assessment*. Unfortunately, the ante-operam method is very difficult to be available for already existing wind farm. When available, it is an expensive and time consuming process since, to get a fair sample size, each test monitoring session lasts for two weeks. In anycase, being performed in a different period, problems related to different vegetation or meteorological conditions can occur.

The *shielding method* involves the use of two microphones, one is exposed to the wind power facility (open monitor), the other is placed behind a shielding mechanism to block sound from the source (shielded monitor). The basic principle of the shielding method is that the open monitor is collecting sound level data that is representative of the wind farm with background sound while the shielded monitor is only collecting sound level data that is representative of background sound. The method presents several clear theoretical flaws: the difficulty of founding an efficient shielding structure and a spatial disadvantage. Indeed, assuming that the background sound levels measured this way are representative of the background sound levels at the open monitor results to be a strong assumption because the shield may also block sound from other sources of background noise that the open monitor may be exposed to. This would result in an overestimation of sound levels attributable to a wind farm. In addition, depending on the location of the source of background sound, it may be possible for the background levels to be amplified by reflections off the shielding mechanism, which would not be an accurate representation of the background sound levels at the open monitor and results in an underestimation of sound levels attributable to a wind farm. Also the shielding mechanism itself may create noise with either wind blowing over the surface or breakout noise from sources located indoors if the shield is a building.

The *shutdown method* is one of the most common methods used to assess background sound levels at an operating wind farm. Wind turbines are shut down to measure background sound levels for a period of time. Depending on the location of the compliance monitor and the wind turbines, some or all of the turbines need to be shut down.

Thus, the most important negative factors with this method are the operational and financial burden it poses on the wind power operator and the potential problems of fluctuating the power supply to the grid at peak power output. Thus, the shutdown method does not allow for continuous compliance monitoring, allowing the background measurement only for discrete time evaluation. This led to another downside of this method related to the possibility that background sound levels change between the operational periods and the shutdown periods.

For these reasons, the validation has been performed in three alternative ways:

1. A comparison of the immission levels of the procedure with the noise prediction models.
2. A method based on the implementation of a computational model for simulated scenarios. An hypothetical set of measured noise level, corresponding to the procedure's measurement period, is simulated summing a theoretical residual noise function of v_{gr}, to a theoretical immission noise as a function of N_{eq} and to a random noise. When applied to this set of data, the procedure should return the two inputs theoretical function.
3. A correlation analysis between the measured 10 min environmental noise level and the environment levels on 10 min predicted by the procedure output curves.

3. Immission levels validation

The procedure has been applied by the authors to several measurement campaigns performed between 2011-2016 along Italy. For all the farms, the maximum immission noise level for both daytime and night-time, corresponding to the maximum N_{eq}, has been compared to the immission noise level predicted by the NORD2000 noise model for the maximum sound production conditions (wind speed 10 m/s). The results are reported in Table 1, together with the uncertainties. The uncertainties on the procedure's immission are estimated with Monte Carlo methods [2], rounded to the first integer, whilst for the noise model they are estimated with a coverage factor 1 (68% L.C.).

Table 1. Comparison between the procedure's maximum immission noise level for daytime and night-time and the immission noise levels predicted by the NORD2000 noise model in the seven measurement sites.

Wind Farm	Receiver alias	$L_{I,max,procedure,day}$ [dB(A)]	$L_{I,max,procedure,night}$ [dB(A)]	$L_{I,max,predicted}$ [dB(A)]
Poggi alti	Poggi alti Scansano	51±3	53±3	53±3
La Miniera	Scapiccioli	43±3	44±3	41±3
La Miniera	Provinca	40±3	36±3	37±3
La Miniera	Palareta	47±3	45±3	42±3
La Miniera	Palareta 2	47±3	47±3	46±3
Poggio Palmorelle	Santa Luce 1	44±3	43±3	44±3
Poggio Malconsiglio	Riparbella	45±3	45±3	45±3
Lucera	Borgo San Giusto	50±3	48±3	47±3

For all the wind farms, the immission noise levels are comparable with the predicted ones within the uncertainties.

4. Numerical validation

The numerical validation [6] is based on the simulation of the immission and residual noise time histories $L_{I,10min}$, $L_{R,10min}$, using independent propagation models for both quantities. The environmental time histories $L_{E,10min}$ are obtained by energetic summation of the simulated immission and residual noise.

The residual and immission noise models simulate in a realistic manner the possible noise levels distribution of a measurement campaign using a semi-empirical approach. This aim is achieved applying propagation models to measured noise source parameters, the blade rotational speed N, wind speed measured at ground near the receiver v_{gr}, wind direction measured at hub w_{dir}. The models include random dispersions with determined distribution calculated by a pseudo-random number generator, and propagation parameters, as residual offset and slope, which can be variated in order to obtain several different scenarios. An example of simulated data is represented in Figure 1.

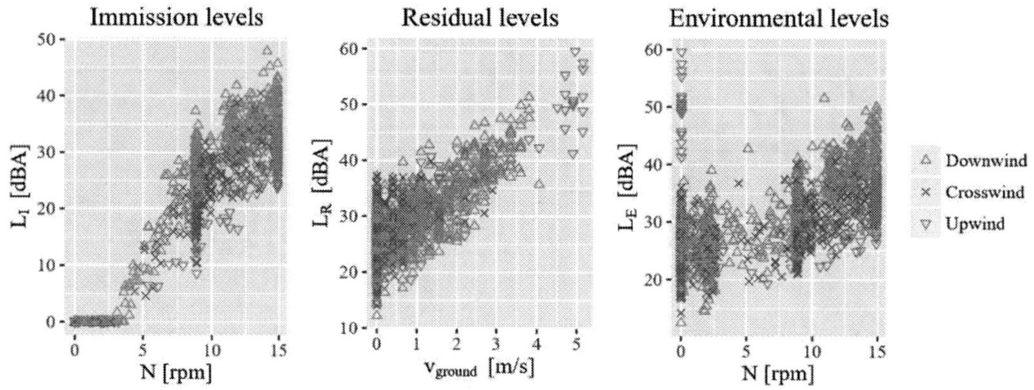

Figure 1. Example of simulated noise levels

The ISPRA-ARPAT procedure is applied to the simulated noise levels, and the obtained processed curves are compared to the theoretical ones calculated directly from the simulated residual and immission levels. The comparison is performed by calculating the uncertainty indicators, root mean square error RMSE, and the bias as difference between simulated and processed average levels. The analysis has been repeated for different models parameters and source data. Results are reported in

Figure 2 as function of the signal to noise ratio SNR, calculated as the difference between the overall theoretical immission and residual. In a range of ±3 dB(A) of SNR, the obtained immission bias values range between ±2 dB(A).

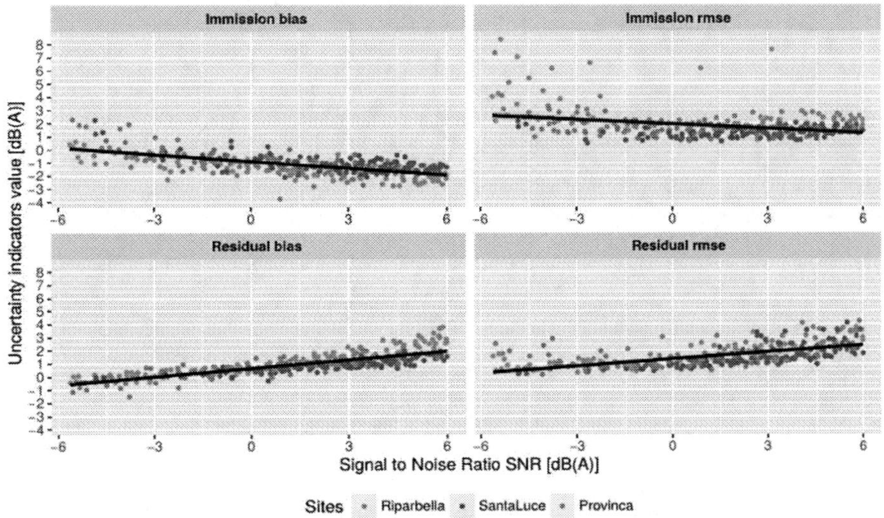

Figure 2. Uncertainty indicators calculated for 3 test wind farm for the validation. Each point is a single results of the numerical validation analysis for a specific simulated data set.

5. Environmental noise levels validation

An indirect validation method consists in comparing the measured environmental noise levels, $L_{Aeq,10min}$, with those calculated by the procedure, which is the energetical sum of the immission and residual noise resulting from the procedure. For each interval of 10 min it is sufficient to apply the output curves function of N_{eq} and v_{gr} to each interval of 10 min in order to have the constructed $L_{Aeq,10min}$. The time histories of measured and calculated environmental noise levels are compared for each of the measurement campaigns, an example is reported in Figure 3. The largest deviations occur when noise levels are very low or when the residual, which is subject to the greater fluctuations, is predominant.

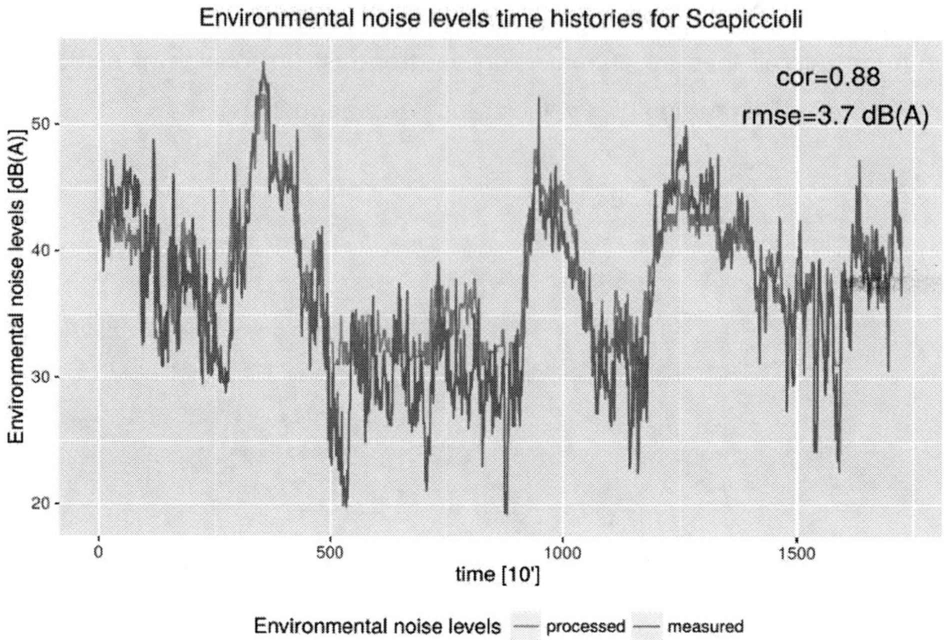

Figure 3. Time histories of measured and calculated (processed) environmental noise levels for "La miniera - Scapiccioli".

In order to quantitatively compare the measured environmental noise levels and the processed levels, the correlation coefficient and the root mean square error RMSE were calculated.

The correlation coefficients resulted quite high for all of measurement campaigns in Table 2. The largest RMSE are for measurement campaigns where the residual is comparable with the immission levels, its fluctuations are then more influential on the environmental levels, as in the case of the Palareta 2 measurements. Figure 4 shows the differences calculated every 10 min between the measured environmental levels and those calculated for all measurements of campaigns.

The resulting distribution is almost symmetrical with average around 1dB, however close to zero. The 90% of the differences lies in a range between -5 and +6 dB(A).

It should be noted that the purpose of the procedure is not to estimate the immission and residual values every 10 min but their average over the measurement period. Therefore, the wide differences shall in part attributable to the strong fluctuations that may occur in the short time intervals of 10 min, which are exactly what the procedure intends to flatten.

Table 2. Correlation and RMSE between the difference of measured and calculated environmental noise levels for all the farm.

	corr	rmse
Scapiccioli	0.88	3.71
Scansano	0.75	3.45
Lucera	0.85	3.94
Riparbella	0.78	4.30
SantaLuce	0.89	3.35
Provinca	0.83	3.04
Palareta1	0.89	3.75
Palareta2	0.85	5.41

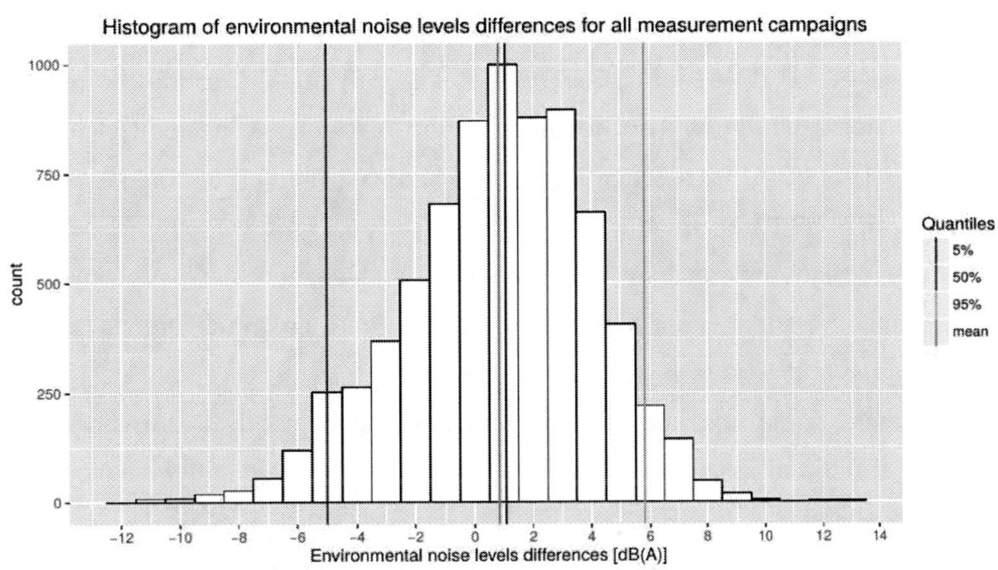

Figure 4. Histogram of environmental noise levels differences for all the wind farms.

6. Conclusions

The Italian ISPRA-ARPAT procedure can assess the WTN by a Data Analysis Procedure (DAP) applied to data from a single measurement campaign, not requiring the wind farm shutdown and avoiding a big economic impact for the farm manager. The DAP extracts the immission and the residual noise from the overall measured environmental levels using a specific filtering analysis.

The immission levels validation using the comparison with NORD 2000 simulations showed that, for all the wind farms analysed, the immission noise levels were comparable with the predicted ones within the uncertainties.

The environmental noise levels validation confirmed the good results, being all the processed time histories highly correlated with the experimental ones. Furthermore, the histogram of the difference between measured and calculated environmental noise levels over all the test sites showed an almost symmetrical distribution with average around 1 dB(A), with the 90% of the differences ranging between -5 and +6 dB(A).

In order to verify and validate the procedure, a numerical validation method based on a semi-empirical simulation of immission and residual noise has also been presented. The validation results are quite good for the immission estimate, being the bias values included in a range between [-2; 2] dB(A) for a SNR range of [-3; 3] dB(A), and [-3; 2] dB(A) for a SNR range of [-6; 6] dB(A). For the the residual estimate the results were also good, being the bias values included in a range between [-1; 2] dB(A) for a SNR range of [-3; 3] dB(A), and [-1; 3] dB(A) for a SNR range of [-6; 6] dB(A).

Generally, a slight underestimation of the immission levels and an overestimation of the residual levels that linearly increases with the SNR, i.e. the difference between the immission and residual levels, are experienced.

Unfortunately, some unexpected case with variation of nearly 6 dB(A) occurred, particularly in the residual noise. These special cases will need further developments of the procedure.

In conclusion the Italian ISPRA-ARPAT has been proved an effective procedure to assess wind turbine noise, with quite good validation results. The procedure is currently being developed to overcome issue in specific cases and to improve the precision of the results.

References

[1] Gallo, P., Fredianelli, L., Palazzuoli, D., Licitra, G., & Fidecaro, F. (2016). *A procedure for the assessment of wind turbine noise. Applied Acoustics, 114,*

213-217.
[2] Fredianelli L , Gallo P , Licitra G, Palazzuoli D, Carpita S (2015). *Sensitivity analysis test on the italian ISPRA-ARPAT methodology to assess noise impact of operational wind farms.* Proceedings of the 6th International Meeting on Wind Turbine Noise. Glasgow 2015.
[3] Licitra G, Gallo P, Palazzuoli D, Fredianelli L, Carpita S (2015). *Sensitivity analysis of modelling parameters in wind turbine noise assessment procedure.* Proceedings of the ICSV 2015, Florence.
[4] Duncan E., (2015). *Methods for assessing background sound levels during post-construction compliance monitoring within a community*, Proceedings of the 6th International meeting on WTN, Glasgow 2015.
[5] Hessler, G.F., Hessler, D.M. and Bay, K. (2008). *Experimental study to determine wind-induced noise and windscreen attenuation effects on microphone response for environmental wind turbine and other applications.* Noise Control Engineering Journal, Vol. 56, No. 4, pp. 300-309.
[6] Carpita S, Fredianelli L, Gallo P, Licitra G, Palazzuoli D. (2016). *Validation of the Italian wind turbine noise assessment procedure through simulated case studies.* Proceedings of the ICSV 2016, Athens.

7th International Conference on Wind Turbine Noise Rotterdam – 2nd to 5th May 2017

Characterizing the acoustic noise from wind turbines by using the divergence of the sound pressure in the ambient

Valentin Buzduga, Silver Spring, MD 20904, USA: buzdugav@yahoo.com
Alexandru Buzduga, London, UK: alexbuzduga@gmail.com

Summary

This paper discusses the possibility of characterizing the dominant noise sources in an ambient, like wind turbines, by simultaneously measuring the sound pressure at certain locations in that ambient and using in calculations the divergence of the sound pressure level between those locations. Unlike the IEC 61400-11 standard method, the technique discussed in this paper does not require stopping the wind turbine for measuring the background noise. The paper shows that the L_{eq} values existing at two locations in the ambient, over the same integration time interval, can be seen as the effect of a single, equivalent sound source. Based on this proof, the authors describe how to use the divergence of the sound pressure between certain locations for determining the region where a presumed dominant sound source may exist. With the new method, all the data collected from a measurement session are used for deciding on the dominant noise source. The paper proposes using statistics methods (the hypothesis test) for deciding if a wind turbine under test is the dominant noise source in the ambient.

1. Introduction

Measuring the acoustic noise from wind turbines (WT) is a challenging task. The variety of the landscapes where the WT are placed, the weather conditions and, ultimately, the presence of other acoustic sources in the field often make this work difficult. A first topic for debates on the WTN 2017 Conference website is the compliance evaluation and ways of measuring the WT noise to exclude background noise. Our paper brings developments to a method of measuring the acoustic noise from a dominant source in the ambient, like WT, by using the divergence of the sound pressure level (SPL) created by sources. This method allows evaluating the noise produced by WT without stopping the WT for measuring the background noise. In our calculations we consider uncorrelated point sources in free field conditions and investigate aspects as regard the SPL produced by the acoustic sources at two locations in the field.

2. Acoustic field properties derived from the constant divergence of the sound pressure at two locations in the field

The theory puts in equation numerous physical characteristics related to the acoustic field originating from various sources under different environmental and propagation conditions. However, interesting aspects can additionally be found from the analysis of the SPL at two locations in the field of a source. The goal of this paper is developing the theory and showing how the simultaneous measurements at two locations allow inferring on the source(s) which originated the SPL at those locations. The main results on this subject are stated below as *properties*.

Let L_s be the SPL produced by a point source at a given location in the field. The SPL is related to the power of the source through the equation:

$$L_s = L_w - 20\log d - 11 \quad [\text{dB}], \qquad (1)$$

where:

L_w = The acoustic power of the source, dB re 10^{-12} watt;

d = The distance from the point source to the considered location, in meters.

If two non-correlated sources produce the individual SPL L_a, L_b, at a given location in the field, then the resultant SPL at that location, L_{a+b}, is calculated from the equations:

$$10^{0.1L_a} + 10^{0.1L_b} = 10^{0.1L_{a+b}}, \text{ or } L_{a+b} = 10\log(10^{0.1L_a} + 10^{0.1L_b}) \quad [\text{dB}] \qquad (2)$$

2.1 The divergence of SPL between two locations in the field of a source

Let us consider a point source S and two locations in the field, M_1 and M_2, represented in a Cartesian system as shown in Figure 1. The coordinates of the point source and the two locations are $S(x, y)$ and $M_1(0,0)$, $M_2(a, 0)$. The distance from S to M_1 and M_2 is d_1, respectively d_2, and the distance from M_1 to M_2 is a. The SPL produced by the point source at the two locations is:

$$L_{s1} = L_w - 20\log d_1 - 11; \quad L_{s2} = L_w - 20\log d_2 - 11; \qquad (3)$$

The difference of the equations in (3) produces:

$$L_{s2} - L_{s1} = 20\log\frac{d_1}{d_2} \quad [\text{dB}] \qquad (4)$$

This equation shows that the divergence of the SPL between any two locations in the field of the source does not depend on the acoustic power of the source. For simplification, this difference will be further written:

$$L_{s2} - L_{s1} = 20\log k \quad [\text{dB}], \quad \text{where } k = \frac{d_1}{d_2} > 0 \qquad (5)$$

If L_{s1} and L_{s2} are known, then the distance ratio k can be calculated as follows:

$$k = 10^{\frac{L_{s2}-L_{s1}}{20}} > 0 \qquad (6)$$

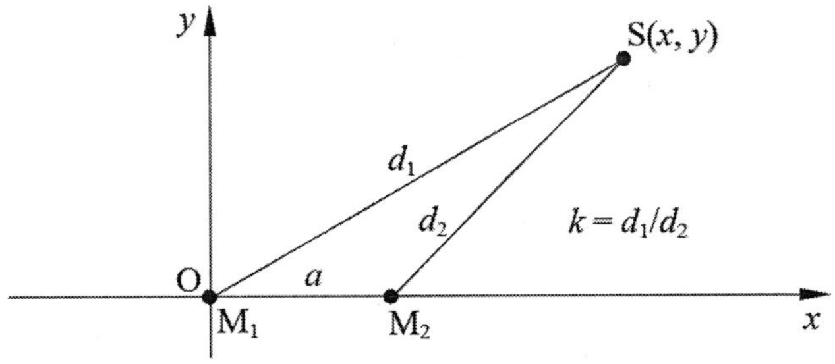

Figure 1: An acoustic point source and two locations in the field

2.2 The locus for the positions of the sources which create the same divergence of SPL between two given locations

It was proved that the acoustic sources at certain positions in the environment can create the same divergence of the SPL between two given locations [2]:

<u>Property 1</u>: *The locus for the positions of the sources which create the same divergence of SPL at two given locations in the field is a spherical surface.*

For the plane problem represented in Figure 1, the locus was found to be a circle. With $d_1 = \mathrm{SM}_1 = \sqrt{x^2 + y^2}$ and $d_2 = \mathrm{SM}_2 = \sqrt{(x-a)^2 + y^2}$, the equality $\frac{d_1}{d_2} = k$ produces:

$$\left(x - \frac{ak^2}{k^2 - 1}\right)^2 + y^2 = \frac{a^2k^2}{(k^2 - 1)^2} \tag{7}$$

The circle defined in (7) has the abscissa of the center point C and the radius:

$$\mathrm{OC} = \frac{ak^2}{k^2 - 1}; \; r = \frac{ak}{|k^2 - 1|}; \tag{8}$$

Depending on *k*, this locus has the particular shapes and positions shown in Figure 2:

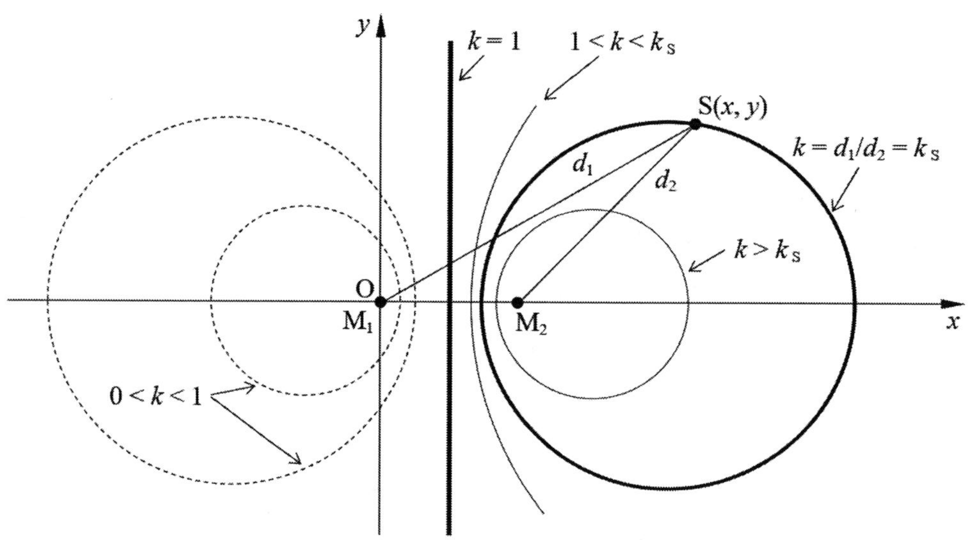

Figure 2: The locus of the points where the ratio of the distances to M_1 and M_2 is constant

- For $k = 1$ the locus is the perpendicular bisector line of M_1M_2 ($k = 1 \leftrightarrow d_1 = d_2$).
- For $k > 1$ the locus is a circle located to the right of the perpendicular bisector of M_1M_2, having the center to the right of M_2. When *k* increases in the interval $(1, \infty)$, the radius of the circle decreases and the center gets closer to M_2.
- For $0 < k < 1$ the locus is a circle located to the left of the perpendicular bisector of M_1M_2, having the center to the left of M1. When *k* decreases in the interval $(0, 1)$, the radius of the circle decreases and the center gets closer to M_1.

For the space problem (in a 3-D system) the locus is obtained by rotating the shapes described above around the horizontal axis. The perpendicular bisector line generates a plane and the circles generate spherical surfaces. Thus we get:

- For $k = 1$ the locus is the perpendicular bisector plane to M_1M_2.

- For $k > 1$ the locus is a spherical surface to the right of the perpendicular bisector, having the radius and center point as described for the circles in Figure 2.
- For $0 < k < 1$ the locus is a spherical surface to the left of the perpendicular bisector plane, having the radius and center point of the left circles in Figure 2.

For the scope of the paper we name the loci described above *circles* or *spheres* of *constant divergence of sound pressure* (CDP) for the locations M_1, M_2, and abbreviate these names *CDP circles* and *CDP spheres*. Each time we will understand that the CDP circles (spheres) regard the discussed locations, M_1 and M_2.

2.3 Special properties associated with the CDP circles and spheres

For simplification, the analysis is made on CDP circles in the Cartesian plane for $k > 1$ as shown in Figure 3. The results can be immediately extended to the corresponding CDP spheres and then extrapolated for $0 < k < 1$ as shown in *Property 3* below.

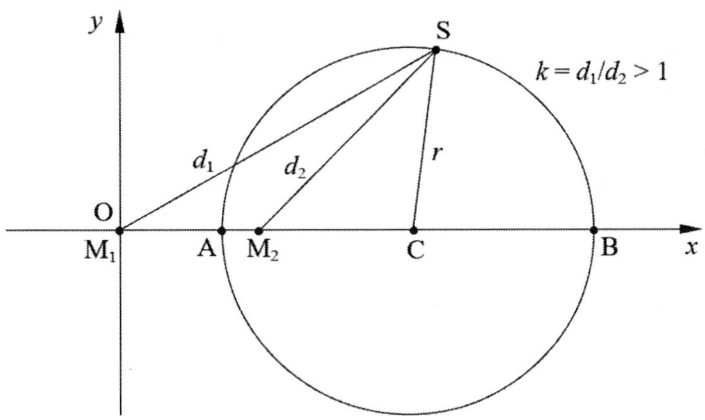

Figure 3: The CDP circle for two locations, M_1 and M_2, in the field of a point source S

Figure 3 shows the center C and the radius *r* of the CDP circle, also the intercept points A and B of the CDP circle with the horizontal axis.
From all possible positions of a source which creates the SPL divergence (4) at M_1, M_2, point A is the closest to and point B is the farthest away from M_1 and M_2.
Obviously, from the definition of the CDP circle we can write:

$$\frac{\mathrm{AM_1}}{\mathrm{AM_2}} = \frac{\mathrm{BM_1}}{\mathrm{BM_2}} = k \tag{9}$$

With $\mathrm{M_1M_2} = a$, from the proportions expressed in (9) we obtain:

$$\mathrm{AM_1} = \frac{ak}{k+1};\ \mathrm{AM_2} = \frac{a}{k+1};\ \mathrm{BM_1} = \frac{ak}{k-1};\ \mathrm{BM_2} = \frac{a}{k-1};\ \frac{\mathrm{BM_1}}{\mathrm{AM_1}} = \frac{k+1}{k-1}; \tag{10}$$

It is interesting to observe the property of a particular sphere in the family of the CDP spheres defined for two given locations by different values of *k*:

*Property 2: If the radius of the CDP circle (sphere) equals the distance between the two locations, then the ratio $k = \frac{d_1}{d_2}$ matches the **golden ratio** value.*

Indeed, the equality of the radius *r* from (8) with the distance M_1M_2 produces:

$$r = a \Rightarrow \frac{ak}{k^2-1} = a \Rightarrow k^2 - k - 1 = 0 \Rightarrow k = \frac{1+\sqrt{5}}{2} = \text{the } \boldsymbol{golden\ ratio}. \quad (11)$$

Further research on this property might reveal interesting aspects related, for example, to the human hearing system.

Property 3: Reciprocal numbers for k produce CDP circles (spheres) symmetrical with respect to the perpendicular bisector of M_1M_2.

Indeed, let us consider $k > 1$ for the CDP circle defined in (7). Then the reciprocal of *k*, $k' = 1/k \in (0,1)$ produces another CDP circle which has the radius *r'* and the abscissa of the center point C' calculated with the equations given in (8) as follows:

$$r' = \frac{ak'}{|k'^2 - 1|} = \frac{ak}{|1 - k^2|} = r; \quad \text{OC}' = \frac{ak'^2}{k'^2 - 1} = -\frac{a}{k^2 - 1}; \quad \frac{\text{OC} + \text{OC}'}{2} = \frac{a}{2}; \quad (12)$$

The first equation in (12) proves that $r' = r$, while the last one shows that the line between the centers of the two circles has the same midpoint as M_1M_2. So, the two CDP circles are symmetrical with respect to the perpendicular bisector of M_1M_2.

Property 4: The SPL values at two locations and the distance between those locations allow calculating a minimum, a maximum, and implicitly the range of the acoustic power for a source which is able to produce those SPL values.

Equation (1) allows calculating the acoustic power of a source if the SPL at a given location and the distance between the source and location are known. In the CDP technique, L_{s1} and L_{s2} allow calculating ratio *k* with (6). Also, Figure 3 shows that the shortest distance from M_1 to the CDP circle is AM_1, while the longest is BM_1. In these conditions, by using (1), (10), and knowing the value of *k*, we can calculate:

$$L_{w(\min)} = L_{s1} + 20\log \text{AM}_1 + 11 = L_{s1} + 20\log\frac{ak}{k+1} + 11 \quad [\text{dB}] \quad (13)$$

$$L_{w(\max)} = L_{s1} + 20\log \text{BM}_1 + 11 = L_{s1} + 20\log\frac{ak}{k-1} + 11 \quad [\text{dB}] \quad (14)$$

$$L_{w(max)} - L_{w(min)} = 20\log\frac{BM_1}{AM_1} = 20\log\frac{k+1}{k-1} \quad [dB] \quad (15)$$

Equations (13) and (14) show that sources of different power levels can produce the same SPL at M_1. Having mandatory positions on the CDP sphere, such sources have the acoustic power between a minimum value (for a source placed at point A) and a maximum value (for a source placed at B).

As an exception, for $k = 1$ (that is $d_1 = d_2$) we only can calculate the minimum power level needed to a source for producing a given SPL, L_{s1}, at M_1. From (13) we get:

$$k = 1 \Rightarrow L_{w(\min)} = L_{s1} + 20\log\frac{a}{2} + 11 = L_{s1} + 20\log a + 5 \quad [\text{dB}] \quad (16)$$

2.4 The addition of SPL from multiple sources in terms of CDP circles (spheres)

First we analyze the properties of the SPL addition at two locations by considering two acoustic sources, S_1 and S_2. We calculate the resulting SPL at M_1, M_2, and identify the characteristics (power and position) of an equivalent source S* which can create at M_1, M_2, the same SPL as the combined effect of S_1 and S_2. Based on these results we will conclude that the procedure can be extrapolated to multiple sources in the ambient.

2.4.1 Two acoustic sources located on the same CDP circle

Figure 4 shows two acoustic sources, S_1 and S_2, on the same CDP circle.

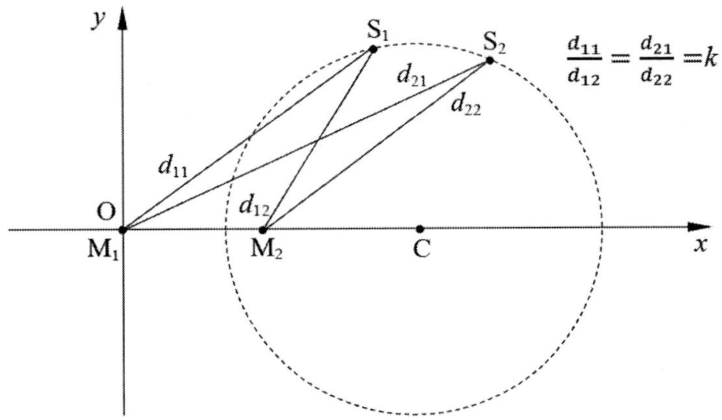

Figure 4: The addition of SPL from two sources placed on the same CDP circle

In the next equations, when two index numbers are used, the first index regards the source while the second is for location. For example, d_{12} is the distance from source S_1 to location M_2, and L_{s21} is the SPL produced by source S_2 at location M_1.
For the situation represented in Figure 4 we can write:

$$L_{s12} - L_{s11} = 20\log k; \quad L_{s22} - L_{s21} = 20\log k \tag{17}$$

The power addition at M_1, M_2, creates the SPL values L_{s^*1}, L_{s^*2}, as follows:

$$L_{s^*1} = 10\log(10^{0.1L_{s11}} + 10^{0.1L_{s21}}); \quad L_{s^*2} = 10\log(10^{0.1L_{s12}} + 10^{0.1L_{s22}}) \tag{18}$$

From (18), and using (17), we get:

$$L_{s^*2} - L_{s^*1} = 10\log\frac{10^{0.1L_{s12}} + 10^{0.1L_{s22}}}{10^{0.1L_{s11}} + 10^{0.1L_{s21}}} = 20\log k \tag{19}$$

We compare the result shown in (19) with the content of (17) and conclude:

Property 5: The SPL created at two locations by two sources which are placed on the same CDP circle (sphere), can be produced by a single source which is situated on the same CDP circle (sphere).

As shown in (13) and (14), such source is not unique. It is reasonable to consider in calculations that the equivalent source S* has the acoustic power L_{w^*} found from the addition of power levels L_{w1}, L_{w2}, of the two sources S_1 and S_2.
Then we calculate the distance d^*_1 from S* to M_1 with equation (1), wherein the SPL L_{s^*1} and power L_{w^*} are known. With this approach, for the case shown in Figure 4 we get:

$$d_1^{*2} = \frac{d_{11}^2 10^{0.1L_{s11}} + d_{21}^2 10^{0.1L_{s21}}}{10^{0.1L_{s11}} + 10^{0.1L_{s21}}}, \qquad \text{which leads to } d_{11} < d_1^* < d_{21}; \tag{20}$$

Consequently, the equivalent source S* is placed on the CDP circle between S_1 and S_2.
When a source creates much higher SPL than the other, for example 6dB higher, it is of interest to evaluate how close to the dominant source is the equivalent source S*. For such situation, from (20) we can calculate:

$$\text{If } \mathbf{S_1} \text{ dominant: } d_{11} < d_{21} \text{ and } L_{s11} \geq L_{s21} + 6 \text{ [dB], then } d_1^* \leq d_{11}\sqrt{0.8 + 0.2\frac{d_{21}^2}{d_{11}^2}} \quad (21)$$

Also,

$$\text{If } \mathbf{S_2} \text{ dominant: } d_{11} < d_{21} \text{ and } L_{s21} \geq L_{s11} + 6 \text{ [dB], then } d_1^* \geq d_{21}\sqrt{0.8 + 0.2\frac{d_{11}^2}{d_{21}^2}} \quad (22)$$

2.4.2 Two acoustic sources located on different CDP circles

Let us consider two sources on distinct CDP circles for M_1, M_2, as shown in Figure 5.

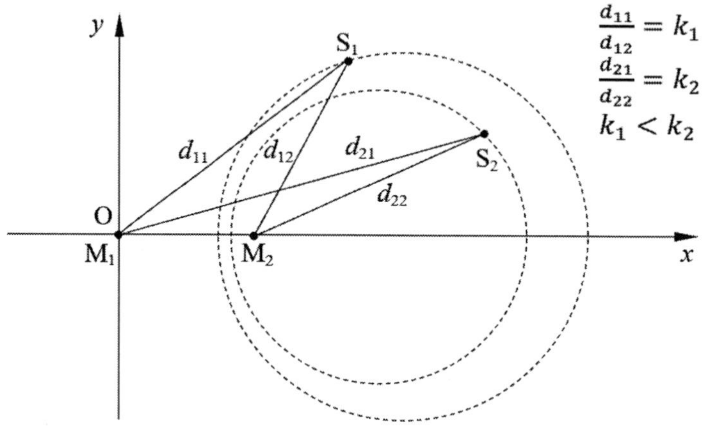

Figure 5: The addition of SPL from two sources placed on different CDP circles

For the SPL produced at the locations represented in Figure 5 we can write:

$$L_{s12} - L_{s11} = 20\log k_1; \quad L_{s22} - L_{s21} = 20\log k_2 \quad (23)$$

The acoustic power addition at M_1, M_2, produces the SPL values as written in (18). With (18), and using (23), we obtain:

$$L_{s^*2} - L_{s^*1} = 10\log\frac{10^{0.1L_{s12}} + 10^{0.1L_{s22}}}{10^{0.1L_{s11}} + 10^{0.1L_{s21}}} = 20\log k^*, \text{which produces } k_1 < k^* < k_2 \quad (24)$$

The double inequality shown in equation (24) can be expressed as follows:

<u>Property 6</u>: The SPL produced at two locations by two sources which are placed on different CDP circles (spheres) can be created by a single source which is situated on an in-between CDP circle (sphere).

For describing an equivalent source S*, we can follow the procedure described above:

- Calculate the SPL values L_{s^*1}, L_{s^*2}, with (18); then calculate k^* from (24);
- Determine the center point and the radius of the CDP circle of S* by using (8);
- Calculate the acoustic power of S* and the distance d^*_1, as indicated at 2.4.1;

2.5 The addition of SPL from multiple sources in terms of CDP circles (spheres)

Based on the properties described above for the CDP circles and spheres in the case of two acoustic sources, we can conclude that the effect of multiple sources in a linear environment can be investigated by using the procedure described in this paper, in successive steps. We can determine finally the position and the acoustic power of an equivalent acoustic source S* which can produce at the given two locations the same SPL values as the set of sources considered in calculations.

3. Localization of the source by using the divergence of SPL

The localization of the acoustic source is not a topic for this paper. However, we want to point out that, in principle, the divergence of SPL between certain locations in the field of an acoustic source can provide cues about the position of the source.
Property 1 and the subsequent proofs related to Figure 2 show that the divergence of SPL between two locations and the distance between those locations allow, in general, finding a sphere (or a plane) on which the acoustic source is placed. Let us consider now an additional location, M_3, positioned in the plane as sketched in Figure 6.

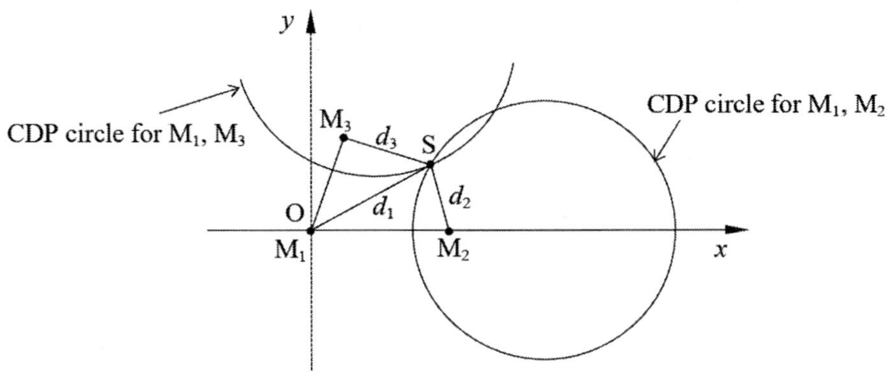

Figure 6: CDP circles for three locations, M_1, M_2, and M_3, in the field of a point source S

As shown before in this paper, we can determine a CDP sphere for M_1, M_2, and another CDP sphere for M_1, M_3. In our calculations we only need the positions of the three locations and the SPL divergence between locations. The intersection of the two CDP spheres is, in general, a circle. It is sufficient now to consider another location, M_4, out of the plane shown in Figure 6, and to visualize the CDP sphere for M_1, M_4. We have to manage the intersection of three spheres having an intersect point in S. With the necessary math we can express the coordinates of source S in terms of the coordinates of four adequately chosen locations and the divergence of SPL between those locations. Details on this topic are revealed in [4].

4. Using the CDP method for evaluating the SPL from a dominant sound source like WT

A method for characterizing the acoustic noise from WT by using the CDP of SPL at two locations in the field was described in [2] and [3]. For the CDP measurements we should consider two locations in the field of the WT, on the same ground line to the WT pole. To facilitate the comparison of the results from the CDP method with those obtained with the IEC 6144-11 standard, the authors proposed placing the measurement location M_1 at the distance $R_0 = H + D/2$ from WT pole as recommended in IEC 61400-11 and outlined in Figure 7.

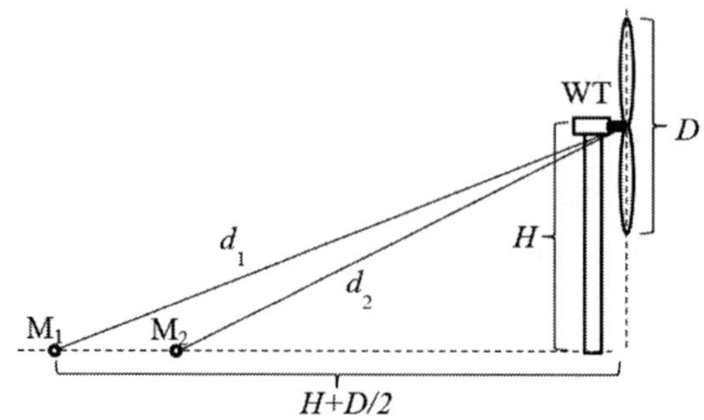

Figure 7: CDP test arrangement for measuring the acoustic noise from a WT

At the two locations, the addition of SPL from WT, L_{s1} and L_{s2}, to the background noise L_{n1} and L_{n2} is described through the following two equations:

$$10^{0.1L_{s1}} + 10^{0.1L_{n1}} = 10^{0.1L_{s+n,1}}; \quad 10^{0.1L_{s2}} + 10^{0.1L_{n2}} = 10^{0.1L_{s+n,2}} \qquad (25)$$

The target is to find the acoustic noise from WT at a given location, for example L_{s1}.
We only can measure combined sound and background noise $L_{s+n,1}$ and $L_{s+n,2}$. Also, we have an additional relation for L_{s1} and L_{s2} from equation (5), $L_{s2} - L_{s1} = 20\log k$, where k is known. The value of k is calculated from the geometry of Figure 7, so that to have adequate divergence of the SPL produced by WT at M_1, M_2 (for example above 1 dB).
We have to manage three equations, two in (25) and one in (5), for getting the values of four unknowns: L_{s1}, L_{s2}, L_{n1} and L_{n2}. This problem has undetermined solutions. In such situations, either an extra relation is found for the unknown quantities, or one unknown quantity is considered to be a variable parameter.
In the referenced papers we considered that the background noise has close values at M_1, M_2, that is $L_{n1} \cong L_{n2}$ [2], [3]. With this agreement, from (25) and (5) we obtained the following equation for calculating the acoustic noise produced by WT at M_1:

$$L_{s1} \cong 10\log(10^{0.1L_{s+n,2}} - 10^{0.1L_{s+n,1}}) - 10\log(k^2 - 1) \ \ [\text{dB}] \qquad (26)$$

For the WT characterization, equation (26) should be considered for each 1/3 OB and for all wind speed brackets. The CDP data $L_{s+n,1}$, $L_{s+n,2}$ obtained from the simultaneous measurements at M_1, M_2, can be treated as shown in [3]. With the results discussed in *property 3*, the $L_{s+n,1} > L_{s+n,2}$ values which are not acceptable for (26), should not be rejected. Such data indicate that dominant noise sources were located to the left of M_1 during certain measurement sessions. We can characterize those equivalent sources by rewriting and using equation (26) for the case $0 < k < 1$.
Moreover, the $L_{s+n,1}$ and $L_{s+n,2}$ data from simultaneous measurements allow determining the CDP sphere for a combined source, equivalent to the WT + background noise. Thus we can decide if the WT is the dominant source by using the procedure described at 2.4.1 and 2.4.2. If the WT noise is at least 6 dB over the background, then the position of the equivalent source should be close to the WT as calculated with (21) and (22).
For further development of the CDP approach, we can consider a new equation so that to link the background noise at the two locations as follows:

$$L_{n2} - L_{n1} = 20\log\delta \ [\text{dB}], \qquad \text{where } \delta > 0 \qquad (27)$$

Equation (27) is similar to (5) and δ is a parameter. In terms of the discussions in this paper, δ indicates the "distance ratio" for the position of an ***equivalent background noise source*** as seen at the two locations. From the equations given in (25), (5) and (27), for the case $\delta < k$ we obtain the following expression for the SPL produced by WT at M_1:

$$L_{s1} = 10\log(10^{0.1L_{s+n,2}} - \delta^2 10^{0.1L_{s+n,1}}) - 10\log(k^2 - \delta^2)\ \ [\mathrm{dB}] \qquad (28)$$

If can be seen that if the background noise has close values at M_1, M_2, that is $L_{n1} \cong L_{n2}$ and $\delta \cong 1$, from (28) we obtain equation (26) discussed above.
Managing equation (28) for different values of δ, both when $\delta < k$ or $\delta \geq k$, is the subject of further research during a project with measurements on WT in real conditions. This topic is the goal of future work.

5. Conclusions

The theoretical aspects developed in this paper provide new tools for managing the noise + background data obtained from the measurements on WT. We consider that the paper provides sufficient basis for an experimental project to determine the acoustical noise from WT by using the CDP method. This approach can be used not only in the case of the wind turbines, but for characterizing the acoustic noise in the ambient in general. Evaluating the background noise through simultaneous measurements at multiple locations as proposed in this paper may develop into a new technology. The analysis of the results of measurements on a WT with this new technique would help improving the CDP methodology.

References

1. *Wind turbines – Part 11: Acoustic noise measurement techniques,* International Standard IEC 61400-11: 2012 (International Electrotechnical Commission, Geneva, Switzerland, 2012).
2. V. Buzduga and A. Buzduga, "Comparison of the divergence of sound pressure method with the IEC 61400-11 method in measuring the wind turbine noise", *Inter-Noise 2015*, San Francisco, California, USA.
3. V. Buzduga and A. Buzduga, "A New Method for Determining the Wind Turbine Noise based on the Constant Divergence of the Sound Pressure Level", *The Sixth International Meeting on Wind Turbine Noise*, Glasgow, Scotland, 20th-23rd April 2015.
4. V. Buzduga, "The localization of the sound sources: a new approach based on the divergence of the sound pressure level", (under development; abstract submitted to *ICBEN 2017,* Zurich, Switzerland).

7th International Conference
on
Wind Turbine Noise
Rotterdam – 2nd to 5th May 2017

Using long term monitoring for noise assessment of wind farms

Eugène de Beer, Peutz bv, e.debeer@peutz.nl

Summary

A relative small wind farm (3 wind turbines) was developed in the center of The Netherlands near a crowded residential area, major highways and waterway. To prevent noise nuisance the local authority applied strict noise limits. These noise limits were based on the background noise (L_{95}) at different wind speeds. However in practice this wind farm had caused a lot of nuisance in the residential area. The local residents had great doubts whether the wind farm would meet its noise limits. Long term monitoring of the noise emission was demanded which had to ensure the local authority and the local residents that the wind farm meets the noise limits.

The local residents demanded that the noise levels were measured at their properties (the location where the nuisance was experienced). From an acoustic point of view such a measurement is less suitable for an assessment due to the disturbing background noise at those locations. The local authority was more interested in accurate measurements with a legally binding assessment. For that reason the noise levels were measured at the nearest dwellings as well at a relative short distance to the wind turbines during a period of a month. Simultaneous the wind speed and wind direction were measured at a height of 10 m. The noise measurements close to the wind turbines were used to calculate the sound power level in accordance with IEC 61400-11. With these sound power levels the noise levels at the dwellings were calculated and assessed. This paper presents the results of the measurements and also gives a consideration whether noise measurements regarding wind turbine noise at locations nearby residential properties can be useful for assessment and to provide these results to local residents.

1. Introduction

A relative small wind farm was developed in the center of The Netherlands. In figure 1 the location of the wind farm is given. The wind farm consists of three wind turbines of the type Vestas V90 with a power of 2 MW each and a hub height of 105 m. The location of the wind farm is located at the border of the municipality. The major noise sources in this area are the motorway A27, the barges on the waterway and the traffic on the major ring roads. Few scattered dwellings are located around the wind turbines. The nearest dwellings are located at a distance of 300 m from a wind turbine. At the east side of the wind farm a residential area is located at a distance of more than 500 m from the nearest wind turbine.

To prevent noise nuisance the local authority applied strict noise limits in the permit. These noise limits were based on the background noise (L_{95}) at different wind speeds.

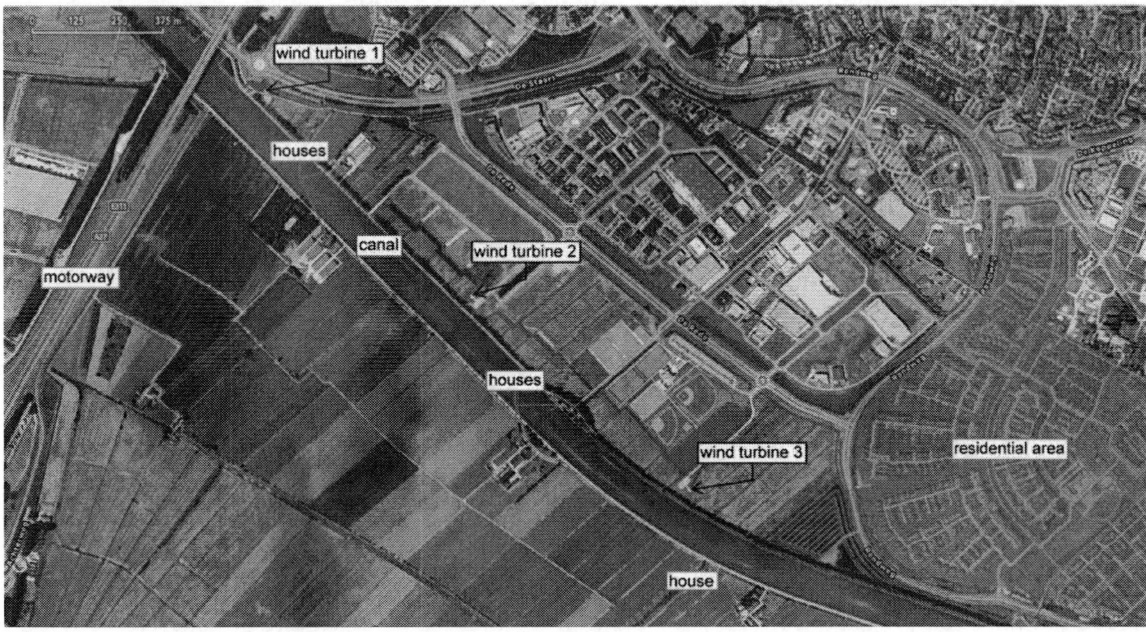

Figure 1: Location of wind farm

2. Reason for noise survey

2.1. Noise limits permit

Prior to the construction of the wind farm there has been much debate about the suitability of the wind farm in this area. The residents feared noise nuisance. Especially in situations when there is low wind speed at ground level (therefore a low background noise level) and a relative high wind speed at hub height (therefore a relevant noise emission from the wind turbine).

On behalf of the permit of the wind farm the authority has extensively investigated the background noise levels in this area. The authority based the noise limits on the background noise (L_{95}) at different wind speeds. The authority wanted to ensure a low noise level due to the wind turbine in all situations. Also in the situation where there is a low wind speed and hence a low background noise level. The noise levels which are included in the permit are given in table 1. The noise limits refer to an equivalent noise level in dB(A) over a 10-minute period and applies at the dwellings. The noise limits are dependent of a measured wind speed at a height of 10 m in the neighborhood of the wind turbines.

	L_{eq} in dB(A)											
Wind speed in m/s	0	1	2	3	4	5	6	7	8	9	10	≥ 11
Day period	34	34	34	34	38/39*	42	43	43	42	42	42	42
Evening period	34	34	34	34	38/39*	42	43	43	42	42	42	42
Night period	34	34	34	34	38	39	40	43	42	42	42	42

* 38/39 dB(A): 38 dB(A) applies when wind direction is between 130 and 210 degrees, 39 dB(A) applies when wind direction is between 210-130 degrees (calculated clockwise; north is 0 degrees)

2.2.Survey questions

After commissioning of the wind farm noise complaints were reported by the local residents. The residents were very doubted whether the wind farm is in compliance with its noise limits of the permit. For the municipality these noise complaints were the reason to investigate it.

In general the compliance checks of wind turbine noise in The Netherlands are done by determining the sound power level of the wind turbine at some occurring wind speeds and comparing these with the manufacturer's specified sound power levels. Such a method of random assessment was insufficient for the municipality as well as the local residents. Long term monitoring of the noise emission of the wind farm was demanded. Only with long term monitoring all different wind speed conditions could be assessed.

The local residents demanded not only long term monitoring of the noise emission but also that the noise levels were measured at their dwellings. The location of their dwellings (facades) was where nuisance was experienced and also the location where the noise limits should be met. From an acoustic point of view such a measurement location is less suitable for an assessment due to the disturbing background noise at these locations. But the local residents had no more confidence in calculation of the noise at their dwellings. So they insisted that the noise levels were measured at their dwellings.

The local authority was more interested in accurate measurements with a legally binding assessment. For that reason it was necessary to measure the noise at a relative short distance from the wind turbines. These measurements were as much as possible carried out in conformity with NEN-EN-IEC 61400-11 by using a measurement board at 150 m (= vertical distance from ground to the rotor center + 0.5 · the diameter of the rotor) of the wind turbine.

As known, background noise measurements should be carried out for more accurate noise measurements on wind turbines. The operator of the wind farm was only willing to stop the wind turbine during some short periods (a few times for 30 minutes). These few short periods were however not enough for an accurate background noise correction for all the measurements during the period of a month. Only the period immediately before and after this background noise measurements are suitable for any background noise correction.

3. Measurements

3.1.Measurement method

In order to satisfy the needs of both the municipality and the local residents both types of measurements (noise measurements at the facades of the dwelling and noise measurements with a measurement board) were carried out simultaneously for a continuous period of one month. In figure 2 a schematic representation of the measurements is shown. For all three wind turbines, the noise is measured at 150 m distance with a measurement board in the field as well as at the facades of the nearest residences.

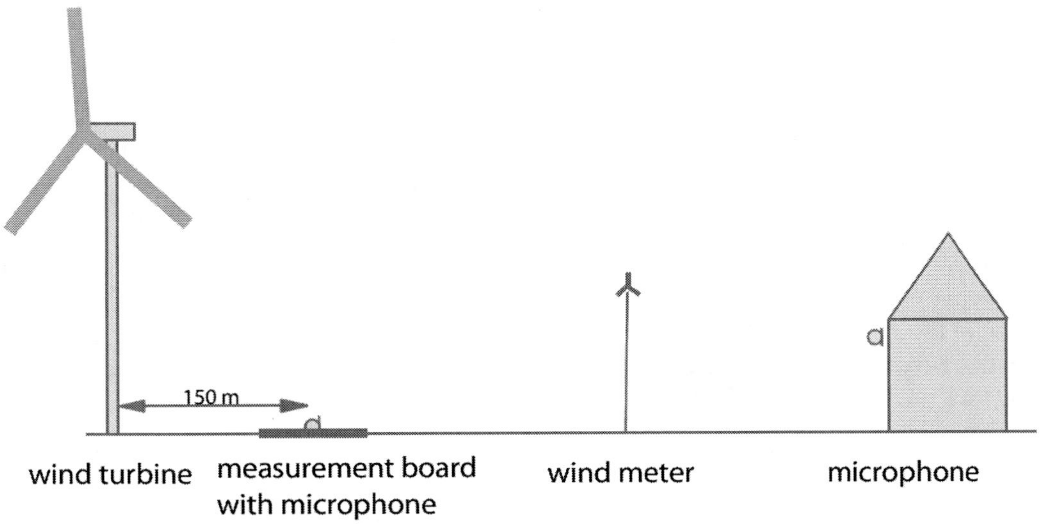

Figure 2: Schematic representation of measurements

The measurement were carried out during a month (February 25th until March 25th). During this period of the year normally it is quite windy in the Netherlands. Beside the noise measurements also weather conditions measurements (wind speed, wind direction, temperature, barometric pressure and rainfall) were carried out. These measurements were carried out at a height of 10 m and averaged every 10-minute period. Also if the wind turbine was in operation during that 10-minute period was registered.

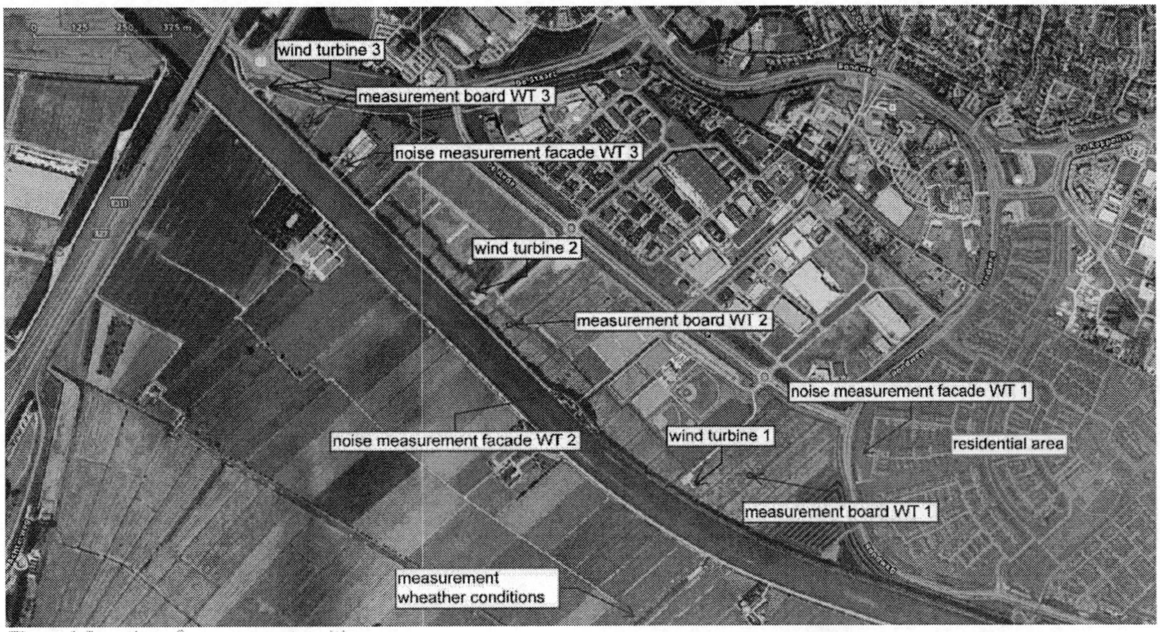

Figure 3:Location of measurement positions

All the measured data has been real time uploaded to an internet database. For the local residents an internet portal was readily available so they could see the measured data and compare it with their own experience at that moment. With this system the confidence of the local residents in the way of assessment has been grown.

3.2. Method of assessment

Except for the noise contribution of the ambient noise, the noise levels as measured at the facades of the nearest dwellings can directly be compared with the noise limits after a correction for the reflection of the facade. For direct assessment of the noise measurements at short distance (150 m) the noise limits at the dwellings was first converted to maximum allowable noise levels at measurement boards.

By using an acoustical model the noise transmission (D) was calculated from the wind turbines to the facades of the nearest dwellings. With this noise transmission and the noise limit at the dwelling a maximum allowable sound power level ($L_{W,j,max}$) of the wind turbine per wind speed was be calculated.

$$L_{W,j,max} = L_{eq,j,noise\ limit} + D_{wind\ turbine\text{-}house}$$

With this maximum allowable sound power level ($L_{W,j,max}$) a maximum noise level at the measurement boards can be calculated using the transmission as defined in IEC 61400-11.

$$L_{eq,j,max} = L_{W,j,max} - 5 - 20\log(R) = L_{W,j,max} - 5 - 20\log(183)$$

Where

$L_{eq,j,max}$ is maximum allowable noise level at measurement board per wind speed j

R is slant distance from rotor center to measurement board

So for each of the measurement boards and the different wind speeds a maximum allowable noise level at measurement board was calculated.

Figure 4: Wind turbine 2 with measurement board in front (inside barrier fence)

4. Measurement results

4.1. Measurements at the facades

In figure 5 the results of the measurements at the facades are given. In this figure just the results of one day (24 hours) are shown. That day (between February, 28th and March , 1st) can be seen as a representative day of the monitoring period.

The orange line represents the measured wind speed at 10 m height and is related to secondary Y-axis (right side). The grey dotted line represents the noise limit which is associated with the determined wind speed. There is only a value given for the noise limit when the wind turbine is running. In this example the wind turbine starts running at 09:30 am and stops running at 01:40 am next day with a few short stops in between.

The red, blue and green lines represent the measured noise levels at the facades of the dwellings which are nearest to respectively wind turbine 1, wind turbine 2 and wind turbine 3 (see figure 3). The noise levels refers to an equivalent noise level over a 10-minute period and is corrected for the facade reflection.

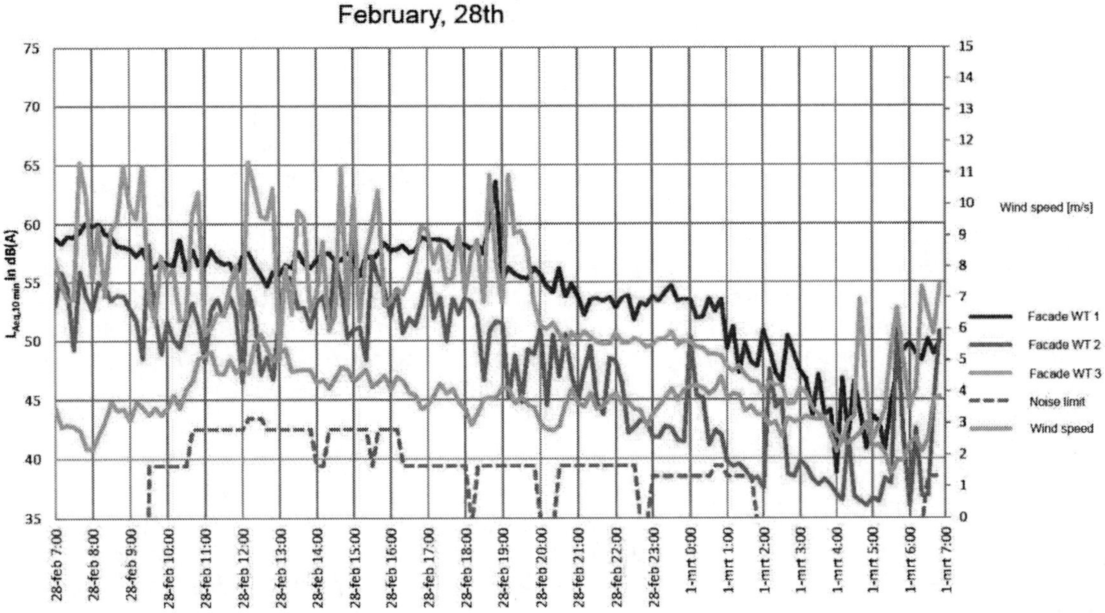

Figure 5: Results of measurements at facades at February, 28th

During the whole day the measured noise levels are significantly higher (up to 20 dB) than the noise limits of the permit. The background noise as traffic of the highway and local roads as well as the shipping on the canal and noise from the dwellings totally dominates the noise of the wind turbine. By listening to the audio files no noise of the wind turbines could be distinguished at any time. This phenomenon occurs throughout the whole month of monitoring. Only in some quiet nights between 01:00 am and 05:00 am a few moments occur that the noise of the wind turbines could be distinguished from the background noise. An example of such a moment is the night period of March, 2nd and 3rd. Figure 6 shows the results of that day.

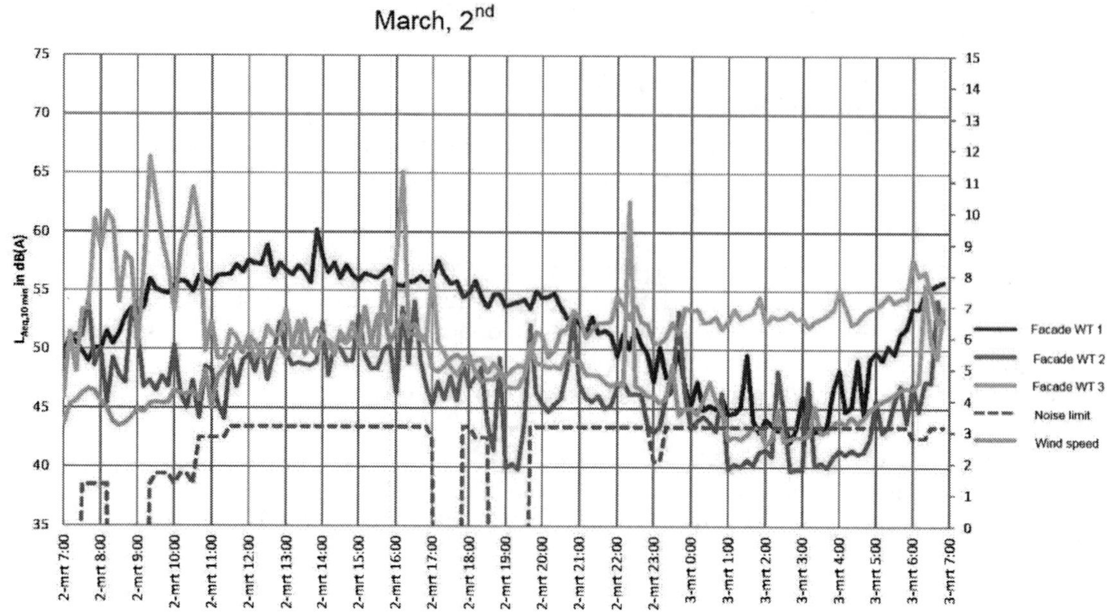

Figure 6: Results of measurements at facades at March, 2nd

Figure 6 shows that during the period between 01:00 am and 05:00 am the measured noise levels at the facade are regularly below the noise limit of the permit. By listening to the audio files of those moments the noise of the wind turbine can be clearly distinguished of the background noise.

Based on the results of the measurement between 02:00 and 04:00 am you should expect that the wind turbines can meet the noise limits of the permit. The transient increases in the noise level in this period of time are caused by background noise sources which are not related to wind turbines. An accurate measurement with a legally binding assessment of the noise of the wind turbines is not possible by these noise measurements at the facades. Even not when long term monitoring is used.

4.2.Noise levels at measurement boards

In figure 7 some measured noise levels at the measurement board are shown. In this figure only the hours between 09:00 pm and 05:00 am of February 28th and March 1st are given. In other hours of the day too many background noise occurs. These moments are not suitable for a comparison with noise limits.

The orange line represents the measured wind speed at 10 m height and is related to secondary Y-axis (right side). The red, blue and green lines represents the measured noise levels at measurement boards of respectively wind turbine 1, wind turbine 2 and wind turbine 3 (see figure 3). The dotted red, blue and green lines represents the maximum allowable noise levels on the measurement board of respectively wind turbine 1, wind turbine 2 and wind turbine 3 to meet the noise limits of the permit. All noise levels refers to an equivalent noise level over a 10-minute period.

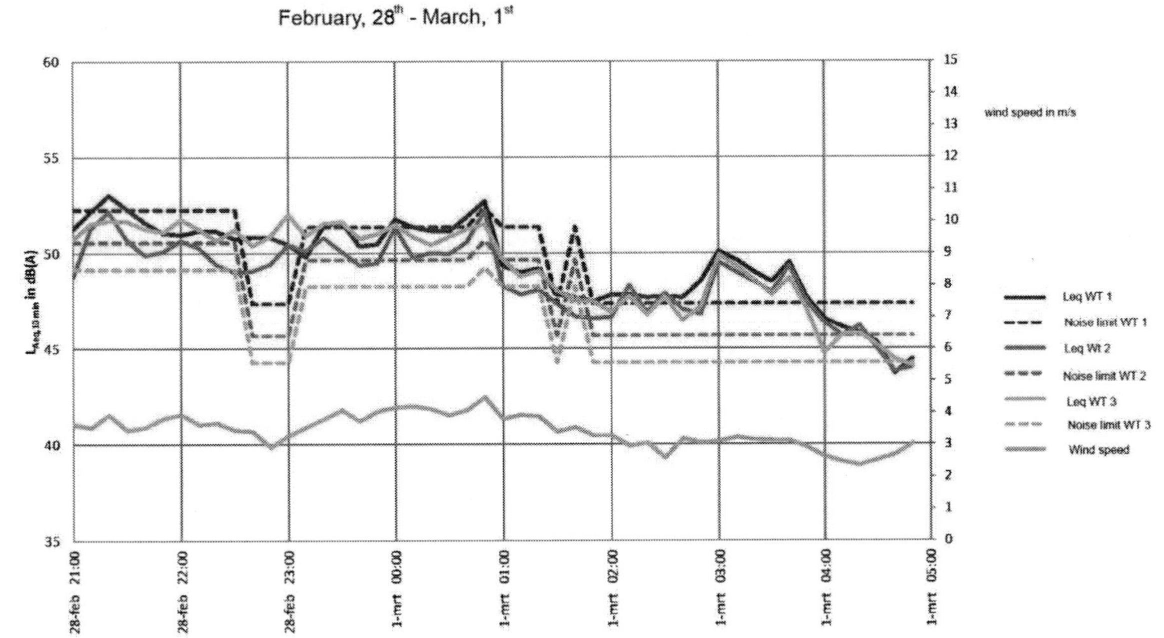

Figure 7: Results of measurements at measurement board in night period February, 28th - March 1st

Figure 7 shows that the noise levels at the measurement boards are often slightly higher than the maximum allowable noise levels. This is mainly caused by all kind of background noise events. By listening to the audio files of this night you can distinguish some wind turbine noise but there is also a lot of disturbing noises. Even as we focus at the relative quiet period of the night.

All the other days during the monitoring period give almost the same results. An accurate measurement with a legally binding assessment of the wind turbine noise seems in general not possible with this kind of measurement, mainly due to dominant disturbing noise.

During the monitoring period there was one relative quiet night (the night of March, 2nd and 3rd) with moments where the wind turbine noise was dominant compared to the background noise. In figure 8 the results of the measurements of this night of March, 2nd and 3rd are given. In the period from 01:00 am until 05:00 am the background noise seems low enough to state that there was no relevant disturbance by the background noise. But there is no legal proof with unmanned measurements that there is no relevant disturbance by the background noise. This is not a problem in situations that the measured noise levels (wind turbine noise and background noise) is lower the noise limits, like the period between 01:00 am and 05:00 am in figure 8. During the time before 01:00 am there are several moments that the measured noise levels exceed the maximum allowable noise levels at the measurement boards. When listening to the audio files of these moments some background noise can be distinguished. In this period the wind turbines meet the noise limits.

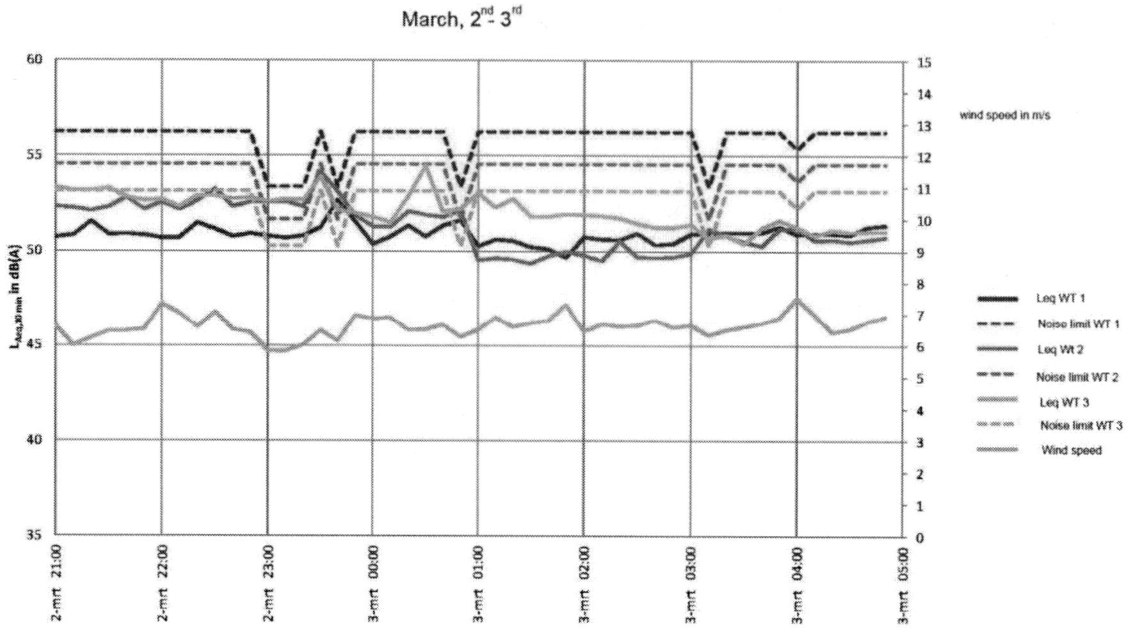

Figure 8: Results of measurements at measurement board in night period March, 2nd-3rd

4.3.Manned measurements

After a few weeks of monitoring it was clear that for legally binding assessment of the wind turbine noise at this wind farm also manned measurements with background noise corrections were necessary. Therefore manned measurements during the night period were carried out wherein the wind turbine was turned off twice for 30 minutes during the measurements.

In figure 9 a part of the measured noise levels at the measurement board 1 as well as the measured wind speed as function of the time is given. The blue line represents the measured noise level (L_{eq}) over 1 minute period, the orange line represents the measured wind speed over 1 and 10 minute period. Any incidental background noise as some geese near the measurement board and a car with the engine running on the other side of the canal are carefully filtered out of the measurement.

Based on these measurements with a background noise correction an accurate noise emission of the wind turbines was established for the wind speeds that occurred during these measurements. These measurements showed that the wind turbines met the noise limits for wind speeds that have occurred.

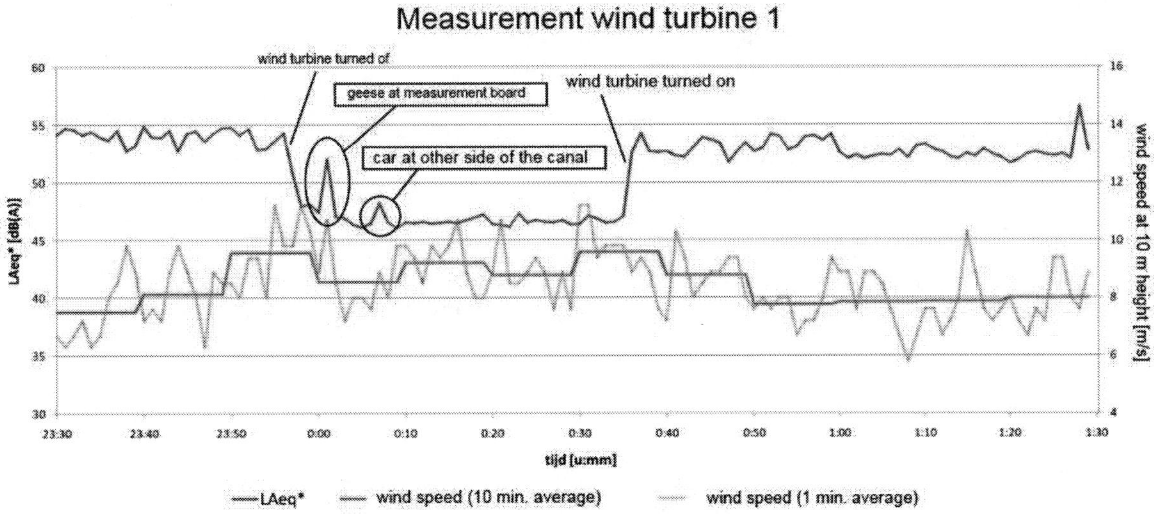

Figure 9: Results of manned noise measurements wind turbine 1

5. The usefulness of long term monitoring

For legally binding assessment of wind turbine noise long term monitoring is limitedly suited. But in many cases where there is distrust between operators of wind farms, local authorities and local residents long term monitoring of wind turbine noise can be very useful, especially when it is combined with an internet portal so anyone has direct access to the measured data. The following aspects are of considerable value for mainly the local residents:

- Transparency. Especially when long term monitoring is used in combination with an internet portal which provides access to the measured data.
- Awareness that many other noise sources in their environment than wind turbine noise determine the acoustic environment in the vicinity of their dwellings
- (Indicative) assessment of wind turbine noise under all kind of circumstances. Although the measurements are indicative the residents have the idea that there has been an assessment under all kind of wind and weather conditions.
- The measured values make the discussions more objective.

The above mentioned aspects may cause that the local residents form a more nuanced representation of the experienced nuisance.

6. Conclusions

Most areas in The Netherlands have significantly high background noise levels. These background noise levels impede an accurate determination of the noise emission of wind turbines. For accurate measurements with a legally binding assessment background noise correction is (almost always) necessary. Background noise corrections are difficult to combine with long term monitoring.

This practice case shows that long term noise monitoring at the dwellings has limited value for legally assessment of the wind turbine noise emission. In this study (monitoring during one month) only one time in a very quiet night an indication of the noise emission of the wind farm is obtained with the noise measurements at the facades of the dwellings.

By using measurement boards at a relative short distance of the wind turbines, the disturbance of ambient noise (background noise) can be reduced. But the results of these measurements even in the relative quiet night periods remain indicative.

For legally binding assessment of the wind turbine noise emission, manned measurements with careful adjustment for background noise are necessary. A major disadvantage of these kind of measurements is that in practice not all wind and weather conditions can be investigated. Only the wind and weather conditions that occur during that manned measurements are considered.

Despite the limitations of long term noise monitoring (without background noise correction) these measurements can be very useful for the local residents to get a more nuanced point of view about the experienced nuisances.

References

Staatscourant 2010 nr. 19592 (December, 23th 2010) *Bijlage 4 bij de regeling Algemene regels voor inrichtingen milieubeheer – Reken- en meetvoorschrift windturbines (Annex 4 of the regulation General rules for environmental management act - Calculation and Measurement Regulations wind turbines)*

IEC 61400-11: 2002 (May 2003) *Wind turbine generator systems Part 11: Acoustic noise measurement techniques*

IEC 61400-11 edition 3.0 2012-11 *Wind turbines – Part 11: Acoustic noise measurement techniques*

7th International Conference
on
Wind Turbine Noise
Rotterdam – 2nd to 5th May 2017

The Challenges and Benefits of Long-Term Sound Monitoring of Wind Farm Sites

Ethan Brush, Acentech Incorporated, Cambridge, MA USA: ebrush@acentech.com
James Barnes, Acentech Incorporated, Cambridge, MA USA: jbarnes@acentech.com
Marc Newmark, Acentech Incorporated, Cambridge, MA USA: mnewmark@acentech.com
Bill Yoder, Acentech Incorporated, Charlottesville, VA USA: byoder@acentech.com

Summary

Turbine noise continues to be a contentious issue in the development and operation for a number of wind farms. Developers seeking to site a wind farm may take every precaution to meet all noise regulations and project goals, yet rigorous ambient sound measurements and extensive modelling to demonstrate regulatory compliance can still be met with intense public skepticism and resistance from some neighbors. Once operational, even with demonstrated project compliance per regulatory rules, wind farms may face heightened scrutiny as nearby community members begin to experience the noise emissions first hand.

The authors of this paper have been involved with a number of long-term noise studies at wind sites in the northeastern United States, lasting months to several years. The long-term studies have been sponsored by government agencies and wind farm operators with input from community groups. Their purpose was to document turbine noise levels through varying seasons, wind environments, temperatures, foliage, and turbine operating conditions. The main goal in each case has been to provide as much objective evidence as possible that compare the turbine noise levels at a site to applicable regulations and to provide more insight into the potential concerns surrounding infrasound, and noise amplitude modulation.

This paper presents the many technical challenges that have been addressed in these projects, such as how to separate the turbine noise from background sounds, remote monitoring system power and communications, gathering appropriate meteorological data, and maintaining reliability even in harsh climates. Political challenges are also a reality when handling such an abundance of data. Consultants representing competing interests can sometimes draw different conclusions. However, this sharing of information and communication, if executed well, can help to strengthen the credibility of the final conclusions to a skeptical audience.

1. Introduction

1.1 State of Wind Turbine Research and Development in the Northeastern United States

The northeastern United States is currently experiencing significant growth in wind projects. This area of the world possesses abundant resources in terms of the annual average wind power experienced at turbine hub heights. The terrain contains numerous hilly areas and coastal zones for potential future wind developments. Offshore wind energy in the northern

Atlantic Ocean is also very attractive and is expected to become a large industry. America's first offshore wind farm just became operational off the coast of Rhode Island in late 2016.

The Massachusetts Clean Energy Center (MassCEC) was formed in 2009 to aid in the growth and success of clean energy technologies, thereby creating an avenue for high quality jobs and a long-term economic growth industry for workers in the state. This agency supports solar and wind power developments through direct grants of funds and services. The MassCEC and others have also funded specific research projects as part of an effort to advance the understanding of wind turbine acoustics. In 2013 and 2014 the MassCEC along with the Massachusetts Department of Environmental Protection (MassDEP) commissioned an extensive study of wind turbine acoustics at five sites throughout the winter, spring, and summer months. The final report of this study was released in February of 2016 and provides new information on amplitude modulation, comparison of measured data to modelling predictions, and infrasound [5]. A few of these results will be discussed in this paper.

Other long-term wind farm measurement programs in the northeastern U.S. have been implemented as a way to continuously monitor for noise compliance. Data from these projects are presented in this paper. Many of the results of the wind farms studies are still under review by agencies and project opponents, therefore the project sites in this paper will remain anonymous.

1.2 Noise Concerns and Regulations

The regulations pertaining to wind farm noise can vary by state and even within communities in the same state. Most regulations specify daytime and night time limits as measured at the nearest receptors or non-participating property line by Leq periods usually one hour in length. The limit may be a fixed overall level or determined by thresholds above existing ambient conditions. Tonal sound may also be prohibited, and incur an overall level penalty. This occurs when a particular octave or 1/3-octave band frequency exceeds its two adjacent bands by a set amount. Tonal sounds are typically evaluated for shorter duration Leq periods, usually 1 to 10 minutes.

Currently, most noise regulations do not cover all of the concerns surrounding wind turbine acoustics. The phenomena of amplitude modulation and infrasound are largely absent. One particular noise regulation in the state of Maine, however, limits the amount of short duration repetitive sounds that are allowed. This is defined as "A sequence of repetitive sounds which occur more than once within an hour, each clearly discernible as an event causing an increase in the sound level of at least 6 dBA on the fast meter response above the sound level observed immediately before and after the event."

Given the variety of regulations and extra scrutiny for other concerns such as amplitude modulation and infrasound, long-term sound monitoring systems for wind farms need to be very versatile.

2. Monitoring System Components and Data Collection

2.1 Visualizing the Noise Monitoring System

A wind farm noise measurement system can vary in complexity from a single sound level meter stuffed in a weatherproof box to something much more complex with multiple sound measurement locations, meteorological logging, waveform recording, and internet connectivity. Systems with higher complexity have become typical due to their flexibility and increased requirements due to regulations and other concerns. Figure 2-1 shows a block diagram of a noise monitoring system the authors have used for wind farm noise monitoring projects.

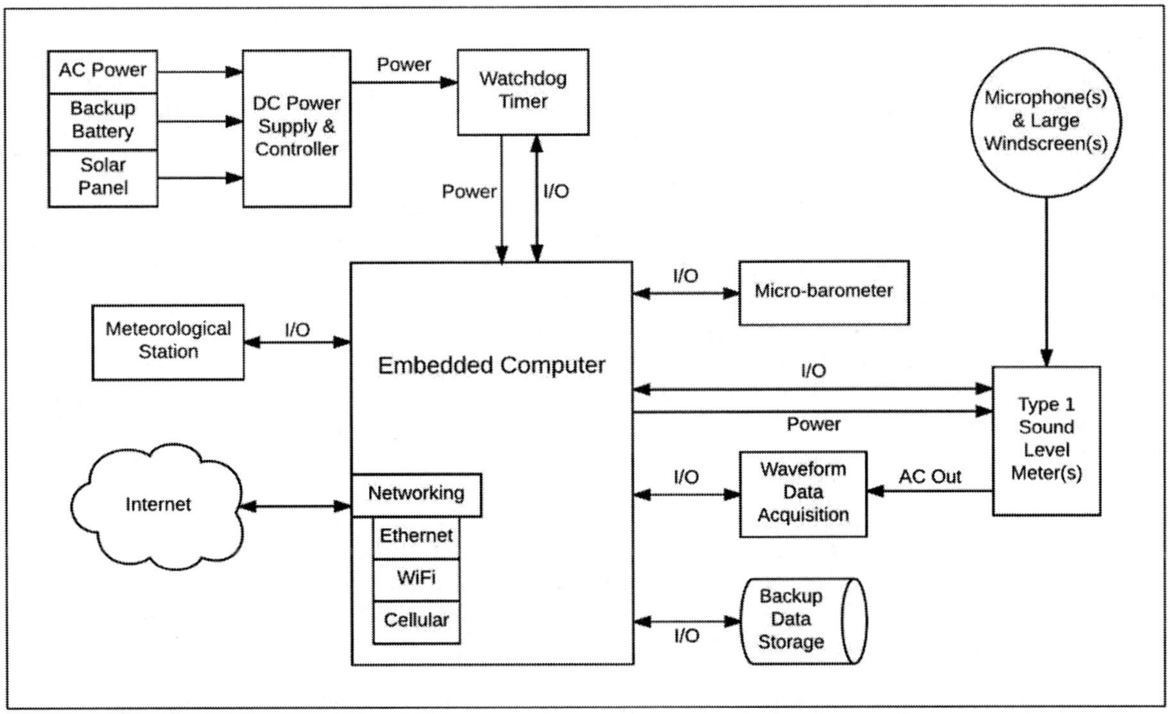

Figure 2-1: Block Diagram of a Typical Wind Farm Noise Monitoring System

2.2 Understanding the Need for Complexity

The block diagram shown in Figure 2-1 is made up of many components, the most important being the computer, which is the hub of the entire system. The computer controls every attached device and archives all collected data. In addition to those primary duties, it acts as the master clock for all devices, ensuring that all data are in near-perfect time sync. A computer allows custom software to be run to perform unique data processing, recording, logging, alarming, streaming, control, and any other needs.

Often an external storage device might be attached to the computer. This is done to provide data redundancy and a simple means of data retrieval when visiting the monitoring site. The presence of an active internet connection provides a way to not only control, update, and maintain the system, but also the ability to stream real-time data, alerts, and recordings to the cloud. However, the remote nature of wind farms often makes the availability of internet quite a challenge. In cases where measurements are made in an area where there is no hardwired internet or Wi-Fi the only options left for communications are cellular and satellite connections. Cellular is an easy choice if accessible due to quick availability of equipment (such as M2M modems, amplifiers, cabling, high gain antennas, etc.), multiple service providers, and easy installation.

Measurements can be made using sound level meters, a micro-barometer, and a meteorological station. The sound level meters used will meet the requirements of the regulation, often ANSI or IEC Type 1. The micro-barometer is used to measure infrasonic levels between 0Hz and 10Hz. The meteorological station will likely measure wind speed, wind direction, temperature, humidity, barometric pressure, and rain. Each of these components communicates directly with the computer for configuration, coordination, and exchange of measured data. Further data capture is performed by the data acquisition module, which is used to digitize the waveform coming from the sound level meter's measurement microphone. This digital waveform can then be saved continuously or as a recording of a specific event.

Providing a robust and reliable system is a major challenge. Many parts of the monitoring system are themselves microcomputers and subject to the same sort of failures often encountered with personal computers. In order to conduct long-term measurements the system must be made resilient enough to continue operating flawlessly and without interruption. To help prevent problems the entire system can be rebooted on a set schedule. However, this does not allow recovery from a component locking up or crashing. In order to fully recover from this the ability to identify that a failure has occurred is required so that the system can be reset accordingly. This process is managed by a watchdog timer, a component that controls the flow of system power to the computer, where power is further distributed. If the computer crashes, power is toggled, rebooting the entire system. If a subcomponent crashes, the computer will request the power to be toggled. This sort of self-healing allows the system to be unattended for very long periods. Note though that this does not allow recovery from a component failure, such as a broken microphone, which would require replacement of the failed component.

Power is also a concern in remote areas. A perfect measurement location would provide AC power and the battery would only be used during brown-outs and short-term power outages. In situations where AC power is unavailable a good option is an adequately sized solar panel and a battery large enough to power the system for an extended period without adequate solar power.

The weather is often extreme during long-term measurements, especially in the northeast United States, ranging from below zero to over 100 degrees Fahrenheit (-18 to 38 degrees Celsius) over the life of the project. Add to that the solar loading provided by the sun hitting the weatherproof enclosure and temperatures inside are even higher. To avoid component failures an enclosure including a ventilation fan or air conditioning should be considered. Winter temperatures are of less concern than the summer, but must be considered if the system contains components with moving parts (e.g. traditional hard disk drives). If possible, it is advantageous to move the electronics indoors and place only those components necessary outside. The picture shown in Figure 2-2 is an example of a noise monitoring installation where the sensors are placed at a property line and the computer and sound level meter are kept indoors.

Figure 2-2: Noise Monitoring Installation (computer and sound level meter are indoors)

3. Technical Challenges and Examples of Monitoring Project Data

3.1 Monitoring Locations and Distance Adjustments for Regulatory Compliance

Measuring for noise compliance typically requires that data be collected at the nearest receptor's property line. However, in remote areas the closest property line to the wind farm site may be in the middle of a dense forest or in an area that is inhospitable for a monitoring system for any number of reasons. The general practice to deal with this challenge is to locate monitoring systems in areas that are feasible and to adjust the results accordingly due to the difference in distance from the nearest turbines and other scientific modelling factors. This should be discussed well in advance of any monitoring program, as all parties concerned with compliance must agree on the proposed protocols.

3.2 Separating Turbine Noise from Background Sounds

Separating turbine noise from the background sounds proves to be very challenging. Background sounds might be primarily wind and transportation in the winter, but the variety of sources explode in the warmer months with insects, lawn mowers, and leaves rustling, among others. Sometimes the regulations or permitting ordinances have very specific testing protocols laid out to isolate wind turbine levels, but this assumes there are no intermittent sound sources and it only works in short term compliance testing where the turbines can be turned on and off on demand. To address the need for long-term unattended monitoring, the ability to identify concerning sounds is very important. The most effective way to do this is to listen to a recording of the concerning period to identify background sound sources and their times of occurrence, and to ignore those time periods. This requires a substantial amount of data storage capacity. The waveforms can be saved in a number of bit-depths, sampling rates, and even compressed file formats as necessary to still allow identification of sources while making it easier to manage disk space.

These techniques to eliminate non-turbine sounds unfortunately don't allow for real-time monitoring or high level analysis of archived data. An approach has been used recently to place a secondary sound monitor behind a nearby large structure such as a barn or house to block the line of sight from the wind farm turbines, creating a shadow zone. Measurements can be made in the shadow zone to provide a real-time estimate of the non-turbine sounds. This approach assumes that all environmental sounds and turbine noise are measured by the line-of-sight primary location and only the environmental sounds are measured at the secondary shadow location, which is a fair assumption above low frequencies. While not perfect, this approach does provide a quick way to view estimates of the turbine-only noise level in real-time or in large sets of data without further processing.

During the summer months, insect and bird related sounds can contaminate wind farm noise data. To combat this challenge faced in all environmental sound studies, the Ai-Weighted Sound Level (dBAi) was proposed. This is a proposed standard based on the A-weighted sound level but limited to the 1/3-octave frequency bands from 10 Hz to 1.25 kHz. The concept behind this weighting is to provide a sound level that accounts for the response of the human ear but without the contribution of insect and bird-related sounds that may emerge during the day or night. This metric is used in environmental sound studies to filter out sounds of local insects and birds, yet keep typical sounds from distant sources such as wind turbines in the measured overall A-weighted sound levels. The rational for employing the Ai-weighted sound level is discussed in reference [6].

3.3 Infrasound

The noise regulations that are typically applicable to wind turbine operation make no reference of infrasound. However, communities and action groups often cite infrasound as a hidden contributor to the impact of wind turbines. While there are no regulations to meet, it is possible to measure infrasound during the course of normal measurements to determine infrasonic sound levels.

To measure infrasound there are common approaches. The simplest is selecting a capable sound level meter. Another approach would be to use a dedicated infrasonic sensor and measurement system. In either case a very large windscreen needs to be implemented to eliminate false infrasound readings created by airflow over the sensor.

Selecting a sound level meter that measures these low frequencies is trivial, but not a given. As an example, a typical sound level meter might only measure down to the 12.5 Hz 1/3-octave band. However, manufacturers are now producing low frequency versions, or standalone models, that can measure down to the 1 Hz band. These sound level meters are readily available but do have a higher purchase price. A significant benefit to this approach is that the infrasonic frequency levels will be archived along with all the other frequency data.

A common dedicated instrument to measure infrasound is a very sensitive differential pressure sensor known as a micro-barometer. The micro-barometer has the potential to measure much lower in frequency (~0 Hz – 10 Hz) than a typical sound level meter and is limited in high frequency response, avoiding contamination due to audible noise sources. The downside of this approach is one of integration; there does not appear to be a sound level meter equivalent using a micro-barometer. This means that data collection, processing, and storage must be done using additional hardware and software. Figure 3-1 shows the sound pressure levels measured below 1 Hz in 0.1 Hz increments versus time for a weeklong period at a wind farm site. These values were measured using a micro-barometer connected to a separate piece of data acquisition hardware and custom spectral analysis software. Also shown is the average electrical power output of the wind farm. The potential link between wind turbines and infrasound is a topic of much debate and continued research, and can be considered a part of long-term noise monitoring programs to continue this industry-wide study.

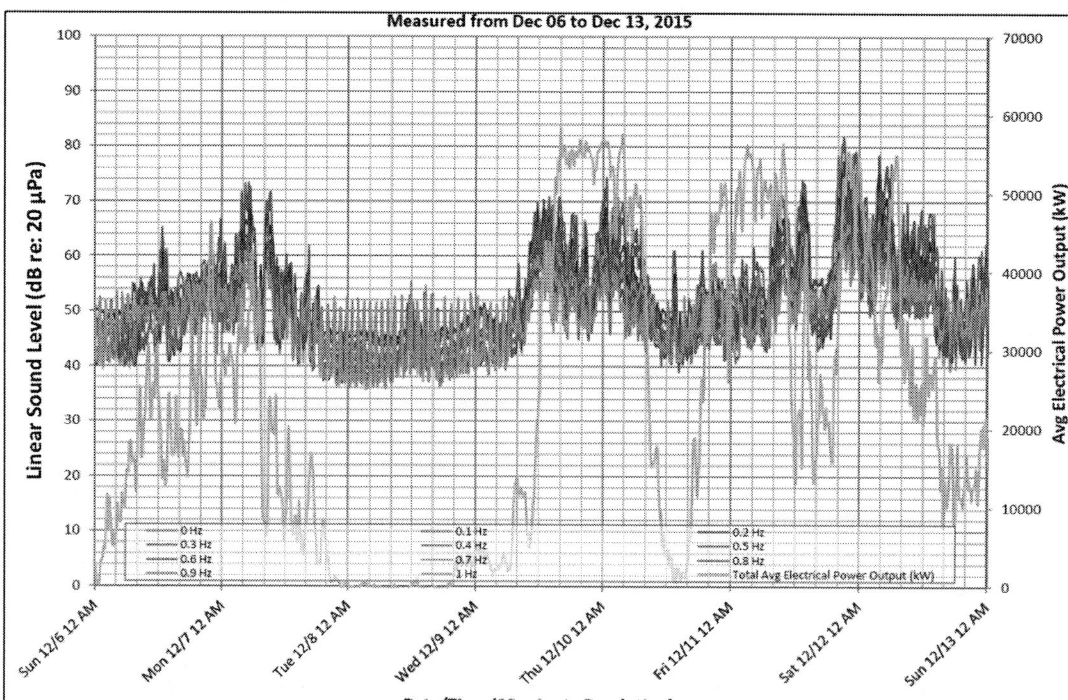

Figure 3-1: Example of Infrasound Compared to Wind Farm Electrical Generation

3.4 Amplitude Modulation

Amplitude modulation (AM) is a recurring periodic change in sound levels over time. It is typically broadband in character and occurs from interactions of the blade with the atmosphere, turbulence, directionality of the broadband sound of the blades, and/or tower interaction with the wake of the blade. Figure 3-2 shows an example of wind turbine amplitude-modulated sound with slow and fast response (100 millisecond) data collection frequencies.

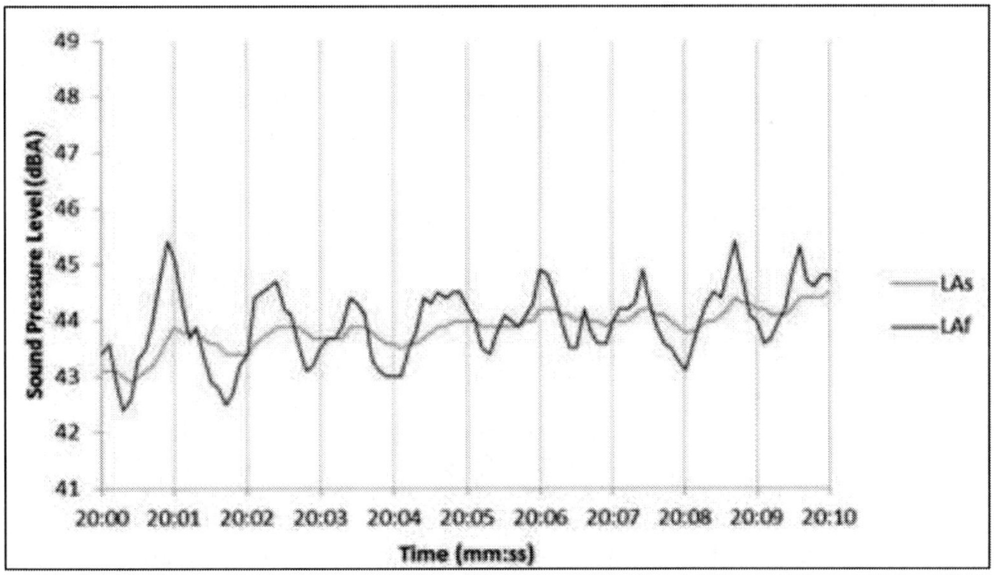

Figure 3-2: Ten Second Example of Wind Turbine AM Sound (Courtesy of MassCEC [5])

The challenge for long-term monitoring of amplitude modulations is that sound level meter data must be saved very frequently. A typical monitoring program may use a sound level meter that reports octave or 1/3-octave spectral data 10 to 20 times per second (every 50 to 100 milliseconds). Monitoring for maximum and minimum oscillations in an automated way is also a challenge, as many naturally occurring noise sources already exhibit AM.

The authors of the MassCEC study [5] developed an approach to characterizing amplitude modulation by calculating a spectrogram of the 1/3-octave sound data collected at 50 millisecond intervals. A spectrogram is a three dimensional graphic showing time on the horizontal axis, frequency on the vertical axis, and the level of the sound represented in a color scale. A Fourier transform was applied to each 1/3-octave band time history to determine the time-varying frequency content of that particular band. Figure 3-3 shows an example of this technique applied to a turbine shutdown test period for the 500 Hz 1/3-octave band. This method was applied to both the overall levels as well as to individual 1/3-octave bands. It was found that AM could be seen much more clearly in individual bands rather than in the overall level, which could be masked by ambient conditions.

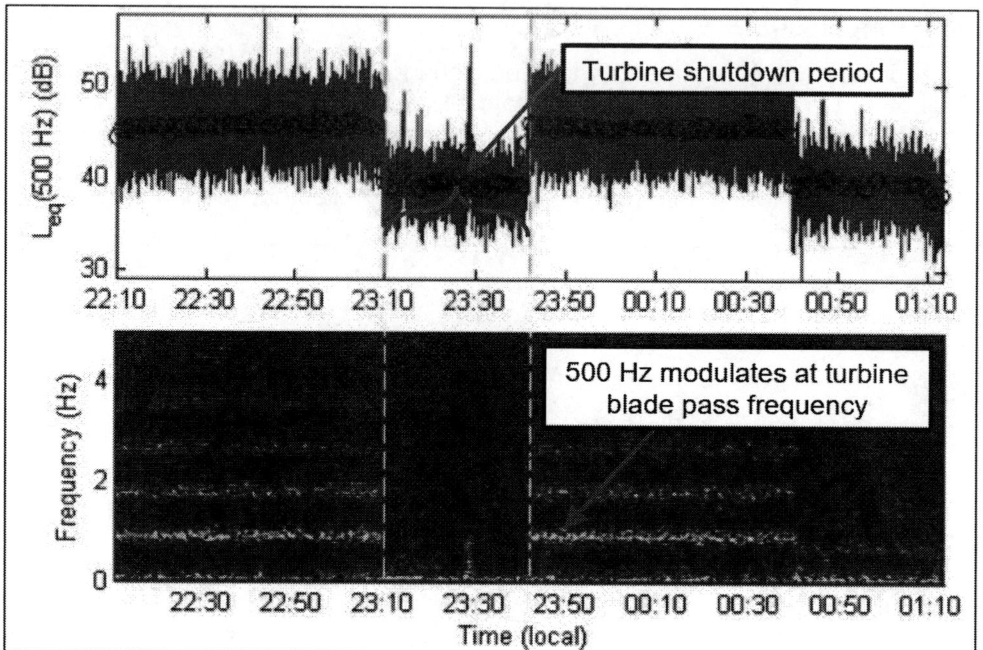

Figure 3-3: Time Series (Top) and Spectrogram (Bottom) For the 500 Hz 1/3-Octave Band During a Turbine Shutdown Test (Courtesy of MassCEC [5])

This technique is effective at isolating frequency-specific amplitude modulated sounds from background measurements. The MassCEC study found amplitude-modulated sounds in the mid-frequency range of about 250 Hz to 2 kHz, but did not find notable amplitude modulation in infrasonic, low, and high frequencies.

3.5 Tonality

Tonal conditions are evaluated for the acceptability of turbine-related sound, and are based on the prominence of the sound pressure levels at one 1/3-octave frequency band with respect to its adjacent frequency bands, including frequencies from 25 Hz to 10 kHz. For many projects, the threshold for an excessive 'tonal condition' is defined according to Annex B of ANSI S12.9-2013 Part 3. For a prominent discrete tone to be deemed excessive, the one-minute Leq sound pressure level in the 1/3-octave band of interest is required to exceed the arithmetic average of the equivalent continuous sound pressure level for the two adjacent 1/3-octave bands by a constant level difference, depending on frequency. For low frequency 1/3-octave bands (25–125 Hz), the difference is 15 dB; in middle frequency bands (160–400 Hz), the difference is 8 dB; and in high frequency bands (500–10,000 Hz), the difference is 5 dB.

Having to monitor for tonal exceedances over extended periods of time means that a great deal of data must be evaluated. This presents a challenge for efficiently communicating weeks or months of data. Figure 3-4 shows one way that has been used to quickly evaluate trends in tonal exceedances over an extended period of time. Times throughout a month-long period where a tonal condition occurred are illustrated by frequency versus time where colors indicate the level above the tonality condition.

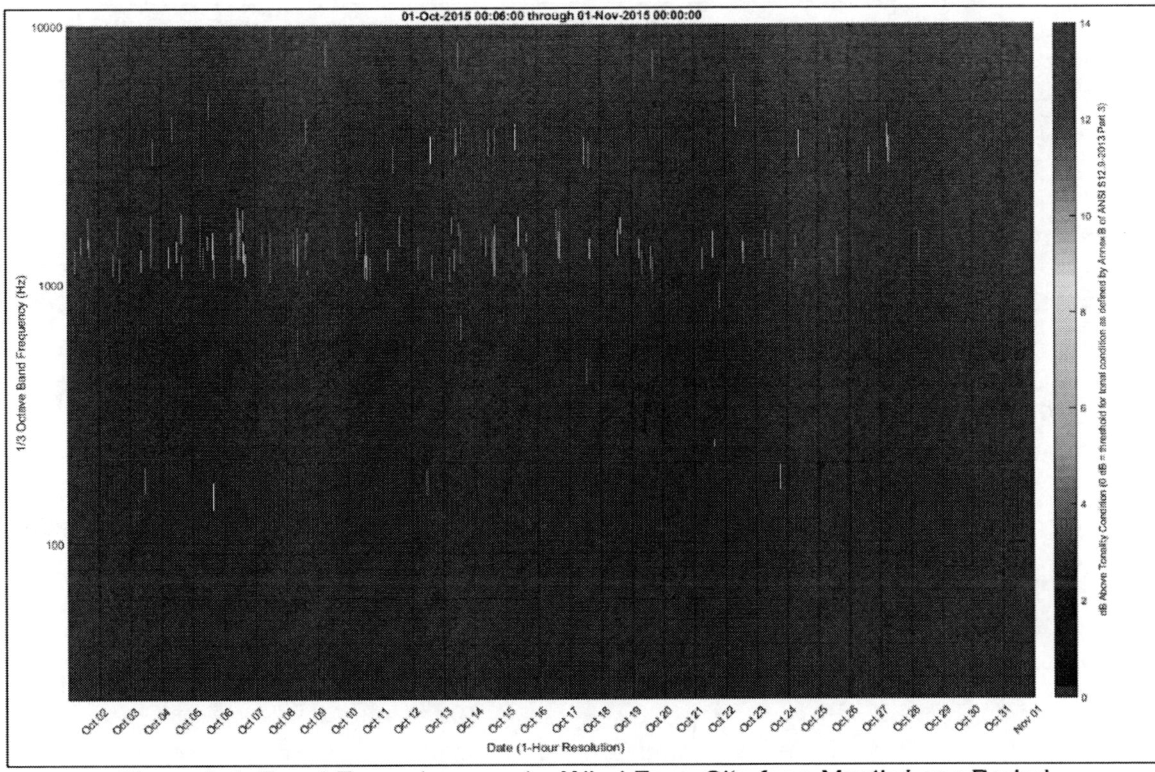

Figure 3-4: Tonal Exceedances at a Wind Farm Site for a Month-Long Period

3.6 Meteorological and Wind Turbine Output Data

Long-term sound monitoring programs often utilize local weather stations to collect meteorological data to better understand how select atmospheric factors affect sound levels at receiver locations. Typical variables measured include rain, relative humidity, temperature, wind speed, and wind direction data. There are a number of commercially available meteorological systems that can be implemented into the sound monitoring diagram in Figure 2-1. A more sophisticated device called a Light Detecting and Ranging (LIDAR) meteorology system is being used in some studies to measure wind speeds and profiles at various heights.

Wind data from anemometers mounted on the nacelles of turbines can also be supplied by the wind farm supervisory control and data acquisition (SCADA) systems to supplement the meteorological data sets. As seen in Figure 3-1 the wind farm SCADA output can also provide other important information such as when and how the turbines were operating and how much energy is being produced at any given moment in time.

Figure 3-5 shows a plot of the hourly average wind speed measured at a turbine nacelle with the corresponding hourly overall Ai-weighted Leq levels. For this project, additional information from the local meteorological station and the wind farm SCADA allowed for categorizing the wind conditions relative to the turbine and microphone. The sound levels are color coded for the downwind, upwind, and crosswind conditions. Also recall that the Ai-weighted number only considers frequency bands between 10 Hz and 1.25 kHz in an effort to eliminate insect and bird noise contamination.

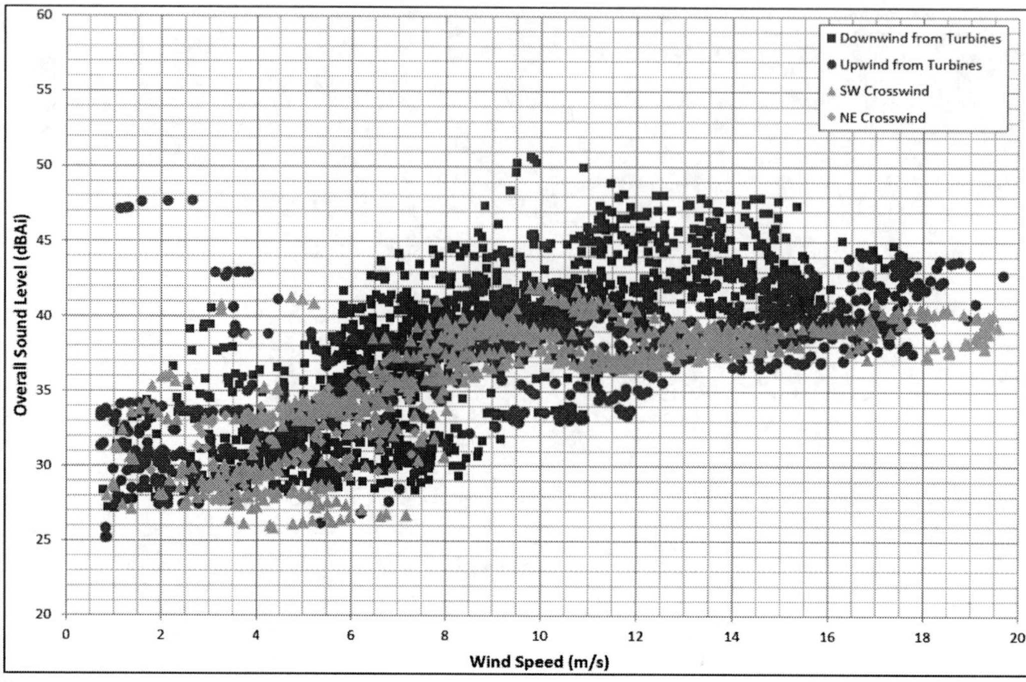

Figure 3-5: One Month of Overall Ai-weighted 1-Hour Leq vs Average Hub Wind Speed

A multitude of atmospheric factors and wind farm operating conditions can be compared to sound levels at receiver locations. A comprehensive sound monitoring solution should be able to collect compatible data sets from all of the local meteorological stations and windfarm operational outputs.

4. Sharing of Information

A wind farm's obligation to conduct noise compliance monitoring is an increasingly significant aspect of its acceptance by a community. The continuous noise monitoring of airports around many cities is an example of successful real-time public sharing of noise data. However, noise from wind farms at neighboring residential properties can often occur at levels close to or lower than the ambient sounds generated by sources such as traffic, birds, rain, blowing leaves and farming activities. That is why there are numerous techniques to isolate the wind turbine sound, but most of these rely on offline data analysis.

An example of a wind farm real-time noise monitoring website is shown in Figure 4-1. For all the reasons outlined in this paper, simply showing overall levels and current spectra would not be beneficial for public viewing, as any sounds above the criteria would not be properly vetted as being solely from the wind farm. For now, some ongoing and long-term wind farm noise studies have made their raw data available to the public, thus allowing others to provide comments or perform their own analyses. Opponents of wind farms may hire their own consultants to review times when complaints have been issued. Having transparent access to the monitoring data may buy goodwill with neighbors to show that their concerns are not being ignored. It can also be used as a means for enforcing operational changes to the wind farm so that noise limits are not exceeded.

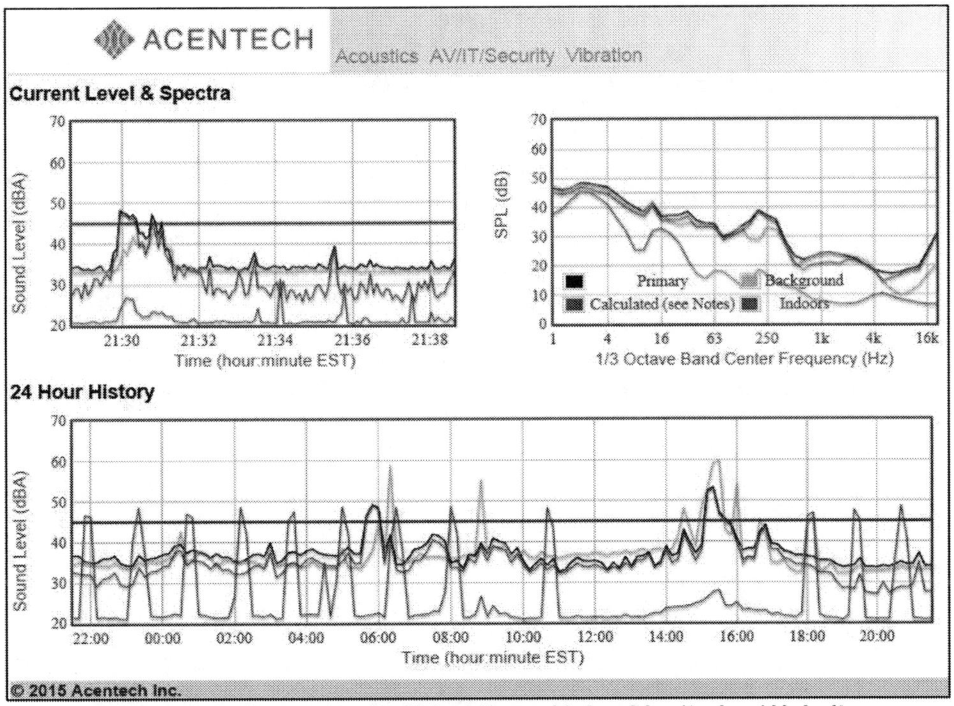

Figure 4-1: Example of a Wind Farm Noise Monitoring Website

5. Conclusions

Turbine noise continues to be a contentious issue for the development and operation of some wind farms. Long-term sound monitoring provides a clear benefit for a project to address the concerns of members in a community that may be upset or worried about noise. The ability to provide as much objective information as possible to compare the ongoing wind farm noise levels to regulations is a valuable tool. Being able to measure and discuss other aspects of wind turbine noise that are not part of most regulations is also important since many detractors focus on these phenomena, and their connection to adverse health effects is still under study.

As has been discussed, there are many challenges faced in long-term sound monitoring of wind farm sites. These include the stress on hardware due to extreme environmental conditions, the sometimes difficult locations for monitoring compliance, along with the multitude of acoustical, meteorological, and wind farm operational data to bring all together. Once all of these initial challenges are met the task of determining whether the noise produced by the wind farm is compliant with regulations is another hurdle to complete. Protocols for this determination should be agreed upon so that data can be shared with interested parties and so that conclusions from the raw data can be corroborated by secondary parties if desired. This sharing of information can help to strengthen the credibility of final conclusions to a skeptical audience. It can also help to advance the understanding and knowledge of wind turbine acoustics and shape how noise regulations are crafted for future wind farms.

References

[1] http://www.nrel.gov/gis/wind.html

[2] Barnes, et. al. (2007) A Variety of Wind Turbine Noise Regulations in the United States – 2007. 2nd International Meeting on Wind Turbine Noise.

[3] Barnes, et. al. (2013) *State of wind turbine developments in northeastern USA – 2013.*

5th International Conference on Wind Turbine Noise.

[4] Delaire, et. al. (2015) *Research into a continuous wind farm noise monitoring system.* 6th International Conference on Wind Turbine Noise.

[5] RSG et al., "Massachusetts Study on Wind Turbine Acoustics," Massachusetts Clean Energy Center and M achusetts Department of Environmental Protection, 2016.

[6] Schomer, et. al. (2010) *Proposed 'Ai'-Weighting; a weighting to remove insect noise from A-weighted field measurements* Proceedings of Inter-Noise 2010, Lisbon, Portugal

[7] Stigwood, et. al.(2015) *Cotton Wind Farm – Long term community noise monitoring project – 2 years on.* 6th International Conference on Wind Turbine Noise.

7th International Conference
on
Wind Turbine Noise
Rotterdam – 2nd to 5th May 2017

Tonal Noise Mitigation on Wind Turbines

Brett Marmo, Xi Engineering Consultants, brettmarmo@xiengineering.com
Jutta Stauber, Xi Engineering Consultants, juttastauber@xiengineering.com
Donald Black, Xi Engineering Consultants, donaldblack@xiengineering.com
Mark-Paul Buckingham, Xi Engineering Consultants, mp@xiengineering.com

Abstract

Problematic tonal noise from wind turbines can be caused by frequency matching between gear meshing and tower resonances. A mitigation technique is to increase the damping of a wind turbine tower thereby reducing the amplification and radiation of tonal noise. Acoustic-structural interaction models using finite element methods have been used to examine the effectiveness of two mitigation strategies; free layer damping and advance particle damping. The models show that both techniques effectively mitigate tonal noise and that engineering constraints and cost benefit analysis should be used to select which technique should be selected for a given turbine. Models were validated with a field test of free layer damping that showed that the audibility of tonal noise is significantly reduced with a commensurate reduction in broadband noise of 3 dB.

1. Introduction

Tonal noise emission by onshore wind turbines can adversely impact neighbouring residential communities leading to loss of sleep, stress and related health problems [1]. For these reasons tonal noise incurs strict regulatory penalties[2-4] which may include running speed restrictions, night time curtailment and, in some cases, the complete closure of wind turbines leading to considerable financial losses. Wind turbine manufacturers and operators are therefore highly motivated to mitigate tonal noise that is emitted by any turbines within their fleet.

Tonal noise emission is commonly associated with noise radiation from the tubular steel towers. Wind turbine towers can become modal if matched closely in frequency with the excitation associated with rotating components in the drive train, such as gearboxes and generators. When these conditions are met, the modal response is greatly amplified due to the very low structural damping of the steel structure resulting in undesired audible tones. Furthermore, the steel structures have large surface areas making them very efficient at radiating tonal noise.

The model response of wind turbine towers can be reduced by increasing the structural damping. This paper investigates the effectiveness of structural damping of towers on the reduction of far field tonal noise. Given that wind turbines are commonly variable speed the rotational excitation and related tones can vary over wide frequency ranges; mitigation strategies must therefore be broadband in nature. Two broadband damping approaches are examined; free layer damping (FLD) and advanced particle damping (APD). The relative effectiveness of these two mitigation strategies is compared using finite element models.

Field tests of tonal noise mitigation were conducted on a medium sized wind turbine with a 40 m hub height. The wind turbine make and model has been withheld for commercial reasons and will henceforth be referred to as the *target turbine*. The wind turbine emitted several tones between 100 and 3500 Hz at wind speeds between 6 and 10 ms^{-1}. The wind turbine tower was treated with FLD and the reduction in far field assessed in accordance with the IEC 61400-11 standard.

2. Technical Background

2.1 Tonal Noise Produced By Wind Turbines

Noise from wind turbines has two main sources: mechanical noise associated with components in the drive train which tends to be tonal in nature, and aerodynamic noise associated with blades slicing through the air, which tends to produce a broadband frequency range [5]. Mechanical vibrations in the drive trains of wind turbines are created by imbalances of the rotating components, for example, through the teeth in the gearbox coming into contact with each other (referred to as gear meshing), or the electro-magnetic (E-M) interaction between the spinning poles and stationary stators in the generator [6]. Each of these vibration sources occurs in discrete frequency bands related to the rotation speed of each component: the vibrations and resultant noise therefore tend to be tonal. Rotational imbalances tend to occur at very low frequencies (< 20 Hz) below the audible range of human hearing. Conversely, gear meshing and E-M interactions tend to occur at low to moderate frequencies (50 Hz to 2 kHz) and are therefore most likely to produce tonal noise that impacts humans.

Drive train vibration becomes problematic when it excites resonances of large surfaces in contact with the air such as the tower, blades and nacelle walls, which the authors refer to as radiating surfaces. Gearbox and generator vibration can move through mounting systems and the drive shafts to the radiating surfaces. In instances where the discrete frequency produced by the gearbox or generator match those of resonances in the blades or tower, tonal noise is amplified and radiated. These radiating surfaces, especially tubular steel towers, tend to have very poor damping characteristics and are, therefore, readily excited by vibration and extremely effective at amplifying tonal noise.

Medium and large scale wind turbines are commonly variable speed devices where the rotor speed varies between ~5 and ~20 rpm. The rotation speed of drive train components and related vibration also vary over a wide range as the rotor changes speeds to accommodate different wind conditions. The frequency of discrete mechanical vibrations will also vary and may run through one or more resonant frequencies resulting in the production of tonal noise with an intermittent nature (e.g. tonality may only occur at particular wind speeds).

2.2 Free Layer Damping

Free layer damping (FLD) involves the bonding of a sheet(s) of visco-elastic material to the substrate, in this case the sheet steel that forms the wind turbine tower. The visco-elastic material must be highly damped and dynamically stiff. The flexural motion of the vibrating substrate results in extension and compression of the visco-elastic material (Figure 1). The visco-elastic material stores strain-energy then dissipates a portion of this energy via hysteresis each vibration cycle, thereby increasing structural damping and reducing the vibration amplitude. The amount of energy dissipated increases with the thickness of the layer of visco-elastic material. FLD is a close relation of constrained layer damping, where a rigid layer of material is fixed to the free surface of the visco-elastic material and energy is dissipated by forcing the damping layer to shear (as opposed to extend and contract).

Free layer damping is well understood technique for increasing structural damping and it's therefore often favoured by vibration and acoustic engineers. It is effective at suppressing resonant motion making it appropriate for wind turbine towers where tonal noise is amplified and radiated by modal behaviour. Retrofitting FLD to existing wind turbines however can be challenging as it requires rope access engineers working in confined spaces with chemical adhesives.

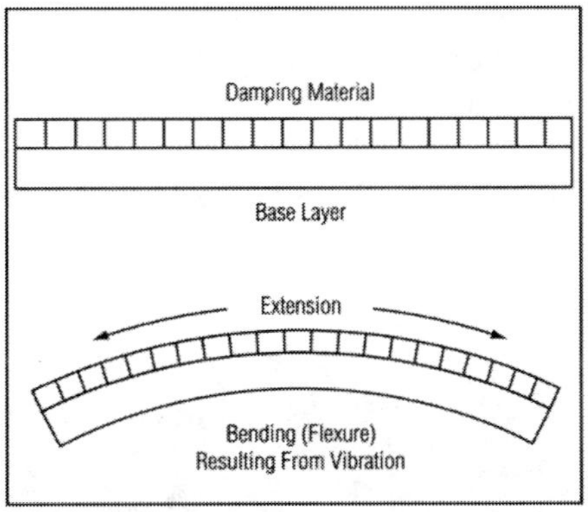

Figure 1. Free layer damping involves attaching a damping material to a base layer (substrate). When the base layer bends during a vibration cycle the damping material is forced to extend and dissipated energy via hysteresis (after [7]).

2.3 APD pods

Advanced particle damping (APD) media is a custom granulated elastomer of various sizes and materials. This media can be housed in a soft-shell or hard-shell container (Figure 2) and applied to a vibrating structure for the purpose of damping structural vibration modes. Prior work [8] to characterize APD has shown that altering the makeup of APD can result in a broadband damping performance. Various APD materials are used having different stiffness properties. The granules have a rough and irregular construction and finish, and granulated sizes are randomized.

APD is effective through two types of damping characteristics. At lower levels of input excitation, the inherent material damping properties are the dominant mechanism for attenuation. At higher input levels, the particle interactions become more dominant as they begin to move relative to one another. This results in a media that is not only effective at various levels of excitation, but also across a broad frequency spectrum.

Since APD is manufactured with elastomeric materials, the inherent structural properties, such as damping ratio and Young's modulus, are compounded to optimize performance. Materials with high levels of damping will absorb higher energy than that of lower damped compounds. Caution needs to be exercised for adding too much damping in the elastomeric materials since there is also a desire to create motion and interaction between particles. Various elastomer hardness, or stiffness's, are used for the purpose of creating individual spring elements within the APD mixture. This results in a randomized body of tuned mass dampers, which will respond over a wide frequency range.

The APD granules are manufactured with rough and textured surfaces. The interaction between particles with rough surfaces creates higher damping levels than that of typical spherical bead damping media with smooth surface conditions. The irregular shapes of the APD media also result in higher number of contact points and interacting surface areas between particles than spherical media. A container of spherical particles can only make 'point' contacts with adjacent spheres, as opposed to ADP's irregular particles which can make 'surface area' contacts in addition to 'point' contacts. Once more, the number of contact points between spherical elements is fixed where APD has the possibility of increased contact points with adjacent particles. The effectiveness of APD is related to the attached structures mass and mobility, which will drive the amount of APD needed to achieve the desired damping. Structures with low mass and high mobility requires less APD media than heavy structures with low mobility.

In wind turbines the APD material is contained within steel containers (referred to as pods) which can be magnetically attached to the inside of the wind turbine wall. The magnetic attachment circumvents the need for chemical adhesives and their application in the confined space of the tower. APD is a novel mitigation strategy resulting in risks and engineering challenges associated with any new technology.

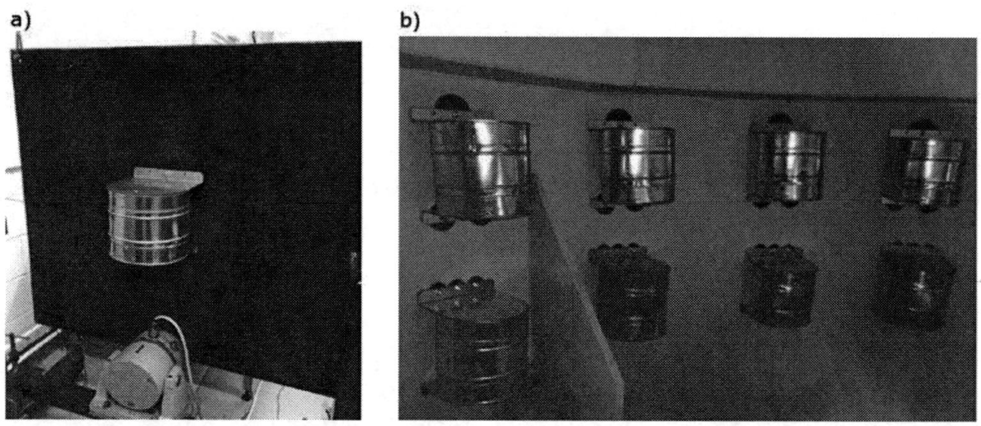

Figure 2. Pods containing advanced particle damping material. a) Experimental set up where an APD pod was magnetically attached to a piece of sheet steel which was excited using a probe shaker (yellow device at bottom of photo). b) APD pods magnetically attached to a wind turbine tower.

3. Finite Element Analysis

A fully-couple structural-acoustic interaction model was created of the target wind turbine using the commercially available software package COMSOL Multiphysics. The unmitigated wind turbine, referred to as the *Native turbine*, was modelled using solid elements to represent the drive train and blades and shell elements to model the tower. The effect of FLD and APD on tonality were modelled by modifying the properties of shell elements that represent the tower of the Native model. Experimental data was used to determine appropriate material properties for FLD and APD attached to the steel tower walls.

3.1 Free Layer Damping

The modelling approach required that the FLD solution is incorporated as a shell element in the solution models of the full turbine (i.e. the sheet steel in the native model is replaced with shell material that represents a composite of sheet steel and damping layer). Three dimensional models were constructed of steel sheets that were 0.9m square and thickness varying between 8 and 14 mm.

Shell models of 0.9 m square plates with appropriate thicknesses and densities to represent the three dimensional plates with polymer and constrained layer were then parameterised for Young's modules and loss factor damping. Results from the shell elements were then compared to the three dimensional models and the best fit results used to construct lookup tables showing appropriate material properties for any given sheet steel thickness with the FLD attached.

3.2 APD Pods

Structural damping characteristics of APD pods on sheet steel were determined using lab-base experiments [9]. A flat 1 m × 1 m industrial steel plate, which was 12 mm thick, was suspended (Figure 2). A force was applied to the plate by a shaker whose output frequency was both stepped from 80 Hz to 600 Hz in 2 Hz increments and applied as white noise. The normal surface acceleration was measured at 12 evenly distributed sensor locations. In order to determine the effect of placing APD pods on the steel plate experiments were run, where a single APD pod was placed in the centre of the plate. The spatially averaged RMS (root mean square) acceleration was obtained by taking the arithmetic average of the RMS acceleration measured at the 12 evenly distributed measuring locations (Figure 3).

The modelling complexity of the mitigated wind turbine model was minimised by representing the APD pods as shell elements. The experimental set up described above was replicated in a COMSOL Multiphysics model that was calibrated by modifying the dynamic and damping properties of the shell elements that represent APD pods (see reference 8). The calibrated properties of the APD pods were then used in the mitigated wind turbine models.

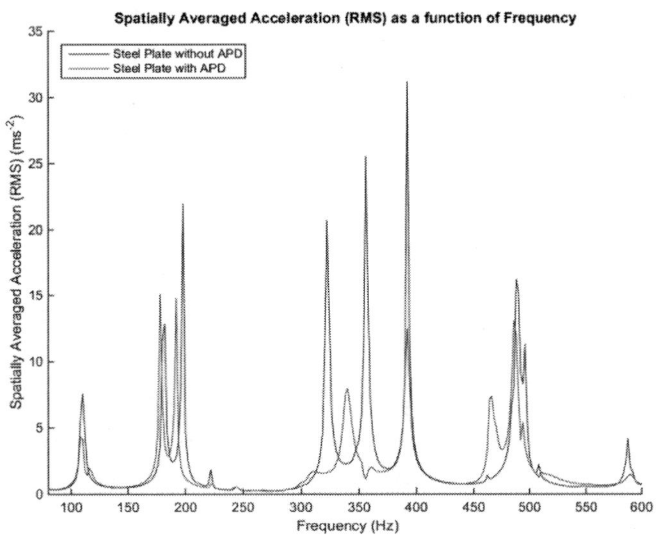

Figure 3. Spatially averaged RMS acceleration as a function of frequency of the plate without and with the APD pod installed as shown in Figure 2. The installation of the APD pod reduces vibrations of the steel plate over a wide range or frequencies, from approximately 100 Hz to 600 Hz.

3.3 Native wind turbine model

The modelled geometry, mass of components and blade stiffness of the target wind turbine was based on data sheets and engineering drawing supplied by the manufacturer (Figure 4a). The drivetrain of the target wind turbine has a two-stage gearbox which was identified as the source of tonal noise using contemporaneous vibration and acoustic surveys. The gear meshing

frequencies of the first step-up stage was 64 Hz and the second step-up stage was 350 Hz. Acoustic surveys of the target turbine found a very prominent tone at 350 Hz. Vibration path analysis indicated that the second stage gear meshing was the source of this tone [9].
The models were excited in the frequency domain between 50 and 750 Hz using a force function applied to the gearbox that represents gear meshing at each step-up stage and their harmonics (Figure 5). The force function was calibrated using a vibration data from accelerometers mounted on the torque arms of the gearbox when the turbine was operating in wind speed of 9 ms^{-1} (Figure 5). The wind turbine was surrounded with an acoustic domain that was coupled to the tower wall surface and the near-field sound field model. A far-field analyser was used to extrapolate the sound field to a position 48 m down-wind that correlates with the measurement position used in IEC 61400 acoustic surveys.

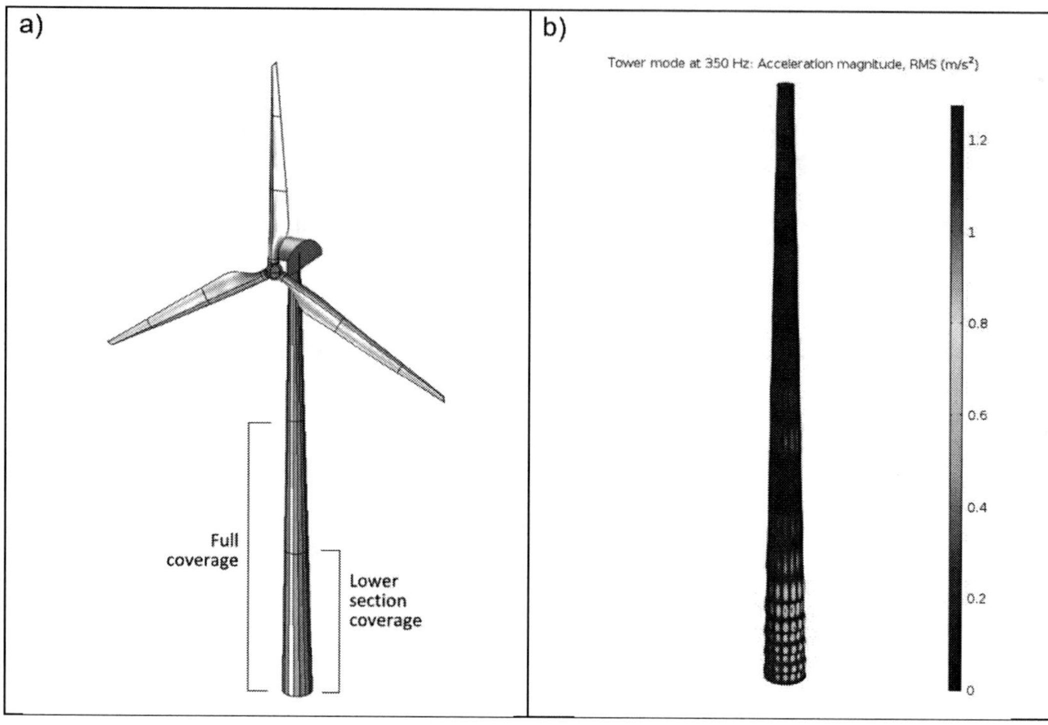

Figure 4. a) Geometry used to model the target turbine. Two strategies involving FLD were modelled; 1) Full coverage and 2) Lower section coverage. APD pods were modelled covering the full coverage area. The density of APD pods covering this area were varied with a model contained 88 APD pods and another model contained 176 pods. b) Modal shape of the tower at 350 Hz which is association with a prominent tonality.

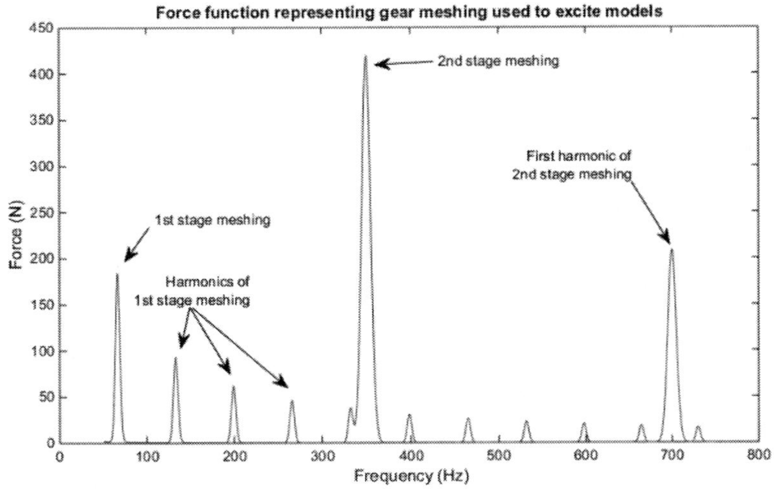

Figure 5. Force function applied to the gearbox and used to excite the models in the frequency domain. The gearbox has two step-up stages at 64 Hz and 350 Hz. The topology of the function was calibrated using gearbox vibration data.

3.4 Mitigation wind turbine models

The wind turbine model was used to determine how different vibration mitigation strategies using FLD and APD affected far-field tonal noise levels. The effect of FLD and APD was implemented in the models using dynamic and damping properties based on experimental data performed on sheet steel with thicknesses comparable to those used in the target wind turbine tower.

Vibration surveys and modelling of the native turbine showed that the model shape associated with the prominent to at 350 Hz was localised in the lower section of the tower (Figure 4b). Mitigation strategies were, therefore, designed to increase the structural damping of the lower and middle tower sections. Two FLD coverage were examined, the first where the lower section of the target turbine was covered and a second were the lower and middle section of the turbine were covered (referred to as full coverage). APD pods were modelled attached to the full coverage area. Two APD coverage strategies were used; one where 88 APD pods covered the full coverage area of the tower and a second where 176 APD pods were used to cover the same area.

Far-field modelled sound pressure levels 48 m downwind from the turbine was added to the background noise measured at the equivalent position in the field to produce realistic sound spectra (Figure 6). The Native turbine has prominent spectral peaks at 64, 350 and 700 Hz. The tonal levels of these peaks are significantly reduced in all four mitigation models. The largest reduction was achieved when 176 APD pods were applied to the full coverage area. The tonality algorithm outlined in IEC 61400 was applied to the modelled spectra and the 350 Hz peak was identified as being a reportable tone for the Native turbine with an audibility of 5.8 dB (Table 1). The audibility of the 350 Hz tone was significantly reduced for each of the mitigation strategies with tone for 176 APD reduced below 0 dB to a level that is defined as inaudible.

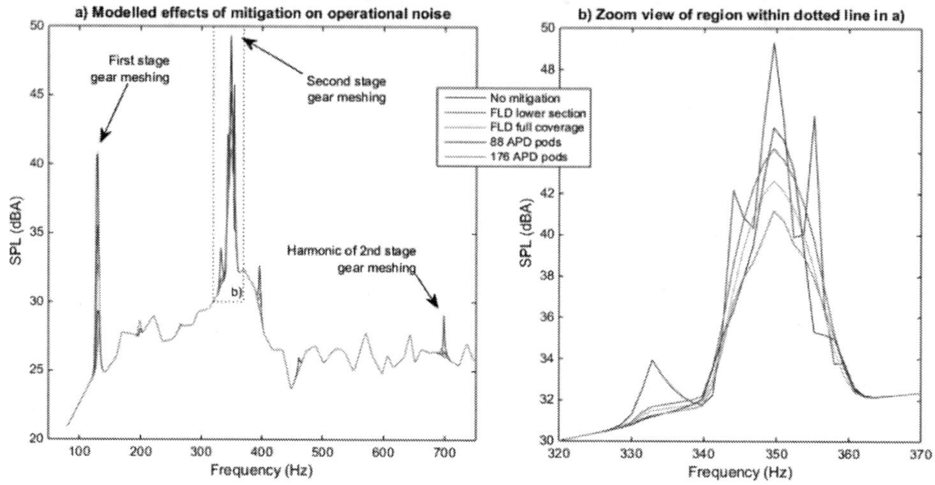

Figure 6. a) Modelled far-field sound 48 m downwind from the Native unmitigated turbine compared to the two FLD and two APD mitigation strategies. The spectra were produced by combining modelled operational noise with background noise level measured in accordance with IEC 61400. b) Close up of the affect of mitigation on the 350 Hz tone associated with the second stage gear meshing.

350 Hz at 9 ms^{-1} wind speed	Model results		Field measurements	
	Tonal Level dB	Audibility dB	Tonal Level	Audibility
Native	49.3	5.8	49.9	6.9
FLD Lower Section	44.2	3.8		
FLD Full Coverage	42.6	2.2	38.2	4.0
88 APD pods	45.2	3.6		
176 APD pods	41.2	-0.6		

Table 1. Modelled and measured tonal level of the 350 Hz tone emitted by the target turbine. The audibility of the modelled acoustic output was calculated by applying the algorithm outlined in IEC 61400 to the modelled spectra shown in Figure 6. Field measurements were conducted on a Native turbine and a turbine with FLD covering middle and lower sections (full coverage). Tonal with audibility less than 0 dB are defined as inaudible.

4. Field Test of FLD

Free layer damping was applied to the target turbine at a site in the United Kingdom to determine the effectiveness of structural damping on mitigating far field tonal noise. FLD tiles were applied to the middle and lower section of the tower (see Figure 4 Full Coverage). An acoustic measurement was conducted and the resulting data processed in accordance with IEC 61400.

The audibility of the tonal noise before and after the installation of the damping tiles was significantly reduced after the installation. An example of a comparison between the audibility of tones for the native and mitigated turbine is shown in Figure 7 for the 9 m/s wind speed bin. The black dashed lines denote the audible tone threshold (0 dB) and the threshold for when to report a tone (-3 dB). The most audible tone, with an audibility of approx. 6.9 dB for the Native turbine, at approximately 350 Hz is reduced to 4.0 dB. The other audible tone, at approximately 700 Hz, is also significantly reduced by 5.5 dB. The application of the damping tiles, not only reduced the audibility of the 350 Hz tone by 2.9 dB and the tonal level by 11 dB (Table 1), but also reduced all other tones in the 50 to 4000 Hz range to inaudible levels (Figure 7). Furthermore, the tones at 3150 Hz and 3300 Hz detected for the native turbine were not detectable when the mitigated turbine was measured (e.g. Figure 7). Note, although not shown here, the audibility of most prominent tone at 350 Hz was reduced by approximately 3 dB across all wind speed bins.

The reduction in tonal noise audibility was accompanied by a commensurate reduction in broadband noise. The broadband sound power level of the mitigated turbine was reduced by ~3 dB across all wind speed bins measured (Figure 8).

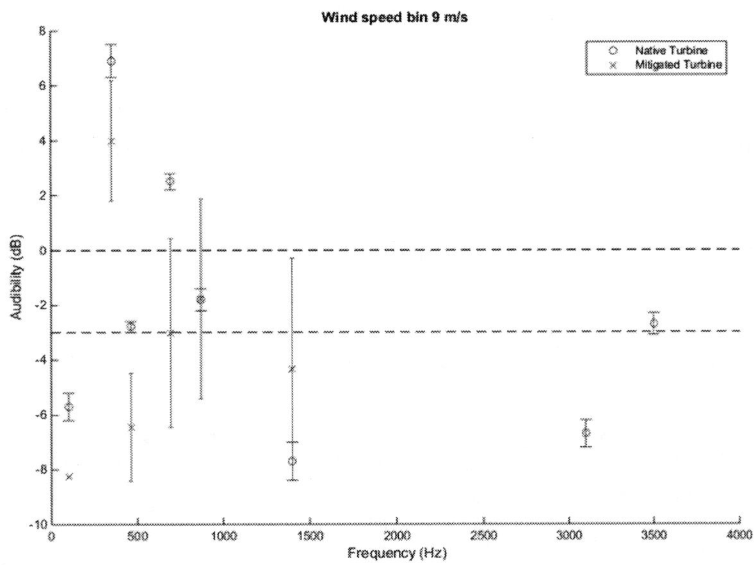

Figure 7. Tonal audibility as a function of frequency for the 9m/s wind speed bin.

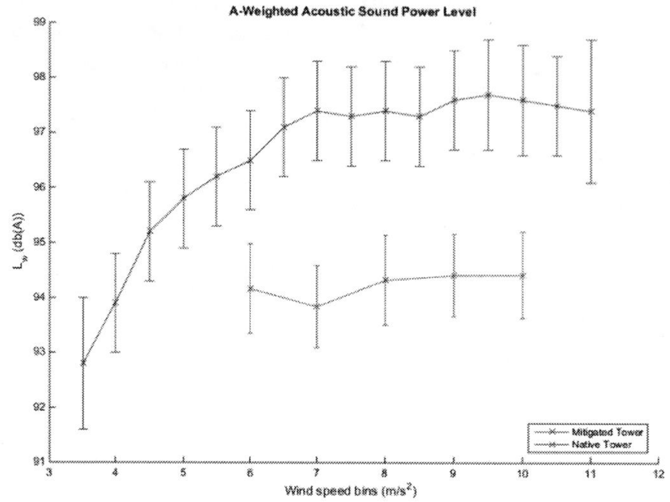

Figure 8. A-weighted sound power level as a function of wind speed for the mitigated and native turbine.

5. Discussion

The modelling approaches presented include assumptions and simplifications that affect the modelled sound field. The acoustic domain was limited to the volume surround the tower wall to limit the mesh size and make convergence of the model possible. Thus the acoustic domain was coupled to the tower only, and any radiation of tonal noise from the blades was not included in the far-field analysis resulting in the possible under estimation of modelled tonal levels. The authors feel the assumption is justified based on previous vibration surveys of the target wind turbine that showed the principal radiation source of the 350 Hz tone was the tower. The model was excited using gearbox meshing only. It is likely that other mechanical vibrations in components such as the generator and bearing would also contribute to the sound field. However, the vibration survey demonstrated that the gearbox was the principal vibration source for tonal noise in the target turbine.

The modelled far-field sound from the Native turbine and the turbine mitigated using full coverage of FLD are in good agreement with similar tonal levels at 250 Hz and reductions in tonal audibility (Table 1). The spectral position and prominence of peaks at 64, 350, 400 and 700 Hz are also in good agreement. The measured ~3 dB reduction in broadband noise will have contributions from the reduction in tones at frequencies greater than 750 Hz (e.g. 1200, 3150 and 3500 Hz see Figure 7). Given that the model analysed frequencies in the 50 to 750 Hz band, it is not possible to conduct an equivalent broadband sound pressure analysis to compare with the broadband measured level. At the time of writing field test of APD pods on a wind turbine tower were commencing; thus comparison of empirical data and APD model results are imminent but cannot be presented here.

The field test demonstrated that structural damping of a wind turbine tower can reduce far field tonal noise levels. The models presented here demonstrate how different damping strategies and their design can be compared with respect to effectiveness of tonal noise mitigation. The application of FLD and APD pods to wind turbine towers reduced tonal noise levels and by increasing the coverage area of the FLD or the number of APD pods the tonal levels were reduced further (Figure 6). It should be noted that comparing full coverage of FLD to 176 APD pods based only on the resultant tonal levels is somewhat meaningless in a commercial and engineering respect. Rather the modelling approach detail should be used in concert with consideration of engineering (geometric constraints, safety on installation, etc.), and commercial factors such as material and installation to allow a cost benefit analysis. The modelling approach also allows optimisation of design by providing a comparison of far field noise for different FLD and APD pod coverage strategies. Cost benefit analysis and design optimisation can provide wind turbine operators with the most favourable tonal noise solutions.

6. Conclusion

Structural damping of wind turbine towers resulted in significant reduction in tonal noise emissions. A field test of free layer damping covering the middle and lower section resulted in reduction in tonal level of the dominant tone at 350 Hz by 11 dB and a reduction of audibility of the tone by 3 dB. All other tones in between 30 and 4000 Hz were made inaudible and there was a commensurate reduction in broadband noise of 3 dB. The field test validated far-field tonal noise levels calculated using a structural-acoustic models of the Native turbine and the turbine modified with FLD. The model was used to show that structural damping with both FLD and advance particle damping are effective methods for mitigating tonal noise radiated by wind turbine towers. The tonal levels from the modelling approach presented can be combined with engineering and commercial factors to design tone mitigations appropriate for specific wind turbines.

References

1. Pedersen E. (2009) *Effects of wind turbines of humans*. Third international meeting on wind turbine noise; 17-19 June 2009; Aalborg, Denmark 2009.
2. ETSU-R-97. (1996). *The assessment and rating of noise from wind farms.*
3. International Standard IEC 61400-11(2012), *Wind turbine generator system - Part 11: Acoustic noise measurement techniques.*
4. British Standards Institution. (2014) *Method for rating and assessing industrial and commercial sound. BS 4142.*
5. Burton, T, Sharpe D, Jenkins, N and Bossanyi, E. (2001). *Wind Energy Handbook.* Chichester, England: John Wiley & Sons Ltd.
6. Marmo, BA, and Carruthers, BJ. (2010) *Modelling and analysis of acoustic emissions and structural vibration in a wind turbine.* COMSOL Conference Proceedings, Paris, France, 2010.
7. Renninger, J. *Understanding Damping Techniques for Noise and Vibration Control.* E-A-R Specialty Composites, Indianapolis, Indiana.
8. Ott, M, Weisbeck, J, Gerges, SNY and Bustamantes M. (2014). *The effectiveness of particle damping for use on vertical surfaces.* Proc INTER-NOISE 2014; 16-19 November 2014; Melbourne, Australia 2014.
9. Stauber, J, Marmo, B, Ott, M, Snaith, M and Puff, P. (2016) *Wind turbine noise attenuation using modal structural damping.* Proc INTER-NOISE 2016; 21-24 August 2016; Hamburg, Germany 2016.

7th International Conference
on
Wind Turbine Noise
Rotterdam – 2nd to 5th May 2017

Origin, Transfer and Reduction of Structure-Borne Noise in Wind Turbines

Dr. Lukas Schneider, ESM Energie- und Schwingungstechnik Mitsch GmbH, Rimbach: Lukas.Schneider@esm-gmbh.de
Dr. Karl-Heinz Hanus, ESM Energie- und Schwingungstechnik Mitsch GmbH, Rimbach: Karl-Heinz.Hanus@esm-gmbh.de

Summary

While producing electricity, wind turbines emit noise that is mainly generated by the aerodynamics of blades, by the meshing of gear-wheels in gearboxes and by interaction of poles in generators. The focus of this contribution is on the structure-borne noise coming from the gearbox and the generator. Depending on the eigenmodes of a wind turbine and the points of force-transmission into the structure, excitation forces can lead to amplified small-band noise in the surrounding of wind turbines. The amplification depends on drive-train eigenmodes and the properties and behaviour of the large scale surfaces, i.e. tower, blades and nacelle. It is highly desired to prevent the emission of amplified structure vibrations into the surrounding. There are several approaches to circumvent tonal issues. A systematic classification of relevant counter measures is: 1. Reduction of excitation forces, 2. Isolation of excitation parts, 3. Isolation of emitting surfaces, 4. Detuning of eigenmodes, 5. Damping of eigenmodes and 6. Introduction of counter forces. This classification is true for basically all machinery. In this contribution, the listed points will be specialised to the case of wind turbines using real life examples such as elastomer couplings in the drive train (isolation of excitation), decoupling of nacelle covers (isolation of emitting surfaces) and changing of suspension properties (detuning of eigenmodes). In some cases retrofit solutions are necessary to solve tonal issues in the field. For low damped eigenmodes the installation of passive tuned mass dampers is a convenient solution, because damping can easily be introduced into the system and vibration amplification can be reduced. In case eigenmodes have higher damping or eigenmodes are not accessible, adaptive or active vibration absorber are applied. They are installed at the point of excitation, i.e. at the gear box or the generator, and they cancel out the excitation forces by introducing counter forces. Various examples of tuned mass dampers and vibration absorbers will be presented.

1. Introduction

The requirements on noise behaviour of wind turbines increase because of changing standardization, competition between manufacturer and political pressure. Therefore, it is important to understand the generation of noise, to generate countermeasures and to avoid critical designs. ESM Energie- und Schwingungstechnik Mitsch GmbH develops, tests and manufactures products that are relevant for the reduction of vibrations in wind turbines. These include

- elastomer supports for gearboxes, generators, cabinets, conductor rails and nacelle covers,
- couplings for the low speed shaft
- tuned mass dampers for low-frequency tower-vibrations and

- mass dampers and vibration absorbers for sound-related vibrations.

Furthermore, ESM supports manufacturer of wind turbines in optimizing the sound-behaviour of wind turbines. To do so, ESM performs measurements on turbines and test rigs, develops customized products and manufactures these innovative products.

2. Origin of vibration causing noise

Under normal operation, noise of wind turbines is caused by airflow around the blades, by meshing of gear-wheels in gearboxes or by interaction of poles and slots in generators. Airborne noise has a broadband frequency spectrum whereas structure-borne noise caused by gearboxes and generators is composed of a few narrowband tones. Additional noise sources can be specific events like turning of the tower-head or activities of fans, pumps or brakes. Also damaged or defective elements can cause a higher noise level. Below, only noise related to normal operation of wind turbines is described.

2.1 Noise emission caused by air flow

There are different mechanisms of noise generation by airflow around the blades. The typical broadband sound spectrum of blades is generated through the interaction of turbulent flow with the trailing edge of blades, the blade tip and front edge during high wind.
Also undesired flow separation due to malposition of blades can cause undesired sound. Low frequency noise is generated by interaction of wake turbulences with the tower during tower-passage of the blades. High frequency noise, on the other hand, is generated when laminar flows propagate to the trailing edge or when there are defects in the blade. Below, airborne noise is not further discussed. More information can be found in reference [1].

2.2 Excitation through gear-meshing

Gearboxes in wind turbines are made of several planetary or spur gearings. Meshing of gears can be regarded as small shocks causing the vibration in surrounding structure. The excitation frequencies of these vibrations are the result of rotor speed and the number of teeth in the individual gears. Because most of the wind turbines have variable rotor speeds, the excitation frequency changes with speed. Due to the impulsive character of the excitation, the multiples (harmonics) of the fundamental frequency are also present in the excitation spectrum. Figure 1 shows an exemplary excitation spectrum of a wind turbine with gearbox in a speed-frequency-chart. It shows that with one speed there are several narrowband excitation frequencies present.

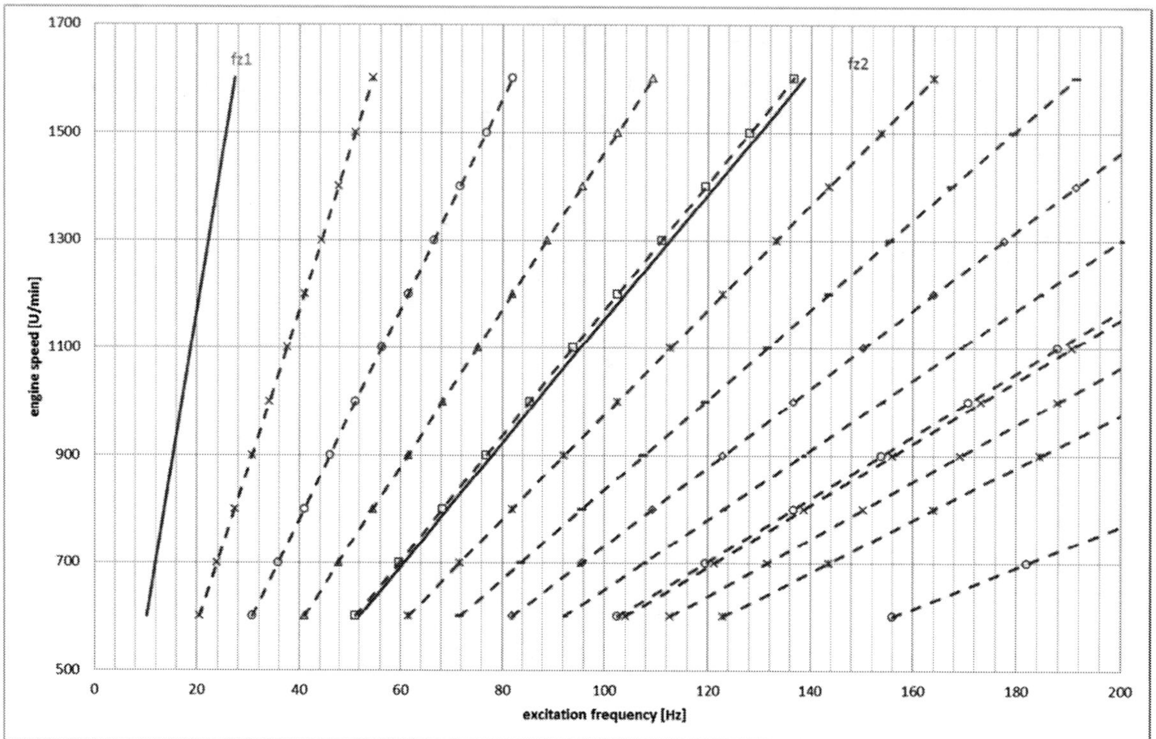

Figure 1: schematic excitation spectrum of a wind turbine out of first and second gear

The amplitude of gearbox vibrations depends on

- the macro geometry of the gearbox, i.e. number of gears, force path, gear ratio and number of teeth,
- the micro geometry of teeth, i.e. the polish of surface,
- the rotor speed and torque,
- the dislocation of teeth due to deformation of shaft and support and
- the surface quality of teeth flank

2.3 Excitation through interaction of poles and slots in generators

Generators consist of poles and slots positioned at the stator and rotor. Depending on the generator concept, the electrical excitation field is either generated at the rotor or stator side. The magnetic field in the air gap is unequally distributed because of the inhomogeneity in magnetic field between the poles. That's why with every crossing poles and slots there are impulsive forces in tangential, rotational and with inclination of poles also in axial direction. The frequencies of these forces are proportional to rotor speed and number of poles. Similar to the excitation in gearboxes also the harmonics of the fundamental frequency are present. Therefore the excitation spectrum direct-drive turbines also looks similar to Figure 1. The amplitude of excitation forces depends on

- number of poles and slots,
- rotation speed and torque
- arrangement and geometry of poles,
- size and precision of air gap,
- stiffness of connected structure.

Knowing these influencing factors countermeasure can be developed to reduce the excitation forces and consequently the emitted noise of wind turbines.

3. Transmission, enhancement and emission of structure vibrations

The forces introduced in the gearbox or the generator cause vibrations in the structure. These vibrations are transmitted through the turbine until they reach the surface. The vibration of surfaces causes pressure variation in the air, which are transmitted with sound velocity through the environment. This, so-called structure-borne sound, is mostly transmitted through the large surfaces of the turbine i.e. blades and tower. Hub and nacelle are secondary because their emitting surfaces are much smaller compared to blade and tower.

3.1 Transfer of vibrations in wind turbines with gearbox

In wind turbines with gearbox the vibration can go through the following path from gearbox to outer surface:

1. Gearbox – elastomer support – mainframe – tower,
2. Gearbox – low speed shaft – main support – mainframe – tower,
3. Gearbox – low speed shaft – hub – blades.

Elastomer supports prevent the transition of vibrations above a certain frequency limit to mainframe and tower by their isolating effect. The path through the low speed shaft into the tower and the blades are predominantly metallic connections hence a good transition paths for structure vibration. The vibrations caused by the generator can go the following path:

4. Generator – elastomer support – mainframe – tower,
5. Generator – coupling – gearbox – see transmission paths above (1.-3.)

Both transmission paths of generator include soft connecting parts, which isolate the vibrations below a relatively low limiting frequency.

3.2 Transfer of vibrations in direct drive wind turbines

Most direct-drive wind turbines have metallic connections between Geno-stator and tower and between geno-rotor and blades. Additionally, vibrations can be transmitted through the rotor bearing, i.e. vibrations from the rotor can migrate to the tower and those of the stator can go to the blades. So, in most direct drive wind turbines no element is present that isolates or damps the vibration on the way to the transmitting surfaces.

3.3 Natural vibration behaviour of wind turbines

Beside the transmitting paths, the natural vibration behaviour of the wind turbine is also relevant. In case excitation frequencies agree with undamped eigenfrequencies of the wind turbine, small excitation amplitudes can cause intense vibrations. This is critical for eigenfrequencies in the drive train because oscillations are superposed on the power transmission. To understand the natural vibration behaviour of wind turbines two important results of modal analysis of linear low damped systems are quoted (see [2]):

- A coupled multibody system with N degrees of freedom has N eigenmodes, which can be represented as eigenfrequencies ω_i, modular masses m_i and modular damping d_i $(i = [1, N])$.
- The vibration answer $w(x,t)$ of this system with sinusoidal excitation consists of the sum of general eigenmodes $\phi_i(x)$, weighted with excitation $P_i(x)$ and an amplification function. For an undamped system this statement reduces to

$$w(x,t) = \phi_1(x)\frac{P_1 \sin(\Omega t)}{s_{gen1}\left(1-\left(\frac{\Omega}{\omega_1}\right)^2\right)} + \phi_2(x)\frac{P_2 \sin(\Omega t)}{s_{gen2}\left(1-\left(\frac{\Omega}{\omega_2}\right)^2\right)} + \phi_3(x)\frac{P_3 \sin(\Omega t)}{s_{gen3}\left(1-\left(\frac{\Omega}{\omega_3}\right)^2\right)} + \cdots,$$

  where Ω is the excitation frequency and $s_{gen\,i}$ the generalized stiffness of the ith eigenmode.

These results of the modal analysis allow the interpretation that a vibration with an excitation frequency Ω is a combination of the neighbouring eigenmodes. This means that the excitation forces alone cannot cause large deflection. It is also necessary that the frequency and shape of the eigenmodes of the wind turbine fit the frequency and direction of the excitation force distribution to cause large vibration on the outer surface. Therefore, it is necessary to know the eigenmodes of the wind turbine to analyse its noise behaviour.

The excitation spectrum of wind turbines with gearbox and direct driven turbines is relatively wide so that during operating a lot of eigenmodes are excited. This is true in particular for the tower and blades. These eigenmodes cause a tonal issue only if the excitation is strong enough, if the damping of the respective eigenmode is low, if the size of the vibrating outer surface area is large and if the energy transfer of the structure vibration to air pressure variations is good (radiation factor).

The eigenfrequencies of the drive-train are relevant because they can be directly excited by the gearbox or the generator. At the same time they have a rigid connection to blade and/or tower and can cause intensive vibrations there. If, in addition, the tower and the blades have eigenmodes at the same frequencies as the drive train, which is very likely, the vibrations are further animated. It is also possible that the intense resonance movement of the drive train has negative impact on the animating mechanisms.

3.4 Emission of noise from the outer surface of a wind turbine

The outer surfaces area of tower and blades are similar, whereas the surfaces of nacelle cover and hub are ten times smaller. So, the latter surfaces emit only a large part of the structure-borne sound if their vibration amplitude is large. Also, the radiation factor is relevant. Surfaces made of thin damped material radiate less than surfaces made of thick undamped materials assuming they are excited at the same frequency.
For analysing the emission of structure-borne sound it is important to know, which emitter is dominant at which frequency. A helpful physical phenomena is the Doppler Effect, i.e. a frequency shift of moving sound sources [3]. If a periodic frequency shift is observed in the sound measurement that goes with the blade rotation, it can be concluded that the sound is emitted by the blades.

4. Countermeasures to reduce structure-borne sound

The measures to reduce structure-borne sound can be classified into the following classes:

1. Reduction of excitation forces,
2. Isolation of excitation parts,
3. Isolation of emitting surfaces,
4. Detuning of eigenmodes,
5. Damping of eigenmodes and
6. Introduction of counter forces

Reducing of excitation (point 1) forces is already discussed in chapter 2.

4.1 Isolation of excitating parts

To prevent the transmission of vibrations from gearbox or generator to surrounding structure the connections to neighbouring components can be design to be elastic coupling components. These components behave like a low-pass filter, i.e. above a specific frequency only small forces can be transferred. The stiffness of the coupling component depends on the transferred loads and on the isolating limit frequency. The generator of a turbine with gearbox, for example, can be suspended in relatively soft elastomer springs because only small moments and forces need to be transferred. So, a good isolation can be achieved for generators. In contrast, on the low speed shaft and the gearbox large torques and forces require stiff elastomer elements and limiting frequencies cannot be as low as for the generator. In three- or four-point-suspensions, for example, the gearboxes usually are mounted in multilayer sandwich mounts and rubber bushings (see Figure 2Figure 2: gearbox supports; left: three-point-support; right: four-point-support). During the design process not only the isolation need to be looked but also the extreme and fatigue loads and the reduction of constraining forces due to manufacturing, mounting and deformation of the main frame. At the same time, these components allow the detuning of eigenfrequencies and introducing damping into the drive train.

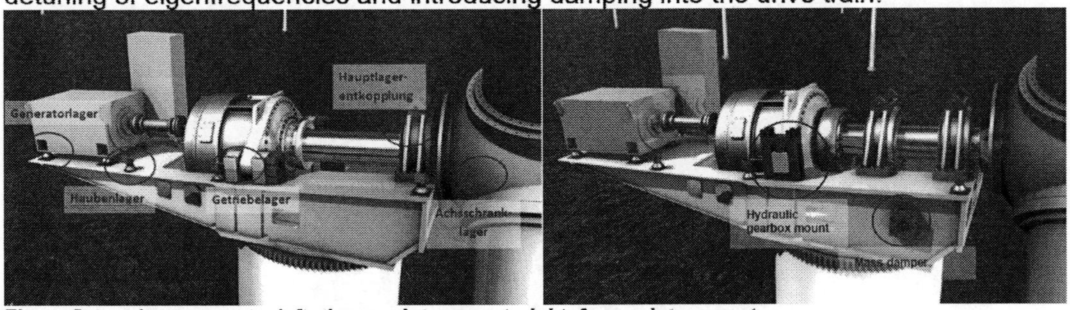

Figure 2: gearbox supports; left: three-point-support; right: four-point-support

The isolation of gearbox and generator in the direction of the blades requires the use of couplings in the low speed shaft (Figure 3). As these couplings require large stiffness in the torsional direction, the limiting frequency in this rotational direction is relatively high.
In the axial, radial and cardanic directions, on the other hand, the coupling can be soft and lower limiting frequencies are possible.

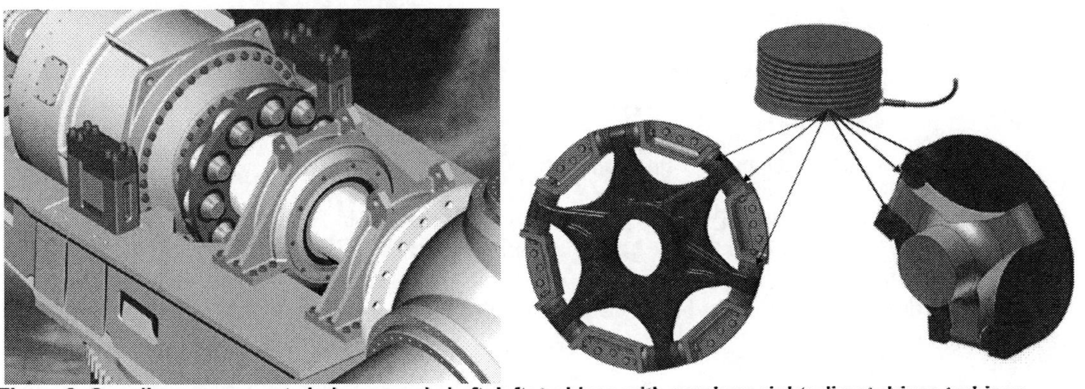
Figure 3: Coupling components in low speed shaft; left: turbines with gearbox; right: direct driven turbines

4.2 Detuning of eignemodes

If critical eigenfrequencies of the wind turbine are inside the excitation frequency range they can be moved up or down. To this end the stiffness or the mass of the affected structure must be altered. Usually these measure require big effort.
Sometimes through the change of the elastomer stiffness of the gearbox mount or the coupling in the low speed shaft the eigenmodes of the drive train can be moved out of the critical range.

4.3 Damping of eigenmodes

Critical eigenmodes that cause tonal issues can be calmed by introducing damping in the respective modes. One effective way in doing so is the use of tuned mass dampers (TMD). This type of damper is made of a mass-spring-damper system that is added to the vibration structure (see figure 4). By accurately tuning the mass damper to the vibrating structure their interaction leads to an increase of damping and therefore a decline of vibration amplitude in the respective structure eigenmode. An external energy supply is not necessary. Figure 4 shows the calculated movement of the structure (mass M1) over excitation frequency for different parameters of the tuned mass damper (mass M2). It is observed that a reduction in vibration of at least one order can be achieved.

The principle of tuned mass damper works for vibrations of any frequency hence ESM builds mass dampers for frequencies from 0.2 Hz to 500 Hz.

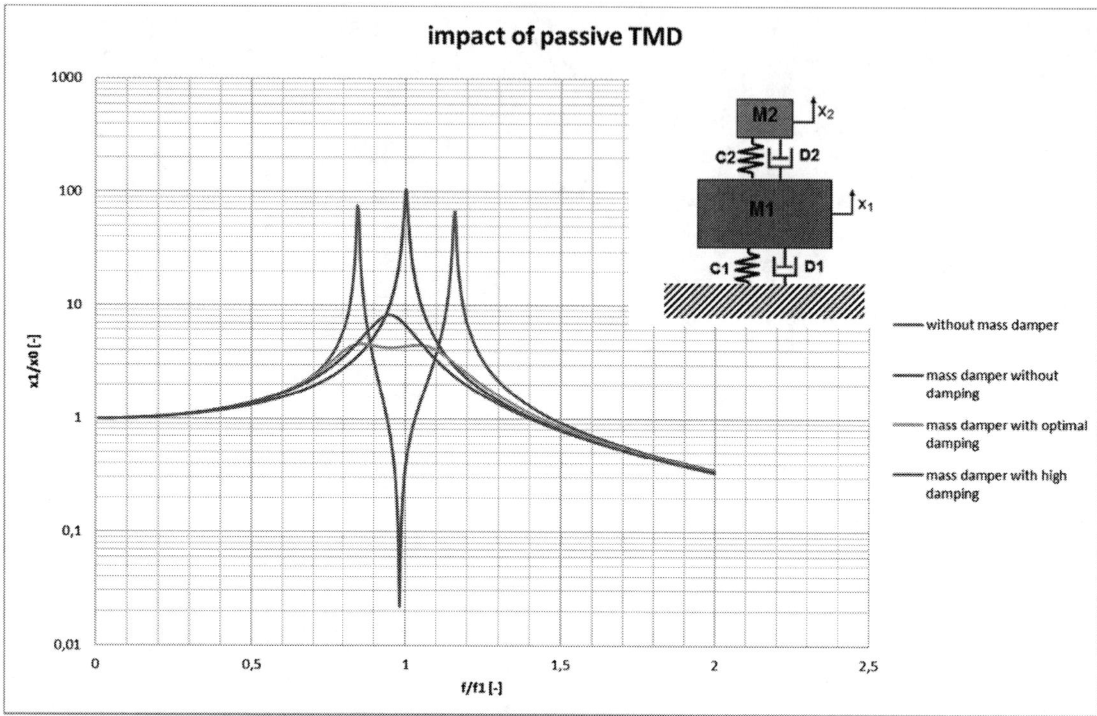

Figure 4: Movement of mass M1 versus the eigenfrequency of force F

Figure 5 shows a typical TMD design for frequencies between 50 Hz and 500 Hz. This design consists of serval mass plates and a tuneable elastomer element. The adaption of frequency and mass can be done on sight. One more feature of the mass damper is, that the frequencies in axial and radial direction can be set independently, i.e. it can work against two eigenmodes at the same time.

Figure 5: passive mass damper of ESM (400 mm x 400 mm x 300 mm; 200 kg)

The frequency of the passive mass damper is temperature-independent. The changing stiffness of the rubber elements with temperature is passively compensated by an appropriate mechanism to increase the pretension of the rubber elements.

4.4 Introduction of counter forces

4.4.1 Adaptive mass damper

The range of use of passive mass damper is limited to weakly damped eigenfrequencies or excitation on one fixed frequency. If the mass damper should operate on damped eigenfrequencies or against a changing force attacking directly at the gearbox or generator there are adaptive mass damper (means variable frequencies). They consist of a mass and a weak damped spring therefore it has a big movement in the point of resonance and big counterforces to transmit in structure of vibration. This effect is useful to calm down the vibrating structure away from the main eigenfrequencies (Figure 6). To follow the changing frequency of excitation the stiffness of the spring has to be balanced by suitable mechanic and control mechanism. The mass damper needs small electric power.

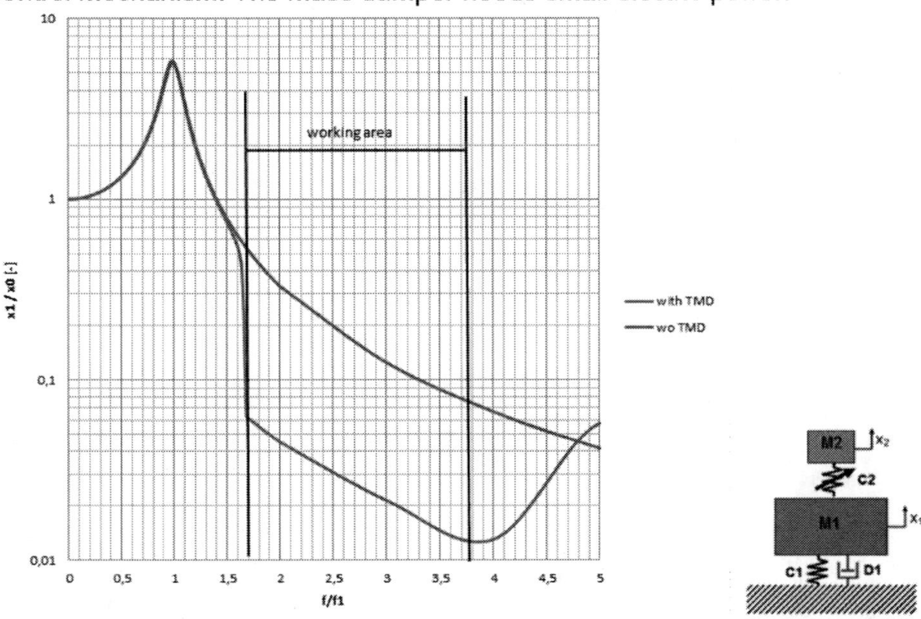

Figure 6: effect of an adaptive mass damper (M2) away from the main eigenfrequency of the vibrating structure (M1)

4.4.2 Active vibration absorber

Another arrangement to fight against vibration in generators or gearboxes is the use of electric actuators like electrodynamic and electromagnetic shaker or coupled unbalance motors. A controller sets the right frequency, phase and amplitude for the actuator and the active mass damper generates counterforces into the structure which erase the excitating forces. The essential components of active vibration absorbers are

- one or more electric actuator,
- power electronics,
- control unit and
- sensors.

Using an electromagnetic shaker forces up to 15 kilo-Newton can be applied by using a power of 500W. The use of active vibration absorber in wind turbines is challenging because the demands on climate conditions, high and low temperatures and lifetime are hard for this type of electric systems.

5. Conclusion

In this contribution the origin of structure-borne sound in wind turbines was explained, how it is transferred through the turbine and emitted from the large surfaces of the turbine. It was shown under which conditions structure-borne sound can be prevented in the design phase, how noise sources can be found and what kind of countermeasures are available to prevent or reduce tonal issues. The countermeasures related to the reduction and isolation of the excitation forces in the gearbox and the generator are related to the specific concept of the wind turbine and can only be influenced during the construction phase of the wind turbine. Detuning of eigenmodes and the use of different mass dampers and vibration absorbers are available for retrofit solutions.

References

[1] Wagner S., Bareiß R. and Guidati G.(1996) *Wind Turbine Noise* Springer

[2] Gasch R., Knothe K., Liebich R. (2012) Strukturdynamik, 2. Auflage, Springer

[3] Falourd, X. et al., Low Frequency Amplitude Modulation related to Doppler frequency shift: an experimental study of a 101m diameter wind turbine in a swiss valley. 6th International meeting on Wind Turbine Noise, Glasgow 2015.

7th International Conference
on
Wind Turbine Noise
Rotterdam – 2nd to 5th May 2017

Low-frequency micro-seismic radiation by wind turbines and it's interaction with acoustic noise emission

Theodore V. Gortsas, Karlsruhe Institute of Technology (KIT), Institute of Soil Mechanics and Rock Mechanics + University of Patras, Department of Mechanical & Aeronautical Engineering
Demosthenes Polyzos, University of Patras, Department of Mechanical & Aeronautical Engineering
Theodore Triantafyllidis and Peter Kudella, Karlsruhe Institute of Technology (KIT), Institute of Soil Mechanics and Rock Mechanics
Toni Zieger and Joachim Ritter, Karlsruhe Institute of Technology (KIT), Geophysical Institute

Summary

Low frequency acoustic noise from WTs is a complaint often brought forward as an argument against wind parks, even though measured levels remain within the legal thresholds. On the other hand, geophysical measuring stations around the world are increasingly disturbed by micro-seismic vibrations of WTs. The joint research project "TremAc" focusses on the interdependence of both wave types which in the past has been widely neglected but could become a key to a better understanding. "Low frequencies" in the sense of this report are emissions below 100 Hz including infrasound.

For a generic WT, an acoustic sound source and dynamic mechanical forces at the foundation were combined and the dynamic soil-structure-interaction was established. The Boundary Element Method was applied to model the propagation in air and ground, finally reaching a nearby building. As the waves interact along the ground surface, the acoustic wave field generates secondary solid-borne noise in the ground while the seismic wave field generates secondary acoustic noise in the air.

Furthermore, geophysical measurements in the neighbourhood of WT are reported. We will present ground motion amplitude and frequency analyses from windfarms in the Upper Rhine Graben, SW Germany where the ground is composed of unconsolidated sediments. There clear signals with discrete frequency peaks between 1 and 7 Hz are observed which increase with wind speed.

1. Introduction

The last 35 years, significant progress has been made on the design of offshore and onshore wind turbines (WT) rendering the exploitation of wind power as one of the fast-growing renewable energy sources (Hau 2006, Al-Bahadly 2011). However, as WTs become numerous, taller and more powerful, many problems dealing with the disturbance of the surrounding environment due to installation and operation of wind farms have been reported (Abbasi et al. 2014, Bakker et al. 2012, van Renterghem et al. 2013, Klæboe and Sundfør 2016). Focusing on onshore WTs, this leads to an ongoing public discussion of requirements for building and operational permission, especially of minimum distances to residential areas.

WTs emit a broadbanded acoustic noise, which may be audible in distances of 300 to 1000 m as a diffuse "wsch-wsch" under certain wind speeds and operational conditions. The sound is annoying particularly because of its pulsating character. Infrasound, which corresponds to frequencies below 20 Hz is inaudible by human ears, but may produce a feeling of static pressure and periodic masking effects (so-called amplitude modulation) for high pressure levels (Lindberg and Backteman 1988).

The main sources responsible for emitting low frequency acoustic noise by a WT are of aerodynamic nature coming from blade vortex interaction, turbulent inflow noise, blade tower interaction and blade tip turbulent flow (Wagner et al. 1996, Oerlemans and Schepers 2009, Mo and Lee 2011). Secondary cause is low-frequency sound generated by the rotor and other rotating mechanical parts in the nacelle of a WT (Adewumi et al. 2015). Low frequency sound has very low absorption in the atmosphere. It can propagate for long distances from a WT and is affected by the near ground surface, the atmospheric refraction and the temperature gradient of the atmosphere. Many published papers and reports can be found on the subject, such as Hubbard and Shepherd 1991, Leventhall 2003, Pedersen et al. 2004, 2009, Jakobsen 2005, Møller and Pederson 2011, Öhlund and Larsson 2015, Keith et al. 2016, 2016a, Michaud et al. 2016, 2016a and Katinas et al. 2016.

Microseismic waves are emitted in a large frequency range, caused by vibrating blades, rotating parts and excited tower structure, where the overall dynamics is also influenced by the foundation structure and the soil stiffness. In the ground, the waves propagate over long distances. Amplitudes are much too small to annoy nearby residents directly, but the waves can still be detected with very sensitive seismic stations within distances of up to 15 km. (Rushforth et al. 1999, 2003, Styles et al. 2005, 2011, Saccorotti et al. 2011, Stammler and Ceranna 2016). The lower the frequency the longer the distance where the microseismic disturbance is detectable.

However, until now acoustic emission through the atmosphere and seismic emission through the ground have been regarded as independent phenomena. When they were measured in the field, each of the wave types was compared to its respective threshold only. The leading question of this contribution, is their interaction along the propagation paths. Are they able to channel or amplify amplitudes under certain conditions? Surveys report observations speaking for an acoustic-seismic interaction (Møller and Pederson 2011). To the authors best knowledge, the contribution of microseismic waves to the perceptible structural vibrations (soil-structure interaction) and the background low frequency noise (soil-air interaction) has not been reported so far.

The first goal of the present work is to simulate the propagation of all kinds of low frequency noise produced by a WT both in soil and air and to compare their influence inside and outside a simple building located at 500 m distance. A short explanation of the numerical tools used is given at first. It follows an analysis of the soil-structure interaction of the WT foundation and the seismic wave propagation using 3D models.

The second goal is to present first results of in-situ measurements of micro-seismic events near WTs. Remote locations far away from any human activity are preferred for seismological stations because of the low anthropogenic noise. However, WTs at some kilometers distance lead to a significant impact of wind dependent signals in a wide frequency range (0.5 Hz to 10 Hz) in seismic recordings (Stammler & Ceranna 2015). This impact reduces the signal-to-noise ratio of earthquake signals or signals of other origin (e.g. nuclear test explosions etc.). Systematic observations of WTs as a seismic source are few and the way of signal propagation is not fully understood yet, also in comparison with meteorological data (e.g. wind speed) or underground conditions (e.g. type of rock). For this reason, we analyze long-term measurements at wind farms near the town of Landau, SW Germany, and compare our seismological results to the wind speed.

2. Advanced ACA/BEM for solving acoustic-structure and soil-structure interaction problems

For the simulations, the Boundary Element Method (BEM) was employed as a robust numerical tool for solving acoustic, elastic and fluid-structure interaction and wave propagation problems like the aforementioned ones. With a reduction of the dimensionality of the problem by one and with a high solution accuracy it offers two remarkable advantages compared to the Finite Element Method, and this is in particular of great advance for large domains under consideration.

Most of the WT signals are either periodic or transient with relatively short duration. They are converted to the frequency domain by means of the Fast Fourier Transform (FFT) and the corresponding boundary value problems are solved for each frequency of the spectrum by the frequency domain ACA/BEM explained in brief below. The obtained results are shifted to time domain again through a standard inverse FFT algorithm (Vavourakis et al. 2008).

For all elastic regions the fundamental solution U for the Navier-Cauchy equation

$$\boldsymbol{\mu \cdot \nabla^2 u(x) + (\lambda + \mu) \cdot \nabla\nabla u(x) + \rho \cdot \omega^2 \cdot u(x) = 0}$$

has to be found and for all air regions the solution G of the Helmholtz equation

$$\boldsymbol{\nabla^2 p(x) + \left(\frac{\omega}{c}\right)^2 \cdot p(x) = 0}$$

where λ, μ represent the Lamé constants of the regions occupied by elastic bodies, ρ the mass density, ω the frequency and c the phase velocity of sound waves propagating in air. Boundary conditions at the interfaces of the regions, require the following conditions to be satisfied:

$$\boldsymbol{u^1 = u^2,\ t^1 = -t^2}$$

$$p^1 = p^2,\ \partial_n p^1 = -\frac{\rho^1}{\rho^2}\partial_n p^2$$

$$\frac{1}{\omega^2 \rho^1}\partial_n p^1 = -\boldsymbol{\hat{n} \cdot u^2},\ p^1 = \boldsymbol{\hat{n} \cdot t^2}$$

for elastic-elastic, acoustic-acoustic, and acoustic-elastic interfaces respectively and with ρ^i being the density of each acoustic medium.

Instead of solving the above differential equations everywhere inside the regions, BEM solves the problem only on their boundaries. The following integral representations for the displacements and pressures of the just described problem can be used within the formalism of the boundary integral equations (Polyzos et al. 1998, Wrobel 2002, Aliabadi 2002)

$$\boldsymbol{k(x) \cdot u(x) + \int T(x,y) \cdot u(y) dS = \int U(x,y) \cdot t(y) dS}$$

$$\boldsymbol{c(x) \cdot p(x) + \int \partial_n G(x,y) \cdot p(y) dS = \int G(x,y) \cdot \partial_n p(y) dS}$$

where *U*, *G* are the fundamental solutions of the Navier-Cauchy Equations and the Helmholtz Equations, respectively, T and t are traction fields corresponding to the elastic fundamental solution U and displacement u, ∂_n denotes derivative across the unit vector normal to the boundary of the corresponding region, while the coefficients $\boldsymbol{k}(\boldsymbol{x})$ = a**I** and c($\boldsymbol{x}$) = a, with a taking the values 0, 1 and 0.5 depending on the position of x.

All the boundaries and interfaces are discretized into four-node linear isoparametric quadrilateral boundary elements. After the application of the integral at the corresponding nodes, one obtains the following linear systems of algebraic equations:

$$T \cdot u = U \cdot t, \qquad \text{respectively} \qquad Q \cdot p = G \cdot q$$

The matrices contain integrals evaluated as explained in (Polyzos et al. 1998, Agnantiaris and Polyzos 2003) and the vectors contain the nodal values of displacements, tractions, air pressures and their fluxes corresponding to the surface elements.

The fully populated and non-symmetric matrices of the equations confines the application of the BEM to problems with no more than 80,000 degrees of freedom (dofs). A very efficient methodology that circumvents that problem and accelerates remarkably the solution process of a BEM code is the use of Hierarchical Matrices along with the Adaptive Cross Approximation (ACA/BEM) technique, which means that the matrices are represented hierarchically by using a block tree structure. Blocks, which correspond to large distances between source and collocation points, are characterized as far field blocks and compressed using low rank matrices found by an ACA algorithm (Gortsas et al. 2015). Only the near field blocks, which are dominated by the singular behavior of the fundamental displacement and traction kernels, are fully calculated as in conventional BEM. Furthermore, a significant reduction of the solution time of the problem is accomplished by utilizing the iterative solver GMRES (Saad and Schultz 1986). More details on the ACA/BEM technique can be found in (Gortsas et al. 2015).

3. Description of the model

3.1 The topography model

The model geometry is shown as a plane-strain section of the half space in Fig. 1. A slope with inclination β was introduced because later simulations shall characterize WTs in mountainous areas. A one-room building with elastic walls and strip foundations represents the "receiver".

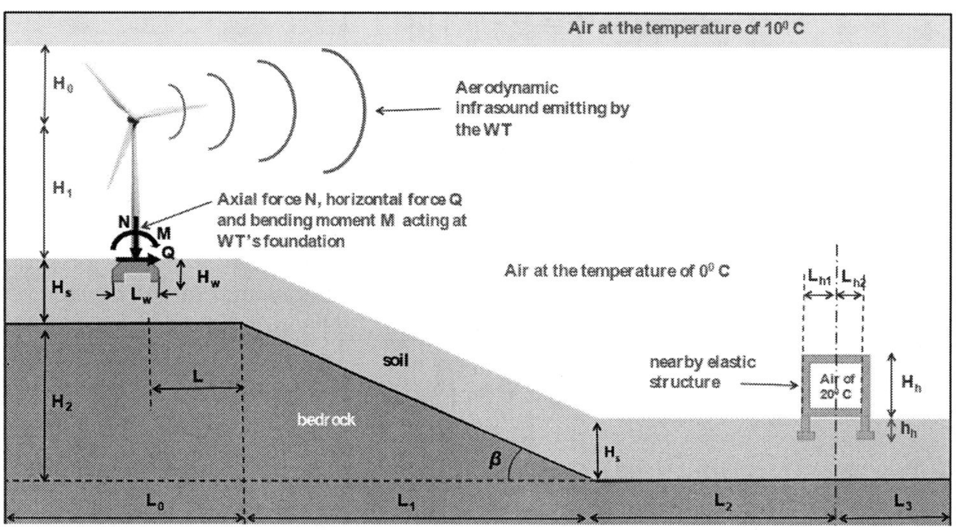

Figure 1: Geometry and materials for the 2D and 3D model (radial section) used in the numerical simulation for the propagation of infrasound and micro-seismic waves generated by a WT.

The simulation considers four different material regions one on the top of another: Rock underlying a granular soil and two acoustical regions consisting of air at two different temperatures. Rock and soil are considered as linear elastic materials, a hypothesis which is also justified by measurements due to the very small amplitudes and strains. The material properties of all regions are given in Tab. 1.

Table 1: Default geometrical details and material properties for the model shown in Fig. 1.

Region	Material Properties	Geometrical Characteristics
Soil	Density: ρ = 1800 kg/m^3 Young Modulus: E = 4.05x10^8 N/m^2 Poisson Ratio: ν = 0.25	H_s = 100 m (default: no bedrock)
Bedrock	ρ = 2600 kg/m^3, E = 32x10^9 N/m^2, ν = 0.23	H_2 = depends on the slope angle β L = 100 m, $L_0 = L_1 = L_2 = L_3$ = 200 m
Air at 0^0 C	Density: ρ = 1.2922 kg/m^3 Sound velocity: c = 331.30 m/s	H_0 = 350 m, H_1 = 150 m
Air at 10^0 C	ρ = 1.2466 kg/m^3 , c = 337.31 m/s	Semi-infinite
Air at 20^0 C	ρ = 1.2041 kg/m^3 , c = 343.21 m/s	Cube of (7 m) x (7 m) x (7 m)
Foundation of WT	ρ = 2500 kg/m^3, E = 39x10^9 N/m^2, ν = 0.2	H_w = 2.97 m, L_w = 19.8 m
Structure at 500m from WT	ρ = 2500 kg/m^3, E = 39x10^9 N/m^2, ν = 0.2	H_h = 7 m, h_h = 1.5 m

The frequency domain fluid-structure interaction problem is solved numerically via the ACA/BEM technique illustrated in section 2. Alltogether 92063 linear elements have been utilized for the discretization of the free surface of the soil.

3.2 The WT foundation

This simulation is based on a generic wind turbine before the authors had access to measurements of real turbines. The underlying model is the generic NREL 5 MW turbine, whose details have been published (Jonkman et al 2009). As the NREL turbine originally was a floated offshore turbine, an appropriate tower and a foundation had to be designed. Load simulations (provided by partner institutes) delivered design loads for various wind speeds and extreme conditions as shown in Tab. 2.

Table 2: Design loads (characteristic mean values and amplitudes) and dynamic soil-structure interaction stiffness for the NREL 5 MW foundation of Fig. 1. For Min and Max values the loads in x and y direction have been added to a vector sum acting in arbitrary direction.

	DLC 1.1 (12 m/s)	DLC 6.1	DLC 6.2	DLC 6.2 + Foundation	Dynamic spring	Dynamic damping
Min. Fz	6929 kN	6389 kN	6674 kN	27711 kN		
Max. Fxy	966 kN	1393 kN	2178 kN	2178 kN		
Max. Mxy	85852 kNm	103040 kNm	173070 kNm	179540 kNm		
Amplit. Fz	1,3 kN				8554 MN/m	240 MNs/m
Amplit. Fx / Fy	11 / 46 kN				7332 MN/m	325 MNs/m
Amplit. Mx / My	819 – 3748 kNm				559000 MNm	4180 MNms
Amplit. Mz	43 kNm				not evaluated	not evaluated

The base reaction eccentricity in load case DLC 6.2 was decisive for the dimensions of the foundation. Assuming a medium-dense sand or gravel (properties according to Tab. 1), a spread foundation forming a ring with an outer diameter of 19.8 m, an inner diameter of 11.88 m and a height of 2.97 m proved to be adequate (Fig. 2). To provide overturning stability its dead weight must be 21.037 MN, three times the weight of turbine and tower. The dynamic stiffness parameters of soil-structure interaction given in Tab. 2 can be calculated using established analytical formula or semi-empirical procedures (Triantafyllidis et al. 1987) but they also result from a BEM calculation and have thus been proved to be of correct order.

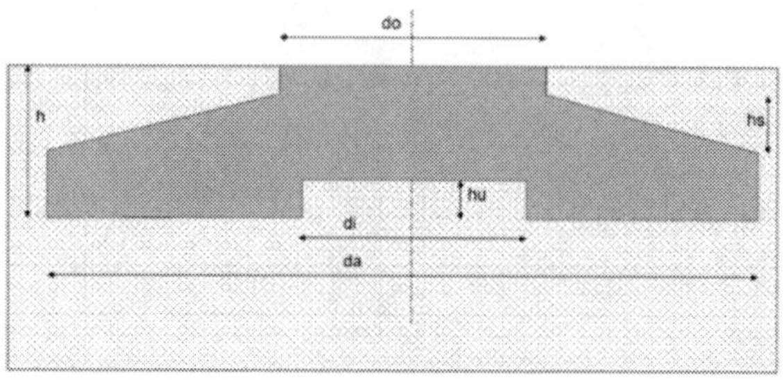

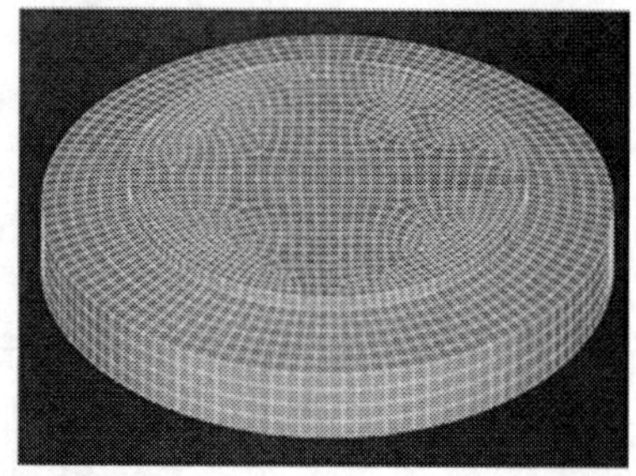

Figure 2: Axisymmetric cross-section and FE model of the circular foundation used in a BEM analysis.

It should be mentioned that the "Maximum" and "Minimum" design loads of Tab. 2 are static values relevant only for the size of the foundation. The characteristic dynamic loads are much smaller and derived with the procedure explained in 4.2.

In the BEM model, the load transfer of the components is realized by a distribution of normal and shear tractions at the soil-structure interface.

4 Simulation of combined 3D acoustic emission and vibration

4.1 **Acoustic source field**

The underlying load set was taken from a combined aero-acoustic and mechanical simulation of the generic NREL 5MW-WT provided by the University of Stuttgart, Institute of Aerodynamics and Gas Dynamics (unpublished).

On one hand, this simulation provides pressure levels of the acoustic noise and flow velocity emitted by the fictitious WT on a cylinder of 100 m radius and 200 m height (Fig. 3). Time series are given on 810 equidistant points on the cylinder surface. On the other hand, the mechanical loading of the foundation is given as time series of the 6 force/moment components. For both time series the simulation period was 41 seconds and the sampling rate 50 resp. 100 Hz. The load simulation covers a detail of a normal WT operation at design wind speed 12 m/s. The procedure is as follows:

- A FFT is performed for each time series. All mean values, either due to constant loads like dead weight or steady wind are subtracted from the signal. If the signals are not mean-value-free, this will result in unrealistic high amplitude ta frequencies towards to zero.
- Non-consideration of irregular signals and overshooting amplitudes at the beginning of the time series. They are believed to be artefacts of the load simulation starting from zero rotation.
- Identification of frequencies dominating from elastic vibration as well as from acoustic sound.
- Application of the selected frequencies to their respective source points at the foundation and the air cylinder as mean-value-free harmonic signals and performance of the BEM simulation in the frequency domain.

The acoustic and elastic time series are uncoupled, so phase shifts between them are not yet considered in this model. In order to exclude load simulation artefacts affecting the amplitude level, the first 20 seconds of the time series were disregarded.

FFT's have been performed for four representative source points (Fig. 3). The dominating frequencies are 0.37 Hz, 0.59 Hz and multiples thereof. The source fields have a characteristic directivity for each frequency (Fig. 4). The sound level maximum is 89 dB (0.56 Pa), radiated with 0.59 Hz to both sides in the rotor plane. Also the 0.37 Hz signal points sidewards in an inclined direction towards the ground. This signal is also present in upwind and downwind direction where it's amplitude is comparable to 0.59 Hz. Other frequencies as 3.52 Hz are emitted in the downwind direction. These characteristics are identical to earlier findings (Oerlemans et. al. 2007). All sound pressure levels on the cylinder surface are below the detection threshold.

4.2 Seismic source field

The dominating vibration frequencies are again 0.37 Hz (predominantly in the F_y and the corresponding M_x signal), 0.41 Hz (only in F_x and M_y) and 0.59 Hz (dominating in M_z but present also in all other components). All time series contain multiples of 0.59 Hz, in particular 3.51 Hz which is strong in F_x and F_z. The 0.59 Hz signal reflects the blades passing the tower, the 0.37 Hz signal is dominating in the first 20 seconds of simulation and is believed to represent a tower bending mode. A 0.046 Hz signal dominating F_z could be a simulation artefact. Fig. 5 shows all components. The dynamic mean amplitudes evaluated from the FFT are given in Tab. 2.

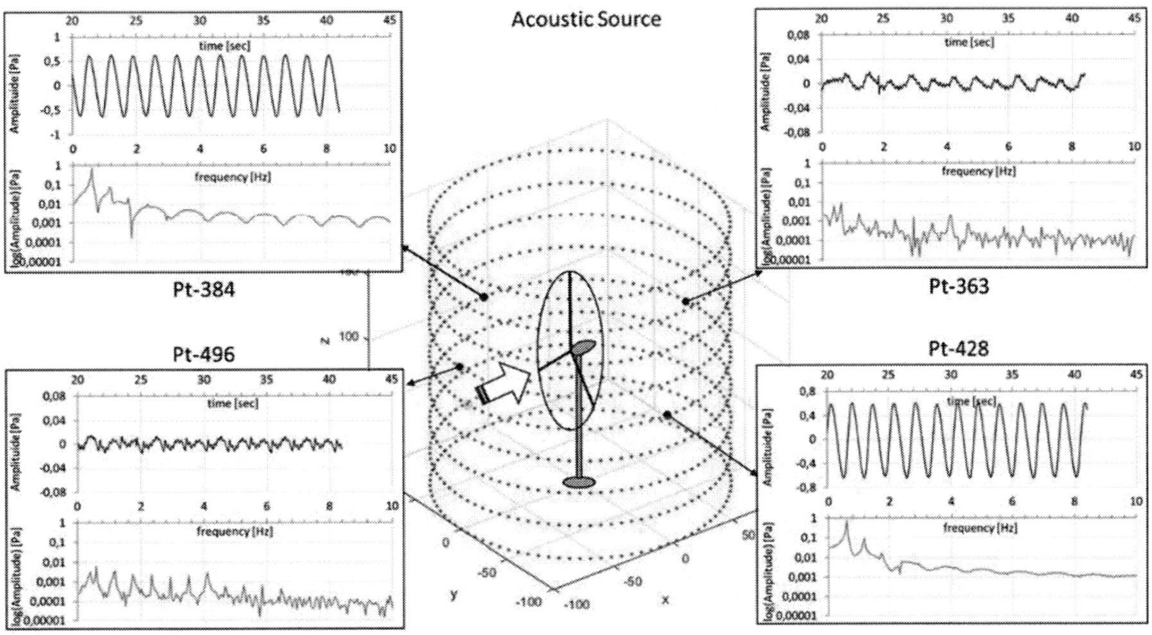

Figure 3: Acoustic source point cylinder around WT with time series and FFT for 4 representative points

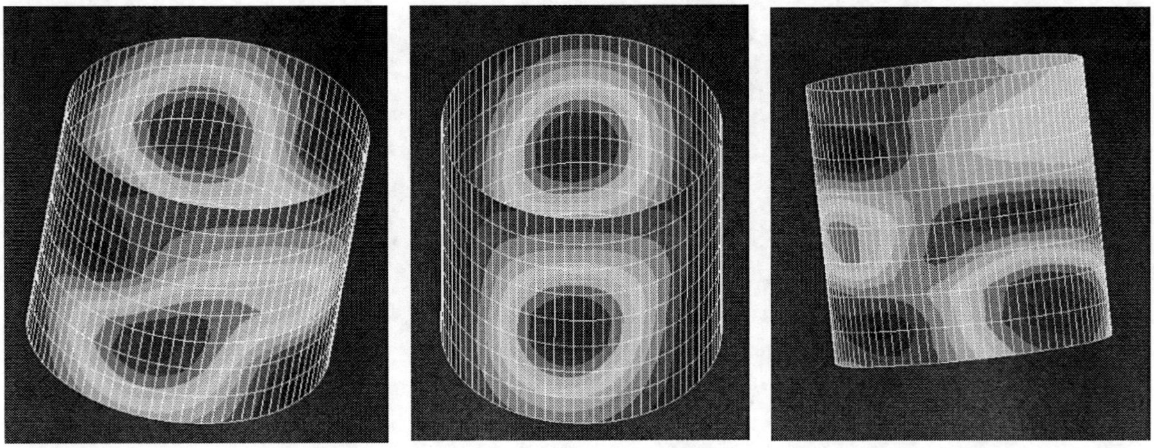

Figure 4: qualitative acoustic pressure distribution on the cylinder seen from the –y direction, a): at 0.37 Hz, b) at 0.59 Hz, c: at 3.52 Hz (cylinder slightly rotated for better visibility)

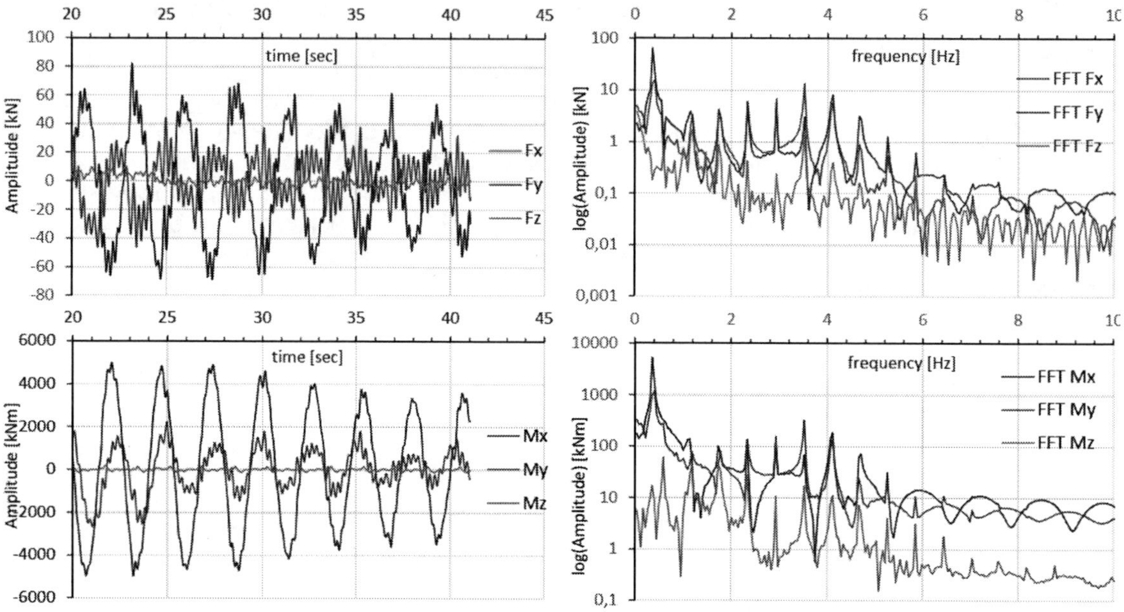

Figure 5: Time series and FFT of the dynamic parts of all 6 load components acting on the foundation

Only the waves radiated by vertical motion of the WT due to F_z spread uniformly to all directions because of the axisymmetry of the problem. The rocking mode of the WT foundation due to the overturning moments M_x/M_y, causes a wave field with a clear dipole directivity perpendicular to the rotation axis. The shear forces F_x/F_y and the rotational component M_z emit longitudinal waves parallel to the acting force and also shear waves in transverse directions. For simplicity, only one force and moment direction was modelled and M_z was ignored. For each frequency, the combination of the largest amplitudes from F_x/F_y, M_x/M_y, and F_z was applied. Fig. 6 demonstrates the dipole character of the wave emission, which is dominated by the moment loading in all frequencies investigated.

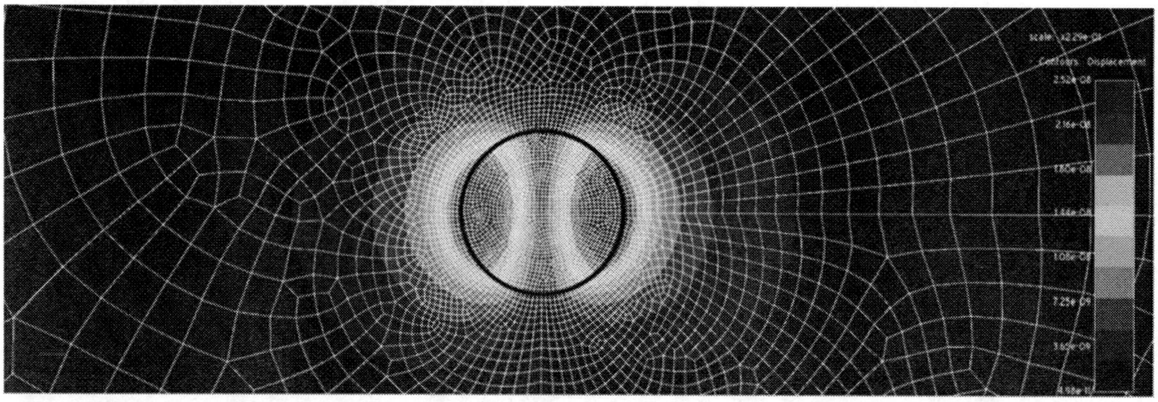

Figure 6: Field of vertical soil displacement amplitudes around the foundation (black circle, rotated by 90°) for combined M_x-F_y-F_z loading at 0.37 Hz

Summarizing, we can say that for the low frequency range, a WT behaves as a frequency-dependent dipole emitter with seismic and acoustic emissions predominantly in the rotor plain.

However, as the problems are linear, the simulation can handle each of the motion modes and frequencies separately, and superimpose the resulting amplitudes.

4.3 Propagation and immission

The BEM simulation of the wave propagation was executed for the frequencies 0.37 Hz, 0.59 Hz and 3.52 Hz. Wavefield directivity, amplitudes and interaction were studied along the ground surface and at a building in 500 m distance from the WT. Because the mesh refinement allowed only one building placed in +x direction, the source pattern was rotatet clockwise by 90° to simulate the emission in -y direction from the WT. Fig. 7 shows the pressure levels and induced displacement fields due to acoustic emission for 0.37 and 0.59 Hz, Tab. 3 shows the results in numbers.

The pressure level distribution is more or less symmetric around the source while the soil vibration amplitude is not (Fig. 7). For 0.37 Hz, the peak soil displacements are in opposite direction to the sound levels.

For low frequencies, the radial attenuation of acoustic sound is much higher than expected. Theoretically the decay should be between $1/r^{0,5}$ (half space) and $1/r$ (full space). In fact, $1/r^{2}$ was found for 0.37 Hz, $1/r^{1,8}$ for 0.59 Hz and $1/r^{0,6}$ for 3.52 Hz.

For the radial decay of seismic amplitudes, about $1/r^{1,5}$ was found for 0.37 Hz and for 0.59 Hz and $1/r^{1,1}$ for 3.52 Hz. Previous systematic studies have shown that for 1 Hz the decay for the vertical force excitation is $1/r^{0,9}$, for shear force $1/r$ and for moments $1/r^{2}$. For higher frequencies of 5 Hz and 10 Hz the exponent tends to smaller numbers.

Secondary soil displacements due to acoustic emission are much smaller than those of direct seismic emission with the exception of 0.59 Hz, where the magnitude is similar. This holds vice versa for secondary sound, but the difference becomes lower for higher frequencies.

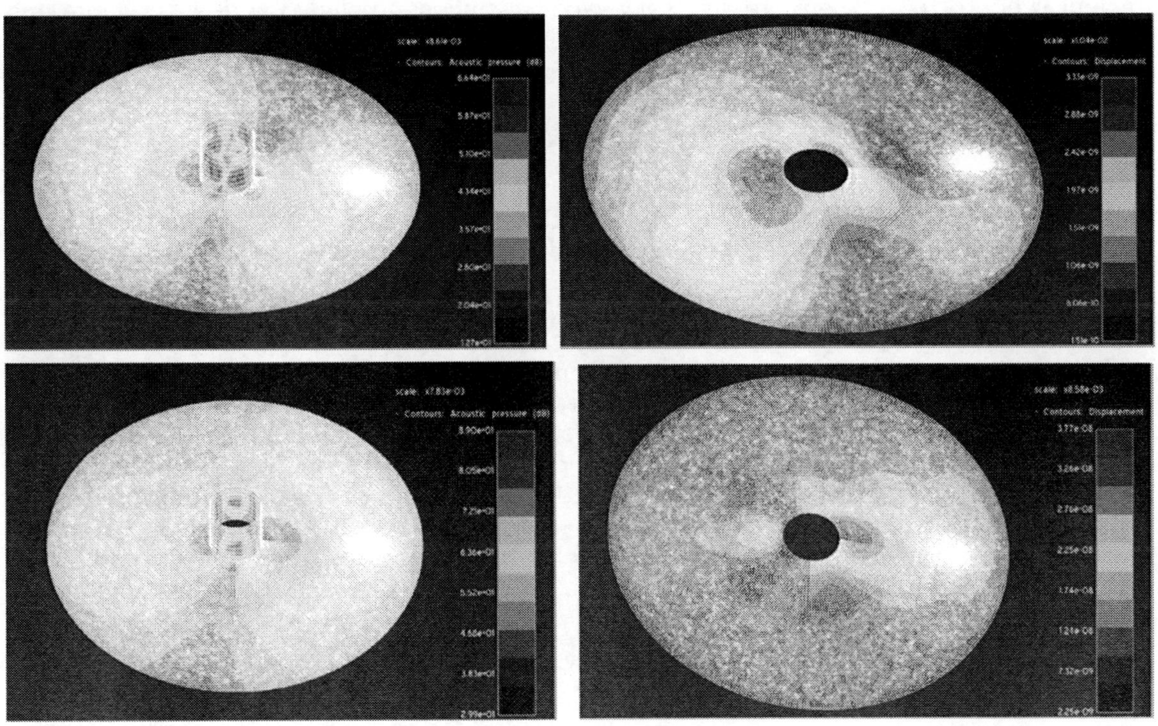

Figure 7: wavefields around the source cylinder to a radius of 700 m, with all source fields rotated clockwise 90°, left: sound pressure levels, right: soil displacements, top: for 0.37 Hz bottom: for 0.59 Hz

Table 3: Acoustic pressure levels and seismic displacement amplitudes due to acoustic and seismic WT emission (separated). Values in italic are back-calculated for comparison using the decay function.

Freq. [Hz]	emission	pressure level at source [dB]		Displacement amplitude at source [m]		pressure level outside building [dB]	Displ. ampl. outside ≈ inside [m]	pressure level inside [dB]
		found.	cylinder	found.	cylinder			
0.37	acoustic		66		3.3E-09	38	7.3E-10	-56
	seismic	30		2.5E-05	*7.2E-07*	10	6.3E-08	7
0.59	acoustic		89		3.8E-08	64	1.4E-08	-22
	seismic	15		1.6E-06	*5.4E-08*	-4	5.2E-09	-7
3.52	acoustic		49		5.2E-10	42	1.5E-10	
	seismic	47		1.8E-06	*1.3E-07*	22	2.2E-08	4

Direct acoustic sound disappears inside the house (-90 dB), while secondary sound does not (-4 dB). Inhouse noise, though far away from perceptibility, is more likely to be caused by seismic effects than by acoustic radiation. This is different, however, if the structure prohibits resonance (as for 3.52 Hz in compare to lower frequencies). Also higher frequencies of 15 and 23 Hz were detectable in the FFT. However, the existing model cannot process them, because wavelengths fall below 30 meters and the used BEM mesh is too coarse to perform simulations above 10 Hz.

5. Influence of Wind Turbines on Seismic Stations

5.1 Setting

The town of Landau is located in the central part of the Upper Rhine Graben, SW Germany. This area is of special interest, because there are two geothermal power plants and therefore a dense seismic network including borehole stations for seismological monitoring were installed (Vasterling et al. 2016). There are also seismological stations to study the deep structure and local earthquakes (Ritter et al. 2008). This allows us to use long-term waveform recordings over a time period of several years. The seismic stations used for this study are shown in Figure 8 as triangles.

We also focus on seismic borehole stations and their interference with WTs. Borehole stations are marked as reversed triangles (Fig. 8). Our study area also includes several wind farms (Offenbach, Bellheim, Herxheimweyher and Rülzheim) with overall 18 WTs of different types until today.

5.2 Results from Surface-Station Recordings

For the data analysis we first remove the instrument response from the acquired continuous vertical-component ground motion data, to achieve the true ground motion velocity. Then we divide each month into one hour-long time segments and calculated the power spectral density spectra (PSD) of the ground motion for each segment by applying a simple taper. After smoothing the spectra with a width of 0.05 Hz, we compared them with the wind speed and averaged the PSD spectra within each wind speed bin (0-3 m/s, 3-4 m/s, 4-5 m/s, 5-6 m/s, 6-7 m/s, 7-8 m/s, 8-9 m/s, 9-10 m/s, >10 m/s) with the 75% percentile, to minimize anomalies. The wind speed measurements were provided by the Institute of Meteorology and Climate Research – Troposphere Research (KIT) in Karlsruhe.

Fig. 9 shows one-sided logarithmic PSD spectra for the time period of one month (December 2014) for the seismic station TMO57 (see Fig. 8 for location). The number of used hourly windows

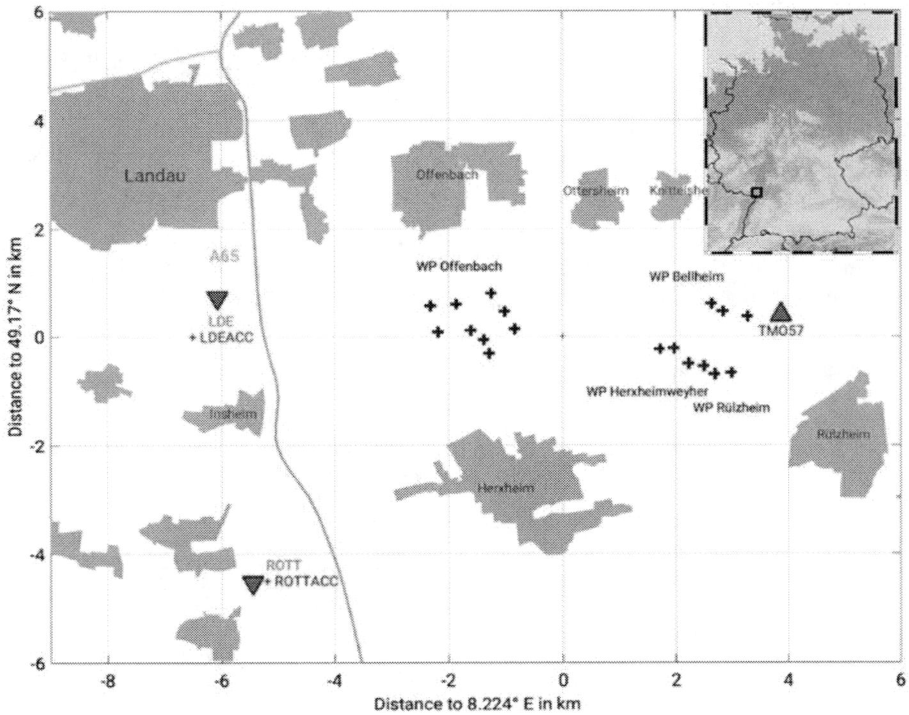

Figure 8: Map of the studied area. Seismic stations are indicated as triangles (surface instruments) and reversed triangles (borehole instruments), WTs and wind parks (WP) are indicated as blue crosses.

to determine the average is shown in the legend, marked with "n". The diagram shows a clear increase of the PSD with increasing wind speed over the entire frequency range. Such an increase of ground motion with wind is well known and occurs in many regions (Withers et al., 1996; Lott et al., 2017). Remarkable in our recordings are the major peaks at 1.08 Hz, 1.48 Hz, 1.85 Hz, 2.76 Hz, 3.67 Hz, 4.59 Hz, 5.5 Hz and 6.41 Hz which are not observed at many other places. The noise level increases at these mentioned discrete frequency peaks by 100 times

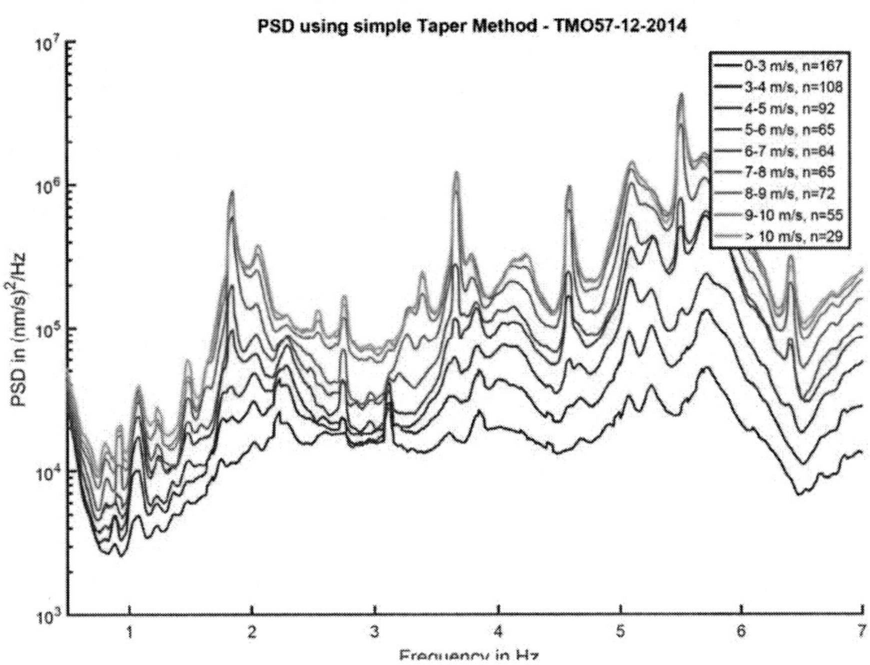

Figure 9: The Power-Spectral-Density (PSD) over the frequency range of 0.5 Hz to 7 Hz. Shown is seismic station TMO57 for December 2014. The station is located at a distance of 600 m to the next wind turbine. The legend indicates the used wind speed bins and the number of PSD spectra of hourly time segments.

(e.g. for 1.85 Hz: 1.2×10^4 $(nm/s)^2/Hz$ at low wind speed to 9×10^5 $(nm/s)^2/Hz$ at high wind speed). Because we observed that the lowest frequency (about 1 Hz) coincides with the blade passing frequency of the WTs, we infer that the origin of the high-peak signals are WTs and that the higher frequencies are multiples of the blade passing frequency.

5.3 Results of Borehole Station Recordings

Seismometers are sometimes installed in boreholes, because tremor in deeper locations can be reduced significantly compared to surface recordings. We use recordings from borehole measurements in order to study a possible attenuation of WT induced seismic signals. For this analysis we use the seismic station LDE in 150 m depth and station ROTT in 305 m depth and a different distance to the next turbine compared to LDE (see Fig. 8). At both locations are also

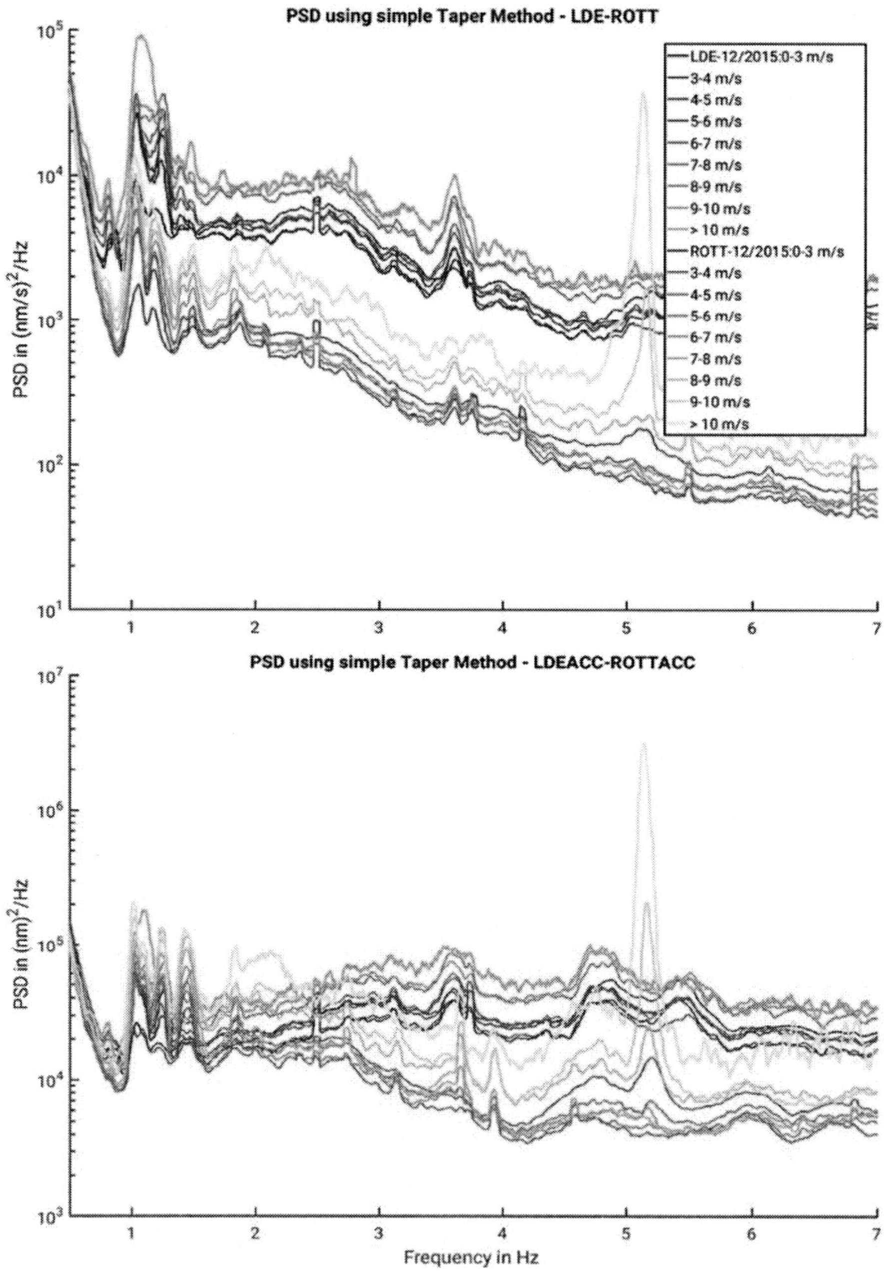

Figure 10: The Power-Spectral-Density (PSD) over the frequency range of 0.5 Hz to 7 Hz. Shown are the borehole stations LDE and ROTT (top panel) as well as the surface stations LDEACC and ROTTACC at the same locations (bottom panel) for December 2015. Station ROTT (305 m under surface) has a much lower noise level than the shallower station LDE (150 m under surface). However, the surface stations indicates no difference between both locations below 2 Hz, and also the frequency peaks coincide to each other.

surface stations installed (LDEACC and ROTTACC); thus we are able to investigate the geometric influence (distance, depth) on seismic impact by disturbances due to WTs.

As shown in Figure 10, the deeper recording at station ROTT (yellow-red colors) has a much lower noise level, most probably because of the larger depth which better shields near-surface noise. But also at this station we can recognize clear wind dependent frequency peaks around 1 Hz, which are also noticeable at station LED (blue-green colors). The large peak at 5.2 Hz recorded at station LDE is not visible during other monthly periods or at other stations, so this is probably due to a specific event at this station. The bottom panel of Figure 10 shows the comparison between the two surface stations LDEACC (yellow-red colors) and ROTTACC (blue-green colors) at the same location. ROTTACC is further away to the next turbine than LDEACC (see Fig. 8). Nevertheless we can observe the same maximum value (1-2 x 10^5 $(nm/s)^2/Hz$) at the frequency peaks in the lower part of the PSD spectrum (between 1 Hz and 2 Hz). This differs from the observation at the borehole stations, in which there is a clear difference between the amplitudes of the peaks. For this reason, we infer that the attenuation effects due to distance is much less than the attenuation effect with depth.

6. Conclusions

The capability of the BEM simulation method has been demonstrated for micro-seismic radiation and acoustic emission in large areas under consideration. It seems that this technique can model seismic and acoustic emission and interaction effects for infrasound frequencies, governed by multiples of the blade-passing-frequency. Secondary noise effects from underground vibrations cannot be neglected even though the main noise contribution seems to result from the acoustic source.

Depending on the frequency the secondary displacement field caused by the acoustic source around the wind turbine can be stronger in front or behind the wind turbine. The main propagation direction is in the rotor plane, for acoustic infrasound as well as for seismic vibrations. Attenuation relations are highly frequency-dependent.

In a building in some hundred meters distance, the calculated displacement amplitudes due to seismic WT emission and the amplitudes produced as a secondary effect by acoustic emission, are of the same order of magnitude. Depending on frequency and structural geometry, secondary acoustic noise may be larger than direct acoustic immission. Contributions from elastic solutions of different wave types and frequencies can be superimposed.

However, it is not yet clear what may happen for higher frequencies. To simulate them, the number of degrees of freedom must be increased dramatically which seems only possible by parallelization for distributed memory systems. We need further investigation to understand whether the combining effects of coupled phenomena can result in stressful situations for the people. At some point hopefully we will perform the transient analysis and not only harmonic.

Based on PSD-spectra at wind farms around the town Landau, SW Germany, we find discrete frequency peaks, which we can associate as multiples of the blade-passing-frequency. The noise level at these frequency peaks increase up to 100 times with rising wind speed.

Effects of WT related signals can be also observed at borehole stations with 100 m to 300 m depths. By comparing with seismic data of surface station, we assume that the attenuation due to distance is much less than the attenuation effect with depth.

Comparison of simulations with field measurements for different topography and geology are also planned at a later stage.

Acknowledgements

This work is part of the joint research project "Objective Criteria for Seismic and Acoustic Emission of Inland Wind Turbines (TremAc), FKZ 0325839A", funded by the German Federal

Ministry for Economic Affairs and Energy (BMWi). The authors are grateful for the financial support.

References

Abbasi, T., Premalatha, M., Abbasi, T., Abbasi, S.A. (2014). "Wind energy: Increasing deployment, rising environmental concerns", Renewable and Sustainable Energy Reviews, 31:270–288.

Adewumi, G., Nyamayoka, L.T and Inambao, F. (2015). "Reducing infrasound and low frequency noise from wind turbine blades and rotors using ANC", Renewable Bioresources 2015.

Agnantiaris, J. P. and Polyzos, D. (2003). "A Boundary Element Method for Acoustic Scattering from Non-Axisymetric and Axisymmetric Elastic Shells", Comp. Modeling in Eng. Sciences, 4(1):197-212.

Al-Bahadly, I. (Ed.) (2011). "Wind Turbines", Published by InTech, Croatia.

Aliabadi, M. H. (2002). "The Boundary Element Method, Volume 2: Applications in Solids and Structures", John Wiley, England.

Bakker, R.H., Pedersen, E., van den Berg, G.P., Stewart, R.E., Lok, W., Bouma, J. (2012). "Impact of wind turbine sound on annoyance, self-reported sleep disturbance and psychological distress", Science of the Total Environment, 425:42–51.

Gortsas, T., Tsinopoulos, S. V. and Polyzos, D. (2015). "An advanced ACA/BEM for solving 2D large-scale problems with multi-connected domains", CMES: Computer Modeling in Engineering & Sciences, 107(4):321-343.

Hau, E. (2006). "Wind Turbines. Fundamentals, Technologies, Application, Economics", Springer.

Hubbard, H. and Shepherd, K. P. (1991). "Aeroacoustics of large wind turbines," J. Acoust. Soc. Am. 89(6):2495–2508.

Jakobsen, J. (2005). "Infrasound Emission from Wind Turbines", Journal of Low Frequency, Vibration and Active Control, 24(3):145-155.

Jonkman, J., Butterfield, S., Musial, W. and Scott, G. (2009). "Definition of a 5-MW Reference Wind Turbine for Offshore System Development", Technical Report NREL/TP-500-38060.

Katinas, V., Marčiukaitis, M. and Tamašauskienė, M. (2016). "Analysis of the wind turbine noise emissions and impact on the environment", Renewable and Sustainable Energy Reviews, 58:825–831.

Keith, S.E. *et al.* (2016). "Wind turbine sound power measurements", J. Acoust. Soc. Am., 139(3):1431-1435.

Keith, S.E. *et al.* (2016a). "Wind turbine sound pressure level calculations at dwellings", J. Acoust. Soc. Am., 139(3):1436-1442.

Klæboe, R. and Sundfør, H.B. (2016). "Windmill Noise Annoyance, Visual Aesthetics, and Attitudes towards Renewable Energy Sources", Int. J. Environ. Res. Public Health, 13:746-765.

Leventhall, G. (2003). "A Review of Published Research on Low Frequency Noise and its Effects", DEFRA report EPG 1/2/50, Defra publications, London.

Lindberg, S. and Backteman, O. (1988). „Loudness of impulse and community noises", journal of low frequency noise and vibration, 7:98-103

Lott, F., Ritter, J., Al-Qaryouti, M. and Corsmeier, U. (2017), On the Analysis of Wind-Induced Noise in Seismological Recordings, Pure Appl. Geophys., doi: 10.1007/s00024-017-1477-2

Michaud, D.S. et al. (2016). "Personal and situational variables associated with wind turbine noise annoyance", J. Acoust. Soc. Am., 139(3):1455-1466.

Michaud, D.S. et al. (2016a). "Exposure to wind turbine noise: Perceptual responses and reported health effects", J. Acoust. Soc. Am., 139(3):1443-1454.

Mo, J-O. and Lee, Y-H. (2011). "Numerical simulation for prediction of aerodynamic noise characteristics on a HAWT of NREL phase VI", Journal of Mechanical Science and Technology, 25 (5):1341-1349.

Møller, H. and Pedersen, C.S. (2011). "Low-frequency noise from large wind turbines", J. Acoust. Soc. Am. 129 (6):3727-3744.

Oerlemans, S., Sijitsma, P. and Mendez Lopez, B. (2007). "Location and quantification of noise sources on a wind turbine", Journal of Sound and Vibration 299:869-883.

Oerlemans, S. and Schepers, J.G. (2009). "Prediction of wind turbine noise and validation against experiment", Report NLR-TP-2009-402.

Öhlund, O. and Larsson, C. (2015). "Meteorological effects on wind turbine sound propagation", Applied Acoustics 89:34–41.

Pedersen, E. and Waye, K. P. (2004). "Perception and annoyance due to wind turbine noise—A dose-response relationship", J. Acoust. Soc. Am., 116(6):3460–3470.

Pedersen, E., van den Berg, F., Bakker, R. and Bouma, J. (2009). "Response to noise from modern wind farms in The Netherlands," J. Acoust. Soc. Am. 126(2):634–643.

Polyzos, D., Tsinopoulos, S. V., Beskos, D. E. (1998). "Static and Dynamic Boundary Element Analysis in Incompressible Linear Elasticity" European Journal of Mechanics A/Solids, 17(3):515-536.

Ritter, J., Wagner, M., Wawerzinek, B. and Wenzel, F. (2008). "Aims and First Results of the TIMO Project – Tiefenstruktur des mittleren Oberrheingrabens", Geotectonic Research, 95:151-154.

Rushforth, I., Styles, P., Manley, D. J. M. P. and Toon, S. (1999). "Microseismic Investigation of Low Frequency Vibrations and their possible effects on Population", Journal of Low-Frequency Noise, Vibration and Active Control, 18(3):111-121.

Rushforth I., Moorhouse A. and Styles P. (2003). "An Integrated Acoustic/Microseismic Approach to Monitoring Low Frequency Noise & Vibration – A Case Study", Building Acoustics, 10(1):77-95.

Saad, Y., Schulz, M.H, (1986). "GMRES: A generalized minimal residual algorithm for solving nonsymmetric linear systems", SIAM J. Sci. Stat. Comput. 7(3):856-869.

Saccorotti, G., Piccinini, D., Cauchie, L. and Fiori, I. (2011). "Seismic Noise by Wind Farms: A Case Study from the Virgo Gravitational Wave Observatory, Italy", Bulletin of the Seismological Society of America, 101(2):568–578.

Stammler, K. and Ceranna, L. (2016). "Influence of wind turbines on seismic records of the Gräfenberg array", Seismological Research Letters, 87(5):1-7.

Styles, P., Stimpson, I., Toon, S., England, R. and Wright, M. (2005). "Microseismic and infrasound monitoring of low frequency noise and vibration from wind farms" https://docs.wind-watch.org/AEG-Eskdalemuir.pdf.

Styles, P., Westwood, R.F., Toon, S. M., Buckingham M.P., Marmo B. and Carruthers B. (2011). "Monitoring and Mitigation of Low Frequency Noise from Wind Turbines to Protect Comprehensive Test Ban Seismic Monitoring Stations", Fourth International Meeting on Wind Turbine Noise, Rome.

Triantafyllidis, Th., Prange, B., Vrettos, C. (1987). "Circular and Rectangular Foundations on Halfspace: Numerical Values of Dynamic Stiffness Functions", in Developments in Geotechn. Eng. Vol. 4.4: Ground Motion and Engineering Seismology (Ed. A.S. Cakmak), Elsevier and CMP, 409-426.

van Renterghem, T., Bockstael, A., De Weirt, V. and Botteldooren, D. (2013). "Annoyance, detection and recognition of wind turbine noise", Science of the Total Environment, 456–457:333–345.

Vasterling, M., Wegler, U., Bruestle, A., and Becker, J. (2016). "Real time monitoring of induced seismicity in the Insheim and Landau deep geothermal reservoirs, Upper Rhine Graben, using the new SeisComP3 cross-correlation detector", Geophysical Research Abstracts, Vol. 18, EGU2016-3007.

Vavourakis, V., Protopappas, V. I. C., Fotiadis, D. I. and Polyzos, D. (2008). "Numerical determination of modal dispersion and AE signal characterization in waveguides through LBIE/BEM and Time-Frequency analysis", Computational Mechanics, 43(3):431-441.

Wagner, S., Bareiβ, R., Guidati, G. (1996). "Wind Turbine Noise", Springer.

Withers, M.M., Aster, R.C., Young, C.J., Chael, E.P. (1996). "High-frequency analysis of seismic background noise as a function of wind speed and shallow depth", Bull Seismol Soc., Am 86(5):1507-1515

Wrobel, L.C. (2002). "The Boundary Element Method, Volume 1: Applications in Thermo-Fluids and Acoustics", John Wiley, England.

7th International Conference
on
Wind Turbine Noise
Rotterdam – 2nd to 5th May 2017

Subjective experiments on the perception of tonal component(s) contained in wind turbine noise

Sakae Yokoyama, Kobayasi Institute of Physical Research: sakae@kobayasi-riken.or.jp.
Tomohiro Kobayashi, Kobayasi Institute of Physical Research: kobayashi@kobayasi-riken.or.jp.
Hideki Tachibana, Professor Emeritus, The University of Tokyo: pon-t@iis.u-tokyo.ac.jp.

Summary

Tonal components in wind turbine noise increase psycho-acoustical annoyance in the areas around wind farms. Therefore, the methods to assess the characteristics of this kind of sound should be investigated in both viewpoints, physically and psycho-acoustically. Regarding the latter problem, the authors performed auditory experiments by using a test facility capable of reproducing low frequency sounds including infrasound. As a first experiment on the effect of tonal components in wind turbine noise, the change of auditory impression was examined using artificially synthesized noises containing tonal components with varying frequencies and levels. For the test stimuli, synthesized noises modelling the frequency characteristics of the general wind turbine noise observed outdoors/indoors in immission areas were used. For the simulation of the transmitted sound from outdoors into a residential room, the proposed house-filter for single-pane windows was applied. As the second experiment, the annoyance sensation due to a tonal component in WTN was examined applying the method of subjective adjustment. For the test stimuli, artificially synthesized sounds modelling outdoor were also used. From the results of these investigations, the method for assessing the tonal components in wind turbine noise in immission areas is discussed.

1. Introduction

Regarding wind turbine noise (WTN) problem, a research project entitled "Research on the evaluation of human impact of low frequency noise from wind turbine generators" has been conducted over the three years from fiscal year 2010, funded by a grant from the Ministry of the Environment, Japan. In this research project, nationwide field measurement [1,2], social survey [3], and auditory experiments [4-8] were performed. Regarding the third topic, in the research project, the hearing thresholds for low frequency pure tones, the audibility of low frequency component in WTN and the effect of amplitude modulated sounds were investigated. Furthermore, tonal components with discrete frequencies are often contained in WTN and they are apt to increase psycho-acoustical annoyance. Therefore, the authors have investigated the effect of tonal components contained in WTN using artificially synthesized noises containing tonal components with varying frequencies and levels. The experimental results was reported at the past conferences [9,10]. The experiments have been continued by increasing the number of the test subjects to improve the experimental reliability and the results are shown in this paper. The auditory experiment consisted of two subjects: one was to examine the difference of auditory impression by a tonal component in WTN and the other was to investigate the effect of a tonal component on annoyance sensation. In the first experiment, method of paired comparison was applied for both of outdoor/indoor scenario and in second experiment the method of subjective adjustment was used for outdoor scenario. From the results of these investigations, the standard method for assessing the tonality of WTN in immission areas is discussed.

2. Subjective Experiments on Audibility of Tonal Components in WTN

As a basic experiment on the effect of a tonal component in WTNs, the change of auditory impression was examined by varying the frequency and "Tonal Audibility" (TA) of a tonal component by using the method of paired comparison.

2.1 Experimental Setting

To investigate the difference threshold of auditory impression when any tonal component was included in WTN, the method of paired comparison was applied. In this study, all of test stimuli were presented diotically through an audio interface (RME, Multiface) and an electrodynamic headphone (SONY, MDR-Z7). This system was set in consideration of the possibility of reproducing low frequency sounds. The experimental setting is shown in Fig.1. The volume controller for Comparison stimulus (Sc) in Fig.1 was used in the following annoyance matching test. The frequency characteristics of reproduction system through the headphone measured using HATS (B&K 4100) is shown in Fig.2.

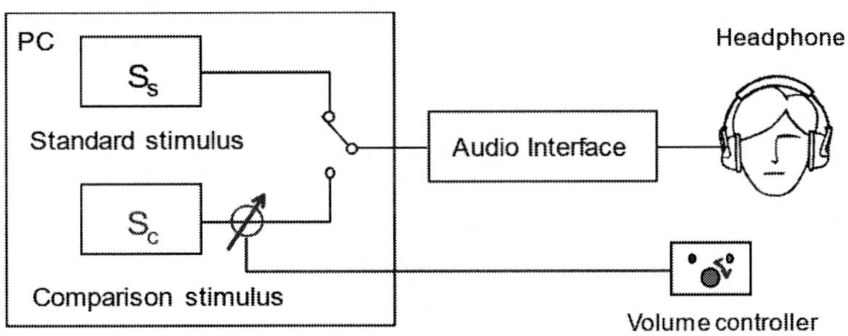

Figure 1 : Experimental setting; Volume controller was used in the following annoyance matching test.

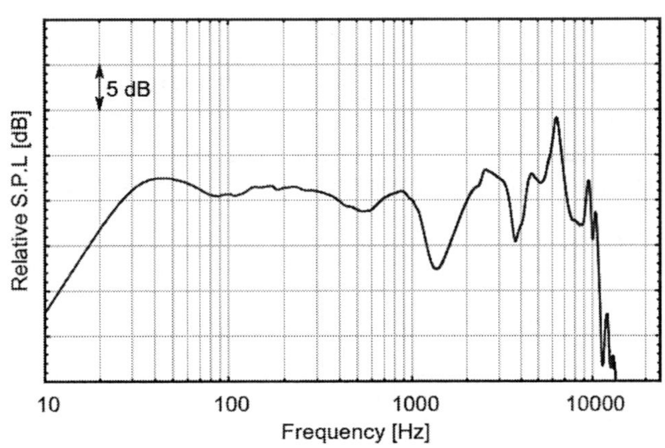

Figure 2 : Frequency characteristics of the reproduction system through a headphone (SONY, MDR-Z7) measured using HATS (B&K4100).

2.2 Test Sounds for Outdoor/Indoor Scenario

In the auditory experiments, both of outdoor and indoor scenarios were assumed.

2.2.1 For Outdoor Scenario

As the standard stimulus (Ss) for the outdoor scenario, an artificially synthesized noise modelling the frequency characteristics of general WTNs observed outdoor (-4 dB/octave in band spectrum [1]) was used and its presentation level was fixed at 45 dB in terms of A-weighted sound pressure level. It should be noted that the stimuli were steady noises. As for the comparison stimulus (Sc), a pure tone (50, 100, 200, 400 or 800 Hz) was superimposed on the model noise so that its TA varies in 7 steps as shown in Table 1 in consideration of results of TAs detected in field measurements. In this study, "Tonal Audibility" (TA) specified in IEC 61400-11: 2012 was used. As an example, the FFT spectrum of a test stimulus containing a tonal component at 200 Hz is shown in Fig. 3(a). As the comparison stimuli Sc, 36 sounds in total (5 different tone frequencies, 7 variations of TA and the same sound as Ss) were synthesized and they were paired with Ss. In the outdoor scenario, A-weighted sound pressure level of Sc ranged from 45.0 dB (without a tonal component) up to 50.5 dB (a 200 Hz tone, 15 dB TA shown in Fig.3(a)). The duration time of each sound was 5 s for both Ss and Sc. To avoid click sounds, the each sound was gradually risen/fallen with a time of 0.5 s.

2.2.2 For Indoor Scenario

In addition, to simulate the situation indoor, all of test sounds mentioned above (36 sounds in total) were convolved with the house filter for the single-pane window proposed in the reference [11]. The 1/3 octave band spectra of the house filter model is shown in Fig.4 (type-B), in which the house filter models for three kinds of window constructions are proposed. As an example, the FFT spectrum of the test sound for indoor scenario containing a tonal component at 200 Hz is shown in Fig. 3(b). Figure 5 shows the comparison between TAs for outdoor/indoor scenarios. In the Fig.5, TAs of test sounds for both of scenarios agreed except for experimental condition of a 50 Hz tone. TAs of test sounds including a 50 Hz tone for indoor scenario ranged from 0.0 dB up to 17.3 dB under the experimental condition of TA for outdoor scenario from -3 dB up to 15 dB. After the convolution with house filter, TAs of test sounds containing a tonal component at 50 Hz became higher than that for the outdoor scenario by about 3 dB, respectively. As the standard stimulus (Ss) for indoor scenario, the convolved model noise without tonal components was fixed at 35.0 dB in terms of A-weighted sound pressure level. The presentation level of Sc ranged from 35.0 dB (without a tonal component) up to 44.6 dB (a 50 Hz tone, 15 dB TA) in terms of A-weighted sound pressure level.

Table 1 : 72 test sounds used in the experiments

Stimuli	Tonal Audibility [dB]
Model noise (-4 dB/octave band)	No tonal components
Model noise + 50 Hz tone	-3, 0, 3, 6, 9, 12, 15
Model noise + 100 Hz tone	-3, 0, 3, 6, 9, 12, 15
Model noise + 200 Hz tone	-3, 0, 3, 6, 9, 12, 15
Model noise + 400 Hz tone	-3, 0, 3, 6, 9, 12, 15
Model noise + 800 Hz tone	-3, 0, 3, 6, 9, 12, 15

* 16 test sounds enclosed by dotted line in Table 1 were also used in the following annoyance matching test.

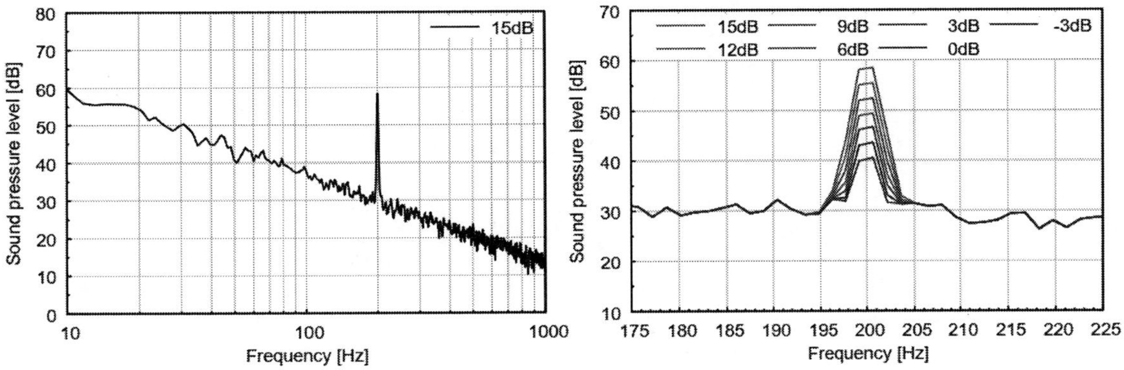

(a) For outdoor scenario (Ss; 45 dB)

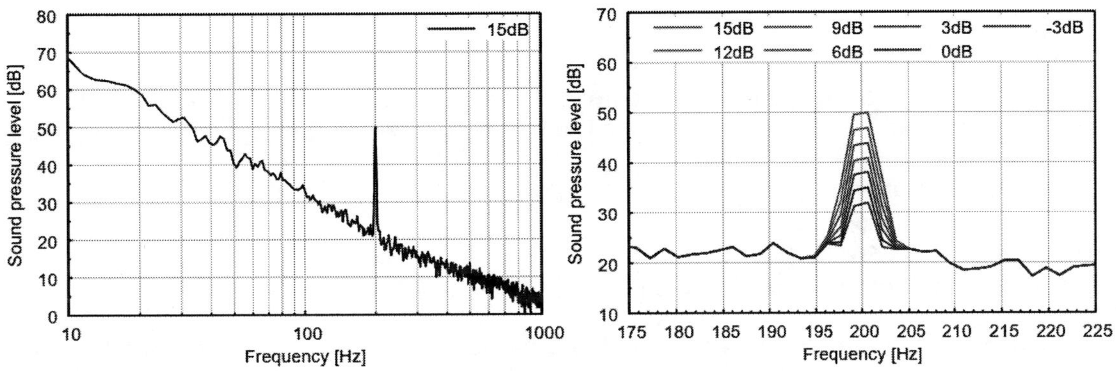

(b) For indoor scenario (Ss; 35 dB)

Figure 3 : An example of FFT spectrum of the test sound with 200 Hz tone frequency (left figure; 15 dB Tonal Audibility, right figure; 7 variations)

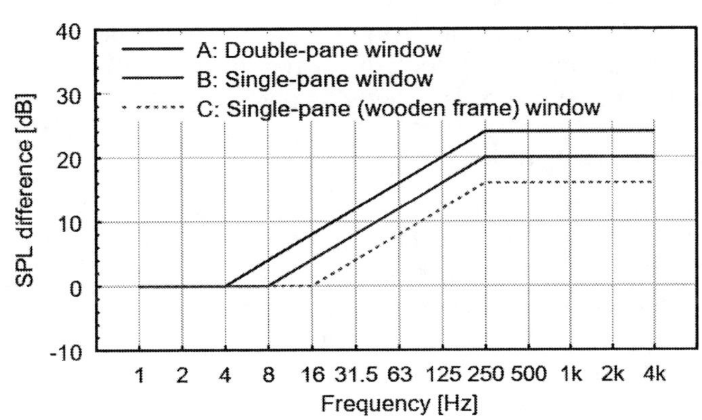

Figure 4 : House Filter Models for three kinds of window constructions proposed in the reference [11].

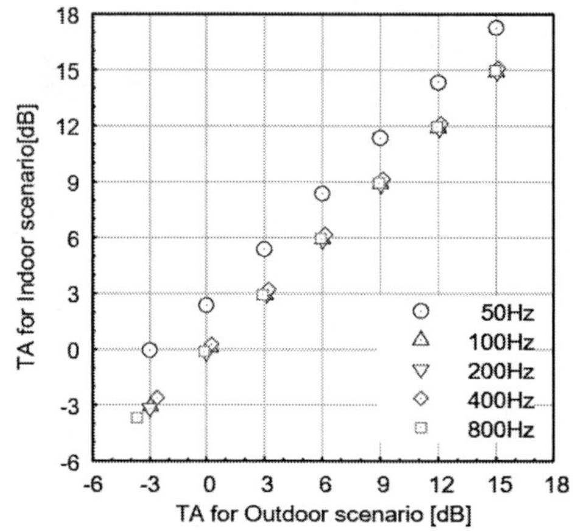

Figure 5 : Comparison between TAs for outdoor/indoor scenarios.

2.3 Experimental Procedure

In the experiment, the subject in a test room was asked to judge the difference of the auditory impression between Ss and Sc according to the following four-category system (in Japanese):

1: They are not different at all.
2: They are slightly different.
3: They are considerably different.
4: They are definitely different.

36 pairs of Ss - Sc were randomly arranged and presented to each test subject twice (72 times pair comparisons in total). The total time needed to complete the test on 72 pairs of the test sounds was about three quarters hours including rest times in between. In this experiment, 19 subjects from twenties to forties (8 males and 11 females) with normal hearing abilities participated.

2.4 Experimental Results of Audibility Tests

From the experimental results, the difference threshold of auditory impression was investigated. In these experiments, Ss (without a tone component) was also included as a Sc in order to check the reliability of participants' judgment. As a result, only 10% or 5% of the participants' response of 2 (slightly different) was found, respectively.

2.4.1 Subjective Response of "Difference"

In Figs. 6 and 7, the percentage of the subjective responses in the four categories for outdoor/indoor scenario was compared with TAs of Sc. For all of the tone frequencies, the tendency that positive response monotonically increased with the increase of tone level was found. Especially in case of a 50 Hz tone for indoor condition, subjective response for difference was higher than other cases clearly. This is caused by the fact that TAs in X-axis in Figs. 6 and 7 indicates TAs set for test stimuli assumed outdoor scenario (see; Fig.5).

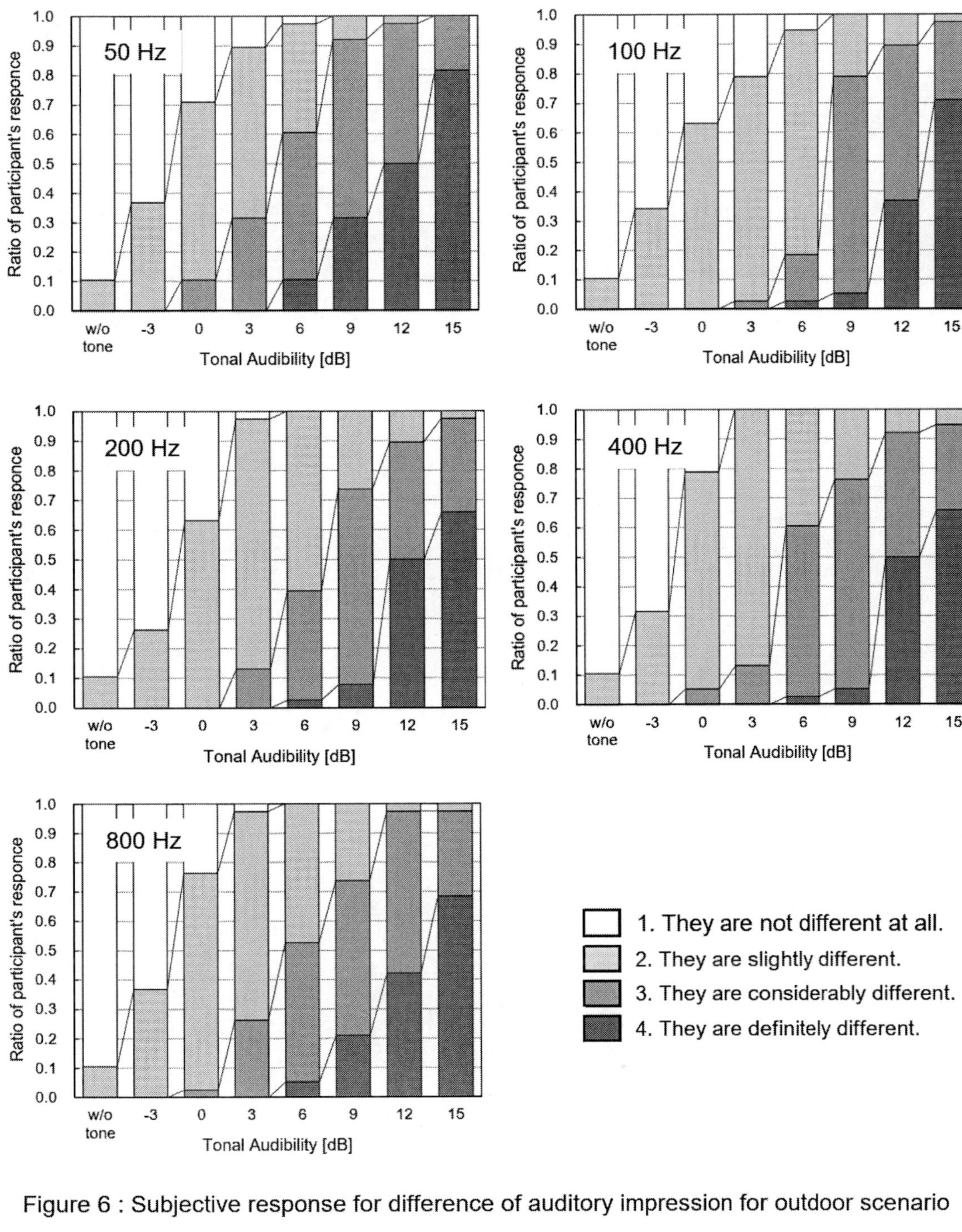

Figure 6 : Subjective response for difference of auditory impression for outdoor scenario

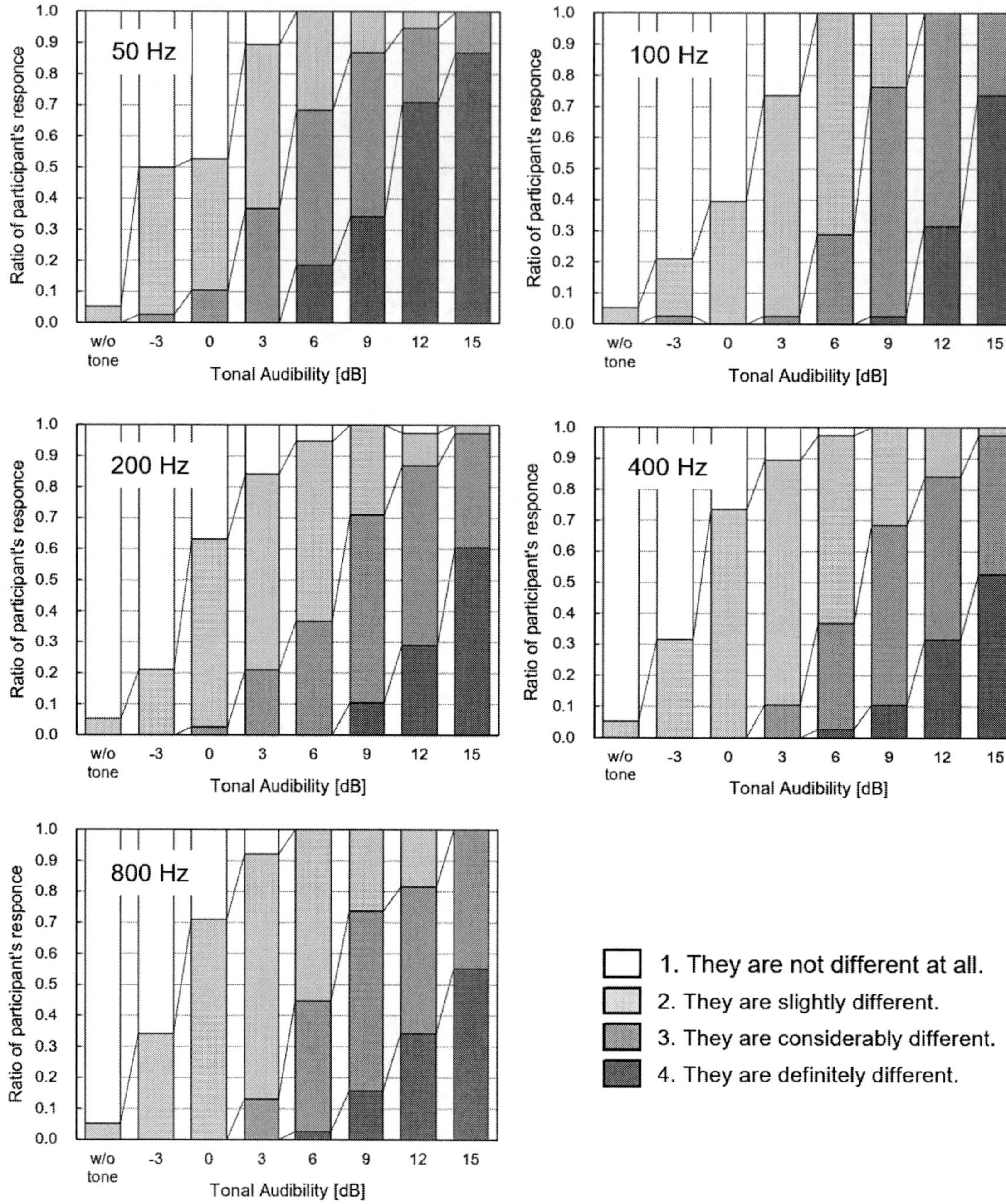

Figure 7 : Subjective response for difference of auditory impression for indoor scenario

2.4.2 Difference Threshold

To quantitatively assess the experimental results, the ratio of positive subjective response for "difference" was calculated by applying the logistic regression analysis. In the analysis, two steps of positive response regarding the difference between Ss and Sc were assumed: one is categories 2+3+4 (case-1) and the other is categories 3+4 (case-2).
Figures 8 and 9 show the relationship between TAs of Sc and the ratio of the results judged as "difference". As for the results for outdoor scenario shown in Fig.8, it is seen that the subjective response for "difference" was caused when TA was higher than around -2 dB for case-1 and between 6 dB and 8 dB according to the frequency of a contained tonal component for case-2, respectively. This tendency is consistent with reference [11]. In the results for indoor scenario shown in Fig.9, we can see that the response of "difference" was caused when TA was higher than between -3 dB and 0 dB according to the frequency of a tonal component for case-1 and around 8 dB for case-2 except for the experimental condition of a 50 Hz tone because of the reason mentioned above. In the Figs. 9(a) and 9(b) for indoor condition, the difference threshold for 50 Hz tone frequency was lower than that for other tone frequency by about 3 dB, respectively. The tendency could be explain by the difference of TAs in both scenarios. From the results, the difference threshold of auditory impression is around -2 dB in case-1 and 8 dB in case-2 in terms of TA.

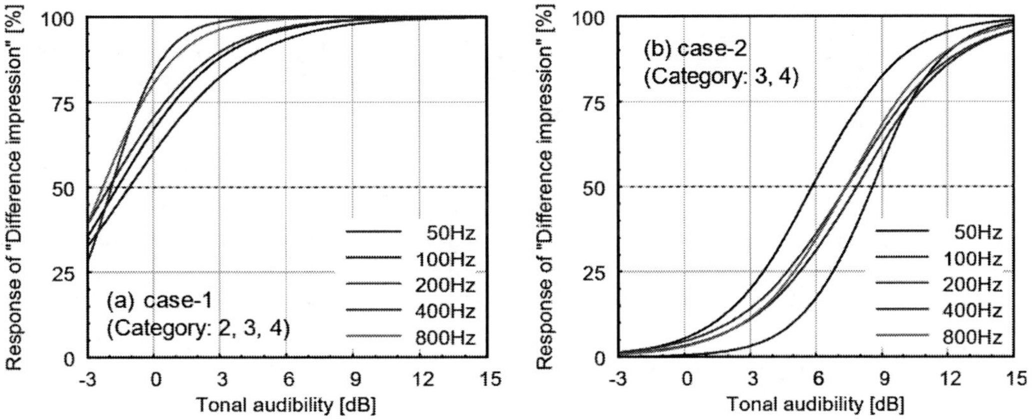

Figure 8 : Difference threshold calculated from the subjective responses for outdoor scenario

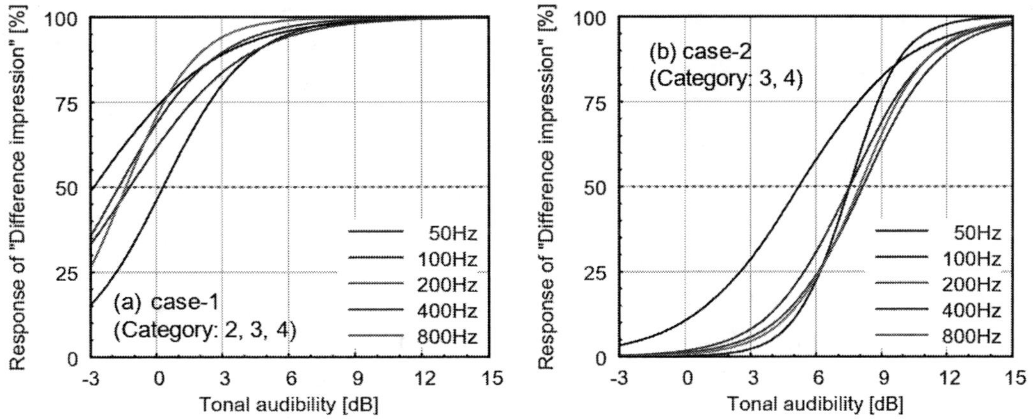

Figure 9 : Difference threshold calculated from the subjective responses for indoor scenario

3. Subjective Experiments on Annoyance of Tonal Components in WTN

To examine the effect of a tonal component in WTN on "annoyance" sensation, auditory experiment was performed applying a method of subjective adjustment.

3.1 Experimental Setting and Test Sounds

In order to investigate the relationship between "annoyance" and the strength of TA, annoyance matching test was performed for outdoor scenario. In this experiment, the reproduction system with a volume controller shown in Fig.1 was used. All test sounds were presented diotically through a headphone system.
As for the standard stimulus (Ss) in this experiment, the model noise for WTN containing a tonal component used in the previous paired comparison test assumed outdoor scenario where the sounds were used as the comparison stimulus (Sc) was again used (see; Table 1). For Ss in the annoyance matching test, 16 test sounds in total including a model noise without tonal component were chosen; a tonal component was set at 100, 200 or 400 Hz and TA was set at -3, 0, 3, 6 or 9 dB. The reproduction level of test sounds was set at the same level as test sounds for outdoor scenario mentioned above, respectively. A-weighted time-averaged SPLs ($L_{Aeq,5s}$) of Ss were listed in Table 2. As the Comparison stimulus (Sc), the model noise for WTN without a tonal component was used.

3.2 Experimental Procedures

As the test procedure, the method of adjustment was applied using the experimental system shown in Fig.3. In each condition, the standard stimulus (Ss) was firstly presented and secondly the comparison stimulus (Sc) was presented. After that, the subject was asked to adjust the "annoyance (impression of unpleasant or harsh)" of Sc so as to be equal to that of Ss by using a volume controller. For the ascending/descending series for outdoor scenario where the reproduction level of Ss was set at the level shown in Table 2, Sc was firstly set at 30/60 dB, respectively. The pair of Ss and Sc was repeated until the subject completed the adjustment. For each experimental condition, two trials (ascending and descending) were performed. The total time needed to complete the test of 16 test sounds was about three quarters hours including rest times in between. For this experiment, 18 subjects from twenties to forties (8 males and 10 females) with normal hearing abilities participated. All subjects have also participated in the previous experiment.

Table 2 – L_{Aeq} of the standard stimuli (Ss).

Tonal Audibility [dB]	100 Hz tone	200 Hz tone	400 Hz tone
No tonal components	(45.0)	(45.0)	(45.0)
-3 dB	45.1	45.2	45.1
0 dB	45.2	45.3	45.3
3 dB	45.5	45.7	45.5
6 dB	45.9	46.2	46.0
9 dB	46.6	47.1	46.8

3.3 Experimental Results of Annoyance matching Test

Figure 10 shows the experimental results of the annoyance matching test. In the figure, X-axis indicates TAs of Ss and Y-axis indicates the each level of test sounds Sc in $L_{Aeq,5s}$ adjusted by all of test subjects. In the figures, the level of Ss which contains a tonal component shown in Table 2 is also represented. In these experiments, Sc (without a tone component) was also included as a Ss in order to check the reliability of subjects' judgment. Looking at the results of investigation for the reliability, we can see that the level difference between Sc and Ss (without a tone component) was sufficiently low. As an experimental result, it is seen that the level of the adjusted Sc increased as the TA of Ss became higher in all cases of the frequency of a tonal component. The difference between the levels in $L_{Aeq,5s}$ of Ss and Sc in each test condition ranged from 0.8 dB to 2.5 dB.

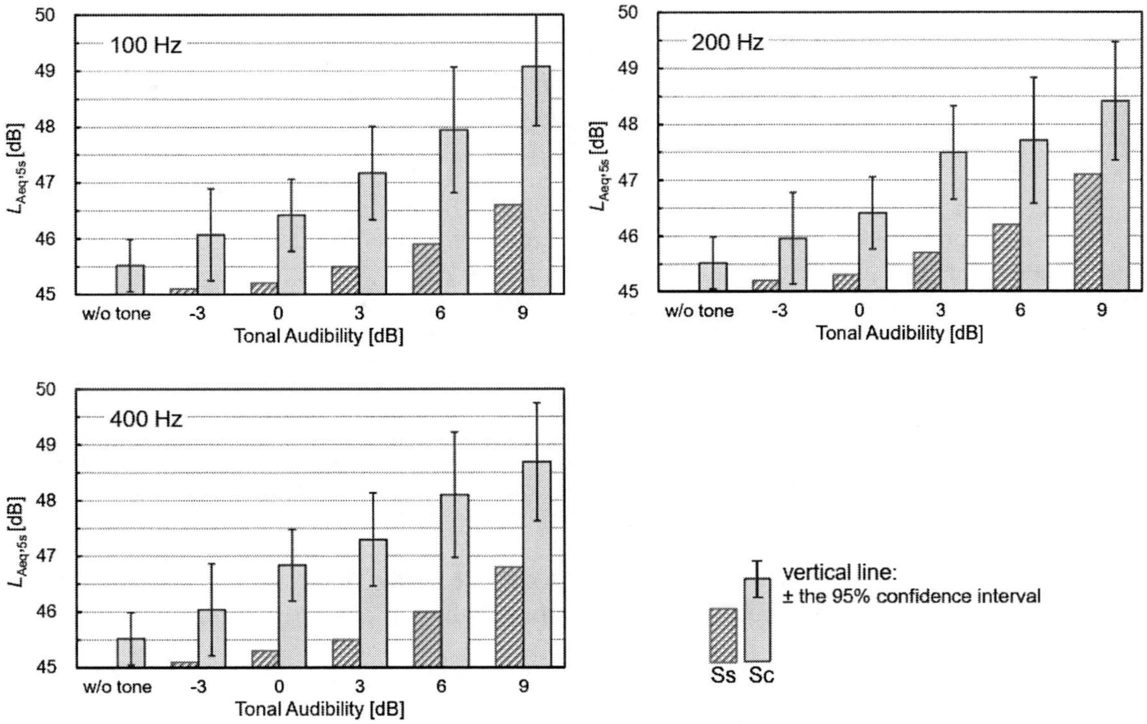

Figure 10 : Experimental results of annoyance matching test by all subjects.

4. Conclusions

Regarding the effect of tonal components contained in wind turbine noise (WTN), two kinds of auditory experiments were performed.

As for the first experiment, to investigate the validity of the numerical assessment method regarding the tonality of WTN, a basic auditory experiment on the difference threshold of auditory impression was tried using an artificially synthesized model noise of WTN and that including a tonal component. As a result, the difference threshold for "considerably different" + "definitely different" was around 8 dB in Tonal Audibility (TA) in all frequency conditions (50, 100, 200, 400 and 800 Hz) for both of outdoor/indoor scenarios.

From the results of the second experiment in which the effect of a tonal component in WTN on "annoyance" sensation was examined, the tendency has been found that the "annoyance" impression increases as the increase of tonal audibility (TA) of model noise with a pure tone at 100, 200, 400 Hz on the whole, whereas there are differences among individuals in "annoyance" sensation for noises with a tonal component.

In our study, the effect of tonal components, amplitude/frequency modulation of tonal component(s) and amplitude modulation sounds (swish sounds) in WTNs have not been considered. Further investigation on the penalty for tonal component(s) in WTNs should be continued.

References

[1] H. Tachibana, H. Yano, A. Fukushima and S. Sueoka (2014) Nationwide field measurements of wind turbine noise in Japan, Noise Control Eng. J., 62(2), 90-101

[2] A. Fukushima, K. Yamamoto, H. Uchida, S. Sueoka, T. Kobayashi and H. Tachibana (2013) Study on the amplitude modulation of wind turbine noise: Part 1 - Physical investigation, Proc INTER-NOISE 13, Paper 0668

[3] S. Kuwano, T. Yano, T. Kageyama, S. Sueoka and H. Tachibana (2014) Social survey on wind turbine noise in Japan, Noise Control Eng. J., 62(5), 503-520

[4] S. Sakamoto, S. Yokoyama and H. Tachibana (2014) Experimental study on hearing thresholds for low frequency pure tones, Acoustical Science and Technology, 35(4), 213-218

[5] S. Yokoyama, S. Sakamoto and H. Tachibana (2014) Perception of low frequency components in wind turbine noise, Noise Control Eng. J., 62(5), 295-305

[6] S. Yokoyama, S. Sakamoto, S. Tsujimura, T. Kobayashi and H. Tachibana (2015) Loudness experiment on general environmental noises by considering low-frequency components down to infrasound, Acoustical Science and Technology, 36(1), 24-30

[7] S. Yokoyama, S. Sakamoto and H. Tachibana (2013) Study on the amplitude modulation of wind turbine noise: part 2- Auditory experiments, Proc. INTER-NOISE 13, Paper 0670

[8] S. Yokoyama, et.al. (2015) Subjective experiments on the auditory impression of the amplitude modulation sound contained in wind turbine noise, Proc. 6th International Meeting on Wind Turbine Noise

[9] T. Kobayashi, et.al. (2015) Assessment of tonal components contained in wind turbine noise in immission areas, Proc. 6th International Meeting on Wind Turbine Noise

[10] S. Yokoyama, et.al. (2016) Perception of tonal components contained in wind turbine noise, Proc. INTER-NOISE 16, 2628-2639

[11] H. Tachibana, A. Fukushima and H. Ochiai (2015) Modelling of house filter for wind turbine noise, Proc. 6th International Meeting on Wind Turbine Noise

7th International Conference
on
Wind Turbine Noise
Rotterdam – 2nd to 5th May 2017

Propagation Through A Turbulent Atmosphere Makes Blade Passage Harmonics Audible

Werner Richarz, Richarz Technical Solutions, werner@richarztechnicalsolutions.com
Harrison Richarz, Iconac, harrison@iconac.co

Summary

Autocorrelations of measured wind-turbine sound exhibit periodic 'infrasonic pulses' at the blade passage frequency. Aero-acoustic models must account for unsteady inflow conditions to capture the observed features. Furthermore, propagation through a turbulent atmosphere randomizes the relative phase of the frequency components of the infrasonic pulse. Whereas the low frequency pulses are virtually inaudible, phase randomization gives rise to an audible swoosh-like sound even though the pressure perturbation pressure levels are about 30 dB below those of the infrasonic pulse.

1. Introduction

Wind turbine noise is often compartmentalized into audible, low frequency and infrasound. The respective boundaries are somewhat elastic. Traditional noise impact assessment is focused on the audible range with the A weighted sound pressure level as a 'universal' metric. C-weighting, tonal corrections and certain low-frequency/infrasound ratings are used by many regulators to address the unique features of wind turbine noise.

Emphasis is on average exposure levels. Other than amplitude modulation, 'real-time' analysis via auralization [1] and similar approaches have received less attention. This need not be so since appropriate instrumentation and signal processing permits one to examine many distinct features of wind turbine noise. This note addresses blade passage tones of a wind turbine. Walker and others [2,3] have shed light on the infrasonic pulses. Usually only the first few harmonics can be identified as broad-band noise tends to mask the overtones. Even though obscured, they are still generated by the forces that are imparted to the air by the rotating blades.

Once generated, these signals propagate through an in-homogeneous medium: the turbulent atmospheric boundary where velocity and temperature gradients interact with the waves. The effect of turbulent eddies on mean square sound pressure levels, a scattering phenomenon, are usually dealt with in a statistical sense.

Real-time changes in sound pressures due to propagation through turbulence have received little attention owing to analytical as well as experimental challenges. The effect of random phase delays on the periodic component of wind turbine sound due to rotating aerodynamic forces are examined in this paper.

2. Sound emitted by rotating forces on wind turbine rotors

Sound generation by wind turbines and aircraft propellers is virtually identical. Whereas propellers impart mechanical energy to the surrounding air, wind turbines extract it. Typical operating parameters for a wind turbine and an aircraft propeller are summarized in table 1.

Table 1. Nominal Operating Parameters of Wind Turbines and Aircraft Propellers

Parameter		WT	Propeller	
Wind or forward speed	V	15	175	m/s
Rotation rate	N	0.4	14	Hz
Rotor diameter	D	90	4	m
Tip Mach number	M_t	0.34	0.73	
Number of blades	B	3	6	
Nominal power	W	2	3.0	MW

Most wind power projects are more than 300m from the closest critical point of reception. At these distances the sound propagation may be considered as 'free-field'. In the far-field the acoustic pressure is proportional to the time rate of change of the force exerted on the medium. From an aerodynamic perspective, wind turbine blades are lightly loaded and operate in the low subsonic speed regime. This permits further simplifications, namely the use of an effective radius R_e, along with an equivalent source which rotates about the wind turbine axis (Figure 1).

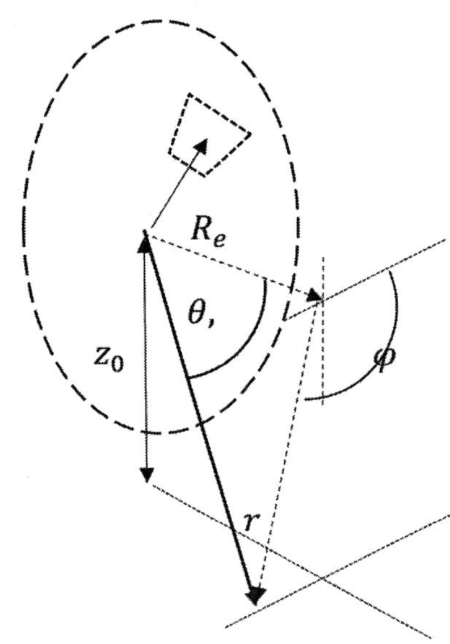

Figure 1. Sound radiation from an open rotor

Following along the lines of Morse and Ingard [4], the far-field sound pressure generated by an equivalent dipole sources on a ring of radius R_e. The radius is about 80% of the blade radius. It is a consequence of applying the mean value theorem to the more exact integral over the source region. The numerical value of R_e is a function of blade loading and geometry. It fails for higher order harmonics as the source region is no longer acoustically compact. With uniform chord-wise blade loading the far field acoustic pressure is

$$p(r,\theta,\varphi,t) = \frac{3\omega_1}{4\pi c r}\frac{W}{V_\infty}\sum_{n=1}^{\infty} P_n \sin\left[n\left(\omega_1\left(\frac{r}{c}-t\right)+B(\varphi+0.5\pi)\right)\right];\ \omega_1 = 2\pi NB$$

$$P_n = n\alpha_n\{\cos\theta + \frac{2}{3}\frac{V_\infty}{\Omega R_e M_e}\}J_{nB}(nBM_e\sin\theta)\ ;\ \alpha_n = \frac{\sin(0.5\,nbB/R_e)}{nbB/R_e}$$

Thrust and torque have been expressed in terms of power, wind-speed and rotation rate for a wind turbine operating at near optimum efficiency. The power (W) extracted from the wind turbine is the product of the force across and the flow speed through the rotor disk. Classical momentum theory shows that $V = \frac{2}{3}V_\infty$. The force is $\frac{3}{2}\frac{W}{V_\infty}$; similarly, the torque is $\frac{W}{\Omega}$.

The magnitudes of the harmonics scale as $\alpha_n J_{nB}(nBM_e \sin\theta)$. For n small α_n is approximately constant so that the Bessel functions set the relative magnitudes. As can be seen in Figure 2 the levels of the higher harmonics relative to the blade passage frequency are substantially lower.

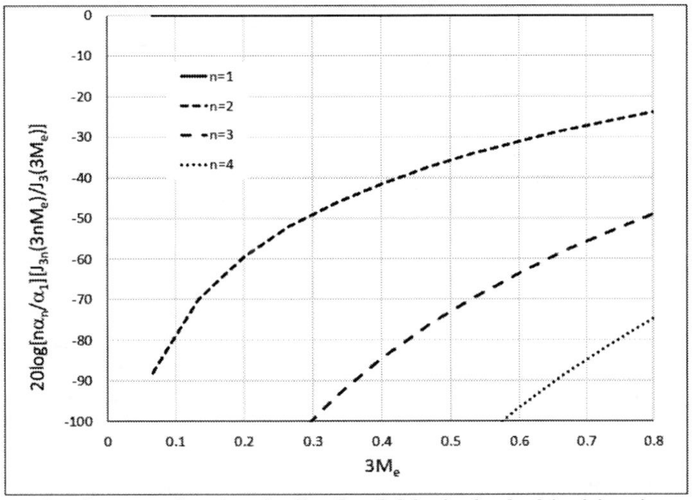

Figure 2. Low order harmonics (n=1..4) of a 3 bladed wind turbine in a uniform wind.

This is not observed in measured low frequency spectra, where the first few harmonics can be readily identified (Figure 3). The sound was measured with a ½" condenser microphone and captured on a digital data recorder. The 3dB point of the low frequency roll-off is near 5Hz so that the absolute spectrum levels are higher than indicated in the figure.

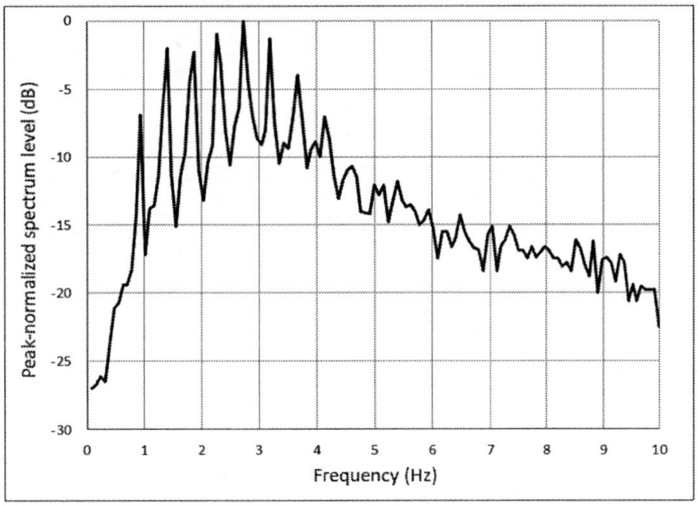

Figure 3. Measured low frequency spectrum of wind turbine sound.

3. Effect of Non-uniform Inflow

The wind turbine does not operate in a homogenous atmosphere. Non-uniform inflows are known to lift the higher harmonics of the blade passage tone by 10 to 40 dB [5]. Even a steady local shear induces a time-varying force as the inflow speed increases with the height above ground.

The approximate description of the far field sound is cumbersome but at low tip Mach numbers reduces to a set of double summations:

$$p(r,\theta,\varphi,t) = \frac{3\omega_1}{4\pi cr}\frac{W}{V_\infty}\sum_{n=1}^{\infty} na_n \sum_{l=0}^{\infty} P_{n,l}\sin\left(n\omega_1\left(\frac{r}{c}-t\right)+(nB-l)(\varphi+0.5\pi)\right)$$

$$P_{n,l} = (\beta_l\cos\theta + \vartheta_l\frac{2}{3}\frac{nB-l}{nB}\frac{V_\infty}{\Omega R_e}\frac{1}{M_e})J_{nB-l}(nBM_e\sin\theta)$$

$$Thrust(t) = \frac{W}{V_\infty}\sum \beta_l\cos l\Omega t\,;\ \ Torque(t) = \frac{W}{\Omega}\sum \vartheta_l\cos l\Omega t$$

If $l > nB$, $J_{nB-l}(z)$ must be replaced with $(-1)^{l-nB}J_{l-nB}(z)$. In general, $\vartheta_l \neq \beta_l$. Equality is assumed here for convenience. The sound field spins with an annular velocity $\frac{nB}{nB-l} \geq \Omega$. Dooley and Metalka [6] applied the concept of spinning modes to rotor-tower interaction, but did not consider the more general problem addressed here.

The wind shear at hub-height is moderate and the blade chord is small compared to the rotor diameter so that the aerodynamic forces scale as $[V^2(t)+(\Omega R_e)^2]$. The atmospheric boundary layer is modelled as $V(z) = V(z_0)(z/z_o)^\eta = V(z_0)(1+R_e/z_0\cos\Omega t)^\eta$ and the relative magnitudes of the Fourier series coefficients are

$$\beta_l = \frac{1}{\pi}\int_{-\pi}^{\pi}[(1+R_e/z_0\cos z)^{2\eta}+(\Omega R_e/V(z_0))^2]\cos(lz)dz$$

The integrals are evaluated numerically. Values for a range of exponents are listed in Table 2.

Table 2. Values of β_l for several shear exponents η; $R_e/z_0 = 0.45$

ℓ \ η	0.1	0.2	0.3	0.4
0	9.71E+00	9.70E+00	9.70E+00	9.71E+00
1	9.36E-02	1.85E-01	2.74E-01	3.62E-01
2	-8.88E-03	-1.31E-02	-1.29E-02	-8.50E-03
3	1.27E-03	1.66E-03	1.43E-03	8.02E-04
4	-2.10E-04	-2.56E-04	-2.03E-04	-1.04E-04
5	3.80E-05	4.37E-05	3.27E-05	1.58E-05
6	-7.22E-06	-7.96E-06	-5.70E-06	-2.63E-06
7	1.42E-06	1.51E-06	1.04E-06	4.64E-07
8	-2.87E-07	-2.96E-07	-1.98E-07	-8.54E-08

The β_l diminish rapidly with increasing index number, so that the summation over l can be terminated at nB. For $\varphi = 1.5\pi$ the pressure is:

$$p(r,\theta,\varphi,t) = \frac{3\omega_1}{4\pi c}\frac{W}{V_\infty}\sum_{n=1}^{\infty} na_n \sum_{l=0}^{nB} \beta_l\,(\cos\theta + \frac{2}{3}\frac{nB-l}{nBM_e}\frac{V_\infty}{\Omega R_e})J_{nB-l}(nBM_e\sin\theta)\sin\left(n\omega_1\left(\frac{r}{c}-t\right)\right)$$

The summation over l, when multiplied by na_n, is directly proportional to the amplitude of the nth harmonic. Table 3 summarizes the estimated levels for the first eight harmonics at various receiver locations and wind shear exponent of 0.4.

Table3. Normalized amplitudes ($P_n(\theta)/P_n(0)$) at selected azimuthal receiver positions

n \ θ	0.0	22.5	45.0	67.5	90.0	112.5	135.0	157.5	180.0
1	-1.00	5.00	1.00	1.00	1.00	1.00	-1.00	1.00	1.00
2	1.08	-8.27	-4.07	-2.30	-1.90	-2.36	4.02	8.49	-1.08
3	-0.85	15.91	2.98	-0.11	-0.69	-0.10	-2.93	-16.29	0.85
4	0.49	-15.71	1.41	2.44	2.20	2.51	-1.39	16.07	-0.49
5	-0.13	8.30	-3.63	-0.98	-0.08	-1.00	3.58	-8.50	0.13
6	-0.12	0.19	1.75	-1.19	-1.38	-1.22	-1.72	-0.20	0.12
7	0.20	-3.82	0.58	0.78	0.31	0.80	-0.58	3.91	-0.20
8	-0.09	1.59	-0.43	0.08	0.19	0.09	0.42	-1.62	0.09

In contrast to the operation in a quiescent medium, the sheared flow lifts the first few harmonics to levels comparable with the fundamental. The dipole nature of the sources is evident for receivers well upstream or downstream of the rotor: $P_n(\theta) \cong -P_n(\pi - \theta)$ for $\theta < 45^o$. Receivers closer to the plane of rotation are in the pressure field that is dominated by sources associated with the torque. Their dipole axes are tangent to the ring with radius R_e. At higher frequencies $n > 10$, the simple ring-source assumption is no longer valid. Observed wind turbine spectra do not appear to contain identifiable blade passage harmonics beyond $n{\sim}10$. It is likely that they are obscured by background noise.

Extraction of the tonal sound from measured signals is difficult. Signal averaging has been used with varying degrees of success []. If a trigger signal is not available some useful information can be obtained from the autocorrelation of the measured wind turbine sound. In general, the sound in the far field is composed of broad noise, machinery noise, which may be tonal, and the low frequency tones generated by the rotating blades. Therefore, the autocorrelation of the sound pressure generated by a wind turbine with an annular speed of Ω is:

$$R_{pp}(\tau) = \frac{1}{2}\sum P_n^2 \cos(n\omega_1\tau) + \frac{1}{2}\sum Q_n^2 \cos(n\omega_2\tau) + R_{nn}(\tau); \;\; \omega_1 = \Omega B$$

The last two terms are assigned to tonal machinery noise and broad-band noise from all other sources. Figure 4 shows a measured autocorrelation. Only values for positive τ are shown as autocorrelations are even functions of delay time.

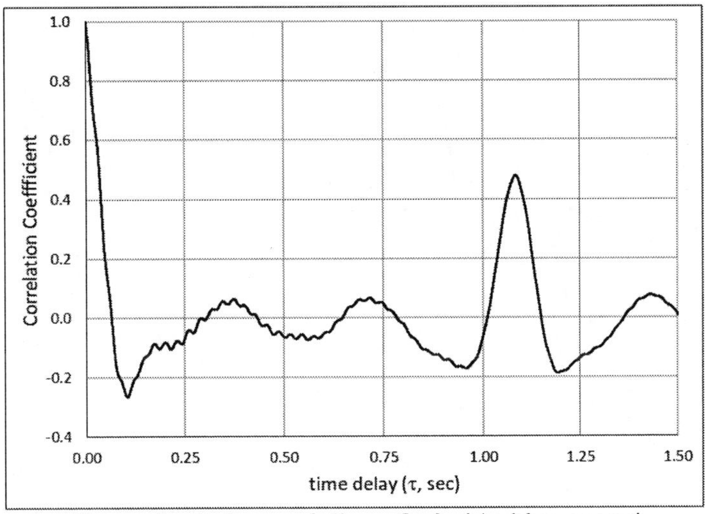

Figure 4. Autocorrelation of wind turbine sound

There is a small contribution from a mechanical tone. The infrasonic pulse can be identified for $|\omega_1\tau| > \pi$. The second prominent maximum corresponds to the period $(2\pi/\omega_1)$ of the blade passage tone. Since $R_{pp}(0) = 1 \approx \frac{1}{2}\sum P_n^2 \cos(0) + R_{nn}(0)$ and $\frac{1}{2}\sum P_n^2 \cos(2\pi n) \sim 0.5$ the overall signal energy is shared equally between the tonal sound and the broadband noise.

Autocorrelations for the model developed above do not have any broadband noise component. The latter could be incorporated if needed but does not figure in the analysis of infrasonic pulses. An autocorrelation as might be measured at a field point 22.5° from WT axis is shown Figure 5.

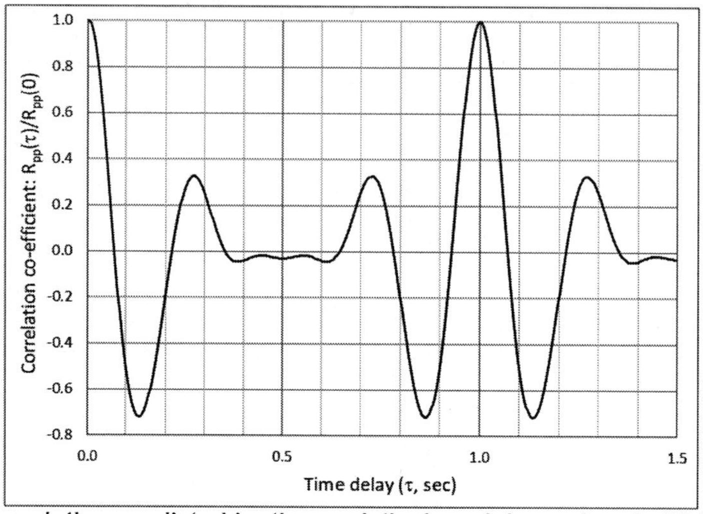

Figure 5. Autocorrelation predicted by the analytical model; receiver is 22.5° from WT axis

The patterns attributed to the rotational noise (Figure 5) are quite similar. The general agreement with measurement suggests that the source model is an appropriate one. The amplitudes P_n^2 are functions azimuthal angle which is indicative of the directivity of each harmonic. So, it is no surprise that the same applies to the autocorrelations (Figure 6). For field points in line with the rotor axis the model predicts major peaks at delay times of $\pm 2m\pi/\Omega B$. For field points close to the rotor plane, there are secondary peaks at $\pm(2m + 1)\pi/\Omega B$.

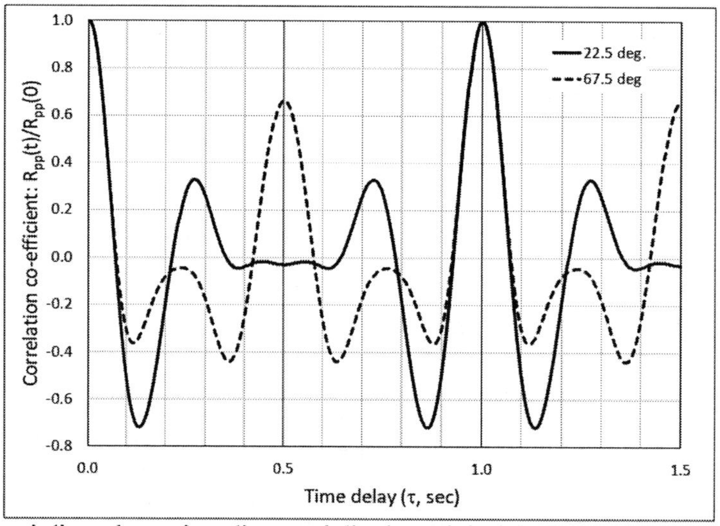

Figure 6. Autocorrelations based on the analytical model at 22.5° and 67.5° from the WT axis.

The simulated pressure-time history is computed directly: $p(\theta,t) = \sum P_n(\theta)\sin(n\omega_1 t)$. For clarity, the propagation time (r/c) is suppressed. The time signals are odd valued functions (Figure 6).

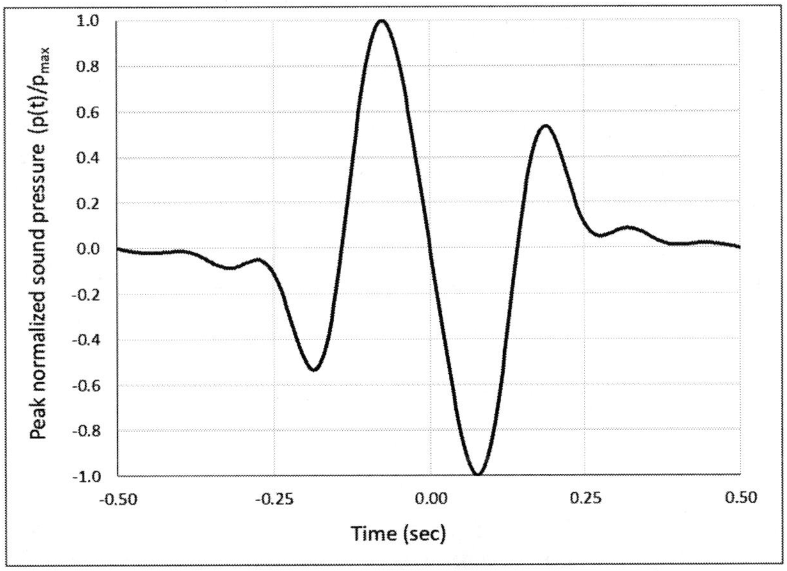

Figure 6. Infrasonic pulse predicted by the analytical model at 22.5° from the WT axis.

This behaviour was suggested by Walker on physical grounds. Vanderkooy and Mann extracted many infrasonic pulses with the aid of an elaborate signal processing algorithm. Their measured signatures (Figure 7) are quite similar to the present model.

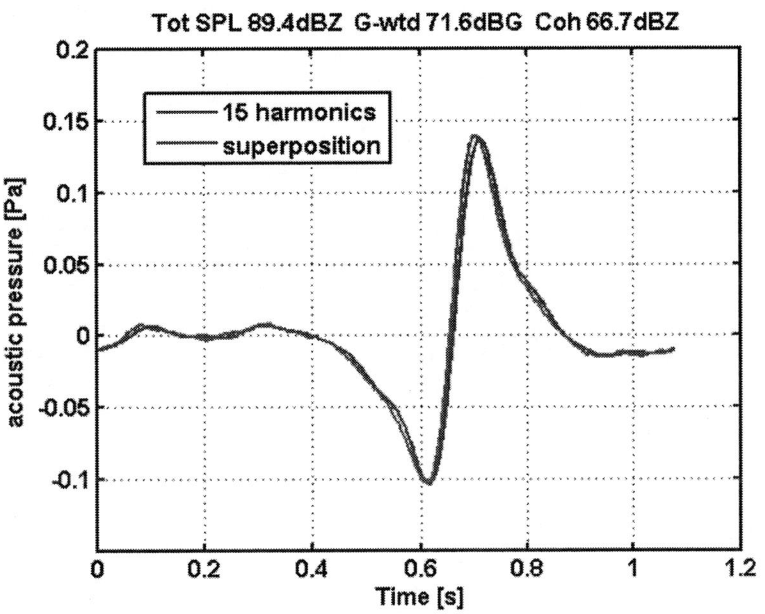

Figure 7. High definition image of infrasonic pulse {Fig. 9 in reference [3]}.

4. Wind turbine sound propagation in a turbulent atmosphere

The analysis suggests that the low frequency tones are due to periodic blade loads induced by sheared inflow or similar aerodynamic phenomena. The simplified analysis used herein does not permit one to predict higher order harmonics with any degree of accuracy. However, it is likely that they diminish in strength at a rate of the order of $1/n$.

As the infrasonic pulse propagates through the atmosphere it interacts with turbulent eddies. For a discussion of the salient aspects of the reader is referred to the work of Ostashew and colleagues [7,8].

Here a simple, empirical approach is used to explore the phenomenon. When the length scale of the eddy is compatible with the acoustic wavelength of a harmonic component, the local acoustic wave speed is altered. The result is a random time shift whose magnitude is a function of frequency.:

$$p(t,Z) = \sum_{n=1}^{\infty} b_n \sin(n\Omega \mathrm{B}(\mathrm{t} + dt_{n+100}))$$

$$\emptyset_j = Z_j \frac{j}{10+j}\,; \quad dt_{j+1} = 0.5(\emptyset_j + dt_j)$$

Daigle et al [9] provide some background for modelling the time delay. The multiplicative factor Z is random variable with zero mean and variance of the order of 0.005. The randomized arrival times have little effect on the low order harmonics so that the overall shape of the infrasonic pulse is not affected. However, as seen in Figure 8, small perturbations are superimposed on the infrasonic pulse.

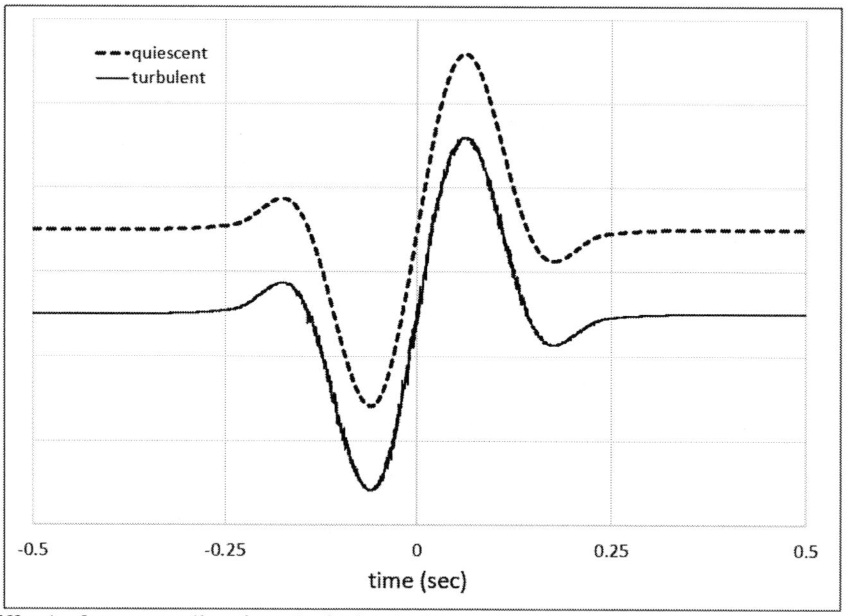

Figure 8. Effect of propagation in a turbulent atmosphere on an infrasonic pulse(simulation).

The perturbations are quite minute. For the parameters used above, the one second averaged pressure levels of the perturbations are 30dB below those of the infrasonic pulse. As the infrasonic levels are of the order of 70 to 90 dB, the perturbations are, in principle, audible. At very large distances conventional atmospheric attenuation will reduce the levels well below the natural background sound.

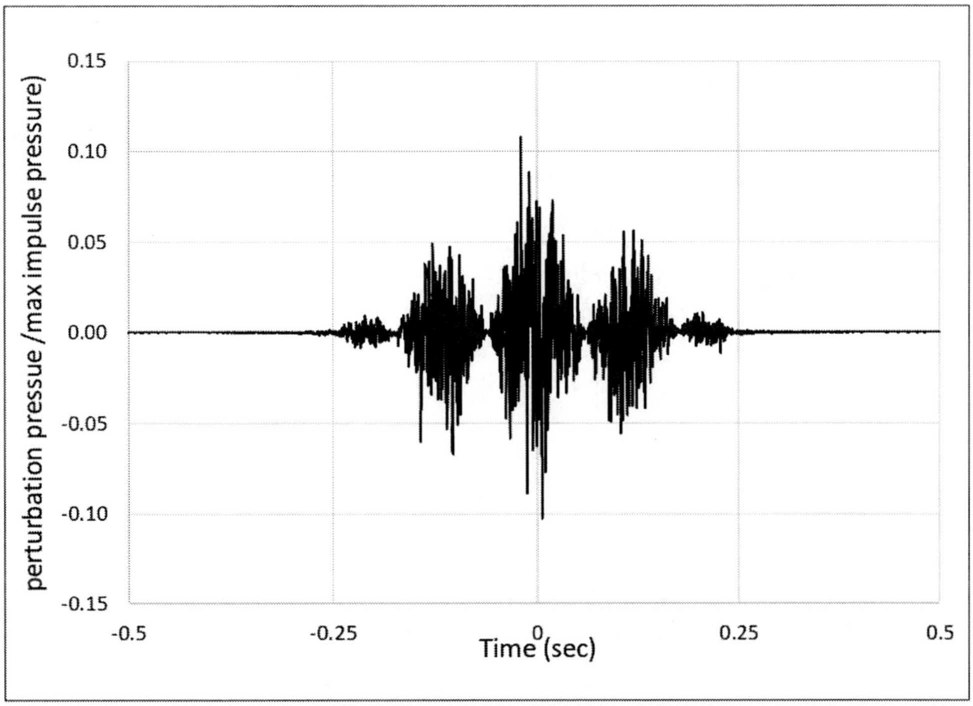

Figure 9. Perturbation to infrasonic pulse due to propagation in a turbulent atmosphere

It is evident in figure 9, the perturbations are most pronounced when the instantaneous amplitude of the infrasonic pulse is large.

5. Concluding remarks

Low frequency sound emitted by a wind-turbine is dominated by the blade passage tone and its harmonics. The higher harmonics have less energy than the fundamental, but do extend well into the audible region. As the frequency spacing is of the order of one Hertz, they are not readily detected with spectral analysis and may also be obscured by other, uncorrelated wind-turbine and background sound. Blade passage tone levels based on steady loading are not in line with observation. Realistic spectra and auto-correlations of far-field pressures are predicted only when unsteady loading, such as induced by a steady sheared inflow, is included.

As the wind-turbine sound propagates through the turbulent atmosphere it is affected by locally unsteady flow and temperature. These eddies induce random phase shifts, especially at higher harmonics where the length and time scales of sound and turbulence are compatible. Most everyone is familiar with the twinkling of stars in a clear night sky or the shimmering image of a distant hot road surface when light waves encounter similar disturbances.

The random phase distorts the smooth infrasonic pulse. The distortions are most pronounced where the amplitude of the pulse is large. The signals are audible and are like the swoosh often associated with wind turbine noise. Broad-band boundary layer/trailing edge noise also possesses a swoosh-like signal. This suggests that substantial reductions of trailing edge and boundary layer noise may not result in any appreciable acoustic benefit.

Nomenclature

b aerodynamic blade chord, (m)
B number of rotor blades
c speed of sound, (m/s)
D rotor diameter, (m)
n, l harmonic indices
r wind turbine hub-receiver distance, (m)
R_e radius of equivalent dipole ring source, (m)
t time, (s)
V wind speed, also V_∞, (m/s)
W wind turbine shaft power, (MW)
z source height above ground, (m)
z_0 hub-height, (m)
Z random variable
α_n blade loading function
β_l unsteady blade loading function
α_n blade loading function
β_l unsteady blade loading function
ω_1 (rad/s) fundamental radian frequency ($2\pi NB$), (rad/s)
Ω rotation rate ($2\pi N$), (rad)
θ polar angle of receiver with respect to wind turbine axis, (rad)
φ azimuthal angle of receiver with respect to wind turbine, (rad)
η wind shear exponent
τ time delay, (s)
$R_{pp}(\tau)$ autocorrelation function of $p(t)$; $R_{pp}(0) = \lim_{T\to\infty} \frac{1}{2T} \int_{-T}^{T} p(t)p(t-\tau)dt$

References

1] Richarz, W., Richarz, H.; (2009) *Wind turbine noise diagnostics.* 3rd Int. Conf. Wind Turbine Noise, Aalborg

2] Walker, B. (2011); *Coherence issues in wind turbine noise assessment.* 4th Int. Conf. Wind Turbine Noise, Rome.

3] VanderKooy, J., Mann, R. (2015); *Measuring wind turbine coherent infrasound.* 6th Int. Conf. Wind Turbine Noise, Glasgow

4] Morse, P.M., Ingard, K.U.; (1968) *Theoretical Acoustics, McGraw Hill,* pp. 738-747.

5] Borchers, I.U., Scholten, R., Gehlhar, B.W.; (1986) *Experimental results of the noise radiation of propellers in non-uniform flows.* AIAA Paper 86-1928

6] Dooley, K.A., Metelka, A.; (2013)*Acoustic Interaction as a primary cause of infrasonic spinning mode generation and propagation from wind turbines*, Proc. Mtg. Acoust. Soc. Amer., JASA Vol 20.

7] Ostashev, V.E., Wilson, D.K., Vecherin, S.N.; (2013) *Frequency decorrelation and pulse propagation in a turbulent atmosphere.* Proc. Mtg. Acoust. Soc. Amer., JASA Vol 20.

8] Ostashev, V.E, Scanlon, M.V., Wilson, D.K., Vecherin, S.N.; (2008) *Source localization from an elevated acoustic sensor in a refractive atmosphere.* JASA, Vol 124 (6) pp. 3413-3420.

9] Daigle, G.A, Piercy, J.E., Embleton, T.F.W.; (1983) *Line of sight propagation through atmospheric turbulence near the ground.* JASA Vol 74 (5) pp. 1505-1513

7th International Conference
on
Wind Turbine Noise
Rotterdam – 2nd to 5th May 2017

The assessment of a hearing thresholds in the presence of infrasound

Branko Zajamsek - Flinders University, Sturt Rd., Bedford Park 5042, SA, Australia.
branko.zajamsek@flinders.edu.au
Kristy L. Hansen - Flinders University, Sturt Rd., Bedford Park 5042, SA, Australia.
kristy.hansen@flinders.edu.au
Gorica Micic - Flinders University, Sturt Rd., Bedford Park 5042, SA, Australia.
branko.zajamsek@flinders.edu.au
Peter Catcheside - Flinders University, Sturt Rd., Bedford Park 5042, SA, Australia.
peter.catcheside@sa.gov.au
Colin H. Hansen - University of Adelaide, North Tce., Adelaide, 5005, SA, Australia.
colin.hansen@adelaide.edu.au

Summary

Preliminary results on hearing threshold at low frequencies in the presence of infrasound at inaudible levels are presented. The measurements were taken under free-field conditions in an anechoic room. The infrasonic source is able to generate pure-tone sound pressure levels up to 68 dB in the 8 Hz one-third octave band and up to 57 dB in the 1 Hz one-third octave band without any significant harmonic distortions. Hearing thresholds were determined in the frequency region from 50 to 400 Hz for 11 otologically normal participants, with and without infrasound present. Although some participants showed altered hearing thresholds in the presence of infrasound, there were no statistically significant group effects. These preliminary results require cautious interpretation, with further data required to clarify if there are any consistent effects of infrasound on low frequency hearing thresholds.

1. Introduction

Wind farm noise measured up to a few kilometres away from a wind farm occupies the frequency range between 0.8 and 200 Hz as shown by Zajamsek et al. (2016). This noise is commonly present during the night time when the wind farm power output is relatively high (>50 %) due to strong wind, and the background noise is low due to stable atmospheric conditions. The infrasound part (0.8 – 20 Hz) of that noise is well below the normal hearing threshold while the low-frequency noise on the other hand can exceed the normal hearing threshold. Although the infrasound is not audible (being perceived as sound), it could nevertheless *a)* have an effect on the physiology of the inner ear or *b)* have an effect on the perception of audible low-frequency noise.

It has been shown by Hensel et al. (2007) that a tone at 6 Hz and 130 dB can alter cochlear processing and hence perception of high frequency sound. Such an infrasonic tone causes the cochlear partition to be displaced at the tone frequency, which consequently causes cochlear amplifier characteristics to vary at that frequency. Thus, cochlear amplifier characteristics are amplitude modulated as they vary at the frequency of the infrasonic tone. However, fundamental mechanisms of this cochlear amplifier system responsible for the wide dynamic and frequency range of human hearing remain poorly understood (Ashmore et al., 2010). Presumably due to this amplitude modulation, most of the subjects in the Hensel et al. (2007) study reported an audible humming-like sound sensation. Furthermore, Lichtenhan and Salt (2013) also propose that audible higher frequency sounds could be modulated by infrasound based on the measurements of a single-auditory nerve in cats. The authors assert that this phenomenon could explain why some people living in the vicinity of wind farms complain about a rumbling type of noise (Zajamsek et al., 2014). If Lichtenhan and Salt (2013) are correct, then this rumbling character of wind farm noise could not be measured with a microphone as it would be a function of infrasound influences on sensory processing, rather than a specific characteristic of wind farm noise itself. Moller and Pedersen (2004) also suggest that infrasound can influence hearing detection thresholds through amplitude modulation of higher frequency noise when measuring the infrasound hearing threshold. Thus, low background noise is essential for determining hearing thresholds of infrasound, and if low-frequency wind farm noise is just below the normal hearing threshold the presence of infrasound could potentially still be perceived by people living in close vicinity to wind farms where background noise is low. Bell (2014) also suggests that perhaps infrasound can be heard as a modulation of higher frequency sounds. According to Bell (2011) the amplitude modulation sensation is caused by middle ear muscles which serve to compensate for pressure variations in the cochlea caused by infrasound. Bell (2014) suggests this process could potentially lead to annoyance and sleep disturbance. However, no experimental evidence is yet available to clarify if wind farm infrasound affects the cochlear amplifier to cause infrasound perception via amplitude modulation of higher frequency noise, or via cochlea pressure effects.

This paper focuses on the effect of inaudible infrasound on the hearing threshold at frequencies below 400 Hz as this scenario is representative of real wind farm noise at a distance of a few kilometres away from a wind farm. At such large distances, inaudible infrasound and low-frequency noise, which is audible or just below the normal hearing threshold, are usually present during the night time (Zajamsek et al., 2016). Consequently, we tested the hypothesis that inaudible infrasound would influence low frequency hearing thresholds below 400 Hz due to amplitude modulation or sensory processing effects.

2. Methodology

2.1 Equipment and anechoic room

For reproducing infrasound and low-frequency (< 400 Hz) pure tones, a Krix Phoenik V2.1 speaker with 6 Ohms input impedance (35 Hz - 40 kHz frequency response) and a Krix KX-4010S commercial cinema subwoofer were used, respectively. The subwoofer has been modified (vent removal) by the manufacturer in order to extend its frequency range into the infrasonic region. The subwoofer consists of a single 18 inch driver, with 8 Ohms input impedance and can handle 700 Watts of input power. Its dimensions are; 1180 (H) × 670 (W) × 410 (D) mm. Both loudspeakers were driven by a dual channel Crown DC-300 power amplifier capable of delivering 155 Watts per channel. The frequency range of the amplifier extends from 0 Hz to 20 kHz.

Hearing threshold tests were carried out in the anechoic room at the University of Adelaide, which has the dimensions of 4.79 (L) × 3.9 (W) × 3.94 (H) m (73.6 m^3). The listener was positioned 3 m away and facing away from the loudspeakers aligned at an angle of incidence of 0°.

Sound pressure level (SPL) measurements were carried out at the listener position (reference point) and 6 adjacent points located 15 cm up, down, back, forward, left or right with the respect to the reference point in order to test the validity of free-field condition assumption in the anechoic room according to ISO 8253-2 (2009). According to this standard, the SPL variation around the reference point must be less than 1 dB. These results are shown in Figure 1 from where it can be seen that the free-field condition is satisfied down to 40 Hz at that particular listener and source position configuration.

The sound pressure at the listener position was measured with a G.R.A.S. 40 AZ low-frequency and G.R.A.S. 40 HL low-noise microphone. Time series data from the microphones were recorded using a PROSIG P8012 24 bit data acquisition system.

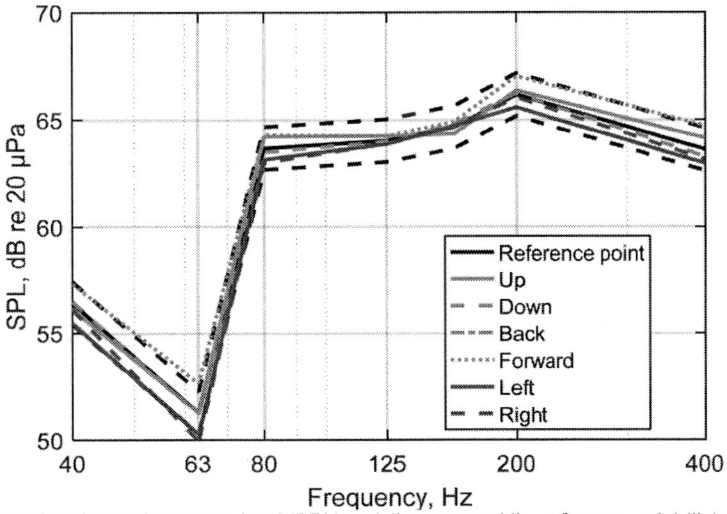

Figure 1 One-third octave band sound pressure level (SPL) variations around the reference point (listener position) at locations specified in ISO 8253-2 (2009). Dotted black lines indicates permissible 1 dB SPL variations around the reference point under the free-field condition.

2.2 Experimental procedure

The hearing threshold was assessed at 400, 200, 100, 80, 63 and 50 Hz with or without the presence of infrasound tones of constant magnitude at 0.8 or 8 Hz. The 0.8 Hz tone was chosen because that is a typical wind turbine blade pass frequency (BPF) and 8 Hz was chosen because this is usually the highest BPF harmonic in the wind farm noise spectrum (Zajamsek et al., 2016). The reproduction of these two tones using a Krix KX-4010S loudspeaker measured at the listener position is shown in Figure 2. As can be seen in Figure 2, the reproduction system reproduces these two tones with minimal total harmonic distortion.

Figure 3 shows the SPL of infrasonic tones in relation to the background noise, normal hearing threshold (ISO 226, 2003), permissible background noise for hearing threshold testing (ISO 8253-2, 2009) and infrasound hearing threshold (Moller and Pedersen, 2004). As can be seen in this figure, the infrasonic tones are well below the normal hearing threshold and the background noise is well below the permissible levels in the frequency range of interest.

Eleven participants (2 females and 9 males), of age between 20 and 70, participated in this study. Each participant underwent 2 sets of testing. Each set consisted of 2 separate listening tests, one without (control) and one with a single infrasonic tone present. The tests were performed separately with a short interruption to mark the end of one and start of another test.

One set took approximately 10 min. to complete. The break between the sets was around 10 minutes. The complete testing thus took approximately 30 min. to complete. After each completed test, the participants were asked if they noticed any difference between the tests. This measurement procedure is summarised in Table 1. The order of measurements in each set was randomly varied between the participants.

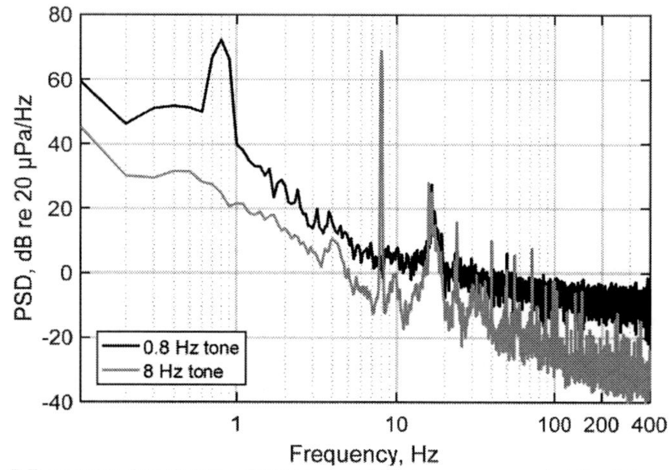

Figure 2 Power spectral density of 0.8 Hz and 8 Hz tone measured at the listener position.

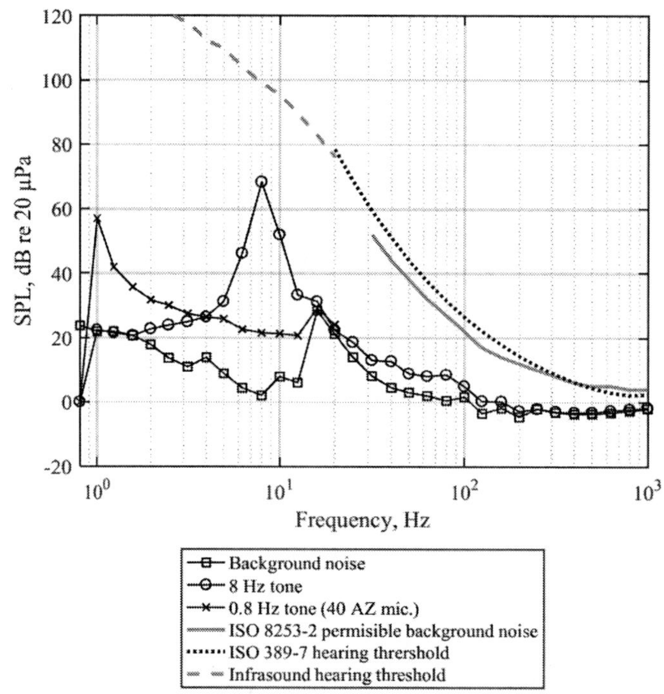

Figure 3 Comparison between background noise, infrasonic tones, normal hearing threshold, permissible background noise and infrasound hearing threshold.

The hearing threshold test tones (50 – 400 Hz) were presented in descending order from 400 to 50 Hz, and for a duration of 2 seconds. The tones were presented with the rise and fall time of

100 ms which is in accordance with the IEC 60645-1 (2012) recommendations. The pause between tone presentations was around 2 seconds, but varied and controlled manually. These characteristics of test tones were also applied to infrasound tones, which were played simultaneously with the test tones.

Table 1 Measurement matrix where HT is short for hearing threshold.

SET 1	SET 2
1. Normal HT 2. HT in the presence of 0.8 Hz (57 dB)	3. Normal HT 4. HT in the presence of 8 Hz (68 dB)

The hearing threshold level determination was done according to ISO 8253-1 (2010) and (ISO 8253-2, 2009) using a combination of a short and long version of the ascending method with a 3 dB step. The SPL was recognised as the threshold level if it was identified by the participant 2 times out of three attempts. If that failed, then the testing continued up to 5 attempts, out of which the SPL had to be recognised as a hearing threshold 3 times. The measurement uncertainty for 95 % coverage probability, U, rounded to the nearest full decibel was estimated to be 5 dB. This estimation is based on;

- measured 0.5 dB standard uncertainty for repeated measurements under identical test conditions and
- 0.1 dB standard uncertainty for audiometric performance based on a 3 dB step size and measured 0.5 dB standard uncertainty for repeated measurements

3. Results

Table 2 presents results for all participants as a difference between the normal hearing threshold (HT) and HT in the presence of infrasound. For the normal HT, a reference HT value was considered to be the lowest HT value of the two tests, if the difference between them did not exceed 2.5 dB or one standard deviation (ISO 8253-1, 2010). The results are given in a number pair where the first number represents the difference between the normal HT and HT in the presence of 0.8 Hz and the second number represents the difference between the normal HT and HT in the presence of 8 Hz. Significant differences (differences larger than measurement uncertainty of 5 dB) are shown in red. At some frequencies the difference could not be established as the reference HT value could not be established due to a difference larger than 2.5 dB between the two normal HT. For example, P5 represents an extreme case where the reference HT value could not be determined for any frequency.

The results in Table 2 show that the presence of infrasound has a limited effect on the HT as significant changes are observed for only 4 participants at selected frequencies. As can be seen, the HT can be either elevated or decreased in the presence of infrasound. An interesting case is seen for P2 at 80 Hz, where the presence of 0.8 Hz increases the HT and the presence of 8 Hz decreases the HT. A mixed model ANOVA analysis showed no statistically significant group effects of infrasound on HT.

It should be reported that three participants after completing normal HT test and HT test with the infrasound noticed a difference between the tests in terms of changed timbre of the test tone and a pressure sensation in the ears. One participant described that some of the test tones appeared to be “wobbly” in the presence of infrasound.

Table 2 The difference between the normal hearing threshold (HT) and the HT in the presence of infrasound in dB. Red numbers indicate differences larger than measurement uncertainty. Negative difference indicates that the HT in the presence of infrasound is higher than without it. Column under P5 is empty because no reference HT value could be established at any frequency.

Freq., Hz	P1	P2	P3	P4	P5	P6	P7	P8	P9	P10	P11
400		3, 6							3, 0		
200	-3, 3	0, 0		3, 3		6, 0	0, 0	3, 0	3, 0	3, 0	3, 3
100	-6, 3			0, 3						0, 0	
80		-7, 7	2, 1			0, 0		-3, -6		0, 0	
63	2, 0	0, 0	3, 0	3, 3				-6, -6			
50	-1, 0			0, 3			0, 0		3, 0	3, 0	

4. Conclusion

This paper presents preliminary results regarding the effects of tonal infrasound on the hearing threshold (HT) at frequencies below 400 Hz. These results suggest that infrasound could effect the HT at some frequencies in some participants, but do not support systematic group effects. However, further data from a larger sample size are needed to clarify infrasound effects on wind farm noise perception.

Acknowledgements

Financial support from the Australian Government National Health and Medical Research Council (NHMRC), Project GNT1113571, is gratefully acknowledged.

References

ASHMORE, J., AVAN, P., BROWNELL, W. E., DALLOS, P., DIERKES, K., FETTIPLACE, R., GROSH, K., HACKNEY, C. M., HUDSPETH, A. J., JÜLICHER, F., LINDNER, B., MARTIN, P., MEAUD, J., PETIT, C., SANTOS SACCHI, J. R. & CANLON, B. 2010. The remarkable cochlear amplifier. *Hearing Research,* 266, 1-17.

BELL, A. 2011. How do middle ear muscles protect the cochlea? Reconsideration of the intralabyrinthine pressure theory. *JHS,* 1, 9-23.

BELL, A. 2014. NHMRC Draft Information Paper: Evidence on Wind Farms and Human Health submission. https://consultations.nhmrc.gov.au/public_consultations/submissions/WIND/2922: NHMRC.

HENSEL, J., SCHOLZ, G., HURTTIG, U., MROWINSKI, D. & JANSSEN, T. 2007. Impact of infrasound on the human cochlea. *Hearing Research,* 233, 67-76.

IEC 60645-1 2012. Electroacoustics - Audiometric equipment - Part 1: Equipment for pure-tone audiometry. Geneva, Switzerland: International Electrotechnical Commission.

ISO 226. 2003. *Acoustics: Normal equal-loudness-level contours* [Online]. International Standards Organisation. [Accessed].

ISO 8253-1 2010. Pure-tone air and bone conduction audiometry. Geneva, Switzerland: International Standards Organisation.

ISO 8253-2 2009. Sound field audiometry with pure-tone and narrow-band test signals. Geneva, Switzerland: International Standards Organisation.

LICHTENHAN, J. & SALT, A. 2013. Amplitude modulation of audible sounds by non-audible sounds: Understanding the effects of wind turbine noise. *Proceedings of Meetings on Acoustics,* 19, 040064.

MOLLER, H. & PEDERSEN, C. 2004. Hearing at low and infrasonic frequencies. *Noise and Health,* 6, 37-57.

ZAJAMSEK, B., HANSEN, K. L., DOOLAN, C. J. & HANSEN, C. H. 2016. Characterisation of wind farm infrasound and low-frequency noise. *Journal of Sound and Vibration,* 370, 176-190.
ZAJAMSEK, B., MOREAU, D. J. & DOOLAN, C. J. 2014. Characterising noise and annoyance in homes near a wind farm. *Acoustics Australia,* 42, 14 - 19.

Seventh International Meeting
on
Wind Turbine Noise
Rotterdam, the Netherlands, 2nd – 5th May 2017

Measurement Techniques for determining Wind Turbine Infrasound Penetration into Homes

Andy Metelka, SVS Canada Inc, Canada ametelka@cogeco.ca

Summary

Previous measurements using advanced measurement instrumentation and Narrowband 3-dimensional FFT (Fast Fourier Transform) based signal processing indicate the presence of wind turbine induced pressures at infrasonic Blade Pass Frequencies (BPF) inside dwellings from multiple active turbines. This has been detected inside homes with new measurements which included homes in both near and extreme far fields, revealing that infrasonic wind turbine noise emissions do penetrate some homes. The role of the home in attenuating the infrasonic wind turbine emissions inside may be described by a transmissibility calculation, proposed here, and applied to simultaneous long-term measurements at three homes. Various locations inside homes are also compared to outside measurements relating wind speed, wind direction and other audible sound level meter (SLM) parameters.

Introduction

Initial measurements of wind turbine BPFs were made in 2008 when colleagues from Canadian universities queried the existence of infrasound at wind turbine installations. At that time, previous reports had indicated that wind turbines created no apparent infrasound and thus the A-weighted SLM Method was used and deemed perfectly adequate for use in describing wind turbine noise.

Since then, thousands of measurement sets have been made throughout North America by the author and others. Yet, NASA 1986 publications confirmed the presence of BPFs and concern for wind turbine infrasound long before. (*DOE/NASA/20320-68*) The measurement of operational wind turbines has consistently found the presence of a distinct pressure signature associated with the BPF and harmonics of the BPF frequency *(166th Meeting ASA San Francisco Dec 2013 Kevin A. Dooley and Andy Metelka)*. The BPF occurs in the lower portion of the infrasound range.

Newer measurements of various soundscapes, not involving wind turbines, have been made. These soundscapes include airports, passenger jet fuselages, train tracks, homes near train tracks, infrasound in automobiles, and infrasound in natural extremely quiet rural Canadian settings. The relative extended octave bands and FFT spectra below 4 Hz were low during steady state conditions compared to wind turbine sites with steady state conditions. This was also apparent both inside and outside nearby homes. Spectral content below 4 Hz was also not apparent in most of these sites, however one produced Blade Pass Frequencies which were detected approximately 120km away from the source wind turbines. *(Dooley and Metelka ASA 166th 2013)*

This was discovered accidentally by performing a 3-month long term system noise test with four infrasound microphones. Originally, the test was to determine the maximum dynamic range of the entire measurement system in a near perfect quiet soundscape. During times where maximum total wind turbine power output in Ontario was greater than 400-500MWatts or more, up to 4 BPF harmonics appeared. Average FFT LEQs(6Hr) were low (30-35dBLin), however still present. This was validated using three different measurement systems over a four-year span. When the power output dropped below 200 MWatts, the BPF became undetectable using the same basic methods for measurement used in following measurement study.

Southern Ontario has been selected for numerous wind turbine projects with 7700 wind turbines announced and to be installed. Currently there are approximately 2700 wind turbines operational. Many homes are available as measurement locations. Noise and infrasound BPF measurements in the subject homes of this wind facility were made with the occupants present and under normal living conditions.

The purpose of this paper is to present measurements inside and outside three homes simultaneously, develop a preliminary penetration criterion under different operating conditions and to develop a basis for an accurate, traceable method for infrasound penetration in homes.

Previous studies report blade-to-tower interaction resulting in BPFs. *(Dooley and Metelka, ASA 166th, 2013) (Vanderkooy and Mann, WTN 2015) (Swinbanks, WTN2015) (Cooper, Cape Bridgewater, 2016)* The measurement techniques used in this report were also used earlier to detect BPFs at over 120km away from the nearest turbine. (Dooley and Metelka, ASA 166th, 2013) These measurements indicate propagation at 3dB/distance doubling for BPFs in the far field and may not exist all the time, but can, especially during specific atmospheric conditions when wind shear and temperature inversion create a waveguide effect. (*Internoise 2014-Kristy Hansen; Branko Zajamsek: Colin Hansen University of Adelaide, Australia)* With propagation in the far field closer to 3dB, BPF infrasound measurements have consistently found propagation rates not indicating spherical 6dB/doubling typical for a point source. This is also valid for infrasound measurements made previously by SVS, confirming that wind turbine infrasound does not propagate as a point source into the extreme far field. BPFs also increased with total turbine power output. It was also observed that wind speed and wind direction (thus, turbine directionality) plays a significant role with propagation and linear weighted pressure penetration inside homes.

Contents

1. Test Sites, location and home construction

The three homes chosen for the study are located in a Class 3 Rural country environment approximately 100 km northwest of Toronto. (*Fig. 1*) Ambient rural levels of 14.8 LAeq and 14 dB L90 were commonly measured at night.

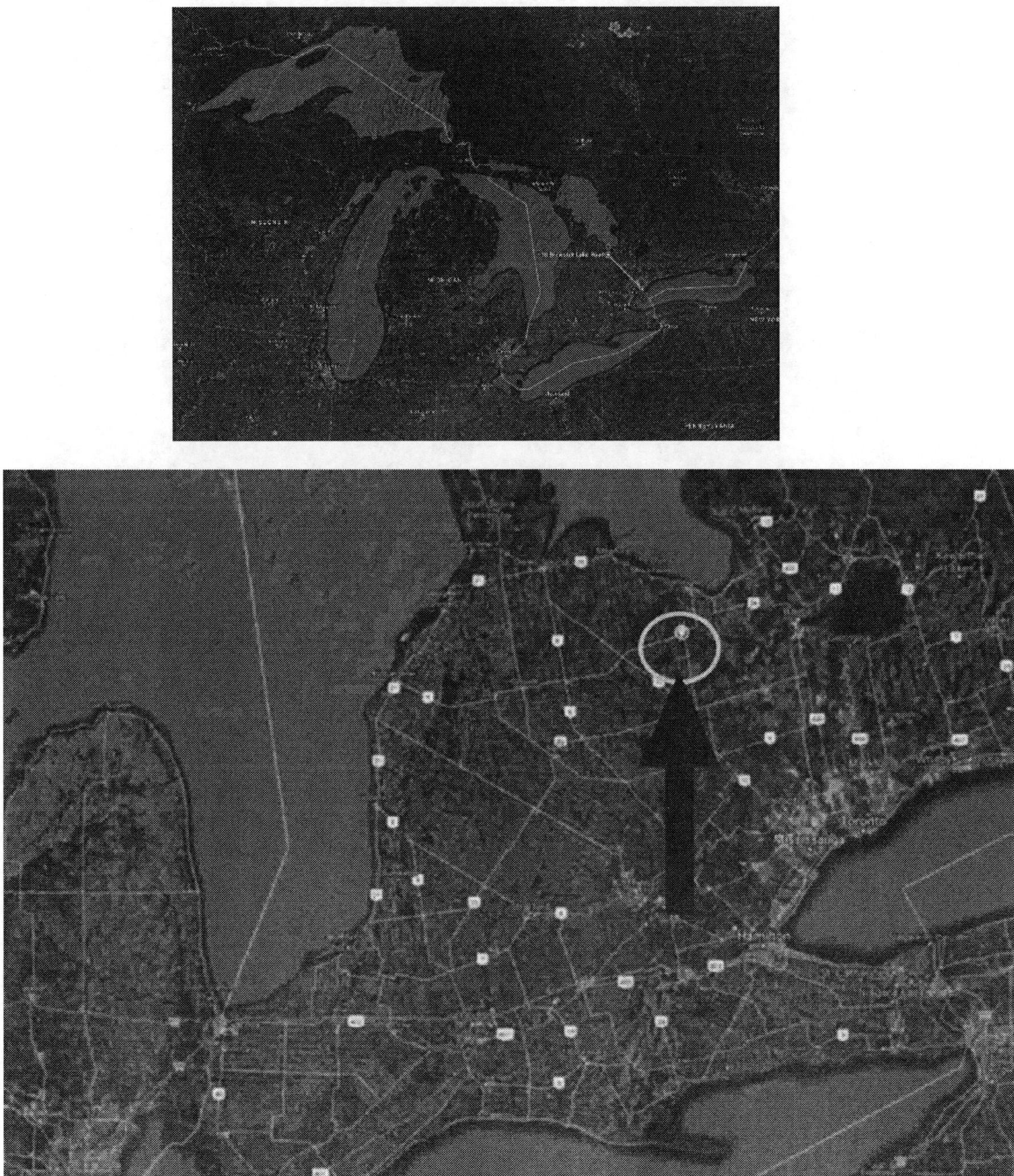

Fig.1: Study area.

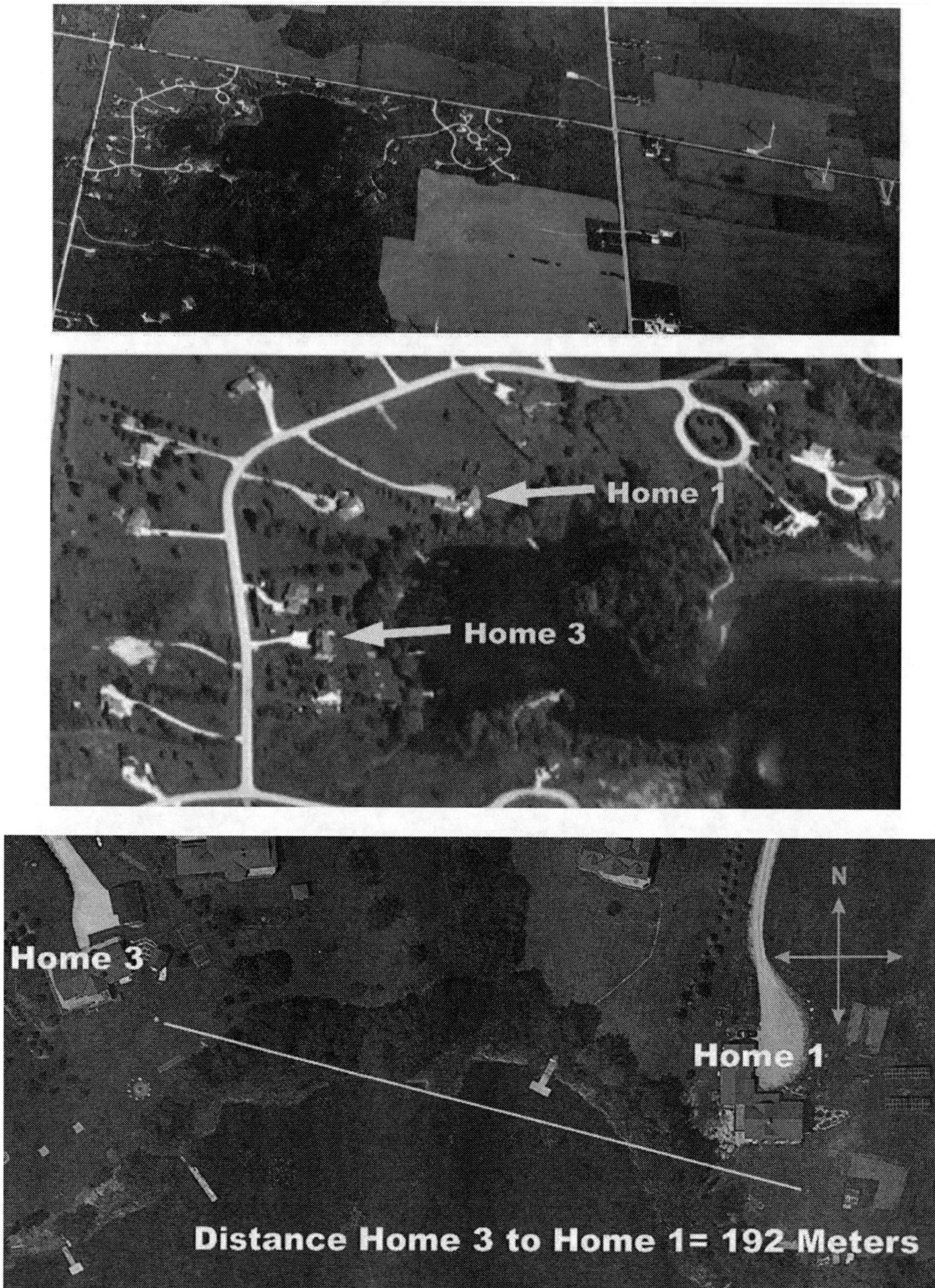

Fig. 2: Homes 1 & 3 are situated close together.
Signatures from these 2 homes are similar and repeatable.

Home 2 was further away from Homes 1 and 3, however, there was a turbine located approximately 600 meters NNW of Home 2. (*Fig. 3)* Home 2 was an earlier farmhouse design over 100 years old. It was not airtight and had vented attics. (*Fig. 3a)* Measurements were made at Home 1 & 3 simultaneously for 3 ½ months as the weather conditions changed from summer to winter. Measurements at Home 2 were made for 2 ½ months and sensors used from Home 2 were transferred inside to Home 1 for the duration of the project to measure multiple rooms simultaneously.

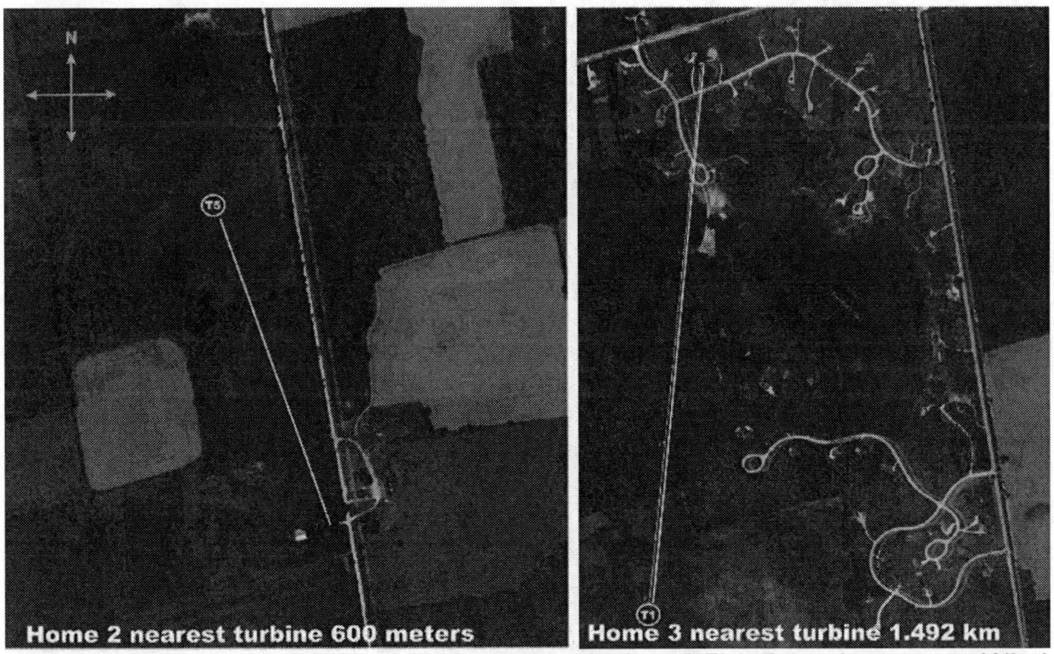

Fig. 3: Homes 1 & 3 exhibit a newer airtight design, 1.5km from the nearest Wind Turbine

Fig. 3a: Home 2, 100 years old, 600 meters from the closest turbine

Fig. 3b: 14 Turbines are S, SE & SW within 5.6 km of Homes 1 & 3. Home 2 has Turbines to the North and West with the nearest turbine approximately 600 meters north of Home 2.

Fig. 4: Wind Turbines visible to South and South West of Home 1

Fig. 5: Home 1 microphone placement and vaulted ceiling with air tight construction.

Home 3 was also modern airtight design with a small attic. (*Fig 6*)

Fig. 6: Home 3 Microphone inside a Secondary windscreen placed in open area.

Fig. 6a: Home 3 Backyard view, with USB Webcam and SINUS Apollo with Laptop

2. Test equipment and layout

Improvements to the general method for environmental acoustics are realized prior to equipment deployment and they include the use of a shielded outdoor infrasound microphone. Outside measurements used a secondary and primary windscreen which has been adopted from the Ontario Ministry of Environment and Climate Change NPC-350 guidelines. NPC-350 is an audible noise guideline and the secondary windscreen from this guideline was adopted to compare audible noise to infrasound at that same location. Currently, this guideline is under review, yet to be released. A wideband pressure sensor, GRAS 40AZ was used to measure both audible noise (Type 1) and infrasound from 0.5Hz to 20kHz. This is a precision laboratory microphone which permits comparison of audible noise and infrasound with the same traceable high dynamic range sensor as well using the same input channel.

Resulting measurements show infrasonic wind disturbance from turbulence around the secondary windscreens in combination with other low frequency artifacts inside and outside homes, yet measurement of infrasound and BPFs from the wind turbines were distinct and accurately measured.

Wind turbine infrasound as an absolute level is deemed irrelevant by most experts, but the debate with human sensory perception is beyond the focus of this paper. Measurement engineering units used to describe the dynamic nature of the wind turbine pressures, specifically the infrasonic pressures below 10Hz, are also in dB which references the 20 micro pascal human hearing threshold. A barometer was also used in combination with an advanced ultrasonic weather station. The weather station outputs entering the Soundbook slow input channels were sampled at 200 SPS. The GRAS 41AC outdoor microphone was originally used as a backup, however, the 40AZ showed the best infrasound response and was used outside and located inside the secondary wind screen. It survived the harshest tests during rain, ice and snow. Calibration checks were done with results that stayed within a 0.5dB variation during the entire testing period which proved it functioned well during the harshest rain, ice and snow. It was specially adapted with waterproof weather screens.

Measurements were made:

1. In Real-time, reducing the potential for human error compared to conventional processes of editing, data file handling, filtering and post processing.
2. With parallel processing in Real-time. No post-processing required creating directly traceable results. (*Dia 1, 2 & 3*)
3. Simultaneously in three homes, to compare results during identical environmental and wind turbine operating conditions.
4. Using the highest grade equipment, traceable to industry measurement standards.
    a. GRAS 40AZ IEC 61094-4-WS2P compliance http://www.gras.dk/.

b. Two Soundbook analyzers and 1 SINUS Apollo Analyzer were used. (*Dia 1 & 3*) Home 1 used a Thies Ultrasonic Weather Station synchronized with the SINUS Samurai Software. No programming was required other than various advanced setups. SINUS Soundbook PTB 21.21/13.05, https://sinus-leipzig.de/en/produkte/messsysteme/soundbook-mk2

Advanced Acoustic, Infra sound, Weather, Vibration, Analysis & Recording System

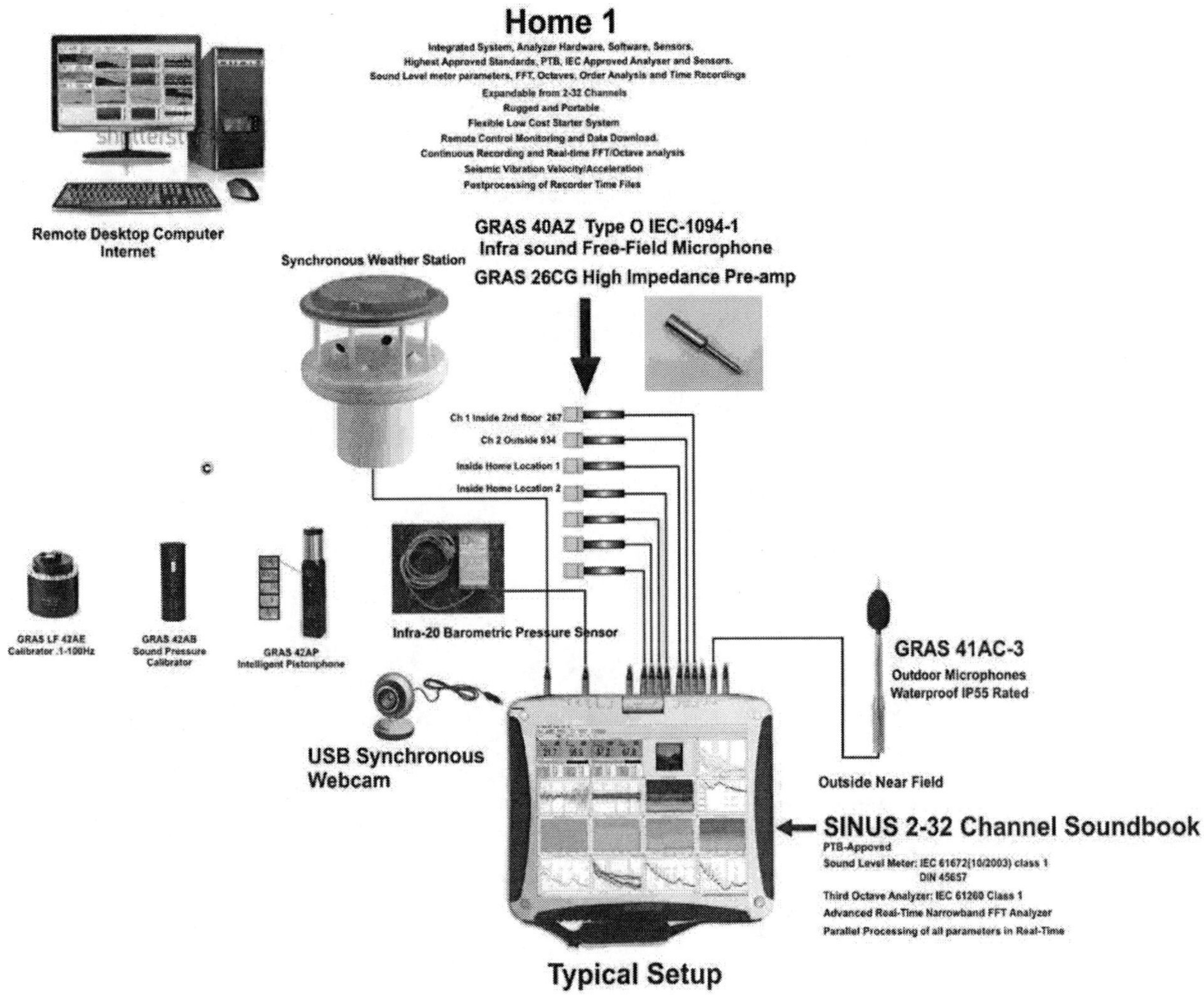

Dia. 1, Home 1: A more complex system was used, including a Thies advanced weather station mounted on the roof.

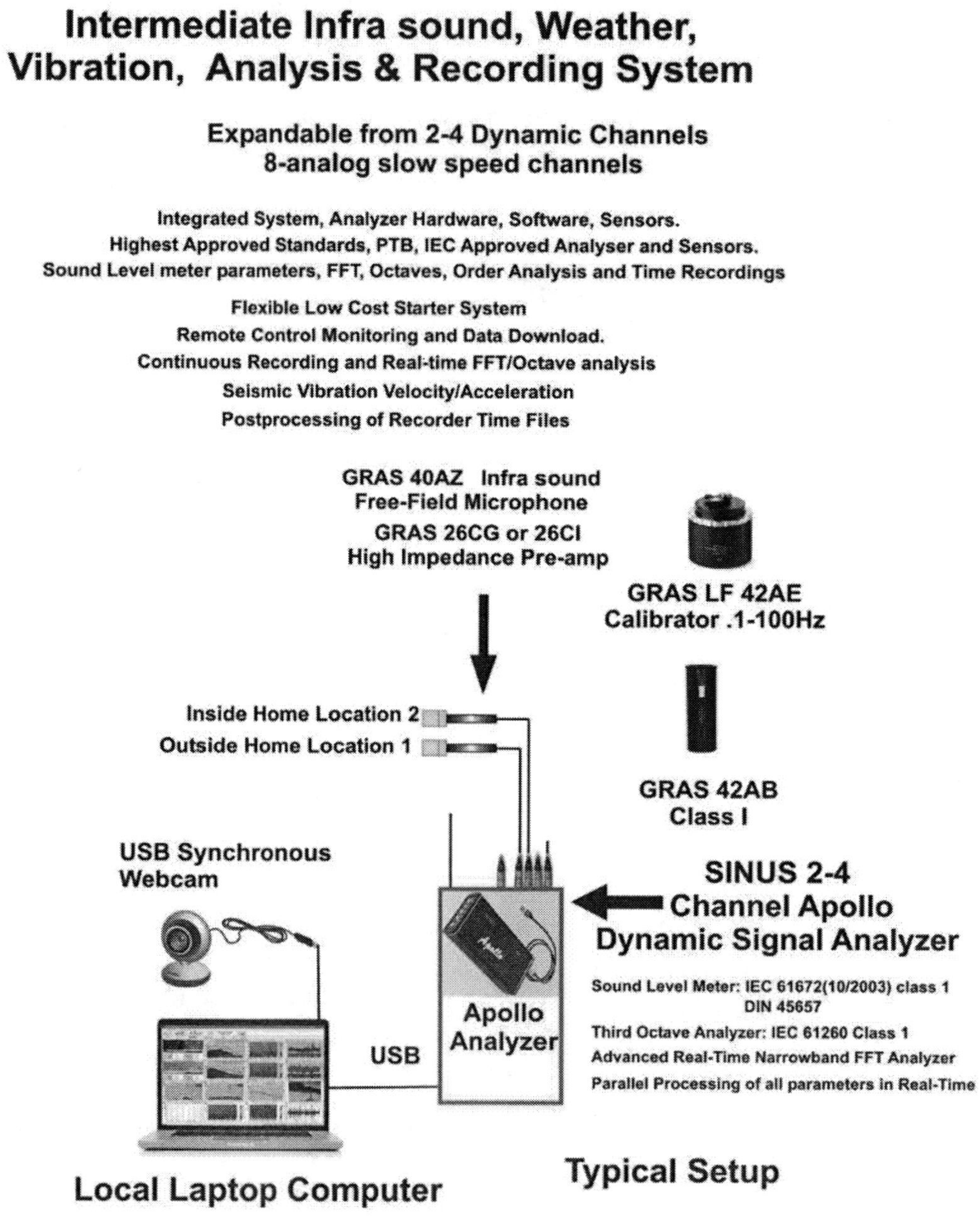

Dia. 2, Home 3: A GRAS 40AZ was placed inside and a second 40AZ was placed outside.

2 channel Acoustic, Infrasound, SLM, Real-time FFT Weather, Vibration, Analysis & Recording System

Home 2

Integrated System, Analyzer Hardware, Software, Sensors.
Highest Approved Standards, PTB, IEC Approved Analyser and Sensors.
Sound Level meter parameters, FFT, Octaves, Order Analysis and Time Recordings
Flexible Low Cost Starter System
Remote Control Monitoring and Data Download.
Continuous Recording and Real-time FFT/Octave analysis
Postprocessing of Recorder Time Files

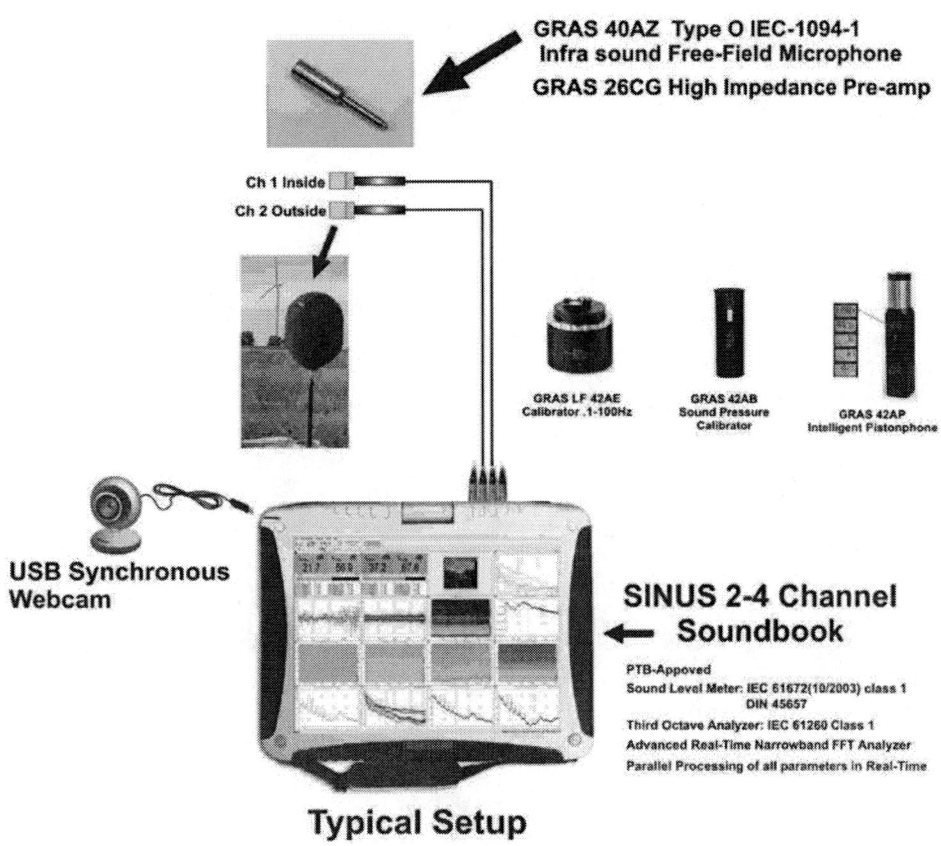

Dia.3, Home 2: A lower cost Apollo system with similar hardware and identical Samurai Software was placed.

3. Measurement methods and results

Several measurement methods were chosen in order to obtain the best results in each situation, mostly for real-time visualization and validation. When combined simultaneously they became useful. The following briefly describes attributes of these measurement methods with turbine measurement examples.

3.1. Basic Time & Pressure Domain Representation

Here, we observed the "Heart Beat" of a Wind Turbine.

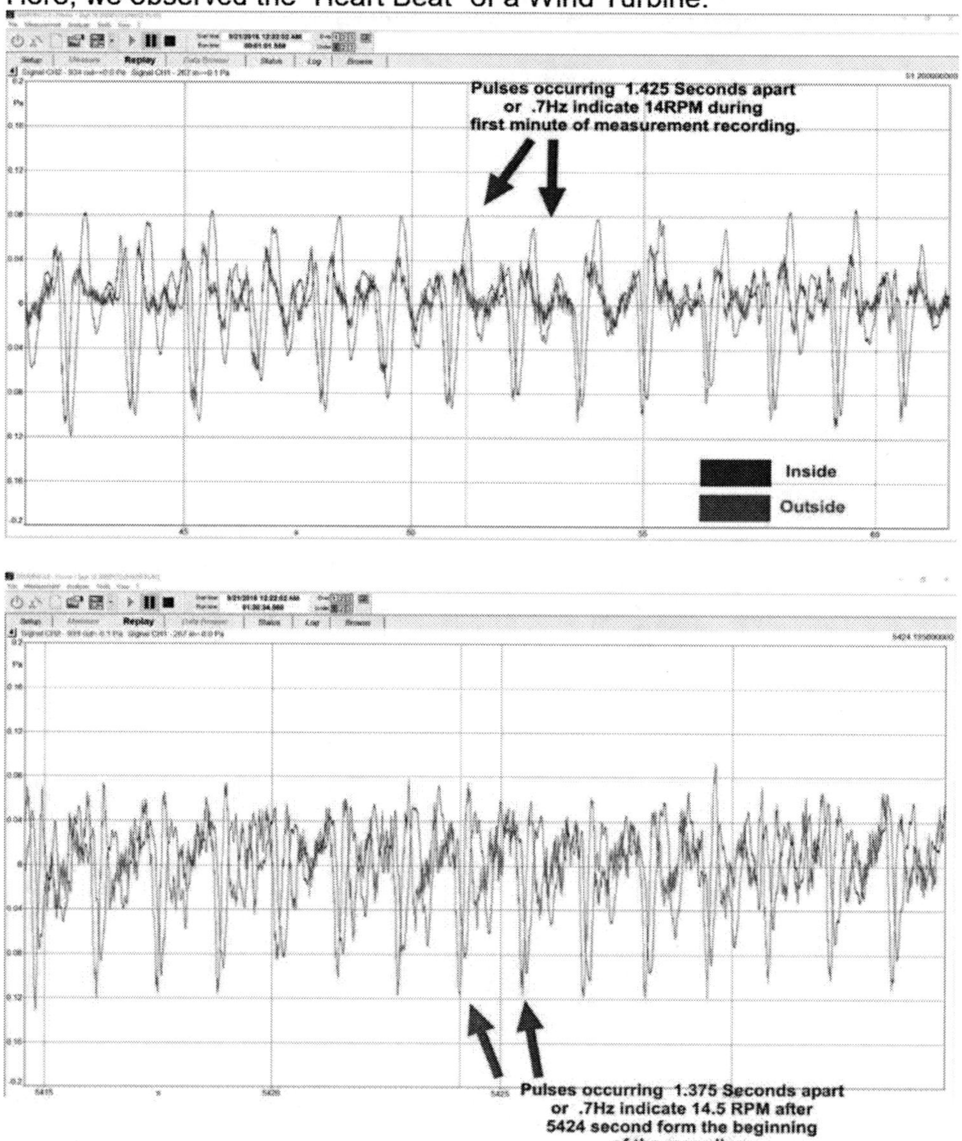

Fig. 7 Time vs. Pressure wave inside and outside a home using GRAS 40AZ infrasound Microphones with a Sinus Soundbook. The cyclical nature of overall pressure captured from the wind turbine is observed.

The basic time domain wave shapes can be observed during ideal conditions when background noise is not present. An absence of such high frequency components inside the home is clearly observed. (*Fig. 7*) Note that the levels in this example indicate large pressure variations inside the home. This is measured more precisely in sections following. The plots in *(Fig. 7)* differ slightly due to a slight difference in rotational speed of each wind turbine as calculated from the noise emission waveform, measured in the time domain. (14RPM vs. 14.5RPM) Both plots indicate the turbines were approaching their 15RPM steady state.

3.2. Sound Level Meter Time History

We observed a blade-to-tower interference with peculiar properties. Since it is a dynamic, cyclical event that comes and goes with changing wind conditions, measurement opportunities are difficult to predict. Common noise measurement tools such as SLM's and logging methods require extensive time and editing of SLM parameters and fail to accurately describe both the cyclical nature of the sound (aerodynamic modulation) as well as BPFs. Extracting turbine BPF's from background infrasound and audible noise cannot be performed using SLMs. Further, SLM metrics for establishing annoyance require an extensive approval process from international standards committees. The simple Time Level History (*Fig. 8*) is useful for comparing SLM relationships over time.

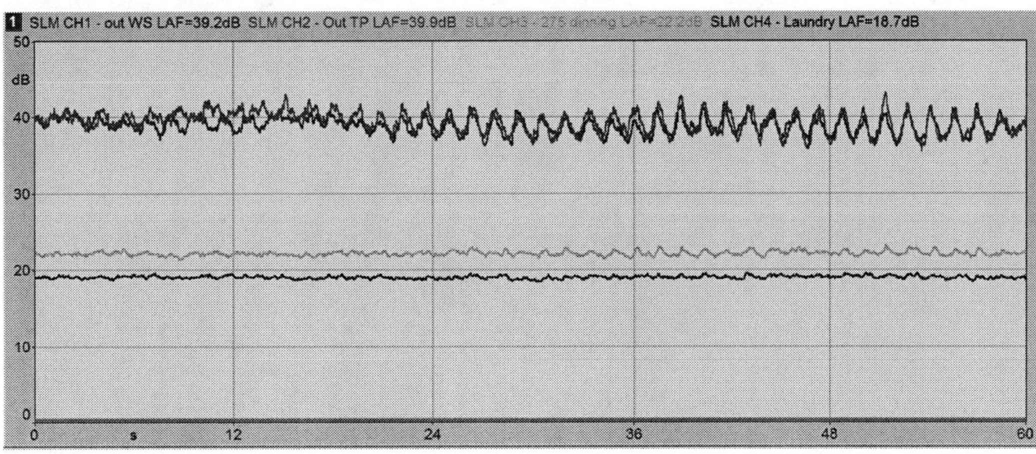

Fig. 8: SLM Time History (60 Seconds) showing the cyclical nature of the LAF inside and outside a home.

3.3. Infrasound BPFs using 1/3-Octave Band Analysis

As a turbine blade passes the tower, it creates a pressure disturbance. (*REF: 1-9*) The complex harmonic nature of this passage is referred to as a Blade Pass Frequency. Fig. 9 shows the blade pass frequencies that show basic errors with 1/1 or 1/3-octave filtering that do not identify blade pass frequencies.

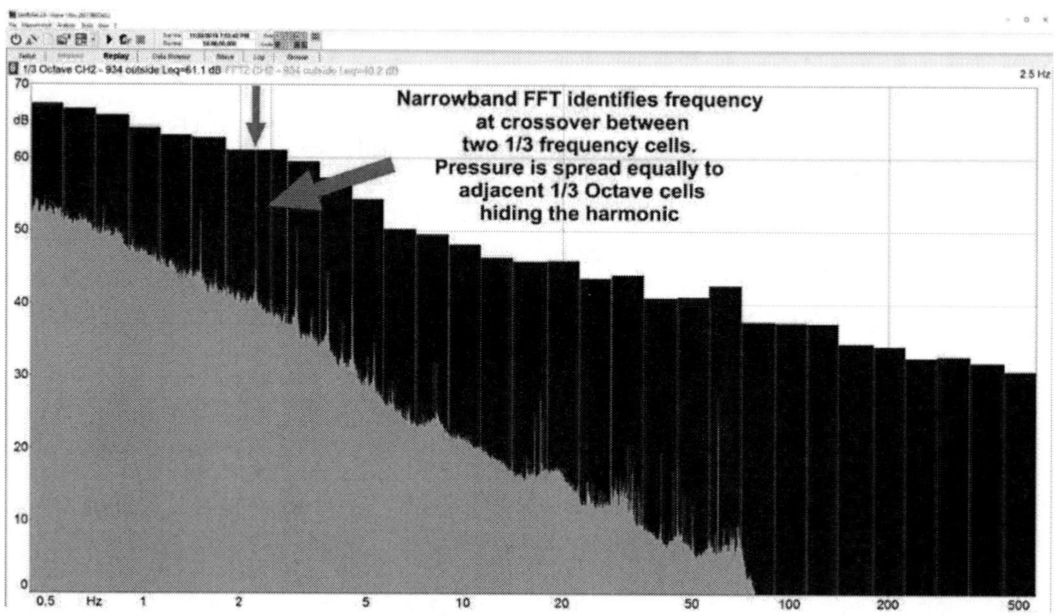

Fig. 9: Example of 1/3 octave bands (extended LF to 0.4Hz) compared to Narrowband FFT.

Only Real-time 1/3 Octaves were measured through-out this study. Octaves from FFT calculations were not considered.

3.4. Infrasound BPFs using Narrowband Real time FFT analysis.

The mathematical analysis of a pulse will produce a fundamental frequency of the occurrence of the pulse and low order harmonics of that pulse. The use of Narrowband analysis permits the identification of the fundamental frequency of the pulsing where the rotor speed of the turbine varies with the wind speed. Over time, there will be a smearing of the discrete frequencies as the RPM changes. A noticeable increase of BPFs inside the home at 0.72Hz, 1.5Hz, 2.25Hz and 6.75Hz is evident in *Figs. 10 and 11*. Another set of wind turbines at a higher RPM has a smaller set of associated harmonics. The comparison of outdoor (input) vs. indoor (output) narrowband pressures can be described as a transmissibly function. This paper refers to the inverse of this ratio as the Grey Highlands Inverse Transmissibility Ratio (GHITR).

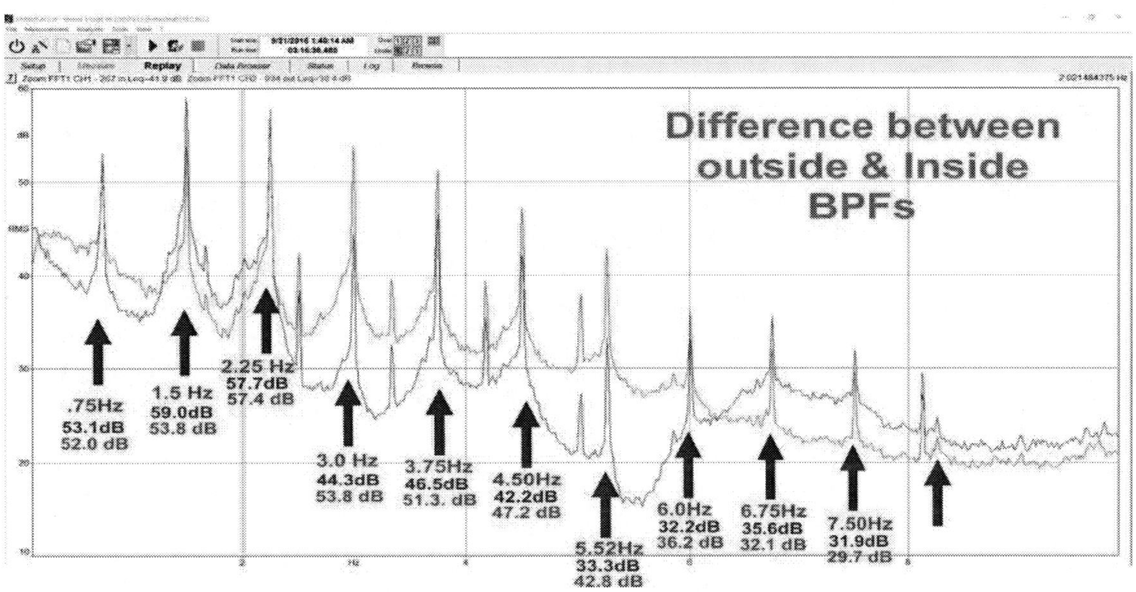

Fig. 10: The turbine with the highest level BPF peaks was selected for calculating the GHITF

				1/GHT Ratio RMS
Sept 21, 2017 1:40 AM7th		Level Lin dB	Home 3	
	Frequency	Outside	Living Room	Inverse Trransmissabilty
BP Fundamental	.750 Hz	52	53.1	1.021
BP 1st Harmonic	1.5 Hz	53.8	59	1.097
BP 2nd Harmonic	2.25 Hz	57.4	57.7	1.005
BP 3Rd Harmonic	3.0 Hz	53.8	44.3	0.823
BP 4th Harmonic	3.75 Hz	51.3	46.5	0.906
BP 5th Harmonic	4.5 Hz	47.2	42.2	0.894
BP 6th Harmonic	5.25 Hz	42.8	33.3	0.778
BP 7th Harmonic	6 Hz	36.2	32.2	0.890
BP 8th Harmonic	6.75 Hz	32.1	35.6	1.109
BP 9th Harmonic	7.5 Hz	29.7	31.9	1.074

Fig. 11 Calculation of Transmissibility from outside to inside Home 3

3.5. 3D SONOGRAMS related to Wind Speed and Direction.

Both wind direction and wind speed are related to the Blade Pass Frequencies as indicated in Fig. 12. When wind speed is below 3m/sec, turbine RPM instability is apparent. As wind speed increases and stabilizes, so does the Wind Turbine rotational speed. (RPM) Wind turbines with constant rotational speed produce stable FFT calculations over constant conditions. After 3 ½ hours of monitoring, we observed that these variable speed turbines behave similarly to fixed speed turbines. The definition of a variable speed turbine may be brought into question here. Some operate at different multiple stationary speeds and are referred to as variable speed.

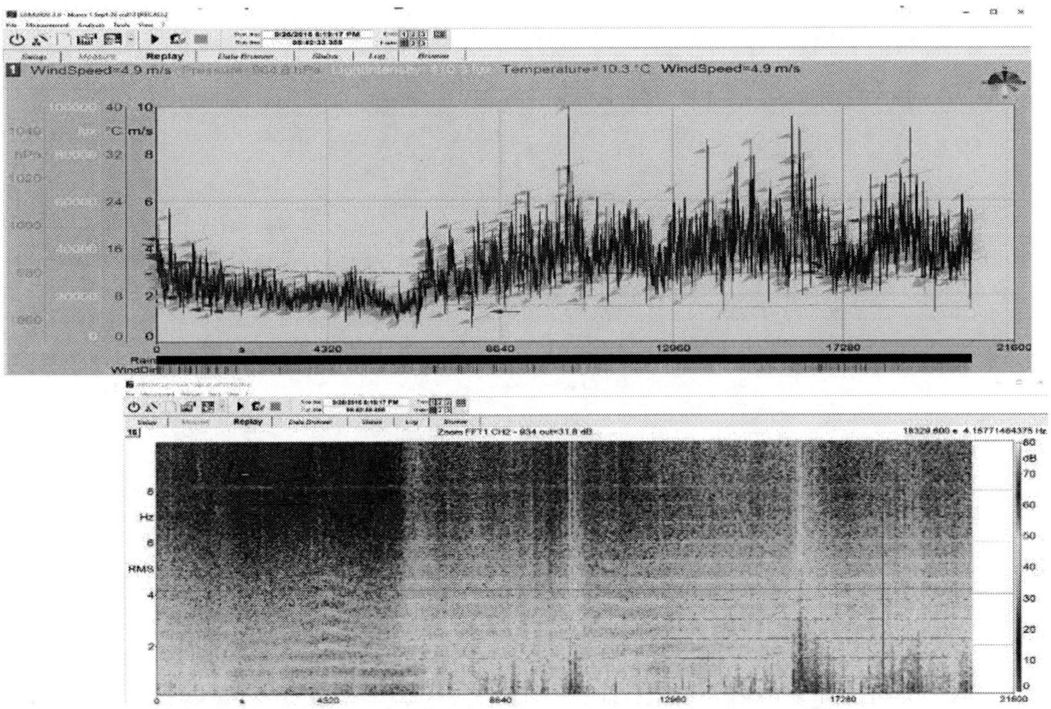

Fig. 12: Home 1, Sept 26. Wind turbine RPM stability is observed despite changes in wind speed above 4m/sec in the SONOGRAM. Red and pink arrows on wind speed plot indicate the wind direction was from the WSW.

Simultaneous measurements in homes allow for revealing comparisons. An overall BPF reduction of approximately 30dB is illustrated in *Fig.13* in Home 3, which is the difference between the green and red horizontal lines in the sonogram. This was also noted during summer conditions where windows and doors opened and closed (*Fig. 29*). There may be several reasons for this, however, this study remains focused on the measurements to identify conditions which impact wind turbine noise transmissibility into homes. Further repeatable measurements validate that air flow in and out of the home correspond to changes in BPF levels measured. Room dynamics can account for this amplification. Further studies are required.

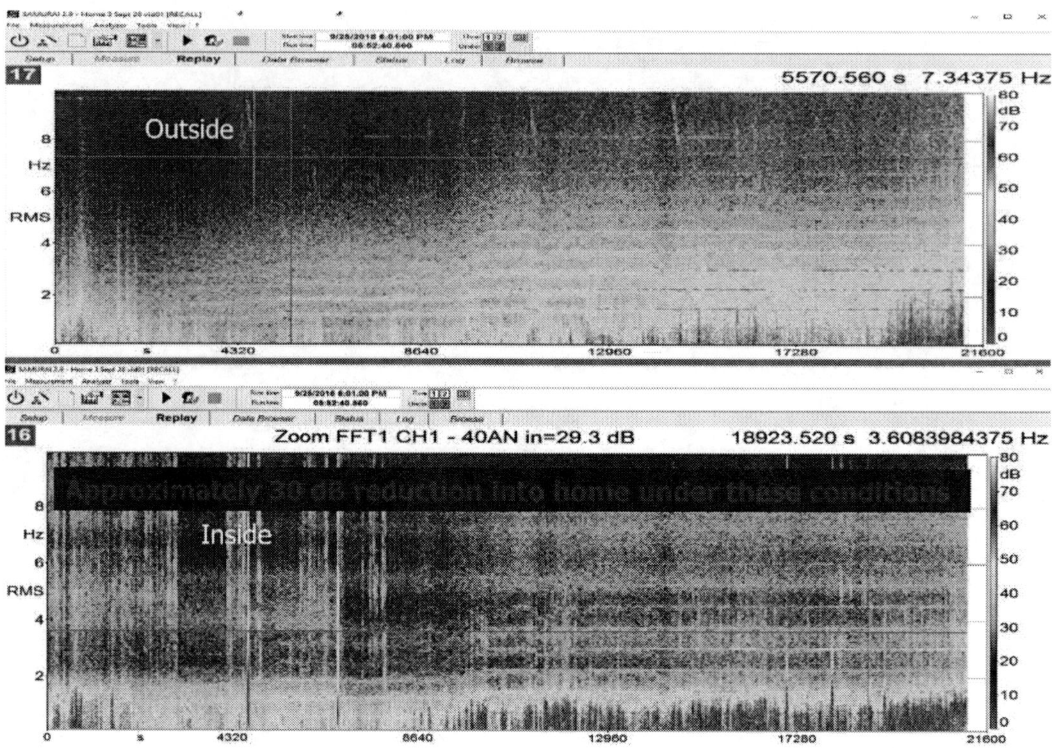

Fig. 13: Sept 26. 6 Hour measurement commencing 12:00 AM indicated relatively low BPF penetration into the home compared to measurements made earlier in the summertime.

SONOGRAMs indicate signatures of multiple wind turbines. Relationships with weather parameters, especially wind direction and wind speed are apparent in *Fig. 13a.*

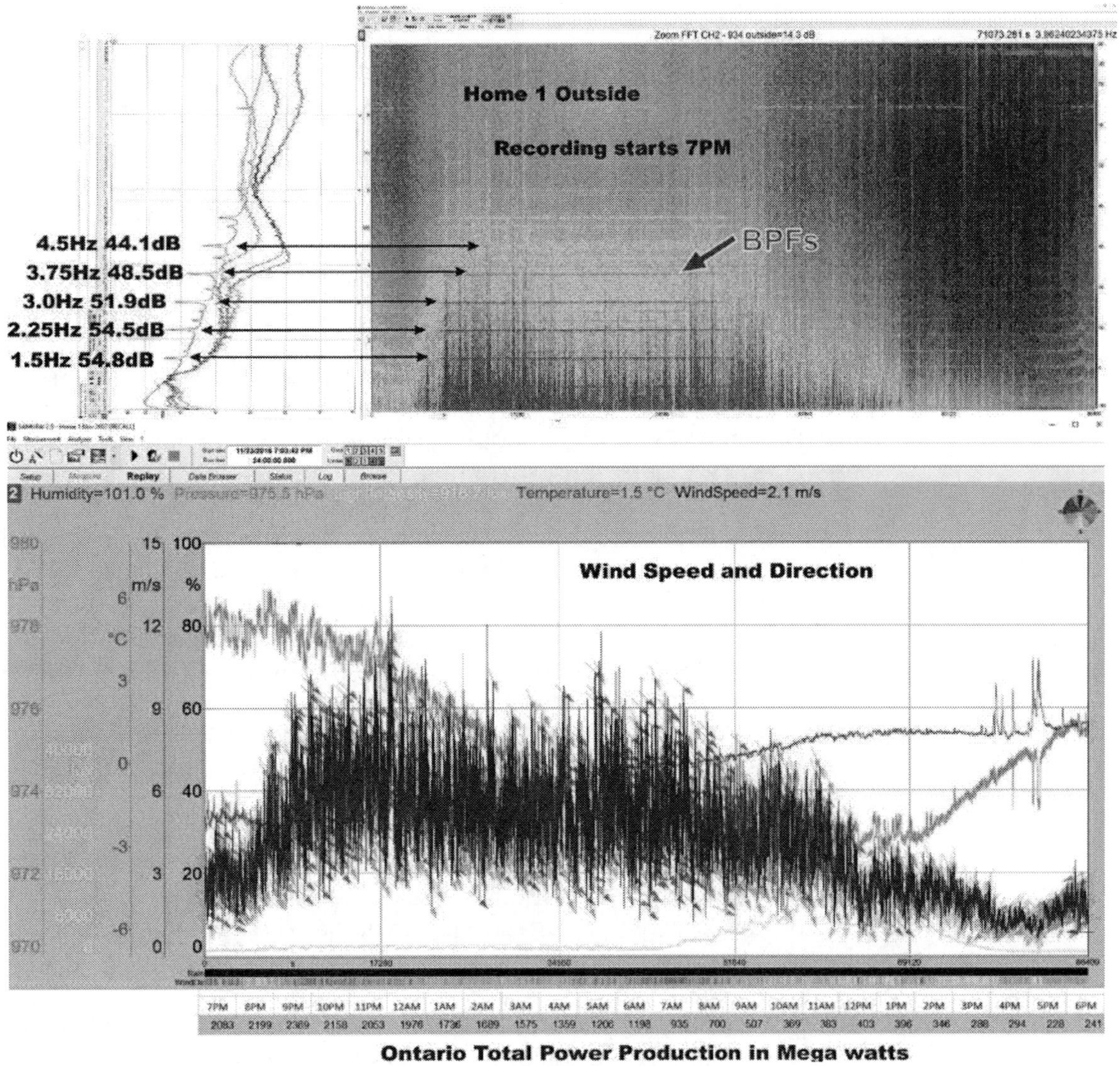

Fig. 13a: FFT Rotation illustrates FFT LEQ over the entire 6-hour duration

Ideal measurements were selected that had no contamination from humans or the sounds of nature. They were selected from continuous Real-time recording and Real-time processing over longer periods of time relative to most studies. Invalid contaminated projects were also quickly recalled and identified. A determination was made to review only valid projects for analysis and invalid measurement sets were rejected. One project can contain up to 300 files for various trace parameters. Project based recording and Real-time Analysis are thus important and provide the following advantages:

1. Continuous recording and SLM, FFT, Octave processing in real-time does not require operator intervention.
2. All computations are processed in real time, therefore the process of recalling and processing many files is eliminated. One can still post-process after the fact for select time recorded files.
3. The amount of data can be millions of files and terabytes of data. Recalling and reviewing processed projects with weather parameter files eliminates confusion with data handling and reduced probability of human error.
4. Earlier methods using time domain recorders required analysis after the data collection phase. One could not validate the data until the analysis phase was recalled in the lab. The "time record only method" and "analyze later" requires much more time to complete and requires knowledge of events to occur in order to minimize recordings to analyze which would require one to search, calculate and repeat. The system we used also recorded raw time data in parallel with processing and allowed more timely analysis while events and conditions were fresh in mind.
5. Real-time analyzers have instrumentation traceable to standards for hardware and combined software traceable to PTB, IEC etc.
6. Synchronous video and weather station parameters are all contained within the project to the same index, thus eliminating the risk of error when importing files that may not be time synchronized to the measurements. As an example: weather and video coming from other software applications. Playback of all data with synchronized video closely replicates witnessed measurements and facilitates greater understanding of the measurements.
7. We received good acceptance and development support from the measurement equipment manufacturers and the whole industry will benefit from specialized future developments.
8. As the turnkey instrumentation is traceable to PTB, IEC etc., confidence in data increases.

Keeping these advantages in mind, analysis proceeded employing these methods with complex conditions which required thousands of data sets. This has allowed for the proposal of this method of how infrasound propagates.

Multi-processing and validation of data

It is evident that this 6-hour measurement is clean from artifacts and contamination within the infrasound range for FFT results as *Fig.14* verifies. FFT peaks are distinct and SONAGRAMs indicate BPFs are present without transient artifacts such as door slams and other common human interference.

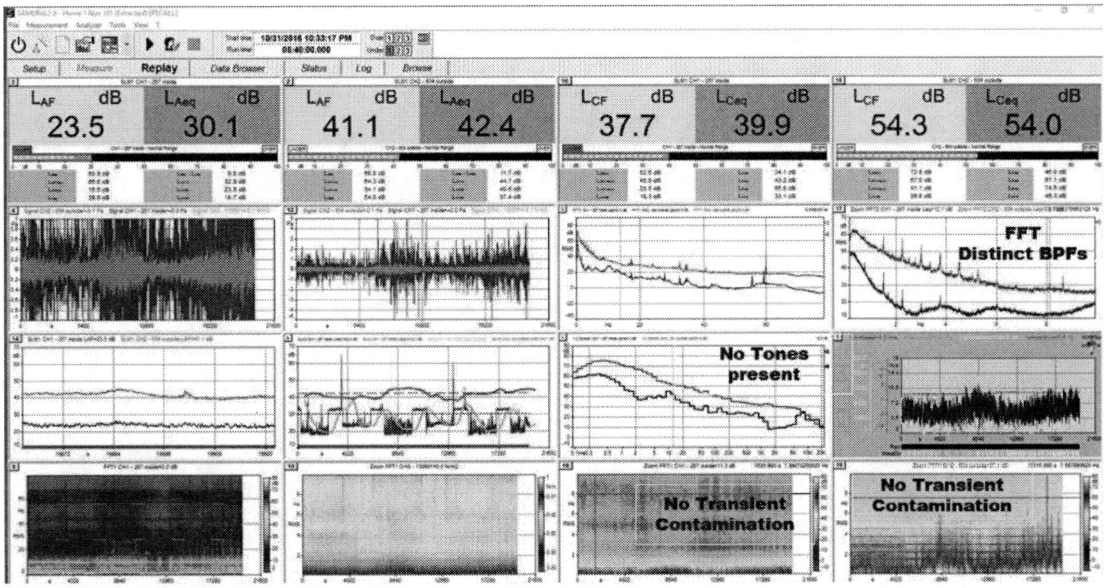

Fig. 14: 16 Graphs produced in Real-time validate a data set.

On/off cycling of a home furnace does not create infrasound in the BPF region. Octave bands show no distinct peaks, however the FFT spectrum clearly indicates peaks even at very low levels inside the home.

Measurements contained BPFs in various rooms inside the home (*Fig.15*) where levels of BPFs vary and fluctuations remain constant over long time intervals. Complex room dynamics and open-concept rooms combined with airflow complicated the amplification effect. To establish the relationship between outside and inside infrasound BPFs, a simple transmissibility calculation was developed. The ratio indicates the home's impedance to BPFs in each room or how much the home attenuates the infrasonic emissions from the wind turbines. The BPF inverse transmissibility factor, similar to a Transfer Function was developed as an indicator of BPF penetration. The process to determine transmissibility begins with reviewing Fig 15 for distinct FFT LEQ spectra calculated over a 6-hour duration. Inside measurements indicate the only contaminant present was low level noise from a furnace. The furnace noise contaminant was above 8 Hz and therefore our ability to discern BPFs were not affected. Once again, the FFT Spectrum is valid leading to a correct transmissibility calculation.

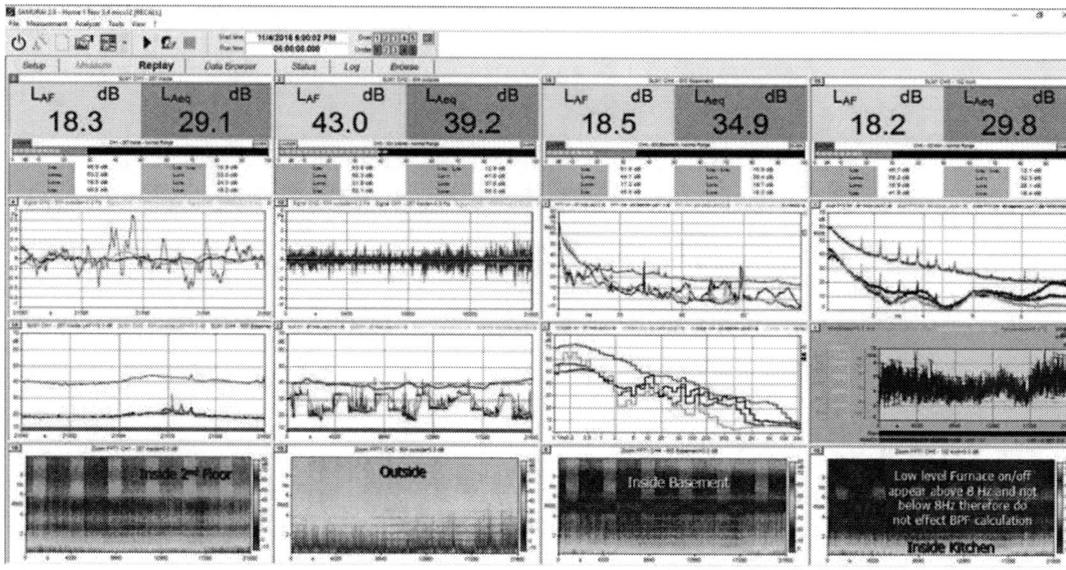

Fig. 15 Home 1, Nov. 4, 6:00PM 6-hour Real-time measurement results.

Outside LAeq(6HR) fell below the nighttime limits of 40dBa in this case (39.2dBA) despite the fact wind speeds varied from 3m/s to 9m/sec. The lack of leaf wind noise, crickets, migrating birds and people outdoors during November were noticed and help account for these low audible levels. Measurements made in August and early September were contaminated with both human interference and the sounds of nature. Audible noise during this time period requires removal and filtering, spending considerable time and careful procedures in editing out. The winter months in Canada have relatively lower audible background noise. Wind was also constant from the west for this measurement. All these factors validate our FFT LEQ(Lin) to be accurate and without contamination. Further, most BPFs were present when wind was from the SW (*Fig.16*) during the recording of the data from *Fig 15*.

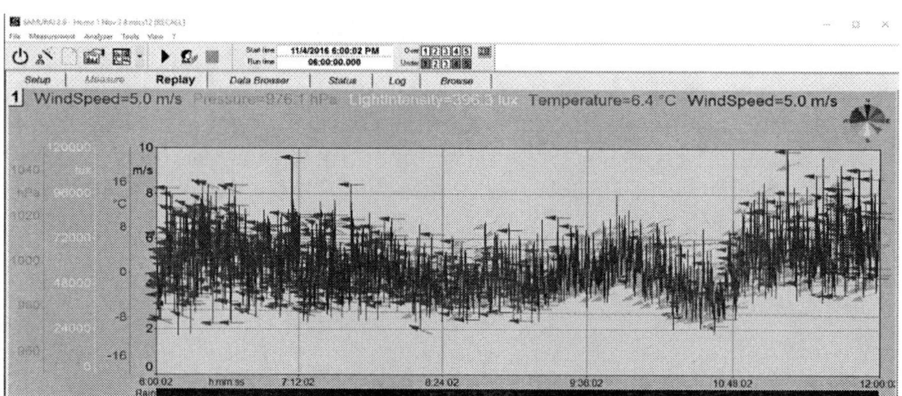

Fig. 16 Home 1, Nov 4, 6PM. Weather Station, Wind Speed & Direction.

Several measurements at Home 1 were made and since human artifacts were not present, such as transient door and window closures, we can conclude that windows remained closed. Validation of artifacts with FFT averaged spectra appear in Fig. 17. Numerical results appear in Fig. 18. Note that as indicated in the SONOGRAM, only furnaces turn on and off and do not contribute significantly to the BPF calculations to follow.

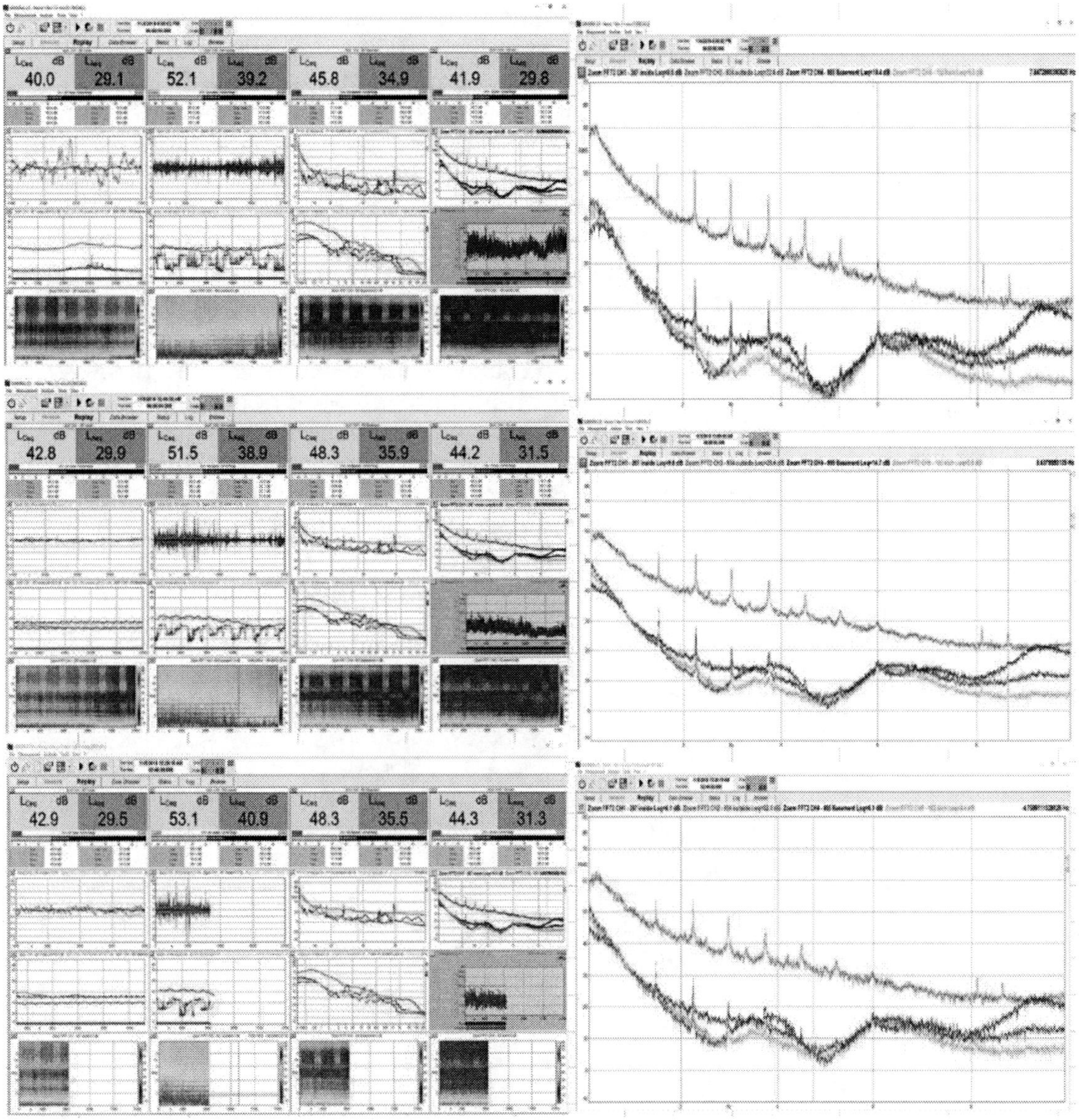

Fig. 17: Mic location indoor kitchen, 2nd floor & basement vs outdoor BPFs

Measurements were made with a 25,600 line resolution and even though these were variable speed turbines, one can see from the discrete peaks in the FFT spectra that there is insignificant RPM variation during the entire 6-hour duration as the SONOGRAM further validates. Transmissibility calculations appear in Fig. 18. The Grey Highlands Inverse Transmissibility Ratio (GHITR) remains repeatable within 2% when Nov 4th to Nov 5th are compared. As the ratio is unitless, it does not indicate amplitude severity, however, levels are also given in the table. *(Fig.18)* The GHITR quickly indicates the location in the home that relative BPF penetration is least and most on a room-by-room basis. It also provides distinctness and clarity which may also be used for human perception independent of levels. Random pressure in the infrasonic range would mask these peaks and perhaps human perception as well. Transient contaminants such as door slams drastically effect the long term FFT Spectrum Leq. Quickly extracting these transients and reprocessing saves portions of the records that may be valuable and therefore they are not discarded or lost. *(Fig. 17a)* Wind turbine rotational speed changes are also immediately recognized in the SONOGRAM. The GHITF technique can also be used in real-time so one can change room dynamics and flow while watching the results in Real-time. The 3rd set of data *Fig. 18* on Nov 5th repeats the 2nd set, however, half the 6-hour record was deleted where turbines were not present. Yet the GHITR is still within 2% at higher harmonics. Higher BPF harmonics exhibited inaccuracies due to dynamic range issues. A GHITR of 1 would indicate identical wind turbine infrasound noise levels inside and outside of a home. A value of > 1 indicates that noise levels are higher inside the home. The dB ratio being logarithmic needs to be realized when comparing higher frequency BPFs close to the noise floor of the entire measurement chain.

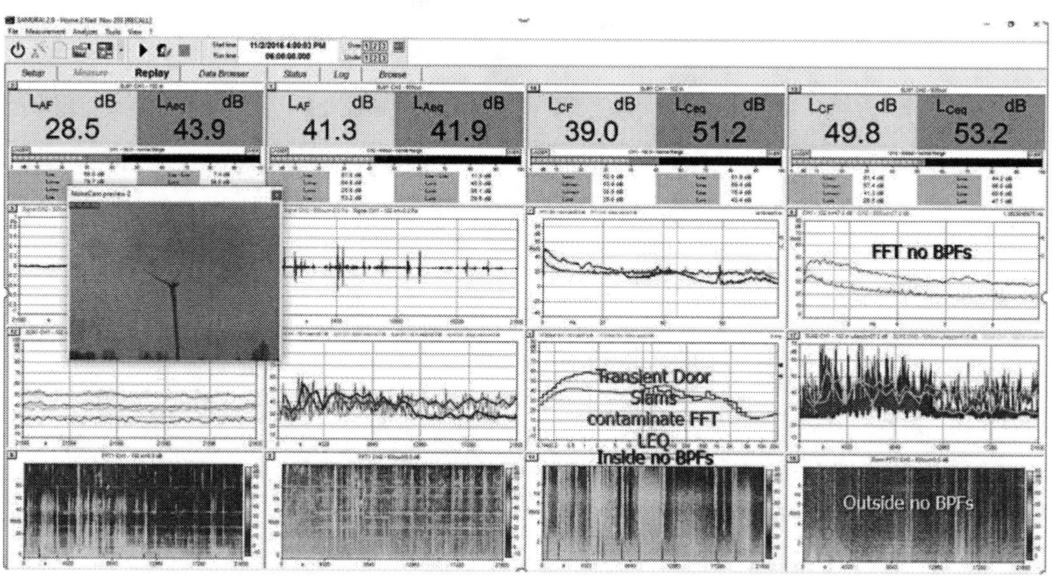

Fig. 17a: Nov 2, 6:00PM. Turbine 600 meters from Home 2 real-time video

verifies turbine was not rotating, and as expected, no BPFs were measured. Transient Door Slams are indicated by vertical lines in the SONOGRAM.

.

RMS Units				Home 1				
Nov 4, 6:00PM	Level Lin dB					1/GHT Ratio RMS		
	Frequency	Outside	Basement	2nd Floor	Kitchen	Basement	2nd Floor	Kitchen
BP Fundamental	.75 Hz	Buried	Buried but noticable	Buried but noticable	Buried but noticable			
BP 1st Harmonic	1.5 Hz	52.3	33.1	31.3	31.8	0.633	0.598	0.608
BP 2nd Harmonic	2.25 Hz	50.7	26.6	20.6	22.8	0.525	0.406	0.450
BP 3Rd Harmonic	3.0 Hz	48.9	21.8	17.4	14.5	0.446	0.356	0.297
BP 4th Harmonic	3.75 Hz	45.1	21.9	17.8	11.5	0.486	0.395	0.255
BP 5th Harmonic	4.5 Hz	40.6	12.3	5.7	3.1	0.303	0.140	0.076
BP 6th Harmonic	5.25 Hz	35.4	7.7	6.8	6.5	0.218	0.192	0.184
RMS Units								
Nov 5, 12:00AM	Level Lin dB			Home 1		1/GHT Ratio RMS		
	Frequency	Outside	Basement	2nd Floor	Kitchen	Basement	2nd Floor	Kitchen
BP Fundamental	.75 Hz	Buried	Buried	Buried	Buried			
BP 1st Harmonic	1.5 Hz	52.7	33.9	32.2	32.7	0.643	0.611	0.620
BP 2nd Harmonic	2.25 Hz	52.4	27.9	22.7	24.4	0.532	0.433	0.466
BP 3Rd Harmonic	3.0 Hz	47.5	20.9	17.3	14.9	0.440	0.364	0.314
BP 4th Harmonic	3.75 Hz	43.3	19.7	16.5	11.3	0.455	0.381	0.261
BP 5th Harmonic	4.5 Hz	39	10.5	7.3	4.5	0.269	0.187	0.115
BP 6th Harmonic	5.25 Hz	Buried	Buried	Buried	Buried	#VALUE!	#VALUE!	#VALUE!
RMS Units								
Nov 5, 12:00AM extracted	Level Lin dB			Home 1		1/GHT Ratio RMS		
	Frequency	Outside	Basement	2nd Floor	Kitchen	Basement	2nd Floor	Kitchen
BP Fundamental	.75 Hz	Buried	Buried	Buried	Buried			
BP 1st Harmonic	1.5 Hz	53.9	34.4	32.7	32.2	0.638	0.607	0.597
BP 2nd Harmonic	2.25 Hz	52.4	29.4	24.3	26	0.561	0.464	0.496
BP 3Rd Harmonic	3.0 Hz	48.7	21.7	18.9	16.2	0.446	0.388	0.333
BP 4th Harmonic	3.75 Hz	39.3	19.9	17.1	12.7	0.506	0.435	0.323
BP 5th Harmonic	4.5 Hz	37.8	13.4	10.7	8.2	0.354	0.283	0.217
BP 6th Harmonic	5.25 Hz	Buried	Buried	Buried	Buried	#VALUE!	#VALUE!	#VALUE!

Fig.18: Measurement location indoor kitchen, 2nd floor and basement vs outdoor BPFs taken on Nov 4 & 5th from data *(Fig.17)*.

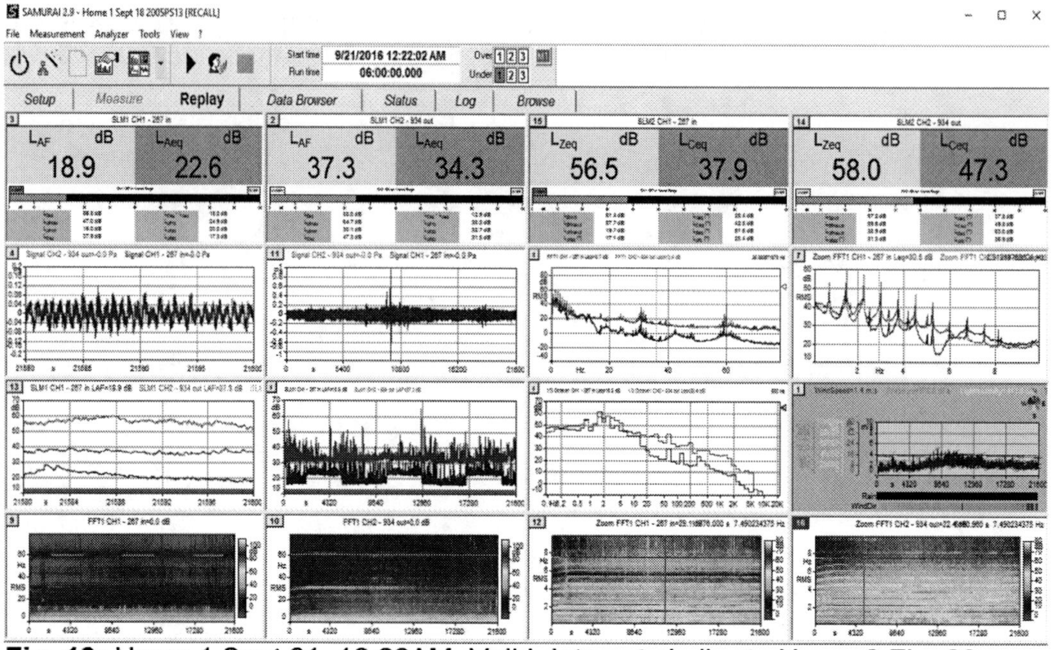

Fig. 19: Home 1 Sept 21, 12:22AM. Valid data set similar to Home 3 Fig. 20

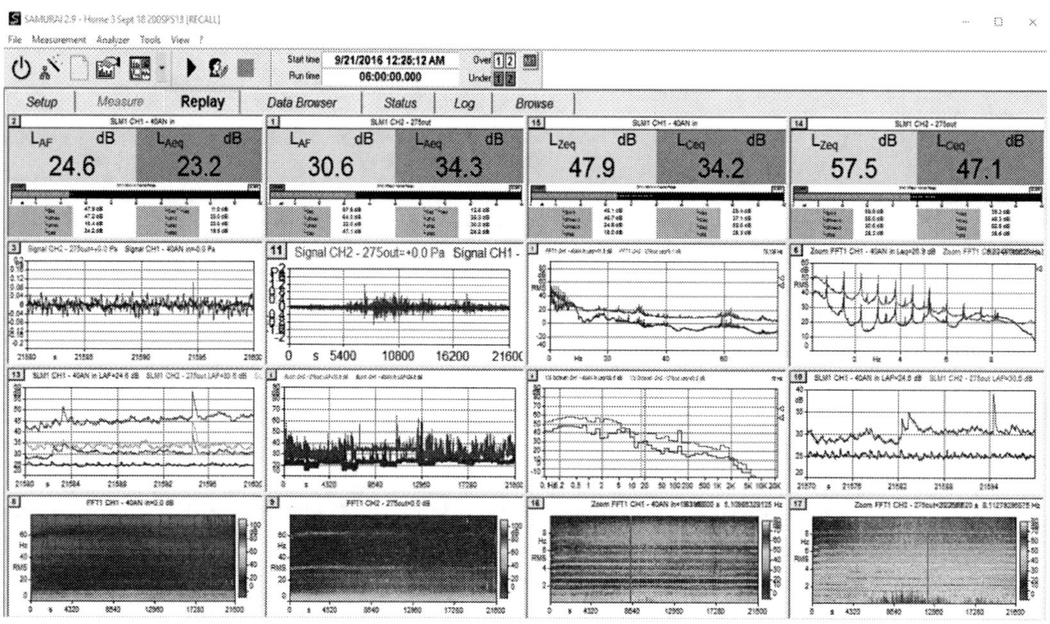

Fig. 20: Home 3 Sept 21, 12:25AM. Valid data set similar to Home 1.

During colder climate conditions in November, infrasound penetration is less, (*Fig.18*) likely due to an increase in air circulation in and out of the home due to increased ventilation. Large attics with roof vents for air circulation also play a role and result in higher GHITR values. (Such as on Sept. 18 *Figs. 21, 22, 23, & 24*) Larger sets of data are required with various home constructions to verify this. (Figs. 19 & 20 made simultaneously in both homes show different results primarily because of different ventilation conditions.)

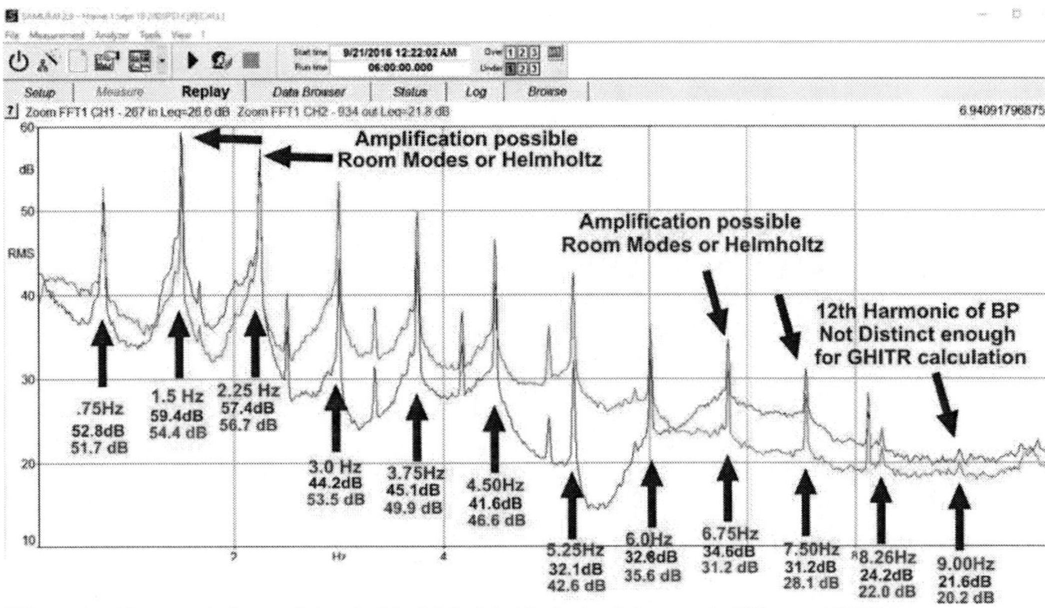

Fig. 21: Home 1 Sept 21 12:22 AM, Multiple turbines at different RPMs

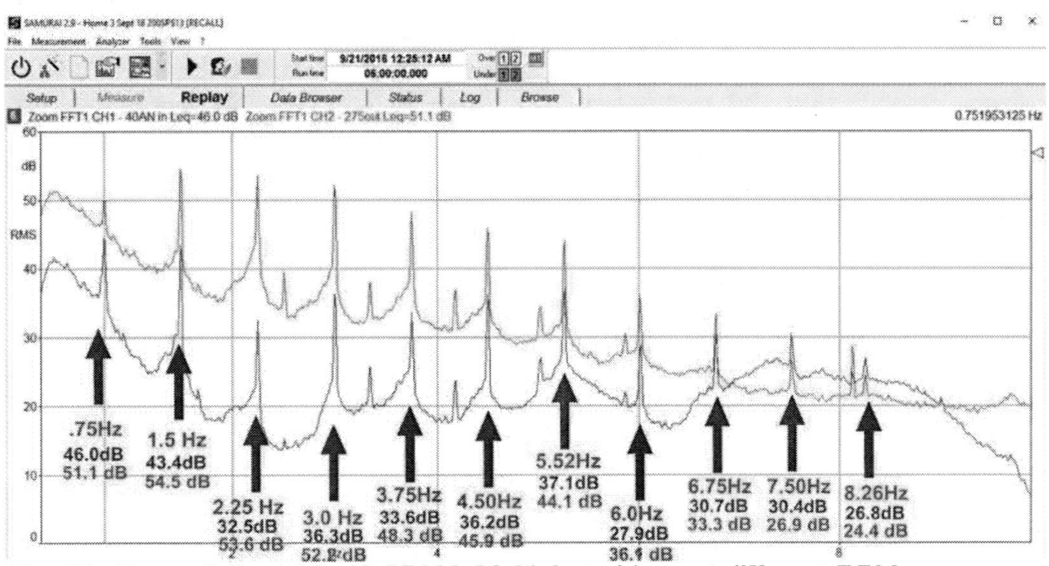

Fig. 22: Home 3 Sept 21, 12:25AM, Multiple turbines at different RPMs

The home with the lowest repeatable GHITR in the winter shows the much higher levels in the *summer*. Fig. 23 shows window and door openings may have played a role with this and more measurements were performed at Home 1 that showed reduced values of GHITR similar to November values.

Home 1, 9/21/2016	12:25:00 AM			
	Frequency	**Outside**	Living Room	GHITR
BP Fundamental	.75 Hz	**51.7**	**52.8**	**1.021**
BP 1st Harmonic	1.50 Hz	54.4	59.4	**1.092**
BP 2nd Harmonic	2.25 Hz	56.7	57.4	**1.012**
BP 3Rd Harmonic	3.00 Hz	53.5	44.2	**0.826**
BP 4th Harmonic	3.75 Hz	49.9	45.1	0.904
BP 5th Harmonic	4.50 Hz	46.6	41.6	0.893
BP 6th Harmonic	5.25 Hz	**42.6**	**32.1**	0.754
BP 7th Harmonic	6.00 Hz	**35.6**	**32.8**	0.921
BP 8th Harmonic	6.75 Hz	**31.2**	**34.6**	1.109
BP 9th Harmonic	7.50 Hz	**28.1**	**31.2**	1.110
BP 10th Harmonic	8.26 Hz	**22**	**24.2**	1.100
BP 10th Harmonic	9.00Hz	**20.2**	**21.6**	1.069

Fig. 23: Home 1 Sept 21, 12:25 Amplification conditions

			1/GHT Ratio RMS	
Home 3, 9/21/2016	12:25:00 AM	**Level Lin dB**	Home 3	
	Frequency	**Outside**	Living Room	GHITR
BP Fundamental	.75 Hz	**51.1**	**46**	0.900
BP 1st Harmonic	1.50 Hz	54.5	43.4	0.796
BP 2nd Harmonic	2.25 Hz	53.6	32.5	0.606
BP 3Rd Harmonic	3.00 Hz	52.2	36.3	0.695
BP 4th Harmonic	3.75 Hz	48.3	33.6	0.696
BP 5th Harmonic	4.50 Hz	45.9	36.2	0.696
BP 6th Harmonic	5.25 Hz	**44.1**	**37.1**	0.841
BP 7th Harmonic	6.00 Hz	**36.1**	**27.9**	0.773
BP 8th Harmonic	6.75 Hz	**33.3**	**30.7**	0.922
BP 9th Harmonic	7.50 Hz	**26.9**	**30.4**	**1.130**
BP 10th Harmonic	8.26 Hz	**24.4**	**26.8**	**1.098**

Fig. 24: Home 3 Sept 21 12:25AM

Home 2 also had amplification inside the home at BPFs from different turbines operating at different speeds. Notice from the spectral data (*Fig 25*). that the frequency decay levels are similar from both groups of turbines.

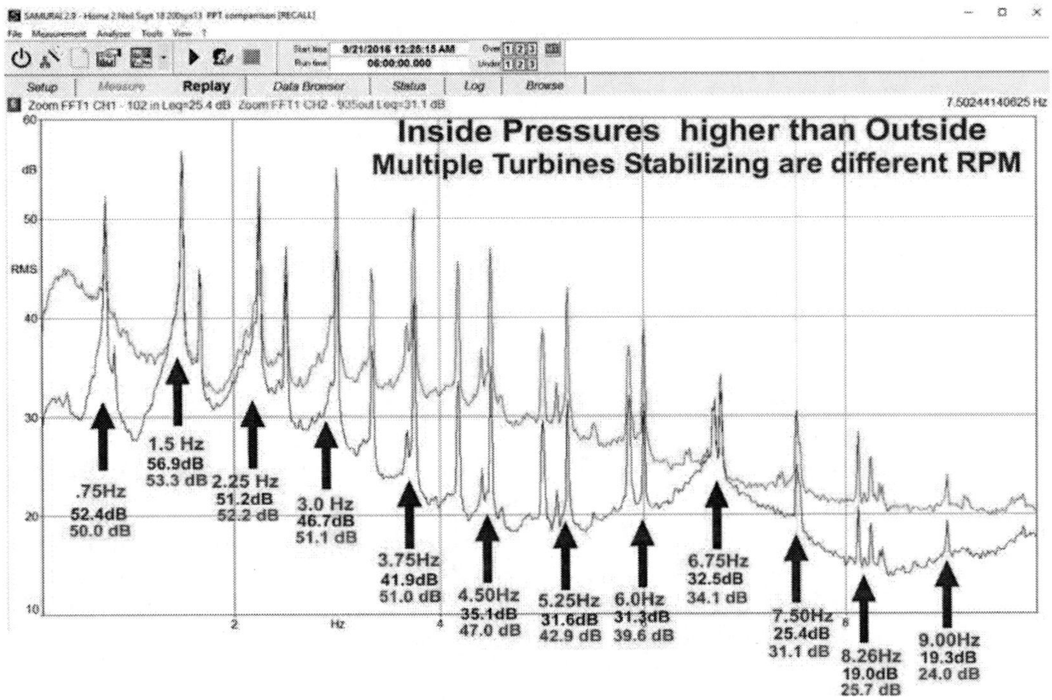

Fig. 25: Home 2, September 21 12:25AM, Multiple Turbines present both indicate similar GHITR

	Frequency	Outside	Living Room	GHITR
BP Fundamental	.75 Hz	50	52.4	1.048
BP 1st Harmonic	1.50 Hz	53.3	56.9	1.068
BP 2nd Harmonic	2.25 Hz	52.2	51.2	0.981
BP 3Rd Harmonic	3.00 Hz	51.1	41.9	0.820
BP 4th Harmonic	3.75 Hz	51	41.9	0.822
BP 5th Harmonic	4.50 Hz	47	35.1	0.747
BP 6th Harmonic	5.25 Hz	42.9	31.6	0.737
BP 7th Harmonic	6.00 Hz	39.6	31.3	0.790
BP 8th Harmonic	6.75 Hz	34.1	32.5	0.953
BP 9th Harmonic	7.50 Hz	31.1	25.4	0.817
BP 10th Harmonic	8.26 Hz	25.7	19	0.739
BP 10th Harmonic	9.00Hz	24	19.3	0.804

Fig. 26: Home 2 September 21, 12:25 AM

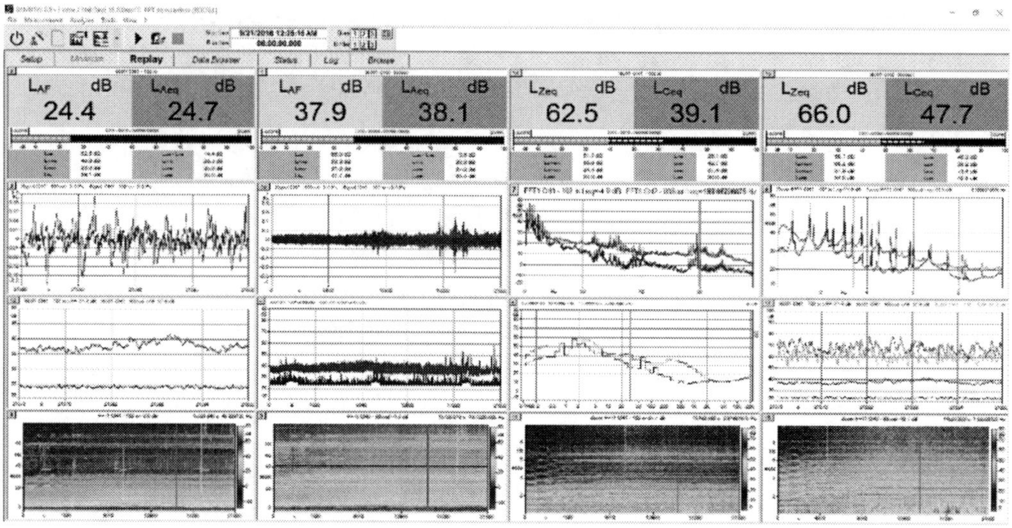

Fig. 27: Home 2 Sept 21, 12:25AM Validation

The Sept 21 measurement data indicates quiet rural levels. Some methods in ANSI/ASA S12.9-2013 were adopted for outside measurements in order to filter insect noise. SLM parameters in Home 1 and 3 being identical at 34.3db(LAeq) while home 2 was higher at 38.1dB. L90 values measured 26.8 dB for Home 3, 35.0 dB for Home 2 and 31.5 dB for Home 1. This level of audible noise is below the MOECC (Ministry of Environment and Climate Change) nighttime guidelines of 40dB set in the Canadian standard NPC-350, yet clear and distinct BPFs are present thought-out the entire 6-hour duration.

Winds speeds were between 1m/sec and 4M/sec at 10 meter heights. Real-time Octaves show no signs of crickets and sonograms have no transient artifacts. FFT Peaks are distinct and clear even at low wind speeds. (As presented at *Internoise 2014 Hansen, University of Adelaide.*)

Vented homes can also have controlled conditions that indicate certain BPF harmonics are reduced with the size of the opening. Figure 28. In this example a door was opened when a person entered, then BPF at 3Hz, 3.75Hz and 4.5Hz are reduced also with some broadband pressure. Fig. 29

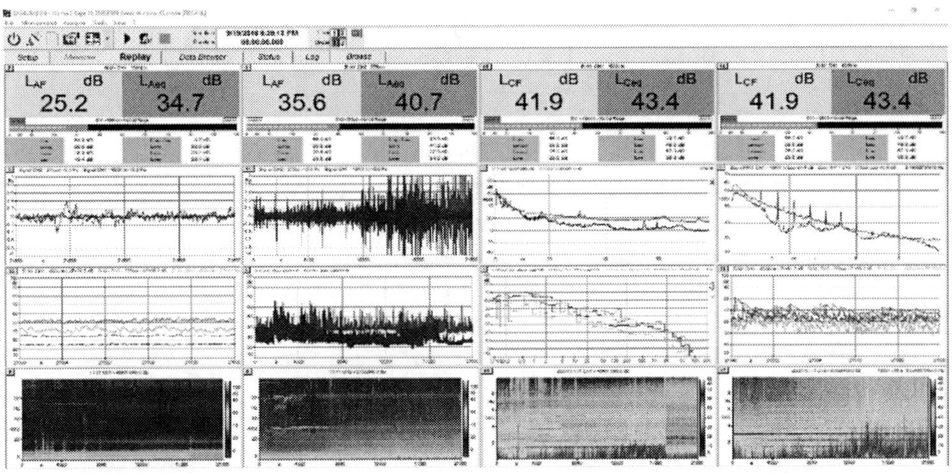

Fig. 28: Sept 19 6:25 PM indoor SONOGRAM indicates window and door closures while outside does not.

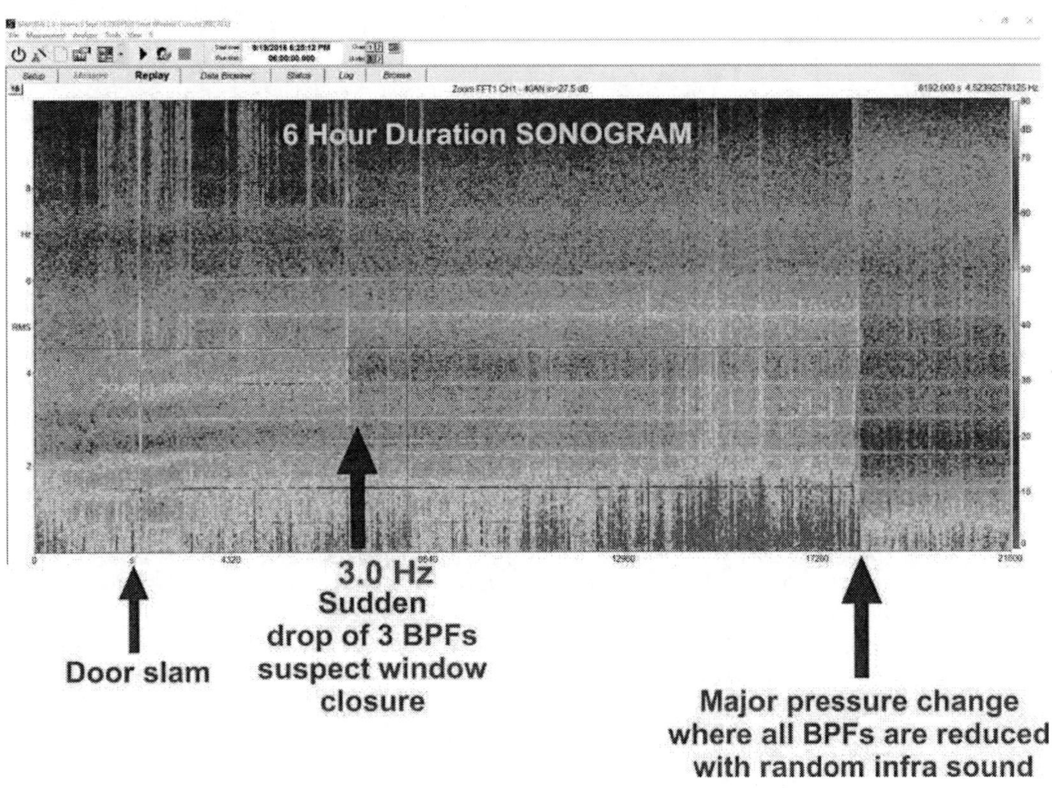

Fig. 29: Sept 19, 6:25PM Sonogram indicates abrupt indoor pressure changes both with Random and BPFs

Temperatures during the evening were 20.2 degrees, so it would be possible the major pressure change was due to the main back deck entrance being closed. These two large doors may have been shut. Occupant believes this to be the case but cannot be certain.

Multiple Turbine Considerations

During times where all turbines stabilize, it would be extremely difficult to separate the individual contribution especially inside when turbines are all rotating at the same speed. Areas with 2 different sets of turbines operating at different speeds do have valid FFT/SONGRAM signatures that separate these harmonic sets. If the Harmonic groups do not overlap, then the total GHTR contribution from each group can be validated and calculated individually. Fig. 30

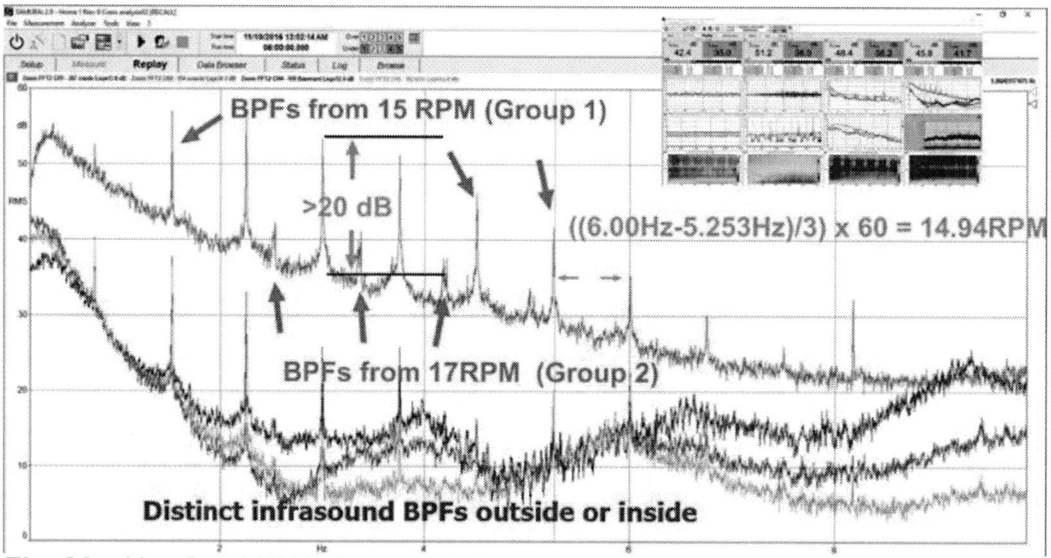

Fig. 30: Nov 2, 6:00PM, Group 2 indicates either a second group of turbines or more likely a singular turbine in this case. Notice FFT peaks are clear and distinct using 25,600-line resolution.

Once again, the inside broadband energy with 3 lobes with all indoor measurements can be seen. These Broadband levels have similar rise and falls. Future measurements will require accurate room dimensions to determine room modes.

Figs. 31, 32 & 33 show measurements made in an abandoned home near Underwood, Ontario. Measurements made there are now compared with the GHITR. This home is surrounded by mainly fixed speed 14.4 RPM turbines. The measurements were made in 2012 using the same instrumentation and sensors. Four locations inside the home are compared to 2 locations outside the home. Note that the comparison is referenced to the microphone in the far field outside approximately 90 meters from the home closer to three turbines directly in-line with the home. The FFT spectrum was also 25600 lines and free of artifacts. Characteristic humps deemed to be room modes only appear inside.

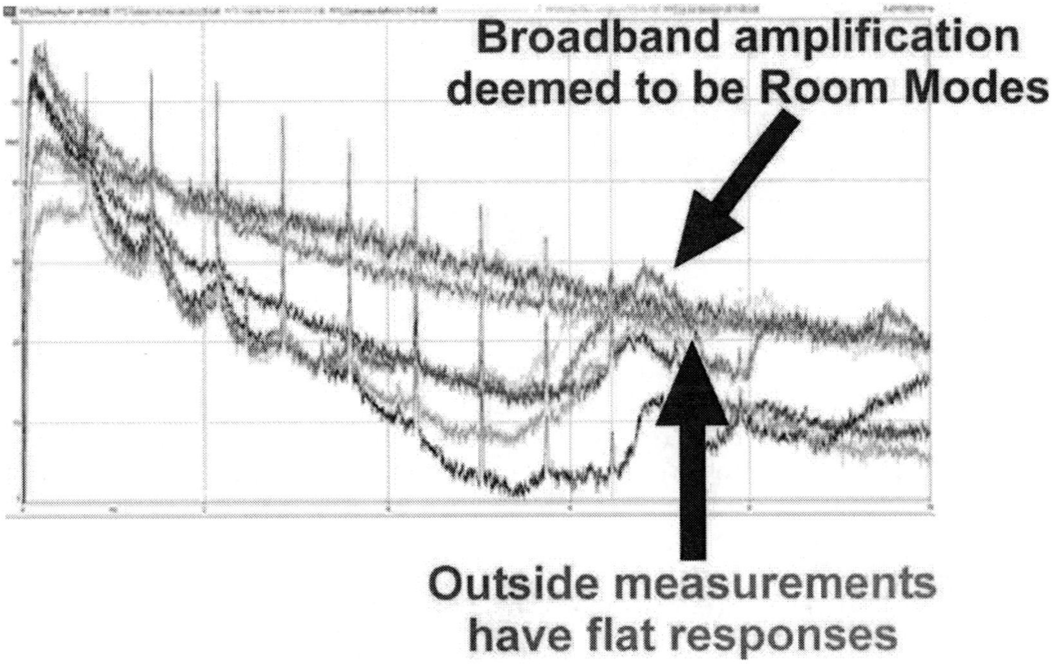

Fig. 31: Infrasound Measurements made in an abandoned home near Underwood Ontario, curtesy of Norma and Ron Schmidt 2012

	Norma and Ron Schmidt				levels in dB (20 micropascal)				GHITR	GHITR	GHITR	GHITR	GHITR	GHITR
Harmonic	Frequency	Dining Room	Outside near Turbine	Outside Front Deck	Downstairs Bathroom	Downstairs Living Room	Upstairs West Bedroom	Upstairs east Bedroom	Dining Room	Outside Front Deck	Downstairs Bathroom	Downstairs Living Room	Upstairs West Bedroom	Upstairs east Bedroom
BP Fundamental	0.717	53.1	53.2	50.8	53.7	47.7	51.2	52.4	0.998	0.955	1.009	0.897	0.962	0.985
2nd	1.435	52.0	54.0	50.6	52.6	50.2	50.9	51.1	0.963	0.937	0.974	0.930	0.943	0.946
3rd	2.152	47.2	52.5	48.2	47.8	46.7	45.4	45.2	0.899	0.918	0.910	0.890	0.865	0.861
4th	2.872	38.7	48.0	40.6	38.7	38.4	36.0	37.2	0.806	0.846	0.806	0.800	0.750	0.775
5th	3.589	32.6	45.4	40.1	32.4	33.4	35.2	35.6	0.718	0.883	0.714	0.736	0.775	0.784
6th	4.306	25.8	40.8	36.5	23.0	25.7	32.7	32.5	0.632	0.895	0.564	0.630	0.801	0.797
7th	5.026	20.1	36.0	33.0	13.5	20.1	27.4	26.6	0.558	0.917	0.375	0.558	0.761	0.739
8th	5.74	18.2	32.7	29.4	11.0	20.5	23.7	22.9	0.557	0.899	0.336	0.627	0.725	0.700
9th	6.46	22.6	29.4	26.2	9.3	24.3	26.7	28.3	0.769	0.891	0.316	0.827	0.908	0.963
10th	7.17	20.4	26.8	24.1	14.1	26.6	24.9	26.5	0.761	0.899	0.526	0.993	0.929	0.989
11th	7.898	11.8	23.6	21.4	12.2	12.5	27.1	17.6	0.500	0.907	0.517	0.530	1.148	0.746
27th	19.34	19.5	15.7	14.0	12.1	6.2	4.9	1.2	1.242	0.892	0.771	0.395	0.312	0.076
28th	20.059	34.6	29.6	26.2	22.2	22.3	21.8	17.6	1.169	0.885	0.750	0.753	0.736	0.595
29th	20.776	20.8	18.3	14.6	9.7	12.8	4.2	4.8	1.137	0.798	0.530	0.699	0.230	0.262

Fig. 32: High GHITRs appear in most locations in this home

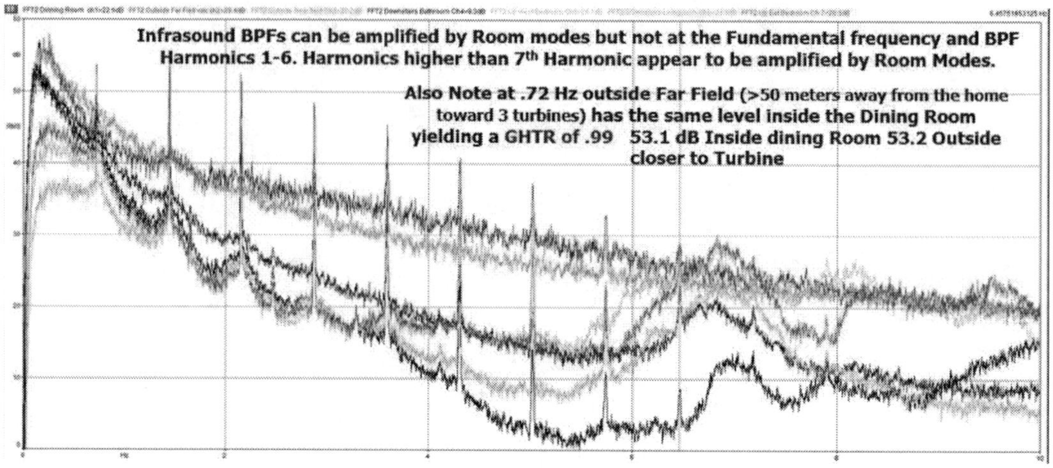

Fig. 33: Rooms modes may amplify BPFs inside, however, one can also see broadband room resonance almost masks the BPF to the right of the broadband room resonances that appear only inside the home. One may conclude that certain BPFs may be masked buy broadband room resonances.

The higher BPF Harmonics decay with distance and also decay similarly inside a home in effect shaping a pulse into a sinusoidal pressure wave.

Narrowband FFT Analyzers have been developed over 40 years ago. Newer Technology has increased reliability, cost, durability, size and most importantly for Wind turbines, the ability to multi-process everything in real time. This parallel processing is a major contributing factor with producing the reported results with this study.

4. Conclusions

- Constant speed wind turbines have stationary and very stable signatures, allowing them to be measured and separated from random, naturally occurring infrasound using Advanced Narrowband FFT analysis and multi-processing.
- Random geophysical, atmospheric and man-made infrasound are present as well and by using several forms of signal processing, one can eliminate these asynchronous fluctuations showing only the synchronous repetitive signals which would include both constant speed and variable speed wind turbines.
- FFT Sonograms show signatures of turbines that appear simultaneously at many locations inside and outside homes and require proper validation before any further calculations are applied.

- Real-time Analysis is a technique to produce valid recordings vs. record collecting and post-processing off-site. This speeds up the entire process and allows for validation through the constant review of multiple measurement results. The technique requires no data editing by third parties if implemented properly and minimizes human error with multiple file handling.

- Wind turbine BPFs which have long wavelengths (Low Frequencies) travel great distances and are present inside homes.

- Audible noise from wind turbines, under normal operating conditions, indicate the noise present in the far field is dominantly below 1kHz. A-weighting attenuates this pressure and is not a true representation of the noise present. C-Weighting would be a better indicator of wind turbine noise in the far field, however, new levels for day and night time may need to be studied for Canada.

- The preliminary use of the GH Transmissibility technique by SVS can be of value in understanding how home construction can impact the wind turbine noise present inside homes, but is purely exploratory at this time and needs to be refined with more measurement data at multiple sights.

- The GH Transmissibility technique is not an indicator of human or animal perception nor does SVS Canada Inc. imply health risk with this measurement study. Further investigation by medical professionals is required to determine this.

- It is apparent that Infrasonic BPFs are present at higher levels in homes with vented attics where the pressure waves enter into the home from soffits, roof vents and chimneys. Modern homes without attics constructed with vaulted air tight ceilings prevent these pressures from entering the home when windows are closed. Testing at additional sites is required to confirm this.

- The GHT Ratio indicates noise levels inside increase during the summer due to windows being opened. Spectral examination revealed that this **is not due** to room resonant modes. Room resonance plays an amplification role for higher BPF harmonics with the room dynamics associated with these 3 homes.

- During colder temperatures where windows are closed, modern energy efficient homes which are more air tight attenuate BPFs as much as 30-40dB

- Wind turbine BPFs are present even in the quietest audible conditions. This can occur at wind speeds less than 2m/sec. They become more distinct from random wind turbulence inside the home.

- Wind turbine BPFs increase in amplitude with higher wind speeds and global power output. Individual electrical power output was not taken into account with these measurements.

- Inaudible pressure waves from turbines can be very complex in nature. The multi-dimensional nature, being time, frequency, amplitude and three-dimensional variant space. Accurate weather and atmospheric conditions also need to be measured and carefully considered, especially in the far field where thermal inversion and wind shear play a role.

- Further measurements need to be made to validate these results in homes such that ideal homes can be constructed and measured for further studies.

- BPFs can be measured under the harshest conditions outside if the proper techniques and equipment are selected.

- Further measurements need to be made with outdoor microphone repositioning. NPC-350 guidelines employ microphone positions at 1.5 and 4.5 meter heights directly in the wind field where 10m/sec wind speeds may be present. Wind turbulence not only effects audible sounds with masking it has a greater effect at lower frequencies especially below 10Hz where BPFs are present.

- Background measurements can also be made when turbines are off due to lack of wind causing them to stop rotating, therefore not inconveniencing the turbine power production. Segments from the SONOGRAM where BPFs do not exist would be processed during these times.

5. Acknowledgements

The measurement, data collection and analysis detailed in this paper was enabled by a research contract from the Municipality of Grey Highlands in the County of Grey in Ontario, Canada. The Municipality has been concerned with the social, economic, environmental and health impacts of industrial scale wind turbines on its citizens from the early 2000s, and this research on infrasound continues a number of initiatives undertaken by its Councils over the last 15 or so years.

6. References

1. Acoustic Interaction a primary cause of infrasonic spinning mode generation and propagation from wind turbines 166th Meeting ASA San Francisco Dec 2013 *Kevin A. Dooley an Andy Metelka*

2. Comparison of the noise levels measured in the vicinity of a wind farm for shutdown and operational conditions *Kristy Hansen; Branko Zajamsek: Colin Hansen University of Adelaide, Australia*

3. Direct Experience of Low Frequency Noise and Infrasound within a Windfarm Community M.A. Swinbanks, MAS Research Ltd. 6th International Meeting on Wind Turbine Noise Glasgow April 2015

4. The Results of an Acoustic Testing Program Cape Bridgewater Wind Farm 44.5100.R7:MSC Prepared for: Pacific Energy Pty Ltd. Steven Cooper

5. Dynamic Measurements of Wind Turbine Acoustic Signals Employing Sound Quality Engineering Methods considering the Time & Frequency Sensitivities of Human Perception Noise-Con 2011, July 25-27, 201 1 Wade Bray and Rick James.

6. A Cooperative Measurement Survey and Analysis of Low Frequency and Infrasound at Shirley Wind Farm in Brown County, Wisconsin Channel Island Acoustics, Camarillo, CA Dr. Bruce Walker. Hessler Associates Inc Haymaket, VA George F and David M. Hessler, Rand Acoustics, Brunswick, ME, Robert Rand; Schomer and Associates, Inc. , Champaign, Il. Dr. Paul Schomer

7. Room Modes & Wind Turbines Developing a basis for why wind turbines bother people, Presented to Multi Municipal Wind Turbine Working Group October 2014 Bill Palmer

8. Wind Turbine Noise and Health Stud Summary of Results, 6th International Meeting on Wind Turbine Noise, Glasgow April 2015 David S. Michaud, PhD, Health Canada, Consumer and Clinical Radiation Protection Bureau, Environmental Health and Risk Assessments Division, Ottawa, Ontario, K1A1C1

9. Measuring Wind Turbine Coherent Infrasound John Vanderkooy and Richard Mann. Department of Physics and Astronomy, Department of Computer Science University of Waterloo, Waterloo, ON, Canada, N2L 3G1 jv@uwaterloo.ca, Date posted: 2 October, 2014

7th International Conference
on
Wind Turbine Noise
Rotterdam – 2nd to 5th May 2017

Computational Aeroacoustics of Small Vertical Axis Wind Turbines by Applying a Hybrid Approach

Johannes Weber, FAU University, Cauerstraße 4, 91058 Erlangen, Germany wb@ipat.uni-erlangen.de
Matthias Tautz, FAU University, Cauerstraße 4, 91058 Erlangen, Germany
Andreas Hüppe, Vienna University of Technology, Getreidemarkt 9, 1060, Vienna, Austria
Stefan Becker, FAU University, Cauerstraße 4, 91058 Erlangen, Germany
Manfred Kaltenbacher, Vienna University of Technology, Getreidemarkt 9, 1060, Vienna, Austria

Summary

The motivation of using small wind turbines in urban and suburban areas for decentralized power generation are grown in the last years. In order to avoid annoyance of residents, it is important to reduce noise emissions of such turbines. In order to design a low-noise small wind turbine, it is necessary to get the physical understanding of the flow physics and the acoustics. Beside experimental measurements, CFD programs have been used so far successfully for the calculation of the flow field of small wind turbines. In this study, a hybrid computational aeroacoustics approach based on perturbation ansatz (APE) will be presented and compared with experimental measurements. As usual for hybrid methods, the CFD simulation is performed firstly and the acoustics are calculated in a second step. The flow around the turbine is computed by scale adaptive simulation of ANSYS CFX using a transient rotor-stator-interface. After that, the acoustic field is computed by application of a modified perturbation ansatz, which results in a convective wave equation based on the acoustic scalar potential and sources computed from the incompressible flow pressure. This computational scheme is implemented as Finite-Element-method. The comparison of the experimental measurements and the acoustic simulation resulted in very good agreement.

1. Introduction

Understanding the physical mechanisms of noise generation of wind turbines is an important factor to design turbines at a low-noise level and therefore for increasing social acceptance. Beside the horizontal wind turbines, small wind turbines are considered as a possible solution for harvesting wind energy especially at small scales in urban areas. Due to the different requirements such as turbulent, complex wind situations in urban environment, vertical axis wind turbines are investigated by research and development.

In this study, a complementary approach consisting of experimental measurements, computational fluid dynamic simulations (CFD) and aeroacoustics simulations (CAA) were performed. The subject of interest is a helical Darrieus turbine. For the CFD the commercial code ANSYS CFX was used. The first code used for CAA was SPySI (sound prediction by surface integration), which is predicated on the porous Ffowcs Williams-Hawkings method [1]. The second code for Computational Aeroacoustics, which is named CFS++, makes use of the Perturbed Convective Wave Equations (PCWE) [2]. It is implemented in a finite-element solver. With the help of this code, it is possible to understand the excitation and propagation of acoustic waves, because the wave equation is solved for the whole computational domain.

2. Aeroacoustic Analogy

2.1 Ffowcs Williams-Hawkings

The SPySI code evaluates an integral solution of Lighthill's aeroacoustic analogy as established by Ffowcs Williams and Hawkings method, which is given in eq. (1).

$$4\pi\{p'H(f)\} = \frac{\partial^2}{\partial x_i \partial x_j}\int_{R^3}\left[\frac{T_{ij}H(f)}{r|M_r-1|}\right]_{\tau_e} d^3\eta_j + \frac{\partial}{\partial t}\int_S\left[\frac{[\rho(u_i - v_i) + \rho_0 v_i]n_i}{r|M_r-1|}\right]_{\tau_e} dS(\eta_j) + \frac{\partial}{\partial x_j}\int_S\left[\frac{[\rho u_j(u_i - v_i) + P_{ij}]n_i}{r|M_r-1|}\right]_{\tau_e} dS(\eta_j) \quad (1)$$

The first term of the equation at the right hand side represents the quadrupole sources, which are generated by the turbulent wakes by the wind turbine. The variable $H(f)$ corresponds to the Heaviside function, M_r equates to the radial Mach number and T_{ij} is the Lighthill tensor. The second term is called thickness noise and is caused by the displacement of the air by the rotor blades. The size u_i describes the fluid velocity and v_i characterizes the velocity of the surface. The last term is named loading noise and corresponds to the sound which is generated by the acceleration of the force distribution on the air around the rotor blade. This term includes the compressive stress tensor.

$$P_{ij}=(p-p_0)\delta_{ij}+\tau_{ij}, \quad (2)$$

For calculation of radiation into free field, the FW-H can be solved with the free-space Green's function,

$$G(x,t)=\frac{\delta(t-|x_i-y_i|-\tau/c_0)}{4\pi|x_i-y_i|}, \quad (3)$$

where c_0 represents the ambient speed of sound and τ is the retarded time [1,3]. An advantage of this method is that only two-dimensional surface integral has to be calculated. Drawbacks are a missing insight into the acoustic sources and the necessity to perform compressible CFD.

2.2 Acoustic Pertubation Equations

The approach to compute the propagated noise are the Acoustic Pertubation Equations (APE-2) by Ewert [3], which is applied in the formulation PCWE previously proposed by Hüppe et al. [2]. APE-2 was developed for low Mach-Number flows (smaller than 0.3) and is valid for incompressible flow simulations [3]. These conditions are fulfilled for the presented study and therefore an incompressible CFD simulation was performed. The advantage of the incompressible CFD in comparison to the compressible CFD simulation, which was used for the subsequently Ffowcs Williams-Hawkings approach, is the less computational time. The starting point of the derivation of PCWE resp. APE-2 is the perturbation ansatz, which means splitting the flow variables in temporal mean and fluctuating quantities. Applying this for the velocity u and pressure p results in

$$u=\bar{u}+u^{ic}+u^{a};\ p=\bar{p}+p^{ic}+p^{a}. \quad (4)$$

The superscript ic is referred to the flow components, which are received from the CFD simulation. The superscript a denotes the acoustic part. As Hüppe et al. derived in [2] the Perturbed Convective Wave Equation (PCWE) can be written as

$$\frac{D^2\phi^a}{Dt^2} - \nabla \cdot \nabla\phi^a = -\frac{1}{c_0^2}\frac{Dp^{ic}}{Dt} \tag{5}$$

Using the acoustic potential ϕ^a. The acoustic pressure p^a is calculated after solving previous inhomogeneous wave equation using

$$p^a = \rho_0\left(\frac{\partial}{\partial t} + \bar{u} \cdot \nabla\right)\phi^a. \tag{6}$$

3. Experimental Set-Up

The experimental measurements were performed in an anechoic wind tunnel at the University of Erlangen-Nürnberg. The subject of these investigations is a generic model scale rotor of a 3-bladed helical Darrieus turbine as illustrated in fig. 1. Due to the small wind tunnel working section, a model of 0.2 m diameter and height is used. The chord length c of the model was chosen as 0.05 m.

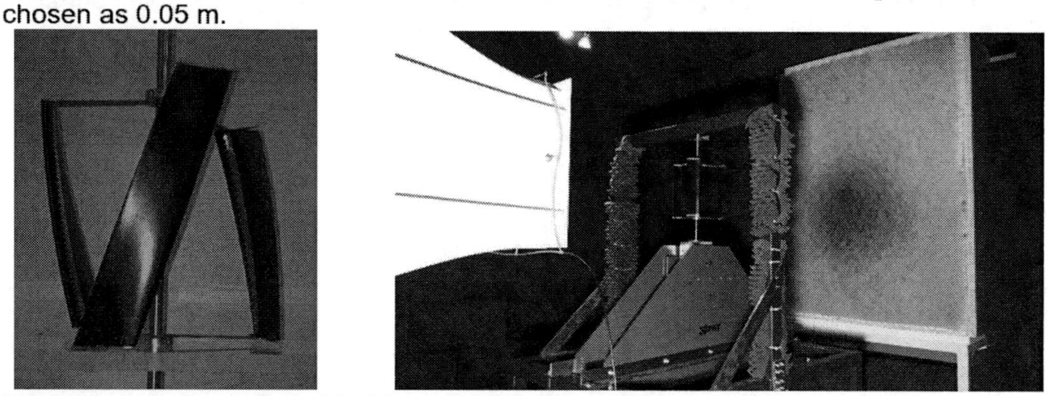

Figure 1: Investigated wind turbine model (left) and experimental set-up (right)

The rotor speed of the model can be adjusted by the servomotor Siemens Simotics S 1Fk7. In order to measure the torque of the model scale, a torque sensor with a measuring range of 1 Nm was used. By measuring the pressure drop along the nozzle and applying Bernoulli's formula the desired wind speed is adjusted. A schematic drawing of the experimental set-up is depicted in figure 2.

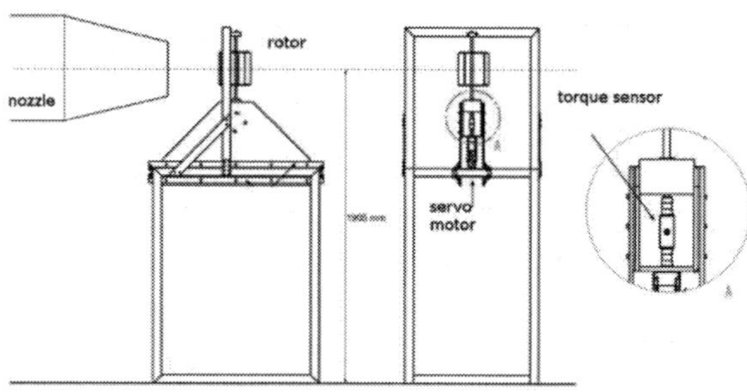

Figure 2: Schematic drawing of the experimental set-up with torque sensor and servomotor

4. Computational Domains

4.1 Aerodynamic Simulations

In case of FW-H a compressible simulation and in case of APE an incompressible simulation were performed. Each simulation was performed on the same grid using the same boundary conditions. Figure 3 depicts the computational fluid domain, which is composed of a rotating and a stationary region. These regions are coupled by a transient rotor stator interface. The inlet is located on the left half of the fluid domain and the inlet boundary condition was specified by v = 15 m/s wind speed. At the outlet, which is located on the right half, an opening boundary condition was defined at relative pressure 0 Pa in order to permit backflow. From this follows that vortices are allowed to pass the outlet boundary. The rotational speed of the airfoils was set to $n = 3500$ rpm, which results in an operating point of $\lambda = 2.44$ (tip speed ratio $\lambda = 2\pi nr/v$). This operating point was the optimum tip-speed ratio of this configuration in the experimental measurements. In this study, the turbulence model SAS-SST (Scale Adaptive Simulation) was used. The mesh of the simulation is depicted in fig. 3 , which was a hybrid mesh composed of an unstructured region in the inner part of rotating domain and a structured grid at the interface and at the stationary domain. The structured grid at the interface was used to get equal cell sizes between the rotating and stationary regions. This should avoid any numerical errors at the interface, which is important for the acoustic simulation. The mesh consists of 17 Mio nodes. The major amount of the nodes (11.2 Mio) are located in the rotating area. About 5.8 Mio nodes are placed in the stationary part and are concentrated in the wake of the rotor. The level of the grid resolution of the boundary layers on the blades and on the shaft was chosen very fine in order to obtain a normalized distance of the wall nearest grid cell of $y^+ < 1$. The time step of the simulation was selected to t=2.5e-05 which leads to a mean CFL number of 3.65. The only part of the grid where the CFL number is higher than 1 are directly located at the leading edge of the airfoil due to the small cells sizes and high velocities at this point. A grid study was performed prior to this work in order to ensure mesh independency.

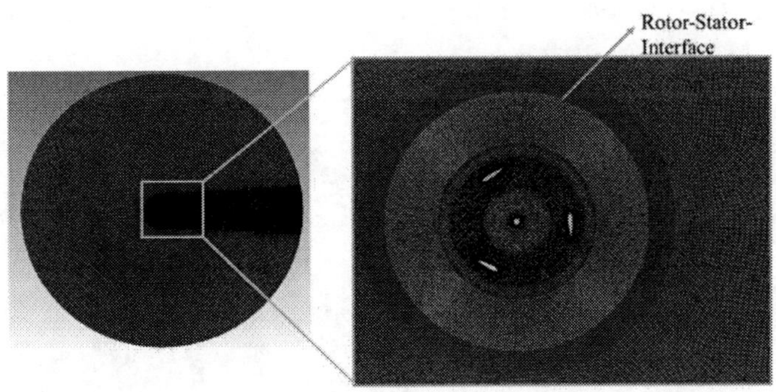

Figure 3: CFD domain (left) and close up of the mesh at the rotating-stationary Interface

4.2 Aeroacoustics

4.2.1 Ffowcs Williams-Hawkings

In order to apply the FW-H analogy using the SPySI code, the three velocity components, the pressure and the density are interpolated onto an integration surface placed around the acoustic sources which are generated by the wind turbine (see Fig. 4). The flow variables of interest are exported from the CFD calculation at every time step. In this study, the integration surface is placed in a distance of 0.4 m from the center. It should be positioned as near as possible to the acoustic sources in order to prevent the influence of numerical damping. If the surface is located too close to the flow field, there is a risk that not only acoustical pressure fluctuations are included, but also hydrodynamic pressure fluctuations, which is unwanted.

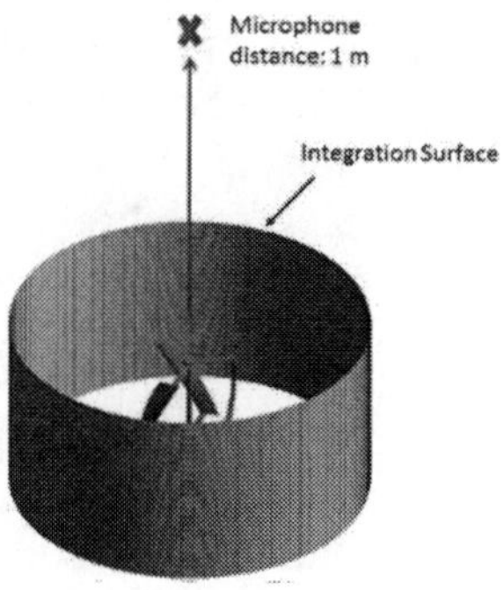

Figure 4: Integration surface on which the flow variables are interpolated

4.2.2 Perturbed Convective Wave Equation

Creating the acoustic domain the properties of the used geometry and the acoustic charactertistic of the wind tunnel should be taken into account. The acoustic domain can be seen in fig. 5. In order to model the characteristic of free field radiation, an absorbing time domain perfectly matched layer is used. This means that acoustic waves cannot be backscattered. There is also an advantage of reducing the size of the acoustic domain and only include region of the monitor point (distance 1 m), which was also used in the experimental measurements. The reduction of size of the computational domain leads to a significant reduction in computational time. The smallest cell size of 4 mm of the acoustic mesh are located around the airfoil and the biggest cells of 3 cm are placed in the stationary and rotating region. The total number of nodes are 470000, of which 360000 elements are in the stationary region and 112000 in the rotating region.

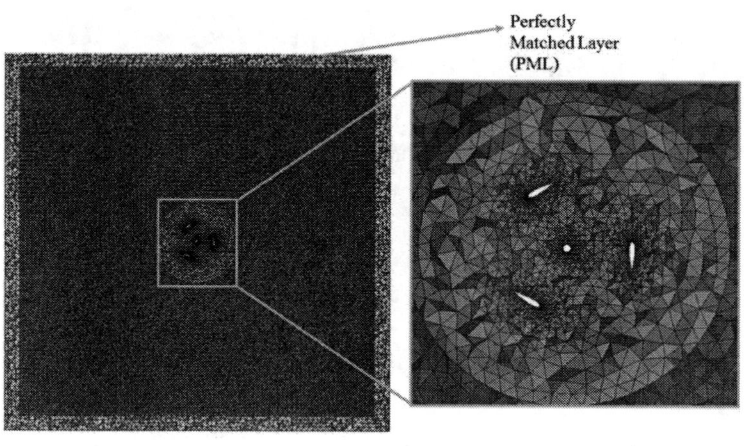

Figure 5: Illustration of the acoustic domain of the stationary (light blue) and rotating part (dark blue)

5. Results and discussion

5.1 CFD Simulation

In this section the CFD results will be discussed. To illustrate the transient flow field the temporal development of the flow of the rotor is presented in fig. 6. Due to the rotational symmetry, a whole revolution can be illustrated by 120°. In order to visualize the vortex structures of the flow, the turbulent kinetic energy is shown. In this illustration, the wind is arriving from the left side and the rotation direction of the wind turbine is counter clockwise. Due to the turbines movement it will cut its own wake during normal operation. That leads to strong blade vortex interaction. Highly turbulent structures can be seen at the blade tip of the turbine. The tip vortices move quickly inwards after being released and are intersected by downwind blade pass. Furthermore, a large vortex structure is separating from the trailing edge, which has an impact of the following blade.

0° 40° 80° 120°

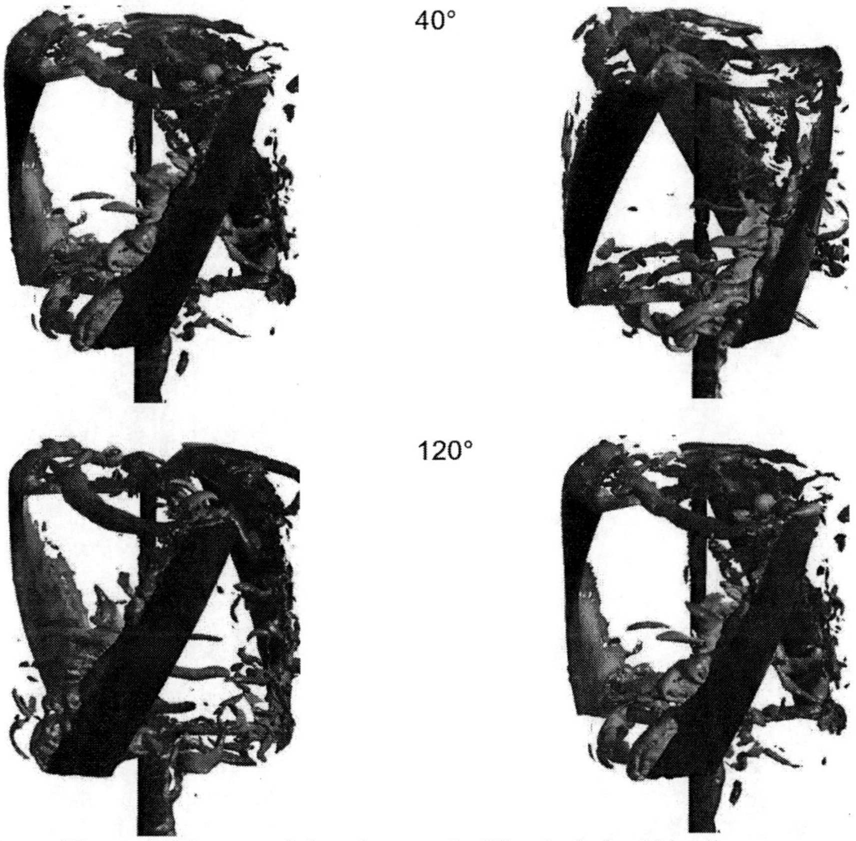

Figure 6: Temporal development of the turbulent kinetic energy

5.2 FW-H

Figures 7 shows the comparison between the measurement and the FW-H simulation. The acoustic pressure of the straightbladed H-Darrieus turbine are computed with the in-house code SPySI. To calculate the sound pressure spectrum of the simulation and measurement, a Fast Fourier Transformation is done. The sound pressure level spectrum can be describe as follows: The first major peak belongs to the blade passing frequency of 175 Hz. Higher Harmonics can be detected up to the 7th one. Between these harmonics small peaks can be seen which refer to the noise of the bearings. Furthermore, broadband noise appears at around 2000 Hz and becomes dominant at frequencies above 3000 Hz. The height of the amplitude of the BPF at 175 Hz will be captured very well by the simulation. Further higher harmonics of the BPF can be resolved by the simulation. In the region of 1000 Hz, the SPySi-code underestimate the noise level of the measurement by a small margin.

At a frequency of 2000 Hz random noise is dominating in the simulation. This is caused by the SAS-Turbulence model, which can only resolve the larger vortices. The smaller vortices, which are responsible for noise generation in the high frequency range could not be resolved.

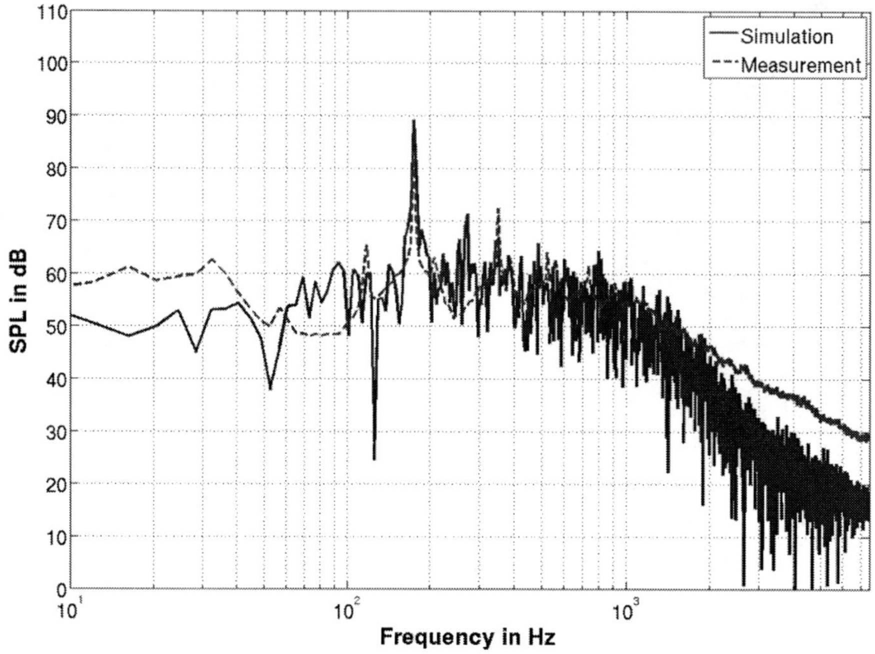

Figure 7: Sound pressure level spectrum of FW-H Simulation (blue) and the experimental measurement (red)

5.3 Acoustic Pertubation Equation

If one consider the acoustic source terms in fig. 8 (left), which are calculated by the FE-Simulation, different acoustic sources can be detected on both ends of the airfoil. At first, one can see a dipole source at the leading edge. Beside this, source terms can be found on the inner side of the airfoil if boundary layer separation occurs, which are increased by large stall angles. This leads to dynamic stall. Furthermore, the vortex separation at the trailing edge will also has an impact on the overall acoustic. In general, there is a high interaction between the different vortices and the blades.

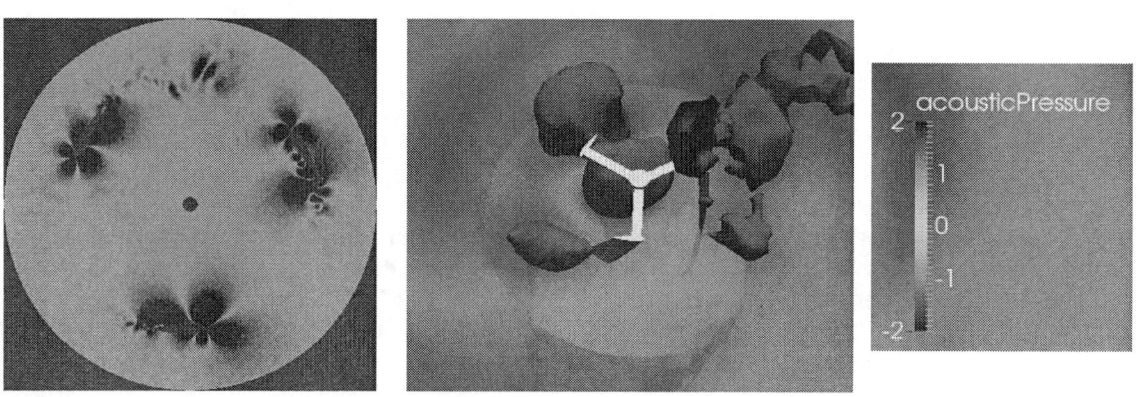

Figure 8: Acoustic source terms (left) and the acoustic wave propagation (right)

In comparison to Ffowcs Williams-Hawkings, the Perturbed Convective Wave Equation shows also very good agreement. At a frequency range of 2000 Hz the simulation underestimates the broadband noise. As describe at the FW-H case, the underestimation refers to the SAS simulation, which cannot resolve the small vortices.

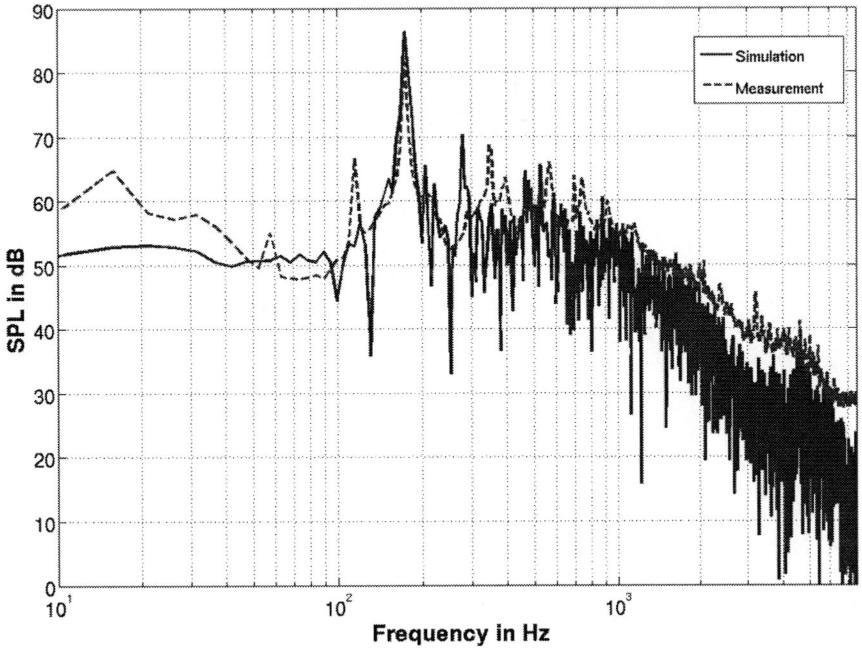

Figure 9: Sound pressure level spectrum of PCWE Simulation (blue) and the experimental measurement (red)

6. Conclusions

CFD and CAA simulations were performed of the helical Darrieus turbine. In this study, two SAS simulations were performed in order to calculate the flow field. Computing the sound propagation of the turbine, a FW-H code and Perturbed Convective Wave Equation method were used. The acoustic computations indicate in case of the tonal components very good agreement. In case of the broadband noise at higher frequencies, the simulation underestimates the noise level, because of the chosen turbulence model SAS. In order to get a better agreement in this frequency range a Large Eddy Simulation can be performed in future. The disadvantage of applying LES is the necessity to use very high computational power, which leads to long simulation time. The acoustic source terms, which are calculated by the theory of Perturbed Convective Wave Equation give an impression about the production of the sound at this operating point. In further investigations filtering of the acoustic source terms should be applied in order to investigate the contribution of the different noise mechanisms to the overall sound emission.

Funding

The authors would like to acknowledge the financial support by the Bavarian research project E|Home-Center [5], which is funded by the Bavarian government.

References

[1] Scheit, C., Karic, B. and Becker, S., 2011, "Effect of blade wrap angle on effiency and noise of small radial fan impellers – A computational end experimental study", Journal of Sound and Vibration, No. 331, pp. 996-1010. DOI: 10.1016/j.jsv.2011.10.014

[2] Hüppe, A., and Kaltenbacher, M., 2015, "Computational Aeroacoustics for ducted FANS", The 22nd International Congress on Sound and Vibration, Florence, Italy

[3] Ffowcs Williams, J. E. and Hawkings, D. L., 1969, "Sound Generation by Turbulence and Surfaces in Arbitrary Motion Philosophical Transactions of the Royal Society", Vol. 264, No. 1151, pp. 321-342

[4] Ewert, R. and Schröder, W., 2003, "Acoustic pertubation equations based on flow decomposition via source filtering", Journal of Computational Physics, Vol. 188, Issue 2, pp 365-398. DOI:10.1016/S0021-9991(03)00168-2

[5] E|Home-Center, www.ehome.center.de

7th International Conference
on
Wind Turbine Noise
Rotterdam - 2nd to -5th May 2017

High Fidelity Airfoil Trailing Edge Noise Predictions via Lattice-Boltzmann Simulations

F. Manegar	University of Siegen, 57068 Siegen, Germany
E-Mail:	farhan.manegar@uni-siegen.de
T. Carolus	University of Siegen , 57068 Siegen, Germany
E-Mail:	thomas.carolus@uni-siegen.de
S. Erbslöh	Senvion SE, 24783 Osterrönfeld, Germany
E-Mail:	sascha.erbsloeh@senvion.com

ABSTRACT

Sound emissions from an isolated airfoil immersed in a free-stream are caused by a number of mechanisms. In some installations such as wind turbines, the turbulent boundary layer interacting with the trailing edge (TE) is, in most instances, dominant. Objective of this study is the validation of numerical simulations with experimentally obtained measurements of TE noise from a Somers S834 airfoil section. Measurements were conducted in the University of Siegen acoustic wind tunnel. The numerical method chosen is the Lattice-Boltzmann scheme as it promises a significant reduction in CPU time compared to Navier-Stokes based LES simulation. Two different setups are investigated. The first setup ("*2D*") is an airfoil section with only a relatively short span-wise extension (7.5% of chord length). Installation effects due to the wind tunnel are not taken into account. In the second setup ("*3D*"), the computational domain covers the wind tunnel and the entire semi-anechoic room. At first the *2D* configuration is analyzed. Compared to experiments the predicted boundary layer induced surface pressure, near-field velocity fluctuations close to the TE and far field sound pressure are in good agreement. Predictions for the configuration "*3D*" show fair agreement with the direct recording of far field sound pressure spectra, but not with near field pressure and velocity fluctuations. Lastly the results of the *2D* configuration are compared with the results of a Navier-Stokes based LES simulation, whose data is available from a previous study of the same setup.

1. INTRODUCTION

The ongoing energy revolution requires installing wind turbines in the vicinity of residential areas, at least in areas with dense population. Wind turbines are known to produce noise, which may be a reason for annoyance. Hence there arises an ongoing effort to reduce wind turbine's noise further. Since half a century numerous studies have been conducted to understand the mechanism by which a flow encountering airfoil emits noise. One early study is by BROOKS et al [1] of a NACA 0012 airfoil section. They identified five airfoil self-noise mechanisms due to boundary layer phenomena. Noise from a full scale wind turbine was analyzed e.g. by OERLEMANS et al [2] employing an experimental and semi-analytical method. They showed that the blade trailing edge (TE) region contributes the most to the overall wind turbine noise. TE noise is caused due to the scattering of the turbulent boundary layer at the TE of an airfoil. A recent study was conducted by GERHARD [3-5] on TE noise and on active/passive ways to reduce it. In this study, the computational aero-acoustic (CAA) method "CURLE's analogy [6] based on a numerical large eddy simulations (LES)" was utilized.

In general CAA methods are of two types: Direct and hybrid. In an hybrid method, an acoustic analogy is required to predict the far field noise. By contrast, in a direct method, the far field acoustics and the near field flow variables are simulated simultaneously. The usage of the Lattice BOLTZMANN Method (LBM) as a direct tool to predict far field noise has been successfully demonstrated by several studies [7 ,8]. LBM has also been used to simulate a section of an airfoil, using both direct and hybrid methods, especially to investigate TE noise [9, 10]. The objective of this study is to use LBM to simulate the TE noise from a Somers S834 airfoil, validate it with the experimental results and draw a comparison with LES/CURLE results from GERHARD's study.

2. Governing Equations

LBM is different from the traditional Computational Fluid Dynamics (CFD) methods. Traditional CFD methods solve partial differential NAVIER-STOKES (N-S) equations, to simulate the fluid. On the other hand, LBM uses discrete BOLTZMANN equations to simulate the flow at kinetic level [11]. Particle distribution functions (PDF), which are defined as the number density of molecules at position x and speed v at a time t, are used by BOLTZMANN equations to capture the kinetic behavior of particles in the lattice world. The basic difference between traditional CFD and LBM lies in this fact, that the LBM approach has much simpler physics to deal with, compared to solving the non-linear PDEs in the N-S approach. LBM is inherently time-dependent. Fluid properties like density and velocity are derived from these PDFs. Such a discretization strategy leads to conservation of mass, momentum and energy. The discrete Lattice BOLTZMANN equations and the associated terms with it are

$$f_i(x+c_i\Delta t, t+\Delta t) - f_i(x,t) = C_i(x,t), \tag{1}$$

where f_i denotes the movement of the distribution of particles in the i-th direction. $c_i\Delta t$ and Δt are the space and time increments.

The right hand side of the eq. (1) consists of the collision term and is called as BHATNAGAR GROSS and KROOK [12] collision operation. Its main function was found out to be that it drives the velocity distribution function towards its equilibrium distribution. The collision term

$$C_i(x,t) = -\frac{\Delta t}{\tau}[f_i(x,t) - f_i^{eq}(x,t)]. \tag{2}$$

consists of relaxation time τ which describes how quickly the velocity distribution function relaxes towards equilibrium and it relates to the fluid viscosity. It uses a

3D cubic lattice D3Q19 to discretize the velocity space into 19 discrete speeds. The usage of D3Q19 enables enough number of velocity components for sufficient lattice symmetry to recover the N-S equations [13]. The f_i^{eq} term is the equilibrium distribution function and in order to recover macroscopic hydrodynamics [14], f_i^{eq} has be chosen in such a way that the conservation laws are satisfied. As mentioned earlier, the fluid properties like density and velocity are obtained by taking the moment summations over the velocity vectors:

$$\rho(x,t) = \sum_i f_i(x,t), \quad \rho u(x,t) = \sum_i c_i f_i(x,t)] \tag{3}$$

In order to recover the compressible N-S equations, CHAPMAN-ENSKOG expansion can be used for small Mach number (*Ma*). The resulting equation of state obeys the ideal gas law: $p = \rho RT$. The relaxation time parameter τ is related to the kinematic viscosity of the fluid as

$$\tau = \frac{v}{RT} + \frac{\Delta t}{2}. \tag{4}$$

[15].

In this study, a turbulence model called very large-eddy simulation (VLES) is used. It consists of a two-equation *k*-□ renormalization group (RNG) [16]. These two equations are further modified to incorporate a swirl correction factor [17]. This enables the resolution of unsteady large-scale vortices in regions, where these can be resolved.

In high Reynolds number (*Re*) applications, a wall function is used to model the effect of the boundary layer on the rest of the flow because fully resolving the near wall region is computationally expensive. Hence the cell closest to a surface is assumed to obey the law of the wall. A hybrid wall function smoothly transitions from a turbulent wall function (i.e. a logarithmic profile) at high *y+* values to a viscous wall function (i.e. a linear profile) at low y+ values as given in eq. (5). Along with the velocity profiles, this hybrid wall function is coupled with a wall model pressure gradient extension to account for the effects of favorable and adverse pressure gradient (APG) on the near-wall boundary layer profile [18].

$$u^+ = u(y)/u_T = \begin{cases} y^+ & for \quad y^+ < 5, \\ g(y^+) & for \quad 5 < y^+ < 35, \\ \frac{1}{k}\log(y^+) + C_1 & for \quad y^+ > 35. \end{cases} \tag{5}$$

In this study, the commercial software Exa PowerFLOW™ 5.0c has been used to set up two different types of simulation case in order to simulate the TE noise emitted.

3. SIMULATION CASE SETUP

The case investigated here is a SOMERS S834 airfoil segment, which has a chord based *Re* of $3.5 \cdot 10^5$ with tripping bands positioned at 17% and 76% of chord length on the suction and pressure side such that it mimics a *Re* of $3.5 \cdot 10^6$. It has a chord length $c = 0.2$ m and an aspect ratio of 1.33. The inlet velocity, $U_{ref} =$ 25.55 m/s and the effective angle of attack (*AOA*) is 4.7°. Since the airfoil is placed in a jet flow, a correction factor is applied to the angle of attack as suggested by BROOKS et al [19]. This leads to a geometrical *AOA* of 12.7°. It has to be noted that the previous study comprising of the LES simulations and experimental measurement also considered this correction factor while calculating the *AOA*.

The first case setup is called *2D*, where only a segment of the span (7.5% chord) is simulated with periodic boundary condition in the span. By simulating only a segment of the span, the advantage lies in the fact that a very fine layer of cells can be used in the near wall region of the airfoil and is still computationally affordable. This leads to surface y^+ of less than 5 on the airfoil segment and hence a wall model is not used. However, a direct acoustic prediction is only possible with a correction factor as introduced by OBERAI [10] for reduced span and cyclic boundary conditions.

As already mentioned, LBM has an advantage compared to traditional CFD approaches, that it can solve the full compressible flow equations for determining both the hydrodynamic and acoustic pressure fluctuations. Hence, in order to utilize this advantage of LBM, a second simulation case called *3D* is set up. In this setup, the entire anechoic room, where the acoustic measurements were held, is simulated. The simulation domain of both setups are shown in Fig. 1. Obviously, with such a large computational domain, the near wall cells can not be as fine as they are in the *2D* setup. They are four times coarser than the *2D* setup. Another advantage of such a configuration is that the installation effects (side plates) are also considered.

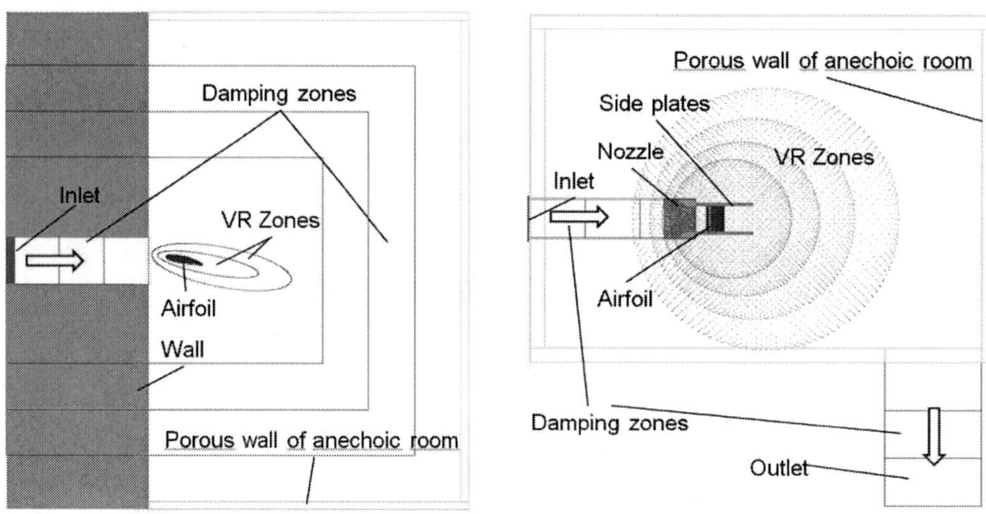

Fig. 1 Left: Simulation domain (top view) of *2D* setup; right: Simulation domain (side view) of *3D* setup.

In LBM, discretization takes place using a strategy called variable resolution (VR) zones. Each VR zone consists of cubical volume cells called voxels. The size of voxels increase by the factor two in adjacent VR zones. In Fig. 2, the various VR zones for the *3D* setup are shown.

In both setups, a velocity is provided at the inlet boundary condition and the outlet boundary condition is set as atmospheric pressure. The inlet and outlet region are modeled as damping zones to avoid acoustic reflections. The simulated Ma (Ma = 0.075) is chosen the same as in experiment, such that the acoustic waves propagate at the same speed as they do in experiment.

Before getting into the results section, there are two interesting aspects to be checked; the wall model that would be used and the percentage of resolved turbulence. The surface y^+ is depicted in Fig. 3. It is observed that due to finer resolution near the wall, *2D* setup has y^+ value less than 5 in most portions of the airfoil surface. This means that the velocity profile would be calculated and a wall model would not be used. Due to coarser mesh, the *3D* setup has surface y^+ considerably greater than 5 in most of the airfoil surface, except for some regions near TE on the suction side (SS). In Fig. 3, only a section of airfoil span in the *3D* setup is shown, whereas the entire span is shown in the *2D* setup.

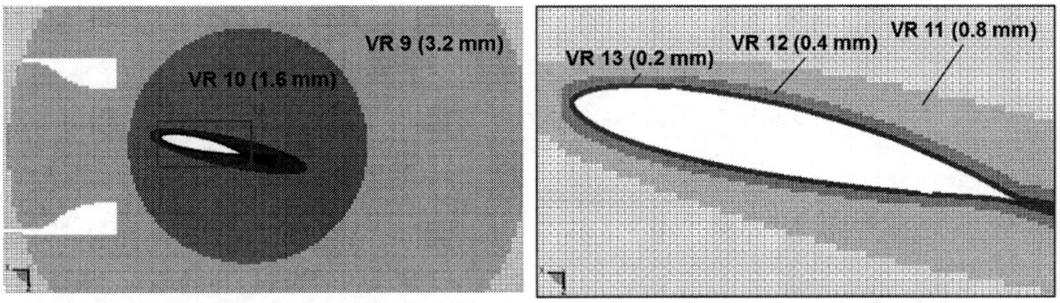

Fig. 2 Left: Top view of VR zones (plane cut at mid span) in the *3D* setup; right: Zoomed view of the same.

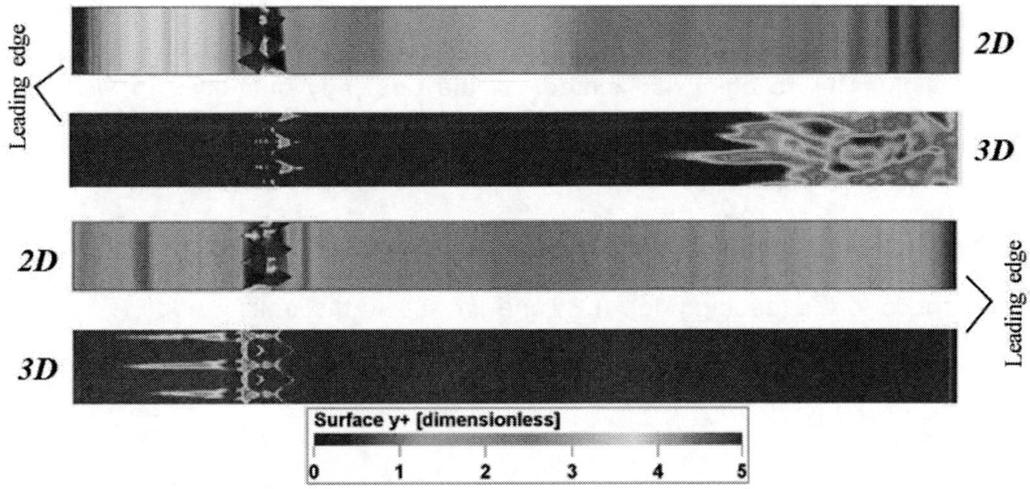

Fig. 3 Top: Surface y^+ on the suction side; bottom: Surface y^+ on the pressure side.

Exa PowerACOUSTICS™ (one of the post processing tools in PowerFLOW™) allows the access to information on the REYNOLD's stresses or fluctuation kinetic energy (FKE) and the total turbulence kinetic energy (TTKE). In order to compute the amount of resolved turbulence, a percentage of FKE over TTKE is taken and shown in Fig. 4. It is noticed that after the trip, *2D* setup has more amount of turbulence resolved than the *3D* setup on both sides of the airfoil. Table 1 summarizes other interesting information regarding the setups.

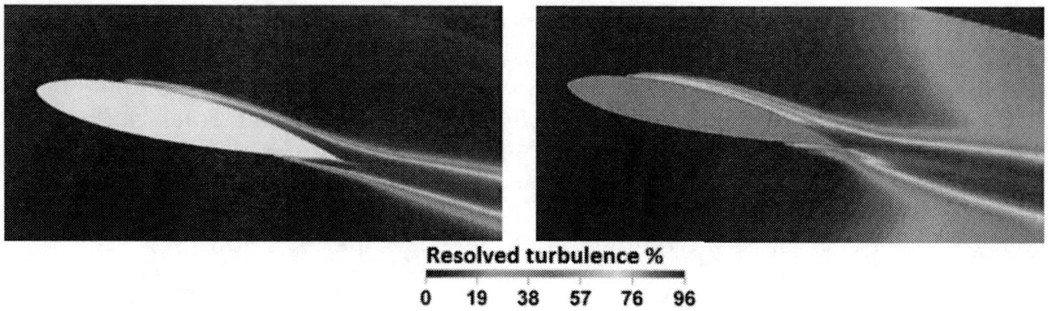

Fig. 4 Left: Resolved turbulence % - plane cut at mid span: *2D* setup; right: *3D* setup.

Table 1 Details of simulation setups in LBM

Parameter	*2D*	*3D*
Time step at finest voxel VR zone	$3.3 \cdot 10^{-7}$ s	$2.66 \cdot 10^{-7}$ s
Finest voxel	0.0586 mm	0.2 mm
Total no. of voxels	28 mio	129 mio
CPU hours (for 0.5 s physical time)	37000	25000

4. FLOW FIELD RESULTS

All results shown here are taken after the flow field had reached a statistically steady state. Unless otherwise stated, all the unsteady quantities in LBM are captured with a sampling frequency f_s = 20 kHz for 36 through-flow times $T_f = c/U_{ref}$. The power spectral density (PSD) shown in the spectral analysis is obtained using the *pwelch* routine in Matlab™ Vers. R2014b (Δf_{ref} = 1 Hz, $p_0 = 2 \cdot 10^{-5}$ Pa).

The mean pressure distribution in terms of the pressure coefficient $c_p = p_{static} / p_{dynamic}$ on the airfoil surface is shown in Fig. 5. The abscissa is such that $x/c = 0$ corresponds to the leading edge (LE) and x/c = 1 to the trailing edge (TE). The values are time averaged for 12 T_f.

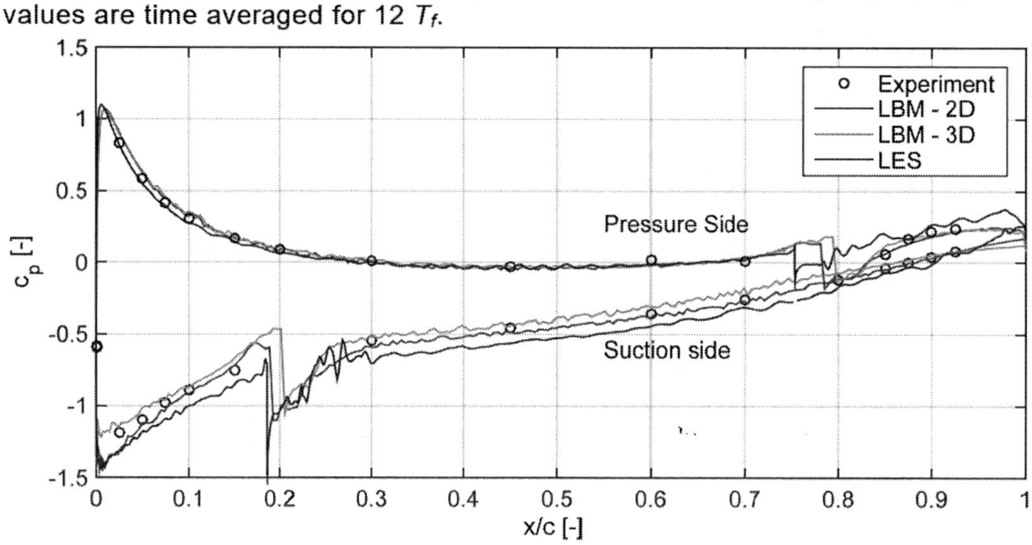

Fig. 5 Mean pressure distribution on the airfoil surface.

It is observed that the results from both LBM setups as well as LES show a good agreement with the experiment. The sudden jumps in the pressure distribution on both sides are due to the tripping.

Fig. 6 shows the root-mean-square (RMS) value of fluctuating surface pressure p'_{rms}, normalized with dynamic free stream pressure p_{dyn} near the trailing edge region on the SS. Since such measurement was not done experimentally, only a comparison between LBM and LES is drawn. The LBM surface pressure fluctuations have higher values compared to LES. But in the trailing edge region (x/c > 0.9), LBM *2D* and LES show the same tendency but not LBM *3D*. Experimentally, the blade pressure fluctuations have been recorded at chord wise position x/c = 0.9.

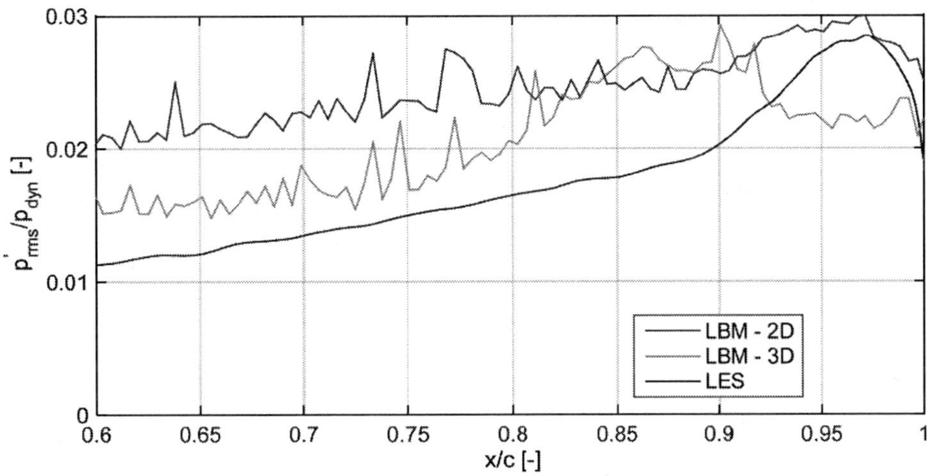

Fig. 6 Simulation predicted pressure fluctuations in the TE region on the SS.

In Fig. 7, the PSD of pressure fluctuations G_{xx}, at this position is compared for all setups. Here the availability of experimental measurement enables us to realize which among the three simulation setups produce similar tendency as experiment and it is apparently the *2D* setup which has the same shape as experimental spectrum. But there is an overprediction in all three simulation setups. However, it is observed that the *3D* setup produces higher PSD values in the lower frequency range (< 1000 Hz) and less in the higher frequency range. This might be due to the coarser voxels in the *3D* setup which were required given the available computational resources.

The next interesting comparison are boundary layer details on the SS. Fig. 8 shows the comparison of boundary layer displacement thickness δ^*, normalized with chord for all setups. It is observed that the LBM 3D overpredicts the boundary layer displacement thickness near the trailing edge whereas LBM 2D has a fair agreement with experiment.

In Fig. 9 (upper row) the velocity profiles perpendicular to the surface near the TE on the SS is plotted for three different chord wise positions, x/c = 0.9, 0.95 and 1 respectively. Note that y is always perpendicular to wall with y/c = 0 at the wall. It is observed that there are some deviations in the velocity profiles compared to experiment in all three simulation setups. Despite the fact that both are LBM simulations (*2D* and *3D*) of the same airfoil, one could observe a clear difference between the velocity profile near the wall. This implies that in *2D* setup a wall model is not used, whereas in *3D* setup it is used.

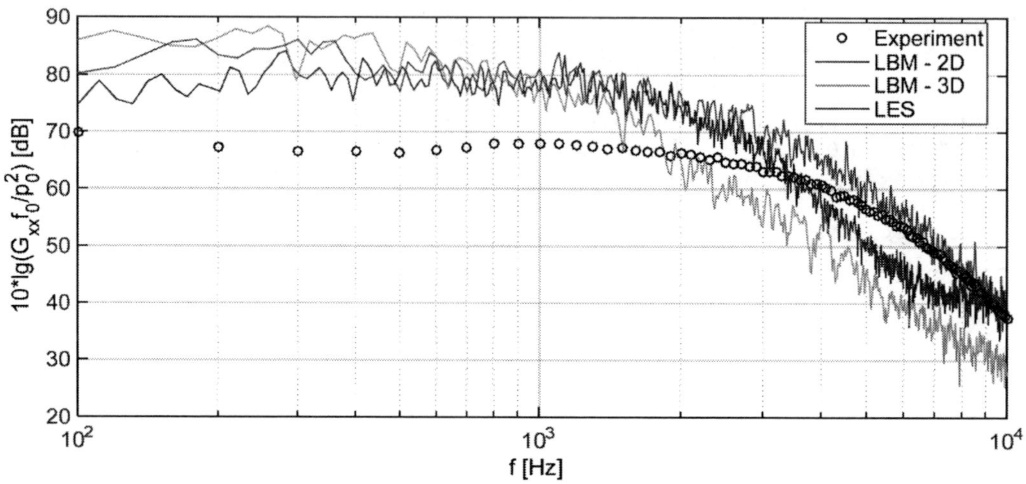

Fig. 7 Experiment and simulation predicted surface pressure spectrum (*x/c* = 0.9 on SS).

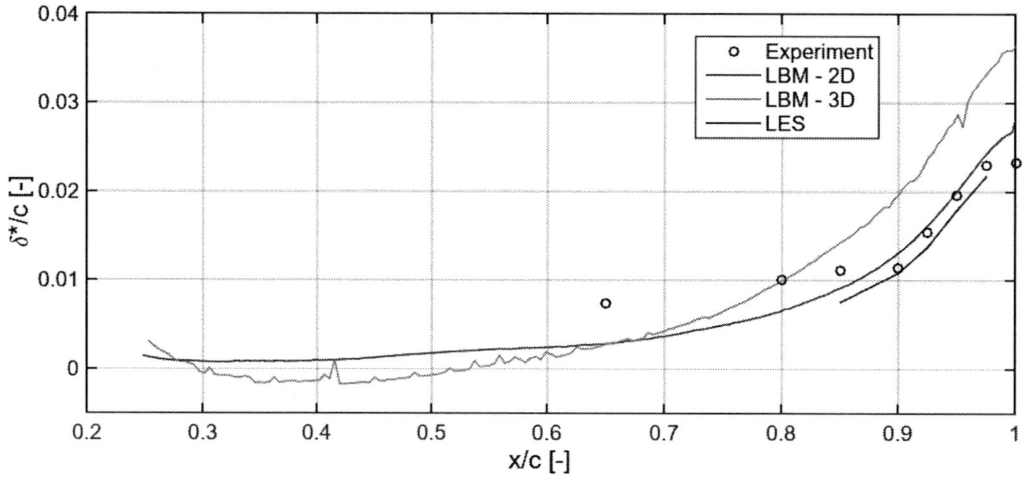

Fig. 8 Boundary layer displacement thickness (SS).

The turbulence intensity

$$TI = u_{rms}' / U_{ref} \quad (6)$$

in the boundary layer is shown in the lower row of Fig. 9 for the same chord wise positions as velocity profiles. In the LBM *2D* it matches quite well with the experiment, but LBM *3D* overpredicts at all three positions, especially at the TE.

Fig. 10 shows the PSD of velocity fluctuations at a point *y/c* = 0.005 on the SS for two chord wise positions near the TE, *x/c* = 0.975 and 1 respectively. The LBM *2D* spectrum shows very good agreement with experiment till 3 kHz and the spectrum falls down after that, owing to the fact that it is modeled and not resolved anymore. And the spectrum of LBM *3D* falls down starting from the low frequency range and doesn't match with the experiment. At the TE (*x/c* = 1), LBM *2D* matches better with experiment than the LES till 3 kHz.

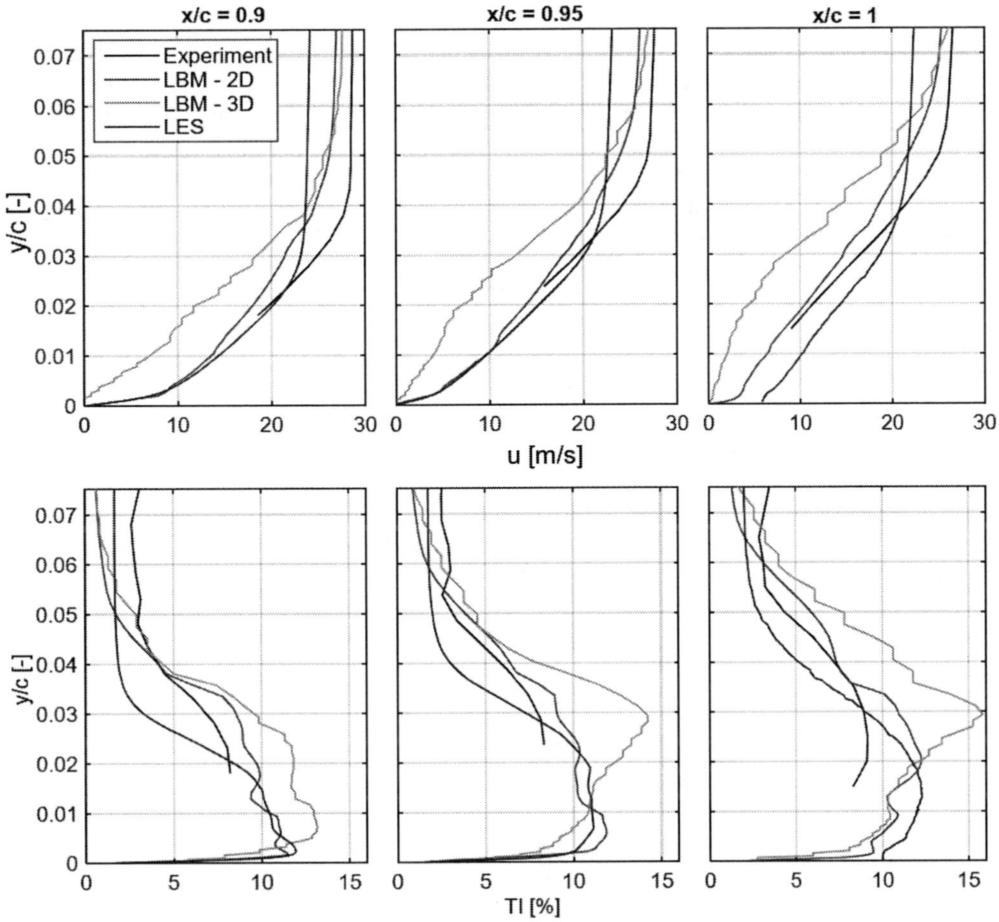

Fig. 9 Top: Velocity distribution; bottom: Turbulence intensity in the turbulent boundary layer on the SS in the TE region.

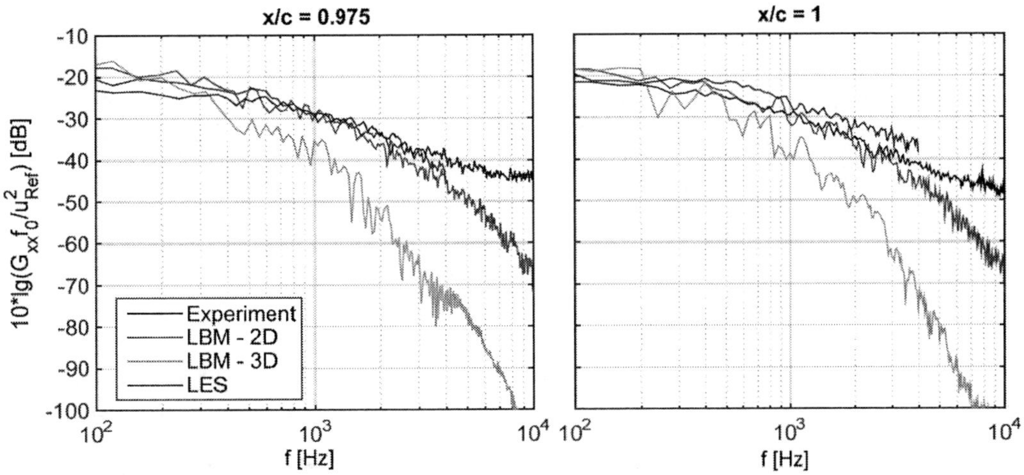

Fig. 10 Experiment and simulation predicted velocity spectrum at y/c = 0.005 on the SS.

5. Acoustic Results

As mentioned earlier, LBM is advantageous because the far field pressure fluctuations are measured directly using probes in the simulation domain. In the *3D* setup, the microphone probes directly capture the far field pressure fluctuations as it is done in experiment, whereas in the *2D* setup, due to reduced span and cyclic boundary conditions, a correction factor has to be added to the direct probe measurements. For low *Ma* flows, OBERAI [10] recommends a frequency dependent correction factor

$$\beta = 10\log_{10}\left(\frac{fb^2}{aR}\right), \tag{7}$$

where b is the segmented span, a is the speed of sound and R is the observer distance. This correction factor has been applied to the direct probes measured in the *2D* setup. In all setups (experiment and simulations), the microphones were placed at 3 locations: 1.5c on either side of airfoil from TE and 1.5c on the SS from LE as shown in Fig. 11.

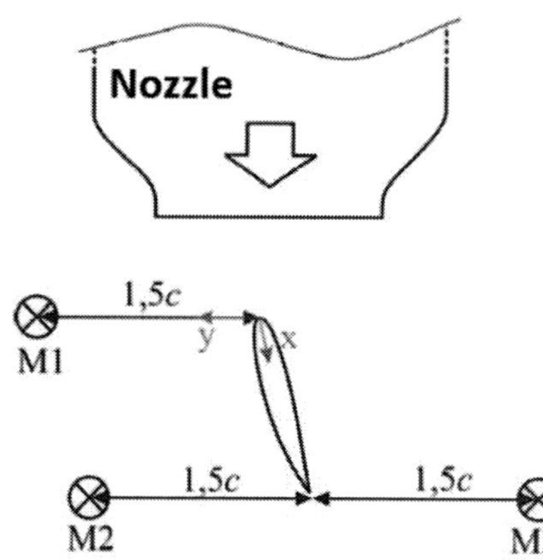

Fig. 11 Schematic diagram of microphone locations.

All recordings were captured with a f_s = 52 kHz. The approach of segmented span in LBM *2D* is valid only when the span wise coherence decays within the simulated span. The coherence function

$$\gamma^2(f) = \frac{|G_{ab}(f)|^2}{G_{aa}(f)G_{bb}(f)} \tag{8}$$

is a function of the power spectral densities G_{aa} and G_{bb} and the cross power spectral density G_{ab} of signals a and b. In this study it is obtained using the *mscohere* routine in Matlab® Vers. R2014b (Δf = 40 Hz). Span wise coherence is calculated at chord wise position x/c = 0.95 on the SS. Fig. 12 shows the contour plot of span wise coherence of the *2D* setup. The *y*-axis covers the simulation domain in spanwise direction (*z* is the distance in span). Only the frequency range between 300 Hz - 2000 Hz has been plotted, since it falls in the interest of TE noise evaluation. It is observed that for frequencies less than 500 Hz, a complete decay of span wise coherence is not observed. Extending the simulation domain in the spanwise direction will capture the largest flow structures correctly. However, the very strong coherent structures (with coherence function > 0.4) decay well in the frequency range less than 500 Hz.

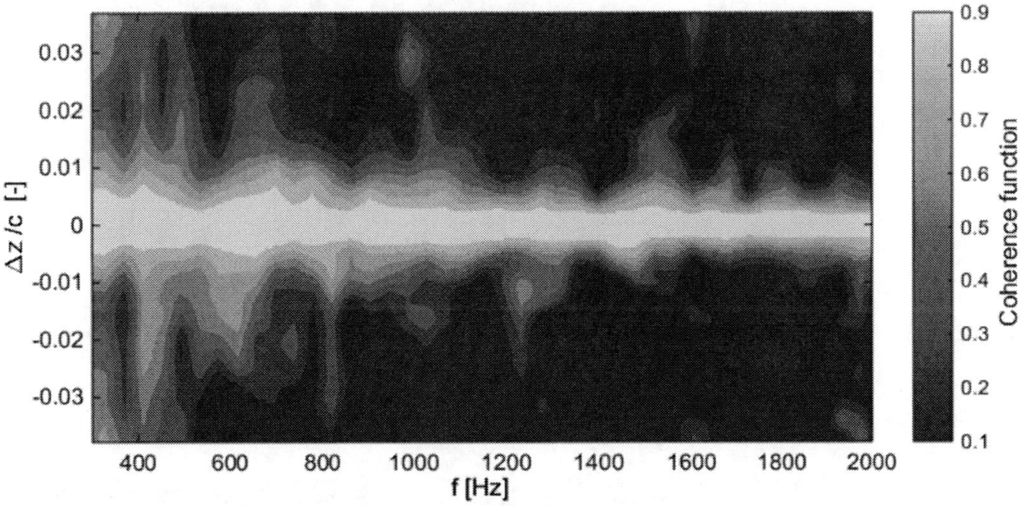

Fig. 12 Span wise coherence of blade pressure for LBM *2D* at *x/c* = 0.95 on the SS.

Fig. 13 depict key results of this study: the spectral sound pressure level (SPL) (average of 3 microphones according to Fig. 11). The signal-to-background noise ratio is large enough only in the frequency range of 300 Hz to 2000 Hz, where the TE noise can be identified in the experiment. As stated by GERHARD [4], the acoustic evaluations in his study of the same setup showed that the airfoil TE noise is the most prominent noise source from frequencies of 160 Hz to 3000 Hz and that one can expect a hump dominating the TE noise spectrum to lie in a frequency range between 350 Hz and 550 Hz. It is observed that the LBM *2D* and experimental measurement match quite well in the frequencies ranging from 400 Hz to 1000 Hz. It is also observed that the direct *SPL* predicted by LBM *3D* is overpredicted in the TE noise region. This again correlates with the results seen earlier in blade pressure fluctuations spectrum due to coarser cells.

Fig. 14 shows the instantaneous dilatation $\frac{1}{\rho}\frac{\partial \rho}{\partial t}$ field in LBM *2D* setup. It is confirmed that the major acoustic source is identified at the TE, where the acoustic waves propagate from the TE region due to diffraction of turbulent eddies on the TE.

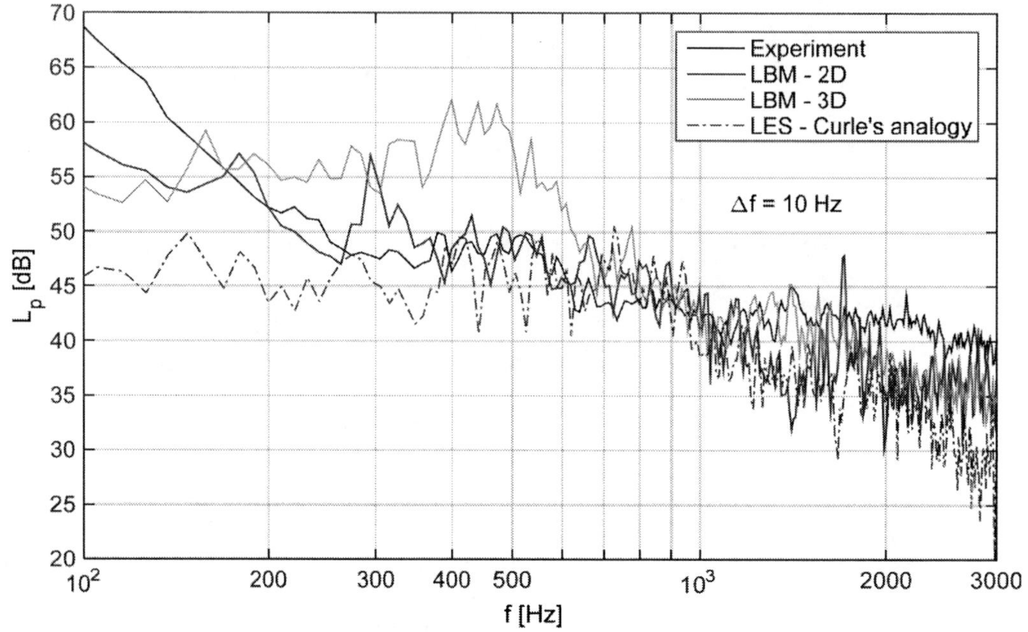

Fig. 13 SPL at observer points according to Fig. 11 (all microphones averaged).

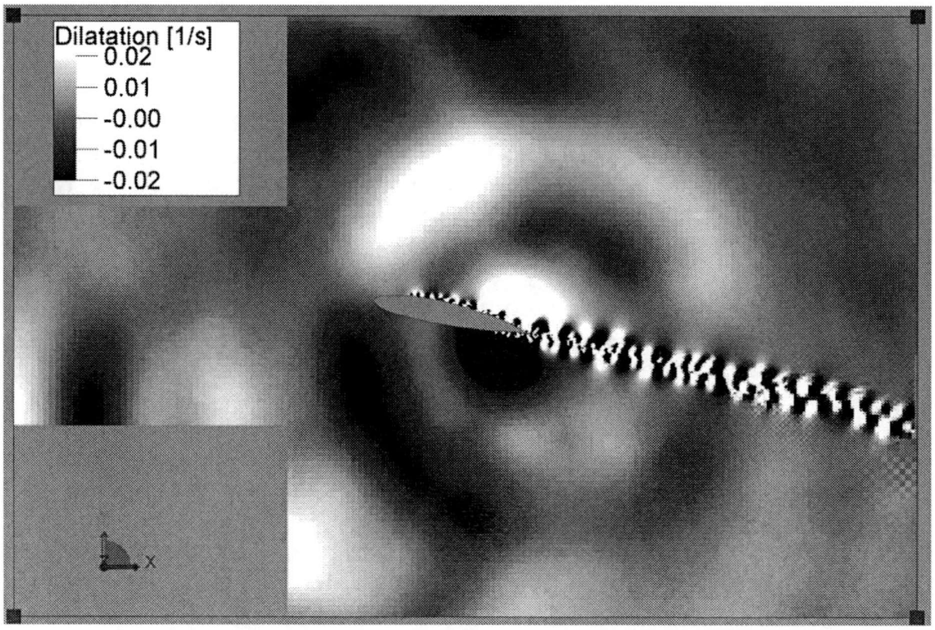

Fig. 14 Instantaneous image of dilatation field in LBM *2D* (plane cut at mid span).

6. Conclusion

Aero-acoustic simulation of a S834 airfoil section has been successfully conducted using a Lattice-Boltzmann method. The important aspect to consider while predicting trailing edge noise lies on the fact that the mesh resolution in the near wall should be fine enough, e.g. such that the surface y^+ is less than 5. As in this work, because of computational cost, this triggered the simulation of an airfoil section with only a short spanwise extension (segmented airfoil) rather an airfoil in a complete wind tunnel. For this case, as compared to experiments, the predicted boundary layer induced surface pressures, near-field velocity fluctuations close to the TE and far field sound pressure are in good agreement. Another important conclusion is that, given a simulation domain of same size, LES requires more computational time and also a finer mesh than a comparable Lattice-Boltzmann method.

ACKNOWLEDGEMENTS

Financial support for this work was partly provided by the Federal Ministry for Economic Affairs and Energy of Germany (BMWi) within the project RENEW (FKZ 0325838B).

The authors would like to thank Franck Perot (Exa Corporation), Marlene Sanjose (University of Sherbrooke) and Tao Zhu (University of Siegen) for their technical assistance in setting up the simulation case as well as Tom Gerhard (Vaillant GmbH) for providing the experimental and LES data.

REFERENCES

[1] BROOKS, T.F., POPE, D.S., MARCOLINI, M.A., 1989, "Airfoil self-noise and prediction", NASA RP-1218, *NASA Langley Research Center, Hampton*, Virginia.

[2] OERLEMANS, S., SIJTSMA, P., MÉNDEZ LÓPEZ, B.M., 2007, "Location and quantification of noise sources on a wind turbine", *Journal of Sound and Vibration*, **299**: 869-883.

[3] GERHARD, T., 2015, "Umströmte Tragflügelsegmente: Auswirkungen des Hinterkantenausblasens auf Strömungsgrenzschicht und Schallemission", *Doctor of Philosophy Thesis*, Universität Siegen.

[4] GERHARD, T., CAROLUS, T., 2014, "Investigation of airfoil trailing edge noise with advanced experimental and numerical methods", *The 21st International Congress on Sound and Vibration*.

[5] GERHARD, T., ERBSLÖH, S., CAROLUS, T., 2014, "Reduction of airfoil trailing edge noise by trailing edge blowing", *Journal of Physics*, **524**: 012123.

[6] CURLE, N., 1955, "The influence of solid boundaries upon aerodynamic sound", *Proceedings of the Royal Society of London. Series A, Mathematical and Physical Sciences*, **231**(1187), pp. 505-514.

[7] MANN, A., PÉROT, F., KIM, M.S., CASALINO, D., FARES, E., 2012, "Advanced noise control fan direct aeroacoustics predictions using a Lattice-Boltzmann method", *18th AIAA/CEAS Aeroacoustics Conference*.

[8] ZHU, T., STURM, M., CAROLUS, T., 2014, "Experimental and numerical investigation of tip clearance noise of an axial fan using a Lattice-Boltzmann method", *The 21st International Congress on Sound and Vibration*.

[9] VAN DER VELDEN, W.C.P., VAN ZUIJLEN, A.H., DE JONG, A.T., BIJL, H., 2015, "On the noise prediction of a serrated DU96 airfoil using the Lattice Boltzmann Method", *6th International Meeting on Wind Turbine Noise*.

[10] VAN DER VELDEN, W.C.P., VAN ZUIJLEN, A.H., RAGNI, D., 2016, "Flow topology and noise emission around straight, serrated and slitted trailing edges using the Lattice Boltzmann methodology", *22nd AIAA/CEAS Aeroacoustics Conference*.

[11] SUCCI, S., 2001, "The Lattice Boltzmann Equation for fluid dynamics and beyond", *Oxford Univesity Press,* Oxford.

[12] BHATNAGAR, P.L., GROSS, E.P., KROOK, M., 1954, "A model for collision processes in gases. I. small amplitude processes in charged and neutral one-component systems", *Physical Review*, **94(3)**: 511-525.

[13] FRISCH, U., HASSLACHER, B., POMEAU, Y., 1986, "Lattice-Gas automata for the Navier-Stokes equation", *Physical Review Letters*, **56**: 1505-1508.

[14] CHEN, H., CHEN, S., MATTHAEUS, W.H., 1992, "Recovery of the Navier-Stokes equations using a lattice-gas Boltzmann method", *Physical Review A*, **45(8)**: R5339-R5342.

[15] CHEN, S., DOOLEN, G., 1998, "Lattice Boltzmann method for fluid flows", *Annual Review of Fluid Mechanics*, 30**(1)**: 329-364.

[16] YAKHOT, V., ORSZAG, S., 1986, "Renormalization group analysis of turbulence. I. basic theory", *Jorunal of Scientific Computing*, **1:** 3-51.

[17] CHEN, H., KANDASAMY, S., ORSZAG, S., SHOCK, S., SUCCI, S., YAKHOT, V., 2003, "Extended Boltzmann kinetic equation for turbulent flows", *Science*, **301**: 307-314.

[18] KOTAPATI, R., SHOCK, R., CHEN, H., 2014, "Lattice-Boltzmann simulations of flows over backward-facing inclined steps", *International Journal of Modern Physics C*, **25**: 1340021.

[19] BROOKS, T.F., MARCOLINI, M.A., POPE, D.S., 1984, "Airfoil trailing edge flow measurements and comparison with theory incorporating open wind tunnel corrections", *Proc. AIAA/NASA 9th Aeroacoustics Conference*, 1-12.

7th International Conference
on
Wind Turbine Noise
Rotterdam – 2nd to 5th May 2017

Small Horizontal Axis Wind Turbine: Aeroacoustic and Aerodynamic Optimization of Airfoil Shape and Blade

K. Volkmer, N. Kaufmann, T. Carolus
University of Siegen, 57076 Siegen, Germany

Corresponding author: kevin.volkmer@uni-siegen.de

Summary
Objective of this contribution is an acoustic and aerodynamic optimization of the full 3D blade geometry for a small horizontal axis wind turbine. Utilizing a refined aerodynamic blade element momentum (BEM) method and a combination of ROZENBERG's wall pressure and AMIET's trailing edge noise model an evolutionary algorithm is implemented. The validity of ROZENBERG and AMIET models was checked by comparison with recent own measurements of the wall pressure fluctuations and trailing edge sound of a small airfoil section. The optimization is subdivided into two independent steps, (i) the airfoil optimization, and (ii) the optimization of the blade twist angle and chord length distributions. To mimic a fully turbulent flow around the blades of a realistic wind turbine - a worst case scenario - tripping was applied close to the leading edge of all airfoils. The airfoil optimization resulted in novel airfoil shapes. As compared to a chosen benchmark airfoil S834 they promise a better lift-to-drag-ratio and/or lower non-dimensional wall pressure fluctuations in the trailing edge region (WPS). The predictions forecast a reduction by more than 10 dB. Utilizing such a low noise airfoil and optimizing spanwise chord length and twist angle distribution in a second step results in new blade designs. As compared to an existing, non-optimized research turbine with SOMERS airfoil shaped blades the predicted sound power of an optimized turbine is substantially lower without degradation of its power coefficient. A detailed analysis shows that the sound reduction is mainly attributed to the improved airfoil sections. Since the optimization and all results presented here are mainly based on models, future experimental validation is indispensable.

Nomenclature

Symbols

A	m^2	swept area of the turbine rotor
C	m	chord length
C_f	-	skin friction coefficient

C_D	-	drag coefficient
C_L	-	lift coefficient
C_P	-	power coefficient
D	-	Dimension
F	N	Force
H	-	boundary layer shape factor
I	-	radiation integral
K	rad/m	convective wavenumber
L	m	span
L_{Spp}	dB	level of power spectral density of far field acoustic pressure
$L_{\Phi pp}$	dB	level of power spectral density of wall pressure
$LPBE$	dB	overall sound power level of a blade element
M	-	free stream Mach number
$OSPL$	dB	overall sound power level of whole wind turbine
P	W	power
$P_{\tilde{\Phi}\,pp}$	-	power of the non-dimensional wall pressure fluctuations
PT	-	penalty term
R	m	total radius
Re	-	Reynolds number
Re_θ	-	Reynolds number based on θ and w_e
R_T	-	Ratio of outer to inner boundary layer timescale
S_O	m	corrected observer distance
S_P	W/Hz	sound power spectral density
S_{PP}	Pa^2/Hz	power spectral density of far field acoustic pressure
c_0	m/s	speed of incoming wind far upstream
c_s	m/s	speed of sound
f	Hz	frequency
f_{obj}	-	objective function
k	rad/m	acoustic wavenumber
l_y	m	spanwise correlation length
n	1/s	rotational speed
n_{BE}	-	number of blade elements
$n_{2D,PT}$, $n_{3D,PT}$	-	number of penalty terms in 2D and 3D
p	Pa	pressure
r	m	radius
u	m/s	circumferential velocity
w	m/s	relative velocity
$x_{1,2,3}$	m	Cartesian co-ordinates
z	-	number of blades

Greek symbols

Δ	-	ZAGAROLA-SMITS' parameter

Φ_{pp}	Pa^2/rad	Power spectral density of surface pressure fluctuations
$\tilde{\Phi}_{pp}$	-	Normalized power spectral density of surface pressure fluctuations
Π	-	COLE's wake strength parameter
α	°	angle of attack
β	°	flow angle
β_c	-	CLAUSER's equilibrium parameter
γ	°	twist angle
δ	m	boundary layer thickness
δ^*	m	boundary layer displacement thickness
ε	-	lift-to-drag-ratio
κ	rad/m	frequency parameter
θ	m	boundary layer momentum thickness
λ	-	tip speed ratio
ν	m2/s	kinematic viscosity
ρ	kg/m^3	air density
τ_{max}	Pa	maximum shear stress
τ_w	Pa	wall shear stress
ω	rad/s	angular frequency
$\tilde{\omega}$	-	Strouhal number based on external variables

Subscripts

0	position far upstream
∞	position in rotor plane
Spp	power spectral density of far field acoustic pressure
e	position at boundary layer edge
obj	objective
ref	reference
sh	shaft
tip	at rotor tip
u	circumferential
w	weighing

Abbreviations

BE	blade element
BEM	blade element momentum theory
WPS	wall pressure spectra
WT	wind turbine

1. Introduction

Several criteria are relevant for designing the twisted, tapered and carefully profiled blades of horizontal axis wind turbines. Naturally, the maximum energy output of a wind turbine is of primary concern and achieved by optimal aerodynamic design. Yet, wind turbine industry is focusing on the analysis and mitigation of flow induced noise – on a par with efficiency, structural health, cost etc.

A classical semi-analytic blade design method is e.g. by GLAUERT and SCHMITZ, see for instance GASCH and TWELE [1]. Its outcome is a 3D blade, where the energy transferred to the shaft by each thought blade element of the segmented blade is theoretically maximal. However, the resulting spanwise geometry and aerodynamic loading distribution may not be optimal with respect to other criteria, e.g. flow induced noise.

A combined acoustic and aerodynamic optimization of the full 3D blade geometry is demanding. LELOUDAS [2] reported an optimization of twist and chord length distribution as well as the spatial distribution of given airfoil shapes, but excluded the optimization of the 2D airfoil shape itself. TIAN et al. [3] created an aeroacoustic prediction methodology for wind turbines by combining the aerodynamic airfoil performance prediction tool XFoil (DRELA [4]), a model for wall pressure spectra (WPS) by ROZENBERG et al. [5], and AMIET's trailing edge noise model [6] with advanced models by ROGER and MOREAU [7]. In addition, AMIET's inflow turbulence noise model [8] was also implemented. Comparison to experimental data from isolated airfoil sections and a complete wind turbine were promising. In principle such a model could be used for optimization, but TIAN et al. did not elaborate on that.

Detailed aerodynamic and aeroacoustic airfoil shape optimization has been tried in the past. SCHEPERS et al. [9] and BERTAGNOLIO [10] utilized the TNO-model (named after the TNO Institute of Applied Physics, Netherlands) with simplified boundary layer characteristics as inputs. GÖCMEN and ÖZERDEM [11] utilized the Brooks-Pope-Marcolini (BPM) model [12] to implement an airfoil optimization tool.

A more integrated optimization, i.e. of twist and chord length distribution with respect to only aerodynamic performance and of airfoil shape with respect to aerodynamic and aeroacoustic performance, has been reported by HAO et al. [13]. Based on the simple BPM model, RODRIGUES and MARTA [14] synchronously optimized twist and chord length distribution as well as airfoil shape with respect to aerodynamic and aeroacoustic performance.

There is no doubt that the acoustic model is an essential and challenging ingredient for any type of a combined acoustic and aerodynamic optimization. According to LOWSON [15] wind turbine sound prediction methods can be subdivided into three classes: (i) empirical single equation models using general design parameters of a turbine (ii) semi-analytical models which model sound source mechanisms and (iii) high fidelity computational aeroacoustic methods.

The objective of this contribution is a two-step acoustic and aerodynamic optimization of the full 3D blade geometry, here exemplary for a small horizontal axis wind turbine. The sub-models should be more advanced as compared to the literature. Thus we focus on a combination of refined blade element momentum (BEM) method and a combined ROZENBERG/AMIET trailing edge noise model. The detailed airfoil boundary layer data relevant for the acoustic source model are taken into account. The optimization, however, is still subdivided into two independent steps, (i) the airfoil optimization, and (ii) the optimization of the blade twist angle and chord length distributions. This two-step approach is thought reducing the optimization time and should enable a separate assessment of the optimization potential in both steps. The optimization itself is based on an in-house evolutionary algorithm as developed recently for axial fans (BAMBERGER [16]) and tidal horizontal axis turbines (KAUFMANN [17]).

2. Sub-models and optimization methodology

2.1 Aerodynamic performance prediction

A widely used representation of aerodynamic wind turbine performance is shown in **Fig. 1** in terms of the non-dimensional power coefficient

$$C_P \equiv \frac{P_{sh}}{0.5 \rho c_0^3 A} \tag{1}$$

as a function of tip-speed ratio

$$\lambda \equiv \frac{u_{tip}}{c_0} . \tag{2}$$

A is the area swept by the rotor and c_0 the speed of the incoming wind far upstream of the turbine.

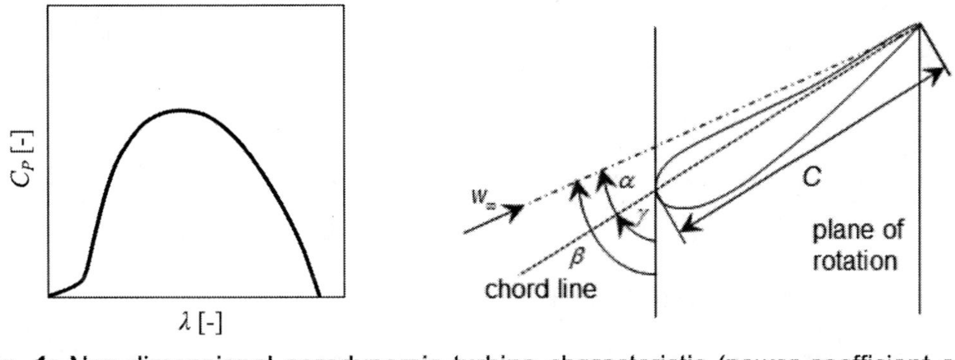

Fig. 1: Non-dimensional aerodynamic turbine characteristic (power coefficient as a function of tip-speed ratio) and relevant quantities at blade element (schematically)

Here we apply an enhanced in-house blade element momentum (BEM) method for performance prediction of a given wind turbine as described by KAUFMANN et al. [18]. In the context of this paper it is relevant, that within the BEM each blade is segmented along its span into a number of blade elements (BE), and that the inflow velocity w_∞, the angle of attack α (see **Fig. 1**) and the local Reynolds number are known at each BE. The local aerodynamic forces on a BE can be used to calculate the power coefficient over all BE by evaluating the local circumferential force δF_u acting on each BE.

$$C_P(\lambda) \equiv \frac{\sum_{all\,BE} \delta F_u r 2\pi / n}{0.5\rho c_0^3 A} = \frac{\sum_{all\,BE} P_{sh,BE}}{0.5\rho c_0^3 A}. \tag{3}$$

n is the rotational speed of the turbine and ρ the fluid density. The lift coefficient C_L of an airfoil can be calculated by the lift force F_L acting on each BE.

$$C_L \equiv \frac{F_L}{0.5\rho w_\infty^2} \tag{4}$$

and local lift to drag ratio

$$\varepsilon \equiv \frac{F_L}{F_D} \tag{5}$$

are determined utilizing the public domain tool XFoil [4]. Hereby incompressibility is assumed as the Mach number is well below 0.3 for all BEs. In addition, to mimic a fully turbulent flow around the blades of a realistic wind turbine - a worst case scenario - tripping is applied at 2% and 5% of the BE chord length on suction and pressure side, respectively (unless specified otherwise).

2.2 Aeroacoustic performance prediction

For the prediction of the aeroacoustic performance we follow TIAN et al. [3], however with some extensions and modifications. Only trailing edge noise is taken into account. The BE is considered to be stationary (i.e. non-rotating) in a flow approaching the BE with w_∞. **Fig. 2** depicts the work flow.

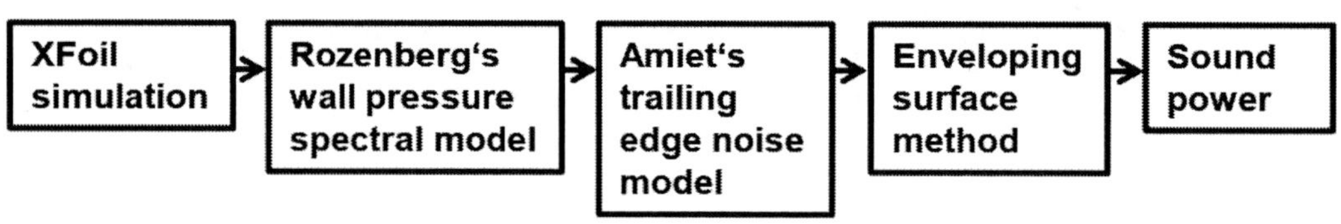

Fig. 2: Flow diagram of acoustic prediction model

On top of lift and drag XFoil yields BE boundary layer parameters which are essential input parameters in ROZENBERG's semi-empirical model (ROZENBERG et al. [5]) for the wall pressure spectrum beneath a turbulent boundary layer. According to ROZENBERG the normalized power spectral density of the wall pressure fluctuations at a given chordwise position is

$$\frac{\Phi_{pp}(\tilde{\omega}) w_e}{\tau^2 \delta^*} = \frac{\left[2.82\Delta^2 \left(6.13\Delta^{-0,75} + F_1\right)^{A1}\right]\left[4.2\left(\frac{\Pi}{\Delta}\right)+1\right]\tilde{\omega}^2}{\left[4.76\tilde{\omega}^{0.75} + F_1\right]^{A1} + \left[8.8R_T^{-0.57}\tilde{\omega}\right]^{A2}}. \tag{6}$$

The normalized frequency, i.e. the Strouhal number, is

$$\tilde{\omega} = \omega\delta / w_e \tag{7}$$

with $\omega = 2\pi f$ being the angular frequency. The fluctuations are normalized with the free stream velocity w_e just outside of the boundary layer and a characteristic shear stress τ in the boundary layer. ROZENBERG sets τ as the maximum of the shear stress in the complete boundary layer. The earlier and well known wall pressure spectral model by GOODY [19] and more recent ones by KAMRUZZAMAN et al. [20] and CATLETT et al. [21] use the *wall* shear stress instead. In the current implementation we also use the wall shear stress which is easily obtained via the skin friction coefficient C_f from XFoil[1]

$$\tau_w = C_f \cdot 0.5 \rho w_\infty^2 \tag{8}$$

δ^* is the boundary layer displacement thickness. Furthermore, the ZAGAROLA-SMITS' parameter $\Delta = \delta / \delta^*$, COLE's wake strength parameter Π and CLAUSER's equilibrium parameter $\beta_c = (\Theta / \tau_w)(dp/dx)$ - see for instance WHITE [22] - are needed. δ and Θ are the boundary layer and momentum thickness, respectively, $R_T = 0.5 \cdot C_f Re \cdot (\delta / C)$ is the ratio of the outer to inner boundary layer timescale. Additional abbreviations are $A_1 = 3.7 + 1.5\beta_c$ and $F_1 = 4.76(1.4/\Delta)^{0.75}(0.375A_1 - 1)$. The parameter A_2 is set to 7, differently to ROZENBERG's suggestion, in order to limit the decrease of the wall pressure fluctuations at high frequencies to $\tilde{\omega}^{-5}$ as suggested by ROGER and MOREAU [23] and SANJOSE [24]. The boundary layer thickness is calculated as in DRELA and GILES [25]. It is convenient for the optimization of the airfoil contour to define a new non-dimensional wall pressure spectrum

$$\tilde{\Phi}_{pp}(\tilde{\omega}) \equiv \frac{\Phi_{pp}(\tilde{\omega}) w_\infty}{\left(\frac{1}{2}\rho w_\infty^2\right)^2 C} = \frac{C_f^2 \left(\dfrac{\delta^*}{C}\right)}{\dfrac{w_e}{w_\infty}} \frac{\left[2.82\Delta^2\left(6.13\Delta^{-0.75} + F_1\right)^{A1}\right]\left[4.2\left(\dfrac{\Pi}{\Delta}\right) + 1\right]\tilde{\omega}^2}{\left[4.76\tilde{\omega}^{0.75} + F_1\right]^{A1} + \left[8.8R_T^{-0.57}\tilde{\omega}\right]^{A2}}, \tag{9}$$

which is normalized with parameters w_∞ and chord length C, given and fixed during an optimization of the airfoil contour (but not necessarily during optimization of the complete blade shape). Integration of $\tilde{\Phi}_{pp}(\tilde{\omega})$ over $\tilde{\omega}$ yields the total power of the wall pressure fluctuations

$$P_{\tilde{\Phi}pp} = \int_{\tilde{\omega}1}^{\tilde{\omega}2} \tilde{\Phi}_{pp}(\tilde{\omega}, \alpha) d\tilde{\omega} \tag{10}$$

The computed wall pressure spectrum is then used as an input for AMIET's trailing edge noise model [6]. The trailing edge is equivalent to the x_2-axis in spanwise direction; x_1 is the chordwise coordinate and x_3 the direction perpendicular to the airfoil surface; the origin of the coordinate system is at mid span, **Fig. 3.**

---

[1] Throughout this study the XFoil parameters are set as follows: *Ncrit* = 9, *M* = 0 (incompressible), tripping at 2% and 5% on suction and pressure side, respectively.

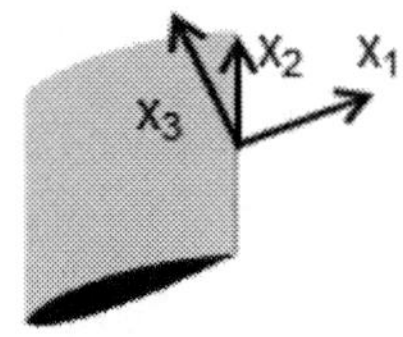

Fig. 3: BE and co-ordinate system; origin at mid span at the trailing edge

ROGER and MOREAU [7] added back scattering at the leading edge and a 3D extension to AMIET's model. Both extensions and the simplified model equation for large aspect ratios are used. The spectral density of the acoustic far field sound pressure is

$$S_{pp}(\mathbf{x},\omega)=\left(\frac{x_3}{S_0^2}\right)^2\left(\frac{\omega C}{4\pi c_s}\right)^2 2\pi L\left|I\left(\frac{K}{C},\frac{kx_2}{CS_0},\kappa\right)\right|^2\frac{1}{\pi}\Phi_{pp}l_y \tag{11}$$

with the speed of sound c_s, the observer distance corrected for convection effects

$$S_0=\sqrt{x_1^2+(1-M^2)(x_2^2+x_3^2)} \tag{12}$$

the chord length C, the span of the blade element L, the wall pressure spectrum Φ_{pp} and the spanwise correlation length l_y. The spanwise correlation length l_y is calculated as in TIAN et al. [3] with the help of CORCOS' model [26]. The airfoil response function I is among others dependent on the acoustic wave number $k = \omega/c_s$ and the convective wave number $K = \omega/U_c$ with U_c being the convection velocity. Also super- and subcritical gusts are included which occur for $\kappa^2 > 0$ and $\kappa^2 < 0$, respectively, where

$$\kappa^2=\left(\frac{KM}{1-M^2}\right)^2-\frac{K_2^2}{1-M^2} \tag{13}$$

and the free stream Mach number $M = w_\infty/c_s$. As the large aspect ratio assumption is applied, the aerodynamic wave number in spanwise direction is $K_2 = kx_2/S_0$.

Eventually, the spectral far field sound power density is calculated on a spherical surface enveloping the BE. The radiation of the spectral density of the acoustic far field sound pressure calculated by eq. (11) is depicted in **Fig. 4**. The geometry and flow parameters of the S834 case which are described in the appendix are used for the calculation.

Integration over the surface area of the sphere yields the sound power spectral density emitted by a BE

$$S_{P,BE}(\omega)=\int_{A_{sp}}\frac{S_{pp}(\mathbf{x},\omega)}{\rho_0 c_s}dA \tag{14}$$

with $\rho_0 c_s$ being the characteristic impedance of the acoustic medium.

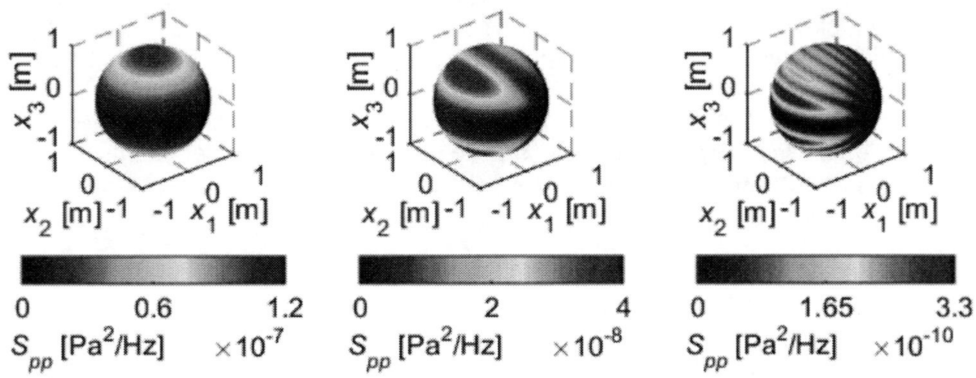

Fig. 4: Calculated spectral density of the acoustic far field sound pressure for different frequencies on a sphere as enveloping surface; frequency: $f = c_s / (C / 4)$ (left), $f = c_s / (C)$ (middle), $f = c_s / (4C)$ (right); values from the example in the appendix, note the different colour scales

A second integration over frequency from ω_1 to ω_2 results in the overall sound power or sound power level of a BE

$$P_{BE} = \int_{\omega_1}^{\omega_2} S_{P,BE}(\omega)d\omega, \quad LPBE = 10\log_{10}\left(P_{BE}/P_{ref}\right). \tag{15a, b}$$

Finally, the overall sound power level *OSWL* of the complete wind turbine is obtained by summation of the contributions from all n_{BE} BEs, the z blades and division by the reference sound power:

$$OSPL = 10\log_{10}\left(2z\sum_{i=1}^{n_{BE}}\left(P_{BE,i}\right)\Big/P_{ref}\right) \tag{16}$$

($P_{ref} = 10^{-12}$ W) with z being the number of blades. The factor 2 accounts for suction and pressure side, a conservative estimate, since the pressure side may contribute less as compared to the suction side. It has to be mentioned that neither an effect of the geometrical airfoil camber on the airfoil's radiation (ROGER and MOREAU [23]) nor the Doppler effect is considered as both do not change the radiated sound power.

2.3 Optimization

Optimization is done in two steps. Firstly, we seek an airfoil contour which provides maximum lift-to-drag-ratio at minimum trailing edge sound of a stationary BE. The trailing edge sound is due to the wall pressure fluctuations; hence, we minimize the wall pressure fluctuations rather the sound. As a representative location on the blade we take a monitoring point close to the trailing edge, here 90% of *C*. In a second step the 3D blade shape, described by the distributions of chord length and twist angle, is optimized for high power coefficient and low overall sound power level.

Optimization of 2D airfoil contour. A design Reynolds number and a range of flow angles of attack are set for airfoil optimization. The airfoil geometry is parameterized utilizing Bezier curves. In total two times seven Bezier points are used to define upper and lower side of the airfoil for the upper and lower airfoil side as in KAUFMANN [17]. This is depicted in **Fig. 5**. During optimization the *y*-coordinates of ten Bezier points can be varied. The two leading and two trailing edge Bezier points are fixed. The trailing edge thickness is fixed to 0.2 % of the chord length.

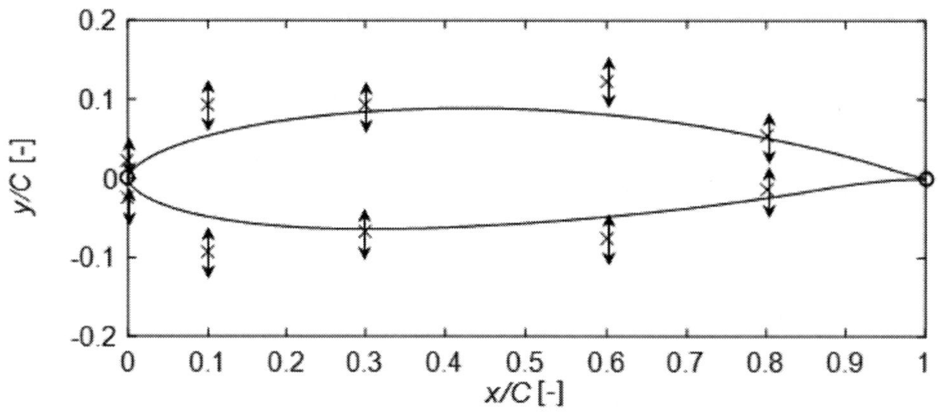

Fig. 5: Bezier points describing the airfoil geometry; the leading edge and the two trailing edge Bezier points denoted with circular markers are fixed

The objectives used for the airfoil optimization are listed in **Tab. 1**.

Tab. 1: Objectives for 2D airfoil optimization

Objective	**Definition**	α_{range}
Lift-to-drag-ratio	$\varepsilon_{obj} = \text{mean}(\varepsilon(\alpha_{range}))$	0° - 7°
Non-dimensional wall pressure fluctuations	$P_{\tilde{\Phi}pp,obj} = \text{mean}\left(P_{\tilde{\Phi}pp}\left(\alpha_{range}\right)\right)$	2° - 6°

An airfoil with good aerodynamics for wind turbine applications offers a high lift-to-drag-ratio ε. ε_{obj} is the arithmetical mean of the lift-to-drag-ratio in a range of angles of attack α_{range} = 0° - 7° to safely avoid stall just outside of the targeted range of 2° - 6°. The non-dimensional wall pressure spectrum $\tilde{\Phi}_{pp}$, obtained by eq. (9), is integrated over a Strouhal range $0.05 \le \tilde{\omega} \le 10$ as described in eq. (10) - this is approximately the Strouhal range of the database ROZENBERG et al. [5] have built the model on and covers the maximum of the spectra. The wall pressure fluctuations are minimized for the targeted range of angles of attack α_{range} = 2° - 6°. The two objectives are combined into the objective function

$$f_{obj,2D}\left(\boldsymbol{x}_{Bez,2D}\right) = \varepsilon_{obj} - g_{\Phi} P_{\tilde{\Phi}pp,obj} - \sum_{i}^{n_{2D,PT}} g_{2D,i} PT_{2D,i} \,. \tag{17}$$

$PT_{2D,i}$ are penalty terms which are weighted by weighing factors $g_{2D,i}$. The $n_{2D,PT} = 5$ penalty terms are formulated such that

- the thickness of the airfoil increases monotonically from leading edge to the maximum thickness and then decreases monotonically towards the trailing edge of the airfoil
- extremely thin airfoils close to the trailing edge are avoided
- negative wall shear stress and hence flow separation along the chord are avoided
- a decrease of lift below a set limit is avoided
- contours yielding favourable pressure gradients at the monitoring point 90% C are disregarded since then the ROZENBERG model is invalid.

For $P_{\Phi\ pp,obj}$ the weighing factor is $g_{\Phi} = g_{w,2D}g_{s,2D}$. With $g_{s,2D}$ the order of magnitudes of both, ε_{obj} and $P_{\Phi\ pp,obj}$ are adjusted to similar levels. With $g_{w,2D}$ the focus of the optimization on either lift-to-drag or wall pressure fluctuations is controlled.

Optimization of 3D blade shape. Starting point is the selection of a design operating point of the wind turbine in terms of tip speed ratio λ (eq. (2)) and the specification of the inflow velocity c_0 far upstream of the turbine.

The distributions of twist angle γ (**Fig. 1**) and chord length C are again defined by Bezier curves. Both distributions are parameterized with four Bezier points each as described by KAUFMANN [17] and depicted in **Fig. 6**. Their position in spanwise direction is fixed while γ and C is varied at those positions during optimization. Hence, the parameter space comprises eight parameters for the blade optimization.

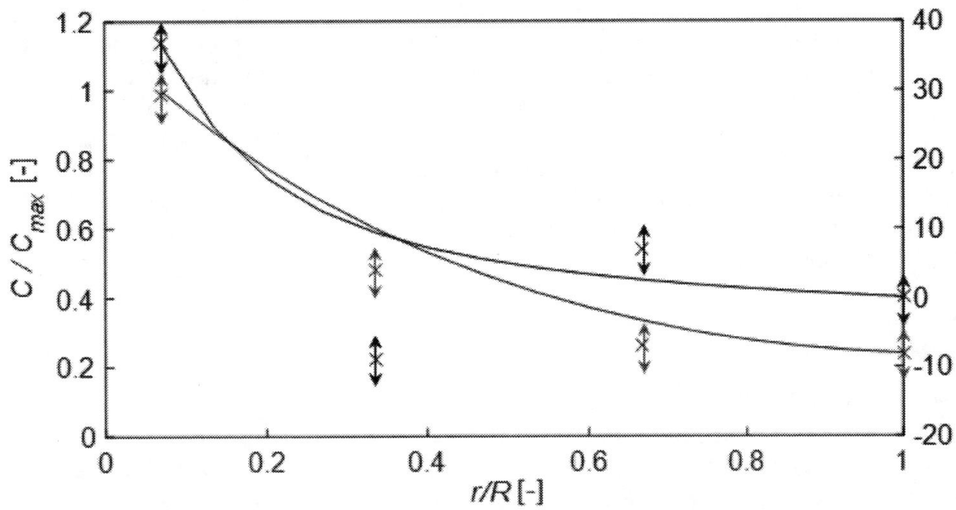

Fig. 6: Bezier points describing the blade distributions of chord length and twist angle

Eventually, with λ, c_0, a given set of airfoils and the distributions of γ and C the turbine's overall sound power and power coefficient can be evaluated. Airfoil data, consisting of lift and drag coefficient and boundary layer information, for the local Reynolds number and angle of attack at each BE is needed for these evaluations. Instead of performing costly XFoil calculations during the blade optimization, these airfoil data are prior to optimization calculated by XFoil for a range of angles of attack and Reynolds numbers. These calculation results then form a characteristic map for each parameter. These maps then are used for interpolating each parameter for the actual Reynolds number and angle of attack of each BE as utilized by KAUFMANN [17].

The objective function comprises the objectives power coefficient according to eq. (3) and the overall turbine sound power level *OSPL* according to eq. (16):

$$f_{obj,3D}\left(\mathbf{x}_{Bez,3D}\right) = g_c C_P - OSPL - \sum_{i}^{n_{3D,PT}} g_{3D,i} PT_{3D,i} \tag{18}$$

The $n_{3D,PT}$ = 2 penalty terms are weighted by the weighing factors $g_{3D,i}$ and formulated such that

- the chord length at the hub of a turbine blade is short enough to avoid overlapping with one of the adjacent blades
- zero or negative wall shear stress and hence flow separation on the BEs is avoided.

For C_P the weighing factor is $g_c = g_{w,3D} g_{s,3D}$. With $g_{s,3D}$ the order of magnitudes of both, C_P and *OSPL* are adjusted to similar levels. With $g_{w,3D}$ the focus of the optimization on either the power coefficient or the overall sound power level is controlled.

2.4 Optimization algorithm and settings

The optimization algorithm employed is an evolutionary algorithm which is described by BAMBERGER [16]. The population sizes are 500 and 200 individuals for airfoil optimization and blade optimization, respectively. Crossover between the parent parameters, i.e. Bezier points, followed by mutation is applied to produce the individuals of the next generation. The mutation range producing the following generation is confined to 50% of the difference between the lowest and highest magnitude of the respective parameter in the previous generation. Only the best individual, the elite, will survive and be transferred to the next generation. Crossover is only performed within the first 40 generations. Convergence is reached if the Bezier points of the best individual vary less than 0.1% of the difference of the initial parameter limits within the last 40 generations.

3. Results

3.1 Benchmark turbine

The University of Siegen operates a small horizontal axis wind turbine USI S83x with 3 m rotor diameter. It was designed utilizing an early in-house design code where the standard blade element momentum method according to GLAUERT and SCHMITZ (i.e. no optimization) is encoded. The design inflow velocity was c_0 = 6 m/s and the design tip speed ratio λ = 7.5. The well-known family of SOMERS airfoil sections S835, S833 and S834 (from hub to tip) has been used to build up each of the three blades. More details are described by GERHARD et al. in [27].

This turbine will serve as a benchmark for the next steps. Therefore, its geometry was used as a first input in the aerodynamic and aeroacoustic performance prediction model eqs. (1) to (16). For that the blade was segmented in 15 BEs. The resulting spanwise distributions of two exemplary quantities are depicted in **Fig. 7**:

The angle of attack - in average it coincides well with the 5° chosen in the initial design -, and the Reynolds number.

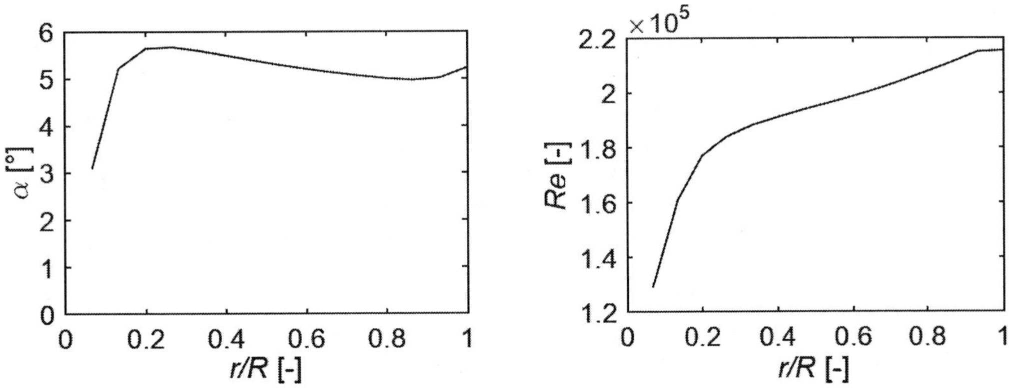

Fig. 7: Benchmark turbine: Predicted spanwise distribution of angle of attack (left) and Reynolds number (right)

Fig. 8 shows the shaft power $P_{sh,BE}$ each BE contributes, and the sound power emitted by each BE. On a trial basis, for the sound prediction the blade was also segmented into eight rather 15 BEs. As a consequence each BE is longer in spanwise direction and hence emits more sound power. But the value of overall sound power level *OSPL* (with 15 BEs: 77.7 dB, with eight BEs: 77.5 dB) essentially remains constant as expected. The *OSPL* is always evaluated within the frequency range 100 Hz to 10 kHz. It is obvious and expected that the inner BEs contribute very little to *OSPL* - roughly speaking, the outer 30% of the blade determine the overall sound power level emitted by the turbine.

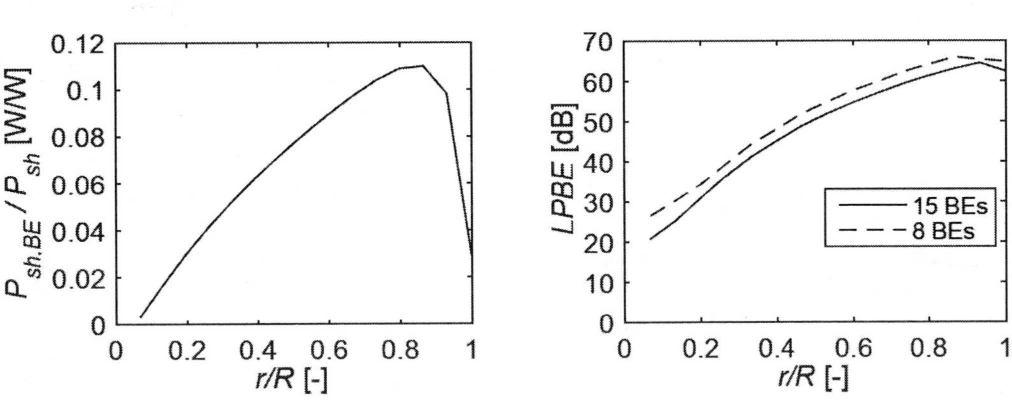

Fig. 8: Normalized spanwise distribution of shaft power from each of the 15 BEs (left); spanwise distribution of overall sound power level (right)

3.2 Optimization

As a potential replacement of the benchmark turbine we now seek an optimized turbine employing the optimization method introduced earlier. As already mentioned optimization is done in two steps, firstly, airfoil optimization, secondly, the 3D blade shape while utilizing the optimized airfoil sections.

3.2.1 Airfoil optimization

The airfoil optimization algorithm is used to find airfoils for a design Reynolds number of 200,000 as for the benchmark turbine. Three different objectives for optimization have been chosen. The resulting airfoils are compiled in **Tab. 2.** Their corresponding lift-to-drag-ratio and the lift coefficient of all airfoils as a function of angle of attack (**Fig. 9**) are obtained utilizing XFoil, their non-dimensional WPS levels (**Fig. 10** and **Fig. 11**) by the aeroacoustic prediction method from section 2.2.

As targeted, both, KV100 and KV102 promise significantly lower non-dimensional WPS levels (more than 10 dB at a Reynolds number of 200,000) as compared to the benchmark airfoil S834. By contrast, KV101 performs at the highest lift-to-drag-ratio as expected, but KV100 is still superior to S834 with respect to lift-to-drag-ratio.

The aeroacoustic prediction method also allows studying the effect of angle of attack and Reynolds number on the near TE wall pressure spectrum WPS, cp. **Figs. 10** and **11**. As a conclusion the exclusively aerodynamically optimized airfoils S834 and KV101 are nearly insensitive to a variation of angle of attack and Reynolds number. By contrast, the acoustically optimized airfoils KV100 and KV102 gain from an increase of both, angle of attack and Reynolds number, at least in the range where the WPS model allows a prediction.

Comparing the lift-to-drag ratios of S834 and the new airfoils, one has to keep in mind that for all airfoils - as stated earlier - tripping was applied very close to the leading edge in order to mimic a fully turbulent boundary layer. Of course, a S834 airfoil with natural transition would have a substantially better lift-to-drag ratio. Finally, it is worth to note, that KV101 has two disadvantages: It is very thin, which may cause structural problems, and its lift polar deviates remarkably from the benchmark.

Tab. 2: Optimized airfoils and objectives

Airfoil	**Objective for optimization**
	none, benchmark
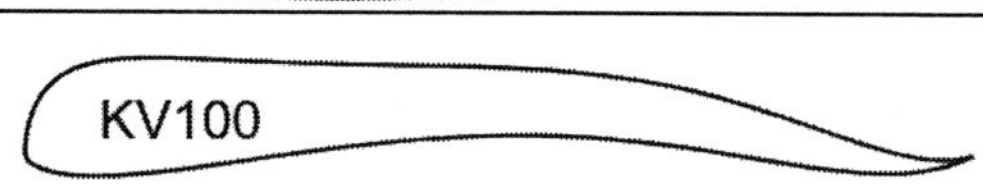	high lift-to-drag-ratio and minimal wall pressure fluctuation near TE
	high lift-to-drag-ratio
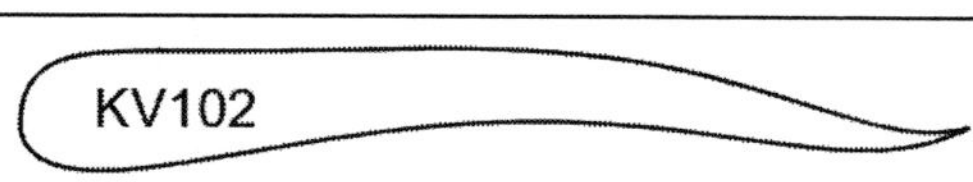	minimal wall pressure fluctuation near TE

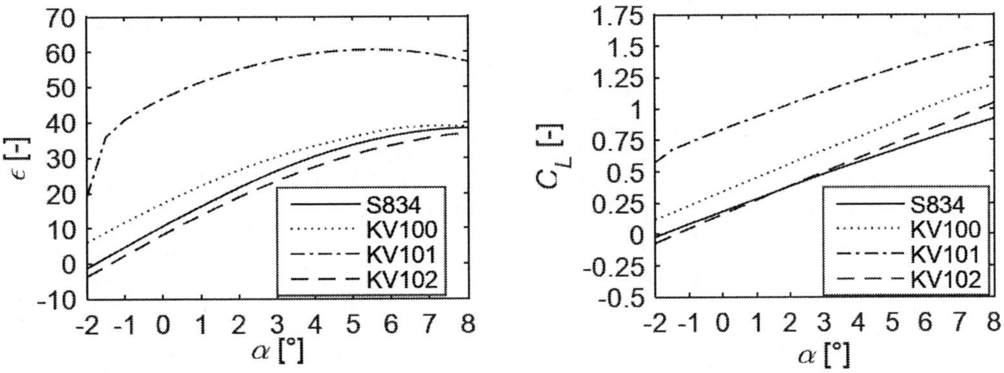

Fig. 9: Lift-to-drag-ratio (left) and lift coefficient (right) of the benchmark and the optimized airfoils; *Re* = 200,000

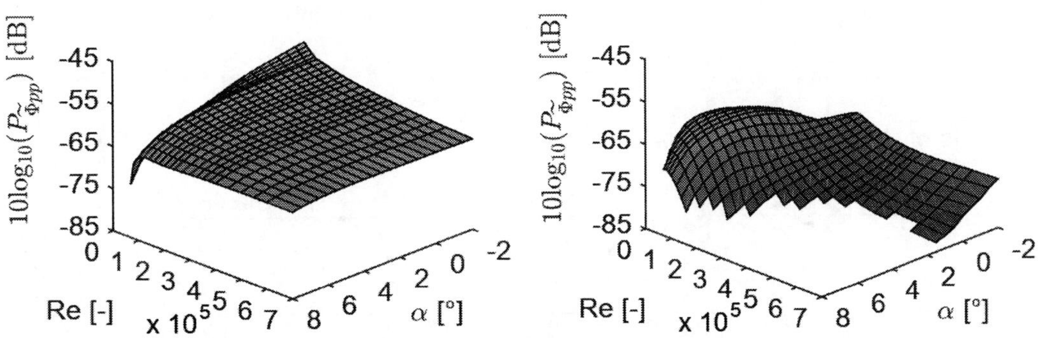

Fig. 10: Power level of the non-dimensional wall pressure fluctuations (eq. (10)) at 90% *C*; left: S834; right: KV100

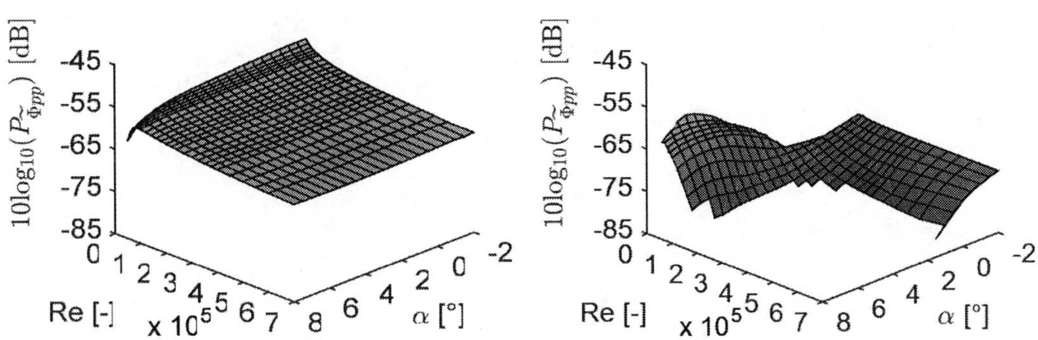

Fig. 11: Power level of the non-dimensional wall pressure fluctuations (eq. (10)) at 90% *C*; left: KV101; right: KV102

In **Tab. 3** most of the boundary layer parameters needed for the WPS calculation in eq. (9) are listed for an exemplary α = 5° and *Re* = 200,000. In addition, the bounds of the empirical database as utilized by ROZENBERG et al. [5] for each parameter are given. ROZENBERG proposed the Reynolds number based on the momentum thickness Re_θ as an important parameter. It can be seen that Re_θ for all airfoils is inside of the bounds. Further examination reveals that for the airfoils S834 and the high lift-to-drag-ratio airfoil KV101 all other parameters are within the database bounds. By contrast, the acoustically optimized airfoils KV100 and KV102 require some extrapolation. Hence, these both airfoils are examined in more detail: As listed, low non-dimensional WPS levels come along with low skin friction coefficients C_f. This coefficient is an important scaling parameter in the applied WPS model eq. (9) as it is even squared. To achieve these low skin frictions the boundary layer shape factor H is large and represents boundary layers which are on the verge of separation at the considered chordwise position. Therefore, also CLAUSER's parameter β_c is very large and outside the database bounds. In addition, the ratio δ^*/C shows that the boundary layer displacement thicknesses is comparatively large. The wake strength parameter Π is also outside from ROZENBERG's empirical database.

Tab. 3: Inner and outer boundary-layer variables of the presented airfoils at an flow angle of 5° at 90% of airfoil chord length on the suction side; the bounds of the database, which ROZENBERG based his model on, are also given for comparison; Re = 200,000

	$H=\delta^*/\theta$	$C_f \cdot 10^4$	β_c	Π	$Re_\theta = w_e\,\theta/\nu$	Δ	δ^*/C
Bounds of ROZENBERG's database	1.33 – 2.55	——	0.19 – 20.9	1.03 – 8.18	564 – 17170	2.23 – 6.39	——
S834	2.26	8.75	14.95	6.03	1364	3.00	0.016
KV100	3.39	0.41	38.59	40.67	2003	2.14	0.035
KV101	1.68	31.0	5.36	1.93	1253	4.37	0.009
KV102	3.51	0.30	60.50	47.90	1812	2.09	0.033

3.2.3 Blade optimization

The blade optimization algorithm is used to find optimal blades for turbines with the same design parameters as the benchmark turbine USI S83x (design inflow velocity c_0 = 6 m/s, design tip speed ratio λ = 7.5, 3 m rotor diameter). Prior to the optimization of spanwise chord length and twist angle (cp. **Fig. 1**), the airfoil sections are chosen according to **Tab. 4**. The objective function comprises both, C_P and *OSPL,* equally weighted.

Tab. 4: Optimized wind turbines and objectives

Wind turbine	**Blade**	**Objective for optimization**
USI S83x	S835, S833 and S834 (from hub to tip)	none; University of Siegen research turbine as described in section 3.1, benchmark
USI KV100	KV100	maximize C_P and minimize *OSPL*
USI KV101	KV101	maximize C_P and minimize *OSPL*
USI KV102	KV102	maximize C_P and minimize *OSPL*

Convergence towards a final geometry is mandatory for optimization results to be acceptable. This is given for all airfoil and blade geometries presented. Additionally, it was checked whether the optimization results are reproducible: Several optimizations with same settings were repeated. The optimization results (i.e. geometries) yielded similar objective function values for each of the repeated optimizations. This indicates that the optimizations tend to find global optima. This is true for airfoil and blade optimization.

Tab. 5 lists the predicted overall performance of the benchmark and optimized turbines. The optimized turbine USI KV102 performs best in terms of sound power emitted. Its *OSPL* is surprising 16 dB less than the benchmark USI S83x. Its power coefficient C_P, however, is 1.5% smaller as compared to USI S83x. USI KV101 performs with the largest power coefficient but the sound emission is slightly increased. USI KV100 is a good compromise: C_P remains as for the benchmark but its *OSPL* some 13 dB less.

Tab. 5: Predicted overall performance of benchmark and optimized turbines

	USI S83x (benchmark)	USI KV100	USI KV101	USI KV102
C_P [-]	0.412	0.413 (+0.2%)	0.456 (+10.7%)	0.406 (-1.5%)
OSPL [dB]	77.7	64.3 (-13.4 dB)	79.7 (+2.0 dB)	61.2 (-16.5 dB)

The corresponding optimized spanwise distributions of twist angle and chord length are presented in **Fig. 12.** The large lift coefficient of airfoil KV101 leads to a reduction of the chord length along the whole span of blade USI KV101 as compared to the benchmark. Otherwise, optimization yields modifications mainly in the hub region. As already pointed out the outer BEs contribute most to the overall performance data of the turbine. Hence, the benefits achieved by optimization are more attributed to the airfoil than to the spanwise distribution.

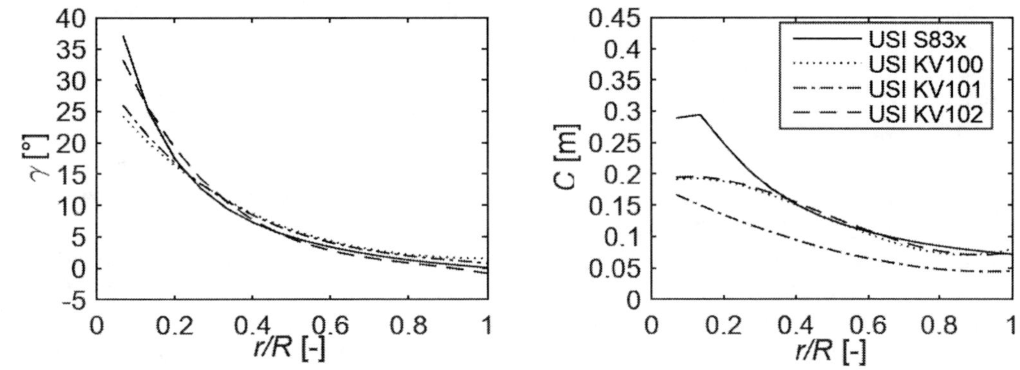

Fig. 12: Benchmark and optimized blades: Spanwise distribution of twist angle (left) and chord length (right) (cp Fig. 1)

It has to be emphasized again that the optimizer is bound to search for geometries only where the underlying models are valid. Utilizing XFoil the chordwise region on the surfaces of each BE, where separation occurs, can be determined, called $C_{separated}$ as the sum from the pressure and suction side. An indicator is the skin friction coefficient with a value smaller than zero. **Fig. 13 (left)** shows a comparsion for all blades. Clearly, at all BEs with the new airfoil sections, the chordwise regions with separation is considerably smaller than for the S83x blade. Again, one has to keep in mind that tripping was applied very close to the leading edge. In general slight separation on the airfoil surface may not harm the quality of the AMIET-based sound prediction (STURM et al. [28]). Nevertheless, as a precaution, regions of separation were avoided by a penalty during airfoil and blade optimization.

Fig. 13 (right) shows the predicted significant reduction of the sound power emitted from each BE of the optimized turbines USI KV100 and USI KV102 as compared to USI S83x.

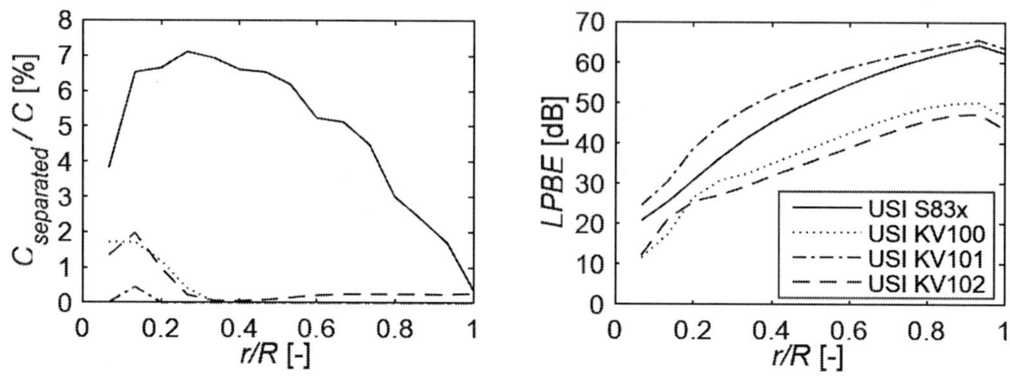

Fig. 13: Benchmark and optimized blades: Spanwise distribution of chordwise length of flow separation in per cent of chord length (left) and sound power emitted (right)

4. Summary and conclusions

Objective of this contribution was an acoustic and aerodynamic optimization of the full 3D blade geometry for a small horizontal axis wind turbine. Utilizing a refined aerodynamic blade element momentum (BEM) method and a combination of ROZENBERG's wall pressure and AMIET's trailing edge noise model (taking into account the adverse pressure gradient developing along a particular airfoil) an evolutionary algorithm was implemented. The optimization was subdivided into two independent steps, (i) the airfoil optimization, and (ii) the optimization of the blade twist angle and chord length distributions.

The validity of ROZENBERG and AMIET models was checked by comparison with recent own measurements of the wall pressure fluctuations and trailing edge sound of a small airfoil section. The agreement was satisfactory.

To mimic a fully turbulent flow around the blades of a realistic wind turbine - a worst case scenario - tripping was applied close to the leading edge of all airfoils. Depending on the set weighting in the objective function the airfoil optimization resulted in various novel airfoil shapes. As compared to a chosen reference airfoil S834 they promise better lift-to-drag-ratio and/or lower non-dimensional wall pressure fluctuations in the trailing edge region. The predictions forecast a possible reduction by more than 10 dB.

Utilizing such a low noise airfoil and optimizing spanwise chord length and twist angle distribution resulted in a new blade design. As compared to the non-optimized benchmark turbine with SOMERS airfoil shaped blades the predicted sound power of a turbine equipped with the most feasible of these new blades is with -13 dB substantially lower, whereas the power coefficient remains constant. The detailed analysis leads to the conclusion that the sound reduction is mainly attributed to the improved airfoil sections.

Although the models implemented may not be very accurate in the complete range of parameters varied during optimization, the combined optimization of aerodynamic and acoustic performance seems to be a promising way for the design of low noise high performance wind turbines. In any case an experimental validation of the results achieved so far is indispensable.

Acknowledgment

The authors thank Julien Christophe from the von Kármán Institute for Fluid Dynamics in Sint-Genesius-Rode, Belgium for discussions regarding AMIET's theory and supplying a code with AMIET's trailing edge noise model. Additionally, the authors highly appreciate the assistance by Marlène Sanjose and Stéphane Moreau from the Université de Sherbrooke, Canada, especially for discussing the ROZENBERG model.

References

[1] Gasch, R., Twele, J.: Wind power plants - Fundamentals, design, construction and operation. Berlin, Heidelberg, Springer, 2012

[2] Leloudas, G.: Optimization of Wind Turbines with respect to Noise, Master Thesis, Technical University of Denmark, 2006

[3] Tian, Y., Cotté, B., Chaigne, A.: Wind Turbine Noise Modelling Based on Amiet's Theory. 5th International Meeting on Wind Turbine Noise. 2013

[4] Drela, M.: XFOIL: An Analysis and Design System for Low Reynolds Number Airfoils. In: Mueller, T. J. (Ed.): Low Reynolds Number Aerodynamics. Proceedings of the Conference Notre Dame, Indiana, USA, 5-7 June 1989. Lecture Notes in Engineering, Vols.54. Berlin, Heidelberg: Springer 1989, p. 1–12

[5] Rozenberg, Y., Robert, G., Moreau, S.: Wall-Pressure Spectral Model Including the Adverse Pressure Gradient Effects. AIAA Journal Vol. 50 (2012) No. 10, p. 2168–2179

[6] Amiet, R. K.: Noise due to turbulent flow past a trailing edge. Journal of Sound and Vibration Vol. 47 (1976) No. 3, p. 387–393

[7] Roger, M., Moreau, S.: Back-scattering correction and further extensions of Amiet's trailing-edge noise model. Part 1 - Theory. Journal of Sound and Vibration Vol. 286 (2005) No. 3, p. 477–506

[8] Amiet, R. K.: Acoustic radiation from an airfoil in a turbulent stream. Journal of Sound and Vibration Vol. 41 (1975) No. 4, p. 407–420

[9] Schepers, J. G., Wuerz, W., Curvers, A., Mantesanz, A., Oerlemans S., Garcillan, L., Braun, K., Fischer, M., Lutz, T., Koegler, K., Herrig, A., Maeder, T.: SIROCCO: Silent rotors by acoustic optimisation, ECN Energy research Centre of the Netherlands, 2007

[10] Bertagnolio, F.: EUDP Project 'Low Noise Airfoil' - Final Report - DTU Wind Energy E-Report, DTU Wind Energy, Technical University of Denmark DTU Wind Energy-E-0004, Roskilde, Denmark
[11] Göçmen, T., Özerdem, B.: Airfoil optimization for noise emission problem and aerodynamic performance criterion on small scale wind turbines. Energy Vol. 46 (2012) No. 1, p. 62–71
[12] Brooks, T. F., Pope, D. S., Marcolini, M. A.: Airfoil Self-Noise and Prediction, NASA Reference Publication 1218 (Ed.), NASA Langley Research Center, 1989
[13] Hao, X., Weimin, Z., Xiao, L., Jieping, L.: Aerodynamic and Aeroacoustic Optimization of Wind Turbine Blade by a Genetic Algorithm. 46th AIAA Aerospace Sciences Meeting and Exhibit
[14] Rodrigues, S. S., Marta, A. C.: On addressing noise constraints in the design of wind turbine blades. Structural and Multidisciplinary Optimization Vol. 50 (2014) No. 3, p. 489–503
[15] Lowson, M. V.: Assessment and Prediction of Wind Turbine Noise, Flow Solutions Report 92/19, ETSU W/13/00284/REP, pp. 1-59, 1992
[16] Bamberger, K.: Aerodynamic Optimization of Low-Pressure Axial Fans. Berichte aus der Strömungstechnik, Ph.D. dissertation at the University of Siegen. Herzogenrath, Shaker, 2015
[17] Kaufmann, N.: Optimierung des hydrodynamischen Entwurfs einer Gezeitenströmungsturbine, Diplomarbeit, Lehrstuhl für Strömungsmechanik und Strömungsmaschinen, Technische Universität Kaiserslautern, 2015
[18] Kaufmann, N., Carolus, T., Starzmann, R.: An Enhanced and Validated Performance and Cavitation Prediction Model for Horizontal Axis Tidal Turbines - submitted to the International Journal of Marine Energy
[19] Goody, M.: Empirical Spectral Model of Surface Pressure Fluctuations. AIAA Journal Vol. 42 (2004) No. 9, p. 1788–1794
[20] Kamruzzaman, M., Bekiropoulos, D., Lutz, T., Würz, W., Krämer, E.: A semi-empirical surface pressure spectrum model for airfoil trailing-edge noise prediction. International Journal of Aeroacoustics Vol. 14 (2015) No. 5-6, p. 833–882
[21] Catlett, M. R., Forest, J., Anderson, J. M., Stewart, D.: Empirical Spectral Model of Surface Pressure Fluctuations beneath Adverse Pressure Gradients. 20th AIAA/CEAS Aeroacoustics Conference. 2014
[22] White, F. M.: Viscous fluid flow. McGraw-Hill series in mechanical engineering. New York, NY, McGraw-Hill Higher Education, 2006
[23] Roger, M., Moreau, S.: Extensions and limitations of analytical airfoil broadband noise models. International Journal of Aeroacoustics Vol. 9 (2010) No. 3, p. 273–306
[24] Sanjose, M.: Rozenberg's wall-pressure spectral model including the adverse pressure gradient effects, private communication, Siegen, Germany 2015
[25] Drela, M., Giles, M. B.: Viscous-inviscid analysis of transonic and low Reynolds number airfoils. AIAA Journal Vol. 25 (1987) No. 10, p. 1347–1355
[26] Corcos, G. M.: Resolution of Pressure in Turbulence. The Journal of the Acoustical Society of America Vol. 35 (1963) No. 2, p. 192–199
[27] Gerhard, T., Sturm, M., Carolus, T. H.: Small Horizontal Axis Wind Turbine: Analytical Blade Design and Comparison With RANS-Prediction and First Experimental Data. Proceedings of the ASME Turbo Expo 2013

[28] Sturm, M., Sanjose, M., Moreau, S., Carolus, T.: Application of Analytical Noise Models Using Numerical and Experimental Fan Data. 11th European Conference on Turbomachinery Fluid dynamics & Thermodynamics. ETC11, 2015

[29] Gerhard, T.: Umströmte Tragflügelsegmente: Auswirkungen des Hinterkanten-ausblasens auf Strömungsgrenzschicht und Schallemission, Ph.D. dissertation at the University of Siegen. Aachen, Shaker Verlag, 2016

Appendix

A validation of the combined ROZENBERG/AMIET sub-model is given in this appendix. GERHARD [29] measured the sound of a SOMERS S834 airfoil section in an aeroacoustic wind tunnel. The Reynolds number $Re = w_\infty C/\nu$ was $3.5 \cdot 10^5$ and the effective angle of attack 4.7°. A special tripping was applied at 17% of the chord length on the suction side and at 76% on the pressure side. The chord length was 0.2 m and the span 0.266 m. The blade's suction side was instrumented with wall pressure sensors. The 90% position, i.e. a location close to trailing edge, is evaluated here. A "far field" microphone was placed at a reference position 0.3 m away from the trailing edge perpendicular to the incoming flow.

Fig. A presents a comparison of the near trailing edge wall pressure at 90% chord length

$$L_{\Phi pp} = 10\log_{10}\frac{\Phi_{pp}}{p_{ref}^2/1\text{ Hz}}\text{ dB}. \qquad \text{(A1)}$$

and the acoustic "far field" spectra with

$$L_{\Phi pp} = 10\log_{10}\frac{\Phi_{pp}}{p_{ref}^2/1\text{ Hz}}\text{ dB}. \qquad \text{(A2)}$$

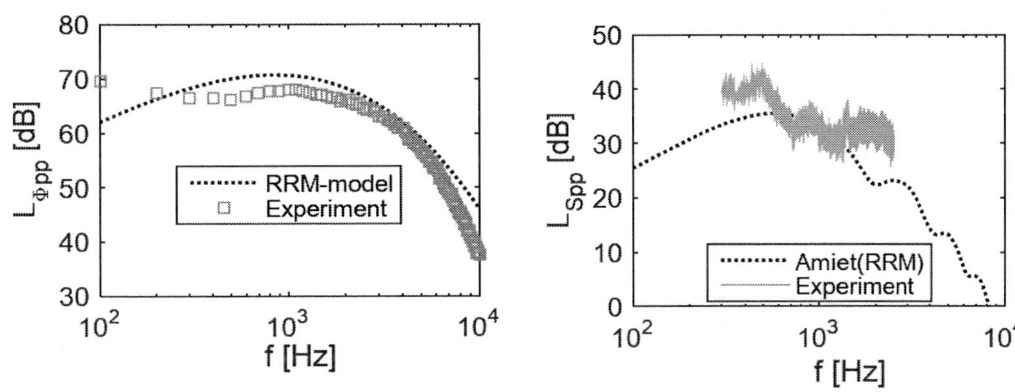

Fig. A: Left: Wall pressure spectral density at 90% chord length; right: "Far field" sound power spectral density at reference measuring position; reference pressure in both diagrams $p_{ref} = 2.5 \cdot 10^{-5}$ Pa; (ROZENBERG-ROBERT-MOREAU: RRM)

The acoustics at the reference position is calculated via AMIET's model fed with the wall pressure spectrum from the ROZENBERG model. Note that the input data for the ROZENBERG model was obtained by XFoil for exactly those transition points as enforced by the above mentioned tripping. Obviously, the agreement of the both, the wall pressure spectrum and the acoustic spectrum based on these data is satisfactory.

7th International Conference on Wind Turbine Noise Rotterdam – 2nd to 5th May 2017

Predicted and Measured Trailing-Edge Noise Emission for a 2.3 MW Wind Turbine

Cordula Hornung: Cordula.Hornung@iag.uni-stuttgart.de
Christoph Scheit: Christoph.Scheit@enercon.de
Christian Napierala: Christian.Napierala@enercon.de
Matthias Arnold: Matthias.Arnold@enercon.de
Dimitrios Bekiropoulos: Dimitrios.Bekiropoulos@enercon.de
Andree Altmikus: Andree.Altmikus@enercon.de
Thorsten Lutz: Thorsten.Lutz@iag.uni-stuttgart.de

Summary

One of the difficulties in predicting wind turbine noise is the transfer from a two dimensional modelling approach to the three dimensional reality. The research is hence based upon the comparison between three different abstraction levels between field measurements and numerical simulations focusing on trailing edge (TE) noise. For the extensive measurement campaign the prototype of an ENERCON onshore 2.3 MW wind turbine was equipped with novel high frequency fibre optic absolute pressure sensors along the chord at three sections in the outer part of the blade. The inflow velocity conditions were monitored at three different heights with a met mast. Additionally the radiated noise spectra were measured at IEC position at the turbine.

For the numerical simulations the semi-analytical TE noise prediction code IAGNoise+ developed at the University of Stuttgart is used. The code uses a 3D RANS FLOWer simulation to determine the turbulence characteristics of the airfoil's boundary layer. They are hence used to deduce the wall pressure fluctuations next to the TE on which a far field model can be applied to account for the diffraction of the fluctuations at the TE and the directivity effects. The underlying model is based on the TNO model as proposed by Parchen, however, extended to be applicable to weakly separated flow. The simulation procedure for the first abstraction levels takes the measured wall pressure fluctuations and applies a far field model. The next level is performed by using steady 3D RANS simulations. As a third approach an evaluation based one three dimensional simulations for different azimuthal positions is used further increasing resolution in inflow conditions. All methods are juxtaposed to evaluate their potential for noise prediction.

1. Introduction

On January first this year, new regulations came into effect in Germany's renewable energy funding, forcing the operators to increase full load time in wind turbine farms. This leads to the development of turbines with larger diameters which will be operated more often at off-design conditions. Obviously noise prediction also for 3D turbines at varying operating conditions is becoming crucial for the whole sector. Depending a little on the observer position in relation to the wind direction trailing edge (TE) noise is the most dominant noise source of modern wind turbines. Turbulent vortices in the boundary layer induce wall pressure fluctuations (WPF) which are scattered at the TE and propagated as sound waves. In research often 2D measurements of airfoils are compared with noise measurements. However, comparisons between 3D measurements and predictions can be found only rarely. To predict TE noise different computational methods, varying in costs and flexibility can be applied. The computational less costly approaches are empirical equations however being restricted to appliances in the range of the model validation cases, only. The next step are semi-analytical methods, providing a reasonable prediction quality of noise also for 3D turbines. The most complex option are CAA methods fully resolving the acoustic length scales. They are computationally expensive for complete blades or rotors and therefore currently not applicable within an industrial design process.

For 2D TE noise prediction a variety of semi analytical models with different model equations for far field and the WPF exist. The basis for these models was laid by Blake (Blake, 1986) and Parchen (Parchen, 1998). Since then many model variations were suggested based on their theory, compare for example (Bertagnolio & Fischer, 2014) and (Kamruzzaman, et.al, 2014). The present model is based on the latter theory, but is extended to be applicable for higher angles of attack. It will be described in more detail in section 3.

There is a variety of literature regarding full rotor measurements. However only few could be found dealing with 3D TE noise prediction, all of them using empirical or semi-empirical models. Grosveld (Grosveld, 1985) also presents models for the remaining noise sources showing that high frequency noise can be related to TE noise whereas low frequency noise sources are mostly located at the leading edge. Oerlemans et al. (Oerlemans, et al., 2007) compared different blade characteristics in measurements against each other. It was found that for the measured positions the TE noise was the dominant one. Field measurement campaigns for wind turbines with respect to aerodynamics were carried out for instance by (Schepers, et al., 1997) for turbines with a rotor diameter up to 27 m. The focus of this research was the collection of data for the assessment and validation of wind turbine design tools. A total of five different turbines, including two and three-bladed rotors, twisted and un-twisted blades, as well as tapered and un-tapered blades were studied. While static and dynamic pressure was measured as well as the angle of attack, no far field acoustic measurements were done. Experimental investigations on multi-megawatt wind turbines have previously been performed within the DAN-AERO campaign (Madsen H. A., et al., 2010) using a fully instrumented rotor blade equipped with four stations for static surface pressure measurements, high-frequency pitot tubes and sixty microphones at the outer blade section. They investigated the influence of atmospheric conditions on local inflow angle and laminar-turbulent transition and compared the measured pressure distribution from the blade with wind tunnel experiments in order to assess the influence of three-dimensional flow patterns on the wind turbine. A detailed analysis of the measurement campaign was presented in (Madsen H. A., et al., 2010) and (Troldborg, et.al, 2013). Another measurement campaign with a multi-megawatt wind turbine has been done by Buck et al. (Buck, Oerlemans, & Palo, 2016) in order determine the effect of inflow turbulence on low frequency noise. They characterized the turbulence by the turbulence dissipation rate. The turbulence dissipation rate was approximated using a correlation between blade-mounted accelerometers and met mast measurements. In addition, acoustic far field measurements were performed using twelve microphones. The data was used to evaluate Amiet's (Amiet, 1975) turbulent inflow noise model.

The noise prediction code used here is IAGNoise+. It is an enhanced version of the code formerly developed by Kamruzzaman (Kamruzzaman, et al., 2014). Based on RANS simulations the turbulent boundary layer characteristics in the vicinity of the TE at distinct sections along the blade are determined. These characteristics are then used to predict the far field noise via a model equation for the WPF. The noise is compared as A-weighted sound power spectrum of an equivalent monopole sound source at hub height derived from an observer at the standardized measurement position according to IEC 61400-11.

Three methods are used for the prediction of TE noise. The first one uses the measured WPF as input to the far field model equation, the second uses a third model CFD simulation to predict the noise and for the third approach three third model simulations are used to take into account the shear of the atmospheric boundary layer (ABL). All methods will be described in detail in section 3.2. The whole research is based upon a 2.3 MW wind turbine with a prototype blade design.

2. Experiments

An extensive measurement campaign was conducted in order to gather further insight on multi-megawatt turbine aerodynamics and aeroacoustics, to validate simulations and to support the further development of numerical algorithms for wind turbine acoustics as presented here. These measurements included the evaluation of static pressure distributions at three radial positions in the aerodynamically and acoustically relevant outer blade section, high-resolution wall-pressure measurements near the TE, and acoustic far field emission measurements according to IEC 61400-11. The nominal pressure-measurement positions are listed in Table 1. Throughout all measurement campaigns, data of the wind turbine's CAN bus has been recorded as well as additional data from a wind met mast installed close to the turbine providing information about wind-speed, shear, and turbulence intensity. Measurements have been conducted for different operating points of the turbine and at different days except for the sound emission measurement, which has been done only once.

The wall pressure sensors used are fiber-optical Fabry-Pérot type sensors (Schmid, et al., 2016) with dimensions 2x3x10 mm³ manufactured by fos4x GmbH and developed together with ENERCON GmbH for pressure measurements under harsh conditions. Due to the small sensor size the natural frequency is above 250 kHz. The sensor is an absolute-pressure sensor with a linear range of approximately ±15 kPa around the design point at approximately 90 kPa and an absolute measurement error of less than 5 % in this range. The sensor is flush-mounted into the blade surface, which is possible due to its small dimensions. Between 100 Hz and approximately 5 kHz the sensor has a constant sensitivity. This is also the main frequency range of interest for wind turbine noise applications. Because of the linear frequency response up to 200 kHz it is possible to use the sensor for quasi-static as well as for acoustic measurements at the same time. Figure 1 shows the sensor compared to a one-cent euro coin as size reference. During the campaign, a total number of 144 pressure sensors have been used.

Radius [%]	Pressure distribution	Near field acoustics
63	x	x
69.1		x
75	x	
87		x
93.3	x	x
95.7		x

Table 1: Measurement positions in radial direction along blade.

Since the acoustic prediction code IAGNoise+ uses a semi-analytical TE noise prediction model based on RANS simulations, a comparison of measured and computed pressure distributions has been conducted. Therefor to ensure approximately matching operating conditions in the RANS simulations, pressure distributions were measured at three radial positions using a total of 42 pressure sensors for each radial position, 25 sensors on the suction side and 17 pressure sensors on the pressure side. In order to capture the high pressure gradient close to the leading edge, a non-equidistant spacing was chosen. In addition, the sensors were aligned downstream of preceding sensors at an angle larger than 20 ° with respect to the chord line in order to avoid disturbed pressure fields due to upstream surface imperfections eventually caused by the implementation of the sensors. Figure 2 shows the distribution of pressure sensors on the suction side for one of the radial positions.

Figure 1: Fiber-optical pressure sensor compared to a one cent euro coin

Figure 2: Distribution of the flush-mounted pressure sensors on the wind turbine blade's suction side.

In order to consider the effect of airfoil shape deviations from the ideal geometry, also a laser scan of the airfoil at each sensor position has been done and analyzed. With respect to the expected lift, no large deviations could be found, i.e. the real profile matches well with the CAD data.

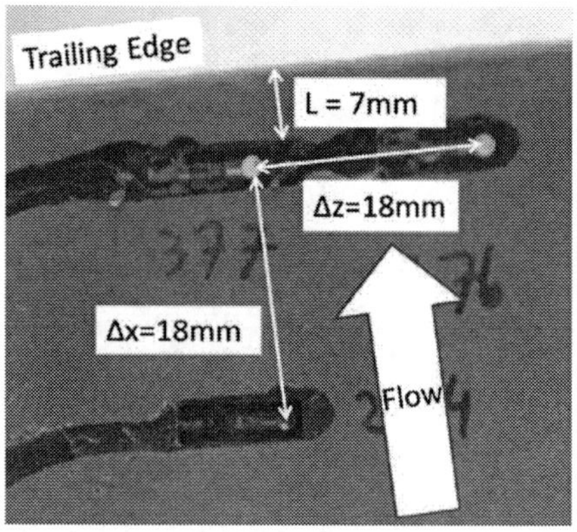

Figure 3: Pressure sensor triplet in order to measure turbulent length scale, convection Mach number and wall-pressure spectral density.

For acoustic analysis the same sensor type was used. L-shaped triplets of the pressure sensor have been integrated close to the TE of the profile such that convection Mach-Number, spanwise turbulent length scale and wall-pressure power spectral density can be evaluated. Sensor triplets were positioned at six different radial positions on both, pressure and suction side. These positions include the same radial positions as for the static pressure distribution plus three additional positions. An example of the pressure sensor setup at the TE is shown in Figure 3.

3. Numerical Methods

As mentioned above, three noise prediction methods were juxtaposed to compare the results. The first is based on measured WPF on which a far field model is applied. The other two use CFD simulations and the TE noise prediction code IAGNoise+. Third-model CFD simulations are used as input. The airfoil noise is evaluated at distinct sections along the blade which are then summed up to the total blade noise using a BEM method. An imaginary rotor is build up using copies of the one evaluated blade. In (Kamruzzaman, et al., 2011) the modelling approach from 2D to 3D is described. For the determination of the directivity of each section pitch and twist angle as well as azimuthal position of the blade on an imaginary rotor are considered. In contrast to the necessary binning of the experiments, only predefined operating points can be simulated. Hence, in order to compare the noise prediction to the measurements for one operating point two operating points were simulated to account for the range driven during measurement. Table 2 gives an overview on the simulated and measured operating points.

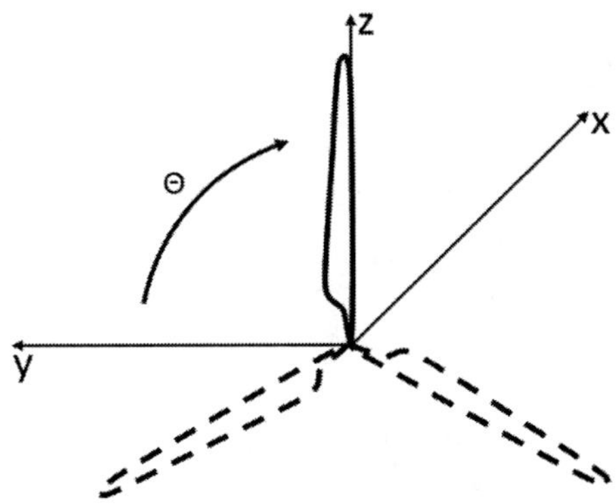

Figure 4: Coordinate system of the turbine, the wind direction is along the x-axis.

	Case 1	***Case 2***
Measurement, binning by wind velocity	14.5 rpm Vhub -1 ° pitch	15.1 rpm Vhub -1 ° pitch
Measurement, binning by rpm	14.4 / 14.6 rpm -1 °pitch	15.0 / 15.2 rpm -1 °pitch
Hybrid Method with WPF	14.5 rpm -1 ° pitch	15.1 rpm -1 ° pitch
IAGNoise+	14.5 rpm Vhub ± 0.35 m/s -1 ° pitch	15.1 rpm Vhub -1 ° pitch
IAGNoise+, resolved ABL	-	15.1 rpm Vhub, $Vtip_{,high}$, $Vtip_{,low}$ -0.9 ° pitch

Table 2: Simulated and measured operating points

3.1 CFD Simulation

In this research the three dimensional flow field around the rotor blades of the wind turbine is simulated with the CFD-solver FLOWer, developed by the German Aerospace Center (DLR). FLOWer is a compressible structured finite volume solver for the steady or unsteady Navier-Stokes-Equations in the RANS formulation. To account for blade pitch and rotational motion the present setup facilitates the Chimera overset grid method. The effects of turbulence are taken into account by the two equation Menter SST k-ω model.

For setting up the grids, the rotor blade surface is assumed to be smooth and identical to the design surface. Surface imperfections due to the sensor assembly, manufacturing accuracy and elastic deformations are not taken into account for the 3D simulations in this research. To evaluate the impact of those, additional 2D simulations comparing the 3D scanned rotor blade surface to the design surface of selected slices are performed. The sensor assembly required a small increase in the TE thickness. However, as will be discussed below, this comparison indicates only local influences of the surface imperfections and confirms the applicability of the surface idealization for the 3D-simulations.

Constricting the simulated domain further to a single blade in a 120°-mesh in order to reduce the computational effort. The mesh setup consists of a total of 14e6 cells in 133 blocks with 6.5e6 cells in blade vicinity as shown in Figure 5. The grid resolution, topology and quality follow the guidelines for CFD-simulations with FLOWer. The boundary layer grid is set up to achieve a y^+ value of approximately 1 on the rotor blade's surface with 30 cell layers across the boundary layer profiles. The measurement conditions during summer where usual surface soiling acting as tripping mechanism, hence all presented simulations were conducted fully turbulent. With this setup each point of operation is simulated for over 30e3 iterations to ensure full convergence.

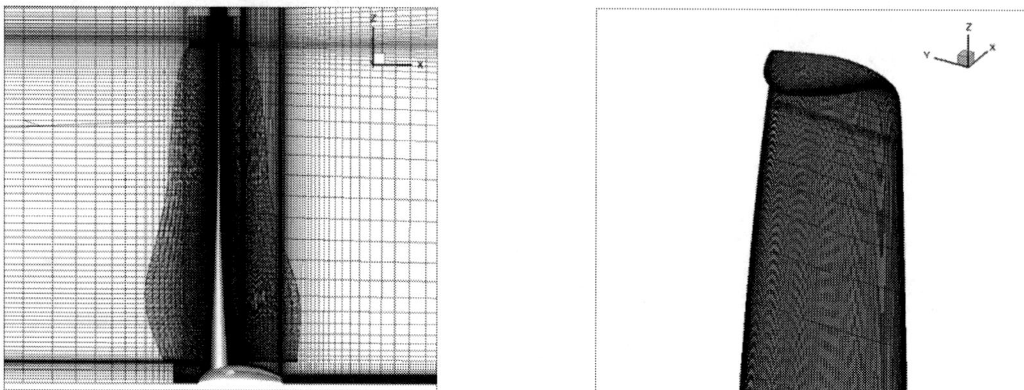

a) Grid in XZ-plane *b) Grid on blade tip surface*

Figure 5: Computational grid for 3D CFD simulations in the XZ-plane and on the blade tip surface.

3.2 Prediction Methods

Hybrid method

The far-field TE noise for the hybrid method can be estimated using the so called "scattered surface pressure prediction" (SSPP) method. Herrig (Herrig, 2012) proposed the SSPP method to analyze the acoustic source characteristics on rotating wind turbine blades. In the present case the far field equation by Blake (Blake, 1986) was employed:

$$G_{p,ff}(\omega,R) \approx \frac{L_{\mathrm{ref}}}{8\pi^2 R^2}\mathrm{Ma}_c(\omega)\Lambda_{p,3}(\omega)G_{pp}(\omega)D \qquad (1)$$

This relation predicts the power density spectrum of the far field sound pressure $G_{p,ff}(\omega,R)$ for an observer at distance R. The directivity is taken into account by the high-frequency directivity function D according to (Schlinker & Amiet, 1981) and depends on the observer position relative to the blade element. L_{ref} is the spanwise extent of the considered blade element. The power density spectrum of wall pressure fluctuations $G_{\mathrm{pp}}(\omega)$ was measured at sensors on the rotor blade in close vicinity to the TE in order to have a good approximation of the hydrodynamic fluctuations at the TE. By considering Taylor's hypothesis valid in between the noise pressure sensor triplets, see also Figure 3, a region considerably smaller than the boundary layer thickness, the frequency dependent convective Mach number $\mathrm{Ma}_c(\omega)$ can be expressed by the average phase difference of the Fourier-transformed wall pressure signals $\varphi(\omega)$ from two sensors aligned in streamwise direction with a known distance Δx (Brooks & Hodgson, 1981) and the speed of sound denoted as c:

$$\mathrm{Ma}_c(\omega) = \frac{\omega\Delta x}{\varphi(\omega)c} \qquad (2)$$

When considering an exponential decay of coherence of wall pressure fluctuations in the spanwise direction the coherence length scale $\Lambda_{p,3}(\omega)$ can be approximated from the coherence of the pressure signals γ_3 of two sensors separated by a distance of Δz from each other parallel to the TE:

$$\Lambda_{p,3}(\omega) \approx -\frac{\Delta z}{\ln(\gamma_3)} \quad (3)$$

In this way a set of six sensors (three on each side of the airfoil) is sufficient to predict the noise emission of a blade element for an observer in the far field. In the current study a total of twelve acoustic triplets were installed on the blade at six stations. Unfortunately some sensors broke during the installation, transport or first operation, therefore the predictions are limited to the five remaining acoustic stations.

Single one-third model prediction

The second method is based on a 120 ° (one-third) model CFD simulation. The CFD simulation is conducted as described above. The inflow velocity is chosen constant across the rotor determined based on the power curve. In contrast to the hybrid method introduced previously the WPF spectra inducing the far field noise need to be modeled when evaluating a RANS simulation. In order to do so the steady boundary layer characteristics in wall normal direction were extracted from the CFD simulation at distinct sections along the blade. To obtain a sufficient resolution 21 sections along the blade are evaluated. Model equation (4) (Parchen, 1998) is used to predict the resulting WPF.

$$P(k_1, k_3, \omega) = 4\rho^2 \left(\frac{1}{k_1^2 + k_3^2}\right) \int_0^\infty (ST) \Lambda_2 \cdot \Phi_{22} \cdot \langle u_2'^2 \rangle \cdot \Phi_m e^{-2|k|x_2} dx_2 \quad (4)$$

In the model $\Lambda_2$is the vertical integral length scale of the wall normal pressure fluctuations. The moving axis spectrum and the spectrum of the vertical velocity fluctuations are described by Φ_m and Φ_{22} respectively. The WPF spectrum is determined integrating the source term ST. As shown in (Lutz, et al., 2015), the prediction quality of the model implemented by Kamruzzaman in IAGNoise+ reduces for higher angles of attack. This is especially due to the anisotropy formulation being only valid for low angles of attack. The modeled parameters are hence modified in comparison to the original model to being able to cover higher loaded boundary layers as well and predict reasonable WPF spectra. The result of equation (4) can then be used in equation (5) to determine the far field noise (Parchen, 1998)

$$S(\omega) = \frac{L}{2\pi R^2} D \int_0^\infty \frac{\omega}{c_0|k_1|} \frac{P(k_1, k_3 = 0, \omega)}{1 - \frac{\omega}{c_0|k_1|}} dk_1 \quad (5)$$

In equation (5) ω represents the frequency and k_1 to $k_3$the wavenumbers in the three spacial directions (compare Figure 4). The sound velocity is denoted by c_0 and the wetted spanwise length by L. The directivity function is the same as in the method above and described by D, and R is the distance to the observer position. Considering the directivity of each blade section with respect to the observer position depending on the azimuthal position of the blade as well as the twist, the rotor noise can then be predicted based on the single sections.

ABL adjusted one-third model prediction

For each simulation of the second method a constant inflow velocity corresponding to the velocity at hub height was chosen. However, the used version of the CFD code is not able to represent shear in an ABL, hence additional simulations were made to increase accuracy. TE noise levels of turbines are dominated by sources located in the outer part of the blade due to a strong Ma dependency. Therefore inflow velocities are chosen corresponding to the velocity at approximately 90% of the blade span.

To evaluate an operation point under sheared inflow conditions three simulations are performed with different inflow velocities but with constant rotational speed. One simulation is conducted using a mean inflow velocity corresponding to the wind velocity in the 12 o'clock rotor position, the second simulation to represent 6 o'clock rotor position and for the third simulation the velocity at hub height is taken (3 and 9 o'clock position). The respective velocities are determined with the velocity profile from the met mast. For the noise determination all three simulations are used, each simulation representing 90° of the rotor gyration, with the simulation for hub height used twice. The same model equation and the same number of sections along each blade are utilized as in the second approach.

4. Results and Discussion

The different methods are applied to up to three operating points (compare Table 2). To ensure the same flow conditions in simulation as well as in measurements the pressure distribution along different blade sections were compared. However, due to the installation of the pressure sensors the blade geometry deviates from the designed one which is used for the 3D simulations. Hence to better understand the effect of the differences in the geometry, two dimensional CFD simulations were carried out for different angles of attack, approximately measured for the real and the design airfoil. In Figure 6 the difference with respect to the pressure distribution of the real airfoil is shown for simulation and measurements for one example. Simulations were carried out assuming a fully turbulent boundary layer flow. In addition, Xfoil (Drela, 1989) simulations were performed for a very similar angle of attack for the ideal profile geometry.

As can be seen, there are differences between pressure distributions based on the real and the ideal airfoil geometry. Regarding the measurements, for most of the sensors the differences are smaller than 10 % of the maximum pressure range in the reference simulation. Comparison is only shown for the profile at position r/R=0.75 but results are similar for the other two stations.

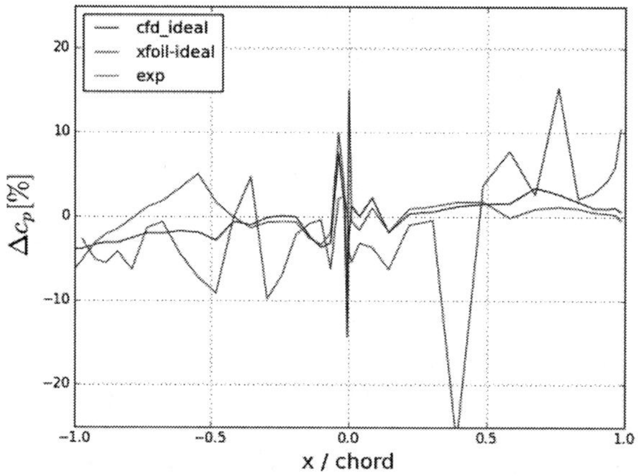

Figure 6: Deviation in pressure distribution with respect to a two-dimensional CFD simulation based on the measured profile geometry. Deviation is normalized by maximum pressure range in the reference CFD simulation. Negative x-coordinates refer to the suction side.

The pressure distributions were also compared with the 3D CFD simulation. In Figure 7 the deviation of the pressure distribution is plotted with respect to Xfoil results for the outermost radial position, showing a reasonable overall agreement. For comparison the 3D CFD results for a clean surface with the AHD transition model has been added to depict the impact of different boundary layer conditions. This comparison confirms the assumption of turbulent boundary layers for the further analysis as discussed above. One should note, that the inflow angles between experiment and simulation are only approximately equal, which results in higher deviation in the pressure level close to the sensitive leading edge area. While the local induction and thus the angle of attack is a part of the solution, the angle of attack needs to be extracted for the measured data from the Xfoil simulations and is therefore optimized in order to achieve a best fit to the measured data. Especially in the acoustically relevant outer part of the blade, cf. Figure 7, a good agreement between all simulations and the measured pressure distribution can be observed, which justifies the comparison of CFD-based acoustic simulations with far-field measurements.

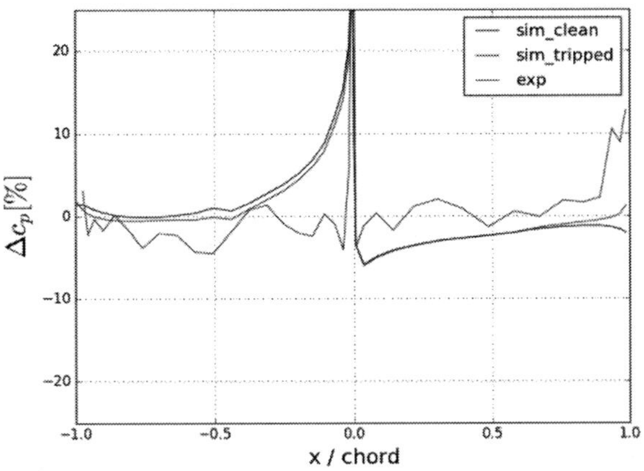

Figure 7: Measured and simulated deviation in pressure distribution at the outermost radial position compared to Xfoil results as reference for 90% of nominal rpm value. Negative x-coordinates refer to the suction side.

In order to approximate the error in noise prediction due to the deviating geometry one of the simulation shown above was also evaluated concerning TE noise. The results are shown in Figure 8 for an exemplary angle of attack. On the pressure side the design geometry underestimates the noise emitted by the real geometry by approximately 1 dB whereas on the suction side it is over predicted.

The three-dimensional noise spectra in comparison to the measured ones are shown in Figure 9 to Figure 13 for all three different evaluation methods. In all figures the A-weighted sound power in third-band levels is plotted over the frequency. When looking at the experimental spectra two local peaks can be perceived. The peak of the real spectra is assumed to be between them since the dent in between is supposed to occur due to the sound absorbing characteristics of the measurement plate (compare Figure 9). Usually the spectra are dominated by the TE noise from the peak towards higher frequencies. It can further be divided into one part related to the suction side – from the last local peak to the inflection point – and the high frequency content originating from on the pressure side.

During measurement in free field constant operating conditions cannot be ensured. Therefore the predicted spectra are always compared to more than one measurement bin adjacent to the simulated operating point. For operating point 1 (case 1 in table 2) noise prediction was conducted with the hybrid (Figure 9) and single one-third IAGNoise+ method (Figure 10). The hybrid method was found to deviate especially in the predicted peak frequency whereas the IAGNoise+ prediction matches the measured spectra very well within the measurement uncertainty of 1 dB. However the contribution of the suction side should be assumed about 1 dB lower and the pressure side contribution about the same amount higher due to the deviation between design and 3D scanned geometry.

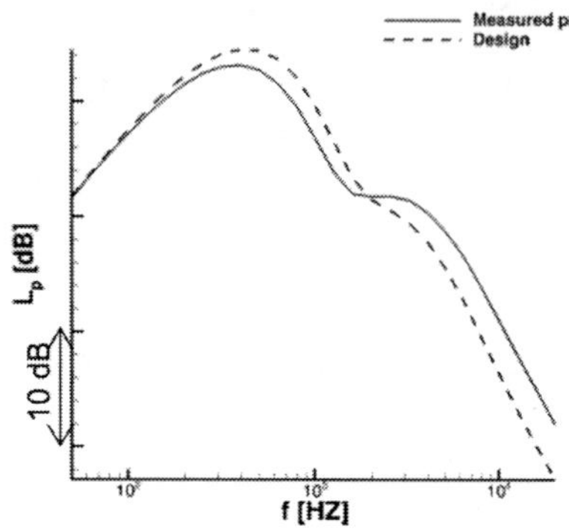

Figure 8: Comparison sound pressure spectra between measured and design profile

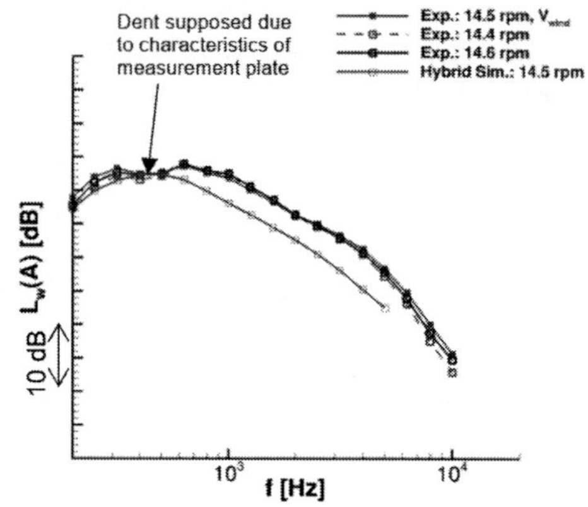

Figure 9: Case 1, prediction with hybrid method

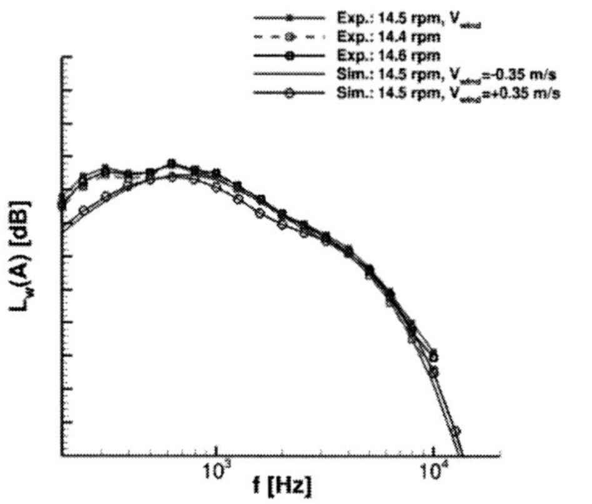

Figure 10: Case 1, prediction with single IAGNoise+ method

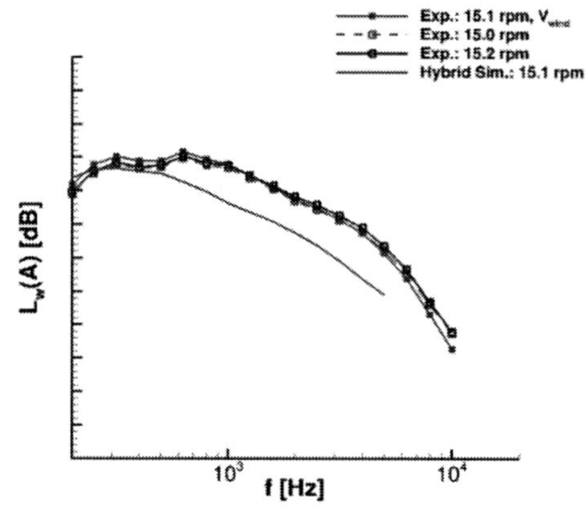

Figure 11: Case 2, prediction with hybrid method

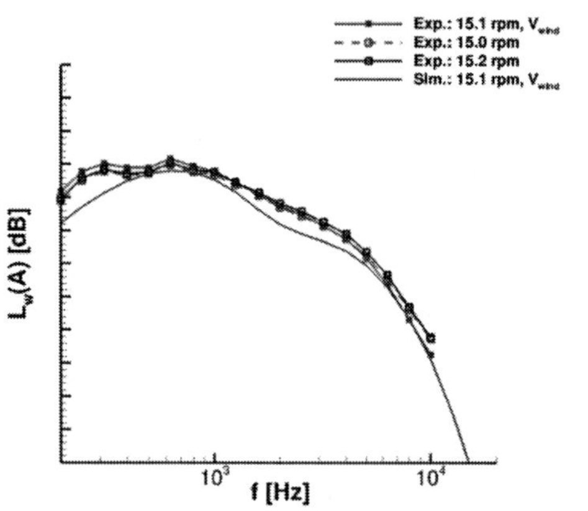

Figure 12: Case 2, prediction with single IAGNoise+ method

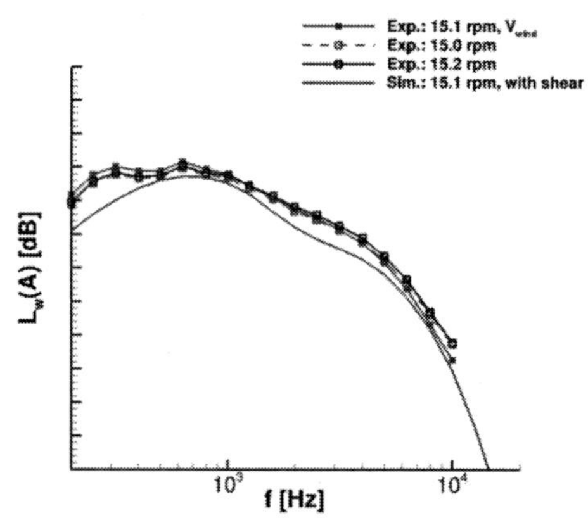

Figure 13: Case 2, prediction with resolved ABL and IAGNoise+

Almost the same behaviour can be observed for case 2 for the hybrid method (compare Figure 11 and Figure 12). This might be due to the fact that the sensors also measure overall pressure fluctuations outside the boundary layer and are hence influenced from the inflow turbulence. The under prediction using the SSPP method was reported in previous studies as well (Herrig, 2012). When comparing Figure 10 and Figure 12 with the single IAGNoise+ method it can be seen that the current implemented method matches the measured spectra for low rpm, however, with increasing rpm and hence higher local inflow Ma-number at the sections the prediction deviates from the measurement. The reason for that might also lie in the directivity since its influence

increases with Ma-number. Up to now the implementation of the far field directivity is determined in relation to a point located 90 ° above the TE of each respective section. In reality the angle to the respective airfoil side should be considered i.e. including the inclination of pressure and suction side to the chord. This modification will be implemented in IAGNoise+ in the future.
For the operating point of case 2 additionally the noise prediction with the third presented method was conducted superposing three complementary CFD simulations to take into account some effects of the shear in the ABL. However, when comparing the predicted noise with ABL consideration in Figure 13 and without it in Figure 12 the influence of the shear was found to be negligible. This can be explained by decreased sound levels on the lower part of the revolution, compensating the increasing sound levels in the upper part of the gyration.

5. Conclusions

This research is the results of an extensive three dimensional aerodynamic and aero-acoustic measurement campaign on a multi-megawatt wind turbine. They were compared to three different TE noise prediction methods. Two in principal different approaches were considered. The first model using measured wall pressure fluctuations as input to a far field model to predict the noise was shown to capture some effects present at the trailing edge probably related to inflow turbulence. In the frequency range normally assigned to TE noise this method under predicts the measured spectra. With the TNO type semi-analytical model, based on the evaluation of the boundary layer data of a one-third model RANS simulation, a very good noise prediction for TE noise could be achieved for small rotational speeds. The performance of the method will be further improved for increased local Ma-number in ongoing research. Consideration of some effects of the shear was not found to have a great influence on the predicted noise.
In general, the presented study proves the validity of the numerical simulations, underlining their applicability in the industrial blade design process. Based on the yielded vast experimental data base, further investigation will be directed towards inflow turbulence, blade deformation and tower effect.

Acknowledgements

Many motivated and highly skilled employees from different ENERCON departments including prototype production, project management, logistics, installation, service and sound emission measurement are responsible for the success of the presented approach. Their due care and personal commitment to our vision of quiet and powerful, high quality wind turbines was crucial in gaining the best results possible.
The authors would also like to thank Mathias Müller, Markus Schmid and Dominik Wetzel from fos4X GmbH, Munich who developed the sensor and data acquisition system in a collaboration project between the two companies. Their assistance and advice throughout the whole project is greatly appreciated.

References

Amiet, R. K. (1975). Acoustic radiation from an airfoil in a turbulent stream. *J Sound Vib, 41*(4), 407-420.

Bertagnolio, F., & Fischer, A. Z. (2014). Tuning of turbunelnt boundary layer anisotropy for improved surface pressure and trailing-edge noise modeling. *Journal of Sound and Vibration 333*, pp. 991-1010.

Blake, W. K. (1986). *Mechanics of flow-induced sound and vibration Volume I&II. Orlando, FL: Academic Press. Inc.*

Brooks, T., & Hodgson, T. (1981). Trailing Edge Noise Prediction from Measured Surface Pressures.

Buck, S., Oerlemans, S., & Palo, S. (2016). Experimental validation of a wind turbine turbulent inflow noise prediction code. *22nd AIAA/CEAS Aeroacoustics Conference, AIAA 2016-2953.* Lyon.

Drela, M. (1989). XFOIL: An Analysis and Design System for Low Reynolds Number Airfoils. *Low Reynolds Number Aerodynamics: Proceedings for.* Notre Dame: Springer.

Grosveld, F. W. (1985). Prediction of Broadband Noise from Horizontal Axis Wind Turbines. *Journal of Propulsion.*

Herrig, A. (2012). *Validation and Application of a Hot-Wire based Method for Trailing-Edge Noise Measurements on Airfoils.* München: Verlag D.Huth.

Kamruzzaman, M., Bekiropoulos, D., Wolf, A., Lutz, T., & Krämer, E. (2014). Rnoise: A RANS based airfoil trailing-edge noise prediction model. *AIAA2014-3305, 20th AIAA/CEAS Aeroacoustics Conference, Atlanta,.*

Kamruzzaman, M., Lutz, T., Nübler, K., & Krämer, E. (2011). Implementation and Verification of an Aeroacoustic Wind Turbine Blade Analysis Tool. *Forth International Meeting on Wind Turbine Noise.*

Lutz, T., Arnold, B., Bekiropoulos, D., & et al. (2015). Prediction of flow-induced noise sources of wind turbines and application examples. *International Journal of aeroacoustics, Volume 14*, pp. 675-714.

Madsen, H. A., Bak, C., Paulsen, U. S., Gaunaa, M., Fuglsang, P., Romblad, J., Jensen, L. (2010). *The DAN-AERO MW Experiments Final Report.* Roskilde: Riso DTU.

Madsen, H. A., Bak, C., Paulsen, U. S., Gaunaa, M., Sorensen, N. N., Fuglsang, P., Jensen, L. (2010). The DAN-AERO MW Experiments. *AIAA Aerospace Sciences Meeting, AIAA 2010-645.* Orlando.

Oerlemans, S., Sijtsma, P., & Méndez López, B. (2007). Location and quatification of noise sources on a wind turbine. *Journal of Sound and Vibrations*, 869-883.

Parchen, R. (1998). Progress report DRAW: a prediction scheme for trailing edge noise based on detailt boundary layer characteristics. *TNO Institute of Applied Physics, The Netherlands.*

Schepers, J. G., Brand, A. J., Bruining, A., Graham, J. M., Hand, M. M., Infield, D. G., Simms, D. A. (1997). *Final report of IEA Annex XIV: Field Rotor Aerodynamics.*

Schlinker, R., & Amiet, R. (1981). Helikopter Rotor Trailing Edge Noise. *NASA CR-3470.*

Schmid, M. J., Müller, M. S., Kuhnle, B. A., Bauer, M. W., Pongratz, R., & Altmikus, A. (2016). Fiber optic accoustic pressure sensor with high dynamic range and low noise. *ETC2016 - 36th European Telemetry and Test Conference.* Nürnberg.

Troldborg, N., Bak, C., Madsen, H. A., & Skrzypinski, W. (2013). *DANAERO MW: Final Report.* Roskilde: DTU Wind Energy.

7th International Conference
on
Wind Turbine Noise
Rotterdam – 2nd to 5th May 2017

Coupled wind turbine noise generation and propagation model: A numerical study

Franck Bertagnolio, DTU Wind Energy, Frederiksborgvej 399, DK-4000 Roskilde, Denmark, E-mail: hama@dtu.dk
Helge Aa. Madsen, DTU Wind Energy, Frederiksborgvej 399, DK-4000 Roskilde, Denmark, E-mail: frba@dtu.dk
Andreas Fischer, DTU Wind Energy, Frederiksborgvej 399, DK-4000 Roskilde, Denmark, E-mail: asfi@dtu.dk

Summary

In this paper, a numerical model consisting of 1) an existing aero-elastic software for wind turbines coupled to a rotor aerodynamic noise source model and 2) a ray tracing algorithm for noise propagation is presented. Several levels of model description for the wind turbine rotor as a noise source are available. Their influence on the immission noise is studied. In addition, the effect of atmospheric conditions on the propagated noise is also investigated, in particular how these conditions influence the perceived amplitude modulation noise in the far-field.

1. Introduction

As a consequence of the rapid development of wind energy in the last few decades and the installation of wind turbines near dwellings, wind turbine noise has become an important concern for neighbours and a driving factor for wind turbine design. In contrast to some other industrial and natural noise sources, wind turbines tend to produce sound with several degrees of unsteadiness and intermittency. The so-called Amplitude Modulation (AM) resulting from the rotation of the blades, sometimes refered to as 'swishing', is one example of such unsteady noise effect (e.g. see the works by Bowdler (2008); Oerlemans et al. (2013)). It is potentially quite annoying as the human perception of sound is particularly sensitive to such rapid changes in noise levels.

The aim of this paper is to present a numerical tool that can simulate wind turbine noise with a sufficient level of accuracy so that AM can be evaluated. Then, different input conditions for the model are used in order to estimate their impact on the resulting immission noise and AM characteristics. The present study will concentrate on the level of accuracy at which the wind turbine rotor and/or rotor blades are modelled as far as noise generation is concerned, as well as the impact of particular atmospheric conditions on the noise propagation.

2. A Wind Turbine Noise Model

The wind turbine noise and noise propagation models presented in this work are constituted of two independent steps. The former is the wind turbine noise generation model as described in this section. It is used to produce noise directivity maps that can be used as input for the latter.

The present wind turbine noise model is built upon the existing aero-elastic code HAWC2 commonly used, both for research and industrial purposes, as a wind turbine loads and energy yield prediction tool (see Larsen & Hansen (2006, 2007)). The blades are discretised along their span into sections (typically of the order of 50 for a typical large wind turbine) of given span lengths. The aerodynamic characteristics of the flow on each of these sections, i.e. mainly the local inflow velocity including rotor induction and associated angle of attack together with the aerodynamic loading in form of lift and drag, are computed using the Blade Element Momentum theory by Glauert (1963).

This aerodynamic data are used as input for aerodynamic noise models, namely turbulent inflow noise, trailing edge noise and stall noise. Note that these models include directivity effects. The noise generated by each blade section can then be calculated with its own directivity pattern. The noise perceived by an observer (or rather a listener in this case) therefore depends on the particular position of the blade section in space relatively to the observer's location. Furthermore, the contributions of all sections along the blade and then of all the blades can be summed op to calculate the noise emitted by each blade and then by the whole wind turbine rotor, respectively, at any given observer's location. Details about the implementation of this wind turbine rotor noise model and an experimental verification can be found in previous publications by Bertagnolio et al. (2016, 2017).

In the present study and in combination with the propagation model introduced in the next section, noise will be evaluated at multiple observer's locations located on a sphere centered at the rotor center with a radius equal to 10 times the wind turbine rotor diameter (this radius is arbitrary and can be freely modified). This step is performed in order to evaluate the noise directivity pattern of the turbine as perceived in the far-field. In this way, the peculiar local directivity patterns that do occur in the vicinity of the turbine itself due to the proximity of the different blades (i.e. the noise sources) relatively to the observer can be gotten rid of. These results are stored and used to produce noise directivity maps representative of each blade noise emission, or for the whole rotor. When using these maps to calculate noise at any location in the far-field, the front vector at which the ray is pointing from a point source, e.g. at the rotor center or on a rotor blade, in order to reach this location (see next section about the ray tracing method) is used to interpolate noise on this spherical directivity map. Through this process, the 10 rotor diameters radius spherical map is transferred back to a point source. For all numerical test cases presented in this paper, 128 observer locations are distributed evenly onto the sphere surface.

Figure 1 shows directivity maps for a typical 2 MW wind turbine. In this case, the computed noise has been averaged over time so that this map is representative of the whole turbine rotor noise emission during a full rotation of the blades. Nevertheless HAWC2 is a time-domain solver, it is therefore possible to compute the noise emitted by each blade at each time-step of the aero-elastic computation. In such case, several noise directivity maps at each time-step of the computation can be established (either for each blade or for the full rotor) yielding an unsteady time-dependant directivity map. This is illustrated in Figs. 2(a) and (b) displaying the noise directivity map of a single blade at two distinct azimuth positions, respectively, during a rotor revolution. Changes of the spatial characteristics of the directivity pattern can be clearly observed. In addition, noise directivity maps are plotted at two different time-steps of a rotor revolution, but this time for the whole rotor noise emission, in Figs. 3(a) and (b), respectively, showing the variation of the rotor noise emission between this two time-steps. Due to the presence of the three blades, the contributions of each individual blade to the directivity pattern are mixed up and the changes are more subtle.

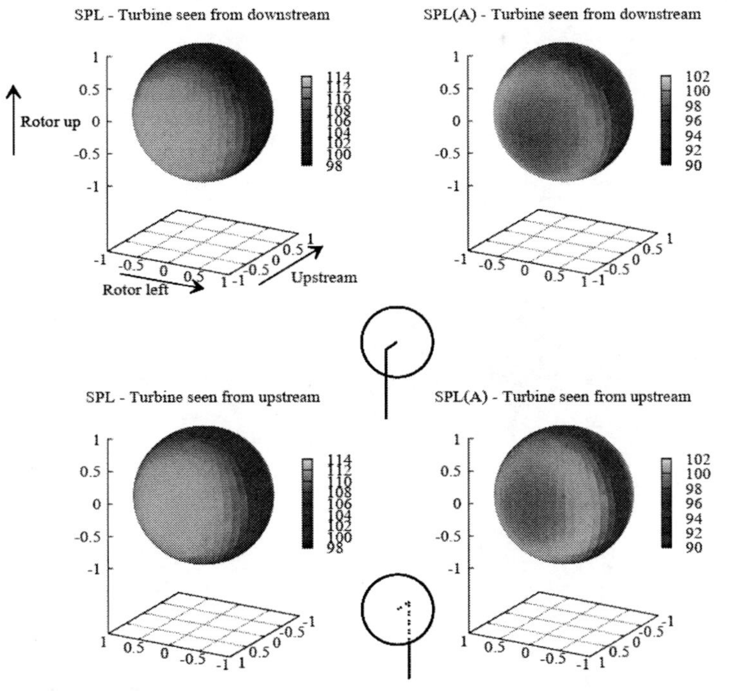

Figure 1: Time-averaged directivity pattern of frequency-integrated SPL for all blades (Top: seen from downstream; Bottom: seen from upstream; Left: SPL; Right: A-weighted SPL)

3. Ray Tracing Method for Noise Propagation

The ray tracing method is used for the modelling of wave phenomena and predicting their propagation paths (see the book by Pierce (1989) for an introduction). It is applied in various contexts ranging from electro-magnetic, seismic to underwater or atmospheric sound waves propagation phenomena. It has also been used by several authors for studying wind turbine noise propagation (e.g. see Zhu et al. (2005); Prospathopoulos & Voutsinas (2007); Heimann et al. (2011) and Bertagnolio (2016)).

The present model is a 3D model of the terrain surrounding the turbine. However, a simple flat terrain is assumed herein. The ray tracing method can handle refraction caused by wind speed and/or temperature gradients, geometric and atmospheric losses. Ground reflection is also included with full reflectivity (i.e. no losses, although it can easily be included in the model).

Given an observer location anywhere in this 3D model (assumed above ground), the ray tracing method finds the possible ray paths between the rotor center, or each

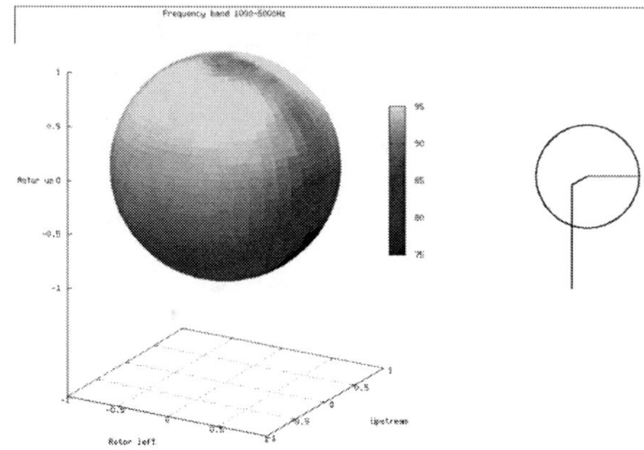

(a) Blade to the left of the rotor disk

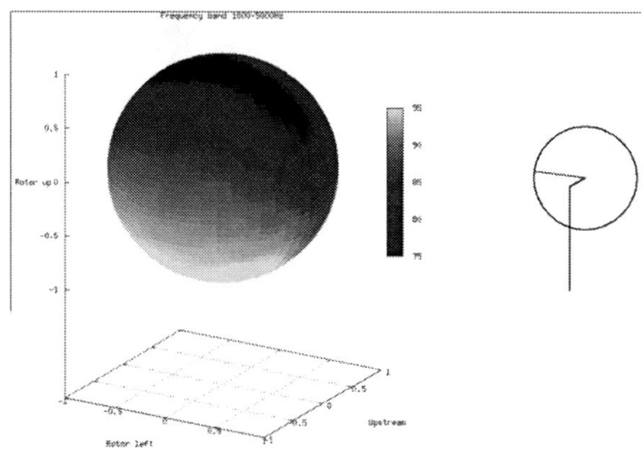

(b) Blade to the right of the rotor disk

Figure 2: Directivity pattern of integrated SPL in 1 kHz to 5 kHz frequency band of one single blade at chosen azimuth positions (Note: the blade is rotating clockwise)

invividual blade, and this location. The noise contribution of each ray paths are added to each other assuming that the noise waves from each path are decorrelated. Note that in the case of a ray path bouncing at a ground location close to the observer and in conjunction with the corresponding neighbouring direct path, then the noise from these two rays is assumed correlated and added up accordingly as described by Piercy et al. (1977). The directions along which these rays are emitted from the rotor center, or from each blade, are used to interpolate the intensity of the emitted noise in these specific directions according to the spherical directivity mapping defined in the previous section. As a result, our ray tracing method is able to simulate the overall directivity pattern of the whole rotor or each individual blades including unsteady features, e.g. related to the rotation of the blades or to some other sources of unsteadiness such as wind speed temporal or spatial variations due to the atmospheric turbulence.

4. Analysis of Model Results

In this section, several aspects of the modelling approaches are investigated looking at the immission noise in the far-field for observers located at 500 m from the turbine, 2 m above ground. In all subsequent plots, time-series of the Sound Pressure Levels (SPL)

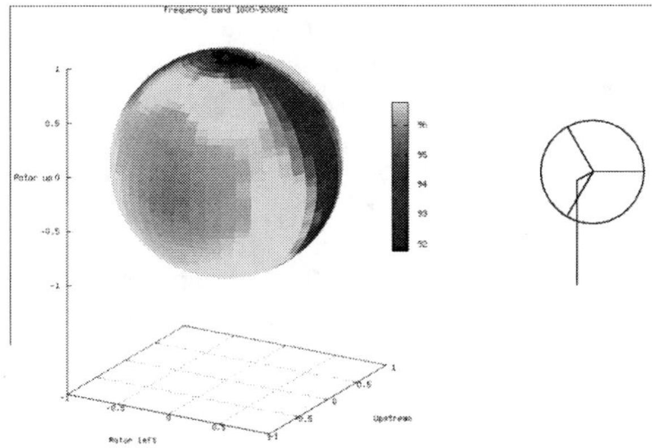

(a) Blade no.1 to the left of the rotor disk

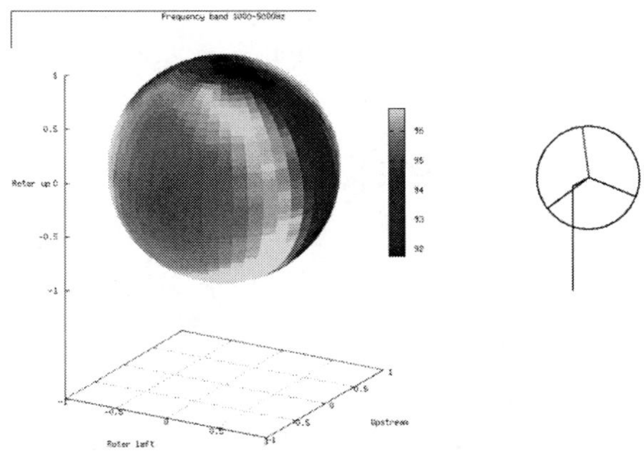

(b) Blade no.1 to the top of the rotor disk

Figure 3: Directivity pattern of integrated SPL in 1 kHz to 5 kHz frequency band of all blades at chosen azimuth positions of one of the blade (Note: the blades are rotating clockwise)

at the observer locations are plotted during one rotor revolution. Both integrated SPL and A-weighted SPL are displayed in order to assess the validity of the conclusions for the human perception.

Note that in this entire article, the wind speed at hub height has been assumed to be equal to 10 m/s.

4.1 Influence of Noise Source Modelling Details

The effect of the level of model description of the rotor as a noise source is investigated. As introduced in Section 2, three levels are available. The rotor noise emission can be modelled as 1) a time-independant point source located at the rotor center (denoted as Steady Rotor), 2) an time-dependent point source at the rotor center (denoted as Unsteady Rotor), or 3) unsteady and moving point sources, one on each of the rotating blades, located at 3/4 of the blades' span (denoted as Unsteady Blades).

The immission noise at an observer location on the left of the rotor (at 500 m from the turbine and 2 m height) with stable atmospheric conditions are plotted in Fig. 4. As expected, the Steady Rotor results are constant in time, while the two other noise modelling approaches show some AM and that this AM is slightly larger when looking at A-weighted SPLs. It appears that the Unsteady Blades model produces more AM than the Unsteady Rotor case (around 1 dB more from peak to peak).

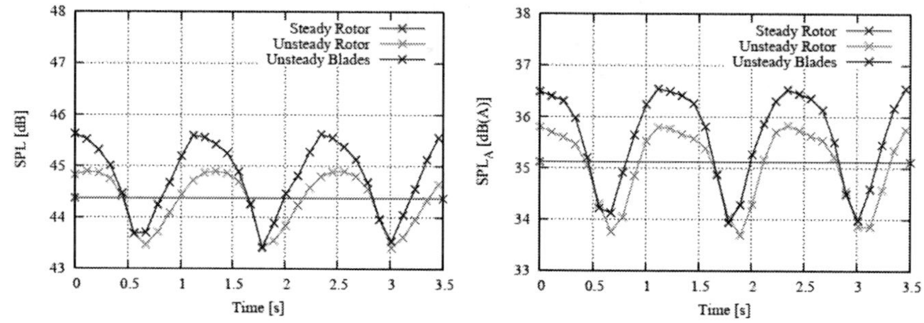

Figure 4: Time-seris of SPL at rotor-left position in atmospheric stable conditions for various types of noise source (Left: SPL; Right: A-weighted SPL)

4.2 Influence of Observer's Position

In order to evaluate the impact of the directivity patterns observed in Section 2, in particular the noise emission deficit in the rotor plane, two positions for the immission noise are considered: one directly downwind of the rotor and one to its left (when facing the rotor from upstream) in the rotor plane, both located at 2 m from the ground and 500 m from the turbine.

The results are displayed in Fig. 5. It can be seen that the observer downwind experiences much larger SPLs (by up to 10 dB and 5 dB(A)) as a consequence of the noise emission deficit in the rotor plane. However, at the same time it appears that the noise AM is considerably reduced downwind to nearly insignificant levels.

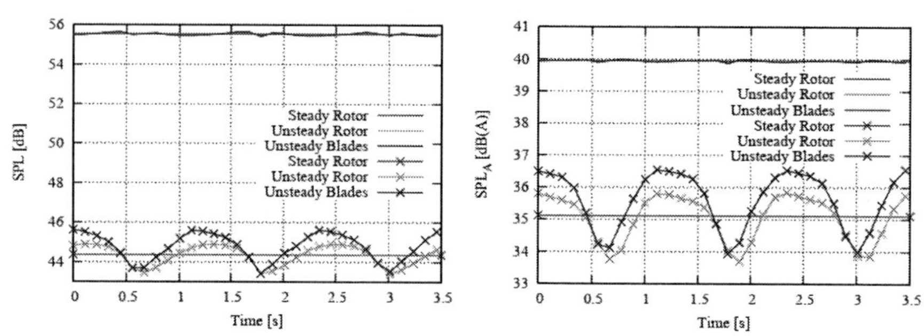

Figure 5: Time-seris of SPL at downwind and left positions in atmospheric stable conditions (Left: SPL; Right: A-weighted SPL; Lines: Observer downwind of the rotor; Lines with cross: Observer located left of the rotor)

4.3 Effect of Atmospheric Parameters on Immission Noise

The last comparison is concerned with the effect of atmospheric stability. In the case of stable conditions, the wind velocity profile is defined by a power law with a coefficient equal to 0.55 and the temperature gradient is +4 K per 100 m elevation. In the case of unstable conditions, the power law coefficient is 0.07 and the temperature gradient is -2 K per 100 m elevation.

The effects of atmospheric stability are displayed in Fig. 6. It can be oberved that these effects are relatively small. Nevertheless, the AM for the Unsteady Blades case is altered and reduced in the case of unstable conditions. This is most probably the result of the rotation of the blades moving up and down in the atmospheric boundary layer.

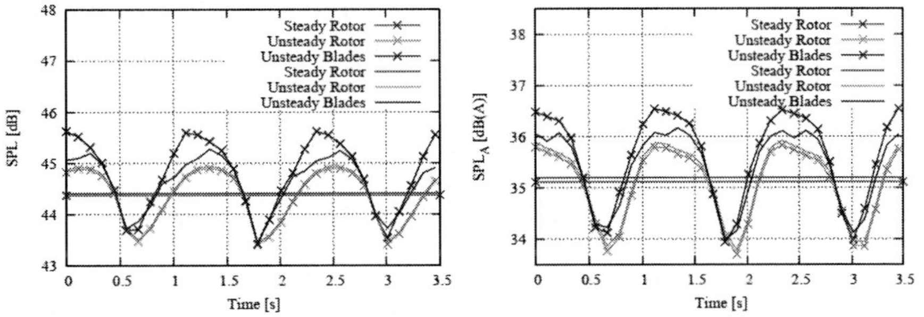

Figure 6: Time-seris of SPL at rotor-left position in stable or unstable atmospheric conditions (Left: SPL; Right: A-weighted SPL; Lines: Unstable; Lines with cross: Stable)

5. Discussion and Future Work

A numerical tool for simulating wind turbine rotor noise in general, and unsteady effects such as AM in particular, has been presented. Investigations showed that the level of accuracy of the model for describing the rotor configuration and geometry have an impact on the computed immission noise. The directivity of the rotor noise are reflected in the far-field. Finally, atmospheric stability effects are shown to have some influence on the immission noise.

However, it should be kept in mind that the present study is very preliminary. Observers located at 500 m from the turbine have been considered only. Two atmospheric conditions, with possibly rather simplistic description of the atmosphere, are considered. Further studies need to be conducted in order to evaluate the influence of the various inputs on the immission noise and a larger variety of conditions must be considered.

Nevertheless, this tool may be prove useful in the future for planning wind farm noise as it can be readily extended to the case of several turbines.

References

Bertagnolio, F. (2016). A noise generation and propagation model for large wind farms. In 22nd *International Congress of Acoustics*, Conference Proceedings. Buenos Aires, Argentina.

Bertagnolio, F., Madsen, H. A., & Fischer, A. (2017). A Combined Aeroelastic-Aeroacoustic Model for Wind Turbine Noise: Validation and Analysis of Field Measurements. *Wind Energy*. (Accepted for publication).

Bertagnolio, F., Madsen, H. A., Fischer, A., & Bak, C. (2016). Validation of an Aero-Acoustic Wind Turbine Noise Model Using Advanced Noise Source Measurements of a 500 kW Turbine. In *ISROMAC 2016*, Conference Proceedings. Honolulu (HI).

Bowdler, D. (2008). Amplitude Modulation of Wind Turbine Noise, A Review of the Evidence. *Institute of Acoustics Bulletin (UK)*, *33*(4), (9 pages).

Glauert, H. (1963). *Airplane Propellers*, vol. Aerodynamic Theory Volume IV. W. F. Durand, The Dover Edition (UK).

Heimann, D., Käsler, Y., & Günter, G. (2011). The Wake of a Wind Turbine and Its Influence on Sound Propagation. *Meteorologische Zeitschrift*, *20*(4), 449–460.

Larsen, T. J., & Hansen, A. M. (2006). Influence of Blade Pitch Loads by Large Blade Deflections and Pitch Actuator Dynamics Using the New Aeroelastic Code HAWC2. In *Proc. of the European Wind Energy Conference and Exhibition*. Athens, Greece.

Larsen, T. J., & Hansen, A. M. (2007). How 2 HAWC2, The User's Manual. Tech. Rep. RISØ-R-1597(ver.3-1), Risø-DTU, Roskilde, Denmark.

Oerlemans, S., Smith, M. G., White, P., von Hünerbein, S., King, A., Piper, B., Cand, M., Bullmore, A., Wilson, B., Madsen, H. A., Fischer, A., & Kragh, K. A. (2013). Wind Turbine Amplitude Modulation: Research to Improve Understanding as to its Cause and Effects. Tech. Rep., RenewableUK, United Kingdom. (Available online: http://www.renewableuk.com/en/publications/reports.cfm/wind-turbine-amplitude-modulation).

Pierce, A. D. (1989). *Acoustics - An Introduction to its Physical Principles and Applications*. The Acoustical Society of America.

Piercy, J. E., Embleton, T., & Sutherland, L. (1977). Review of noise propagation in the atmosphere. *Journal of the Acoustical Society of America*, *61*(6), 1403–1418.

Prospathopoulos, J. M., & Voutsinas, S. G. (2007). Application of a Ray Theory Model to the Prediction of Noise Emissions from Isolated Wind Turbines and Wind Parks. *Wind Energy*, *10*, 103–119.

Zhu, W. J., Sørensen, J. N., & Shen, W. Z. (2005). Airfoil Trailing Edge Noise Generation and Its Surface Pressure Fluctuation. *Wind Engineering*, *29*(2), 129–143.

7th International Conference
on
Wind Turbine Noise
Rotterdam – 2nd to 5th May 2017

A Comprehensive Hamiltonian Ray Tracing Technique for Wind Turbine Noise Propagation Under Arbitrary Weather Conditions

Sterling McBride and Ricardo Burdisso - Virginia Tech Department of Mechanical Engineering, Durham Hall, Blacksburg, Virginia 24060, USA.
msterl6@vt.edu, rburdiss@vt.edu

Summary

Physically accurate prediction of noise produced by wind turbines is crucial for their environmental impact assessment, including their amplitude modulation behavior. Current commercial noise codes for wind turbines are incapable of accurately handling arbitrary weather conditions. Assumptions including a negligible vertical wind component, absence of significant spatial speed of sound gradients, and acoustically modeling the wind turbine as a monopole source are among their principal limitations. The proposed wind turbine noise model incorporates many of the physics needed for accurate noise prediction over long distances. The state-of-the-art approach to model the aerodynamic noise from wind turbine is to divide the blades into a number of radial segments. A noise source is then associated to each element. From the wind and temperature profiles, blade geometry and aerodynamic parameters, the strength and directivity of each of the sources are estimated. Finally, the noise from each source is coupled to a propagation code to account for weather conditions. A Hamiltonian Ray Tracing (HRT) propagation method is used in this model and it is the main focus of the paper. The HRT technique averts many of the setbacks presented by other common approaches (FFP, parabolic equation, Eikonal ray tracing). The HRT method computes curved ray-paths by numerically solving a non-linear system of coupled first order differential equations. Energy conservation along tubes formed by the rays permit the computation of sound pressure levels in 3D-space. The number of rays and integration time dictates the developed tool's computational efficiency. Noise propagation results from a 5MW wind turbine noise over acoustically hard flat terrain are shown in order to illustrate the code's capabilities.

1. Introduction

Wind turbine cumulative installed capacity is projected to rise globally by 40% until the year 2020 [1]. Furthermore, manufacturers are constantly increasing their size to achieve greater power outputs, higher efficiency and profitability. However, this also means that their environmental impact is set to become more significant. Specifically, noise production has become one of the major concerns. For this reason, most onshore wind turbine farms around the world must follow strict noise regulations. Dominant wind turbine noise is aerodynamically produced and is responsible for the observed amplitude modulation (AM) phenomenon. Broadband noise levels fluctuate significantly along crosswind directions, which is usually referred to as "swishing", "whooshing" or "pulsating noise"[2,3]. This distinctive type of noise is propagated towards neighbouring residential areas and is considered to be the principal cause of reported annoyance [4,5].

There is a need for accurate and cost effective tools that improve wind turbine noise predictions over relatively large distances (a few kilometres) around wind turbine farms. Thus, accurate turbine noise models that include comprehensive propagation methods must be developed. Currently, noise propagation models are classified as having an engineering, semi-analytical, or numerical approach [3]. The first one refers to standards such as ISO 9613-2 and commercial software models like Nord2000, Concawe, Harmonoise, and so forth. Even though their degree of sophistication has risen on recent years, they still rely on simplified models where neither actual turbine radiation characteristics, nor real meteorological conditions are taken into account [3, 6]. Semi-analytical approaches consist of acoustic wave analytic solutions limited to 2D media, where the speed of sound has a linear variation. In this case, acoustic rays follow a circular path under upward or downward refraction [7-9]. Finally, numerical methods are based on solutions to multiple forms of the Helmholtz equation. These methods have proven considerable capability for 3D noise propagation, however they still have some simplifications and are computationally very intensive [10].

In this work, a state-of-the-art wind turbine noise prediction tool is presented. Aerodynamic 2D flow field over the blades is used to predict the aerodynamic noise using semi-empirical models [11]. Noise sources over blade elements are constructed and coupled to a newly developed Hamiltonian ray tracing (HRT) code. This propagation technique is based on the acoustic wave's phase behaviour as they travel through the atmospheric media [12]. It avoids the drawbacks presented by other numerical approaches such as parabolic equation solutions, Fast Field Program, and the traditional Eikonal ray tracing [3, 10]. Furthermore, it is capable of taking into account real temperature, wind speed, and humidity variations in a three-dimensional gird over the atmosphere. The simplicity of the equations that have to be solved in order to capture ray bending due to refraction is one of the major advantages from this method. The HRT technique is also computationally efficient.

2. Turbine Noise Model

As shown in Fig. 1, the wind turbine noise (WTN) model consists of five modules. The input to the code consists of the turbine and blade geometry, operating conditions, atmospheric data, ground impedance, and execution control parameters.

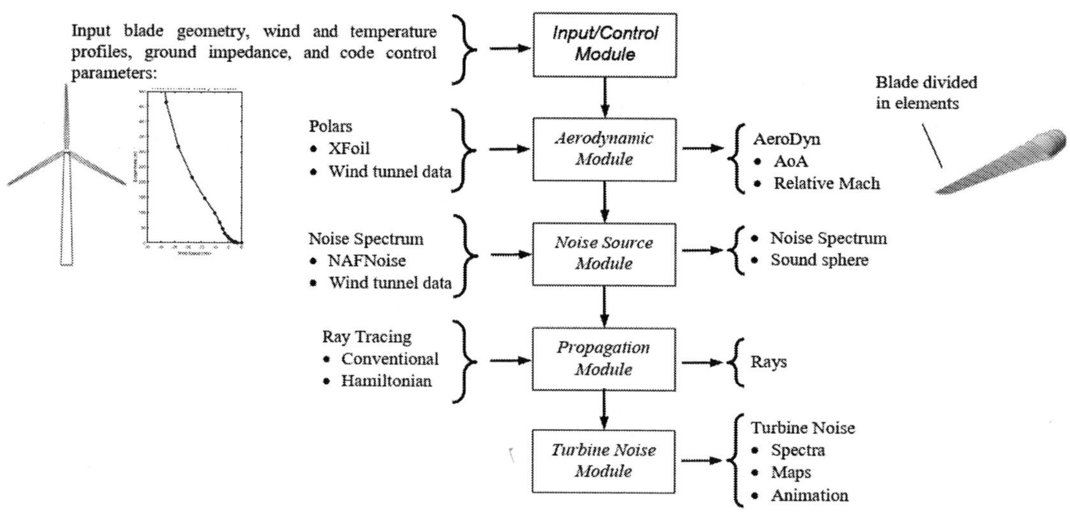

Figure 1: Turbine noise modelling approach.

The turbine blades are assumed rigid, the terrain flat, and the atmospheric conditions uniform over the domain but arbitrary with height. The blades are then split in span-wise direction elements and the blade rotation approximated as a discrete set of azimuth positions. Thus, this approach defines a finite number of positions on the rotor plane to perform aerodynamic and noise calculations as shown in Fig. 2a. The sound sources characterizing the turbine noise radiation will be defined at these points. The second module is the *Aerodynamic*

Module, which uses a blade element momentum method (BEM) to compute the aerodynamic parameters needed for noise calculations [13]. To this end, the airfoil section polars are either computed using XFoil [14] or taken from wind tunnel data [15]. Turbine yaw, tilt, and conning angles are accounted for in the calculations. Fig. 2b illustrates the resulting AoAs for all the positions shown in Fig. 2a for a particular wind profile.

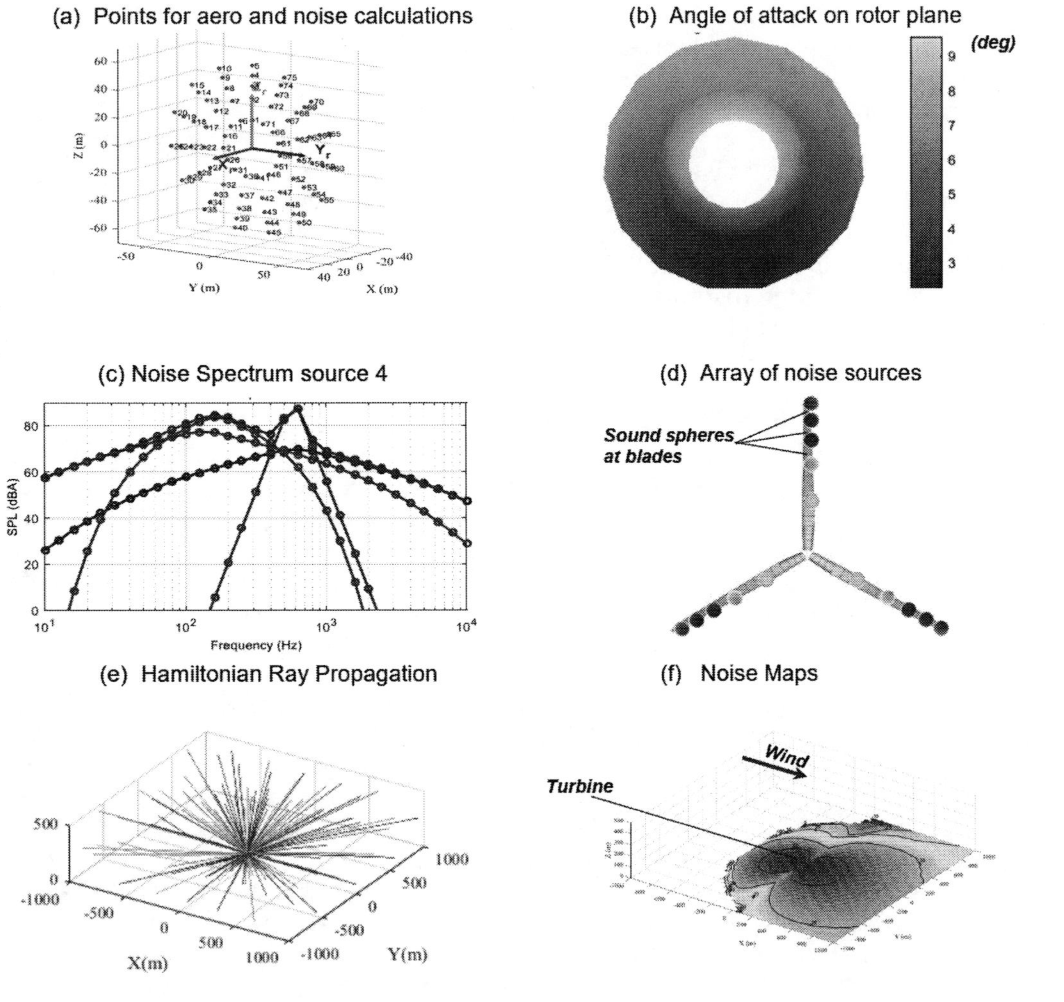

Figure 2: (a) Points on rotor plane for aerodynamic and noise calculation, (b) AoA for full rotation in a non-uniform flow, (c) noise source 4 spectrum computed by NAFNoise, (d) sound spheres array for the wind turbine blades, (e) Hamiltonian ray propagation (f) resulting OASPL noise map due to turbine at 48° azimuth position.

The flow conditions around the blades of a wind turbine govern wind turbine aerodynamic noise generation mechanisms. In the *Noise Source Module*, the aerodynamic noise sources (leading and trailing edge noise) are computed for the selected blade elements and the set of azimuth blade position. This module uses the code NAFNoise [16] or wind tunnel data [15,17] to predict the aerodynamic noise in 1/3rd octave bands at a single point in the direction normal to the airfoil chord line at a distance of 1 meter. The radiation directivity of the sources proposed by Brooks et al. [11] are applied to define sound spheres to couple with the propagation module. An example of the resulting noise spectrum for position 4 in Fig. 2a as computed by NAFNoise is shown in Fig. 2c. Upon implementing the radiation directivity, the resulting sound spheres centred at the trailing (or leading) edge of the airfoil elements are obtained, as shown in Fig. 2d. The sound spheres have varying strength and directivity from the hub to the tip. This is a consequence of the changing inflow, blade twist and airfoil geometry along the blade.

The next step is the *Propagation Module* that implements a Hamiltonian ray tracing propagation technique from the sound spheres located at the wind turbine blades. For this module, it is compelling to implement all atmospheric meteorological conditions and their variability in order to simulate real physical phenomena. The most important parameters to include are temperature, wind, and relative humidity distributions. They affect the propagation behaviour of sound in terms of energy attenuation and absorption, as well as sound propagation paths. Moreover, the terrain type and shape affect wind behaviour, especially near the surface (induced turbulent flows). Here the terrain is assumed flat. The HRT code's output includes the ray paths starting from the wind turbine noise sources and propagating through the atmosphere (see Fig. 2e), as well as noise levels distribution in space.

The final module, *Turbine Noise*, concatenates ground level noise produced by the wind turbine sound sources. For the aerodynamic broadband signals, the noise from the different sources is summed incoherently at the receiver locations and binned according to arrival time. The noise spectrum at the receiver is computed over a uniform time sequence. Fig. 2f shows a typical resulting equivalent noise map.

3. Hamiltonian Ray Propagation Model

The propagation module is the key contribution of this paper and it is presented here. Modelling and simulating atmospheric sound propagation is challenging due to the complexity of atmospheric conditions. Usually, computational time increases proportionally with accuracy. In this case, a Hamiltonian Ray tracing technique has been developed. It is highly capable of handling complicated physical phenomena that lead to high accuracy, while maintaining a relatively high computational efficiency. The proposed model is based on the work by Lighthill [12]. Acoustic wave refraction due to spatial speed of sound gradients, a full Doppler effect formulation resulting from wind velocities in any arbitrary direction, proper acoustic energy dissipation during propagation, and ground reflections are the fundamental issues that the code addresses. For the moment, only a flat soft terrain is considered. However, future work includes different types of terrain in terms of absorption properties and elevation distribution.

The general Hamiltonian propagation problem is divided into two steps. The first one corresponds to the computation of the acoustic ray paths by taking into account the contributions of wind and temperature distributions over the atmosphere. A set of 3D coupled non-linear first order differential equations are solved in order to find the acoustic rays' location in space and local wave number vector. Both are updated as time progresses and the rays propagate. Bent ray paths and wavefront patterns (areas of constant phase) are constructed from the output. The results allow a clear visualization of highly dense acoustic energy zones where a large amount of rays converge, and areas where ray bending is more likely to happen.

The second component of the analysis corresponds to the characterization of the acoustic wave energy associated to the rays. In this case, tubes bounded by a bundle of rays are constructed. Energy conservation laws along the tubes are used to compute the intensity and sound pressure distributions in space. Atmospheric attenuation is also taken into account. Noise over microphones located on the ground is calculated solely from ray tubes that have reached the ground. Finally, ground level noise maps corresponding to noise produced by a single wind turbine at various frequencies are constructed.

3.1 Hamiltonian Ray Path Formulation

The path followed by a ray depends if the media is considered to be homogeneous or inhomogeneous. The first one corresponds to the case where temperature is uniformly distributed in the atmosphere i.e. temperature is constant in space. The second one corresponds to a non-uniform spatial temperature distribution resulting in speed of sound gradients in space. If an inhomogeneous media is considered, rays refract and follow a curved path. However, temperature changes are not significant in the lower atmosphere corresponding to the altitude range of interest (within 500 meters above the ground). In this case, the principal cause of ray bending is mostly influenced by wind velocity.

The Hamiltonian ray tracing approach for ray path computations is based on the definition of a frequency dispersion relationship and an acoustic wave's phase. The dispersion relationship defines the parameters of dependency of the acoustic wave's frequency, and changes depending on the media's homogeneity. For a physically realistic inhomogeneous media, it is given by $\omega = f(k_x, k_y, k_z, x, y, z)$, where (k_x, k_y, k_z) are the spatial components of the local wave number vector at the location (x, y, z). The acoustic wave's phase is formulated as $\alpha = \omega t - (k_x x + k_y y + k_z z)$ and its time and spatial derivatives are

$$\begin{aligned} \partial\alpha/\partial t &= \omega \\ \partial\alpha/\partial x &= -k_x \\ \partial\alpha/\partial y &= -k_y \\ \partial\alpha/\partial z &= -k_z \end{aligned} \tag{1}$$

The combination of both the dispersion relationship and the acoustic wave's phase formulation results in

$$\begin{aligned} \frac{\partial\omega}{\partial k_x}\frac{\partial k_x}{\partial x} + \frac{\partial k_x}{\partial t} &= -\frac{\partial\omega}{\partial x} \\ \frac{\partial\omega}{\partial k_y}\frac{\partial k_y}{\partial y} + \frac{\partial k_y}{\partial t} &= -\frac{\partial\omega}{\partial y} \\ \frac{\partial\omega}{\partial k_z}\frac{\partial k_z}{\partial z} + \frac{\partial k_z}{\partial t} &= -\frac{\partial\omega}{\partial z} \end{aligned} \tag{2}$$

The quasilinear PDEs in (2) must be simplified so that they are easily solved with numerical methods. To do so, the PDEs are expressed as the inner product of two vectors in space-time-wavenumber domain, as follows

$$\begin{aligned} \left(\vec{t} + \frac{\partial\omega}{\partial k_x}\vec{x} - \frac{\partial\omega}{\partial x}\vec{k}_x\right)\cdot\left(\frac{\partial k_x}{\partial t}\vec{t} + \frac{\partial k_x}{\partial x}\vec{x} - \vec{k}_x\right) &= 0 \\ \left(\vec{t} + \frac{\partial\omega}{\partial k_y}\vec{y} - \frac{\partial\omega}{\partial y}\vec{k}_y\right)\cdot\left(\frac{\partial k_y}{\partial t}\vec{t} + \frac{\partial k_y}{\partial y}\vec{y} - \vec{k}_y\right) &= 0 \\ \left(\vec{t} + \frac{\partial\omega}{\partial k_z}\vec{z} - \frac{\partial\omega}{\partial z}\vec{k}_z\right)\cdot\left(\frac{\partial k_z}{\partial t}\vec{t} + \frac{\partial k_z}{\partial z}\vec{z} - \vec{k}_z\right) &= 0 \end{aligned} \tag{3}$$

The vector to the left of each of the three expressions in (3) corresponds to the tangent to a characteristic line over an integral surface constructed in the space-time-wavenumber domain as shown in Figure 3. Only the spatial components associated with the first expression in (3) are shown, i.e. the integral surface is defined only on the (x, t, k_x) domain.

A characteristic curve starting at a specified initial condition is highlighted in red over the integral surface. It corresponds to an assumed solution for the first PDE in (2) under a specific initial condition x_0. This curve is formulated by parametrizing it about s and acknowledging that its derivative about s is both tangent to the characteristic line and to the integral surface. That is,

$$\frac{d\vec{C}}{ds} = \frac{dt}{ds}\vec{t} + \frac{dx}{ds}\vec{x} + \frac{dk_x}{ds}\vec{k}_x \tag{4}$$

This means that it is possible to equate the vector $d\vec{C}/ds$ with the vector to the left on the first expression in (3). Mathematically this results in

$$\frac{dt}{ds} = 1 \quad , \quad \frac{dx}{ds} = \frac{\partial\omega}{\partial k_x} \quad , \quad \frac{dk_x}{ds} = -\frac{\partial\omega}{\partial x} \tag{5}$$

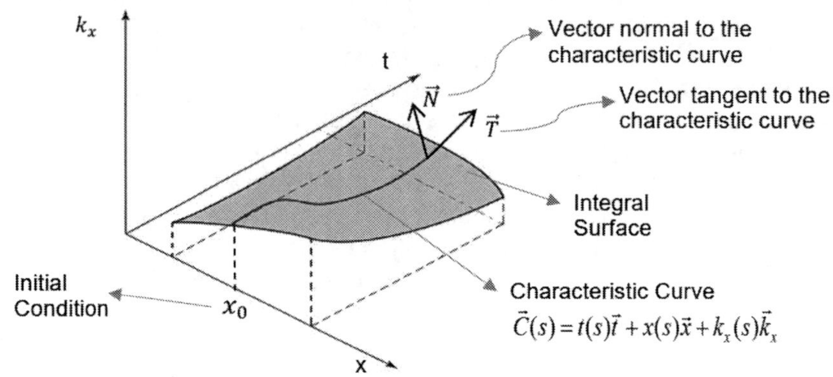

Figure 3: Integral surface and characteristic curve over $(\boldsymbol{x}, \boldsymbol{t}, \boldsymbol{k_x})$ space.

Note that the first PDE in (2) is reduced to 3 equations shown in (5) [19]. They can be further simplified by using $dt = ds$. In this case, time *t* is the parametrizing variable. If the same procedure is followed for all equations in (2), then the following set of coupled first PDEs is obtained

$$\frac{dx}{dt}=\frac{\partial\omega}{\partial k_x} \quad \text{and} \quad \frac{dk_x}{dt}=-\frac{\partial\omega}{\partial x}$$
$$\frac{dy}{dt}=\frac{\partial\omega}{\partial k_y} \quad \text{and} \quad \frac{dk_y}{dt}=-\frac{\partial\omega}{\partial y} \qquad (6)$$
$$\frac{dz}{dt}=\frac{\partial\omega}{\partial k_z} \quad \text{and} \quad \frac{dk_z}{dt}=-\frac{\partial\omega}{\partial z}$$

The set of PDEs in (6) are analogous to Hamilton's equations for dynamical systems. Thus, the denomination of "*Hamiltonian ray tracing*" to this method. This type of equations can be easily solved by using numerical methods. However, an expression for frequency is still required in order to proceed. A Doppler effect between a stationary point of view and a relative one moving with the local wind velocity $\vec{V}=(V_x,V_y,V_z)$ alters the frequency dispersion relationship. The relative frequency ω_r and the absolute one ω are related by $\omega=\omega_r+(V_xk_x+V_yk_y+V_zk_z)$. It is observed that by using a full Doppler formulation all components of wind (including the vertical one) are taken into account. The resultant modified equations after incorporating the Doppler Effect are

$$\frac{dx}{dt}=\frac{\partial\omega_r}{\partial k_x}+V_x \quad \text{and} \quad \frac{dk_x}{dt}=-\frac{\partial\omega_r}{\partial x}-k_x\frac{\partial V_x}{\partial x}-k_y\frac{\partial V_y}{\partial x}-k_z\frac{\partial V_z}{\partial x}$$
$$\frac{dy}{dt}=\frac{\partial\omega_r}{\partial k_y}+V_y \quad \text{and} \quad \frac{dk_x}{dt}=-\frac{\partial\omega_r}{\partial y}-k_x\frac{\partial V_x}{\partial y}-k_y\frac{\partial V_y}{\partial y}-k_z\frac{\partial V_z}{\partial y} \qquad (7)$$
$$\frac{dz}{dt}=\frac{\partial\omega_r}{\partial k_z}+V_z \quad \text{and} \quad \frac{dk_z}{dt}=-\frac{\partial\omega_r}{\partial z}-k_x\frac{\partial V_x}{\partial z}-k_y\frac{\partial V_y}{\partial z}-k_z\frac{\partial V_z}{\partial z}$$

The relative frequency is defined as the speed of sound multiplied by the wavenumber vector's magnitude, this is $\omega_r=f\left(k_x,k_y,k_z,x,y,z\right)=c\left(x,y,z\right)\left|\vec{k}\right|=c\left(x,y,z\right)\sqrt{k_x^2+k_y^2+k_z^2}$. After replacing into (7) and some mathematical manipulations, the result is

$$\frac{dx}{dt}=c\frac{k_x}{\sqrt{k_x^2+k_y^2+k_z^2}}+V_x \quad \text{and} \quad \frac{dk_x}{dt}=-\sqrt{k_x^2+k_y^2+k_z^2}\frac{\partial c}{\partial x}-k_x\frac{\partial V_x}{\partial x}-k_y\frac{\partial V_y}{\partial x}-k_z\frac{\partial V_z}{\partial x}$$

$$\frac{dy}{dt}=c\frac{k_y}{\sqrt{k_x^2+k_y^2+k_z^2}}+V_y \quad \text{and} \quad \frac{dk_y}{dt}=-\sqrt{k_x^2+k_y^2+k_z^2}\frac{\partial c}{\partial y}-k_x\frac{\partial V_x}{\partial y}-k_y\frac{\partial V_y}{\partial y}-k_z\frac{\partial V_z}{\partial y} \tag{8}$$

$$\frac{dz}{dt}=c\frac{k_z}{\sqrt{k_x^2+k_y^2+k_z^2}}+V_z \quad \text{and} \quad \frac{dk_z}{dt}=-\sqrt{k_x^2+k_y^2+k_z^2}\frac{\partial c}{\partial z}-k_x\frac{\partial V_x}{\partial z}-k_y\frac{\partial V_y}{\partial z}-k_z\frac{\partial V_z}{\partial z}$$

Equations in (8) correspond to the final form of the system that must be solved for the spatial components (x,y,z) that define a single ray's location during propagation, and its corresponding acoustic wavenumber components (k_x,k_y,k_z). Thus, in order to propagate a single ray, its initial conditions must be known (initial wavenumber and location in space). This model is physically very accurate because it accepts all velocity components of wind, as well as variations on the speed of sound on any direction. Therefore, simulations that seek considerable accuracy require a complete 3D wind field over the desired propagation volume. Finally, Runge-Kutta (RK) methods are used to solve the system of equations presented in (8). Its numerical accuracy depends on the selected order of the RK and time step taken during propagation. Furthermore, these two parameters and the total simulation time are the prime factors that affect the computational efficiency of this technique.

The more common Eikonal ray tracing method is based on an asymptotic series solution to the acoustic Helmholtz equation. In this case, a complex nonlinear PDE must be solved and moving media is usually taken into account by assuming an effective speed of sound i.e. adding to the speed of sound to the horizontal component of wind velocity in the direction of propagation [10, 26]. Whereas, the Hamiltonian formulation results from acoustic wave frequency analysis during propagation and incorporates wind velocity into the model in an accurate manner.

3.2 Hamiltonian Ray Tracing Energy Analysis

Computing noise levels associated to the ray path equations presented in the previous section is done via acoustic energy analysis. It assumes that acoustic waves propagating in the vicinity of a ray path have a nearly constant phase. Thus, all waves within a ray-tube coherently contribute to the total enclosed pressure fluctuations [12]. A ray-tube corresponds to a bundle of acoustic rays enclosing a cross-sectional area that change depending on the paths taken by the bundling rays. This is a standard approach for computing sound pressure levels as rays propagate in space. If a single ray tube is considered as the system for analysis, the following equation is obtained

$$\frac{\partial E_T}{\partial t}+\nabla\cdot\vec{I}=0 \tag{9}$$

where, E_T corresponds to the total acoustic energy density contained in a tube and the vector $\vec{I}$ is the corresponding acoustic intensity vector. If the media is time-independent and the total energy density can only have variation in space, then the first term in (9) is equal to zero. Furthermore, this also means that there are no sources or sinks of energy flux in space. Thus, the second term in (9) must also be equal to zero [12, 21]. Under these conditions, the average flow of energy per cycle of the acoustic waves must be constant along a propagating tube. That is, $|\vec{I}|\times A$ is constant along a ray tube. *A* corresponds to the cross sectional area of the tube at any arbitrary position in space. Therefore, given an initial ray tube area as well as the corresponding initial intensity, it is possible to compute new intensities for each time step during propagation.

3.3 Turbine Noise Propagation

The Hamiltonian formulation applied to wind turbine noise propagation require a starting noise field. In this case, the sound spheres computed at *Noise Source Module* provide the starting field. Each sphere is composed of an evenly distributed spherical grid (icosahedron), as shown in Fig. 4. Every point over the grid contains sound pressure level data. Additionally, an initial wavenumber is assigned to each grid point. Therefore, an initial triangular area between points in the spherical grid is defined, where the intensity is known. The ray paths propagate according to the system of equations (8) and tubes consisting of three-ray bundles are constructed. For every propagation time-step, a new area and intensity are calculated. Thus, sound pressure levels can be computed in space for any specified simulation time. Furthermore, since the impact of wind turbine noise near the ground is of concern, ground-reflecting bundles of rays provide the necessary noise levels to build noise maps around a wind turbine.

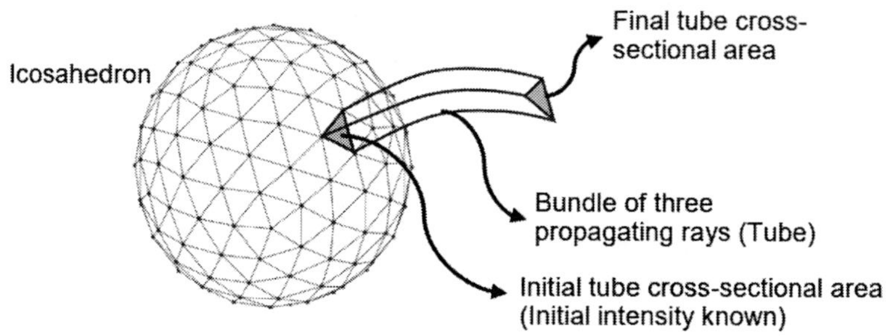

Figure 4: Starting ray propagation grid from sound source at one of the wind turbine blade components.

4. Validation and Numerical Examples

This section presents a limited number of validation and example cases for the HRT propagation method.

4.1 Ray Path Validation

Current analytical solutions for outdoor sound propagation are very limited. However, Mo et al. [22, 23] developed analytical expressions for the Eikonal ray path. It computes refracted ray paths in 2D planes defined parallel to a defined initial propagation direction. The analytical solution to the Eikonal equation is given by

$$x(z)=\frac{\sqrt{1-\xi_0^{'2}c_0^2}-\sqrt{1-\xi_0^{'2}\left(c_0+\alpha z\right)^2}}{\xi_0^{'}\alpha} \tag{10}$$

where, x corresponds to the horizontal component of the ray path and z to the ray's altitude, $\xi_0' = \cos\theta_0/c_0$, θ_0 is the angle between the initial wavenumber vector $\vec{k}_0$ and a line parallel to the x-axis, $c_0$is the speed of sound at the noise source's height, and α is the gradient of the speed linear profile. Thus, the speed of sound must be a 2D profile that follows the linear expression $c(z) = c_0 + \alpha z$. The Hamiltonian ray path numerical solutions have been validated against this analytical approach.

Fig. 5 shows three rays propagated from a noise source located at 100 meters height in a stationary media (negligible wind), where a 2D speed of sound profile is $c(z) = 332 + 0.6z$. Results show that the HRT technique produces ray paths that are very close to those analytically computed. Given that the Hamiltonian ray tracing is solved numerically, these results provide confidence in the HRT method.

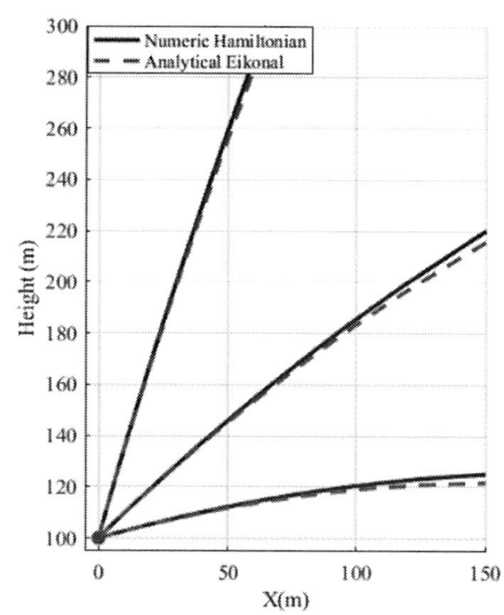

Figure 5: Three rays propagated using a numerical Hamiltonian and Analytical Eikonal approach.

4.2 HRT vs FFP Validation

The next validation case is comparing the HRT method against a Fast Field Program (FFP) output. The FFP implemented for validation is based on a NASA code developed for prediction of noise from fixed wind aircrafts and helicopters [27]. FFP numerically solves a Helmholtz equation transformed to the horizontal wave number domain. It does so over a stratified media where the wave number depends exclusively on height [10]. This is one of the major limitations of this code, in addition to the vast amount of computational time required for simulations. Nevertheless, this method is widely accepted for atmospheric noise prediction purposes.

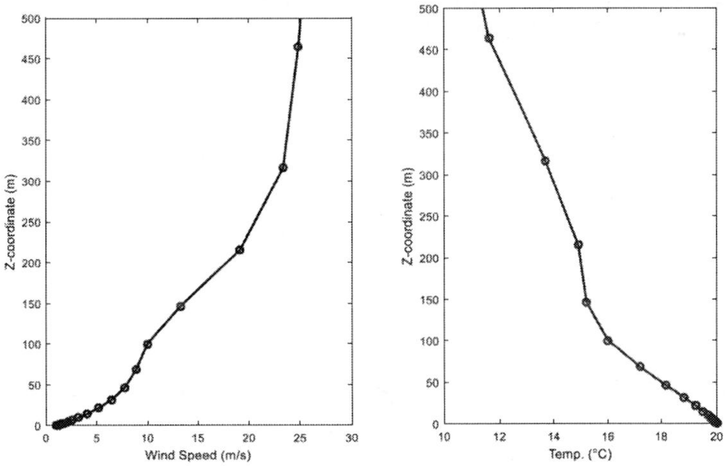

Figure 6: Wind and temperature profile used in the simulations.

The selected validation problem consists of a monopole source located at 100 m height. The weather condition consist of the non-uniform wind and temperature profiles shown in Fig. 6. They were generated by modifying experimentally measured data [25]. There is no vertical wind component in the simulations. The terrain was assumed flat and acoustically hard, e.g. very high uniform flow resistivity. In this case, the HRT code emitted 2,562 rays from the monopole source.

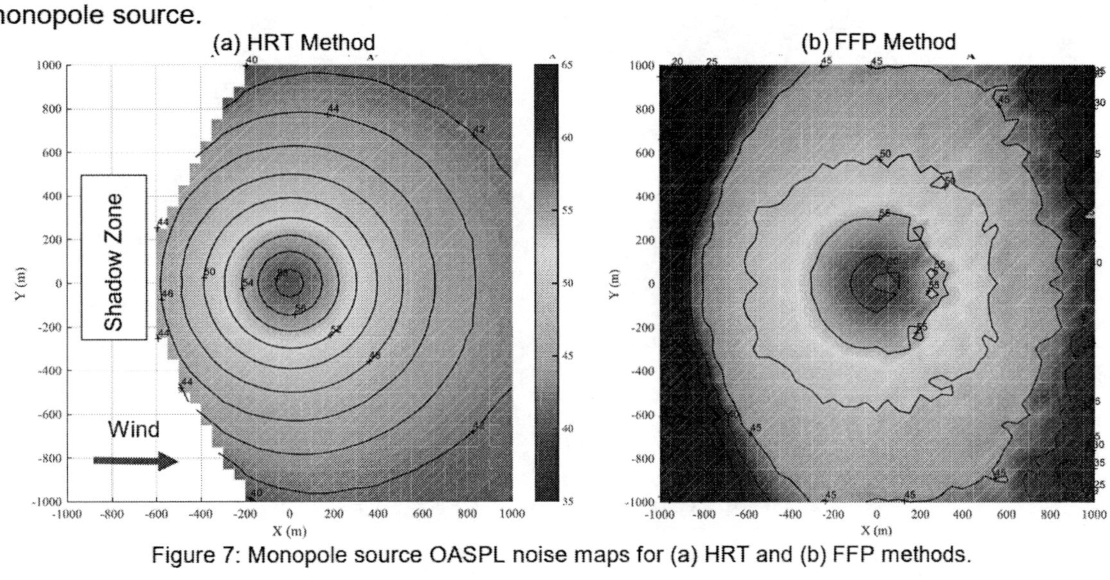

Figure 7: Monopole source OASPL noise maps for (a) HRT and (b) FFP methods.

Fig. 7 shows two simulated monopole noise maps for both the HRT and FFP methods. The noise maps consist of a 2 Km square grid. The monopoles are located at the centre of the grid. Smooth lines with decreasing noise levels surrounding the source characterize the HRT results. The FFP map on the other hand shows significant noise level fluctuations around the noise source. The HRT method is not capable of predicting noise on areas where there are no rays reflecting the ground (commonly referred to as the shadow zone). This constraint is not limited to HRT but to all ray tracing propagation methods. However, there are refraction solution methods to compute noise levels on the shadow zone, as described by L'Esperance et al. [8]. Future work will include shadow zone prediction embed within the HRT code.

Noise for both HRT and FFP noise levels at different locations on the noise maps are shown in Table 1. Sound pressure levels show good agreement between the methods. Yet, computational time for the FFP method was of 3.6 hours using a Fortran code, while the HRT method took only 6 minutes in Matlab. All simulations were performed on a 3.42-GHz quad-core personal computer with 16 GB of RAM.

Table 1: Sound Pressure Level results over noise map grid for HRT and FFP methods.

Noise Map Coordinates	Sound Pressure Levels [dBA]	
	HRT	FFP
(800,0,0)	43.8	44.9
(600,800,0)	41.9	44.4
(-550,0,0)	47.7	50.0

4.3 Wind Turbine Noise Map Results

HRT noise propagation results are presented for a NREL 5MW reference turbine [24]. The reason for selecting this turbine is that the blade geometry and other parameters are available in the open literature. The rated rotor speed is 12.1 rpm. The length of the blades is 61.5 meters and its maximum chord is 4.65 meters. The blade airfoil sections are composed of a series of circular, DU and NACA airfoils. For the simulations, it is assumed the hub height is 100 m and the turbine operates at 12 rpm with an inflow of 10 m/s at the hub. The turbine yaw and tilt angles are set to zero and the rotor is not conned either. The blades were divided in 5 span-wise elements and the rotation accounted for by taking 15 azimuth positions for a total of 75 sound sources distributed on the rotor plane. NAFNoise was used to predict the trailing edge noise for the 75 sound sources, e.g. leading edge noise was not modelled.

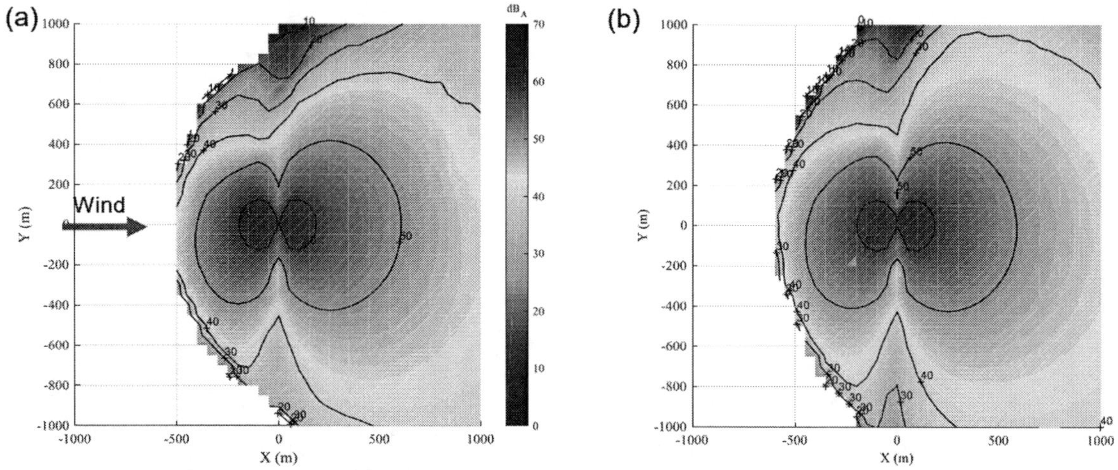

Figure 8: Equivalent Overall A-weighted sound pressure level noise map for one rotor revolution: (a) 2,562 and (b) 10,242 rays emitted by each of the 75 sound source.

The code computes the 1/3rd octave band as well as overall A-weighted SPL (OASPL) spectrum for an array of microphones at each azimuth position of the rotor. In these simulations, a square grid of 1600 microphones was placed on the ground over an area of 2 km by 2 km. The turbine is at the centre in the domain. Background noise was not added to the turbine noise results. The resulting equivalent OASPL noise maps are shown Fig. 8. In this case, the same weather conditions as in Fig. 7 are used, as well as the ground characteristics. The HRT code was used to propagate 2,562 rays from the 75 noise spheres distributed along the three blades (see Fig. 2d). Thus, 187,500 rays were emitted from the wind turbine for propagation with a computational time of approximately 2 hours for this case using Matlab (Fig. 8a). Results show the characteristic directivity pattern of wind turbine noise i.e. dipole resembling. The noise map also shows higher levels towards one side (right of downwind direction). Even though sound pressure levels over the shadow zone in the upwind direction are not known, it is safe to assume that they are very low i.e. as observed in Fig. 8., below the 30 dBA line in the upwind direction. To assess convergence and computational time, a second simulation was performed with 10,242 rays emitted per sound source (a total of 768,150 rays emitted from the wind turbine). The result in Fig. 8b shows nearly the same results, particularly close to the turbine. The main differences are in a reduction of the extent of the shadow region and the levels in the upper right region of the domain. The computational time for this case was approximately 12 hours.

5. Conclusions

The Hamiltonian ray tracing (HRT) approach presented in this work provides accurate and efficient wind turbine noise predictions over large distances. It is a physically thorough model capable of taking into account 3D wind and speed of sound patterns for simulations. It avoids generalized assumptions from other numerical propagation methods such as the commonly used effective speed of sound in the Eikonal ray tracing method. Additionally, contrary to other approaches, it is not limited to linear speed of sound gradients. The model was validated against commonly used methods. It presented good agreement when compared to an analytical Eikonal ray tracing and Fast Field Program solution. The results show accurate predictions in the far field with a significant computation advantage over the FFP method. The HRT approach was integrated with an accurate turbine noise model. A NREL 5MW reference turbine was used to demonstrate the tool's capabilities. HRT was able to capture the characteristic directivity pattern of noise produced by wind turbines. The current HRT formulation is limited to predictions outside the shadow zone. However, future work includes the implementation of known solutions that predict noise over these areas.

References

1. Global Wind Report-Annual Market Update 2015. Global Wind Energy Council.
2. Stigwood M., Large S., Stigwood D. Audible amplitude modulation – results of field measurements and investigations compared to psychoacoustical assessment and theoretical research, *5th International Conference on Wind Turbine Noise,* Denver, Colorado, USA, 28-30 August, (2013).
3. Bowdler, D. & Leventhall, G. *Wind Turbine Noise.* Multi-Science Pub. Brentwood, Essex, United Kingdom (2011).
4. Van den Berg, G. The Beat is Getting Stronger: The Effect of Atmospheric Stability on Low Frequency Modulated Sound of Wind Turbines, *Journal of Low Frequency Noise, Vibration and Active Control.* 24 (1), 1–24, (2005).
5. Bolin K., Bluhm G., Eriksson G. Nilsson M. E. Infrasound and low frequency noise from wind turbines: exposure and health effects. *Environmental Research Letters*, 6 (3), 035103, (2011).
6. Plovsing, B. DELTA acoustics report AV 1851/100, Nord2000. Comprehensive Outdoor Sound Propagation Model. Part 2: Propagation in an Atmosphere with Refraction, (2001).
7. Berengier, M., Gauvreau, B., Blanc-Benon, P., & Juve, D. *Outdoor Sound Propagation: Ashort Review on Analytical and Numerical Approaches.* (2003) *Acta Acustica United with Acustica, 89*(6), 980-991.

8. L'Esperance, A., Nicolas, J., Herzog, J., Daigle G.,*Heuristic model for outdoor sound propagation baseD on an extension of the geometrical ray theory in the case of a linear speed profile.* Applied Acoustics (1992), 37(2), 111-139.
9. Hidaka, T., Kageyama, K., Masuda, S., *Sound propagation in the rest atmosphere with linear sound velocity profile. Journal of the Acoustical Society of Japan (1985), 6*(2), 117-125. doi:10.1250/ast.6.117
10. Salomons, E. *Computational Atmospheric Acoustics.* Springer Science+Media Buisnesss, B.V. (2001).
11. Brooks, t., Pope, S., & Marcolini, M. National Aeronautics and Space Administration, NASA Reference Publication 1218. Airfoil Self-Noise and Prediction, (1989).
12. Lighthill, J. *Waves in Fluids.* Cambridge University Press, Cambridge, Great Britain (1980).
13. Laino, D. User's Guide to the Wind Turbine Aerodynamics Computer Software, AeroDyn 12.50 (2002). National Renewable Energy Laboratory.
14. Drela, M. and Youngren, H. *XFOIL 6.94 User Guide*, Massachusetts Institute of Technology, Cambridge, Massachusetts, (2001).
15. Devenport W., Burdisso R.A., Camargo H., Crede E., Remillieux M., Rasnick M., and Van Seeters P. National Renewable Energy Laboratory, Technical Report NREL/SR-500-4347, Aeroacoustic Testing of Wind Turbine Airfoils, (2010).
16. Moriarty, P. *NAFNoise User's Guide.* National Wind Technology Centre, National Renewable Energy Laboratory. Golden, Colorado, (2005).
17. Migliore P.J. & Oerlemans S. Wind Tunnel Aeroacoustic Tests of Six Airfoils for Use on Small Wind Turbines. *AIAA Wind Energy Symposium*, Reno, Nevada, (2004).
18. Oerlemans, S. *Detection of aeroacoustic sound sources on aircraft and wind turbines.* Thesis Dissertation, University of Twente, (2009).
19. Zuaderer, E. *Partial Differential Equations of Applied Mathematics.* Wiley-Interscience, Hoboken, N.J. (2006).
20. Amiet, R. Acoustic Radiation from an Airfoil in a Turbulent Stream*, Journal of Sound and Vibration*. 41 (4) 407-420, (1975).
21. Dieminger, W. Hartmann, G. & Leitinger, R. *The Upper Atmosphere: Data Analysis and Interpretation.* Springer-Verlag, New York (1996)
22. Mo, Q., Mo, Q., Yeh, H., Lin, M., & Manocha, D. *Analytic ray curve tracing for outdoor sound propagation,* Applied Acoustics 104 (2016) pp.142-151, doi:10.1016/j.apacoust.2015.10.011
23. Mo, Qi, Hengchin Yeh, and Dinesh Manocha. *Tracing Analytic Ray Curves for Light and Sound Propagation in Non-Linear Media*. IEEE Transactions on Visualization and Computer Graphics 22 (2016) pp. 2493-2506, doi:10.1109/TVCG.2015.2509996
24. Jonkman, J., Butterfield, S., Musial, W., Scott, G. National Renewable Energy Laboratory, Technical Report NREL/TP-500-38060, Definition of a 5-MW Reference Wind Turbine for Offshore System Development, (2009).
25. Slawsky L., Zhou L., Roy S., Xia G., Vuille M., and Harris R., Observed Thermal Impacts of Wind Farms Over Northern Illinois, *Sensors*, 15 (7), 14981-15005, (2015).
26. Jensen, N., Kuperman, W., Porter, M. & Schmidt, H. *Computational Ocean Acoustics.* Springer New York, 2nd edition (2011).
27. Burley, C., Pope, S. *APET User Guide*. NASA-LaRC, AS&M, Inc, (2014).

International Conference
on
Wind Turbine Noise
Rotterdam – 2nd to 5th May 2017

Investigation of Amplitude Modulation Noise with a Fully Coupled Noise Source and Propagation Model

Emre Barlas, Technical University of Denmark, Wind Energy Dept., ebarlas@dtu.dk
Wei Jun Zhu, Technical University of Denmark, Wind Energy Dept., wjzh@dtu.dk
Wen Zhong Shen, Technical University of Denmark, Wind Energy Dept., wzsh@dtu.dk
Kaya Onur Dag, Technical University of Denmark, Wind Energy Dept., kdag@dtu.dk
Patrick Moriarty, National Renewable Energy Lab., USA, patrick.moriarty@nrel.gov

Summary

Wind turbines operate in turbulent winds and the sound generation from the blades is affected by their dynamic response. To take these phenomena into account, an advanced modelling technique is developed which couples a wind turbine noise generation and propagation model in a quasi-three-dimensional and unsteady manner with high fidelity flow forcing. The focus of the present study is the amplitude modulation (AM) detection and quantification under various wind directions and ground conditions. The results show that the propagation effects such as refraction, absorption and reflection mostly reduce the time averaged sound pressure level (SPL) in the far field. However, there is a considerable increase of the AM levels, especially in the downwind and upwind areas of a turbine.

1. Introduction

The amplitude modulation (AM) of wind turbine noise is considered as a major cause for annoyance. From the comprehensive and collective research project funded by Renewable UK, two types of amplitude modulation exist, namely Normal AM (NAM) and Other/Enhanced AM (OAM/EAM) [1]. NAM is relatively well understood and explained to be caused by the directivity and convective amplification of the noise emitted from the airfoils' trailing edges as they rotate with the rotor. This is inherent for all wind turbines and the peak-to-through ratios vary between 3 dB and 9 dB depending on the observer location and reflective surfaces around the wind turbine. NAM is observed in the mid frequency range (500-1000 Hz) which is in line with the claimed source spectral content. It is expected that these levels decrease with increasing distance as the distinct directivity loses its effect. However, OAM has an intermittent nature and it is usually observed in the far field with a higher modulation depth than NAM. The spectral content is shifted towards lower frequencies. The complexity of OAM under various weather conditions and its intermittent nature makes it difficult to detect, quantify and subsequently mitigate.

The state-of-the-art work on AM has been focused on sound source modelling [2][3], noise field experiments [3] and listening room experiments [4]. The present paper investigates the AM levels in far field by using the recently developed high-fidelity model in which the wind turbine source and propagation models are coupled dynamically for a wind turbine under realistic atmospheric conditions.

2. Computational Models

For this study a three bladed, stall regulated Nordtank NTK 500/37 wind turbine with a 500 kW nominal power is used. The hub height is 35 m and the rotor diameter is 41 m. Further technical details about the turbine can be found in [5]. In the next three subsections the source, propagation and the flow models are described, followed by the fourth subsection that describes the coupling procedure.

2.1 Wind Turbine Noise Source Model

For the wind turbine noise source modelling we use the recent implementation of aeroacoustics module based on NREL's NAFNoise [6]) within the FAST8 modular framework [7]. This integration allows modelling aerodynamic noise generated by the blades with consideration for wind turbine structural dynamics and its interaction with the incoming turbulent flow. Using in the blade element theory, each blade is divided into a number of two dimensional airfoil elements. The total noise level at a given receiver location is predicted as the sum of the contributions from all blade elements. In the present study, only two types of aerodynamic noise have been included; turbulent boundary layer trailing-edge and turbulent inflow noise. The models and methods used for obtaining the necessary inputs are explained below.

Trailing Edge Noise: The total sound pressure level of noise generated from the interaction of the turbulent boundary layer with the trailing edge is calculated through the summation of the different contributions of noise on the pressure side, the suction side, and from flow separation. These three noise mechanisms can be modelled semi-empirically using scaling laws (see Brooks, Pope, and Marcolini (BPM) [8] for more details). Different to the classical BPM noise generation model, the boundary layer characteristics in this study are obtained from Q^3UIC (DTU's integral boundary layer flow solver) which is shown to predict the boundary layer characteristics more accurately than XFOIL [9]. Additionally, the angle of attack flag, a conditional switch used in the classical BPM model to turn on the separation noise is modified with the values according to the lift coefficient curves of the airfoils used in the blade.

Turbulent Inflow Noise: The turbulent inflow noise model originally proposed by Amiet [10] and further developed by Lowson [11] is used for inflow noise estimation, including a correction for airfoil thickness proposed by Moriarty and Guidati [12]. The turbulence intensity (TI) is calculated by taking into account the inflow characteristics obtained from the flow simulations (Section 2.3) and instantaneous airfoil locations. The rotor area is divided into 24 regions obtained with 6 azimuthal and 4 radial divisions. As the blades rotate, in each region, the wind speed is sampled, and the standard deviation and local mean value are calculated. The sampling time-duration is chosen to be 2 seconds, which corresponds to 400 samples. The TI value in each region is calculated using to the standard deviation and average wind speed (see Figure 1). At each time-instant the TI value in the region where an airfoil is located is assigned to the airfoil.And, a turbulent length scale is calculated using the relationship in [13] that can be expressed as a function of surface roughness zo and height z: $L = 25 * z^{0.35} * z_0^{-0.063}$.

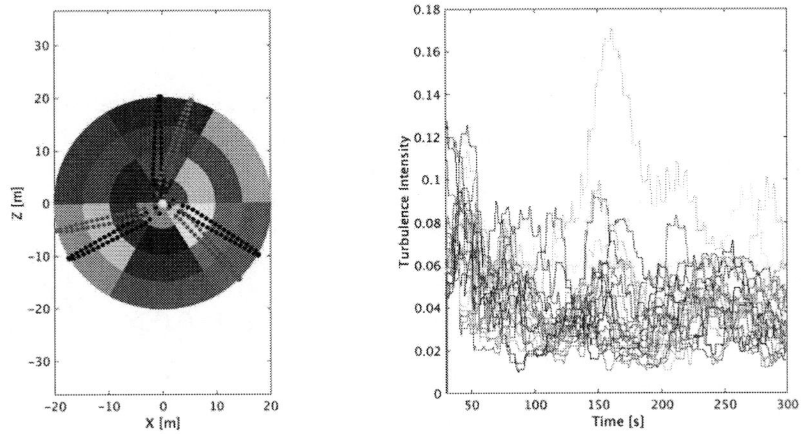

Figure 1: Colored 24 regions of the rotor swept area for turbulence intensity calculations. Right: Time-dependent turbulence intensity signals in the 24 regions; lines with the same colors.

2.2 Sound Propagation Model

DTU's Parabolic Equation (PE) based sound propagation tool WindSTAR-Pro (Wind turbine Simulation Tool for AeRodynamic Noise Propagation) is used. The tool implements the two-dimensional Generalized Terrain PE model [14] with full parallelization for various frequencies and realizations/time steps. The PE method solves the wave equation for harmonic waves from near to far field, with finite angle approximation and forward propagation. The conventional method uses the effective speed of sound approach where the moving atmosphere is replaced by a hypothetical motionless medium with an effective speed of sound $C_{eff}(x,z) = c(x,z) + V_x(x,z)$ where V_x is the wind velocity component projected along the direction of propagation between source and receiver (obtained from the flow solver (see also the next section). For this study, the speed of sound is kept constant ($c(x,z) = 340\frac{m}{s}$) in the whole domain, and the refraction phenomenon dominated by the wind speed deficit and its fluctuations around a wind turbine at the distances of interest. The spatial resolution in both directions is set to one eighth of the wavelength ($\Delta x = \Delta z = \frac{\lambda}{8}$). Only flat terrain is considered and the ground impedance is characterized using the Delany-Bazley model [15] with effective flow resistivity of 200 $kPas/m^2$ and 2000000 $kPas/m^2$ that are representative for grassland and hard ground, respectively. All simulations are carried out for 1/3- octave band centre frequencies up to 2500 Hz and the corresponding sound pressure levels are summed logarithmically to obtain the overall SPL:

$$L_{p_{sum}} = 10\log_{10}\sum_{i=1:N} 10^{Lp(f_i)} \quad (1)$$

where N is the number of frequencies considered. $L_p(f_i)$ is the sound pressure level:

$$L_p(f_i) = L_W(f_i) - 10\log_{10} 4\pi r^2 - \alpha_p(f_i)r + \Delta L_p(f_i) \quad (2)$$

where the source power level for a wind turbine ($L_W(f_i)$) and the second term on the right hand side (geometrical spreading) is obtained from the source model explained in the previous subsection with r being the distance from the rotor plane. The third term represents the atmospheric absorption where the absorption coefficient, $\alpha_p(f_i)$, is calculated according to ISO 9613-1 for air at 20°C with 80% relative humidity. The last term, $\Delta L_p(f_i)$, is the relative sound pressure level $\Delta L = 20\log_{10}(p'/p_f)$ that represents the deviation from the free field of a source due to ground effect, atmospheric refraction, turbulence, etc. This last term is calculated using the PE propagation method.

2.3 Flow Field Model

DTU's pseudo-spectral 3D Navier-Stokes solver is used, which uses the Smagorinsky subgrid scale model for large eddy simulation (LES) [16]. First a precursor simulation is run to establish the developed boundary layer with a surface roughness height of 0.6 m and a mean wind speed of 12 m/s at hub height (36m). This solution is then fed into the computational domain with the wind turbine. The wind turbine rotor is modelled via the actuator line technique, in which the body force is distributed radially along each rotor blade. The domain of [1500 m X 400 m X 400 m] is discretised with a resolution of [600 X 160 X 320 points] in the streamwise, lateral and vertical directions, respectively. The time step is set to 0.007 seconds.

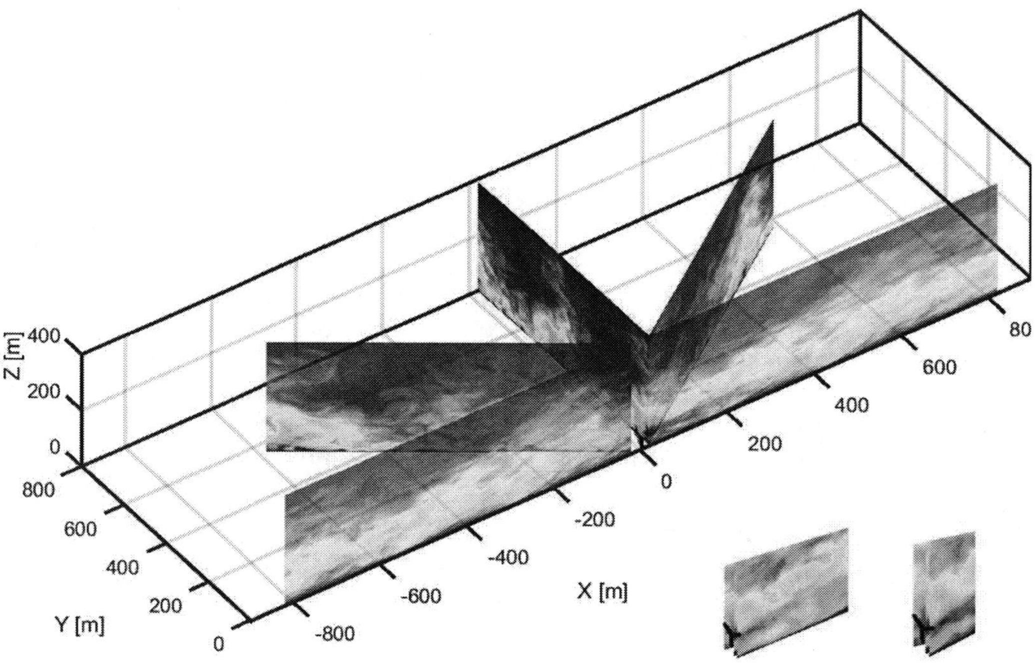

Figure 2: 3D view of the sample flow field and the interpolated flow fields between the three blades and two different receiver locations at one time instant.

2.4 Coupling Procedure

Figure 3 depicts a schematic of the coupling procedure. The different steps are listed below;

- First, the flow field is simulated using LES (see Section 2.3 for the flow model). A two-dimensional (Y-Z) slice located 2D upstream of the wind turbine is stored at each time step (0.07 s). Additionally, the whole three-dimensional flow fields are stored with a sampling frequency of 0.1 s to be used for propagation simulations.
- The flow field, sampled upstream of the turbine, is used as an input to FAST8, creating a model of a fully aero-elastic turbine exposed to a realistic atmospheric flow.
- Frequency dependent sound pressure levels at given receiver locations are calculated and stored via the integrated aeroacoustics module in FAST8 (see Section 2.1 - source model). 25 receiver locations are chosen and shown in Figure 3.
- For these receiver locations, highest SPL contributions along each blade at each time step and for each frequency are detected and their coordinates are stored.

- A two-dimensional PE domain (see Section 2.2 – propagation model) is constructed between each blade and each receiver.
- At each time step, two-dimensional PE simulations are carried out for 3 blades, 25 frequencies and 25 receivers (for a total of 1875 simulations).
- This procedure is repeated for each time step by updating the flow field, source strength and locations. In this study, 2400 realizations are simulated with a time step of 0.1 second.

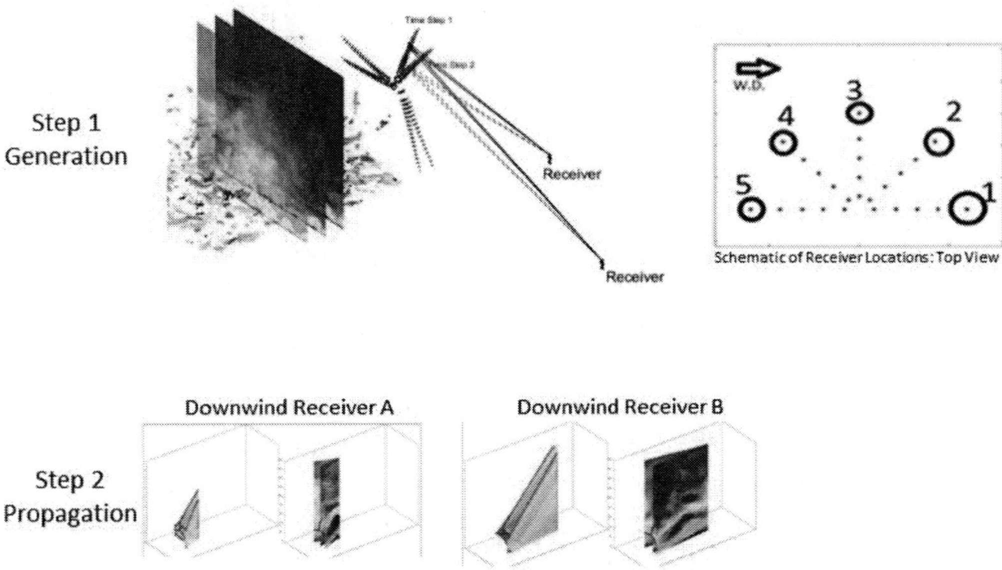

Figure 3: Overview of the coupling steps. Step 1: Noise generation via FAST8+NAFNoise. Step 2: Propagation via 2D PE simulations from each blade to selected receivers. The procedure is repeated at each time-step.

3. Results

The results section is composed of three subsections to investigation: the averaged SPL, time dependent SPL and AM levels. The main focus is the SPL comparison from various cases. Three sets of simulations are carried out as described below:

- Only source simulations ($\mathrm{SPL_S}$): Output of FAST8+NAFNoise at various receiver locations. This means that the last two terms in Equation (2) are neglected, e.g. the atmospheric absorption, and the ground reflection and refraction due to atmosphere.
- Propagation over a grassland ($\mathrm{SPL_{S+P_G}}$): Coupled source and propagation simulations where the ground impedance value is representative of a grassland and the full three dimensional flow field is taken into account as explained in Section 2.4.
- Propagation over a hard ground ($\mathrm{SPL_{S+P_{HG}}}$): Coupled source and propagation simulations where the ground impedance value is infinity for a fully reflective hard ground and the fully three-dimensional flow field is taken into account as explained in Section 2.4.

In order to investigate the effect of propagation at various receiver locations, the focus is on the differences of SPL values, namely $\Delta\mathrm{SPL}$;

$$\Delta_{\mathrm{SPL_G}} = \mathrm{SPL_{S+P_G}} - \mathrm{SPL_S} \quad \text{and} \quad \Delta_{\mathrm{SPL_{HG}}} = \mathrm{SPL_{S+P_{HG}}} - \mathrm{SPL_S}$$

25 centre frequencies are simulated using the PE (from 10 Hz to 2500 Hz) and 1/3-octave bands. For each frequency and each blade, 2400 realizations are simulated with a time step of 0.1 second.

3.1 Comparison of Time Averaged SPLs

Figure 4 shows the time averaged Δ_{SPL_G} in the upper half domain and the time averaged $\Delta_{SPL_{HG}}$ in the lower half domain. From the figure, it is clear that far field noise at upwind locations has lower levels than downwind locations. This is an expected result from the upward/downward refraction due to the flow field and it shows that the results obtained from the propagation coupled simulations (SPL_{S+P_G} and $SPL_{S+P_{HG}}$) can capture this phenomenon as opposed to the source only simulations. (SPL_S). Considering the 41 m turbine diameter and 36 m hub height, it is expected that the atmospheric conditions will affect the propagation over a shorter range than for a taller turbine. Nevertheless, increased sound pressure levels ($\Delta_{SPL} > 0$) are observed up to an upwind distance of 260 m for hard ground, and 230 m for grass-covered ground. The increased SPL upwind is a combination of the initial refraction and the constructive interference of the direct and the reflected waves. It is worthwhile mentioning that since the grass-covered ground has a more absorbing character; the overall sound pressure levels are always lower than those in the hard ground case. This is valid for all directions.

Downwind simulations show that there are increased levels ($\Delta_{SPL_G} > 0$) up to 150 m from the rotor for the grassland and beyond this distance the differential levels are negative ($\Delta_{SPL_G} < 0$). On the other hand, the level difference for the hard ground case ($\Delta_{SPL_{HG}}$) are either higher or equal to zero. Since the same flow fields and atmospheric absorption values are used for both cases this comparison shows the effect of ground absorption.

Furthermore, in the crosswind direction, the increased levels reach up to 6 dB and 4 dB for the hard ground and grassland, respectively and then decrease with increasing distance. It is also observed that the distances where Δ_{SPL} values are greater than or equal to zero, are longer than in the downwind cases for both ground covers. The main difference between two directions is the flow field. Sound waves propagate through the wake induced flow field for the downwind case while for the crosswind case the waves are affected only by the turbulent perturbations since the mean lateral velocity is close to zero and we assume a neutral atmosphere where there is no temperature effect. This means that the refraction caused propagation effects play a bigger role in favour of SPL attenuation for the downwind case than for the crosswind case (valid up to the simulated distance, 600 m). This result does not necessarily mean that the neutral atmosphere causes less attenuation. On the contrary, it highlights the complexity of the refraction phenomenon with respect to the wind turbine induced flow fields. Nevertheless, since the wind turbine has a dipole noise emission the overall levels are significantly lower in the crosswind direction than in the downwind direction. Hence, the attenuation might not be of paramount interest in terms of wind turbine noise annoyance.

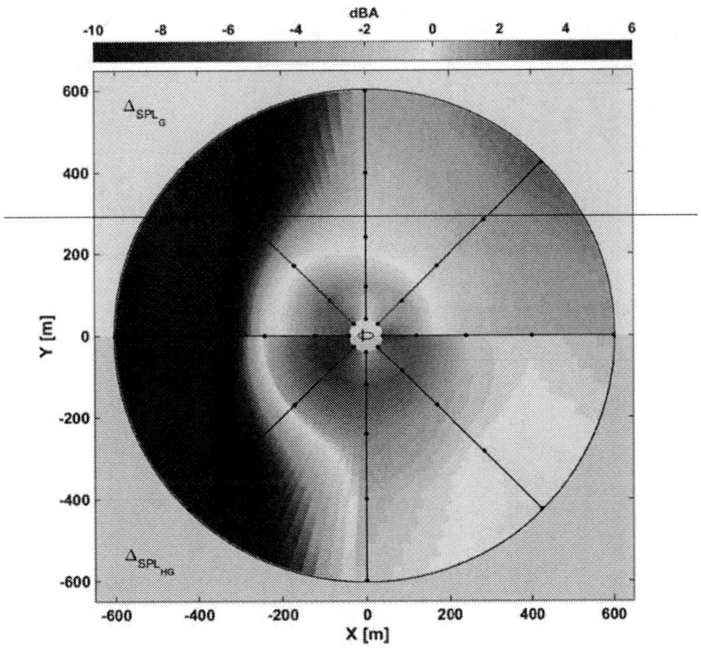

Figure 4: Top view of the time averaged OASPL difference. Upper half domain for a grass-covered land: $\Delta_{SPL_G} = SPL_{S+P_G} - SPL_S$; Lower half domain for a hard ground: $\Delta_{SPL_{HG}} = SPL_{S+P_{HG}} - SPL_S$. Wind direction is from left to right. Black dots represent the receiver locations.

Figure 5 shows frequency dependent results of the time averaged SPL difference. It is observed that it takes more time for the atmospheric effects to impact the lower frequencies than the higher frequencies. This can be seen by comparing the plots of 40 Hz and 2500 Hz. While the increased values due to propagation are seen up to 600 m from the turbine rotor for 40 Hz in the upwind direction, this value is only 100 m for 2500 Hz.

An interesting observation, that is mostly valid for lower frequencies, is that for a certain distance the upwind levels are higher than those in the downwind case. The reason is similar to that for the crosswind-downwind comparison; the initial refraction and the constructive interference of the direct and the reflected waves. The difference between the upwind and downwind propagation is in line with the conclusions deduced in [17] [18].

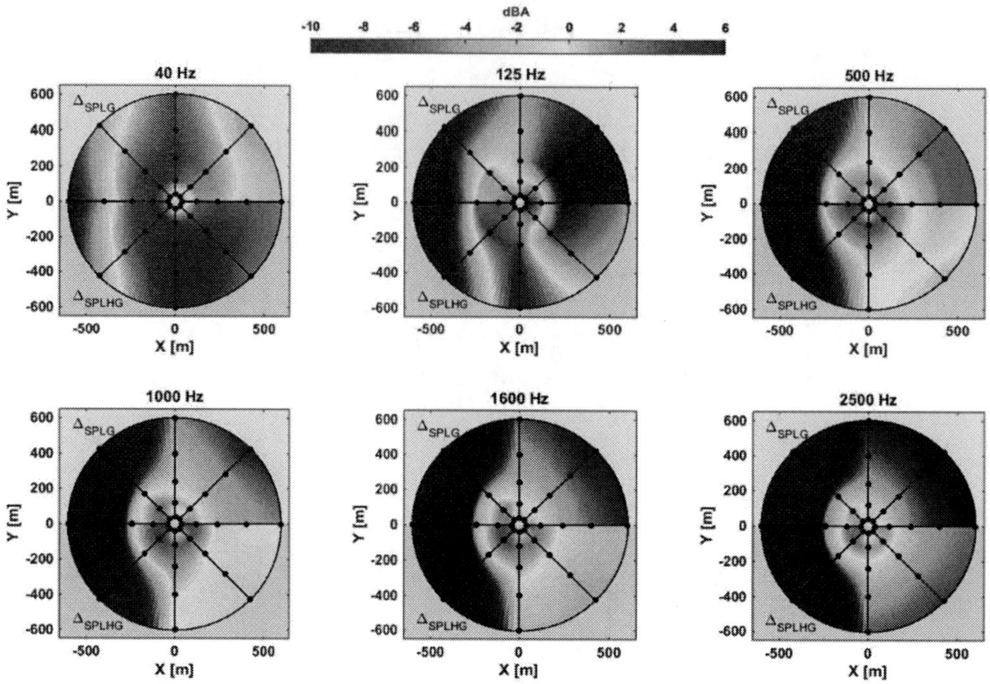

Figure 5: Top view of the frequency dependent time averaged SPL difference. Upper half domain for a grass-covered land: $\Delta_{\mathrm{SPL}_G} = \mathrm{SPL}_{\mathrm{S+P}_G} - \mathrm{SPL}_\mathrm{S}$. Lower half domain for a hard ground: $\Delta_{\mathrm{SPL}_{HG}} = \mathrm{SPL}_{\mathrm{S+P}_{HG}} - \mathrm{SPL}_\mathrm{S}$. Wind direction is from left to right. Black dots represent the receiver locations.

3.2 Comparison of Time Dependent SPLs

As the source-propagation coupling is unsteady, we can investigate the time dependent SPL and look into amplitude modulation. Figure 6 shows spectrograms for a receiver at 2 m height and 45 m downwind of the turbine. The spectrogram difference ($\boldsymbol{\Delta_{\mathrm{SPL}_G} = \mathrm{SPL}_{\mathrm{S+P}_G} - \mathrm{SPL}_\mathrm{S}}$) gives some insight into propagation caused fluctuations, where the propagation effects increase the amplitude, are investigated for all receivers. Some further observations can be deduced (due to the limited length of the paper, the plots are not shown here) as follows:

- At 240 m and 400 m, the downwind locations (receivers 1 and 2, see Fig. 3) have increased levels in the 300-500 Hz band for the hard ground and in the 300-400 Hz band for the grassland.
- Crosswind simulations (in the direction of receiver 3 in Fig 3) show increased noise levels at distances of 240 m, 400 m and 600 m for frequencies less than 300 Hz. The reason for this persistent increase in this frequency range is not fully clear; however it is observed that these levels are much higher for the hard ground.
- Upwind simulations (receivers 4 and 5 in Fig 3) show that there are increased levels at distances up to 400 m, but beyond this distance, there is no increased level. Up to 400 m the increased levels are fairly random and there is not a distinct frequency band that shows consistently higher levels ($\boldsymbol{\Delta_{\mathrm{SPL}}} > 0$).

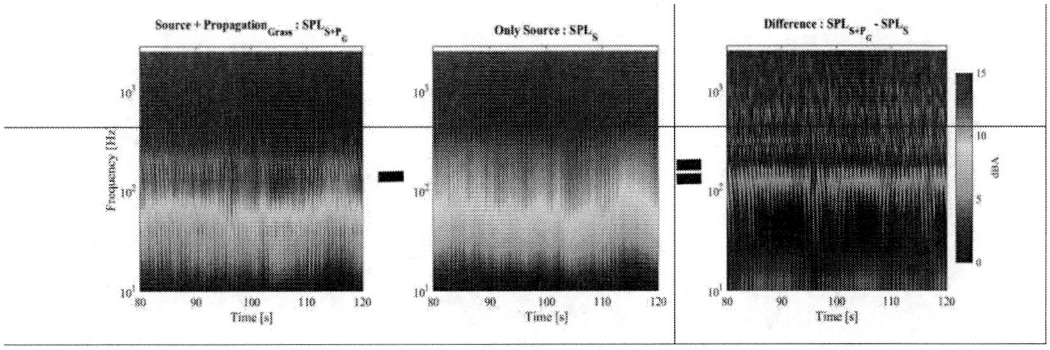

Figure 6: Spectrograms of sound pressure level for Source + Grass propagation (left), Only Source (mid) and Difference (right).

3.3 Comparison of AM Levels

In this study rating AM is carried out via the method proposed by the IOA Noise Working Group on (Wind Turbine Noise) Amplitude Modulation [19]. The method is based on transforming a SPL time series of 10 seconds blocks into the frequency domain in order to detect the fundamental frequency peak and its next two harmonics. After certain checks and filtering, an inverse Fourier transform is applied and the modulation depth is determined from the reconstructed signal.

This method is applied on the three cases described above ($SPL_S, SPL_{S+P_G}, SPL_{S+P_{HG}}$). Detected AM levels in each ten second block are averaged over a four minute simulation. Similar to the differential spectrogram plots, the differences in AM levels between only source and coupled source-propagation simulations are shown in Fig 7. The results indicate a significant increase in AM levels when the propagation effects are taken into account in the far field downwind and upwind of the turbine, and in the near field crosswind. The increase in the near field is believed to be caused by the existence of the ground reflection and destructive/constructive interference. And the increased far field AM levels result from SPL fluctuations caused by turbulence. If we look closely at some of the time series (see Fig 8) the near field crosswind sound pressure of both grassland and hard ground has a much higher AM depth than that of the only source simulations. These distances should be investigated more carefully than the far field sound because the PE method has a limited accuracy in the near field. However, interesting conclusions can be deduced by looking at the time series of downwind and upwind noise in the far field. The time series obtained from receiver number 1 (downwind – 600 m) shows that the relatively small modulations seen for the source only simulations are enhanced with the propagation effects. Most of the time, the modulation depth is higher for both ground types. Additionally there are certain instances the sound pressure levels are higher than those from the source only simulations. Even though this is common for the hard ground, similar observations can also be made for grass-covered ground. The time series obtained from receiver number 2 (crosswind – 40 m) shows a very distinct and periodic increase in the modulation depth. As aforementioned this is essentially the result of ground reflection. The time series obtained from receiver number 3 (upwind – 400 m) shows SPL attenuation even though the modulation depths are considerably higher than in source only propagations.

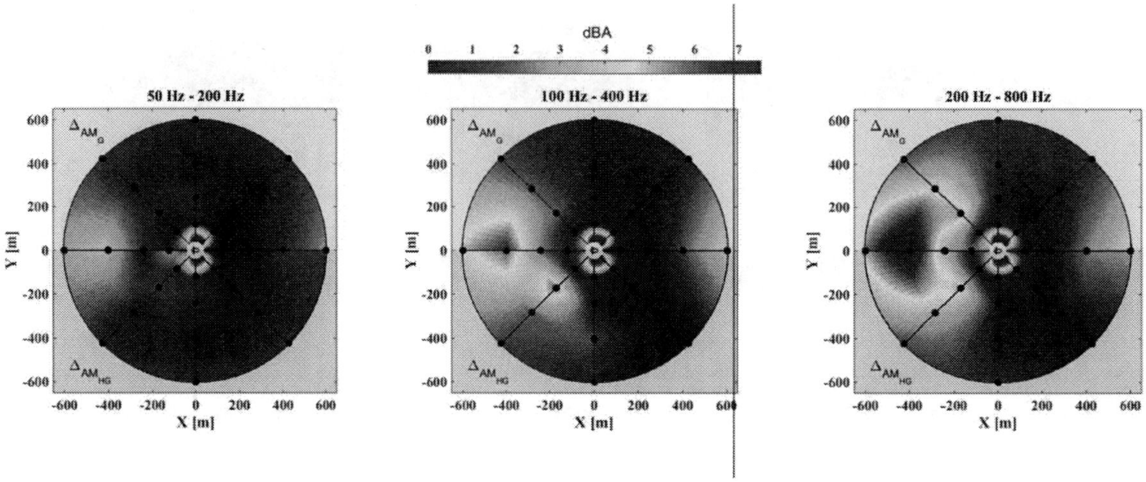

Figure 7: Top view of AM level difference identified for three frequency bandwidths. Upper half domain for the grass-covered land: $\Delta_{\mathrm{AM}_{\mathrm{G}}} = \mathrm{AM}_{\mathrm{S+P_G}} - \mathrm{AM}_{\mathrm{S}}$. Lower half domain for the hard ground: $\Delta_{\mathrm{AM}_{\mathrm{HG}}} = \mathrm{AM}_{\mathrm{S+P_{HG}}} - \mathrm{AM}_{\mathrm{S}}$

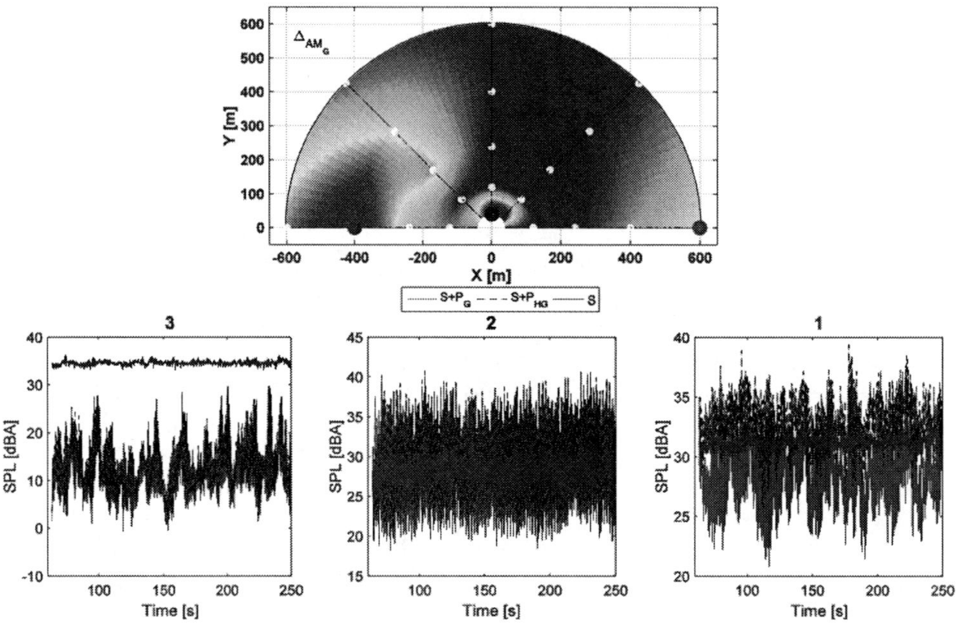

Figure 8: Top: Top view of AM level difference. Bottom: Time series of SPL for three selected receivers enumerated and colour coded in the top plot.

4. Conclusions and Future Work

In this study, an unsteady coupling model has been developed for studying wind turbine noise generation and propagation. Using the solutions from a high fidelity flow solver, the turbine is modelled using an aero-elastic model exposed to a realistic atmospheric flow and the wind turbine noise levels at various receivers are calculated based on a semi-empirical noise model. Subsequently simulations of propagation based on parabolic equation method are conducted with the corresponding flow fields. The conclusion of the present study is that the propagation effects such as refraction, absorption and reflection mostly decrease the time averaged SPL in the far field. However there is a considerable increase of AM levels, especially in the downwind and upwind of the turbine. Future work will focus on the effects of hub height wind speed and incoming flow wind shear on AM of a larger turbine and at longer distances using the developed modelling technique.

References

[1] Renewable UK (2016) *Wind Turbine Amplitude Modulation Research to Improve Understanding as to its Cause and Effect*

[2] Oerlemans S (2011) An explanation for enhanced amplitude modulation of wind turbine noise, RenewableUK contract report. NLR-CR-2011-071.

[3] Fischer A, Aagaard Madsen H,Kragh K A and Bertagnolio F (2014) Analyses of the mechanisms of amplitude modulation of aero-acoustic wind turbine sound, Proceedings of EWEA 2014

[4] Lee S, Kim, W Choi and Lee S (2010) *Annoyance caused by amplitude modulation of wind turbine noise, 2011 Institute of Noise Control Engineering*

[5] S. Markkilde Petersen, Wind turbine test NORDTANK NTK 500/37 system test, Annual report year: 1994, Risø National Laboratory for Sustainable Energy, Technical University of Denmark

[6] P. Moriarty, NAFNoise User's Guide (National Wind Technology Center, National Renewable Energy Laboratory, USA,Golden, Colorado, 2005).

[7] NWTC Information Portal (FAST v8). https://nwtc.nrel.gov/FAST8

[8] N Ramos-García, J N Sørensen, W Z Shen (2014) A strong viscous–inviscid interaction model for rotating airfoils, Wind Energy Volume 17, Issue 12, Pages 1957–1984

[9] Brooks TF, Pope, DS, Marcolini MA. Airfoil Self-Noise and Prediction. NASA Reference Publication 1218, National Aeronautics and Space Administration, USA, 1989.

[10] R K Amiet, "Acoustic radiation from an airfoil in a turbulent stream," Journal of Sound and Vibration, vol. 41, pp. 407-420, 8/22/ 1975.

[11] M V Lowson, (1993) Assessment and prediction of wind turbine noise," Flow Solutions Ltd., Bristol

[12] P J Moriarty, G Gianfranco, P Migliore (2005) Prediction of Turbulent Inflow and Trailing-Edge Noise for Wind Turbines, 11th AIAA/CEAS Aeroacoustics Conference Monterey, California

[13] J Couniham, 1975, "Adiabatic Atmospheric Boundary Layer: A Reviewed and Analysis of Data from the Period 1880-1972," Atmos. Environ., 9, pp.871–905.

[14] R.A Sack and M. West, A Parabolic Equation for Sound Propagation in Two Dimensions over Any Smooth Terrain Profile, 1994, Applied Acoustics

[15] M.E. Delany and E.N. Bazley, Acoustical properties of fibrous absorbent materials

[16] Kaya O Dag (2017) Development of Large Eddy Simulation Tools for Simulation of Atmospheric Boundary Layer in Wind Farms, PhD Thesis, Technical University of Denmark

[17] Barlas E, Zhu WJ, Shen WZ, Andersen SJ. (2016) Wind turbine noise propagation modelling: An unsteady approach. The Science of Making Torque from Wind. Journal of Physics: Conference Series,

[18] Barlas E, Zhu WJ, Shen WZ, Sørensen JN, Kelly M, Andersen SJ.(2017) Effect of wind turbine wake on atmospheric sound propagation. Applied Acoustics 2017

[19] IOA Noise Working Group (Wind Turbine Noise) Amplitude Modulation Working Group
Final Report A Method for Rating Amplitude Modulation in Wind Turbine Noise 9 Aug 2016 Version 1

7th International Conference
on
Wind Turbine Noise
Rotterdam – 2nd to 5th May 2017

Aeroacoustic simulation of multiple wind turbine source interaction

Xavier Robin & César Legendre
{xavier.robin,cesar.legendre}@fft.be

Free Field Technologies S.A. Rue Emile Franqui 9,
Mont Saint-Guibert 1435 Belgium.

Summary

In this paper, the Linearized Euler Equations (LEE) are used to determine the acoustic propagation under complex conditions such as vertical wind variations due to steep terrains or atmospheric temperature gradients. This set of equations is solved by the high-order adaptive Discontinuous Galerkin (DG) scheme in the time domain due to its ability to accurately represent the phenomena involved and its high parallel scalability allowing the analysis of large/high frequency problems at acceptable computational costs from an industrial perspective. Moreover, the analysis is carried out by means of the Actran DGM code using GPU acceleration to further speed-up the solution. With respect to the aeroacoustic noise source, a simplified methodology is used to synthetize the source using an analytical propeller source model based on thickness (monopolar) and loading (dipolar) noise contributions. Despite its simplicity, this methodology allows rapidly comparing the acoustic performance of different wind turbine designs. Thereafter, in this work the solution of multiple turbine sources interaction is presented, including additional effects both in the near field and in the far field: (i) representative topological characteristics of the terrain; and (ii) realistic atmospheric flow conditions. Finally, the noise mitigation effects, due the desynchronization of wind turbines intended to prevent high amplitude modulation (AM) peaks, are assessed.

1. Introduction

Wind energy has become the second largest form of power generation capacity in Europe with almost 300 TWh generated in 2016 covering 10.4 % of the EU's electricity demand [**1**]. This sector is growing creating new jobs and new market opportunities. With this expansion wind farms began to be erected near populated areas and complains have emerged of negative effects on health [**2**]. Therefore, predictions methods to assess the noise of wind farms in the near areas could help improving the environmental impact of this technology. On the other hand, with the progresses in numerical methods, computer science and modeling techniques, it is now possible to obtain acoustic predictions in large domains at relatively high frequencies [**3**].

The present article describes an efficient implementation of the Linearized Euler Equations (LEE) using a high-order adaptive Discontinuous Galerkin (DG) scheme in time domain [**3**] with GPU acceleration. Rotating clouds of point sources are imposed on the right-hand side (r.h.s) of the momentum (dipoles) and mass (monopoles) equations. The cloud of point sources is synthesized using the propeller noise theory from the aerodynamic data computed using a vortex panel method [**4**] and a semi-empirical model of wall pressure function (WPF) [**5**,**6**] while ActranDGM is used as the acoustic solver. Since the DG method operates with large-size elements of high p-order interpolation, the point sources are sub-sampled inside the elements to take into account the large variations due to the higher harmonics of the propeller. The advantages that represent the implementation of the LEE in a DG context are the following: (i) the p-adaptivity allows to use the same mesh for different wave frequencies adjusting dynamically the elements' order; (ii) to explore new physics in propagation and generation of sound using the LEE, i.e. complex weather conditions such as strong cross-wind or temperature gradients; and (iii) high parallel scalability suited to large acoustic problems and high frequency, characteristic of wind turbine noise in exterior conditions. The case selected for this study consists of two generic three blades wind turbines as described by NREL [**10**] in exterior conditions over a realistic land with hills and valleys. Realistic flow conditions are taken into account. The calculation focuses on the blade passing frequency and its harmonics which are responsible for the amplitude modulation phenomenon commonly recognized as a source of annoyance even at large distance [**11**]. The results of the numerical calculation are discussed followed by some concluding remarks and possible further investigations.

2. Theory

2.1 Propeller source model

Noise from wind turbines can be described by three types of noise: (i) trailing edge noise due to the free turbulence and flow unsteadiness generated by the airfoil downstream; (ii) broadband noise due to the turbulent flow field generating a pressure load over the blades; and (iii) tonal sources due to complex interactions of the flow at the blade passing frequency (BPF) and its harmonics. The two latter are characterized by monopole and dipole contributions. In this section, a simplified model of propeller noise is presented. The method defines a series of rotating point sources, i.e. monopoles and dipoles, which serve to characterize the thickness and loading noise respectively [**7**,**8**,**9**]. To determine the amplitude and phase of the thickness and loading noise, the model makes use of: (i) the geometric characteristics of the blade to compute the mass flow rate of the fan (thickness noise); (ii) the aerodynamic pressure over the blade (mean and fluctuating) to determine the momentum sources (loading noise). Finally, the mean and fluctuating pressures are computed by two different methods. The mean pressure over the blade is computed using a vortex panel method [**4**] while the fluctuating pressure is estimated by a semi-empirical method (WPF) to model a realistic turbulent boundary layer excitation using only a few aerodynamic parameters of the blade [**5**,**6**].

2.1.1 Description of the method

The method consists in a cloud of rotating point sources for the acoustic contributions of the individual propeller blades with angular velocity $\mathbf{\Omega}$ located at a certain distance $\mathbf{r}$ from the rotation centre 0^*, as depicted in Figure 1a. Two types of point sources are considered: (i) mass (or monopole) contributions for modelling the thickness noise; (ii) momentum (or dipole) contributions for modelling the loading noise.

For the thickness noise, given a propeller in axial configuration as depicted in Figure 1a with N blades rotating at angular speed $\mathbf{\Omega}$. The acoustic response of the mass displaced by the N blades can be modelled as a cloud of n rotating monopoles (Figure 1b) in the following form:

$$Q_\rho = \frac{N\,M_b}{n} \sum_{i=1}^{n} A_i(t, \mathbf{\Omega}, \mathbf{x}_i)\ \delta(\mathbf{x}_i), \tag{1}$$

to be applied to the r.h.s. of the mass conservation equation, in which, $\delta(\mathbf{x}_i)$ is a generic point monopole source placed at the position $\mathbf{x}_i$, M_b is the volume displaced by the blade.

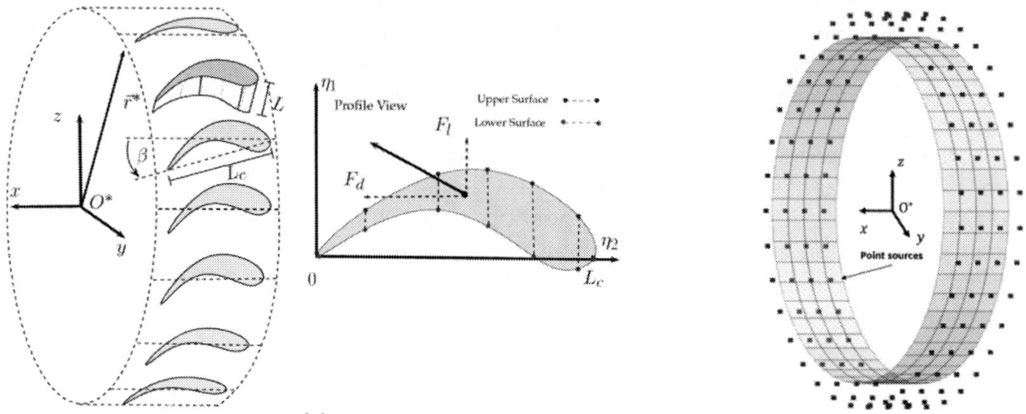

(a) (b)

Figure 1: (**a**) Scheme of an axial propeller, (**b**) cylindrical distribution of points sources.

The mass amplitudes $A_i(t, \mathbf{\Omega}, \mathbf{x}_i)$ of each source depend on time and space to properly take into account: (i) the phase shift between the N blades in the propeller; and (ii) the correct amplitude at the BPF (Blade Passing Frequency) and harmonics in terms of the angular frequency of the propeller $\mathbf{\Omega}$ and the local pitch angle of the blade β. The modelling of $A_i(t, \mathbf{\Omega}, \mathbf{x}_i)$ intends to take into account the rotation of points sources due to the movement of the propeller.

Concerning the loading noise, let's consider the same propeller in axial configuration with N blades rotating at angular speed $\mathbf{\Omega}$ as previously presented. The force induced by the N blades in the fluid can be modelled as a cloud of n rotating dipoles in the following form:

$$\boldsymbol{Q}_m = \frac{N}{n}\sum_{i=1}^{n} \boldsymbol{F}_i(t, \mathbf{\Omega}, \mathbf{x}_i)\, \delta(\mathbf{x}_i), \tag{2}$$

to be applied to the r.h.s. of the momentum conservation equation, in which, $\delta(\mathbf{x}_i)$ is a generic point source placed at the position $\mathbf{x}_i$, $\boldsymbol{F}_i(t, \mathbf{\Omega}, \mathbf{x}_i)$ are the local force amplitudes of each source depending on time and space. The force amplitudes $\boldsymbol{F}_i(t, \mathbf{\Omega}, \mathbf{x}_i)$ are modelled to take into account the rotation of dipole sources due to the movement of the propeller. Moreover, the local force $\boldsymbol{F}_i(t, \mathbf{\Omega}, \mathbf{x}_i)$ may be defined as the local surface over the blade $\Delta \boldsymbol{a}_i$ times the local aerodynamic wall pressure $P_i(t, \mathbf{\Omega}, \mathbf{x}_i)$ as follows:

$$\boldsymbol{F}_i(t, \mathbf{\Omega}, \mathbf{x}_i) = P_i(t, \mathbf{\Omega}, \mathbf{x}_i)\, \Delta \boldsymbol{a}_i = F_l\,(t, \mathbf{\Omega}, \mathbf{x}_i)\frac{\boldsymbol{r}^*}{|\,\boldsymbol{r}^*\,|} + F_d\,(t, \mathbf{\Omega}, \mathbf{x}_i)\frac{\boldsymbol{r}^* \times \mathbf{\Omega}}{|\,\boldsymbol{r}^*\,||\,\mathbf{\Omega}\,|}, \tag{3}$$

decomposed in a lift force F_l in the radial direction $\boldsymbol{r}^*/\,|\,\boldsymbol{r}^*\,|$ and a drag force F_d in the tangential direction. Furthermore, the local pressure over the blades $P_i(t, \mathbf{\Omega}, \mathbf{x}_i)$ can be decomposed as following:

$$P_i(t, \mathbf{\Omega}, \mathbf{x}_i) = P_i^0(\mathbf{\Omega}, \mathbf{x}_i) + P_i''(t, \mathbf{\Omega}, \mathbf{x}_i), \tag{4}$$

a mean pressure depending only in space P_i^0 and a fluctuating pressure P_i'' depending in both time and space. The choice of computing the aerodynamic pressure over the blades will condition the accuracy of the method. For instance, CFD LES[1] or DNS[2] simulations can be used to estimate the wall pressure over the blade. The propeller source model is independent of the

[1] Large Eddy Simulation.
[2] Direct numerical simulation.

computational method used to estimate the pressure over the blade's surface. In this work, the mean pressure is computed using XFOIL [4], a vortex panel method code for subsonic airfoils. On the other hand, the fluctuating pressure is estimated using Goody's semi-empirical model for the wall pressure function. Goody's model consists in a cross-correlation function in space combined with scaling laws for the frequency spectrum to estimate the pressure fluctuations over the wall due to a turbulent boundary layer [5,6].

2.2 Wave operator

For computing the acoustic response of the propeller source model proposed in section §2.1, in this work the linearized Euler equations (LEE) in non-conservative and homentropic form are considered. In addition, notice that index notation is used for the sake of compactness. Therefore, imposing homentropy in the whole domain ($\nabla s = 0 \rightarrow c_0^2 \rho = p$) the mass and momentum equations with generic mass and momentum source terms are written as follow:

$$\frac{\partial \rho'}{\partial t} + \frac{\partial}{\partial x_i}\left(\rho' v_i^0 + \rho_0 v_i'\right) = Q_\rho, \tag{5}$$

$$\frac{\partial v_i'}{\partial t} + \frac{\partial}{\partial x_j}\left(v_j^0 v_i' + c_0^2 \delta_{ij}\frac{\rho'}{\rho_0}\right) + v_j'\frac{\partial v_i^0}{\partial x_j} - v_i'\frac{\partial v_j^0}{\partial x_j} + \frac{\rho'}{\rho_0^2}\left[c_0^2\frac{\partial \rho_0}{\partial x_i} - \frac{\partial p_0}{\partial x_i}\right] = \boldsymbol{Q}_m, \tag{6}$$

where δ_{ij} is Kronecker's delta, c_0 is the speed of sound, s is the entropy and Q_ρ and $\boldsymbol{Q}_m$ are defined by equations (1) and (2) respectively. In addition ρ, p and v_i are the density, pressure and velocity respectively. Prime terms are used for the acoustic fluctuations and zero (sub-zero or over-zero) terms for the mean flow values. It is worth noticing that equations (5) and (6) include complex mean flow effects that may be present in the acoustic propagation at large distances, for example, atmospheric temperature gradients and strong cross-wind conditions over hills and valleys.

2.2.1 Discontinuous Galerkin (DG) method

To establish the description of the DG method, let define the array of variables $\mathbf{q}$, flux matrix $\bar{\bar{\mathbf{F}}}$, array of volume terms $\mathbf{s}$ and the array of sources $\boldsymbol{R}$ the following form:

$$\boldsymbol{q} = \begin{bmatrix} \rho \\ u' \\ v' \\ w' \end{bmatrix}, \quad \boldsymbol{R} = \begin{bmatrix} Q_\rho \\ \boldsymbol{Q}_m^x \\ \boldsymbol{Q}_m^y \\ \boldsymbol{Q}_m^z \end{bmatrix}, \quad \bar{\bar{\mathbf{F}}}_j = \begin{bmatrix} v_j^0 & \rho_0\delta_{1j} & \rho_0\delta_{2j} & \rho_0\delta_{3j} \\ c_0^2\dfrac{\delta_{1j}}{\rho_0} & v_j^0 & 0 & 0 \\ c_0^2\dfrac{\delta_{2j}}{\rho_0} & 0 & v_j^0 & 0 \\ c_0^2\dfrac{\delta_{3j}}{\rho_0} & 0 & 0 & v_j^0 \end{bmatrix}, \tag{7}$$

$$\mathbf{s} = \begin{bmatrix} 0 & 0 & 0 & 0 \\ \dfrac{1}{\rho_0^2}\left(\dfrac{\partial p_0}{\partial x} - c_0^2\dfrac{\partial \rho_0}{\partial x}\right) & \dfrac{\partial v_j^0}{\partial x_j} - \dfrac{\partial u_0}{\partial x} & -\dfrac{\partial u_0}{\partial y} & -\dfrac{\partial u_0}{\partial z} \\ \dfrac{1}{\rho_0^2}\left(\dfrac{\partial p_0}{\partial y} - c_0^2\dfrac{\partial \rho_0}{\partial y}\right) & -\dfrac{\partial v_0}{\partial x} & \dfrac{\partial v_j^0}{\partial x_j} - \dfrac{\partial v_0}{\partial y} & -\dfrac{\partial v_0}{\partial z} \\ \dfrac{1}{\rho_0^2}\left(\dfrac{\partial p_0}{\partial z} - c_0^2\dfrac{\partial \rho_0}{\partial z}\right) & -\dfrac{\partial w_0}{\partial x} & -\dfrac{\partial w_0}{\partial y} & \dfrac{\partial v_j^0}{\partial x_j} - \dfrac{\partial w_0}{\partial z} \end{bmatrix}, \tag{8}$$

so that equations (5,6) may be written in a compact form as follows:

$$\frac{\partial \boldsymbol{q}}{\partial t} + \frac{\partial}{\partial x_j}\left(\bar{\bar{\mathbf{F}}}_j . \boldsymbol{q}\right) - \boldsymbol{s}.\boldsymbol{q} = \boldsymbol{R}. \tag{9}$$

For the variational formulation, considering equation (9) being multiplied by the shape functions $\boldsymbol{N}_\alpha$ and integrated over the volume Ω, this results in:

$$\int_\Omega N_\alpha \frac{\partial \boldsymbol{q}}{\partial t} d\Omega + \int_\Omega N_\alpha \frac{\partial}{\partial x_j}(\overline{\overline{\mathbf{F}}}_j . \boldsymbol{q}) d\Omega - \int_\Omega N_\alpha \boldsymbol{s}.\boldsymbol{q} d\Omega = \int_\Omega N_\alpha \boldsymbol{R}\, d\Omega. \tag{10}$$

Integrating by parts the second term of equation (10) to transfer the derivative in space $\partial/\partial x_j$ to the shape functions N_α, equation (10) results in:

$$\int_\Omega N_\alpha \frac{\partial \boldsymbol{q}}{\partial t} d\Omega = \int_\Omega \frac{\partial N_\alpha}{\partial x_j} \overline{\overline{\mathbf{F}}}_j . \boldsymbol{q} d\Omega + \int_\Omega N_\alpha \boldsymbol{s}.\boldsymbol{q} d\Omega - \oint_{\partial\Omega} N_\alpha \overline{\overline{\mathbf{F}}}_j . \boldsymbol{q}\, n_j\, d\Gamma + \int_\Omega N_\alpha \boldsymbol{R}\, d\Omega\ . \tag{11}$$

In DG methods the solution can be discontinuous from one element to the other. Moreover, the surface integral of equation (11) is computed from the solutions on both sides of the element ensuring a good stability of the numerical method. One of the strengths of the variational formulation presented in equation (11) is that using an explicit time discretization, the linear system becomes block diagonal. Therefore, the linear system inversions at each time step may be avoided and it can be replaced by matrix multiplications. In addition, the last term of equation (11) represents the cloud sources applied to the momentum and mass equations. Since the acoustics (11-l.h.s) and propeller sources (11-r.h.s) are very disparate in terms of characteristic length, the variability of the propeller source terms is taken into account inside element by the high-order shape functions N_α of the numerical algorithm.

3. Wind Turbine and source modelling

A realistic model, the "NREL offshore 5-MW baseline wind turbine" [**10**] has been chosen because it is well described and used in the literature. The following describes the wind turbine geometry, operating conditions and aerodynamic properties. Based on these characteristics the noise sources are calculated according to the theoretical development presented in section §2.1, including thickness noise and loading noise components. The acoustic model is then presented with its characteristics: ground profile, finite element model and outputs.

3.1 Wind turbine specifications and aerodynamic properties

The "NREL offshore 5-MW baseline wind turbine" is described in detail in reference [**10**]. The blade structural and aerodynamic characteristics, hub and nacelle properties, drivetrain, tower and baseline control system properties of the wind turbine are presented. Main characteristics useful for the current acoustic study are underlined in Table 1 including geometry, blade profiles and operating conditions.

Rating	5 MW
Rotor Orientation, Configuration	Upwind, 3 Blades
Control	Variable Speed, Collective Pitch
Drivetrain	High Speed, Multiple-Stage Gearbox
Rotor, Hub Diameter	126 m, 3 m
Hub Height	90 m
Cut-In, Rated, Cut-Out Wind Speed	3 m/s, **11.4 m/s**, 25 m/s
Cut-In, Rated Rotor Speed	6.9 rpm, **12.1 rpm**
Rated Tip Speed	**80 m/s**
Overhang, Shaft Tilt, Precone	5 m, 5°, 2.5°
Rotor Mass	110,000 kg
Nacelle Mass	240,000 kg
Tower Mass	347,460 kg
Coordinate Location of Overall CM	(-0.2 m, 0.0 m, 64.0 m)

Table 1: NREL 5-MW general characteristics from [**10**]

The hub height of the wind turbine is 90m and the blade radius is equal to 63m. The rated operating condition is considered in the study: rated wind speed is 11.4 m/s, rotor speed is 12.1 RPM and tip speed 80 m/s.

Seventeen blade elements are used to describe the profiles of the blade (Table 2). The three inboard and three outboard elements are two-thirds the size of the eleven equally spaced mid-span elements. The chord of each individual blade element is reported as well as their shape listed as profile names. In Figure 2, the topology of the blade with the highlighted sections are presented.

Section	Span mid-location	Span size	Chord	Airfoil Table (see annex 1)	Span rotating speed	Span visible speed	Re	Attack angle
-	m	m	m	(-)	m/s	m/s [Mach]	-	°
1	2.8667	2.7333	3.542	Cylinder1	3.6	12.0 [0.04]	2.81E+06	59.0
2	5.6	2.7333	3.854	Cylinder1	7.1	13.4 [0.04]	3.43E+06	44.8
3	8.3333	2.7333	4.167	Cylinder2	10.6	15.5 [0.05]	4.29E+06	33.9
4	11.75	4.1	4.557	DU40_A17	14.9	18.8 [0.06]	5.66E+06	24.1
5	15.85	4.1	4.652	DU35_A17	20.1	23.1 [0.07]	7.11E+06	18.1
6	19.95	4.1	4.458	DU35_A17	25.3	27.7 [0.08]	8.19E+06	14.1
7	24.05	4.1	4.249	DU30_A17	30.5	32.5 [0.10]	9.16E+06	11.5
8	28.15	4.1	4.007	DU25_A17	35.7	37.4 [0.11]	9.94E+06	9.9
9	32.25	4.1	3.748	DU25_A17	40.9	42.4 [0.12]	1.05E+07	9.0
10	36.35	4.1	3.502	DU21_A17	46.1	47.4 [0.14]	1.10E+07	8.5
11	40.45	4.1	3.256	DU21_A17	51.3	52.5 [0.15]	1.13E+07	8.4
12	44.55	4.1	3.01	NACA64_A17	56.4	57.6 [0.17]	1.15E+07	8.3
13	48.65	4.1	2.764	NACA64_A17	61.6	62.7 [0.18]	1.15E+07	8.2
14	52.75	4.1	2.518	NACA64_A17	66.8	67.8 [0.20]	1.13E+07	8.2
15	56.1667	2.7333	2.313	NACA64_A17	71.2	72.1 [0.21]	1.10E+07	8.2
16	58.9	2.7333	2.086	NACA64_A17	74.6	75.5 [0.22]	1.04E+07	8.3
17	61.6333	2.7333	1.419	NACA64_A17	78.1	78.9 [0.23]	7.42E+06	8.2

Table 2: NREL 5-MW blade characteristics from **[10]** and current paper authors

Due to the rotation of the blade, the local speed of the flow is computed depending on the span radius. For each portion of the blade, the local Reynolds number is obtained based on the chord length, with local flow speed and kinematic viscosity of the air at 20°C. Finally, the angle of attack of each of the seventeen blade elements is also reported in Table 2.

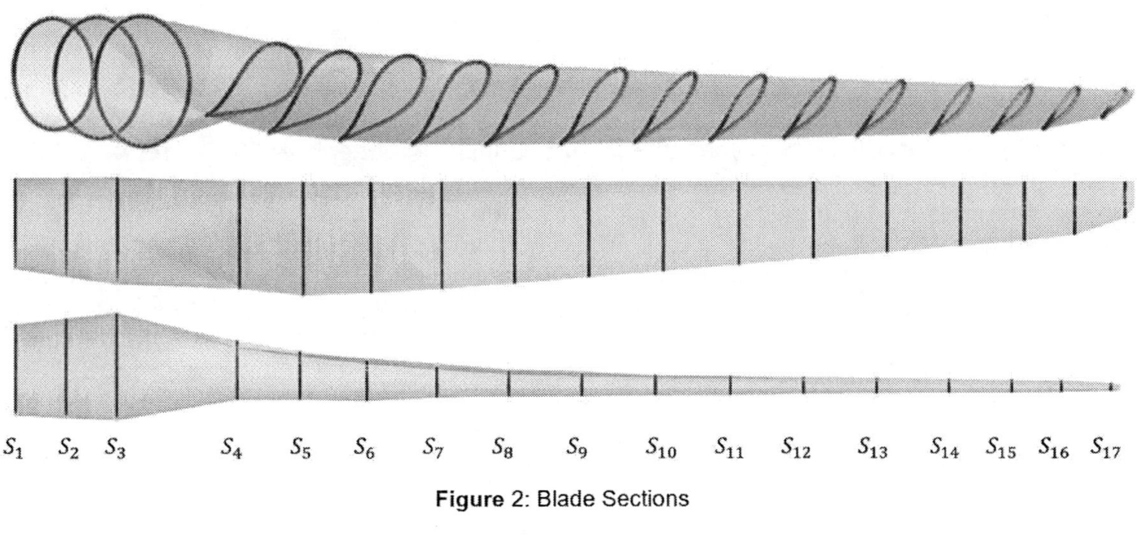

Figure 2: Blade Sections

3.2 Calculation of the noise sources

Both thickness noise and loading noise are handled in the calculation. Thickness noise is obtained thanks to the blade profile (coordinates over the chord). Loading noise requires additional information as it is based on the difference of pressure on each side of the blade. The airfoil profiles, Reynolds and Mach numbers and angle of attack are used to compute the pressure coefficient C_p for each of the seventeen sections. The computation is performed using XFOIL v 6.99, airfoil design tool based on the vortex panel method [**4**]. Moreover, the pressure on blade surface is derived from the pressure coefficient C_p by the following relation:

$$C_p = \frac{p - p_\infty}{\frac{1}{2}\rho V_\infty^2}, \qquad (12)$$

where C_p is the pressure coefficient, p is the aerodynamic pressure over the blade surface and p_∞, V_∞ are the pressure and velocity at upstream conditions. Note that the first three sections have a cylindrical and therefore a symmetric shape; as a consequence from an aerodynamic point-of-view, the pressure conditions in the upper and lower surfaces are symmetric. Besides, since these sections are placed near the centre of rotation, their contribution to the aerodynamic noise may be neglected. For the aforementioned reasons, these sections are not considered in the computation of thickness and loading noise.

The noise sources are represented as a distribution of point sources in space (Figure 1b). The number of points plays an important role in the source representation and can be arbitrarily chosen for the span direction, profile thickness and azimuthal direction. For the current application, the same number of points is chosen for each airfoil section: (i) 5 points are used in span width; (ii) 50 points in profile thickness; and (iii) 1800 in azimuth direction. For a total of 15.3×10^6 sources. Point cloud sources at the three first harmonics are displayed in Figure 3. It can be observed the source's variation over space, for instance, the first harmonic (BPF) contains six lobes corresponding to positive and negative pulses of each blade; the second and third harmonics contains 12 and 24 lobes respectively in a similar configuration.

4. Acoustic Modelling

The NREL 5-MW wind turbine is an offshore model, representative of typical utility-scale land- and sea-based multi-megawatt turbines. For the sake of simplicity, the acoustic model is represented by a realistic topography in onshore conditions as depicted in Figure 4a, although this wind turbine model is not precisely designed for such installations. The topography of the ground describes hills and valleys on a circular region of 1.5 km of diameter and 1.77 km² of surface, Figure 4c,d. Some buildings are added to increase the realism of the simulation. The noise is propagated in free field conditions besides the ground and the acoustic pressure is recorded at building locations. Finally, a flow field over the ground is also taken into account to assess convective effects, wind gusts of 11.4m/s (41 km/h) from south to north are considered (Figure 4b) . In this case, the flow is computed using a potential flow analysis and depends only in space as depicted in Figure 4b.[3] The propagation domain is composed of air at standard conditions $\rho_0 = 1.225\ kg/m^3$ and speed of sound $c_0 = 340\ m/s$.
The size of the finite elements is designed to capture the BPF and nine next harmonics (Table 3). Two additional frequencies are added between each harmonic to handle a broadband component leading to 28 computational frequencies.

---

[3] The method proposed in this work is not restricted to a potential mean flow. A RANS or LES CFD analysis could be used to estimate more realistic flow conditions and atmospheric temperature gradients representative of earth flow profiles.

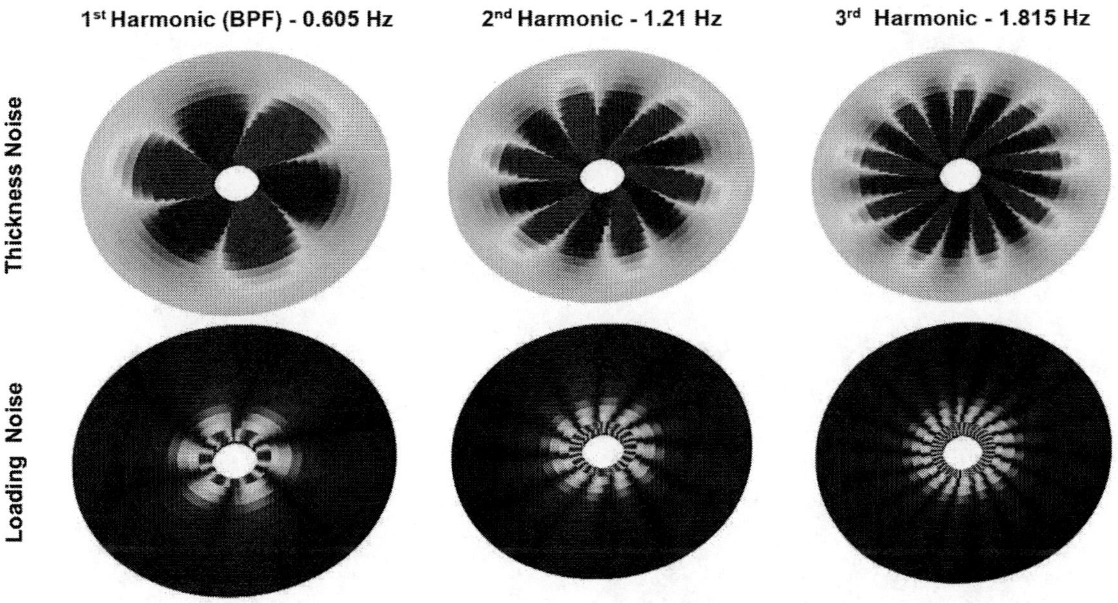

Figure 3: Point cloud source for thickness and loading components at the BPF and the 2 next harmonics. Arbitrary and different scale are used for both types of noise.

The broadband noise is computed using the wall pressure function (WFP) and the cross-correlated space dependency of Goody's model [**5**,**6**]. These estimation give a background pink noise 20 to 30 dB lower than the tonal components and rapidly decaying at higher frequencies.

Harmonic	**1 (BPF)**	**2**	**3**	**4**	**5**	**6**	**7**	**8**	**9**	**10**
Frequency [Hz]	0.605	1.21	1.815	2.42	3.025	3.63	4.235	4.84	5.445	6.05

Table 3: Tonal components used in the simulation.

For the acoustic numerical prediction a time domain model solved by means of ActranDGM solver is used. The *ActranDGM* model solves the Linearized Euler Equations (LEE) with source terms (point sources) in the momentum and mass equations (sections §2.1-2.2). The mesh used in the ActranDGM computation consists in (Figure 5a): (i) an acoustic domain corresponding to a semi-ellipsoid volume covering the ground surface (dark blue domain, Figure 5a); (ii) a buffer zone to damp the sound waves propagated out the physical domain (turquoise domain, Figure 5a); and (iii) non-reflecting boundary conditions surrounding the buffer zone. All the domains necessary for the above model form a mesh of 135,641 3D-Tetrahedra with 4632 2D-Triangular elements.

To be able to handle large domain computations at relatively high frequency, an implementation of the LEE using DG method in time domain with p-adaptivity is considered. This present three main advantages: (i) since the numerical method is explicit, a low RAM consumption is necessary; (ii) in time domain, several frequencies components of the source can be injected simultaneously; and (iii) the current implementation of the solver benefits of a GPU acceleration reaching high parallelism and therefore reduction in computational time. In addition, p-adaptivity allows using the same mesh for different wave frequencies adjusting dynamically the elements' order to capture the smallest acoustic wavelength. In the present case, the mesh size target is 50m with element's orders ranging from 1 to 16. Finally, the sources injected at the r.h.s of the momentum and mass equation contain the 28 frequency components, all injected simultaneously in time domain. The physical duration of the simulation is 15s and then frequency results are obtained from the time solution using a DFT least square method.

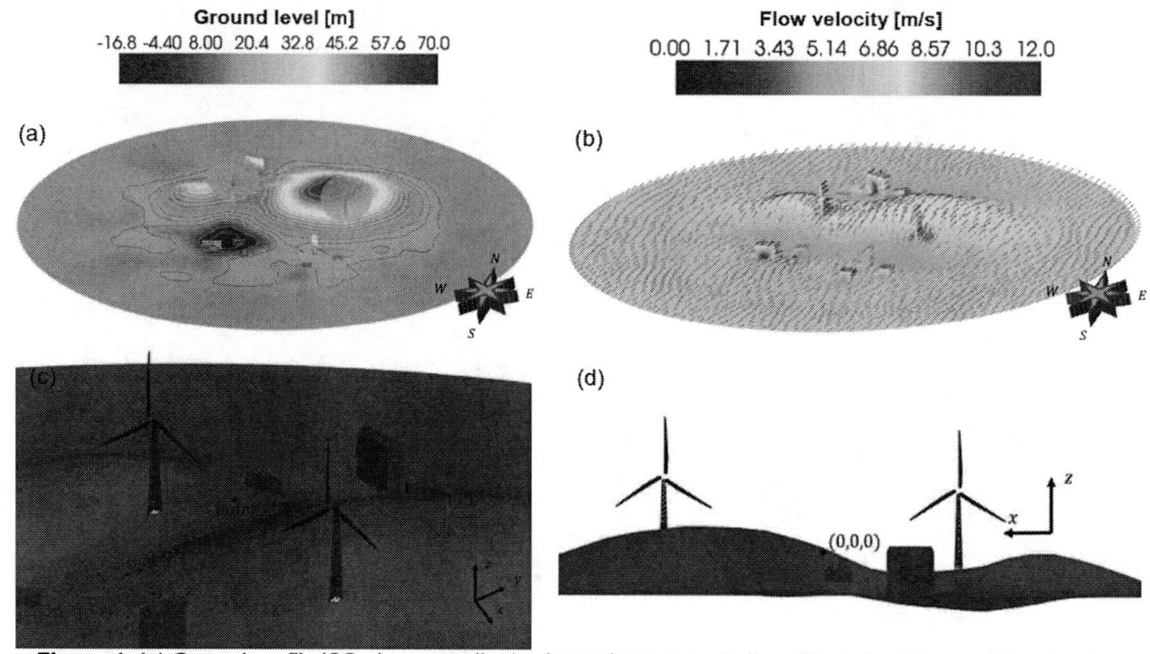

Figure 4: (**a**) Ground profile ISO view, coordinates legends represents the altitude in meters and in grey the area of the point cloud sources (two turbines). (**b**) Mean flow velocity over the ground in m/s. (**c,d**) Detail on the ground and turbines, the origin of the computation domain is depicted. A compass has been added to (**a,b**) as reference.

The analysis is performed in three configurations:

- First configuration consists in the noise propagation of the wind turbine referred as **WT1** and centred at the position (205.5, -71.0, -122.0) m from the origin (Figure 5b).
- Second configuration consists in the noise propagation of the wind turbine referred as **WT2** centred at the position (-126.8, 95.8, 108.57)m from the origin. A shift in phase of 90° is applied with respect to **WT1** (Figure 5b).
- Third configuration consists in the noise propagation of **WT1** and **WT2** combined.

Both wind turbines are aligned for winds south-north as depicted in Figure 5b. Sound pressure level is recorded at two locations near buildings referred **Mic1** and **Mic2** located at (180.08, -243.77, 7.98) m and (-92.89,-107.43,-8.28) m from de origin respectively as depicted in Figure 5b.

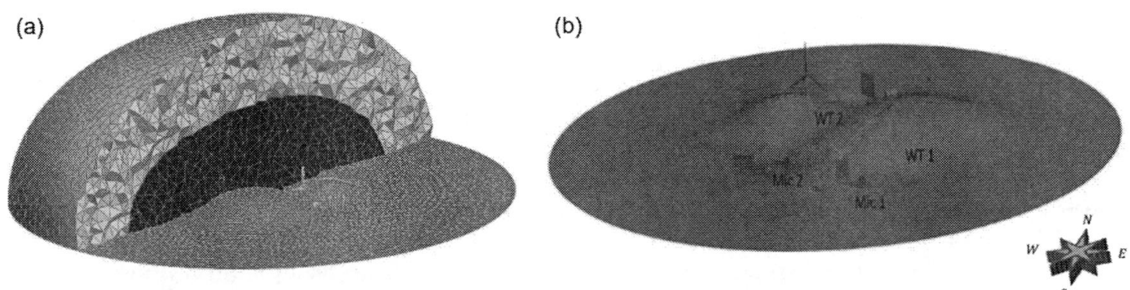

Figure 5: (**a**) Mesh used in the DG computation. The Blue zone represents the computational (acoustic) domains and the turquoise zone the buffer zone for emulating a non-reflecting boundary condition. (**b**) Identification of the wind turbines and location of the microphones. A compass has been added as reference.

5. Results and discussions

The time domain results corresponding to the acoustic pressure in [Pa] recorded at **Mic1** and **Mic2** for all configurations are presented in Figures 6a,b. It can be observed that all the curves start with a transient low amplitude until they reach an established periodic regime. The acoustic pressure recorded at location **Mic1** shows a predominant presence of the noise produced by **WT1** (blue line, Figure 6a) with a little contribution of **WT2** (green line, Figure 6a) to the combined noise perceived from both **WT1** and **WT2** (red line, Figure 6a). This is a consequence of the recording position of **Mic1**, which is closer to **WT1** (217m) than **WT2** (448m). In addition, the combined noise perceived at **Mic1** (red line, Figure 6a) shows the acoustic signature of the noise coming from the wind turbines, the signal present a period of 1.652s with an asymmetric non-sinusoidal waveform rapidly increasing in amplitude during the first 1/5 of the period by ending with a slower decay for the rest (similar to a sawtooth waveform).

(a)

(b)

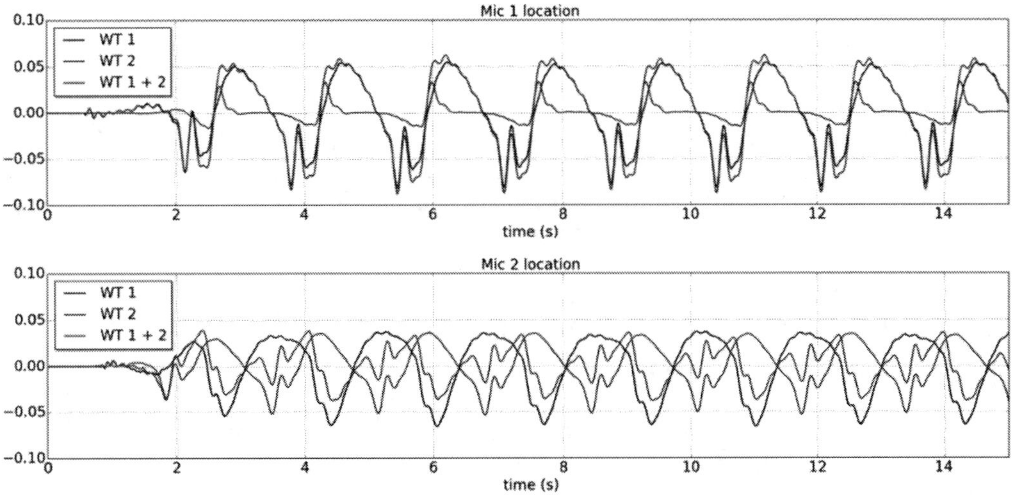

Figure 6: Pressure (Pa) vs. time (s) recorded at **Mic1** (**a**) and **Mic2** (**b**) locations.

In addition, Figure 6b shows the acoustic pressure recorded at location **Mic2**; both **WT1** (blue line) and **WT2** (green line) present a similar participation in amplitude with a notable difference in phase, this is because the more equidistant position of **Mic2** (in comparison with **Mic1**), i.e. 321m from **WT1** and 236m from **WT2**. Furthermore, the combined pressure signal recorded at **Mic2** (red line Figure 6b) present also an asymmetric non-sinusoidal waveform. It might be noticed that both combined pressure signals recorded at **Mic1** and **Mic2** (red lines Figures 6a,b) present small oscillations at higher frequencies typically characteristic of amplitude modulation (AM) noise. This is due by the presence of higher harmonics and broadband components (inter harmonic frequencies) in the modelling of the noise sources.

Following the same line or reasoning, Figures 7a,b shows the frequency results for the sound pressure level (SPL) in [dB] (P_{ref} = 20μPa) recorded at **Mic1** and **Mic2** for all configurations. As in time results (Figures 6a), it may be observed in Figures 7a,b that **WT1** (blue line) predominantly contributes to the overall SPL in **Mic1** with respect to **WT2** (green line) with a large difference at the first harmonic (0.605 Hz) of 10dB. Then, the combined SPL (red line, Figure 7a) has a dominant participation of the first harmonic (75 dB, 0.605 Hz) to subsequently decay on the last harmonic (43 dB, 6.05 Hz).

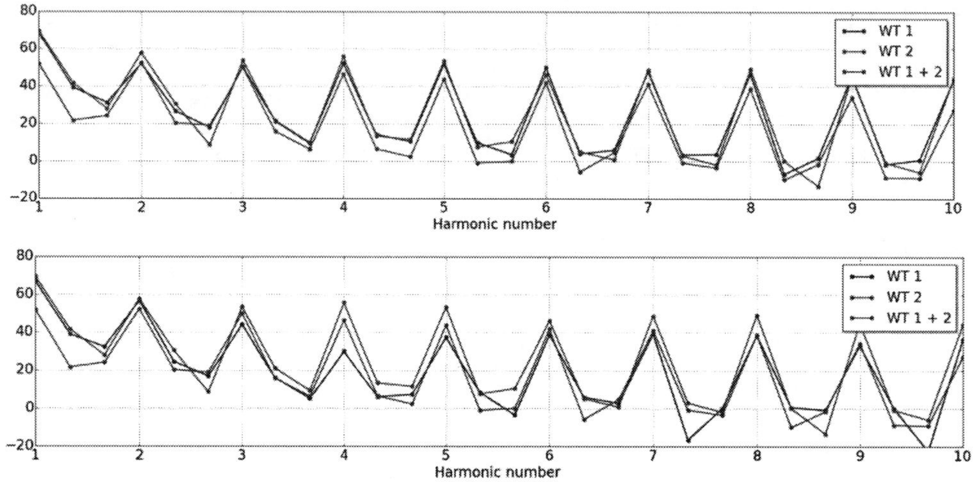

Figure 7: Sound Pressure Level [dB, Pref-2µPa] vs Harmonic number [-] recorded at Mic 1 (a) and Mic 2 (b) locations. The reader may refer to the Table 3 for obtaining the frequencies in Hz.

In addition, the inter-harmonics or broadband have a little contribution to the overall SPL as depicted in Figures 7a,b. The inter-harmonics amplitudes decay in a similar way to the harmonics (1 to 10). For instance, the inter-harmonics in the combined SPL from **WT1** and **WT2** in both Figures 7a,b (red lines) start with amplitude of 41dB (between 1st and 2nd harmonics) to rapidly decay below 0 dB (between 9th and 10th harmonics).

Noise maps help determine the most polluted areas and therefore allow an optimal urban planning reducing the noise impact on people. Consequently, pressure level maps are presented next in Figures 8a-d;9a-f.

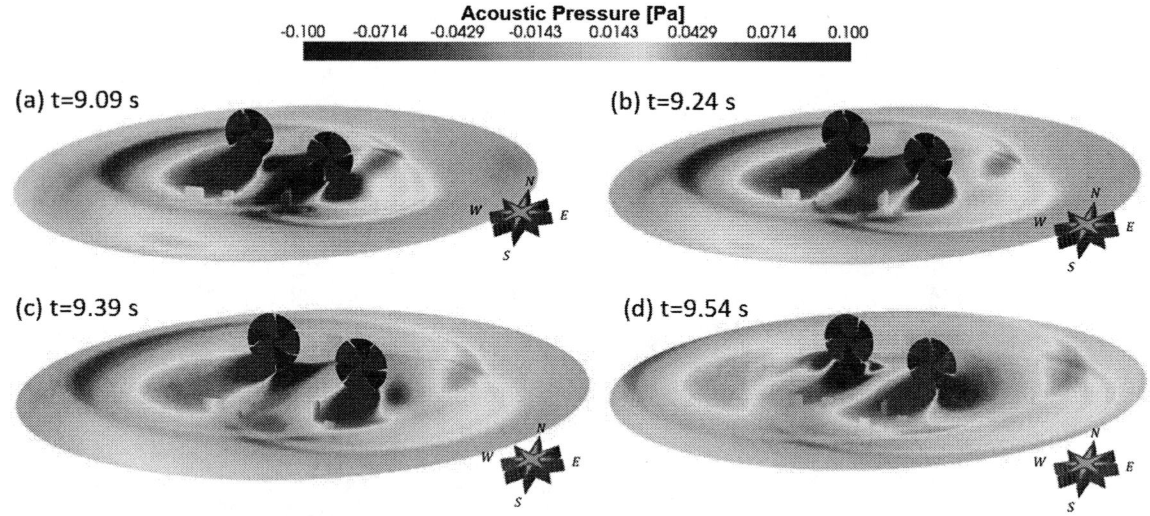

Figure 8: Instantaneous acoustic pressure [Pa] of the time domain simulation. All the frequency content of the sources (28 frequencies) were injected. A compass has been added to each figure as reference.

The pressure maps in time domain are plotted in Figures 8a-d. The values are presented over the ground and in a circular plane at the turbine's locations. It can be observed how the sources generate strong pressure fluctuations near the wind turbines' locations. As time goes by (from t=9.09s in Figure 8a to t=9.54s in Figure 8d) is observed the phase shift between the two sources besides a progressive rotation of 45° over 0.43s over the circular planes near the sources.

Furthermore, it is noticed the wavefront generated at the turbines' location that gradually propagates towards the west and east flanks. On the other hand, in the north and south flanks, the pressure amplitude is notably lower, this is probably due to: (i) the irregularities of the ground; and (ii) the destructive interference of sound emerged from the turbines.

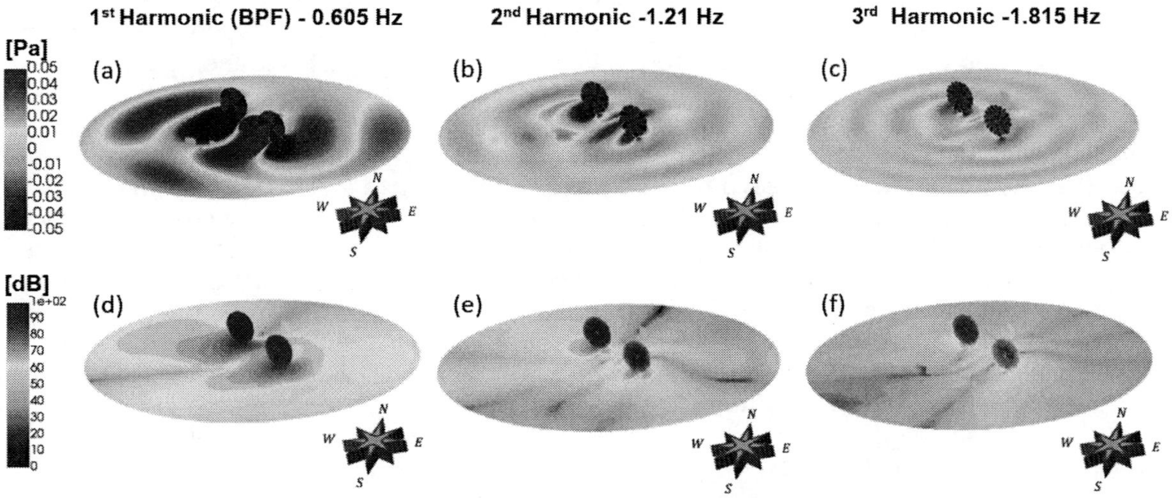

Figure 9: Real part of acoustic pressure [Pa] at different harmonics: (a) 0.605 Hz, (b) 1.21 Hz and (c) 1.815 Hz. Amplitude of acoustic pressure in [dB, Pref 2 μPa]: (d) 0.605 Hz, (e) 1.21 Hz and (f) 1.815 Hz. A compass has been added to each figure as reference.

Frequency maps of the 1st, 2nd and 3rd harmonics corresponding to 0.605, 1.21 and 1.815 Hz respectively are plotted in Figures 9a-f. Concerning the 1st harmonic, the real part of the acoustic pressure (Figure 9a) shows strong variations between the turbines. Similarly, the pressure amplitude in dB depicted in Figure 9d shows a high amplitude area in the vicinity of the two turbines with a narrow zone of silence at the south-west and north flanks. The 2nd harmonic shows large variations on the real part of acoustic pressure on the vicinity of the turbines (Figure 9b). Besides, the pressure amplitude (Figure 9e) shows a reduced zone of high amplitude near the source in comparison to the 1st harmonic, with a wide zone of silence in the south flank with two narrow ones at the north and east. Finally, the 3rd harmonic shows symmetric variations of the real part of acoustic pressure (Figure 9c) with respect to the symmetry axis south-west to north east. The pressure amplitude (Figure 9f) reveals a wide zone of silence in the south-west and north-east flanks.

Some comments about the computational time and memory consumption are finally addressed. The acoustic computations corresponding to the ActranDGM model were performed in 1 GPU accelerators Nvidia Tesla K80 (Kepler architecture) taking in 5h59min with 18.5 Gb of memory consumption in each cases **WT1** and **WT2** (independent wind turbines). When both sources are applied together (**WT1**+**WT2**), the time to process the sources (r.h.s.) in the computational domain increases and computation takes 8h26min with 27.0 Gb of memory consumption for 30.8×10^6 degrees of freedom. It is noticeable the relatively low memory consumption of the DG method to solve the LEE in time domain, this is one of the features that renders DG methods suitable to acoustic problems in large domains and high frequency. Additionally, the current implementation of ActranDGM solver allows parallel computations in multiple CPUs using MPI communicators (not presented here). Loading several sources at different frequencies is demonstrated here and leads to a significant reduction of the total computational time compared to sources injected frequency by frequency.

6. Conclusions

A novel methodology for the prediction of wind turbine noise in a high-order DG method context has been presented. The methodology, although demonstrated in a simplified case, seems promising and it is a first foray into the technology of wind turbine noise prediction in three dimensions and realistic flow and topographic conditions. The modelling of sources was performed using the propeller noise theory presented in section §2.1 and then applied to the r.h.s. of the mass and momentum equations as a cloud of point sources. Moreover, one may mention the advantages of the present method: (i) the numerical method used (DG) has high parallel scalability suited to large acoustic problems and high frequency; (ii) the physical model used (LEE combined with sources) paves the wave to new phenomena to be taken into account, for instance, damping effects by inclusion of viscosity, effects of sound reflections by density or temperature irregularities in the atmosphere (clouds), impedance boundary conditions to assess the ground absorption effects on acoustics. Furthermore, it is worth mentioning the prediction of amplitude modulation noise with the present method; more noticeable due to the use of broadband components besides the tonal noise of the wind turbines. Finally, as future investigations and perspectives, one can mention: (i) to use CFD solutions (pressure over the blades) as input of the propeller noise theory; (ii) increase in frequency on the simulations to better observe the amplitude modulation noise; and (iii) inclusion of damping effects on sound propagation over large distances.

References

[1] **Wind Europe** (2017), Wind in Power 2016 European statistics. (Published in February 2016).

[2] **Phillips**, C. V. (2011). Properly interpreting the epidemiologic evidence about the health effects of industrial wind turbines on nearby residents. *Bulletin of Science, Technology & Society, 31*(4), 303-315.

[3] **B. de Brye et. al** (2015), *Efficient Aeroacoustic Simulations of Jet Engine Noise using GPU acceleration*, DAGA 2015

[4] **Drela, M.** (1989). *XFOIL: An analysis and design system for low Reynolds number airfoils. In Low Reynolds number aerodynamics.* (pp. 1-12). Springer Berlin Heidelberg.

[5] **Corcos G.M.** *Resolution of pressure in turbulence*. Journal of the Acoustical Society of America, 35:192–199, 1963.

[6] **Goody M.C**. *Empirical spectral model of surface pressure fluctuations*. AIAA Journal, 42(9):1788–1794, 2004.

[7] **Dierke, J et al.** *Installation effects of a propeller mounted on a wing with coanda flap. Part II: Numerical investigation and experimental validation*, 2014. AIAA 2014 conference.

[8] **Clair V., Polacsek C., Le Garrec T., Reboul G.,** and **Jacob M.** *Numerical predictions of turbulence/cascade-interaction noise using computational aeroacoustics with a stochastic model*, 2015. AIAA Journal, 1-16.

[9] **Free Field Technologies**, *Actran V.18, User Manual.* Vol. 1.

[10] **J. Jonkman, S. Butterfield, W. Musial, and G. Scott**. "Definition of a 5-MW Reference Wind Turbine for Offshore System Development". 2009.

[11] **G.P. van den Berg**. "Effect of the wind profile at night on wind turbine sound", 2004. J. Sound Vibr., 277(4), 955-970

7th International Conference
on
Wind Turbine Noise
Rotterdam – 2nd to 5th May 2017

Acoustic measurements of a DU96-W-180 airfoil with flow-misaligned serrations at a high Reynolds number in a closed-section wind tunnel

Roberto Merino-Martinez, Wouter van der Velden, Francesco Avallone & Daniele Ragni, Faculty of Aerospace Engineering, Delft University of Technology, Kluyverweg 1, 2629 HS, Delft, the Netherlands.
Email: r.merinomartinez@tudelft.nl

Summary

Trailing-edge serrations are passive noise reduction add-ons widely used in wind-turbine applications. This study presents acoustic beamforming results of microphone array measurements of a cambered airfoil (DU96-W-180) in a closed-section wind-tunnel at a Reynolds number of industrial interest. Two different serration geometries with different lengths were tested and compared with the straight-edge baseline airfoil. The serrations were set at a flap angle of 6 degrees. Several flow velocities and angles of attack were tested at three chord-based Reynolds numbers ranging from 5×10^5 to 1.5×10^6. A phased microphone array was used to obtain source maps of the trailing-edge noise; Particle Image Velocimetry (PIV) was employed to obtain information about the turbulent boundary layer approaching the trailing edge; further, a numerical simulation using the Lattice Boltzmann Method (LBM) was performed for comparison. Far-field noise from the experimental data and computations shows a satisfactory agreement. Noise reductions of several dB were obtained, especially at lower frequencies. An increase in high-frequency noise is observed after a crossover frequency, which is assumed to be due to the set flap angle.

1. Introduction

Noise emission from wind turbines is one of the main issues that the wind energy industry currently must deal with. The power production of a single wind turbine is limited by strict noise regulations: a decrease of 1 dB of the sound pressure level (SPL) is expected to raise the energy production by 2 to 4% (Oerlemans and Fuglsang, 2012).

Trailing-edge serration is the most used passive noise reduction device to reduce turbulent boundary layer trailing-edge noise of wind turbine blades (Oerlemans 2016). The most common design features thin, solid, sawtooth patterns attached to the trailing edge (Gruber et al., 2011).

Previous researches focused on symmetric airfoils in an open-jet wind tunnel, such as NACA 0018 wings with retrofitted serrations with and without flap angle (Arce León et al., 2016a; 2016b; Avallone et al. 2016). Noise reductions of about 6 dB were measured for different flow speeds. In addition, a comparison with numerical simulations showed satisfactory agreement (van der Velden and Oerlemans, 2017). The low Reynolds numbers usually obtainable in open jet facilities and ambiguities in the definition of the angle of attack, due to the flow expansion, limit their industrial utilization. On the other hand, closed-section wind tunnels provide a well-characterized aerodynamic flow, further improving the comparison with numerical simulations (Pagani et al., 2016), but less accurate acoustic measurements. The latter are affected by the typical high background noise levels, the convection of the sound waves due to the air flow, reflections on the tunnel walls, and the interaction of the turbulent boundary layer of the wind tunnel with the microphones, if these are installed flush-mounted on the wall of the tunnel. In order to alleviate these effects, phased microphone arrays and beamforming algorithms are usually adopted to estimate the location and strength of sound sources (Mueller, 2002).

This study investigates the noise emissions of a cambered airfoil (DU96-W-180) in a closed-section wind tunnel at Reynolds numbers of industrial interest. A straight trailing-edge baseline case and two different serration geometries with flap angle are studied.

Microphone array measurements were performed to measure the noise emissions at the trailing edge of each configuration. Particle Image Velocimetry (PIV) measurements provided information about the boundary layer characteristics at the trailing edge. In addition, computational simulations using the Lattice Boltzmann method were performed to compare and assess the accuracy of the experimental far-field acoustic measurements.

The current paper is structured as follows: Section 2 describes the methodology, i.e., the beamforming algorithm and the computational method. The experimental setup and the model geometry are introduced in Section 3. Section 4 shows the experimental results and the comparison with computations. Finally, Section 5 summarizes the manuscript.

2. Methodology

2.1 Beamforming method

The application of beamforming algorithms to the acoustic data recorded by a phased microphone array allows for the estimation of the location and strength of sound sources (Johnson and Dudgeon, 1993; Merino-Martinez et al., 2016; Malgoezar et al., 2017). A scan grid needs to be defined, where each grid point is considered as a potential sound source.

Conventional frequency domain beamforming (Johnson and Dudgeon, 1993) was employed in this study, since it is a robust, simple and fast algorithm. The convection of the sound waves was considered in the formulation (Sijtsma, 2010). Since this method is based on single point sources, an additional integration method, more suitable for distributed noise sources, is used. It consists in integrating over an area of interest and normalizing the result by the integrated array response function (Sijtsma and Stoker, 2004; Sijtsma, 2010). In this paper, an integration method similar to the Source Power Integration technique (Sijtsma, 2010), which considers a covariance matrix fitting (Yardibi et al., 2010) based on the assumption of a line source, was applied over a region of the source map obtained by conventional beamforming (see section 3.2). This method was recently proposed by Sijtsma (Sijtsma, 2016) and was proven to provide very accurate results in the microphone array methods benchmark (Sarradj, 2017) for a simulated linear sound source, heavily contaminated with background noise, which resembles the measurement of trailing-edge noise in a closed-section wind tunnel.

In order to obtain the sound frequency spectrum of the integration area with this method, the following formula is applied for each frequency of interest:

$$A(f)=\frac{\sum_{j=1}^{K}\left(g_j^{*}C_{\exp}g_j\right)}{\sum_{j=1}^{K}\left(g_j^{*}C_{sim}g_j\right)} \tag{1}$$

where an asterisk (*) denotes the complex conjugate transpose, f is the frequency, g_j is the steering vector for the jth grid point in the integration area (Sijtsma, 2010), $C_{\exp}$ is the cross-spectral matrix obtained from the experimental measurements and C_{sim} is the simulated cross-spectral matrix due to the considered line source (using the same microphone distribution). In this case, the line source was assumed to be in the airfoil trailing-edge position. Both summations in Equation (1) apply to the K grid points within the integration area considered.

Performing acoustic measurements in closed wind tunnel test sections is a challenging task (Mueller, 2002; Pagani et al., 2016) as described earlier. Therefore, the main diagonal of the cross-spectral matrix of the Fourier-transformed microphone signals was removed to suppress the effect of incoherent noise (mostly due to the wind tunnel boundary layer interaction with the microphones) and improve the beamforming results (Sijtsma, 2010). Since this technique might cause inaccuracies in the absolute source strength, in the following only the relative differences between values corresponding to different configurations are reported because of their higher reliability (Oerlemans and Sijtsma, 2004).

A similar aeroacoustic experiment is presented in (Pagani et al., 2016) where slat noise was measured, instead of trailing-edge noise. The advanced deconvolution method DAMAS (Brooks and Humphreys, 2006) provided similar source distributions and integrated sound spectra as conventional frequency domain beamforming.

2.2 Lattice Boltzmann method (LBM)

The commercial software package Exa PowerFLOW 5.3c was used to solve the discrete Lattice-Boltzmann equations for a finite number of directions. For a detailed description of the equations used for the source field computations the reader can refer to (Succi, 2001) or (van der Velden et al., 2016). Here, only summary is presented regarding the computational method.

The discretization used for this application consisted of 19 discrete velocities in three dimensions (D3Q19) involving a third-order truncation of the Chapman-Enskog expansion, which is suitable and give accurate results for low Mach number flows. The distribution of particles was solved using the kinetic equations on a Cartesian mesh, with the Bhatnagar-Gross-Krook (BGK) collision term operator (Bhatnagar et al., 1954). A Very Large Eddy Simulation (VLES) was implemented as viscosity model to locally adjust the numerical viscosity of the scheme (Chen et al., 2003). The model consists of a two-equation k-ε Renormalization Group (RNG) modified to incorporate a swirl based correction that reduces the modeled turbulence in presence of large vortical structures. A turbulent wall-model was used to resolve the near-wall region (Chen and Doolen, 1998). The choice of the model allowed to obtain a reliable estimate of the boundary layer till the viscous sub-layer, with feasible turn-around times. The surface itself is modelled by a cut-cell approach, which avoids meshing complex geometry. Especially for the current application, where the flow around complex sawtooth trailing edges have to be solved, this is a huge advantage over other computational methods.

Since the LBM is inherently compressible and it provides a time-dependent solution, the sound pressure field was extracted directly from the computational domain. Sufficient accuracy is obtained when considering at least 16 cells per wavelength for the LBM (Habibi et al., 2013). The obtained far-field noise was further compared with noise estimated by using an acoustic analogy. For this purpose, the Ffowcs-Williams and Hawkings (FWH) (Williams and Hawkings, 1969) equation was employed. The time-domain FWH formulation developed by Farassat (Farassat and Succi, 1980) was used to predict the far-field sound radiation of the serrated trailing edge in a uniformly moving medium (Brès et al., 2010). The input to the FWH solver is the time-dependent pressure field of a surface mesh provided by the transient LB simulations.

3. Experimental setup

3.1 Model geometry

The experiments were performed at the Delft University of Technology Low Turbulence Wind Tunnel (LTT). This wind tunnel has a contraction ratio of 17.8 and the freestream turbulence level in the test section varies from 0.04% at 20 m/s to 0.1% at 75 m/s.

The tunnel has an octagonal closed test section 1.8 m wide, 1.25 m high and 2.6 m long. A DU96-W-180 airfoil with a span of 1.25 m and a chord of 0.6 m was vertically installed flush-mounted to the tunnel section (Figure 1) and tested at different flow speeds (U_∞ = 12.4, 24.8 and 37.4 m/s) and angles of attack, (α = -6, -2, 0, 2, 6, 9 and 12 degrees). These flow speeds were chosen to result in a chord-based Reynolds numbers of 5×10^5 ,10^6 and 1.5×10^6 and Mach numbers of 0.037, 0.073, and 0.11, respectively.

Three different trailing-edge geometries were investigated: a straight edge (which is considered as a baseline configuration), and two sawtooth serrations of different lengths (l): the Sr05 configuration with l equal to 5% of the airfoil chord (3 cm) and the Sr15 with l equal to 15% of the airfoil chord (9 cm). The width of both sawtooth serrations is half of the length, i.e., 1.5 cm and 4.5 cm, respectively (Figure 2). These serration geometries are based on a boundary layer thickness of 3 cm at the suction side, obtained with XFOIL calculations under similar flow conditions, and, in addition, confirmed by Devenport et al. (2010). The serrations were manufactured by laser cutting a steel flat plate with a constant thickness of 1.5 mm and retrofitted to the trailing edge keeping the surface free from irregularities. Both serrations were set at a flap angle of $\phi = 6$ degrees.

The coordinate system adopted in the manuscript is reported in Figure 1. The x axis is oriented in the downwind streamwise direction, the z axis in the vertical direction and the y axis perpendicular to the xz plane and it points at the microphone array. The origin of this coordinate system is located at the midspan of the airfoil.

Computations were performed on the same model geometry under similar flow characteristics. The same coordinate system is used. For the sake of conciseness, only results for chord-based Reynolds number equal to 1.5×10^6 (U_∞ = 37.4 m/s) and angle of attacks of -2 degrees (zero-lift angle for this airfoil) and 6 degrees are presented in this paper. Transition from laminar to turbulent in these conditions is natural as in the experiments. The computational domain is 20 chord lengths long (12 m) in the x and y directions while it is equal to the span of the model (1.25 m) in the z direction. The outer region holds an anechoic layer to damp out the acoustic waves near the far-field boundaries. Spanwise cyclic boundary conditions are applied at the edge of the model span. At the inlet a fixed velocity is described, and the outlet is modeled by fixing the static pressure, while maintaining a free flow direction. The simulated Mach number is identical to the real Mach number, i.e., 0.11. The grid used in this study has 60,000,000 voxels, with 8 different refinement regions located around the airfoil. The finest voxels, around the straight and serrated trailing edges, are cubes of size 3.52 x 10^{-4} m considered sufficient to correctly capture the most relevant features of the boundary layer and the near wake. The boundary layer was modeled using the inbuilt wall model, with the closest cell located around y^+ = 50. The Courant-Friedrichs-Lewy (CFL) stability number was set to unity by the solver, to ensure stable conditions. Simulations were run for a physical time of 0.3 s (approximately 20 flow passes). A total of 10 flow passes were used for detailed analysis. Statistical data was recorded at a frequency of 26 kHz, and used for a prediction of the acoustic far-field noise. For each 0.1 s of physical time (6.5 flow passes), 450 CPU hours were necessary on a Linux Xeon E5-2690 2.9 GHz platform with 20 cores.

Figure 1 - Wind tunnel setup and airfoil model.

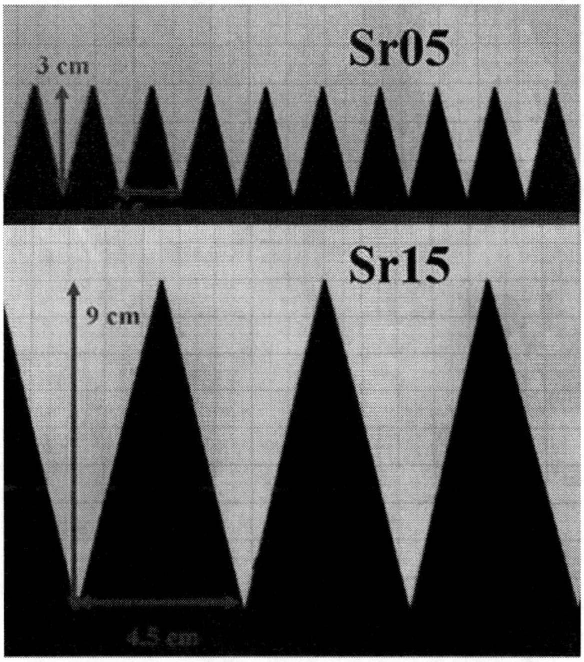

Figure 2 – Trailing-edge serration geometries.

3.2 Acoustic measurements

A phased microphone array was installed with recessed microphones. Recession was about 2 cm deep and had an opening angle of 60 degrees. Microphones were installed along one of the walls of the wind tunnel behind an acoustically-transparent flat Kevlar window (see Figure 3). The setup configuration allows to keep the closed test section configuration alleviating the effect of the turbulent boundary layer convecting along the wall.

The array consisted of 64 microphones in a multi-arm logarithmic spiral distribution with an elliptical shape with a mayor axis of 0.93 m, see Figure 4. The distance from the array to the scan plane (i.e., the airfoil trailing-edge position) was 0.9 m and it was facing the suction side of the airfoil. The center microphone at the array was aligned with the middle point of the trailing edge (for the case with $\alpha = 0°$).

Data was acquired for 30 s at a sampling frequency of 50 kHz. The acoustic data was averaged using time blocks of 2048 samples (Δt = 40.96 ms) for each Fourier transform and windowed using a Hanning weighting function with 50% data overlap. With these parameters, the frequency resolution is 24.4 Hz and the expected error (Brandt, 2011) in the cross-spectrum estimate is 3.7%. Unfortunately, no background noise measurements with the empty tunnel could be performed, so the signal to noise ratio is not known.

The scan grid for beamforming covered the expected area of noise generation, ranging from $x = -1\,m$ to $x = 0.5\,m$ and from $z = -0.65\,m$ to $z = 0.65\,m$ with a separation between grid points of 1 mm, see Figures 5 and 6. Therefore, the considered grid size was 1501 × 1301.

Figure 3 - Kevlar window (in yellow) in one of the LTT wind tunnel walls.

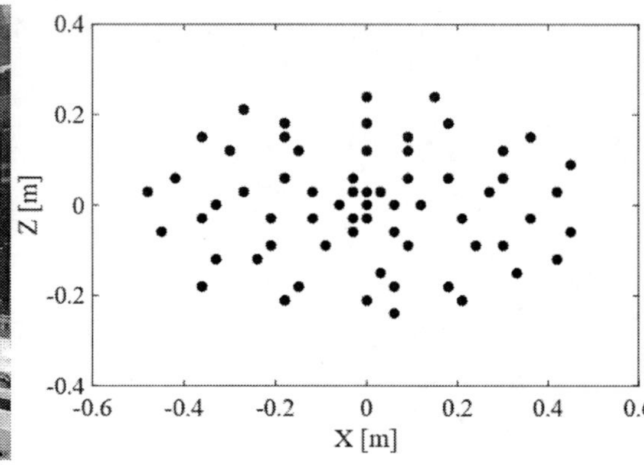

Figure 4 - Microphone array distribution. Coordinates are given in the airfoil system.

The integration area for the application of the method described in section 2.1 extended from $x = -0.1\ m$ to $x = 0.1\ m$ and from $z = -0.5\ m$ to $z = 0.5\ m$ (see the dashed lines in Figures 5 and 6). This region covered the whole serration length for both geometries and prevented possible contaminations from wind tunnel boundary layer interactions with the model ends (corner sources) (Tuinstra and Sijtsma, 2015) and other noise sources while still providing spanwise statistically meaningful results of the trailing-edge noise (Pagani et al., 2016).

3.3 PIV measurements

Stereoscopic PIV experiments were conducted to measure the three-component velocity fields in planes perpendicular to the serration surface at the trailing edge of the wing. The required illumination was provided by a Quantel Evergreen Nd:YAG laser system with an average output of 200 mJ/pulse. The laser light was conveyed to form a 2 mm laser sheet of about 0.3 m width at the field of view. Two LaVision Imager Pro LX 16 Mpix (4870 × 3246 px, 12 bits, pixel pitch of 7.4 µm/px) were used. They were equipped with two Nikon lenses of focal length f = 200 mm and set at an aperture f# = 2.8 – 4. They were set at about 40 degrees angle at about 1 m distance from the model. The resulting field of view was 100 × 140 mm^2. The magnification factor was M = 0.25 resulting in a digital resolution of approximately 34 px/mm. The focusing plane was slightly offset with respect to the laser plane (defocusing), to obtain an image of the particle of about 2.3 px. Therefore, no bias error due to peak-locking is expected (Westerweel, 1997). Seeding was provided in the test section by a SAFEX smoke generator with SAFEX MIX, able to produce liquid droplets of less than 1 µm. Ensembles of 1000 uncorrelated double-frame recordings per dataset were acquired and processed by LaVision Davis 8.1.4. Particle images were processed using final interrogation windows of 24 × 24 px with 75% overlap resulting in a vector spacing of 0.18 mm. The main PIV parameters are gathered in Table 1.

4. Results

4.1 Beamforming source plots

Examples of the beamforming source plots are presented in Figures 5 and 6, where the acoustic images for the three airfoil configurations (straight trailing edge, short serrations Sr05 and long serrations Sr15) are shown for the one-third octave bands of 2 and 4 kHz, respectively. Both figures refer to an angle of attack of 6 degrees and U_∞ = 37.4 m/s corresponding to a Reynolds number of 1.5×10^6.

Table 1 – PIV parameters

Parameters		Stereoscopic PIV setup	
Measurement field of view	FOV	100 × 140 mm^2	4870 × 3246 px^2
Interrogation window size	Iw	0.72 × 0.72 mm^2	24 × 24 px^2
Vector spacing	S	0.18 mm	6 px
Digital resolution	DR	~34 px/mm	
Magnification	M	0.25	
Vectors	NV	538 × 769	

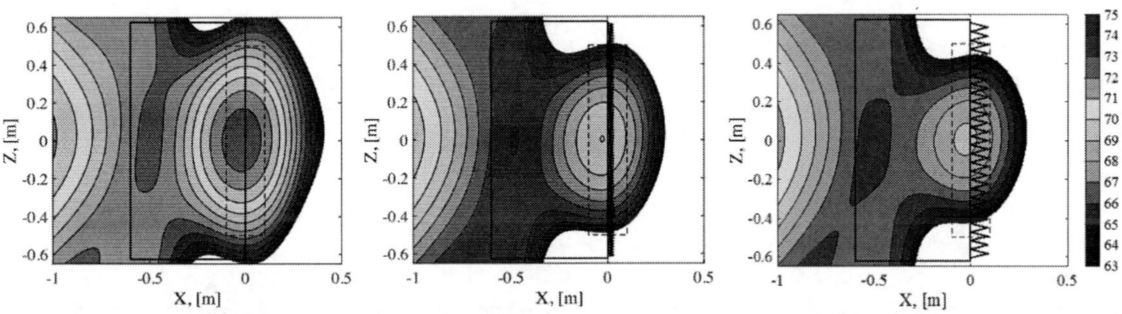

Figure 5 – One-third octave band (2 kHz) beamforming source plot for the straight trailing edge (left), short serrations (center) and long serrations (right) with U$_\infty$ = 37.4 m/s, α = 6 degrees and Re = 1.5×10^6. The solid black line represents the airfoil position and the dashed black line, the integration area.

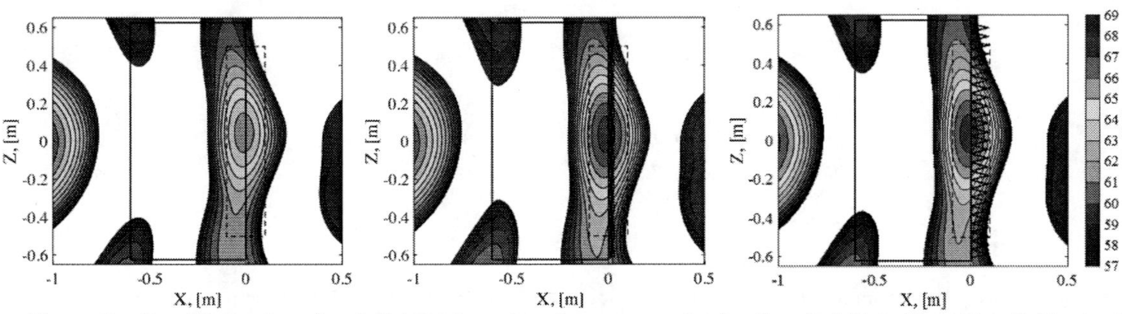

Figure 6 - One-third octave band (4 kHz) beamforming source plot for the straight trailing edge (left), short serrations (center) and long serrations (right) with U$_\infty$ = 37.4 m/s, α = 6 degrees and Re = 1.5×10^6. The solid black line represents the airfoil position and the dashed black line, the integration area.

In Figure 5, trailing-edge noise reductions ($\Delta\phi_{aa}$ in Figures 8 and 9) (considering the peak values in the image) with respect to the straight edge baseline of around 3 dB and of 4 dB are measured for the short and long serrations respectively. As previously mentioned, the integration of the source map over an area (marked with a dashed black line) prevents to some extent the inclusion of unwanted noise sources, such as leading edge noise or noise sources present in the wind tunnel itself. This type of sources can also be seen in Figures 5 and 6.

In Figure 6, on the other hand, it is seen that the short serrations show noise increase (considering peak values) of around 3 dB and 2 dB for the short and long serrations, respectively. The three plots in Figure 6 show two corner sources on the junctions of the leading edge with the tunnel wall, most probably due to the interaction of the boundary layer of the wind tunnel with the model.

The noise reductions observed by the serrated trailing edges agree with those obtained in Oerlemans et al., 2009 in field measurements on full-scale wind turbines, where a noise increase after a crossover frequency was also noticed.

4.2 PIV results

In order to compare computations and experiments, boundary layer profiles at 95% of trailing-edge suction side are investigated for the baseline straight trailing edge case. In Figure 7, both mean ($\bar{u}$) and rms ($\overline{u'}$) streamwise velocity are depicted for $\alpha = -2$ degrees and $\alpha = 6$ degrees. For $\alpha = -2$ degrees (Figure 7 (a) and (b)), the turbulent boundary layer results are very similar for the two methods, both in terms of mean and fluctuating velocity. Excellent agreement is found between PIV and LBM. On average, a boundary layer thickness of δ = 13 mm is found, with a shape factor of H = 1.5, indicating a fully developed turbulent boundary layer.

The mean boundary layer of the $\alpha = 6$ degrees case (Figure 7 (c)) is slightly different close to wall, when comparing PIV and LBM. Due to the cambering of the DU96-W180 airfoil, and therefore the stronger adverse pressure gradient, transition seems to be delayed in the experiment, resulting in a less turbulent boundary layer. The boundary layer thickness was measured to be δ = 21 mm, with a shape factor of 1.9. In the simulated results, the boundary layer behaves differently, although the fluctuations are captured adequately. This deviation could also explain the larger differences between the far-field reduction results for this case, presented in subsection 4.3.

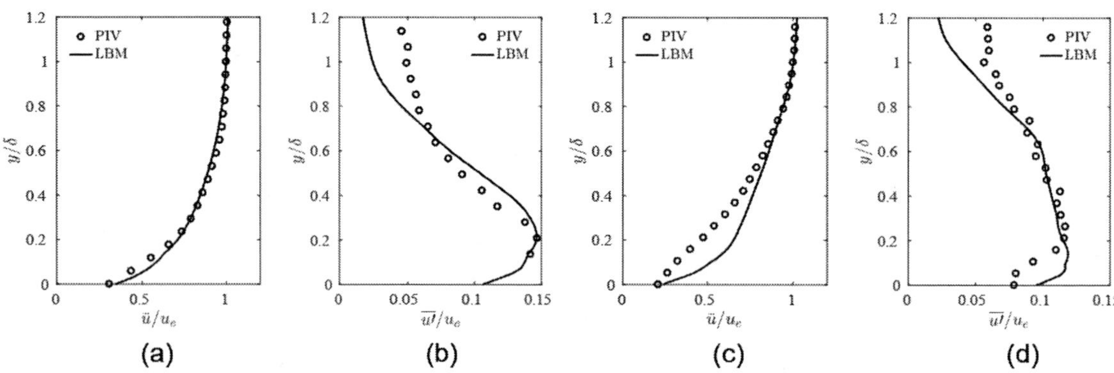

Figure 7 – Boundary layer characteristics from both PIV and LBM obtained at 95% of the baseline trailing-edge suction side. Normalized streamwise mean and rms velocity for $\alpha = -2$ degrees (a and b) and $\alpha = 6$ degrees (c and d).

4.3 Noise reduction comparison

The noise reductions ($\Delta\phi_{aa}$) achieved by both serrated geometries with respect to the straight trailing-edge baseline case are presented in Figure 8 for $\alpha = -2$ degrees and in Figure 9 for $\alpha = 6$ degrees, for a chord-based Reynolds number of 1.5×10^6 (U_∞ = 37.4 m/s). Positive values mean noise reduction. Both the results obtained by the integrated beamforming source plots and the LBM simulations are plotted in the same figure and show similar levels and trends for the selected frequency range (500 – 5,000 Hz). The reason for limiting the analysis to this frequency range is it is the region of highest confidence for both the acoustic measurements and LBM.

It can be observed in Figure 8 that reductions up to 5 dB are obtained for the short serrations and approximately of 4 dB for the long serrations between 1 and 2 kHz approximately. Since the long serrations have length approximately equal to 5 times the boundary layer thickness, minor increase in noise reduction are expected (Gruber et al., 2011).

The general trend shows that the noise reduction performance of the serrations worsens with increasing frequency, leading even to some noise increase after a crossover frequency of approximately 3,800 Hz for both serration geometries. This is supposed to be due the flap angle of the serrations ($\phi = 6$ degrees) (Arce León et al., 2016b). Good agreement is found between experiment and simulation, giving confidence to both proposed methodologies in the paper. In addition, the noise reduction values obtained are of the same order of magnitude as those observed in (Gruber et al., 2011) for a similar experiment.

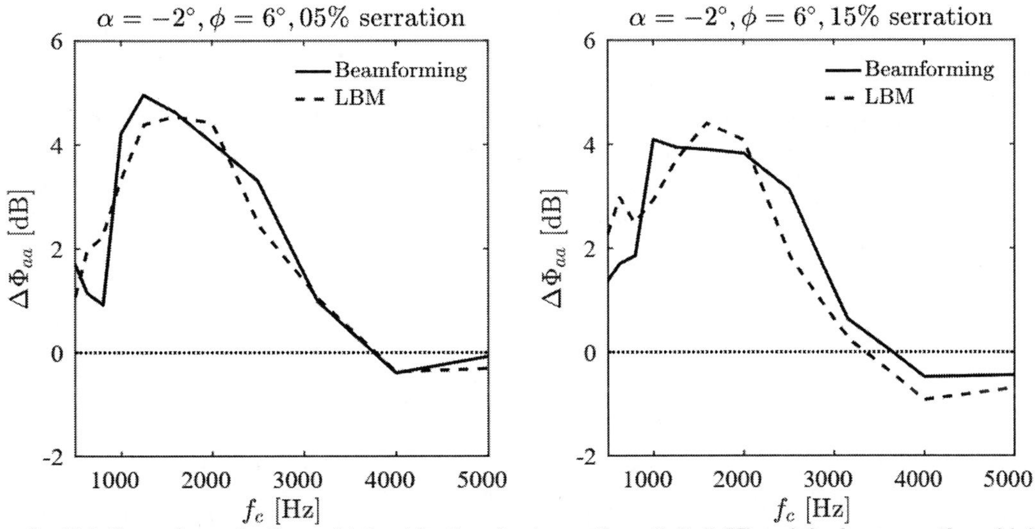

Figure 8 – Relative noise reductions obtained by the short serrations (left, Sr05) and the long serrations (right, Sr15) with respect to the baseline case for U_∞ = 37.4 m/s, α = −2 degrees. Both the experimental (solid line) and LBM (dashed line) results are compared. Positive values represent noise reductions.

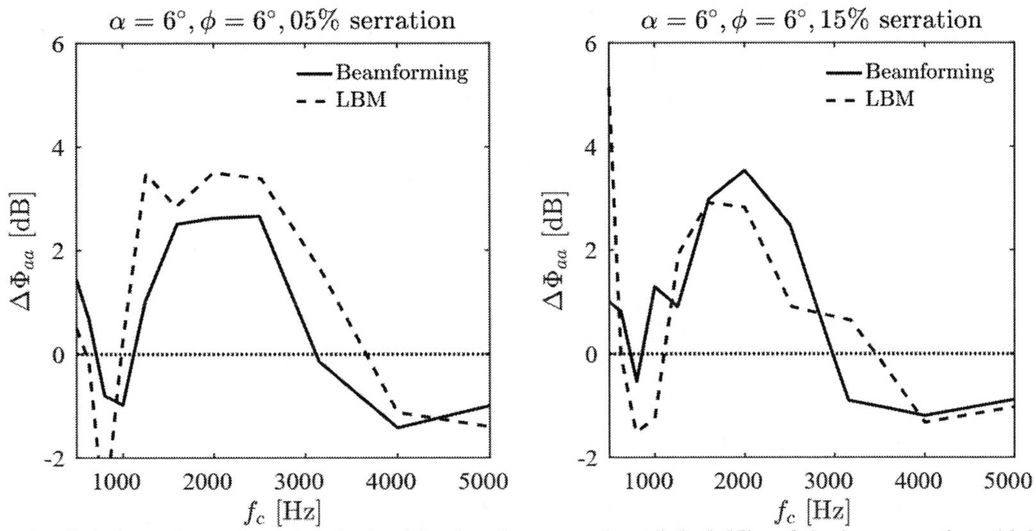

Figure 9 – Relative noise reductions obtained by the short serrations (left, Sr05) and the long serrations (right, Sr15) with respect to the baseline case for U_∞ = 37.4 m/s, α = 6 degrees. Both the experimental (solid line) and LBM (dashed line) results are compared. Positive values represent noise reductions.

In Figure 9 ($\alpha = 6$ degrees), similar trends as in Figure 8 can be observed, although with lower values for the whole spectrum, reaching maximum reductions of around 3 dB for both serrated geometries. Once again, a noise increase of around 1 dB is observed after a crossover frequency. This time the threshold frequency shows lower values, closer to 3 kHz. For $\alpha = 6$ degrees, also a cross-over frequency at the lower frequency is detected, around 800-1000 Hz. This behaviour is different to the behaviour observed in previous studies with symmetric airfoils

(Arce León et al., 2016a). The general agreement between experiment and simulation for $\alpha = 6$ degrees is worse, which could indicate a variation in the flow behaviour.

The noise reduction values obtained from the integrated beamforming results differ from those obtained by simply considering the peak values of the source plots in section 4.1. This confirms that the selected integration method is more suitable for line sources.

4.4 Directivity plots

In addition, the LBM computations provide the sound radiation directivity of the trailing-edge noise. Figure 10 presents the radiation directivity plots for the three configurations at ten chords distance for the case of $\alpha = -2$ degrees and Re = 1.5×10^6, banded in three different frequency ranges: 500-1,000 Hz, 1,000-2,000 Hz and 2,000-5,000 Hz.

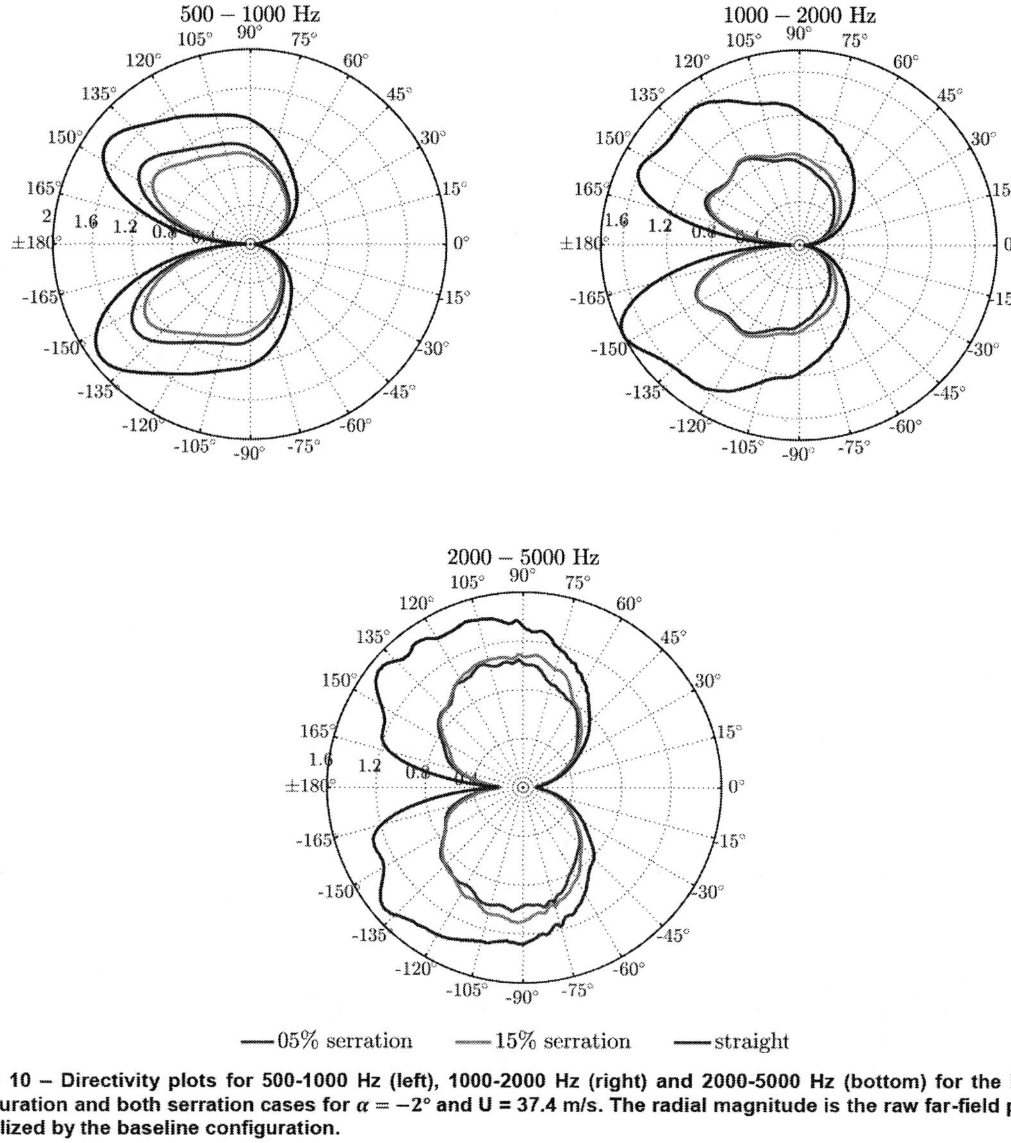

Figure 10 – Directivity plots for 500-1000 Hz (left), 1000-2000 Hz (right) and 2000-5000 Hz (bottom) for the baseline configuration and both serration cases for $\alpha = -2°$ and U = 37.4 m/s. The radial magnitude is the raw far-field pressure normalized by the baseline configuration.

In general, it can be observed that both serrated geometries provide considerable noise reductions at all angles, especially in the upstream direction (120-150°). The pattern of the directivity exhibits a convective dipole, oriented towards the leading edge. This is in line with diffraction pattern of trailing-edge noise cases. At the highest frequency band, lobes start to appear, indicating the change from compact to non-compact acoustic sources.

5. Conclusions

Acoustic beamforming and computational simulations using the Lattice Boltzmann Method are used to investigate the noise emissions of a DU96-W-180 wind turbine profile. Two different trailing-edge serration geometries are studied and compared with the straight trailing-edge case (baseline). Both serrations are set at a flap angle of 6 degrees. The experimental campaign was performed in a closed-section wind tunnel with a microphone array and chord-based Reynolds numbers up to 1.5×10^6. The experimental trailing-edge noise spectra integrated over an area were compared with the simulations results, showing a satisfactory agreement between both and noise reductions up to 5 dB for the lower frequencies. A slight increase in noise is observed after a crossover frequency, which is supposed to be due to the serrations flap angle. The radiation directivity plots show a more dipole-like pattern for the serrated cases and a larger noise reduction in the upwind direction. This contribution also serves as a cross validation of both experimental and numerical approaches.

Acknowledgements

The authors would like to thank Carlos Arce León and Stefan Pröbsting for their support and work during the experimental campaign. The research of the PhD candidate W.C.P. van der Velden is funded and supported by Siemens Wind Power A/S, Brande, Denmark.

References

Arce León, C., Merino-Martinez, R., Ragni, D., Avallone, F. and Snellen, M. (2016a) *Boundary layer characterization and acoustic measurements of flow-aligned trailing edge serrations.* Experiments in Fluids 57(12):182. DOI: 10.1007/s00348-016-2272-z.

Arce León, C., Merino-Martinez, R., Ragni, D., Avallone, F. , Scarano, F., Pröbsting, S., Snellen, M., Simons, D.G. and Madsen, J. (2016b) *Effect of trailing edge serration-flow misalignment on airfoil noise emissions.* Journal of Sound and Vibration. Accepted for publication.

Avallone, F., Pröbsting, S. and Ragni D. (2016). *Three-dimensional flow field over a trailing-edge serration and implications on broadband noise.* Physics of Fluids 28:117101. DOI: 10.1063/1.4966633.

Bhatnagar, P., Gross, E. and Krook, M. (1954) *A model for collision processes in gases. Small amplitude processes in charged and neutral one-component systems.* Physical Review 94(3), 511-525. DOI: 10.1103/PhysRev.94.511.

Brandt, A. (2011) *Noise and vibration analysis: signal analysis and experimental procedures.* 2nd Edition. Wiley, Hoboken. ISBN 978-0-470-74644-8

Brès, G., Pérot, F. and Freed, D. (2010) *A Ffowcs Williams-Hawkings Solver for Lattice-Boltzmann Based Computational Aeroacoustics*, 16th AIAA/CEAS Aeroacoustics Conference, AIAA-2010-3711. DOI: 10.2514/6.2010-3711

Brooks, T. and Humphreys, W. (2006) *A deconvolution approach for the mapping of acoustic sources (DAMAS) determined from phased microphone arrays*. Journal of Sound and Vibration, Vol. 294. Nos. 4-5, 2016, pp 856-879. DOI: 10.1016/j.jsv.2005.12.046.

Chen, S. and Doolen, G. (1998) *Lattice Boltzmann method for fluid flows* Annual Review of Fluid Mechanics 30, 329-364.

Chen, H., Kandasamy, S., Orszag, S., Shock, R., Succi, S. and Yakhot, V. (2003) *Extended Boltzmann Kinetic Equation for Turbulent Flows*, Science (80-.). 301. 633–636. DOI: 10.1126/science.1085048.

Devenport, W., Burdisso, R.A., Camargo, H., Crede, E., Remillieux, M., Rasnick, M. and Van Seeters, P. (2010) *Aeroacoustic testing of wind turbine airfoils*. Virginia Polytechnic Institute and State University Blacksburg, Virginia, USA. May 2010. NREL/SR-500-43471.

Farassat, F., Succi, G.P. (1980) *A review of propeller discrete frequency noise prediction technology with emphasis on two current methods for time domain calculations*, Journal of Sound and Vibration 71, 399–419. DOI: 10.1016/0022-460X(80)90422-8.

Gruber, M., Joseph, P. and Chong, T. (2011) *On the mechanisms of serrated airfoil trailing edge noise reduction*. 17th AIAA/CEAS Aeroacoustics Conference, Portland, Oregon, USA. DOI: 10.2514/6.2011-2781.

Habibi, K., Gong, H., Nayafiyazdi, A. and Mongeau, L.G. (2013) *Prediction of the Sound radiated from Low-Mach Internal Mixing Nozzles with Forced Mixers using the Lattice Boltzmann Method*. 19th AIAA/CEAS Aeroacoustics Conference, Berlin, Germany: pp. 1–15. DOI: 10.2514/6.2013-2143.

Johnson, D.H. and Dudgeon D.E. (1993) *Array signal processing. Concepts and techniques*. PTR Prentice Hall. Englewood Cliffs. ISBN: 978-0-130-48513-7.

Malgoezar, A.M.N., Snellen, M., Merino-Martinez, R., Simons, D.G. and Sijtsma, P. (2017) On the use of global optimization methods for acoustic source mapping. Journal of the Acoustical Society of America. Vol. 141 No. 1. pp. 453-465. DOI: 10.1121/1.4973915.

Merino-Martinez, R., Snellen, M. and Simons, D.G. (2006) *Functional beamforming applied to imaging of flyover noise on landing aircraft*. Journal of Aircraft, Vol. 53 No. 6. pp. 1830-1843. DOI: 10.2514/1.C033691.

Mueller, T. (2002) *Aeroacoustic measurements*. Springer, Berlin. ISBN-978-3-642-07514-8.
Oerlemans, S. and Sijtsma, P. (2004) *Acoustic array measurements of a 1:10.6 scaled Airbus A340 model*. 10th AIAA/CEAS Aeroacoustics Conference, Manchester, United Kingdom. DOI: 10.2514/6.2004-2924.

Oerlemans, S., Fisher, M., Maeder, T. and Kögler, K. (2009) *Reduction of wind turbine noise using optimized airfoils and trailing-edge serrations*. AIAA Journal, Vol. 47, No. 6. 1470 – 1481. DOI: 10.2514/1.38888.

Oerlemans, S. and Fuglsang, P. (2012) *Low-noise wind turbine design* EWEA Workshop Oxford Palumbo, D (2012) Determining correlation and coherence lengths in turbulent boundary layer flight data Journal of Sound and Vibration 331(16), 3721-3737.

Oerlemans, S. (2016) Reduction of wind turbine noise using blade trailing edge devices. 22nd AIAA/CEAS Aeroacoustics Conference. May 30 – June 1 2016. Lyon, France. DOI: 10.2514/6.2016-3018.

Pagani, C.C.Jr., Souza, D.S. and Medeiros, M.A.F. (2016) *Slat noise: Aeroacoustic beamforming in closed-section wind tunnel with numerical comparison*. AIAA Journal 54 (7) 2100 – 2115. DOI: 10.2514/1.J054042.

Sarradj, E., Herold, G., Sijtsma, P., Merino-Martinez, R., Geyer, T.F., Bahr, C.J., Porteus, R., Moreau, D.J. and Doolan C.J. (2017) *A microphone array method benchmarking exercise using synthesized input data*. 23rd AIAA/CEAS Aeroacoustics Conference. June 5-9 2017. Denver, CO, USA. (Paper accepted for the conference).

Sijtsma, P. and Stoker, R.W. (2004) *Determination of absolute contributions of aircraft noise components using fly-over array measurements*. 10th AIAA/CEAS Aeroacoustics Conference. May 10 – 12 2004. Manchester, United Kingdom. DOI: 10.2514/6.2004-2958.

Sijtsma, P. (2010) *Phased array beamforming applied to wind tunnel and flyover tests*. Tech. Rep. NLR-TP-2010-549, National Aerospace Laboratory (NLR), The Netherlands (December 2010).

Sijtsma, P. (2016) *Analytical Benchmark 1*. 22nd AIAA/CEAS Aeroacoustics Conference. May 30 – June 1 2016. Lyon, France. Presentation in Microphone Array Methods Discussion Panel.

Succi, S. (2001) *The lattice Boltzmann equation for fluid dynamics and beyond*. 1st edition. Clarendon Press, Oxford.

Tuinstra, M. and Sijtsma, P. (2015) *Suppression of spurious noise sources in airfoil self-noise measurements*. 21st AIAA/CEAS Aeroacoustics Conference. June 22 – 26 2015. Dallas, TX, USA. DOI: 10.2514/6.2015-2689.

van der Velden, W.C.P., Pröbsting, S., van Zuijlen, A.H., de Jong, A.T., Guan, Y. and Morris, S.C. (2016) *Numerical and experimental investigation of a beveled trailing edge flow and noise field*. Journal of Sound and Vibration, 384 (2016) 113–129. DOI: 10.1016/j.jsv.2016.08.005.

van der Velden, W.C.P. and Oerlemans S. (2017) *Numerical analysis of noise reduction mechanisms on improved trailing edge serrations using the Lattice Boltzmann method*. 35th Wind Energy Symposium (AIAA SciTech Forum) January 9 – 13 2017. Grapevine, Texas, USA. DOI: 10.2514/6.2017-1379.

Westerweel, J. (1997) *Fundamentals of digital particle image velocimetry*. Measurement Science and Technology. 8, 1379–1392. DOI: 10.1088/0957-0233/8/12/002.

Williams, J.E.F., Hawkings, D.L. (1969) *Sound Generation by Turbulence and Surfaces in Arbitrary Motion*. Philos. Trans. R. Soc. London A Math. Phys. Eng. Sci. 264. DOI: 10.1098/rsta.1969.0031.

Yardibi, T., Li, J., Stoica, P., Zawodny, N. S. and Cattafesta III L. N. (2010) A covariance fitting approach for correlated acoustic source mapping. Journal of the Acoustic Society of America 127(5): 2920 – 2931. DOI: 10.1121/1.3365260.

7th International Conference
on
Wind Turbine Noise
Rotterdam – 2nd to 5th May 2017

Trailing Edge Serrations — Effect of Their Flap Angle on Flow and Acoustics

Carlos Arce León, LM Wind Power, the Netherlands: caar@lmwindpower.com
Roberto Merino-Martínez, TU Delft, the Netherlands: r.merinomartinez@tudelft.nl
Daniele Ragni, TU Delft, the Netherlands: d.ragni@tudelft.nl
Stefan Pröbsting, TU Delft, the Netherlands: s.probsting@tudelft.nl
Francesco Avallone, TU Delft, the Netherlands: f.avallone@tudelft.nl
Ashish Singh, LM Wind Power, India: asi@lmwindpower.com
Jesper Madsen, LM Wind Power, Denmark: jema@lmwindpower.com

Summary
Trailing edge serrations have been proven to work as a passive noise reduction device. Nevertheless, they have also previously been found to increase noise in a particular frequency range, argued in earlier research to be due to the misalignment of the serrations with the direction of the flow in the wake. It emerges as a high-frequency noise increase in a broadband region of the spectrum. This study investigates the effect of serration-flow misalignment on the noise emissions using acoustic beamforming, and finds a correlation with observations made on the flow using particle image velocimetry (PIV). The hydrodynamic source of the noise increase is hereby identified, and a Strouhal number relation for the high-frequency noise increase is proposed.

1. Introduction
Trailing edge serrations have become the prevailing device for turbulent boundary layer-trailing edge noise (TBL-TE noise) reduction on wind turbine blades (Hurault et al., 2015; Mathew et al., 2016; Oerlemans, 2016). Their performance nevertheless depends on their correct design and installation. Tooth length and width recommendations with respect to the local airfoil boundary layer thickness, δ, have been proposed by Gruber, Joseph, and Chong, 2011. It is furthermore widely accepted that their edge should be thin with respect to δ to avoid the introduction of additional noise mechanisms, such as vortex shedding[1]. But until recently (Arce León et al., 2016b; Arce León et al., 2016c; Vathylakis et al., 2016), the effect of their flap angle has not been thoroughly addressed.

This parameter is nevertheless greatly important in the application of trailing edge

[1]A similar but distinct device, the trailing edge serrations of cutout type, follows a different design approach, and to avoid tonal noise issues caused by the thick trailing edge, the application of meshes has been proposed by Vathylakis, Chong, and Joseph, 2015.

serrations as a high-volume product in the wind turbine industry. Many careful considerations in their design must be made (Mathew et al., 2016), and the flap angle, φ, is one of them. One needs to contemplate its placement on the blade, and a set of its operating conditions (twist, the range of pitch, and the local inflow velocities, for example). It needs to adapt to the local airfoil shape and the way it will be attached. One should consider and ease the procedural implications of its installation, something that is not easy, especially if it is to be done by a technician dangling from a rope. These considerations should be made such that they minimize the risk of serration-flow misalignment—that is, how much the serrations are misaligned with respect to the direction that the flow in the wake would have if the serrations were not present.

Other than for aerodynamic reasons (minimizing the serration impact on the airfoil performance, unless that is desired), there is good aeroacoustic sense in pursuing serration-flow alignment. It has been previously hypothesized that serration-flow misalignment is the cause of an observed increase in noise with respect to the straight trailing edge. This phenomenon has been observed in wind tunnel measurements, as in Dassen et al., 1996; Gruber, Joseph, and Chong, 2010; Finez et al., 2011, and also in field measurements of serrated wind turbine blades, for example Oerlemans et al., 2009, and independent measurements by LM Wind Power. The conjecture argues that the high frequency noise originates from the regions between the serration teeth, where the flapwise tooth misalignment with the mean flow increases the turbulence intensity.

The noise increase is known to happen after a certain frequency, f_c, entitled as *crossover frequency*. Gruber, Joseph, and Chong, 2010 proposed that it is related to a constant Strouhal number which is based on the boundary layer thickness and the freestream velocity,

$$\mathrm{St_c} = \frac{f_c \delta}{U_\infty}. \quad (1)$$

Using a serrated NACA 6512-10 airfoil and measuring f_c at different freestream velocities, it was established that $\mathrm{St_c} \approx 1$. A variance of around 30% was found, attributed to potential variations in the boundary layer thickness introduced by the serrations. It was further explained by the fact that the boundary layer was not measured for all the given freestream velocities, but was instead approximated using the software XFOIL. A specific relation of $\mathrm{St_c}$ to the airfoil angle of attack, α, is not clear from the research, but results for different angles of attack are provided in Gruber, 2012.

The research here presented aims to review the proposed constant Strouhal number approximation by combined acoustic and aerodynamic measurements of a serrated NACA 0018 airfoil with regard to its original straight trailing edge shape. Several freestream velocities are investigated, 30, 35 and 40 m/s, which at the highest yields a Reynolds number of approximately 533,000. Angles of attack of 0°, 6°, 12° (geometric) and serration flap angles of 0° and 6° are prescribed. The boundary layer thicknesses at all the investigated conditions have been measured on both the pressure and suction sides for the straight edge using PIV.

The hydrodynamic source of this high frequency noise increase is further explored by means of high-speed stereoscopic particle image velocimetry (S-PIV). The time-

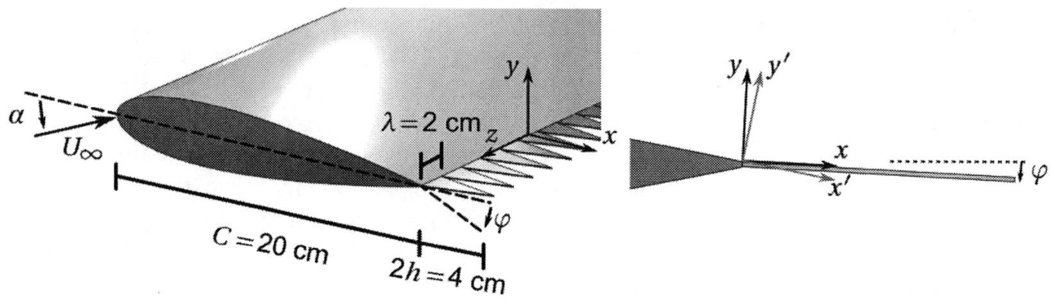

Figure 1: Airfoil and serration dimensions (left), and convention used for the coordinate system rotation over the airfoil and serration surfaces (right).

resolved boundary layer flow is thereby described on both sides, allowing an inspection of mean flow parameters and turbulence characteristics that was previously incomplete. An extended study of the research here presented has been compiled in Arce León et al., 2017.

2. Experimental Setup

The experiment was conducted at the vertical wind tunnel facility (V-Tunnel) of the Delft University of Technology. This is a low turbulence tunnel ($< 1\%$ turbulence intensity, Ghaemi, Ragni, and Scarano, 2012) with an open test section. The two side-plates in this setup form a structural support for the airfoil, and help to maintain a quasi-two dimensional flow behavior over the airfoil, offering a relatively low aspect ratio of 2. This setup further allows an acoustic direct line of sight to the airfoil trailing edge. The wind tunnel nozzle has a dimension of $0.4 \times 0.4\,\mathrm{m}^2$.

The airfoil used was a NACA 0018 profile of 0.4 m span and 0.2 m chord. It was machined from aluminium and has a modular trailing edge that allows converting the straight trailing edge into a serrated edge by the insertion of laser-cut serrations of flat-plate type. The latter had a length of $2h = 4\,\mathrm{cm}$, a width of $\lambda = 2h/2 = 2\,\mathrm{cm}$, and a constant thickness of 1 mm (the same as the baseline airfoil trailing edge thickness). A schematic of of the serrated airfoil and its dimensions can be seen in figure 1. The coordinate system definition is also shown indicated. For consistency, when boundary layer results are shown, it is rotated about the z axis to keep y wall-normal in accordance with the definition of boundary layer. The prime nomenclature will be omitted later for conciseness.

The boundary layer transition was forced using a narrow (2 cm) tape of three-dimensional roughness elements (carborundum, 0.6 mm nominal size) that spanned the whole airfoil at both the suction and pressure sides. It was placed at 20% chord. By doing this it is ensured that laminar boundary layer instability noise is avoided and only TBL-TE noise is produced. The boundary layer was confirmed to remain turbulent at the trailing edge using a microphone probe.

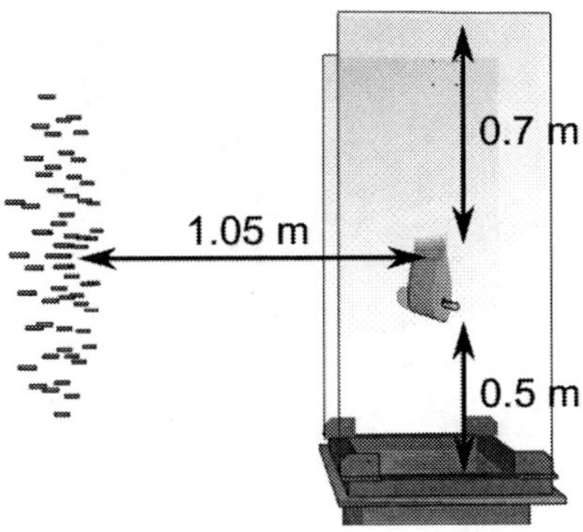

Figure 2: Schematic of the setup used for the acoustic measurements. The side plates are rendered here transparent.

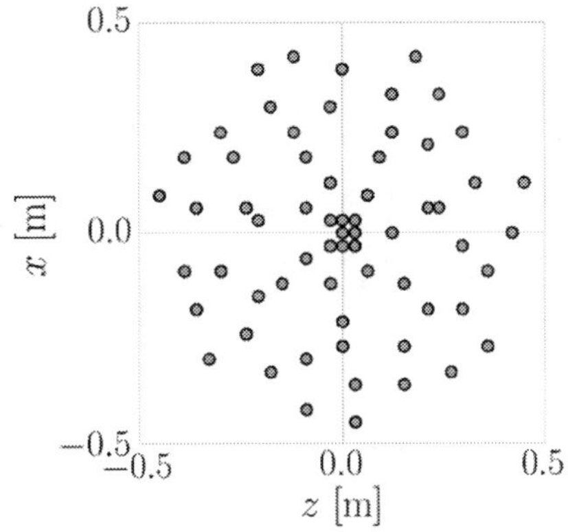

Figure 3: Microphone distribution within the array.

2.1 Acoustic measurements

Acoustic phased array measurements were used to acquire the trailing edge noise source emissions at 30, 35 and 40 m/s. This method was preferred as it allows the inspection of only the region of interest, avoiding unwanted noise sources. Examples of the latter are the noise of the tunnel nozzle and the side-plate edges, and that produced by the side-plate boundary layer interaction with the airfoil. The velocities at which the measurements were taken were chosen because they offered a good signal-to-noise ratio (SNR). A schematic of the setup is shown in figure 2.

Recordings were made over 60 s total time with a pistonphone-calibrated multi-arm logarithmic spirally distributed 64 microphone array (figure 3, Mueller, 2002; Pröbsting et al., 2016). A sampling frequency of 50 kHz was used. The data was averaged using time blocks of 2048 samples of 40.96 ms each. These were windowed by a Hanning weighing function with 50% overlap, to which a fast Fourier transform was applied.

Conventional frequency domain beamforming was used (Johnson and Dudgeon, 1993; Sijtsma, 2010), for which a scan grid of potential sources is defined. The grid covered a trailing edge-centered rectangle from z = -0.22 m to 0.22 m in the airfoil spanwise direction, and z = -0.3 m to 0.3 m in its streamwise direction. The results were integrated, using the Source Power Integration method (Sijtsma, 2010), around the trailing edge region, in the area between z = -0.1 m and 0.1 m, and x = -0.06 m and 0.06 m. The beamforming results can be seen in the source maps of figure 4, where the integration region is indicated with a dashed rectangle.

Results will be presented in one-third octave bands between 1 and 5 kHz. Respectively, the limits are established by the effective array aperture (0.9 m) and a sufficiently good SNR (larger than 10 dB).

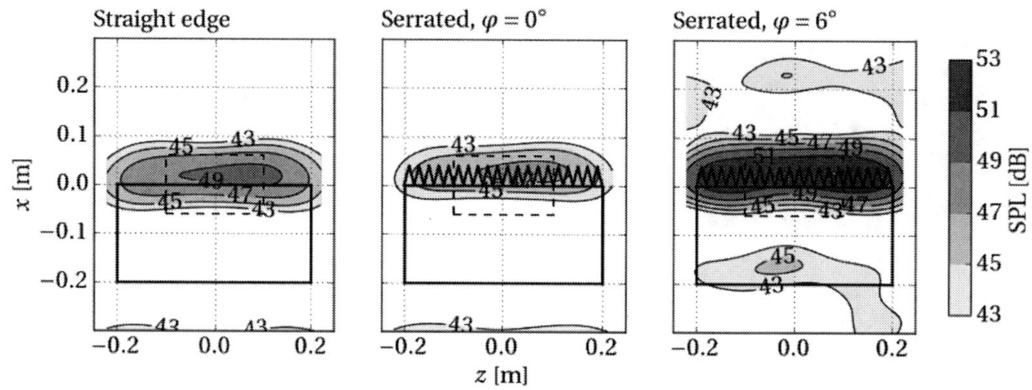

Figure 4: Source map results of the frequency domain beamforming at a one-third octave band of 4 kHz for the straight and serrated trailing edges. The airfoil and serrations are indicated, and the integration region is marked with the dashed rectangle.

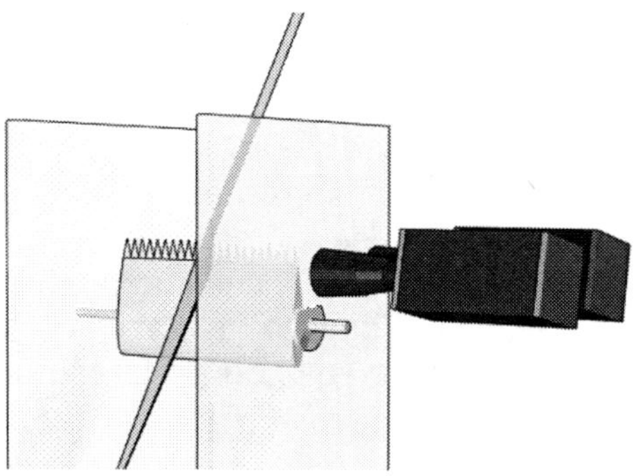

Figure 5: Planar PIV measurement setup.

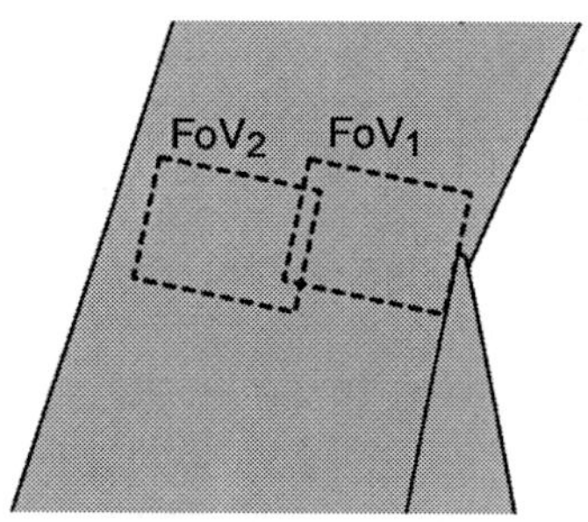

Figure 6: FoV configuration.

2.2 Flow measurements

Two sets of PIV measurements were conducted to characterize the flow. The first aimed at obtaining the boundary layer thickness from the airfoil with the straight trailing edge. Boundary layer properties were measured over both the suction and pressure sides with two-component planar particle image velocimetry (2C-PIV) for all the freestream velocities and angles of attack at which acoustic and S-PIV flow measurements were carried out. The second set of measurements were performed with S-PIV and were used to characterize the boundary layer flow statistics, both time-averaged and time-resolved. Both systems are described below.

The 2C-PIV system (schematic in figure 5) used a low-repetition rate configuration. Tracer particles in the flow (SAFEX, mean diameter $\sim 1\,\mu$m) were illuminated with an Nd:YAG 200 mJ/pulse laser (Quantel Twin BSL 200) with an acquisition frequency of 5 Hz. 300 uncorrelated image pairs were obtained with two CCD cameras (PCO Sensicam QE) with their respective field of view (FoV) overlapping, as shown in figure 6. The

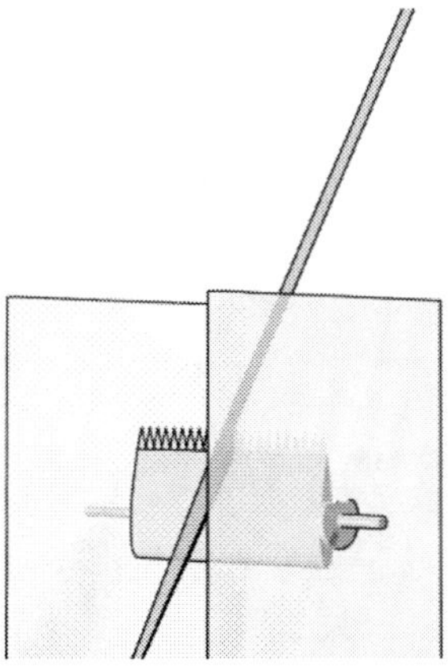

Figure 7: Stereoscopic PIV measurement setup.

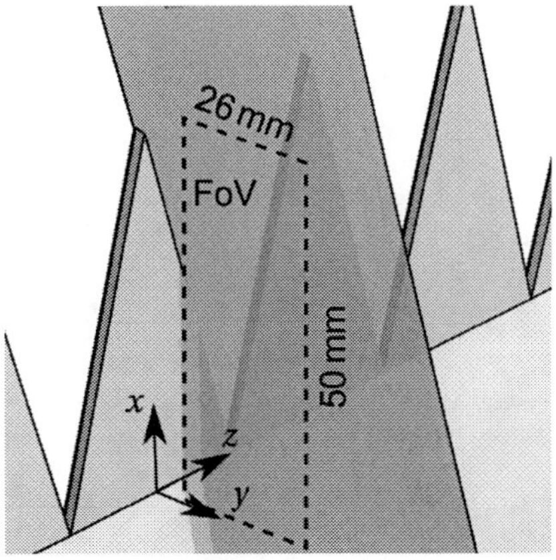

Figure 8: Location of the field of view for the stereoscopic PIV measurements in the $z/\lambda = 0.25$ measurement plane.

combined FoV yielded an effective measurement region of 36×16 mm. The cameras were arranged as shown in figure 5 with their optical axes perpendicular to the flow direction. They have a sensor size of 1376×1040 px with a pixel pitch of 6.7 µm/px, and were equipped with 105 mm focal length Nikon NIKKOR objectives operated at an F-number of $f/8$. The resulting digital image resolution was approximately 65 px/mm. At a pulse separation time of 15 µs, a freestream particle displacement of about 20 px was obtained for 20 m/s.

The S-PIV system (figure 7) was used in two configurations; one for the acquisition of time-averaged data, and another for time-resolved data. In both cases, the tracer particles (same as in the 2C-PIV system) were illuminated with a dual-cavity Nd:YLF high-speed laser (Quantronix Darwin Duo, 2×25 mJ/pulse at 500 Hz frequency) capable of 2×25 mJ/pulse. Two CMOS cameras (Photron Fastcam SA1.1, 1024×1024 px, 20 µm/px pixel-pitch, 12 bit resolution) were set up in stereoscopic configuration (35° angle offset between the optical axes) in order to resolve the out-of-plane (spanwise) velocity component w. The same camera optics were used as before, with an F-number of $f/5.6$. A digital image resolution of 20 px/mm was obtained over an effective FoV of $26 \times 50\,\text{mm}^2$, as indicated in figure 8 (the sensor was cropped to 512×1024 px to reach the desired high-frequency acquisition rate).

For the time-averaged measurements, a 250 Hz acquisition frequency was used to acquire 2000 image pairs at a pulse separation of 50 µs, resulting in a freestream particle displacement of around 12.5 px. The time-resolved flow information was obtained with 10,100 particle images per case at 10 kHz acquisition frequency, such that time separation between image pairs of 100 µs was achieved, resulting in a freestream particle displacement of around 25 px. Due to technical restrictions imposed by the highest achievable acquisition frequency of the current S-PIV system (10 kHz), the freestream

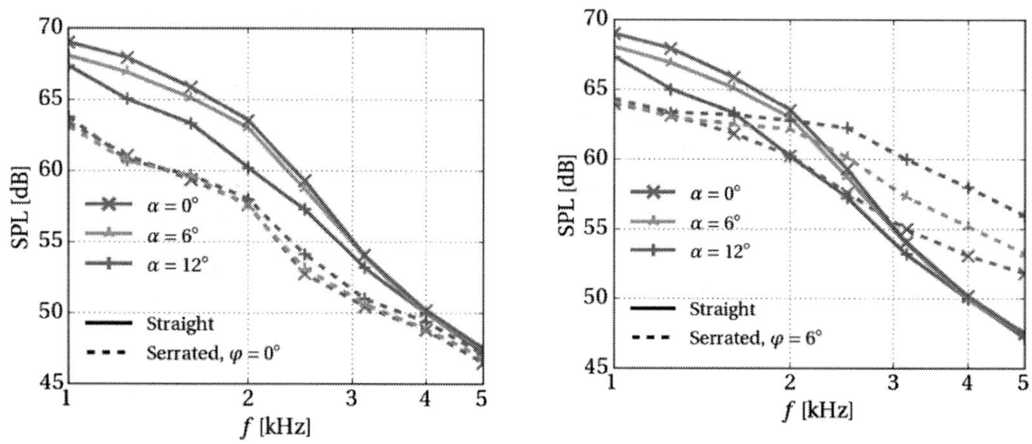

Figure 9: Third-octave band SPL for the straight and serrated trailing edges for various α. $U_{\infty} = 35$ m/s, $\varphi = 0°$ (left) and $\varphi = 6°$ (right).

velocity was limited to 20 m/s.

In all cases, the LaVision software DaVis 8 was used for acquisition and processing. Multi-pass stereoscopic cross-correlation was applied, resulting in a final window size of 16×16 px overlapped by 75% (50% in the case of the 2C-PIV). For the S-PIV, a spatial resolution of 0.8 mm and vector spacing of 0.2 mm were obtained, while for the 2C-PIV it was respectively 0.24 mm and 0.12 mm. An error analysis of the measurements is omitted for conciseness and the reader is referred to Arce León et al., 2016a for further details. The random error is found to be approximately 1% in the freestream velocity and around 3% in the inner boundary layer. Considering the 2000 acquired images, the resulting error in the mean velocity is within 0.05%, and 2% for the root-mean-square (rms).

3. Results

3.1 Acoustic Emissions and Strouhal Number Evaluation

The third-octave band spectra of the trailing edge radiated sound of the straight and serrated trailing edges are shown in figure 9, for $U_{\infty} = 35$ m/s, at the different investigated angles of attack. The case of $\varphi = 0°$ is shown on the left, and $\varphi = 6°$ on the right.

The noise reduction capability of the serrations is evident. For $\varphi = 0°$ serrations perform reasonably well for all investigated angles of attack, reducing noise in the investigated frequency range by up to 7 dB. The application of a flap angle, $\varphi = 6°$, severely degrades their performance.

The level of noise reduction is weaker when a flap angle is present, or equivalently, the serrations appear to become a source of noise themselves beyond a certain frequency, as indicated in Gruber, Joseph, and Chong, 2011.

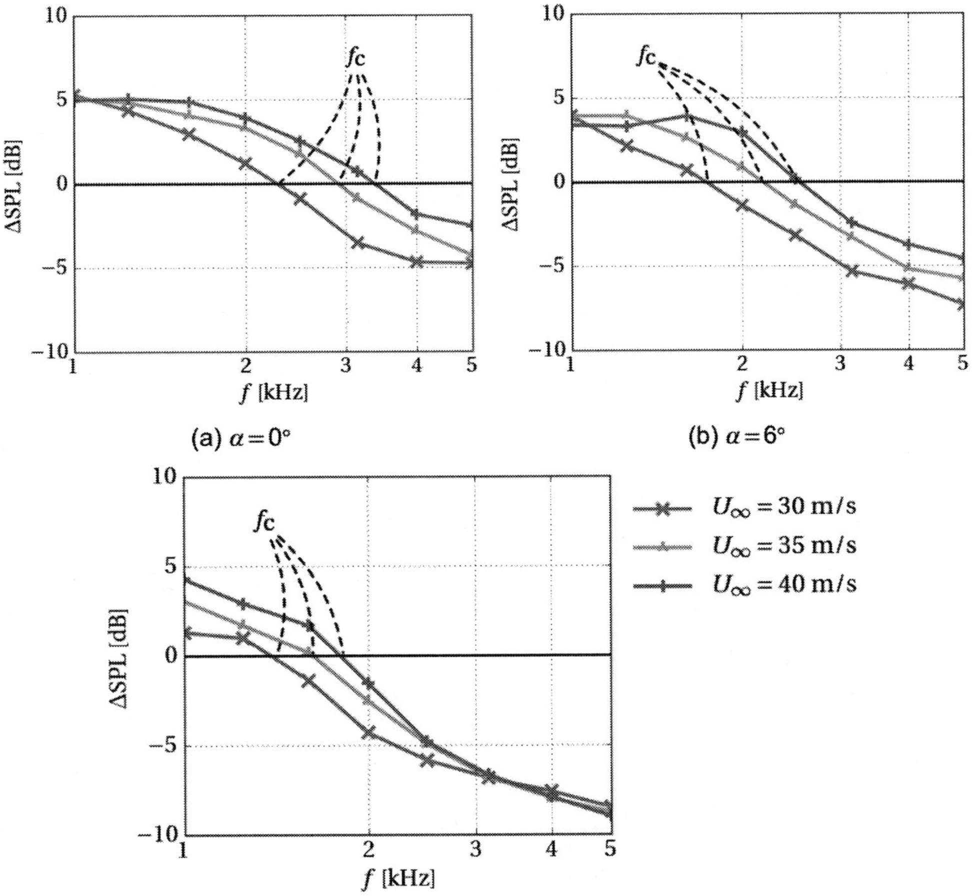

(a) $\alpha = 0°$

(b) $\alpha = 6°$

(c) $\alpha = 12°$

Figure 10: Noise reduction for the serrated trailing edge relative to the baseline airfoil for different α values and $\varphi = 6°$. The crossover frequency f_c is indicated for each U_∞.

To facilitate the analysis, third-octave spectra relative to the noise emissions of the straight trailing edge are presented in figure 10. Here, a positive value indicates noise reduction. The three freestream velocities are shown, at the three angles of attack, for $\varphi = 6°$. The crossover frequency, f_c, is indicated.

The results provide further proof that serration-flow misalignment leads to the production of noise after a certain frequency. Focus will now be paid to whether this frequency can be established using a constant Strouhal number, as suggested by Gruber, Joseph, and Chong, 2011.

The results of the calculated Strouhal number, St_c, for $\varphi = 6°$ are shown in figures 11 and 12. Here f_c is obtained from the results shown in figure 10, and the boundary layer thickness parameters are retrieved from the measured boundary layers of the straight

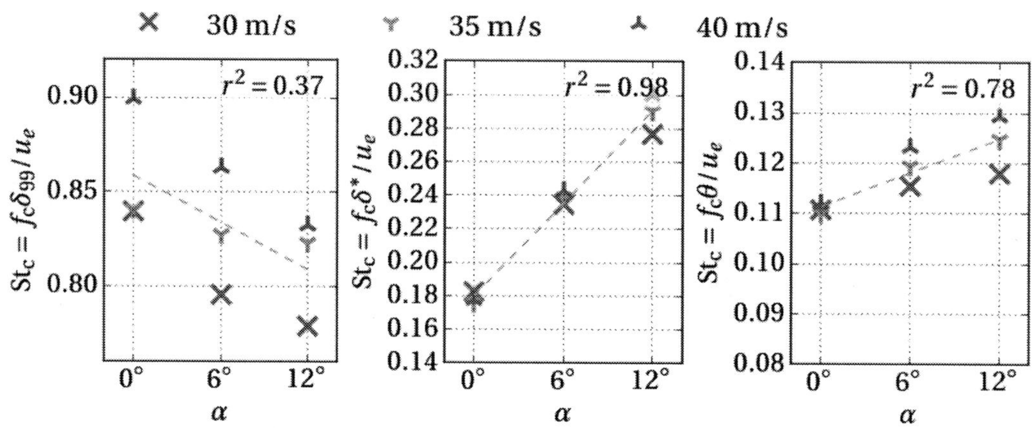

Figure 11: $\mathrm{St_c}$ for different boundary layer thickness parameters measured on the suction side. $U_\infty = 30$ m/s: ×, $U_\infty = 35$ m/s: ⋎, $U_\infty = 40$ m/s: ⅄.

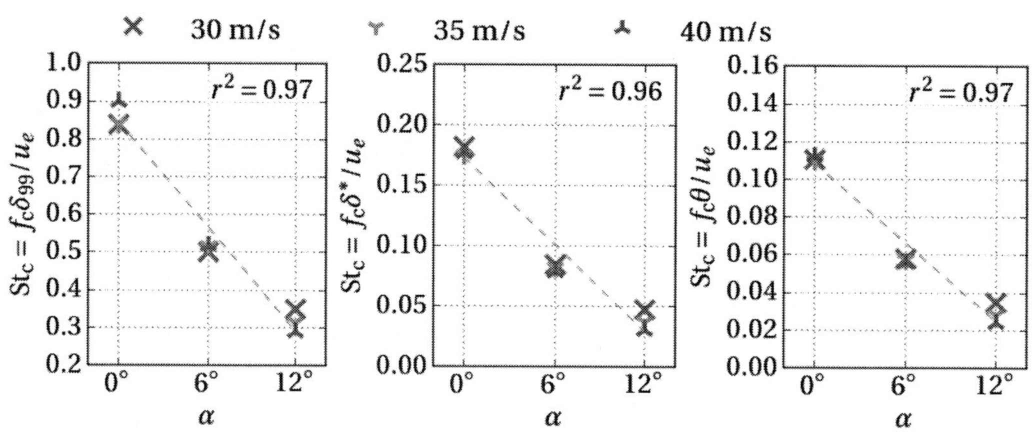

Figure 12: $\mathrm{St_c}$ for different boundary layer thickness parameters measured on the pressure side. $U_\infty = 30$ m/s: ×, $U_\infty = 35$ m/s: ⋎, $U_\infty = 40$ m/s: ⅄.

edge. Three thickness parameters are investigated in order to verify the trend behavior of each: the y location of the 99% edge velocity, δ_{99}, the displacement thickness, δ^*, and the momentum thickness, θ. The edge velocity, u_e, is taken as the velocity measure for $\mathrm{St_c}$. The results are presented for both the suction (figure 11) and the pressure sides (figure 12).

While Gruber, Joseph, and Chong, 2011 suggests that a constant Strouhal number that defines f_c can be established, this is not evident from the present results. Although $\mathrm{St_c}$ exhibits certain level of constancy across different freestream velocities for the same angle of attack, there is a general tendency for it to vary linearly for different values of α.

In the plots, the coefficient of determination, r^2, is indicated. It serves to quantify the quality of a linear fit over the different values of $\mathrm{St_c}$ for the various freestream velocities and angles of attack. It also helps to evaluate which boundary layer parameter, and which boundary layer side (pressure or suction side) would serve to better establish the behavior of $\mathrm{St_c}$. Based on this criterion, the pressure side boundary layer thickness values appear to offer a more robust trend behavior, with higher overall values of r^2.

Despite not finding a universal Strouhal number related to the crossover frequency, the observed linear behavior over the different angles of attack, and its moderate uniformity over different freestream velocities indicate that f_c is indeed related to the boundary layer thickness and u_e.

Nevertheless, developing a more general model to evaluate $f_c(\mathrm{St_c})$ is unlikely to be possible based on the available information. The exhibited trends are presumably dependent on the airfoil shape, and serration flap angle, variations of which are not thoroughly investigated here.

The results do however serve to suggest that the hydrodynamic source causing the increase in noise may reside on the pressure side of the serrations. This assertion is investigated in the following section.

3.2 Near-Edge Flow

Several features of the boundary layer flow are investigated. Focus is laid on the near-edge flow over the straight trailing edge, and over the serrated edge at $z/\lambda = 0.25$ (see figure 8). Comparisons are made between the latter at $\alpha = 12°$, and the former at $\alpha = 0°, \varphi = 0°$ and $\alpha = 12°, \varphi = 6°$. Where omitted, δ refers to δ_{99}.

As indicated earlier, the measurements are taken at 20 m/s freestream velocity. This is the highest velocity at which time-resolved data can be extracted given the hardware limitations of the S-PIV system. On the other hand, at this velocity, the TBL-TE noise is too weak to allow an accurate acoustic measurement in this tunnel, and results therefore in a low SNR, with respect to the background noise. The disparity between the flow and acoustic measurement velocities will be consolidated later.

The mean flow values of the three flow components are shown in figures 13 (stream-

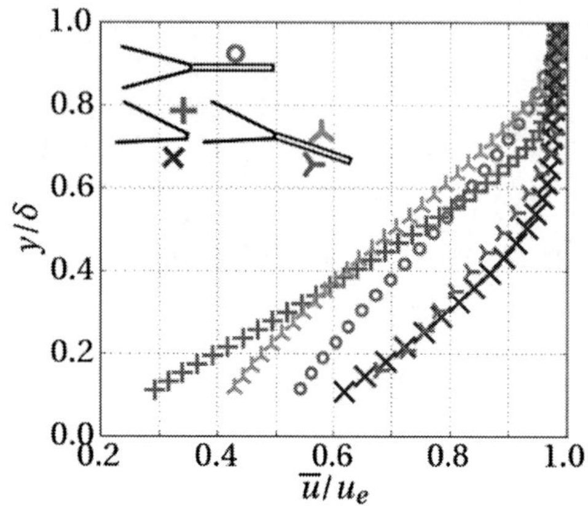

Figure 13: Streamwise mean flow component, $\overline{u}$.

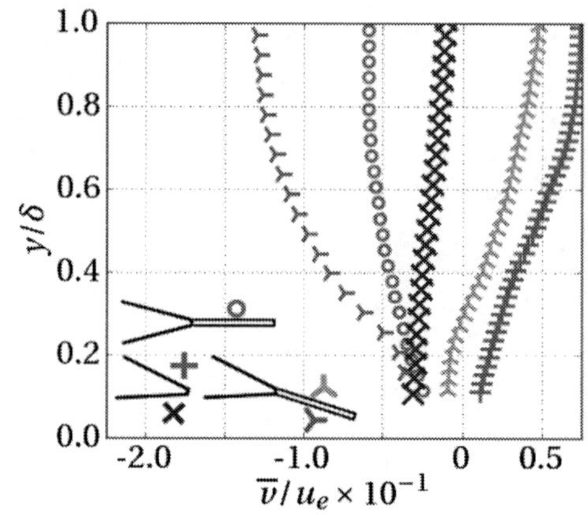

Figure 14: Wall-normal mean flow component, $\overline{v}$.

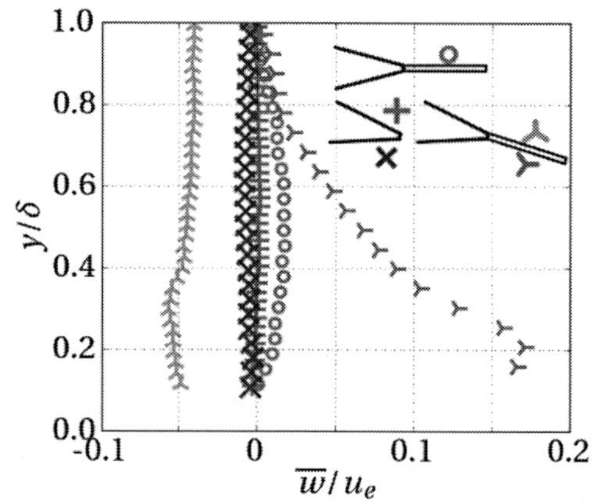

Figure 15: Spanwise mean flow component, $\overline{w}$.

wise, $\overline{u}$), 14 (wall-normal, $\overline{v}$), and 15 (spanwise, $\overline{w}$).

Noticeably higher values of $\overline{u}$ are present at the pressure side of both the straight and serrated edges at $\alpha = 12°$. On the suction side, lower values are observed with regard to the $\alpha = 0°$, $\varphi = 0°$ case, indicative of a higher adverse pressure gradient near the edge.

The mean wall-normal velocity component for the $\alpha = 12°$, $\varphi = 6°$ serrated edge configuration shows flow with an orientation towards the surface (negative values), as would be expected due to the serration misalignment. It further exhibits the largest wall-normal component magnitude between the investigated configurations. Along the suction side, the flow is oriented away from the wall, for both the serrated and straight edge cases, as expected.

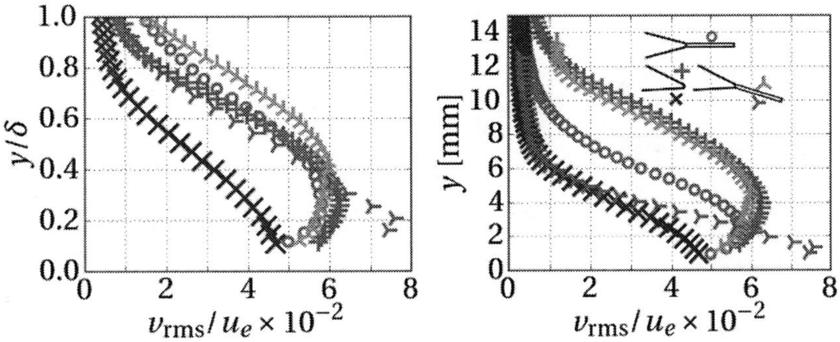

Figure 16: Wall-normal fluctuations, v_{rms}.

A very notable feature in the mean spanwise flow component, $\overline{w}$, is found on the pressure side of the serrated $\alpha = 12°, \varphi = 6°$ case, (figure 15). This large deviation from the other cases is a reflection of the significant spanwise flow deflection found on the pressure side in serration-flow misaligned situations, as detailed in Arce et al., 2015; Arce León et al., 2016b. The negative values observed over the suction side of the same case also relate to this condition, as flow becomes slightly deflected towards the center of the serration.

This feature correlates with a high rms observed in the v component, as shown in figure 16. Here both the δ-normalized (left) and absolute (right) distances from the wall are indicated. The closeness of the high v_{rms} value region to the surface and the edge is evident, at about 1 mm, sustaining the assumption that it could play a role in increasing acoustic emissions. It is important to note that this feature is not present in the non-serrated airfoil, committing it to the effect that the serration misalignment has on the flow.

The measurements of $10\log_{10}\Phi_{vv}$, for Φ_{vv} the wall-normal flow component spectra, are presented in figure 17. The serrated $\alpha = 0°, \varphi = 0°$ case is omitted for brevity, but all the other cases of the previous figures are shown. The results are given for four wall-normal locations. The closest (at $y = 1.5$ mm) relates to the location of high v_{rms} seen in figure 16 for the serrated case, and the furthest (12 mm) is near the edge of the boundary layer of the suction side flow. The Kolmogorov law (dashed line) is shown for reference.

The most distinct feature related to the topic investigated here is the higher turbulent energy observed beyond a frequency of about 1.1 kHz for the pressure side serrated case. The energy increases with increasing frequency, reaching over 5 dB at around 5 kHz. Its relation to the high frequency noise increase is likely, but in order to consolidate this argument, the velocity difference between the flow and acoustic measurements must first be conciliated. In order to do this, the Strouhal number relation discussed in section 2.1 is reintroduced here.

Using the observations relating to figure 12, the expected value of the acous-

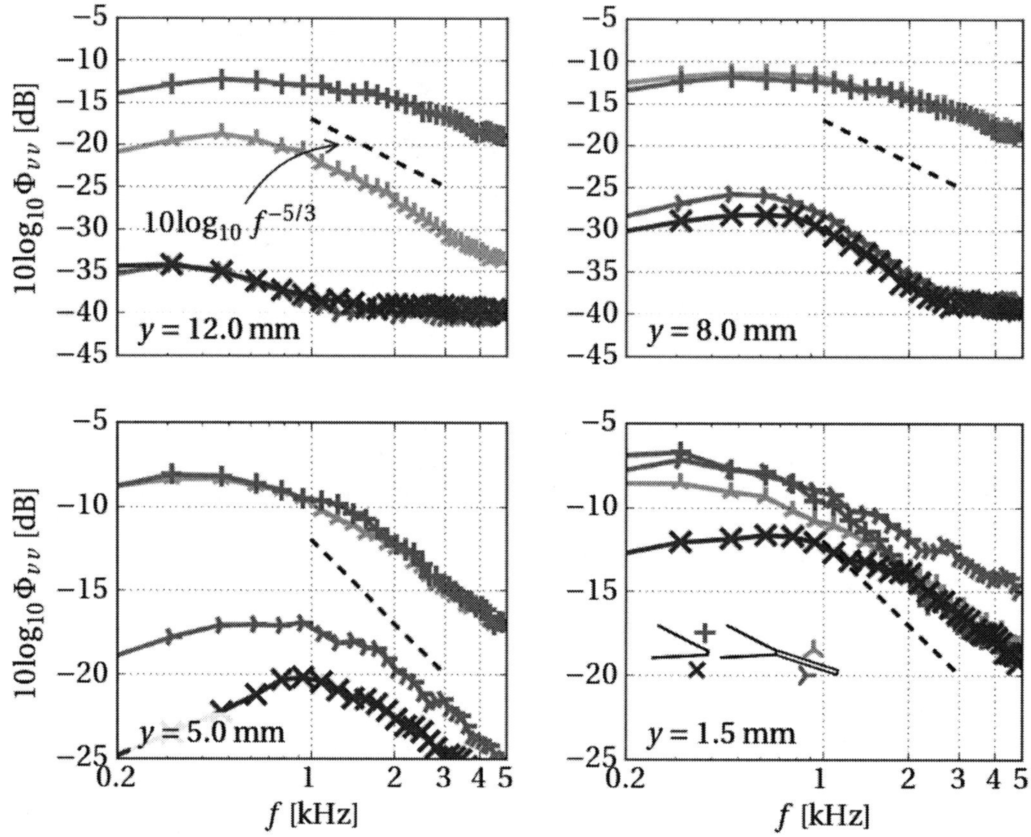

Figure 17: Spectrum of the wall-normal velocity component for different wall-normal locations.

tic crossover frequency, f_c, can be established for comparison with the turbulence crossover frequency at 20 m/s. This analysis is represented in figure 18.

The crossover frequency at 20 m/s is here estimated using the Strouhal numbers found above in figure 12. That is,

$$f_c = \mathrm{St}_c u_e/\delta \tag{2}$$

for $u_e = 20$ m/s and δ measured with 2C-PIV. The results using the St_c values found for the freestream velocities 30, 35 and 40 m/s are indicated by the symbols ×, Y and ⅄ respectively. Since there was no perfect collapse of St_c found for these freestream velocities, the result of St_c for 20 m/s is linearly extrapolated (+ symbol) and used to establish f_c.

While the values differ depending on the boundary layer thickness parameter that is used, the results collapse to around 1.1 kHz. The mean is indicated in the figure. A correlation is then hereby established between the acoustic and turbulence crossovers.

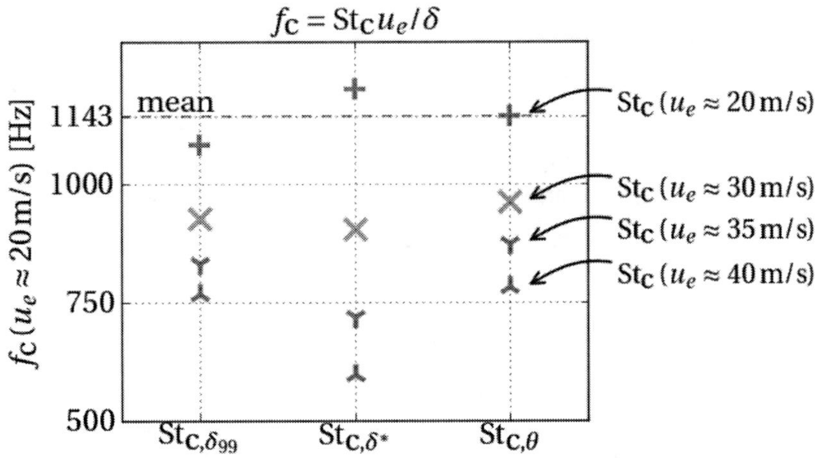

Figure 18: Predicted value of f_C based on the pressure side measurements of St_C based on $U_\infty = 30$ m/s (×), $U_\infty = 35$ m/s (Y), $U_\infty = 40$ m/s (⅄), and the linear extrapolation to $U_\infty = 20$ m/s (+).

4. Conclusions

The effects of serration-flow misalignment have been investigated based on its modification of the acoustic emissions and boundary layer flow.

Measurements using acoustic beamforming confirm that an increase in noise is experienced after a certain crossover frequency, and that it is related to serration-flow misalignment.

Based on the pressure side boundary layer thickness and its edge velocity, this crossover frequency has been found to follow a linear Strouhal number behavior dependent on the angle of attack of the airfoil. While somewhat consistent Strouhal numbers were found for varying edge velocities, the idea that a single Strouhal number can be used to establish the crossover frequency is disputed, and it is more likely to be heavily dependent on angle of attack, serration flap angles, airfoil shape, and a number of other omitted parameters.

Flow measurements of the boundary layer indicate the presence of some notable features on the pressure side. A high sideways flow deflection indicated by $\overline{w}$, and a high level of v_{rms} that correlates to it, occur very near the surface and edge of the serration pressure side. At this location, an increase in the higher frequency turbulence energy is also observed.

The Strouhal number analysis is used to establish a relationship between this increase in the turbulence energy and that of the acoustic emissions. The crossover frequency correlation found between the two, in addition to the mean flow observations, establish that the increased source of noise is of a hydrodynamic nature, and occurs on the pressure side of flow-misaligned serrations.

References

Arce, Carlos et al. (2015). "Flow Field Around a Serrated Trailing Edge at Incidence". In: *33rd Wind Energy Symposium*. Kissimmee, Florida: American Institute of Aeronautics and Astronautics. ISBN: 978-1-62410-344-5. DOI: 10.2514/6.2015-0991.

Arce León, Carlos et al. (2016a). "Boundary layer characterization and acoustic measurements of flow-aligned trailing edge serrations". In: *Experiments in Fluids* 57.12, p. 182. ISSN: 0723-4864. DOI: 10.1007/s00348-016-2272-z.

Arce León, Carlos et al. (2016b). "Flow topology and acoustic emissions of trailing edge serrations at incidence". In: *Experiments in Fluids* 57.5, p. 91. ISSN: 0723-4864. DOI: 10.1007/s00348-016-2181-1.

Arce León, Carlos et al. (2016c). "PIV Investigation of the Flow Past Solid and Slitted Sawtooth Serrated Trailing Edges". In: *54th AIAA Aerospace Sciences Meeting*. January. San Diego, California: American Institute of Aeronautics and Astronautics, pp. 1–15. ISBN: 978-1-62410-393-3. DOI: 10.2514/6.2016-1014.

Arce León, Carlos et al. (2017). "Effect of trailing edge serration-flow misalignment on airfoil noise emissions". In: *Journal of Sound and Vibration* In review.

Dassen, T et al. (1996). *Results of a wind tunnel study on the reduction of airfoil self-noise by the application of serrated blade trailing edges*. Tech. rep. National Aerospace Laboratory, NLR.

Finez, A et al. (2011). "Broadband Noise Reduction of a Linear Cascade With Trailing Edge Serrations". In: *17th AIAA/CEAS Aeroacoustics Conference (32nd AIAA Aeroacoustics Conference)*. Portland, Oregon, USA. DOI: 10.2514/6.2011-2874.

Ghaemi, Sina, Daniele Ragni, and Fulvio Scarano (2012). "PIV-based pressure fluctuations in the turbulent boundary layer". In: *Experiments in Fluids* 53.6, pp. 1823–1840. ISSN: 0723-4864. DOI: 10.1007/s00348-012-1391-4.

Gruber, M (2012). "Airfoil noise reduction by edge treatments". PhD thesis. University of Southampton.

Gruber, M, P Joseph, and TP Chong (2010). "Experimental investigation of airfoil self noise and turbulent wake reduction by the use of trailing edge serrations". In: *16th AIAA/CEAS Aeroacoustics Conference*, pp. 1–23.

Gruber, Mathieu, Phillip Joseph, and Tze Chong (2011). "On the mechanisms of serrated airfoil trailing edge noise reduction". In: *17th AIAA/CEAS Aeroacoustics Conference (32nd AIAA Aeroacoustics Conference)*. Vol. 2781. June. Portland, Oregon, USA: American Institute of Aeronautics and Astronautics, pp. 5–8. ISBN: 978-1-60086-943-3. DOI: 10.2514/6.2011-2781.

Hurault, J et al. (2015). "Aeroacoustic wind tunnel experiment for serration design optimisation and its application to a wind turbine rotor". In: *6th International Meeting on Wind Turbine Noise*. Glasgow.

Johnson, Don H. and Dan E. Dudgeon (1993). *Array Signal Processing, Concepts and Techniques*. ISBN: 9780130485137. P T R Prentice Hall, Englewood Cliffs.

Mathew, J et al. (2016). “Serration Design Methodology for Wind Turbine Noise Reduction”. In: *Journal of Physics: Conference Series* 753, p. 022019. ISSN: 1742-6588. DOI: 10.1088/1742-6596/753/2/022019.

Mueller, TJ (2002). *Aeroacoustic Measurements*. ISBN-978-3-642-07514-8. Springer Science & Business Media, p. 313. ISBN: 3540417575.

Oerlemans, S et al. (2009). “Reduction of wind turbine noise using optimized airfoils and trailing-edge serrations”. In: *AIAA Journal* 47.6, pp. 1470–1481. ISSN: 0001-1452. DOI: 10.2514/1.38888.

Oerlemans, Stefan (2016). “Reduction of wind turbine noise using blade trailing edge devices”. In: *22nd AIAA/CEAS Aeroacoustics Conference*, pp. 1–18. DOI: 10.2514/6.2016-3018.

Pröbsting, S. et al. (2016). “Vortex shedding noise from a beveled trailing edge”. In: *International Journal of Aeroacoustics* 15.8, pp. 712–733. ISSN: 1475-472X. DOI: 10.1177/1475472x16666633.

Sijtsma, P. (2010). *Phased array beamforming applied to wind tunnel and fly-over tests*. Tech. rep. NLR-TP-2010-549. Anthony Fokkerweg 2, 1059 CM Amsterdam, The Netherlands: National Aerospace Laboratory (NLR).

Vathylakis, Alexandros, Tze Pei Chong, and Phillip F. Joseph (2015). “Poro-Serrated Trailing-Edge Devices for Airfoil Self-Noise Reduction”. In: *AIAA Journal*, pp. 1–16. ISSN: 0001-1452. DOI: 10.2514/1.J053983.

Vathylakis, Alexandros et al. (2016). “Sensitivity of aerofoil self-noise reductions to serration flap angles”. In: *22nd AIAA/CEAS Aeroacoustics Conference*, pp. 1–17. DOI: 10.2514/6.2016-2837.

7th International Conference
on
Wind Turbine Noise
Rotterdam - 2nd to -5th May 2017

An Experimental Parametric Study of Airfoil Trailing Edge Serrations

T. Carolus University of Siegen , 57068 Siegen, Germany
E-Mail: thomas.carolus@uni-siegen.de

F. Manegar University of Siegen, 57068 Siegen, Germany
E-Mail: farhan.manegar@uni-siegen.de

E. Thouant École Centrale de Lyon, Lyon, France
E-Mail: elodie.thouant@gmail.com

K. Volkmer University of Siegen , 57068 Siegen, Germany
E-Mail: kevin.volkmer@uni-siegen.de

I. Schmich-Yamane Électricité de France, Grenoble, France
E-Mail: isabelle.schmich-yamane@edf.fr

ABSTRACT
Self-noise of an airfoil arises due to different mechanisms. In applications such as wind turbines the turbulent boundary layer interacting with the trailing edge is thought to be the dominant mechanism. In this study, a parametric study of add-on sawtooth type trailing edge serrations is carried out for trailing edge noise reduction. The main parameters which are known to control the performance of this trailing edge serrations type are the height and width of the serration teeth (i.e., amplitude and wavelength, respectively). Several parameters have been investigated experimentally in an aero-acoustic wind tunnel at the University of Siegen on a Somers S834 airfoil section. These parameters include the two main parameters mentioned above as well as orientation, thickness and the side on the airfoil, where serrations are attached. The experimental results show a maximum noise reduction for thin, long and narrow serration teeth, attached on the pressure side and oriented along the wake of the airfoil.

1. INTRODUCTION

Wind energy is a renewable source of energy and wind turbines convert energy without causing substantial adverse effects to the environment. However, the wide spread of wind turbines near densely populated regions is restricted mainly due to the emitted noise. Studies conducted to identify the noise sources from wind turbines have revealed that the dominant noise source is the flow induced trailing edge noise (TEN) [1]. An airfoil immerged in a flow emits TEN when the turbulent structures within the turbulent boundary layer (TBL) convect past the trailing edge (TE). These turbulent structures are scattered and radiated into the free field because of the edge discontinuity. TEN is broadband in nature [1].

Noise reduction techniques for TEN have been investigated since almost half a century now. They are classified into two types, active and passive. Some active techniques involve trailing edge blowing [2] and trailing edge suction [3]. Examples of passive techniques are the inclusion of porous trailing edges [4], serrations [5] and brushes [6] at the TE. Noise reduction techniques for wind turbines considering the practical constraints, converge towards trailing edge serrations (TES).

HOWE [7] contributed to the understanding of the effect of TES by deriving an analytical noise radiation model for low Mach number (Ma) flows to predict the noise reduction caused by sawtooth serrations on an infinite flat plate airfoil at zero angle of attack. More about HOWE's theory will be discussed in chapter 2. Many experimental works reported large deviations between the noise reduction experimentally achieved with TES and the noise reduction predicted with HOWE's model [5, 8-10]. LYU et al. [11] developed a new theoretical model and compared it with HOWE's model. In LYU's model the convection effects are incorporated in the convected wave equation, thereby making the model valid for all Ma numbers. In addition, by taking the effects of incident wall pressure gust into consideration the model tends to narrow down the deviation with the experiments. It was also enlightened that noise reduction is caused due to the destructive interference, i.e., out-of-phased scattered pressure in the vicinity of the TE.

JONES et al. [12] conducted direct numerical simulations (DNS) of the flow around a NACA-0012 airfoil with and without TES to understand the physical mechanism of TES. Two serration geometries were investigated, one with an amplitude $2h$ (see **Fig. 1**) equal to the boundary layer thickness δ at the TE (named as short) and another one with an amplitude of 2 δ (named as long), with wavelength λ being constant for both designs. Compared to the short TES the long TES provided higher noise reduction for all frequencies. The short TES provided less noise reduction and this only in a finite frequency range, above which the noise level increased. An interesting observation in this study was that the aerodynamic properties remained unchanged upstream of the TE for both cases, with and without TES. Hence the noise reduction is attributed to be caused due to the changes in scattering process and also due to the changes in the aerodynamic behavior in the direct vicinity of the serrations.

GRUBER [5] who did an extensive study of TES showed that the noise reduction is due to the reduction of phase speed at which the turbulence is convected near the sawtooth edges. In addition to that, a reduction of coherence of pressure measured along the sawtooth is also reported in the frequency band where noise reduction happens. OERLEMANS et al. [10] measured noise emission of a large scale wind turbine with one optimized blade, with serrations on another blade and the third blade remaining unaltered. His findings were that serrations resulted in a noise reduction of 2 – 3 dB; at higher frequencies a noise increase was observed, which was explained to be due to misalignment of the serrations with the flow. VATHYLAKIS et al. [13] investigated the effect of flap angles on the noise reduction on a NACA65(12)-10 airfoil, where flap angles from -15° to + 15° in steps of 5° have been studied. The flap-up position (+5°), which is oriented towards the suction side, resulted in the best noise reduction, the flap-down position (-5°) was the worst configuration.

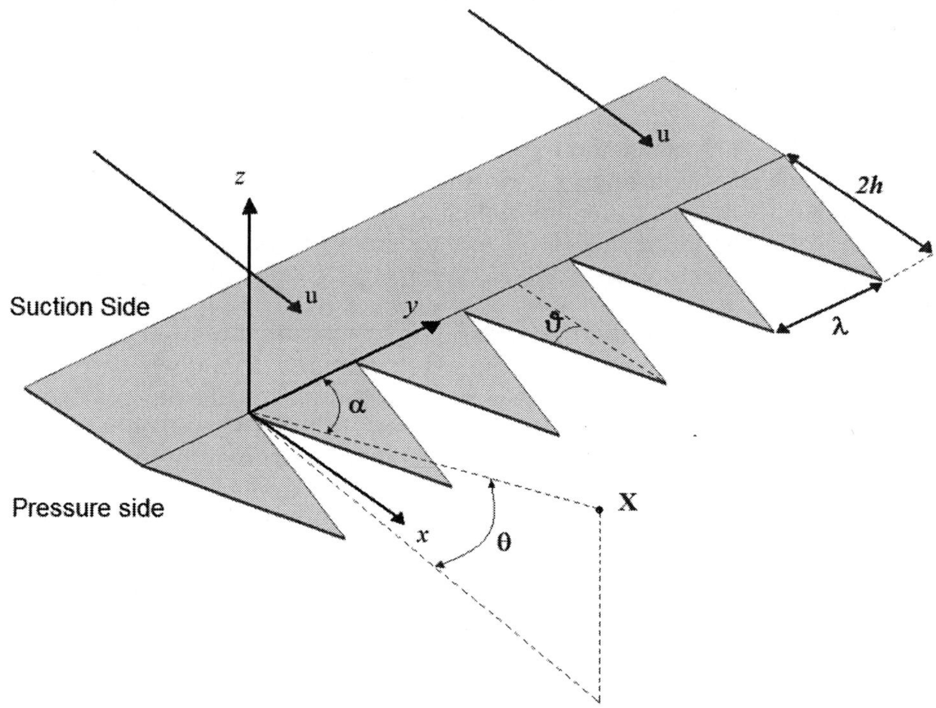

Fig. 1 Geometrical parameters of a serrated trailing edge and the coordinate system.

This paper presents a preliminary experimental model scale study on the acoustic effect of various parameters apart from the classical serration geometry "amplitude" and "wavelength", which could influence the noise reduction capability of TES. Additionally, a discussion on some findings of other researchers like deviation between experiment and HOWE's model, effects of flap angle and effects of serration amplitude.

2. HOWE's theory

The serrations which are investigated in this work are designed based on HOWE's theory leading to a specific design of TES. A brief description of his theory is presented in this section.

HOWE [7] derived an analytical model for predicting the noise from a flat plate at zero angle of attack with sawtooth serrations attached to TE. The sawtooth serrations have a spatial periodicity called as wavelength λ and a root-to-tip distance called as amplitude *2h* as shown in **Fig 1**. Green's function is used to calculate the pressure radiated to the observer's location. The use of modified Green's function in the case of serrated TE is argued by LYU et al. [11], who instead proposes to calculate scattered sound using the convected wave equation.

This is the reason why HOWE's model is valid only for low Ma numbers, because it neglects the effects of convection (the direct radiation to the far field due to the quadrupole sources in the boundary layer). HOWE also assumes that the properties of turbulence remain the same before and after the TE within the boundary layer, in other words, he assumes a frozen turbulence.

Using CHASE's [14] model for wavenumber frequency spectrum of the blocked wall pressure inside the turbulent boundary layer, the far-field sound power spectrum at the observer point at a distance |X| from TE is given as in eq. (1)

$$\frac{\Phi(X,\omega)}{\left(\rho \upsilon_*^2\right)^2\left({l}/{c_0}\right)\left({\delta}/{|X|}\right)^2}=\left(\frac{C_m}{\pi}\right)\sin^2\left(\frac{\theta}{2}\right)\sin(\alpha)\Psi(\omega), \tag{1}$$

where ρ is the fluid density, l is the span, c_0 is the speed of sound, δ is the turbulent boundary layer thickness, θ and α are respectively polar and azimuthal observer angles, $\omega = 2\pi f$, where f is the frequency, skin friction coefficient $\upsilon_* \approx 0.03u$; $C_m \approx 0.1533$, $\varepsilon \approx 0.133$ are constants and the convection velocity is $U_c \approx 0.7u$. The non-dimensional edge noise spectrum is

$$\Psi(\omega)=\left(1+\frac{1}{2}\varepsilon\frac{\partial}{\partial\varepsilon}\right)f\left(\frac{\omega\delta}{U_c},\frac{h}{\lambda},\frac{h}{\delta};\varepsilon\right),$$

where

$$f\left(\frac{\omega\delta}{U_c},\frac{h}{\lambda},\frac{h}{\delta};\varepsilon\right)=\frac{1}{ab+\varepsilon^2}\left(1+\frac{64\left({h}/{\lambda}\right)^3\left({\delta}/{h}\right)\left(\cosh\left(c\sqrt{a+\varepsilon^2}\right)-\cos\left({2\omega h}/{U_c}\right)\right)}{\left(\sqrt{a+\varepsilon^2}\right)\left(ab+\varepsilon^2\right)\sinh\left(c\sqrt{a+\varepsilon^2}\right)}\right)$$

$$a=\left({\omega\delta}/{U_c}\right)^2,\ b=1+\left(4h/\lambda\right)^2,\ c={\lambda}/{2\delta},\ \varepsilon=1.33$$

The normalized spectra $10\log\left[\Psi(\omega)\right]$ (dB)for λ/h = 0.25, 1, 5 and h/δ = 1 and h/δ = 3, along with the spectrum for unserrated edge (not defined here, refer [7]) are plotted in **Fig 2**.

The important factors which control the noise reduction mechanism according to HOWE are as follows:

1) Noise reduction takes place when the non-dimensional frequency $\omega\delta/u$ is larger than 1.
2) The angle ϑ between mean flow and the local tangent to the wetted region should be less than 45°, i.e., the sharper the serrations, the higher the noise reduction.
3) Higher noise reductions occur when the dimensions of the serrations are of the order of the turbulent boundary layer thickness δ and more.

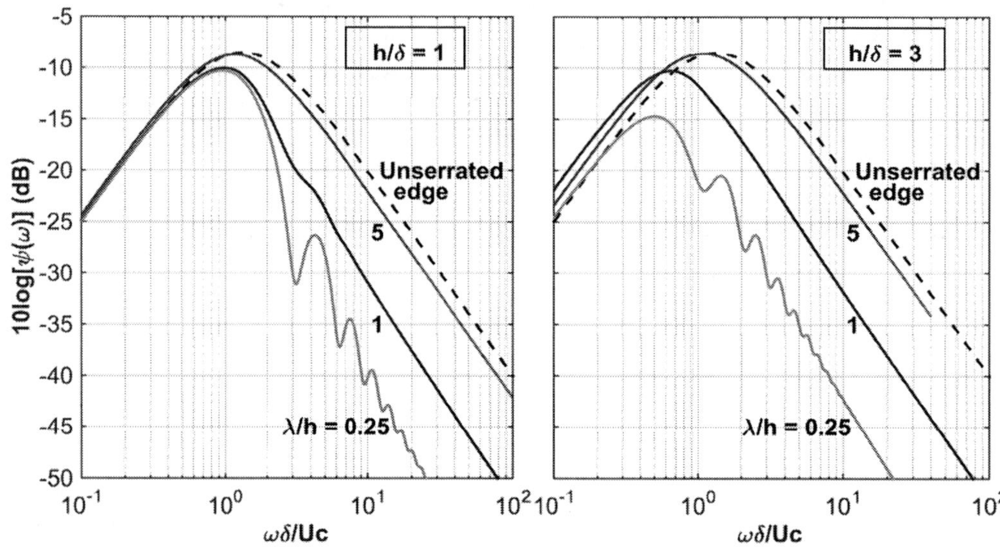

Fig. 2 HOWE's model: Normalized acoustic pressure frequency spectrum for unserrated and serrated edge as a function of λ/h and h/δ.

3. Experimental setup

3.1 Airfoil investigated and aero-acoustic wind tunnel

The experimental investigation was carried out with SOMERS S834 airfoil segment, with a chord length of 0.2 m and a span of 0.266 m in an aero-acoustic wind tunnel facility at the University of Siegen shown in **Fig. 3.** For an airfoil in a confined jet, it is essential to apply a correction factor for the angle of attack (AOA) as stated by BROOKS et al. [15]. This is due to the fact that the lifts produced by an airfoil in a free stream and in a confined jet are not the same, as the finiteness of the jet leads to significant flow deflection. The effective AOA for this study has been chosen such that an infinitely long span of the airfoil has maximum glide ratio. It has been calculated using XFOIL. The result is 4.7°. On applying the correction factor, the geometric AOA is obtained as 12.7°.

The airfoil segment is mounted between the side plates at the end of the contracting nozzle. A centrifugal fan is used to create the desired flow rate and the air is passed through a series of screens, honeycomb and silencers. The aero-acoustic wind tunnel provides a maximum flow velocity u = 25.55 m/s. The chord based Reynolds number (Re) is 350000. In order to replicate or mimic the real conditions in a large wind turbine, a ten times higher Re has to be achieved. Hence the airfoil is tripped at the natural transition position that occur at $3.5{\cdot}10^6$. The tripping positions are calculated using XFOIL and a zig zag trip is applied along the complete span at 34 mm from the leading edge (LE) on the suction side (SS) and 152 mm from the LE on the pressure side (PS).

3.2 Acoustic measurement

The wind tunnel exhausts in a semi-anechoic chamber which allows the acoustic measurements according to ISO 3745. The cut-off frequency of the chamber is 125 Hz. The dimensions of the semi-anechoic chamber are 4.5 m x 3.23 m x

2.9 m. The turbulence intensity of the jet is 0.2 % at a plane 0.01 m downstream the exit of the nozzle. The floor in the chamber is reflective and has an opening covered with grid, through which the flow recirculates. More details about the semi-anechoic chamber can be read in [16].

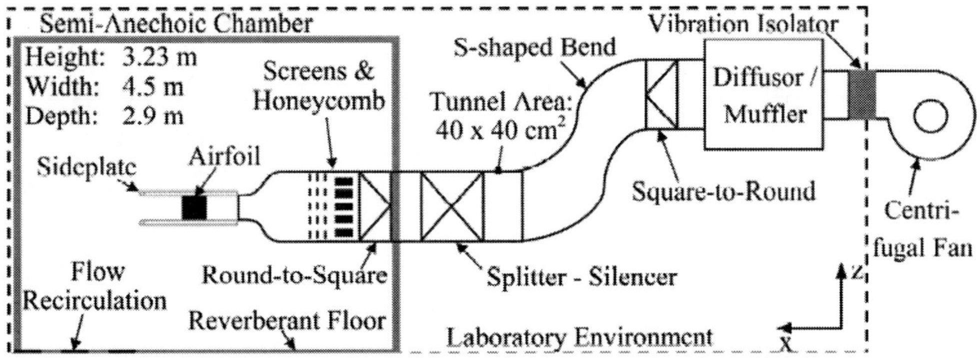

Fig. 3 Schematic diagram of aero-acoustic wind tunnel (not to scale)

Three microphones (1/2" Brüel & Kjaer TM, type 4190) are used to measure synchronously the noise being emitted from the airfoil. The microphones are covered with wind screens to avoid any flow induced pseudo sound and are located at a distance of 2.5 times chord (500 mm) from the TE as shown in **Fig 4**. All measurements are captured with a sampling rate of f_s = 51.2 kHz. The spectral analysis is based on the power spectral density S_{pp} obtained using the *pwelch* routine in Matlab™ Vers. R2014b (Δf_{ref} = 1 Hz, $p_0 = 2 \cdot 10^{-5}$ Pa). L_{Spp} is defined as shown in eq. (2)

$$L_{Spp}(f) = 10 \cdot \lg\left(\frac{S_{pp} \cdot \Delta f_{ref}}{p_0^2}\right) \text{ [dB]} \tag{2}$$

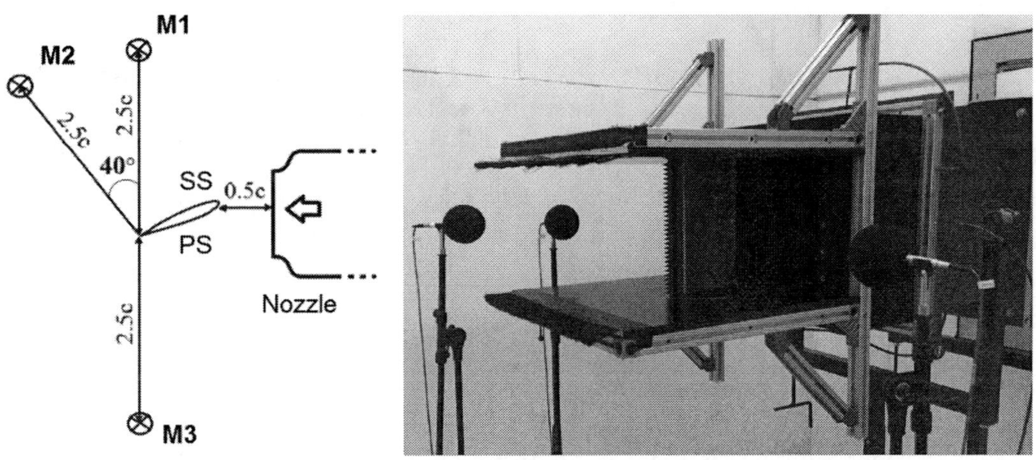

Fig. 4 Left: Schematic diagram of microphone locations, right: Microphones in semi anechoic chamber.

During TEN measurements the background noise due to the wind tunnel contaminates the acoustic signature of the airfoil Therefore, the measurements were done once with the reference airfoil and once without, to estimate the influence of the background noise. The L_{spp} spectrum of all three microphones averaged is shown in **Fig 5**. It has to be noticed that the signal to noise ratio is bad in frequency ranges above 3000 Hz, i.e., the background noise is too high in these frequency to separate airfoil from background noise. Hence in this study, the investigations of TES will be only displayed in a frequency range from 200 Hz to 3000 Hz.

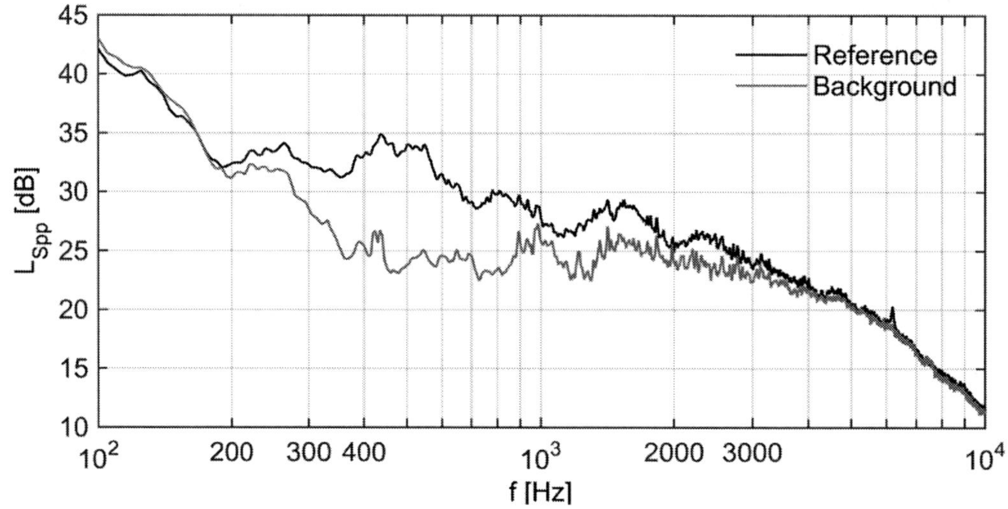

Fig. 5 L_{spp} of reference airfoil and background noise at microphone locations as shown in **Fig. 4** (average of 3 microphones).

3.3 Serrations investigated

As mentioned earlier, the serrations in this study are designed based on HOWE's theory and the following two parameters are defined to design the serration:

- The relative length h/δ, where h is half of serration's amplitude and δ is the boundary layer thickness at the TE, which is obtained from the experimental results of reference airfoil for the same flow characteristics conducted by GERHARD [17]. The here used value of δ is 9 mm.
- The relative wavelength λ/h, where λ being serration's wavelength.

To achieve noise reduction, the designed serrations must satisfy the following two conditions stated by HOWE:

- $1 < h/\delta < 10$
- $\lambda/h < 4$ or $\vartheta < 45°$

Before the various designs were investigated, a parametric study of three following important parameters was carried out:

- Fixation side: The side on which serrations are glued.
- Orientation: The angle φ the serration makes with respect to the camber line of the airfoil. The serrations oriented towards the suction side (SS) have positive values of φ and the serrations oriented towards the pressure side (PS) have negative values of φ. The serration aligned to camber line has φ = 0°. Orientation angles between +15° and -15° in steps of -5° have been investigated.

- Relative thickness t/δ, where t is the thickness of the serration and δ is the boundary layer thickness: the values of t/δ = 0.56 %, 3.3 %, 5.6 % and 11 % have been investigated.

After evaluating the results from this parametric study, the following three sets of serrations incorporated with the best fixation side, orientation and relative thickness are investigated:

Set 1: λ/h = 0.25, 1.00 , 5.00 with h/δ = 1

Set 2: λ/h = 0.25, 1.00, 1.50, 2.00, 2.50, 5.00 with h/δ = 3

Set 3: h/δ = 1.00, 1.50, 2.00, 2.50, 3.00 with λ/h = 1

All the parameters are shown in the schematic diagram in **Fig. 6** (left). A photograph of the serrations of Set 2 is shown in the right side of **Fig 6**.

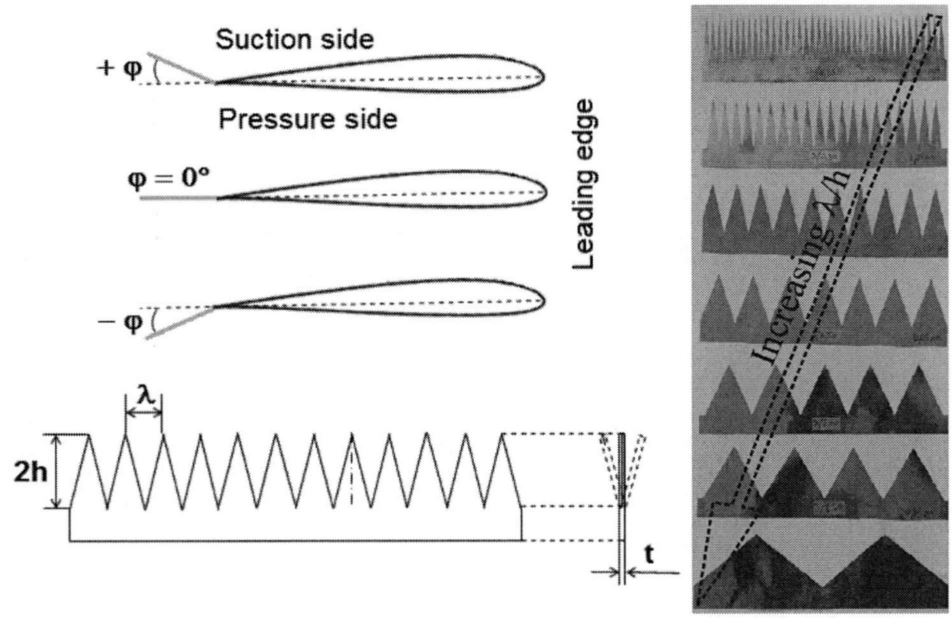

Fig. 6 Left: Schematic diagram of serration parameters, right: Serrations of Set 2.

4. Results

Three main parameters are analyzed one after the other. The best result in each analysis is kept for the next parameter's analysis.

4.1 Fixation side

For this investigation, a serration having h/δ = 1 and λ/h = 1 with a relative thickness t/δ = 0.67 % and the orientation angle φ = +5° is chosen. Measurements are carried out with this serration glued on the SS and PS respectively. It is to be noted that, when the serration is glued on the PS, the orientation angle is naturally φ = +5°, but when it is glued on the SS, the serration has to be oriented towards the SS to yield the same orientation angle φ = +5°(**Fig 7**).

The L_{SPP} spectra for both cases are plotted in **Fig. 8** along with the L_{SPP} spectrum for the reference case without TES. It is observed that the serration when glued on SS results in less noise reduction in comparison to the serration glued on PS. Hence for the following investigations, the serration is glued on the PS.

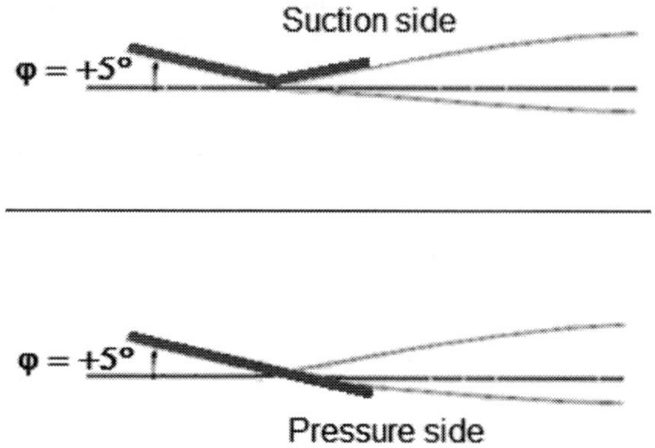

Fig. 7 Schematic diagram of different fixation sides of TES.

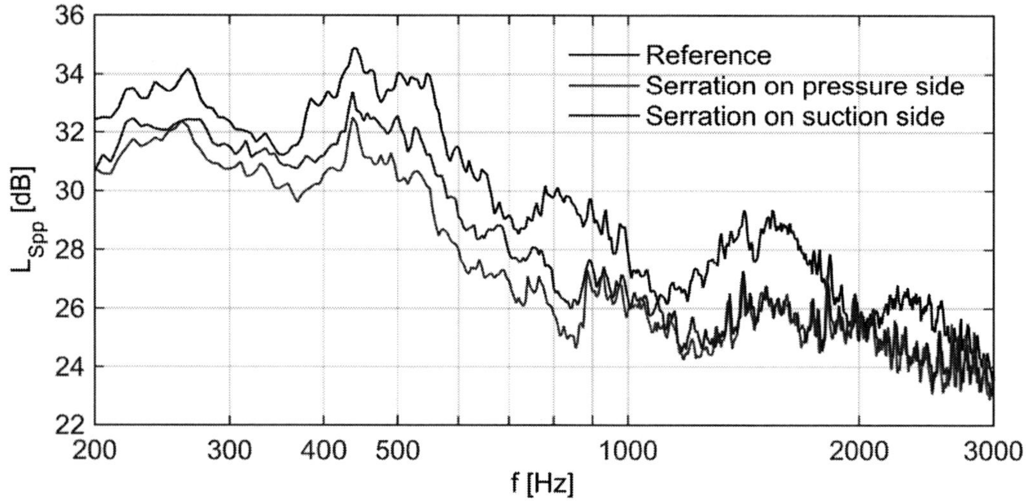

Fig. 8 Effect of fixation side of serrations.

4.2 Orientation

For this investigation, a serration having h/δ = 1 and λ/h = 1 with a relative thickness of t/δ = 3.3 % is chosen and glued on the PS. Measurements are carried out for orientation angles φ = +15° to -15° in steps of -5°. As mentioned earlier, the serrations orienting towards SS have positive values of φ and the serrations orienting towards PS have negative values of φ. The serration aligned to the camber line has an angle φ = 0°. The results are presented in three figures. First a comparison of φ = +15°, 0° and -15° is shown in **Fig 9a**. It is to be noted that the serration oriented towards SS (φ = +15°) brings a potential reduction only beyond 600 Hz. On the contrary, the serration oriented towards PS (φ = -15°) increases noise emission beyond 600 Hz. The serration aligned with camber line (φ = 0°) reduces the TEN almost in the entire frequency range displayed. In **Fig. 9b**, the comparison of φ = -10°, 0° and 10° is shown. The same tendency is observed here except for the fact that the frequency at which the change of behavior happens is shifted to 750 Hz, which was 600 Hz in the previous case.

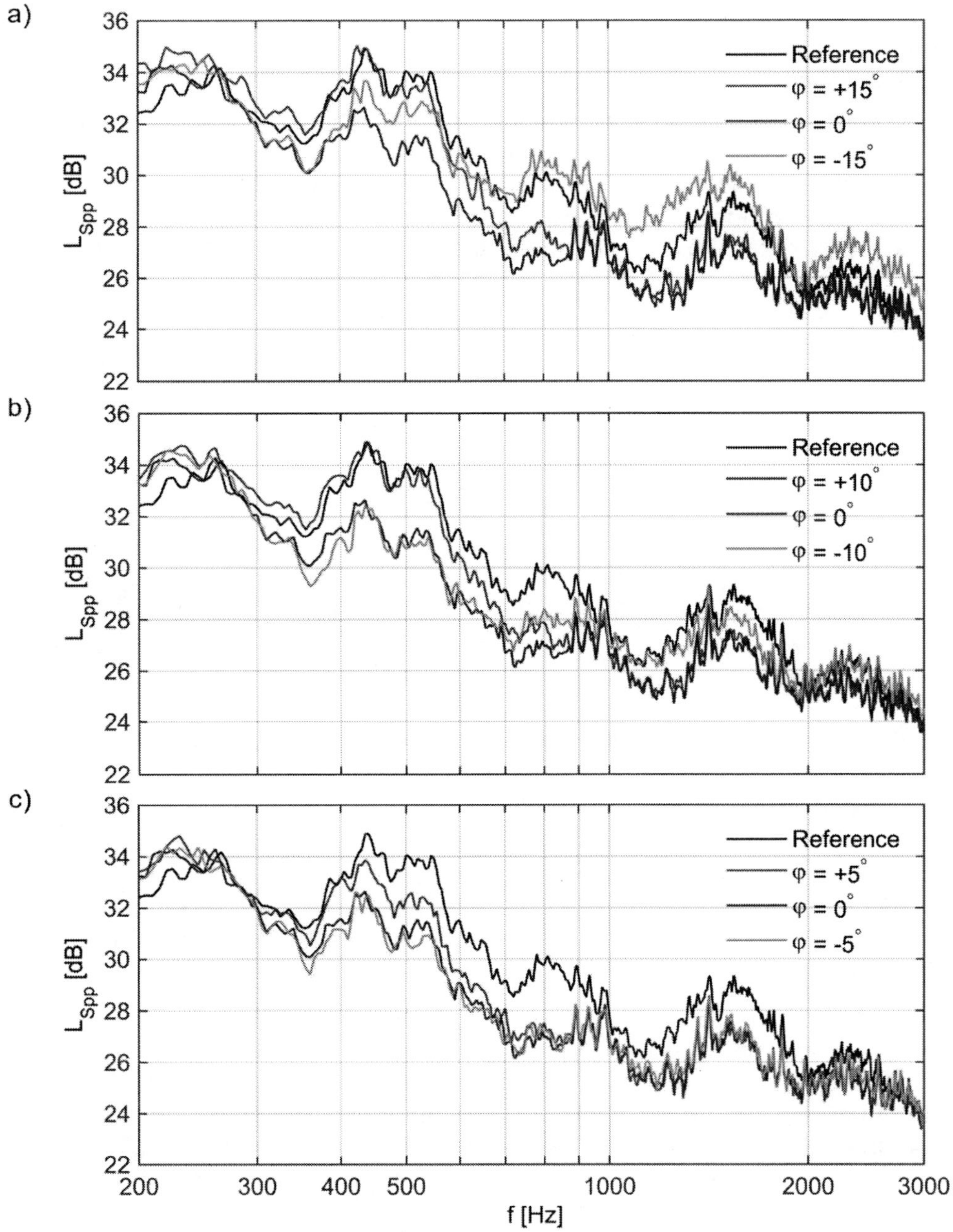

Fig. 9 Effect of orientation angle φ.

The serration aligned with the camber line (φ = 0°) seems to be still the best configuration. Finally the comparison of φ = -5°, 0° and +5° is shown in **Fig 9c.** Here it is observed that, all three orientation angle reduce the noise emission and the best noise reduction is provided by the serration with φ = -5°. This is contradicting with the results presented by VATHYLAKIS et al. [13]. However, for the next investigations, the orientation angle φ = +5° will be implemented, because to achieve φ = -5°, the serrations have to be bent manually towards the PS and the risks of misaligning some of the tooth is high, whereas with φ = +5° the serrations need to be glued on the PS and requires no further bending.

4.3 Relative thickness

Another interesting parameter to be investigated is the relative thickness of the serration. For this investigation, a serration having h/δ = 1 and λ/h = 1 with the orientation angle φ = +5° is chosen and is glued on the PS. Four different relative thicknesses of serration were investigated, t/δ = 0.56 %, 3.3 %, 5.6 % and 11 %. The measurement results are shown in **Fig 10** for only three of them.

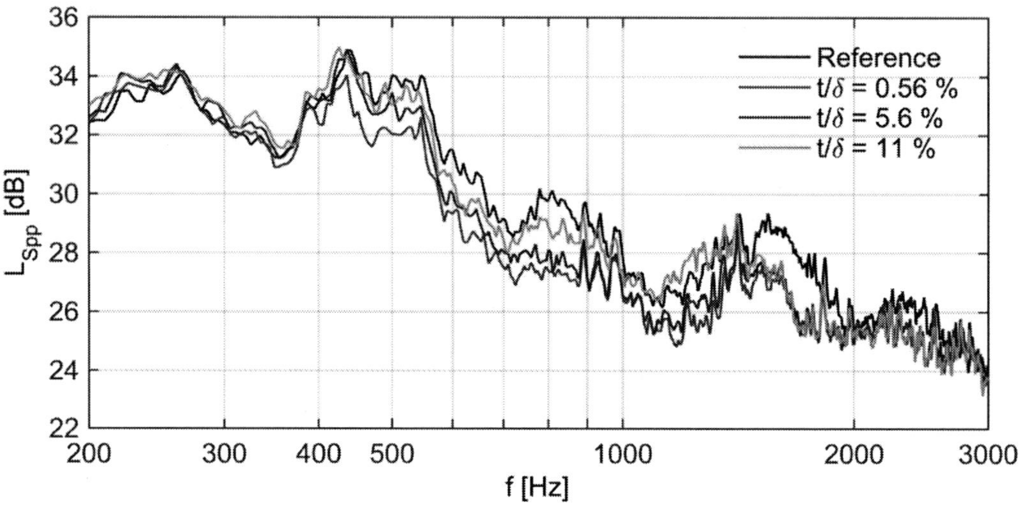

Fig. 10 Effect of relative thickness of serrations t/δ.

The results show that the variation of thickness does not have a big influence in noise reduction. However it is observed that the thinner the serrations are the higher is the noise reduction. For further investigation, the serration with t/δ = 0.56 % is chosen.

4.4 Serrations based on HOWE's theory

As already stated, all the serrations investigated in this section are glued on the PS with φ = +5° and t/δ = 0.56 %. At first an analysis is made between serrations which have a constant h/δ (1 and 3) but varying λ/h (0.25, 1, 5), i.e., Set 1 and Set 2. The measurement results are shown in **Fig 11.a** and **Fig 11.b**. The following two observations are made:

- The longer the serrations, the higher the noise reduction. This is in agreement with the observations in simulations by JONES et al. [12].
- According to HOWE, the sharper the serrations, the higher is the noise reduction. But in both cases presented below, the serration with λ/h = 0.25 has lesser noise reduction compared to λ/h = 1.

Another interesting comparison is presented in **Fig 11.c**, where the serrations are increased in amplitude from h/δ = 1 to 3 in steps of 1 and the wavelength is kept the same of half of amplitude in each case. A clear increase in noise reduction can be seen for the frequency range 300 Hz to 800 Hz, as the serrations become longer and broader.

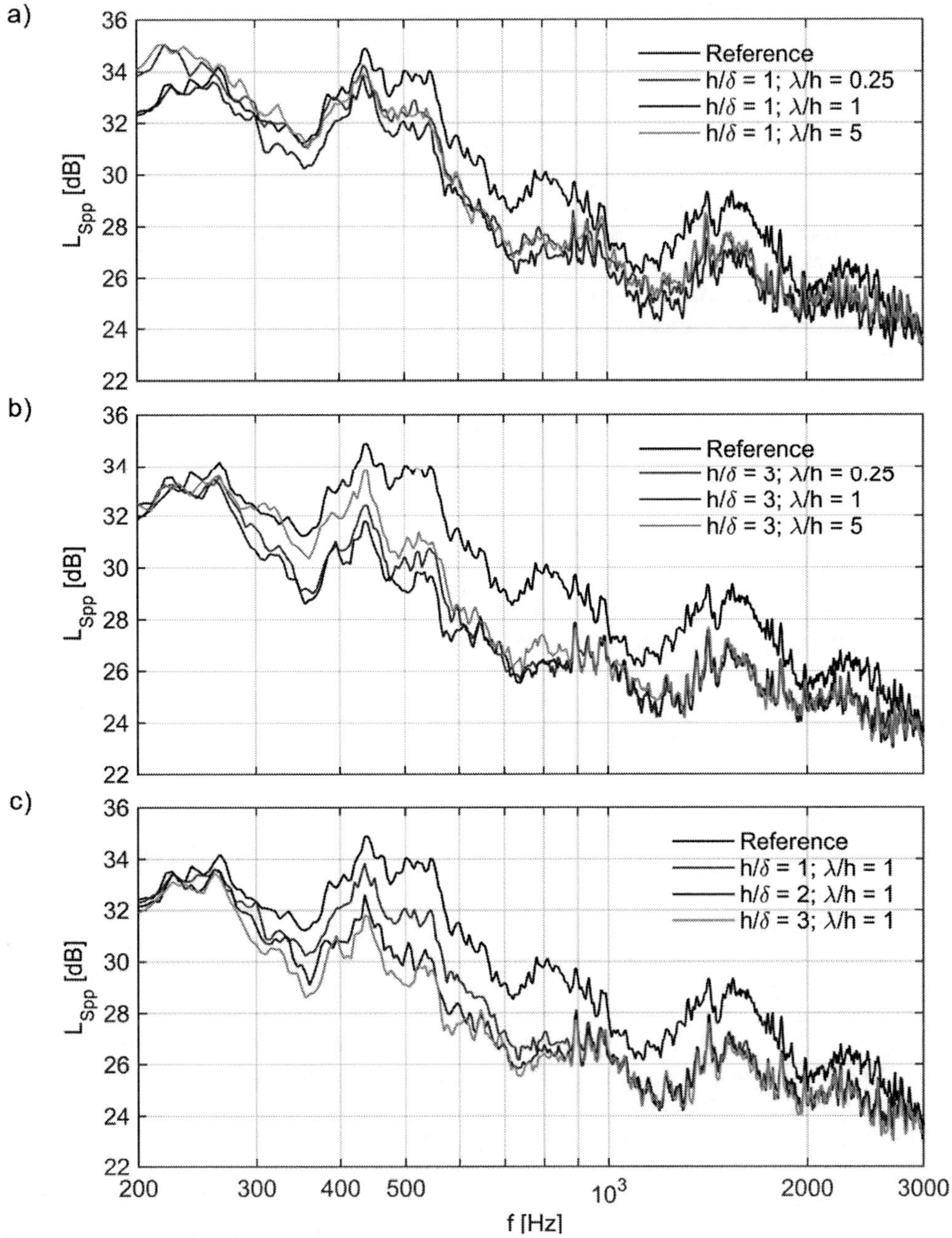

Fig. 11 Effect of "amplitude" and "wave length" of serrations; a) set 1, b) set 2, c) set 3.

The results of other investigated serrations are not shown here. The serration which brought about the least noise reduction was $h/\delta = 1$ and $\lambda/h = 0.25$. The best serration is the one with $h/\delta = 3$ and $\lambda/h = 1$. With this geometry a noise reduction of 3 dB was achieved in the frequency range from 300 Hz to 900 Hz with a maximum of 5 dB noise attenuation at the TE peak around 500 Hz. This finding corresponds to GRUBER's [5] who also reported that the serrations with $h/\delta > 2$

yield the maximum noise reduction. However, the increase in noise beyond $f\delta/u > 1$, as reported by Gruber, was not observed here due to the insufficient signal-to-noise ratio.

5. Conclusion

In this study a variety of triangular type trailing edge serrations have been designed based on HOWE's theory and eventually investigated in model scale experiments. On top of the classical design parameters h/δ and λ/h, more technological parameters have been investigated. The outcome was that it is beneficial to glue the serration on the pressure side rather than the suction side and to keep the thickness at 0.05 mm (i.e., $t/\delta = 0.56$ %). The best angle of orientation is $\varphi = -5°$, which means the serration is oriented towards the pressure side, aligning most probably with the wake, which has not been quantified in this study. The investigation of various serrations based on HOWE's theory showed that the serrations having $h/\delta = 3$ and $\lambda/h = 1$ bring best results. Previous predictions that smaller values of λ/h result in better noise reduction could not be confirmed within this study.

ACKNOWLEDGEMENTS

The authors would like to thank Michel Roger (École Centrale de Lyon) and Marlene Sanjose (University of Sherbrooke) for their technical advice and also the workshop at University of Siegen for manufacturing the serrations. Part of this work has been funded by the Federal Ministry for Economic Affairs and Energy of Germany (BMWi) within the project RENEW (FKZ 0325838B).

REFERENCES

[1] OERLEMANS, S., SIJTSMA, P., MÉNDEZ LÓPEZ, B.M., 2007, "Location and quantification of noise sources on a wind turbine", *Journal of Sound and Vibration*, **299**: 869-883.

[2] GERHARD, T., ERBSLÖH, S., CAROLUS, T., 2014, "Reduction of airfoil trailing edge noise by trailing edge blowing", *Journal of Physics*, **524**: 012123.

[3] WOLF, A., LUTZ, T.H., WURZ, W., KRAMER, E., STALNOV, O., SEIFERT, A., 2015, "Trailing edge noise reduction of wind turbine blades by active flow control", *Wind Energy 18*, **909**.

[4] SARRADJ, E., GEYER, T., 2007, "Noise generation by porous airfoils", *28th AIAA Aeroacoustics Conference*. 2007-3719.

[5] GRUBER, M., 2012, "Airfoil noise reduction by edge treatments", *Ph. D. thesis.*

[6] HERR, M., 2013, "Trailing edge noise - Reduction concepts and scaling laws", *Ph. D. thesis.*

[7] HOWE, M.S., 1991, "Noise produced by a sawtooth trailing edge", *The Journal of the Acoustical Society of America*, **90**: 482-487.

[8] DASSEN, A.G.M., PARCHEN, R., BRUGGEMAN, J., HAGG, F., 1996, "Results of a wind tunnel study on the reduction of airfoil self-noise by the application of serrated blade trailing edges", *Proceeding of the European Union Wind Energy Conference and Exhibition*, pp. 800-803.

[9] PARCHEN, R., HOFFMANS, W., GORDNER, A., BRAUN, K., 1999, "Reduction of airfoil self-noise at low Mach number with a serrated trailing edge", *6th International*

Congress on Sound and Vibration, pp. 3433-3440.

[10] OERLEMANS, S., FISHER, M., MAEDER, T., KÖGLER, K., 2009, "Reduction of wind turbine noise using optimized airfoils and trailing-edge serrations", *AIAA Journal*, Vol. **47**, pp. 1470-1481.

[11] LYU, B., AZARPEYVAND, M., SINAYOKO, S., 2015, "A trailing-edge noise model for serrated edges", *21st AIAA/CEAS Aeroacoustics Conference*, pp. 1-24.

[12] JONES, L.E., SANDBERG, R.D., 2012, "Acoustic and hydrodynamic analysis of the flow around an aerofoil with trailing-edge serrations", *Journal of Fluid Mechanics*, Vol. **706**, pp. 295-322.

[13] VATHYLAKIS, A., CHONG, T.P., PARUCHURI, C., JOSEPH, P. F., 2016, "Acoustic and hydrodynamic analysis of the flow around an aerofoil with trailing-edge serrations", *22st AIAA/CEAS Aeroacoustics Conference*.

[14] CHASE, D.M., 1987, "The character of the turbulent wall pressure spectrum at subconvective wavenumbers and a suggested comprehensive model", *Journal of Sound and Vibration*, Vol. **112**, pp. 125-147

[15] BROOKS, T.F., MARCOLINI, M.A., POPE, D.S., 1984, "Airfoil trailing edge flow measurements and comparison with theory incorporating open wind tunnel corrections", *Proc. AIAA/NASA 9th Aeroacoustics Conference*, 1-12.

[16] WINKLER, J., TEMEL, F.Z., CAROLUS, T., 2007, "Concept, design and characterization of a small aeroacoustic wind tunnel facility with application to fan blade measurements", *Fan Noise*, pp. 1-12.

[17] GERHARD, T., 2015, "Umströmte Tragflügelsegmente: Auswirkungen des Hinterkantenausblasens auf Strömungsgrenzschicht und Schallemission", *Ph. D. thesis*, Universität Siegen.

7th International Conference
on
Wind Turbine Noise
Rotterdam – 2nd to 5th May 2017

Wind Turbine Rotor Noise Prediction & Reduction for Low Noise Rotor Design

M. Kamruzzaman, Specialist Aeroacoustics, Vestas Technology UK Ltd., West Medina Mills, Stag Lane, Newport, Isle of Wight, PO30 5TR, UK. Email: MOKAM@vestas.com

Jeremy Hurault, Lead Engineer, Vestas Technology UK Ltd., West Medina Mills, Stag Lane, Newport, Isle of Wight, PO30 5TR, UK. Email: JEHUR@vestas.com

Kaj Dam Madsen, Senior Specialist, Vestas Wind Systems A/S., Hedeager 42, 8200 Aarhus N, Denmark. Email: KADMA@vestas.com

Summary

Modelling of wind turbine rotor noise sources and development of noise reduction technologies are two key issues for the future generation low noise turbine design. A high-fidelity rotor noise prediction tool combined with in-house aerodynamic and aeroelastic code has been developed to simulate different wind turbine aerodynamic noise sources. Simulation results are validated with dedicated field test data of MW class turbines consistent with different operation conditions. Good agreement between simulation and measurement were found. Noise reduction by serrated trailing-edge (STE) has been successfully applied for Vestas turbines. An efficient serration design and optimization method consistent with turbine operation condition has been developed to improve STE noise reduction performance. Enhanced method is validated with dedicated wind tunnel measurement and full scale turbine field test data. Encouraging results were found. For different wind class turbines, approximately 2 to 3dB noise reduction were found without any power/annual energy production (AEP) loss.

1. Introduction

Acoustic emissions of wind plants have a negative impact on social acceptance of wind energy and can be a barrier for the future spread of wind energy. To comply with local regulations governing community noise, wind turbines are often designed to curtail their operation, degrading efficiency, reducing energy capture, and effectively increasing the cost of wind energy. Thus, development of high quality noise prediction tools and innovative noise reduction technologies are key objectives for wind turbine manufacturers.

Aerodynamic i.e. flow-induced noise from the rotor is generally considered to be the most dominant noise source for a modern large wind turbine, provided that mechanical noise is adequately treated [1]. There are three main aerodynamic noise sources for wind turbines, turbulence inflow noise, trailing-edge noise and separation noise. Blade tip noise can also be a problem, but for modern wind turbines tip noise is not contributing to the overall noise as it can be well controlled by tip shape design. These aeroacoustics noise generation results from the interaction between turbulent flow vortices and the blade surfaces. Acoustic field test of a full scale turbine showed that trailing-edge noise is the most dominant noise source for large wind turbines [2] [3]. Thus, accurate prediction and reduction of this noise source is the main focus for future generation wind turbine blade design. The design process principally depends on the development of accurate theoretical model in which correct evaluation of turbulent boundary-layer (BL) structure plays a key role [4] [5] [6] [7] [8, 9, 10]. Turbulence inflow noise is also important in low frequency noise generation, and mainly depends on the incoming wind conditions (wind shear, turbulence scale, intensity etc.). This paper present a wind turbine rotor noise prediction tool that model above two noise sources (trailing-edge and inflow noise), validate these models against dedicated acoustic field measurements, and further analyse the results to understand the sensitivity of various acoustic parameters.

Second topic of the present paper is noise reduction and mitigation by serrated trailing-edge (STE). Howe [11] [12] discussed the production of sound by low Mach number turbulent flow over the trailing edge of a serrated airfoil (semi-2D) at zero angle-of-attack (AOA). The simplified analytical treatment in these papers and his textbook still remain effective guides to understand the primary mechanisms of noise reduction and design drivers. For serrations of spanwise wavelength λ, amplitude h, and at frequencies f satisfying $fh/U>>1$ (U being the free stream velocity), trailing-edge noise is reduced relative to that for a straight edge by $10\log(10h/\lambda)$.

Researchers at National Aerospace Laboratory (NLR) in the JOULE III project "Investigation of Serrated Trailing Edge Noise (STENO)" investigated the application of STE to reduce the turbulent boundary-layer trailing-edge (TBL-TE) noise of wind turbine blades by wind tunnel tests, numerical prediction methods and free field measurements [13] [14] [15]. Wind tunnel measurements using 2D airfoil sections showed that the STE reduced the level of TBL-TE noise significantly. However, strong indications were found that the noise reduction mechanism may be less effective in case of strong 3D flow (e.g. tip region) and the existence of perpendicular pressure gradient across the serrations. The work described led to the conclusion that it is worthwhile to investigate the optimal application of STE for real wind turbines.

In the STENO project [14] a reduction in the total noise level of about 2 dB in the free-field experiments for the range of operational incidence angles using the STE on the UNIWEX turbine was found. The reduction is much less than the theory, numerical calculations and wind tunnel tests predicted. They could not explain this behaviour, but they provided two possible effects that could play a certain role: first, the alignment of the serrations; second, the boundary layer influence caused by serrations. Outcome of the SIROCCO project [2] [24] shows very encouraging results on noise reduction with STE blades. Roughly 3.2dB overall sound power level (OASPL) reduction is observed compared to the noise of baseline blades. It is also shown in the STENO report that the serration cross section profile shape has a strong impact on both airfoil aeroacoustic and aerodynamic performance.

The work described in present paper aimed to develop noise-reduction by serrations, and the design capability and technology database to apply them to Vestas rotors. As it can be seen in section 4 of the present paper, 2-3 dB(A) (OASPL) noise reduction has been validated for a contemporary Vestas rotor like the V117 3.3MW. Various aspects related to STE noise reduction technologies and an overview of the in-house noise prediction tool are described in Sections 2 to 3. Field measurements are performed in order to validate the enhanced methods, Section 4 focuses on details of the experimental setup and post processing methods. Results, discussions and a conclusion are described in Section 4 and 5 respectively.

2. Rotor Noise Modelling and Simulations

Theory and key steps of the present rotor noise prediction methods are based on Ref. [16]. Similar types of method have been also applied by other researchers [17] [3] [18] [19]. A short overview is described below.

The blades of the wind turbine are non-uniformly divided into a number of airfoil sections or blade elements. The two-dimensional noise prediction model is applied for each blade section and the total noise level is determined by summing up all the noise sources. For the ith blade element,

$$L_{p,i}^{blade}(f) = 10\log_{10}\left[\sum_{j=1}^{S} 10^{0.1.L_{p,j}^{pm}(f)}\right] \qquad 0.1$$

where j denotes the different noise sources such as Turbulent Boundary-Layer Trailing-Edge (TBL-TE) noise, turbulence inflow noise etc. The noise propagation effects are calculated separately for each noise source element to receiver position. The actual turbine geometry (i.e. location, orientation and velocity of each element) is determined from specific radial location, hub height, pitch, tilt and cone angles, azimuth position and rotational speed. In the next step, all the contributions are incorporated to obtain the immission of an element at the receiver position by the following equation

$$L_{p,j}^{pe}(f) = L_{p,j}(f) + L_{di,j}(f) + L_{ca,j}(f) + L_{spre,j} + L_{att,j}(f) + L_{met,j}(f) \qquad 0.2$$

where $L_{p,j}^{pe}$ is the sound pressure level of jth type noise source including propagation effects. $L_{p,j}$ is the corresponding Sound Pressure Level (SPL) without propagation model. $L_{di,j}$ represents directivity, $L_{ca,j}$ convection amplification, $L_{spre,j}$ geometrical spreading, $L_{att,j}$ atmospheric attenuation and $L_{met,j}$ is refraction and ground effect of the corresponding j^{th} source and *f* is the frequency of interest.
The sound pressure level radiated from the single blade number *n is*

$$L_{p,n}^{blade}(f) = 10\log_{10}\left[\sum_{i=1}^{N} 10^{0.1.L_{p,i}^{blade}(f)}\right] \qquad 0.3$$

Finally, the sound pressure level radiated from a wind turbine relative to an observer position *k* is the sum of the sound pressure levels from all elements of all blades, e.g. for a three bladed turbine it is

$$L_{p,k}^{rot}(f) = 10\log_{10}\left[\sum_{n=1}^{3} 10^{0.1.L_{p,n}^{blade}}\right] \qquad 0.4$$

Furthermore, the sound pressure level of various observer positions (varying *k* from 1 to *P*) can be simulated by a polar grid around the turbine. A/B/C-weighting filter of the frequency dependent SPL can be applied after the propagation effects have been incorporated. Calculation of the sound power level L_w(IEC 61400-11: ed.3 standard) from the sound pressure level is straight forward, i.e.

$$L_{w,k}^{rot}(f) = L_{p,k}^{rot}(f) + 10\log_{10}\left[\frac{4\,\pi\,R_1^2}{S}\right] - B_1 \qquad 0.5$$

R_1 is the distance from the observer (measurement microphone position) position to the rotor centre, see Figure 1, B_1 is a correction for reflection from the hard board (where microphone is set up, $B_1 = 6.0$), and S= $1m^2$ is a reference area.

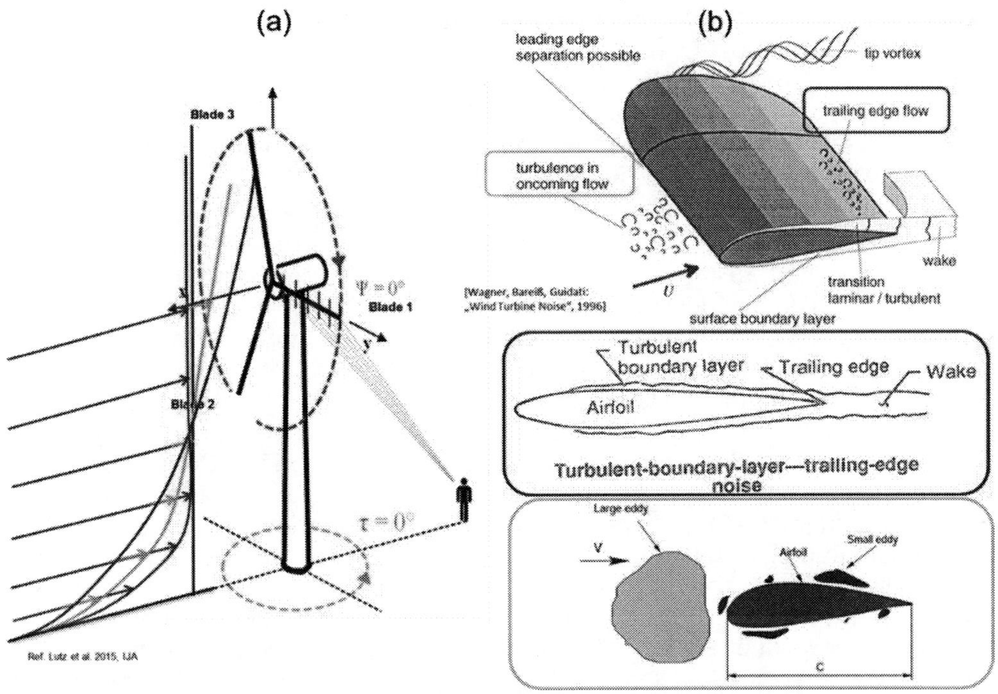

Figure 1: Overview of Vestas rotor noise prediction tool and different wind turbine aerodynamic noise sources, (a) left plot [19], (b) right plots [1].

Finally, the total noise level (sound pressure) over all frequency can be evaluated by

$$L_p^{tot} = 10\log_{10}\left[\sum_{m=1}^{M} 10^{0.1\,L_{p,k}^{rot}(f_m)}\right] \qquad 0.6$$

in case of 1/3-band analysis f_m, with $m = 1,2,3 \ldots . M$ is the corresponding centre frequencies. It is very important to note that during rotation the positions of each blade (each noise source component as well) relative to a fixed observer is also changing. Thus, one needs to find the source to receiver distance appropriately and consider the propagation effect during rotation by Equation 0.2. Calculation of the total rotor noise by above procedures consists of a set of coordinate system transformations and numerical discretization.

In a broad sense, the calculation can be divided into three key steps: i) blade aerodynamics (evaluation of local flow Re, Ma and AOAs) simulation, ii) 2D aerodynamic noise source strength (calculation of $L_{p,j}$ in Equation 0.2 by a 2D model) modelling and iii) noise propagation (calculation of other parameters in Equation 0.2).

Two different aerodynamic noise sources, namely trailing-edge noise and turbulence inflow noise are modelled in the Vestas rotor noise tool. For the trailing-edge noise modelling, different semi-empirical [20] [21] and simplified theoretical [9] [10] models are implemented. Turbulence noise source parameters are modelled based on different class of aerodynamic simulation codes depending on the types of noise prediction models being considered. A modified version of Amiet-Lowson [1] [17] model has been used for inflow noise simulation. Noise propagation and directivity effects are considered applying Howe-BPM theory [22] [23]. Final output is 1/3-band, A-weighted IEC61400-11: ed. 3 standard Sound Power Level. Noise source spectrum related to various rotor azimuth positions relative to a fixed observer is also available. Moreover, rotor noise tool is coupled with the Vestas in-house aerodynamic-aeroelastic tool in order to perform detail simulation consistent with a specific turbine operation condition.

3. Rotor Noise Reduction by Serrated Trailing-edge (STE)

The use of serrated trailing edges for wind turbine noise reduction has now become a mature technology, academic/research institutions and wind turbine manufacturers demonstrating its effectiveness in wind tunnel and turbine tests leading to commercial products. Researchers from NLR, Energy Research Centre of the Netherlands (ECN) and Institute of Aerodynamics and Gas Dynamics (IAG) at University of Stuttgart tried to reduce TBL-TE noise by modifying the airfoil shape and/or implementing STE, during the European project SIROCCO [2] [24]. In this project, acoustic field measurements on a 94 m diameter, three-bladed wind turbine has been conducted. One standard blade, one blade with acoustically optimized airfoil shape, and one standard blade with STE were fitted on a HAWT. Test results for the baseline blade showed that the dominant source was TBL-TE noise from the outer 25% of the blade. Both optimized blades showed a significant TBL-TE noise reduction at low frequencies. For clean blade at normal operation conditions, average overall noise reduction of 0.5 dB for the blade with optimized airfoil shape and 3.2 dB for the blade with STE were observed. For both blades, the noise reduction increased with increasing wind speed on the pitch-regulated test turbine. This motivates turbine manufacturers to use serration as a noise reduction technology for future generation low noise rotor design.

An efficient STE design and optimization method has been developed that enhanced 2 to 3 dB(A) noise reduction for Vestas rotor blades. The STE design procedure relies on dedicated wind tunnel test outcome, and a correlation between local aerodynamics characteristics of 2D airfoil and 3D rotating blade [25]. The ratio of serration length and local turbulence length scale near the blade trailing-edge region is one of the key parameters for appropriate STE geometry design. To achieve best noise reduction for a given turbine and operation condition, serration geometry is optimized consistent to the local turbulent flow characteristic based on the rotor noise simulation outcome and related wind tunnel data. The most important parameters that govern STE noise reduction effectiveness are set to be consistent with local turbulence parameters near the blade trailing-edge. Moreover, clean and rough flow conditions and other off design situation are investigated to consider design uncertainty. In addition to that, detailed geometric refinements were added to the design to ensure manufacturability and reduce fatigue issues for 20 years lifetime requirement. The resulting serration design performance has been validated on several turbines and the results are depicted in the next section. Figure 2 shows few STE geometry examples as tested in the wind tunnel and a Vestas blade with serrated trailing-edge (STE).

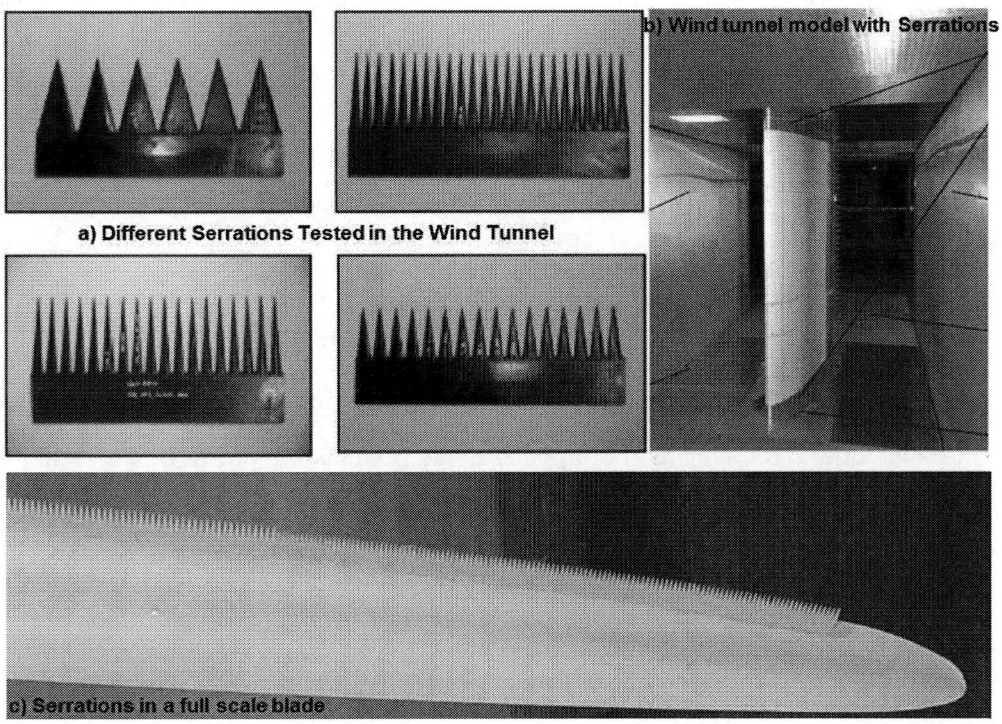

Figure 2: Overview of trailing-edge serrations as investigated in wind tunnel test (a). Wind tunnel airfoil model with STE (b). Vestas blade with serrations (c).

4. Results & Discussions

Detail validation of the Vestas rotor noise simulation tool, and assessment of the optimized STE noise reduction performance are conducted in the following section. Three different MW class turbines are considered for the validation study. Table below shows an overview of the selected test cases.

Case #	Turbine	Rotor Diameter [m]	Hub height, [m]	Wind Class & Other Info
A	V126-3.3MW	126m	116m	IEC 3A
B	V117-3.3MW	117m	92m	IEC 2B
C	V112-3.3MW	112m	116m	IEC 2A

4.1 Rotor Noise Prediction & Validation

Noise simulations are performed with Vestas rotor noise tool as described in Section 2 for three turbines, see Table 1. Predicted total noise is the sum of two aerodynamic noise sources i.e. trailing-edge noise and turbulence inflow noise. For the present simulation, trailing-edge noise is modelled by improved BPM type [20] semi-empirical model (calibrated for Vestas wind tunnel data), and inflow noise is modelled by modified Amiet-Lowson [1] approach. Necessary turbulence noise source parameters are estimated based on the Vestas in-house aerodynamic-aeroelastic tool consistent with turbine operation and geometric conditions. Noise propagation and directivity effects are considered applying Howe-BPM theory [22] [23].

Note that for the present simulation no atmospheric attenuation, refraction and ground effect modelling are included, as shown in Equation 0.2. For one to one comparison with measurement, simulation is performed at IEC61400-11 standard observer distance which provides 1/3-band, A-weighted, Sound Power Levels.

Figures 3 to 5 shows simulation vs measurement overall sound power levels (OASPL) as a function of wind speeds. In general, very good agreement with measurement has been observed near the rated power wind speed region. For Case C (V112-3.3MW), at lower wind speeds, predicted OASPL does not fit well with field test as depicted in Figure 5. In some cases, large scattering has been observed at high wind speeds too. This implies, at very low and/or very high wind speeds predicted noise levels are not as good as medium/design wind speed range. This behaviour can be due to the numerical inaccuracy of the different simulation tools, or can be also due to the limitation of applied noise prediction models. Because at low wind condition, local flow AOAs in a pitch regulated turbine is very high. Thus, aerodynamic tool may needs extra care to converge flow simulation results that can increase numerical errors in the subsequent noise calculation steps. A part of the above inaccuracy can be also due to the higher uncertainties in field test data at low/high wind speed range. Site dependent measurement uncertainty should be also noted, as shown in Figure 5. More validation and further research is necessary to understand this phenomenon.

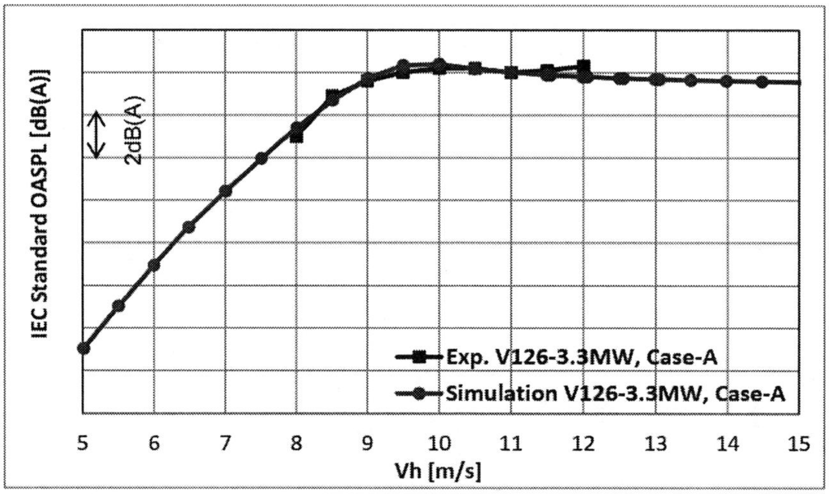

Figure 3: Simulation vs Field test data validation Case-A, V126-3.3MW, H=116m.

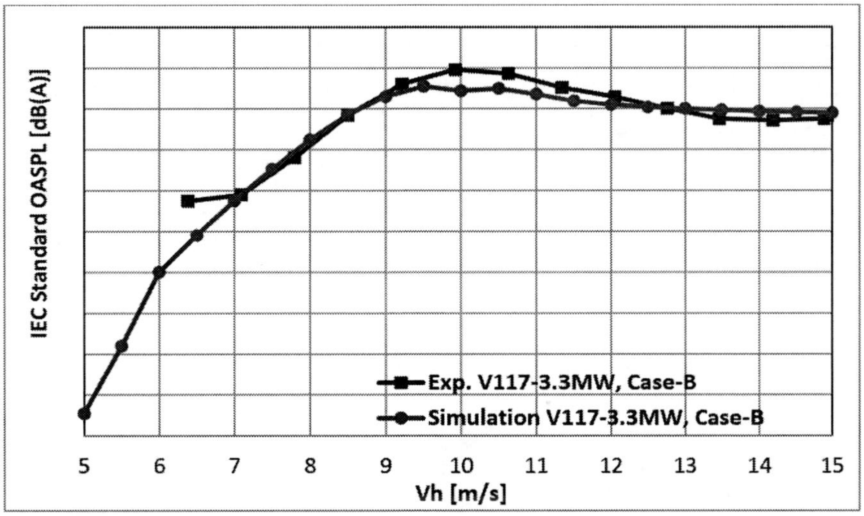

Figure 4: Simulation vs Field test data validation Case-B, V117-3.3MW, H=92m.

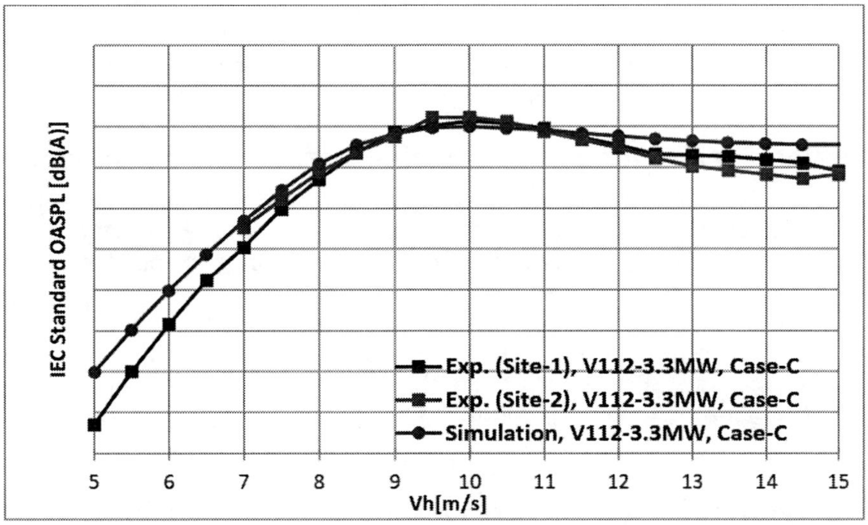

Figure 5: Simulation vs Field test data validation Case-C, V112-3.3MW, H=116m.

Simulated noise spectrum for V112-3.3MW at 9m/s is compared with measurement data and shown in Figure 6. Peak frequency and spectral shape are in good agreement with measurement, except offset in the high frequency region. However, applying an atmospheric attenuation model improve this high frequency offset. Moreover, high frequency spectrum amplitude will not contribute that much on OASPL as spectrum amplitude is dominated by the low frequency region.

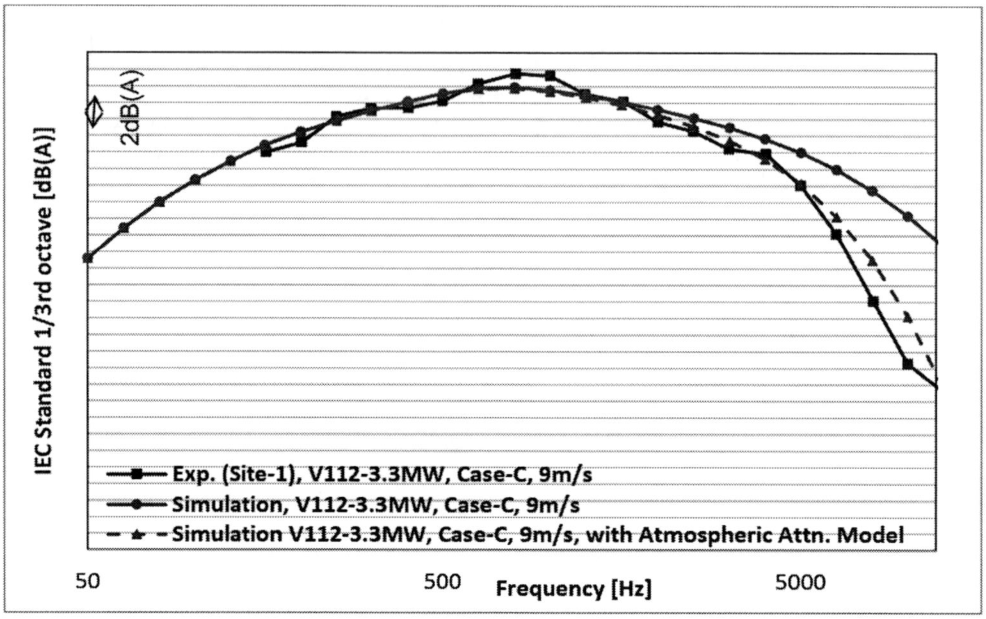

Figure 6: V112 3.3MW, 1/3rd octave noise spectrum at 9 m/s hub height wind speed. Simulation vs measurement.

4.2 Serrated Trailing-edge (STE) Noise Performance Validation

Noise reduction performance of the optimized STE design approach is conducted in the following section. As discussed in Section 3, serration geometry is optimized consistent to the local turbulent flow characteristic based on the rotor noise simulation outcome and corresponding wind tunnel data. IEC 61400-11 ed.3 sound power measurement has been performed for three Vestas turbines as shown in Table 1, with and without STE. Turbines are pitch regulated and thus the angle of attack (AoA) will vary significantly with the wind speed. More information about Vestas 3MW platform can be found in Ref. [26].

Figures 7 to 10 shows field test measurements data comparison plots of the baseline vs STE turbines. STE are effective at reducing the OASPL on a large wind speed range. The overall noise reduction is between 1.5 to 3.1 dB(A) in wind speed range of 7-15 m/s. The noise reduction is overall very good and in agreement to the wind tunnel measurement [25]. One can expect that the other noise sources generated by the wind turbine generator (WTG) that are not tackled by STE would make STE less effective on a full rotor, but these results confirm that TBL-TE noise is the major WTG noise source.

In order to assess the robustness of the STE, it has been applied on a wind turbine of a different wind class, a V126-3.3MW turbine. V117 is an IEC2b class wind turbine whereas V126 is an IEC3a. It is clear on Figure 8 that the STE are also very effective for overall noise reduction on V126-3.3MW. The noise reduction varies function of the wind speed bin from 2.0 to 3.0 dB(A). Figure 9 show a comparison of the 1/3rd octave spectrum with and without serrations for V126-3.3MW. The STE are effective at noise reducing across a wide range of frequencies. At the peak 1/3rd octave band spectrum level, more than 3 dB(A) reduction is achieved by Vestas serrations.

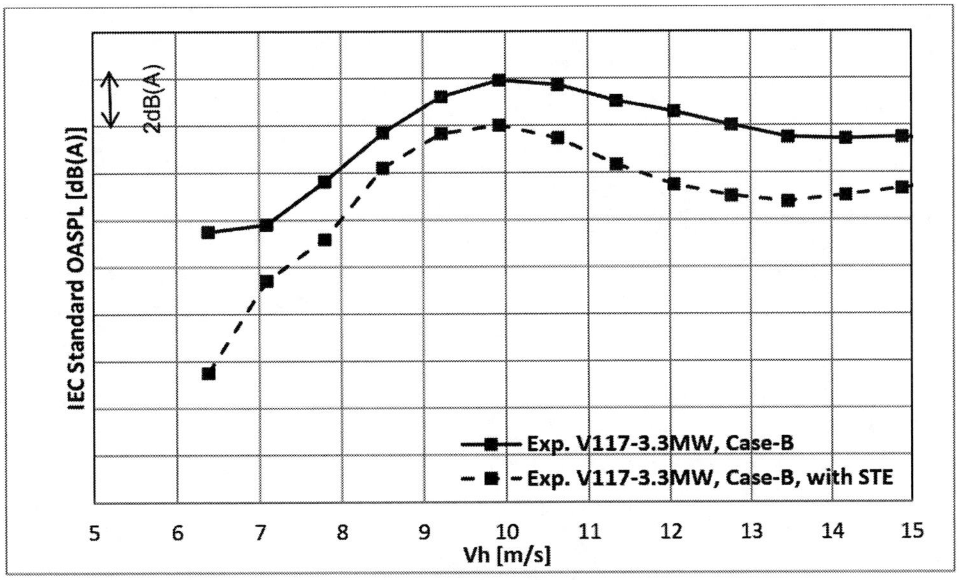

Figure 7: V117-3.3MW noise performance with and without serration based on IEC 61400-11 ed.3 measurement.

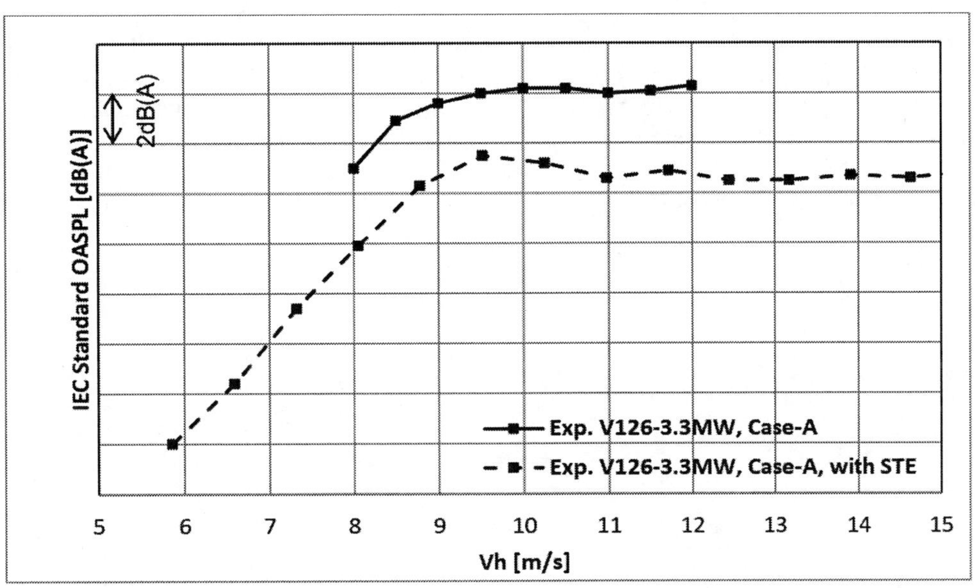

Figure 8: V126-3.3MW noise performance with and without serrations. IEC 61400-11 ed.3 measurement.

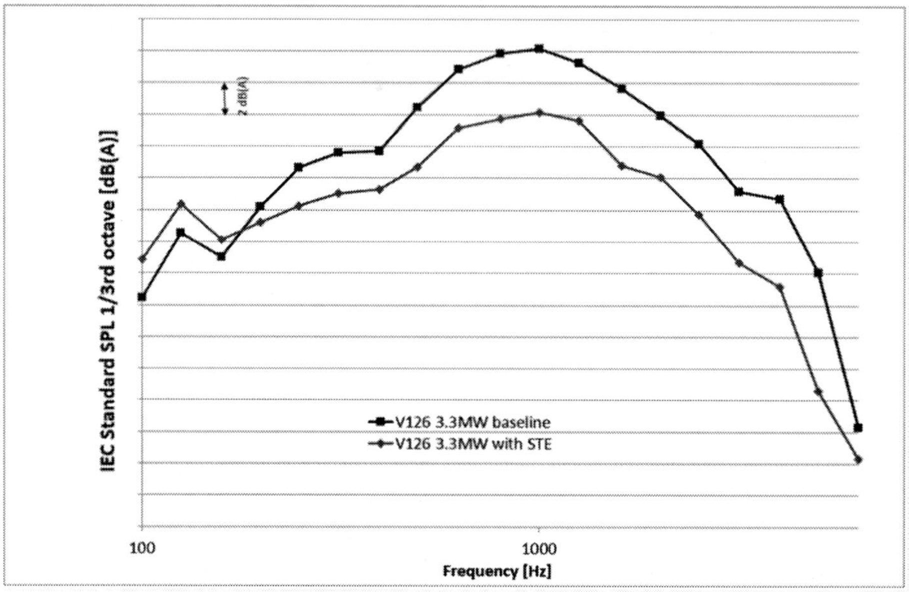

Figure 9: V126 3.3MW, 1/3rd octave noise spectrum at 10 m/s hub height wind speed. Comparison with and without serrations field test.

In order to demonstrate the influence of enhanced serration shape optimisation related to the local turbulent flow properties, the noise level of V112 3.3MW turbine is shown in Figure 10. The comparison of two serrations design show that the new optimized serrations are performing better on a large wind speed range as expected. Noise reduction at the maximum noise level has been improved by 0.8-1.0 dB(A) to reach close to 3.0 dB(A) reduction. This confirms the advantage of enhanced local flow dependent STE geometry design and optimization method.

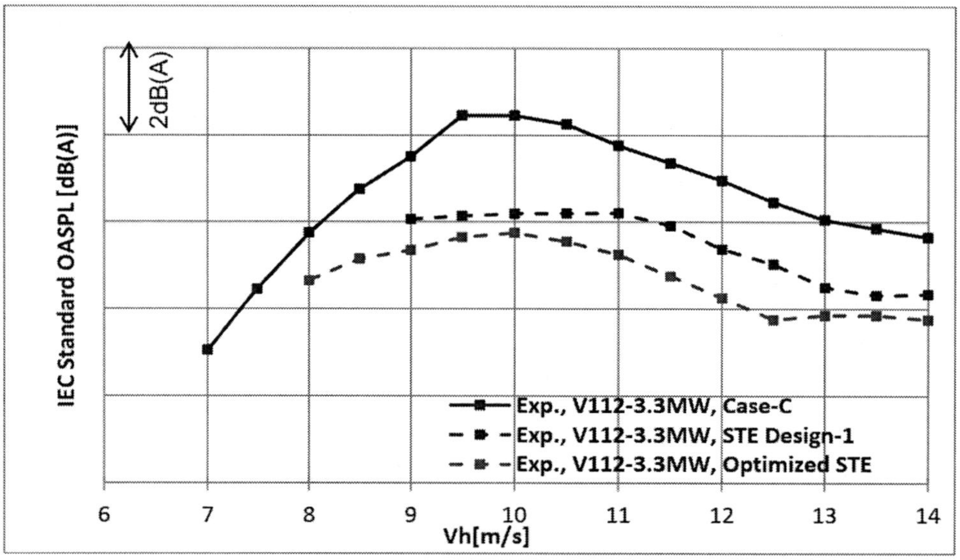

Figure 10: V112-3.3MW noise performance with and without serration. IEC 61400-11 ed.3 measurement.

5. Conclusions

A noise prediction tool has been developed to simulate different wind turbine aerodynamic noise sources consistent with turbine geometry and operation condition. Prediction method is extensively validated with IEC standard noise field test data and encouraging results were found. In general, predicted overall sound power level (OASPL) at the rated power region is in good agreement with measurement, within ±1.0dB uncertainty range. The accuracy at very low and very high wind speed range is not as good as medium/rated power wind speeds. More validation and analysis is necessary to understand this behaviour.

The noise reduction potential of the enhanced trailing-edge serration design and optimization method for full scale turbine is very good. Roughly 3dB noise reduction at the rated power has been achieved without any power/annual energy production (AEP) loss. It has been demonstrated that trailing edge serrations are a very effective way of reducing modern pitch regulated wind turbine noise. Serrations teeth shape and size is critical to achieve best performance and design geometry should be consistent with local turbulent flow characteristics. All Vestas turbines could be supplied with serration add-ons in order to reduce the noise and annual energy production can be increase in noise restrictive sites.

Bibliography

[1] S. Wagner, R. Bareiss and G. Guidati, "Wind turbine noise", Springer Verlag, 1996.

[2] J. Schepers, "SIROCCO: Silent rotors by acoustic optimisation," in *Second International Meeting on Wind Turbine Noise, Lyon, France. Also ECN report: ECN-M--07-064.*, 2007.

[3] S. Oerlemans, "Detection of Aeroacoustic Sound Sources on Aircraft and Wind Turbines," PhD Thesis, University of Twente, ISBN 978-90-806343-9-8, 2009.

[4] W. Blake, Mechanics of Flow-Induced Sound and Vibration Vol.~I and II, Academic Press, INC., 1986.

[5] T. Lutz, A. Herrig, W. Wuerz, M. Kamruzzaman and E. Kraemer, "Design and Wind-Tunnel Verification of Low-Noise Airfoils for Wind Turbines," *AIAA Journal,* vol. 45, no. 4, pp. 779-792, April 2007.

[6] M. Kamruzzaman, "Study of Turbulence Anisotropy and Its Impact on Flow Induced Noise Emission," Verlag Dr. Hut, Muenchen, ISBN. 9783843904902 , 2011.

[7] A. Herrig, "Validation and Application of a Hot-Wire based Method for Trailing-Edge Noise Measurements on Airfoils," Verlag Dr. Hut, ISBN: 9783843905787, 2010.

[8] M. Kamruzzaman, T. Lutz, W. Wuerz, W. Shen, W. J. Zhu, M. O. L. Hansen, F. Bertagnolio and H. A. Madsen, "Validations and Improvements of Airfoil Trailing-edge Noise Prediction Models Using Detailed Experimental Data," *Wind Energy,* vol. 15 (1), pp. 45-61, January 2012.

[9] M. Kamruzzaman, D. Bekiropoulos, A. Wolf, T. Lutz, W. Wuerz and E. Kraemer, "Rnoise: A RANS Based Airfoil Trailing-Edge Noise Prediction Model," in *AIAA/CEAS Aeroacoustics Conference, AIAA-2014-3305, Atlanta, USA*, 2014.

[10] M. Herr, "Broadband Trailing-Edge Noise Predictions: Overview of BANC-III Results," in *21st AIAA/CEAS Aeroacoustics Conference, AIAA-2015-2847, Dallas, TX*, 2015.

[11] M. S. Howe, "A Review of the Theory of Trailing Edge Noise," *Journal of Sound and Vibration,* vol. 61, pp. 437-465, 1978.

[12] M. S. Howe, "Aerodynamic noise of a serrated trailing-edge," *Journal of Fluids and Structures,* vol. 5(1), pp. 33-45, 1991.

[13] T. Dassen and F. Hagg, *Results of a Wind Tunnel Study on the Reduction of Airfoil Self-noise by the Application of Serrated Blade Trailing Edges. Report, NLR-TP-96350.*

[14] K. A. Braun and R. Parchen, "Noise Reduction by using Serrated Trailing Edges," in *EWEC'97 Proceedings, Dublin*, 1997.

[15] T. Dassen, *Comparison of Measured and Predicted Airfoil Self-noise with Application to Wind Turbine Noise Reduction. Report, NLR-TP-97564*

[16] M. Kamruzzaman, T. Lutz, K. Nuebler and E. Kraemer, "Implementation and Verification of an Aeroacoustic Wind Turbine Blade Analysis Tool," in *Wind Turbine Noise*, 2011.

[17] W. Zhu, N. Heilskov, W. Shen and J. S. rensen, "Modelling of aerodynamically generated noise from wind turbines," *Journal of Solar Energy Engineering,* vol.

127, pp. 517-528, 2005.

[18] K. Boorsma and J. G. Schepers, "Enhanced wind turbine noise prediction tool SILANT," in *Fourth International Conference on Wind Turbine Noise, Rome, Italy, 12-14 April*, 2011.

[19] T. Lutz and M. Kamruzzaman, "Prediction of flow-induced noise sources of wind turbines and application examples," *Wind Turbine Noise Special Issue, Int. Journal of Aeroacoustics,* 2015, Vol.14: 5-6.

[20] T. F. Brooks, D. S. Pope and M. A. Marcolini, "Airfoil Self-Noise and Prediction," NASA Report, NAS 1.61:1218, 1989.

[21] M. Kamruzzaman, D. Bekiropoulos, A. Wolf, Th. Lutz, W. Würz and E. Krämer, "A Semi-empirical Wall Pressure Fluctuation Spectrum and Far-field Trailing-edge Noise Prediction Model," *Wind Turbine Noise Special Issue, Int. Journal of Aeroacoustics, Vol. 14: 5-6, page. 833-882,* 2015.

[22] T. F. Brooks and T. H. Hodgson, "Trailing edge noise prediction from measured surface pressures," *Jornal of Sound and Vibration,* vol. 78(1), no. 69, p. 117, 1981.

[23] M. S. Howe, Acoustics of Fluid-Structure Interactions, Online ISBN: 9780511662898, Hardback ISBN: 9780521633208, Paperback ISBN: 9780521054287 ed., Cambridge University Press, 1998.

[24] S. Oerlemans, M. Fisher, T. Maeder and K. Koegler, "Reduction of Wind Turbine Noise using Optimized Airfoils and Trailing-Edge Serrations," *AIAA Journal,* vol. 47(6), 2009.

[25] J. Hurault and P. Ravetta, "Aeroacoustic wind tunnel experiment for serration design optimisation and its application to a wind turbine rotor," in *6th International Meeting on Wind Turbine Noise, Glasgow*, 2015.

[26] Vestas Product Catalogue, 2016 (https://www.vestas.com/en/products/turbines#!).

7th International Conference
on
Wind Turbine Noise
Rotterdam – 2nd to 5th May 2017

Modelling and localizing low frequency noise of a wind turbine using an array of acoustic vector sensors

Daniel Fernandez Comesaña, Microflown Technologies, 6824BV, Arnhem, Netherlands. fernandez@microflown.com

Krishnaprasad N. Ramamohan, Delft University of Technology, 2628 CD Delft, the Netherlands. ramamohan@microflown.com

David Pérez Cabo, ETSI de Telecomunicación, 36310, Vigo, Spain. cabo@microflown.com

Graciano Carrillo Pousa, Microflown Technologies, 6824BV, Arnhem, Netherlands. carrillo@microflown.com

Summary
The large size and low rotational speed of modern wind turbines are often linked to the generation of low frequency noise. This paper proposes a simplified approach to model the sound produced by a wind turbine based on moving monopole sources. Time-dependent Green functions are used to account for the Doppler effect introduced by the relative changes in position between the moving elements and the fixed sensors. The proposed model can be used for understanding how different mechanical defects have an impact on the perceived sound. The sound field is hereby studied through an array of acoustic vector sensors (AVSs) since it enables locating low frequency sound sources with a relatively small aperture. A beamforming method is applied upon the synthetic data for locating the noise emission points along the moving blades. An experimental investigation is also presented introducing a novel in-situ calibration procedure for adjusting the AVS orientation. Both numerical and experimental results show that the proposed approach is suitable for modelling and localizing the sources of noise emission with a low number of acoustic vector sensors.

1. Introduction

Wind power has become an important source of renewable energy, which is significantly helping to reduce the global carbon emission levels. The success of this technology is leading to increase the amount of wind turbines installed every year. Besides energy efficiency and cost, noise emission is one of the key design criterion. As reported by several authors such as Rogers and Manwell (2004), Bass et al. (2011), Doolan et al. (2012), Zajamšek et al. (2016) and Hansen et al. (2017), one of the current concerns of wind farm neighbors is the annoyance caused by the emitted noise.

Wind turbine manufacturers seek solutions to localize and rank noise sources effectively. Understanding the foundations of the problem is crucial to design appropriate noise control strategies. Several methods are available for visualizing the sound field produced by complex structures. Among them, beamforming techniques are often applied employing large microphone arrays (Oerlemans, 2009). Due to the resolution limit and spatial sampling principle such methods require the usage of multiple sensors spread over a large area in order to localize low frequency sound sources.

Alternatively, an Acoustic Vector Sensor (AVS) array has been proven to lead to similar performance using smaller array apertures with less elements (Nehorai and Paldi, 1994 ; Hawkes and Nehorai, 1998; Kitchens, 2010). An AVS consist of a sound pressure microphone and three orthogonally placed particle velocity sensors. Each AVS provides vector information about the sound propagation at the measured point. As a result, an AVS array shows advantages over a traditional microphone array since it combines the information extracted from the spatial interference between sensors and the intrinsic directivity of the particle velocity elements.

The present work introduces a framework to model the sound pressure and particle velocity field produced by a set of arbitrary moving monopole sources. The proposed data model enables to predict the performance of an AVS array in combination with beamforming techniques. Numerical and experimental results are included to evaluate the results obtained accounting for different source conditions and demonstrate the feasibility of using an AVS array for outdoor measurements in a wind turbine field.

2. Data model

Rogers and Manwell (2004) suggest that noise produced by wind turbines is mostly induced by either the motion of mechanical components or aerodynamic effects. Owing to vibration damping and improved mechanical designs, wind turbines have become quieter over the years, especially in terms of the tonal noise produced by gearboxes, generators or yaw drives. On the other hand, Doolan et al. (2012) claim that aerodynamic interactions are one of the dominant sources of noise of large wind turbines. Consequently, this papers is mainly focused on modelling only aerodynamic noise created by flow-blade interactions.

The forward model hereby presented is based on the Equivalent Source Method introduced by Verheij (1997), which is extended to take into account the rotating movement of the blade. Therefore, it is assumed that noise produced by flow-blade interactions can be modelled by a set of moving monopole sources distributed along the structure. Three primary elements are considered: moving sound sources, time-dependent propagation paths and a static sensor array.

The sound sources are used to resemble the aerodynamic noise generated by interactions between the air flow and the moving blades. Considering the random nature of these phenomena, the sources were assumed to be statistically independent, i.e. uncorrelated. Consequently, each source signal $q_i(t)$ is modelled with a unique white noise signal $s_i(t)$ that is filtered and modulated accounting for load variations during each blade cycle, i.e.

$$q_i(t) = s_i(t) * H_i(t)\,, \tag{1}$$

where $H_i(t)$ is the impulse response of an arbitrary filter which may change over time depending on the source conditions, $\mathbf{x}_o$ is the position of the source and the operator $*$ denotes convolution. It is assumed that each source $q_i(t)$ moves along the trajectory $\mathbf{E}_i(t)$. The source strength density of each source can then be defined as

$$Q_i(\mathbf{x},t) = \frac{1}{\rho} q_i(t)\delta\big(\mathbf{x} - \mathbf{E}_i(t)\big)\,, \tag{2}$$

where δ is the delta function and ρ is the air density. The relative movement of the sound sources implies that the propagation path which relates the emission to the reception point will change over time. For the particular case hereby evaluated, the distance between sources and receivers decreases during the down-stroke phase and it increases as soon as the source passes the tower and starts the up-stroke movement. The position variations cause a Doppler effect that modifies the pitch of the original emitted signal. This variable behavior can be modelled using a time-dependent free-field Green function which is a solution of the inhomogeneous wave equation for an arbitrary excitation, defined as

$$G(\mathbf{x},\mathbf{x}_0,t) = \frac{\delta(t-T)}{4\,\pi r}, \tag{3}$$

where $G(\mathbf{x},\mathbf{x}_0,t)$ is the Green function that relates the point $\mathbf{x}$ and $\mathbf{x}_0$ at the time instant t; r is the distance between the points ($\|\mathbf{x}-\mathbf{x}_0\|$); T is the time delay between the two points (r/c) and $\|.\|$ indicates the Eucledian norm of the vector between brackets. It should be noted that Equation 3 models sound propagation imposing a free-field assumption. However, it is also possible to model a more complex propagation channel that includes air flow variations and reflection by redefining this expression. For the sake of simplicity, the present work is solely focused on studying the sound field radiated using the definition provided above.

According to Camier et al. (2012), the velocity potential $\Psi_i(\mathbf{x},t)$ can be defined as the convolution of the source signal and the time-dependent Green function at the observation point $\mathbf{x}$ as

$$\Psi_i(\mathbf{x},t) = Q_i(\mathbf{x}_0,t) * G(\mathbf{x},\mathbf{x}_0,t) = \frac{q_i(t-T)}{4\pi\rho(\|\mathbf{r}_i(t)\| - \mathbf{v}(t-T)\cdot\mathbf{r}_i(t)/c)}, \tag{4}$$

where $\mathbf{r}_i(t) = \mathbf{x} - \mathbf{E}_i(t-T)$, i.e. is the vector from the sound source to observation point at the time instant when sound is emitted; $\mathbf{v}(t-T)$ is the source velocity vector at the time of emission and c is the sound speed. The temporal and spatial derivatives of the velocity potential $\Psi_i(\mathbf{x},t)$ yield the sound pressure $p(\mathbf{x},t)$ and the particle velocity vector $\mathbf{u}(\mathbf{x},t)$. The sound field perceived at the measurement point $\mathbf{x}$ can then be described by the linear superposition of the sound produced by N sources as

$$p(\mathbf{x},t) = -\rho\sum_{i=1}^{N}\frac{\partial\Psi_i(\mathbf{x},t)}{\partial t} + n(t), \tag{5}$$

$$\mathbf{u}(\mathbf{x},t) = \sum_{i=1}^{N}\nabla\Psi_i(\mathbf{x},t) + \mathbf{n}(t), \tag{6}$$

where ∇ denotes the spatial gradient operator and $n(t)$ can be used to model additional noise signals introduced by the measuring instrumentation.

3. Beamforming using an AVS array

One common application for acoustic sensor arrays is the Direction of Arrival (DOA) estimation of propagating wavefronts for the localization of sound sources. Generally, array geometry information is used in combination with the signals recorded in order to create spatially discriminating filters that can be steered to a particular direction. This spatial filtering operation is also known as beamforming. Traditional beamforming techniques steer a beam to a particular direction by computing a weighted sum of the individual sensor signals. This procedure results in the addition of signals coming from the direction of focus, maximizing the energy of the beamformer output whilst sound waves from other directions are attenuated. A set of time delays $\tau_i(\boldsymbol{\kappa})$ can be computed from the scalar product between the sensor position $\mathbf{x}_i$ and a unitary vector $\boldsymbol{\kappa}$ which is aligned with the direction of interest, i.e.

$$\tau_i(\boldsymbol{\kappa}) = \frac{\boldsymbol{\kappa}\cdot\mathbf{x}_i}{c}. \tag{7}$$

The vector $\boldsymbol{\kappa}$ is related to the angle of azimuth ϕ and elevation φ of the propagating wavefronts as follows:

$$\boldsymbol{\kappa} = [\cos(\phi)\cos(\varphi), \sin(\phi)\cos(\varphi), \sin(\varphi)]^T. \tag{8}$$

As suggested by Krishnaprasad (2016), converting the time signals to the frequency domain and treating each frequency beam separately, the steering vector of an array of M acoustic vector sensors can be expressed as

$$\mathbf{a}(\boldsymbol{\theta}, f) = \mathbf{a}_p(\boldsymbol{\theta}, f) \otimes \mathbf{h}(\boldsymbol{\theta}), \tag{9}$$

with

$$\mathbf{a}_p(\boldsymbol{\theta}, f) = \left[e^{\mathrm{j}2\pi f \tau_1}\ ,\ e^{\mathrm{j}2\pi f \tau_2}, \dots, e^{\mathrm{j}2\pi f \tau_M}\right]^T \tag{10}$$

$$\mathbf{h}(\boldsymbol{\theta}) = \begin{bmatrix} 1 \\ \boldsymbol{\kappa} \end{bmatrix} \tag{11}$$

$$\boldsymbol{\theta} = [\phi, \varphi], \tag{12}$$

where $\otimes$ represents the Kronecker product and f is the frequency evaluated. The output of classical Delay-And-Sum (DAS) and Capon beamforming (also known as MVDR) are obtained by maximizing or minimizing the following generalized expression:

$$B(\boldsymbol{\theta}, f) = \mathbf{a}^H(\boldsymbol{\theta}, f)\, \mathbf{R}(f)\, \mathbf{a}(\boldsymbol{\theta}, f), \tag{13}$$

where for classical beamforming the DOA is obtained by maximizing Eq. 13 with $\mathbf{R}(f)$ being the covariance matrix of the measurement data. On the other hand, the DOA for Capon beamforming is obtained by minimizing Eq. 13 with $\mathbf{R}(f)$ being the inverse of the covariance matrix of the measurement data.

4. Numerical investigation

A numerical investigation has been conducted to study the data model proposed. Time signals of sound pressure and particle velocity were computed at multiple location in order to synthesize the data recorded by an array of acoustic vector sensors. After the time data is generated, it is then possible to apply beamforming techniques to locate the sound sources. For the sake of brevity, the numerical study presented in this section is focused on evaluating the sound field perceived by an array of 12 AVSs for different source configurations.

The sound field produced by a single source with narrow band excitation is first evaluated. A white noise signal was filtered with a band-pass filter centered at 200 Hz with a bandwidth of 50 Hz. The source was moved along a circular trajectory of 50 m radius with a speed of 36 RPM. A sketch of the geometry is presented on the left hand side of Figure 1, along with the spectrogram of the synthesized pressure signal at the center of the array and the resulting beamforming map using DAS. As can be seen, despite the narrow band nature of the original source signal, the continuously changing propagation path from the emitting point to the static sensor array introduces a significant Doppler shift. The beamforming map is also affected by the source movement, showing some smearing along the source trajectory.

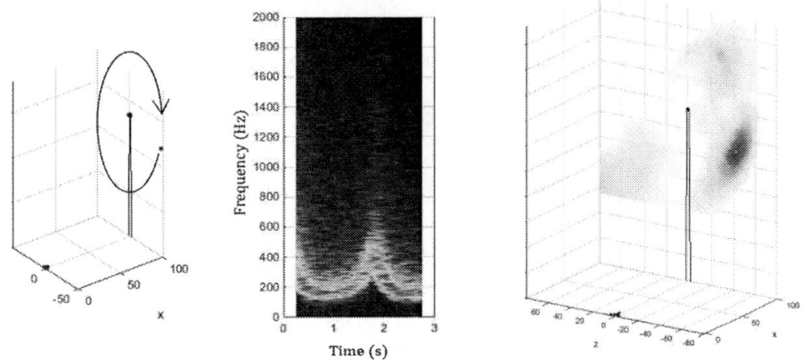

Figure 1: Single source case moving along a circular trajectory (left), spectrogram of the signal perceived at the center of the array (middle) and beamforming map (right).

Secondly, a numerical examples with multiple moving sources is shown in Figure 2. In this case the sound sources are linearly distributed along a straight line, resembling a wind turbine blade. As mentioned above, uncorrelated signals were used for all the sources due to the random nature of aero-acoustic flow interactions around the blade. The beamforming map produced shows that the small array used is not capable of resolving all the sources individually (illustrated with green dots) but it is still possible to detect the main sound emission area.

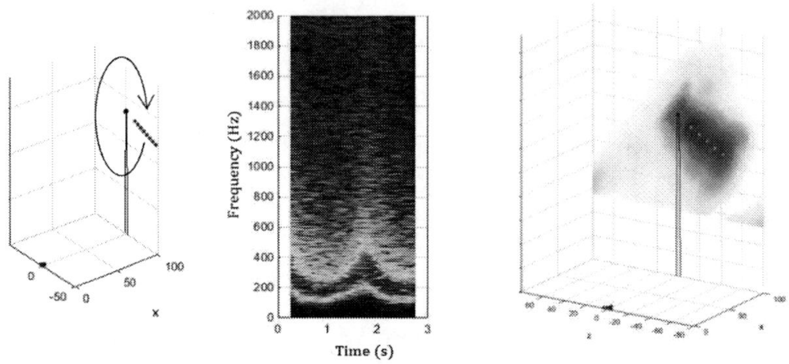

Figure 2: Multiple aligned sources moving along a circular trajectory (left), spectrogram of the signal perceived at the center of the array (middle) and beamforming map (right).

In addition, a scenario with three sets of linearly distributed sources was also studied. Results are presented in Figure 3. Assessing the spectrogram of the measured signals, the strong Doppler effect introduced is no longer apparent. The superposition of the sound generated during down stroke and up stroke phases creates an apparently broad-banded acoustic excitation which seems fairly stationary from 150 Hz to 350 Hz.

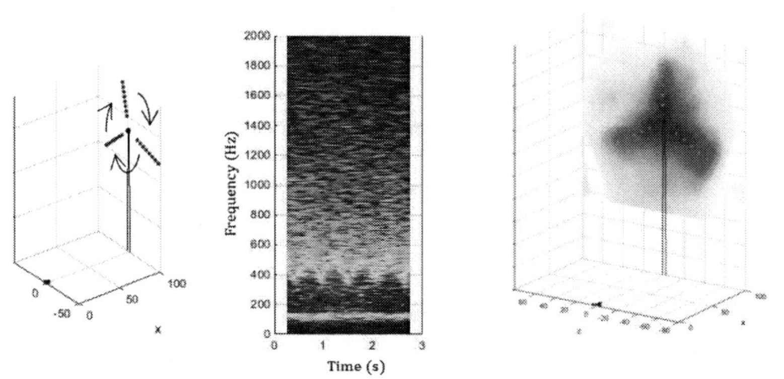

Figure 3: Multiple aligned sources moving along a circular trajectory (left), spectrogram of the signal perceived at the center of the array (middle) and beamforming map (right).

5. Experimental evaluation

An experimental study was conducted for assessing the sound field produced by a large wind turbine with an array of acoustic vector sensors. Measurements were performed about 130 meters away from a wind turbine operating in regular conditions. The measurement campaign was carried out in collaboration with a wind turbine manufacturer in order to gain understanding on the sound radiation mechanism at mid and low frequencies. The following sections provide information about the measurement setup, in-situ calibration and some time-averaged results obtained. However, several details are omitted due to a confidentiality agreement with the wind turbine manufacturer.

5.1 Measurement setup

The sensor array comprised 9 AVSs and 3 microphones deployed over an area of 8 by 8 meters. A nested array configuration was used in order to apply virtual sensor reconstruction techniques in future works. The resulting 39 sensor signals were recorded using a 48 channel data acquisition system connected to a regular laptop. In addition, a video of the wind turbine movement was synchronously recorded in order to match the blade movement with the beamforming results and track the rotating speed. All equipment was powered with a 12V battery which was converted to 220 AC using a sinusoidal power inverter. A picture of the full setup is shown in Figure 4.

Figure 4: Nested array used in the measurement campaign.

Each sensor was covered with a multi-layer wind screen designed for outdoors usage in a wind turbine farm. This element is particularly critical for the usage of particle velocity sensors, since airflow interactions around the sensing element may mask the acoustic response. However, the good performance achieved with the multi-layer screen used in this dataset yield negligible noise induced by wind.

5.2 Calibration of AVS orientation

The frequency response of each sensor was calibrated prior the outdoor measurement campaign. However, the use of an array containing AVSs requires adjusting the sensor orientation after the array is deployed. The large aperture size use to localize low frequency sound sources prevents from building a fix structure which could be used to control sensor positioning and alignment. In order to introduce a calibration step at the post-processing stage, a novel procedure is hereby proposed based on the use of impulse excitation created at known locations.

Misalignments of the particle velocity elements contained in an AVS can lead to significant errors in the beamforming results and therefore they must be corrected beforehand. Since the vector transducers are orthogonally placed, alignment errors can be corrected using a simple rotation matrix. Finding such matrix resembles the "*Procrustes problem*" postulated in linear algebra where it is asked to find an orthogonal matrix $\mathbf{C}$ which most closely maps the raw data matrix $\mathbf{Y}$ to the calibrated data $\mathbf{Y}_c$ such as

$$\mathbf{C} = \arg\min_{\Omega} \|\mathbf{\Omega Y} - \mathbf{Y}_c\|_F \quad \text{subject to } \mathbf{\Omega}^T\mathbf{\Omega} = \mathbf{I}, \tag{14}$$

The problem was originally solved by Schönemann (1966) using the normalized singular value decomposition of the matrix $\mathbf{M}$ resulting from multiplying both raw and expected data, i.e.

$$\mathbf{M} = \mathbf{Y_c}\,\mathbf{Y}^T = \mathbf{U\Sigma V}^T \tag{15}$$

$$\mathbf{C} = \mathbf{UV}^T \tag{16}$$

Each AVS was calibrated individually using the 3D sound intensity vectors measured while the sound field was excited by an impulsive source of known location. Figure 5 shows a picture of the array during the orientation calibration stage along with time signal recorded by one of the sensor elements.

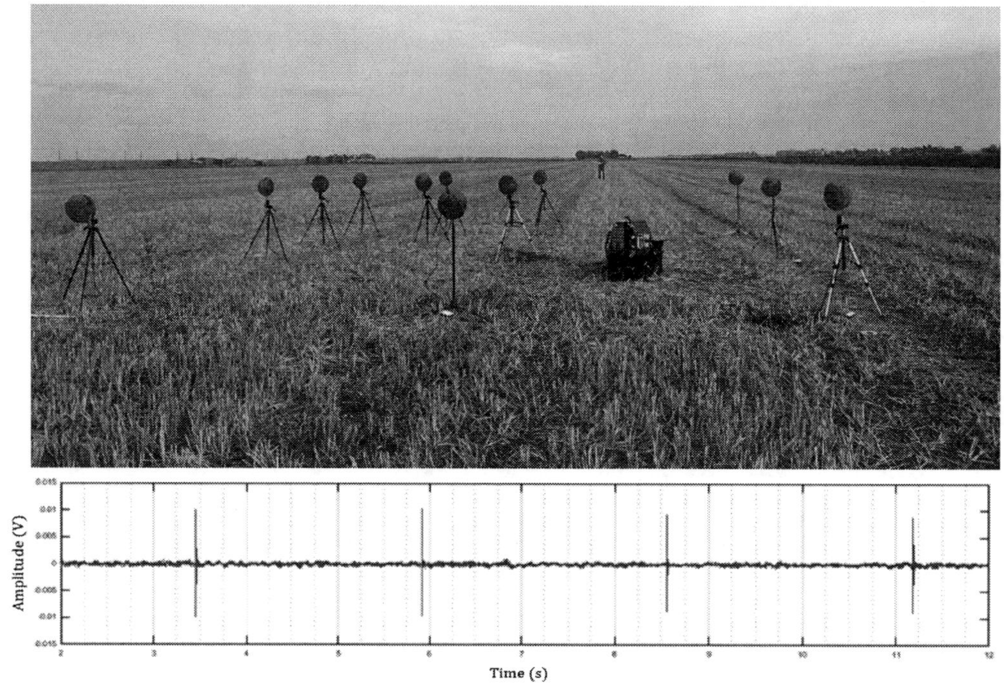

Figure 5: Picture of the array during the calibration procedure while impulsive signals are generated 40 m away from the array (up) and impulsive signal recorded by one of the particle velocity sensors (down).

An impulsive source was repeatability played 40 meters away from the center of the array at two locations that differed 90 degrees in azimuth. The raw sound intensity vectors measured by each sensor for the first source location (green) and the second one (blue) are displayed on the left hand side of Figure 6. The source location was then used to calculate the expected unitary vectors (middle graph of Figure 6). By using Equation 15 and Equation 16, a rotation calibration matrix was obtained and applied to the raw data. Results obtained after applying the rotation matrix to the raw data are shown on the right hand side of Figure 6. As it is shown, results obtained with the calibrated signals match accurately the reference vectors.

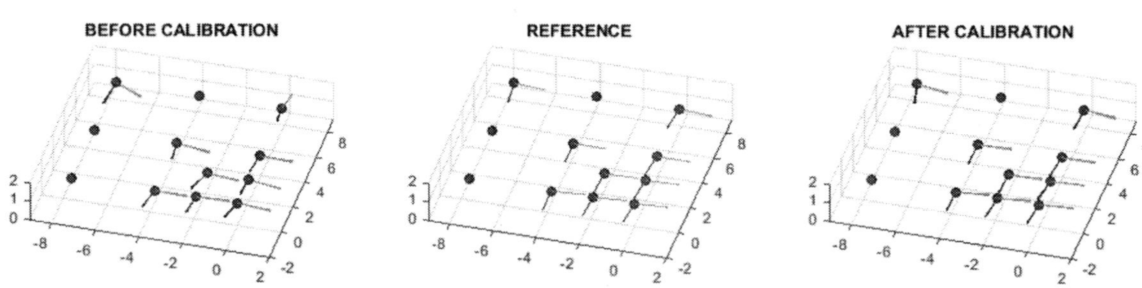

Figure 6: DOA of each AVS measured separately during the reference measurements before (left) and after (right) calibrating the sensors orientations obtained while the source was in the first (green) and second (blue) location.

5.3 Measurement results

Normalized beamforming results are presented in Figure 7 for frequency bands of 50 Hz centered at 200 Hz, 300 Hz, 400 Hz and 500 Hz with a dynamic range of 3 dB. The location of the array is plotted with blue dots whereas the tower and rotating blade area are represented with solid and discontinuous black lines, respectively.

The sound maps were calculated using Capon beamforming with a covariance matrix linearly averaged over 20 seconds. As it is shown, for this particular wind turbine most of the sound is produced during the down stroke movement of the blades, which is in line with previous results published in the literature. It should be noted that despite the large spacing between contiguous sensors, spatial aliasing is very low even at 500 Hz, for which the sampling interval is 5 times over the Nyquist rate.

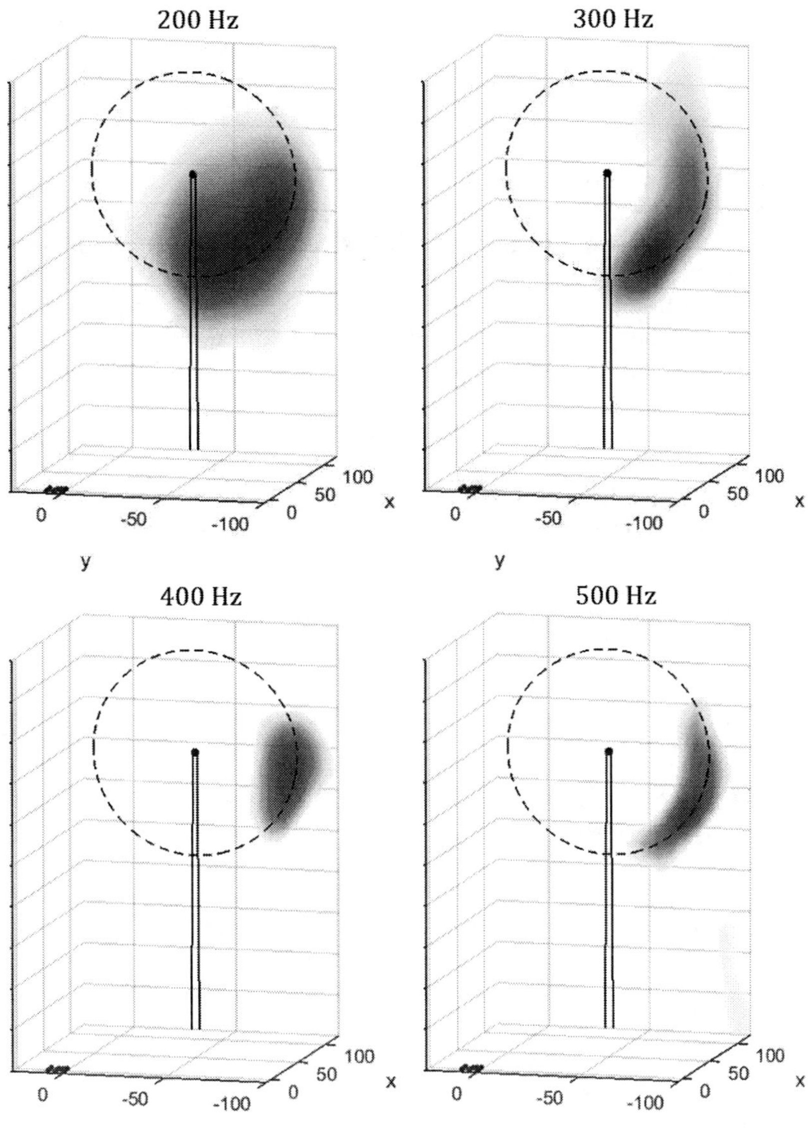

Figure 7: Averaged beamforming results after 20 seconds at 200 Hz (top left), 300 Hz (top right), 400 Hz (bottom left) and 500 Hz (bottom right).

6. Conclusions

This paper proposes a framework to model the sound field produced by a wind turbine based on a set of arbitrary moving monopole sources. Both sound pressure and particle velocity can be computed, therefore this approach is suitable for synthesizing the data of a microphone array or even an acoustic vector sensor array. The data model is formulated in the time domain, allowing to render directly the sensors' output signals. Intrinsic frequency changes due to the relative motion between the sources and receivers are accounted for by using time-dependent Green functions. As a result, Doppler shifts can be predicted and be use to gain a better understanding of the impact of certain defects or noise control measures in the sound field produced.

A numerical study has been presented, illustrating the impact of different source configurations on the sound perceived by a static sensor. Furthermore, the computed data is also used in combination with sound localization maps obtained via beamforming techniques. The application of sound localization algorithms to a complex sound field such as the one produced by a wind turbine may lead to ambiguous results induced by the sensor distribution. The ability to create synthetic data can be very helpful in order to optimize an array configuration for a frequency range of interest while studying the performance obtained with different array geometries.

An experimental study was conducted to verify the feasibility of assessing the sound field produced by a wind turbine with an array of acoustic vector sensors. An array containing acoustic vector sensors and sound pressure microphones have been designed, deployed and calibrated for an outdoor measurement campaign in a wind turbine field. A novel procedure to calibrate the AVS orientation has also been proposed. In conclusion, numerical and experimental evidence demonstrate that it is possible to model and measure the sound field produced by a large wind turbine using an array of acoustic vector sensors.

References

Bass, J., Bowdler, D., McCaffery, M. and Grimes, G. (2011), *Fundamental research in Amplitude Modulation – a project by RenewableUK*, Proceedings of Wind Turbine Noise.

Camier, C., Blais, J. F., Lapointe, R., and Berry, A. (2012). *A time-domain analysis of 3D non-uniform moving acoustic sources: application to source identification and absolute quantification via beamforming*. In Proceedings of the 4th Beamforming Berlin Conference.

Doolan, C. J., Moreau, D. J., Brooks, L. A., Hessler Jr, G. F., Schomer, et al. (2012). *Wind turbine noise mechanisms and some concepts for its control*. Acoustics Australia, 40(1), 7-13.

Hansen, C. H., Doolan, C. J., & Hansen, K. L. (2017). *Wind Farm Noise: Measurement, Assessment, and Control*. John Wiley & Sons.

Hawkes, M., and Nehorai, A. (1998). *Acoustic vector-sensor beamforming and Capon direction estimation*. IEEE Transactions on Signal Processing, 46(9), 2291-2304.

Kitchens, J. P. (2010). *Acoustic vector-sensor array processing*. PhD thesis, Massachusetts Institute of Technology, USA.

Krishnaprasad, N.R. (2016). *Acoustic Vector Sensor Based Source Localization*. MSc thesis, TU Delft, the Netherlands.

Nehorai, A., and Paldi, E. (1994). *Acoustic vector-sensor array processing*. IEEE Transactions on signal processing, 42(9), 2481-2491.

Oerlemans, S. (2009). Detection of aeroacoustic sound sources on aircraft and wind turbines, PhD thesis, University of Twente, The Netherlands.

Rogers, A. L., and Manwell, J. F. (2004). *Wind turbine noise issues*. White Paper by RERLC.

Schönemann, P. H. (1966). *A generalized solution of the orthogonal Procrustes problem*. Psychometrika, 31(1), 1-10.

Verheij, J. W. (1997). *Inverse and reciprocity methods for machinery noise source characterization and sound path quantification*, International Journal of Acoustics and Vibration, 2(1), 11-20.

Zajamšek, B., Hansen, K. L., Doolan, C. J., & Hansen, C. H. (2016). *Characterisation of wind farm infrasound and low-frequency noise.* Journal of Sound and Vibration, 370, 176-190.

7th International Conference
on
Wind Turbine Noise
Rotterdam – 2nd to 5th May 2017

Use of the Acoustic Camera to accurately localise wind turbine noise sources and determine their Doppler shift

Stuart Bradley, Physics Department, University of Auckland, New Zealand: s.bradley@auckland.ac.nz

Michael Kerscher, gfaitech GmbH, Volmerstrasse 3, 12489 Berlin, Germany: kerscher@gfaitech.de

Torben Mikkelsen, Technical University of Denmark, Department of Wind Energy, Roskilde, Denmark: tomi@dtu.dk

Summary

The Acoustic Camera is frequently used to visualize the aero-acoustic noise from wind turbines, but 'ground truth' validation of accuracy has not generally been available. We describe an experiment in which eight small piezoelectric speakers were placed at positions along a turbine blade from near the hub to at the very tip, and on the leading and trailing edge of the blade. Six sources were attached on the suction (downwind) side of the blade, and two on the pressure (upwind) side. Each source generated a loud stable narrow-band tone. All sources were emitting simultaneously at slightly different frequencies. This multi-source configuration was recorded with a star-shaped 48-microphone Acoustic Camera of diameter 3.4 m, and a 120-microphone spiral array of diameter 4 m. Measurements were made with the turbine not rotating, as the turbine rotation speed increased, at full constant rotation speed, and as the turbine slowed down to stop. Cameras were placed on the ground downwind of the turbine along the turbine axis direction. Measurements were also conducted upwind of the turbine at hub height.
All eight sources were clearly identified even though they were close in frequency. The leading edge and trailing edge sources, which experienced very similar Doppler shift, were also clearly resolved. The location of each source predicted acoustically also agreed very closely with the location on the optical image obtained by the reference camera located at the centre of each microphone array. This was true even when the turbine was rotating at full speed.
The Doppler effects were dramatic, with the sources at varying radial distances along the blade experiencing different Doppler shift amplitudes. Particularly interesting effects will be described during speed up and slowing down of the turbine.
Overall, this experiment provides, for the first time, quantitative evidence of the source location accuracy of the Acoustic Camera for wind turbine noise estimation, as well as a quantitative assessment of the ability of the Acoustic Camera to characterise the time-dependent spectral changes along a blade due to Doppler shift.

1. Introduction

Wind turbines generate aero-acoustic noise from the blades which, because of periodicities, can cause annoyance at distances of hundreds of meters (Chen et al. 2016; Michaud et al, 2016; Dröes and Koster , 2016). The typical path length from turbine source to listener is such that meteorological influences are significant (Mittal et al. 2017; Gallo et al, 2016). In the absence of very dense 3D meteorological measurements and an accompanying, *highly reliable*, sound propagation model, characterisation of the sources needs to be done close to the turbine. Measurements can be made on the blade itself (Bertagnolio et al, 2017), but this leaves a challenge to translate these measurements to what is heard in the far field. An alternative is to use a microphone array some tens of meters from the turbine, so that measurements are effectively far-field but close enough that intervening meteorological influences are not significant (Buck et al, 2016).

The Acoustic Camera is frequently used to visualize the aero acoustic noise from wind turbines, but 'ground truth' validation of accuracy has not generally been available. There are two main reasons why such validation is required. The first is that, because of the relatively slow speed of sound, the turbine rotates considerably during the emission of sound and its reception at the Acoustic Camera microphone array. This means that *registration* of the detected sound pattern on the blade requires assumptions about wind speed and direction, generally in addition to any wind vector measurements (Mo and Jiang, 2017; Zhang et al. 2017). The second reason that validation of Acoustic Camera data is required is that the sound from the blade undergoes huge Doppler shift due to the rapid rotation and the relatively close distance to the Acoustic Camera. Placing an Acoustic Camera further from the turbine will reduce Doppler shift, but also reduces the spatial resolution achieved by the Acoustic Camera and increases the uncertainties about meteorological effects.

Notwithstanding these difficulties, there is no doubt that Acoustic Camera images, such as that recorded by Oerlemans et al. (2007) and shown in Fig. 1, provide valuable insights into turbine noise generation, showing source location on the blade, spectral characteristics, and some measure of directionality, as the blade rotates.

Validation of Acoustic Camera methodology using the aero-acoustic noise generated by the turbine is not a realistic approach, because it is necessary to make an assumption about the sound which is being used for validation. For this reason, we have set up a *control experiment*, in which known sound sources are placed in known locations on a rotating turbine, and are then recorded and analysed using an Acoustic Camera.

2. Experimental Design

In this experiment, multiple small piezo-electric tonal sources were placed on one blade of an operational turbine, which was then allowed to rotate. Two models of Acoustic Camera arrays were used to make measurements of the sound received 50 m from the turbine.

2.1 Acoustic Sources

Eight small piezoelectric speakers were placed at positions along a turbine blade from near the hub to at the very tip, and on the leading and trailing edge of the blade. Six sources were attached on the suction (downwind) side of the blade, and two on the pressure (upwind) side. Each source generated a loud stable narrow-band tone. All sources were emitting simultaneously at slightly different frequencies. The sources, each weighing 55 g, were each powered by a light battery (the dominant weight) and taped to the blade at measured locations as shown in Fig. 2.

The tone frequencies, in ascending order, are 3068, 3069, 3090, 3101, 3308, 3405, 3409, and 3453 Hz for sources 4, 5, 2, 3, 1, 7, 8, and 6 respectively. The variation is due to the manufacturer's tolerance for these inexpensive piezo buzzers. Note that sources 4 and 5 are very close in frequency, as are sources 1 and 7 on the tip.

2.2 Acoustic Camera

This multi-source configuration was recorded with a star-shaped 48-microphone Acoustic Camera of diameter 3.4 m (a Star48), and a 120-microphone spiral array of diameter 4 m (a FlexStar120). The Acoustic Cameras were placed on the ground 48.5 m downwind of the turbine along the turbine axis direction, as shown in Fig. 3. Measurements were also conducted upwind of the turbine at hub height using the Star48 from a hoist.

Figure 1. Use of an array of microphones to image turbine noise sources (from by Oerlemans et al. 2007) .

2.3 Turbine Operation

The sources were mounted on a Nordtank NTK 500/41 500 kW turbine of diameter 41 m, swept area 1325 m^2, a maximum rotor speed of 27 rpm, a tip speed of 58 m/s, and hub height 50 m. Measurements were made with the turbine not rotating, as the turbine rotation speed increased, at full constant rotation speed, and as the turbine slowed down to stop. The turbine has an air-brake by turning the tip of one of the blades by 90°, as seen on the left in Fig. 2.

3. Location of Sources

Fig. 4 shows Acoustic Camera resolution of sources on the blade face which is toward the Star 48 using a spectral filter which encompasses all sources (the turbine is static) Sources 3 and 4 show as a double-sized contour, and are not quite individually resolved. Fig. 5 shows source location based on tighter bandwidth filters (as indicated on the figure).

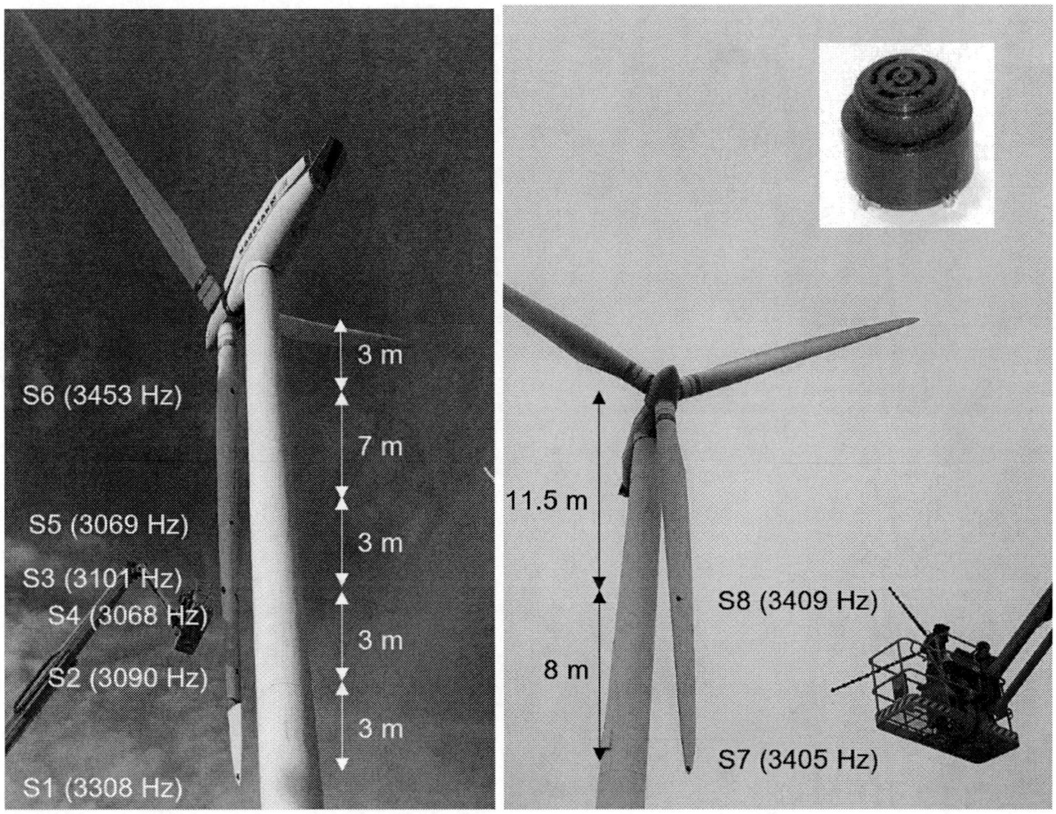

Figure 2. The location and tone frequency of the 8 speakers. Inset: one of the speakers.

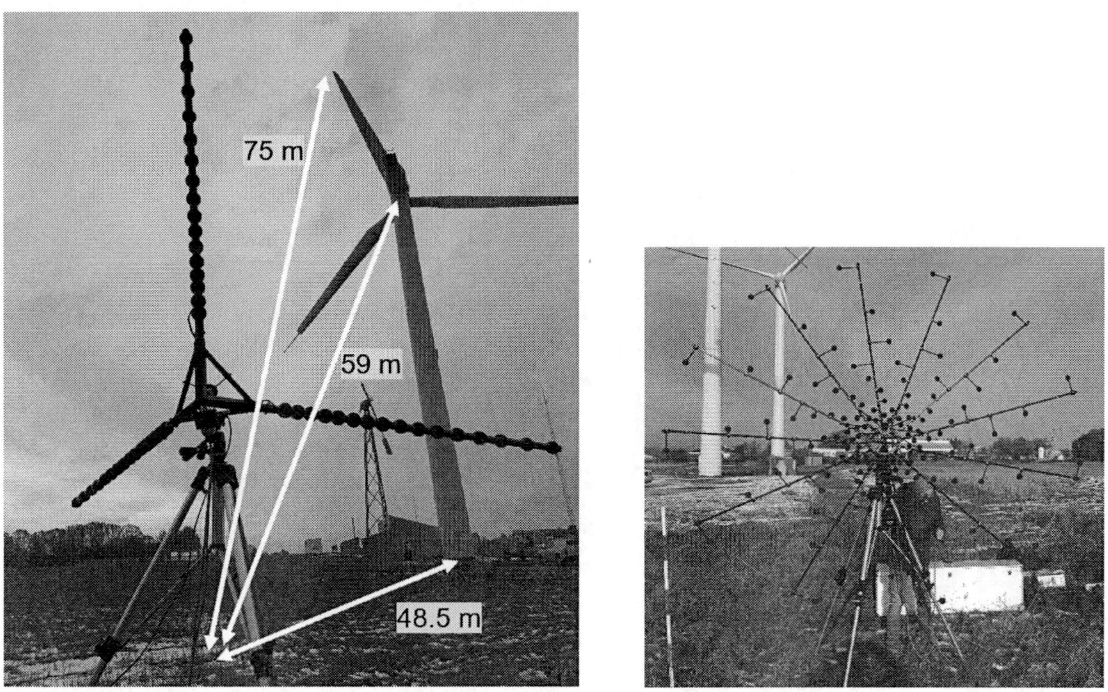

Figure 3. The turbine and Star48 array (left), and the FlexStar120 (right).

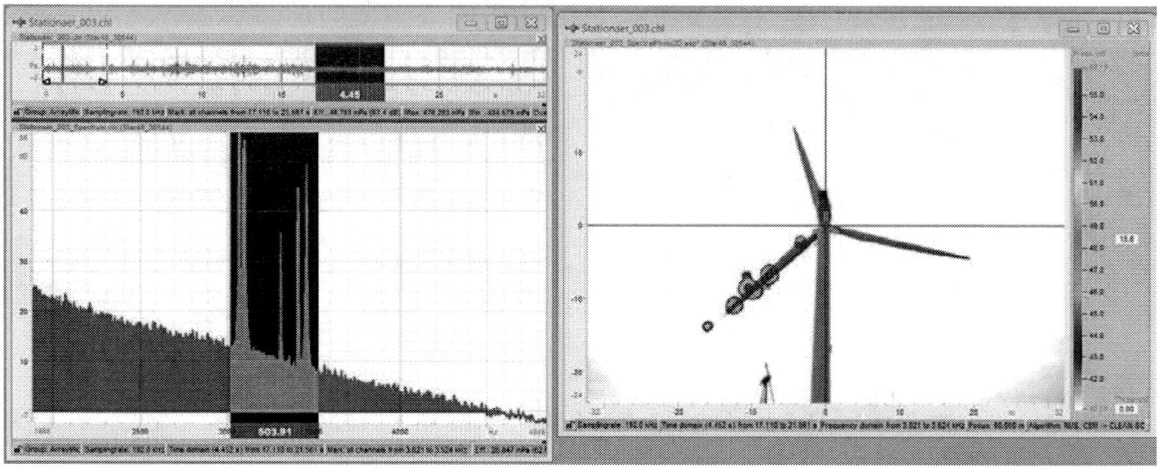

Figure 4. Time record (top, spectrum (left) and sources (right) for the static turbine (Star48).

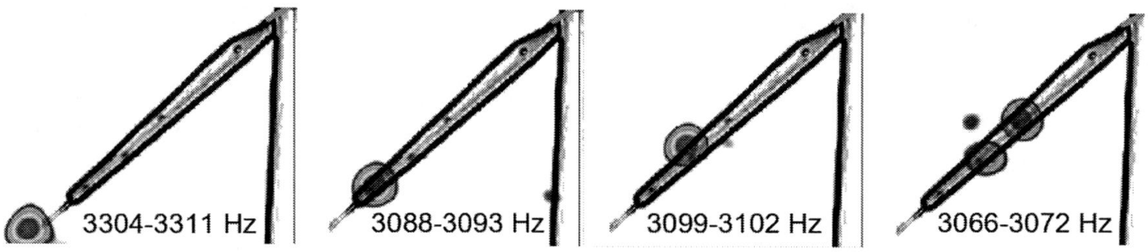

Figure 5. Source location using Star48.

The tighter bandwidth allows source 3 to be clearly separated from the nearby source 4. The rightmost of these frames shows sources 4 and 5 clearly resolved (they are spatially well separated but have frequencies within 1 Hz). In three of these frames minor reflections can also be seen, possibly from the tower, giving a spurious blue-coloured point. Similar results are obtained for source 6.

Figure 6. FlexStar120 image with the turbine rotating at constant speed (left), and Star48 images from the hoist at hub height with filters for source 8 (centre) and 7 (right).

When the turbine is fully rotating at constant speed, source location appears to be not nearly as good (left frame, Fig. 6). Three of the sources are resolved and possibly with further frequency selection 4 could be resolved. It may be possible here that the bandwidth chosen did reflect the Doppler shifts fully, and more work needs to be done on that (this figure was obtained in a 'quick look' at the data, immediately after the experiment). There are also spurious 'sources' from reflections, most likely off the tower.
Measurements were also made of the sources on the upwind face of the blade (also in Fig. 6) when the turbine was static, with similar results to those for the downwind face.

4. Doppler Shift

The Doppler shift is captured in Fig. 7 as the turbine speeds up from rest. The entire 8 sources can be identified, although the two on the blade face away from the Acoustic camera have much lower amplitude. In Fig. 7, from the top down, sources are 6, 7 and 8 (fainter), 1, 3 and 2, and 4 and 5. The relative amplitudes of the frequency variation are expected to increase with radial distance from the hub in the order 6, 5, 8, 3 and 4, 2, 1 and 7. It is difficult to separate those sources which are close in frequency, but it is clear that source 6 has the smallest FM amplitude and sources 1 and 7 have the highest amplitude.

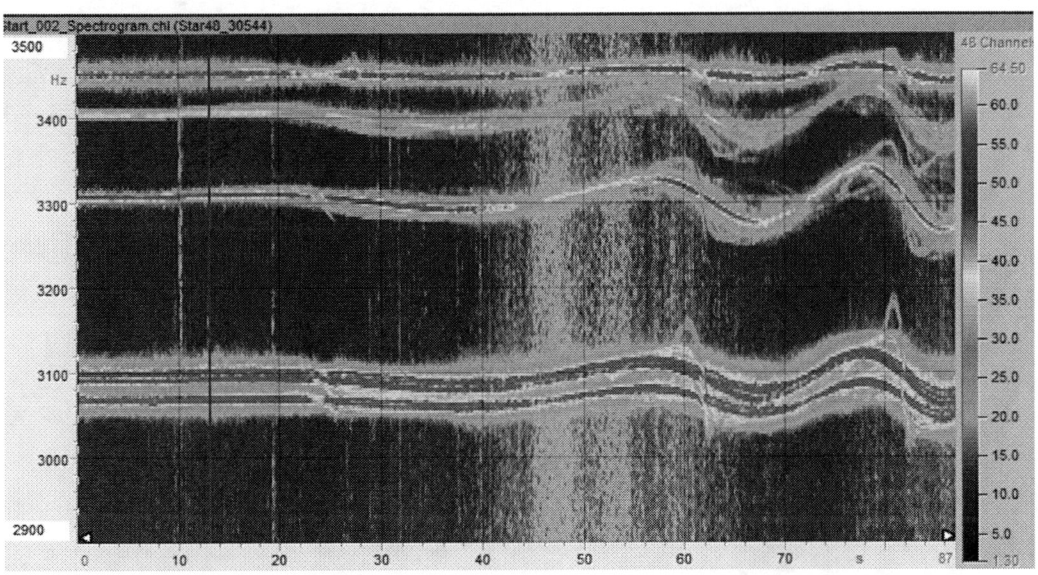

Figure 8.Doppler shift commencing as the turbine spins up.

Sequences are shown in Fig. 9 at constant turbine rotation and as the turbine slows down to stop.

5. Conclusions

Overall, this experiment provides, for the first time, quantitative evidence of the source location accuracy of the Acoustic Camera for wind turbine noise estimation, as well as a quantitative assessment of the ability of the Acoustic Camera to characterise the time-dependent spectral changes along a blade due to Doppler shift. The ability of an Acoustic Camera to localise individual point tonal 3 kHz sources is exceptional! In Fig. 10 the optical and acoustic source separation is compared with a 1 m circle, showing that the acoustic spatial resolution is around ½ m.

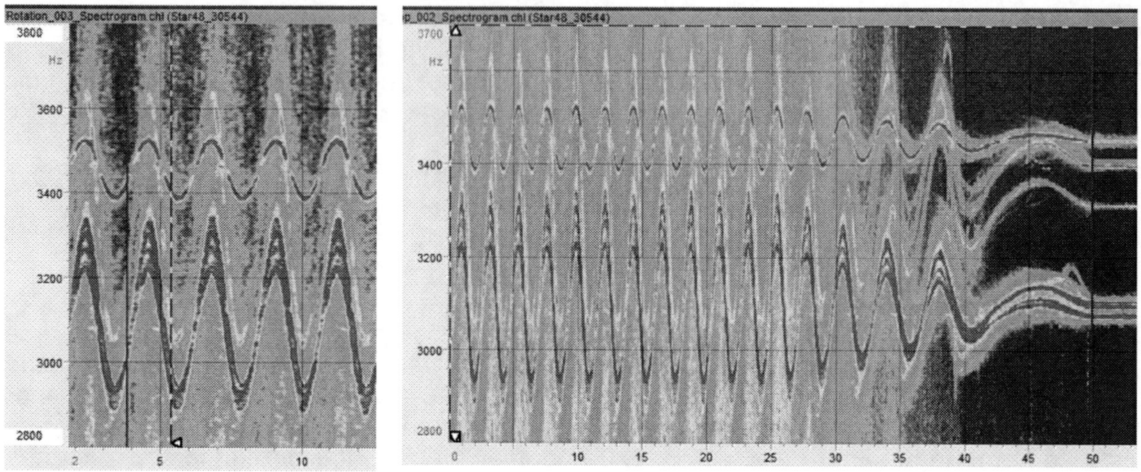

Figure 9. Doppler shift at constant rotation (left) and as the turbine slows down (right) for selected sources.

This work gives confidence in the current ability of Acoustic Cameras to adequately register sources with sufficient precision, and also to measure the Doppler shift as a function of radial distance (i.e. as a function of speed of the sources). This should allow the Doppler effects to be removed so that true source characterisation can be obtained.

Figure 10. An Acoustic camera optical image with edge detection, showing one of the sources, together with the acoustic source location. Just above this point is drawn a 1 m – dimeter circle.

References

Bertagnolio F, Madsen H A, Fischer A and Bak C (2017) *A semi-empirical airfoil stall noise model based on surface pressure measurements*. J Sound Vibration 387, 127–162.

Bower, T, Villeneuve, P J, Russell, E, Koren, G, and van Den Berg, F (2016) *Self-reported and measured stress related responses associated with exposure to wind turbine noise*. JASA 139(3), 1467-1479.

Buck S, Oerlemans S. and Palo S (2016) *Experimental characterization of turbulent inflow noise on a full-scale wind turbine*. J Sound Vibration 385, 219–238.

Chen L, Harding C, Sharma A and MacDonald E (2016). *Modeling noise and lease soft costs improves wind farm design and cost-of-energy predictions*. Renewable Energy, 97, 849-859.

Dröes M I and Koster H R A (2016). *Renewable energy and negative externalities: The effect of wind turbines on house prices*. J Urb Econ, 96, 121–141.

Michaud, D S; Feder, K, Keith, S E, Voicescu, S A, Marro, L, Than, J, Guay, M, Denning, A, Gallo, P, Fredianelli, L, Palazzuoli, D, Licitra, G and Fidecaro F (2016) *A procedure for the assessment of wind turbine noise*. Appl Acoustics 114, 213–217.

Mittal P, Mitra K, and Kulkarni K (2017). *Optimizing the number and locations of turbines in a wind farm addressing energy-noise trade-off: A hybrid approach*. Energy Conversion and Management, 132 147–160.

Mo P, and Jiang W (2017) *A hybrid deconvolution approach to separate static and moving single-tone acoustic sources by phased microphone array measurements*. Mech Systems Sig Proc 84, 399–413.

Oerlemans, S, Sijtsma, P and Méndez López B (2007) *Location and quantification of noise sources on a wind turbine*. J Sound Vibration 299, 869-883.

Zhang X-Z, Bi C-X, Zhang Y-B and Xu L (2017). *A time-domain inverse technique for the localization and quantification of rotating sound sources*. Mech Systems Sig Proc 90,15–29.

7th International Conference
on
Wind Turbine Noise
Rotterdam 2-5 May 2017

Wind Turbine Sound Prediction: Modelling and Case Study on the Effect of Blade Elasticity

L. Schorle University of Siegen, 57068 Siegen, Germany
E-Mail: leonard.schorle@uni-siegen.de

T. Carolus University of Siegen, 57068 Siegen, Germany
E-Mail: thomas.carolus@uni-siegen.de

S. Erbslöh Senvion SE, 24783 Osterrönfeld, Germany
E-Mail: sascha.erbsloeh@senvion.com

Summary

Sound prediction tools for wind turbines are essential in the initial phase of planning in order to meet regulatory immission standards. They are further used for estimating the effect of modifications during turbine development on the sound immitted at relevant monitoring points. They can be classified in three methodologically different categories: Empirical one-equation models, semi-analytical models and computational aeroacoustics methods. Objectives of this contribution are to compile and compare selected aeroacoustic semi-analytical wind turbine sound prediction models available in the open literature, and conduct a first case study.

The general structure of all semi-analytical models is found to be similar: A set of sub-models for the elementary sound sources on the blades is combined with a sound propagation model from the sources to the listener. Trailing edge sound is identified as the dominant source in state-of-the-art wind turbines. However, the sub-models for elementary sound sources found in the literature vary substantially. Combining selected sub-models, a preliminary own wind turbine sound prediction model was compiled and encoded, yielding the acoustic footprint of the turbine on ground level and the swishing character of the wind turbine sound.

HOWE's sound source directivity function seems to yield results, that match best with experimental results from literature. Also, from principle considerations ROZENBERG's approach for the convective amplification factor, taking into account the motion of the elementary sound sources, seems to be the most consistent.

In a first case study, the effect of delocalisation of elementary sound sources along the blades and a modification of the local angle of attack due to the flexibility of the blade has been studied. As a preliminary result, the effect of blade elasticity on the acoustic emission of a complete wind turbine rotor seems to be small when comparing to a rigid rotor.

It is important to note that so far the atmospheric attenuation, refraction and ground effects have not been taken into account - this remains for future studies and may affect these results.

Nomenclature

Symbols

A, B	correction factors	dB
C	chord length of the airfoil	m
CA	convective amplification factor	
D	directivity function of the trailing edge sound	
f	acoustic frequency of the source	Hz
f'	shifted acoustic frequency at the observer	Hz
K_1, K_2, ΔK_1	correction factors	dB
L	spanwise length of the blade segment	m
M	free Mach number	
m	total amount of sound sources	
n	exponent in the convective amplification equation	
$OASPL$	overall sound pressure level (over all frequencies)	dB
p_{ref}	reference value for the sound pressure ($2 \cdot 10^{-5}$)	Pa
r	distance between the source and the observer	m
r_{Obs}	radial distance between the observer and the WT foot	m
R	radial position of a sound source on the rotor blade	m
SPL	sound pressure level	dB
S_{pp}	power spectral density (sound pressure) at the source	Pa^2/Hz
S'_{pp}	power spectral density (sound pressure) at the observer	Pa^2/Hz
St	Strouhal number	
t	source time	s
t'	observer time	s
T	time period of one rotor revolution	s
U	free-stream velocity	m/s
U_W	inflow velocity of the wind in the rotor plane	m/s
v_r	circumferential speed in the rotation plane	m/s
W_∞	relative inflow speed in the rotation plane	m/s
α	angle of attack	°
β_∞	free inflow angle	°
β_{cone}	cone angle	°
γ	stagger angle	°
δ^*	boundary layer displacement thickness	m
ζ	observer angle (ζ = 0° in front of the wind turbine)	°
λ	acoustic wave length	m
σ_R	rotor twist angle	°
σ_T	elastic torsion angle	°
σ_P	pitch angle	°
Γ_W	tilt angle	°
Ψ	azimuth angle	°
β, γ_R, Θ_H, Φ_A, Θ_A, Φ_H	directivity function angles	°

Further indices

am	amplitude modulation
eq	equivalent

norm	normalized with respect to the position directly in front of the WT
p	pressure side of the airfoil
ref	reference case with r = 1 m and D = 1
s	suction side of the airfoil
SO	source and observer
TOT	contributions from suction and pressure side (for an elementary source)
total	total sound pressure (summation over all elementary sources)
1/3	1/3 octave band spectrum

1. Introduction

The sound emitted by wind turbines is a relevant criterion for onshore and potentially even offshore installations. Sound prediction tools are essential in the initial phase of planning a wind turbine installation in order to meet regulatory immission standards and, more recently, for an early assessment of perceptual reaction from listeners as addressed e.g. by EGGENSCHWILER et al. [1]. They also may allow for estimating the effect of modifications of the turbine during the phase of technical development. For instance, how does a delocalisation of sound radiating portions of the blade due to the elasticity of the blades affect the turbine's overall far field sound pressure at relevant observer points? Or modification of the details of the flow around the blade e.g. by means of trailing edge suction [2] or blowing [3]. Or the increasingly popular edge serration [4] which intervene in the sound scattering process.

There is a variety of sound sources potentially relevant in modern wind turbines. Two principal categories may be distinguished: (i) mechanic sound sources like the gear box or the generator, and (ii) aeroacoustic sound sources like the trailing edge, inflow or tip sound source. According to OERLEMANS et al. [5] who did an extensive acoustic field measurement on a wind turbine as well as numerous other researchers in the past, trailing edge sound is regarded as the dominant aeroacoustic sound source in modern wind turbines.

The phenomena relevant for an observer perceiving turbine sound are illustrated schematically in **Fig. 1**: Starting points are a number of elementary sound sources of given strength and directivity distributed at the blade surfaces. The superposition of all elementary sound sources with their instantaneous orientation and motion with respect to an observer forms the time dependent far field sound pressure. While propagating, sound rays from these sources are refracted and damped in the atmosphere and reflected on the ground.

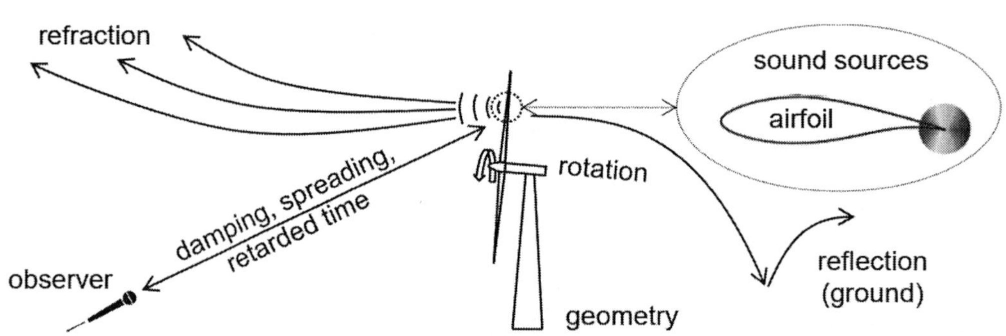

Figure 1: Wind turbine sound: Possible influencing factors on the sound immission at an observer-position

According to LOWSON [6], sound prediction methods for wind turbines can be classified in three methodologically different categories: (i) empirical one equation models as e.g. from LIPS [7] or in the German standard DIN EN 61400-11 [8], that are merely based on basic parameters of the wind turbine, (ii) semi-analytical models, which model different sound source mechanisms, and (iii) methods which are based on a full description of the turbine's geometry and flow, so-called computational aeroacoustics methods (CAA). State-of-the-art semi-analytical sound prediction models contain a number of steps:

- Segmentation of blades into a number of blade elements
- Based on sound source models: computation of sound source data (source strength and directivity for each blade element)
- Calculation of propagation of each elementary sound source to listener position, taking into account (i) convective amplification due to the motion of the elementary sound sources, (ii) geometrical spreading, (iii) atmospheric attenuation, (iv) refraction and ground effects
- Energetic summation of the sound pressures from all elementary sources at listener position

Overall objective of this work is assessing the effect of aeroacoustically relevant blade modifications on a wind turbine's far field acoustics. As a first step, in this paper a few semi-analytical sub-models from literature for the sources, their directivity and the effect of source motion are compiled and integrated into a wind turbine sound prediction tool. Due to complexity, atmospheric damping, ground-effects and refraction are neglected for the time being. Eventually, in a preliminary case study, the tool is applied to tackle the problem, how the sound from wind turbine blades is modified by source delocalisation due to blade elasticity.

2. Sub-models from literature and synthesis

2.1 Trailing edge sound source model

The so-called trailing edge sound results from the interaction of the turbulent boundary layer in the trailing edge region of an airfoil with the trailing edge. It has been subject of many investigations. BROOKS et al. [9] performed an extensive measurement campaign on NACA-0012 airfoils in an aeroacoustic wind tunnel and developed a semi-analytical model for the trailing edge sound, in principle based on the theory by FFOWCS WILLIAMS and HALL [10]. Their model predicts the 1/3 octave band spectrum of the sound pressure level SPL_{TOT} around the trailing edge as a function of boundary layer displacement thickness δ^*, angle of attack α, free mach number M of the airfoil, the spanwise length L of the blade segment, the distance r between the source and the observer, the directivity function $D_{SCHLINKER}$ according to SCHLINKER and AMIET [11], and various correction factors A, B, K_1, K_2 and ΔK_1. The total turbulent boundary layer trailing edge sound comprises three terms

$$SPL_{TOT} = 10\log\left(10^{SPL_\alpha/10} + 10^{SPL_s/10} + 10^{SPL_p/10}\right) \text{ [dB]}, \qquad (1)$$

where SPL_α contains the scaling model for the angle of attack α

$$SPL_\alpha = 10\log\left(\frac{\delta_S^* M^5 L D_{SCHLINKER}}{r^2}\right) + B\left(\frac{St_s}{St_2(\alpha)}\right) + K_2(\alpha) \text{ [dB]}, \quad (2)$$

SPL_p and SPL_S are the contributions from the pressure (*p*) and suction (*s*) side of the airfoil:

$$SPL_p = 10\log\left(\frac{\delta_p^* M^5 L D_{SCHLINKER}}{r^2}\right) + A\left(\frac{St_p}{St_1}\right) + (K_1 - 3) + \Delta K_1 \text{ [dB]} \quad (3)$$

$$SPL_s = 10\log\left(\frac{\delta_s^* M^5 L D_{SCHLINKER}}{r^2}\right) + A\left(\frac{St_s}{St_1}\right) + (K_1 - 3) \text{ [dB]} \quad (4)$$

The Strouhal numbers for the pressure and suction side are

$$St_p = \frac{f\delta_p^*}{U}, \; St_s = \frac{f\delta_s^*}{U} \quad (5, 6)$$

where *f* is the acoustic frequency and *U* is the free-stream velocity. St_1 is a function of Mach number *M* only:

$$St_1 = 0{,}02 \cdot M^{-0{,}6}. \quad (7)$$

St_2 depends on St_1 and α:

$$St_2 = St_1 \cdot \begin{cases} 1 & (\alpha < 1{,}33^\circ) \\ 10^{0{,}0054(\alpha - 1{,}33)^2} & (1{,}33^\circ \le \alpha \le 12{,}5^\circ) \\ 4{,}72 & (12{,}5^\circ < \alpha) \end{cases} \quad (8)$$

The resulting point source is located in the centre of the trailing edge of the blade segment.

The directivity function of the point source is a key component in the model. Originally it was derived by SCHLINKER and AMIET [11] as

$$D_{SCHLINKER} = 2 \cdot \sin^2\left(\frac{\Theta_A}{2}\right) \cdot \sin^2(\Phi_A). \quad (9)$$

The definitions of the angles between the trailing edge sound source and the observer are given in **Fig. 2**. The distance between the observer and the sound source is named *r*. This function is frequency-independent and normalized such that $D_{SCHLINKER}(\Phi_A = \Theta_A = 90^\circ) = 1$.

Two alternative frequency-independent directivity functions for trailing edge sound have been reported:

$$D_{HOWE} = 2 \cdot \sin^2\left(\frac{\Theta_H}{2}\right) \cdot \sin(\Phi_H) \quad (10)$$

(HOWE [12] or KAMBE et al. [13], D_{HOWE} ($\Phi_H = \Theta_H = 90°$) = 1), and

$$D_{OERLEMANS} = \left[2 \cdot \sin^2\left(\frac{\Theta_A}{2}\right) \cdot \sin^2(\Phi_A) \right]_{\Delta\beta,\Delta\gamma-averaged} \quad (11)$$

(OERLEMANS and SCHEPERS [14], $D_{OERLEMANS}$ ($\Phi_A = \Theta_A = 90°$) = 1). The latter is basically a modification of SCHLINKER's directivity function. Based on experimental results OERLEMANS and SCHEPERS applied an averaging over β and γ_R: "($d\beta$, $d\gamma_R$) is chosen to be (π/12, 2π/3) for β = 0°, and is reduced to (0,0) for $\beta = \pi/2$ (using the error function)" [14]. All those directivity functions are visualized in **Fig. 3**.

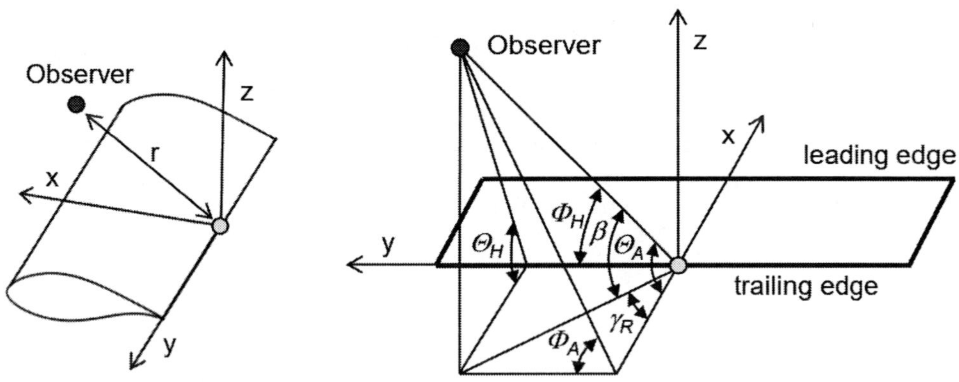

Figure 2: Definition of angles between trailing edge source and observer according to KAMBE et al. [13] , HOWE [12] and SCHLINKER and AMIET [11]

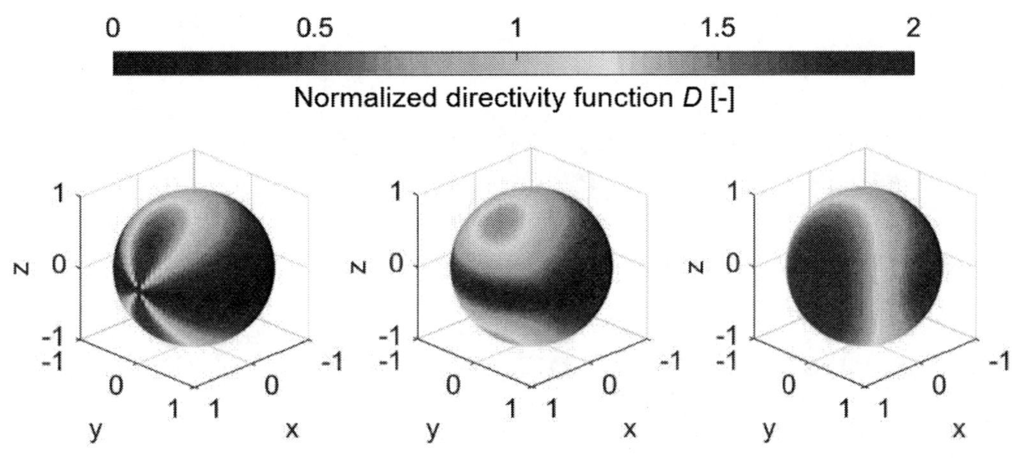

Figure 3: Visualisation of the directivity functions (from left to right): $D_{SCHLINKER}$, $D_{OERLEMANS}$ and D_{HOWE}; adapted from OERLEMANS and SCHEPERS [14].

Another popular trailing edge sound source model for the far-field sound is from AMIET [15]. In contrast to the previous models it is frequency-dependent and requires the surface pressure fluctuations on the airfoil as input. However, MOREAU and ROGER [16] have shown that for low mach numbers (M < 0.3) and acoustic compact radiators (C/λ >> 1) AMIET's directivity function tends to HOWE's [12]. These criterions are met at the outer part of the blades of modern wind turbines where the mach number is around M = 0.2 according to OERLEMANS [17] and the trailing edge sound has its spectral peak at C/λ > 10 according to BROOKS et al. [9].

2.2 Moving sound sources

As described by many authors such as ROZENBERG et al. [18], SINAYOKO et al. [19] or CRIGHTON et al. [20], the motion of a sound source relative to the observer has an influence on the perceived mean square value of the sound pressure and the spectral shape of the sound (in the far-field). The phenomenon of the so-called convective amplification is according to ROZENBERG et al. [18] explainable by the following: When the sound source moves towards the observer, the distance and therefore the travelling time of the acoustic energy to the observer is reduced. Hence, the observer receives the same acoustic energy from the source but in a shorter time which is equivalent to an increase of acoustic power (mean square value of the sound pressure).The convective amplification factor

$$CA = \frac{S'_{pp}}{S_{pp}} = \frac{1}{\left(1 - M_{SO}\right)^n} \tag{12}$$

represents the relationship between the power spectral density of the sound pressure at the source S_{pp} and at the observer S'_{pp}. Assuming CA being independent of frequency, this convective amplification factor also holds true for the overall sound pressure level ($OASPL$). The various amplification-factors found in the literature differ exclusively in the value of the exponent n. The derivation by ROZENBERG et al. [18] is based on the principle of energy conservation and yields an exponent n = 1. Starting with a different approach, SINAYOKO et al. [19] came up with an exponent of n = 2. Other derivations of the convective amplification are based on the inhomogeneous wave equation, which describes the spreading of waves in fluids. By inserting different source-terms, it is possible to investigate the effect of motion for different sound sources. GOLDSTEIN [21] as well CRIGHTON et al. [20] inserted volume and mass point sources and came up with an exponent n = 4.

2.3 Geometrical considerations

The directivity function angles Θ_A, Θ_H, Φ_A, Φ_H are depicted in **Fig. 4**. The source S, observer O and hub H are shown and the angles of the directivity functions are marked according to HOWE [12], KAMBE et al. [13] and SCHLINKER and AMIET [11]. Therefore, an oblique pyramid is spanned between the source and observer with its base point on an edge of the base area. The base area is defined by the span-wise and chord-wise direction vectors at the point S which depend on the stagger and cone angles. After that, the points P_F, P_2 and P_3 can be calculated. The scalar product is then used to determine the desired angles.

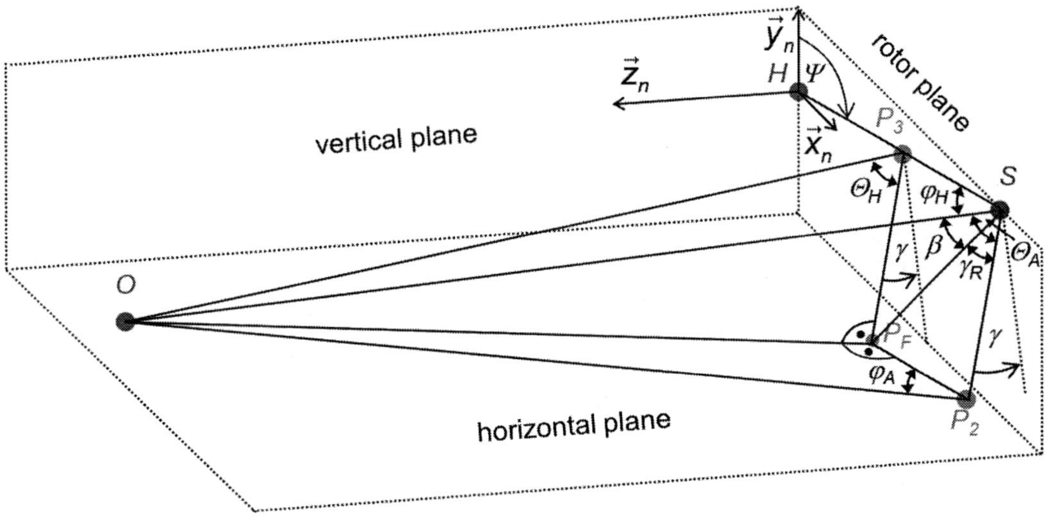

Figure 4: Definition of the directivity function angles Θ_A, Θ_H, Φ_A, Φ_H according to HOWE [12], KAMBE et al. [13] and SCHLINKER and AMIET [11].

The cross section of a blade element is shown in **Fig. 5** which moves in the plane of rotation with the circumferential speed of $-v_r$ from right to left. The direction of the wind is from bottom to top and therefore the relative inflow velocity W_∞ points from bottom left to top right. The definitions of the angle of attack α, the free inflow angle β_∞ and the stagger angle γ are depicted in **Fig. 5** as well. The stagger angle γ is the sum of the rotor twist angle σ_R, the pitch angle σ_P and the elastic torsion angle σ_T.

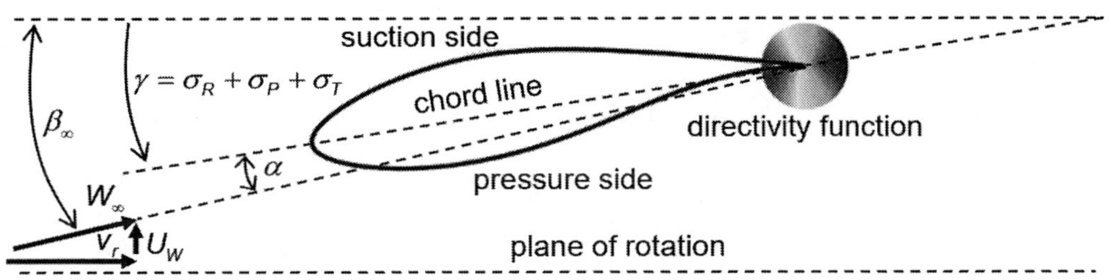

Figure 5: Cross section of a blade element with speed triangle and resulting angles.

The rotor twist angle σ_R is the manufactured twist of the blades and therefore fixed during operation. In contrast, the pitch angle σ_P may be adjusted during operation in order to control the power output of the wind turbine. Finally, the elastic torsion angle σ_T describes the elastic deformation of the rotor blades during the operation and depends on the wind load. In **Fig. 6** a schematic diagram of a sound source on the wind turbine and additionally relevant angles can be seen. The azimuth angle Ψ describes the circumferential position of the source, the stagger angle γ represents the overall rotational angle of the source around the blade radius, while the tilt angle Γ_W shows the tilt of the nacelle relative to the horizon. The cone angle β_{cone} describes the angle between the blade radius and the rotation plane.

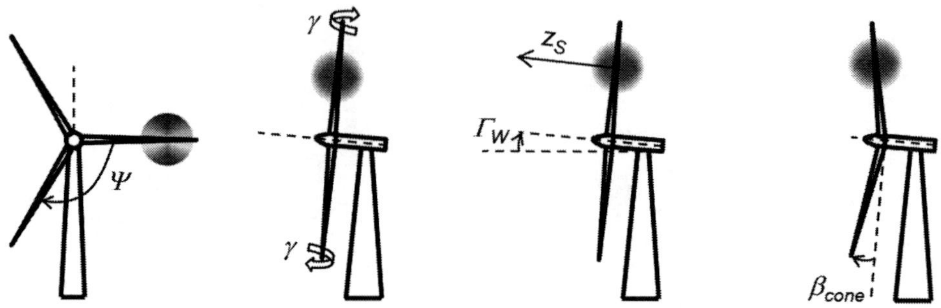

Figure 6: Schematic illustration of a point sound source on a wind turbine blade and the resulting angles.

2.4 Synthesis

In analogy to BOORSMA and SCHEPERS [22] and OERLEMANS and SCHEPERS [14] the sub-models are compiled into one tool. The sound pressure level due to sound radiated by a particular point source and received by an observer at observer time instance t', becomes

$$SPL_{1/3}(r,\ \Phi_A,\ \Theta_A;\ f',\ t') = \left\{ SPL_{TOT,ref}(f) + 10\log\left(D(\Phi_A,\ \Theta_A)\right) + 10\log\left(CA(M_{SO})\right) - 10\log\left(\frac{r}{r_{ref}}\right)^2 \right\}\Bigg|_t \quad \text{[dB]} \tag{13}$$

(here in terms of 1/3 octave bands), with

$$SPL_{TOT,ref}(f) = SPL_{TOT}\left(r_{ref} = 1\ \text{m},\ \Phi_A = \Theta_A = 90°;\ f\right) \quad \text{[dB]} \tag{14}$$

and the Doppler shifted frequency [23]

$$f' = f \cdot \left(1 - M_{SO}\right)^{-1}. \tag{15}$$

The sound pressure spectrum $SPL_{1/3}(r,\ \Phi_A,\ \Theta_A;\ f',\ t')$ belongs to a small time interval in which the motion of the sound source can be regarded as approximately linear with respect to the observer. (When using HOWE's directivity function, the angles Φ_A, Θ_A are replaced by Φ_H, Θ_H.) Having substituted the sound radiating wind turbine blades by m sound sources, each sound source has its own M_{SO} (in CA) as well as r, $SPL_{TOT,ref}$ and D. The overall sound pressure level from one sound source as seen by the observer then becomes

$$OASPL_{source}(t') = 10\log\left(\int_0^\infty 10^{\frac{SPL_{1/3}(t',f')}{10}}\,df'\right) \quad \text{[dB]}. \tag{16}$$

Since each $OASPL_{source}$ arrives at the observer at its individual retarded time instance

$$t' = t + \frac{r}{c}, \tag{17}$$

a resampling with respect to t' within one rotor revolution is required to obtain values at a common time base. Eventually, the overall sound pressure level *OASPL* from all m moving sources as a function of t' is obtained as

$$OASPL(t') = 10\log\sum_{i=1}^{m} 10^{\frac{OASPL_{source,i}(t')}{10}} \quad [\text{dB}]. \tag{18}$$

Hereby we assumed that the sound sources (here trailing edge sound sources) radiate incoherently.

To further analyse the time dependent *OASPL*, the equivalent overall sound pressure level

$$OASPL_{eq} = 10\log\left(\frac{1}{T}\int_{t_0}^{t_0+T} 10^{\frac{OASPL(t')}{10}} dt'\right) \quad [\text{dB}] \tag{19}$$

and the amplitude modulation

$$OASPL_{am} = \max\left(OASPL(t')\right)\Big|_T - \min\left(OASPL(t')\right)\Big|_T \quad [\text{dB}] \tag{20}$$

are introduced. Both are calculated over one period of revolution T.

The steps of the combined sound prediction model, as implemented in a MatLab™ code, are shown in **Fig. 7**. Inputs are the wind turbine parameters, the positions of the sound sources along the blades and the observer points. Then, the angles Φ_A, Φ_H, Θ_A, Θ_H, the Mach number M_{SO} and the retarded time t' for every source-observer-combination are derived. In a next step, the directivity factor D and convective amplification factor CA for every source-observer-combination and the boundary layer displacement thickness on the suction δ^*_s and the pressure side δ^*_p for every blade segment are calculated with XFOIL. Based on the boundary layer information, the model by BROOKS et al. [9] (BPM-Model) is used to calculate the $SPL_{TOT,ref}$ for every sound source. Then, at discrete time instances the sound immission on the observer points are calculated. Finally, the immission is resampled with respect to t' within one rotor revolution to obtain values at a common time base and all sound sources are superpositioned at the observer points.

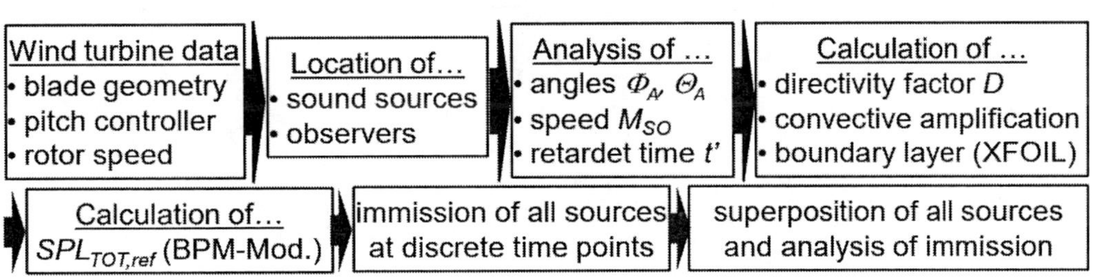

Figure 7: Structure of the sound prediction model.

In contrast to BOORSMA and SCHEPERS [22] and OERLEMANS and SCHEPERS [14] we here implement all three different directivity functions and the three amplification factors in order to assess the impact of the model choice on the results.

3. Case study

3.1 Benchmark turbine

Since published data is rare, we take - as a first benchmark - the horizontal axis wind turbine as investigated by OERLEMANS and SCHEPERS [14]. It has a rotor half diameter of 47 m and a hub height of 100 m. In OERLEMAN et al.'s paper the free field wind speed varies between 6 - 8 m/s. Here we assume a fixed value of 7 m/s.

Since not all parameters of this turbine have been published we more or less arbitrarily estimate other required parameters: The blade varies linearly from 3.5 m at the hub to 0.8 m at the tip. A DU93W210TET03 blade profile is assumed where TET03 indicates a trailing edge thickness of 0.3% of chord length. The rotational speed is set to 11.6 rpm. Within the case study, the angle of attack is fixed to $\alpha = 6° =$ const. along the complete span. Tilt and pitch angle (Γ_{W}, σ_{P}) are set to 0° as well as torsion and cone (σ_T, β_{cone}) angle in the case of rigid blades.

With this basic setup and with the known free inflow angle β_∞, the rotor twist angle σ_R and the stagger angle γ can be calculated for all blade segments. In the case of elastic deformed rotor blades, the rotor twist angle of the rigid blades combined with an assumed elastic torsion according to **Fig. 8** is used to calculate the new stagger angle and new angle of attack.

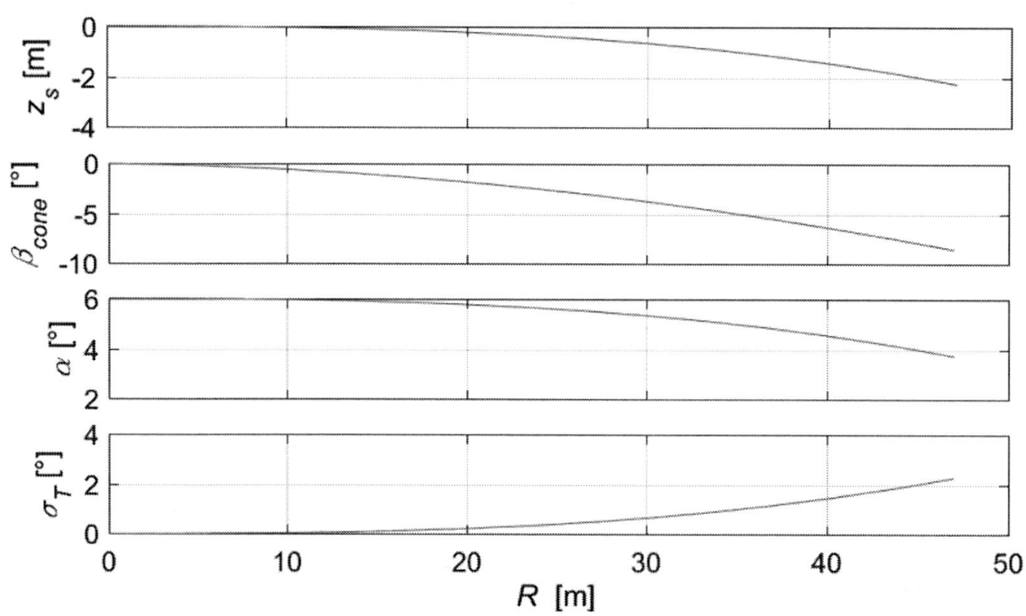

Figure 8: Assumed bending line z_S, cone angle β_{cone}, angle of attack α and elastic torsion angle σ_T in the case of elastic deformation drawn as a function of the blade radius.

For considering elasticity of the blades, we assume a bending line z_S, a cone angle β_{cone}, a resulting angle of attack α and the elastic torsion angle σ_T as in **Fig. 8**. The assumed deformation of the wind turbine blade is shown in **Fig. 9**; the location of the sound sources is also indicated.

In this case study, the time resolutions obtained is due to 40 spatial rotor-positions (i.e. time instances) considered within one rotor revolution.

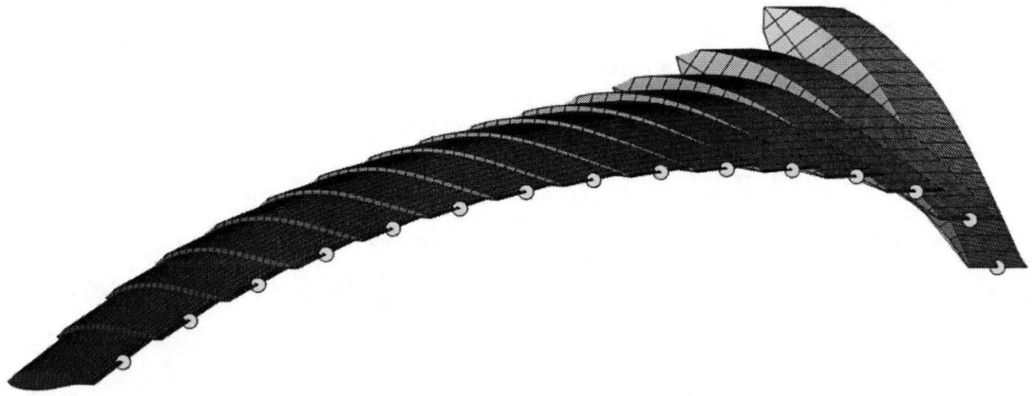

Figure 9: Segmentation of the elastically deformed wind turbine blade and positions of the trailing edge sound sources (yellow spheres).

3.2 Results

3.2.1 Turbine with rigid blades: General radiation characteristic

As an example, we present the radiation characteristic of the benchmark turbine with completely rigid blades as predicted utilizing directivity function and convective amplification factor D_{HOWE} and $CA_{Rozenberg}$ respectively. In **Fig. 10**, the observer positions on a circle around the wind turbine are schematically shown expressed in terms of observer angle ζ and distance from foot of the wind turbine to observer r_{Obs} (left) and the *OASPL* as a function of the observer time t' within one rotor revolution for three different observer positions ζ = 0°, 63°, 90° is shown schematically. The circle has a diameter of 240 m on the ground around the foot of the wind turbine, i.e. r_{Obs} = 120 m. The observer angle ζ is 0° in front and 180° behind the turbine. The swishing character of the wind turbine sound is well distinguishable - particularly at the side of the wind turbine (ζ = 90°).

In **Fig. 11**, the equivalent overall sound pressure level $OASPL_{eq}(\zeta)$ (left) and the amplitude modulation $OASPL_{am}(\zeta)$ (right) for 81 observers around the wind turbine are shown. The highest equivalent $OASPL_{eq}$ are found at ζ = 0° and ζ = 180°, the lowest at ζ = 90° and ζ = 270°. The amplitude modulation is very significant at ζ = 90° and ζ = 270° and close to zero at ζ = 0° and ζ = 180°. This is only a qualitative analysis, in which the absolute levels can not be validated at this time of the project.

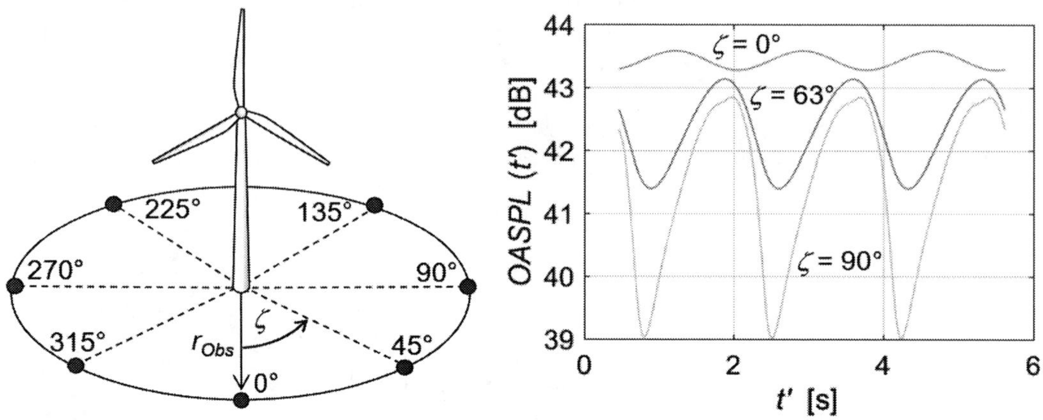

Figure 10: Observer positions in a circle around the wind turbine as a function of the observer angle ζ (left) and $OASPL(t')$ within one rotor revolution for three different observer positions $\zeta = 0°, 63°, 90°$ at r_{Obs} = 120 m - utilizing D_{HOWE} and $CA_{Rozenberg}$

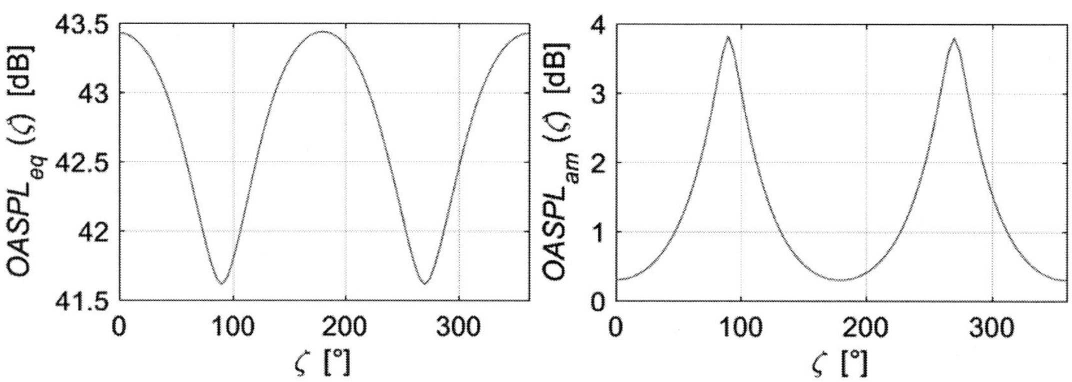

Figure 11: Equivalent overall sound pressure level $OASPL_{eq}(\zeta)$ (left) and amplitude modulation $OASPL_{am}(\zeta)$ (right) at r_{Obs} = 120 m around the wind turbine - utilizing D_{HOWE} and $CA_{Rozenberg}$

3.2.2 Effects of sub-model choice

The different trailing edge sound directivity functions have a major impact on the predicted overall sound pressure level (*OASPL*) at the observer points. Because OERLEMANS and SCHEPERS [14] documented normalized measurement data, all simulation results are also normalized with respect to the observer position directly in front of the wind turbine:

$$OASPL_{eq,norm}(\zeta) = OASPL_{eq}(\zeta) - OASPL_{eq}(\zeta = 0°) \quad [\text{dB}] \tag{21}$$

$$OASPL_{am,norm}(\zeta) = OASPL_{am}(\zeta) - OASPL_{am}(\zeta = 0°) \quad [\text{dB}] \tag{22}$$

For the same case as in section 3.2.1 it can be seen in **Fig. 12** that the immission for all directivity functions is quite similar directly in front of ($\zeta = 0°$) and behind ($\zeta = 180°$) the wind turbine. Moreover, it becomes clear that the highest immission is directly up- and downstream of the wind turbine. The major differences due to the directivity functions chosen are observed at the sides of the wind turbine ($\zeta = 90°$ and 270°). The directivity functions $D_{OERLEMANS}$ and $D_{SCHLINKER}$ cause a pronounced directivity pattern of $OASPL_{eq,norm}$ (with a variation up to 11 dB and 18 dB, respectively) whereas with D_{HOWE} the directivity is nearly level.

The normalized amplitude modulation $OASPL_{am,norm}$ (**Fig. 12**, lower diagram) is minimal directly in front of and behind the turbine. At the turbine sides, the highest normalized amplitude modulation and the largest impact of the directivity functions can be seen. The directivity function $D_{SCHLINKER}$ predicts over 50 dB of normalized amplitude modulation, whereas with $D_{OERLEMANS}$ and D_{HOWE} approximately 4 dB are predicted. As initially mentioned OERLEMANS and SCHEPERS [14] published experimental data, which is plotted in **Fig. 12** as well. Obviously, utilizing the directivity function D_{HOWE} in our prediction scheme yields results which come closest to OERLEMANS' experimental data.

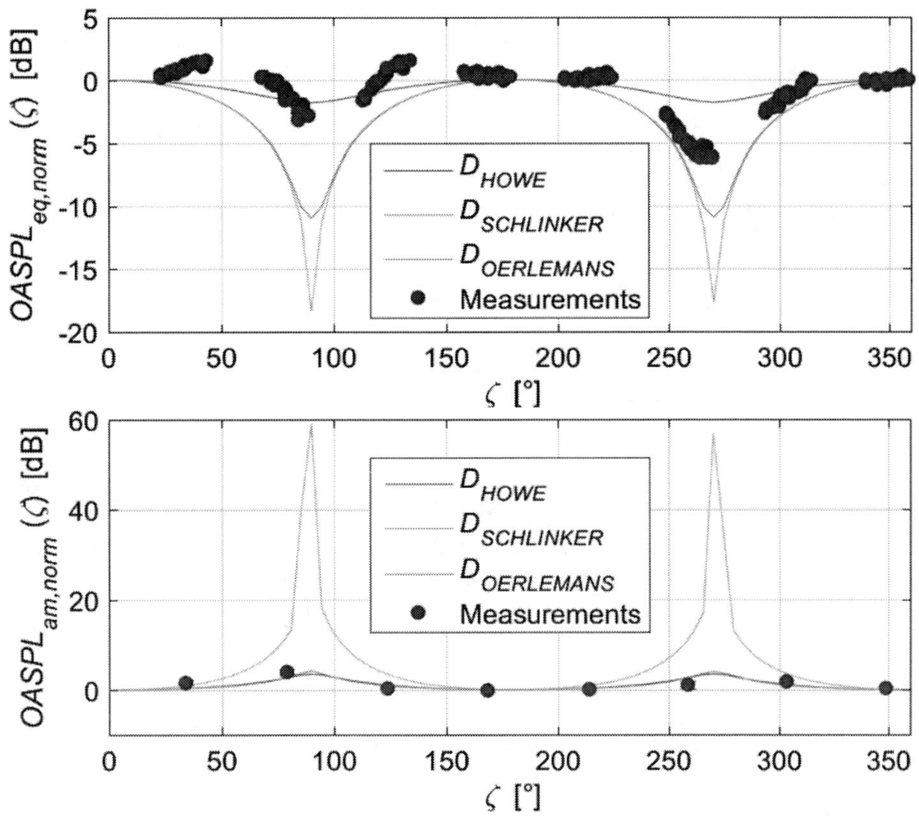

Figure 12: Effect of the directivity sub-models on the normalized equivalent $OASPL_{eq,norm}$ (upper diagram) and the normalized amplitude modulation $OASPL_{am,norm}$ (lower diagram) for observer positions in a circle around the wind turbine at r_{Obs} = 120 m. Experimental data from OERLEMANS and SCHEPERS [14].

Another more fundamental experiment reported in the literature confirmed these findings. KAMBE et al. [13] validated experimentally HOWE's directivity function for a flat plate rather than an airfoil. They generated eddies with a radius of 4.7 mm and sent them over the edge of a plate (M = 0.1 to 0.2, C = 1 m). The major advantage of this experiment was its reproducibility and the absence of any superimposed flow. In consequence, KAMBE et al. [13] were able to measure the sound pressure at 70 positions in a plane normal to the surface and along the cord line and in a plane parallel to the surface in an angular range from 0° to 360°. **Fig. 13** depicts the measured sound pressure amplitudes and the predicted sound pressure amplitudes with HOWE's model in the plane normal to the surface of the plate and along the cord line (left) and in a plane parallel to the surface (right). It can be seen that HOWE's directivity function agrees very well with the measurements and only small differences were found.

The directivity functions $D_{SCHLINKER}$ and $D_{OERLEMANS}$ are inconsistent with the measurement results from KAMBE et al. [13] because they are zero on the surface of the plate. Therefore, again HOWE's directivity function seems to be the most plausible.

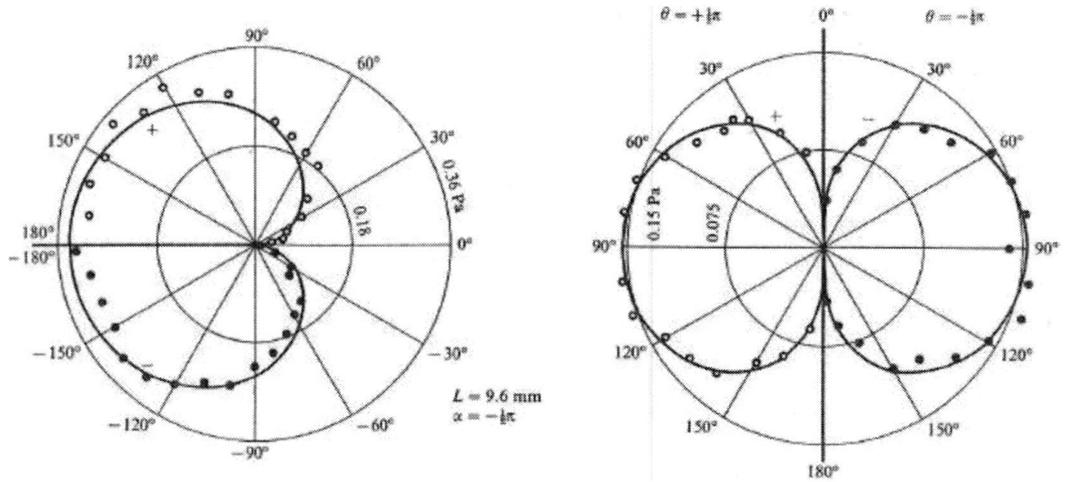

Figure 13: Trailing edge sound: Directivity of sound pressure amplitudes in a plane normal to the surface of the flat plate and along the cord line (left) and in a plane parallel to the surface (right); HOWE's prediction (solid line) and experimental data (circles) by KAMBE et al.; from KAMBE et al. [13].

The impact of the different convective amplification sub-models on the predicted $OASPL_{eq,norm}$ and $OASPL_{am,norm}$ can be seen in **Fig. 14.** Here we use the sub-model D_{HOWE}. The greatest impact is found at ζ = 90° and 270° but compared to the choice of D its impact is comparably small. Hence, the convective amplification sub-model $CA_{Rozenberg}$ is chosen for a consistent prediction model since it satisfies the conservation of acoustic energy.

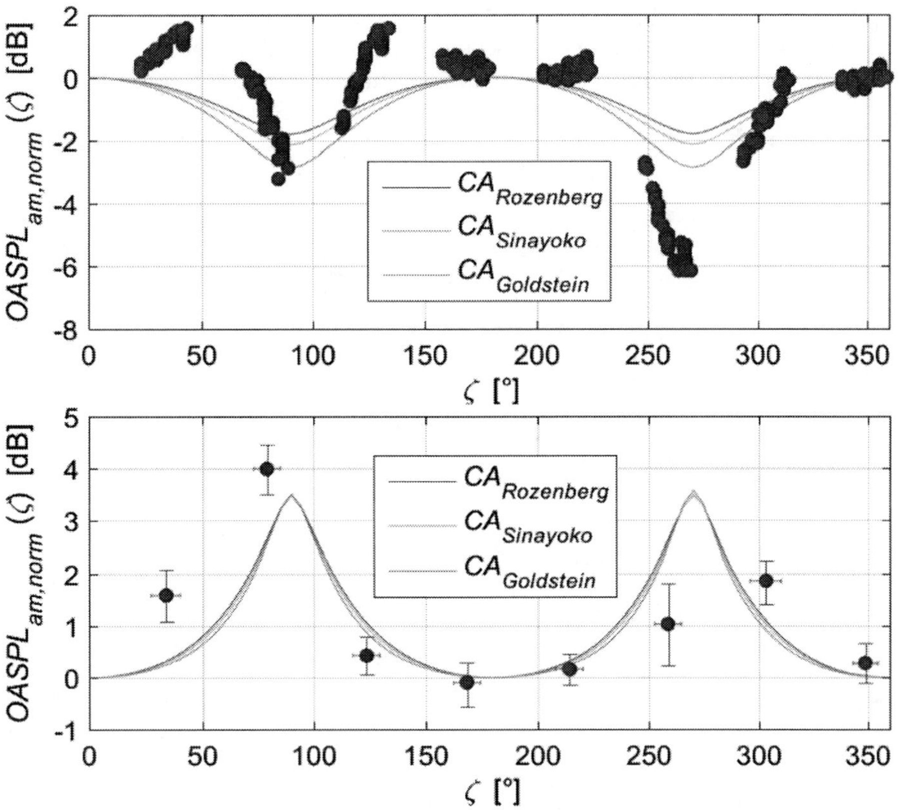

Figure 14: Effect of the convective amplification sub-models on the normalized equivalent $OASPL_{eq,norm}$ (upper diagram) and the normalized amplitude modulation $OASPL_{am,norm}$ (lower diagram) for observer positions in a circle around the wind turbine at r_{Obs} = 120 m. Experimental data from OERLEMANS and SCHEPERS [14].

3.2.3 Effect of elastic blade deformation

To investigate the influence of the elastic deformation of the rotor blades we exclusively utilize the sub-models D_{HOWE} and $CA_{Rozenberg}$. **Fig. 15** compares the sound immitted at the obervers points from the rigid and elastic blade. It can be seen that the differences are smaller than 0.5 dB, hence negligible.

4. Summary and Conclusions

The objective of this paper was to compile and compare selected aeroacoustic semi-analytical wind turbine sound prediction models available in the open literature.

The general structure of those models is found to be similar: A set of sub-models for the elementary sound sources on the blades is combined with a sound propagation model from the sources to the listener. However, the sub-models found in the literature vary and a thorough assessment based on experiments is not an easy task.

The predicted *OASPL* around a wind turbine is shown to be affected substantially by the choice of the modelled directivity function of the elementary point sources representing the full blades. Given three simple, frequency-independent directivity functions ($D_{SCHLINKER}$, $D_{OERLEMANS}$ and D_{HOWE}) HOWE's source directivity function seems to yield results, that match best experimental results from literature.

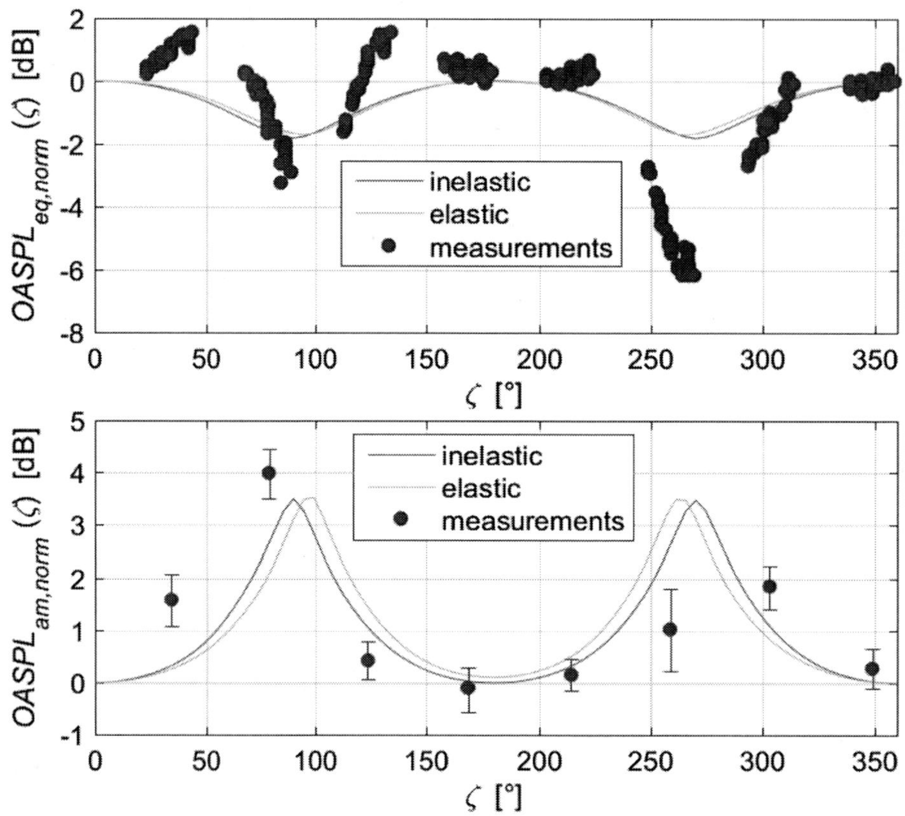

Figure 15: Effect of the elastic deformation of the rotor blades on $OASPL_{eq,norm}$ and $OASPL_{am,norm}$ at r_{Obs} = 120 m - sub-models from $CA_{Rozenberg}$ and D_{HOWE}. Experimental data from OERLEMANS and SCHEPERS [14].

The effect of the models which take into account the motion of the sound sources is significantly smaller in comparison to the source models. Three models for the convective amplification factors ($CA_{ROZENBERG}$, $CA_{SINAYOKO}$ and $CA_{CRIGHTON}$) where compared. From principle considerations ROZENBERG's approach seems to be the most consisting.

Eventually, combining selected sub-models, a preliminary own wind turbine sound prediction model was compiled and encoded, yielding the acoustic footprint of the turbine on ground level and the swishing character of the wind turbine sound.

In a first case study, the effect of delocalisation of elementary sound sources along the blades and a modification of the local angle of attack due to the flexibility of the blade has been studied. As a preliminary result, the effect of blade elasticity on the acoustic emission of a complete wind turbine rotor seems to be small when comparing to a rigid rotor. It is important to note that so far the atmospheric attenuation, refraction and ground effects have not been taken into account - this remains for future studies and may affect these results.

Acknowledgement

This work has been funded by the Bundesministerium für Wirtschaft und Energie (Federal Ministry for Economic Affairs and Energy) of Germany within the project RENEW (FKZ 0325838B). The authors greatly appreciate the ministry's support. We also want to thank B.Eng. M. Schüssler for his support in this work's editing process.

References

[1] EGGENSCHWILER, K.; HEUTSCHI, K.;SCHÄFFER, B.; PIEREN, R.; BÖGLI, H.; BÄRLOCHER, M.; Wirkung und Beurteilung des Lärms von Windenergieanlagen, Lärmbekämpfung, Springer VDI Verlag, 2016 (Nr. 5, pp. 159-167)

[2] LUTZ, T. ; ARNOLD, B. ; WOLF, A. ; KRÄMER, E.: *Numerical Studies on a Rotor with Distributed Suction for Noise Reduction,* TORQUE 2014, Journal of Physics: Conference Series 524, 2014

[3] GERHARD, T. ; ERBSLÖH, S. ; CAROLUS, T.: *An Experimental and Numerical Parameter Study on Trailing Edge Blowing for Reduced Trailing Edge Noise,* 6th International Conference on Wind Turbine Noise, Glasgow, 2015

[4] VATHYLAKIS, A. ; CHONG, T.P. ; PARUCHURI, C. ; JOSEPH, P.F.: *Sensitivity of Aerofoil Self Noise Reductions to Serration Flap Angles,* 22nd AIAA/CEAS Aeroacoustics Conference, France, 2016

[5] OERLEMANS, S.;SIJTSMA, P.; MÉNDEZ LÓPEZ, B.: Location and quantification of noise sources on a wind turbine, Journal of Sound and Vibration 299, 869-883, 2007

[6] LOWSON, M.V.: Assessment and Prediction of Wind Turbine Noise. Flow Solutions Report 92/19, ETSU W/13/00284/REP, pp. 1-59, Dec. 1992

[7] LIPS, W.:"Strömungsakustik in Theorie und Praxis: Anleitung zur lärmarmen Projektierung von Maschinen und Anlagen", Expert-*Verlag*, Renningen, 2008.

[8] DIN EN 61400-11 Windenergieanlagen - Teil 11: Schallmessverfahren (IEC 61400-11:2012); Deutsche Fassung EN 61400-11:2013

[9] BROOKS, T. F. ; POPE, D. S. ; MARCOLINI, M. A.: *Airfoil Self-Noise and Prediction*: *NASA RP-1218.* Hampton, Virginia : NASA Langley Research Center, 1989

[10] FFOWCS WILLIAMS, J. E., HALL, L. H.: Aerodynamic Sound Generation by Turbulent Flow in the Vicinity of a Scattering Half Plane, *Journal of Fluid Mechanics*, 1970 (40(4), pp. 657-670)

[11] SCHLINKER, R. H. ; AMIET, R.K.: *Helicopter rotor trailing edge noise*: *NASA Contractor Report 3470*, 1981

[12] HOWE, M. S.: *A Review of the Theory of Trailing Edge Noise*: Journal of Sound and Vibration, 1978 (61(3), pp. 1-145)

[13] KAMBE, T. ; MINOTA, T. ; IKUSHIMA, Y.: *Acoustic wave emitted by a vortex ring passing near the edge of a half-plane* : Journal of Fluid Mechanics, 1985 (Volume 155, pp. 77 - 103)

[14] OERLEMANS, S. ; SCHEPERS, J. G.: *Prediction of wind turbine noise and validation against experiment*. National Aerospace Laboratory NLR-TP-2009-402, 2009

[15] AMIET, R. K.: Noise due to turbulent flow past a trailing edge, Journal of Sound and Vibration, Vol. 47, No. 3, 1976, pp. 387-393.
[16] MOREAU, S. ; ROGER, M.: Back-scattering correction and further extensions of Amiert's trailing-edge noise model. Part II: Application, Journal of Sound and Vibration, 2009 (323, pp. 397-425)
[17] OERLEMANS, S.: Wind turbine noise: primary noise sources, National Aerospace Laboratory NLR-TP-2011-066, 2011
[18] ROZENBERG, Y. ; ROGER, M. ; MOREAU, S.: *Rotating blade trailing-edge noise: experimental validation of analytical model* : AIAA Journal, American Institute of Aeronautics and Astronautics, 2010 (48, pp. 951 - 962)
[19] SINAYOKO, S. ; KINGAN, M. ; AGARWAL, A.: *Trailing edge noise prediction for rotating blades: analysis and comparison of two classical approaches* : 33rd AIAA Aeroacoustics Conference, American Institute of Aeronautics and Astronautics, 2012
[20] CHRIGHTON, D. G. ; DOWNLING, A. P. ; FFOWCS WILLIAMS, J. E. ; HECKL, M. ; LEPPINGTON, F. G.: *Modern Methods in Analytical Acoustics.* 1. Auflage. Springer-Verlag, 1992
[21] GOLDSTEIN, M.: *Aeroacoustics.* USA: McGraw-Hill, 1976
[22] BOORSMA, K., SCHEPERS, J.G.: *Enhanced wind turbine noise prediction tool SILANT*: Wind Turbine Noise, 2011
[23] KUTTRUFF, H.: Akustik - Eine Einführung. Stuttgart : S. Hirzel Verlag, 2004

7th International Conference
on
Wind Turbine Noise
Rotterdam – 2nd to 5th May 2017

The Occurrence of Nocturnal Wind Farm Rumbling Noise

Kristy L. Hansen - Flinders University, Sturt Rd., Bedford Park 5042, SA, Australia.
kristy.hansen@flinders.edu.au
Branko Zajamsek - Flinders University, Sturt Rd., Bedford Park 5042, SA, Australia.
branko.zajamsek@flinders.edu.au
Colin H. Hansen - University of Adelaide, North Tce., Adelaide, 5005, SA, Australia.
colin.hansen@adelaide.edu.au

Summary
Nocturnal low-frequency noise from a wind farm was analysed in terms of its magnitude and amplitude modulation (AM) content. Acoustic and meteorological measurements were conducted at an unoccupied dwelling located 3.3 km from a wind farm. At times when the wind farm was operating at a capacity factor greater than 40% and the residence was downwind from the wind farm, low-frequency wind farm noise consistently exceeded the normal hearing threshold at 1/3-octave frequencies of 50 Hz and above. During these times, AM at the blade-pass frequency was found to be present at low frequencies. The AM was evaluated using the 'Reference' method that was recently introduced by the Institute of Acoustics in 2016. This method successfully identified AM most consistently when compared to other detection algorithms. The presence of AM was verified through reference to narrowband plots, which show AM as side-bands adjacent to the tonal frequency that is modulated. Based on the fact that the noise is shown to be both audible and contain potentially annoying AM, the potential for sleep disturbance is established. The implications arising from repeated arousals are discussed and include daytime sleepiness and impaired mood and function.

1. Introduction
Wind farm low-frequency noise (commonly defined between 20 - 200 Hz, although sometimes defined between 20 and 160 Hz) is often reported to cause annoyance (Doolan, 2013, Pedersen and Waye, 2008, Pedersen et al., 2007, Leventhall, 2009) and affect sleep (Smith et al., 2016, Bakker et al., 2012, Nissenbaum et al., 2012, Shepherd et al., 2011). It has been shown that this noise can exceed the normal hearing threshold at locations ranging from a few hundred metres (Sondergaard, 2014) to a few kilometres away from a wind farm (Zajamšek et al., 2016b, Marcillo et al., 2015), and be amplitude modulated (Hansen et al., 2015b, Perkins et al., 2016). During the night-time, wind farm noise has often been found to be present when atmospheric conditions are stable (Zajamšek et al., 2016b, Van den Berg, 2004), with strong high level winds allowing ongoing wind turbine operation while winds at typical receiver heights may be negligible. Thus, the associated background noise at a typical residence can be very low, particularly in rural areas, resulting in wind farm noise being the dominating noise source. No information has been found about how likely it is for wind farm low-frequency noise to exceed the normal hearing threshold and be amplitude modulated during the night-time, when it will be more intrusive and noticeable than during the day.

To evaluate the sleep disturbance potential of wind farm noise, it is first necessary to understand the nature of the noise source. For instance, it has been found that at comparable levels, wind farm noise is more annoying than other sources such as transportation noise and industrial noise (Pedersen & Waye, 2004). This suggests that the character of the noise that is perceived by nearby residents is just as important as the overall sound pressure level. The character of a noise source can be described by its frequency content and time-varying nature. Wind turbine noise spectra are dominated by low frequency energy and the associated noise levels are highly time dependent due to meteorological factors, blade loading variations, directivity and interaction between the signals from two or more turbines. Periodic variations in the level of wind farm noise with time are commonly referred to as amplitude modulation (AM) and these variations typically occur at the blade-pass frequency (0.5 – 1 Hz). Our measurements have shown that at distances up to 3.5 km, wind farm noise can be audible and contain significant AM, tonal components, low-frequency noise and infrasound (Hansen et al., 2014), all of which have the potential to adversely affect sleep, health and wellbeing.

Modern wind turbines produce various types of amplitude modulated noise including "swishing," "thumping" and "rumbling". "Swishing" is caused by the directivity characteristics of trailing edge noise on rotating blades and convective amplification of that noise (Oerlemans and Schepers, 2009). Since wind turbine trailing edge noise spans between 400 Hz to 1.4 kHz, it is significantly attenuated with distance from the wind turbine and is therefore not significant at distances of a few kilometres from a wind farm (Zajamšek et al., 2016b). "Thumping" and "rumbling" noise, on the other hand, contain lower frequencies and can thus propagate further. Therefore, at locations up to a few kilometres away from a wind farm, this can be the dominant contribution of wind farm noise at a residence (Hansen et al., 2015a).

It is well established, that environmental noise disturbs sleep (Muzet, 2007). During sleep the auditory system remains fully functional and thus recognises, evaluates and reacts to environmental noise (Muzet, 2007). The sleeper response to noise depends on several factors such as the type of noise (continuous or intermittent), noise intensity, frequency content, noise interval (duration, regularity, expectation), noise significance and the difference between the background noise level and the maximum amplitude of the noise stimulus (Muzet, 2007). Since wind farm low-frequency noise is amplitude modulated (intermittent) and is usually present in a rural environment where the background noise levels are low, it is likely to have an effect on sleep even at low to moderate levels. This hypothesis is supported by some recent preliminary results by Smith et al. (2016) which show that low-frequency AM has an impact on sleep. The purpose of this paper is to show that amplitude modulated low-frequency wind farm noise can exceed the normal hearing threshold for a significant amount of time during the night at a distance of 3.3 km away from a wind farm. This type of wind farm noise could therefore have an effect on annoyance and sleep.

2. Methodology

2.1 Field Measurements

Continuous outdoor acoustic measurements were taken at a residence (R) located 3.3 km from the nearest wind turbine in the Waterloo wind farm as shown in Figure 1. The measurements took place over one week during the Australian winter. The wind farm is located in South Australia, which has a "Mediterranean" type of weather with mild wet winters and hot summers. The wind farm consists of 37 Vestas v90, 3 MW wind turbines, which are located on a ridge. The average height of the wind turbine base is 200 m above the measurement location. The rural environment surrounding the wind farm is relatively flat with sparse low hills and farm fields. Measurements were undertaken when the wind farm was operating and shut down. The wind farm operational state was determined from the capacity factor, which is calculated as the ratio between the maximum and actual wind farm power output, expressed as a percentage. There were no significant extraneous noise sources nearby the measurement location, except countryside roads and farming activity. The influence of these noise sources is assumed to be minimal during the night time.

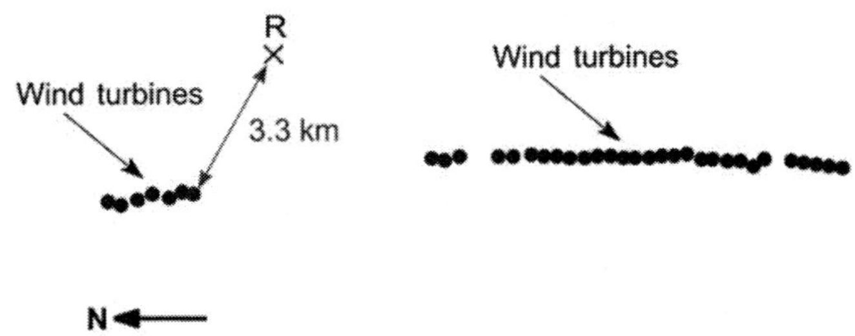

Figure 1 - Measurement location with respect to the wind farm.

Measurements were taken from 24/07/2013 to 1/08/2013, during which time the wind farm was non-operational for 2 nights and operational for 5 nights. This preliminary analysis considers one operational night only, during which the residence was located downwind (±45°) from the nearest wind turbine for the majority of the night.

Table 1 – Range of capacity factors and wind speeds that occurred for measurements taken during operational conditions. The wind speed at heights of 1.5 m and 10 m was measured within 30 m of the residence, and the wind speed at hub height was measured using a SODAR that was positioned on the same ridge-top as the wind turbines.

	Capacity factor (%)	Wind speed (m/s)			Start Date
		1.5 m	10 m	hub	
Operational	2 - 63	0 - 2.2	0 - 4.3	0.2 - 10.4	28/7/13

The outdoor acoustic measurements were taken using a B&K type 4955 1/2 inch microphone. This microphone has a 6.5 dBA noise floor and flat frequency response from 6 Hz to 20 kHz. The microphone was connected to a B&K type 3050 module with 24-bit dynamic resolution, which was operated using Pulse software. The data on this device were recorded at a sampling rate of 8192 Hz. The sampling frequency was chosen to be relatively low, since wind farm noise is biased towards the lower frequencies and its contribution at frequencies above 2 kHz is negligible at the large propagation distances considered here. Measurements were recorded in 10-minute long time blocks. Wind speed and direction were measured at heights of 1.5 m and 10 m using Davis Vantage Vue and Vantage Pro weather stations, respectively. Wind speed and direction at hub height were measured using a SODAR system, which was mounted on the same ridgetop as the wind turbines.

The outdoor microphone was positioned more than 20 m away from the residence and greater than 10 m away from surrounding vegetation in order to minimise sound reflections from the façade and vegetation noise, respectively. To protect the microphone from wind-induced noise, a primary 90 mm wind screen and a spherical secondary wind screen were used. The spherical wind screen consists of a 16 mm layer of acoustic foam covered by a layer of SoundMaster acoustic fur, supported by a 450 mm diameter steel frame of spherical shape, made out of thin-wire steel. The microphone and wind screen were mounted on a star-dropper using a custom-made connector.

The narrowband power spectral density (PSD) estimation was calculated using Welch's averaged modified periodogram method of spectral estimation with a Hanning window of length 81920 (10x the sampling frequency) points and 50% overlap, which gives a 0.1 Hz frequency resolution. To obtain the power spectrum (PS), the PSD was corrected for windowing and frequency resolution effects in the following way according to (Randall, 1977):

$$PS = PSD + \log_{10}(B_{en} \times \Delta f) \quad (1)$$

where B_{en} is the normalised noise bandwidth (1.5 for a Hanning window) and Δf is the frequency resolution.

2.2 Algorithm for detection of AM

Recently, the Amplitude Modulation Working Group (AMWG), on behalf of the UK Institute of Acoustics, has devoted considerable effort to develop an effective and reliable method of detecting and quantifying the AM content of wind farm noise. After conducting an extensive literature review, the group identified that AM detection/quantification methods fall into three main categories; namely, time-series methods, frequency-domain methods and hybrid methods (Bass et al., 2015). The group then identified one method from each of these categories for further consideration. After a period of consultation, during which comments, observations and criticisms were received, the group developed a robust method called the 'Reference Method' (Bass et al., 2016b). This is a hybrid approach based on the AMWG Method 3 (Bass et al., 2015) and incorporating concepts from other methods developed by Tachibana's group (Fukushima et al., 2013) and Renewable UK (Renewable UK, 2013b). The resulting method is rather complex and implementation requires the development of specific computer programming routines. Conveniently, the AMWG have provided a code in the Python language which is available at (Bass et al., 2016a). This code has been used to detect and quantify the AM in the analysis presented in this paper. A brief description of the method is provided below and interested readers are encouraged to consult the AMWG documentation (Bass et al., 2016b) for more detailed information.

The 'Reference Method' involves dividing the signal into 10-minute long periods, which are then further divided into non-overlapping 10-second segments. These 10-second segments consist of $L_{Aeq,\,100ms}$ data, which have been band-limited in one of four frequency ranges. Although not explicitly stated in the final AMWG report, the sampling time to acquire the $L_{Aeq,\,100ms}$ data is 100 ms, resulting in 100 non-overlapping data points for each 10-second period. Since the data considered in this paper contain AM at frequencies below 100 Hz, the lowest frequency range of 50 – 200 Hz is used in this analysis. Narrowband analysis has revealed that the most significant AM occurs at approximately 46 Hz (Hansen et al., 2014) and therefore the lower bound of the range has been reduced to 40 Hz. Since third-octave analysis indicates that the noise in the 50 Hz 1/3-octave band is audible according to the normal threshold of hearing (ISO389-7, 2005), and A-weighting applies heavy penalties at this frequency, it is considered that A-weighting is not appropriate in this case. In fact, for distances greater than 1-1.5 km from a wind farm, it is believed that use of the A-weighting when applying the 'Reference' method will result in inaccuracies. It is well-known that the A-weighting is not appropriate for characterising signals with dominant low frequency components (Hellman and Zwicker, 1987).

The band-limited $L_{Aeq,\,100ms}$ data are transformed to the frequency domain using a Fast Fourier transform. From the resulting AM spectrum, the blade-pass frequency can be identified either using SCADA (System Control and Data Acquisition) data or by determining the applicable range of blade-pass frequencies at a given location. According to the latter method, the maximum value of the modulation spectrum that lies within the expected range is the blade-pass frequency (Bass et al., 2016b). The blade-pass peak amplitude is compared to four surrounding (but not adjacent) peaks, which are averaged to determine the masking level. The ratio of the peak and masking level is referred to as the 'prominence' and this ratio must be greater than or equal to 4 for the 10-second period to be considered 'valid'. For each 10-minute period, at least 50% of 10-second periods must be 'valid' for the AM in that period to be identified as associated with wind farm operation. The aim of these criteria is to minimise false positives and to ensure that the identified periods contain relatively continuous AM.

Once the blade-pass frequency has been detected and classified as 'prominent', a window is applied around the corresponding frequency. Windows are also applied around the next two harmonics, provided specific criteria are met. The next step in the analysis is to perform an inverse Fourier transform to reconstruct the filtered time series. The modulation depth for each 10-second period is determined by finding the 95th percentile minus the 5th percentile (L_5 - L_{95}) of the reconstructed time-series data. The modulation depth for a 10-minute period is determined by finding the 90th percentile of the valid 10-second values.

3. Results

The presence of wind farm noise at this residence, which is located 3.3 km from the nearest wind turbine, can be clearly seen in the narrowband spectra. A frequency resolution of 0.1 Hz has been chosen, as this is sufficiently lower than the blade-pass frequency of 0.8 Hz. The peaks occur at frequencies below 100 Hz, as wind farm noise in the mid- to high-frequency ranges has been attenuated by atmospheric and ground absorption. The peaks are significantly higher than the background noise level which is measured when the wind farm is non-operational (Zajamšek et al., 2016a, Hansen et al., 2014).

Figure 3 shows a typical plot of the outdoor narrowband spectra over the infrasonic frequency range when the wind farm is operational and the wind speed at the residence is very low. The wind farm noise signature contains 10 harmonics of the blade-pass frequency of 0.8 Hz and the peaks are as high as 20 dB above adjacent troughs.

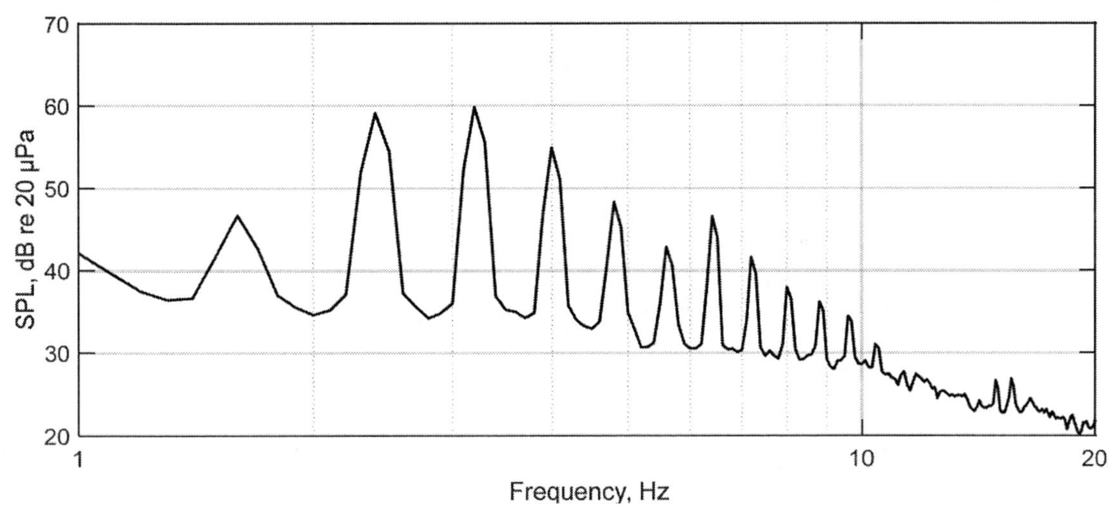

Figure 2 – Narrowband plot showing infrasonic frequencies (f ≤ 20 Hz) only. Harmonics of the blade-pass frequency of 0.8 Hz are clearly visible. Data were measured on 29/7/2013 at 3:30 am.

Figure 3 shows the wind farm noise signature that is typically measured over the low frequency range. The most significant AM occurs below 100 Hz and therefore the narrowband spectra above this frequency are not shown. There are 5 main tonal peaks visible in Figure 3 and these peaks are surrounded by side-bands spaced at the blade-pass frequency. The presence of these side-bands indicates that the tones are amplitude modulated. Adjacent side-bands show the occurrence of AM at the blade pass frequency and each subsequent pair of side-bands reveal that AM also occurs at harmonics of the blade-pass frequency.

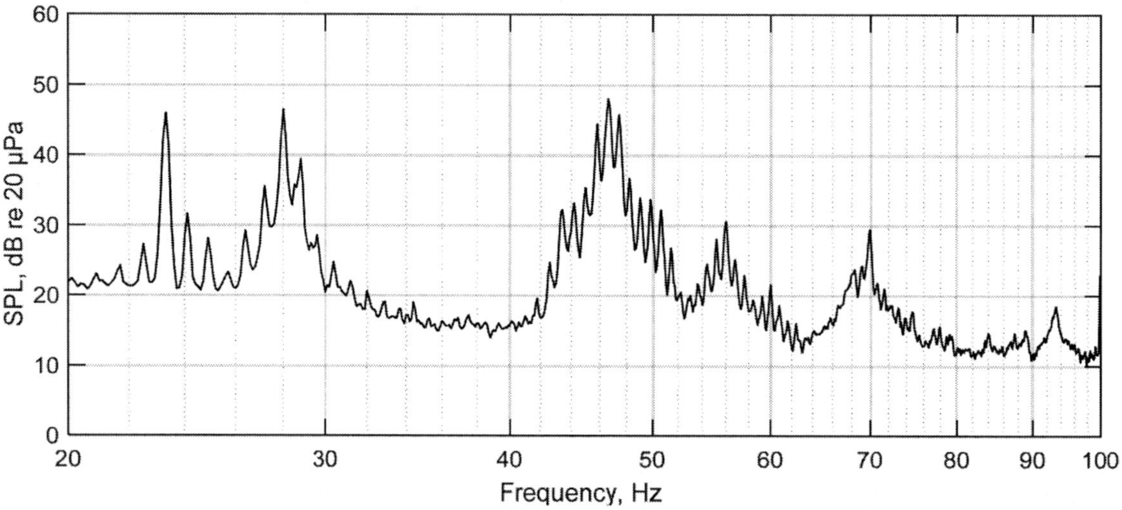

Figure 3 – Power spectral density plot showing the presence of several tones with side-bands spaced at the blade-pass frequency, which is indicative of AM. Data were measured on 29/7/2013 at 3:30 am.

The most convenient method of assessing the audibility of wind farm noise is to compare 1/3-octave spectra with the ISO389-7 (2005) hearing threshold. This threshold represents the median value of the normal hearing threshold of young adults (18-25 years) for pure tones and binaural listening. For operational conditions where the power output is greater than 40%, the mean wind farm low-frequency noise spectra shown in Figure 4 exceed the normal hearing threshold curve at 50 Hz and remain above the normal hearing threshold curve over the entire low-frequency region up to 200 Hz. This indicates that the outdoor wind farm noise would be audible to most people over this frequency range.

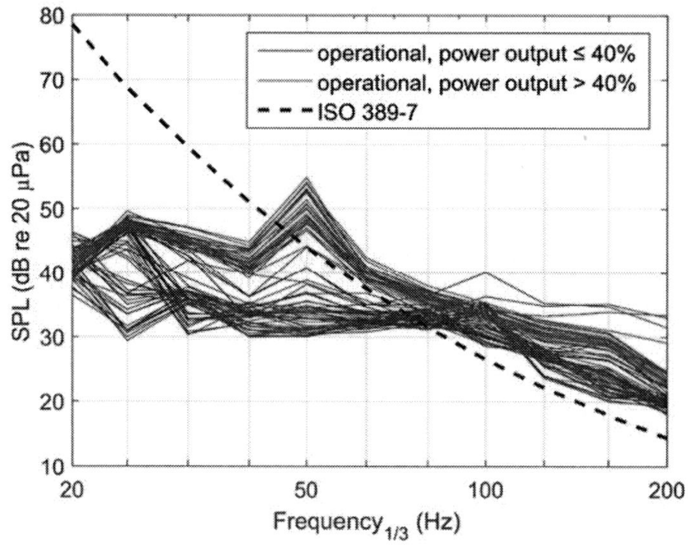

Figure 4 – Low-frequency 1/3-octave levels measured between 7 pm and 7 am.

The modulation depth associated with the outdoor wind farm noise is shown for the period between 7 pm and 7 am in Figure 5. In this figure, the small and large dots represent the modulation depth for 10-second and 10-minute periods, respectively. The 10-minute values are obtained through finding the 90th percentile of the valid (according to the prominence criterion)

10-second values. The large red dots represent 10-minute periods where the prominence criterion was exceeded for at least half of the 10-second periods, indicating that the AM can be attributed to the wind farm. Comparison with the power output data (not shown here), indicates that the wind farm was operating at a capacity factor greater than 40% for each of the 10-minute periods containing wind farm AM. Moreover, during the night selected for analysis, the wind farm was operating at a capacity factor greater than 40% for only 37% of the time. Wind farm AM occurred 29% of the time during the night; however, during the times that the wind farm was operating at a capacity factor greater than 40%, wind farm AM occurred 80% of the time.

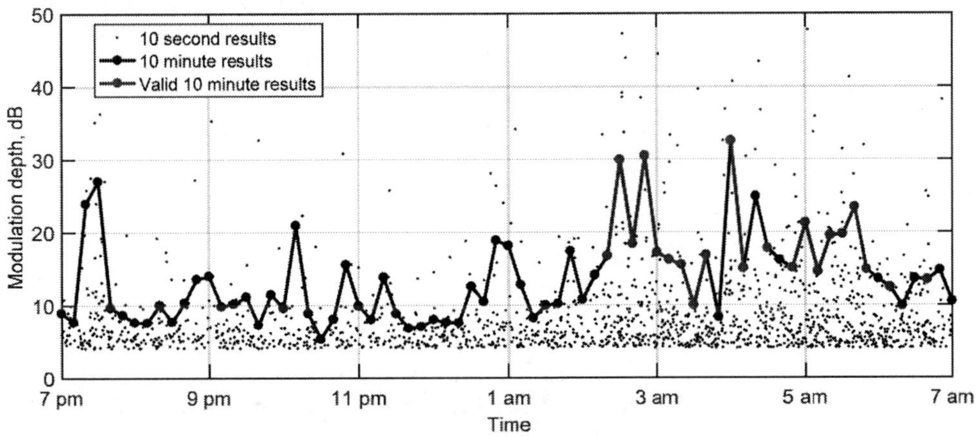

Figure 5 – Amplitude modulation depth spectra obtained using the 'Reference' method proposed by the AMWG with unweighted data filtered between 40 and 200 Hz. Data points representing 10-second periods where the modulation depth was below 4 dB (prominence cutoff) are not shown.

In summary, it has been shown that nocturnal low-frequency wind farm noise can exceed the normal hearing threshold and be amplitude modulated at the blade pass frequency of 0.8 Hz. For the data analysed here, wind farm AM occurred 29% of the time; however, this value is dependent on the power output of the wind farm. On an occasion where the power output was consistently greater than 40% during the night-time, wind farm AM could occur as much as 80% of the time. In particular, it has been shown that the 50 Hz 1/3-octave band exceeds the normal hearing threshold by the largest amount. This 1/3-octave band also contains the most significant amount of AM.

4. Discussion

When evaluating the impact of low-frequency wind farm noise on residents living near a wind farm, it is important to consider the time variability of the noise for a number of reasons. According to listening tests conducted by Lee et al. (2011), AM of wind turbine noise significantly contributes to annoyance. Moreover, for a given AM depth, annoyance is greater when the overall noise level is lower. According to Renewable UK (2013a), an equivalent demodulated sound would need to be adjusted on average by 3.5 dB(A) to be equally annoying to an amplitude modulated sound at 30 dB(A). On the other hand, this adjustment would only need to be 1.7 dB(A) for a 40 dB(A) test sound. These results suggest that overall noise levels may not need to be high for a noise to be annoying.

For amplitude modulated noise, the hearing threshold can only give an approximate indication of the potential audibility. The reason for this is that the 10-minute averaged 1/3-octave sound pressure levels are not very representative of the actual levels that occur as a function of time, as the peaks in sound pressure level are much higher (Zajamšek et al., 2016b). Therefore if a low-frequency noise is amplitude modulated as well as being above the normal hearing threshold, it is likely to be annoying to many people, especially as the detection and annoyance thresholds are close together for low-frequency noise. Even if the noise is slightly below the hearing threshold, it will still be heard by some people due to the natural spread in hearing thresholds (Møller and Pedersen, 2011). In this case, the noise could manifest as inaudible or soft to some people, yet could be loud and annoying to others. In more recent work, Yoon et al. (2015) showed that the detection threshold for an individual's response to wind turbine noise generally decreases as the extent of variation of the noise increases. In addition, the equal loudness contours at lower frequencies are closer together, meaning that low-frequency noise that is only moderately above the audibility threshold may be perceived as loud or even annoying (Møller and Pedersen 2011).

Therefore, the fact that noise in the 50 Hz 1/3-octave band can be as high as 10 dB above the normal hearing threshold indicates that this noise would be audible to most people, particularly when it is amplitude modulated. Results from a recent EPA study (SA Environmental Protection Authority, 2013) indicated that residents living in close proximity to the Waterloo wind farm complained about a "rumbling" noise, which is consistent with the presence of a 50 Hz amplitude modulated tone. Whether or not "rumbling" is equivalent to "thumping" is unclear at this moment, although both descriptors indicate a noise which is low in frequency content with a magnitude that varies with time.

To evaluate the relationship between AM and annoyance, listening tests have been conducted by Yokoyama et al. (2013). The researchers used real wind farm noise data recorded closer than 1 km from the nearest wind turbine. This means that recordings were presumably dominated by trailing edge noise and thus the "swishing" type of AM character (Baath, 2013). The researchers report very little or no "fluctuation sensation" for wind farm noise low-pass filtered at 100 Hz or lower. Due to the expected difference between wind farm noise presented in this paper (biased towards lower frequencies) and wind farm samples used by Yokoyama et al. (2013) (biased towards mid- to high-frequency trailing edge noise), the findings cannot be extrapolated to the present data. For a more comprehensive review of human response (listening tests) to AM see (Perkins et al., 2016).

Opinions regarding sleep disturbance and health problems from wind farm noise remain strongly divided, and are confounded by an ongoing lack of adequate data. The Health Canada Wind Turbine Noise and Health Study (Michaud et al., 2015, Michaud et al., 2016), the largest systematic study to date, do not support any significant associations between wind farm noise exposure and self-reported or objectively measured impacts on sleep, illnesses, perceived stress or quality of life. However, only indirect assessments of sleep were evaluated and wind farm noise was modelled rather than measured and thus the data is inevitably limited. Jalali et al. (2016) conducted a "before-after" study in which they monitored sleep and noise before and after wind farm installation. The study found no significant change in objective measures of sleep after wind farm installation. However, subjective sleep quality was significantly worsened after the wind farm installation (Jalali et al., 2016). The outcome of this research may have been affected by the fact that the authors considered whole-night sleep parameters only, rather than focussing on specific noise events. Smith et al. (2016) recently presented some preliminary results showing that synthesized low-frequency band, amplitude modulated noise has an impact on sleep. The noise samples used in the study were based on real recordings of wind farm noise. The researchers also noted that AM was a particularly important constituent contributing to sleep disruption. The noise levels that were used in this study are higher than the allowable limits in Australia and this combined with the small sample size used in the investigation indicates that further work is needed in this area.

Epidemiological studies suggest that nocturnal noise exposure might be more relevant for the creation of long-term health issues such as cardiovascular disease than daytime noise exposure (Katsouyanni et al., 2008). Pre-existing stress and/or extraneous noises can impair sleep initiation, maintenance and the return to sleep after awakenings and thus contribute to insomnia. Once sleep is initiated, sensory processing is markedly diminished but remains preserved, even in very deep or "slow wave" sleep. Throughout sleep, the primitive brainstem mechanisms continue to "gate" physiological responses according to the stimulus salience and intensity to either; (a) ignore the stimulus in favour of ongoing sleep, (b) give rise to a brief arousal or (c) trigger a full-awakening. Arousal is an abrupt change in the pattern of brain wave activity, as measured by an electroencephalogram (EEG). It has been shown by Basner et al. (2011) that even at low noise levels, physiological reactions to noises such as traffic disturbances can be reliably measured from EEG micro-arousals, cardiovascular activation responses and body movements without necessarily frequent or prolonged full awakenings. Another recent study found some evidence for EEG changes during short wind farm noise exposure periods during wake (Inagaki et al., 2015). However, with the current lack of conventional sleep measures (Inagaki et al., 2015), meaningful extrapolation of these findings to wind farm noise effects on sleep is not possible.

Even a single night of acoustic disturbance (such as analysed in this paper) can produce frequent cardiovascular activation responses without increasing awakenings or discernible EEG arousals. Despite the lack of discernible awakenings, this type of disturbance can result in daytime sleepiness and impaired mood and function (Martin et al., 1997). It has been found by Catcheside et al. (2002) that a 5-second long 500 Hz acoustic tone between 54-90 dB(A), present during non-rapid-eye movement sleep can produce a clear reflex cardiovascular response.

5. Conclusions

This paper has shown that low-frequency wind farm noise can be above the normal hearing threshold and contain significant levels of amplitude modulation for a large proportion of the night-time. This persistent low-frequency noise from wind farms is a plausible cause of chronic poor sleep and adverse health impacts to nearby residents, perhaps more so than other noise in a normally quiet rural environment. However, opinions about the effects of wind farm noise on human health and well-being remain divided and thus further research is warranted. Future work will involve quantifying the annoyance and sleep disturbance potential of the low-frequency amplitude modulated noise analysed in this paper.

Acknowledgements

Financial support from the Australian Research Council, Project DP130103136, and the Australian Government National Health and Medical Research Council (NHMRC), Project GNT1113571, is gratefully acknowledged.

References

BAATH, L. B. 2013. Noise spectra from wind turbines. *Renewable Energy,* 57, 512-519.

BAKKER, R. H., PEDERSEN, E., VAN DEN BERG, G. P., STEWART, R. E., LOK, W. & BOUMA, J. 2012. Impact of wind turbine sound on annoyance, self-reported sleep disturbance and psychological distress. *The Science of the total environment,* 425, 42-51.

BASNER, M., MULLER, U. & ELMENHORST, E.-M. 2011. Single and combined effects of air, road, and rail traffic noise on sleep and recuperation. *Sleep,* 34, 11-23.

BASS, J., CAND, M. & COLES, D. 2016a. *Institute of Acoustics AM code* [Online]. Available: https://sourceforge.net/projects/ioa-am-code/ [Accessed].

BASS, J., CAND, M., COLES, D., DAVIS, R., IRVINE, G., LEVENTHALL, G., LEVET, T., MILLER, S., SEXTON, D. & SHELTON, J. 2015. Discussion Document: Methods for Rating Amplitude Modulation in Wind Turbine Noise. London: Institute of Acoustics.

BASS, J., CAND, M., COLES, D., DAVIS, R., IRVINE, G., LEVENTHALL, G., LEVET, T., MILLER, S., SEXTON, D. & SHELTON, J. 2016b. A method for rating amplitude modulation in wind turbine noise. London, UK: Institute of Acoustics.

CATCHESIDE, P. G., CHIONG, S., MERCER, J., SAUNDERS, N. & MCEVOY, R. 2002. Noninvasive cardiovascular markers of acoustically induced arousal from non-rapid-eye-movement sleep. *Sleep,* 25, 797-804.

DOOLAN, C. 2013. *Advanced Topics in Aerospace Engineering.*

FUKUSHIMA, A., YAMAMOTO, K., UCHIDA, H., SUEOKA, S., KOBAYASHI, T. & TACHIBANA, H. 2013. Study on the amplitude modulation of wind turbine noise: Part 1-Physical investigation. *Internoise 2013.*

HANSEN, K., ZAJAMŠEK, B. & HANSEN, C. 2014. Comparison of the noise levels measured in the vicinity of a wind farm for shutdown and operational conditions. *Internoise2014.* Melbourne, Australia.

HANSEN, K. L., ZAJAMŠEK, B. & HANSEN, C. H. 2015a. Quantifying the character of wind farm noise. *In Proc. of ICSV22, Florence, Italy.*

HANSEN, K. L., ZAJAMŠEK, B. & HANSEN, C. H. 2015b. Quantifying the character of wind farm noise. *ICSV 22.* Florence, Italy: International Institute of Acoustics and Vibration.

HELLMAN, R. & ZWICKER, E. 1987. Why can a decrease in dB (A) produce an increase in loudness? *The Journal of the Acoustical Society of America,* 82, 1700-1705.

INAGAKI, T., LI, Y. & NISHI, Y. 2015. Analysis of aerodynamic sound noise generated by a large-scaled wind turbine and its physiological evaluation. *International Journal of Environmental Science and Technology,* 12, 1933-1944.

ISO389-7. 2005. *Acoustics: Reference zero for the calibration of audiometric equipment - Part 7: Reference threshold of hearing under free-field and diffuse-field listening conditions* [Online]. International Standards Organisation. [Accessed].

JALALI, L., BIGELOW, P., NEZHAD-AHMADI, M.-R., GOHARI, M., WILLIAMS, D. & MCCOLL, V. 2016. Before–after field study of effects of wind turbine noise on polysomnographic sleep parameters. *Noise and Health,* 18, 194-205.

KATSOUYANNI, K., CADUM, E., DUDLEY, M., SAVIGNY, P., SEIFFERT, I., SWART, W., BREUGELMANS, O., BLUHM, G., SELANDER, J., HARALABIDIS, A. & DIMAKOPOULOU, K. 2008. Hypertension and exposure to noise near airports: the HYENA study. *Environmental health perspectives,* 116, 329-333.

LEE, S., KIM, K., CHOI, W. & LEE, S. 2011. Annoyance caused by amplitude modulation of wind turbine noise. *Noise Control Engineering Journal,* 59, 38-46.

LEVENTHALL, G. 2009. Low Frequency Noise. What we know, what we do not know, and what we would like to know. *Noise Notes,* 8, 26.

MARCILLO, O., ARROWSMITH, S., BLOM, P. & JONES, K. 2015. On infrasound generated by wind farms and its propagation in low-altitude tropospheric waveguides. *Journal of Geophysical Research: Atmospheres,* 120, 9855-9868.

MARTIN, S. E., WRAITH, P. K., DEARY, I. J. & DOUGLAS, N. J. 1997. The effect of nonvisible sleep fragmentation on daytime function. *American journal of respiratory and critical care medicine,* 155, 1596-601.

MICHAUD, D., FEDER, K., KEITH, S., VOICESCU, S., MARRO, L., THAN, J., GUAY, M., DENNING, A., MURRAY, B., WEISS, S. & OTHERS 2015. Effects of Wind Turbine Noise on Self-Reported and Objective Measures of Sleep. *Sleep,* 39, 97-109.

MICHAUD, D. S., FEDER, K., KEITH, S. E., VOICESCU, S. A., MARRO, L., THAN, J., GUAY, M., DENNING, A., BOWER, T., LAVIGNE, E. & OTHERS 2016. Exposure to wind turbine noise: Perceptual responses and reported health effects. *The Journal of the Acoustical Society of America,* 139, 1443-1454.

MØLLER, H. & PEDERSEN, C. S. 2011. Low-frequency noise from large wind turbines. *Journal of the Acoustical Society of America,* 129, 3727-3744.

MUZET, A. 2007. Environmental noise, sleep and health. *Sleep medicine reviews,* 11, 135-142.

NISSENBAUM, M. A., ARAMINI, J. J., HANNING, C. D. & OTHERS 2012. Effects of industrial wind turbine noise on sleep and health. *Noise and Health,* 14, 237-243.
OERLEMANS, S. & SCHEPERS, J. 2009. Prediction of wind turbine noise and validation against experiment. *International Journal of Aeroacoustics,* 8, 555-584.
PEDERSEN, E. & WAYE, K. P. 2008. Wind turbines—low level noise sources interfering with restoration? *Environmental Research Letters,* 3, 1-5.
PEDERSEN, S., MØLLER, H. & WAYE, K. P. 2007. Indoor Measurements of Noise at Low Frequencies-Problems and Solutions. *Journal of Low Frequency Noise, Vibration \& Active Control,* 26, 249-270.
PERKINS, A. R., LOTINGA, J. M., BERRY, B., GRIMWOOD, J. C. & STANSFELD, A. S. 2016. A review of research into the human response to amplitude-modulated wind turbine noise and development of a planning control method. *In proc. of Internoise, Hamburg.*
RANDALL, R. 1977. Application of B&K Equipment to Frequency Analysis, 2nd Edition. B&K.
RENEWABLE UK 2013a. The Development of a Penalty Scheme for Amplitude Modulated Wind Farm Noise: Description and Justification. Renewable UK.
RENEWABLE UK 2013b. Wind turbine amplitude modulation: research to improve understanding as to its cause & effect. Renewable UK.
SA ENVIRONMENTAL PROTECTION AUTHORITY 2013. Waterloo Wind Farm - Environmental noise study. EPA.
SHEPHERD, D., MCBRIDE, D., WELCH, D., DIRKS, K. N. & HILL, E. M. 2011. Evaluating the impact of wind turbine noise on health-related quality of life. *Noise and Health,* 13, 333.
SMITH, G. M., OGREN, M., THORSSON, P., PEDERSEN, E. & PERSSON WAYE, K. 2016. Physiological effects of wind turbine noise on sleep. *In Proc. of ICA2016, Buenos Aires.*
SONDERGAARD, B. 2014. Low frequency noise from large wind turbines. *In proc. of InterNoise 2014, Melbourne.*
VAN DEN BERG, G. 2004. Effects of the wind profile at night on wind turbine sound. *Journal of Sound and Vibration,* 277, 955-970.
YOKOYAMA, S., SAKAMOTO, S. & TACHIBANA, H. 2013. Perception of low frequency components contained in wind turbine noise. *5th International Meeting on Wind Turbine Noise.* Denver, Colorado.
YOON, K., GWAK, D. Y., LEE, S. & LEE, S. 2015. Experimental study of relationship between amplitude modulation and detection threshold of wind turbine noise. *6th International Meeting on Wind Turbine Noise.* Glasgow, UK.
ZAJAMŠEK, B., DOOLAN, C. J., HANSEN, K. L. & HANSEN, C. H. 2016a. Characterisation of wind farm infrasound and low-frequency noise. *Journal of Sound and Vibration,* 370, 176-190.
ZAJAMŠEK, B., HANSEN, K. L., DOOLAN, C. J. & HANSEN, C. H. 2016b. Characterisation of wind farm infrasound and low-frequency noise. *Journal of Sound and Vibration,* 370, 176-190.

7th International Conference
on
Wind Turbine Noise
Rotterdam – 2nd to 5th May 2017

Pre-construction Site Prediction Tool for Wind Farm AM - Do We Now Know Enough?

Andrew Birchby, Renewable Energy Systems: andrew.birchby@res-group.com
Fraser Tough, Renewable Energy Systems: fraser.tough@res-group.com
Jeremy Bass, Renewable Energy Systems: jeremy.bass@res-group.com

Summary

Amplitude Modulation (AM) refers to variations in the level of audible noise from a wind farm at the frequency with which a blade passes a fixed point in space. As acoustic character is a component of how noise from wind turbines is perceived, it is important for wind farm developers to understand the circumstances in which AM can occur at residential distances. At some sites, mitigation may be required in order to reduce the level of AM under certain meteorological conditions. Rather than waiting until a site is operational, and responding reactively, a more proactive approach has advantages. To this end, a statistical model has been developed with which the probability of AM occurring can be estimated at the pre-construction stage. The variables found to be significant to the likelihood of AM occurring add weight to previous research identifying the mechanisms responsible.

1. Introduction

Noise character, in addition to noise level, is a factor in the perception of sound. In the context of wind farm noise, Amplitude Modulation (AM) is defined as "periodic fluctuations in the level of audible noise from a wind turbine (or wind turbines), the frequency of the fluctuations being related to the blade passing frequency of the turbine rotor(s)" (IOA, 2016). Levels of AM beyond those anticipated by the guidance relevant to the assessment and rating of noise from wind farms, ETSU-R-97 (DTI, 1996), have been found to occur in the far-field at certain sites under specific meteorological conditions. Whilst mitigation options exist (HLA, 2015) such that the level of AM can be reduced should action, whether on a good neighbour basis (RES, 2015) or to fulfil the obligations of a Planning Condition (WSP, 2016), be required post-construction, a proactive approach has advantages. Being able to predict levels of AM at the pre-construction stage would be beneficial in the following ways:

- more accurate project valuations could be made, taking into account any energy loss that could potentially result from AM mitigation, should it be required
- the results could be fed into negotiations with turbine manufacturers, informing the turbine selection process, allowing improved warranties to be sought and making mitigation options available from the outset where necessary.

To this end, a study has been undertaken into the incidences of AM occurring at a residential location neighbouring an operational wind farm site. The presence of AM has been shown to be associated with prevailing meteorological conditions and turbine operational information. A statistical model has been used to determine the significance of the different variables. The model has then been used to estimate the probability of AM occurring under a given set of conditions.

2. Methodology

A noise survey was undertaken at an operational wind farm site and data over a two month period has been analysed. The level of AM was calculated using the metric recommended by the Institute of Acoustics AM Working Group (IOA, 2016). The IOA metric was specifically designed to better distinguish between AM generated by wind turbines and variations in the noise level due to other sources when compared to alternative methods. No value of AM is returned when fluctuations in the noise level within a 10 minute period are deemed to be due to other sources. The metric is therefore useful not only in determining the level of AM but also whether AM due to wind turbine noise is present or not.

A database was created containing the result provided by the IOA metric for each 10 minute period along with data measured by the on-site meteorological mast and the turbines themselves. Filtering was undertaken to remove variables that are proxies for others e.g. time of day, highly correlated with others e.g. rotational speed of the rotor (due to its correlation with wind speed), or site dependent e.g. wind direction. Statistical tests were applied in order to assess the relative significance of each variable in determining whether AM was present or not. Interaction variables and powers of variables were not tested. The most significant variables were used to create a logistic regression model with which the probability of AM occurring can be predicted. The performance of the model was evaluated and some example probabilities calculated to illustrate the concept.

3. Results

3.1 Model Formulation

To obtain the log odds of AM, compute:

$$\log\left(\frac{\phi}{1-\phi}\right) = X\beta = \beta_0 + \beta_1 x_1 + \cdots + \beta_n x_n$$

Where X represents the model design matrix and β represents the vector of coefficients. Utilising Maximum Likelihood Estimation (MLE) to estimate β, the odds of AM can be computed as:

$$\frac{\phi}{1-\phi} = e^{X\beta}$$

The probability of AM can be found by rearranging such that:

$$\phi = \frac{e^{X\beta}}{1+e^{X\beta}}$$

3.2 Variable Importance and Effect

The most significant variables for the data considered, in order, were found to be:

1. shear exponent (reflecting the change in wind speed with height above ground)
2. wind veer (the change in wind direction with height above ground)
3. air temperature
4. angle of attack (angle between the blade chord and the relative wind)
5. barometric pressure
6. pitch angle (angle between the blade chord and the plane of rotation)
7. wind speed at hub height
8. standard deviation of wind speed at hub height.

Z scores were computed as the ratio of the parameter estimates divided by their standard errors. The importance of the variable with the largest Z score was set to 100 and the importance of other variables set proportionally as the ratio of their score vs. the highest score. The results, demonstrating the relative importance of the variables, are shown in Figure 1.

Figure 1 - Variable Importance (based on relative Z scores)

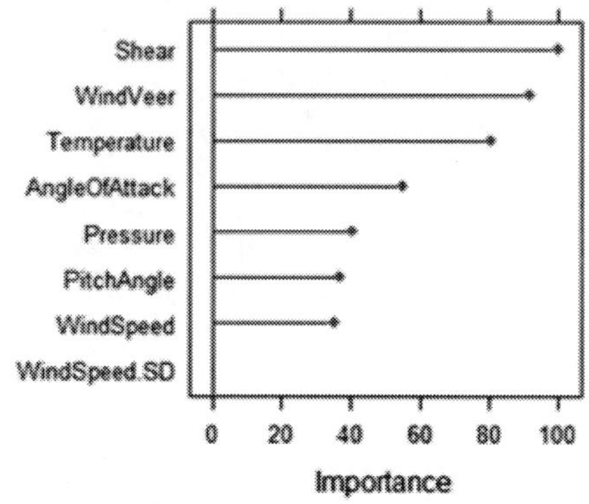

The effect of each variable, in terms of whether increases in the variable result in increases or decreases in the probability of AM occurring, are outlined in Table 1 below. Increases in shear, temperature, angle of attack and wind speed increase the likelihood of AM occurring, whereas increases in wind veer, pressure, pitch angle and the standard deviation of wind speed decrease the likelihood. The variables identified as significant support the results of previous research identifying transient blade stall due to asymmetric flow into the turbine rotor as the causal mechanism of AM (RUK, 2013). The increased probability of AM for high shear exponents, temperatures and angles of attack, along with low turbulence (of which the standard deviation of wind speed is a proxy), suggests that periods of high shear associated with stable atmospheric conditions are the primary driver, rather than other causes of asymmetric flow into the rotor such as turbulence, at this specific location.

Table 1 - Direction of Variable Effect

Variable	Effect on Probability of AM
Shear	Increases
Wind veer	Decreases
Temperature	Increases
Angle of attack	Increases
Pressure	Decreases
Pitch angle	Decreases
Wind speed	Increases
Wind speed standard deviation	Decreases

3.3 Model Performance

A Receiver Operating Characteristic (ROC) curve was computed from the fitted data to illustrate the performance of the model (see Figure 2). The sensitivity (y-axis) describes the proportion of true positives (when AM would be predicted by the model and occurs according to the measured data), and the specificity (x-axis) describes the proportion of false positives (when AM would be predicted by the model but does not occur according to the measured data). The blue line shows the effect of changing the classification threshold probability. Setting the classification threshold probability to 1 would result in a true positive rate of 0 and a false positive rate of 0, as every point would be classified as "no AM". Setting the classification threshold to 0 would result in a true positive rate of 1 and a false positive rate of 1, as every point would be classified as AM. Setting the threshold at 0.5 would result in a true positive rate of 0.9 and a false positive rate of 0.2. The Area Under the Curve (AUC) is 0.9, indicating that the ability of the model to discriminate between periods when AM does and does not occur in the dataset is good-to-excellent, with anything greater than 0.75 suggesting that the model performs well (1 being perfect and 0.5 being worthless).

Figure 2 - Model Performance

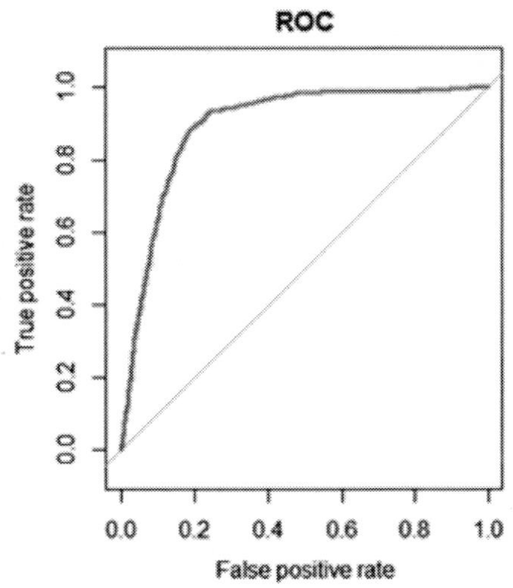

3.4 Example Probabilities

Holding all other inputs constant (at average levels), the probability of AM occurring for various combinations of the variables determined to be significant can be calculated. Figures 3 and 4 illustrate how the impact of different variables on the probability of AM can be visualised. Figure 3 demonstrates that AM is more likely to occur in weather conditions which result in high shear exponents and low levels of wind veer. Figure 4 shows that AM is more likely to occur when high shear exponents coincide with high temperatures.

Charts such as those detailed in Figure 5 illustrate how sensitive the probability of AM is to changes in two given variables, for those shown:

a) the probability of AM is equally sensitive to changes in shear exponent for different levels of wind veer, although the range of shear exponents to which it is sensitive differ. The probability of AM is most sensitive to wind veer for a specific range of shear exponents and insensitive when the shear exponent is below or above a certain value
b) the probability of AM is equally sensitive to changes in shear exponent for different angles of attack, although the range of shear exponents to which it is sensitive differ. The probability of AM is most sensitive to angle of attack for a specific range of shear exponents and insensitive when the shear exponent is below or above a certain value
c) the higher the shear exponent, the more sensitive the probability of AM is to changes in temperature and vice versa
d) changes in wind speed appear to have the greatest impact on the probability of AM for a certain range of temperatures and changes in temperature appear to have a slightly greater impact on the probability of AM the lower the wind speed

Now that a model has been established based on data from an operational site, it can be applied to a time series containing wind climate variables, along with information on how the proposed turbines would be expected to respond, for potential future sites allowing the probability of AM occurring at any point over a typical year to be ascertained.

Figure 3 - Probability of AM by Shear and Veer

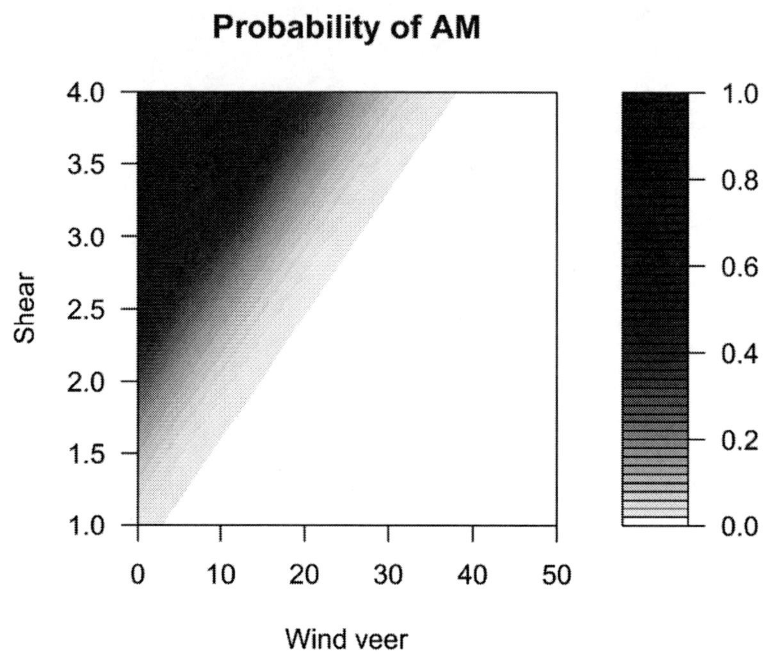

Figure 4 - Probability of AM by Shear and Temperature

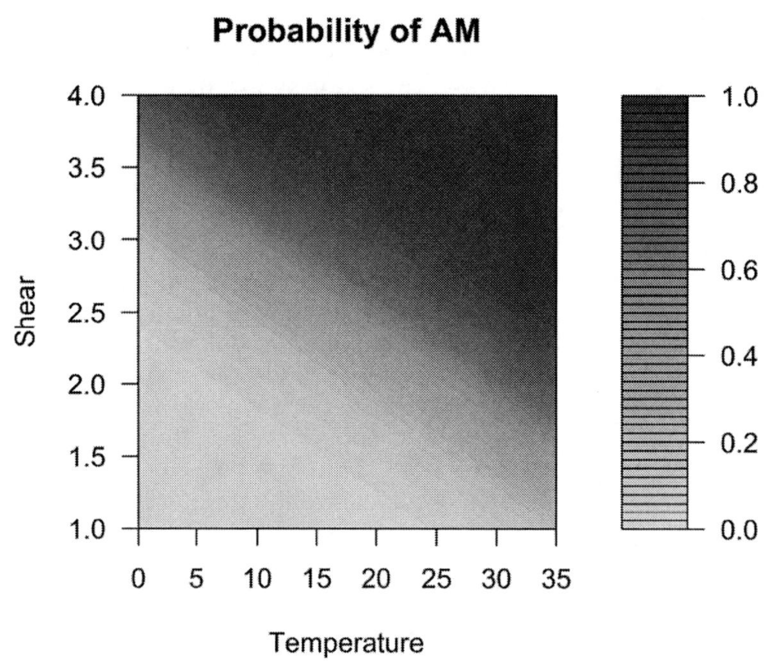

Figure 5 - Probability of AM for Given Variables

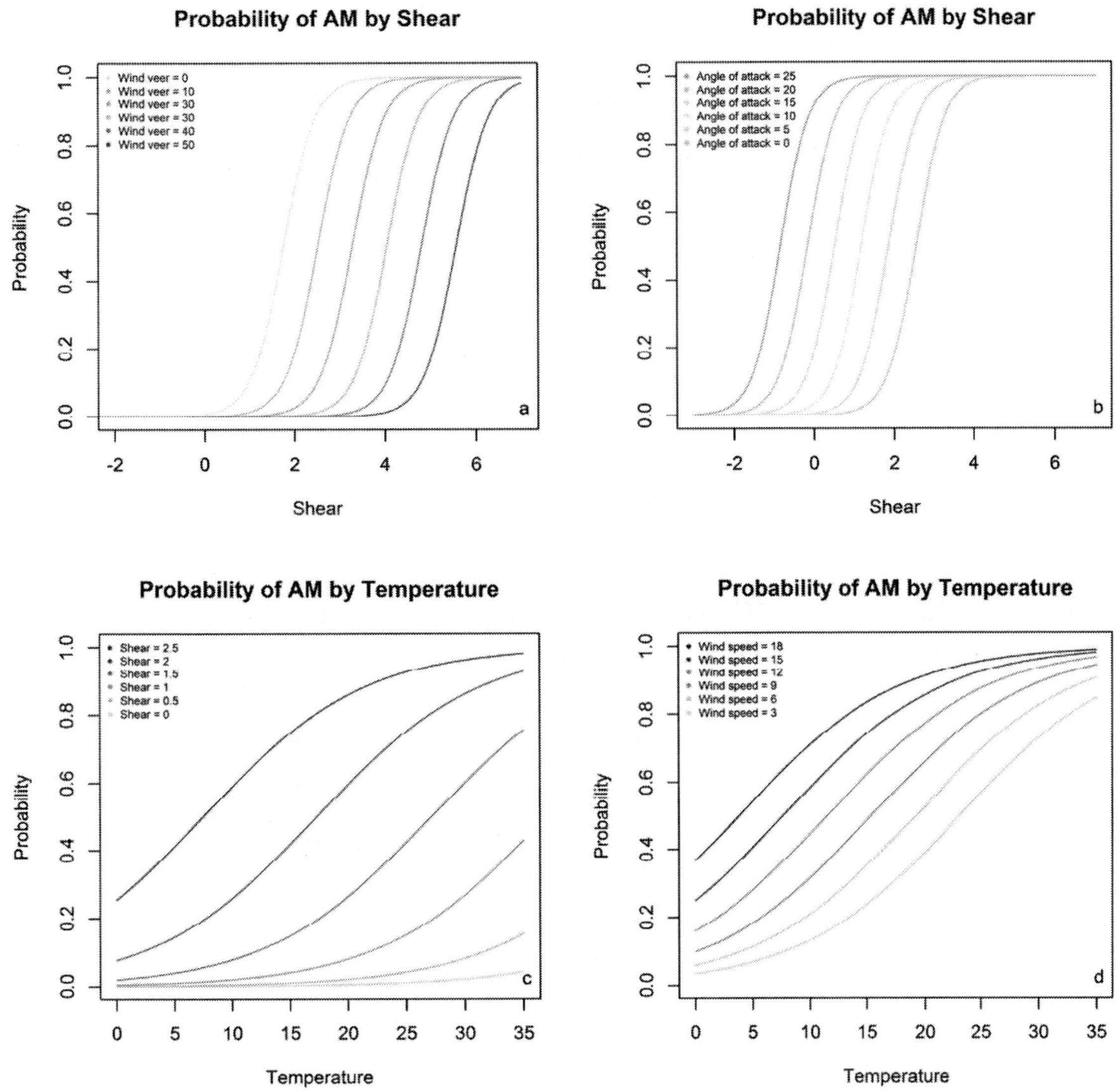

4. Conclusions

A statistical model has been created which allows the probability of AM to be estimated based on a combination of meteorological conditions and turbine operational parameters. The variables discovered to be significant in the likelihood of AM occurring add weight to previous research identifying the mechanisms responsible. The tool provides a first step towards enabling AM to be predicted at the pre-construction stage, suggesting that it should be possible to ascertain the need for mitigation and any potential impact on project value in advance. The results could also be fed into the turbine selection process, allow improved warranties to be obtained and enable mitigation options to be available from the outset where necessary.

5. Further Work

Suggestions for further investigation are as follows:

- data from other sites can be added to the database in order to identify whether the significant variables for this site are consistent across others
- a statistical model could be developed that is capable of predicting not only the likelihood of occurrence, but also the level of AM, for a range of sites. A semi-continuous regression model could be utilised, including interactions and powers of variables to model any non-linear relationships between the log odds of response and the inputs
- rather than basing the model on all of the data provided by the IOA metric, the model could be focused on higher levels of AM
- the dataset could be filtered to determine whether evidence of other drivers, responsible for a smaller number of instances of AM, also exists.

References

IOA (2016), *A Method for Rating Amplitude Modulation in Wind Turbine Noise,* Institute of Acoustics Noise Working Group (Wind Turbine Noise) Amplitude Modulation Working Group Final Report, August 2016

DTI (1996), *The Assessment and Rating of Noise from Wind Farms*, The Working Group on Noise from Wind Turbines, ETSU Report for the DTI, ETSU-R-97, September 1996

HLA (2015), *Measurements demonstrating mitigation of far-field AM from wind turbines*, Hoare Lea Acoustics, 6th International Meeting on Wind Turbine Noise, Glasgow, April 2015

RES (2015), *Addressing the Issue of Amplitude Modulation: A Developer's Perspective*, Renewable Energy Systems, 6th International Meeting on Wind Turbine Noise, Glasgow, April 2015

RUK (2013), *Wind Turbine Amplitude Modulation: Research to Improve Understanding as to its Cause and Effect*, Renewable UK, December 2013

WSP (2016), *Wind Turbine AM Review*, WSP Parsons Brinckerhoff, Phase 2 Report, August 2016

7th International Conference
on
Wind Turbine Noise
Rotterdam, Netherlands, 2 – 5 May 2017

Title: An Investigation into Short-Term Fluctuations in Amplitude Modulation of Wind Turbine Noise. Preliminary Results.

Authors: **Ian Bonsma, BASc, PEng (ibonsma@hgcengineering.com)**
Nathan Gara, CET (ngara@hgcengineering.com)
Brian Howe, MEng, MBA, PEng (bhowe@hgcengineering.com)
Nick McCabe, MESc, PEng (nmccabe@hgcengineering.com

The degree of amplitude modulation present in wind turbine noise can vary rapidly, possibly in response to rapid fluctuations in inflowing air.

The goal of this work was to investigate whether correlations between properties of inflowing air and the degree of amplitude modulation could be observed on a short time scale of less than a minute. In order to conduct the work, continuous recordings of noise were made at three locations, selected to be in predominantly downwind and crosswind directions, and simultaneous measurements of wind speed and direction were made at various heights spanning the turbine's rotor diameter using a LIDAR system.

The degree of amplitude modulation was calculated using the refined method published by the Institute of Acoustics in 2016.

This paper is based on the results of research work carried out by HGC Engineering, a consultancy located near Toronto, Ontario, Canada.

Summary

This paper has adopted the method proposed by Renewable UK for the quantification of amplitude modulation in wind turbine noise, and has sought to extend past work by others which searched for potential correlations between amplitude modulation and various characteristics of the airflow entering a wind turbine.

Several recent papers have attempted to find such correlations. This paper is based on short averaging times (10-seconds) and on a detailed measurement regime examining inflowing air, obtained using a LIDAR system located near the turbine in question.

While some trends are suggested by the data, no clear correlations were found between short time period AM ratings and various wind-derived meteorological quantities.

1.0 Introduction

There appears to be a strengthening consensus that the amplitude modulation of Wind Turbine Noise (WTN), and in particular the "enhanced amplitude modulation" (EAM) of wind turbine noise, which occurs under certain situations, possibly certain atmospheric or meteorological conditions, is a significant contributor to the annoyance which many experience in response to audible WTN [1].

Past work by many, including work by contributors to the Wind Turbine Noise conferences, has explored and developed methods for the quantification of AM, and also looked for correlations between properties of inflowing air and the degree of AM observed.

Thoroughly exploring the connection between conditions of inflowing air and AM is important, since if a reliable correlation can be found, it may be possible to eliminate or reduce EAM through sensing technology and modification to wind turbine control systems.

HGC Engineering has been involved with the measurements and analysis of WTN in Canada, and particularly in the province of Ontario since 2005, and is a regular participant at the WTN conferences, having presented on several occasions. HGC Engineering holds ISO 17025 accreditation for IEC 61400-11 testing. Through our contacts with industry groups, wind farm developers and wind farm operators, HGC Engineering has opportunities to collect acoustic data in and around wind energy projects.

2.0 Past Studies

2.1 Quantification of AM

Various metrics have been proposed in recent years for the quantification of AM in WTN [2][3], including Fluctuation Strength [4], direct determination of peak to trough ratios [5], autocorrelation [6], the Fast and Slow sound level weighting metrics [7], and various envelope analysis techniques [8], [9], [10].

At the 2015 WTN Conference, a paper was presented summarizing work by the UK's Institute of Acoustics (IOA), intended as progress toward an agreed metric for AM [11]. Since that time, a final IOA report describing a proposed AM metric was published [12].

This method proposed by the IOA is based on sequential $L_{AEQ\ 100ms}$ data, calculated over three separate band-limited frequency ranges together spanning the range from 50 to 800 Hz. Each 10s segment of data is de-trended by subtracting a 3rd order polynomial best fit curve. A Fourier transform is used to calculate a power spectrum, and the highest peak in a range of possible blade passing frequencies is found. The energy represented by this peak, and its possible harmonics, is used to calculate an inverse Fourier transform. Finally, the modulation depth is calculated by subtracting the L_{95} from the L_5 of the resulting time series. Python language code implementing the method was released by the IOA in 2016.

The method results in a series of 10 second data, as well as a series of 10 minute averaged results.

2.2 Quantification of Meteorological Properties Derived from Wind Data

Air flowing into a turbine is often simplified for visualization purposes as an ideal laminar flow, or a consistent profile of smoothly moving air displaying an exact profile of increasing speed with height. In reality, real-world winds can vary over small areas in many ways. With modern sensing techniques such as LIDAR and SODAR, it is possible to gain a detailed picture of inflowing air arriving at a turbine rotor.

Various papers in the literature discuss the quantification of different characteristics of the inflowing air. The meteorological metrics used in this paper are based on short-duration (10-second average) measurements made using a LIDAR system and include the (vertical) wind speed shear, the wind speed shear residual sum of squares (RSS; a measure of how closely the data fits a best-fit idealized wind shear profile), equivalent wind speed (a measure of the "average" wind speed flowing into a rotor disk, taking into account the wind speed at different heights and the fraction of the disk which any given measurement height represents), wind veer (horizontal wind direction shear), and turbulence intensity.

A method for quantifying wind speed shear when multiple wind speed measurements over the rotor swept area are available is described in [13]. The measured wind speed u at multiple heights z is fit to a power law curve with an exponent of α_{fit}. The resulting curve is forced through the coordinate (u_{hub}, z_{hub}) using Equation 1.

$$u_{fit}(z) = u_{hub}\left(\frac{z}{z_{hub}}\right)^{\alpha_{fit}} \qquad [1]$$

The goodness of fit between the resulting power law equation and the actual measured wind speed data is described by the residual sum of squares (RSS), where *n* is number of measurement heights, and may be calculated according to Equation 2.

$$RSS = \sum_{i=1}^{n}\left(u_{fit}\,(z_i) - u_i)\right)^2 \quad [2]$$

Another quantity used to describe the inflowing air which is obtainable when wind speed measurements are available at several heights is the equivalent wind speed, described in [14] and [13]. The equivalent wind speed formula (Equation 3) is derived from the equation for kinetic energy flux and describes the equivalent wind speed u_{eq} as a function of the measured wind speeds *u* at heights *i* and the corresponding segment areas A_i of the rotor swept area *A* as illustrated in Figure 1.

$$U_{eq} = \left(\sum_i {u_i}^3 \frac{A_i}{A}\right)^{\frac{1}{3}} \quad [3]$$

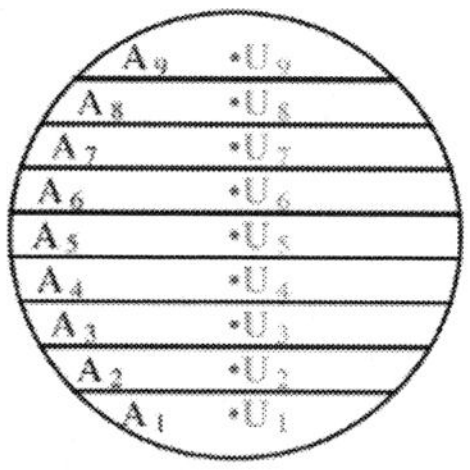

Figure 1: Diagram showing fractional segment areas (A_i) of rotor disk and corresponding wind speed measurement heights (U_i).

Equivalent wind speed can also be expressed as a percentage of the hub height speed, as a measure of how closely the equivalent wind speed is estimated by an anemometer at hub height (see Equation 4).

$$\%\,Difference = \frac{U_{eq} - u_{hub}}{u_{hub}} * 100 \quad [4]$$

Wind veer can be quantified using the standard deviation of the measured wind directions measured at several heights [15]. As described in that paper, the standard deviation of wind direction, σ_ϑ, has been calculated using Equation 5.

$$\sigma_\theta = \sin^{-1}(\epsilon)[1.0 + (\tfrac{2}{\sqrt{3}} - 1)\,\epsilon^3] \quad [5]$$

Where

$$\epsilon = \sqrt{1 - [(n^{-1}\textstyle\sum_n \sin\theta_i)^2 + (n^{-1}\sum_n \cos\theta_i)^2]} \quad [6]$$

n is the number of wind speed measurements, and ϑ_i is the wind direction at height *i*.

Another common metric used to describe inflowing air is turbulence intensity, *TI*, defined as the ratio of the standard deviation of the horizontal wind speed σ_u, at a given height, and the mean of the horizontal wind speed $\bar{u}$ at that height. Various recent papers have examined potential relationships between TI and AM [10], [12]. In this paper, a single-number descriptor for the weighted TI, denoted here as TI_w, has been calculated using Equation 7, which takes into account the relative area represented by the TI data obtained at each measurement height.

$$TI_w = \frac{\sum_i\left[\left(\frac{\sigma u_i}{\bar{u}_i}\right)A_i\right]}{\sum_i(A_i)} \qquad [7]$$

2.3 Correlation

Although many papers discuss possible factors contributing to EAM (several are summarized in [16]) and various parameters of inflowing air have been discussed, demonstration of a clear relationship between EAM and such parameters is elusive. Various recent papers have looked for correlations between airflow data and AM. Past related work by HGC Engineering tried to correlate EAM to ten metre height wind data [2][8], but this is now known to be insufficient to show a reliable correlation. Possible correlations between a limited number of windspeed-derived meteorological metrics and AM were examined in [10] and [17], based on a sophisticated campaign involving 15-second AM ratings and meteorological observations at multiple heights on tall met masts located within several kilometres of turbines. Again, no strong correlations with wind-speed derived quantities were described, but some connection between positive sound speed gradients and positive temperature gradient and AM was found. Two papers, [18] and [19], describe mitigation of wind turbine AM based on a suggested relationship between AM and local blade stall associated with high wind shear conditions, but no clear measurement data were presented.

Another sophisticated paper examining possible correlations between meteorological quantities and AM was presented in [16]. This study looked at short duration (10 second) measurements of AM, and correlated them to meteorological properties obtained using a LIDAR system and to wind turbine properties, all over a 6-week period. Figures comparing a short time history (a few hours) of temporal variation of various meteorological properties (presented as a fairly course average over time) were shown, although no statistical analysis was presented. No strong correlations with meteorological properties were found.

3.0 Current Methodology

It has been hypothesized that EAM may be related to a wide variety of meteorological or geometrical conditions. The authors' practical experience suggests that the subjective degree of EAM can change rapidly over periods of only a few seconds, an observation also described in [10], [18], and thus a detailed knowledge of changes in the inflowing air over relatively short periods is likely critical for the prediction of EAM based on airflow data. It is also hypothesized that because airflow properties can vary quickly and vary over small areas, the use of multiple wind speed measurement heights is important, measured at a location as close to a wind turbine as possible.

In order to further explore the potential to detect an airflow property or properties which may correlate to AM, this current work focuses on a statistical analysis between various airflow-derived properties and AM. The methodology is similar to that described in [16], but examines additional windspeed-derived properties, examines shorter-duration airflow observations, and was extended for a longer time period.

In this work, the AM metric proposed by the IOA has been applied to the collected data. The results are discussed herein.

3.1 Measurement Setup

Measurements were completed at a wind project in Ontario, Canada. The wind power facility comprises a small number of turbines, each with electrical power ratings greater than 2 MW, hub heights greater than 90 metres and rotor diameters greater than 100 metres.

With the cooperation of the facility operator a test turbine was selected. To the extent possible, the selection of the test turbine was made to a) minimize background sound by maximising the distance from roadways or forested areas, b) allow unobstructed airflow from the turbine, to downwind and crosswind measurement locations considering the dominant wind direction in the area, c) provide a stable and accessible area at which to deploy a trailer-mounted LIDAR system in the upwind direction of the turbine, d) allow clear air upwind of the turbine.

This paper is based on data collected at three sound level monitoring stations: one at a location relatively close to the turbine (approximately 150 metres in the prevailing downwind direction); one at a distance of 440 meteres from the base of the turbine in the prevailing downwind direction; and one at a distance of 440 metres in a crosswind direction. The 440 metre distance was selected as it was in the acoustic "mid-field" range discussed in [16] (roughly equivalent to a few rotor diameters) and provided a convenient installation location in both downwind and crosswind directions. Each station comprised a Norsonic Nor140 integrating sound level meter connected to a ½" microphone. The meters were configured to continuously produce 12 kHz 24 bit audio recordings. 1/3 octave L_{AEQ} data was also collected and logged on a 1-minute basis.

Each microphone was set at 1.5 metres above the local grade, and equipped with a 175 mm diameter windscreen to reduce wind-induced microphone self-noise. Although equipped with solar cells and high-capacity batteries to allow continuous operation, the stations were visited periodically to inspect the equipment, verify correct calibration, and to retrieve the audio recordings. Figure 2 illustrates a typical monitoring station.

Figure 2: Typical Sound Level Monitor Deployment Used For this Work

A Windcube V2 LIDAR system was installed 165 metres from the test turbine in the prevailing upwind direction. The unit was configured to collect data at 10 selected heights, varying between 40 metres and 170 metres, at a fast temporal resolution (on the order of one reading per second).

In addition to the LIDAR system, 1.5 metre and 10 metre anemometers were installed 185 metres west of the turbine. Wind speed and direction, temperature, humidity, pressure, and rainfall data were collected at this location. Figure 1 shows the test turbine and equipment locations.

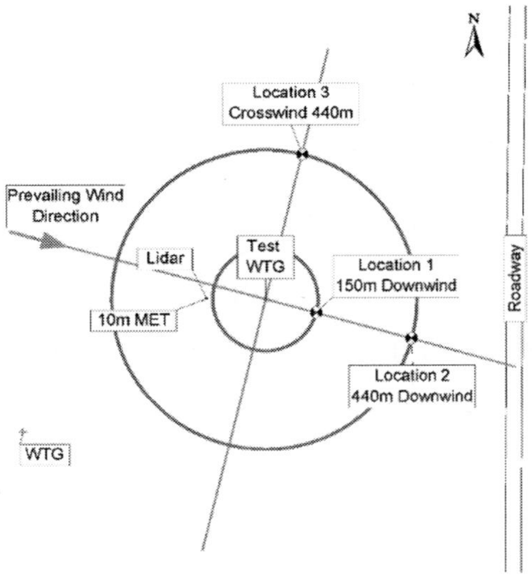

Figure 3: Test Turbine and Measurement Locations

Data for the test turbine were provided by the operator and the manufacturer in 10 minute resolution.

4.0 Analysis

4.1 Computation AM Ratings

The audio recordings were used to determine the AM rating of sequential 10-second periods. A Python language computer program was written to calculate 100 ms L_{AEQ} sound levels in the three frequency bands specified by the IOA method, using the audio recordings as input. Once data were available, the AM rating was calculated using the Python language computer code released by the IOA in 2016, which implements the IOA method [12].

4.2 Wind-derived Meteorological Quantities

The data recorded by the LIDAR comprises one set of measurements at each height from each of five beams every few seconds. A second data set is produced by calculating the meteorological properties averaged over sequential 10-minute periods. Based on information from the manufacturer, the same calculations were performed to yield a set of data averaged over sequential 10-second periods.

This 10-second data was used to compute the wind speed shear, the wind speed shear RSS, the equivalent wind speed, the equivalent wind speed percent difference, the standard deviation of the wind direction, and the weighted turbulence intensity. These quantities were calculated using the equations described in Section **2.2**.

4.3 Data Reduction and Filtering

The data collection extended over a 10 week monitoring campaign. Numerous criteria were used to reduce the size of the data set:

a) data collected during rainfall was excluded;

b) data collected during periods of no or low winds were excluded (low wind was taken as hub height wind speeds, u_{hub}, of 5 m/s or less);

c) data obtained when the turbine was operating at speeds of less than 2 rpm was excluded.

d) during portions of the measurement campaign, the turbine was prevented from operating within its normal programme in response to restrictions from the electrical system operator. Data obtained during these periods is also excluded from the analysis discussed here due to uncertainty as to the effects of these restrictions and the additional variables they may represent.

e) a manual process of identifying contaminating sounds was employed. This involved a review of the third-octave data to identify unusual spectral content (e.g., loud birds noises, tractor noise, etc.) and a review of selected audio recordings to validate and eliminate the periods of interference.

f) periods of low quality data were also excluded. This was done in response to listening tests of many of the 10-second periods where the IOA method identified significant AM ratings, where sounds other than the wind turbine noise were clearly dominant. To exclude this data, periods where the IOA method determined a low Prominence score were eliminated. A Prominence score of 10 was selected as the cut-off, which appeared to eliminate many of these calculated "false positive" data points.

Once filtered, the remaining data was divided into categories to arrive at six sets of data: data obtained under prevailing downwind and crosswind conditions at each of the three monitoring locations.

Downwind conditions are defined here as periods where the wind direction was within +/- 45° of the line of sight between the turbine and the monitor location in question. Crosswind is defined as a period where the wind direction was +/- 45° of the line normal to the line of sight between the turbine and the monitor location. The average wind direction from the LIDAR system was utilized for this filter.

In accordance with the IOA method, the frequency band showing the strongest AM was selected as shown in Figure 4. For this purpose, one of the data sets identified above was used. The 100-400 Hz band showed somewhat larger AM ratings when the AM rating was greater than about 5 and according to the method is the most appropriate. In addition, subjectively, the listening tests suggested that the lower frequency band (50-200 Hz) tends to contain more wind noise, and the higher frequency band (200-800 Hz) tends to contain more background interference such as bird calls or distant vehicle noise. As a result, the 100-400 Hz band was selected for the subsequent analysis.

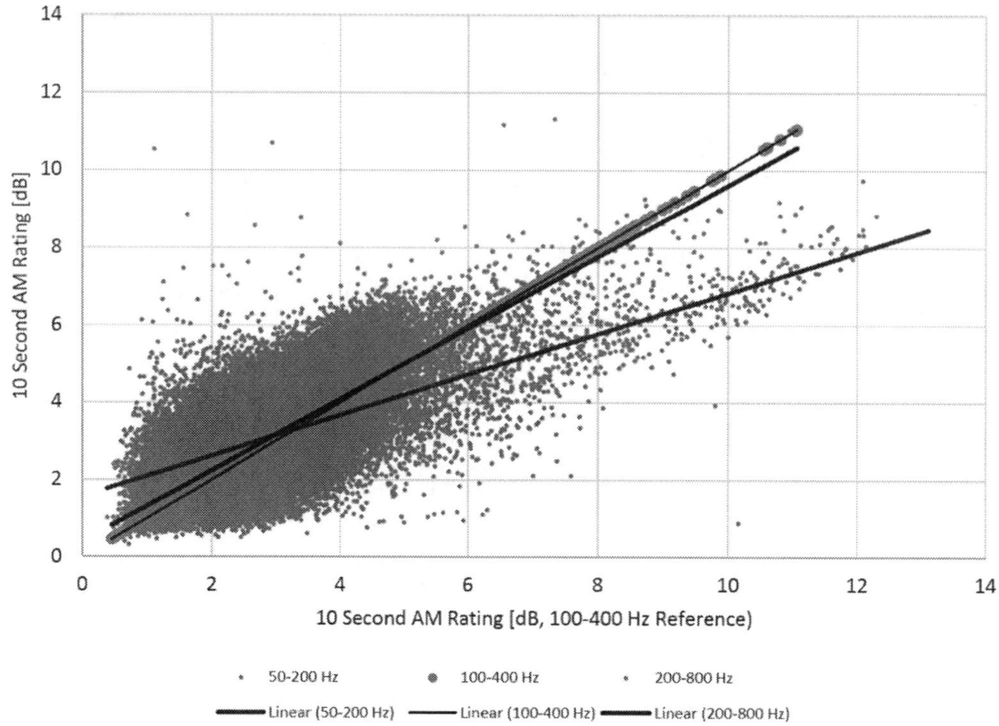

Figure 4: Identification of Frequency Band with Strongest AM

5.0 Results

The 10-second AM rating calculated according to the IOA method [12] was compared to the various 10-second windspeed-derived meteorological properties described above, for each of the six data sets described in Section **4.3**.

Scatter charts illustrating the relationship between the AM rating and the meteorological properties based on downwind and crosswind data for Measurement Location 1 are shown in Figures 5 and 6, respectively.

The downwind and crosswind data for Location 2 are shown in Figures 7 and 8, and for Location 3, Figures 9 and 10.

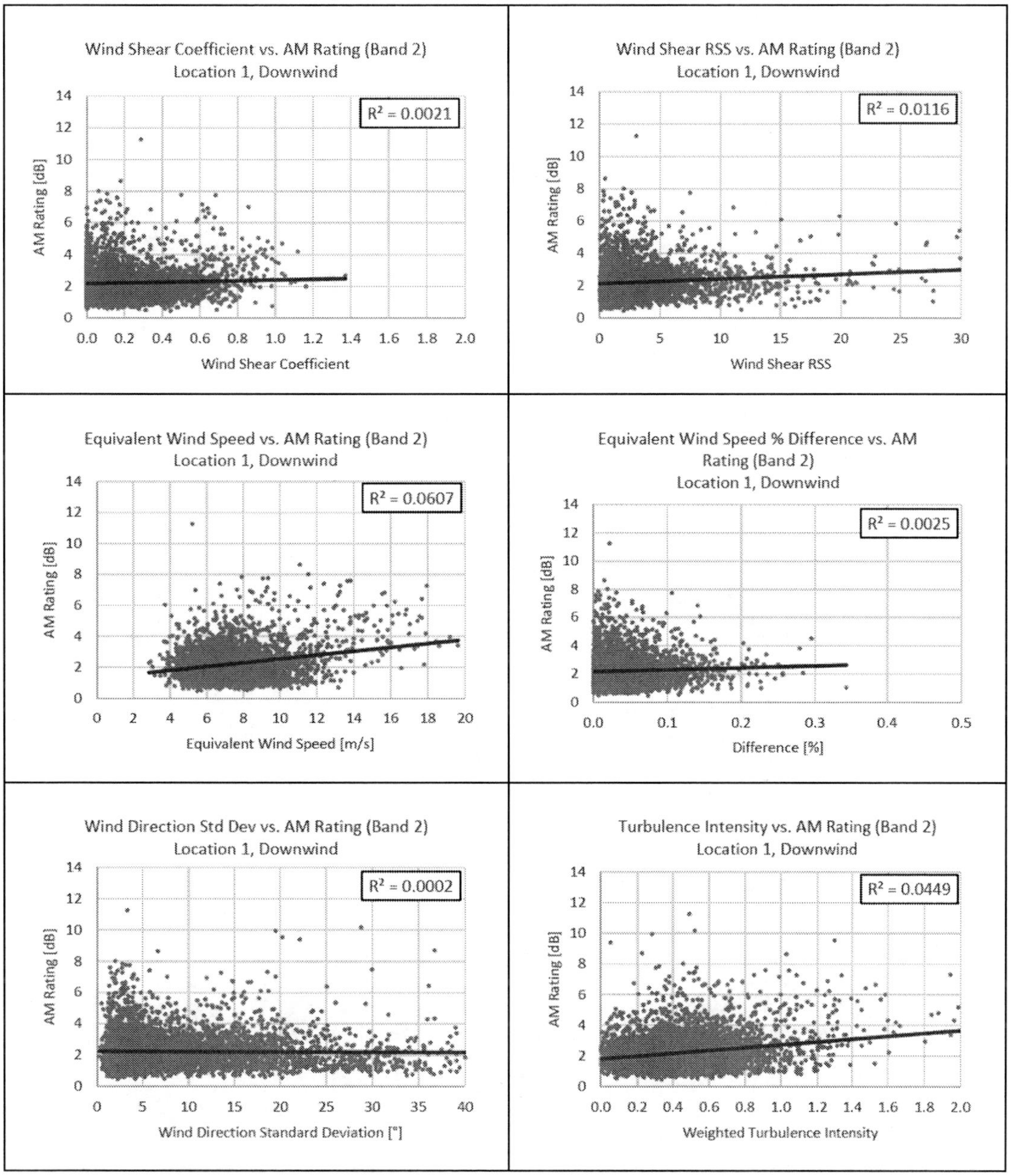

Figure 5: Correlation Between AM Rating and Meteorological Properties. Measurement Location 1, 150 metres from Turbine, Downwind Conditions

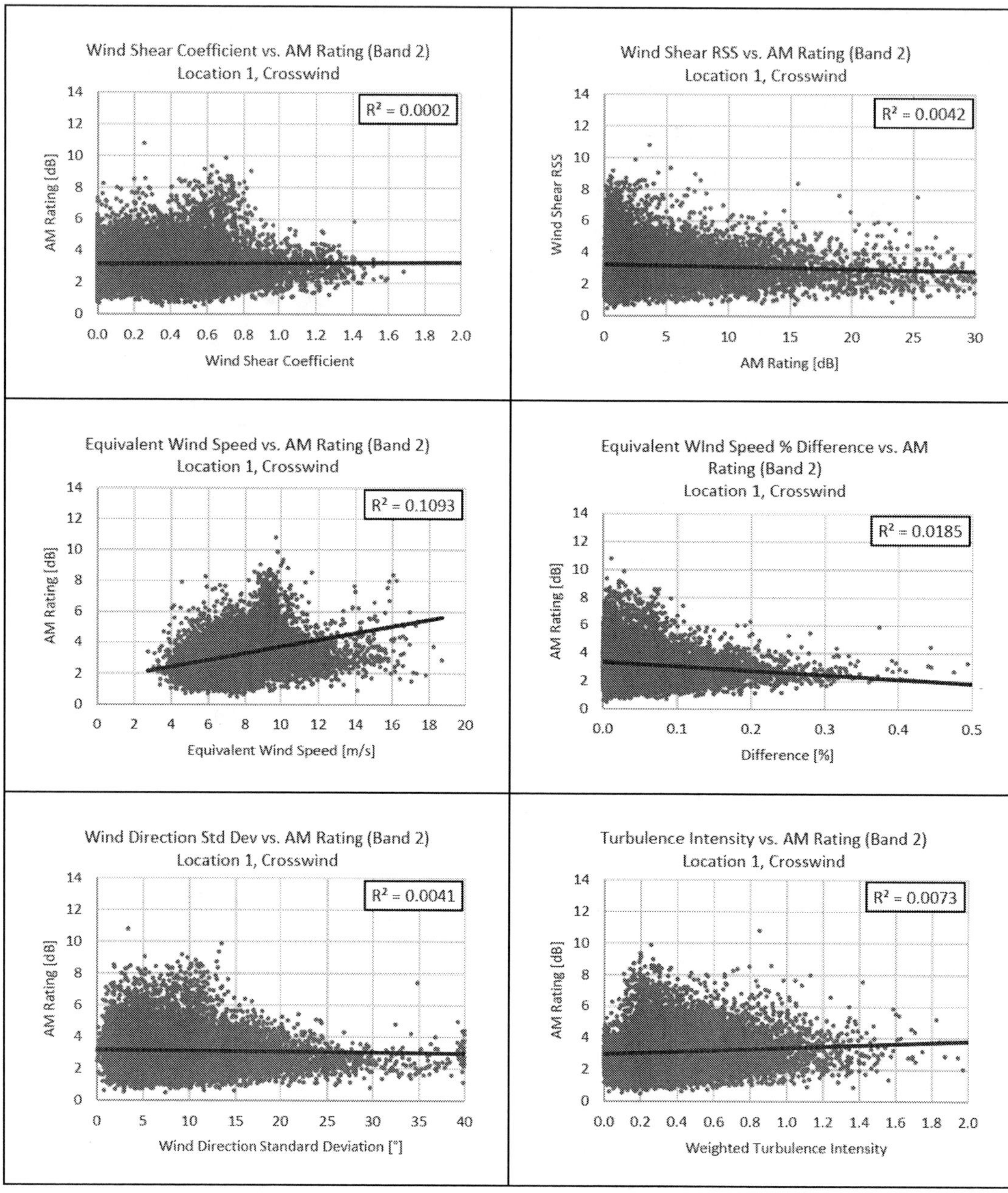

Figure 6: Correlation Between AM Rating and Meteorological Properties. Measurement Location 1, 150 metres from Turbine, Crosswind Conditions

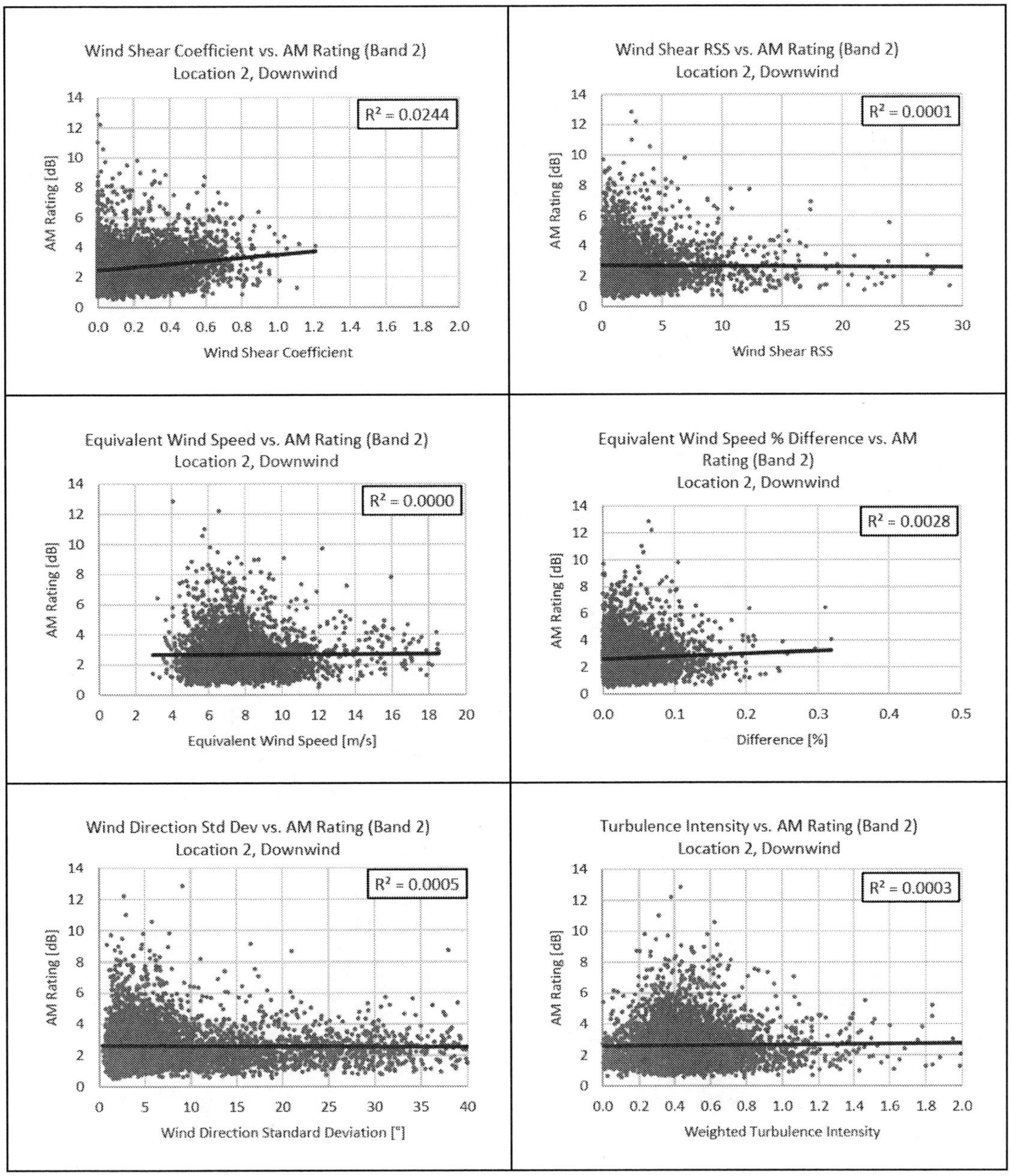

Figure 7: Correlation Between AM Rating and Meteorological Properties. Measurement Location 2, 440 metres from Turbine, Downwind Conditions

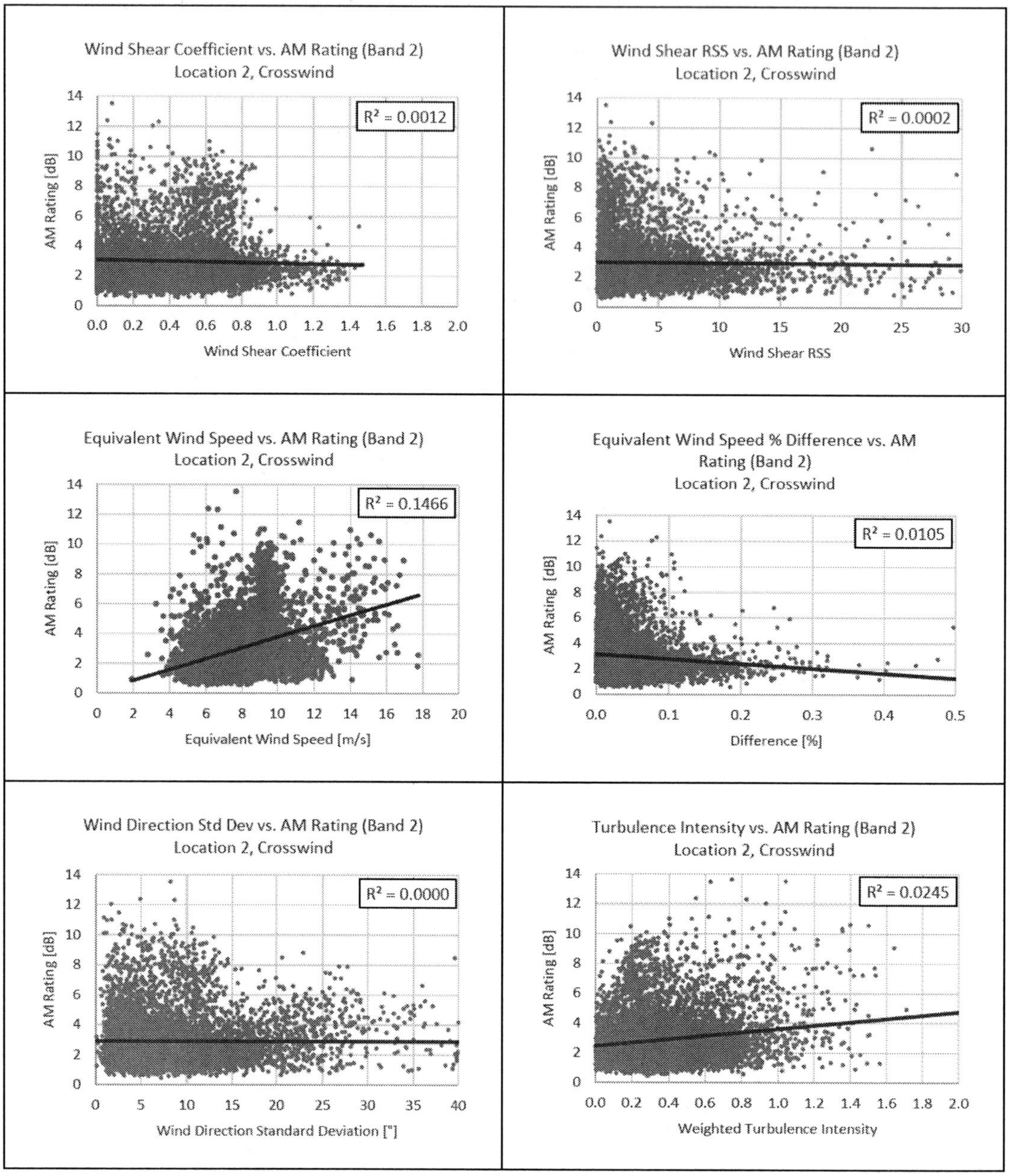

Figure 8: Correlation Between AM Rating and Meteorological Properties. Measurement Location 2, 440 metres from Turbine, Crosswind Conditions

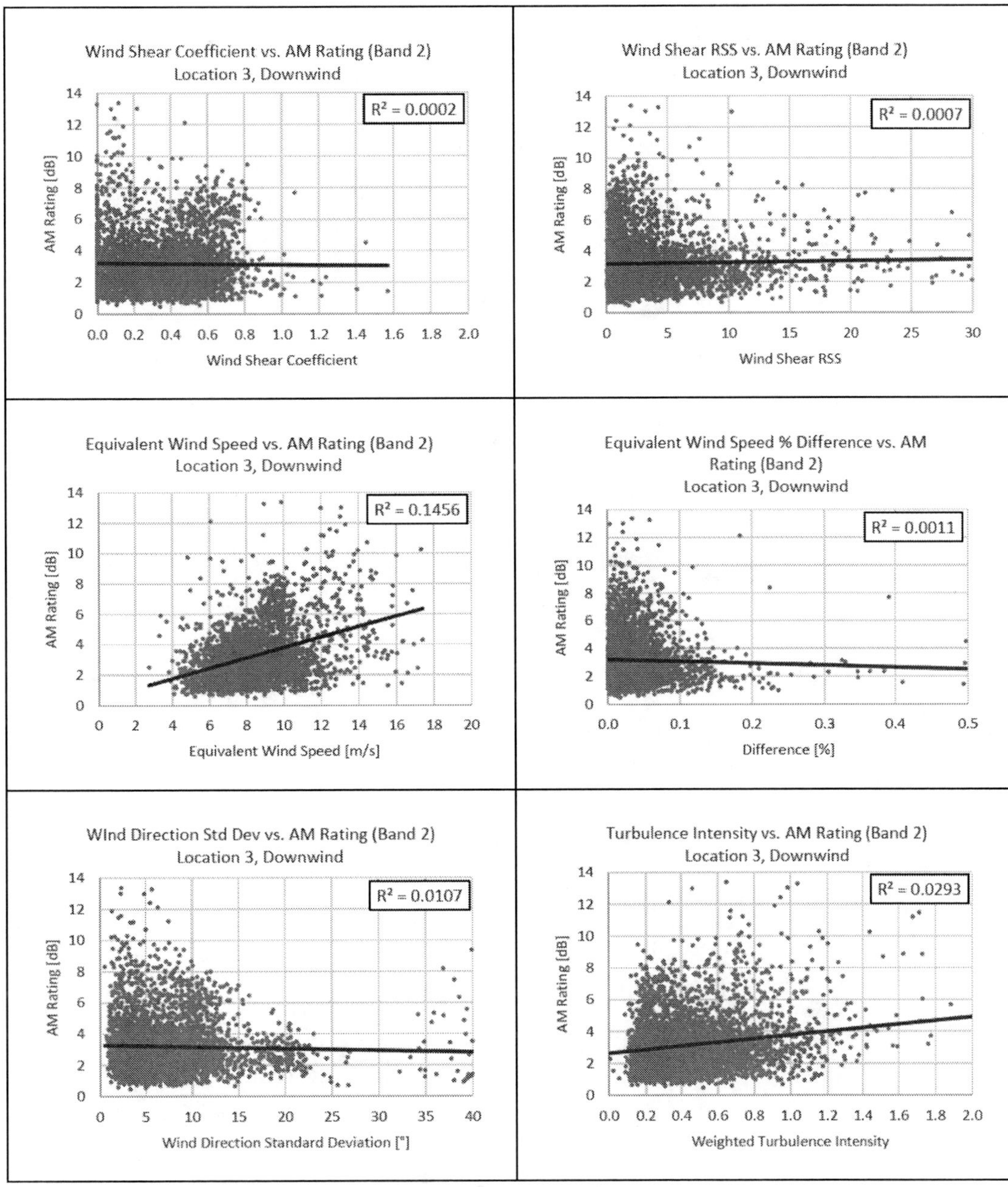

Figure 9: Correlation Between AM Rating and Meteorological Properties. Measurement Location 3, 440 metres from Turbine, Downwind Conditions

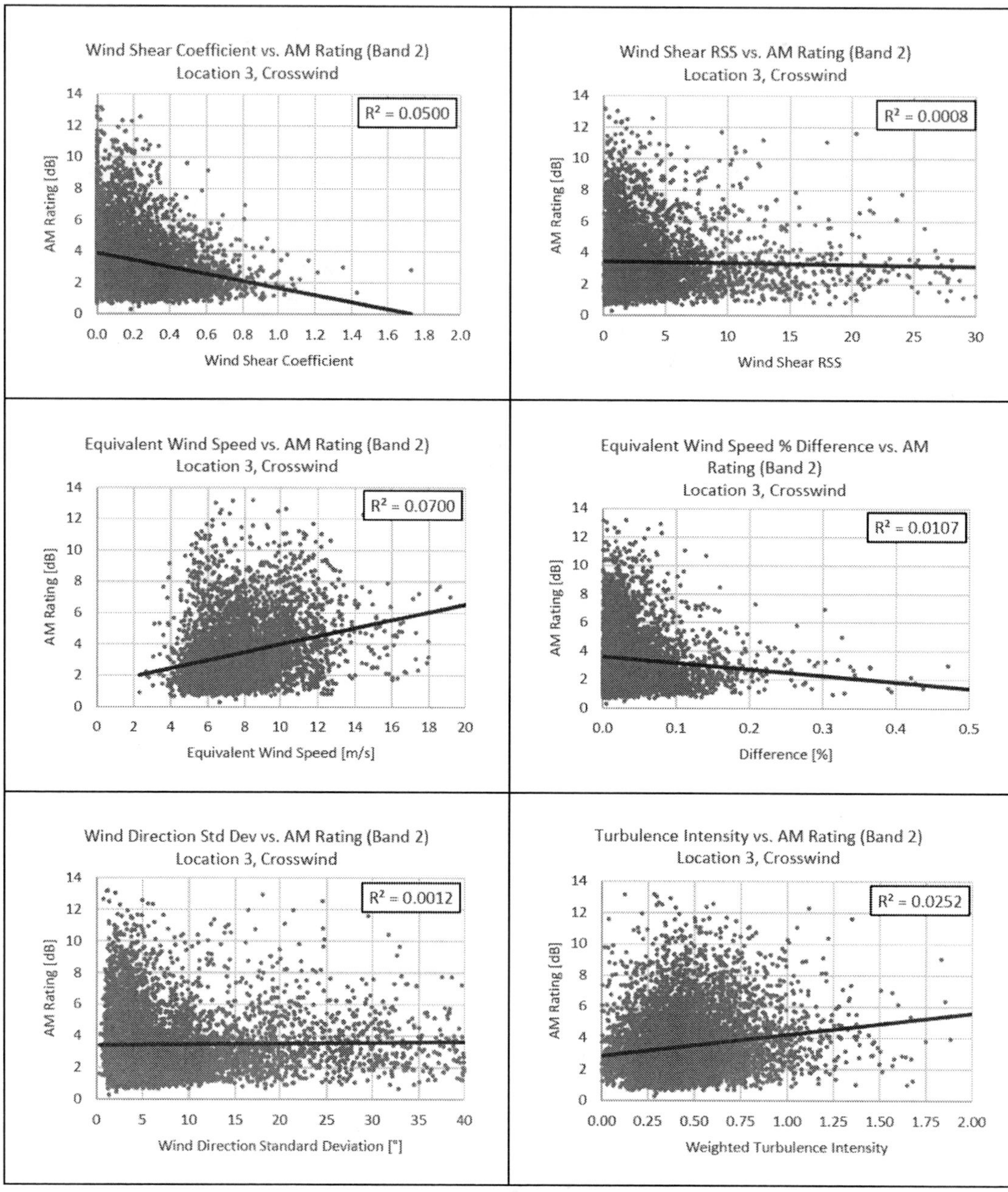

Figure 10: Correlation Between AM Rating and Meteorological Properties. Measurement Location 3, 440 metres from Turbine, Crosswind Conditions

As shown in the scatter charts (Figures 5 through 10), a wide range of short-duration (10 second) amplitude modulation ratings were found, up to about 10 dB in most cases.

No strong correlations between any of the considered meteorological metrics (wind shear coefficient, wind shear RSS, equivalent wind speed, equivalent wind speed percent difference, standard deviation of wind direction, and weighted turbulence intensity) were found. The correlation coefficients were all very low, with the strongest reaching only about 0.15.

Although subjective listening tests showed that most of the highest AM ratings were determined when subjectively strong modulation is apparent in the audio recordings, it is recognized that there are likely other sounds being identified by the IOA method as AM despite the introduction of various filters. This has the effect of contaminating the data to some extent, an inherent problem for this type of measurement.

It is noted that there appears to be a continuum from little or no AM to higher levels of AM. This observation reflects the difficulty in establishing a clear differentiation between AM and EAM. However, it is not clear to what extent the range of modulation ratings produced by the IOA method (or any other quantification method) is able to represent the range of subjective impressions that different wind turbine sounds produce.

The data is shown in Figures 5 to 10 is represented in a different way in Figures 11, 12, and 13, for Locations 1, 2, and 3, respectively, for both downwind and crosswind conditions. These figures illustrate the percentage of the data which passed the filters described above, having an AM rating greater than 3 dB, subdivided into sequential meteorological data bins.

It is recognised that the LIDAR data record is not perfect in that it contains gaps as a result inherent limitations of the technology, and the filters discussed in Section **4.3** result in other gaps in the data, and thus the "percentage of data" values must be interpreted as a fraction of the available data, rather than as a fraction of the total elapsed time.

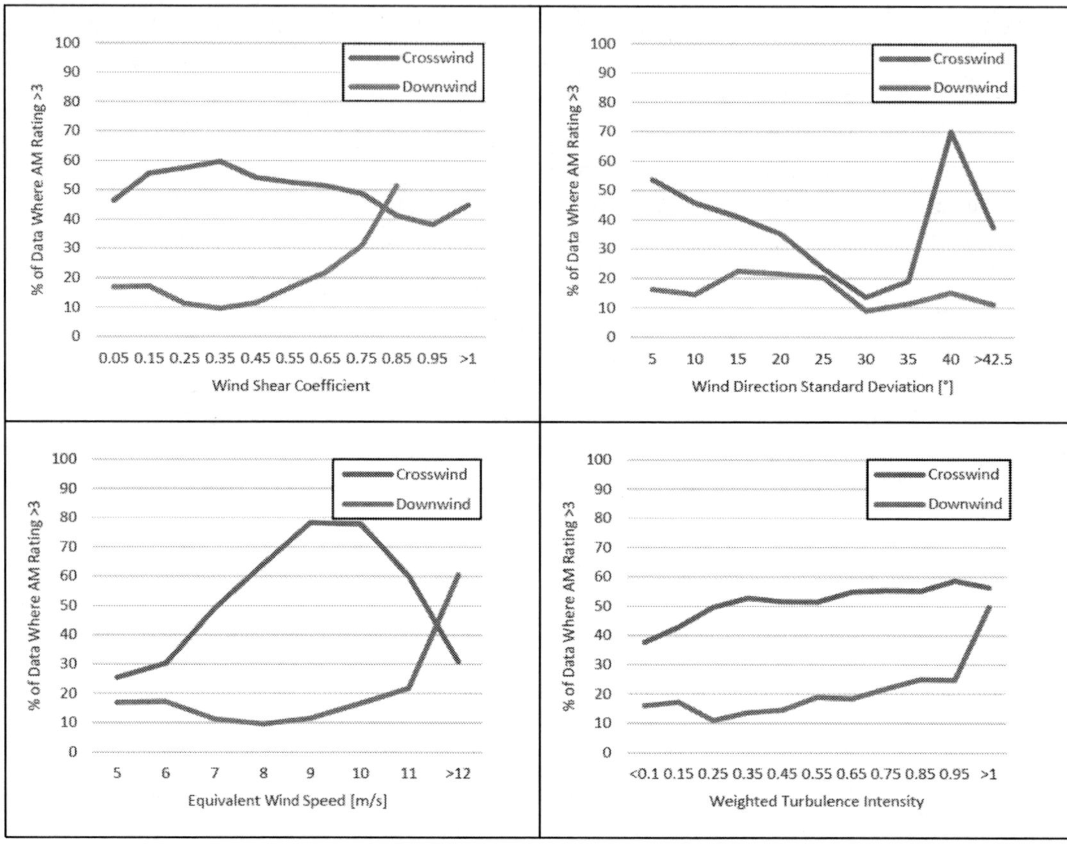

Figure 11: Comparison of High AM Ratings Crosswind versus Downwind and Meteorological Properties.
Measurement Location 1, 150 metres from Turbine

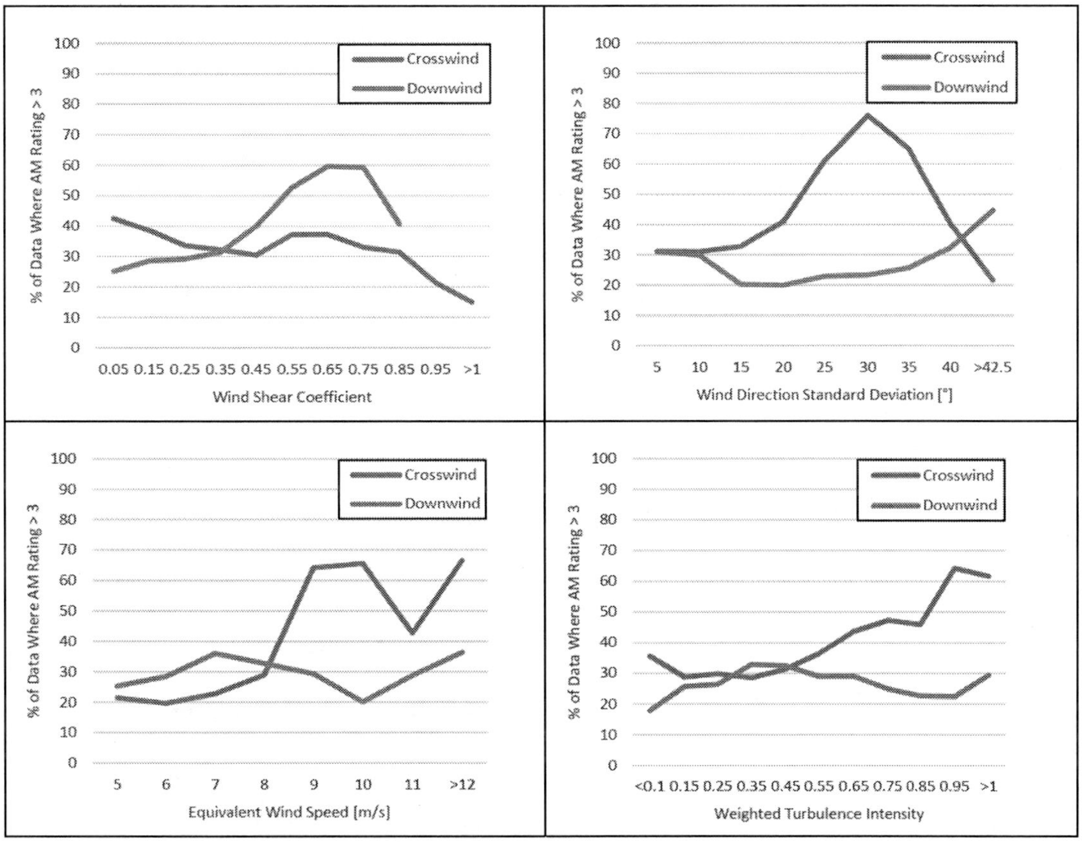

Figure 12: Comparison of High AM Ratings Crosswind versus Downwind and Meteorological Properties.
Measurement Location 2, 440 metres from Turbine

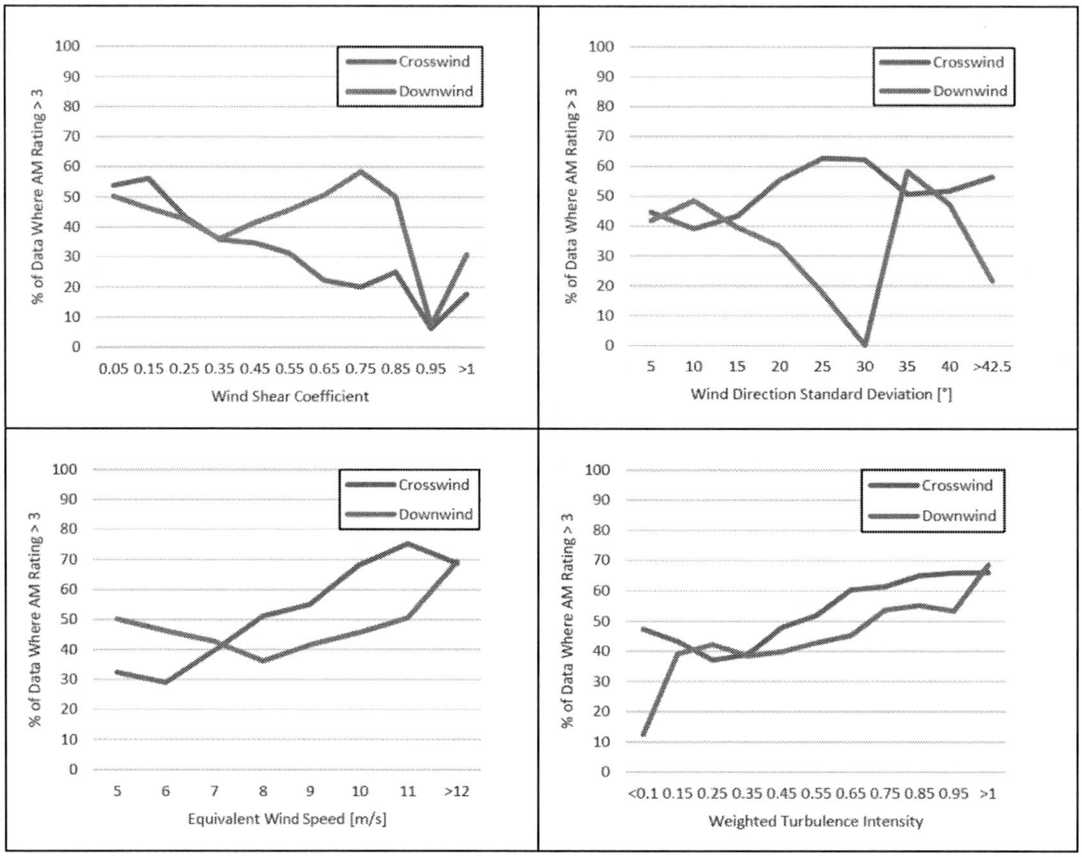

Figure 13: Comparison of High AM Ratings Crosswind versus Downwind and Meteorological Properties.
Measurement Location 3, 440 metres from Turbine

As shown in Figures 11 through 13, the data indicate a fairly consistent trend to greater AM ratings under crosswind conditions. There also appears to be an indication of some relationship between increasing weighted turbulence intensity and the prevalence of higher AM ratings.

6.0 Conclusions

No clear statistical relationships were found between the examined wind-derived meteorological properties and the AM rating as calculated by the IOA method.

Elevated AM ratings were found to be more common under crosswind conditions, and when examining at least the higher AM ratings, there appears to be some increase in the prevalence of higher AM rated periods with increasing weighted turbulence intensity.

Further work towards identifying conditions under which EAM is more prevalent could include a comparison between high resolution turbine parameters and the

AM ratings, and re-evaluating the wind-derived data with sensitivity to this turbine data.

References

[1] Pedersen, E. and Persson Waye K., *Perception and annoyance due to wind turbine noise - a dose-response relationship*, JASA, 116(6). 2004

[2] McCabe, N., *Detection and Quantification of Amplitude Modulation in Wind Turbine Noise through Envelope Analysis*, Research Thesis. 2010

[3] Institute of Acoustics, *Amplitude Modulation Working Group Discussion Document Methods for Rating Amplitude Modulation in Wind Turbine Noise*, version 7. 2015. Available at http://www.ioa.org.uk/sites/default/files/AMWG%20Discussion%20Document.pdf

[4] Legarth, S.V., *Auralization and Assessments of Annoyance from Wind Turbines*, Proceedings of the 2nd International Meeting on Wind Turbine Noise. 2007

[5] Cooper, J. and Evans, T., *Automated detection and analysis of amplitude modulation at a residence and wind turbine*, Proceedings of Acoustics 2013. 2013

[6] Cheong, C., Stalnov, O. and Joseph, P., *On the Measurement and Prediction of Wind-Turbine Swishing Noise*, Proceedings of the 6th International Conference on Wind Turbine Noise. 2015

[7] Fukushima, A., Kobayashi, T. and Tachibana, H., *Practical measurement method of wind turbine noise*, Proceedings of the 6th International Meeting on Wind Turbine Noise. 2015

[8] McCabe, N., *Detection and Quantification of Amplitude Modulation in Wind Turbine Noise,* Proceedings of the 4th International Meeting on Wind Turbine Noise. 2011

[9] Lee, S., Kim, K., Lee, S., Kim H. and Lee, S., *An estimation method of the amplitude modulation in wind turbine noise for community response assessment*, Proceedings of the 3rd Meeting on Wind Turbine Noise. 2009

[10] Larsson, C. and Öhlund, O., *Amplitude modulation of sound from wind turbines under various meteorological conditions*. JASA 135(1). 2014

[11] Irvine, G., *The Institute of Acoustics' Working Group on Amplitude Modulation in Wind Turbine Noise - Progress towards an agreed rating and assessment metric*, Proceedings of the 6th International Meeting on Wind Turbine Noise. 2015

[12] Institute of Acoustics, *IOA Noise Working Group (Wind Turbine Noise) Amplitude Modulation Working Group Final Report A Method for Rating*

Amplitude Modulation in Wind Turbine Noise. 2016. Available at http://ioa.org.uk/sites/default/files/AMWG%20Final%20Report-09-08-2016_1.pdf

[13] Wagner, R., Courtney, M., Gottschall, J. and Lindelöw-Marsden, P., *Accounting for the speed shear and in wind turbine power performance measurement*, Wind Energy 14(8). 2011

[14] Choukulkar, A., Pichugina, Y., Clack, C. T. M., Calhoun, R., Banta, R., Brewer, A. and Hardesty M., *A new formulation for rotor equivalent wind speed for wind resource assessment and wind power forecasting*. Wind Energy 19(8). 2015

[15] Yamartino, R. J., *A Comparison of Several "Single-Pass" Estimators of the Standard Deviation of Wind Direction*, Journal of Climate and Applied Meteorology, 23. 1984

[16] M. Cand, A. Bullmore, B. Willson, *Wind Turbine Amplitude Modulation: Research to Improve Understanding as to its Cause & Effect. Work Package D (WPD) – Measurement and Analysis of New Acoustic Recordings*, part of the Renewable UK report available at http://www.renewableuk.com/resource/collection/4e7cc744-fef2-473b-af2b-135ff2aa3a43/ruk_wind_turbine_amplitude_modulation_dec_2013_v2_(1).pdf?hhSearchTerms=%22%22work+package+d%22%22

[17] Öhlund, O. and Larsson, C. *Sound propagation from wind turbines under various weather conditions*, Proceedings of the 5th International Conference on Wind Turbine Noise. 2013

[18] Cand, M. and Bullmore, A., *Measurements demonstrating mitigation of far-field AM from wind turbines*, Proceedings of the 6th International Meeting on Wind Turbine Noise. 2015

[19] Cassidy, M. and Bass, J. *Addressing the Issue of Amplitude Modulation: A Developer's Perspective*, Proceedings of the 6th International Meeting on Wind Turbine Noise. 2015

7th International Conference
on
Wind Turbine Noise
Rotterdam – 2nd to 5th May 2017

Application of the UK IOA Method for Rating Amplitude Modulation

David Coles, 24 Acoustics Ltd, davecoles@24acoustics.co.uk
Tom Levet, Hayes McKenzie Partnership Ltd, tom@hayesmckenzie.co.uk
Matthew Cand, Hoare Lea, matthewcand@hoarelea.com

Summary

Following the work of the Amplitude Modulation Working Group (AMWG) for the UK Institute of Acoustics (IOA), a method for the quantification of amplitude modulation from wind turbines has been proposed. The method was developed to obtain a consistent and repeatable measure of the modulation depth characteristics of wind farm noise, which can be related to the psycho-acoustic response people experience. Details of the method are discussed and pertinent aspects highlighted.

Results are presented, discussing the analysis of noise measurements undertaken at residential receptor locations near wind farm sites.

1. Introduction

The Amplitude Modulation Working Group (AMWG) was established by the UK Institute of Acoustics (IOA) to derive a method for measuring and rating amplitude modulation (AM) in wind turbine noise.

Amplitude modulation (in this context) is a regular fluctuation in the level of noise, the period of fluctuation being related to the rotational speed of the turbine. AM is considered an inherent characteristic of wind turbine noise. However, a number of factors can give rise to an increase in modulation depth, which can cause specific complaints from residents neighbouring a wind farm. The characteristic of the sound might be described by a listener as a regular 'swish', 'whoomph' or 'thump', depending on the cause and severity of the modulation.

Given the varying severity and perceived annoyance of AM, it is vital to be able to rate AM in a robust and repeatable manner. This in turn allows policy makers to consider penalising levels of AM that are considered unacceptable.

The AMWG has developed a method to reliably identify the presence of amplitude modulated wind turbine noise within a sample of data and rate its magnitude.

The AMWG published a Discussion Document in April 2015 (IOA AMWG, 2015). Following publication, comments, observations and criticisms were received from interested parties. Taking input from the responses, a final 'Reference Method' was developed for adoption.

2. Method Strengths

The method proposed by the AMWG addresses a number of key issues that should be considered when assessing AM. The primary strengths of the method are described below.

2.1 Quantification of Amplitude Modulation

The ability to quantify the magnitude of AM is crucial. The metric obtained should be meaningful and representative. Where it is considered that levels of AM at a site are unacceptable, quantification of AM also allows the effect of any mitigation to be measured and therefore determine whether sufficient mitigation has been applied.

2.2 Repeatability – Minimal User Input

The method proposed by the AMWG is repeatable as it requires very little input from the practitioner prior to running the algorithm. The only required input for the processing algorithm is to define an allowable range for the fundamental frequency of modulation. This is straightforward to determine, as it is directly related to the rotor speed, and can therefore be calculated from the turbine specification.

2.3 Resistant to Extraneous Noise Sources

The method implements a number of techniques to minimise the effect of extraneous noise sources. These range from band-filtering the input data, to assessing the prominence of spectral peaks in the frequency domain. As such, many samples that are corrupted by extraneous noise (and would usually result in large false-positive values) are rejected by the method and not assigned a value for AM. This significant and effective reduction in false-positives comes with minimal introduction of false-negatives.

2.4 Meaningful Results Can be Obtained Quickly

The assessment of noise from wind turbine sites usually involves analysis of large datasets, spanning weeks or months. Since the AMWG method rejects corrupted noise samples (along with those samples containing no sustained modulation), it allows the practitioner to process large datasets and obtain meaningful results quickly. This enables issues to be addressed and resolved more efficiently.

Notwithstanding the strengths outlined above, it is still essential for the practitioner to exercise professional judgement and review any dataset with an appropriate level of scrutiny. Following the processing of the data, user input is required in the form of a verification process, to ensure identified periods are wind farm related and not affected by other modulating sources. It is possible that other sources in the local environment may be modulating at frequencies similar to the blade passing frequency, and in the same acoustic range, e.g. a dog barking, or a pigeon cooing.

3. Method Overview

The proposed method is a 'hybrid' approach. The modulation depth is calculated in the time domain, while the frequency domain is used to discriminate wind turbine AM and reject samples corrupted by extraneous noise sources (or those containing no obvious modulation).

An overview of the method is presented here, and some key aspects are highlighted. Full details of the method are described in the report published by the IOA AMWG (2016), which should be read by anyone considering implementing this method.

The principal output from the method is a series of 10-minute values representing modulation depth. The 10-minute values are calculated from a sequence of 10-second results. Analysis of each 10-second block comprises the following:

- Band-filtering the input data to focus the analysis on frequencies associated with wind turbine AM;
- Using Fourier analysis to assess the power spectrum and remove energy not associated with the fundamental modulation frequency (which itself should be related to the wind turbine(s));
- Performing an inverse Fourier transform to provide a 'clean' time-series containing energy only at the fundamental modulation frequency (and associated harmonics);
- Calculating the modulation depth by subtracting the L_{95} from the L_5 of the reconstructed time-series.

A key strength of this method is its ability to reject samples corrupted by extraneous noise sources. The techniques used to achieve this, along with other pertinent details, are described below in Section 4.

4. Pertinent Details of the Method

Section 3 provides an overview of the method proposed by the AMWG and highlights the simple principles upon which the method is based. However, the sophistication of the method is contained within the details. Key aspects of the method are highlighted below, however, as mentioned above, the AMWG report should be referred to for full details of the procedure. The reader will note that 10-second samples can be rejected at various stages of the analysis, as described below. The effect of these rejections is realised when calculating the 10-minute values, and forms a fundamental role in the method's ability to discriminate genuine AM. This is detailed further in Section 4.6.

4.1 Band-Filtering Input Data

The input signal for the method is a time-series of band-limited, A-weighted, 1/3 octave L_{eq} data in 100 millisecond samples. The following three frequency ranges (which each encompass seven 1/3 octave bands) are defined:

- 50 to 200 Hz
- 100 to 400 Hz (reference)
- 200 to 800 Hz

The seven 1/3 octave bands should be A-weighted and then summed logarithmically into a single band-passed stream of data for input to the method.

Focussing on a limited frequency range dominated by modulation, assists in both the identification of AM and in excluding spurious data. It also results in higher levels of AM compared to those obtained from broadband (A-weighted) analysis. In fact, the band-limited data can detect AM which might have been masked using a broadband analysis based on overall L_{Aeq} values.

4.2 Fourier Transform

A standard Fourier transform is applied to the input time-series to transform the data into the frequency domain and obtain a modulation spectrum. An important distinction from some frequency-domain based methods, such as that proposed by RUK (2013), is that both the real and imaginary parts of the Fourier output are retained. The full complex output contains phase information and is used later in the analysis to transform the data back into the time-domain.

The input to the Fourier transform is a 10 second block of 100 ms L_{eq} samples. This results in a frequency resolution of 0.1 Hz, and a maximum resolved frequency of 5 Hz. This places a limit on the maximum modulation frequency that can be assessed using this method – since three harmonics are considered, a maximum fundamental frequency of 1.6 Hz can be assessed. This translates to a rotor speed of 32 RPM for a 3-bladed turbine.

The output of the Fourier transform is converted to a power spectrum using equation 1:

$$S = \frac{|F\{x\}|^2}{n^2} \tag{1}$$

where $F\{x\}$ is the output from the Fourier transform, and n is the length of input data (100, in this case). Analysis of the power spectrum is performed to determine whether the sample contains valid AM. Pertinent details of this analysis are described below in Section 4.3.

4.3 Analysis of the Power Spectrum

A typical power spectrum for a sample containing AM is shown below in Figure 1.

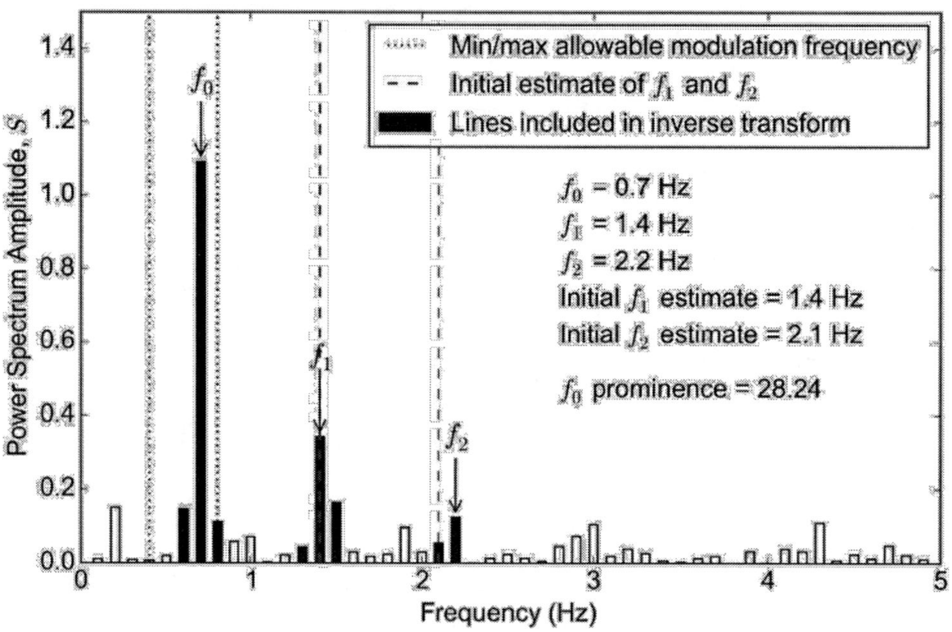

Figure 1 – Power spectrum of a sample containing AM. The positions of the first three harmonics (f_0, f_1, and f_2) are shown, along with the prominence of the fundamental peak.

The first stage of analysis is to find the highest peak within the allowable range for the fundamental modulation frequency (as set by the practitioner). A peak is simply defined here as a local maximum. If a peak is not found within the allowable range, this is a clear indication that the sample has been corrupted (or doesn't contain any notable modulation) and the 10-second sample is rejected from the analysis.

Once a fundamental frequency of modulation has been found, the location of associated harmonics is determined close to the multiples of the fundamental frequency. The method for doing this is described in the AMWG report.

The identification of a peak in the allowable range is not necessarily an indication that the sample contains genuine wind turbine AM. It is possible to greatly reduce the number of false positives by assessing the prominence of the peaks in the power spectrum. This exploits the fact that genuine wind turbine AM produces pronounced peaks in the power spectrum. Figure 1 shows a sample containing high modulation, which produces a very clear peak at the fundamental frequency of modulation (0.7 Hz). Figure 2 shows the power spectrum of a sample containing no notable modulation. There are clearly local maxima within the allowable range of fundamental modulation frequencies. However, none of the identified peaks identified are 'prominent' relative to the neighbouring spectral frequencies and this sample should not be considered further in the analysis.

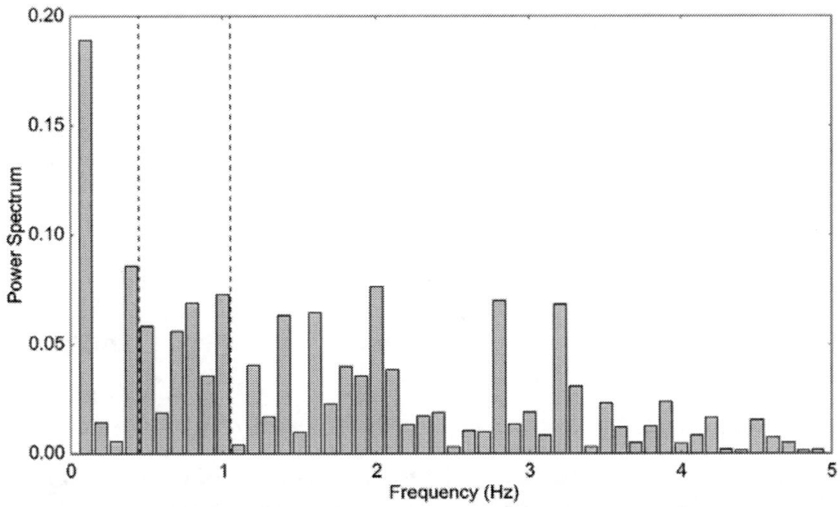

Figure 2 – Power spectrum of a sample containing no AM. Although local maxima are found within the allowable range of fundamental modulation frequencies (marked by the dashed vertical lines), none of the peaks are considered prominent.

The AMWG has proposed a means of determining the prominence of a peak within a power spectrum. This forms a critical part of the analysis and greatly reduces the number of false positives. The method is described below:

1. The magnitude of the fundamental peak, L_{pk}, is taken as the amplitude of a single line in the power spectrum at the frequency of the peak;
2. The two lines either side of the peak are ignored;
3. The masking level, L_m, is taken as the linear average of two lines each side of the peak (beyond those lines immediately adjacent to the peak);
4. The prominence, p, of the peak is calculated using:

$$p = \frac{L_{pk}}{L_m} \quad [2]$$

An example clarifying the classification of masking lines in the power spectrum is shown below in Figure 3. The lines adjacent to the peak are ignored. The masking lines are the two lines beyond the adjacent lines either side of the peak.

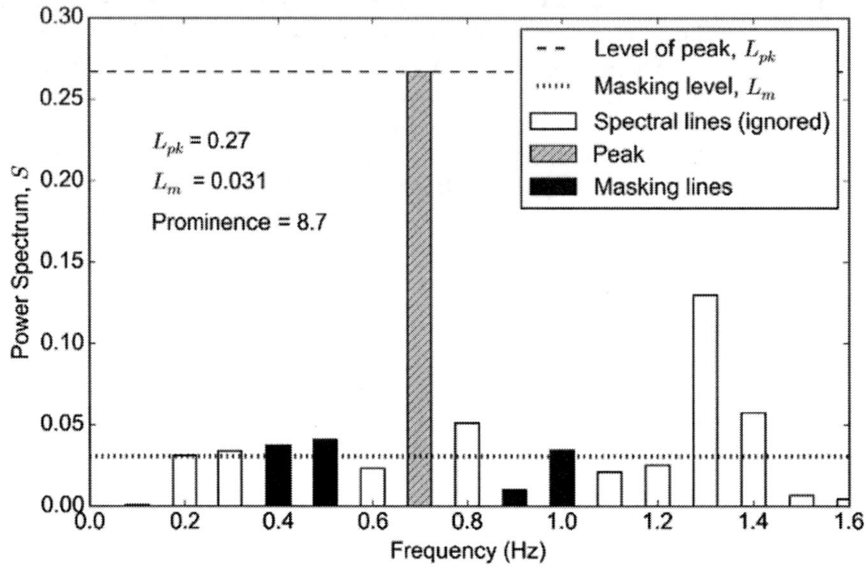

Figure 3 – An example calculation of the peak prominence.

If the prominence of the peak is less than a value of four, the 10-second sample is rejected from the analysis.

4.4 Inverse Fourier Transform

Analysis of the power spectrum is used to identify the frequencies of interest (namely the fundamental and the next two harmonics). However, after these frequencies have been identified, the rest of the analysis is performed on the original output of the Fourier transform (containing real and imaginary components) rather than the power spectrum. For each harmonic identified, three lines in the Fourier output are retained (the centre line, and one line either side). Lines for the corresponding negative frequencies are also retained. All other values in the Fourier output are set to zero. The inverse Fourier transform is then performed on this array (which should only contain energy associated with the fundamental frequency of modulation and its main harmonics). This is clarified in Figure 4 below.

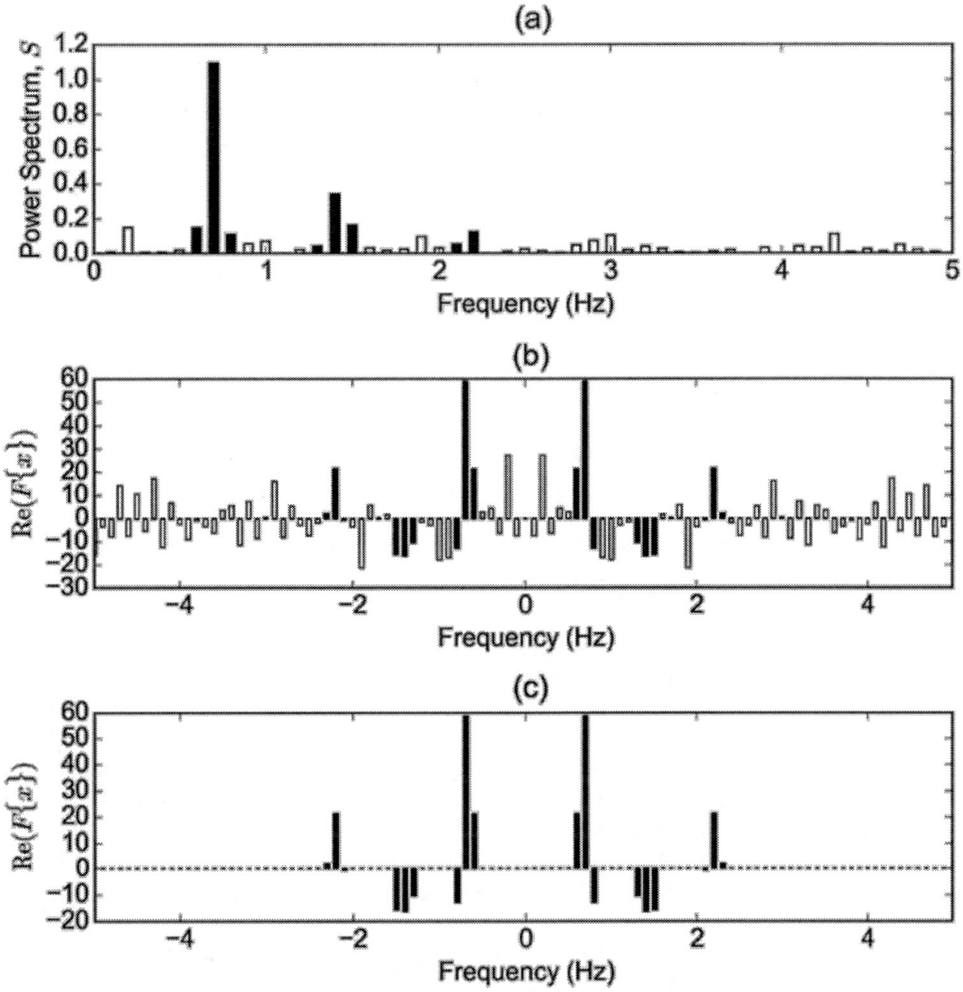

Figure 4 – A clarification on indices to include in the inverse Fourier transform. Panel (a) shows the power spectrum and the identification of indices to include. Panel (b) shows the original output from the Fourier transform (only the real part is shown here) with the identified indices shown as black lines. Note that the complex conjugates are also shown as black lines (the negative frequency components). Panel (c) shows the array with the identified indices included, and zeros at all other values. The inverse Fourier transform is performed on this array (note that the full array, including imaginary components, should be used – only the real part is shown here).

The result of the inverse Fourier transform should be a 'clean' version of the original time-series, containing only energy related to the fundamental frequency of modulation (and its main harmonics). An example is shown in Figure 5.

4.5 Determination of Modulation Depth

Once the reconstructed time-series has been generated, the modulation depth for the 10-second period is calculated simply by subtracting the L_{95} from the L_5, in a similar manner to Fukushima, Yamamoto et al. (2013). Calculating the modulation depth in this manner has the effect of weighting the value towards the highest modulation within the 10-second period.

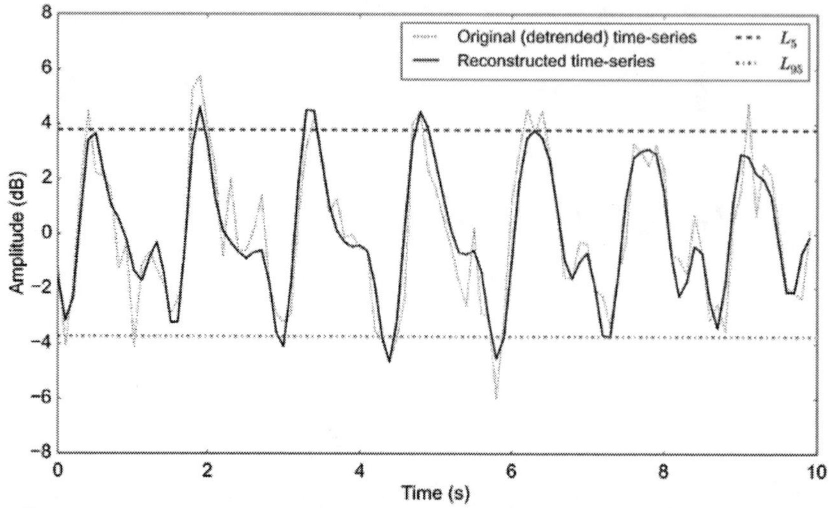

Figure 5 – The reconstructed time-series compared to the original (detrended) time-series. The modulation depth is calculated from the difference between the L_5 and L_{95} (both shown on the chart).

4.6 Calculation of 10-Minute Value

There are a number of possible ways to calculate a value for a 10-minute period from a sequence of (up to 60) 10-second results. One method would be to take the linear average, however, since the modulation within each 10-second period is averaged, averaging these results may undervalue the impact of AM within a 10-minute period. Another option is to take the maximum 10-second result within a 10-minute period. However, this would be very prone to spurious results and result in a value that is not robust from one 10-minute period to another. The AMWG method uses the 90th percentile (L_{10}) of the valid 10-second results. This is considered to represent the typical worst-case instances of AM within a 10-minute interval, without being excessively sensitive to possibly spurious extreme values. Figure 6 shows 10-second and 10-minute results for a 100 minute period, and gives an indication of where the 10-minute values sit within the spread of 10-second results.

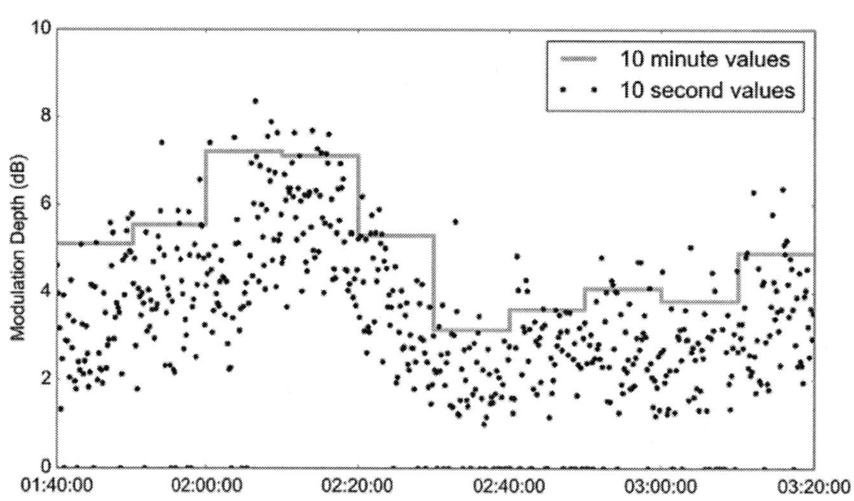

Figure 6 – A series of 10-second results and the corresponding 10-minute values.

It is important to note that only valid 10-second samples are used in the determination of the 10-minute value (some will have been discarded as detailed above). Furthermore, and critically, a value for AM is only assigned to a 10-minute period if there are at least 30 (i.e. 50%) valid 10-second results within that period. This criterion has been found to be a very effective indicator to exclude spurious data where little continuous AM attributable to wind turbines could be detected. In other words, this is an objective indicator of the presence of sustained wind turbine AM with varying magnitude. This criterion was chosen to be conservative, to minimise the risk of false exclusion of valid data, and so it is possible that some samples, i.e. 10-minute periods with more than 50% valid 10-second blocks still represent erroneous data (i.e. false positives). Conversely the 50% criterion will exclude isolated periods of sporadic/brief AM.

The effectiveness of the method to identify and quantify wind turbine AM (even in the presence of extraneous noise sources) is demonstrated in Section 5.

5. Application to Real-World Data

A method for rating wind turbine AM can only be considered fit for purpose if it produces meaningful results when applied to real-world data. The examples presented below demonstrate the effectiveness of the AMWG method in quantifying wind turbine AM and enabling a meaningful assessment to be undertaken efficiently.

5.1 Detecting Amplitude Modulation in the Presence of Noise

The 50% criterion, described above, is a very effective means of suppressing extraneous noise in a dataset. This is demonstrated in Figure 7, which shows a 24-hour dataset corrupted by sources of extraneous noise. Panels (b) and (c) show the difference made by applying the 50% criterion – in Panel (c), the extraneous noise is suppressed (no 10-minute AM values are reported) and ratings are assigned to 10-minute samples within the only period of the day in which the turbines were operational. This illustrates the effectiveness of the method.

5.2 Determining Prevalence of Amplitude Modulation

Determining the prevalence of AM is simplified by the AMWG method since results are not reported for periods which do not contain sustained modulation. It is possible to review data from longer noise surveys quickly and ascertain whether AM has occurred. Figures 8 and 9 demonstrate this – Figure 8 shows a one week period with a relatively high occurrence of AM, while Figure 9 shows a one week period with a relatively low occurrence of AM.

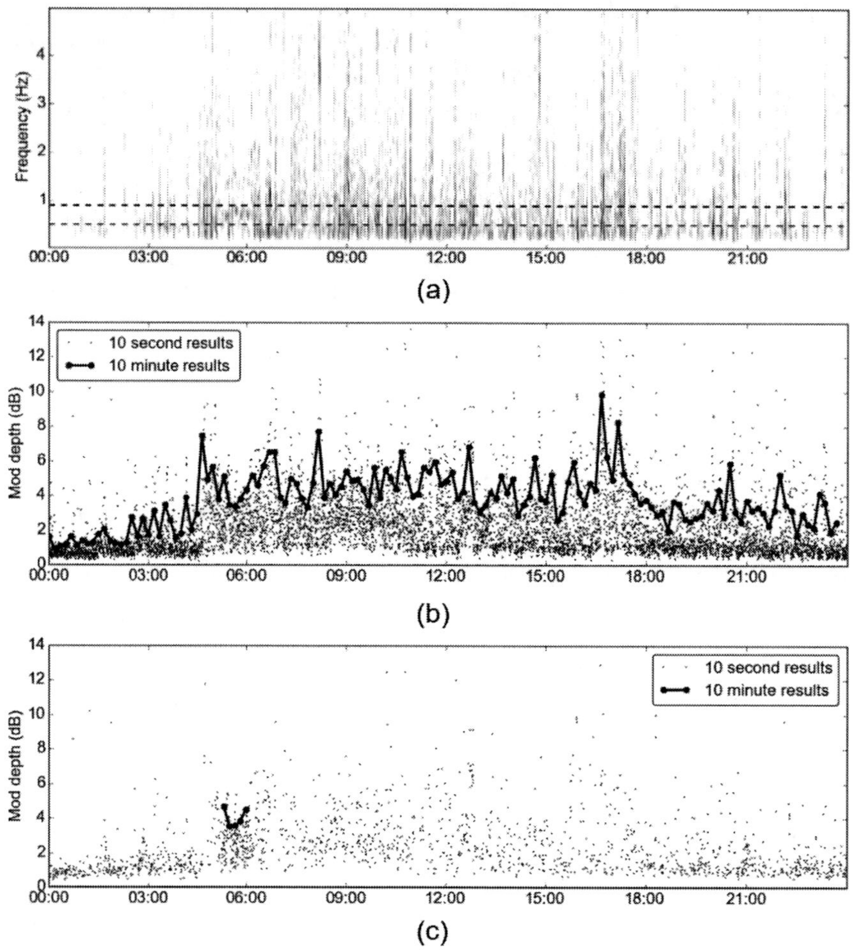

Figure 7 – Example of 50% criterion applied to data with a relatively large amount of corruption from non-turbine sources (birds, trees, etc.). Panel (a) shows a waterfall plot, which shows that there is only a consistent trend of modulation apparent in the expected modulation frequency range (shown by dashed lines) around 06:00. The 10-minute results are shown both without (b) and with (c) the 50% criterion applied. It was verified in this case that the only valid period in which 10-minute results are presented in (c) corresponds to the only period in which the turbines operated on that day.

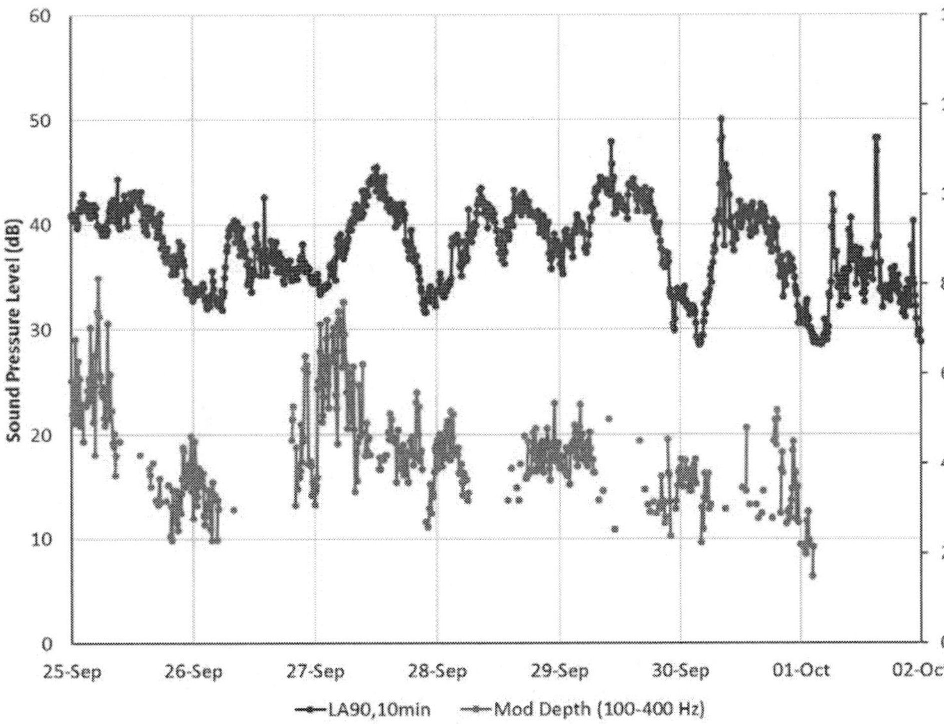

Figure 8 – A one week period with a relatively high occurrence of wind turbine AM. The chart shows data from the same survey location as Figure 9.

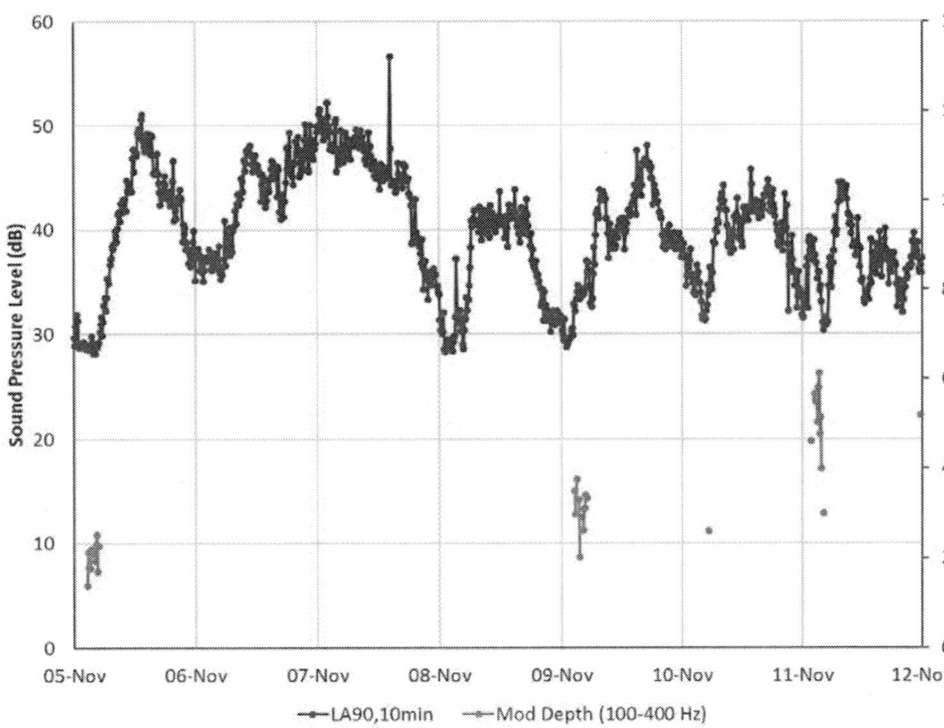

Figure 9 – A one week period with a relatively low occurrence of wind turbine AM. The chart shows data from the same survey location as Figure 8.

5.3 Determining Conditions For Mitigation

The full potential of the AMWG method is realised when noise data are combined with meteorological data to identify conditions under which AM occurs at any given site. This could form the basis of a mitigation scheme to reduce AM levels at relevant times. Figure 10 shows data analysed from a site where AM is found to occur. Panels (a) and (b) show the drastic effect of applying the 50% data filtering criterion – a large number of false positives are removed and two distinct regions are identified showing the conditions under which significant AM is occurring. This provides a substantial increase in efficiency when analysing large datasets, and enables conditions under which AM occurs to be identified quickly.

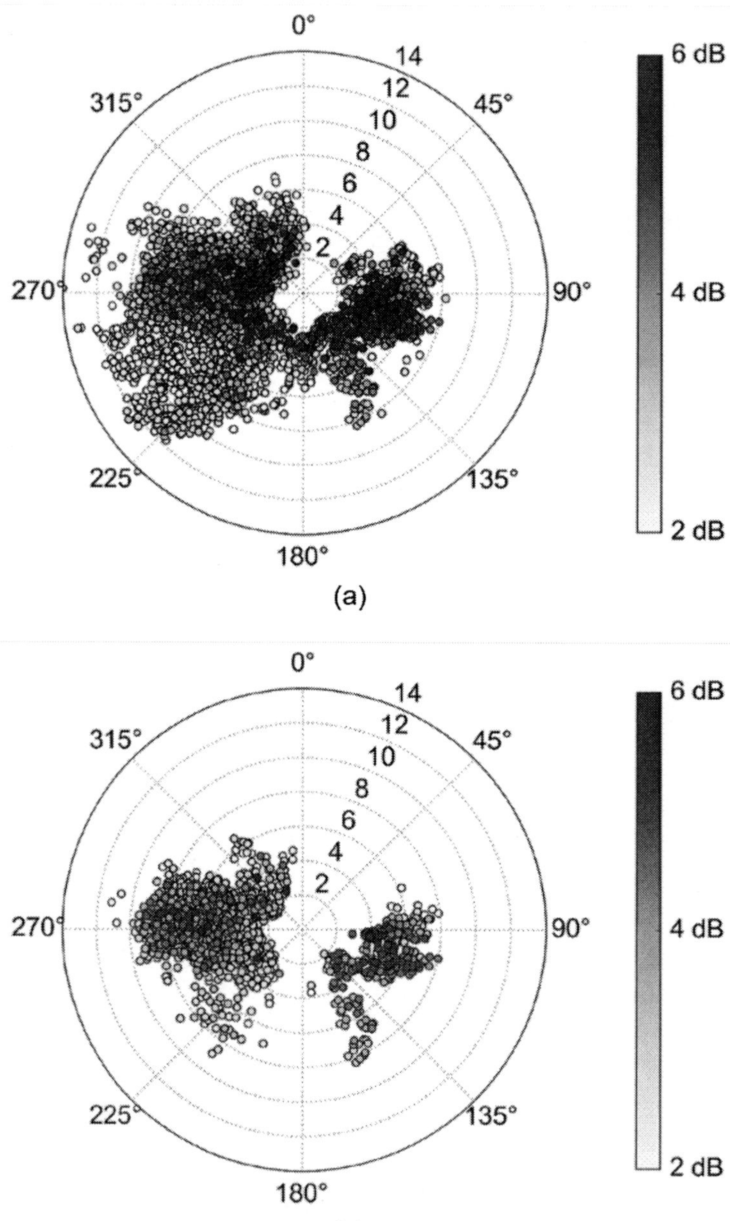

Figure 10 – A demonstration of the 50% criterion when applied to a dataset comprising noise and meteorological data. Panels (a) and (b) show the data with and without the 50% criterion respectively.

6. Conclusions

The IOA AMWG has published a method for rating wind turbine AM. It provides a meaningful and representative value of the modulation in measured signals, but requires 1/3 octave band measurements at 100 ms resolution. The method utilises a 'hybrid' approach, with the modulation depth being calculated in the time-domain, while filtering of extraneous noise sources is conducted in the frequency-domain. Numerous techniques are employed to minimise false positives and remove samples that are either corrupted or do not contain sustained modulation. The result is a robust and repeatable method for rating AM, which performs well when applied to real-world data and has the potential to significantly increase the efficiency of analysing large datasets.

References

Fukushima, A., Yamamoto et al. (2013) *Study on the amplitude modulation of wind turbine noise: Part 1 – Physical investigation.* Proc. Internoise 2013

IOA Amplitude Modulation Working Group (2015) *Discussion Document, Methods for Rating Amplitude Modulation in Wind Turbine Noise.*

IOA Amplitude Modulation Working Group (2016) *A Method for Rating Amplitude Modulation in Wind Turbine Noise.*

RenewableUK (2013) *Wind Turbine Amplitude Modulation: Research to Improve Understanding as to its Cause and Effect.*

7th International Conference
on
Wind Turbine Noise
Rotterdam – 2nd to 5th May 2017

Comparison of the IOA method and Japanese F-S method for quantitative assessment of amplitude modulation of wind turbine noise – A study based on the field measurement results in Japan

Akinori Fukushima, NEWS Environmental Design Inc.: fuku-new@wonder.ocn.ne.jp
Hideki Tachibana, Professor Emeritus, The University of Tokyo, Japan.

Summary

Amplitude modulation sound is generally contained in wind turbine noise (WTN), and often causes serious annoyance in the neighboring areas around wind farms. Therefore, it is very important to establish the method to evaluate the extent of amplitude modulation. In actual conditions, WTN varies slowly being affected by meteorological conditions and the depth of the amplitude modulation (peak-trough difference) varies and its time-averaged level also fluctuates slowly as a trend. To deal with amplitude modulation sound with such temporal characteristics, the authors proposed the "F-S method" in which the level difference between the A-weighted sound pressure levels obtained by Fast time-weighting and Slow time-weighting is firstly calculated and the amplitude modulation depth (D_{AM}) is evaluated as the 90% range of the level difference signal. On the other hand, the Institute of Acoustics (IOA) has published the final report regarding the assessment of amplitude modulation sound, in which rather sophisticated signal processing techniques are applied to assess the extent of amplitude modulation (the IOA reference method). By applying the F-S method and the IOA reference method to the WTN data obtained at 66 measurement points around 11 wind farms in Japan, the correspondence of the results obtained by the two methods were examined.

1. Introduction

Wind turbine noise (WTN , hereafter) generally contains amplitude modulation (AM, hereafter) sound, and it tends to make WTN more annoying. Therefore, it is necessary to assess the extent of AM quantitatively from both of physical and psycho-acoustical viewpoints [1,2,3].

The magnitude of WTN observed in the residential areas around wind farms fluctuates slowly caused by the change of meteorological conditions and it is a key point to eliminate such temporal fluctuation (trend) and to extract instantaneous alteration of sound pressure when assessing the strength of AM. As a simple method, the authors proposed a method (F-S method) in which the trend is assessed by using the Slow time-weighting of sound level meter and the AM component is detected as the level difference $\Delta L_A(t)$ between the sound pressure level (SPL, hereafter) by Fast time-weighting and that by Slow time-weighting [1, 2]. As the 90% range of $\Delta L_A(t)$, the AM depth (D_{AM}), is statistically estimated. As another method for rating AM in WTN, the Institute of Acoustics (IOA) has published a reference method to assess the AM rating (R_{AM}) [4], which is similar to the F-S method but much more sophisticated in signal processing and judgment procedures. To examine the compatibility between the F-S method and the IOA reference method, the D_{AM} and R_{AM} were examined by applying the two methods to the WTN data obtained at wind farm sites in Japan and they were compared.

2. WTN data used in this study

The authors have conducted a series of field measurements of WTN at 34 wind farm sites across Japan in 2010 to 2012 [5]. In the measurements, prototype wide-frequency-range sound level meters were used by attaching double-skin type wind screen sets to the measurement microphones. From the data obtained in these field measurements, 66 data recorded at around 11 wind farms were used in this study. Table 1 shows the wind farms under measurement, the rated output power, the number of wind turbines and the measurement distances. The recordings of one hour duration time which contained AM components were chosen from the data and were analysed by the F-S method and the IOA reference method. Figure 1 shows the A-weighted 1/3 octave band sound pressure level (SPL) spectra of the WTNs measured at the 66 points which are used in this study. On the whole, the A-weighted spectra are highest at the middle frequency range around 125 Hz – 1 kHz bands.

Table 1 - The specifications of wind turbines and measurement points

Wind farm	Rated output power [kW]	Numbers of WT	Measurement points	Horizontal distance from the nearest WT [m]
W01	1,980	1	6	250 - 470
W02	2,500	6	7	240 - 560
W03	2,300	8	6	370 - 1,150
W04	2,400	21	7	790 - 1,160
W05	1,500	9	3	210 - 370
W06	1,500	1	6	320 - 530
W07	1,980	1	6	660 - 1,240
W08	1,950	1	7	170 - 1,150
W09	1,995	1	6	270 - 720
W10	1,300	10	5	310 - 670
W11	1,300	7	7	680 - 900
11 wind farms	1,300 - 2,500	1 - 21	66 points	170 - 1,240

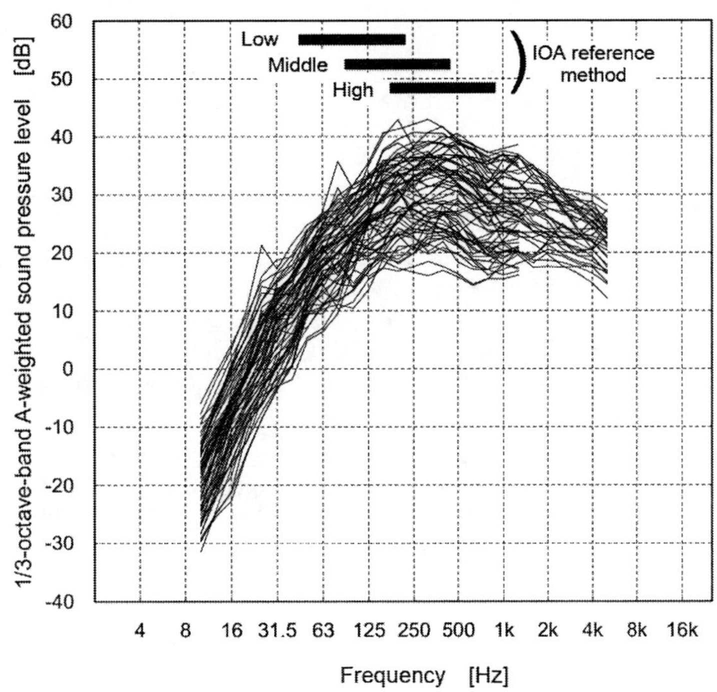

Figure 1 - The A-weighted 1/3 octave band SPL spectra of the WTNs measured at the 66 measurement points in residential areas around 11 wind farms in Japan.

3. Data processing

Table 2 shows the specifications of the data processing of the F-S method and the IOA reference method applied in this study. The survey period was one hour and data processing was conducted for every 10 seconds during the survey period in both of the two methods.

Figure 2 shows the comparison of the flow of the data processing by the two methods in detail. In the F-S method, data processing is very simple; the AM component $\Delta L_A(t)$ is simply obtained as the level difference between the instantaneous A-weighted SPL through F time-weighting $L_{A,F}(t)$ and that obtained through S time-weighting $L_{A,S}(t)$, and the Amplitude Modulation Depth (D_{AM}) is obtained as the 90% range of $\Delta L_A(t)$. On the other hand, in the IOA reference method, WTN signal is divided into three different frequency ranges (low: 50 – 200 Hz, middle: 100 – 400 Hz, and high: 200 – 800 Hz), and the A-weighted SPL time-series is obtained as L_{Aeq} for 100 ms, $L_{Aeq,100ms}(t)$ for each 7/3 octave band. To eliminate the effect of trend ("detrend"), the 3rd order polynomial of $L_{Aeq,100ms}(t)$, $L_{trend}(t)$, is calculated for every 10 s and the AM component $\Delta L_A(t)$ is obtained as the level difference between $L_{Aeq,100ms}(t)$ and $L_{trend}(t)$. Further, to make $\Delta L_A(t)$ more definite, "recreate" processing is performed through DFT (Discrete Fourier Transform) and IDFT (Inverse Discrete Fourier Transform), and finally, the AM rating (R_{AM}) is obtained as the 90% range of $\Delta L_{A,recreat}(t)$. In the IOA reference method, it is specified that the AM rating is performed for 10 minutes (the major time interval), but it was made for 10 s (the minor time interval) in this study in order to make simple comparison between the F-S method and the IOA reference method.

Table 2 - Comparison of the F-S method and the IOA reference method.

Items	F-S method	IOA reference method
Survey period	1 hour	
Minor time interval	10 seconds	
Frequency weighting	A	
Frequency range	Over-all	low : 50 Hz - 200 Hz middle : 100 - 400 Hz high : 200 - 800 Hz
A-weighted SPL	$L_{A,F}(t)$, $L_{A,S}(t)$	$L_{Aeq,100\,ms}(t)$ for each frequency range
Sampling interval of SPL	50 ms	100 ms
Presence of WTN	Aural check	Automatically
Elimination of BGN	Aural check	Automatically
Trend	$L_{A,S}(t)$	$L_{trend}(t)$: 3rd order polynomial of $L_{Aeq,100\,ms}(t)$
AM components $\Delta L_{AM}(t)$	$L_{A,F}(t)$ - $L_{A,S}(t)$	$L_{Aeq,100\,ms}(t)$ - L_{trend}
Frequency components of $\Delta L_{AM}(t)$	Over-all	Fundamental (BPF)$\pm\Delta f$ (2nd harmonic$\pm\Delta f$) (3rd harmonic$\pm\Delta f$)
AM rating index	D_{AM} : 90% range of $\Delta L_{AM}(t)$	R_{AM} : 90% range of recreate filtered time series of frequency components

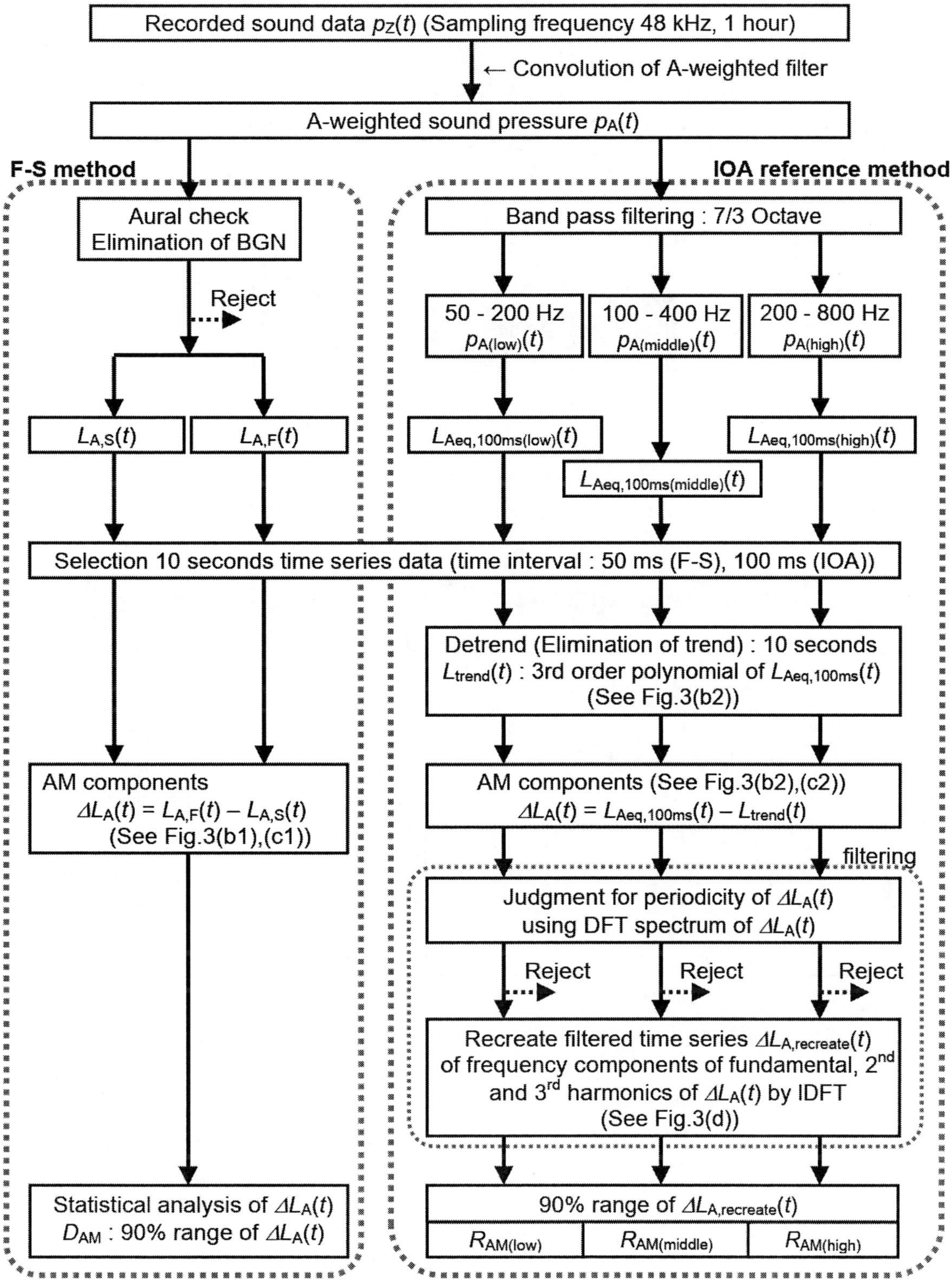

Figure 2 - Procedures of determining AM depth D_{AM} by the F-S method and AM rating R_{AM} by the IOA reference method of AM sound.

4. Results

4.1 Detection of AM component from WTN

Figure 3 shows an example of the process for detecting AM components by the two methods. Figure 3(a) shows the A-weighted sound pressure of a WTN for 80 seconds recorded at a measurement point. Figure 3(b1) shows $L_{A,F}(t)$ (blue) and $L_{A,S}(t)$ (red). In the F-S method, the AM component $\Delta L_A(t)$ is simply obtained as $L_{A,F}(t) - L_{A,S}(t)$ as shown in Figure 3(c1), and the

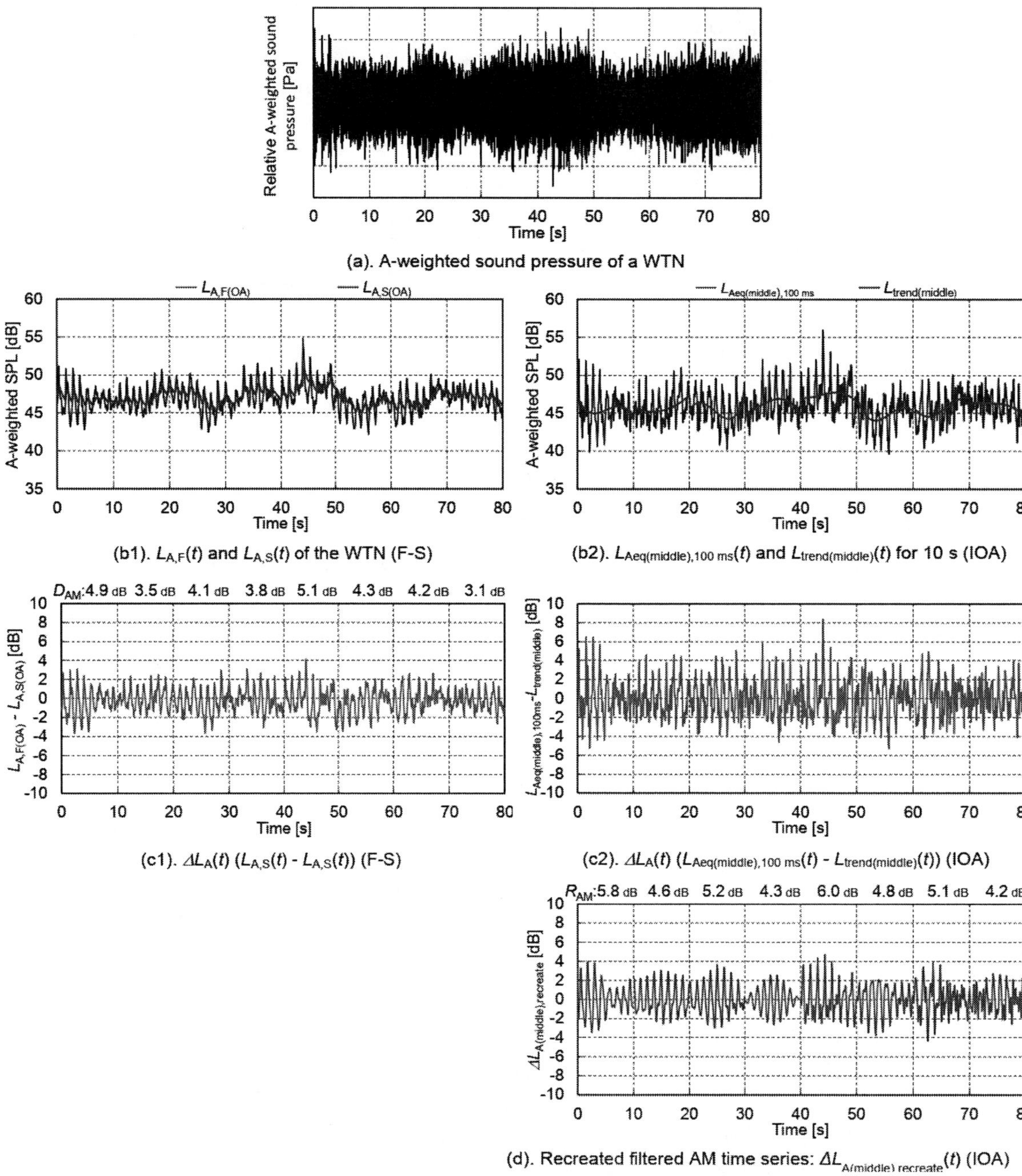

Figure 3 - Examples of WTN and AM components.

AM depth (D_{AM}) is estimated as the 90% range of the $\Delta L_A(t)$. On the other hand, Figure 3(b2) shows the time history the A-weighted equivalent continuous SPL for 100 ms for middle frequency range as an example, $L_{Aeq(middle),100ms}(t)$ (blue) and the trend $L_{trend(middle)}(t)$ (red) obtained by the IOA method. As the difference between $L_{Aeq(middle),100ms}(t)$ and $L_{trend(middle)}(t)$, the raw AM component $\Delta L_{A(middle)}(t)$ is obtained as shown in Figure 3(c2). Further, to make $\Delta L_{A(middle)}(t)$ more definite based on the blade passing frequency (BPF), the fundamental component and 2nd and 3rd harmonics of $\Delta L_{A(middle)}(t)$ are detected by DFT processing and $\Delta L_{A(middle),recreate}(t)$ shown in Figure 3(d) is obtained through IDFT using the fundamental and 2nd and 3rd harmonic components. Finally, the AM rating (R_{AM}), is obtained as the 90% range of $\Delta L_{A(middle),recreate}(t)$.

4.2 Correspondence between D_{AM} and R_{AM}

For all of the data, D_{AM} (by the F-S method) and R_{AM} (by the IOA reference method) for every 10 s were calculated according to the procedures shown in Figure 2. Although R_{AM} analysis for over-all A-weighted SPL is not included in the IOA reference method, it was also performed in this study for a reference. The correspondences between D_{AM} ($D_{AM,10s}$: only for over-all A-weighted SPL) and R_{AM} ($R_{AM(low),10s}$ for low frequency range, $R_{AM(middle),10s}$ for middle frequency range, $R_{AM(high),10s}$ for high frequency range, and $R_{AM(OA),10s}$ for over-all frequency range) are compared in Figure 4(a), (b), (c), and (d), respectively. Among these results, Figure (a) is relatively poor in correspondence, whereas the other cases Figures (b), (c) and (d) are in fairly

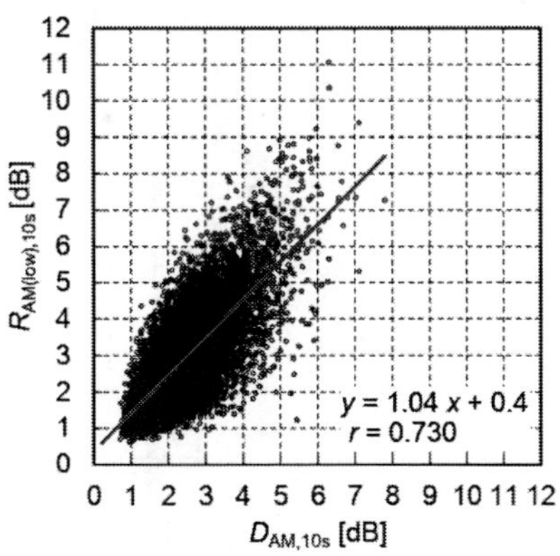

(a). R_{AM} for low frequency range vs. D_{AM}

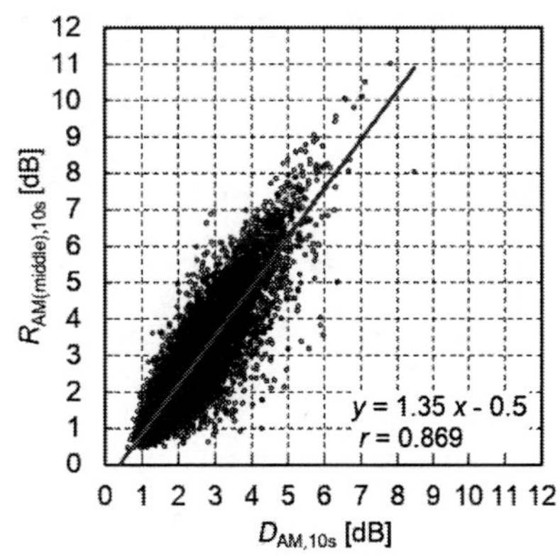

(b). R_{AM} for middle frequency range vs. D_{AM}

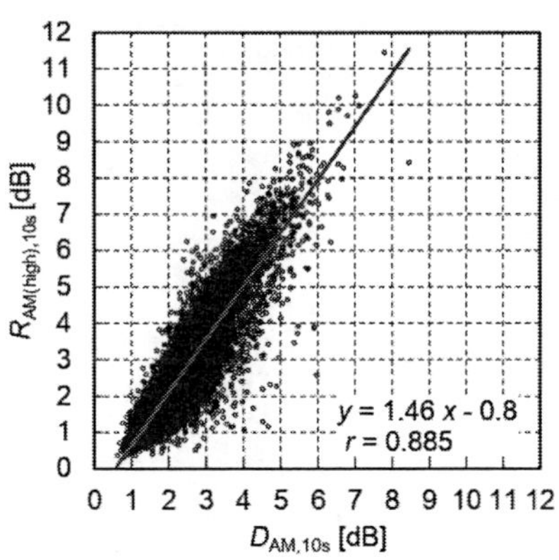

(c). R_{AM} for high frequency range vs. D_{AM}

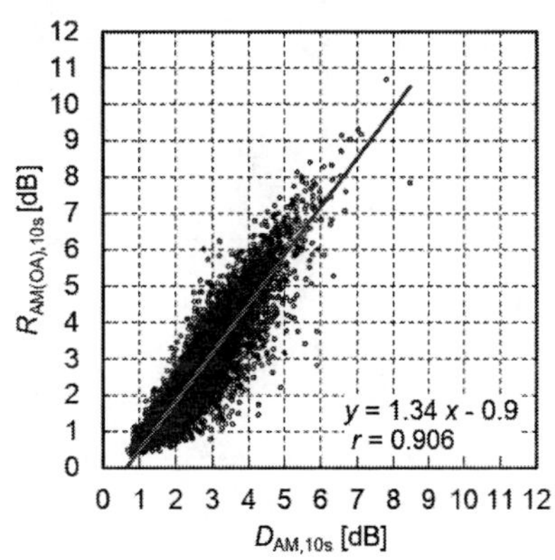

(d). R_{AM} for over-all frequency range vs. D_{AM}

Figure 4 - Comparison of D_{AM} (F-S method) and R_{AM} (IOA reference method) for 10 seconds.

high correspondence between D_{AM} and R_{AM}.

From the raw data shown above, the highest 10% values of D_{AM} and R_{AM} were obtained according to the procedure specified in the IOA reference method and compared as shown in Figure 5. In the cases of $D_{AM,10s}$ vs. $R_{AM(middle),10s}$, $D_{AM,10s}$ vs. $R_{AM(high),10s}$, and $D_{AM,10s}$ vs. $R_{AM(OA),10s}$, the correlation coefficients are more than 0.9 and standard deviations are less than 0.58 dB, whereas the correspondence between $D_{AM,10s}$ and $R_{AM(low),10s}$ is relatively poor compared to other three cases.

Next, the arithmetic mean values of D_{AM} and R_{AM} were obtained and compared as shown in Figure 6. Also in this case, $D_{AM,10s}$ vs. $R_{AM(middle),10s}$, $D_{AM,10s}$ vs. $R_{AM(high),10s}$, and $D_{AM,10s}$ vs. $R_{AM(OA),10s}$ are highly correlated with correlation coefficient of around 0.9 and standard deviation less than 0.5 dB. The correspondence between $D_{AM,10s}$ and $R_{AM(low),10s}$ is relatively poor in this case, too.

4.3 Comparison of AM components obtained by the two methods

The way to detect the AM component of WTN is very much different in the F-S method and the IOA reference method as mentioned earlier. To see this difference, the cross-correlation between $\Delta L_A(t)$ by the F-S method and $\Delta L_{A,recreate}(t)$ by the IOA reference method was analysed. As an example the cross-correlation analysis was performed using the WTN sound shown in Figure 3. The cross correlation functions between $\Delta L_A(t)$ by the F-S method (Figure 3(c)) and $\Delta L_{A,recreate}(t)$ by the IOA reference method for each 10 s for low frequency range, middle

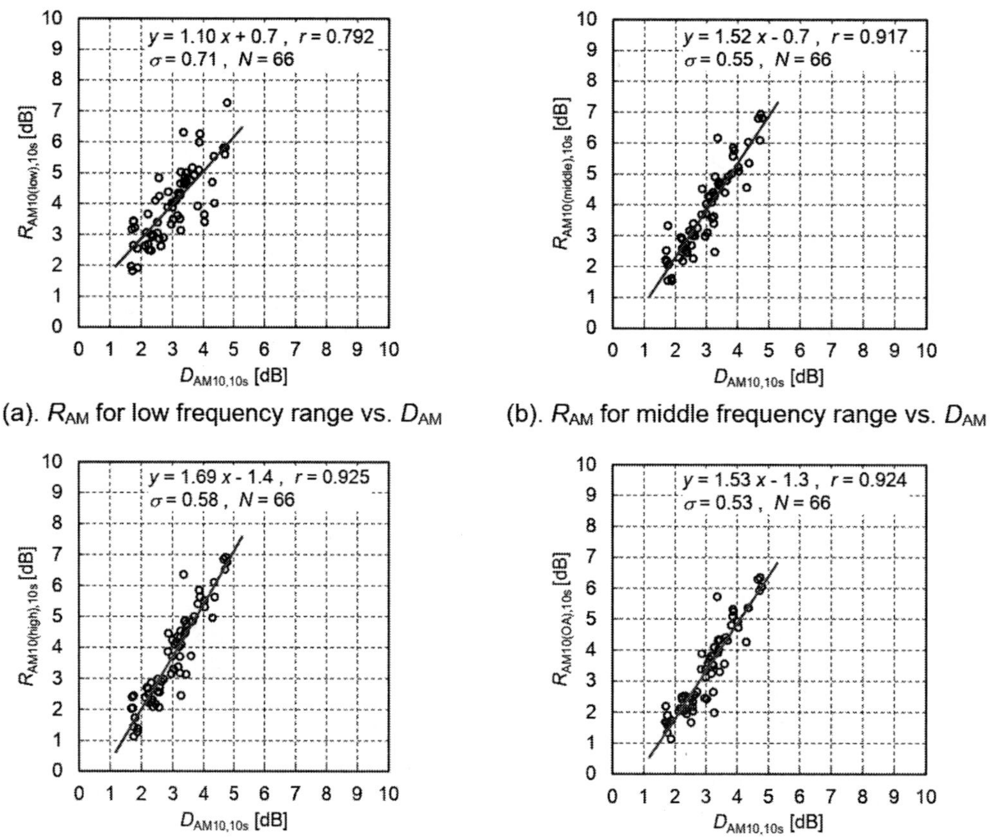

(a). R_{AM} for low frequency range vs. D_{AM}

(b). R_{AM} for middle frequency range vs. D_{AM}

(c). R_{AM} for high frequency range vs. D_{AM}

(d). R_{AM} for over-all frequency range vs. D_{AM}

Figure 5 - Comparison of the highest 10% values of the 10 s D_{AM} and R_{AM} for 1 hour.

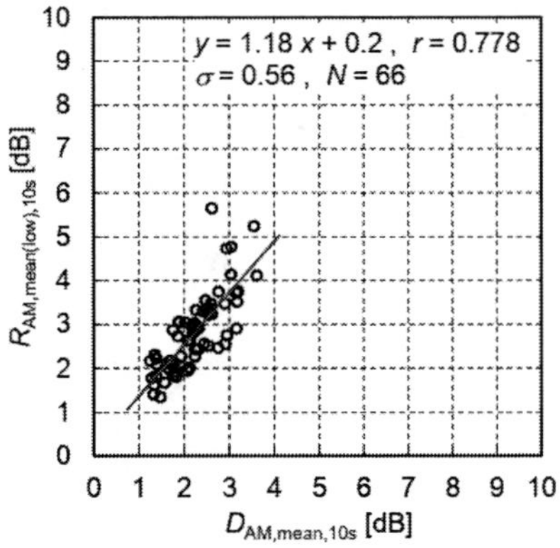

(a). R_{AM} for low frequency range vs. D_{AM}

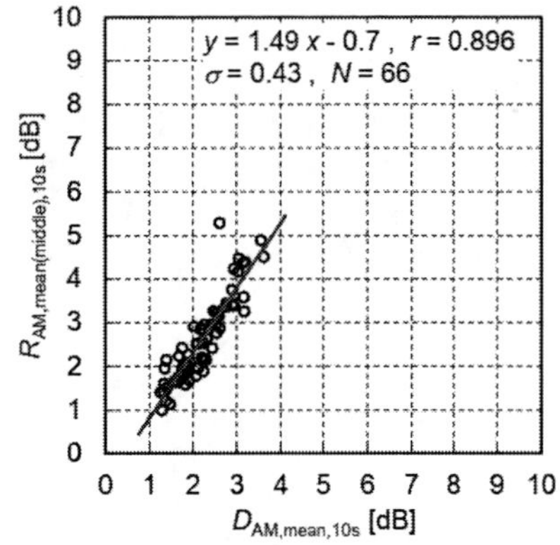

(b). R_{AM} for middle frequency range vs. D_{AM}

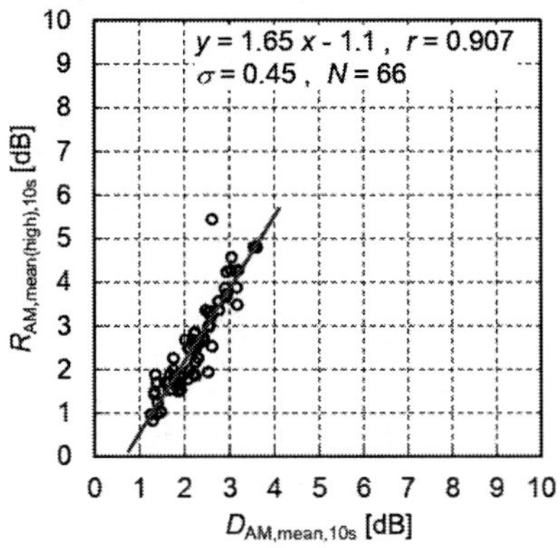

(c). R_{AM} for high frequency range vs. D_{AM}

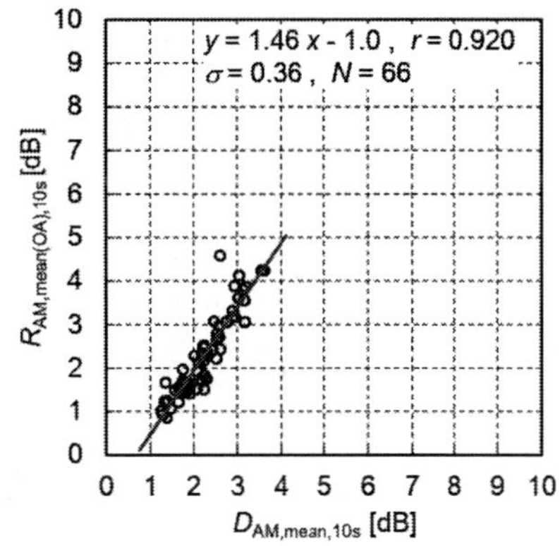

(d). R_{AM} for over-all frequency range vs. D_{AM}

Figure 6 - Comparison of the arithmetic mean values of the 10 s D_{AM} and R_{AM} for 1 hour.

frequency range, high frequency range and over-all frequency range are shown in Figure 7. As is seen in this figure, the peak is observed very near the time delay τ =0; the maximum cross-correlation coefficients are almost 0.7 - 0.8 for the cases of middle (b), high (c) and over-all (d) frequency ranges, whereas it is a bit lower in the case of low frequency range (a).

Such a cross-correlation analysis was performed for all of the 10 s data chosen by aural check (the F-S method) and by automatic judgement algorithm (the IOA reference method) and the results were arranged in the form of histogram as shown in Figure 8 for low frequency range (a), middle frequency range (b), high frequency range (c), and over-all frequency range (d), respectively. In the case of low frequency range shown in Figure 8(a), the maximum cross-correlation coefficient appeared in the class 0.55 - 0.60 and the rate that the cross-correlation coefficient Is lower than 0.5 is 33%. On the other hand, in the cases of middle frequency range (Figure 8(b)), high frequency range (Figure 8(c)) and over-all frequency range (Figure 8(d)), cross-correlation coefficient appeared in the class of 0.65 – 0.70 and the rate that the cross-correlation coefficient is lower than 0.5 is 25%, 23% and 20%, respectively. The results of this analysis indicates that the AM components detected by the F-S method and those by the IOA reference method are in fairly high correlation especially in the cases where middle, high or over-all frequency ranges are assessed in the IOA reference method, nevertheless the two methods are very different.

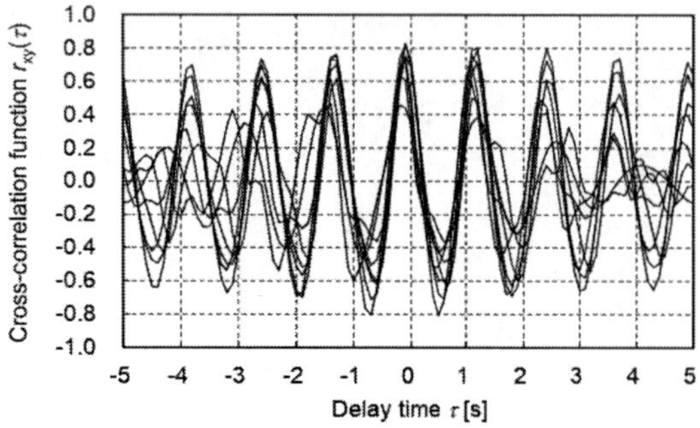

(a). low frequency range

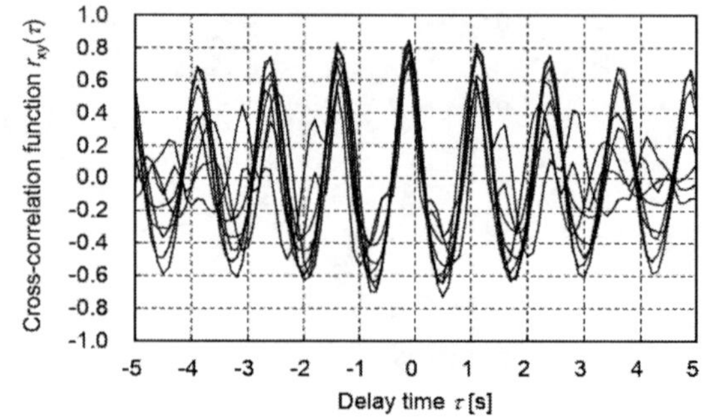

(b). middle frequency range

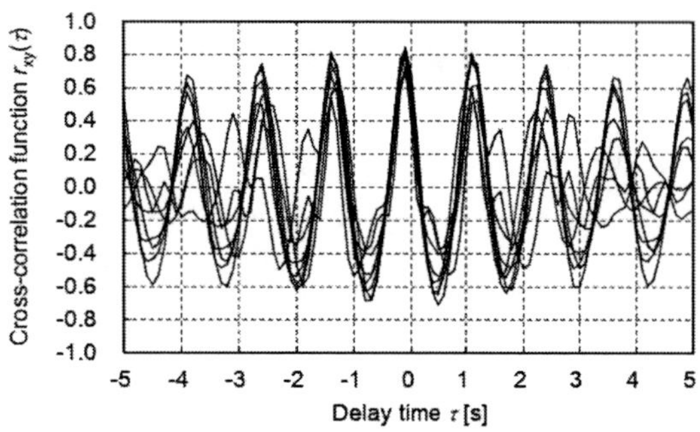

(c). high frequency range

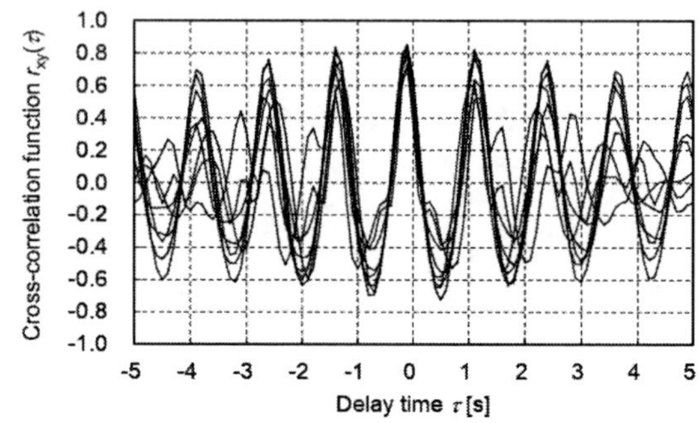

(d). over-all frequency range

Figure 7 - Examples of cross-correlation coefficients between AM components $\Delta L_A(t)$ obtained by F-S method and $\Delta L_{A,recreate}(t)$ obtained by the IOA reference method in Figure 3.

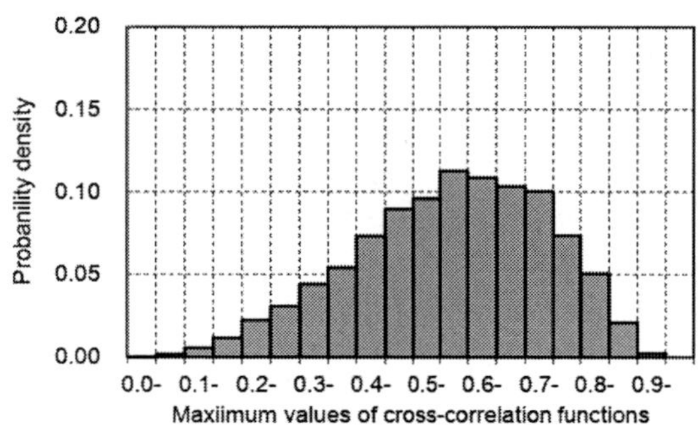

(a). low frequency range

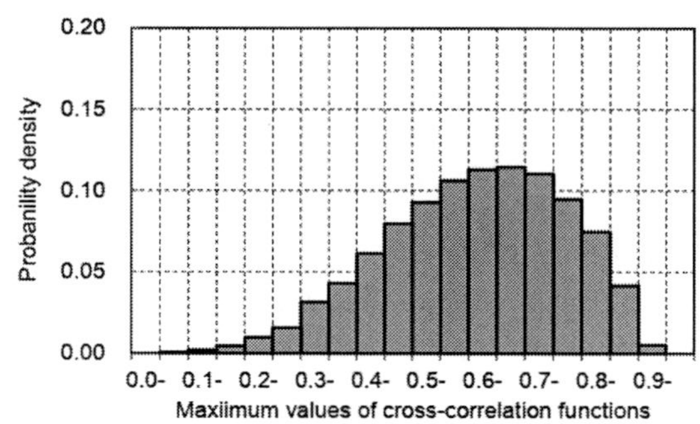

(b). middle frequency range

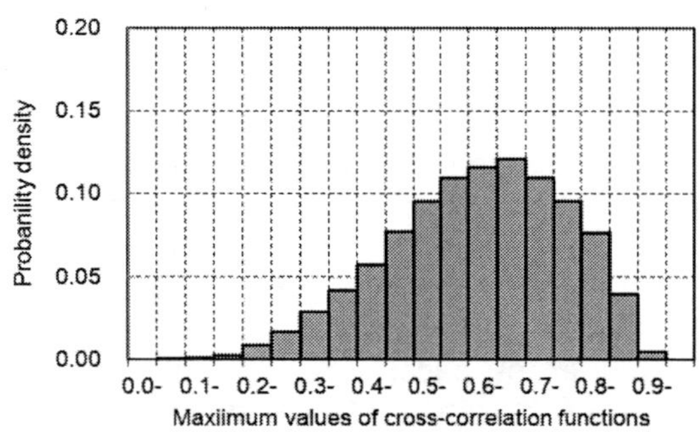

(c). high frequency range

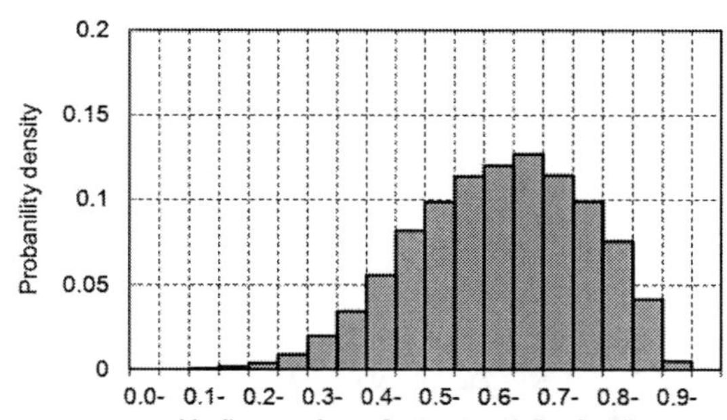

(d). over-all frequency range

Figure 8 - Distribution of cross-correlation coefficients of $\Delta L_A(t)$ obtained by F-S method and $\Delta L_{A,recreate}(t)$ obtained by the IOA reference method.

5. Conclusions

Regarding the assessment of Amplitude Modulation in WTN, two methods, the F-S method proposed by the authors and the IOA reference method proposed by the Institute of Acoustics were compared using the WTNs measured at 11 wind farm sites in Japan. The results of the study are as follows.

- In the comparison between D_{AM} and R_{AM} for the fundamental time interval, 10 s, considerably high correlation was found between D_{AM} and R_{AM} in the middle frequency range $R_{AM(middle)}$ (correlation coefficient : r=0.87) and between D_{AM} and R_{AM} in high frequency range $R_{AM(high)}$ (r=0.89). Further, if R_{AM} is assessed for over-all frequency, the correlation between D_{AM} and R_{AM} becomes a bit higher. Compared to these frequency ranges, the correlation between D_{AM} and R_{AM} in low frequency range $R_{AM(low)}$ is relatively low (r=0.73).
- Next, when comparing D_{AM} and R_{AM} of the highest 10% values in the survey period (1 h), considerably high correlation was found between D_{AM} and R_{AM} in the middle frequency range $R_{AM(middle)}$ (r=0.92), between D_{AM} and R_{AM} in high frequency range $R_{AM(high)}$ (r=0.93), and between D_{AM} and R_{AM} in over-all frequency range $R_{AM(OA)}$ (r=0.92). Compared to these frequency ranges, the correlation between D_{AM} and R_{AM} in low frequency range $R_{AM(low)}$ is relatively low (r=0.79) in this case, too.
- Further, when comparing DAM and RAM of the arithmetic mean values in the survey period, considerably high correlation was again found between D_{AM} and R_{AM} in the middle frequency range $R_{AM(middle)}$ (r=0.90), between D_{AM} and R_{AM} in high frequency range $R_{AM(high)}$ (r=0.91), and between D_{AM} and R_{AM} in over-all frequency range $R_{AM(OA)}$ (r=0.92). Compared to these frequency ranges, the correlation between D_{AM} and R_{AM} in low frequency range $R_{AM(low)}$ is relatively low (r=0.78) in this case, too.
- Regarding the difference of the way of detecting AM components, the cross-correlation between $\Delta L_A(t)$ simply obtained by the F-S method and $\Delta L_{A,recreate}(t)$ obtained by the IOA method applying DFT and IDFT processing was examined. As a result, it was found that they are in fairly high correlation when assessing the middle, high or over-all frequency ranges.
- From the results mentioned above, it can be concluded that D_{AM} obtained by the F-S method and R_{AM} obtained by the IOA reference method are highly correlated and they are compatible when assessing the middle and high frequency range defined in the IOA reference method. As shown in Figure1, the A-weighted spectra of WTN are apt to be highest in the frequency range from about 125 Hz to 1 kHz in 1/3 octave bands and it is easily supposed that the effect of AM is determined by the frequency components in this frequency range not only physically but also psycho-acoustically.

References

[1] A.Fukushima, K.Yamamoto, H.Uchida, S.Sueoka, T.Kobayashi and H.Tachibana, *Study on the amplitude modulation of wind turbine noise : Part 1 – Physical investigation*, Proc. INTER-NOISE 2013.
[2] A.Fukushima, T.Kobayashi and H.Tachibana. *Practical measurement method of wind turbine noise,* Proc. Wind Turbine Noise 2015.
[3] S. Yokoyama, S. Sakamoto and H. Tachibana, *Study on the amplitude modulation of wind turbine noise: part 2- Auditory experiments*, Proc. INTER-NOISE 2013.
[4] Institute of Acoustics IOA Noise Working Group (Wind Turbine Noise) Amplitude Modulation Working Group Final Report, *A Method for Rating Amplitude Modulation in Wind Turbine Noise*, Version 1, 2016.8.
[5] H.Tachibana, H.Yano, A.Fukushima and S.Sueoka. *Nationwide field measurements of wind turbine noise in Japan.* Noise Control Eng. J. 2014, 62 (2), 99-101.

7th International Conference
on
Wind Turbine Noise
Rotterdam – 2nd to 5th May 2017

Putting the IOA preferred AM assessment method and the proposed penalty scheme into practice – an outlook for future developments of wind farms in the UK.

Krispian Lowe, Innogy Renewables UK Limited
Sylvia Broneske, Innogy Renewables UK Limited

Summary

Introduction: The methodology to measure and assess amplitude modulation, assign a level that is acceptable and to provide a suitable planning condition, has been a source of discussion in the UK and amongst other countries for many years. In 2016 a methodology to measure the level of AM has been published by a working group of the Institute of Acoustics. The research paper 'Review of the evidence on the response to amplitude modulation from wind turbines' has been published by the Department of Business, Energy and Industrial Strategy on 25th October 2017. This paper also includes a proposal for a penalty scheme, but is expressively not planning guidance.

Aim: Determine the level of AM, subsequent penalty and the overall rating level for two wind farms, for which data is available.

Methods: To analyse data from existing immission measurements of the wind farms in operation utilising the Institute of Acoustics adopted method. To assess the potential impact on future developments by calculating the rated sound pressure level at properties using the proposed penalty scheme described in the Department of Energy, Business and Industrial Strategy's research report.

Results: AM was found at both wind farms, with the amount of time that AM occurs ranging from 7.7 to 36.8 per cent of the operational time, where AM that would invoke a penalty occurred 3.8 to 13.7 per cent of the operational time.

Conclusion: A property that is downwind of a wind farm may experience AM, but this level does not necessarily give rise to a penalty. The penalty may be related to the meteorological conditions at the time of measurement.

Further work: It has been recommend to carry out further measurements on wind farm sites to link site conditions, before construction of the wind farm to the level of AM after construction.

1. Introduction

Amplitude modulation (AM) is *"periodic fluctuations in the level of audible noise from a wind turbine (or wind turbines), the frequency of the fluctuations being related to the blade passing frequency of the turbine rotor(s)."* (IOAAMWG, 2016). The effect of AM is apparent close to the turbine and manifests itself as a swishing sound; it sometimes has a thumping quality at longer residential distances (RenewableUK, 2013a). The AM at residential distances is not a common occurrence, can be transient and occurs during specific conditions (RenewableUK, 2013a). There have been several attempts to define a methodology, penalty scheme and planning conditions (RenewableUK, 2013b; West Devon Council, 2005; Planning Inspectorate, 2009a; Planning Inspectorate, 2009b).

The endorsed guidance on wind farm noise assessment in the United Kingdom is "The Assessment and Rating of Noise from Wind Farms", which is henceforth referred to as ETSU-R-97 (NWG, 1996). ETSU-R-97 sets out the basic methodology for background noise monitoring, establishing noise limits, tonal assessment and example planning conditions. In addition, there is a brief mention of amplitude modulation within ETSU-R-97 (NWG, 1996, p. 12), which states:

This modulation might be expected to be clearly apparent when performing noise measurements close to wind turbines. However, the modulation of the A-weighted noise level is of the order 2-3 dB(A) for typical wind turbine configurations. Measurements performed in Denmark and at some locations in the UK indicate that this level of amplitude modulation may be greater if analysis is performed using third octave or narrow band analysis of the radiated noise from a wind turbine.

ETSU-R-97 refers only to the level of AM at the turbine location, but also recommends that AM was to be pursued under further work. However, it took until 2010 that Renewable UK first invited to tender for research work on AM. No actual research on AM was commissioned by the Government, in particular the very important impact study. The recently published paper is merely a review of existing research.

With respect to AM, very few planning conditions that apply to UK wind farms either state: "*AM shall be assessed using the best available method*" or the *Den Brook planning condition* (West Devon Council, 2005; Planning Inspectorate, 2009a; Planning Inspectorate, 2009b). The *Den Brook condition*, developed by Mr Stigwood of MASENV, uses a time domain signal to analyse the variation of the equivalent continuous sound pressure level, $L_{Aeq,125ms}$, subject to additional criteria, to determine if an AM event has occurred. Bass J., (Nov/Dec 2011) analysed the *Den Brook condition* with data from background noise measurements, without the presence of a wind farm, and found that AM false positives occurred 88 to 92 per cent of the time. Therefore, the *Den Brook condition* is considered insufficient in determining AM, but adequate in measuring the variation of noise level with respect to time. The condition does not discriminate with respect to a presence of a wind farm or its absence. The *Den Brook condition* is thoroughly inadequate to determine and rate the level of AM in wind farm noise in an automated assessment.

The subsequently published "Good Practice Guide to the Application of ETSU-R-97 for the Assessment and Rating of Wind Turbine Noise", henceforth referred to as GPG, states that "*The evidence in relation to 'Excess' or 'Other' Amplitude Modulation (AM) is still in development. At the time of writing, current practice is not to assign a planning condition to deal with AM.*" (Wind Turbine Noise Working Group, 2013, p. 29).

RenewableUK performed research on AM using a literature review, performing measurements to ascertain the conditions in which AM occurs, a human response acoustic study and an example planning condition (RenewableUK, 2013a). AM seems to occur when the blade of a turbine reaches partial stall at the top of the rotor. This generates a lower frequency thumping noise, which is heard at residential distances. The human response study showed that people are less sensitive to the change of the level of AM greater than 2 dB; however the underlying sound pressure level generates more clearer dose response. The suggested example planning condition performs a frequency domain analysis of the measurements (RenewableUK, 2013b). However, this approach was not adopted by the wind industry or the Institute of Acoustics' expert group.

The Institute of Acoustics Amplitude Modulation Working Group (IOAAMWG) investigated the best method of determining AM from wind turbine sound. The IOAAMWG requested consultation responses for the following three methods (IOAAMWG, 2015a).

1. method 1 - time series method (after TACHIBANA),
2. method 2 - Fourier Analysis Method (FFT) and
3. method 3 - hybrid reconstruction method.

Method 1 measures the difference between the Fast and the Slow time weighted L_{Aeq} levels for a 10 second period. The subsequent depth of modulation is calculated from the 5 minus 95 percentile for the 10 second period. Method 2 is based upon the methodology stated in RenewableUK (2013a); this method uses an FFT for a 10 second period to determine the AM at a frequency within the operational range of the turbine. Method 3 uses method 2 to determine the frequency at which AM is occurring, then filters the time domain signal with respect to this frequency, which is then used to determine the peak-to-trough value.

A response to this discussion document was submitted by Dr Krispian Lowe on behalf of RWE Innogy UK Limited[1] (IOAAMWG, 2015b, pp. 7-30) which was praised for its detailed analysis and diligence. The susceptibility of each method to the variation of turbine RPM, overall background level, level of AM and the level of random noise, to mimic true background noise, was investigated. Lowe et al found that Method 1 was wholly inadequate to measure AM, if random noise is introduced into the signal. Method 2 had systematic errors in the methodology, but did determine the level of AM. Method 3 also determine the level of AM. The preferred method, to determine the level of AM, recommended by Innogy Renewables UK Limited was Method 2, but with some modification.

IOAAMWG (2016) recommends the hybrid method, but with modifications from the initial discussion document. The hybrid method uses a FFT to determine the fundamental frequency where AM occurs for each 10 second period, in three frequency bands; extracts the fundamental frequency and its harmonics to a time domain signal; determines the modulation depth of the time domain signal; combines all 10 second modulation depth levels for a 10 minute value and calculates the 90th percentile to give the overall AM depth of the 10 minute period; each band is compared to the others and the greatest AM depth values are used.

The review of response to amplitude modulation was assessed by Department of Business, Energy and Industrial Strategy (DECC, 2016). This review does not represent planning guidance. An example penalty scheme would apply a penalty of 3 dB at an AM depth of 3 dB, as determined in IOAAMWG (2016) and increase linearly up to a penalty of 5 dB for an AM depth of 10 dB. As described in BS4142, BSi (2014), penalties are accumulated to an overall penalty, i.e. the AM penalty is added to the tonal penalty value. Therefore, the maximum penalty for a wind farm exhibiting tones[2] and AM would be 10 dB, which may make a wind farm

[1] As of 1st September 2016 Innogy Renewables UK Limited

[2] tonal penalty ranges from 1.54 dB to 5 dB

financially nonviable if the rated noise level (measured/predicted sound pressure level plus all applicable penalties) exceeds the noise limits. As the typical levels of AM for a wind farm are not known, this would introduce a significant amount of uncertainty in future developments. (DECC, 2016) states:

"This method is by necessity an interim recommendation based on the available evidence to date, and supplemented with professional experience. It is suggested that any planning condition derived from this report would be subject to a period of testing and review. The period should cover a number of sites where the condition has been implemented, and would be typically in the order of 2-5 years from planning approval being granted. The review would involve the analysis of any new AM research at the time, and case studies from sites where a condition has been implemented."

2. Methodology

The methodology to analyse AM used in this paper is described in IOAAMWG (2016). The analysis was conducted using MATLAB code and has been verified with the example data and the example python script[3] published by the IOAAMWG. Our methodology differs by constraining the minimum and maximum blade passing frequency (BPF), for each 10 minute period, with the values reported via SCADA. Please note that the minimum and maximum BPF values are those that occur during the 10 minute period. An example result for a single 12 hour period is shown in Figure 1. In the case where AM has been detected a 10 minute in one or more of the three frequency bands, then the greatest value is recorded.

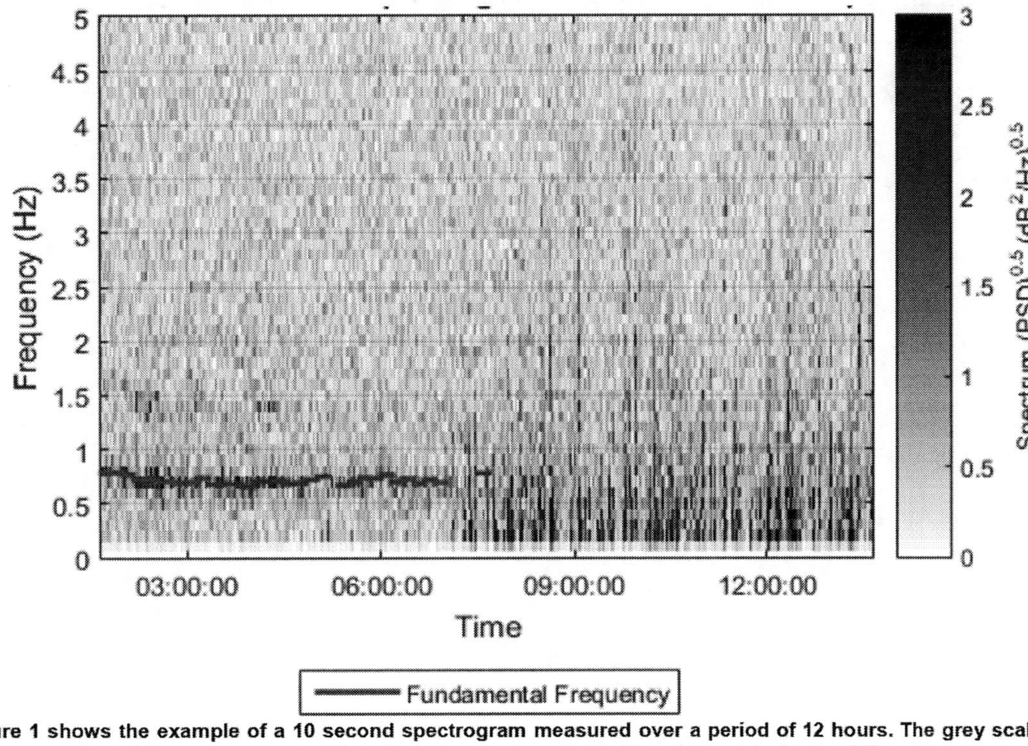

Figure 1 shows the example of a 10 second spectrogram measured over a period of 12 hours. The grey scale part is the square root of the power spectral density to improve contrast. The red bars indicates AM occurs greater or equal to 50 per cent of the time during a 10 minute period. The results are shown for the 50 Hz to 200 Hz range.

[3] https://sourceforge.net/projects/ioa-am-code/

3. The Proposed AM Penalty Scheme

The AM values derived in Section 2 can be converted to an AM penalty as proposed in DECC (2016) using Figure 2. The penalty value would be added to the corresponding $L_{A90,10min}$, to give the rating level of each 10 minute period. It should be noted that for the purposes of this paper tonality has not been assessed and therefore the rating level only consitsts of the measured $L_{A90,10min}$ plus the potential AM penalty.

The rating levels are divided between day-time, 0700-2300, and night-time, 2300-0700. The night time rating level will include the difference between the day-time and night-time limit, if the night-time limit is higher. The average rating noise level for each integer wind speed shall be compared to the respective limit for each property. It is assumed that the rating level is the arithmetic mean for a given integer wind speed. An example for a single 10 minute period is given in Table 1.

It should be noted that the authors do not endorse, support or approve of this penalty scheme.

Table 1. Example of the application of the penalty scheme for a single wind speed for a single $L_{A90,10min}$ value.

	Day-time	**Night-time**
Period	0700-2300	2300-0700
Wind Speed at 10 m standardised (m/s)	6	6
Limit [dB(A)]	35	43
Measured $L_{A90,10min}$ (dB)	30	30
AM Depth using IOA method (dB)	3	3
AM penalty (dB)	3	3
'Night-time' correction (dB)	0	8
Rating Level (dB)	33	41
Difference between Limit and Rating Level	2	2

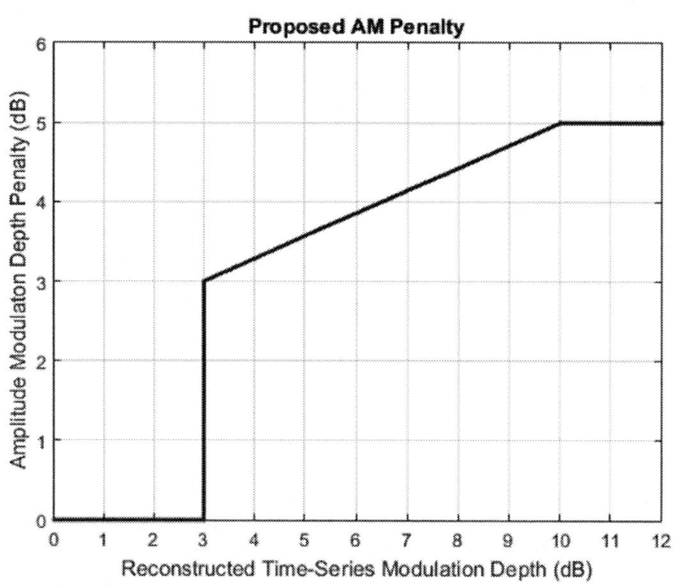

Figure 2 showing the conversion of the measured AM value to the AM Penalty. (Source: (DECC, 2016)

4. Measurements

The measurements have been taken for two sites: Site A with stall regulated turbines and Site B with pitch regulated turbines. Site A consists of turbines of a hub height of less than 50 m and Site B consists of turbines of more than 50 m hub height. The sound pressure level measurements for Site A are broadband only, owing to the measurements being taken prior to the publication of IOAAMWG (2016). Therefore, Site A results can only be seen as indicative.

Two measurement positions (A-up) and (A-down) are in the prevailing upwind and downwind positions and are 500 and 400 m from the nearest turbine, respectively. The total time for monitoring was 42 days.

The sound pressure level measurements for Site B are in third octaves. A total of two locations were selected that are in downwind (B-down) and upwind (B-up) position. The properties range from 600 m to 900 m from the nearest turbine. The total time of monitoring was 116 days.

5. Results

5.1 Site A

The measurements for Site A are not separated into narrow band frequencies as per the methodology stated in Section 2. The levels are solely broadband and have a frequency range between 20 Hz and 20,000 Hz; therefore the level of AM may be underestimated, owing to masking of high frequencies, or overestimated owing to contamination at these high frequencies by other sources such as bird song.

A-up is to the west of the site and is not in the prevailing wind direction. The majority of the sound emanating from the wind farm would be in a upwind direction. A-down is to the east of the site and would be downwind of the wind farm for most of the time. Table 2 summarises the measurement statistics.

The wind farm was operating normally 71.6 per cent of the total monitoring time. AM in general was detected 14.8 per cent and 23.4 per cent of the time for A-up and A-down, respectively. The level of AM that would constitute a penalty occurred 9.1 and 3.8 per cent of time for A-up and A-down, respectively.

Table 2. Summary statistics of data points and percentage for AM measurements at Site A. The percentage of AM detected and when AM would generate a penalty are respect to the wind farm is operating.

A-up	Number	Time (days)	Percentage
Number of Data points	5976	41.5	-
Wind Farm Operating	4281	29.7	71.6
AM Detected	635	4.4	14.8
AM Penalty	389	2.7	9.1
A-down			
Number of Data points	5972	41.5	-
Wind Farm Operating	4277	29.7	71.6
AM Detected	1002	7.0	23.4
AM Penalty	164	1.1	3.8

For A-up, the level of AM ranges from 1 dB to 10 dB, with most of the AM in the 3 dB bin with a fraction of occurrence >20 per cent (Figure 3). The shape of the measured AM is Gaussian up to the peak at 3 dB and then exponentially decays. The AM that has been measured occurs at two distinct fundamental frequencies of 1 Hz and 1.4 Hz, which related to a rotor rpm of 20 rpm and 28 rpm (Figure 4). The level of AM that would invoke a penalty occurs mostly between

1900 to 0800 hrs (Figure 5). It should be noted that the percentage of occurrence is 9.1 and is a total of 2.7 days over a 41.5 day period. The difference between the rating level and the overall rating ranges from 0 dB to 0.5 dB for day-time periods (Table 3; Figure 9). The difference between the rating level and the overall rating ranges from 0 dB to 3.6 dB for night-time periods (Table 3; Figure 10). The majority of the contribution for the AM penalty at night-time is the difference between the day-time and night-time limits, which is equal to 3 dB. Therefore, the AM component part of the penalty ranges from 0 dB to 1.9 dB. It should be noted that background noise has not been removed from these data.

For A-down, the level of AM ranges from 1 dB to 10 dB, with most of the AM in the 2 dB bin with a fraction of occurrence >35 per cent (Figure 6). The shape of the measured AM is Gaussian up to the peak at 2 dB and then exponentially decays. The AM that has been measured occurs at two distinct fundamental frequencies of 1 Hz and 1.4 Hz, which related to a rotor rpm of 20 rpm and 28 rpm (Figure 7). The level of AM that would invoke a penalty occurs mostly between 0100 to 1700 hrs (Figure 8). It should be noted that the numbers of occurrence is low and is a total of 1.1 days over a 41.5 day period. The difference between the rating level and the overall rating ranges from 0 dB to 0.5 dB for day-time periods (Table 4; Figure 11). The difference between the rating level and the overall rating ranges from 0 dB to 0.6 dB for night-time periods (Table 4; Figure 12). The majority of the contribution for the AM penalty at night-time is the difference between the day-time and night-time limits, which is equal to 3 dB. Therefore, the AM component part of the penalty ranges from 0 dB to 0.3 dB.

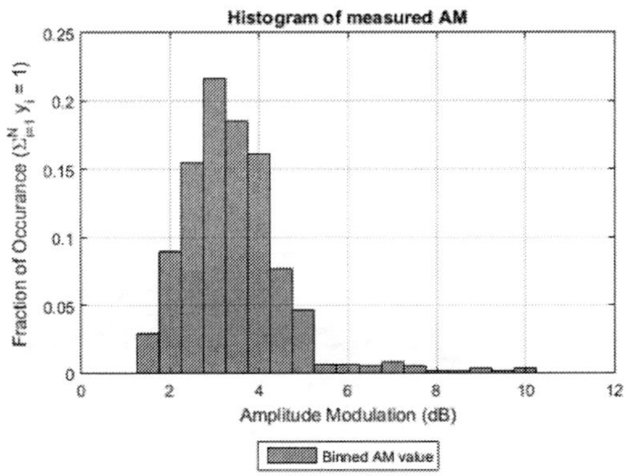

Figure 3 The histogram of the measured level of AM, 10 minute rating, at A-up for the entire monitoring period. The x-axis is the amplitude modulation in 0.5 dB bins and the y-axis is the fraction of occurrence.

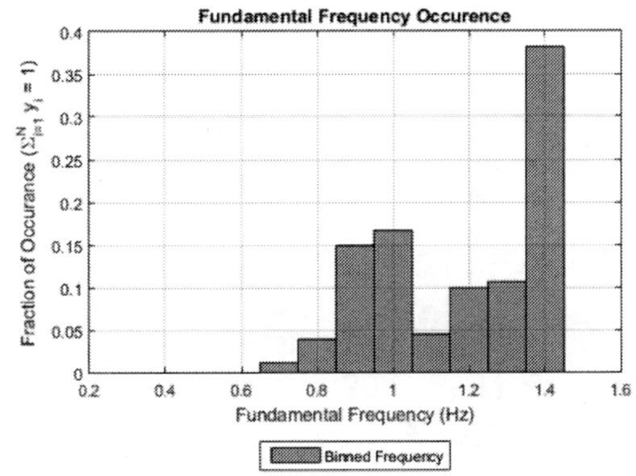

Figure 4 The histogram of the mean fundamental frequency for a 10 minute period, at A-up for the entire monitoring period. The x-axis is the frequency that AM occurs in 0.1 Hz bins and the y-axis is the fraction of occurrence.

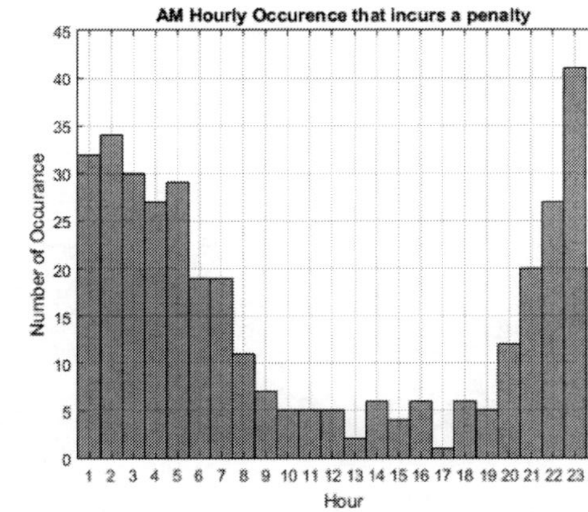

Figure 5 The number of occurrence for AM for each hour over the monitoring period at A-up.

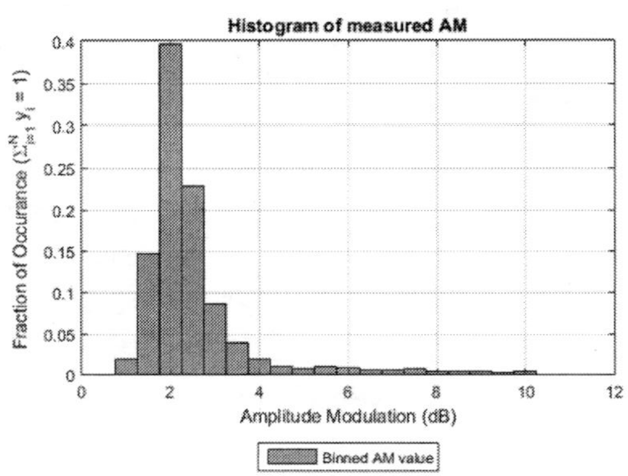

Figure 6. The histogram of the measured level of AM, 10 minute rating, at A-down for the entire monitoring period. The x-axis is the amplitude modulation in 0.5 dB bins and the y-axis is the fraction of occurrence.

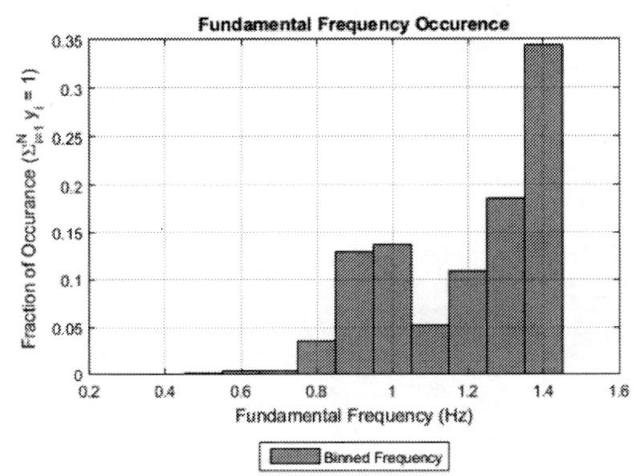

Figure 7 The histogram of the mean fundamental frequency for a 10 minute period, at A-down for the entire monitoring period. The x-axis is the frequency that AM occurs in 0.1 Hz bins and the y-axis is the fraction of occurrence.

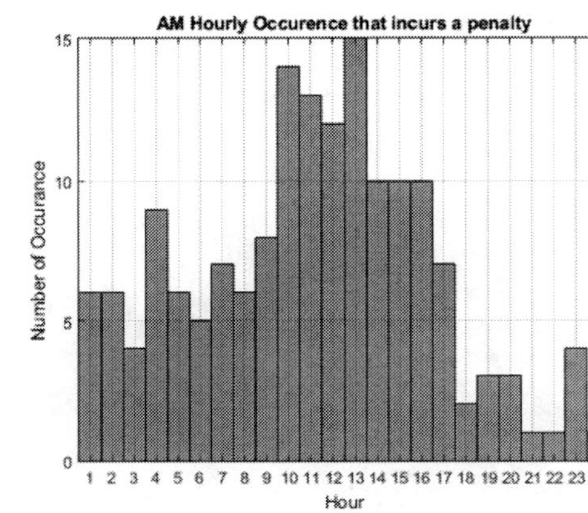

Figure 8 The number of occurrence for AM for each hour over the monitoring period at A-down.

Table 3 The results for A-up for day-time and night-time periods for the wind speed range from 3 m/s to 12 m/s, binned to 1 m/s. The mean $L_{A90,10min}$ values, subsequent Rating level, which includes the AM penalty, and the difference between the rating level and the mean $L_{A90,10min}$ values are displayed for day-time and night-time periods. In the case of night-time periods the difference between the rating level and the mean $L_{A90,10min}$ values without the night-time correction. All values, excluding wind speed are in dB.

Wind Speed (m/s)	**3**	**4**	**5**	**6**	**7**	**8**	**9**	**10**	**11**	**12**
Day-time Binned $L_{A90,10min}$ (dB)	30.2	32.0	33.9	35.4	37.1	39.2	39.8	42.1	41.6	44.3
Rating Level including AM penalty (dB(A))	30.3	32.1	34.2	35.7	37.3	39.7	39.9	42.1	41.6	44.3
Difference Between Rating and $L_{A90,10min}$	0.1	0.1	0.3	0.3	0.2	0.5	0.1	0.0	0.0	0.0
Night-time Binned $L_{A90,10min}$ (dB)	24.9	23.4	29.7	33	34	36.9	37.9	41.4	42.8	45.5
Rating Level including AM penalty (dB(A))	25.4	23.8	31.9	36.6	35.9	37.7	37.9	41.5	42.8	45.5
Difference Between Rating and $L_{A90,10min}$	0.5	0.4	2.2	3.6	1.9	0.8	0	0.1	0	0
Difference Between Rating and $L_{A90,10min}$ without Night-time correction	0.3	0.2	1.2	1.9	1	0.4	0	0	0	0

Table 4 The results for A-down for day-time and night-time periods for the wind speed range from 3 m/s to 12 m/s, binned to 1 m/s. The mean $L_{A90,10min}$ values, subsequent Rating level, which includes the AM penalty, and the difference between the rating level and the mean $L_{A90,10min}$ values are displayed for day-time and night-time periods. In the case of night-time periods the difference between the rating level and the mean $L_{A90,10min}$ values without the night-time correction. All values, excluding wind speed are in dB.

Wind Speed (m/s)	**3**	**4**	**5**	**6**	**7**	**8**	**9**	**10**	**11**	**12**
Day-time Binned $L_{A90,10min}$ (dB)	31.3	32.6	34.5	37.4	39.7	42.6	45	47	47.2	50.3
Rating Level including AM penalty (dB(A))	31.3	32.7	34.7	37.7	39.9	43	45.2	47	47.7	50.3
Difference Between Rating and $L_{A90,10min}$	0	0.1	0.2	0.3	0.2	0.4	0.2	0	0.5	0
Night-time Binned $L_{A90,10min}$ (dB)	30.4	30	32.8	34.5	37.9	41.1	44.1	45.7	47.2	48.2
Rating Level including AM penalty (dB(A))	30.4	30.6	33.2	34.6	37.9	41.1	44.1	45.7	47.2	48.2
Difference Between Rating and $L_{A90,10min}$	0	0.6	0.4	0.1	0	0	0	0	0	0
Difference Between Rating and $L_{A90,10min}$ without Night-time correction	0	0.3	0.2	0.1	0	0	0	0	0	0

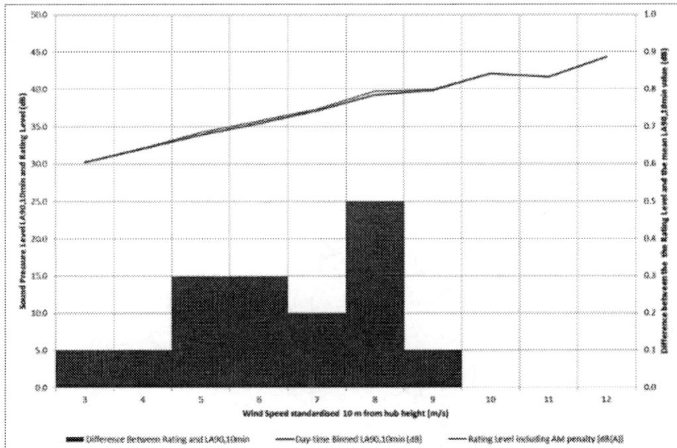

Figure 9 The mean sound pressure level (magenta line) and rating level inclusive of the AM penalty (green line) for day-time period for A-up. The difference between the sound pressure level and the rating level is displayed as blue bars.

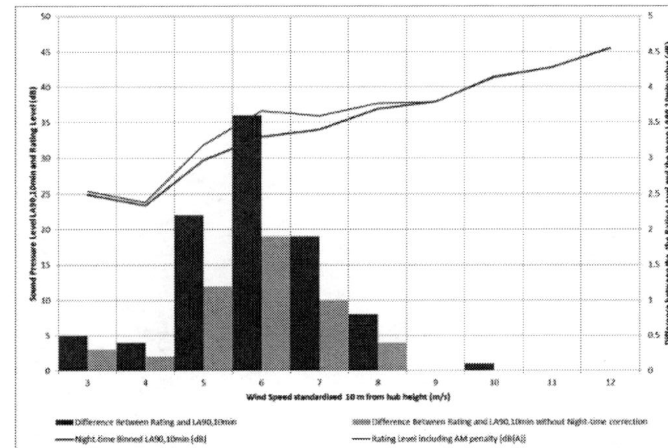

Figure 10 The mean sound pressure level (magenta line) and rating level inclusive of the AM penalty (green line) for night-time period for A-up. The difference between the sound pressure level and the rating level is displayed as blue bars. The difference between the sound pressure level and the rating level, without the night-time correction, as yellow bars.

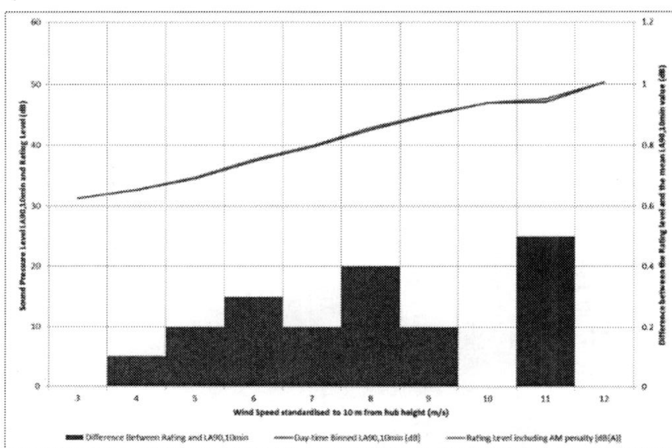

Figure 11 The mean sound pressure level (magenta line) and rating level inclusive of the AM penalty (green line) for day-time period for A-down. The difference between the sound pressure level and the rating level is displayed as blue bars.

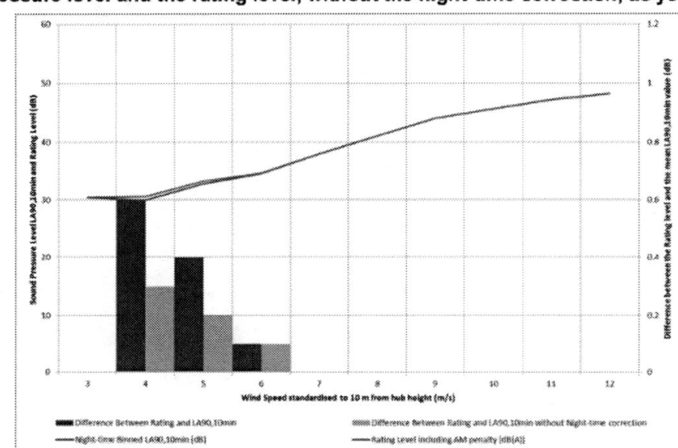

Figure 12 The mean sound pressure level (magenta line) and rating level inclusive of the AM penalty (green line) for night-time period for A-down. The difference between the sound pressure level and the rating level is displayed as blue bars. The difference between the sound pressure level and the rating level, without the night-time correction, as yellow bars.

5.2 Site B

For the B-up position, the wind farm was operating for 59.6 per cent of the time and AM was recorded 7.7 per cent of this operational time (Table 5). It was found that 65 per cent of the AM detected would constitute a penalty, which corresponds to 5 per cent of the operational time. The level of AM ranged from 1 dB to 9.5 dB with the most common level located at 3.5 dB (Figure 13). The fundamental frequency peaks at 0.6 Hz (Figure 14), which equates to a rotational speed of 12 rpm. The fundamental frequency range detected was from 0.4 Hz to 0.9 Hz, which equates to 8 to 18 rpm. The majority of AM that would incur a penalty occurs during late day-time and night-time periods of 1700 to 0900 hrs (Figure 15). The potential penalty for AM was determined for day-time and night-time periods (Table 6; Figure 19). The day-time AM penalty would range from 0 to 0.1 dB with the greatest levels of AM occurring at wind speeds 5 ms^{-1} to 6 ms^{-1}. The night-time periods showed an increase in the AM penalty ranging from 0 to 1.3 dB (Table 6; Figure 20). The greatest penalty would occur at the low wind speeds and decrease at the higher wind speeds. The AM component, which excludes the day-night-time limit correction, shows a range of 0 to 0.7 dB, with the larger component at the mid-range wind speeds of 5 to 6 ms^{-1}.

For the B-down position, the wind farm was operating for 59.6 per cent of the time and AM was recorded 36.8 per cent of operational time (Table 5). It was found that 37 per cent of the AM detected would constitute a penalty, which corresponds to 13.7 per cent of the operational time. The level of AM ranged from 1 dB to 7.5 dB with the most common level located at 2.5 dB (Figure 16). The fundamental frequency peaks at 0.8 Hz (Figure 17), which is 18 rpm. The fundamental frequency range detected was from 0.4 Hz to 0.9 Hz, which equates to 8 to 18 rpm. The majority of AM that would incur a penalty occurs during late day-time and night-time periods of 1500 to 0800 hrs (Figure 18). The potential penalty for AM was determined for day-time and night-time periods (Table 7). The day-time AM penalty would range from 0 to 0.4 dB with the greatest levels of AM occurring at wind speeds 7 ms^{-1} to 10 ms^{-1} (Table 7; Figure 21). The night-time periods showed an increase in the AM penalty ranging from 0 to 2.6 dB (Table 7; Figure 22). The penalty and AM component increases from low wind speeds to high wind speeds. The AM component ranges from 0 dB to 1.6 dB (Table 7; Figure 22).

Table 5. Summary statistics of data points and percentage for AM measurements at Site B. The percentage of AM detected and when AM would generate a penalty are respect to the wind farm is operating.

B-up	Number	Days	Percentage
Number of Data Points	16773	116.5	
Wind Farm Operating	9989	69.4	59.6
AM detected	766	5.3	7.7
AM Penalty	495	3.4	5
B-down			
Number of Data Points	16788	116.6	
Wind Farm Operating	10010	69.5	59.6
AM detected	3683	25.6	36.8
AM Penalty	1374	9.5	13.7

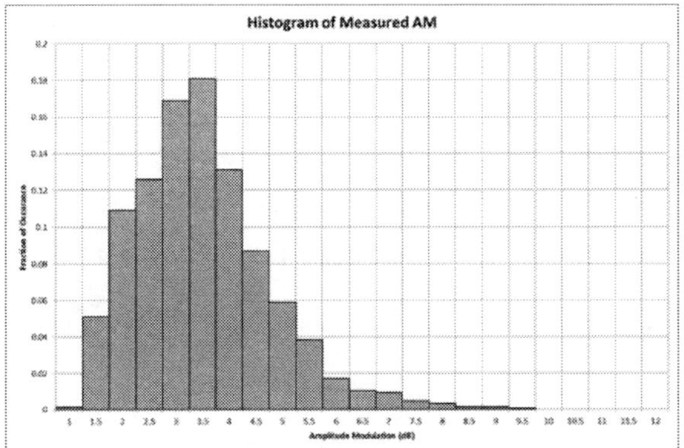

Figure 13 The histogram of the measured level of AM, 10 minute rating, at B-up for the entire monitoring period. The x-axis is the amplitude modulation in 0.5 dB bins and the y-axis is the fraction of occurrence.

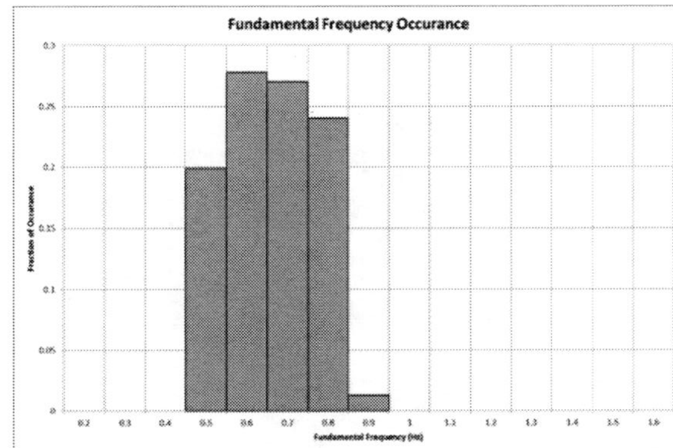

Figure 14 The histogram of the mean fundamental frequency for a 10 minute period, at B-up for the entire monitoring period. The x-axis is the frequency that AM occurs in 0.1 Hz bins and the y-axis is the fraction of occurrence.

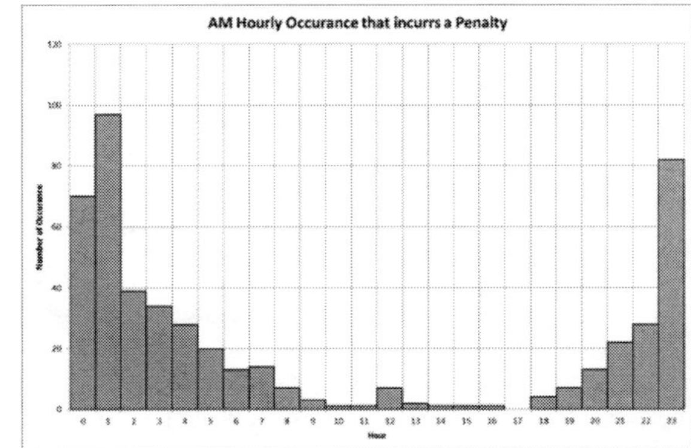

Figure 15 The number of occurrence for AM for each hour over the monitoring period at B-up.

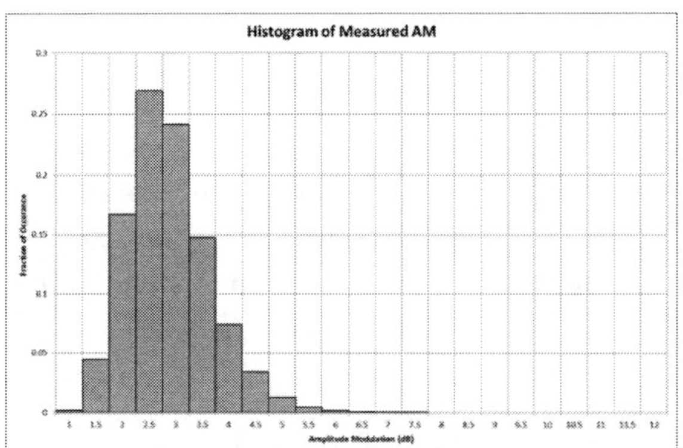

Figure 16. The histogram of the measured level of AM, 10 minute rating, at B-down for the entire monitoring period. The x-axis is the amplitude modulation in 0.5 dB bins and the y-axis is the fraction of occurrence.

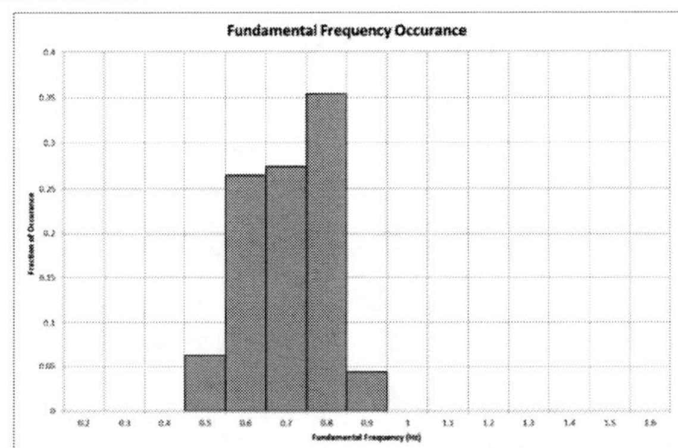

Figure 17 The histogram of the mean fundamental frequency for a 10 minute period, at B-down for the entire monitoring period. The x-axis is the frequency that AM occurs in 0.1 Hz bins and the y-axis is the fraction of occurrence.

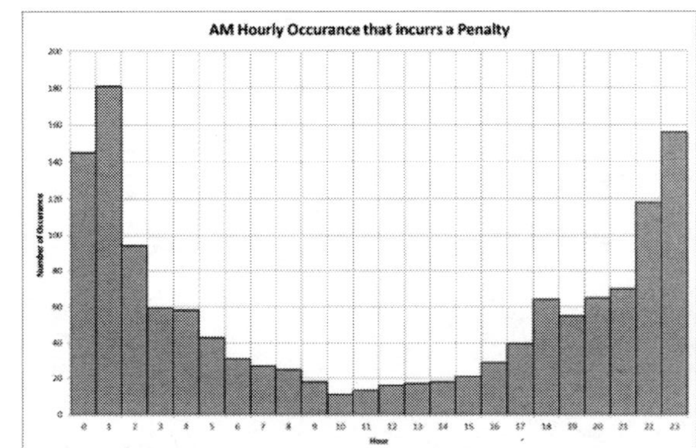

Figure 18 The number of occurrence for AM for each hour over the monitoring period at B-down.

Table 6 The results for B-up for day-time and night-time periods for the wind speed range from 3 m/s to 12 m/s, binned to 1 m/s. The mean $L_{A90,10min}$ values, subsequent Rating level, which includes the AM penalty, and the difference between the rating level and the mean $L_{A90,10min}$ values are displayed for day-time and night-time periods. In the case of night-time periods the difference between the rating level and the mean $L_{A90,10min}$ values without the night-time correction. All values, excluding wind speed are in dB.

Wind Speed (m/s)	1	2	3	4	5	6	7	8	9	10	11	12
Day-time Binned $L_{A90,10min}$ (dB)	26.1	31.4	30.8	30	30.3	31.9	32.8	34.1	35.7	37	38.6	38.7
Rating Level including AM penalty (dB(A))	26.1	31.4	30.8	30	30.4	32	32.8	34.1	35.7	37	38.6	38.7
Difference Between Rating and $L_{A90,10min}$	0	0	0	0	0.1	0.1	0	0	0	0	0	0
Night-time Binned $L_{A90,10min}$ (dB)	27.9	27	25.7	25.4	26.4	28.5	29.2	30.9	33.4	34.4	30.7	29.6
Rating Level including AM penalty (dB(A))	29.2	27.4	26.4	26	27.1	29.1	29.6	31.2	33.6	34.4	30.7	29.6
Difference Between Rating and $L_{A90,10min}$	1.3	0.4	0.7	0.6	0.7	0.6	0.4	0.3	0.2	0	0	0
Difference Between Rating and $L_{A90,10min}$ without Night-time correction	0.7	0.2	0.4	0.4	0.6	0.6	0.4	0.3	0.2	0	0	0

Table 7 The results for B-down for day-time and night-time periods for the wind speed range from 3 m/s to 12 m/s, binned to 1 m/s. The mean $L_{A90,10min}$ values, subsequent Rating level, which includes the AM penalty, and the difference between the rating level and the mean $L_{A90,10min}$ values are displayed for day-time and night-time periods. In the case of night-time periods the difference between the rating level and the mean $L_{A90,10min}$ values without the night-time correction. All values, excluding wind speed are in dB.

Wind Speed (m/s)	**1**	**2**	**3**	**4**	**5**	**6**	**7**	**8**	**9**	**10**	**11**	**12**
Day-time Binned $L_{A90,10min}$ (dB)	31.1	34.9	33.6	33.6	34.7	36.2	37.6	39.1	40.5	41.6	42.2	42.9
Rating Level including AM penalty (dB(A))	31.1	34.9	33.7	33.7	35	36.6	38	39.5	40.9	42	42.3	43.2
Difference Between Rating and $L_{A90,10min}$	0	0	0.1	0.1	0.3	0.4	0.4	0.4	0.4	0.4	0.1	0.3
Night-time Binned $L_{A90,10min}$ (dB)	32.1	31.7	30.6	30.8	32.1	34.3	35.7	37.4	39.4	40.2	39.4	37.2
Rating Level including AM penalty (dB(A))	32.6	32	31.1	31.5	32.9	35.4	36.6	38.4	40.6	41.6	42	39.5
Difference Between Rating and $L_{A90,10min}$	0.5	0.3	0.5	0.7	0.8	1.1	0.9	1	1.2	1.4	2.6	2.3
Difference Between Rating and $L_{A90,10min}$ without Night-time correction	0.5	0.3	0.5	0.7	0.8	1.1	0.9	1	1.2	1.4	1.6	1.5

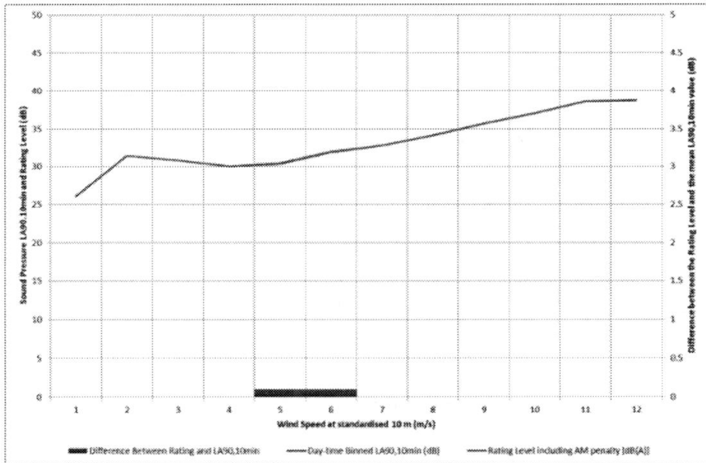

Figure 19 The mean sound pressure level (magenta line) and rating level inclusive of the AM penalty (green line) for day-time period for B-up. The difference between the sound pressure level and the rating level is displayed as blue bars.

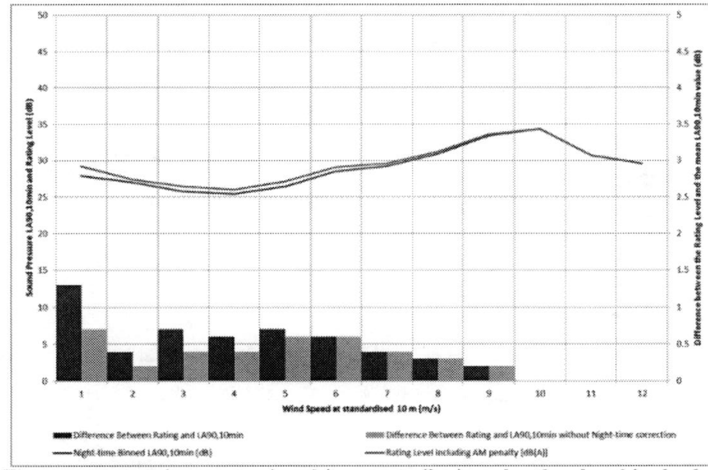

Figure 20 The mean sound pressure level (magenta line) and rating level inclusive of the AM penalty (green line) for night-time period for B-up. The difference between the sound pressure level and the rating level is displayed as blue bars. The difference between the sound pressure level and the rating level, without the night-time correction, as yellow bars.

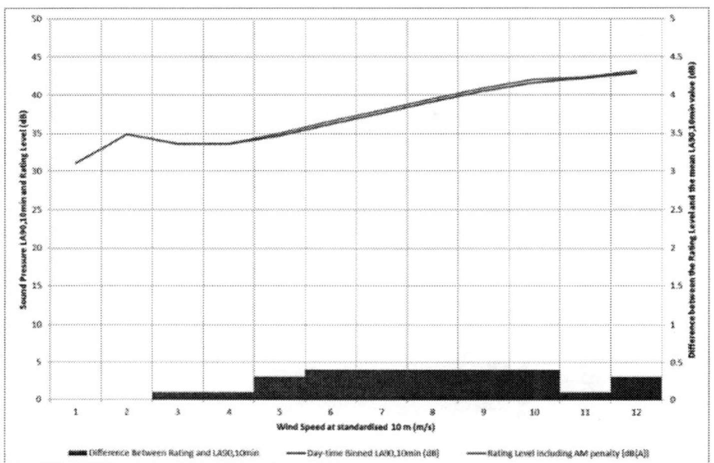

Figure 21 The mean sound pressure level (magenta line) and rating level inclusive of the AM penalty (green line) for day-time period for B-down. The difference between the sound pressure level and the rating level is displayed as blue bars.

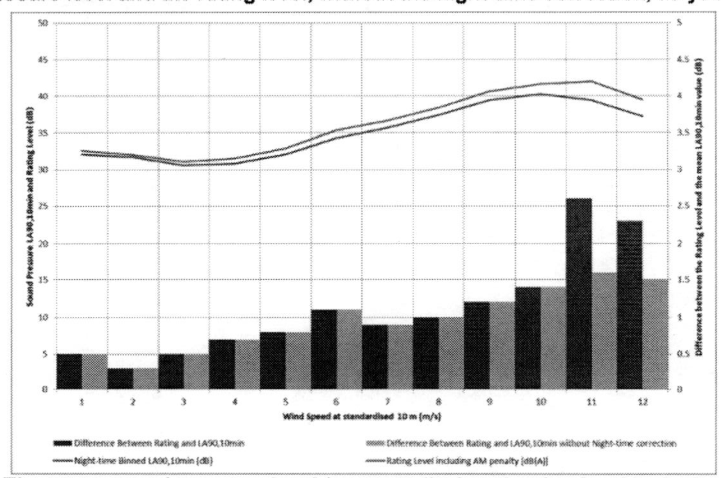

Figure 22 The mean sound pressure level (magenta line) and rating level inclusive of the AM penalty (green line) for night-time period for B-down. The difference between the sound pressure level and the rating level is displayed as blue bars. The difference between the sound pressure level and the rating level, without the night-time correction, as yellow bars.

6. Discussion

A total of two wind farms have been assessed for this paper and measurements taken at 4 positions, two at each wind farm. Site A consists of stall regulated and Site B of pitch regulated turbines. The hub heights are 65 m and do not reflect the hub and tip heights for future developments, where the tip heights are expected to be 90 m or greater.

Broadband sound measurements were made at Site A and third octave measurements were made at Site B. The length of time for each measurement campaign varied from 41 days to 116 days. The results at Site A can be interpreted as indicative. Site B measurements comply with the procedure set out in IOAAMWG (2016). In addition to the preferred method in (IOAAMWG, 2016), Innogy also uses SCADA data to complement the assessment where possible.

Overall, AM detected using the IOA's method ranged from 7.7 to 36.8 per cent of the operational time. The percentage of operational time, that the AM would invoke a penalty ranged from 3.8 to 13.7 per cent.

For each position of Site A, AM does occur with differing durations with a Mode average AM level of 3 dB and 2 dB respectively. At A-up AM occurred 14.8 per cent of the operational time and would invoke a penalty for 9.1 per cent of the time. In addition, majority of AM occurred during the evening and night-time periods. At A-up, the maximum penalty owing to AM, for the day-time and night-time periods would be 0.5 dB and 1.9 dB, respectively.

Conversely, A-down experienced AM for a greater amount of time of 23.4 per cent, but lesser penalty with 3.8 per cent of the time, with AM occurring mostly through the morning and day-time periods. In addition, the maximum penalty owing to AM, for the day-time and night-time periods would be 0.5 dB and 0.2 dB, respectively.

Site B generates level of AM ranging from 7.7 to 36.8 per cent of the operational time with a potential penalty being generated from 5 to 13.7 per cent. The most of the detected levels of AM occurs at B-down, which is in the prominent downwind of the wind farm. The greatest level of AM occurs at B-down at all wind speeds at night-time.

The AM methodology described in Section 2 provides suitable detection of AM originating from wind turbines. The occurrence of AM, that would give rise to a penalty in accordance to the proposed penalty scheme in (DECC, 2016), during the day-time periods 0700 to 2300 is significantly lower than those during the night-time period 2300 to 0700, for the majority of monitoring positions. Overall, for all sites, the level of AM increases during evening and night-time periods. These are periods of time that wind shear typically increases. The apparent trend is that levels of AM penalty increases at the mid-range wind speeds. This may be caused by the background noise climbing slowly as a function of wind speed, at mid-range wind speed, whilst turbine noise increases quickly at mid-range wind speeds. Therefore, the background noise masks AM at low and high wind speeds. In this report we have not looked at the level of wind shear for each wind farm.

The level of AM varies for upwind and downwind locations. The histogram of the occurrence of the level of AM shows that the Mode is greater for upwind locations rather than downwind locations. The duration of AM that would invoke a penalty in accordance to the proposed scheme differs between upwind and downwind conditions, but is not consistant between Site A and Site B. For Site A, the applied penalty is greater for upwind locations rather than downwind locations. For Site B, the applied penalty is greater for downwind locations rather than the upwind location. Therefore, based on these two data sets, it is not possible to determine that the penalty for AM is greater for any particular downwind or upwind location.

7. Conclusions

The effect of AM from wind turbines is noticeable near to the turbine(s) and could be masked at residential conditions. However, AM in certain conditions can be audible and change in character of the sound. The definition and the methodology of measuring of AM, owing to wind turbines, is defined in IOAAMWG (2016). This AM methodology has been adopted by (DECC, 2016) and planning conditions including a penalty scheme has been proposed for further review, but not endorsed. The methodology and subsequent penalty scheme proposal is the latest in the discussion of AM from the Den Brook condition and RUK condition.

Innogy looked at two wind farms where monitoring has taken place and the data has met sufficient quality, quantity and indices, to analyse AM. One stall regulated turbine site and one pitch regular turbines site have been examined, Site A and Site B. The tip heights are below 110 m and are unlikely to reflect the future developments for wind turbine sites, but are suitable to provide an indication on the level of AM and the impact of the proposed penalty scheme as interpreted by Innogy. It should be noted that the measurements taken at Site A do not conform to the IOA methodology, but could provide an indicative assessment of AM. The measurements at Site B do conform to the IOA methodology.

AM was found at all sites and occurred from 7.7 to 36.8 per cent of the time, when turbines where operating. The percentage of AM that would invoke a penalty ranged from 3.8 to 13.7 per cent of the operational time. The majority of AM occured during evening and night-time periods. The amount of time AM occurs is greatest for downwind properties, but the level of AM that would generate a penalty is not associated with these downwind properties. This indicates that other factors such as background noise, shear or meteorological conditions may contribute to levels of AM that would generate a penalty.

Further investigation of AM should be conducted at sites that experience a range of wind conditions. A link between the site conditions prior to the wind farm construction and the level of AM at the wind farm should be established.

It should be noted that the authors do not endorse, support or approve of this proposed penalty scheme.

References

Bass J., Nov/Dec 2011. *Acoustics Bulletin,* St. Albans: Institute of Acoustics.
BSi, 2014. *Methods for rating and Assessing Industrial and Commerical Sound,* London: BSi Standards Limited.
DECC, 2016. *Wind Turbine AM Review,* London: Department for Business, Energy & Industrial Strategy.
IOAAMWG, 2015a. *Discussion Document: Methods for Rating Amplitude Modulation in Wind Turbine Noise,* St. Albans: Insitute of Acoustics.
IOAAMWG, 2015b. *Responses to the Discussion Document: Methods for Rating Amplitude Modulation in Wind Turbine Noise,* St. Albans: Institute of Acoustics.
IOAAMWG, 2016. *Final Report: A Method for Rating Amplitude Modulation in Wind Turbine Noise,* St Albans: Institute of Acoustics.
NWG, 1996. *ETSU-R-97: The Assessment and Rating of Nosie from Wind Farms.* s.l.:Department of Trade and Industry.
Planning Inspectorate, 2009a. *APP/F2415/A/09/2096369,* London: Planning Inspectorate.
Planning Inspectorate, 2009b. *APP/Q1153/A/06/2017162,* London: Planning Inspectorate.

RenewableUK, 2013a. *Wind Turbine Amplitude Modulation: Research to Improve Understanding as to its Cause and Effect,* London: Renewable UK.

RenewableUK, 2013b. *RenewablesUK AM Noise Guidance Notes,* London: RenewableUK.

West Devon Council, 2005. *Planning Application 8250/2005/OKE,* Barnstaple: West Devon Council.

Wind Turbine Noise Working Group, 2013. *A Good Practice Guide to the Application of ETSU-R-97 for the Assessment and Rating of Wind Turbine Noise,* St. Albans: Institute of Acoustics.

7th International Conference
on
Wind Turbine Noise
Rotterdam – 2nd to 5th May 2017

AN INVESTIGATION INTO THE EFFECT OF WIND SHEAR ON THE SOUND EMISSION OF WIND TURBINES

Payam Ashtiani, Aercoustics Engineering Ltd.: PayamA@aercoustics.com
Duncan Halstead, Aercoustics Engineering Ltd.: DuncanH@aercoustics.com

Summary

There has been some debate regarding the effect of changing wind profiles on the noise emission of a wind turbine, particularly in conditions of high wind shear. With modern wind turbines having rotor diameters in excess of 90m, the difference in wind velocity from the bottom to the top of the swept rotor area has the potential to be dramatic. This paper presents a range of wind turbine acoustic test data collected in differing wind shear profiles to determine if there are any noticeable differences in the sound emissions. The measurement data used in this study was collected per the IEC 61400-11 test standard for measuring acoustic emissions from a wind turbine.

As part of the IEC 61400-11 test standard, synchronous wind speed measurements are collected at 10 metre and hub heights in 10 second averages. This allows for an approximation of the degree of wind shear at a given time. Test data is split between conditions with lower wind shear and higher wind shear, and the calculated noise emissions are compared.

1. Introduction

Wind shear, or wind gradient, refers to the difference in wind speeds with vertical height, typically measured from the Earth's surface. Friction at the surface results in slower wind speeds closer to the ground. Increasing in height away from the ground results in higher observed wind speeds, up to a point. A greater degree of wind shear represents a greater difference in wind speeds measured at two different heights. In a stable atmosphere, with little inter-layer mixing, this wind speed gradient is smooth and pronounced. As atmospheric turbulence increases, more mixing occurs and the vertical wind speed gradient decreases.

Modern wind turbines, which can have rotor diameters of over 80 meters in length, could experience large differences in wind speeds from the bottom of the rotor swing to the top in cases of high wind shear.

2. Aerodynamic Noise from a Wind Turbine

Wind turbine noise can be attributed broadly to one of two mechanisms: mechanical noise and aerodynamic noise. Mechanical noise arises from the rotating machinery of the turbine – such as gearboxes, fans, and pumps – and is usually generated in the nacelle of the wind turbine. Aerodynamic noise, on the other hand, is generated by the airflow interacting with the blades as they rotate.

Aerodynamic noise accounts for much of the broadband noise emitted by a wind turbine; the strength of which is highly dependent on flow speed. Trailing edge noise is considered to be the most important noise source for modern turbines [1] and has a fifth-power relationship to wind speed, as derived in the following equation by Ffowcs Williams and Hall [2].

$$p^2 \sim U^5 \frac{L\delta^*}{r^2} \cos^3 \gamma \sin^2(\theta/_2) \sin \varphi$$

p = sound pressure [Pa]
U = mean flow velocity [m/s]

With this high sensitivity to flow speed, it could be hypothesised that increased wind shear could influence the noise emission of a wind turbine, due to the difference in wind speeds between the bottom and the top of the rotor swept area. However, it is important to account for the impact of rotational speed of the turbine on the flow velocity incident on a turbine blade. Modern wind turbines rotate around 15 rpm which, for a blade length of 50m would correspond to a tangential speed – and consequently a flow velocity – of roughly 40m/s at the midpoint of the blade.

3. Measurements

The analysis data used in this paper was acquired using the IEC 61400-11 test methodology. The test method utilizes a microphone mounted on a flat reflective plane roughly 130-150 meters downwind of the wind turbine. Wind speed measurements are taken simultaneously from anemometers located at the turbine nacelle as well as a 10m tower erected 150-200 meters from the turbine. Acoustic measurements are sorted based on the wind speeds acquired at the two measurement points. Turbine operational parameters – including electrical power, yaw angle, and rotor rotational speed – are also acquired as part of the test method. All measurement data is acquired synchronously, averaged in 10 second intervals. Measurements are taken with the wind turbine running and with the wind turbine parked to separate the wind turbine noise from the ambient noise [3].

The degree of wind shear during is characterized using the difference in wind speed measured at hub height and 10 meters for a given measurement interval. Using these two measurement points, the wind shear coefficient was calculated using formula below.

$$\frac{v_{10}}{v_H} = \left(\frac{h_{10}}{h_H}\right)^{\alpha}$$

V_{10} = *wind speed measured at 10-meter height*
V_H = *wind speed measured at hub height*
h_{10} = *10 meters*
h_H = *hub height*
α = *wind shear coefficient*

Higher values of α correspond to greater differences between 10-meter and hub height wind speeds, and therefore indicate a greater degree of wind shear.

Using this method, the wind shear coefficient for each measurement interval was calculated for each 10 second interval over the course of an entire test. Tests whose wind shear coefficients changed appreciably were selected for further analysis. Most often, the selected tests were those where the measurement periods spanned both day and night time periods. This is not unexpected; the absence of solar heating at night allows for a more stable atmosphere and, consequently, higher wind shear profiles are created [4].

For this study, four tests on four different turbines were selected for analysis. The results of which are detailed in the following section.

4. Results

Plots of wind speed, wind shear coefficient, overall sound level, and 1/3rd octave sound level have been created for each of the four tests used in the analysis. Sound level plots have been separated by colour into high and low shear periods. Average calculated wind shear exponents for each measurement period are provided in the plot legends. Measured ambient sound levels are also provided to show the signal-to-noise ratio observed during the measurements. Wind speeds used in the sound pressure level plots are all measured at hub height.

Each dataset represents a test completed on a single turbine, sometimes spanning multiple test days. The tests span multiple turbine manufacturers, locations, and blade lengths. The turbines tested all had rotor diameters at least 100 meters in length.

4.1 Overall Sound Level Comparison

The following plots compare the measured overall sound levels in high and low wind shear periods. High wind shear data points are identified in green, low wind shear points in blue, and ambient (turbine parked) points in brown. Each data point represents one 10-second measurement interval.

4.1.1 Test #1

Roughly the first half of the test sees wind speeds steadily increasing in an atmosphere with low wind shear (little difference between hub height and 10-meter wind). The second half has the wind speeds increasing through the same range, but this time in an environment with higher wind shear.

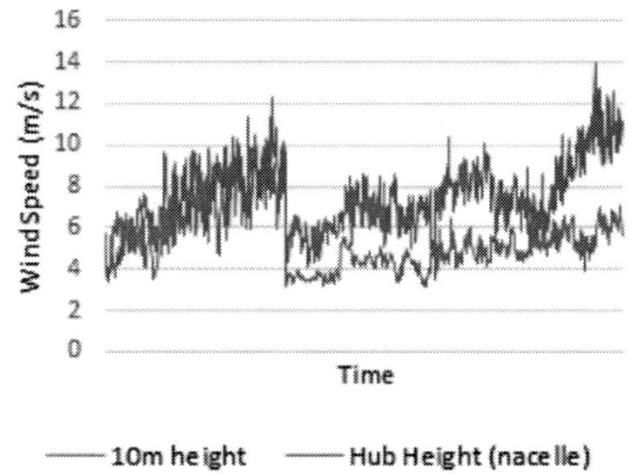

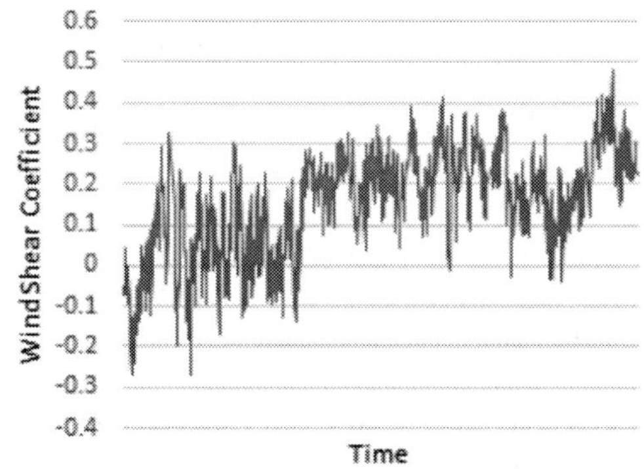

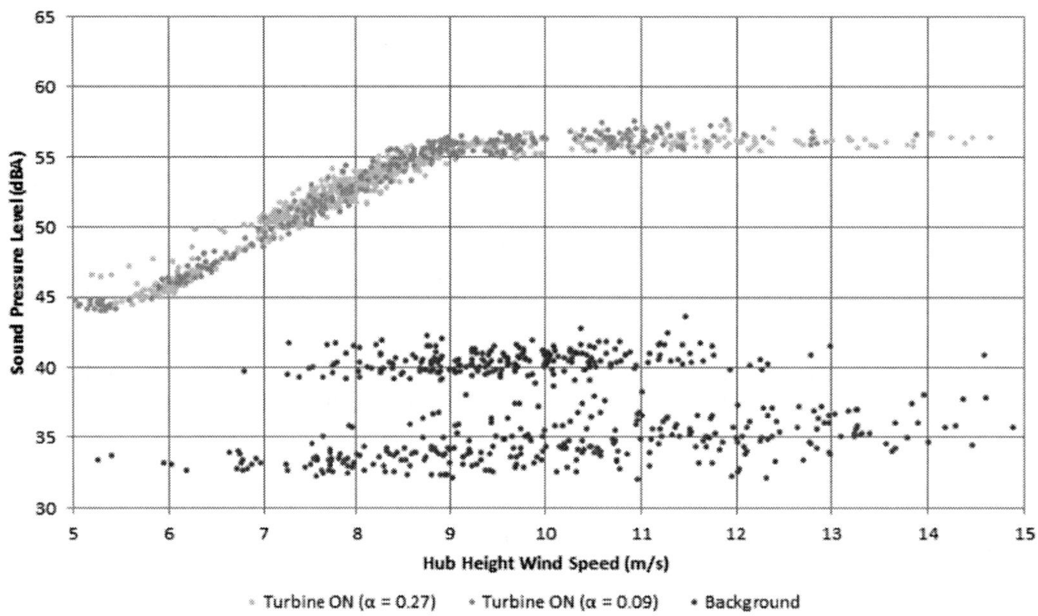

In this test, the measured wind speeds between high and low wind shear periods overlap consistently throughout almost the entire wind speed range. The two data sets show a very similar relationship between sound pressure level and wind speed.

4.1.2 Test #2

This test has alternating periods of low and high wind shear with little overlap in wind speeds between wind shear regimes. Low wind shear periods are observed at high and low wind speeds, with a high wind shear cluster of data in between.

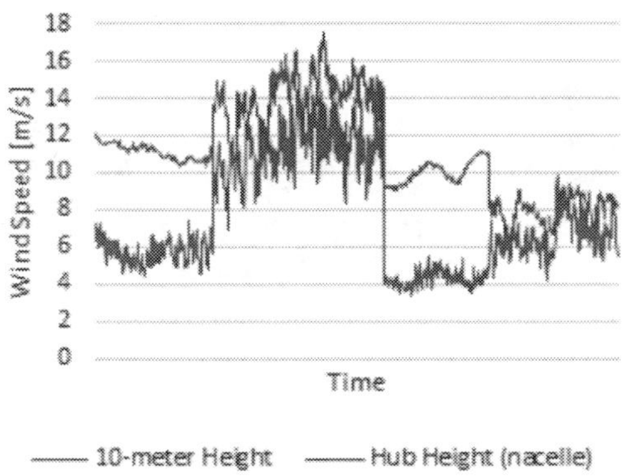

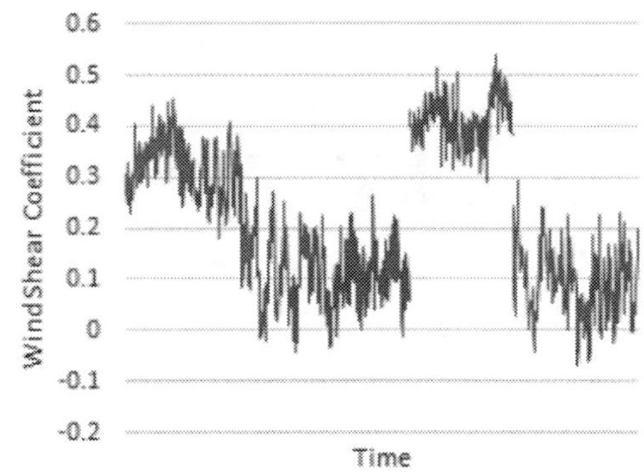

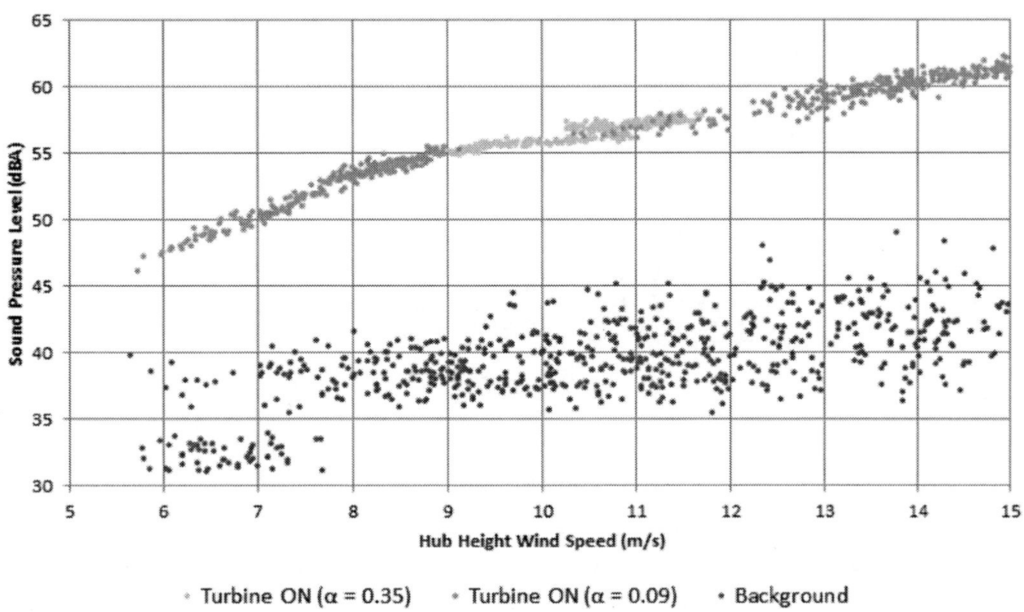

There is a gap evident in this plot between high and low shear points, with a little overlap in the datasets around 11 m/s. The high shear points in the middle of the wind speed range appear to follow the same linear trend as the low wind shear clusters on either side.

4.1.3 Test #3

The beginning of this test shows a high shear environment, after which the wind shear drops significantly. Some overlap in hub height wind speeds is observed.

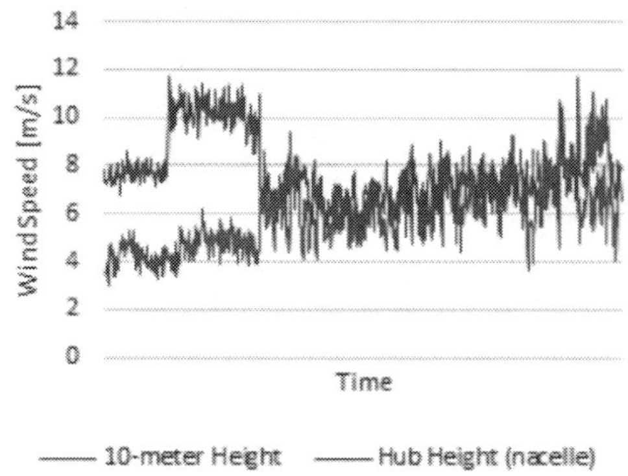

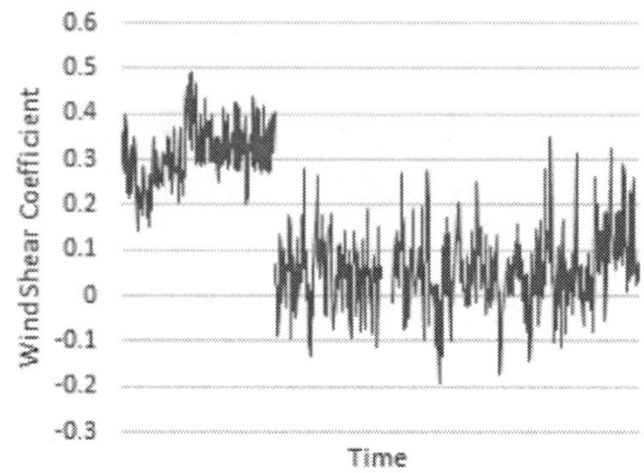

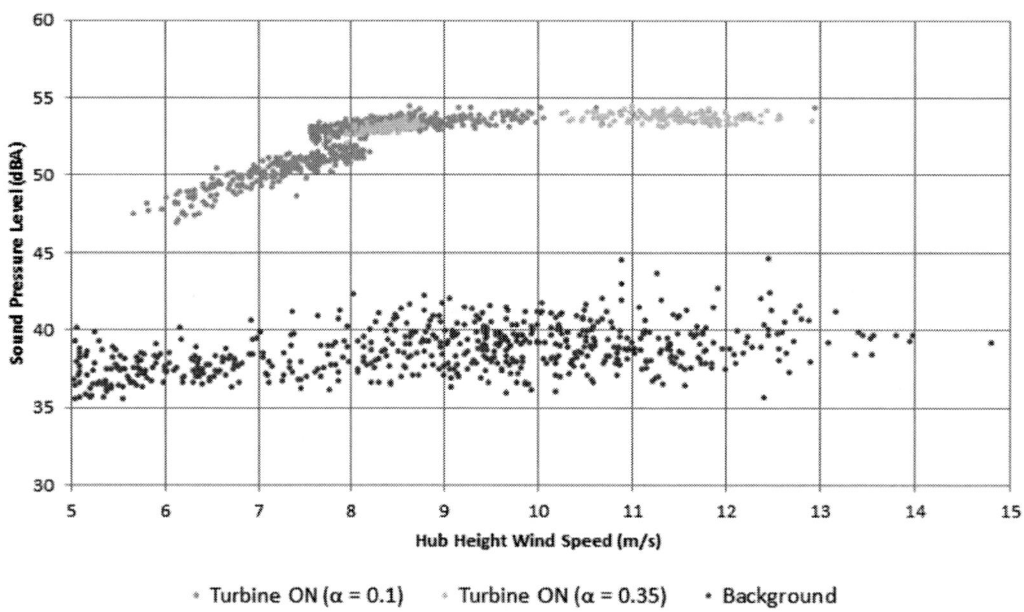

The low wind shear data in this set seems to show two trends around 8m/s for the low wind shear set. The high wind shear points follow the higher of the two, but do not show any levels that fall higher than the low wind shear points.

4.1.4 Test #4

In the beginning of this test there is degree of wind shear. This is followed by a relatively low shear period for the rest of the test. Overlap in the wind speed ranges between high and low shear periods is observed between 7 and 9 m/s hub height winds.

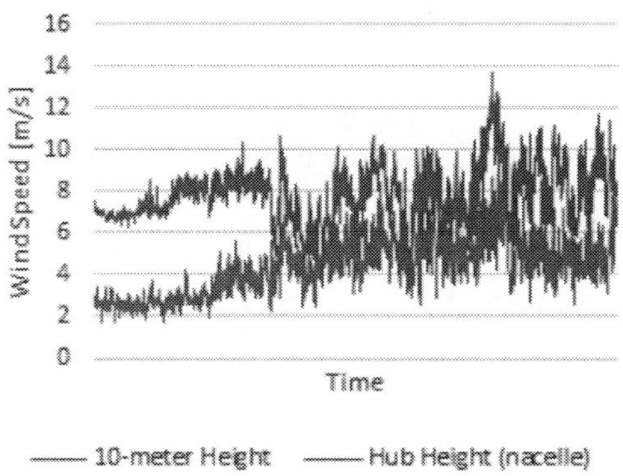

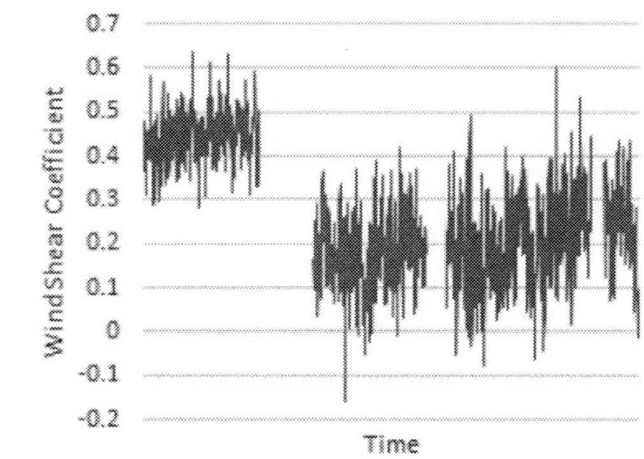

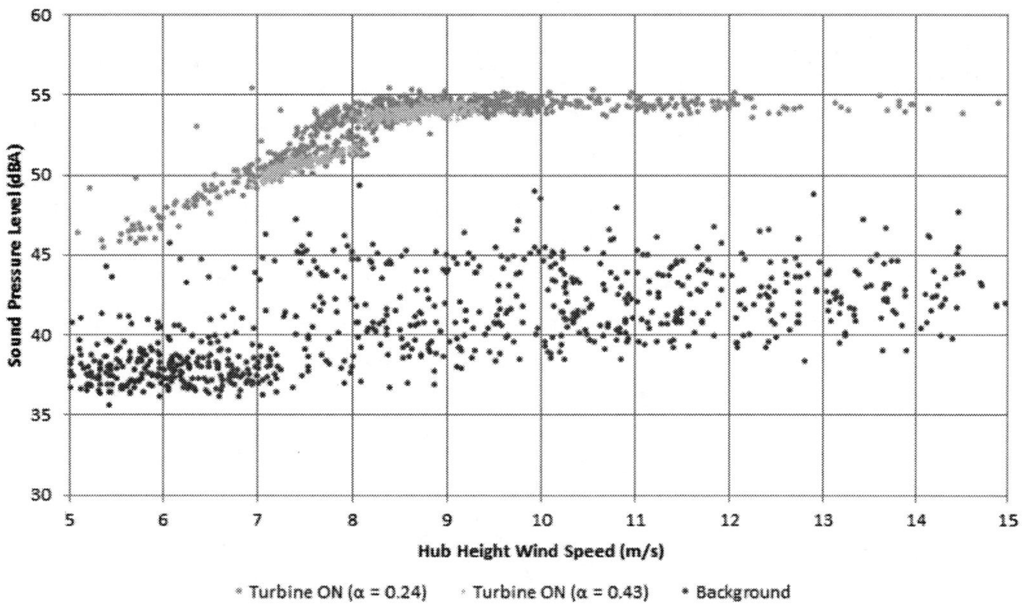

This data set has a great deal of overlap in datasets between 7-9m/s. The high wind shear spread appears to have a lower deviation in sound level compared to the low wind shear data from the same wind speeds.

1/3 Octave Band Comparison

The following plots compare the measured 1/3rd octave band spectra between high and low shear periods. Background levels are also provided to show the overall signal to noise ratio at each frequency band. The analysis range has been limited to 20Hz-3150Hz; measured noise levels above this range are usually driven by the ambient environment, rather that the wind turbine.

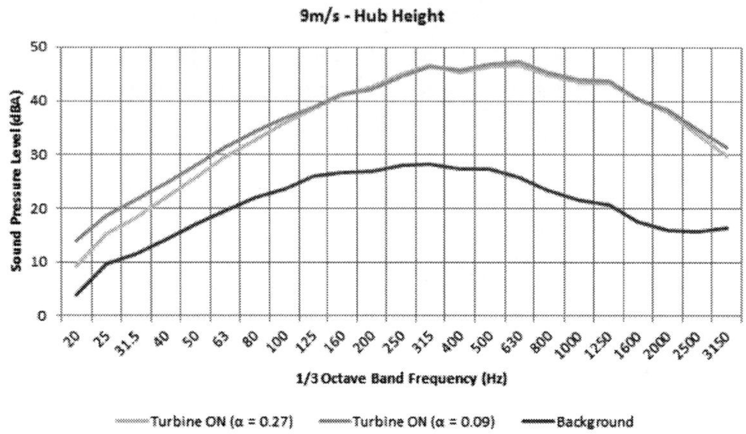

4.1.5 Test #1

Largely the same sound levels are measured at frequencies between 160Hz-2000Hz. High and low frequencies show a slightly increased sound level from the low wind shear dataset.

4.1.6 Test #2

Test #2 shows increased levels at frequencies below 400Hz during periods of low shear. A slight increase in level at high shear is apparent in the 1600Hz frequency band.

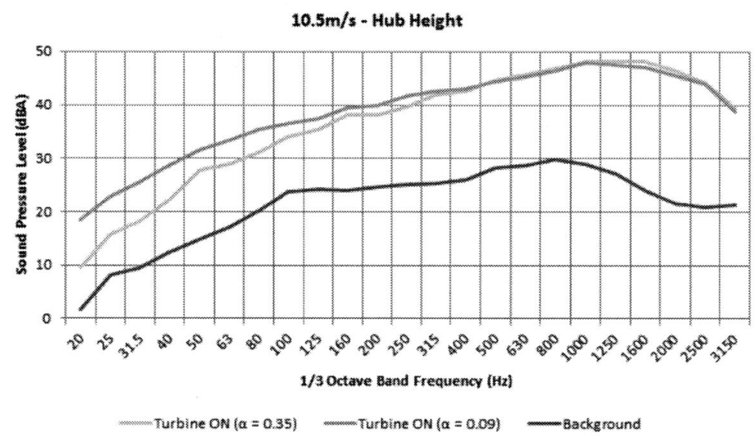

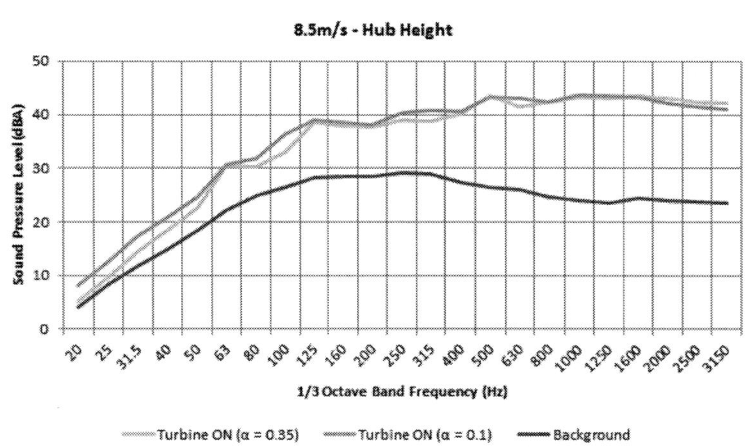

4.1.7 Test #3

Low wind shear levels are generally higher their high shear counterparts in most frequencies up to 1600Hz. Local maxima at 63Hz, 125Hz, and 500Hz have near identical levels between high and low shear datasets.

4.1.8 Test #4

Here the levels are nearly identical above 250Hz. High wind shear levels below 250Hz are equal or lower than the low wind shear data.

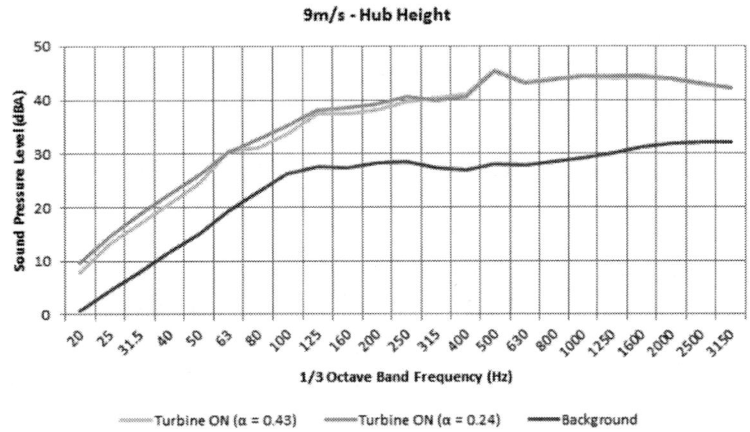

5. Discussion

The measurement data presented in the previous sections do not show any consistent difference between the measured sound level between periods of low wind shear and periods of high wind shear. The overall sound levels were largely identical, and the sound levels in 1/3rd octave bands were often higher in periods of low wind shear.

There are three things to note regarding the differences in sound levels between high and low wind shear.

1. Periods of high wind shear most often occur at night, when the ambient sound levels are generally lower due to reduced human activity.
2. Low wind shear is a side effect of turbulence causing vertical mixing in the airflow. Local eddies and turbulent effect can contribute to a higher sound level due to wind induced noise in the ambient environment surrounding the measurement location.
3. In almost all cases, the differences in measured levels between high and low wind shear conditions are less than the measurement tolerance of the test method.

It should be noted that airflow stall has not been considered in this study. All the datasets presented in this paper were taken from turbines with active pitch control systems on each blade. Older generation stall-controlled turbines are not included in this study. Stall occurring along a portion of the blade length (also known as dynamic stall) has also not been addressed in this study.

6. Conclusion

Based on the data presented in this paper, there does not appear to be a link between higher wind shear and increased noise generation of a wind turbine. In fact, the sound emission level even dropped in some tests as shear coefficient increased although the changes are within the general range of standard error of the test. The turbine acoustic signatures did not appear to change dramatically between high and low shear conditions.

Furthermore, the difference in incident wind speed on the turbine blades due to wind shear is small when compared to the tangential flow velocity caused by the rotation of the blades.

7. References

[1] D. Bowdler and G. Leventhall, Wind Turbine Noise, Essex: Multi-Science Publishing Co. Ltd., 2011.

[2] J. E. Ffowcs Williams and L. H. Hall, "Aerodynamic sound generation by turbuent flow in the vicinity of a scattering half plane," *Journal of Fluid Mechanics,* no. 40, pp. 657-670, 1970.

[3] IEC 61400-11 , *Wind Turbines - Part 11: Acoustic noise measurement techniques,* International Electrotechnical Commission, 2012.

[4] K. Attenborough, K. M. Li and K. Horoshenkov, Predicting Outdoor Sound, New York: Taylor & Francis, 2007.

7th International Conference on Wind Turbine Noise
Rotterdam – 2nd to 5th May 2017

Variations in measured noise emission of wind turbines due to local circumstances

Ing. W. van der Maarl, Peutz bv, w.vandermaarl@peutz.nl
Ing. E.H.A. de Beer, Peutz bv, e.debeer@peutz.nl.

Summary

When measurements are performed to determine the noise emission of onshore wind turbines, it is often found that the noise emission not only depends on the wind speed at hub height but also varies due to other aspects. Meteorological conditions and location specific circumstances like nearby objects and terrain conditions influencing the wind profile appear to be relevant. In this paper these situational factors are assessed.

Multiple close distance measurements in conformity with IEC61400-11 are performed at several wind turbine sites. From analysis of these measurements it is found that the occurring sound power levels deviate significantly day-by-day and per site, even with identical wind turbines. In the paper these deviations are related to the measurement method, changing atmospheric conditions and roughness length. Also the configuration of the wind turbine and location specific circumstances have a contribution in the found deviations.

Assessment of the sound power levels of wind turbines at project sites should be done taking into account the found spread of measured sound power levels. This spread is of concern when e.g. sample compliance checks are performed by the authorities by measuring the sound power level at dominant wind speeds. In the paper the practical implications are discussed.

1. Introduction

When nuisance is reported by residents due to the noise of a wind park local authorities may order a survey to check whether the noise limits at the nearby dwellings are met. With the European standard based on year equivalent noise limits (L_{den} and L_{night}) it is not possible to check these limits directly by measuring the sound pressure level at the nearby dwellings. In year round measurements disturbing noise would be dominant. Therefore the L_{den} and L_{night} are calculated. To check if the sound power levels provided by the manufacturer are also applicable for that specific situation, measurement at the project site of the occurring sound power levels is sometimes demanded.

Results of measurements at project sites show differences between the sound power levels provided by the manufacturer and the sound power levels measured at the project site. In paragraph 3.1 of this paper several measurement results are presented which are performed at project sites during varying conditions. In paragraph 3.2 measurement results performed at one wind turbine during varying conditions are given. In chapter 4 the found deviations are assessed and possible explanations are given.

2. Situation and method

2.1 Situation

The area in the north-west of The Netherlands is very suitable for wind parks because it is an windy area due to flat land and the nearby North Sea. This rural area is according to Dutch standards an area with low density of population and low background noise. However, still many sites are nearby dwellings. Nuisance due to solitary wind turbines and wind parks is reported frequently.

The local authorities often demand (driven by complaints of local residents) a survey to check whether the noise limits are met in practice.

In a time span of several years (2011 up until now) the sound power levels of many wind turbines were validated in practice by our company.

In figure 1 a typical situation of a solitary wind turbine in the aforementioned area is given.

Figure 1. Typical situation of wind turbine in rural area of The Netherlands

For this paper we made a selection of project sites where compliant checks were ordered by the local authorities ranging from solitary wind turbines of 800 kW up to more than 7 MW and wind parks. The surface area can be qualified as flat with occasional embankments and water channels. The surface is covered with seasonal agricultural crops or grass lands.

2.2 Method

Since 2011 the general noise limits in The Netherlands for wind turbines are 47 dB L_{den} and 41 L_{night} at the facade of dwellings, independent of the background noise levels.

In the Dutch law a Calculation and measurement method for wind turbines is prescribed for prognosis studies and compliance checks [1]. This method is based on the IEC 61400-11 edition 2 standard as well as the Dutch 'Guide for measuring and calculating industrial noise'. Contrary to the IEC61400-11 standard a 1 minute equivalent sound pressure level is used and the sound power level is related to V_{hub} (wind speed at hub height), calculated from the actual power output of the wind turbine and the applicable power curve. To obtain a year average noise emission the sound power levels are corrected with the statistical occurring wind speed at hub height.

For the selected wind turbines two comparisons on the datasets are made:
1) Comparison of guaranteed sound power levels with sound power levels determined at project sites
2) Comparison sound power levels of one specific wind turbine measured during different meteorological conditions

Dataset 1) consists measurements at project sites with an identical wind turbine type (800 kW, blade length 27 m). The hub height is 60 m for measurement 2 – 5 and 73 m for measurement 1.The sound power levels for dataset 1) are determined in conformity with the Dutch method for determination of the sound power levels. Per wind speed bin at least 6 valid measurements (with and without wind turbine in operation) are mandatory.

Dataset 2) consists measurements at one project site with a wind turbine of 2.5 MW, hub height 85 m and a blade length of 60 m. Measurements on this site were done in conformity with IEC 61400-11 edition 3 (with wind speed bins of 0.5 m/s, equivalent sound pressure level of 10 s etc.). Measurement no.6 en 7 (see table 2) are performed with the same wind turbine configuration. Measurements 8 – 10 are performed with another configuration but this configuration was unaltered during the measurements.

2.3 Measurement conditions

At every project site noise measurements were done to determine the sound power levels per wind speed bin. Measurements were done for at least several hours in the day or evening period during changing wind speeds in the range of 5 to 11 m/s (at hub height). The wind speed at hub height was derived from the actual power output (1 minute average for measurement 1-4 and 10 s for measurements 5 - 9) and the applicable power curve.

In table 1 the meteorological conditions during the measurements of dataset 1) are summarized.

Tabel 1 – Meteorological conditions at project sites with identical wind turbine types

Parameter	Measurement no.		
	1+2	3	4
Timespan	19:00 - 22:30	16:45 - 20:45	17.00-22.00
General wind direction	0 degrees	210-240 degrees	210-240 degrees
Temperature	7 °C	17 °C	14 °C
Pressure	1021 hPa	1008 hPa	1011 hPa
Relative humidity	79%	90%	80%
Cloud cover	Clear sky	Heavily cloudy	Heavily cloudy

The wind speed is standardized to 15 °C and 1013 hPa for measurements 1 to 5 (in conformity with the Dutch standard) although the majority of the measurements were performed already around these standard conditions.

In table 2 the meteorological conditions during the measurements of dataset 2) are summarized.

Table 2 - Meteorological conditions at one wind turbine at different days

Parameter	Measurement no.				
	5	6	7	8	9
Timespan	18:00 - 23:45	12:00 - 18:00	10:30 - 16:30	11:00 - 17:30	09:30 - 12:00
General wind direction	210-300 degrees	317 degrees (North West)	281 degrees (West)	200-240 degrees	210-220 degrees
Temperature	4,6-5,7 °C	5.4 - 11.0°C	4.3 - 12.1°C	8,6 -11,1°C	10,3-11,1 °C
Pressure	1011 hPa	1027 hPa	1024 hPa	1021 hPa	1023 hPa
Relative humidity	71-82%	75%	81%	80-97 %	84-90 %
Cloud cover	Partly cloudy	Half cloudy	Few clouds (2/8)	Heavily clouded	Heavily clouded/foggy

3. Measurement results

3.1 Wind turbines of same type at different locations

In figure 3 the measured sound power levels of four different measurements at three different project sites (and the same production model wind turbine) are given.

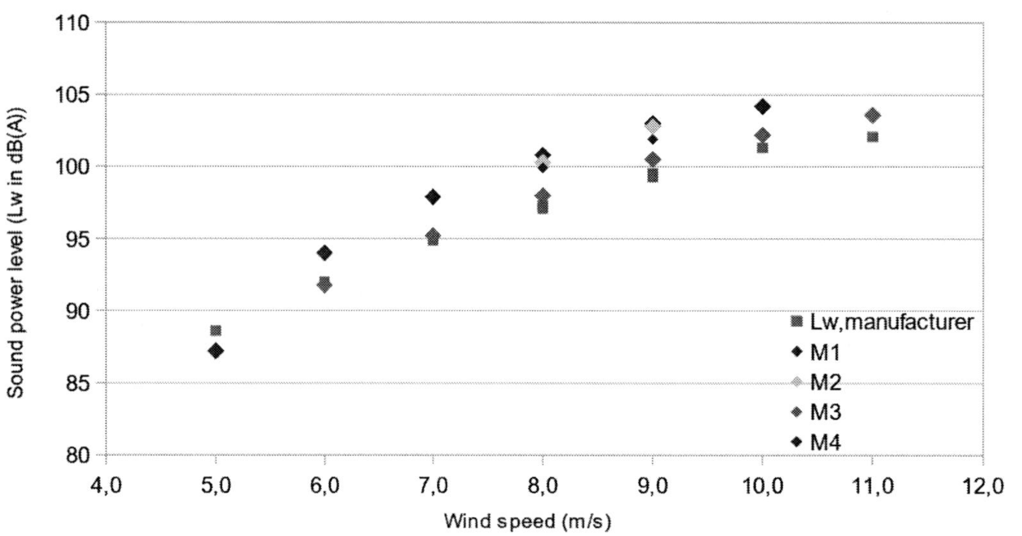

Figure 3 Sound power levels based on measurements at the project sites (measurements 1 to 4) compared to sound power levels supplied by the manufacturer.

The difference between the guaranteed sound power levels of the manufacturer and the measured sound power levels range from -1.4 to +4.5 dB(A) per wind speed bin. The standard deviation of the found spread is +1.8 dB(A). Figure 3 shows that at wind speed bins 5 and 6 m/s at hub height the deviation from the guaranteed sound power level is quite small (minus 1 to plus 2.0 dB(A)). At the wind speeds of 7 - 11 m/s the deviation is significantly higher. Since this range of wind speeds occurs during approximately 50% of the time these sound power levels contribute for a great part to the year average emission.

In figure 4 the measured A-weighted third octave band spectra are given for measurements 1 to 4.

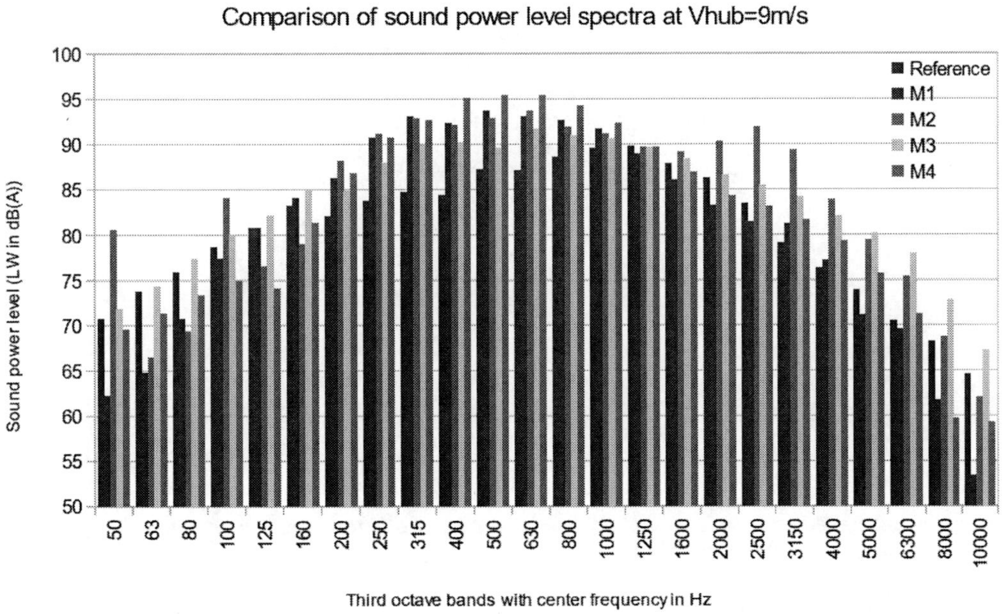

Figure 4 Comparison of measured third octave band spectra of measurements 1 - 4

From figure 4 can be concluded that in the frequency range 160 to 800 Hz and 2,5 – 10 kHz a significant increase occurs compared to the reference sound power level (supplied by the manufacturer). Analysis of all measurement data shows that within measurements M2 en M3 significant disturbing noise occur in the frequency range of 2,5 – 10 kHz, since the difference between the measured sound pressure levels with turbine in operation and with turbine shut down is less than 3 dB. This disturbing noise is caused by a row of trees close to the wind turbine and measurement position. However, the energy in these octave band is small. The broadband value (dB(A)) is at maximum 0.5 dB(A) lower if the third octave band from 2,5 to 10 kHz are neglected.

In the frequency range of 160 – 800 Hz also a significant increase occurs. This frequency range contribute greatly to the total sound power levels in dB(A). In chapter 4 possible causes are further investigated.

3.2 Comparison of sound power levels day-to-day

In figure 4 the average sound power levels measured at one wind turbine at different days and different meteorological conditions are given. The occurring wind speeds during these measurement days were approximately the same. The configuration of the wind turbine was during the different measurement days unaltered (same blade configuration and management). The sound power levels are normalized (and anonymised) to 100 dB(A) at 8 m/s for the dataset with lowest sound power level at 8 m/s. The underlying dataset consists of at least 10 measurements of 10 s per wind speed bin for total noise (wind turbine in operation) and background noise (wind turbine shut down).

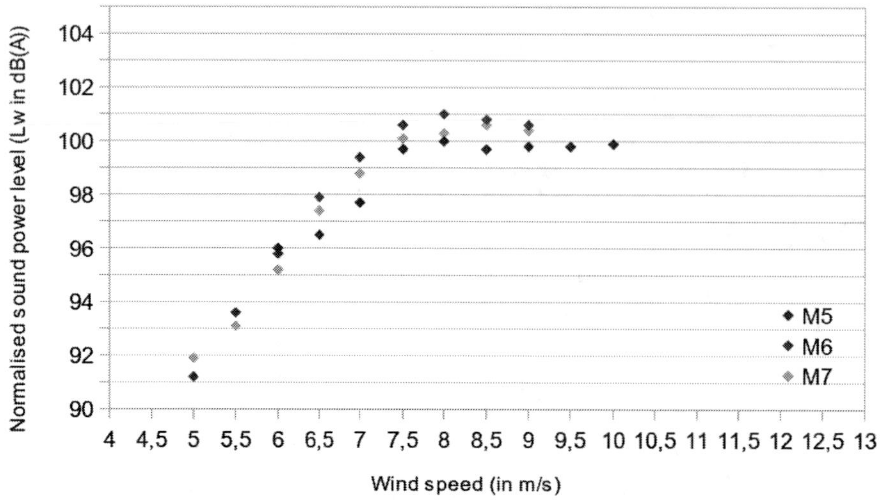

Figure 5 Sound power levels based on measurements at a wind turbine at different days and unaltered configuration of the wind turbine (measurements 5-7)

The deviation of the measured sound power levels at different days range from 0,5 to +1,7 dB(A). The standard deviation of the found spread is +0,4 dB(A).The deviation in this dataset is at the lower wind speed bins of 4,5 to 6 m/s smaller than at the higher wind speeds between 6,5 and 9 m/s although in this wind speed range more data points are collected.

In figure 6 the A-weighted third octave band spectra are given for the wind speed bin at 7,0 m/s for measurements 5 - 7.

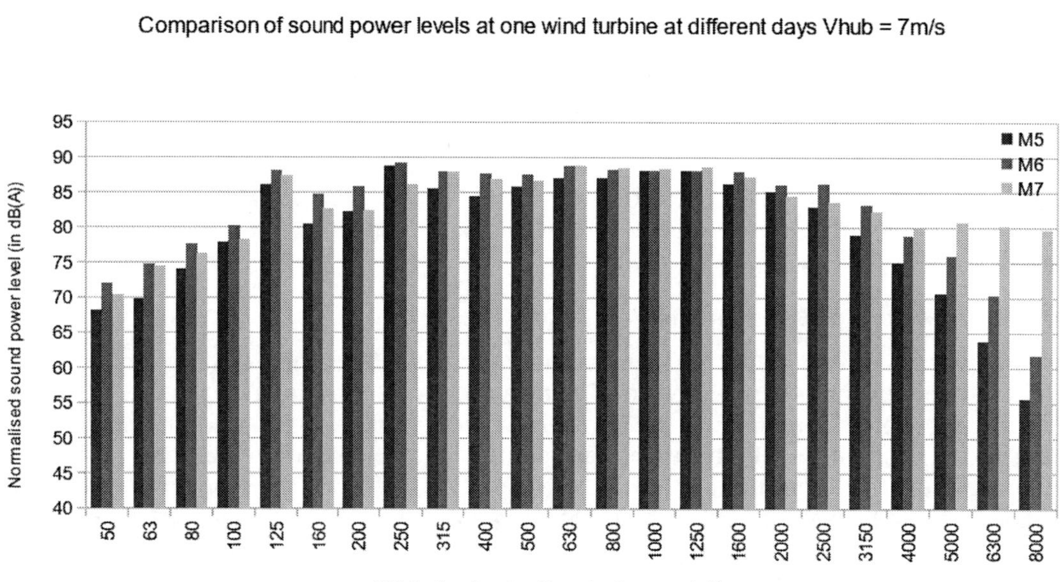

Figure 6 Measured sound power levels in third octave band spectra of one wind turbine during different days at V_{hub} = 7 m/s

Figure 6 shows that the spread of the sound power levels at one wind speed ranges from 1 to 5 dB (excluding the deviations at 5 kHz and higher since these sound power levels are determined without the necessary 3 dB difference between turbine in operation and turbine shut down). The contribution of the third octave bands of 5 kHz and higher to the total sound power level in dB(A) is however limited to 0,3 dB(A) at maximum, which is negligible. Therefore the found spread in dB(A) occurs mainly due to the energy in the 50 Hz to 800 Hz third octave bands.

Figure 7 shows the measured sound power levels from one wind turbine at two different days.

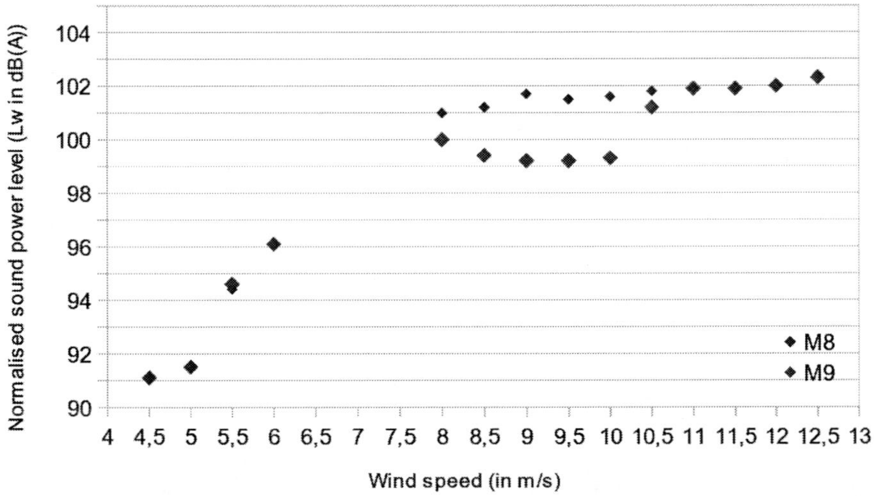

Figure 7 Sound power levels based on measurements at a wind turbine at different days and unaltered configuration of the wind turbine (measurements 8 and 9)

In figure 7 a remarkable pattern is shown whereby lower wind speeds (4,5-6 m/s) are almost identical, while 7,5 – 10 m/s deviates significantly (up to 3 dB(A)). Even more remarkable is that measurement 7 is performed the following day (even only 12 hours later) after measurement 6, with verified unchanged settings of the wind turbine.

In figure 8 the A-weighted third octave band spectra are given for the wind speed at 8,5 m/s for both measurements.

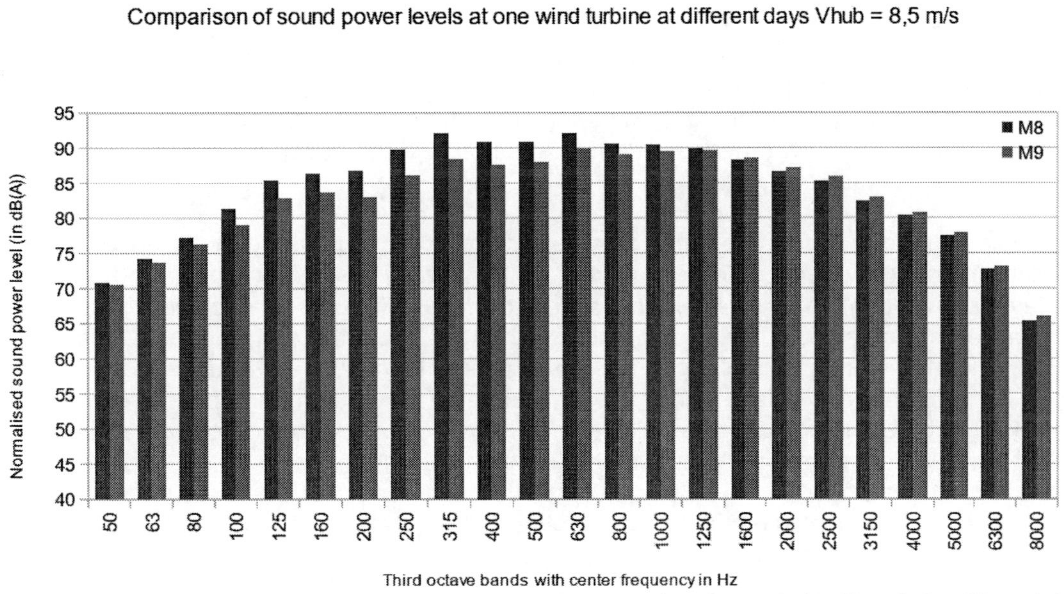

Figure 8 Measured sound power levels in third octave band spectra of one wind turbine during different days at V_{hub} = 8,5 m/s

Figure 8 shows that the found difference in sound power levels occurs mainly in the 100 Hz to 1 kHz third octave bands.

Possible causes of the found deviations are further discussed in chapter 4.

4. Assessment of effects contributing to deviations

4.1 General

In general the following aspects may contribute to the found spread of measured sound power levels:

- Accuracy of the measurements (method, disturbances);
- Wind turbine conditions;
- Meteorological conditions;
- Objects and terrain influencing disturbances and wind profile.

In the following paragraphs possible causes for the found deviations are considered.

4.2 Accuracy of measurements

The sound power levels determined in practice at project sites deviate at some sites and some wind speeds significantly from the sound power levels of the manufacturer. The sound power levels are determined exactly the same measurement procedure as they are determined at the test field. Therefore comparison is possible, and several variables can be excluded as main cause. Close field measurements have in this regard advantages over far field measurements because deviations due to the effects of noise propagation are (as much as possible) eliminated.

It is remarkable that in almost all cases the sound power levels determined at the project sites are higher than the sound power levels delivered by the manufacturer.

The accuracy of measurements in conformity with IEC 61400-11 edition 2 and 3depends on several aspects but is typically due to site effects ('type B' uncertainty) between 1 and 3 dB(A). In appendix D of the standard (edition 2) the accuracy of the measurements are given. Since the wind speed at hub height is derived from the power output of the windturbine and the power curve, the accuracy of the power measurement equipment and the power curve contribute for the most part the actual accuracy.

Disturbing noise during the measurements may contribute to the found spread. However in these measurements the effects of foreground noise (temporary elevation of the sound pressure level due to local noise sources not related to the wind turbine) is excluded from the measurements as much as possible. Incidental disturbances could therefore not be a suitable explanation. The third octave band spectra show a significant increase at the octave bands higher than 2,5 kHz. In these octave bands the difference between the sound pressure level due to the wind turbine and background noise is less than 3 dB(A). At these octave bands background noise is dominant. The total contribution of these octave bands to the total sound power level is however limited to 0,5 dB(A) maximum. The increases at the lower third octave bands (200 Hz to 1 kHz) have the highest contribution to the total sound power levels. At these third octave bands the background noise is at least more than 6 dB(A) lower than the noise due to the wind turbine.

For the measurements at one wind turbine (Measurements 5 to 9) the aforementioned considerations apply as well, although in these measurements comparison is only made between the different measurement results. The actual accuracy of these measurements is even higher because the measurements were done with exactly the same instruments leading to a smaller margin for the measurement accuracy.

4.3 Wind turbine conditions

The deviations may be caused by wind turbine configuration due to:
1) Dimension tolerances;
2) Cleanliness of blades;
3) Management of the pitch angles of the blades.

Measurements 1 to 5 are done at the same fabricate and type wind turbine. Due to tolerances in the production process the dimensions of the blades may deviate from each other leading to possible increase of the sound power level. Measurements 6 to 10 are done at one specific wind turbine where a comparable spread of sound power levels is found. The effect due to dimension tolerances is therefore assumed to have a minor contribution.

It is known that blade cleanliness has an influence on the noise emission [Oerlemans, 2]. At the measured wind turbines no dust or dirt was visible at the blades. Unknown is if the blade roughness was increased by prolonged operation. The measured wind turbines were operational for less than a year and were not operating in specific dusty areas since the soil was during the measurements covered with crops or grass.

Rotational speeds were observed during the measurements and checked with the range of the specifications. The rotational speeds were in accordance with the specification of the manufacturer so the measurements performed at the project site were done without reduced (or increased rotational speeds). Therefore is a significant contribution to the increased sound power levels not likely.

Management of the pitch angles is for measurements 1 to 5 considered to be exactly identical. However this is not confirmed by the operator of manufacturer so it cannot be excluded as a possible cause. Measurements 6, 7 and 8 to 10 were done at one wind turbine with confirmed unchanged management of the blade pitch, as it was done at consecutive days.

4.4 Meteorological conditions

During the measurements several meteorological conditions varied significantly. The relative humidity (influencing air density) varied 70-97% (RH) and cloud cover (influencing atmospheric stability) varied from clear sky to fully clouded.

Inflow turbulence may occur due to specific meteorological conditions, for instance sunny days with heated ground surface leading to a buoyancy effect of air mixing with higher layers. Since most measurements were done with (almost) full cloud cover or were done in the evening at sunset or after sunset the effect of 'turbulent mixing in the boundary layer' [YY2] is assumed to be of lesser influence. Inflow turbulence is however considered as one of the most relevant aspects contributing to the increased sound power levels (see paragraph 4.5).

4.5 Objects and terrain disturbances

Upwind nearby objects and terrain disturbances like existing channels, rows of trees, embankments lead to increased (inflow) turbulence, and decrease the stability of the atmosphere. The roughness length of the surface near the areas of the project sites (measurement no. 1 to 4) ranges from 0.03 to 0.3, which deviates at some sites significantly from the reference roughness length of 0.05 from the IEC 61400-11 standard used to determine the sound power levels of the wind turbine by the manufacturer.

The nearby objects and terrain disturbances also may change the inflow angle and increase inflow turbulence. All wind turbines of the considered project sites have upwind objects, like homesteads, rows of tree, channels and soil bodies at a distance of less than 150 m.
The influence of objects near to the ground is greater when they are more near the wind turbine but may have an influence on the inflow turbulence and wind profile up to several kilometers [Ragheb, 3]. Increased inflow turbulence is found to increase the sound power levels [Søndergaard, 4], although others did not find a significant influence [Evans and Cooper, 5].

At the project site of measurement no. 2 and 3 the wind turbine was located near the homestead surrounded by a row of trees. During measurement 3 the row of trees were at a distance of less than 50 m from the wind turbine and the measurement location. At the project site of measurement no. 4 an embankment and homestead was present most probably leading to inflow turbulence. This inflow turbulence may even occur during specific meteorological conditions (wind direction and speed).

Measurements 5 – 9 were done at a site where upstream at a distance of 400 m a row of trees was present, while no other nearby objects (of considerable size) were present. This site was therefore without nearby disturbing objects. Nearby objects and terrain disturbances on these measurements is considered to be of minor influence.

4.6 Implications on compliance

The found increased sound power levels based on measurements at the project sites are between 0 and 4 dB(A) per wind speed bin. In the researched sites the found increase of sound power levels did not lead to exceedance of the noise limits at the nearby dwellings. However the found increase of sound power levels lead to an increase of the year average emission factor (L_E) of maximum 2 dB. Since the initial calculated L_{den} and L_{night} was about 1 to 2 dB lower than the noise limits, no substantial exceedance was found.

The found increase is however higher than is expected only on measurement uncertainty. In situations where neighbors are complaining about noise due to the wind turbine measurement of the sound power levels in practice may be a good starting point to check whether increased sound power levels occur.

In the planning phase it may in critical situations be necessary to avoid compliance issues and discussion afterwards. To be more certain that the noise limits in practice will be met a surplus of 1 to 3 dB(A) per wind speed bin could be considered in the calculation results before assessment with the noise limits is done.

5. Conclusions

From the measurement results of 9 measurements at different sites and meteorological conditions a significant spread is found in the occurring sound power levels. The found spread is higher than measurement accuracy only, therefore other relevant influences occur.

The wind turbine configuration was probably the same for 1-4 and confirmed the same for M5-9 so the found spread is probably not caused by different wind turbine configurations

Meteorological conditions varied quite heavily during several conditions which is a probable cause for relevant part of the found spread. Nearby objects and terrain influences were at some project sites present, probably leading to an increased inflow turbulence, because of nearby rows of trees and buildings. For measurements 1-4 inflow turbulence could well be the most relevant cause for the found increased sound power levels.

The found elevated sound power levels due to local circumstances can be between 1 and 3 dB(A). To avoid compliance issues one may consider to add a surplus of 2 or 3 dB(A) in the planning phase in the development of a wind turbine site before assessment with noise limits is done.

Since the contribution of the considered possible causes is yet unknown differentiation between specific local circumstances (size and type of nearby objects, predicted inflow turbulence) is not possible based on the datasets used in this paper. Further investigation to the contribution of these effects is necessary.

References

[1] Reken en meetvoorschrift windturbines 2011, Activiteitenregeling bijlage 4

[2] Acoustic Array Measurements on a Full Scale Wind Turbine, S. Oerlemans and B. Méndez López, 2005

[3] Wind shear, roughness classes and turbine energy production, M. Ragheb, 2015

[4] Noise from wind turbines under non-standard conditions, L. Søndergaard, 2012

[5] Effects of different meteorological conditions on wind turbine noise, T. Evans and J. Cooper, 2013

7th International Conference
on
Wind Turbine Noise
Rotterdam – 2nd to 5th May 2017

Vertical directivity observations based on statistics of low frequency tonal components measured at downwind and upwind locations.

X. Falourd, P. Marmaroli, R. Feuz, PRONA SA, Yverdon-les-Bains, Switzerland
D. Bollinger, HEIG-VD, Yverdon-les-Bains, Switzerland

E-mails: x.falourd@prona.ch , dominique.bollinger@heig-vd.ch

Summary

The time dependent noise footprints of a 101-m diameter wind turbine are measured over four seasonal periods with collection of data from four sound level meters, one meteorological station and one LIDAR. Correlated to the turbine operational parameters, the time series and statistics of the tonal low-frequency components (presenting Doppler shift) are significant of one vertical directivity related on the rotor speed and probably the pitch of blades. This study helps to deepen the knowledge on vertical directivity and how to consider the sound power level based on IEC61400-11 in the assessment of wind turbine noise.

1 Introduction

In order to answer questions about the modelling and measurement methodology of noise from a wind turbine, the study is based on a detailed analysis of acoustic data collected at different weather and wind conditions during the four seasons of the year, specifically with the methodology of the Swiss noise abatement regulation for this type of noise source in mind.
In the previous study presented in conference WTN2015 [1] the time series and statistics of noise levels and amplitude modulations in the bands of ⅓-octave (50 Hz to 125 Hz) were presented and compared with Doppler frequency shift analyses of recorded sounds. It has been showed that spectro-temporal sound analysis related to turbine operation parameters presents strong correlations validating the Doppler low-frequency shift hypothesis of three tonal acoustic sources located on the blades. The result characterized the amplitude modulation depths of low-frequency sound from the wind turbine and allows extracting a generic proposal for low-frequency noise mapping in azimuth around the wind turbine.

With the aim of qualifying and quantifying the propagation of wind turbine noise, the study presents results on the emergency of the WTN and focus on the analysis of statistics of low frequency tonal components presenting noticeable emergency at 500 m range for different wind speed conditions. The main goal is to better understand how low-frequency noise is generated and propagated for various wind classes at longer distance. The statistics on the propagation loss between distant sound level meters (range 150 m to 500 m) shows that the radiated sound power between the four sound level meters is not constant and is dependent to the rpm of the rotor. Based on these observations, the hypothesis relies on that the blades pitch regulation is the main factor in the observed vertical directivity.

2 Wind turbine site and descriptions

The studied wind site is representative of an alpine Valley (villages, roads, highway and railways) and it allows studying the acoustic propagation over a flat ground for two opposite regimes of wind.
The time dependent noise footprints of a 101-m diameter wind turbine are measured over four periods lasting at least 10 days for four seasons with collection of data from four sound level meters, one meteorological station, one wind profiler LIDAR and an array of microphones (see [1] for details). Turbine operations parameters are recorded with mean values (10 min.) for the entire campaign. Sound level meters were arranged on both sides of wind (two upstream, two downstream) as shown in figure 1.

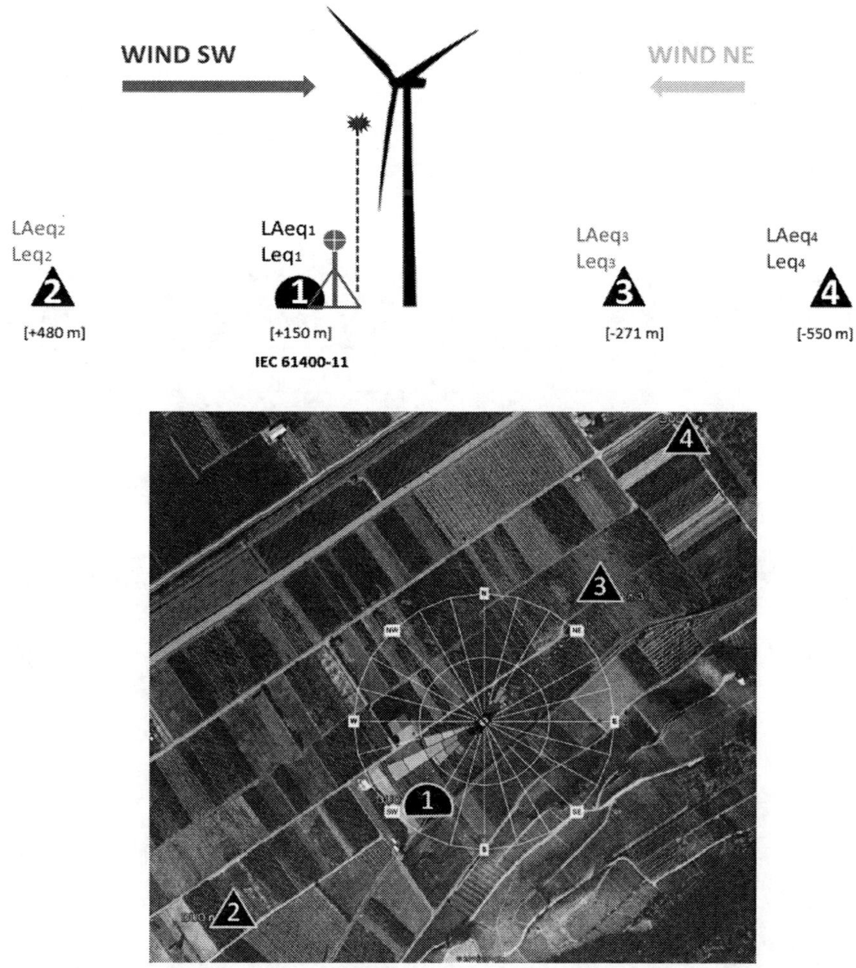

Figure 1 : Locations of the wind turbine (red line) and of the four sound level meters and one characterical wind rose of the site. (credit Google Earth)

This minimum configuration is used to collect noise data for the two principal wind regimes. Sound level meters synchronized by GPS clock were programmed to collect noise data (125 ms recordings) completed with intermittent sound recordings for subsequent acoustic analysis. The wind profiler was located upwind of the turbine for controlling the wind direction near the SLM n°1 (Sound Level Meter) positioned according to standard IEC 61400-11 for the downward wind. SLM n°1 is placed on a rigid circular board laid on the ground and is equipped with a secondary windscreen.

3 Analysis of an episode of stop of the wind turbine

The control of WTN levels in dB(A) at immission points requires that the WTN is measurable and if possible the noise levels are corrected using coefficients qualifying the audibility of the WTN (tonal and/or impulsive). The characterization of the WTN at average distances (between 250 m and 1000 m) and at long distances (beyond 1000 m) is thus subjected to the criterion of emergence or the discrimination of the WTN on the ambient noise. The simplest method is thus to carry out measurements of the noise levels LAeq and Leq in bands of third of octave for stable conditions and to discriminate these noise levels according to whether the wind turbine is under operation or not. This method is usually used during measurements of reception but requires carrying out series of "stop and go" for each class of wind and each angular sector to measure around the wind turbine.

In order to estimate the emergence of the WTN for longer distance than the locations defined in IEC 61400-11, the event of a stop of the WT is analyzed to figure out if it possible to measure WTN levels at immission points. Figure 2 presents one daily report of the SLM n°1, located at +150 m upstream of the wind turbine for an event of stopping of the wind turbine during favorable wind conditions to the wind energy production.

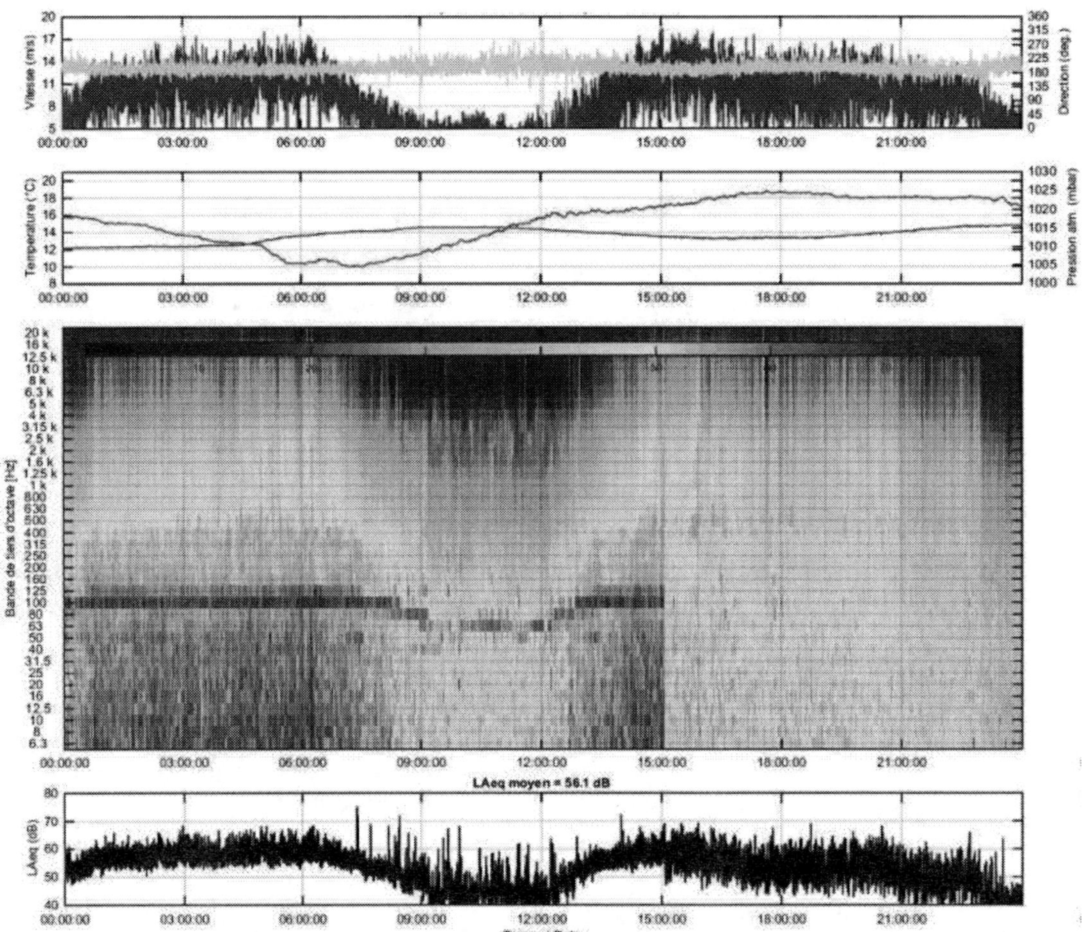

Figure 2 : Meteorological and acoustic daily report at location SLM n°1 figuring the stop of the wind turbine at 15:00:00. From top to bottom: wind speed and direction; weather conditions, ⅓ octave spectrogram, LAeq time serie.

The ⅓ octave band spectrogram clearly documents the stopping of the wind turbine starting at 15:00:00. The noise levels of the first ⅓ octave bands up to 800 Hz decrease suddenly after the stop of the WT; on the contrary ⅓ octave bands of above 800 Hz are only slightly or not affected due to the ambient noise of the same order of the WTN.

3.1 A-Weighting noise level assessment

Figure 3 presents the statistics of the noise levels LAeq measured at the SLM n°1 according to the speed of wind collected at 10m height as well as the weak models of regression of the wind noise for the four positions of SLM for this same day of measurement.

The data in the figure are in:

- light blue points: LAeq (SLM n°1, r =150 m) during the operation of the WT
- gray points: LAeq (SLM n°1) related to the ambient noise only (WT stopped)
- blue line: weak LAeq regression of the LAeq the operation of the WT
- gray Line: weak LAeq regression of the ambient noise
- cyan line: transposition of the LAeq regression of SLM n°1 to SLM n°3 (r = 270 m)
- pink line: transposition of the LAeq regression of the SLM n°1 to SLM n°2 (r = 480 m)
- red line: transposition of the LAeq regression of SLM n°1 to SLM n°4 (r = 550 m)
- green pointed curve: LAeq levels (SLM n°1) according to ISO9613-2 calculations based on the WTN certificate.
- black pointed curve: LAeq levels sum of LAeq (WTN) and LAeq (wind weak model).

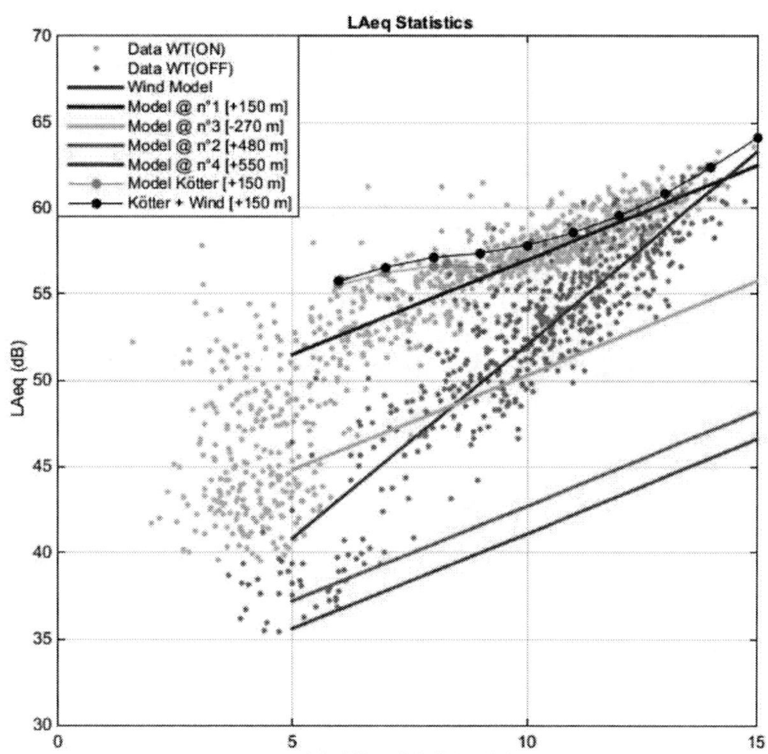

Figure 3: Raw WTN and wind noise statistics (LAeq), comparisons with simplified modelisations

Observations

a) For the range of wind speeds described in the WTN certificate (6 to 9 m/s), measured LAeq levels at SLM n°1 are in accordance to the certified noise levels (green curve).

b) The comparison between the weak regressions of WTN and the wind noise model makes it possible to estimate the emergence of the WTN on the ambient noise for various speeds of wind at the four measuring locations (SLM n°1 to 4) under the assumption that the ambient noise is identical on the four sites.

c) For wind speeds greater than 9 m/s, the global noise level LAeq (black pointed curve) is composed of the WTN and the wind noise. By applying propagation loss terms of the WTN models (between 5 and 10 m/s) to greater ranges (SLM n°2,3 and 4), it is clear that the WTN noise is of the order of the wind noise (SLM n°3, r = 270 m) and is far below the wind noise (SLM n°2 and 4, ranges greater than 500 m).

In regards of these statistics, the WTN assessment at immission points for ranges greater than 150 m based on LAeq measurement is limited by the ambient noise levels.

3.2 Zero-Weighting noise level assessment

An identical approach of this stop event is proposed considering the Leq noise levels (zero weighting). Figure 4 presents the statistics of the noise levels Leq measured at SLM n°1 according to the speed of wind as well as the models of regression of the wind noise for the four locations of SLMs. The representations are identical to those of the preceding figure.

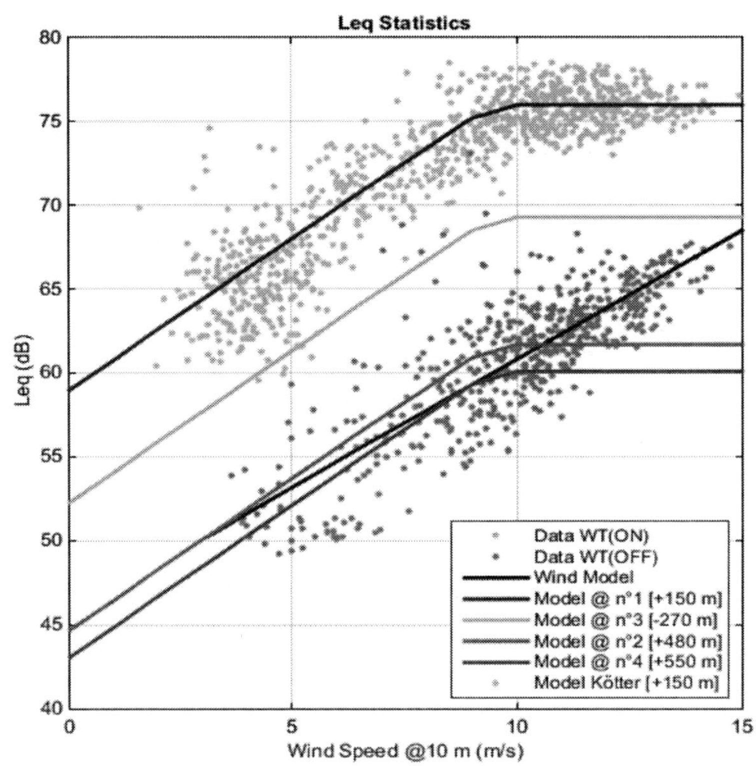

Figure 4 : Raw WTN and wind noise statistics (Leq), comparisons with weak modelisations

Observations

The emergence of the WTN Leq in dB(Z) at SLM n°1 is obtained for all speeds. Beyond 10 m/s wind speeds, the Leq noise level is considered as constant as for the turbine maximum power. For the wind speeds lower than 10m/s, the measurement of the WTN can be undertaken on the four sites according to favorable conditions' of instantaneous emergence of the WTN. The wind turbine noise is mainly composed of low frequencies below 1 kHz as depicted in the following figures showing the event stop for the ⅓ octave bands between 8 Hz and 1 kHz.

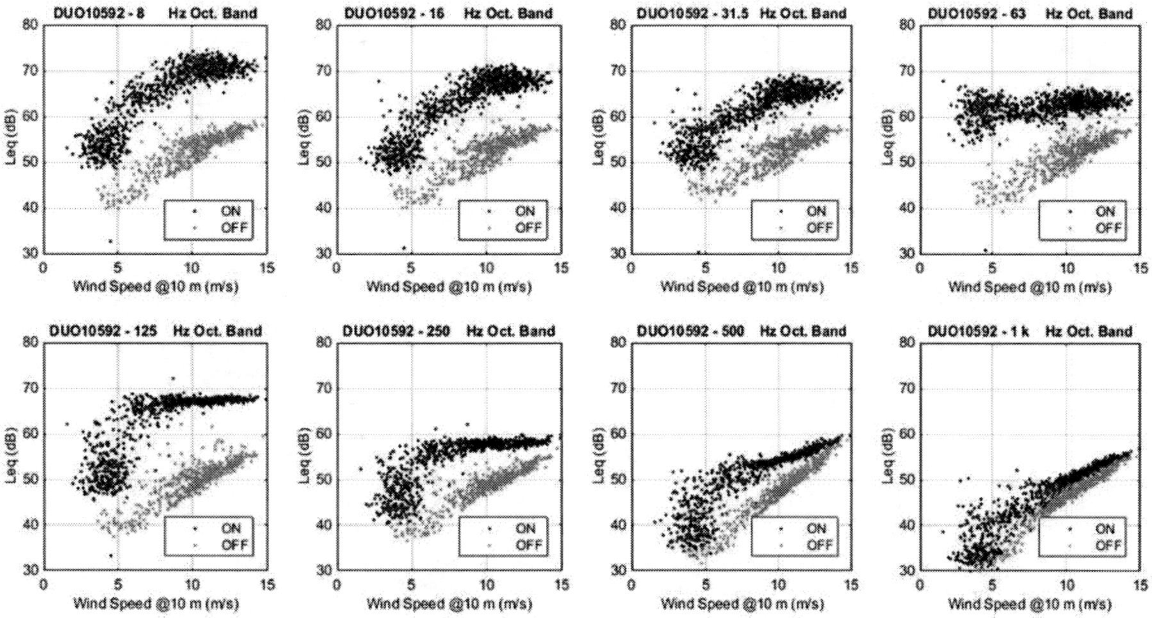

Figure 5 : Raw WTN and wind noise statistics (Leq) in ⅓ octave bands at SLM n°1 (IEC 61400-11)

The emergence of the wind noise (black) on the background noise (blue) is clearly visible for all frequencies up to 1 kHz. It is remarkable that best emergencies obtained are those associated the very low frequencies (of 8 Hz with 125 Hz). A-Weighting function strongly attenuates the contribution of the first frequency bands lower than 1 kHz in the total level LAeq. On the basis of these results, the measurement of the WTN using LAeq for ranges greater than the IEC61400-11 distance is not recommended; measurements at greater distances are subjected to errors dependent on the considerations of ambient noise caused by the wind.

In order to qualify the long range acoustic propagation of the WTN, it is required to collect statistics on low frequencies transmission losses between the four SLMs. With the benefit of the low frequency tonal components presenting emergencies on the background noise, it is possible to measure the propagation loss by comparison of sound levels from the four sound level meters.

4 Measurement of WTN low-frequencies at long distances

For this site, the project profits of emergencies of varying low frequency tonal components as depicted in the following figure. As described in [1], knowing the rpm of the WT (given or estimated) it is possible to track the varying tonal components related to three acoustic sound sources located on the three blades. These lasts are dominating the 50 Hz - 125 Hz bands of frequencies and are directly correlated with the wind speed i.e. the rotation speed of the wind turbine.

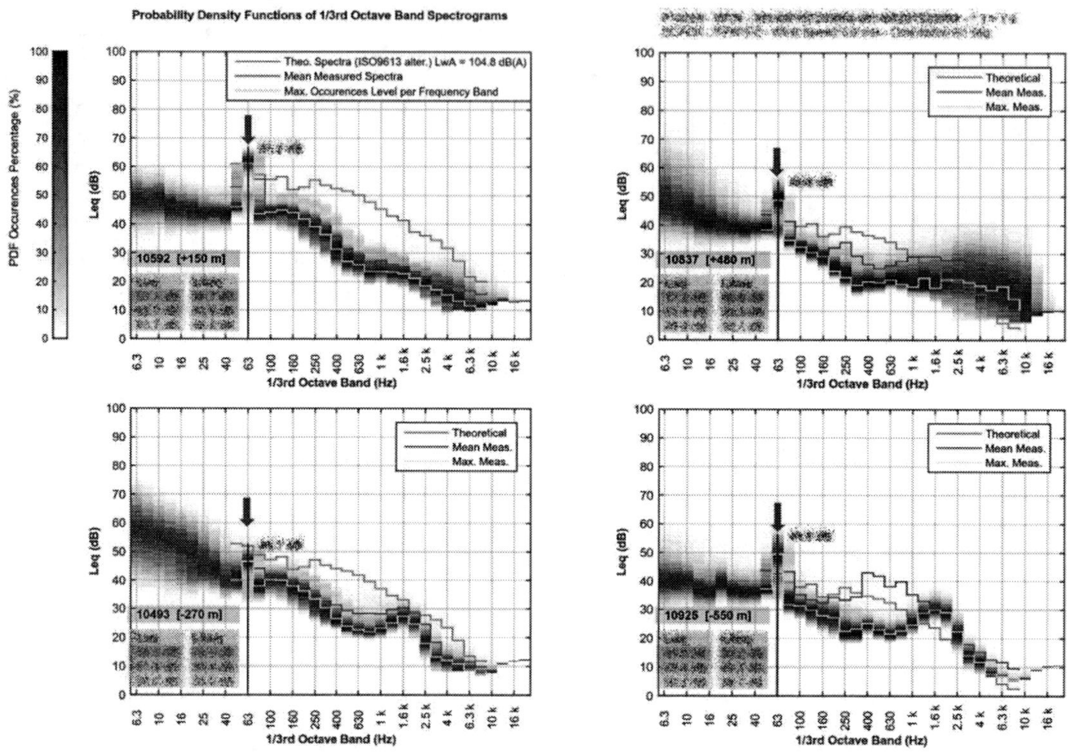

Figure 6: Probability density functions of ⅓ octave bands of the SLMs n°1 to 4.

Under the assumption that the tonal low frequencies radiated by the wind turbine have the same characteristics of broad band acoustic sources located on the blades, the analysis of the propagation of the noise is based on the comparison of transmission losses (TL) measured between the positions from the SLMs n°2, 3 and 4 and the position of the reference SLM n°1 (Figure 7).

The analysis of the time series and statistics of transmission losses (TL_{21}, TL_{31}, TL_{41}, TL_{42}) for different classes of wind speed authorizes the qualification of propagation effects on the WTN levels for distances up to 500 m. The relative Transmission Loss of the instantaneous tonal component is calculated by:

$$TL_{21} = Leq(2)_{,\ tonal} - Leq(1)_{,\ tonal.}$$

The process to extract instantaneous tonal components is described in [1].

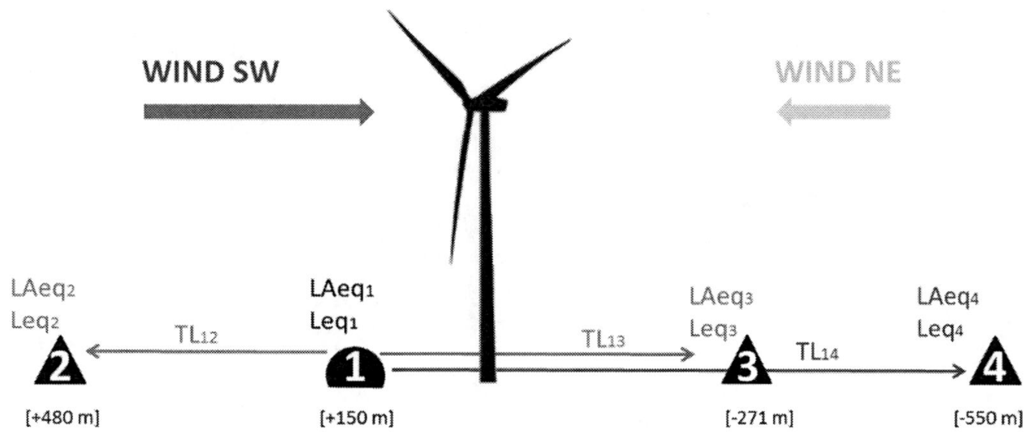

Figure 7: Transmission Loss (TL) calculation principle.

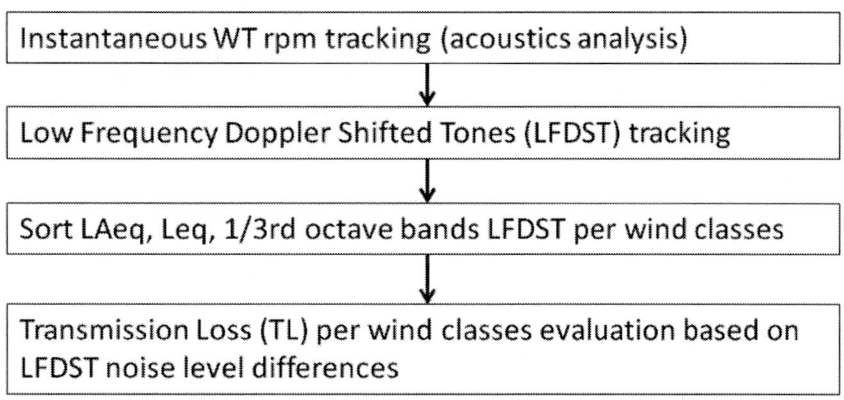

Figure 8: Principle of calculation of wind dependent transmission losses of low tonal frequencies

Notes on the variability of the tonal components

1. As explained in [1], the tracked central frequencies are Doppler shifted tonal sound sources located on the blades. The three rotating sound sources are received by SLM with Doppler shifts and it creates an interference pattern explaining amplitude modulations relative to rpm and depending on the location of the SLM. It means that depending on the relative position of the SLM facing the WT, the cyclic Doppler shift changes the central frequency drastically (up to more than 10 Hz) and causes amplitude modulations which might be distributed over two ⅓ octave bands.
2. SLM n°2, 3 and 4 are not equipped with secondary windscreen and might be exposed to wind. However an overview of the probability density functions of ⅓-octave bands for each wind bin allows us to consider these measurements as relevant.

Keeping in mind that this tracking and this evaluation of TL might be blurred by Doppler effect, the seasonal statistics of TL are produced for each wind speed bin. The following figures present:

1. Top Left: LAeq statistics for SLM n°1 (IEC 61400-11 location, r = 150 m)
2. Top right: TL_{21} statistics (SLM n°2, r = 480 m)
3. Bot. left: TL_{42} statistics
4. Bot. right: TL_{41} statistics (SLM n°4, r = 550 m)

In yellow are depicted weak regressions of the scatter points. In blue lines are represented expected transmission losses considered as constants (independent of the wind speed).

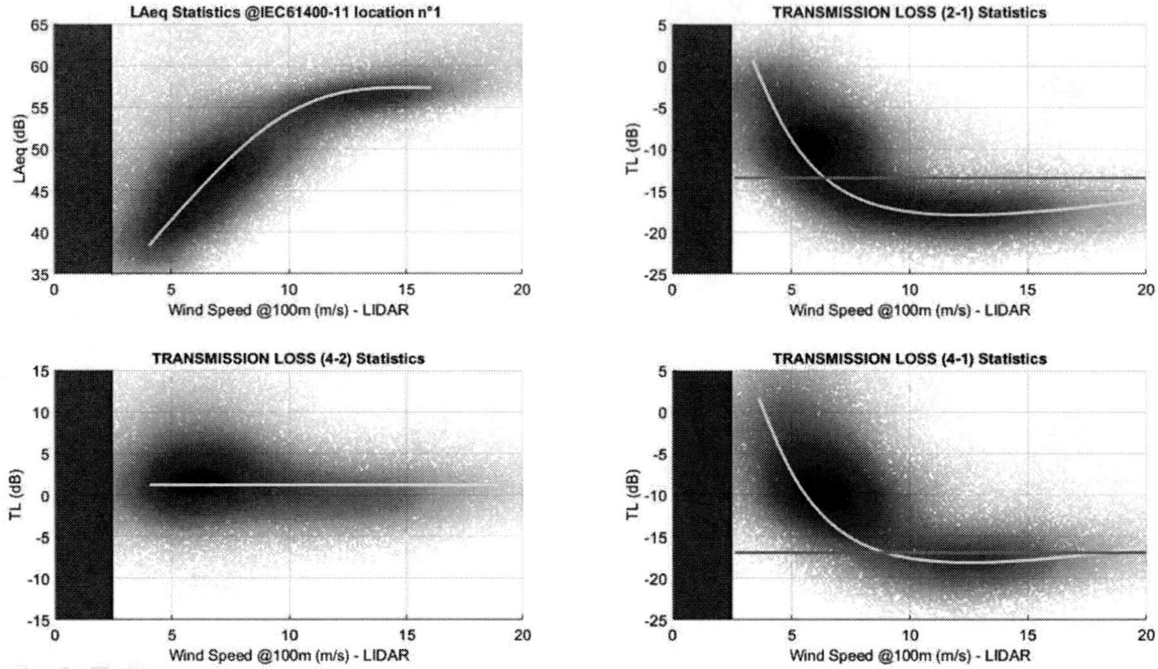

Figure 9: LAeq statistics (SLM n°1) and Transmission Loss (TL) estimated at SLM n°2 and n°4.

Observations:

TL_{21} statistics and TL_{41} statistics figure out similar curves which are not constant. For low wind speed the WTN differences of tonal levels at SLMs n°2 and n°4 with the tonal levels at SLM n°1 is less than expected considering ISO9613-2 for an omnidirectional sound source.

The transmission losses evolve towards an asymptotic limit predicted by ISO9613-2. The effect of the wind on the propagation could be at the origin of this loss of sound level but one discovers by analyzing the statistics downstream and upstream of the wind turbine of the same evolutions of the losses by transmission. TL_{42} statistics figure out that TL_{42} is almost constant (independent of the wind speed) and the WTN level at SLM n°4 is greater than at SLM n°2. (SLM n°4 is mostly downwind).

The vertical directivity due to the blade pitch regulation might be responsible of lower sound levels of tonal components at SLM n°1 (IEC 61400-11) due the directivity of the three sound sources located on the blades (supposed as dipoles [1]). Based on these observations, one hypothesis is that the blade pitch regulation is the main factor in the observed vertical directivity. Under the assumption that tonal components are radiating the same way that broadband WTN, the modeling of the sound levels at long distances might be underestimated by taking into account of the measurements according to the IEC61400-11 due to the supposed vertical directivity of blade pitch regulation minimizing the WTN at SLM n°1 for low wind speeds.

The transmission losses per wind classes have been evaluated for the four seasonal periods of measurements (May, August, December and February). As depicted in the following graphs, one can see that the transmission loss curves are all very similar with a deficit of TL for low wind speeds. More precisely one can observe that for warm months, TL are greater than for cold months figuring the effect of the ground on the propagation.

Considering the previous notes on the way these statistics were done, a further investigation shall merit to discriminate downwind and upwind data with an adequate processing of the acoustic signals. However the data collection does not permit to investigate deeply this hypothesis since it requires having continuous recordings of WTN with sound level meters equipped with secondary windscreen laid on a flat board.

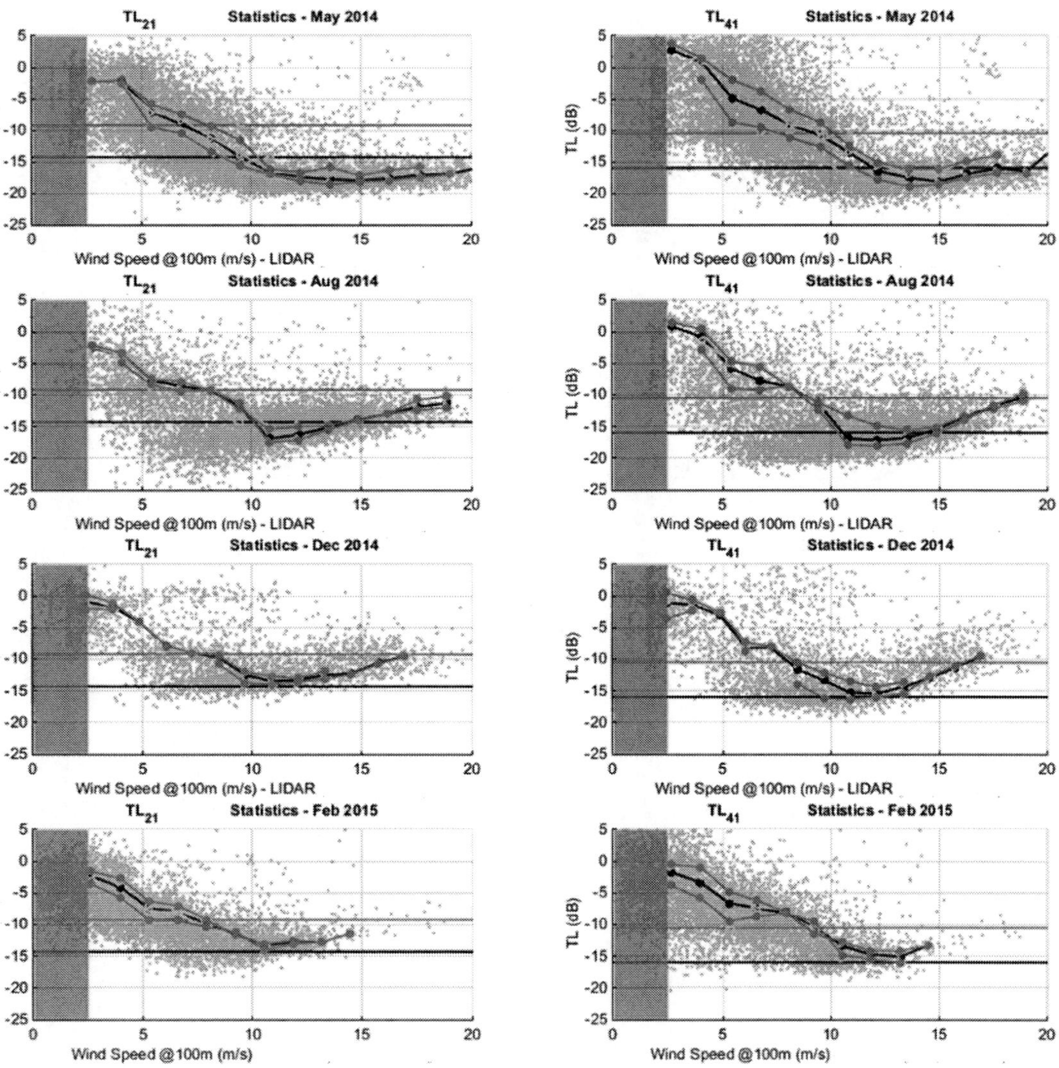

Figure 10: Seasonal statistics of Transmission Loss (TL_{21} and TL_{41}).

5 Conclusions

The time dependent noise footprints of a 101-m diameter wind turbine were measured over four seasonal periods with collection of data from four sound level meters, one meteorological station and one LIDAR. By tracking specific tonal components related to the rpm of the WT, it has been shown that relevant statistics of WTN at distant locations (up to 500 m) permit to reveal that the acoustic radiation of this sound sources contribute to a vertical directivity which is rpm dependent. Correlated to the turbine operational parameters, the time series and statistics of the tonal low-frequency components (presenting Doppler shift) are significant of one vertical directivity related to the blade pitch regulation.

6 Acknowledgements

The project has been possible through the support of the Federal Office for the Environment (FOEN) and the Environment Department of Canton Vaud (DGE-Vaud). The authors want to thank both FOEN and DGE for their support, advices and help for the realization of this project, research and papers.

7 References

[1] X.Falourd & al., *Low frequency amplitude modulation related to Doppler frequency shift: an experimental study of a 101m diameter wind turbine in a Swiss valley*. Wind Turbine Noise Conference 2015, Glasgow.

[2] D. Bowdler, Geoff Leventhall. *Wind turbine noise*, Brentwood, Essex: Multi-Science Publishing, 2011.

7th International Conference
on
Wind Turbine Noise
Rotterdam - 2nd to 5th May 2017

Wind turbine noise measurements in controlled conditions

K. Boorsma, J.G. Schepers
ECN Wind Energy, Westerduinweg 3, 1755 LE Petten, The Netherlands
E-mail: boorsma@ecn.nl

Abstract To validate and reduce the uncertainty associated with noise prediction models for wind turbines, there is a need for detailed noise measurements on wind turbines in controlled conditions. However, high quality wind tunnel campaigns on horizontal axis wind turbine models are scarce due to the large wind tunnel size needed and consequently high associated costs. To serve this purpose an experiment using the 4.5 meter diameter Mexico turbine was set-up in the large low speed facility of the DNW wind tunnel.

An overview of the experiments is given including a selection of results. Both far field microphone as well as microphone array measurements have been performed, together with unsteady force measurements on five instrumented blade sections. This allows a unique insight in the relation between acoustics and the underlying aerodynamics. Overall noise characteristics of the turbine have been determined for a variety of operational conditions by varying tip speed ratio and blade pitch angle. Scaling of the noise is studied by comparing similar combinations of tip speed ratio and pitch angle for different tip speeds. The effect of blade soiling on the noise is evaluated using roughness strips, as well as the influence of yawed or misaligned inflow on the rotor noise. A comparison to calculations using a BPM model is given indicating for which operational conditions this model suffices and for which parts the noise prediction can be improved.

In summary, after years of preparation, ECN and partners have performed very successful aero-acoustic experiments in the largest wind tunnel in Europe. The comprehensive high quality database that has been obtained will be used in the international Mexnext consortium to further improve wind turbine acoustic modeling.

1. Introduction

Uncertainty in aerodynamic noise (and load) prediction is an important parameter driving the price of wind energy [1, 2]. An accurate prediction of noise can aid the design of more quiet blades but also noise mitigation strategies. Validation by experiments is the most plausible route to model improvement. Although many field measurements on wind turbines exist [3], the uncertainty in inflow conditions (turbulence, shear, gusts) and turbine specifications complicates progress. To reduce this uncertainty, experiments in controlled conditions as featured in wind tunnels are a prerequisite. However, high quality wind tunnel campaigns on horizontal axis wind turbine models are scarce due to the large wind tunnel size needed and consequently high associated costs [4, 5]. To serve this purpose, measurements on the Mexico wind turbine were carried out in the Large Scale Low Speed Facility (LLF) of the German Dutch Wind Tunnels (DNW) in 2014 as a follow up of its previous campaign in 2006 [6]. One of the special features of this experiment is that in addition to the loads and flow velocities also noise sources were measured using a phased microphone array.

An illustration of this experiment, called New Mexico, is given in Figure 1. Readily published results featuring loads and flow field measurements plus their comparison against simulations can be found in [7, 8, 9, 10, 11]. The present paper gives an overview of the acoustic part of the experiment, including selected results together with a comparison to noise predictions.

Figure 1. Test set-up of the experiment

2. Test set-up

The set-up of the experiment was largely identical to the first Mexico campaign, featuring an open jet configuration. A picture of the set-up is added in Figure 1. The model features a three-bladed 4.5 m diameter upwind rotor, including a speed controller and pitch actuator. The blade features the DU-91-W2-250 profile for the inboard sections, RISØ-A2-21 midboard and NACA-64418 for the outboard sections. The model is instrumented with unsteady pressure sensors at five sections (25%R, 35%R, 60%R, 82%R, and92%R), distributed over the three blades. Strain gauges were added to the root of the blades to measure flap- and edgewise bending moments. Several model related sensors were installed in the nacelle (e.g. generator torque, 1p sensor, accelerometer and inclinometer) to track turbine performance. The model was suspended to a 6 component balance at the tower foot to measure forces and moments. Phase locked stereo PIV measurements were performed a the 9 o'clock plane of the rotor for a variety of configurations and locations.

A difference with the previous campaign lies in the fact that acoustic measurements were performed. Thereto an acoustic array was positioned between nozzle exit and the model, below the jet (depicted in red in Figure 1). As can be observed the array could not be placed directly upstream of the model, but was positioned slightly sideways due to the restricted space available between the nozzle (depicted in orange) and external balance (depicted in blue). The 4m x 4m phased array consisted of 140 electret microphones (circular arrangement) sampled at a frequency of 51.2 kHz over a period of up to 60 s for each data point. In addition to that 48 far field microphones, arranged in three horizontal rows on the side wall of the test chamber (covering directivity from about 40° to 140° with respect to the rotor center, where 90° denotes sideways propagation) were used, featuring the same data acquisition parameters as the array.

Although all microphones are positioned outside of the open jet, they are protected by so-called foam 'wind balls' against wind noise from secondary flows in the test hall. All microphones were calibrated using a certified pistonphone, providing a pure tone of 94dB at 1 kHz. Acoustic lining was applied to the test chamber side walls, floor and ceiling wherever possible to prevent reverberations. Foam padding was applied to the top side of the balance (depicted in grey in Figure 1) to prevent noise disturbance from the impingement of the jet shear layer onto the sharp objects of the model support frame. More details about model, test set-up and instrumentation can be found in [12, 7, 13].

2.1. Post-processing and uncertainty

To reduce the enormous amount of acoustic data resulting from the raw time series of the far field microphones, DNW has applied a fast Fourier transform using a block size of 4096 yielding a narrow band frequency resolution of 12.5Hz. Since the model was initially not designed to perform aero-acoustic research with, the motor/generator and gearbox appeared to be rather noisy. Unfortunately this noise over shadowed the aerodynamic rotor noise, which is the subject of research, for all conditions. The background noise caused by the wind tunnel itself was shown to be lower than the model noise for all conditions. The first observation makes it rather difficult to separate rotor noise from the motor/generator/gearbox noise. As such the results from the farfield noise measurements are not discussed in the current paper.

The array data are processed by DNW using beamforming with the same block size as the far field microphones. The CLEAN-SC enhanced beamforming algorithm [14] has been used to separate the rotor noise from the motor/generator and gearbox noise by defining the scan grid as the rotor plane. Integrating over the scan grid then yields the narrow band and 1/3-Octave band power spectra. The resulting spectra have been corrected for convective amplification (due to the source moving with the blade) and shear layer diffraction of the open jet. A distance correction was employed using the 1/r law. It was found that after transformation of all array results to 0.28 m relative to the scan grid location, the measured Sound Pressure Level (SPL) would equal the Sound Power Level (PWL). Therefore this this distance has been used throughout the processing. Weighting has not been applied.

It has been shown by former tests at DNW that the absolute accuracy of the sound levels, calculated from the integrated scan areas by enhanced CLEAN-SC beam forming processing, is about ± 3 dB. However, for the relative (delta) accuracy of the sound levels, calculated from the integrated scan areas by enhanced CLEAN-SC beam forming processing, this figure improves to about ± 1.5 dB (or better). More details about the DNW applied post-processing can be found in [13].

Sectional forces (normal and tangential to the local chord) at the five sections have been obtained by linearly integrating the measured pressure distribution along suction and pressure side. The rotor axial force was determined by decomposing these forces in the axial direction and integrating them linearly over the span from blade root to tip, assuming zero loading at the ends. The axial force coefficient Cdax was obtained by dividing this quantity by the freestream dynamic pressure and rotor disk area.

3. Test matrix and configurations

An overview of the configurations relevant for the acoustic analysis is given in Table 1 and some of them are depicted in Figure 2. The first configuration featured roughness strips along the full blade span at a variety of operational conditions in axial flow. After that the roughness strips were removed from the outboard part of the blades housing the NACA profile. A full sweep through the operational regime was performed in axial flow conditions, which included lambda traverses by varying tunnel speed for both 325 rpm and 425 rpm at various pitch angles. In addition to that, the performance was assessed for various yaw angles.

Next several blade add-ons were tested out on the turbine. All of them featured a full sweep through the rotating operational regime, just as was performed for the partly clean configuration. Firstly Guerney flaps were applied to the blades up to 60%R (Figure 2(a)), later they were cut off to extend to 46%R. The Guerney flap consisted of a 0.5 mm thick L-shape strip from thin sheet metal. The shorter side was non-uniform, tailored to the local chord length (2%c). The longer side was kept to 20 mm and mounted to the pressure side of the blades (aligned with the trailing edge) using adhesive tape.

Inspired by IEC pitch fault load cases, a pitch misalignment run was performed. The pitch angle of blade 2 was reduced by 20° in comparison to the other blades. The rotational speed was limited to 325 rpm to keep the instability due to the aerodynamic imbalance low (the nose cone could be observed to 'wiggle' around a bit). A full sweep through the operational regime was performed, featuring the standard pitch angles for blade 1 and 3. In addition to that, lambda sweeps at 15° and 20° (referring to the blade 1 and 3 pitch angle) were performed.

4. Experimental results

An example of resulting beamformning plots is given in Figure 3 for the rough and clean blade configuration. The results clearly show the dominance of turbulent boundary layer trailing edge noise at the outboard part of the blade which features the highest incoming flow speeds. Acknowledging the clockwise rotation of the blade, the most noisy part of the revolution is the downgoing motion of the blade in agreement with previous research on this topic [15, 3]. However it is noted that the peak is observed slightly before the 9 o'clock position, which is attributed to the off-axis location of the array.

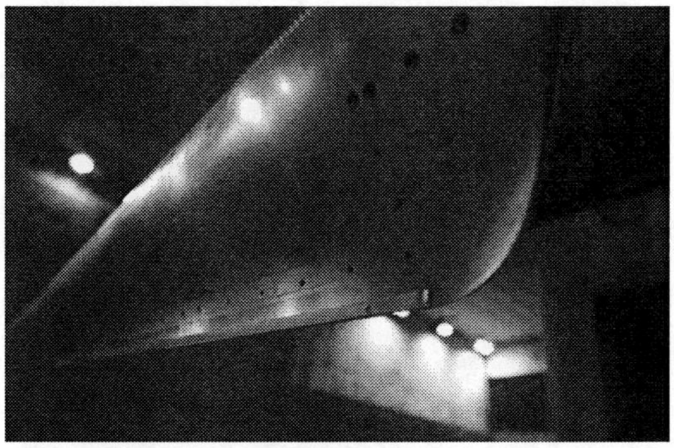

(a) Guerney flap

(b) Smoke visualization

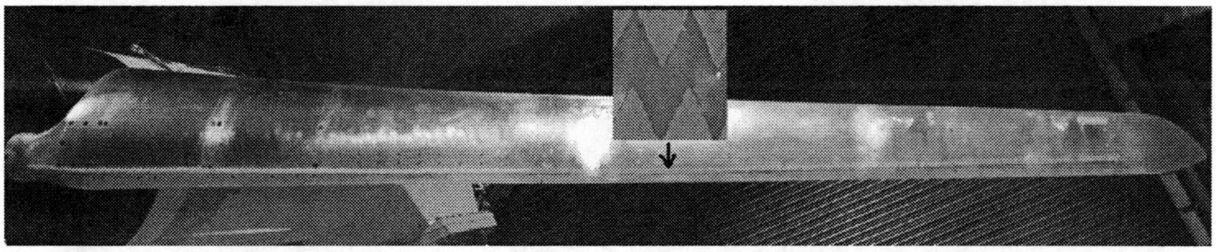

(c) Roughness (zigzag) strips along the full blade span (pressure side) plus inset showing detail

Figure 2. Pictures of different New Mexico configurations

Table 1. Blade configuration legend for New Mexico

Id	Configuration	Roughness
B0		Roughness on full blade
B1	Guerney flaps long (r/R<0.60)	Outboard blade clean (r/R>0.7)
B2	Guerney flaps short (r/R<0.46)	Outboard blade clean (r/R>0.7)
B3		Outboard blade clean (r/R>0.7)
B6	Pitch misalignment blade 2 (-20°)	Outboard blade clean (r/R>0.7)

Resulting 1/3-Octave band spectra are depicted in Figure 4(a) for several operational conditions. A haystack shaped spectrum is observed which is common for this noise source type. A lower tip speed ratio λ will result in a higher angle of attack increasing the noise level but also lower the frequency for which the peak occurs. From these spectra, overall noise levels in terms of Overall Power Watt Level (OAPWL) are obtained which can be plotted and compared for a variety of conditions as shown in Figure 4(b). The same variation with tip speed ratio can be observed as was noted from the spectra.

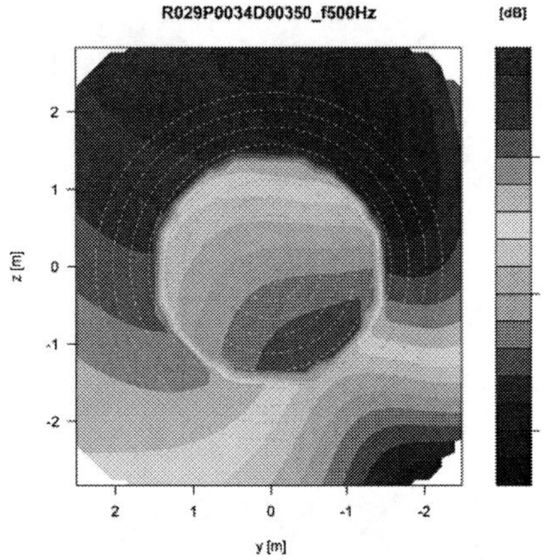

(a) B0, f=500 Hz

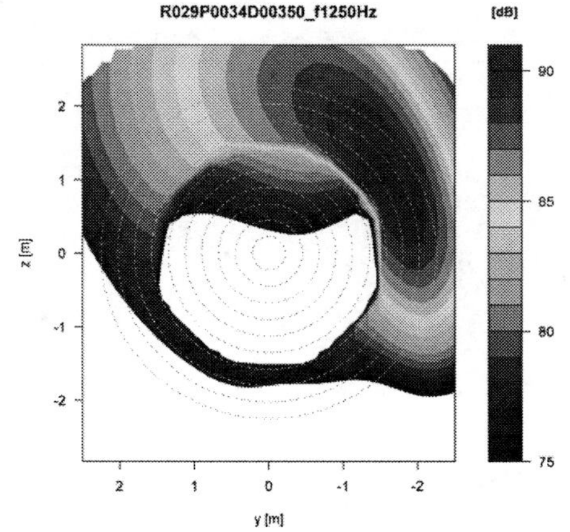

(b) B0, f=1250 Hz

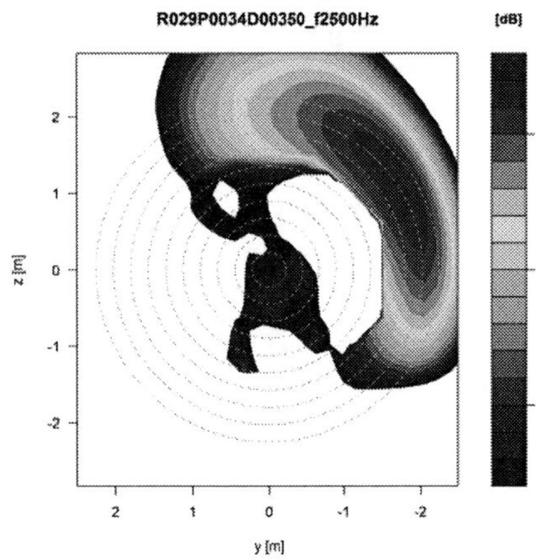

(c) B0, f=2500 Hz

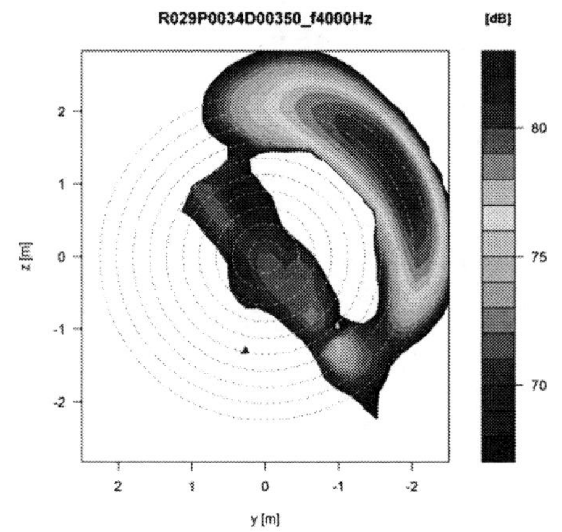

(d) B0, f=4000 Hz

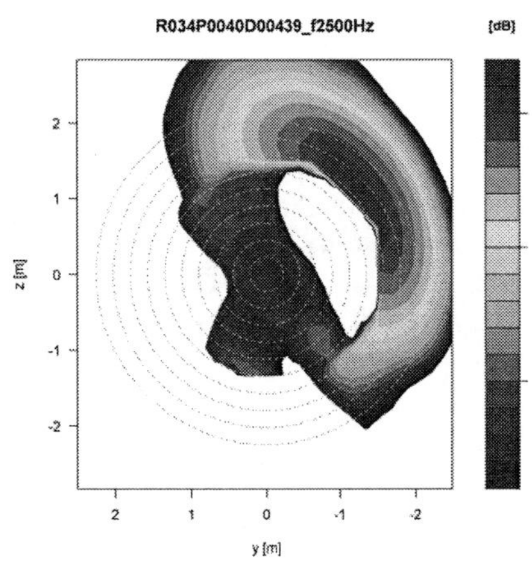

(e) B3, f=2500 Hz

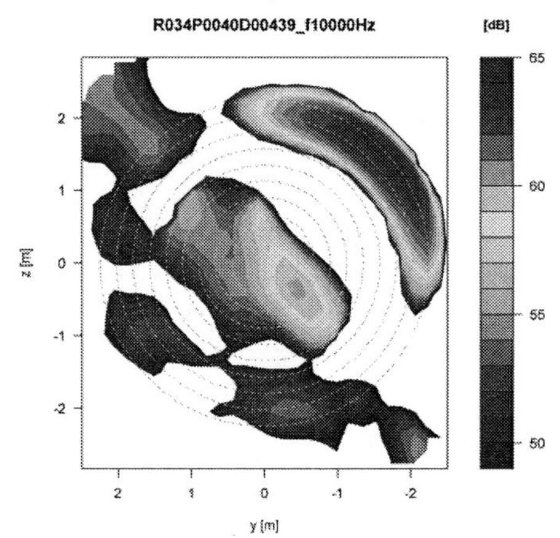

(f) B3, f=10000 Hz

Figure 3. Selected beamforming plots for rough (B0) and partially clean blade configuration (B3), λ=6.7, 425 rpm, pitch=-2.3°

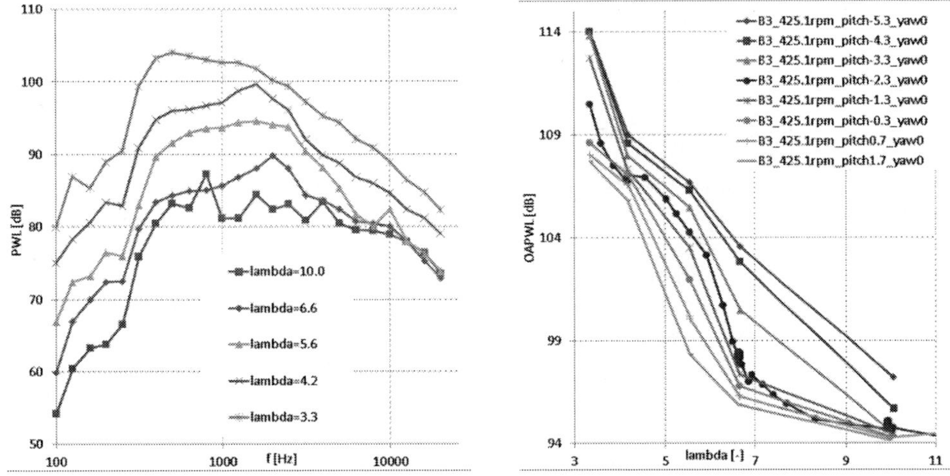

(a) 1/3 Octave band PWL spectra for pitch=-1.3°

(b) OAPWL levels versus tip speed ratio

Figure 4. Noise plots for partially clean configuration B3, 425 rpm

The sharp increase from λ=7 to 5 originates from turbulent transition moving rapidly forward causing more turbulent conditions at the trailing edge. For lower tip speed ratios also trailing edge separation occurs, increasing the noise levels even further. Some of the operational conditions at -2.3° pitch angle were repeated many times (>5) giving an indication of the repeatability. The dependency on pitch angle is also clearly illustrated, confirming that lowering the pitch will increase noise levels due to the higher local angles of attack and consequently more turbulence at the trailing edge. A rule of thumb originating from field test [16] was established in the past estimating 1dB noise reduction per degree increase of pitch angle. Judging by the graph this rule of thumb is not far off, although the amount seems to depend on the tip speed ratio (or local aerodynamic state) under consideration.

4.1. Effect of roughness

The effect of removing the strips from the outboard region is illustrated in Figure 3(e) and 3(f) for two frequencies at design conditions. Comparing around the spectrum peak frequency (f=2500 Hz) to the rough configuration from Figure 3(c) clearly shows the highest noise levels to move further inboard which still has the roughness strips applied. Also it can be observed that at f=10000 Hz the dominant noise source moves further outboard to the tip again indicating the dominance of tip noise at these high frequencies.

The effect on the noise levels can be observed in Figure 9. For the 425 rpm case, a rather large noise increase due to roughness of about 5dB is observed above λ=6.7 in attached flow conditions. The larger than expected increase could be related to the thickness effect of the roughness strip which adds extra to the boundary layer thickness.

Because the boundary layer has now already been triggered to a turbulent state and hence transition does not creep up, there is no steep increase in noise levels below λ=7 as for the partially clean case. For separated flow conditions at very low tip speed ratios the noise levels converge because at these high angles of attack natural transition occurs prior to the trigger position. Similar differences between partially clean and rough conditions are observed for the other pitch angles.

It was also observed that for partially clean blades the repeatability of the noise levels for identical operating conditions was not always perfect. For attached flow conditions differences larger than 1 dB were observed between identical runs which were performed on different days. Although great care was taken to clean the blades each morning, dust particles or similar could have triggered early transition in some cases influencing the noise level. It indicates how sensitive emitted noise levels are to soiling.

4.2. Special configurations and conditions

The influence of **yaw**ing the turbine on the overall noise levels is shown in Figure 5(a). Small yaw angles up to 15° hardly influence the noise levels. Exceeding this misalignment, which is not very common for regular wind turbine operation, can offset the noise levels to about 4dB at 45°. Due to the skewed wake and advancing and retreating effect, local angle of attack and apparent velocity at the blade sections vary with azimuth angle, causing fluctuations during a rotor revolution. As such the azimuth position for which the maximum noise levels are perceived changes, together with a decrease of the time averaged noise level. This is illustrated by the source plot in Figure 5(b) in comparison to the corresponding plot in axial flow conditions in Figure 3(e). This effect also explains why positive and negative yaw misalignment $\pm$ 30° result in different perceived noise levels. So although source levels principally are the same between positive and negative yaw, the perceived noise levels on the ground differ due to this effect.

The **Guerney flaps** were not installed to have an impact on the noise signature but it is worthwhile to investigate whether a noise penalty exists. As indicated in Table 1, the spanwise extension of the flaps was varied with a short (up to r/R=0.46) and long configuration (up to r/R=0.6). As reported previously a clear benefit in terms of power production was measured for the short configuration together with a change in spanwise load distribution for both configuration [8]. Figure 6(a) indicates that extending the flaps to 60%R results in a significant noise increase for tip speed ratios exceeding design conditions (λ >6.7). The corresponding frequency spectra at λ=10 in Figure 6(b) indicate a clear noise increase for frequencies that are normally dominated by trailing edge noise. Apparently the separated flow from the flap interacting with its sharp edges results in a relatively strong noise source, even though apparent velocities are below the velocities in the outboard region.

Introducing a **pitch misalignment** of -20° for blade 2 to mimic a pitch fault condition will result in high angles of attack for this blade. Previous research on the loads in this condition showed, for high tip speed ratios, the blade 2 stalled wake impinging on the following blade 3 [11]. Figure 7(a) shows that this off-pitching results in a noise increase between 5 to 10 dB, due to the extra separation-stall noise on blade 2. It is noted that

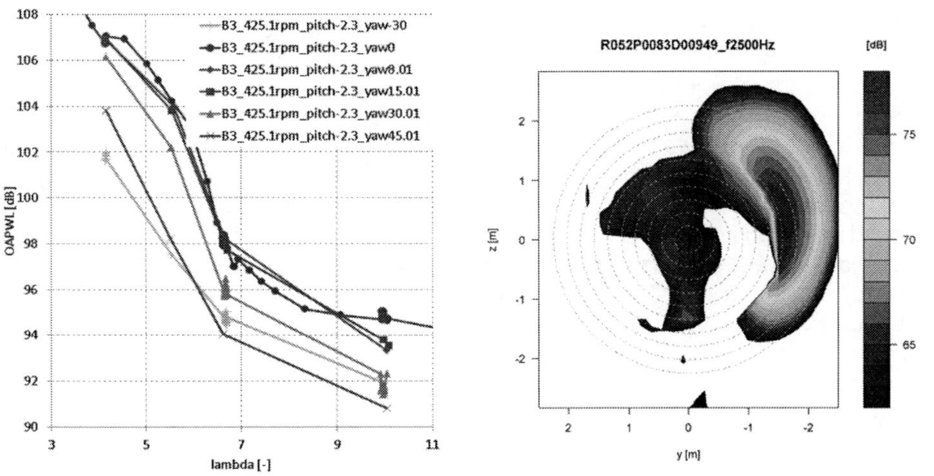

(a) 1/3 Octave band PWL spectra

(b) Source plot at f=2500Hz for 45° yaw, λ=6.7

Figure 5. Influence of yawing the turbine for partially clean configuration B3, pitch=-2.3°, 425 rpm

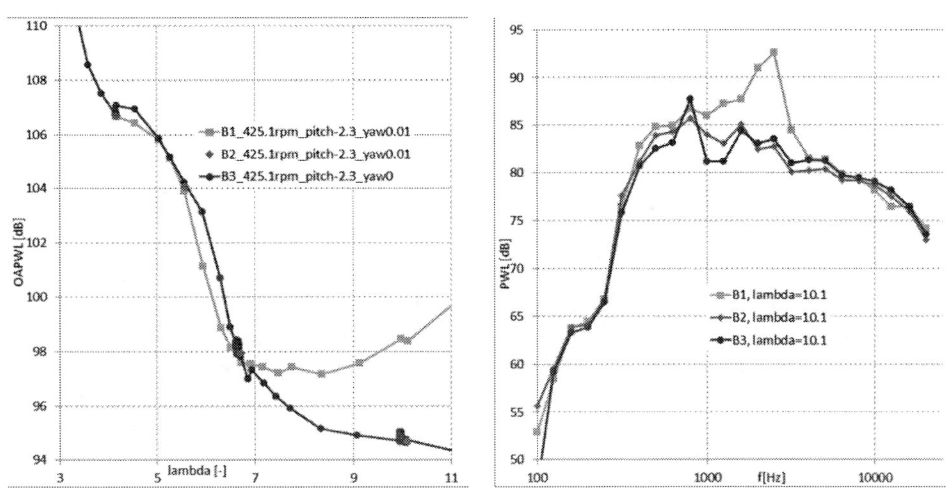

(a) OAPWL levels versus tip speed ratio

(b) 1/3 Octave band PWL spectra at λ=10

Figure 6. Influence of the two different Guerney flaps configurations (B1, B2) compared to the reference (B3), pitch=-2.3°, 425 rpm the indicated pitch angle in the legend corresponds to the pitch angle on blade 1 and 3. The noise trend with tip speed ratio is different in comparison to the reference due to the fact for the featured tip speed ratios, the angle of attack range covered remains in the separated flow region. Also featured in the plots is a configuration with a large pitch angle of blade 1 and 3 (20°), which results in more conventional angles of attack for blade 2. However the rather negative angles of attack for blade 1 and 3 at moderate to high tip speed ratios induce stall for these blades together with a noise increase. For low tip speed ratios, angles of attack will increase to operation in the attached flow region for this configuration, which explains the lower noise levels below λ=7. The corresponding frequency spectra at λ=6.7 in Figure 7(b) show the peak level to slowly shift to lower frequencies below 1kHz in case of stalled flow.

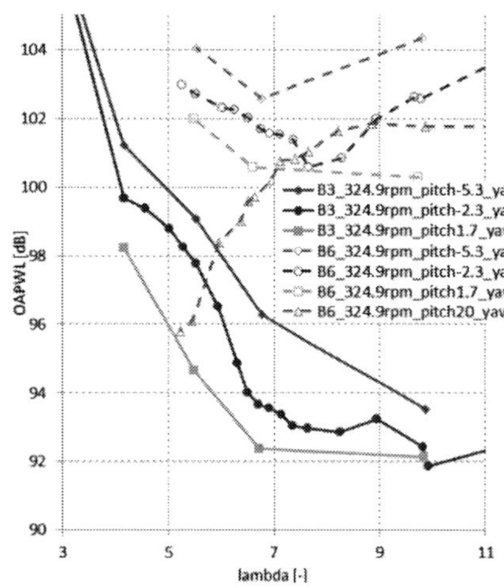

(a) OAPWL levels versus tip speed ratio

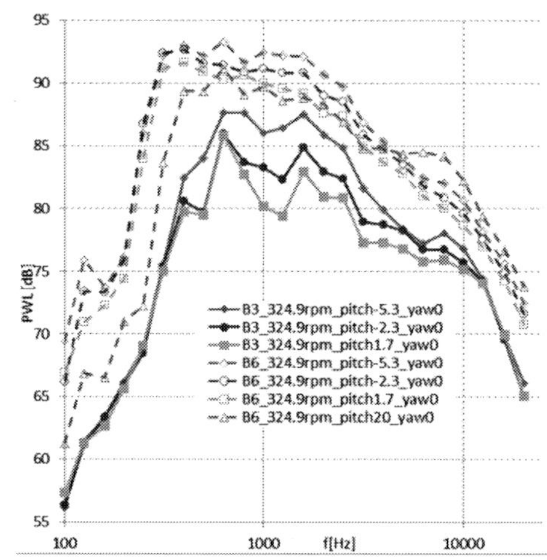

(b) 1/3 Octave band PWL spectra at λ=6.7

Figure 7. Influence of off-setting blade 2 pitch angle with -20° (B6, dashed line) compared to the reference (B3, solid line), 325 rpm

The above results are only a small portion of the available data, as there still is a large amount of data which is untouched. It is recommended to analyze the data using blade tracking as well to analyse sectional noise levels better as a function of azimuth angle. Since the acoustic and aerodynamic measurements were synchronized in the time domain, it is possible to perform cross correlations between microphone and unsteady pressure sensors.

5. Comparison to predictions

A comparison has been made to predictions from an engineering model widely used throughout the wind industry. First the model implementation named Silant is described together with the applied settings. Acknowledging that the noise is driven by the underlying aerodynamics a comparison of measured and prediction forces is given first, after which the noise levels are studied.

5.1. Silant model

Silant originated in 1996 from a Dutch consortium consisting of Stork Product Engineering BV, the Netherlands Organisation for Applied Scientific Research (TNO) and the Dutch Aerospace Laboratory (NLR). The model was designed to calculate noise emission of wind turbines, based on the sources that are considered most important: trailing edge noise (including separation-stall noise) and inflow noise. After ECN became the manager of the tool several improvements have been made, partly in cooperation with NLR.

Silant divides the rotor blades into a number of segments, usually in the order of 10 to 20 per blade. For each element, the trailing edge and inflow noise source are calculated. For the tip element, the contribution of tip noise is added. To determine the total emission, the element contributions are acoustically summed, assuming the sources to be incoherent.

The BPM turbulent boundary layer trailing edge noise model [17] is implemented to model the first noise source. This model necessitates the input of boundary layer parameters at the trailing edge for both pressure and suction side of the airfoil. In this case these are obtained from an a priori created database generated by RFOIL [18], which is based on XFOIL [19], and essentially is a 2D panel code featuring a viscous inviscid interaction scheme. Airfoil coordinates of the profiles used in the blade serve as input to this code. The roughness strips were mimicked by prescribing the laminar to turbulent transition location. The critical amplification factor needed for transition as used in the underlying e^n model was set at 9, corresponding to smooth, low turbulence inflow conditions. In addition to the boundary layer variables this model needs several rotor aerodynamic variables (sectional angle of attack and apparent velocity), which are estimated by a BEM based code [20] after feeding the operational conditions. Here the relevant airfoil polars originate from dedicated airfoil wind tunnel tests, with the exception of the midboard RISØ profile. The tip noise model for rounded tips from [17] is also implemented, where level and spectral content of the tip noise are determined using the spanwise extent of separation at the trailing edge due to the tip vortex. Here the spanwise extent is estimated using a representative angle of attack in the tip region, obtained from the BEM code.

Although inflow noise due to the interaction of the airfoil with turbulence in the oncoming flow is predicted by Silant, this noise source is discarded from the current comparison due to the low turbulence levels in the tunnel and the expected dominance of turbulent boundary layer trailing edge noise. For more details about Silant please consult the relevant publication [21].

5.2. Load verification

To assess the validity of the BEM simulations of which the results are used as input to the Silant code, a comparison is made in terms of axial force coefficient Cdax obtained from the pressure sensors for pitch=-2.3° and 425 rpm in Figure 8(a). To prevent differences due to the limited number of sensors, the experimental resolution in spanwise direction is used to obtain the axial force from the simulations. The agreement is quite good for a variety of operational conditions, although results seem to slightly diverge for high tip speed ratios towards the turbulent wake state. It is noted here that these conditions feature relatively low tunnel speeds and consequently low dynamic pressures utilizing only a small fraction of the measurement range (plus the fact that absolute differences are non-dimensionalized with a lower velocity enlarging differences in Cdax). Because a good agreement in axial force can also be a result of compensating errors along the blade span and since the outboard part of the blade is mostly responsible for the noise, a comparison of chord normal force at 82%R is given in Figure 8(b). Except for the kink due to stall just below λ=6, the trend is well captured. In attached flow conditions (λ >6) predicted loads are roughly 20 N/m lower, which is a satisfactory agreement.

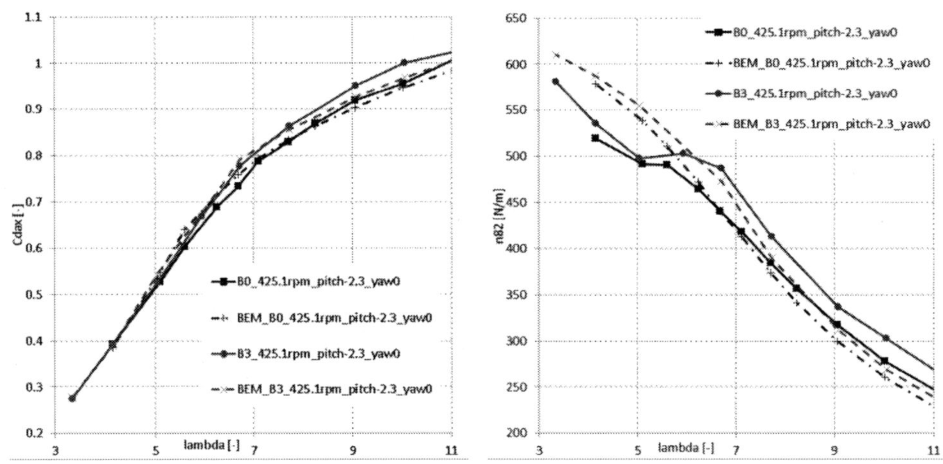

(a) Axial force coefficient

(b) Normal force at 82%R

Figure 8. Comparison to predicted loads (dashed) at pitch=-2.3° and 425 rpm for partially clean (B3) and rough configuration (B0)

5.3. Noise validation

The comparison of predicted noise levels to the measurements is shown in Figure 9 for the two different rotational speeds, while a selection of underlying spectra at 425 rpm is given in Figure 10. For the clean configuration the agreement is within 1 dB below $\lambda = 7$. The implemented switch to separation-stall noise seems to yields a good

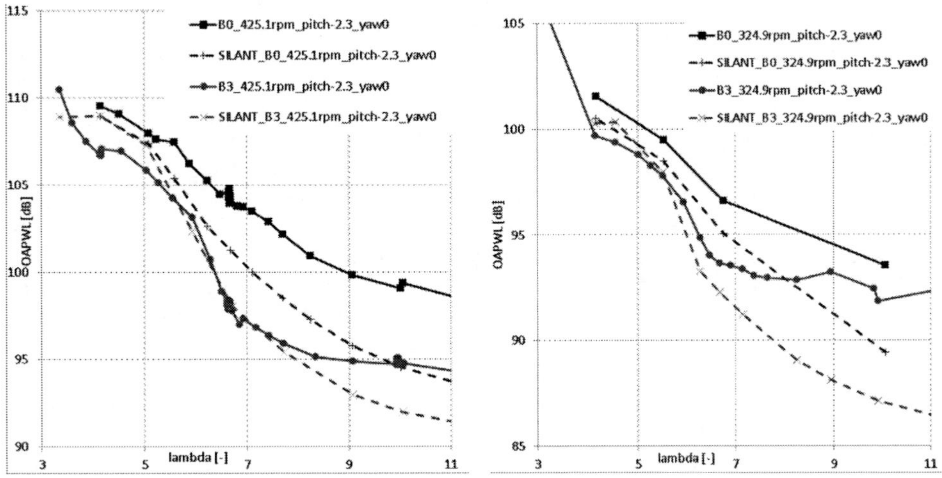

(a) 425 rpm (b) 325 rpm

Figure 9. Comparison to predicted OAPWL (dashed) at two different rotational speeds for partially clean (B3) and rough configuration (B0), pitch=-2.3°

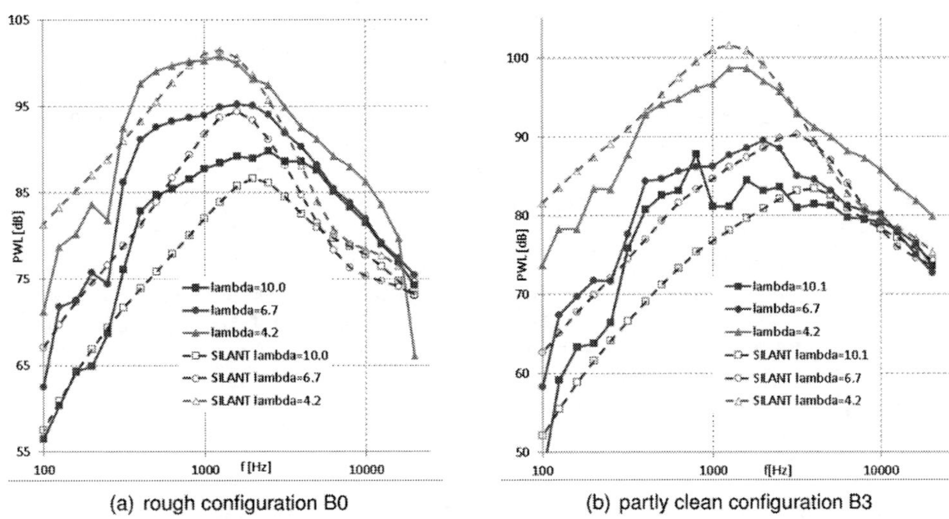

(a) rough configuration B0 (b) partly clean configuration B3

Figure 10. Comparison to predicted spectra (dashed) for partially clean (B0) and rough conditions (B3), pitch=-2.3°, 425 rpm agreement for low tip speed ratios. However the measured sharp noise increase trend towards λ=3 in massively separated flow is not captured by the calculations. For higher tip speed ratios the results seem to diverge slightly, similar to what was shown for the loads. Hence the question that can be asked is whether this discrepancy arises from a shortcoming of the BPM model or the aerodynamic input to this model, of which the last option seems to be more likely in this case. The underlying spectra show a surprisingly good agreement. The maximum levels and their corresponding frequency are well approximated and the shape agreement is also fair. Generally speaking the measured peaks are slightly more broad than the predictions.

The effect of the roughness strip on the noise appears to be underestimated by the predictions. In addition to enforcing laminar to turbulent transition, turbulators have a finite thickness which is known to induce an increase in boundary layer thickness. This thickness effect is not modeled in the RFOIL code which was used to create the airfoil database. Possibly this effect is responsible for the larger discrepancy between measured and predicted noise levels for the rough configuration. The scaling of the noise between the two different rotational (or tip) speeds is well predicted by the code, confirming the validity of the underlying model for this purpose.

6. Conclusions

Wind turbine noise measurements have successfully been performed in the wind tunnel for a variety of operational conditions and model configurations. The rule of thumb stating that per degree of pitch angle increase the noise roughly reduces with 1dB was assessed as a reasonable approximation. Addition of roughness by means of zigzag strips increased the overall noise levels up to 5 dB for the experiment under investigation. Yawing the turbine is found to significantly decrease the noise above 15° misalignment. The influence of various configurations on the noise signature has been assessed. The predictions of the BPM model were found to agree well within the specified uncertainty band of the experiment and trends are well captured. An excellent agreement was obtained for design conditions. Here it must be stated that the BPM model is dependent on the accuracy of the inputted airfoil data and rotor aerodynamic state.

In summary, ECN and partners have performed very successful aero-acoustic experiments in the largest wind tunnel in Europe. A comprehensive high quality database has been obtained which is shared in the wind energy R&D community. A sample of results has been shown in the current paper, however a large portion of the measurement data is still untouched. The database will be analyzed further in the international IEA Wind context in order to validate and advance the aero-acoustic modeling of wind turbines. With the results future large wind turbines will be designed reducing annoyance and optimizing noise reduction potential.

Acknowledgements

Hermann Holthusen (DNW) is acknowledged for data reduction of the raw time series into beamforming plots and resulting spectra plus the corresponding explanations. Financial support for this experiment was given in part by the EU INNWIND project.

In addition to that the European ESWIRP project has been responsible for sponsoring the tunnel time in the DNW. The support of the the steering committee (Delft University of Technology, Technion Israel Institute of Technology and DTU Technical University of Denmark), which advised during the preparation of the tunnel test, was greatly appreciated. The IEA Wind Executive Committee is acknowledged for enabling the valuable cooperation within IEA Wind Task 29 Mexnext.

References

[1] J.G. Schepers and K. Boorsma et al. Final report of IEA Task 29, Mexnext (Phase 2). ECN-E-14-060, Energy Research Center of the Netherlands, December 2014.

[2] J.G. Schepers and K. Boorsma et al. Final report of IEA Task 29, Mexnext (Phase 1): Analysis of MEXICO wind tunnel measurements. ECN-E-12-004, Energy Research Center of the Netherlands, February 2012.

[3] J.G. Schepers et al. Sirocco: Silent rotors by acoustic optimisation. In *Conference proceedings European Wind Energy Conference*, Athens, Greece, March 2006.

[4] S. Wagner, G. Guidati et al. Design and Testing of Acoustically Optimized Airfoils for Wind Turbines (DATA). Technical Report Publishable Final Report - Contract JOR3-CT98-0248, , 2001.

[5] T. Cho, C. Kim and D. Lee. Acoustic measurement for 12% scaled model of NREL Phase VI wind turbine by using beamforming. *Current Applied Physics*, 10:5320–5325, 2010.

[6] J.G. Schepers and H. Snel. MEXICO, Model experiments in controlled conditions. ECN-E-07-042, Energy Research Center of the Netherlands, 2007.

[7] K. Boorsma and J.G. Schepers. New MEXICO Experiment, Preliminary overview with initial validation. Technical Report ECN-E–14-048, ECN, September 2014.

[8] K. Boorsma and J.G. Schepers. Rotor experiments in controlled conditions continued: New Mexico. Technical Report ECN-M–16-063, presented at Torque 2016, October 2016.

[9] E.A. Parra, K. Boorsma et al. Momentum considerations on the New MEXICO experiment. Technical Report ECN-M–16-064, presented at Torque 2016, October 2016.

[10] N.N. Sørensen, F. Zahle et al. CFD computations of the second round of MEXICO rotor measurements. *Journal of Physics, Conference Series*, 753, A. Aerodyanmics and noise, 2016.

[11] L. Oggiano, K. Boorsma et al. Comparison of simulations on the New Mexico rotor operating in pitch fault conditions. *Journal of Physics, Conference Series*, 753, A. Aerodyanmics and noise, 2016.

[12] K. Boorsma and J.G. Schepers. Description of Experimental Setup, New Mexico Experiment. Technical Report ECN-X–15-093, ECN, August 2015.

[13] I. Philipsen, S. Heinrich, K. Pengel, and H. Holthusen. Test Report for Measurements on the New Mexico Wind Turbine Model in DNW-LLF. LLF-2014-19, August 2015.

[14] P. Sijtsma. CLEAN based on spatial coherence. Technical Report AIAA paper 2007-3436, presented at the 13th AIAA/CEAS aeroacoustics conference, Rome, Italy, May 2007.

[15] S. Oerlemans. *Detection of Aeroacoustic Sound Sources on Aircraft and Wind Turbines.* PhD thesis, University of Twente, 2009.

[16] N.J.C.M. van der Borg and P.M. Vink. Acoustic Noise Measurement on Wind Turbines Performed in the Frame of JOU2-CT92-0233. Technical Report ECN-C–95-112, ECN, November 1995.

[17] T.F. Brooks, D.S. Pope, and M.A. Marcolini. Airfoil self noise and prediction. Technical Report Reference publication 1218, NASA, 1989.

[18] B.O.G. Montgomerie, A.J. Brand, J. Bosschers, and R.P.J.O.M Van Rooij. Three-dimensional effects in stall. Technical Report ECN-C–96-079, ECN, 1996.

[19] M. Drela. Xfoil: An analysis and design system for low reynolds number airfoils. In *Conference on Low Reynolds Number Airfoil Aerodynamics.* University of Notre Dame, June 1989.

[20] C. Lindenburg. Bladmode, program for rotor blade mode analysis. Technical Report ECN-C–02-050-r2, ECN, 2002.

[21] K. Boorsma and J.G. Schepers. Enhanced wind turbine noise prediction tool SILANT. Technical Report ECN-M–12-004, presented at the Fourth International Meeting on Wind Turbine Noise, Rome, Italy, April 2011.

Institute of Noise Control Engineering/Europe (INCE/Europe)
Riverside House
4 Oakland Vale
New Brighton, Merseyside CH45 1LQ
United Kingdom

ISBN 978-1-5108-5210-5

CURRAN ASSOCIATES INC.
proceedings.com